S. CHAND'S SUCCESS GUIDES
(Questions and Answers)

PHYSICAL CHEMISTRY

For the Students of B.Sc. I, II and III of All Indian Universities

[Including Solved Long Answer Type, Short Answer Type Questions and Numerical Problems]

REVISED AND UPDATED EDITION

Dr. R.L. MADAN
M.Sc., Ph.D.
Former Principal, Govt. College Panchkula
and Former Head of Chemistry Department
Govt. Post-Graduate College
Faridabad

G.D. TULI
Former Principal
Shyam Lal College
Delhi University, Delhi

S. CHAND
PUBLISHING

S Chand And Company Limited
(ISO 9001 Certified Company)

S Chand And Company Limited

(ISO 9001 Certified Company)

Head Office: D-92, Sector–2, Noida – 201301, U.P. (India), Ph. 91-120-4682700

Registered Office: A-27, 2nd Floor, Mohan Co-operative Industrial Estate, New Delhi – 110 044, Phone: 011-49731800

www.schandpublishing.com; e-mail: **info@schandpublishing.com**

Marketing Offices:

Chennai	:	Ph: 23632120; chennai@schandpublishing.com
Guwahati	:	Ph: 2738811, 2735640; guwahati@schandpublishing.com
Hyderabad	:	Ph: 40186018; hyderabad@schandpublishing.com
Jalandhar	:	Ph: 4645630; jalandhar@schandpublishing.com
Kolkata	:	Ph: 23357458, 23353914; kolkata@schandpublishing.com
Lucknow	:	Ph: 4003633; lucknow@schandpublishing.com
Mumbai	:	Ph: 25000297; mumbai@schandpublishing.com
Patna	:	Ph: 2260011; patna@schandpublishing.com

First Edition 1999
Subsequent Editions and Reprints 2001, 2003, 2005, 2007
Fifth Revised and Enlarged Edition 2009
Sixth Revised Edition 2012
Reprint with corrections 2014 (Twice), Reprint 2016, 2018 (Twice), 2020, 2022

Reprint 2023

ISBN: 978-81-219-1881-7 **Product Code:** H6PCQ68CHEM20ENAF12O

PRINTED IN INDIA

By Vikas Publishing House Private Limited, Plot 20/4, Site-IV, Industrial Area Sahibabad, Ghaziabad – 201 010 and Published by S Chand And Company Limited, A-27, 2nd Floor, Mohan Co-operative Industrial Estate, New Delhi – 110 044.

PREFACE TO THE SIXTH EDITION

The author is pleased to place the fifth enlarged and updated edition of the book at the hands of the students and teachers. Overwhelming response to the earlier editions has encouraged me to undertake thorough revision of the text. Some of the salient features of this edition are :

1. A number of questions that appeared in the papers of different universities during the last couple of years have been solved in different chapters.
2. Physical chemistry in a Nutshell, which is a special feature of the book, has been provided in the beginning. It will prove to be very useful to the students in last minute revision for the examination.
3. Collision frequency, mean free path, viscosity of a gas and barometric formula have been added to the chapter on gaseous state.
4. Structural differences between solids, liquids and gases, various intermolecular forces in liquids, hydrogen bonding and Pioseulle's equation have been discussed in the chapter on liquid state.
5. Powder method and Laue's method for the determination of crystal structure have been added to chapter on solid state.
6. Various kinds of heats of reaction, Hess's law of constant heat summation and bond energy have been discussed in detail in the chapter on thermodynamics I.
7. Different techniques for the determination of transition point have been provided. Carbon dioxide, zinc-cadmium and Magnesium-zinc systems have been discussed in the chapter on phase rule.
8. Dipping calomel electrode, glass electrode, quinhydrone electrodes and $Fe^{2+} - K_2Cr_2O_7$ and $Fe^{2+} - KMnO_4$ titrations have been discussed in the chapter on electrochemical cells. Also various kinds of corrosion, theory and methods of prevention have been treated in this chapter.
9. In the chapter on physical properties and molecular structure, electronic spectra, width and intensity of spectral lines, Born-Oppenheimer approximation, Frank-condon principle and electronic transitions in σ, π and *n*-molecular orbitals have been included.
10. Chapters on solutions and dilute solutions, quantum mechanics and statistical thermodynamics have been expanded to include several new concepts.
11. Appendix 1 on laws of logarithms and Appendix 2 on mathematical relations, which are quite useful, have been provided at the end but before some useful tables.

It is hoped that the book in the present form will be found to be even more useful to the reader.

Any suggestions for further improvement of the book are welcome.

DR. R.L. MADAN

Mobile 09971775666

PREFACE TO THE FIRST EDITION

I feel immense pleasure in presenting the Simplified Course in Physical chemistry for B.Sc. I, II and III (Pass and Honours) students of Indian universities. I have written this book keeping in view the difficulties faced by the students, as observed by me during the last three decades. I have presented the matter in a manner that will inspire the students to read physical chemistry with interest.

This book has been written in question-answer style. This will apprise the students of the style and nature of questions that are asked in the examination and the manner of answering them appropriately. Every effort has been made to present the subject matter lucidly by using a simple and clear language. If the student makes an earnest effort, he can follow the subject-matter readily.

A large number of solved numericals have been included for a clear understanding of the subject. Numerous well-labelled structures and diagrams have been given to explain the points. Important principles, laws, concepts and results have been printed in italics or bold to help the students grasp the subject easily. Questions from university papers of various Indian Universities have been included in the text.

I thank Shri Ravindra Kumar Gupta, Director, S. Chand & Co. Ltd., for the inspiration that I received from him. I also thank the staff of this Company for their co-operation and efforts in bringing out the book in a nice form.

I sincerely hope that the book will go a long way to satisfy the long-felt need of students for a friendly book on Physical Chemistry. Any suggestions for improvement of the book are welcome and will be gratefully acknowledged.

DR. R.L. MADAN

CONTENTS

Note : For topics on structure of atom (Classical and wave-mechanical approach), nuclear chemistry, chemical bonding and acids and bases, consult author's success guide Inorganic Chemistry.

For Rapid Revision

PHYSICAL CHEMISTRY IN A NUTSHELL
EQUATIONS, RELATIONS AND FORMULAE

I. GASEOUS STATE

1. Collision number (z_1) $= \sqrt{2}\,\pi\sigma^2\bar{c}\,\rho$

where σ = Diameter of the molecule, $\bar{c}$ = Average velocity

ρ = No. of molecule per unit volume of the gas

2. Collision frequency (involving two molecules of the same type)

$$z_{11} = \frac{1}{\sqrt{2}}\pi\sigma^2\bar{c}\,\rho^2$$

3. Collision frequency (involving molecules of two different types)

$$z_{12} = \frac{1}{\sqrt{2}}\pi\sigma^2\bar{c}\,\rho_1\rho_2$$

where ρ_1 and ρ_2 are densities of molecules of type 1 and 2 respectively.

4. Mean Free path $\lambda = \dfrac{\bar{c}}{z_1} = \dfrac{\bar{c}}{\sqrt{2}\,\pi\sigma^2\bar{c}\,\rho}$

Also $\lambda = \dfrac{kT}{\sqrt{2}\pi\sigma^2\rho}$

where k = Boltzmann constant

P = Pressure of the gas

5. Compressibility factor, Z $= \dfrac{PV}{nRT}$

6. Van der Waals equation for real gases

$$\left(P + \frac{a}{V^2}\right)(V - b) = RT \quad \text{(for one mole of the gas)}$$

$$\left(P + \frac{an^2}{V^2}\right)(V - nb) = nRT \quad \text{(for } n \text{ moles of the gas)}$$

7. Kammerlingh-Onnes (virial) equation of state for real gases

$$PV = RT\left[1 + \frac{B}{V} + \frac{C}{V^2} + ...\right]$$

where B, C, etc. are second, third, ... virial coefficients.

8. Relation between critical constants and Van der Waals constants

$$V_C = 3b, \qquad P_C = \frac{a}{27b^2} \qquad T_C = \frac{8a}{27Rb}$$

9. **Relation between critical constants**

$$P_C V_C = \frac{3}{8} R T_C$$

10. **Reduced Equation of State**

$$\left(\pi + \frac{3}{\phi^2}\right)(3\phi - 1) = 8\theta$$

where $\pi = \dfrac{P}{P_C}$, $\phi = \dfrac{V}{V_C}$ and $\theta = \dfrac{T}{T_C}$

11. **Kinetic Gas Equation**

$$PV = \frac{1}{3} mnu^2$$

where n = Total no. of molecules in the gas

m = Mass of each molecule

u = Root mean square velocity

12. $PV = \frac{2}{3} K.E = \frac{2}{3} kT$

13. $u = \sqrt{\dfrac{3P}{d}}$

14. **Most Probable velocity (α)** $= \sqrt{\dfrac{2RT}{M}}$

15. **Average velocity (v)** $= \sqrt{\dfrac{8RT}{\pi M}}$

16. **Root mean square velocity (u)** $= \sqrt{\dfrac{3RT}{M}}$

17. $\alpha : v : u$ **= 1 : 1.128 : 1.224**

18. **Degree of freedom for a molecule containing *N* atoms**

	Translational	Rotational	Vibrational
Linear molecule	3	2	$3N-5$
Non-linear molecule	3	3	$3N-6$

II LIQUID STATE

1. **Relationship between boiling point T_b and critical temperature, T_c (Guldberg's Rule)**

$$T_b \approx \frac{2}{3} T_C$$

2. **Relationship between boiling point and heat of veporisation ΔH_v (Trouton's rule)**

$$\frac{\Delta H_v}{T_b} = 21$$

Here ΔH_v is in calories and T_b is in degrees absolute

3. **Relationship between temperature and surface tension**

$$\gamma\left(\frac{M}{D}\right)^{\frac{2}{3}} = k(t_c - t)$$

where M = Molecular Mass, D = Density of the liquid at t° C.
t_c = critical temperature, γ = surface tension of the liquid at t° C
k = A constant

4. $\gamma = \frac{rhdg}{2}$

5. **Surface Tension Determination (by stalgmometer)**

$$\frac{\gamma_1}{\gamma_2} = \frac{m_1}{m_2}$$ (Drop weight method)

$$\frac{\gamma_1}{\gamma_2} = \frac{n_2 d_1}{n_1 d_2}$$ (Drop number method)

6. $\frac{\gamma^{\frac{1}{4}}}{D-d} = C$

$$\frac{M\gamma^{\frac{1}{4}}}{D-d} = MC = [P] \text{ where } [P] \text{ is the parachor}$$

Neglecting d compound to D, we have

$$\frac{M\gamma^{\frac{1}{4}}}{D} = [P]$$

7. **Flow of liquid Through a tube is governed by**

$$f = \eta\frac{Av}{x}$$

8. **Determination of viscosity (By viscometer)**

$$\frac{\eta_1}{\eta_2} = \frac{d_1 t_1}{d_2 t_2}$$

9. $\frac{d}{M} \cdot n \times 10^6 \approx 40 \text{ to } 80$

10. **Rheochor [*R*]** $= \frac{M}{D} (\eta)^{\frac{1}{8}}$

11. $[\eta] = k\overline{M}^a$

where $[\eta]$ = Intruisic viscosity, $\overline{M}$ = Average molecular weight. k and a are constants

12. **Refractive Index, μ** $= \frac{\text{Velocity of light in vacuum}}{\text{Velocity of light in the medium}}$

13. **Specific Refractivity,** $r = \left(\frac{n^2-1}{n^2+2}\right) \cdot \frac{1}{d}$

where n = Refractive Index and d is the density of the liquid at the same temperature

Molar Refractivity R $= \left(\frac{n^2-1}{n^2+2}\right) \cdot \frac{M}{d}$

Molar Refractivity R_s of the solution $= \left(\frac{n^2-1}{n^2+2}\right)\left(\frac{N_1M_1+N_2M_2}{d}\right)$

III SOLUTIONS AND DILUTE SOLUTIONS

1. **Modes of expressing concentration**

(*a*) **Percentage by weight** $= \frac{\text{Weight of the solute}}{\text{Volume of the solution}} \times 100$

(*b*) **Strength** $= \frac{\text{Weight of the solute in grams}}{\text{Volume of the solution in litres}}$

(*c*) **Molarity** $= \frac{\text{Strength of the solution in grams/litre}}{\text{Molecular weight of the solute}}$

(*d*) **Molality** $= \frac{\text{No. of grams of the solute per 1000 g of solvent}}{\text{Molecular weight of the solute}}$

(*e*) **Normality** $= \frac{\text{Strength of the solution in grams per litre}}{\text{Equivalent weight of the solute}}$

(*f*) **Mole fraction of the solute in the solution** $= \frac{\text{No. of moles of the solute}}{\text{No. of moles of the solute + No. of moles of solvent}}$

2. **Relative lowering of vapour pressure (Raoult's Law)**

$$\frac{p^\circ - p_s}{p^\circ} = \frac{n_2}{n_1+n_2}$$

where n_2 and n_1 are the no. of moles of the solute and solvent respectively.

3. **Henry's Law** $m = kp$

where m = mass of the gas dissolved per unit volume of the solvent

p = pressure of the gas in equilibrium with the solvent

4. **Duhem-Margules Equation**

$$x_1 \frac{d \ln p_1}{d x_1} = x_2 \frac{d \ln p_2}{d x_1} \quad \text{(one form)}$$

$$\frac{d \ln p_1}{d \ln x_1} = \frac{d \ln p_2}{d \ln x_2} \quad \text{(Another form)}$$

5. **Composition of the distillate (mixture of two immiscible liquids)**

$$\frac{W_A}{W_B} = \frac{p_A^\circ \times M_A}{p_B^\circ \times M_B}$$

where W, p° and M stand for the weight, vapour pressure and molecular weight respectively

6. **Lowering of vapour pressure**

$$\frac{p_A^\circ - p_A}{p_A^\circ} = \frac{wM}{mW}$$

where w = Mass of the solute

W = Mass of the solvent

m = Mol. mass of the solute

M = Mol. mass of the solvent

p_A° and p_A represent the vapour pressures of the pure solvent and solution respectively.

7. **Elevation in boiling point**

$$\Delta T_b = k_b \times \text{molality}$$

where ΔT_b = Elevation in *b.p.*

k_b = Molal elevation constant

$$m = \frac{1000 \times k_b \times w}{W \times \Delta T_b}$$

8. **Depression in freezing point**

$$\Delta T_f = k_f \times \text{molality}$$

where ΔT_f = Depression in *f. p.*

k_f = Molal depression constant

$$m = \frac{1000 \times k_f \times w}{W \times \Delta T_f}$$

9. Osmotic pressure (Vant Hoff's equation)

$$\pi V = nRT$$

where π = Osmotic pressure R = Solution constant (equal to gas constant)
V = Volume of the solution in litres n = No. of moles of the solute

$$m = \frac{w.R.T}{\pi V}$$

Here m is the molecular mass of the solute

10. Relation between lowering of vapour pressure and osmotic pressure

$$\frac{p^\circ - p}{p^\circ} = \frac{M\pi}{\rho.RT}$$

where M = Mol. mass of the solvent

ρ = Density of the solution

11. Association and dissociation of solutes (Vant Hoff's factor)

$$\text{Vant Hoff factor, } i = \frac{\text{Observed colligative property}}{\text{Calculated (normal) colligative properly}}$$

$$\text{Also,} \quad i = \frac{\text{Calculated (normal) molecular mass}}{\text{Observed molecular mass}}$$

12. Relation between degree of association and vant Hoff's factor

$$\text{Vant Hoff's factor } (i) = 1 - \alpha + \frac{\alpha}{n}$$

where α is the degree of association

13. Relation between degree of dissociation and Vant Hoff's factor

Vant Hoff's factor, $i = 1 - \alpha + n\alpha$

where α is the degree of dissociation

IV. CHEMICAL KINETICS

1. Rate of reaction

For a general reaction : $aA + bB \rightarrow pP + qQ$

$$\text{Rate of reaction} = -\frac{1}{a}\frac{d[A]}{dt} = -\frac{1}{b}\frac{d[B]}{dt} = \frac{1}{p}\frac{d[P]}{dt} = \frac{1}{q}\frac{d[Q]}{dt}$$

2. Rate law and rate constant

For the reaction $\alpha A + \beta B \rightarrow$ Products

$$\text{Rate} = k[A]^\alpha[B]^\beta$$

where k is rate constant

3. First order reactions

$$k = \frac{2.303}{t}\log\frac{C_0}{C_t}$$

Where C_0 = Initial concentration

C_t = Concentration after the time t.

Also

$$k = \frac{2.303}{t} \log \frac{a}{a-x}$$

Where a = Initial concentration

x = Change in concentration after the time t.

4. Second order reactions (one reactant only)

$$k = \frac{1}{t} \frac{x}{a(a-x)} \quad \text{or} \quad k = \frac{1}{t}\left[\frac{1}{C_t} - \frac{1}{C_0}\right]$$

5. Second order reaction (two different reactants)

$$A + B \longrightarrow \text{Products}$$

$$k = \frac{2.303}{t(a-b)} \log \frac{b(a-x)}{a(b-x)}$$

where a is the initial concentration of A and b that of B. x is the change is concentration of A and B that takes place after time t.

6. Third order reaction

$$3\,A \longrightarrow \text{Products}$$

$$k = \frac{1}{2t}\left[\frac{1}{C_t^2} - \frac{1}{C_0^2}\right] \quad \text{or} \quad k = \frac{1}{2t}\left[\frac{x(2a-x)}{a^2(a-x)^2}\right]$$

7. Relation between half-life period and order of the reaction

$$t_{0.5} \propto \frac{1}{a^{n-1}}$$

8. Order of the reaction using differential rate equation

$$r_1 = k\,C_1^n \qquad r_2 = k\,C_2^n$$

$$n = \frac{\log r_1 - \log r_2}{\log C_1 - \log C_2}$$

9. Order of the reaction using fractional change (say half change)

$$\frac{(t_{0.5})_1}{(t_{0.5})_2} = \left(\frac{a_2}{a_1}\right)^{n-1}$$

$$n = 1 + \frac{\log(t_{0.5})_1 - \log(t_{0.5})_2}{\log a_2 - \log a_1}$$

10. Effect of temperature on rate of reaction (Arrhenius equation)

$$\log \frac{k_2}{k_1} = \frac{E_a}{2.303\,R}\left[\frac{T_2 - T_1}{T_1 T_2}\right]$$

k_1 and k_2 are rate constants at temperatures T_1 and T_2 respectively. E_a is activation energy of the reaction.

11. Collision Theory (Lindemann's theory)

$$k = z\,e^{-E_a/RT}$$

12. Transition State theory (Theory of activated complex formation)

$$k = \frac{RT}{Nh} e^{\Delta S^*/R} \cdot e^{-\Delta H^*/RT}$$

V. SURFACE CHEMISTRY (ADSORPTION AND CATALYSIS)

1. Freundlich Adsorption Isotherm

$$\frac{x}{m} = k\,P^{\frac{1}{n}} \qquad \log \frac{x}{m} = \log k + \frac{1}{n} \log P$$

where x = Mass of gas adsorbed

m = Mass of the adsorbent

P = Pressure of the gas

k and n are constants

2. Langmuir adsorption isotherm

$$\frac{x}{m} = \frac{a\,P}{1 + b\,P}$$

a and b are constants

3. B.E.T. (Brunauer, Emmett and Teller) equation

$$\frac{P}{v(P^\circ - p)} = \frac{1}{V_m C} + \left[\frac{C-1}{V_m C}\right]\frac{P}{P^\circ}$$

v = volume of the gas (reduced to STP) adsorbed at pressure P and constant temperature T

V_m = volume of the gas (reduced to STP) adsorbed when the surface is completely covered by a unimolecular layer

P° = vapour pressure of the liquefied gas at temperature T.

C = Constant

4. Surface area from B.E.T. equation

$$\text{Specific surface area} = \frac{6.023 \times 10^{23}}{w \times 22400} \times V_m \times S$$

S = surface area of a single adsorbate molecule

5. Adsorption of solids from solutions

$\frac{x}{m} = kC^{\frac{1}{n}}$ where C is the equilibrium concentration of the solution in moles per litre.

6. **Gibbs adsorption equation**

$$S = -\frac{C}{RT}\frac{d\gamma}{dC}$$

7. **Rate of an acid-catalysed reaction**

For the reaction : $S + HA \underset{k'_1}{\overset{k_1}{\rightleftharpoons}} SH^+ + A^-$

$$\frac{-d[S]}{dt} = \frac{k_1 k_2 [S][HA]}{k'_1[A^-]}$$

8. **Michaelis-Menten equation**

$$v = -\frac{d[S]}{dt} = k_2[ES] = \frac{k_2[E_0][S]}{k_m + [S]}$$

E_0 = Total concentration of the enzyme
E = Concentration of free enzyme.

VI. THERMODYNAMICS - I

1. **Work done in the expansion of gas against constant pressure *P*.**

$$w = P(V_2 - V_1)$$

where V_2 and V_2 are initial and final volumes

2. **First law of Thermodynamics**

$$q = \Delta E + w \qquad \text{... }(i)$$

q = Heat supplied to the system

ΔE = Increase in internal energy

w = work done by system

According to latest S.I. sign convertions, as w is negative if the work is done by the system, we write eq. (i) as

$$q = \Delta E - w \qquad \text{or} \qquad \Delta E = q + w$$

3. **Relation between enthalpy change and internal energy change**

$$\Delta H = \Delta E + P\Delta V \quad \text{or} \quad \Delta H = \Delta E + \Delta nRT$$

4. **Heat capacity at constant volume, C_v**

$$C_v = \left(\frac{\partial E}{\partial T}\right)_v \quad \text{or} \quad C_v = \frac{dE}{dT}$$

5. **Heat capacity at constant pressure, C_p**

$$C_p = \left(\frac{\partial H}{\partial T}\right)_P \quad \text{or} \quad C_p = \frac{dH}{dT}$$

6. **Relation between C_p and C_v**

$$C_p - C_v = R \text{ where } R \text{ is gas constant}$$

7. **Joule-Thomson effect, μ**

$$\mu = \left[\frac{\partial T}{\partial P}\right]_H$$

8. **Work done is isothermal reversible expansion of a gas**

$$w = 2.303\, n\, RT \log\frac{V_2}{V_1} = 2.303\, n\, RT \log\frac{P_1}{P_2}$$

V_1 and V_2 are initial and final volumes respectively
P_1 and P_2 are initial and final pressures respectively

9. **Relation between temperature and volume in reversible adiabatic expansion.**

$T_1V_1^{\gamma-1} = T_2V_2^{\gamma-1}$ or $TV^{\gamma-1}$ = constant

V_1 and V_2 are initial and final volumes

T_1 and T_2 are initial and final volumes

$$\gamma = \frac{C_p}{C_v}$$

10. **Relation between temperature and pressure in adiabatic expansion of an ideal gas**

$T_1P_1^{(1-\gamma)/\gamma} = T_2P_2^{(1-\gamma)/\gamma}$ or $T\,P^{(1-\gamma)/\gamma}$ = constant

P_1 and P_2 are initial and final pressures respectively

11. **Relation between pressure and volume in adiabatic expansion of an ideal gas**

$P_1V_1^{\gamma} = P_2V_2^{\gamma}$ or PV^{γ} = constant

12. **Work done is adiabatic reversible expansion of an ideal gas**

$$w = \frac{P_2V_2 - P_1V_1}{1-\gamma}$$

13. **w, q, ΔE and ΔH in the expansion of a real gas in isothermal reversible expansion.**

$$w = nRT \ln\frac{V_2 - nb}{V_1 - nb} + an^2\left(\frac{1}{V_2} - \frac{1}{V_1}\right)$$

$$\Delta E = -\,an^2\left(\frac{1}{V_2} - \frac{1}{V_1}\right)$$

$$q = nRT \ln\frac{V_2 - nb}{V_1 - nb}$$

$$\Delta H = -2an^2\left(\frac{1}{V_2} - \frac{1}{V_1}\right) + n^2bRT\left(\frac{1}{V_2 - nb} - \frac{1}{V_1 - nb}\right)$$

14. **q, w, ΔE and ΔH in adiabatic reversible expansion of a real gas**

$q = 0$

$$\Delta E = nC_v(T_2-T_1) - an^2\left(\frac{1}{V_2}-\frac{1}{V_1}\right)$$

$$w = an^2\left(\frac{1}{V_2}-\frac{1}{V_1}\right) - nC_v(T_2-T_1)$$

$$\Delta H = nC_v(T_2-T_1) - 2an^2\left(\frac{1}{V_2}-\frac{1}{V_1}\right) + nR\left(\frac{V_2T_2}{V_2-nb}-\frac{V_1T_1}{V_1-nb}\right)$$

15. Kirchoff's equation

$$\frac{\Delta H_2 - \Delta H_1}{T_2 - T_1} = \Delta C_p \quad \text{or} \quad \left[\frac{\partial(\Delta H)}{\partial T}\right]_P = \Delta C_p$$

when the reaction is carried out at constant volume

$$\frac{\Delta E_2 - \Delta E_1}{T_1 - T_1} = \Delta C_v$$

VII THERMODYNAMICS - II

1. Efficiency of an engine, η

$$\eta = \frac{W}{Q}$$

W = Work done

Q = Heat supplied

2. Efficiency of a carnot engine

$$\frac{w}{q_2} = \frac{T_2-T_1}{T_2}$$

T_1 = Temperature of the sink

T_2 = Temperature of the source

q_2 = Heat absorbed at T_2

3. Carnot Theorem

$$\eta = \frac{w}{q_2} = \frac{q_2-q_1}{q_2} = \frac{T_2-T_1}{T_2}$$

4. Relation between entropy change and heat change

$$\Delta S = \frac{q_{rev}}{T}$$

5. For an irreversible process

$$\Delta S_{system} + \Delta S_{surrounding} > 0$$

6. Entropy change for an ideal gas under different conditions

$$\Delta S = C_v \ln\frac{T_2}{T_1} + R\ln\frac{V_2}{V_1} \quad \text{and} \quad \Delta S = C_p \ln\frac{T_2}{T_1} + R\ln\frac{P_1}{P_2}$$

$$\Delta S = R \ln \frac{V_2}{V_1} + R \ln \frac{P_1}{P_2} \qquad \text{(Isothermal process)}$$

$$\Delta S = C_p \ln \frac{T_1}{T_2} \qquad \text{(Isobaric process)}$$

$$\Delta S = C_v \ln \frac{T_2}{T_1} \qquad \text{(Isochoric process)}$$

7. Entropy of melting and entropy of vaporisation

$$\Delta S_m = \frac{\Delta H_m}{T_m}$$

ΔH_m = Enthalpy of melting

T_m = Melting temperature

$$\Delta S_v = \frac{\Delta H_v}{T_b}$$

ΔH_v = Enthalpy of vaporisation

T_b = Boiling temperature

8. Work function A (Helmholtz free energy)

A = E – TS

$$\Delta A = \Delta E - T\Delta S$$

9. Gibb's free energy, *G*

G = H – TS

$$\Delta G = \Delta H - T\Delta S$$

$$\Delta S = \frac{q_{rev}}{T}$$

$$T\Delta S = q_{rev}$$

$$-\Delta G = w_{\max} - P\Delta V$$

10. Change of work function with temperature and volume

$$\left(\frac{\partial A}{\partial v}\right)_T = -P \qquad \left(\frac{\partial A}{\partial T}\right)_V = -S$$

11. Change of free energy with temperature and pressure

dG = V dP – S dT

$$\left(\frac{\partial G}{\partial P}\right)_T = V \qquad \left(\frac{\partial G}{\partial T}\right)_P = -S$$

12. Change in free energy for a process under isothermal condition

$$(\Delta G)_T = nRT \ln \frac{P_2}{P_1} = nRT \ln \frac{V_1}{V_2}$$

13. Maxwell relations

$$\left[\frac{\partial T}{\partial V}\right]_s = -\left[\frac{\partial P}{\partial S}\right]_v \quad \left[\frac{\partial T}{\partial P}\right]_s = \left[\frac{\partial V}{\partial S}\right]_P$$

$$\left[\frac{\partial S}{\partial V}\right]_T = \left[\frac{\partial P}{\partial T}\right]_v \quad -\left[\frac{\partial S}{\partial P}\right]_T = \left[\frac{\partial V}{\partial T}\right]_P$$

14. Thermodynamic relations

$$\left(\frac{\partial E}{\partial S}\right)_V = T \quad \left(\frac{\partial E}{\partial V}\right)_S = -P$$

$$\left(\frac{\partial H}{\partial S}\right)_P = T \quad \left(\frac{\partial H}{\partial P}\right)_S = V$$

15. Gibbs-Helmholtz equation

$$\Delta G = \Delta H + T\left[\frac{\partial(\Delta G)}{\partial T}\right]_P$$

$$\left[\frac{\partial(\Delta G/T)}{\partial T}\right]_P = -\frac{\Delta H}{T^2}$$ The symbols have usual meanings

16. Relation between e.m.f. and enthalpy change

$$nFE = -\Delta H + TnF\left(\frac{\partial E}{\partial T}\right)_P$$

F = One faraday of electricity

n = No. of electron involved in the redox reaction

17. Clausius-Clapeyron equation

$$\frac{dP}{dT} = \frac{\Delta H}{T\Delta V}$$

$$\frac{dT}{dP} = \frac{T_m(V_l - V_s)}{\Delta H_m}$$ (For melting equilibrium)

$$\frac{d\ln P}{dT} = \frac{\Delta H_v}{RT^2}$$ (For vaporisation equilibrium)

$$\log\frac{P_2}{P_1} = \frac{\Delta H_v}{2.303R}\left(\frac{T_2 - T_1}{T_1 T_2}\right)$$ (Integrated form of Clausius-Clapeyron equation)

18. Nernst heat theorem

$$\lim_{T\to 0}\frac{d(\Delta G)}{dT} = \lim_{T\to 0}\frac{d(\Delta H)}{dT} = 0$$

19. Expression for entropy at temperature T(Third law of thermodynamics)

$$S = \int_0^{T_m} C_p(s)d\ln T + \frac{\Delta H_m}{T_m} + \int_{T_m}^{T_b} C_p(l)d\ln T + \frac{\Delta H_v}{T_b} + \int_{T_b}^{T} C_p(g)d\ln T$$

VIII THERMODYNAMICS - III

1. Partial molar quantity

$$\left[\frac{\partial X}{\partial n_i}\right]_{T,P,n_1,n_2,......} = \bar{X}_i$$

2. Chemical potential or partial molar free energy μ_i

$$\mu_i = \left[\frac{\partial G}{\partial n_i}\right]_{T,P,n_1,n_2...} = \bar{G}_i$$

3. Relation between activity coefficient γ and fugacity f

$$\gamma = \frac{f}{p} \qquad p = \text{pressure of the gas}$$

4. Gibbs-Duhem equation

$$\Sigma n_i\, d\mu_i = 0$$

5. Variation of chemical potential with temperature and pressure

$$\left(\frac{\partial G}{\partial T}\right)_{P,N} = -S \qquad \left(\frac{\partial G}{\partial P}\right)_{T,N} = V$$

6. $\boldsymbol{\mu_i = \mu_i^\circ + RT\ \ln p_i}$

7. Phase equilibrium for multiple component

$$(\mu_i)_{P_1} - (\mu_i)_{P_2} = (\mu_i)_{P_3} = ...$$

8. Criterion for two-phase equilibrium for one-component system

$$S = -\frac{a}{RT}\frac{d\gamma}{da} \qquad S = -\frac{C}{RT}\frac{d\gamma}{dC}$$

where a and c stand for activity and concentration respectively and S stands for excess concentration of surface phase.

9. Excess concentration versus pressure

$$S_2 = -\frac{p}{RT}\cdot\frac{d\gamma}{dp}$$

IX POLYMERS OR MACROMOLECULES

1. Number average molecular weight

$$\overline{M}_n = \frac{\Sigma N_i M_i}{\Sigma N_i}$$

N_i is the number of molecules having molecular weight M_i

2. **Weight average molecular weight**

$$\overline{M}_w = \frac{\Sigma N_i M_i^2}{\Sigma N_i M_i}$$

3. **Osmotic pressure method for mol. weight determination**

$$M = \frac{RT}{(P/C)_{c\to 0}}$$

4. **Ultracentifuge method for mol. weight determination**

$$x_2 - x_1 = \frac{2r^2 g(d-d')(t_2-t_1)}{9\eta}$$

η = coefficient of viscosity of the medium

Mass of the particle, $m = \frac{4}{3}\pi r^3 d$

Molecular weight of the polymer = Nm

N is the Avogadro's number

5. **Viscosity method for the molecular weight determination**

$$[\eta] = \left[\frac{\eta_{sp}}{C}\right]_{C\to 0}$$

$$[\eta] = k\,M^a$$

$[\eta]$ = Intrinsic viscosity. k and a are constants

6. **Light scattering method for molecular weight determination**

$$\lim \left[\frac{HC}{\tau}\right]_{C\to 0} = \frac{1}{M'} \qquad \tau = \frac{1}{l}\log\frac{I_0}{I}$$

Molecular weight $M = M' \times \alpha \times \beta$ $\qquad H = \frac{32\pi^3 n_0^2 (dn/dc)^2}{3N\lambda_0^4}$

where α and β are constants $\qquad n_0$ = Refractive index of the solvent

λ_0 = Wavelength of light used in vacuum

X. ELECTROCHEMISTRY

1. **Degree of dissociation, α**

$$\alpha = \frac{\text{No. of molecules dissociated}}{\text{No. of molecules taken}}$$

$$\alpha = \frac{\wedge_C}{\wedge_0} = \frac{\text{Equivalent conductance at concentration C}}{\text{Equivalent conductance at zero concentration}}$$

2. Ohm's law

$$R = \frac{E}{I} \qquad \text{or Resistance} = \frac{\text{Voltage}}{\text{Current}}$$

$$\text{Conductance} = \frac{1}{\text{Resistance}}$$

3. Specific conductance K

$$K = C \times \frac{l}{a} \qquad \text{where } C \text{ is conductance}$$

$$\text{Resistivity or specific resistance } (\rho) = R\frac{a}{l}$$

a = Area of cross section
l = Length of the conductor

4. Equivalent conductance $\wedge_{eq}$

$\wedge_{eq} = K \times V$ where K = Specific conductance

V = Volume of the solution containing one gram equivalent of the electrolyte

5. Molar conductance (conductivity), $\wedge_m$

$$\wedge_m = K \times V$$

Here V is the volume containing one mole of the electrolyte

6. Cell constant

Cell constant $= \frac{l}{a}$ where l is the distance between the electrodes a is the area of cross-section of the electrode

$$\text{Specific conductivity} = C \times \frac{l}{a} = \text{conductance} \times \text{cell constant}$$

7. Relation between equivalent conductivity $\wedge_{eq}$ and specific conductivity K

$$\wedge_{eq} = K \times \frac{1000}{C_{eq}}$$

where C_{eq} = concentration of the solution in gram-equivalents per litre.

8. Ostwald's delution law

$$k = \frac{C\alpha^2}{1-\alpha} \quad \textbf{(Exact formula)} \qquad K = C\alpha^2 \quad \textbf{(Approximate formula)}$$

C = concentration of the solution
α = Degree of dissociation

9. Debye-Huckel-Ousagar equation or variation of conductivity of a strong electolyte with dilution

$$\wedge_m^C = \wedge_m^0 - (A + B\wedge_m^0)\sqrt{C}$$

$\wedge_m^C$ = Molar conductance of the solution at the concentration C

$\wedge_m^0$ = Molar conductance of the solution at infinite dilution

10. Kohlrausch law

$$\wedge_{eq}^0 = \lambda_c^0 + \lambda_a^0$$

where $\wedge_{eq}^0$, λ_c^0 and λ_a^0 stand for the equivalent conductance of the electrolyte, cation and anion respectively at infinite dilution.

11. Relation between solubility and molar conductance

$$\text{Solubility} = \frac{K \times 1000}{\wedge_m^0}$$

12. Transport number

$$\frac{\text{Fall in conc. around anode}}{\text{Fall in conc. around cathode}} = \frac{\text{Speed of the cation}}{\text{Speed of the anion}}$$

Transport no. of the cation $n_c = \dfrac{u_c}{u_a + u_c}$

where u_a and u_c are speeds of anion and cation respectively.

Transport no. of the anion. $n_a = \dfrac{u_a}{u_a + u_c}$

$$n_a + n_c = 1$$

In moving boundary method

$$n_c = \frac{aa'}{aa' + bb'} \qquad n_a = \frac{bb'}{aa' + bb'}$$

13. Relation between ionic conductance and transport number

$$\lambda_a = n_a \times \wedge_0$$

$$\lambda_c = n_c \times \wedge_0$$

14. Relation between ionic mobility and ionic conductance

$$\text{Ionic mobility} = \frac{\text{Ionic conductance}}{96500}$$

15. Solubility of a sparingly soluble salt by conductance measurement

$S = \dfrac{1000 \times K}{\wedge_0}$ S is solubility in gram-equivalents per litre

$\wedge_0$ = Equivalent conductance

16. Henderson equation for *pH* of a buffer mixture

$$pH = pK_a + \log\frac{[Salt]}{[Acid]} \quad \text{(For Acid buffer)}$$

$$pOH = pK_b + \log\frac{[Salt]}{[Base]} \quad \text{(For Basic buffer)}$$

17. Hydrolysis constant, degree of hydrolysis and *pH* of solution of a salt of weak acid and strong base

Hydrolysis constant $K_h = \frac{K_w}{K_a}$ $\quad K_w$ = Ionic product of water

Degree of hydrolysis $h \simeq \sqrt{\frac{K_w}{K_a.C}}$ $\quad K_a$ = Dissociation constant of the acid

$$pH = -\frac{1}{2}\left[\log K_w + \log K_a - \log C\right]$$

18. Hydrolysis constant, degree of hydrolysis and *pH* of a solution of a salt of strong acid and weak base

$K_h = \frac{K_w}{K_b}$ $\quad K_b$ = Dissociation constant of the base

$$h = \sqrt{\frac{K_w}{K_b.C}}$$

$$pH = -\frac{1}{2}\left[\log K_w - \log K_b + \log C\right]$$

19. Hydrolysis constant, degree of hydrolysis and pH of a solution of a salt of weak acid and weak base

$$K_h = \frac{K_w}{K_a K_b}$$

$$h \simeq \sqrt{\frac{K_w}{K_a.K_b}}$$

$$pH = -\frac{1}{2}\left[\log K_a + \log K_w - \log K_b\right]$$

20. Faraday's first law of electrolysis

$W = Z \times C \times t \qquad Z = \frac{\text{Eq. wt. of the metal}}{96500}$

W = Weight of the metal deposited

Z = Electrochemical equivalent
C = Current in amperes
t = Time in seconds

21. Faraday's second law of electrolysis

$$\frac{\text{Weight of A deposited}}{\text{Weight of B deposited}} = \frac{\text{Eq. weight of A}}{\text{Eq. weight of B}}$$

XI ELECTROCHEMICAL CELLS

1. Relation between $\Delta G, \Delta H, \Delta S$ and EMF of the cell

$$\Delta G = -nFE$$

$$\Delta H = nF\left[T\left(\frac{\partial E}{\partial T}\right)_P - E\right]$$

$$\Delta G = \Delta H - T\Delta S$$

2. Standard EMF of the cell

Standard EMF of the cell = [Standard reduction potential of RHS electrode] – [Standard reduction in potential of LHS electrode]

3. Nernst equation for the electrode potential of metal-metal ion electrode

$$E_{M^{n+},M} = E^0_{M^{n+},M} + \frac{RT}{nF}\ln\left[M^{n+}\right]$$

n = No. of electrons participating in the redox reaction
F = One faraday of electricity
M^{n+} = Concentration of metal ion
T = Temperature

$$E = E^\circ - \frac{RT}{nF}\ln\frac{[\text{Reduced State}]}{[\text{Oxidised State}]}$$ E° = Standard electrode potential

$$E = E^\circ + \frac{RT}{nF}\ln\frac{[\text{Oxidised State}]}{\text{Reduced State}}$$ E = Electrode potential under the experimental conditions

4. Nernst equation for measuring EMF of a cell

For the cell reaction $aA + bB \rightleftharpoons xX + yY$

$$E = E^\circ - \frac{RT}{nF}\ln\frac{[X]^x[Y]^y}{[A]^a[B]^b}$$ E° = Standard EMF of the cell

E = EMF under the experimental conditions.

5. Nernst equation for the potential of hydrogen-electrode

$$E_{H^+,H_2} = \frac{RT}{F}\ln[H^+] = 0.0591\log[H^+] \text{ at } 25^\circ C$$

6. EMF of electrode concentration cell without transference

For the cell $H_2(P_{H_2} = P_1)\,|\,H^+(a_{H^+})\,|\,H\,(P \quad = P_2)$

$$E = \frac{RT}{2F} \ln \frac{P_1}{P_2}$$

7. EMF of electrolyte concentration cell with transference

For the cell $H_2(1\,atm) | H^+(a_1) || H^+(a_2) | H_2(1\,atm)$

$$E = \frac{RT}{F} \ln \frac{a_2}{a_1}$$

8. EMF of the concentration cell

H_2 (1 atm) | HCl (a_1) || HCl (a_2) | H_2 (1 atm)

$$E = \frac{t_- RT}{F} \ln \frac{m_2^2 \gamma_2^2}{m_1^2 \gamma_1^2}$$

where t_- = Transport no. of Cl^-

m_1 and m_2 stand for molalities

γ_1 and γ_2 represent activity coefficients of the ions

9. Liquid junction potential, E_j

$$E_j = (t_- - t_+) \frac{RT}{F} \ln \frac{m_2 \gamma_2}{m_1 \gamma_1}$$

t_- and t_+ represent the transport numbers of anion and cation respectively.

XII PHOTOCHEMISTRY

1. Lambert's law of transmission of light

$I = I_0 e^{-kx}$

I_0 = original intensity of light

I = Intensity after crossing through a thickness x of the medium

k = Absorption coefficient

2. Extinction coefficient or absorptivity, k'

$$k' = \frac{k}{2.303} = \frac{1}{x} \log \frac{I_0}{I}$$

$-\log \frac{I}{I_0}$ is called absorbance of the medium

3. Beer's law

$I = I_0 \times 10^{-\varepsilon' cx}$

c = concentration of the solution

x = Thickness of the solution

ε = molar absorption coefficient

$$\varepsilon' = \frac{\varepsilon}{2.303}$$

4. Quantum efficiency or quantum yield, ϕ

$$\phi = \frac{\text{No. of moles reacting in a given time}}{\text{No. of Einsteins of light absorbed in the same time}}$$

XIII QUANTUM MECHANICS

1. Kirchoff's law of black body radiation

$$\frac{E_S}{A_S} = E_B$$

E_S = Emissive power of a substance
A_S = Absorptive power of the substance
E_B = Emessive power of a perfectly black body

2. Stefan's fourth power law

$$E = \sigma T^4$$

E = Energy radiated by a perfectly black body per unit area per unit time at temperature T

σ = Stefan's constant

3. Wien's displacement law

$$\lambda_m \propto \frac{1}{T}$$

λ_m = Wavelength for which the emittance of a black body is maximum

$$E_\lambda = \frac{8\pi hc}{\lambda^5} e^{-hc/kT_\lambda} \quad \text{(Wien's radiation law)}$$

$$E_\lambda = \frac{8\pi}{\lambda^4} kT \quad \text{(Rayleigh-Jeans radiation law)}$$

4. Planck's radiation law

$$E_\lambda = \frac{8\pi hc}{\lambda^5} \cdot \frac{1}{e^{hc/\lambda kT} - 1}$$

E_λ = Energy radiated by a black body at wavelength λ

5. Schrodinger wave equation

$$\frac{\partial^2 \psi}{\partial x^2} + \frac{\partial^2 \psi}{\partial y^2} + \frac{\partial^2 \psi}{\partial z^2} + \frac{8\pi^2 m(E-V)}{h^2}\psi = 0$$

6. Eigen value of energy for a particle in one dimensional box

$$E = \frac{n^2 h^2}{8ma^2}$$

7. Energy of a particle in a three-dimensional box

$$E = \frac{h^2}{8m}\left(\frac{n_x^2}{a^2} + \frac{n_y^2}{b^2} + \frac{n_z^2}{c^2}\right)$$

XIV STATISTICAL MECHANICS

1. Striling formula for a large number N of molecules

$\ln N! = N \ln N - N$

2. Probability of an event

$$\text{Probability} = \frac{\text{Number of ways which lead to that event}}{\text{Total number of equally likely ways possible}}$$

3. Maxwell-Boltzmann distribution law

$$n_i = \frac{1}{e^{\alpha+\beta\varepsilon_i}} \qquad \text{(one form)}$$

$$n_i = \frac{g_i}{e^{\alpha+\beta\varepsilon_i}} \qquad \text{(second form)}$$

4. Boltzmann distribution law

$$\frac{n_i}{N} = \frac{e^{-\varepsilon_i/kT}}{\Sigma e^{-\varepsilon_i/kT}} \qquad \text{(one form)}$$

$$\frac{n_i}{N} = \frac{g_i e^{-\varepsilon_1/kT}}{\Sigma g_i e^{-\varepsilon_1/kT}} \qquad \text{(second form)}$$

5. Relationship between entropy S and thermodynamic probability W

$$S = k \ln W$$

6. Partition function

Partition function $Q = \Sigma g_i e^{-\varepsilon_1/kT}$

7. Translational partition function

$$Q_t = \frac{(2\pi m kT)^{\frac{3}{2}}}{h^3} \cdot V$$

8. Rotational partition function

$$Q_r = \frac{8\pi^2 I kT}{\sigma h^2}$$

9. Vibrational partition function

$$Q_v = \frac{1}{1 - e^{-h\nu_0/kT}}$$

10. Electronic partition function

$$Q_e = g_0 + g_i e^{-\varepsilon_1/kT} + \ldots$$

XV DISTRIBUTION LAW

1. Nernst Distribution Law

$$\frac{\text{Concentration of X in solvent A}}{\text{Concentration of X in solvent B}} = \frac{C_1}{C_2} = K \text{ (Constant)}$$

Provided *X* remains in the same molecular species in both the solvents

2. Modified distribution law when association takes place in one of the solvents

$$\frac{C_1}{\sqrt[n]{C_2}} = K$$

3. Modified distribution law when dissociation takes place in one of the solvents

$$\frac{C_1(1-\alpha)}{C_2} = K \qquad \alpha = \text{Degree of dissociation}$$

4. Amount of substance left unextracted after *n*th operation

$$w_n = \left(\frac{KV}{v+KV}\right)^n W$$

XVI PHASE RULE

1. Gibb's phase rule

$F = C - P + 2$

F = No. of degrees of freedom

C = No. of components

P = No. of phases

2. Reduced phase rule

$F = C - P + 1$

XVII SOLID STATE

1. Radius of ratio

$$\text{Radius ratio} = \frac{\text{Radius of Cation } (r^+)}{\text{Radius of anion } (r^-)}$$

2. Relation between radius ratio, coordination number and arrangement of atoms.

Radius ratio	*Coordination no.*	*Arrangement*
1.0 — 0.732	8	cubic
0.732 — 0.414	6	octahedral
0.414 — 0.225	4	tetrahedral
0.225 — 0.155	3	triangular

3. Density of the unit cell of a solid

$$d = \frac{Z \times \text{Atomic mass}}{6.02 \times 10^{23} \times a^3 \times 10^{-30}} \; g/cm^3$$

4. Interplanar spacing d_{hkl}

$$d_{hkl} = \frac{a}{\sqrt{h^2+k^2+l^2}} \qquad a = \text{side of the cube}$$

5. **Bragg's equation**

$$2d\sin\theta = n\lambda$$

λ = Wavelength of X-rays used

θ = The angle at which maxima occurs

d = Distance between two successive planes

XVIII PHYSICAL PROPERTIES AND MOLECULAR STRUCTURE

1. **Specific rotation, $[\alpha]_\lambda^t$**

$$[\alpha]_\lambda^t = \frac{v \times \alpha}{l \times m}$$

m = Mass of the substance

l = Length of the solution

v = Volume of the solution

α = Angle of rotation produced

λ = Wavelength of light used

t = Temperature of observation

2. **Dipole moment, μ**

$$\mu = q \times d$$

where q = Charge on the +ve or –ve end

d = Distance between the changes

3. **Induced molar polarization, P_i**

$$P_i = \left[\frac{D-1}{D+2}\right]\frac{M}{\rho}$$ where D = Dielectric constant

ρ_i = Density of the molecule

M = Molecular mass of the molecule

4. **Magnetic permeability**

$$F = \frac{1}{\mu}\cdot\frac{m_1 \times m_2}{r^2}$$ μ = Magnetic permeability of the medium

m_1 and m_2 are magnetic poles

r = Distance between the poles

5. **Molar magnetic susceptibility**

$$\chi_M = \frac{M}{P}\frac{I}{H}$$

$$\mu = 1 + 4\pi\chi$$

6. **Relation between magnetic moment and number of unpaired electrons**

$$\mu_m = 2\sqrt{S(S+1)} \quad B.M$$

Here S represents the total spin quantum number. For numpaired electrons, $S = \frac{n}{2}$,

Hence

$$\mu_m = \sqrt{n(n+2)} \quad B.M.$$

7. **Relation between molar susceptibility and the number of unpaired electrons**

$$\chi_M = \frac{N}{3kT}[n(n+2)]$$

8. **Rotational energy of a diatomic molecule taking it as a rigid rotator**

$$E_r = \frac{h^2}{8\pi^2 I} J(J+1)$$

J is rotational quantum number

$$I = \left(\frac{m_1 m_2}{m_1 + m_2}\right) r_0^2$$

9. **Wave number of lines in the rotational spectrum**

$$\bar{\nu} = \frac{h}{8\pi^2 IC} \times 2(J+1)$$

XIX CHEMICAL EQUILIBRIUM

1. **Law of mass action**

For the reaction $A + B \rightleftharpoons C + D$

$$K_C = \frac{[C][D]}{[A][B]}$$

2. **For the reaction $aA + bB + cC + ... \rightleftharpoons lL + mM + nN + ...$**

$$K_c = \frac{[L]^l [M]^m [N]^n ....}{[A]^a [B]^b [C]^c ....}$$

3. **For the above reaction in gaseous phase**

$$K_p = \frac{[p_L]^l [p_M]^m ....}{[p_A]^a [p_B]^b ....}$$

4. **Vant Hoff reaction isotherm**

$$\Delta G = \Delta G° + RT \ln K$$

where $K = \frac{(p_L)^l (p_M)^m ....}{(p_A)^a (p_B)^b ....}$

5. **Relation between K_p and K_c**

$$K_p = K_c (RT)^{\Delta n}$$

Δn = No. of moles of the products – No. of moles of the reactants

6. **Vant Hoff's equation of temperature dependence of equilibrium constant**

$$\frac{d(\ln k_p)}{dT} = \frac{\Delta H^\circ}{RT^2}$$

$$\log K''_p - \log K'_p = \log \frac{K''_p}{K'_p} = \frac{\Delta H}{2.303R} \left[\frac{T_2 - T_1}{T_1 T_2} \right]$$

7. Vant Hoff equation of temperature dependence of equilibrium constant in terms of K_c

$$\frac{d \ln K_c}{dT} = \frac{\Delta E^\circ}{RT^2}$$

XX COLLOIDAL STATE

1. Size of the particle from Brownian Movement

$$\frac{RT}{N_0} \ln \frac{n_1}{n_2} = \frac{4}{3} \pi r^3 (h_2 - h_1)(\rho - \rho')$$

r_0 = Radius of the particles

N_0 = Avogadro's number

n_1 = Number of particles at the depth h_1

n_2 = Number of particles at the depth h_2

ρ = Density of the particle

ρ' = Density of the medium

2. Size of particles from light scattering using ultramicroscope

$$r = \left(\frac{3m}{4\pi \rho n} \right)^{\frac{1}{3}}$$

r = Radius of the particles

n = Number of particles in unit volume

ρ = Density of the particles

m = Mass of the particles per unit volume

1

GASEOUS STATE

Q. 1. Describe important characteristics of gases.

Ans. Some important characteristics of gases are given below:

(*i*) They fill up the whole vessel in which they are placed.

(*ii*) They exert pressure on the walls of the vessel in which they are taken.

(*iii*) They can be compressed by applying the pressure.

(*iv*) Cooling and compression of the gas brings about its liquefaction.

(*v*) They exhibit the property of intermixing with one another—a property known as *diffusion*.

(*vi*) At ordinary conditions of temperature and pressure, their densities are low.

Q.2. Discuss Boyle's law and Charles' law of gases. *(Kumaon 2000; Himachal 2000)*

Ans. Boyle's law: It states as follows:

Temperature remaining constant, the volume of a given mass of a gas is inversely proportional to its pressure.

Thus, if V is the volume of the gas at pressure P, then

$$V \propto \frac{1}{P} \quad \text{(At constant temperature)}$$

or

$$P\,V = \text{constant}$$

If V_1 is the volume of the gas at pressure P_1 and V_2 is the volume at pressure P_2, then

$$P_1 \times V_1 = P_2 \times V_2$$

Charles law:

Pressure remaining constant, the volume of a given mass of a gas increases or decreases by 1/273 of its volume at 0°C for every degree rise or fall of temperature.

Thus if V_0 is the volume of a gas at 0°C and V_t is the volume at t°C, then mathematically, the law may be written as

$$V_t = V_0 + \frac{V_0}{273} \times t$$

$$= V_0\left(1 + \frac{t}{273}\right)$$

$$= V_0\left(\frac{273+t}{273}\right) \qquad ...(1.1)$$

Thus at $t = -273°C$

$$V_t = V_0\left(\frac{273 - 273}{273}\right)$$

$$= 0$$

This means that at –273°C, the volume of the gas is reduced to zero *i.e.*, the gas ceases to exist; in fact, the gas gets liquefied. *This temperature (–273°C) at which the gas ceases to exist is called* **Absolute Zero**. It is represented by 0 K. Thus

$$-273°C = 0\ K$$

or

$$0°C = 273\ K$$

This means that to convert $t°$ C into K, we should add 273

$$\therefore \quad t°\ C = (t + 273)\ K = T\ K$$

Putting $273 + t°\ C = T\ K$ in equation (1.1), we get

$$V_t = V_0 \frac{T}{273}$$

This implies that

$V_t \propto T$ if pressure remains constant

or simply, $V \propto T$ if pressure remains constant.

Thus the Charles law may also be defined as follows:

Pressure remaining constant, the volume of a given mass of a gas is directly proportional to its absolute temperature.

Mathematically, the law may also be written as

$$\frac{V}{T} = \text{constant, if pressure is kept constant}$$

or

$$\frac{V_1}{T_1} = \frac{V_2}{T_2} \text{ under constant pressure}$$

where V_1 *is* the volume of the gas at pressure P_1 and V_2 is the volume at pressure P_2.

(Meerut 2000)

Q.3. State and explain Avogadro's law.

Ans. Avogadro's law may be defined as under:

Under similar conditions of temperature and pressure, equal volumes of all gases contain the same number of molecules.

In other words, if temperature and pressure are kept constant, the volume (V) of a gas is directly proportional to the number of molecules (N) present in the gas. Mathematically,

$$V \propto N \text{ (At constant } P \text{ and } T)$$

It has been found that one mole of every gas (*i.e.*, 22.4 litres at *N.T.P.*) contains 6.023×10^{23} molecules. This number is known as **Avogadro's Number** and is usually represented by N_0.

Obviously, the number of moles (n) of a gas may be calculated as follows:

$$n = \frac{N}{N_0}$$

As N_0 is a constant quantity, we can say that

$$n \propto N$$

Hence the equation, $V \propto N$, may also be written as:

$V \propto n$, where n is the number of moles of the gas.

Q. 4. How is combined gas equation obtained?

Ans. It is obtained by the combination of Boyle's law, Charles' law and Avogadro's law.
According to *Boyle's law,*

$$V \propto \frac{1}{P} \quad \text{if } T \text{ is kept constant} \qquad ...(1.2)$$

According to *Charles law,*

$$V \propto T \text{ if } P \text{ is kept constant} \qquad ...(1.3)$$

According to *Avogadro's law,*

$$V \propto n \text{ if } T \text{ and } P \text{ are kept constant} \qquad ...(1.4)$$

Combining equations (1·2), (1·3) and (1· 4), we get

$$V \propto \frac{1}{P} \times T \times n$$

or $$V = R \times \frac{1}{P} \times T \times n$$

where R is a constant of proportionality and is called *Gas Constant.*

or $$PV = nRT \text{ for } n \text{ moles of the gas.}$$

This equation is known as *Gas Equation.*

Q. 5. Give the postulates of kinetic theory of gases.

(G. Nanakdev 2000; Nagpur 2002; Lucknow 2002; Guwahati 2006; Vidyasagar 2007; Purvanchal 2007)

Ans. (*i*) Every gas is made up of a large number of extremely small particles called molecules. All the molecules of a particular gas are identical in mass and size and differ from gas to gas.

(*ii*) The molecules of a gas are separated from each other by large distances so that the actual volume of the molecules is negligible as compared to the total volume of the gas.

(*iii*) The distances of separation between the molecules are so large that the forces of attraction or repulsion between them are negligible.

(*iv*) The molecules are supposed to be moving continuously in different directions with different velocities. Hence, they keep on colliding with one another (called molecular collisions) as well as on the walls of the containing vessel.

(*v*) The pressure exerted on the walls of the containing vessel is due to the bombardment of the molecules on the walls of the containing vessel.

(*vi*) Since the molecules are moving with different velocities, they possess different kinetic energies. However, the average kinetic energy of the molecules of a gas is directly proportional to the absolute temperature of the gas.

(*vii*) Molecules are supposed to be perfectly elastic hard spheres so that no energy is lost when the molecules collide with one another or with the walls of the vessel.

(*viii*) The force of gravitation on the molecules is also supposed to be negligible.

Q. 6. Derive kinetic gas equation $PV = \frac{1}{3} mnu^2$.

(G. Nanakdev, 2000; Lucknow 2002; Garhwal 2010)

Ans. Consider a cubical vessel containing the gas (Fig. 1.1). Suppose,
Total number of molecules present in the gas = n,
Mass of each molecule of the gas = m,
Length of each side of the cube = l cm.
According to the kinetic theory of gases, the molecules of the gas are moving in different

directions with different velocities. Consider any one molecule, moving with a velocity ϑ_1 cm/sec in any direction. The velocity can be resolved into three mutually perpendicular components ϑ_x, ϑ_y and ϑ_z parallel to X-axis, Y-axis, and Z-axis respectively as shown in Fig. 1.1. Then mathematically,

$$\vartheta_1^2 = \vartheta_x^2 + \vartheta_y^2 + \vartheta_z^2 \qquad ...(i)$$

Now, consider the motion of the molecule due to component ϑ_x only. Due to this component, the molecule will strike the opposite faces *A* and *B* which are at right angles to the X-axis.

Momentum of molecule due to component ϑ_x just before it strikes the face *A*

$$= m\vartheta_x$$

As the molecules are supposed to be perfectly elastic, after colliding the face *A*, the molecule will return with the same velocity. As the direction has changed, the velocity will now be taken as $-\vartheta_x$. Hence

Momentum of the molecule due to component ϑ_x just after it strikes the face *A*

$$= m \times (-\vartheta_x) = -m\vartheta_x$$

∴ Change of the momentum of the molecule in *one collision* on the face *A* due to component ϑ_x

$$= m\vartheta_x - (-m\vartheta_x) = 2\,m\vartheta_x$$

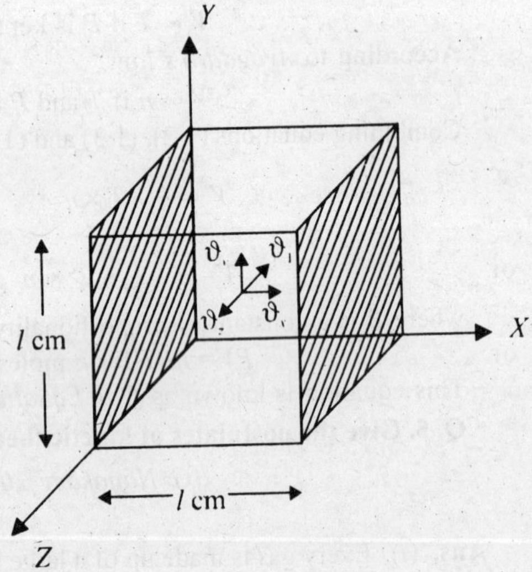

Fig. 1.1. A cubical vessel containing *n* molecules of the gas.

Now let us calculate the number of collisions which take place on the face *A* in one second due to component ϑ_x.

For next collision on the face *A*, the molecule has to travel from face *A* to *B* and then back from *B* to *A i.e.*, a total distance of

$$l + l = 2l \text{ cm}$$

As the component being considered is ϑ_x cm/sec, this means that the distance travelled by the molecule in one second is ϑ_x cm.

∴ Number of collisions which take place on the face A in one second due to component ϑ_x

$$= \frac{\vartheta_x}{2l}$$

As the change of momentum in one collision is $2m\vartheta_x$

∴ Change of momentum of the molecule in one second due to component ϑ_x for collisions on the face *A*

$$= 2m\vartheta_x \times \frac{\vartheta_x}{2l} = \frac{m\vartheta_x^2}{l}$$

Similarly, change of momentum of the molecule in one second due to component ϑ_x for collisions on the face *B*

$$= \frac{m\vartheta_x^2}{l}$$

∴ Change of momentum of the molecule in one second due to component ϑ_x for collisions on the opposite faces perpendicular to X-axis.

$$= \frac{m\vartheta_x^2}{l} + \frac{m\vartheta_x^2}{l} = \frac{2m\vartheta_x^2}{l}$$

Similarly, change of momentum of the molecule in one second due to component ϑ_y, for collisions on the opposite faces perpendicular to Y-axis and due to component ϑ_z for collisions on the opposite faces perpendicular to Z-axis will be respectively.

$$\frac{2m\vartheta_y^2}{l} \text{ and } \frac{2m\vartheta_z^2}{l}$$

∴ Change of momentum of the molecule in one second due to all the components for collisions on all the six faces

$$= \frac{2m\vartheta_x^2}{l} + \frac{2m\vartheta_y^2}{l} + \frac{2m\vartheta_z^2}{l}$$

$$= \frac{2m}{l}(\vartheta_x^2 + \vartheta_y^2 + \vartheta_z^2) = \frac{2m\vartheta_1^2}{l} \qquad \text{from eqn. } (i)$$

Similarly, the change of momentum in one second, for molecules having velocities ϑ_2, ϑ_3, ϑ_4, ...ϑ_n will be respectively.

$$\frac{2m\vartheta_2^2}{l}, \frac{2m\vartheta_3^2}{l}, \frac{2m\vartheta_4^2}{l} \ldots\ldots \frac{2m\vartheta_n^2}{l}$$

∴ Total change of momentum in one second for all n molecules for collisions on all the six faces will be

$$= \frac{2m\vartheta_1^2}{l} + \frac{2m\vartheta_2^2}{l} + \frac{2m\vartheta_3^2}{l} + \ldots + \frac{2m\vartheta_n^2}{l}$$

$$= \frac{2m}{l}(\vartheta_1^2 + \vartheta_2^2 + \vartheta_3^2 + \ldots\ldots + \vartheta_n^2)$$

$$= \frac{2m}{l} \times n \times \frac{\vartheta_1^2 + \vartheta_2^2 + \vartheta_3^2 + \ldots\ldots + \vartheta_n^2}{n} = \frac{2mn}{l} \times u^2$$

(where $u = \sqrt{\frac{\vartheta_1^2 + \vartheta_2^2 + \vartheta_3^2 + \ldots\ldots + \vartheta_n^2}{n}}$ is called root mean square velocity.)

But the change of momentum in one second according to Newton's second law is equal to the force. Hence, the force (F) exerted by all the 'n' molecules on all the six faces is

$$F = \frac{2mnu^2}{l}$$

As Pressure = $\frac{\text{Force}}{\text{Area}}$

and area of all the faces = $6\,l^2$

∴ Pressure (P) exerted on the walls of the cubical vessel will be

$$P = \frac{2mnu^2}{l} \times \frac{l}{6l^2} = \frac{1}{3}\frac{mnu^2}{l^3}$$

But l^3 is the volume of the cubical vessel and hence is equal to the volume (V) of the gas

$$\therefore \qquad P = \frac{1}{3}\frac{mnu^2}{V}$$

or $$PV = \frac{1}{3} mnu^2$$

This is called *kinetic gas equation.*

Q. 7. Deduce (*i*) Boyle's law (*ii*) Charles law (*iii*) Avogadro's law from kinetic gas equation.

(G. Nanakdev, 2000; Kumaon, 2000; Punjabi, 2002; Arunachal 2002; Nagpur 2003)

Ans. (*i*) Boyle's law

According to kinetic gas equation

$$PV = \frac{1}{3} mnu^2 = \frac{2}{3} \cdot \frac{1}{2} mnu^2 = \frac{2}{3} \cdot \frac{1}{2} Mu^2$$

(where $m \times n = M$ is the total mass of the gas)

But $$\frac{1}{2} Mu^2 = \text{Kinetic Energy of the gas}$$

$$\therefore \quad PV = \frac{2}{3} \text{ K.E.} \qquad ...(i)$$

Further, according to one of the postulates of the kinetic theory of gases,

$$\text{K.E.} \propto \text{Absolute Temperature } (T)$$

or $$\text{K.E.} = kT$$

where k is a constant of proportionality.

Putting this value in equation (*i*), we get

$$PV = \frac{2}{3} kT \qquad ...(ii)$$

As $\frac{2}{3}$ is a constant quantity, k is also a constant, therefore, if T is kept constant, $\frac{2}{3} kT$ will be constant. Hence

$$PV = \text{constant, if } T \text{ is kept constant.}$$

Boyle's law is thus deduced.

(*ii*) **Charles Law**

Consider eqn. (*ii*) deduced under Boyle's law

$$PV = \frac{2}{3} kT$$

This may be rewritten as

$$\frac{V}{T} = \frac{2}{3} \frac{k}{P}$$

Now $\frac{2}{3}$ is a constant quantity, k is also constant, therefore, if P is kept constant, the whole quantity on the right hand side becomes constant. Hence

$$\frac{V}{T} = \text{constant, if } P \text{ is kept constant}$$

Charles law is thus deduced.

(*iii*) **Avogadro's Law**

Consider any two gases. According to kinetic gas equation,

For 1st gas $$P_1 V_1 = \frac{1}{3} m_1 n_1 u_1^2 \qquad ...(i)$$

For 2nd gas $$P_2 V_2 = \frac{1}{3} m_2 n_2 u_2^2 \quad ...(ii)$$

where P_1 = Pressure of the 1st gas
V_1 = Volume of the 1st gas
m_1 = Mass of each molecule of the 1st gas
n_1 = Total number of molecules of the 1st gas
u_1 = Root mean square velocity of the molecules of 1st gas.

P_2, V_2, m_2 and u_2 are the corresponding values for the 2nd gas.

If conditions of pressure are similar for both the gases, then

$$P_1 = P_2 \quad ...(iii)$$

If equal volumes of both the gases are taken, then

$$V_1 = V_2 \quad ...(iv)$$

Combining equations (*iii*) and (*iv*), we get

$$P_1 V_1 = P_2 V_2 \quad ...(v)$$

Thus left hand sides of equations (*i*) and (*ii*) are equal. Hence right hand sides must also be equal, *i.e.*,

$$\frac{1}{3} m_1 n_1 u_1^2 = \frac{1}{3} m_2 n_2 u_2^2$$

or $$m_1 n_1 u_1^2 = m_2 n_2 u_2^2 \quad ...(vi)$$

Further, according to one of the postulates of the kinetic theory of gases,

Average K.E. of a gas ∝ absolute temperature.

Since conditions of temperature are similar in both the cases,

∴ Average K.E. of 1st gas = Average K.E. of 2nd gas

i.e., $$\frac{1}{2} m_1 u_1^2 = \frac{1}{2} m_2 u_2^2$$

or $$m_1 u_1^2 = m_2 u_2^2 \quad ...(vii)$$

Dividing equation (*vi*) by equation (*vii*), we get

$$n_1 = n_2$$

Thus Avogadro's law is proved.

Q. 8. Deduce (*i*) Ideal gas equation (*ii*) Graham's law of diffusion (*iii*) Dalton's law of partial pressure from kinetic gas equation. *(Guwahati 2006; Nagpur, 2003, 2008)*

Ans. (*i*) Ideal gas equation

According to kinetic gas equation

$$PV = \frac{1}{3} mnu^2 = \frac{2}{3} \cdot \frac{1}{2} mnu^2$$
$$= \frac{2}{3} \text{K.E.} \left(\because \frac{1}{2} mnu^2 = \text{K.E. of the gas} \right)$$

But K.E. ∝ T (one of the postulates of Kinetic theory)

or K.E. = kT

where k is constant of proportionality.

$$\therefore \quad PV = \frac{2}{3}kT$$

$\frac{2}{3}$ is a constant quantity, k is also constant

$\therefore \ \frac{2}{3}k = R$, another constant, called gas constant

Hence $\quad PV = RT$

which is the equation of state for 1 mol of the ideal gas.

(*ii*) **Graham's Law of Diffusion**

According to kinetic gas equation

$$PV = \frac{1}{3}mnu^2 = \frac{1}{3}Mu^2 \qquad (\therefore\ mn = M, \text{ total mass of the gas})$$

or $$u^2 = \frac{3PV}{M} = \frac{3P}{M/V} = \frac{3P}{d} \qquad \left(\because \frac{M}{V} = d \text{ is the density of the gas}\right)$$

or $$u = \sqrt{\frac{3P}{d}} \qquad ...(i)$$

But greater the velocity (u) of the molecules, greater is the rate of diffusion (r) of the gas

i.e., $$r \propto u \qquad ...(ii)$$

Combining equations (*i*) and (*ii*), we get

$$r \propto \sqrt{\frac{3P}{d}}$$

Hence, if pressure (P) is kept constant, then

$$r \propto \sqrt{\frac{1}{d}}$$

Thus, Graham's law follows from the kinetic gas eqation.

(*iii*) **Dalton's Law of Partial Pressures**

Let us consider two gases. According to kinetic gas equation,

$$PV = \frac{1}{3}mnu^2$$

or $$P = \frac{1}{3}\frac{mnu^2}{V}$$

Now, if only the first gas is enclosed in the vessel of volume V, the pressure exerted would be

$$p_1 = \frac{1}{3}\frac{m_1 n_1 u_1^2}{V}$$

Again,if only the second gas is enclosed in the same vessel (so that V *is* onstant), then the pressure exerted would be

$$p_2 = \frac{1}{3}\frac{m_2 n_2 u_2^2}{V}$$

Lastly, if both the gases are enclosed together in the same vessel then since the gases do not react with each other, their molecules behave independent of each other. Hence, the total pressure exerted would be

$$P = \frac{1}{3}\frac{m_1 n_1 u_1^2}{V} + \frac{1}{3}\frac{m_2 n_2 u_2^2}{V}$$

$$= P_1 + P_2$$

Similarly, if more than two gases are present, then it can be proved that

$$P = p_1 + p_2 + p_3 + \ldots$$

Q. 9. Starting from kinetic theory of gases (*i*) Derive relations between molecular velocity, pressure, temperature, volume and density of the gas (*ii*) Derive the relation

$$\textbf{K.E.} = \frac{3}{2}\textbf{RT}$$ *(Arunachal 2003)*

Ans. (*i*) Calculation of Molecular Velocities. According to kinetic gas equation

$$PV = \frac{1}{3} mnu^2$$

$$u^2 = \frac{3PV}{mn} = \frac{3PV}{M} \qquad (\because mn = M \text{ is the mass of the gas})$$

If 1 mol of the gas is taken, M will be equal to the molecular weight of the gas,

$$u = \sqrt{\frac{3PV}{M}} \qquad \ldots(i)$$

Further, for 1 mole of the gas

$$PV = RT$$

$$\therefore \quad u = \sqrt{\frac{3RT}{M}} \qquad \ldots(ii)$$

Equation (*i*) can be rewritten as

$$u = \sqrt{\frac{3P}{M/V}}$$

But $\frac{M}{V} = d$ is the density of the gas

$$u = \sqrt{\frac{3P}{d}}$$

(*ii*) **Calculation of the Kinetic Energy of the Molecules.** For 1 mol of the gas containing n molecules, each of the mass m, the kinetic energy will be

$$\text{K.E.} = \frac{1}{2} mnu^2$$

According to kinetic gas equation

$$PV = \frac{1}{3} mnu^2$$

or $$mnu^2 = 3\,PV$$

$$\therefore \quad \frac{1}{2} mnu^2 = 3/2\,PV$$

$$\text{K.E.} = 3/2\,PV$$

Further for 1 mol of the ideal gas,

$$PV = RT$$

$$\therefore \quad \boxed{\text{K.E.} = 3/2\,RT}$$

Q. 10. Give an account of Maxwell's distribution of velocities. Explain graphically how the velocities change with temperature. *(Calcutta 2000; Lucknow 2000; Awadh 2000; G. Nanakdev, 2003; Nagpur, 2003, Raipur 2010)*

Ans. Molecules of a gas are moving continuously in different directions with different velocities. Hence they keep on colliding with one another. However, the collisions are supposed to be perfectly elastic and there is no net loss of kinetic energy. In other words, whereas some molecules are speeded up, the others are slowed down after collision. Maxwell suggested that at a particular temperature, the fractions of molecules possessing particular velocities remain almost constant. Such a state is called a *steady state.* On the basis of the laws of probability, he calculated that the fraction of molecules having velocities between c and $c + dc$ is given by

$$\frac{dn_c}{n} = 4\pi \left(\frac{M}{2\pi RT} \right)^{3/2} e^{-Mc^2/2RT} c^2 \, dc \qquad ...(i)$$

where dn_c is the number of molecules having velocities between c and $c + dc$, out of a total of n molecules of the gas and M and T are the molecular weight and absolute temperature of the gas respectively.

The above eqn. may be rewritten as

$$\frac{1}{n} \cdot \frac{dn_c}{dc} = 4\pi \left(\frac{M}{2\pi RT} \right)^{3/2} e^{-Mc^2/2RT} c^2 \qquad ...(ii)$$

The quantity $\frac{1}{n} \cdot \frac{dn_c}{dc}$ is called the probability (p) of finding molecules with velocity c.

The significance of the above equation may be seen by plotting the probability $\left(p = \frac{1}{n} \frac{dn_c}{dc} \right)$ versus the velocity c at a particular temperature T. Such a plot is known as Maxwell's distribution of velocities and is shown in Fig. 1.2. α represents the most probable velocity.

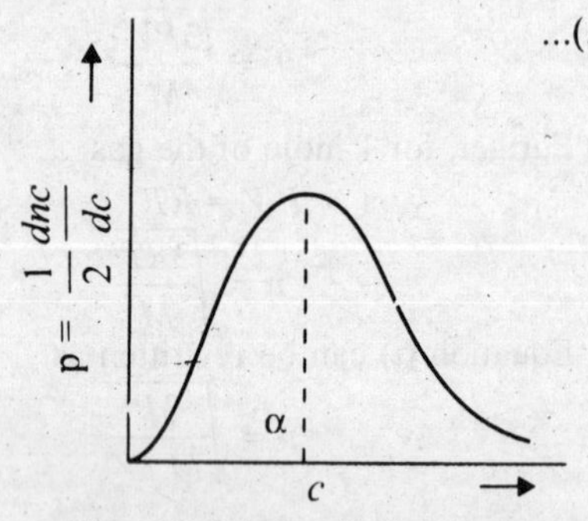

Fig. 1.2. Maxwell distribution of velocities.

The area between two ordinates separated by dc is equal to dn_c/n *i.e.*, the fraction of the total number of molecules possessing velocities between c and $c + dc$ (Fig.1.3) Hence the area under the whole curve is equal to the total number of molecules and the height of the ordinate

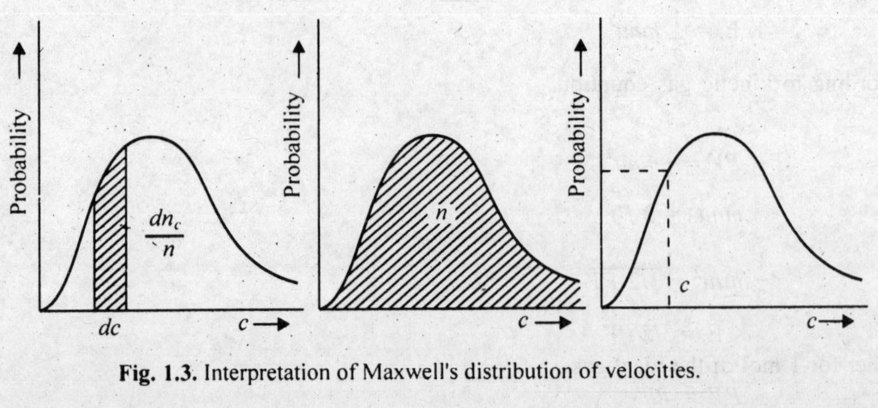

Fig. 1.3. Interpretation of Maxwell's distribution of velocities.

corresponding to any velocity is a measure of the fraction of the molecules which have that velocity. From the Maxwell's distribution of velocities as given in Fig. 1.2, it may be observed that the fraction of molecules having zero velocity *i.e.*, of molecules which are motionless is zero and the fractions of molecules having velocities greater than zero increase with increase of velocities, reach a maximum value corresponding to a particular velocity and then begin to decrease. Thus it may be concluded that

(*i*) The fractions of molecules having very low velocities or very high velocities are very small.

(*ii*) There is a velocity (α) corresponding to which the fraction of molecules is maximum.

In other words, α is the velocity possessed by the maximum fraction of the molecules. *This velocity* is called the *most probable velocity.* This *most probable velocity may be defined as the velocity possessed by the maximum fraction of the molecules of the gas at a particular temperature.*

Effect of Temperature on Maxwell's Distribution of Velocities. With increase of temperature, the curve shifts as shown in Fig. 1.4. Two important changes may be noticed:

(*i*) The peak of the curve shifts forward.

(*ii*) The peak of the curve shifts downward and is flattened.

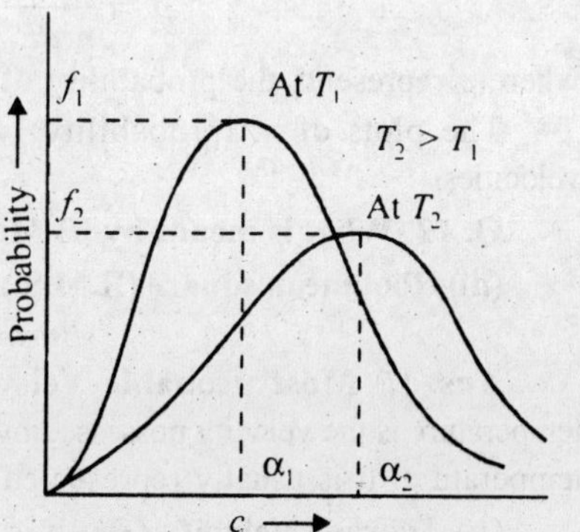

Fig. 1.4. Distribution of velocities at two different temperatures.

Following two important conclusions may be drawn from the above observations.

(*i*) With increase of temperature, the value of the most probable velocity increases.

(*ii*) The fraction of molecules possessing the most probable velocity decreases with increase of temperature ($f_2 < f_1$).

Q. 11. Derive mathematical equation to represent Maxwell's distribution of energies.

(G. Nanakdev 2000; Vidyasagar 2002; Aligarh 2002)

Ans. Eqn. (*i*) and (*ii*) in Q. 10 representing the distribution of velocities can be converted to represent the distribution of energies as follows:

$\frac{1}{2}Mc^2$ can be replaced by *E*, the kinetic energy of the gas per mole

$$\frac{1}{2}M\bar{c^2} = E \qquad ...(iii)$$

To change $c^2\,dc$ into the energy terms, Eqn. (*iii*) can be rewritten as

$$Mc^2 = 2E \qquad ...(iv)$$

Differentiating both sides, we get

$$2Mc\,dc = 2dE$$

or

$$Mc\,dc = dE \qquad ...(v)$$

Taking square roots of **both** sides of Eqn. (*iv*), we get

$$M^{1/2}c = (2E)^{1/2} \qquad ...(vi)$$

Multiplying Eqns. (*v*) and (*vi*), we get

$$M^{3/2}c^2\,dc = (2E)^{1/2}\,dE$$

$$c^2 dc = \frac{(2E)^{1/2}\,dE}{M^{3/2}} \qquad ...(vii)$$

Substituting the value of $\frac{1}{2}Mc^2$ from Eqn. (*iii*) and that of $c^2\, dc$ from Eqn. (*vii*) in Eqn. (*i*), we get

$$\frac{dn_c}{n} = 4\pi\left(\frac{M}{2\pi RT}\right)^{3/2} e^{-E/RT} \frac{(2E)^{1/2}\, dE}{M^{3/2}}$$

or
$$\frac{dn_c}{n} = \frac{2\pi}{(\pi RT)^{3/2}} e^{-E/RT} E^{1/2} dE \quad \text{... (viii)}$$

where dn_c/n now represents the fraction of the total number of molecules having kinetic energies between E and dE.

Eqn. (*viii*) can be rewritten as

$$p' = \frac{1}{n}\frac{dn_c}{dE} = \frac{2\pi}{(\pi RT)^{3/2}} e^{E/RT} E^{1/2} \quad \text{...(ix)}$$

where p' represents the probability of molecules having energy E.

The plots of p' (probability) vs E obtained are similar to those for the distribution of velocities.

Q. 12. What is meant by (*i*) Most probable velocity (α) (*ii*) Average velocity (ϑ) (iii) Root mean square (R.M.S.) velocity (u)? *(Sambalpur 2003; Madras, 2004; Burdwan 2004; Calcutta 2007; Nagpur 2008; Himachal 2010)*

Ans. (*i*) **Most probable Velocity (α).** Most probable velocity of a gas at a particular temperature is the velocity possessed by the maximum fraction of the total number of molecules at that temperature. It is usually represented by '**α**'.

(*ii*) **Average Velocity (ϑ).** It is defined as the arithmetic mean of the velocities possessed by the different molecules of a gas at a particular temperature. It is usually represented by ϑ. It is expressed in two different ways.

(*a*) If all the n molecules of the gas are supposed to have different velocities, say $\vartheta_1, \vartheta_2, \vartheta_3 ...... \vartheta_n$ cm /sec, then

$$\vartheta = \frac{\vartheta_1, \vartheta_2, \vartheta_3 ...... \vartheta_n}{n}$$

(*b*) If n_1 molecules of the gas have velocity ϑ_1 cm/sec, n_2 have velocity ϑ_2 cm/sec, n_3 have velocity ϑ_3 cm/sec and so on, then

$$\vartheta = \frac{n_1\vartheta_1 + n_2\vartheta_2 + n_3\vartheta_3 + ...}{n_1 + n_2 + n_3 + ...(= n)}$$

(*iii*) **Root Mean Square Velocity (*u*).** It is defined as the square root of the mean of the squares of the different velocities. It is usually represented by '*u*'. Like the average velocity, it can be expressed in two different ways.

(*a*) Assuming all the n molecules to have different velocities *viz.* $\vartheta_1, \vartheta_2, ... \vartheta_n$ cm/sec, we have

$$u = \sqrt{\frac{\vartheta_1^2 + \vartheta_2^2 + \vartheta_3^2 + ... + \vartheta_n^2}{n}}$$

(*b*) Assuming n_1 molecules have velocity ϑ_1, n_2 molecules have velocity ϑ_2, n_3, have ϑ_3 and so on, we have

$$u = \sqrt{\frac{n_1\vartheta_1^2 + n_2\vartheta_2^2 + n_3\vartheta_3^2 + ..}{n_1 + n_2 + n_3 .....(= n)}}$$

Q. 13. Derive expression for (*a*) most probable velocity (α) (*b*) average velocity (ϑ) (*c*) root mean square velocity (*u*). from Maxwell distribution of velocities. *(Lucknow 2000)*

Ans. (*a*) **Most Probable Velocity** (α). Maxwell's distribution law equation can be treated mathematically to get the condition for the maximum in the probability distribution curve. The condition obtained is

$$c = \sqrt{\frac{2RT}{M}}$$

Representing the most probable velocity by a in place of *c*, we have

$$\alpha = \sqrt{\frac{2RT}{M}} \qquad ...(i)$$

This result is obtained by differentiating right hand side of Eqn. (*ii*) (See Q. 10) w.r.t. *c* and equating the result to zero *i.e.*, we get

$$\left(1 - \frac{Mc^2}{2RT}\right) ce^{-Mc^2/2RT} = 0$$

Taking $1 - \frac{Mc^2}{2RT} = 0$ we get $\frac{Mc^2}{2RT} = 1$

$$c = \sqrt{\frac{2RT}{M}}$$

(*b*) **Average Velocity** (ϑ). By using Maxwell's distribution law equation, it can be derived mathematically that

$$\vartheta = \sqrt{\frac{8RT}{\pi M}} \qquad ...(ii)$$

[This result is obtained by making use of the fact that

$$\vartheta = \frac{\int_0^\infty c\,dn_c}{n}$$

and then substituting the value of dn_c from Eqn. (*i*) and solving it].

(*c*) **Root Mean Square Velocity** (*u*). From the kinetic gas equation,

$$PV = \frac{1}{3} mNu^2 = \frac{1}{3} Mu^2 \qquad (mN = M \text{ gram molecular mass})$$

or $$u^2 = \frac{3PV}{M} = \frac{3RT}{M} \qquad (\because PV = RT \text{ for 1 mol of an ideal gas})$$

$$\therefore \quad u = \sqrt{\frac{3RT}{M}} \qquad ...(iii)$$

Q. 14.Derive relationships between (*i*) Most probable velocity (*ii*) Average velocity (*iii*) Root mean square velocity. *(M. Dayanand 2000; Kurukshetra 2002)*

Ans. Expressions for α, ϑ and *u* are

$$(i)\ \alpha = \sqrt{\frac{2RT}{M}},\ (ii)\ \vartheta = \sqrt{\frac{8RT}{\pi M}},\quad (iii)\ u = \sqrt{\frac{3RT}{M}}$$

Combining (*i*) with (*iii*), we get $\alpha = \sqrt{\frac{2}{3}}\, u$

i.e., $\alpha = 0.816\,u$ ...(*i*)

Combining (*ii*) with (*iii*), we get

$$\vartheta = \sqrt{\frac{8}{3\pi}}\,u$$

i.e., $\vartheta = 0.921\,u$ ...(*ii*)

Alternatively, the relationship between α, ϑ and u will be

$$\alpha : \vartheta : u = \sqrt{\frac{2RT}{M}} : \sqrt{\frac{8RT}{\pi M}} : \sqrt{\frac{3RT}{M}}$$

$$= \sqrt{2} : \sqrt{\frac{8}{\pi}} : \sqrt{3}$$

$$= 1{\cdot}414 : 1{\cdot}595 : 1{\cdot}732$$

or $\alpha : \vartheta : u = 1 : 1{\cdot}128 : 1{\cdot}224$ ...(*iii*)

Q. 15. Calculate the temperature at which the hydrogen molecules will have an average speed of 176400 cm s^{-1}. **(*Kurukshetra, 2001*)**

Ans. We are given that, ϑ = 176400 cm s^{-1}

Applying the formula $\vartheta = \sqrt{\frac{8RT}{\pi M}}$

Putting $R = 8{\cdot}314 \times 10^7$ ergs/degree/mol

$M = 2$ (for H_2)

$$176400 = \sqrt{\frac{8 \times 8.314 \times 10^7 \times T}{22/7 \times 2}}$$

or $$T = \frac{176400^2 \times 22 \times 2}{7 \times 8 \times 8.314 \times 10^7} = 300\text{ K or }27°\text{C}$$

Q. 16. Calculate the root mean square velocity, average velocity and most probable velocity of sulphur dioxide molecules at 427°C.

Ans. (*a*) **Calculation of root mean square velocity**

$$u = \sqrt{\frac{3RT}{M}}$$

Here, T = 427°C = 427+ 273 K = 700 K

M = Mol. wt. of SO_2 = 32 + 32 = 64

$R = 8.314 \times 10^7$ ergs/degree/mol (C.G.S. Units)

Putting the values in the above equation, we get

$$u = \sqrt{\frac{3 \times 8{\cdot}314 \times 10^7 \times 700}{64}}$$

$$= \sqrt{\frac{3 \times 8{\cdot}314 \times 10^8 \times 70}{64}}$$

$$= \sqrt{\frac{3 \times 8{\cdot}314 \times 70}{64}} \times 10^4 = 5.224 \times 10^4\text{ cm s}^{-1}$$

= 52240 cm s^{-1}

(*b*) *Calculation of average velocity* (ϑ)

$\vartheta = 0{\cdot}9213 \times u = 0{\cdot}9213 \times 5.2240 = \mathbf{48128\ cm\ s^{-1}}$

(*c*) *Calculation of most probable velocity* (α)

$= 0{\cdot}816 \times u = 0.816 \times 52240 \text{ cm s}^{-1} = \mathbf{42628\ cm\ s^{-1}}$

Q. 17. At what temperature the root mean square velocity of chlorine gas will be equal to that of SO_2 at N.T.P. ?

Ans. The root mean square velocity is given by the equation

$$u = \sqrt{\frac{3RT}{M}}$$

For SO_2 $T = 273$ K

$M = 64$

$$\therefore \quad u_{SO_2} = \sqrt{\frac{3R \times 273}{64}}$$

For Cl_2 $T = ?$

$M = 71$

$$\therefore \quad u_{Cl_2} = \sqrt{\frac{3R \times T}{71}}$$

But $u_{SO_2} = u_{Cl_2}$

$$\therefore \quad \sqrt{\frac{3R \times T}{71}} = \sqrt{\frac{3R \times 273}{64}}$$

or
$$\frac{3R \times T}{71} = \frac{3R \times 273}{64}$$

or
$$T = \frac{273 \times 71}{64} = 302{\cdot}9 \text{ K}$$

$$= 302{\cdot}9 - 273 = \mathbf{29{\cdot}0^\circ\ C}$$

Q. 18. Oxygen has a density of 1.429 g per litre at N.T.P. Calculate the R.M.S. and average velocity of oxygen molecules. ***(Delhi 2003; Himachal 2010)***

Ans. (*a*) *Calculation of R.M.S. velocity*

R.M.S. velocity (u) is given by the formula

$$u = \sqrt{\frac{3P}{d}}$$

Here $P = 1$ atmosphere $= 76 \times 13{\cdot}6 \times 981$ dynes/sq cm

$$d = 1{\cdot}429\ gL^{-1} = \frac{1{\cdot}429}{1000} \text{ g cm}^{-3} \quad \text{(C.G.S. Units)}$$

$= 0.001429 \text{ g cm}^{-3}$

Putting these values in the above formula, we get

$$u = \sqrt{\frac{3 \times 76 \times 13.6 \times 981}{0.001429}}$$

$$= \sqrt{\frac{30.42}{1{\cdot}429} \times 10^8}$$

$$= 4{\cdot}61 \times 10^4 \text{ cm s}^{-1} = \mathbf{46100\ cm\ s^{-1}}$$

(*b*) *Calculation of Average Velocity* (ϑ)

$$v = 0{\cdot}9213 \times u = 0{\cdot}9213 \times 46100 = \mathbf{42461\ cm\ s^{-1}}$$

Q. 19. Calculate the root mean square velocity of chlorine molecules at 17°C and 800 mm pressure.

Ans. This numerical problem involves two steps as follows:

Step 1. *To calculate the volume under given conditions* (for one mole of gas) one mole of a gas at N.T.P. occupies 22400 cm^3.

Applying the equation

$$\frac{P_1V_1}{T_1} = \frac{P_2V_2}{T_2}$$

$$\frac{760\times22400}{273} = \frac{800\times V_2}{290}$$

$$\therefore V_2 = \frac{760\times22400\times290}{273\times800} = 22600\ \text{cm}^3$$

Step 2. *To calculate the R.M.S. velocity using the equation*

$$u = \sqrt{\frac{3PV}{M}}$$

Here, P = 800 mm = 80 cm = 80 × 13·6 × 981 dynes per sq cm (C.G.S. Units)

V = 22600 cm^3

M = 71 (for Cl_2)

$$u = \sqrt{\frac{3\times80\times13{\cdot}6\times981\times22600}{71}}$$

$$= \sqrt{10{\cdot}18\times10^8} = \mathbf{3{\cdot}19 \times 10^4\ cm\ s^{-1}}$$

Q. 20. Calculate translational K.E. of 2 moles of a gas at 27°C

Ans. For 1 mol of a gas, K.E. $= \frac{3}{2}RT$

$\therefore$ For 2 moles of the gas,

$$\text{K.E.} = \left(\frac{3}{2}RT\right)\times2$$

Putting R = 8·314 × 10^7 ergs per degree per mol

and T = 27°C = (27 + 273) K = 300 K

we get K.E. = 3 × 8·314 × 10^7 × 300 ergs (C.G.S. Units)

= 7482·5 × 10^7 ergs = **7482·6 joules.**

Q. 21. Calculate the root mean square velocity of nitrogen at N.T.P. (R = 8·314 × 10^7 ergs/degree/mol). ***(Vidyasagar 2002)***

Ans. For N_2 at N.T.P.

P = 760 mm = 76 cm = 76 × 13·6 × 981 dynes cm^{-2} (C.G.S Units)

V = 22400 cm^3

M = 28

Putting these values in the equation

$$u = \sqrt{\frac{3PV}{M}}$$

we get $$u = \sqrt{\frac{3 \times 76 \times 13.6 \times 981 \times 22400}{28}}$$

$$= \mathbf{4{\cdot}92 \times 10^4\ cm\ s^{-1}}$$

Q. 22. What is meant by degree of freedom of a molecule? How is this classified into different types? *(G. Nanakdev 2000; Kumaon, 2000; Indore, 2001; M. Dayananad 2002. Mizoram 2003; Punjabi 2002; Nagpur 2008)*

Ans. The number of coordinates required to specify the position of all the mass points *i.e.*, atoms in a molecule is called the number of **degrees of freedom.**

To represent one atom, we need three coordinates.

Thus a molecule made up of N atoms has 3 N degrees of freedom..

According to kinetic theory of matter, when thermal energy is absorbed by a molecule, it is stored within the molecule in the form of:

(*i*) translational motion of the molecule

(*ii*) internal movement of the atoms within the molecule, *i.e.*, rotational motion and vibrational motion.

The translational motion of the molecule means the motion of the centre of mass of the molecule as a whole. The centre of mass of the molecule can be represented by three coordinates. Hence there are ***three translational degrees of freedom.*** **The remaining (3*N*– 3) coordinates** represent the *internal degree of freedom*.

However, as already mentioned, the internal degrees of freedom may be subdivided into

(*i*) rotational degrees of freedom, and

(*ii*) vibrational degrees of freedom.

For a rotational motion, there are two degrees of freedom for a linear molecule and three for a non-linear molecule as shown in Fig. 1.5.

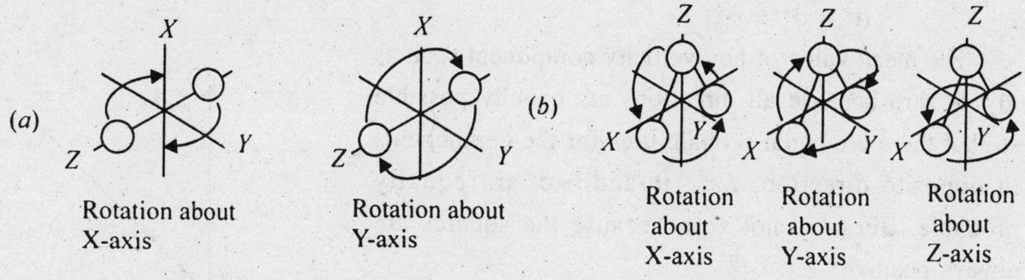

Fig. 1-5. (*a*) Rotations of a linear molecule about two mutually perpendicular axes (*b*) Rotations of a non-linear molecule about three mutually perpendicular axes.

Remaining degrees of freedom for the linear molecules are equal to (3*N*– 5) and for the non-linear molecules they are equal to (3*N*– 6). These describe the motion of the nuclei with respect to one another and thus represent the number of *vibrational degrees of freedom*. Hence,

Vibrational degrees of freedom of a linear molecule containing N atoms = $3N - 5$.

Vibrational degrees of freedom of a non-linear molecule containing N atoms = $3N - 6$.

The following examples illustrate calculation of vibrational degree of freedom of molecules:

(*i*) As a diatomic molecule is always linear, its vibrational degrees of freedom $= 3N - 5 = 3 \times 2 - 5 = 1$.

(*ii*) For the linear polyatomic molecule, $HC \equiv CH$, (N= 4), the vibrational degrees of freedom $= 3N - 5 = 3 \times 4 - 5 = 7$.

(*iii*) For the non-linear polyatomic molecule, HCHO, (N = 4) the vibrational degrees of freedom $= 3N - 6 = 3 \times 4 - 6 = 6$.

It may be summarised that for a molecule consisting of N atoms

	Translational	*Rotational*	*Vibrational*
For linear molecule :	3	2	$3N-5$
For non-linear molecule:	3	3	$3N-6$

Thus the total degrees of freedom in each case = $3N$

Q. 52. Discuss the principle of equipartition of energy.

(Vidyasagar 2002; Delhi 1997; M. Dayanand 2000; Garhwal 2000; Kumaon 2000; Calcutta 2007; G. Nanakdev 2003; Calcutta 2007)

Ans. According to kinetic theory of gases, the molecules of a gas are moving in different directions with different velocities. Consider any one gas molecule moving in any direction with velocity ϑ. The velocity ϑ can be resolved into three components ϑ_x, ϑ_y, and ϑ_z along the three mutually perpendicular axes (Fig. 1.6) such that

$$\vartheta^2 = \vartheta_x^2 + \vartheta_y^2 + \vartheta_z^2 \quad \text{... (i)}$$

Multiplying both sides of the equation by $\frac{1}{2}m$, we get

$$\frac{1}{2}m\vartheta^2 = \frac{1}{2}m\vartheta_x^2 + \frac{1}{2}m\vartheta_y^2 + \frac{1}{2}m\vartheta_z^2 \quad \text{...(ii)}$$

i.e., K.E. of the gas molecule $= (\text{K.E.})_x + (\text{K.E.})_y + (\text{K.E})_z$ ...(*iii*)

Considering all the molecules of the gas and taking their average value, we can say,

$$\overline{\text{K.E.}} = (\overline{\text{K.E.}})_x + (\overline{\text{K.E.}})_y + (\overline{\text{K.E.}})_z \quad \text{...(iv)}$$

Since the molecules of the gas move in a random manner and do not have any preference for a particular direction, therefore, it follows that

$$\overline{\vartheta_x^2} = \overline{\vartheta_y^2} = \overline{\vartheta_z^2} \quad \text{...(v)}$$

The mean value of any velocity component such as $\overline{\vartheta_x}$ is zero because all directions are equally possible so that there are equal probabilities for the components in opposite direction, *i.e.*, ϑ_x and $-\vartheta_x$ are equally probable. But $\overline{\vartheta_x^2}$ is not zero because the squares are always positive.

Multiplying throughout by $\frac{1}{2}m$, we get

$$\frac{1}{2}m\overline{\vartheta_x^2} = \frac{1}{2}m\overline{\vartheta_y^2} = \frac{1}{2}m\overline{\vartheta_x^2} \quad \text{...(vi)}$$

i.e $(\text{K.E.})_x = (\text{K.E})_y = (\text{K.E.})_z$ ...(*vii*)

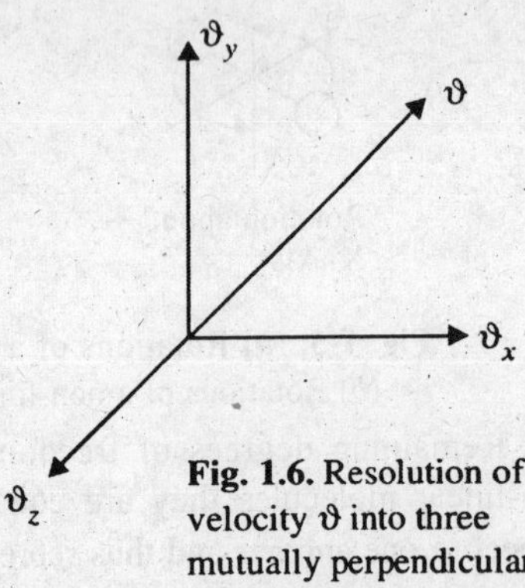

Fig. 1.6. Resolution of a velocity ϑ into three mutually perpendicular components.

Combining eqn. (*v*) and (*vii*), it becomes clear that the average translational kinetic energy of a molecule is the sum of their components along the three mutually perpendicular axes and these three components are equal to one another.

From kinetic theory of gases, as the total is translational kinetic energy of a molecule is $\frac{3}{2}kT$, therefore, each component will be $\frac{1}{2}kT$ where k is Boltzmann constant and T is the absolute temperature.

Now let us consider the rotational and vibrational energies associated with the gas molecules. Just as translational kinetic energy of a molecule is the sum of the three squared terms, each squared term contributes $\frac{1}{2}kT$ to the total energy of the molecules, the rotational energy is kinetic in nature and is represented by one squared term ($1/2I\omega^2$ where, I is the moment of inertia and ω is the angular velocity of rotation). Each type of rotation (*i.e.*, each squared term) contributes $\frac{1}{2}kT$ to the energy of the molecule. However, vibrational energy is partly kinetic and partly potential and is represented by the sum of two squared terms. Each squared term contributes $\frac{1}{2}kT$ to the energy of the molecule so that each mode of vibration contributes kT to the energy of the molecule.

The above discussion leads to the following generalization known as the "principle of equipartition of energy".

The total energy of a molecule is divided equally among the different degrees of freedom, the amount of energy associated with each degree of freedom being equal to $\frac{1}{2}kT$ *where the term 'degree of freedom' refers to the number of independent squared terms (i.e., terms involving the square of a coordinate or momentum/velocity) required to express the total energy of the molecule.*

It may be summarised as :

Energy contributed by each translational degree of freedom = $\frac{1}{2}kT$.

Energy contributed by each rotational degree of freedom = $\frac{1}{2}kT$.

Energy contributed by each vibrational degree of freedom (or vibrational mode) = kT.

Q. 24. Calculate the average energy of the molecule of C_2H_6 as well as the molar energy at 300 K as predicted by the law of equipartition of energy.

Ans. As C_2H_6 is a non-linear molecule, and $N = 8$

$\therefore$ Translational degrees of freedom = 3

Rotational degrees of freedom = 3

Vibrational degrees of freedom = $(3N - 6) = 18$

$$\therefore \quad \bar{\varepsilon} = \bar{\varepsilon}_t + \bar{\varepsilon}_r + \bar{\varepsilon}_\vartheta$$

$$= \left(3\times\frac{1}{2}kT\right) + \left(3\times\frac{1}{2}kT\right) + (18\times kT)$$

$$= \frac{3}{2}kT + \frac{3}{2}kT + 18\,kT = 21\,kT$$

$$= 21\times\frac{8{\cdot}314}{6{\cdot}022\times10^{23}}\times300$$

$$= 8.70\times10^{-20}\text{ Joules}$$

Average energy per mole will be given by

$$\overline{E} = \overline{\varepsilon} \times N$$
$$= 8.70 \times 10^{-20} \times 6.022 \times 10^{23} \text{ Joules}$$
$$= 52.39 \text{ kJ}$$

Q. 25. Calculate the average internal energy of a diatomic molecule at 300 K using law of equipartition of energy. *(Panjab 2002)*

Ans. For a diatomic molecule, as it is linear, and $N = 2$.

Average translational energy

$$(\overline{\varepsilon}_t) = 3 \times \frac{1}{2}kT = \frac{3}{2}kT$$

Average rotational energy

$$(\overline{\varepsilon}_r) = 2 \times \frac{1}{2}kT = kT$$

Average vibrational energy

$$(\overline{\varepsilon}_9) = (3N - 5) \times kT = (3 \times 2 - 5)kT = kT$$

Hence the average internal energy of the molecule is

$$\overline{\varepsilon} = \overline{\varepsilon}_t + \overline{\varepsilon}_r + \overline{\varepsilon}_9$$
$$= \frac{3}{2}kT + kT + kT = \frac{7}{2}kT$$
$$= \frac{7}{2} \times \frac{8.314}{6.022 \times 10^{23}} \times 300 \text{ Joules}$$
$$= \mathbf{1.45 \times 10^{-20}} \textbf{ Joules}$$

Q.26. At what temperature would ethane molecules have the same r.m.s. velocity as methane molecules at 27°C ? *(Delhi 2003)*

Ans. r.m.s. velocity $= \sqrt{\frac{3RT}{M}}$

For methane, *r.m.s.* velocity at 27°C (300 K) $= \sqrt{\frac{3 \times R \times 300}{16}}$

Let ethane possess the same velocity at TK. Then,

$$\sqrt{\frac{3RT}{30}} = \sqrt{\frac{3 \times R \times 300}{16}} \quad or \quad \frac{T}{30} = \frac{300}{16} \quad or \quad T = \frac{30 \times 300}{16}$$

or $$T = 562.5 \text{ K or } 289.5°\text{C}$$

Q. 27. Write notes on

***(i)* Collision diameter** *(Lucknow 2001; Punjab 2002; ; Nagpur 2008, G. Nanakdev 2003)*

***(ii)* Collision number** *(Purvanchal 2007; Nagpur 2008; Himachal 2010)*

***(iii)* Collision frequency** *(Mangalore 2004; Purvanchal 2007; Himachal 2010)*

***(iv)* Mean free-path.** *(Purvanchal 2002; Nagpur 2003; Mangalore 2004; Punjab 2002; Kalyani 2003, Garhwal 2010)*

Ans. (*i*) **Collision diameter**

Collision is an event in which the centres of two identical molecules come within a distance σ from one another as shown in the figure below.

Sigma (σ) which is the distance between centres of the molecules at the point of closest approach is called the **collision diameter.**

(*ii*) **Collision number**

The number of molecules with which a single molecule will collide per unit time per unit volume of the gas is called *collision number.* It is given by

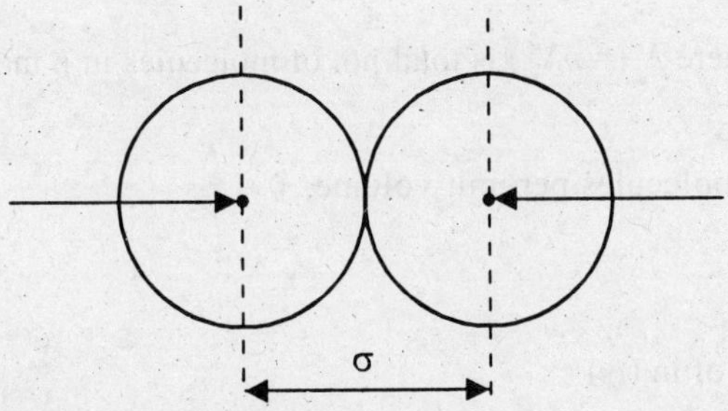

$$Z_1 = \sqrt{2}\,\pi\,\sigma^2\,\overline{c}\,\rho \qquad ...(i)$$

where Z_1 = Collision number

σ = Diameter of molecule

$\overline{c}$ = Average velocity

ρ = No. of molecules per unit volume of the gas

(*iii*) **Collision frequency**

It is the number of molecular collisions occurring per unit time per unit volume of the gas.

Total number of molecules colliding per unit volume per unit time is obtained by multiplying collision number by number density ρ. Thus total number of colliding molecules $= \sqrt{2}\,\pi\,\sigma^2\,\overline{c}\,\rho^2$. Since each collision involves two molecules of the same type, number of collisions is given by one half of this

$$Z_{11} = \frac{1}{2}\cdot\sqrt{2}\,\pi\sigma^2\,\overline{c}\,\rho^2$$

$$= \frac{1}{\sqrt{2}}\pi\,\sigma^2\,\overline{c}\,\rho^2 \qquad ...(ii)$$

It readily follows that number of collisions of molecules of type 1 with those of type 2 would be given by

$$Z_{12} = \frac{1}{\sqrt{2}}\pi\,\sigma^2\,\overline{c}\,\rho_1\,\rho_2 \qquad ...(iii)$$

where ρ_1 and ρ_2 are densities of molecules of type 1 and 2 respectively.

(*iv*) **Mean Free Path**

Mean free path (λ) is defined as the mean distance travelled by a gas molecule between two successive collisions. Thus

$$\lambda = \frac{\overline{c}}{Z_1} = \frac{\overline{c}}{\sqrt{2}\pi\sigma^2\,\overline{c}\,\rho^2} \qquad \text{[Substituting from } (i)] \qquad ...(iv)$$

Also $PV = nRT = nN_A\ kT$

where A_A = Avogadro's number

and k = Boltzmann constant

[Gas constant (R) = Avogadro's no. (N_A) × Boltzmann constant] or $$P = \frac{nN_A kT}{V} = \frac{NkT}{V}$$ where $N (= nN_A)$ is total no. of molecules in n moles of the gas

Since ρ is the number of molecules per unit volume, $\rho = \frac{N}{V}$

Thus $P = \rho\, kT$ or $\rho = \frac{P}{kT}$ ...(*v*)

Substituting the value of ρ of in (*iv*)

$$\lambda = \frac{\bar{c}}{\sqrt{2}\,\pi\sigma^2\,\bar{c}\,P/kt} = \frac{kT}{\sqrt{2}\,\pi\sigma^2 P}$$

Mean free path is also related to the viscosity of the gas. It is given by the equation

$$\lambda = \eta\sqrt{\frac{3}{Pd}}$$

where P = Pressure of the gas

d = Density of the gas

η = coefficient of viscosity of the gas

Q. 28. Discuss the effect of temperature and pressure on (*i*) mean free path (*ii*) collision frequency.

Ans. Let us first take up *mean free path*. It is given by the equation

$$\lambda = \frac{kT}{\sqrt{2}\,\pi\sigma^2 P}$$ Refer to Q. 27.

Clearly, the variables on the R.H.S. of the equation are P and T. Other quantities are constants. T appears in the numerator and P in the denominator. We can say that

$$\lambda \propto \frac{T}{P}$$

or λ varies directly as the temperature and varies inversely as the pressure.

Taking the case of *collision frequency*:

Collision frequency $$Z_{11} = \frac{1}{\sqrt{2}}\pi\sigma^2\,\bar{c}\,\rho^2$$ ...(*i*)

Effect of temperature

From the above equation $Z_{11} \propto \bar{c}$

Also $\bar{c} \propto \sqrt{T}$ (R.M.S. velocity and average velocity vary directly as the square root of absolute temperature)

$$\therefore \quad Z_{11} \propto \sqrt{T}$$

Thus collision frequency varies directly as the square root of absolute temperature

Effect of pressure

From equation (*i*) above

$$Z_{11} \propto \rho^2$$

But $P \propto \rho$ (Pressure of a gas is directly proportional to P *i.e.*, number of molecules present in per unit volume of the gas)

$\therefore \quad Z_{11} \propto P^2$

or collision frequency varies directly as the pressure of the gas.

Q. 29. RMS velocity of hydrogen at STP is 1·85 × 10⁵ cm sec⁻¹ and its mean free path is 1·75 × 10⁻⁵ cm. Calculate the collision number.

Ans. RMS velocity, $\mu = 1{\cdot}85 \times 10^5 \text{ cm sec}^{-1}$

Average velocity $\bar{c} = 0{\cdot}9213 \times \text{RHS Velocity}$

$= 0{\cdot}9213 \times 1{\cdot}85 \times 10^5 \text{ cm sec}^{-1}$

$$\text{Mean free path} = \frac{\text{Average velocity}}{\text{Collision number}}$$

$$1{\cdot}75 \times 10^{-5} = \frac{0{\cdot}9213 \times 1{\cdot}85 \times 10^5}{\text{Collision number}}$$

$$\text{or} \quad \text{Collision number} = \frac{0{\cdot}9213 \times 1{\cdot}85 \times 10^5}{1{\cdot}75 \times 10^{-5}} = \mathbf{9{\cdot}734 \times 10^9 \ sec^{-1}}.$$

Q. 30. At 0°C and 1 atmosphere pressure, molecular diameter of a gas is 4 × 10⁻⁸ cm. Calculate the mean free path of its molecule.

Ans. Mean free path is given by the equation

$$\lambda = \frac{1}{\sqrt{2}\,\pi\sigma^2\rho}$$ See Eq. (*iv*) Q. 5

where σ is molecular diameter.

22400 ml of a gas at 0°C and 1 atm. pressure contains $6{\cdot}02 \times 10^{23}$ molecules.

Hence, number of molecules per c.c,

$$\rho = \frac{6{\cdot}02 \times 10^{23}}{22400} = 2{\cdot}689 \times 10^{19}.$$

Substituting the values in the above equation

$$\lambda = \frac{1}{1{\cdot}414 \times 3{\cdot}14 (4 \times 10^{-8})^2 \times 2.689 \times 10^{19}} = \mathbf{0{\cdot}524 \times 10^{-5} \ cm.}$$

Q.31. What is (*i*) an ideal gas (*ii*) a real gas?

(Madras 1994, Magadh 1996, Nagpur 1997; G. Nanakdev 2003)

Ans. Ideal gas

A gas which obeys Boyle's law, Charles law etc., under all conditions of temperature and pressure is known as an ideal gas or a perfect gas.

An ideal gas may also be defined as:

A gas which obeys the gas equation (PV = nRT) under all conditions of temperature and pressure is called an ideal gas.

It is observed that there is no gas which obeys the gas laws or the gas equation under all conditions of temperature and pressure. The gases are found to obey the gas laws fairly well if the pressure is low or the temperature is high.

Real Gas.

A real gas is one which obeys the gas laws fairly well under low pressure or high temperature.

All gases are real gases. They show more and more deviations from the gas laws as the pressure is increased or the temperature is decreased.

Q.32. Discuss how real gases deviate from Boyle's law.

(Jamia Millia 2002; Purvanchal 2002)

Ans. According to Boyle's law, $PV=$ constant at constant temperature. This means that the product PV should remain constant at all pressures provided temperature is kept constant. Thus a plot of PV *vs* P should be a horizontal straight line parallel to X-axis. However, in actual practice, it is found that no gas gives a straight line plot. Two types of curves are generally obtained.

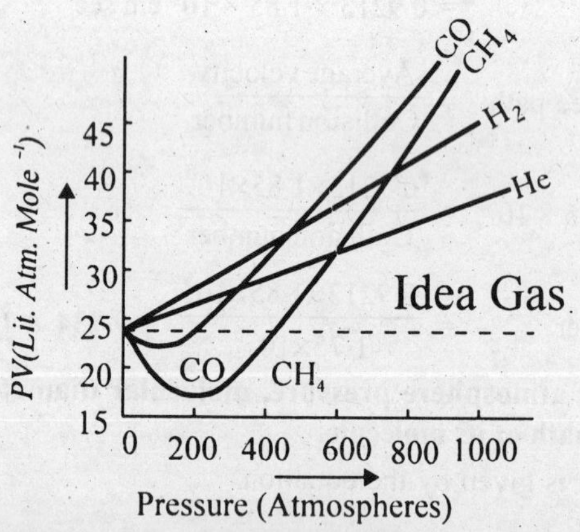

Fig. 1.7. Plots of PV *vs* P for different gases at 0°C (273 K).

(*i*) For gases like hydrogen and helium, the product PV continuously increases with increase of pressure (see fig. 1.7).

(*ii*) For gases like CO and CH_4, the product PV first decreases with increase of pressure, reaches a minimum value and then begins to increase (see Fig. 1.7).

Q.33. Discuss deviation of real gases from Boyle's and Charles's law in terms of compressibility factor, z. *(Aligarh 2002; Sambalpur 2003; Burdwan 2004) G. Nanakdev, 2000; Kumaon 2000)*

OR

What do you mean by compressibility factor 'Z' of a gas ? show graphically the variation of 'Z' with pressure of a real gas at a constant temperature. *(Calcutta 2007)*

Ans. The effect of temperature and pressure on the behaviour of a gas may be studied in terms of quantity 'z' called **compressibility factor,** which is defined as

$$z = \frac{PV}{nRT}$$

For an ideal gas, $z = 1$, at all temperatures and pressures. In case of real gases, *the factor* z varies from values less than 1 to values greater than 1, with changes of temperature and pressure. For example, the plots of compressibility factor z *vs.* pressure P at different constant temperatures for nitrogen gas are shown in Fig.1.8. From these plots it is observed that as the temperature increases, the minimum in the curves shifts upwards. Ultimately, a temperature is reached at which the value of z remains close to 1 over an appreciable range of pressure. For example, in case of nitrogen gas, at 50°C, the value of z remains close to 1 upto nearly 100 atmospheres. *This temperature at which a real gas behaves like an ideal gas over an appreciable pressure range is called* **Boyle temperature** *or* **Boyle point** *because at this temperature Boyle's law is obeyed over a range of pressures.* Obviously, above the Boyle temperature, a gas shows positive deviations only.

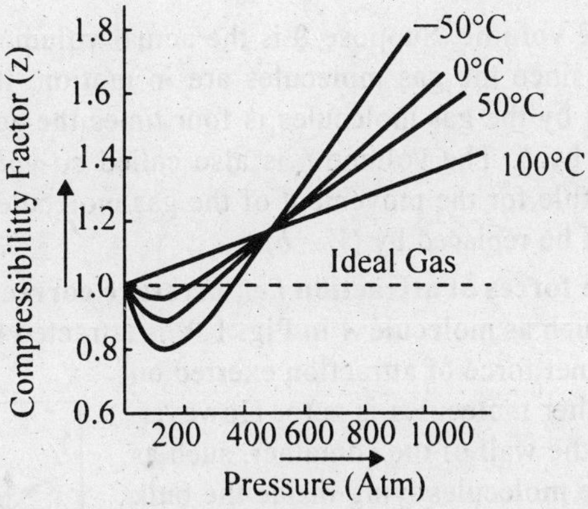

Fig. 1·8. Plot of Z *vs* P for N_2 at different temperatures.

Deviations from Charles-Gay Lussac's Law. According to Charles-Gay Lussac's law, the coefficient of volume expansion of every gas should be same [*i.e.*, 1/273 of its volume at 0°C] independent of pressure. However, experiments have shown that this is true only at low pressures. Appreciable deviations are observed at high pressures.

Q.34. Calculate the volume of 10 moles of methane at 100 atm pressure and 0°C. At this temperature and pressure z = 0.75.

Ans.
$$z = \frac{PV}{nRT}$$

$$\therefore \quad V = \frac{znRT}{P}$$

$$= \frac{0.75 \times 10 \times 0.0821 \times 273}{100}$$

$$= 1.68 \text{ litre}$$

Q.35. What are the causes of deviation of real gas from ideal behaviour?

(Kerala 2001; M Dayanand 2002; Allahabad 2002; Nagpur, 2003; G. Nanakdev 2003; Anna 2004; Guwahati, 2006; Vidyasagar 2007)

Ans. The derivation of the gas laws is based upon the Kinetic Theory of Gases which in turn is based upon certain assumptions. Thus there must be something wrong with certain assumptions. A careful study shows that at high pressure or low temperature, the following two assumptions of Kinetic Theory of Gases become invalid.

(*i*) *the volume occupied by the gas molecules is negligible as compared to the total volume of the gas.*

(*ii*) *The forces of attraction or repulsion between the gas molecules are negligible.*

The above two assumptions are true only if the pressure is low or the temperature is high so that the distance between the molecules is large. However, if the pressure is high or the temperature is low, the gas molecules come close together. Under these conditions:

(*i*) the forces of attraction or replusion between the molecules are not negligible.

(*ii*) the volume of the gas is so small that the volume occupied by the gas molecules cannot be neglected.

Q.36. Derive van der Waal's equation for real gases.

(Anna 2004; Mangalore 2004; M.. Dayanand 2002; Purvanchal 02; Nagpur, 2002; Guwahati 2006; Vidyasagar 2007; Calcutta 2007)

Ans. van der Waal obtained the equation for real gases by applying the corrections for volume and pressure as follows:

1. Correction for the volume. Suppose ϑ is the actual volume occupied by one mole of the gas molecules. Then since the gas molecules are in motion, it has been found that the *effective volume* occupied by the gas molecules is four times the actual volume *i.e.*, equal to 4ϑ. Let it be represented by b. The volume b is also called *co-volume* or *excluded volume.* Thus the free space available for the movement of the gas molecules is $(V - b)$. Hence in the real gas equation V should be replaced by $(V - b)$.

2. Correction for the forces of attraction *i.e.*, pressure correction. A gas molecule lying in the interior of the gas such as molecule A in Fig. 1.9, is attracted by all other gas molecules surrounding it. Hence the net force of attraction exerted on such a molecule by the other molecules is zero. However, for a molecule lying near the wall of the container, such as molecule B in Fig. 1.9, the molecules lying inside the bulk of the gas exert some net inward pull. Thus the effect of such an inward pull is 'dragging back' of the molecule. Consequently, the pressure with which the gas molecule strikes the wall of the vessel is less than the pressure that would have been exerted if there were no such inward pull. Thus the ideal pressure would be greater than the observed pressure by a factor 'p'. Hence

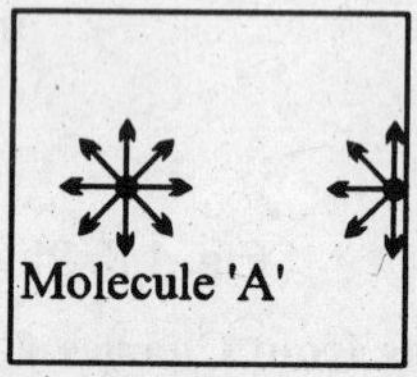

Fig. 1.9. Intermolecular forces of attraction.

Corrected pressure = $P + p$

where p is the 'correction factor' due to the inward pull.

This inward pull on the gas molecules lying near the wall depends upon:

(*i*) Number of molecules surrounding a molecule,

(*ii*) Total number of molecules.

Each of these factors in turn depends upon the density (ρ) of the gas. Hence

correction factor, $p \propto \rho^2$

But for a given mass of the gas

$$\rho \propto \frac{1}{V}$$

where V is the volume of the gas

$$\therefore \quad p \propto \frac{1}{V^2}$$

or

$$p = \frac{a}{V^2}$$

where a is a constant, depending upon the nature of the gas. Thus

$$\text{Corrected pressure} = P + \frac{a}{V^2}$$

Putting the corrected values of volume and pressure in the ideal gas equation, $PV = RT$, for 1mole of the gas, the equation is modified to

$$\left(p + \frac{a}{V^2}\right)(V - b) = RT$$

This equation is known as van der Waal's equation. The constants a and b are called van der Waal's constants whose values depend upon the nature of the gas and are independent of the temperature and pressure.

For n moles of the gas, $p \propto \frac{n}{V}$ and effective volume = $n\ b$. Hence the van der Waal's equation takes the form

$$\left(P+\frac{an^2}{V^2}\right)(V-nb)=nRT$$

Q.37. What is the significance of van der Waals constants *a* and *b* ?

(Punjab 2002; Vidyasagar 2007)

Ans. It is observed that the values of a for the easily liquefiable gases are greater than those for the so called permanent gases like H_2 and He. Moreover, the value of a increases with the ease of liquefaction of the gas. We know that a gas which is more easily liquefiable has greater intermolecular forces of attraction. Hence *'a' is a measure of the intermolecular forces of attraction in a gas.*

'*b*' is the 'effective volume' of the gas molecules. The constancy in the value of b for any gas over a wide range of temperature and pressure confirms that the *gas molecules are incompressible.*

Q. 38. Explain the behaviour of real gases by van der Waal's equation (*i*) At low pressure, (*ii*) At moderate pressure, (*iii*) At high pressure, (*iv*) At high temperature, (*v*) Exceptional behaviour of hydrogen and helium.

(M. Dayanand 2002; Delhi, 2002; Islamia, 2002; Himachal 2010)

Ans. (*i*) **At low pressure.** At extremely low pressure, V is very large. Hence the correction term $\frac{a}{V^2}$ is very small. Similarly, the correction term b is also very small as compared to V. Thus both the correction terms can be neglected so that the van der Waal's equation reduces to $PV= RT$. This is why at extremely low pressures, the gases obey the ideal gas equation.

(*ii*) **At moderate pressure.** As the pressure is increased, the volume decreases and hence the factor $\frac{a}{V^2}$ increases. Thus the factor $\frac{a}{V^2}$ can no longer be neglected. However, if the pressure is not too high, the volume V is still sufficiently large so that b can be neglected in comparison with V. Thus the van der Waal's equation reduces to

$$\left(P+\frac{a}{V^2}\right)V=RT$$

or

$$PV+\frac{a}{V}=RT$$

or

$$PV=RT-\frac{a}{V}$$

Thus PV is less than RT by a factor $\frac{a}{V}$. As the pressure increases, V decreases, so that the factor $\frac{a}{V}$ increases. Thus PV decreases as the pressure is increased. This explains why a dip the plots of PV vs P of real gases is obtained (see fig. 1.7).

(*iii*) **At high pressure.** As the pressure is increased further so that it is fairly high, V is so small that b can no longer be neglected in comparison with V. Although under these conditions the factor $\frac{a}{V^2}$ is quite large but since P is very high so $\frac{a}{V^2}$ can be neglected in comparison with P. Thus the van der Waal's equation reduces to

$$P(V-b)=RT$$

or $$PV-Pb=RT$$

or $$PV=RT+Pb$$

Thus PV is greater than RT by a factor Pb. Now as the pressure is increased, the factor Pb increases more and more. This explains why after the minima in the curves, the product PV increases continuously as the pressure is increased more and more (Fig. 1.7).

(*iv*) **At high temperature.** At any given pressure, if the temperature is sufficiently high, V is very large so that as in case (*i*), the van der Waal's equation reduces to $PV=RT$. Hence at high temperature, real gas behaves like the ideal gas.

(*v*) **Exceptional behaviour of hydrogen and helium.** In case of hydrogen and helium, their molecules have very small masses. Hence in the case of these gases, intermolecular forces a of attraction are extremely small even at high pressures. In other words, the factor $\frac{a}{V^2}$ is negligible at all pressures. Hence the van der Waal's equation is applicable in the following form at *all pressures and ordinary temperatures.*

$$P(V-b)=RT \text{ or } PV=RT+Pb$$

This explains why hydrogen and helium show positive deviation only which increases with increase in the value of P.

Q. 39. Calculate from the van der Waal's equation the temperature at which 3 moles of SO_2 would occupy a volume of 0.01 m³ at 1519875 Nm⁻² pressure (a = 0.679 Nm⁴ mol⁻², b = 5.64 × 10⁻⁵ m³ mol⁻¹, R = 8.314 JK⁻¹ mol⁻¹).

Ans. van der Waal's equation can be written as :

$$T=\frac{1}{nR}\left(P+\frac{an^2}{V^2}\right)(V-nb)$$

Substituting the given values, we get,

$$\therefore \quad T=\frac{1}{3\times 8.314}\left(1519875+\frac{0.679\times 3^2}{(0.01)^2}\right)(0.01-3\times 564\times 10^{5})$$

$$=\frac{1}{24.942}\times 1580985\times 0.0098308$$

$$=623.1 \text{ K.}$$

Q.40. Two moles of ammonia gas are enclosed in a vessel of 5-litre capacity at 27°C. Calculate the pressure exerted by the gas, assuming that

(*i*) the gas behaves like an ideal gas (using ideal gas equation);

(*ii*) the gas behaves like a real gas (using van der Waal's equation).

Given that for ammonia a = 4.17 atm litre² mol⁻² (or 4.17 atm dm⁶ mol⁻²) and b = 0.037 litre mol⁻¹ (or 0.037 dm³ mol⁻¹)

Ans. We are given that

n = 2 moles

V = 5 litres or dm³

T = 27°C = (27 + 273) K = 300 K

a = 4.17 atm litre² mol⁻² or atm dm⁶ mole⁻²

b = 0.037 litre m or ¹ or dm³ mol ⁻¹

We know that

$R = 0.0821$ litre atm K^{-1} mol^{-1} atm or dm^3 atm K^{-1} mol^{-1}

(*i*) If the gas behaves like an ideal gas, we have

$$PV = nRT$$

$$\therefore \quad P = \frac{nRT}{V}$$

$$= \frac{2 \times 0.0821 \times 300}{5} = \mathbf{9.85\,atm}$$

(*ii*) If the gas behaves like a real gas, we apply van der Waal's equation *i.e.,*

$$\left(P + \frac{an^2}{V^2}\right)(V - nb) = nRT$$

or
$$P = \frac{nRT}{V - nb} - \frac{an^2}{V^2}$$

$$= \frac{2 \times 0.0821 \times 300}{5 - 2 \times 0.037} = \frac{4.17 \times (2)^2}{(5)^2} = \mathbf{9.33\ atm}$$

Q.41. Describe Kammerlingh-Onnes (virial) equation of state for real gases.

(Calcutta, 2000)

Ans. It is an *empirical equation* which expresses PV as a power series of pressure P at any given temperature. It is written as

$$PV = A + BP + CP^2 + DP^3 + \ldots$$

where P is the pressure in atmospheres and V is the *molar* volume in litres. The coefficients A, B, C etc., are known respectively as the first, second, third etc., *virial coefficients.* The following points may be noted about the above equation:

(*i*) The virial coefficients are different for different gases.

(*ii*) At very low pressure, only the first virial coefficient is significant and it is equal to RT *i.e.,*

$$A = RT$$

(*iii*) At higher pressures, the other virial coefficients also become important and must also be considered.

(*iv*) For any particular gas, the values of A, B, C, etc., are constant at constant temperature. Their values change with change of temperature. The first virial coefficient A is always positive and increases with temperature. On the other hand, the second virial coefficient B is negative at low temperatures. With increase of temperature it increases to zero and then becomes more and more positive. The temperature at which $B = 0$ is called *Boyle temperature.* At this temperature, Boyle's law is valid over a fairly wide pressure range.

The other form in which Onnes equation is often written is by expressing the compressibility factor (z) as a power series in $1/V$ *i.e.,*

$$Z = \frac{PV}{RT} = 1 + \frac{B}{V} + \frac{C}{V^2} + \cdots$$

or
$$PV = RT\left[1 + \frac{B}{V} + \frac{C}{V^2} + \cdots\right]$$

where the coefficients B, C etc., (called second, third virial coefficients etc.) for a particular gas depend only on the temperature and not on the pressure.

Q.42. What is the molar volume of N_2 (*g*) at 500 K and 600 atm according to (*a*) the perfect gas law and (*b*) the virial equation? The virial coefficient B of N_2 (g) at 500 K is

0.0169 litre mol^{-1}.

Ans. (*a*) According to perfect gas law

$$V = \frac{RT}{P} = \frac{0.0821 \times 500}{600} = \textbf{0.0684 litre.}$$

(*b*) $$PV = RT\left(1 + \frac{B}{V}\right) = RT + \frac{RT}{V}B$$

or $$PV = RT + PB$$

or $$V = \frac{RT + PB}{P}$$

or $$V = \frac{RT}{P} + B = \left(\frac{0.0821 \times 500}{600}\right) + 0.0169$$

$$= 0.0684 + 0.0169 = \textbf{0.0853 litre mol}^{-1}$$

Q. 43. What is Boyle's temperature ? How can be obtain it from van der Waal's equation?

(M. Dayanand 2002; G. Nanakdev, 2003; Burdwan 2004; Raipur 2010)

Ans. It is a temperature at which a gas behaves like an ideal gas. This temperature is given by

$$T_B = \frac{a}{Rb}$$

It may be derived from the van der Waal's equation as follows :

$$\left(p + \frac{a}{V^2}\right)(V - b) = RT$$

It may be written in the form

$$PV = RT - \frac{a}{V} + bP + \frac{ab}{V^2} \quad ...(i)$$

As both a and b are small, and if the pressure is not too high so that V is not so small, $\frac{ab}{V^2}$ can be neglected.

Further V in the correction term a/V may be replaced by RT/P.

Then equation (*i*) reduces to

$$PV = RT - \frac{aP}{RT} + bP$$

$$= RT + P\left(b - \frac{a}{RT}\right)$$

Since the gas behaves ideally at Boyle's temperature, $PV = RT$

Hence the second term on R.H.S. should be zero.

Since P has a finite value

$$b - \frac{a}{RT} = 0 \quad \text{or} \quad T = \frac{a}{Rb}$$

or Boyles temp., $T_b = \dfrac{a}{Rb}$ ... (*ii*)

Q. 44. How can you calculate second virial coefficient using van der Waal's equation?

(Himachal 2000)

Ans. The general equation of state as given by Onnes is

$$PV = RT\left[1 + \frac{B}{V} + \frac{C}{V^2} + ...\right] \quad ...(i)$$

The second virial coefficient B of this equation can be obtained by comparing it with van der Waal's equation

$$\left(P+\frac{a}{V^2}\right)(V-b)=RT$$

which may be rewritten in the form

$$PV = RT + bP + \frac{ab}{V^2} - \frac{a}{V}$$

or
$$PV = RT + bP - \frac{a}{V} + \frac{ab}{V^2}$$

As a and b are very small and if the pressure is not too high, $\frac{ab}{V^2}$ can be neglected. Further P may be replaced by $\frac{RT}{V}$. Hence we get

$$PV = RT + b\frac{RT}{V} - \frac{a}{V}$$

or
$$PV = RT\left[1+\frac{1}{V}\left(b-\frac{a}{RT}\right)\right] \qquad ...(ii)$$

Comparing eqns. (*i*) and (*ii*), we get

$$\boxed{B = b - \frac{a}{RT}}$$

Thus second virial coefficient can be calculated from a knowledge of the van der Waal's constants a and b of the gas at temperature T.

Q. 45. The van der Waal's constant b for CO_2 is 4.28×10^{-2} $dm^3\ mol^{-1}$. Assuming CO_2 molecule to be spherical, calculate its molecular diameter.

Ans. For 1 mole of the gas

$$b = 4N\vartheta_m$$

where N is Avogadro's number and ϑ_m is the actual volume of the gas molecule.

$$\vartheta_m = \frac{n}{4N} = \frac{0.0428}{4\times 6.022\times 10^{23}} \text{ litres}$$

$$= 1{\cdot}777 \times 10^{-26} \text{ litres} = 1{\cdot}777 \times 10^{-23} \text{ cm}^3$$

But
$$\vartheta_m = \frac{4}{3}\pi r^3 = \frac{4}{3}\pi\left(\frac{\sigma}{2}\right)^3$$

$\therefore$
$$\sigma^3 = \frac{6\vartheta_m}{\pi} = \frac{6\times 1.777\times 10^{-23}}{3.143} = 33.92\times 10^{-24}$$

which gives $\sigma = 3.24 \times 10^{-8}$ cm = **3.24 Å.**

Q. 46. What is meant by critical temperature, critical pressure and critical volume ?

(Lucknow 2001; Sri Vankateswara 2002; Jamia Millia 2002; M. Dayanand, 2002; G. Nanakdev, 2003; Vidyasagar 2007)

Ans. Critical temperature (T_o) *of a gas may be defined as that temperature above which the gas cannot be liquefied, howsoever high pressure we may apply on the gas. The minimum pressure required to liquefy the gas at the critical temperature is called the* **critical pressure**

(P_c). *The volume occupied by one mol of the gas at the critical temperature and pressure is called the* **critical volume (V_c)**.

The critical temperature, critical pressure and critical volume of a gas are collectively called as the **critical constants** of the gas. For example, the critical constants of carbon dioxide (CO_2) are

$$T_c = 31.1°\text{C} \quad \text{or} \quad 304.2\text{ K}$$
$$P_c = 72.9 \text{ atmospheres}$$
$$V_c = 95.65 \text{ cm}^3 \text{ mol}^{-1}$$

Q. 47. Describe the measurement of critical constants (T_c, P_c and V_c)

(Lucknow 2001; Venkateswara 2002 G. Nanakdev, 2003)

Ans. Measurement of T_c and P_c. The measurement of critical temperature (T_c) is based upon the observation that when a liquid *e.g.* ether or carbon disulphide, is heated in a closed space the surface of separation between the liquid and the vapour disappears at a definite temperature and on cooling the surface of separation reappears at the same temperature. The temperature at which this occurs is the **critical temperature.**

The apparatus used for the measurement of the critical temperature is shown in Fig. 1.10. It helps to measure critical pressure simultaneously. It consists of a glass bulb *B* containing only the liquid and its vapour sealed over mercury. The bulb is connected to a manometer *M* containing air above the mercury.

Further, the bulb *B* is surrounded by a heating jacket as shown in the Fig. 1.10. The bulb *B* is heated gently by circulating hot water in the heating jacket and the temperature is noted at which the meniscus disappears. The bulb is then allowed to cool till the meniscus reappears or the cloudiness appears on the surface of the bulb due to condensation of vapour. The mean of the two temperatures gives the critical temperature of the liquid/gas enclosed in the bulb *B*.

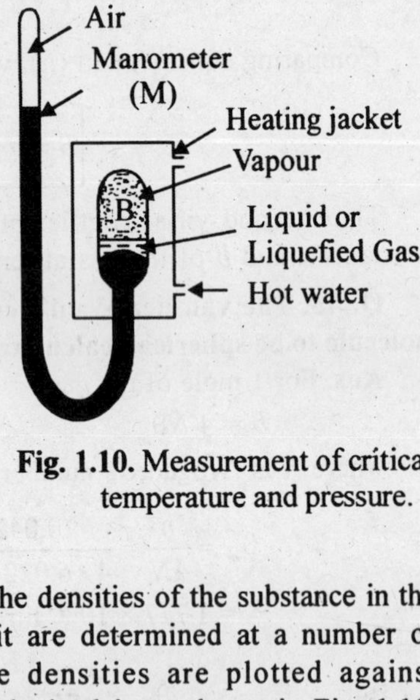

Fig. 1.10. Measurement of critical temperature and pressure.

Corresponding pressure from the manometer at the two stages is noted and taking their mean, gives the critical pressure (P_c).

Measurement of V_c. The measurement of critical volume (V_c) is based upon a rule which states that '*the mean of the densities of any substance in the state of liquid and saturated vapour at the same temperature is a linear function of the temperature.*'

Thus to determine the critical volume of any substance, the densities of the substance in the liquid state and those of the vapour in equilibrium with it are determined at a number of temperatures. If these densities are plotted against temperature, the curve obtained is as shown in Fig. 1.11. The densities of the vapour lie along the curve *AC* while those of the liquid lie along the curve *BC*. The two curves meet at the point *C* which corresponds to the critical temperature. If the mean values of the densities $\left(\frac{d_l + d_g}{2}\right)$ are plotted against temperature, a straight line *CD* is obtained. The line on extrapolation meets the density axis at *D* which corresponds to the critical density. Dividing the molecular weight of the substance by the critical density, ***critical volume*** is obtained.

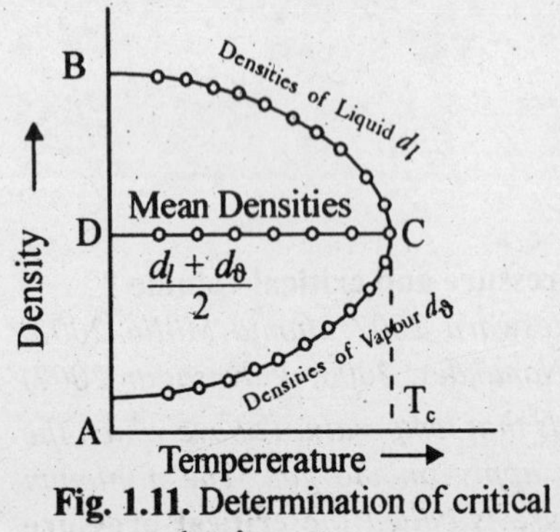

Fig. 1.11. Determination of critical density

Q. 23. Describe isotherms of carbon dioxide as studied by Andrews.

OR

Describe Andrews experiments on critical phenomenon.

(Bangalore, 2002; Nagpur, 2008; Raipur 2010)

Ans. Andrews, in 1861, was the first to study the effect of temperature and pressure on the volume of carbon dioxide. Each time, keeping the temperature constant at a particular value, he measured the volume of carbon dioxide at different pressures. He then plotted the volumes against pressures, at the constant temperature. Such a plot of P vs V at constant T is called an *isotherm.* A number of such isotherms were obtained at different temperatures as shown in Fig. 1.12.

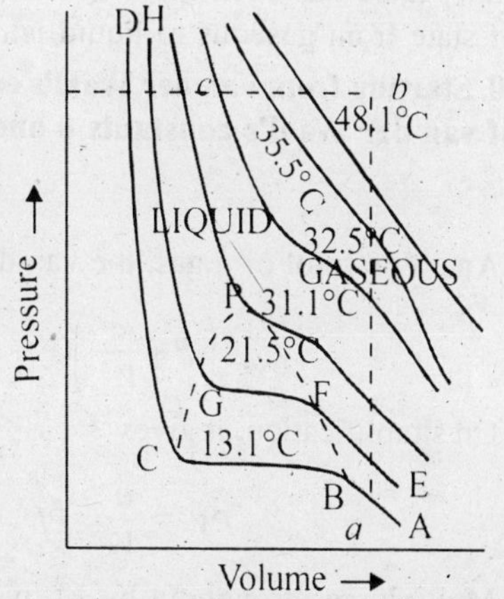

Fig. 1.12. Isotherms of carbon dioxide.

At the lowest temperature *i.e.*, 13.1°C (286.2 K), at low pressure, carbon dioxide exists as a gas, as shown at the point *A.* As the pressure is increased, the volume of the gas decreases along the curve *AB.* At the point *B,* the liquefaction of the gas starts. Hence the volume decreases rapidly along *BC* because the liquid has much less volume than the gas. At the point C, the liquefaction is complete. Now the increase of pressure has very little effect upon the volume because liquids are hardly compressible. Hence a steep curve *CD* is obtained as shown in Fig. 1.12. Thus along the portion *AB,* carbon dioxide exists only as gas, along the portion *BC,* the liquid and the vapour are in equilibrium with each other and along *CD,* carbon dioxide exists only as liquid.

The isotherm *EFGH* at 21.5°C is similar to that at 13.1°C except that the horizontal portion over which the liquefaction occurs is shorter. In fact, as the temperature is raised, this portion becomes smaller and smaller and finally at 31.1°C it is reduced only to a point. In the isotherms above 31.1°C, the horizontal portion completely disappears. This shows that above 31.1°C, the liquefaction of the gas does not take place at all howsoever high pressure may be applied. Thus 31.1°C is the critical temperature of carbon dioxide. The pressure required to liquefy the CO_2 gas at 31.1°C (304.2 K) is found to be 72.9 atmospheres, which is, therefore, the critical pressure for CO_2.

It has been found that in case of all other gases, similar isotherms are obtained. However, lengths of the horizontal portions and the peak of the dotted parabola are different, depending upon the nature of the gas.

Q. 49. What is meant by continuity of state? *(Aligarh 2002; Kurukshetra 2002)*

Ans. If the ends of the horizontal portions of the different isotherms are joined together, a parabolic curve (shown by dotted lines in Fig. 1.12) is obtained. The peak of this parabola represents the critical point. Within the parabola, the liquid and the vapour exist together in equilibrium. Outside the parabola, the substance exists only as gaseous under the conditions represented on the right side and only as liquid under the conditions represented on the left side. A careful thinking reveals that below the critical temperature it is possible to change a substance from gaseous state to liquid state without having both the phases present together at any stage. This is explained as under :

Suppose the substance under consideration is CO_2 so that Fig. 1.12 may be referred again. At 13.1°C, if pressure and volume correspond to the point *a*, the substance is gaseous. Keeping the volume constant, if the gaseous CO_2 is heated till it has pressure corresponding to the point *b*, the path *ab* is followed. Now keeping the pressure constant, suppose the gaseous CO_2 is cooled the volume decreases along the path *bc*. At *c*, CO_2 will exist as liquid. Thus nowhere along the path *abc*, there has been more than one phase present. This is expressed by saying that the change of state from gaseous to liquid is *continuous* or it is referred as the *continuity of state*.

Q. 50. Starting from van der Waal's equation, obtain an expression for critical constants in terms of van der Waal's constants *a* and *b*. *(Meerut, 2002; Lucknow, 2002; Sambalpur 2003; Vidyasagar 2007; Guwahati 2007; Calcutta 2007; Himachal 2010)*

Ans. For 1mol of a gas, the van der Waal's equation is

$$\left(P+\frac{a}{V^2}\right)(V-b)=RT$$

On simplification, it gives

$$PV+\frac{a}{V}-Pb-\frac{ab}{V^2}=RT$$

Multiplying throughout by V^2, we get

$$PV^3+aV-PbV^2-ab=RTV^2$$

Dividing throughout by P, we get

$$V^3+\frac{aV}{P}-bV^2-\frac{ab}{P}=\frac{RT}{P}V^2$$

Rearranging in the descending powers of V, we get

$$V^3-\left(b+\frac{RT}{P}\right)V^2-\frac{a}{P}V-\frac{ab}{P}=0$$

Using the appropriate values of the van der Waal's constants a and b and the gas constant R, for any constant value of T (used by Andrews in his experiments), the values of V can be obtained for different values of P. The *theoretical isotherms* thus obtained are shown in Fig. 1.13. These curves are similar to those obtained experimentally (Fig. 1.12) except that in the curves below the critical temperature, the horizontal portions are replaced by curved portions. For example, in the isotherms at 13.1°C the horizontal portion BC has been replaced by the curved portion $BXYZC$. This is expected because the van der Waal's equation is cubic in V, as written above. Hence for every value of pressure P it has three roots (*i.e.* three values of V), all of which may be real or one may be real and the other two may be imaginary. *Below the critical temperature*, all the three roots are real in a certain pressure range, and these values come closer as the isotherm moves upwards. At the *critical temperature*, all the roots become equal for a particular pressure. *Above the critical temperature*, for any value of P; there is only one real root *i.e.*, only one value of V.

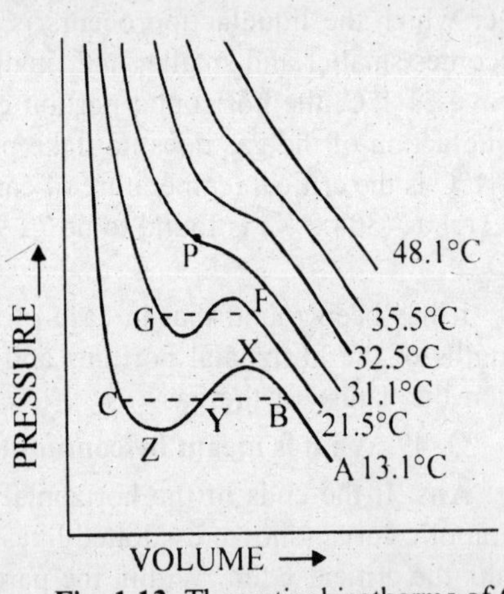

Fig. 1.13. Theoretical isotherms of CO_2

It has been observed that the portions BX, XYZ and ZC represent metastable conditions and can be actually obtained in practice by careful experiments.

As stated above, at the critical temperature the three values of V become identical for a particular value of pressure P. If we represent critical volume by V_c, then

$$V = V_c \text{ or } (V - V_c)^3 = 0$$

Expanding this, we get

$$V^3 - 3V_cV^2 + 3V_c^2V - V_c^3 = 0 \quad ...(i)$$

Further, as already explained, the van der Waal's equation may be written in the form

$$V^3 - \left(b + \frac{RT}{P}\right)V^2 + \frac{a}{P}V - \frac{ab}{P} = 0$$

At the critical point, the terms P and T may be replaced by the critical values P_c and T_c. Hence the above equation may be written as

$$V^3 - \left(b + \frac{RT_c}{P_c}\right)V^2 + \frac{a}{P_c}V - \frac{ab}{P_c} = 0 \quad ...(ii)$$

At the critical point, equation (*i*) and (*ii*) must be identical. Hence the coefficients of similar powers of V must be equal in both the equations. Equating the coefficients of terms of similar powers of V, we have

$$3V_c = b + \frac{RT_c}{P_c} \quad ...(iii)$$

$$3V_c^2 = \frac{a}{P_c} \quad ...(iv)$$

and

$$V_c^3 = \frac{ab}{P_c} \quad ...(v)$$

These equations can be solved to calculate the critical constants V_c, P_c and T_c in terms of van der Waal's constants a and b. Dividing equation (*v*) by (*iv*), we get

$$\frac{V_c^2}{3V_c^2} = \frac{ab}{P_c} \times \frac{P_c}{a}$$

or

$$\boldsymbol{V_c = 3b} \quad ...(vi)$$

Putting this value in equation (*iv*), we get

$$3 \times (3b)^2 = \frac{a}{P_c}$$

or

$$\boldsymbol{P_c = \frac{a}{27b^2}} \quad ..(vii)$$

Putting the values of V_c and P_c in equation (*iii*), we get

$$3 \times 3b = b + \frac{RT}{a/27b^2}$$

or

$$9b = b + \frac{RT_c \times 27b^2}{a}$$

or

$$\frac{RT_c \times 27b^2}{a} = 8b$$

or

$$\boldsymbol{T_c = \frac{8a}{27Rb}} \quad ...(viii)$$

Q. 51. Derive the relation $P_cV_c = \frac{3}{8}RT_c$

(Kurukshetra 2002; Lucknow 2002; Meerut 2002; Nagpur 2008)

Ans. Writing again the equations between critical constants and van der Waal's constants

$$V_c = 3\,b \qquad ...(i)$$

$$P_c = \frac{a}{27b^2} \qquad ...(ii)$$

$$T_c = \frac{8a}{27Rb} \qquad ...(iii)$$

To obtain a relationship between P_c, V_c and T_c we have to eliminate a and b from equations (*i*), (*ii*) and (*iii*). This may be done as follows :

From eqn. (*i*)

$$b = \frac{V_c}{3} \qquad ...(iv)$$

Substituting this value of b in eqn. (*ii*), we get

$$P_c = \frac{a}{27(V_c/3)^2} = \frac{a}{3V_c^2}$$

or

$$a = 3\,P_cV_c^2 \qquad ...(v)$$

Substituting the values of a and b from eqns. (*iv*) and (*v*) in eqn. (*iii*), we get

$$T_c = \frac{8 \times 3P_cV_c^2}{27R \times V_c/3}$$

or

$$T_c = \frac{8}{3}\frac{P_cV_c}{R}$$

or

$$P_cV_c = \frac{3}{8}RT_c \qquad ...(vi)$$

Q. 52. Express van der Waal's constants *a* and b in terms of critical constants.

(Indore 2001; Sambalpur 2003)

Ans. Expressions for a and b in terms of P_c and V_c are

$$b = \frac{V_c}{3} \text{ and } a = 3\,P_c^3\,V_c^3$$

Expressions in terms of P_c and T_c are

$$a = \frac{27R^2T_c^2}{64P_c} \quad \text{and} \quad b = \frac{RT_c}{8P_c}$$

These equations can be obtained by eliminating a and b separately using the equations

$$P_c = \frac{a}{27b^2} \text{ and } T_c = \frac{8a}{27Rb^2}$$

Q. 28. The critical temperature of carbon dioxide gas is 31.1°C and its critical density is 0.455 g/cm^3. Find out the values of *a*, *b* and P_c for the gas (*R*=0082 litre/atm/degree/mol).

(M. Dayanand 2000; Kurukshetra 2001)

Ans. $T_c = 31.1°C = 304.1$ K

$$V_c = \frac{\text{mol. wt.}}{\text{critical density}} = \frac{M}{d} = \frac{44}{0.455} = 96.70\,\text{cm}^3 = 0.0967\,\text{litre}$$

$$b = \frac{V_c}{3} = \mathbf{0.0322\ litre\ mol^{-1}} \quad ...(1)$$

$$P_c = \frac{a}{27b^2} \quad ...(2)$$

$$T_c = \frac{8a}{27Rb^2} \quad ...(3)$$

Dividing (2) by (3)

$$\frac{P_c}{T_c} = \frac{R}{8b}$$

or $$P_c = \frac{RT_c}{8b} = \frac{0.082 \times 304.1}{8 \times 0.0322} = \mathbf{96.80\ atm.}$$

From $$P_c = \frac{a}{27b^2}$$

$$\mathbf{a = 27\ P_c\ b^2}$$

Substitute the value of b from (1) in the above equation

$$a = 27 \times 96.8 \times 0.0322^2$$
$$= 2.71\ \text{atm litre}^2\text{mol}^{-2}$$

Q. 54. The critical temperature and critical pressure of chlorine are 146°C and 93.5 atmosphere respectively. Calculate the values of the van der Waal's constants *a* and *b*.

Ans. Here we are given that

$$T_c = (146 + 273)\text{K} = 419\ \text{K}$$

and $$P_c = 93.5\ \text{atmospheres}$$

The relations between critical constants and van der Waal's constants are

$$P_c = \frac{a}{27b^2} \quad ...(i)$$

and $$T_c = \frac{8a}{27Rb} \quad ...(ii)$$

Dividing equation (*ii*) by equation (*i*), we get

$$\frac{T_c}{P_c} = \frac{8a}{27Rb} \times \frac{27b^2}{a} = \frac{8b}{R}$$

or $$b = \frac{RT_c}{8P_c} \quad ...(iii)$$

From equation (*i*), $a = 27\ P_c\ b^2$

$$= 27P_c\left(\frac{RT_c}{8P_c}\right)^2$$

$$= \frac{27R^2T_c^2}{64P_c} \quad ...(iv)$$

Taking $R = 0.0821$ litre atm degree^{-1} and mol^{-1} and putting the given values of T_c and P_c in equations (*iii*) and (*iv*), we get

$$a = \frac{27R^2T_c^2}{64P_c} = \frac{27 \times (0.0821)^2 \times (419)^2}{64 \times 93.5}$$

$$= \mathbf{5.34\ atm\ litre^2\ mol^{-2}}$$

and
$$b = \frac{RT}{8P_c} = \frac{0.0821 \times 419}{8 \times 93.5}$$

$$= \mathbf{0.046\ litre\ mol^{-1}}$$

Solution in terms of SI units:

$T_c = 419$ K

$P_c = 93.5$ atm $= 93.5 \times 101325$ Nm^{-2}

($\because$ 1 atm = 101325 Nm^{-2})

$R = 8.314$ JK^{-1} mol^{-1}

$$\therefore \quad a = \frac{27R^2T_c^2}{64P_c}$$

$$= \frac{27 \times (8.314)^2 \times (419)^2}{64 \times 93.5 \times 101325} \text{ Nm}^4 \text{ mol}^{-2}$$

$$= \mathbf{0.54\ Nm^4\ mol^{-2}}$$

$$b = \frac{RT_c}{8P_c} = \frac{8.314 \times 419}{8 \times 93.5 \times 101325} \text{ m}^3 \text{ mol}^{-1}$$

$$= \mathbf{4.6 \times 10^{-5}\ m^3\ mol^{-1}}$$

Q. 55. Derive the reduced equation of state and state the law of corresponding state.

(Kurukshetra 2001; Lucknow 2001; M. Dayanand 2002; Nagpur 2002; Arunachal 2002; Vidyasagar 2002; Delhi 2003; Purvanchal 2007)

Ans. If the pressure, volume and temperature of a gas are expressed in terms of the critical pressure, volume and temperature respectively, then we may write

$$P = \pi P_c,\ V = \phi\ V_c,\ T = \theta\ T_c \qquad ...(i)$$

Thus the quantities π, ϕ and θ may be defined as

$$\pi = \frac{P}{P_c},\ \phi = \frac{V}{V_c} \text{ and } \theta = \frac{T}{T_c}$$

The quantities π, ϕ and θ are called reduced pressure, reduced volume and reduced temperature respectively.

Substituting the values of P, V and T given by equation (*i*), in van der Waal's equation

$$\left(P + \frac{a}{V^2}\right)(V - b) = RT$$

we get
$$\left(\pi P_c + \frac{a}{\phi^2 V_c^2}\right)(\phi V_c - b) = R\theta T_c.$$

Further, substituting the values of P_c, T_c and V_c in terms of van der Waal's constant *viz.*

$V_c = 3\ b$

$$P_c = \frac{a}{27b^2}$$

$$T_c = \frac{8a}{27Rb}$$

we get
$$\left(\pi\frac{a}{27b^2} + \frac{a}{\phi^2.9b^2}\right)(\phi.3b - b) = R\theta\frac{8a}{27Rb}$$

or
$$\left(\pi a + \frac{3a}{\phi^2}\right)(3b\phi - b) = 8\,\theta ab$$

or
$$\boxed{\left(\pi + \frac{3}{\phi^2}\right)(3\phi - 1) = 8\,\theta}$$

This equation which relates the reduced pressure, volume and temperature (*i.e.*, π, ϕ and θ) is called the **reduced equation of state**. The most important point about this equation is that this does not involve the constants *a, b* and *R* and hence is applicable to all substances in the liquid and the gaseous state. Another important conclusion drawn from the above equation is as follows :

If two substances have the same reduced pressure and the same reduced temperature, they must have the same reduced volume.

The above statement is called the **law of corresponding states** and the substances having the same reduced pressure and same reduced temperature and hence the same reduced volume are said to be in the corresponding states.

Q. 56. Calculate the pressure exerted by one mol of CO_2 at 40°C when confined to a volume of 0.17 dm^3 by making use of the law of corresponding states. The values of the critical pressure (P_c), critical volume (V_c) and critical temperature (T_c) are 73 atmospheres, 0.0957 dm^3 and 30.42°C respectively.

Ans. $T = 40 + 273 = 313$ K, $V = 0{\cdot}7$ dm^3

$P_c = 73$ atm, $V_c = 0.0957$ dm^3, $T_c = (30.42 + 273)$ K or 303.42 K

$$\theta = \frac{T}{T_c} = \frac{313}{303.42} = 1.032$$

and
$$\phi = \frac{V}{V_c} = \frac{0.7}{0.0957} = 7.315$$

Substitute in the reduced equation of state and calculate π.

$$\left[\pi + \frac{3}{\phi^2}\right](3\phi - 1) = 8\theta$$

$$\left[\pi + \frac{3}{7.315^2}\right](3 \times 7.315 - 1) = 8 \times 1.032$$

$$(\pi + 0.056) \times 20.945 = 8.256$$

$$\pi = \frac{8.256}{20.945} - 0.056 = 0.338$$

∴ $P = \pi P_c = 0.338 \times 73 =$ **24.674 atm**

Q. 57. The reduced volume and temperature of a gas are 10.2 and 0.7. What will be its pressure if its critical pressure is 42 atm ?

Ans. It is given that

Reduced volume, $\phi = 10.2$, Reduced temperature, $\theta = 0.7$

Critical pressure, $P_c = 42$ atm

According to the reduced equation of state

$$\left(\pi + \frac{3}{\phi^2}\right)(3\phi - 1) = 8\theta$$

Substituting the values of ϕ and θ, we get

$$\left(\pi + \frac{3}{(10.2)^2}\right)(3 \times 10.2 - 1) = 8 \times 0.7$$

or $$\left(\pi + \frac{3}{104.4}\right)(30.6 - 1) = 5.6$$

or $$(\pi + 0.0288)(29.6) = 5.6$$

or $$\pi + 0.0288 = \frac{5.6}{29.6} = 0.1892$$

or $$\pi = 0.1604$$

i.e. $$\pi = \frac{P}{P_c} = 0.1604$$

or $$P = 0.1604 \times P_c \quad (P_c = 42 \text{ atm})$$

$$= 0.164 \times 42 = 6.7368 \text{ atm.}$$

Q. 58. Describe (*i*) Adiabatic expansion (Linde's) method (*ii*) Claude's method (*iii*) Adiabatic demagnetisation method for liquefaction of gases.

(H.S. Gaur 2002; M. Dayanand 2002; Lucknow 2002; Raipur 2010)

Ans. (*i*) **By the adiabatic expansion of compressed gas—Linde's Process.** This process is based upon Joule-Thomson effect which states as follows :

When a gas under high pressure is allowed to expand adiabatically through a fine orifice (hole) into a region of low pressure, it is accompanied by cooling.

An obvious reason for the cooling is that when the gas expands from a region of high pressure to the region of low pressure, the intermolecular distances increase. Some work is done in separating the molecules apart. This is done at the cost of kinetic energy of the gas. Thus the kinetic energy of the gas decreases and therefore, the temperature falls.

It may be mentioned here that in case of hydrogen and helium, instead of cooling, these gases are warmed up if allowed to expand as above, under ordinary temperature. However, if these gases are first cooled to some particular temperature and then allowed to expand, they behave like other gases. The temperature to which a gas should thus be cooled to give normal Joule-Thomson effect is called the *inversion temperature.* The inversion temperatures for hydrogen and helium are – 80°C and – 240°C respectively.

The process based upon the above principle is known as **Linde's process.** The apparatus used is shown in Fig. 1.14.

The above apparatus was used by Linde for the liquefaction of air. The procedure adopted is as follows:

The air is first compressed to about 200 atmospheres in the compressor and then passed through the condensing coils. As a result, the air is cooled and the water vapours present condense to form the liquid which is removed. The dry air is then passed through the copper spiral which terminates into a jet. The air expands through the jet into the chamber where the pressure of the air falls down to about 50 atmospheres. Thus as a result of Joule-Thomson effect, the air is cooled.

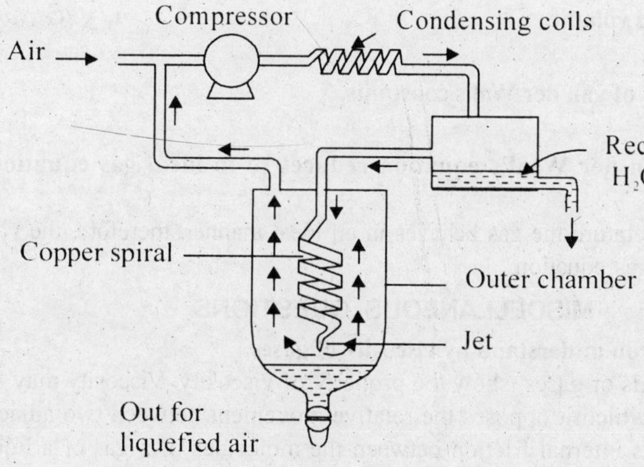

Fig. 1.14. Linde's process for liquefaction of air.

As this cooled air moves up, it further cools the incoming air. The cooled air is sent to the compressor again and the process is repeated a number of times till ultimately the air is cooled to such an extent that it liquefies. The liquefied air collects at the bottom of the chamber and can be drawn off Any uncondensed air is recirculated.

(*ii*) **By the adiabatic expansion of compressed gas involving mechanical work–Claude's process**. This process is based upon the principle that when a gas expands adiabatically against a piston in an engine, it does some external work, hence its internal energy falls and consequently the temperature of the gas falls. This principle, combined with the Joule-Thomson effect, has been applied in the Claude's process for the liquefaction of air. The apparatus used is shown in Fig. 1.15.

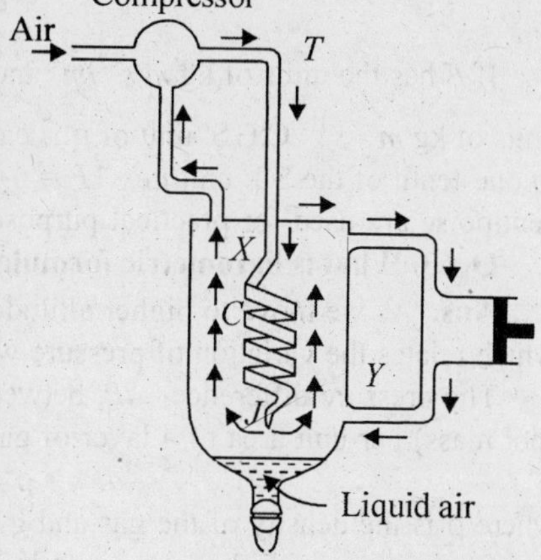

Fig. 1.15. Claude's process.

The purified air *i.e.*, air freed from carbon dioxide and water vapour is first compressed to about 200 atmospheres. The compressed air then passes through the tube T which bifurcates at the point X. A part of the air goes into the cylinder fitted with an airtight piston and the rest of the air passes through the coil C having a jet J at its end. The air that goes into the cylinder pushes the piston and thus does some external work. As a result, the internal energy of the air falls and hence the temperature falls. The cooled air then enters the chamber D at the point Y. The air then passes through the coil C is cooled by Joule-Thomson effect as it comes out of the jet J into the chamber D. The cooled air going upwards in the chamber D further cools the incoming air in the coil C. The cooled air is led to the compressor and the process is repeated a number of times till the air is cooled to such an extent that it liquefies. The liquefied air collects in the chamber and is withdrawn.

(3) **By adiabatic demagnetisation.** The process is based upon the principle that when a magnetised body is demagnetised adiabatically (*i.e.*, in such a way that heat can neither enter nor leave the system) the temperature of the body must fall. When a suitable solid is magnetised, the order is increased, since various magnetic entities are aligned. When demagne-tisation is allowed to occur, the disorder returns and, therefore, the entropy increases. Hence heat must be absorbed and the cooling of the substance and its surroundings results. Now-a-days, the lower temperatures are commonly attained by this methods.

Q59. What is excluded volume. *(G. Nanakdev 2000)*

Ans. See Q. 36.

Q. 60. Write the units of van der Walls constants. *(Meerut 2000)*

Ans. See Q. 39.

Q.61. Prove that van der Waal's equation reduces to to ideal gas equation at Boyle's temperature. *(Meerut 2000)*

Ans. At Boyle's temperature the gas behaves in an ideal manner, therefore, the van der Waal's equation reduces to ideal gas equation.

MISCELLANEOUS QUESTIONS

Q. 62. Explain what you understand by viscosity of gases.

Ans. All fluids, liquids or gases, show the property of viscosity. Viscosity may be defined as the property by virtue of which it opposes the relative movement between two adjacent layers of the substance. There exists internal friction between the molecules of a gas or a liquid, which is manifested as viscosity. In the case of gases, transfer of momentum between two adjacent layers of the gas takes place, which gives rise to viscosity. Coefficient of viscosity, η, is given by the relation

$$F = -\eta \frac{dv_y}{dz}$$

Coefficient of viscosity may be defined as **force per unit area required to establish a unit gradient between two adjacent layers of a fluid (liquid or gas), a unit distance apart.**

The negative sign in the equation signifies that if F is in the positive y-direction, the velocity v_y decreases in successive layers and $\frac{dv_y}{dz}$ is negative.

If F has the units of kg $m\ s^{-2}/m^2$ and $\frac{dv_y}{dz}$ has the units of $m\ s^{-1}/m$ (S.I. units), then η has the units of kg $m^{-1}\ s^{-1}$. C.G.S. unit of η is called poise which is still very much in use. A poise is equal to one-tenth of the S.I. unit *i.e.,* $1P = 0{\cdot}1$ kg $m^{-1}\ s^{-1}$. Generally, smaller units like micropoise and centipoise are used for practical purposes,

Q. 63. What is barometric formula? What is the utility of this formula?

Ans. As we move to higher altitudes, the pressure of the atmosphere decreases. The formula which relates the variation of pressure with altitude is known as barometric formula.

The pressure difference $-dP$, between the altitude x and altitude $x + dx$ is equal to the weight (not mass) per unit area of a layer of gas of thickness dx. It is given by

$$-dP = \rho\, g\, dx \qquad ...(i)$$

where ρ is the density of the gas and g is the acceleration.

Density of an ideal gas is given by

$$\rho = \frac{PM}{RT} \qquad ...(ii)$$

where M is the molecular mass.

From (*i*) and (*ii*)

$$-dP = \left(\frac{PMg}{RT}\right) dx$$

or

$$-\frac{dP}{P} = -\left(\frac{Mg}{RT}\right) dx$$

Integrating the equation, we get

$$ln \frac{P}{P_0} = -Mg\, x/RT$$

This is the equation for barometric formula.

Q. 64. The values of van der waal constant '*a*' for the two gases are *x* and *y* respectively (x >> y). Which of these can be liquefied easily ? *(Calcutta 2007)*

Ans. Force of attraction between the molecules of a gas is one factor that governs the liquefaction process. The quantity '*a*' in van der waal equation for real gases is related to force of attraction between the molecules. A greater value of '*a*' means greater force of attraction, which in turn means greater chance of liquefaction thus the gas having '*x*' as the value of '*a*' can be liquafied more easily.

Q. 65. An ideal gas can never be liquefied. Justify. *(Vidyasagar 2007)*

Ans. According to kinetic theory of gases which is applicable to ideal gases, there are no forces of attraction between the molecules of a gas. In the absence of force of attraction, it is not possible to bring them together and obtain them in the liquefied state.

Q. 66. Can a gas of the type P (V – b) = RT liquefied ? Explain *(Calcutta 2007)*

Ans. In the above equation, the quantity '*a*' which refers to the force of attraction between gaseous molecules, does not occur. In the absence of force of attraction, it is not possible to liquefy the gas.

Q. 67. Show that the excluded volume of a gas is four times the volume of the gas molecules. *(Garhwal 2010)*

Ans. '*b*' in the van der Waals equation is volume correction that must be applied. This is the volume that is not available for the movement of the molecules. However, this is not the actual volume of the molecules. The molecules of a gas are not static. They are in a state of continuous movement in all directions. We may observe that a particle which is in a particular position at one moment may be to the left or right or upwards or downwards next moment. Thus, we may roughly compute that a volume equal to four times the actual volume of the molecules is not available for the movement of other molecules.

2

LIQUID STATE

Q. 1. Discuss structural differences between solids, liquids and gases.

Ans. We shall describe the structures of solids, liquids and gases in terms of kinetic molecular theory.

Molecules in a *gas* are in a state of constant random motion. The space between the molecules is large and therefore, there is a negligible force of attraction between the molecules.

In *liquids*, the force of attraction between the molecules are sufficiently strong to hold them in contact with one another. But, the molecules can move past one another through the vacant spaces or 'holes' here and there.

A liquid may be defined as **a collection of molecules held close to one another but showing random motion through the intervening empty spaces.**

In *solids*, there are strong intermolecular forces, thus fixing the position of each molecule. The molecules in a solid are not able to leave their position. However, small oscillation around the mean position is possible. Structures of gases, liquids and solids may be represented with the help of Fig. 2.1.

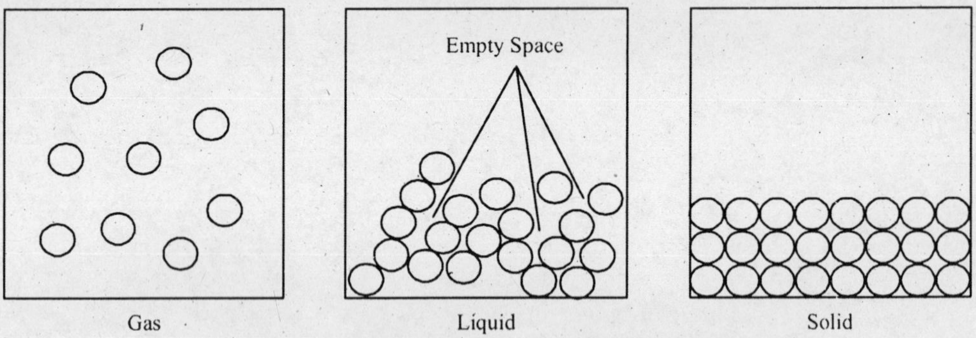

Fig. 2.1. Molecular structures of gas, liquid and solid.

Q. 2. Discuss various intermolecular forces in liquids.

Ans. The intermolecular forces in liquids are of three kinds:

(*a*) *Dipole-dipole interactions.* Such interactions occur in polar molecules. Take the example of HCl. It is a polar molecule with one end having a partial positive charge and the other end having a partial negative charge. Such molecules are called *dipoles.* The positive end of one molecule attracts the negative end of another molecule. There is a net attractive force between the polar molecules. Such attractive forces are called dipole-dipole attraction. The strength of such forces is generally 1% of that of a covalent bond. It may be observed that attractive forces between the opposite poles are stronger than the repulsive forces between like poles. Thus, there is a net force of attraction. It is illustrated in the Fig. below.

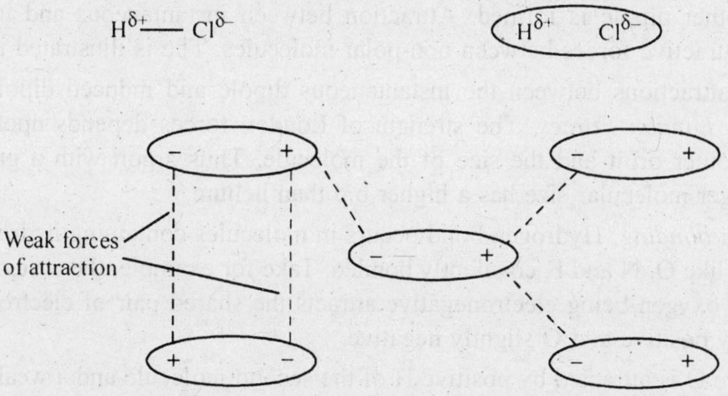

Fig. 2.2. Weak forces of attraction between polar HCl molecules.

(*b*) *London Forces*. Fritz London in 1930 explained the force of attraction between non-polar molecules. A non-polar molecule is formed between the atoms of similar electronegativity by sharing of electrons. The shared electron pairs lie exactly in the middle.

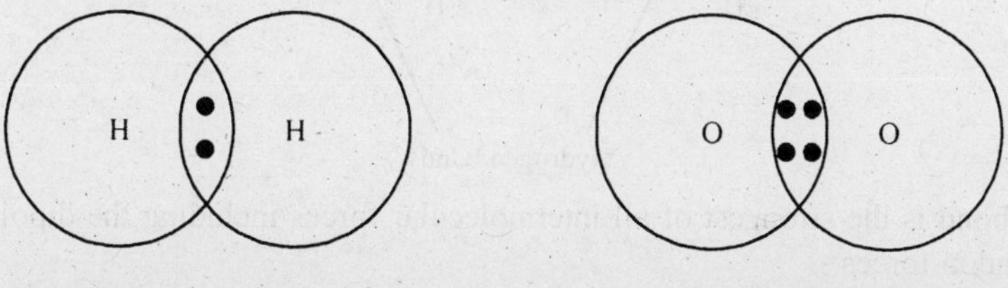

Even these molecules exhibit weak attractive forces. London explained this by saying that according to principle of probability, at a particular instant, the electrons may shift to one side of the molecule making that side negative and the other side positive, thus creating a dipole.

Such a dipole is called *instantaneous dipole.* The negative pole of such a molecule repels the electrons of neighbouring non polar molecule making that end positive and the other end negative.

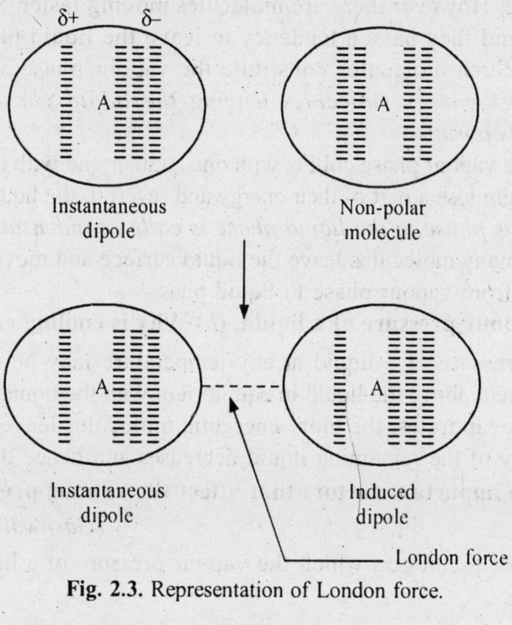

Fig. 2.3. Representation of London force.

Thus, an *induced dipole* is created. The instantaneous dipole and induced dipole will now attract each other. Because of the mobile nature of electrons, instantaneous dipole may disappear the next moment and another dipole is formed. Attraction between instantaneous and induced dipole is responsible for attractive forces between non-polar molecules. The is illustrated in Fig. 2.3.

Momentary attractions between the instantaneous dipole and induced dipole in liquid molecules are called *London Forces.* The strength of London forces depends upon the number of electrons in the outer orbit and the size of the molecule. Thus argon with a greater number of electrons and larger molecular size has a higher b.p than helium.

(*c*) *Hydrogen bonding*. Hydrogen bond occurs in molecules containing hydrogen and an electronegative atom like O, N and F, covalently bonded. Take for example, the case of H_2O molecule. In this molecule, oxygen being electronegative attracts the shared pair of electrons towards itself making H slightly positive and O slightly negative.

Now negative O is attracted by positive H of the second molecule and a weak bond is created between the two, which is called *hydrogen bond.*

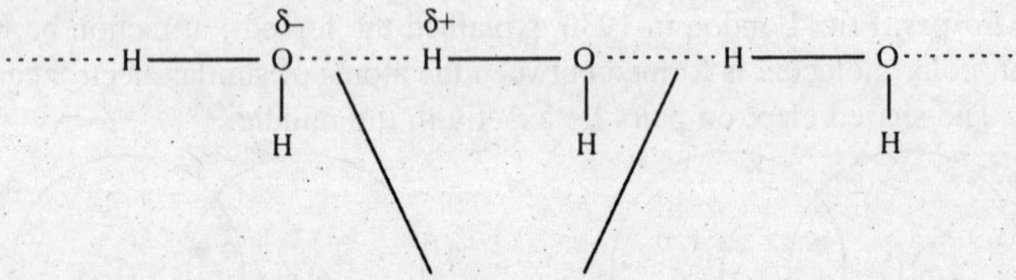

Hydrogen bonds

Hydrogen bond is the strongest of all intermolecular forces including the dipole-dipole interactions and London forces.

It may be mentioned that all the three intermolecular forces mentioned above are collectively called van der walls' forces.

Q. 3. Discuss the phenomenon of evaporation and condensation.

Ans. According to kinetic theory of gases and liquids, the molecules are constantly moving in all possible directions and with all possible velocities. Some of the liquid molecules are moving with small velocities. They are associated with smaller kinetic energies. Such molecules continue moving in the liquid state. However there are molecules moving faster. Such molecules possess higher kinetic energies and they have a tendency to leave the liquid phase and move into the space above the liquid. Such molecules constitute the vapour phase above the liquid phase. *This phenomenon of high-energy molecules leaving the liquid phase and moving into the vapour phase is called evaporation.*

The molecules in the vapour phase collide with one another and with the walls of the container. In this process, some of them lose a part of their energy and revert to the liquid state. *Transference of molecules from the vapour phase to the liquid phase is called condensation.*

At equilibrium, as many molecules leave the liquid surface and move to the vapour phase as the molecules that move from vapour phase to liquid phase.

Q. 4.(*a*) Define vapour pressure of a liquid. (*b*) Why is cooling caused by evaporation?

Ans. (*a*) Vapour pressure of a liquid at any temperature may be defined as the pressure exerted by the vapour present above the liquid in equilibrium with the liquid at that temperature.

(*b*) When a liquid evaporates, the more energetic molecules leave the liquid. As a result, the average kinetic energy of the remaining liquid decreases and hence the temperature falls.

Q. 5.What are the important factors that affect the vapour pressure ?

(*Garhwal 2000; Guwahati 2006*)

Ans. Two important factors on which the vapour pressure of a liquid depends are:

(*a*) *Nature of the liquid:* If the intermolecular forces of attraction in the liquid are weak, the molecules can easily leave the liquid and come into the vapour phase and hence the vapour pressure is higher. For example, the vapour pressures of ether, acetone, benzene etc. are higher than that of water at the same temperature. This is illustrated in Fig. 2.4. Acetic acid gives lower vapour pressure than water as the former has greater intermolecular forces caused by hydrogen bonding.

(*b*) *Effect of temperature :* As the temperature of a liquid is increased the number of molecules with higher kinetic energy increases. Hence the vapour pressure of the liquid increases. This is again evident from each curve in Fig. 2.4.

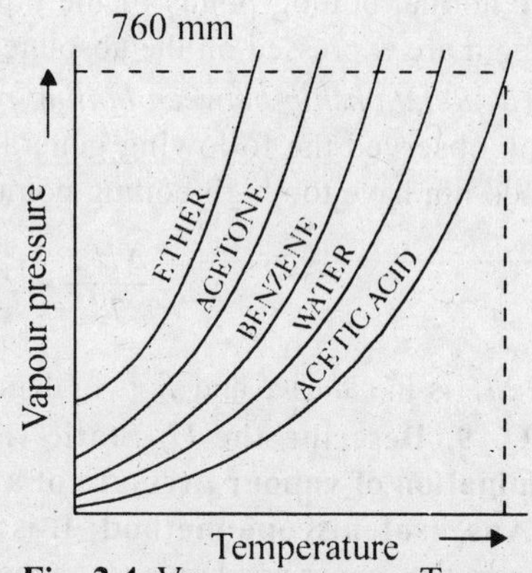

Fig. 2.4. Vapour pressure- Temperature curves for some liquids.

Q. 6. What is meant by boiling point of a liquid ?

Ans. *The temperature at which the vapour pressure of the liquid becomes equal to the atmospheric pressure is called the boiling point of the liquid.* For example, boiling point of water is 100°C. The boiling point for any liquid is obtained from the vapour pressure temperature curves, as shown for acetic acid in Fig. 2.4. If the atmospheric pressure is 760 mm, the boiling point is called normal boiling point.

Obviously if the external pressure is higher, more heat will be required to make the vapour pressure equal to the external pressure and hence higher will be the boiling point. Similarly, if the external pressure is decreased, the boiling point is lowered. This is the reason that a liquid boils at a lower temperature on the top of a mountain (where pressure is low) than on the seashore.

Q. 7. What is meant by heat of vaporisation (ΔH_v) of a liquid ?

(Garhwal 2000)

Ans. After the liquid starts boiling, the extra heat supplied to the liquid is used up not in increasing the temperature of the liquid but to overcome the intermolecular forces of attraction existing in the liquid and thus changing the liquid into vapour. Hence the temperature remains constant till whole of the liquid changes into vapour.

The amount of heat required to change 1 mole of the liquid into its vapour at the boiling point is called the heat of vaporisation of the liquid.

Obviously, greater the intermolecular forces of attraction present in a liquid, greater is the heat of vaporisation. For example, the heat of vaporisation of water is more than that of ether, acetone or benzene.

Q. 8. Give the relationship between (*i*) Boiling point and critical temperature (*ii*) Boiling point and heat of vaporisation (Trouton's rule).

Ans. (*i*) *Relationship between boiling point and critical temperature* (*Guldberg's Rule*): Guldberg gave the following general relationship between the normal boiling point (T_b) and the critical temperature (T_c) of the liquids :

$$T_b \approx \frac{2}{3} T_c$$

i.e., the normal boiling points of the liquids are nearly two -third of their critical temperatures when both are expressed on the absolute scale.

(*ii*) *Relationship between boiling point and heat of a vaporisation* (*Trouton's Rule*): Trouton observed the following generalization about liquids which are non-associated and which do not have too high boiling points.

$$\frac{\Delta H_\vartheta}{T_b} \approx 21$$

where ΔH_ϑ is in calories and T_b is in degrees absolute. This is known as Trouton's Rule.

Q. 9. Describe the (*i*) Static (isoteniscope) method (*ii*) Dynamic method for the determination of vapour pressure of a liquid. *(Gauhati 2006)*

Ans. Isoteniscope method: It is a convenient and accurate method which can be used to measure the vapour pressure over a range of temperatures. The apparatus is shown in Fig. 2.5. The main part of the apparatus is the isoteniscope which comprises a bulb *B* and U-shaped portion of the tube *C*. The bulb *B* is nearly half-filled with the liquid under investigation and the same liquid is placed in the U-shaped portion with its level about 2-3 cm below that in the bulb. The isoteniscope is placed in the thermostat *A* whose temperature can be noted from the thermometer *T*. The isoteniscope is then connected to the rest of the apparatus which comprises a barometric tube *D* and a big bottle *E* as shown in the Fig. 2.5. The bottle *E* can be connected to the atmosphere or to the vacuum pump as desired.

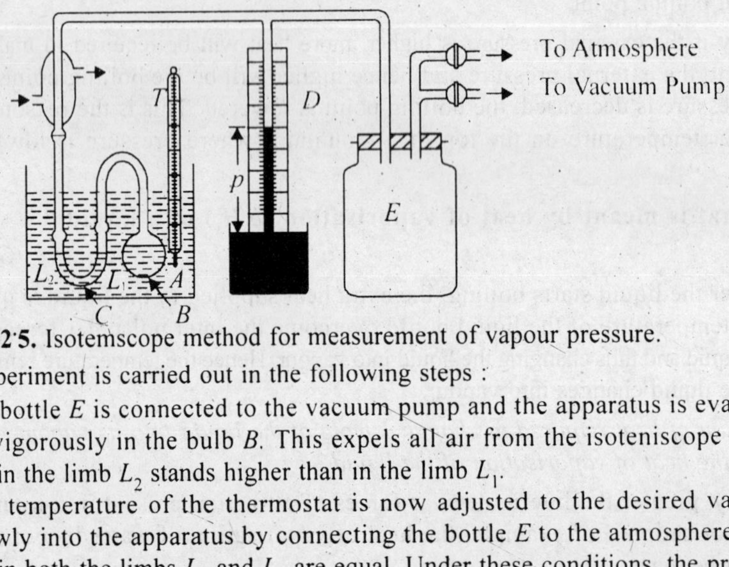

Fig. 2·5. Isotemscope method for measurement of vapour pressure.

The experiment is carried out in the following steps :

(*i*) The bottle *E* is connected to the vacuum pump and the apparatus is evacuated till the liquid boils vigorously in the bulb *B*. This expels all air from the isoteniscope *BC*. The level of the liquid in the limb L_2 stands higher than in the limb L_1.

(*ii*) The temperature of the thermostat is now adjusted to the desired value and air is admitted slowly into the apparatus by connecting the bottle *E* to the atmosphere till the levels of the liquid in both the limbs L_1 and L_2 are equal. Under these conditions, the pressure in both the limbs L_1 and L_2 will be equal. Pressure in the limb L_1 is the vapour pressure whereas pressure in the limb L_2 is equal to the pressure in the rest of the apparatus and is equal to atmospheric pressure minus the height of the mercury column in the tube *D*. Hence this difference is the vapour pressure of the liquid at the temperature of the thermostat.

2. Dynamic methods: The dynamic method is based upon the principle that a liquid boils at the temperature at which the vapour pressure becomes equal to the external pressure over the liquid. Thus by fixing the external pressure over the liquid at different values, the boiling point of the liquid is determined at each external pressure. Suppose the boiling points are t_1, t_2, t_3 etc. at the external pressure p_1, p_2, p_3 etc. respectively. Then obviously, the pressures p_1, p_2, p_3 etc. will be the vapour pressures at the temperatures, t_1, t_2, t_3 etc. respectively. By plotting the pressures against the corresponding temperatures for any particular liquid, a curve of the type shown in Fig. 2·4 is obtained. From this curve, the vapour pressure at any temperature can be found.

Q. 10. What is meant by surface tension of a liquid ?

(Indore 2001; Purvanchal 2002; Jammu 2002; G.Nanak Dev 2003; Arunachal 2003; Guwahati 2007)

Ans. Molecules of the liquid at the surface are in a different situation than those in the interior of the liquid. For example, a molecule lying inside the liquid is surrounded by other molecules and so is attracted equally in all directions (Fig. 2·6). Hence the net force of attraction acting on the molecule is zero. But, a molecule lying at the surface is attracted downwards by the molecules lying in the bulk of the liquid. Thus a molecule lying at the surface experiences a net inward attraction. As a result of this inward pull on all molecules lying at the surface, the surface behaves as if it were under tension. This property of liquids is called surface tension.

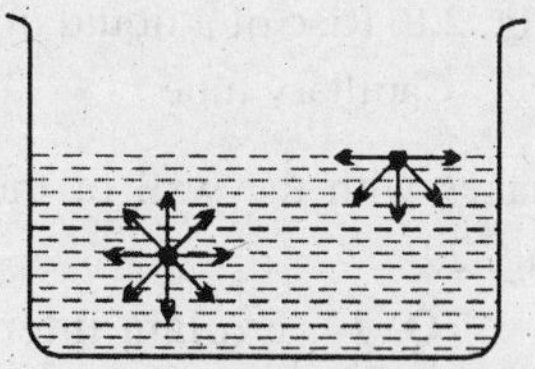

Fig. 2.6. Surface tension

The surface tension of a liquid is defined as the force acting at right angles to the surface along one centimetre length of the surface. Thus the units of surface tension are dynes per cm (or Newtons per metre *i.e.*, Nm^{-1} in the S.I. system).

Q. 11. Why is the drop of a liquid spherical ? *(Arunachal 2003; Guwahati 2007)*

Ans. As a result of the inward pull on the molecules at the surface, the surface of the liquid tends to contract to the smallest possible area for a given volume of the liquid. It is for this reason that the drops of a liquid are spherical because for a given volume, a sphere has minimum surface area.

Q. 12. What is meant by surface energy? *(Garhwal 2000)*

Ans. To increase the area of surface, some work has to be done against the inward pull. Consider a soap solution film contained in rectangular wire frame *ABCD* in which the side *CD* is movable (Fig. 2·7). In order to extend the surface area of the film, the movable wire has to be pulled from position *CD* to position *EF.* Thus some work has to be done against surface tension.

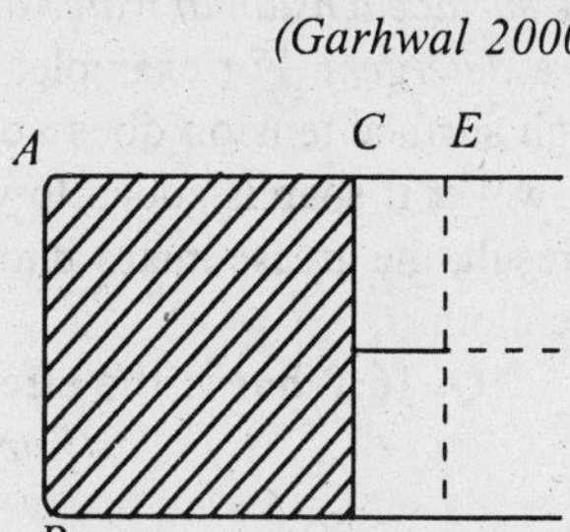

Fig. 2.7. Concept of surface energy.

The work in ergs required to be done to increase or extend the surface area by 1 sq. cm is called surface energy. The *units* of surface energy are, therefore, ergs per sq. cm (or joules per sq. metre in S.I. system).

In terms of dimensions,

Surface energy = work per sq. cm. = (Force × length) per sq. cm.

$$= \frac{\text{dynes} \times \text{cm}}{\text{cm}^2} = \text{dynes cm}^{-1}$$

These units are the same as those of surface tension. Thus the surface energy is same thing as surface tension.

Q. 13. Why does a liquid rise in a capillary tube ? What is meant by meniscus and contact angle ? *(Delhi 2002)*

Ans. When one end of a capillary tube is put into a liquid that *wets glass,* the liquid rises into the capillary tube to a certain height. This rise is obviously due to the inward pull of surface which

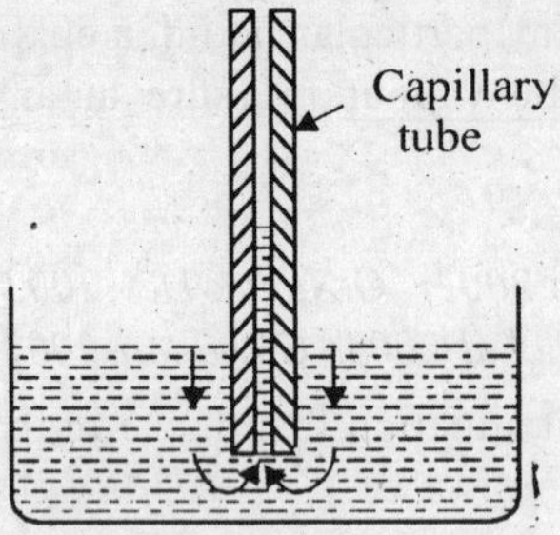

Fig. 2.8. Rise of a liquid in a Capillary tube

pushes the liquid into the capillary tube (Fig. 2.8). In case of liquids which do not wet glass *e.g.* mercury, the level inside the capillary falls below the level outside. Further whereas the upper surface of a liquid that wets glass is concave, that of mercury is convex. Such a surface is called **Meniscus** (Greek : a little moon). The angle which the curved surface (*i.e.* meniscus) makes with the wall of the tube is called **contact angle**, Fig. 2.9.

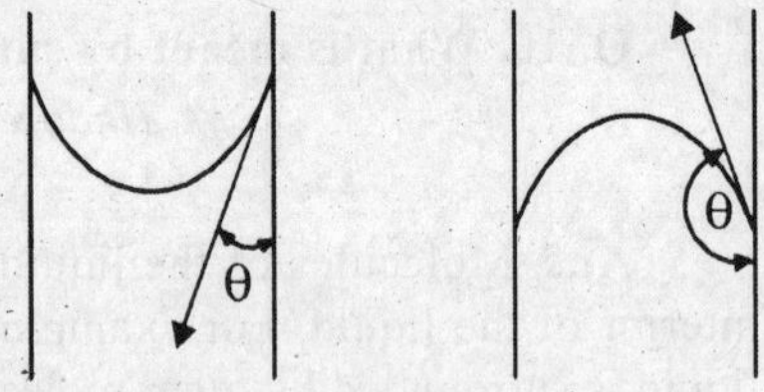

Fig. 2.9. Contact angle for a liquid (*a*) that wets glass (*b*) that does not wet glass.

Q. 14. Explain interfacial tension.

Ans. If two immiscible or partially miscible liquids are taken in a vessel, then the surface tension acting along their surface of separation *i.e.*, along the interface is called the **interfacial tension.** Its value is generally intermediate between the surface tension of the two liquids but sometimes it is less than both. This is due to the fact that at the interface, the molecules of one liquid are attracted by the molecules of the other.

In fact, the surface tension of a liquid is also an interfacial tension, the two phases involved being the liquid and the vapour above it.

Q. 15. What are surface active agents ? *(Lucknow 2000)*

Ans. There are certain substances like soaps, sulphonic acids and organic compounds like methyl alcohol, ethyl alcohol, acetone which when added to water even in small amount decrease the surface tension of water to a considerable extent. *Such substances which lower the surface tension of water are called surface active agents.* It is for this reason that the soap acts as a *detergent.* For example, if grease is sticking on the surface of a cloth, water because of high surface tension does not wet the grease and hence grease is not removed by water alone. However if soap is added to water, it lowers the interfacial tension between water and grease. As a result, the grease mixes into the water (soap solution) and hence is removed from the surface of the cloth.

Q. 16.What is the effect of temperature on surface tension ?
(Kumaon 2000;Madurai 2000; Kurukshetra 2002; M. Dayanand 2002)

or

Surface tension of a liquid vanishes at its critical temperature.
(G. Nanakdev 2003; Calcutta 2007)

Ans. The surface tension of liquids generally decreases with increase of temperature and becomes zero at the critical temperature. The decrease in surface tension with increase of temperature is obviously due to the fact that the kinetic energy of the molecules increases and, therefore, the intermolecular attraction decreases.

An important relationship between temperature and surface tension is as follows:

$$\gamma\left(\frac{M}{D}\right)^{2/3} = k(t_c - t)$$

where M = molecular mass of the liquid

D = density of the liquid at temperature t°C

t_c = critical temperature of the liquid

γ = surface tension (or surface energy) of the liquid at temperature t°C

k= a constant

Katayama suggested a more correct relation *viz.*

$$4\gamma\left(\frac{M}{D-d}\right)^{2/3} = k'(t_c - t)$$

where d is density of the vapour at temperature t. In this equation $\gamma = 0$ when $t = t_c$.

Q. 17. Describe capillary rise method for determining surface tension of a liquid.

(Lucknow 2000; Purvanchal 2002)

Ans. The principle of the method is as follows :

Suppose one end of a capillary tube is put into a liquid which *wets* glass. Then the force of surface tension pushes the liquid into the capillary to a certain height (as shown in the Fig. 2.10) till the force of surface tension acting upwards balances the weight of the liquid column acting downwards.

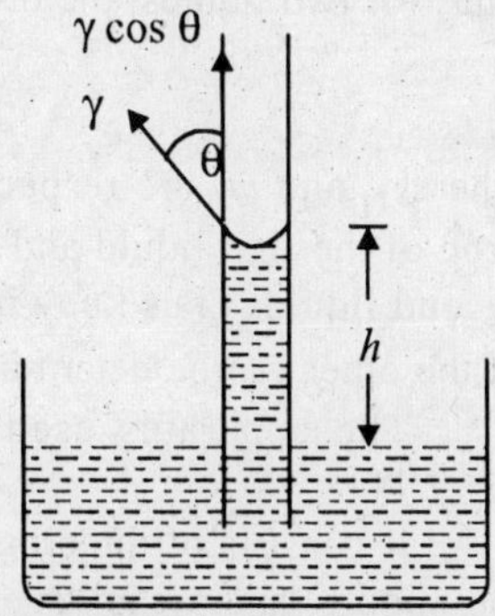

Fig. 2.10. Capillary rise method for measurement of surface tension.

Suppose, height to which liquid rises in the capillary = h cm, density of the liquid $= d$ g/ml, inner radius of the capillary = r cm, surface tension of the liquid = γ dynes/cm and contact angle = θ. Then component of surface tension acting in the vertical direction = $\gamma \cos \theta$.

Inner circumference of the capillary = $2\pi r$

$\therefore$ Force of surface tension acting upwards all along the circumference = $2\pi r .\gamma \cos \theta$ Weight of the liquid acting downwards = $\pi r^2 hdg$ where g is acceleration due to gravity, and $\pi r^2 h$ is the volume of the liquid column.

Hence at equilibrium

$$2\pi r.\gamma\cos\theta = \pi r^2 hdg$$

or

$$\gamma = \frac{rhdg}{2\cos\theta}$$

For most of the liquids which wet glass,

$$\theta = 0$$

so that $\cos \theta = 1$

Hence $\boxed{\gamma = \frac{rhdg}{2}}$

Thus the only two parameters that need to be measured are the inner radius (r) of the capillary tube and the height (h) to which the liquid rises.

Q. 18. At 20°C, toluene rises 1·95 cm in a capillary tube of radius 0·3412 mm. Calculate the surface tension of toluene. The density of toluene at 20°C is 0·866 g per c.c.

(Venkateswara 2004)

Ans. Surface tension is given by the relation

$$\gamma = \frac{rhdg}{2}$$

$r = 0{\cdot}3412$ mm $= 0{\cdot}03412$ cm

$h = 1{\cdot}95$ cm

$d = 0{\cdot}866$ g cm^{-3}

$g = 981$ cm s^{-2}

Substituting the values in the equation above

$$\gamma = \frac{0{\cdot}03412 \times 1{\cdot}95 \times 0{\cdot}866 \times 981}{2} = 28{\cdot}26 \text{ dynes cm}^{-1}$$

Q. 19. Describe (*i*) Drop weight method (*ii*) Drop number method for the determination of surface tension of a liquid using stalgmometer.

(G. Nanakdev 2003; Jammu 2004; H. S. Gaur 2003; Delhi 2004)

Ans. (*a*) **Drop weight method:** This method is based upon the *principle* that the weight of drop falling slowly out of a capillary held vertically is directly proportional to its surface tension. Thus for two liquids, the drops of which fall almost at the same rate out of the same capillary,

$$\frac{\gamma_1}{\gamma_2} = \frac{m_1}{m_2},$$

where γ_1 and m_1 are respectively the surface tension and weight of each drop of the first liquid and γ_2 and m_2 are the corresponding values for the second liquid. Thus knowing the surface tension of one of the liquids, that of the other can be determined.

The apparatus used is called a stalgmometer and is shown in Fig.2.11.

It consists of three parts, a glass tube in the upper portion, a bulb in the middle and a capillary tube as the lower portion. The lowest point of the capillary is flattened to help formation of spherical drops and smooth flow of liquid. There is an etched mark M on the glass tube portion.

Fig. 2.11. Stalgomemter

The procedure consists of the following steps :

(*i*) Stalgmometer is clamped vertically in a clamp stand and a rubber tube with a pinch cock is attached to the upper portion.

(*ii*) Distilled water is sucked into the apparatus and then the pressure adjusted with the help of pinch cock so that the rate of fall of drops is about 15-20 drops per minute.

(*iii*) 10 drops of water are allowed to fall into a clean weighing bottle which has already been weighed when empty. The weight of the weighing bottle alongwith the drops is taken.

(*iv*) The apparatus is now rinsed with acetone and then allowed to dry.

(*v*) Now, the experimental liquid is sucked into the stalgmometer, the rate of fall of drops adjusted as before and 10 drops of the liquid taken into the empty weighing bottle and the bottle weighed.

Calculations :

Suppose weight of empty bottle $= w_1$ g

Weight of bottle + 10 drops of water $= w_2$ g

Weight of bottle + 10 drops of liquid $= w_3$ g

$\therefore$ Weight of 10 drops of water $= (w_2 - w_1)$ g

Weight of 10 drops of liquid $= (w_3 - w_1)$ g

If γ_w is surface tension of water, then the surface tension of the experimental liquid, γ is given by

$$\frac{\gamma_l}{\gamma_w} = \frac{w_3 - w_1}{w_2 - w_1} \quad \text{or} \quad \gamma_l = \frac{w_3 - w_1}{w_2 - w_1} \times \gamma_w$$

Knowing γ_w, γ_l can be calculated.

(*b*) **Drop number method:** As it is more convenient to count the number of drops formed from a certain volume of the liquid than to take the weight of a drop of a liquid, therefore the above method may be carried out in an alternative manner as follows :

Suppose the number of drops formed from v ml of a liquid = n.

Then weight of one drop of the liquid

$$= \frac{v \times d}{n}$$

where d is the density of the liquid. Thus if from the same volume (say v ml) of the two liquids, the number of drops formed are n_1, n_2 and d_1, d_2 are their respective densities, then

$$\frac{\gamma_l}{\gamma_w} = \frac{m_1}{m_2} = \frac{\frac{v \times d_1}{n_1}}{\frac{v \times d_2}{n_2}} = \frac{n_2 \times d_1}{n_1 \times d_2}$$

Hence knowing the number of drops (n_1 and n_2) formed from the same volume of two liquids, knowing their densities (d_1 and d_2) and knowing the surface tension of one of them, that of the other can be calculated.

In the procedure, the apparatus is cleaned and clamped vertically. Distilled water is first sucked into the apparatus upto mark M. The number of drops, while water flows from the mark M to the mark N, is noted. The apparatus is rinsed with acetone and dried. The experimental liquid is now sucked in upto the mark M and the number of drops formed, while the liquid flows from M to N, is counted. The result is then calculated using the formula derived above.

Q. 20. Describe du Nouy torsion balance method for the measurement of surface tension.

Ans. The method is based upon the principle that *the force required to take out a horizontal platinum ring dipping just below the surface of a liquid is directly proportional to the surface tension of the liquid.* The platinum ring used is usually of about 4 cm diameter and the liquid is usually taken in a watch glass. The force required is measured either by attaching the platinum ring to one end of the beam of a sensitive balance and placing weights on the other side or by using a special balance called *torsion balance.* In this balance, the platinum ring is attached to one end of the beam and the other end of the beam is fixed to a wire which can be given a torsion by suitable device. The amount of torsion just sufficient to take out the ring can be recorded on a circular scale attached to one end of the wire as shown in Fig. 2·12. The experiment is first performed with water and then with the experimental liquid. Suppose the torsion angles are θ_1 and θ_2 and their surface tensions are γ_1 and γ_2 respectively.

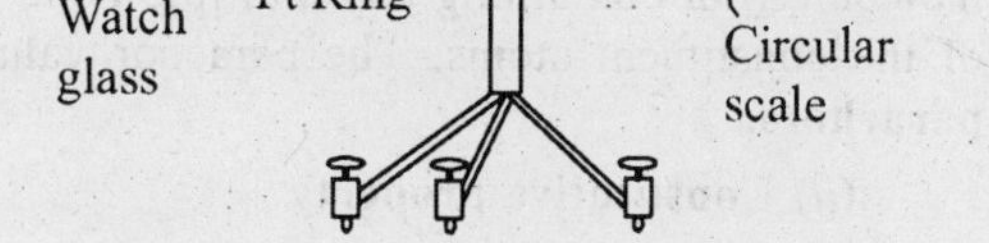

Fig. 2.12. Principle of torsion balance.

Then $$\frac{\gamma_1}{\gamma_2} = \frac{\theta_1}{\theta_2}$$

Thus knowing the surface tension of water, that of the experimental liquid can be found.

Q. 21. Explain the term parachor. *(M. Dayanand, 2000 Garhwal 2000; Indore 2001; Meerut 2002; Lucknow 2002; Purvanchal 2003; Arunachal 2003; Madras 2004;)*

Ans. The following relationship between surface tension and density exists

$$\frac{\gamma^{1/4}}{D-d} = C \qquad ...(i)$$

where γ = surface tension of the liquid

D = density of the liquid

d = density of the vapour at the same temperature

and C = a constant, independent of temperature

Multiplying both sides of equation (*i*) with the molecular weight (M) of the liquid, we get

$$\frac{M\gamma^{1/4}}{D-d} = MC = [P] \qquad ...(ii)$$

The constant $[P]$ thus obtained is called **Parachor.**

At ordinary temperature, d is very small in comparison to D and hence can be neglected in the denominator. Hence equation (*ii*) becomes

$$\frac{M\gamma^{1/4}}{D} = [P] \qquad ...(iii)$$

Putting $\frac{M}{D} = V$, the molar volume of the liquid, equation (*iii*) become

$$V.\gamma^{1/4} = [P]$$

In this equation if $\gamma = 1$, then

$$[P] = V$$

Hence *parachor of a liquid is the molar volume of the liquid at a temperature at which the surface tension of the liquid is unity.*

Q. 22. What is meant by (*i*) additive property (*ii*) constitutive property ?

(Kerala 2001, Garhwal 2000)

Ans. (*i*) Additive property

The additive property of a molecule is defined as that property, the total value of which is the sum of the values of its constituent atoms. Parachor is an example of such a property. Thus each atom has a definite value of parachor and the total parachor value of a simple molecule (not containing any multiple bond or a ring etc.) is the sum of the parachor values of the constituent atoms. The parachor values associated with the atoms are called **atomic parachors.**

(*ii*) Constitutive property

A constitutive property of a molecule is defined as that property which depends upon the constitution of the molecule *i.e.*, the arrangement of atoms within the molecule. In case of parachor, it is found that the total parachor value of a molecule containing some multiple bond or a closed ring etc. depends not only upon the atomic parachors but also upon the multiple bonds or rings present. The parachor values associated with structure of the molecule *i.e.*, the multiple bonds, closed rings etc. are called **structural parachors.**

Values of some atomic and structural parachors are given below in Table 2.1

Table 2-1. Some atomic and structural parachors

Atomic Parachors		*Structural Parachors*	
Atom	*Parachor*	*Structure*	*Parachor*
C	4.8	Single bond	0.0
H	17.1	Double bond	23.2
O	20.0	Triple bond	46.6
Cl	53.8	3-membered ring	16.7
Br	68.0	4-membered ring	11.6
I	90.0	5-membered ring	8.5
N	12.5	6-membered ring	6.1
P	37.7	Co-ordinate bond	– 1.6
S	48.5	O_2 in esters	60.0

Q. 23. Establish the structure of quinone based upon parachor values.

(*Meerut 2002*)

Ans. The following two structures for quinone are possible, the molecular formula for quinone being $C_6H_4O_2$.

I

And

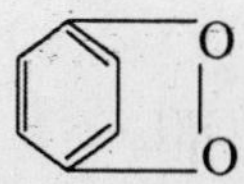

II

Parachor of quinone according to structure I (using values from Table 2.1)

$= 6\ [P]_c + 4\ [P]_H + 2\ [P]_O + 4\ [P]_{\text{Double bond}} + [P]_{\text{6-ring}}$

$= (6 \times 4.8) + (4 \times 17.1) + (2 \times 20.0) + (4 \times 23.2) + 6.1$

$=$ **236.1**

Parachor of quinone according to structure II (using values from Table 2.1)

$= 6\ [P]_c + 4\ [P]_H + 2\ [P]_O + 3[P]_{\text{Double bond}} + 2[P]_{\text{6-ring}}$

(∵ there are two 6-membered rings in II)

$= (6 \times 4.8) + (4 \times 17.1) + (2 \times 20.0) + (3 \times 23.2) + (2 \times 6.1) =$ **219.0**

Observed value of parachor for quinone (From experiment) = **236.8**

This value is close to the value calculated for structure I.

Hence I must be the correct structure for quinone.

Q. 24. Establish the structure of benzene based on parachor values.

Ans. The structure of benzene is confirmed by the parachor values as follows :

Kekule's structure for benzene (C_6H_6) is 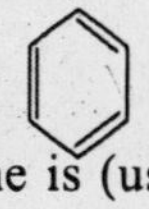

Calculated value of parachor for benzene is (using values from Table 2.1)

$= 6\ [P]_c + 6\ [P]_H + 3\ [P]_{\text{Double bond}} + [P]_{\text{6-ring}}$

$= (6 \times 4.8) + (4 \times 17.1) + (3 \times 23.2) + (6.1)$
$=$ **207.1**

Density of benzene at 20°C, (D)

$= 0.878$ g/ml

Surface tension of benzene at 20°C, (γ)

$= 29.3$ dynes/cm

Molecular weight of benzene (C_6H_6)

$= 78$

∴ Experimental value of parachor, (P)

$$= \frac{M\gamma^{1/4}}{D}$$

$$= \frac{78 \times (29.3)^{1/4}}{0.878} = \mathbf{206.7}$$

Thus the experimental value is in close agreement with the calculated value. Hence Kekule's structure is confirmed.

Q. 25. The parachors of ethane and propane are 110.5 and 150.8 respectively. What values of parachor do you expect for hexane? *(Kurukshetra 2001; Delhi 2003)*

Ans. $[P]\ C_2H_6 = 110.5$

$[P]\ C_3H_8 = 150.8$

$[P]\ CH_2 = [P]\ C_3H_8 - [P]\ C_2H_6$

$= 150.8 - 110.5 = 40.3$

$[P]\ C_6H_{14} = [P]\ C_3H_8 + 3[P]CH_2$

$= 150.8 + 3 \times 40.3 = 271.7$

Q. 26. How high will a liquid rise in a capillary tube of 0.1 mm diameter, if its density be 1.3 g cm^{-3} and surface tension 65 dynes cm^{-1} ? (g= 980 cm s^{-2})

Ans. Diameter of the capillary ($2r$)

$= 0.1$ mm $= 0.01$ cm

∴ Radius, $r = 0.005$ cm

Density of the liquid (d) $= 1.3$ g cm^{-3}

Surface tension of the liquid (γ)

$= 65$ dynes cm^{-1}

The formula for the rise of liquid in the capillary is

$$\gamma = \frac{rhdg}{2}$$

$$h = \frac{2\gamma}{rdg}$$

$$= \frac{2 \times 65}{0.005 \times 1.3 \times 980}$$

$= \mathbf{20.41\ cm}$

Q. 27. The density of acetone at 20°C is 0.7910 g/ml. Calculate the surface tension of acetone, given that the parachor values of C, H, O and double bond are 7.2, 16.2, 20.0 and 23.2 respectively.

Ans. Density of acetone at 20°C (D)

$$= 0.791 \text{ g/ml}$$

Parachor values : $[P]_C = 7.2$

$$[P]_H = 16.2$$

$$[P]_O = 20.2$$

$$[P]_{\text{Doublebond}} = 23.2$$

$\therefore$ Parachor for acetone $(CH_3 - \overset{\overset{\displaystyle O}{\|}}{C} - CH_3)$

$$= 3[P]_C + 6[P]_H + [P]_O + [P]_{\text{Double bond}}$$

$$= 3 \times 7.2 + 6 \times 16.2 + 20.0 + 23.2 = 162$$

Molecular weight of acetone (M) = 58

Surface tension is related to parachor according to the formula

$$\frac{M\gamma^{1/4}}{D} = [P]$$

$$\therefore \quad \gamma^{1/4} = \frac{[P]D}{M}$$

or

$$\gamma = \left[\frac{[P]D}{M}\right]^4$$

Substituting the values for $[P]$, D and M

$$\gamma = \left[\frac{162 \times 0.791}{58}\right]^4$$

$$= \mathbf{23.83 \text{ dynes/cm}}$$

Q. 28. Explain the terms (*i*) Viscosity (*ii*) Coefficient of viscosity.

(Indore,2001;Meerut, Kurukshetra 2002; Lucknow, 2002; Nanakdev, 2003; Kalyani 2003; Delhi ,2003;Madras,2004; Calcutta 2007)

Ans. Consider a liquid flowing through a narrow tube, Fig. 2.13. All parts of the liquid do not move through the tube with the same velocity. Imagine the liquid to be made up of a large number of thin cylindrical coaxial layers. The layer which is in contact with the walls of the tube is almost stationary.

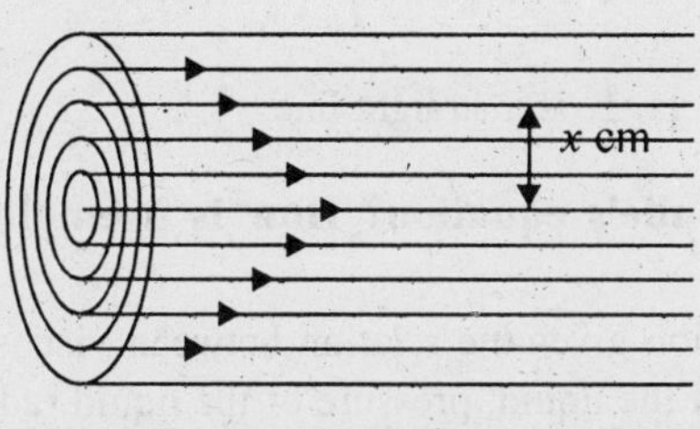

Fig. 2.13. Flow of a liquid through a narrow tube.

As we move from the walls towards the centre of the tube, the velocity of the cylindrical layers goes on increasing. It is maximum at the centre. Conversely, we may say that as we move from the centre towards the walls, the velocity of the layers keeps on decreasing. In other words, every layer offers some resistance or friction to the layer immediately below it. *This force of friction which one part of the liquid offers to another part of the liquid is called viscosity.*

It has been found that the *force of friction* (*f*) between two cylindrical layers each having area *A* sq. cm, separated by a distance *x* cm, and having a velocity difference of *v* cm/sec., is given by

$$f \propto Av/x$$

or

$$f = \eta \frac{Av}{x}$$

where η is a constant known as coefficient of viscosity.

If $x = 1$ cm, $A = 1$ sq. cm and $v = 1$ cm/sec, then $f = \eta$

Hence *coefficient of viscosity may be defined as the force of friction* (*in dynes*) *required to maintain a velocity difference of 1 cm/sec between two parallel layers, 1 cm apart and each having an area of 1 sq. cm.*

The unit of viscosity is poise.

The unit of viscosity in SI units is $Ns\ m^{-2}$ or $kg\ m^{-1}s^{-1}$ It can be shown that

$$1 \text{ poise} = \frac{1}{10} kg\ m^{-1} s^{-1}$$

Q. 29. What is fluidity ? What are the units of fluidity ?

(Kurukshetra 2000; M. Dayanand 2002)

Ans. Reciprocal of viscosity is fluidity represented by ϕ. Thus $\phi = \frac{1}{\eta}$ Unit of fluidity is poise^{-1}.

Q. 30. How does the viscosity of a liquid change with temperature ?

(Lucknow 2000;Nagpur 2002; Delhi 2002; Calcutta 2007)

Ans. With increase of temperature, the kinetic energy of the molecules of the liquid increases. Hence the liquid starts flowing faster. In other words, *the viscosity of a liquid decreases with increase in temperature.* It has been found that the decrease is about 2% per degree rise of temperature.

Arrhenius gave the following relationship between viscosity and temperature:

$$\eta = Ae^{B/RT}$$

where *A* and *B* are the constants for the liquid and *T* is the absolute temperature. The above equation may be written as

$$\log \eta = \log A + \frac{B}{RT}$$

or

$$\log \eta = A' + \frac{B'}{T}$$

Thus the plot of log η vs $\frac{1}{T}$ is a straight line.

Q. 31. What is Poiseulle's equation? How is it employed to find the viscosity of a liquid?

Ans. Poiseulle's equation gives the relation between viscosity of a liquid flowing through a capillary tube and volume of the liquid, pressure of the liquid radius and length of the tube and the time taken for the flow

$$n = \frac{\pi P r^4 t}{8 l v} \quad \text{(Poiseulle's equation)}$$

where v = volume of the liquid flowing through the capillary in time *t*

P = Pressure-head

r = Radius of the capillary tube

l = Length of the tube through which the liquid flows.

It is quite cumbersome to determine experimentally the values of P, r, l and v. Hence, it may not be practical to use Poiseulle's equation for the determination of viscosity of a liquid.

Viscosity is generally determined w.r.t. water. This value is called *relative viscosity*. We note down the time for flow of water from the point M to the point M' on the viscometer (see Fig. 2.14). We repeat the experiment with the unknown liquid. Let t_l and t_{H_2O} be the times for the flow of liquid and water in the same viscometer. As the viscometer is the same, the value of r, l and v will remain the same. We can write the Poisecuille's equation for the liquid and water as under:

$$n_l = \frac{\pi P_l r^4 t_l}{8 l v} \quad ...(i)$$

$$n_{H_2O} = \frac{\pi P_{H_2O} r^4 t_{H_2O}}{8 l v} \quad ...(ii)$$

Divide eq. (i) by eq. (ii)

$$\frac{n_l}{n_{H_2O}} = \frac{P_l t_l}{P_{H_2O} t_{H_2O}} \quad ...(iii)$$

Pressure is proportional to the density. Thus $P_l \propto d_l$ and $P_{H_2O} \propto d_{H_2O}$.

Therefore, equation (iii) is reduced to

$$\frac{n_l}{n_{H_2O}} = \frac{d_l t_l}{d_{H_2O} t_{H_2O}}$$

t_l and t_{H_2O} are determined from the experiment.

Substituting the values of d_l, d_{H_2O} and n_{H_2O}, we can find out the n_l, the viscosity of the liquid.

Q. 32. What is a viscometer ? Describe Ostwald's viscometer method for the determination of viscosity of a liquid.

(Nagpur 2002; Lucknow 2002; Aligarh 2002; Delhi 2003: Calcutta 2007; Garhwal 2010)

Ans. The apparatus used for the measurement of relative viscosities is called a *viscometer*. The simplest is the one given by Ostwald and is called Ostwald's viscometer. It is shown in Fig. 2.14. It consists of a bulb B, to which is sealed a tube above and a capillary below. There is a mark M on the tube and mark M' on the capillary as shown. The lower end of the capillary is sealed to a bigger bulb C and a wider tube CD.

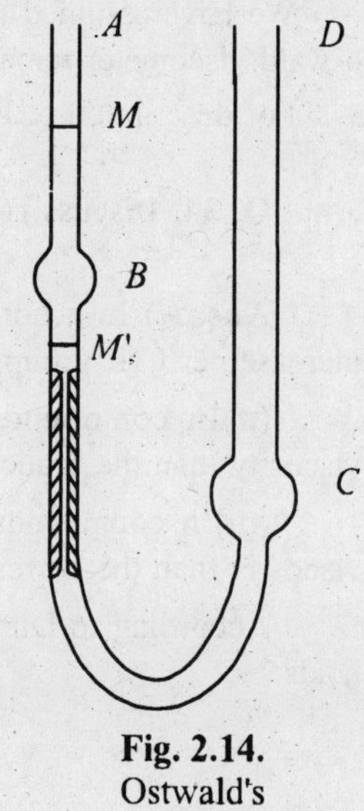

Fig. 2.14. Ostwald's viscometer.

The procedure for viscosity determination consists of the following steps :

(i) The apparatus is first washed with chromic acid and then thoroughly with water. It is then clamped vertically and a rubber tube is attached to the end A.

(ii) A definite volume of water is introduced into the bulb C through the end D.

(*iii*)Through the rubber tube attached to the end *A*, distilled water is sucked up into the bulb *B* so that the level of water is a little above the mark *M*. The water is then allowed to flow back and the time taken for the water to flow from the mark *M* to *M'* is noted.

(*iv*)The apparatus is then dried and the experiment is repeated as above taking the same volume of the experimental liquid. The result is then calculated using the formula as given below:

$$\frac{\eta_l}{\eta_{H_2O}} = \frac{d_l t_l}{d_{H_2O} t_{H_2O}}$$

where η_l = coefficient of viscosity of liquid

η_{H_2O} = Coefficient of viscosity of water

d_l = Density of liquid

d_{H_2O} = Density of water

t_i = Time of flow with the liquid

t_{H_2O} = Time of flow with water

knowing the values of all other terms, η_l can be calculated.

Q. 33. Why do we use the same viscometer for the liquid and water during the experimental determination of viscosity by ostwald viscometer? Describe the experiment.

(Delhi 2000)

Ans. For determination of viscosity of the liquid by ostwald viscometer, we use the equation

$$\frac{n_l}{n_{H_2O}} = \frac{d_l\, t_l}{d_{H_2O}\, t_{H_2O}}$$

This equation is derived using Poisecuille's equation

$$n = \frac{\pi P r^4 t}{8 l v}$$

We have assumed that the values of *r, l* and *v* remain the same. This can happen when the oswald viscometer remains the same for measurement with the liquid and water.

For description of the experiment, refer to *Q*. 32

Q. 34. Discuss relationship between viscosity and chemical constitution.

(Madurai 2000;Purvanchal 2002)

Ans. (*i*) In a homologous series, the viscosity increases as we go up a series and the increase per CH_2 group is almost constant.

(*ii*)In compounds involving chain isomerism, the n-isomers have generally greater viscosity than the branched chain isomers.

(*iii*)In compounds involving geometrical isomerism, trans-isomers have higher value of viscosity than the corresponding cis-isomers.

According to Dunstan's rule, for non-associated or normal liquids, the following equation holds

$$\frac{d}{M}.\eta\times10^6 \approx 40\text{ to }80$$

where d = density of the liquid

η = coefficient of viscosity

M= Molecular mass

For associated liquids, the value is much greater than 80.

Q. 35. Explain the term rheochor. *(Meerut, 2000; Indore, 2001)*

Ans. The following relation exists between rheochor $[R]$, mol wt, density and coefficient of viscosity of the liquid.

$$[R]=\frac{M}{D}=(\eta)^{1/8}$$

where M, D and η are the molecular mass, density and coefficient of viscosity of the liquid respectively.

Thus if $\eta = 1$,

then $[R] = \frac{M}{D} = V$, molar volume

Hence *rheochor may be defined as the molar volume of the liquid at a iemperature at which its coefficient of viscosity is unity.*

Q. 36.How is the molecular weight of polymers determined using coefficient of viscosity? *(Lucknow 2000; Burdwan 2003)*

Ans. If the viscosity of a solution is η at concentration c, and the viscosity of pure solvent is η_O, then the specific viscosity of a solution is defined by the relationship

$$\frac{\eta-\eta_0}{\eta_0} = \eta_{sp}$$

The plot of $\frac{\eta_{sp}}{c}$ vs c is found to be linear. The value of $\frac{\eta_{sp}}{c}$ obtained by extrapolating the plot to zero concentration gives what is called **intrinsic viscosity**, $[\eta]$. It is found to be related to the average molecular weight of the polymer, $\overline{M}$,by the equation

$$[\eta] = k\overline{M^a}$$

where k and a are constants for a given type of polymer and a solvent. The values of these constants are found by first performing the experiments with two samples of the given polymer whose molecular masses have already been found by some other method.

Q. 37. At 20°C, the density of water is 0.9983 g/ml and its viscosity is 0.010087 poise. Explain how these figures indicate that water is an associated liquid.

Ans. Here we are given that

d = 0.9983 g/ml

η = 0.010087 poise

For H_2O

M = 18.

According to Dunstan's rule, $\frac{d}{M}\eta\times10^6$ for normal liquids is from 40 to 60

In this case $\frac{d}{M}\eta\times10^6 = \frac{0.9983}{18}\times0.010087\times10^6 = 559.4$

which is much greater than 80. Hence water is an associated liquid.

Q. 38. The water flow time for an Ostwald viscometer is 59.2 sec at 25°C. If 46.2 seconds are required for the same volume of ethyl benzene (density = 0.867 g cm^{-3}) to flow through the capillary, calculate its absolute viscosity at 25°C, that of water being 0.00895 poise at the same temperature.

Ans. *For water :* t_1 = 59.2 sec

η_1 = 0.00895 poise

d_1 = 1.0 g/cc

For ethyl benzene : t_2 = 46.2 sec

d_2 = 0.867 g /cc

η_2 = ?

Substituting the values in the formula

$$\frac{\eta_1}{\eta_2} = \frac{d_1 t_1}{d_2 t_2}$$

$$\eta_2 = \frac{\eta_1 \times d_2 \times t_2}{d_1 t_1}$$

$$= \frac{0.00895 \times 0.867 \times 46.2}{1 \times 59.2} = \mathbf{0.00606\ poise}$$

Q. 39. Explain the terms (i) refractive index (ii) Snell's law of refraction. *(Kerala 2001; Garhwal 2010)*

When a ray of light travels from one medium to another, it undergoes a change in direction, as shown in Fig. 2.15. This phenomenon is called *refraction*.

If a ray of light travels from air or vacuum to a more dense medium (*i.e.*, any liquid or solid), then the ratio of the sine of the angle of incidence, *i*, to the sine of the angle of refraction, *r*, is constant depending upon the nature of the medium. This constant is called the **refractive index** of the medium. It is usually represented by μ. Thus

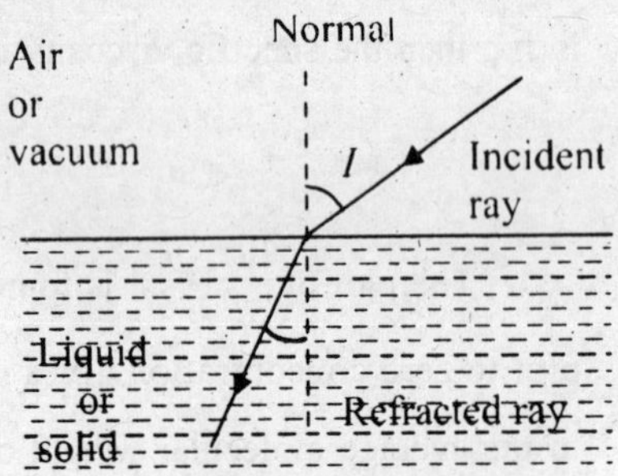

Fig. 2.15. Refraction of light from air to a more dense medium.

$$\mu = \frac{\sin i}{\sin r}$$

This is called **Snell's law.**

The value of refractive index is also found to depend upon the wavelength of the light used and the temperature. These are usually specified as subscripts and superscripts respectively.

Thus $\mu_D^{20°}$ indicates a refractive index taken at 20°C with light corresponding to yellow *D*-line of the sodium spectrum obtained from sodium vapour lamp.

Alternatively, *the refractive index of a medium is also defined as the ratio of the velocity of light in vacuum to the velocity of light in the medium, i.e.,*

$$\mu = \frac{\text{Velocity of light in vacuum}}{\text{Velocity of light in the medium}}$$

Q. 40. (*i*) What is a refractometer (*ii*) What are different types of refractometer (*iii*) Explain the principle of Abbe's refractometer.

Ans. The instrument used for the measurement of refractive index is called a *refractometer.* The two most common refractometers are :

(*i*) Pulfrich refractometer (*ii*) Abbe's refractometer.

The principle of the Abbe's refractometer is explained in Fig. 2.16. The actual shape of the instrument is shown in Fig. 2.17. It consists of two Thin layer of right angled prisms P and Q with their hypotenuse liquid faces together. These can be clamped together with a clamp X in a metal box hinged at Y, as shown in Fig. 2.17. The box is attached to an arm R with the help of which the prism system can be rotated and the value of refractive index can be read directly on the circular scale fixed at the top of the arm R. The hypotenuse face of the lower prism (P) is ground whereas that of the upper prism (Q) is polished. A drop of the liquid is placed on the hypotenuse face of the prism P which spreads between the two faces when the two prisms are fixed together. The light from a suitable source is then reflected by the mirror M on to the liquid layer. On striking the ground surface of the prism P, the light gets diffused in all directions. Thus the rays are incident to the polished face of the prism Q at angles less than 90° (in the liquid medium) and at angles greater than 90° (in the prism Q). The rays incident at angles greater than 90° undergo total

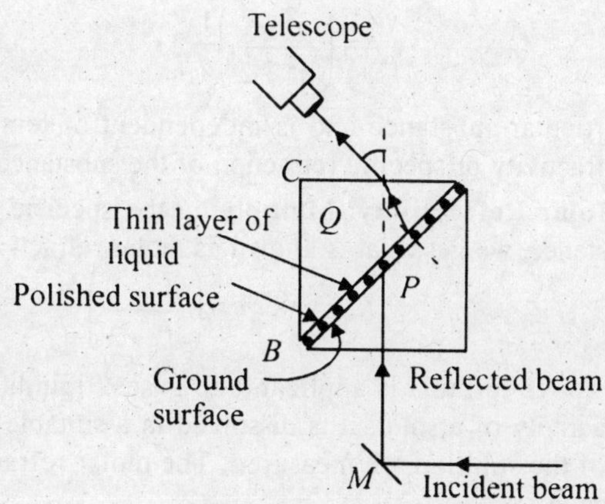

Fig. 2.16. Principle of Abbe's Refractometer.

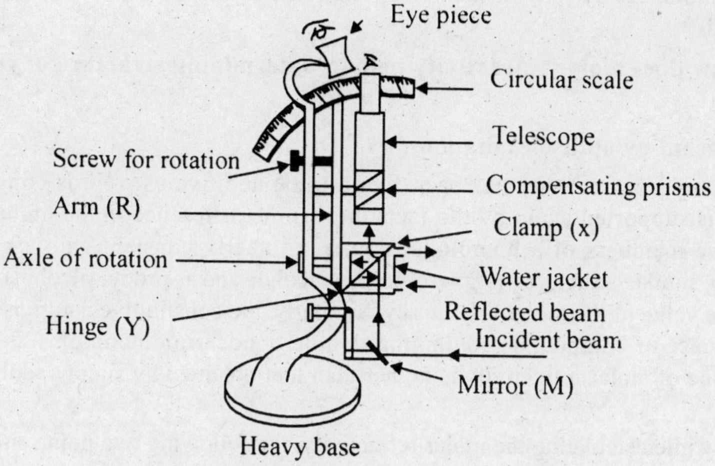

Fig. 2.17. Design of the instrument.

internal reflection within the prism Q and do not come out of the prism Q. The rays incident at angles less than 90° undergo refraction and come out of the prism Q in the form of a bright band of light. To observe the edge of the bright band corresponding to the 'grazing incidence' a telescope is fixed above the prism system and the box containing the prism system is rotated till the edge of the bright band coincides with the cross-wire of the telescope. This setting corresponds to the refracted ray, refracted at the critical angle, e. However, instead of measuring the angles, the circular scale is graduated directly for reading the refractive indices.

Q. 41. Explain the terms (*i*) Specific refractivity (*ii*) Molar refractivity.

(G. Nanakdev, 2000; Kurukshetra, 2001; M. Dayanand, 2002)

Ans. (*i*) **Specific Refractivity:** It has been found that for any substance having refractive index n and density d at the same temperature, the following equation holds good

$$r = \left(\frac{n^2-1}{n^2+2}\right)\frac{1}{d}$$

for the particular substance and is independent of temperature. The quantity r is called the specific refractivity or specific refraction of the substance.

2. Molar Refractivity: Multiplying the specific refractivity by the molecular mass M of the substance, we get what is known as molar refractivity or molar refraction R. Thus

$$R_m = r \times M = \left(\frac{n^2-1}{n^2+2}\right)\frac{M}{d}$$

The above formula is applicable to gases, liquids as well as to solids. To determine the molar refractivity of a solid, it is dissolved in a suitable solvent and the refractive index n and density d of the solution are measured. The molar refractivity R_s of the solution is then given by the formula

$$R_s = \left(\frac{n^2-1}{n^2+2}\right)\left(\frac{N_1M_1+N_2M_2}{d}\right)$$

where N_l and N_2 are the mole fractions of the solvent and solute respectively and M_1 and M_2 are their molecular masses. But the total molar refractivity of the solution is the sum of the contributions made by the solute (solid) and the solvent. Hence

$$R_s = N_1R_1 + N_2R_2$$

where R_1 is the molar refractivity of the solvent and R_2 that of the solid. Thus knowing R_1, R_2 can be calculated.

Q. 42. How does molar refractivity help in determining structure of compounds ?

OR

What is meant by optical exaltation ? *(M. Dayanand 2002)*

Ans. The molar refractivity has been found to be additive as well as constitutive. The additive nature is supported again by the facts that (*i*) the difference in the molar refractivity of the successive members of a homologous series is nearly constant and (*ii*) the isomeric compounds with similar structures (*e.g.* iso-propyl alcohol and *n*-propyl alcohol) are found to have nearly same value of the molar refractivity. Similarly, the constitutive nature is supported by the fact that in case of compounds containing double bond, triple bond or a closed ring etc, the observed value of molar refractivity is higher than that obtained by simply adding the atomic refractivities.

However, while calculating the molar refractivity, the following two points are noteworthy.

(*i*) *The atomic refractivity of oxygen atom is different in alcohols, ethers and ketones.*

(*ii*) *When a compound contains more than one double or triple bonds, the molar refractivity depends not only on the number of double or triple bonds but also on their relative position in the molecule.* When the double bonds or triple bonds are present in conjugated position *e.g.* —CH = CH—CH = CH—, the molar refractivity is higher than the calculated value. This is known as **optical exaltation.**

Thus for hexatriene,

$$CH_2 = CH — CH = CH— CH = CH_2$$

Refractivities for C and H are 2.418 and 1.1 respectively. Refractivity for carbon-carbon double bond is 1.733.

Calculated value of $[R] = 28.52$

Experimental value of $[R] = 30.58$

Optical exaltation $= 30.58 - 28.52 = 2.06$

Q. 43. Density of acetic acid at 25°C is 1.046 g/ml. Calculate the refractive index of acetic acid. Values of R_m (Molecular refractivity) for C = 2.42, H = 1.10, O in OH = 1.50, O in CO = 2.21.

Ans. 1*st step* : To calculate the value of R_m for CH_3COOH

Here we are given

$$R_C = 2.42,\ R_H = 1.10,$$

$$R_O \text{ in} - OH = 1.50,\ R_O \text{ in} > CO = 2.21$$

$$R_m \text{ for } CH_3-\overset{\overset{\displaystyle O}{\|}}{C}-OH = 2R_c + 4R_H + R_O \text{ in—OH} + R_O \text{ in} > CO$$

$$= 2 \times 2.42 + 4 \times 1.10 + 1.50 + 2.21$$

$$= 12.95 \text{ cm}^3 \text{ mol}^{-1}$$

2*nd step* : To calculate the value of n

Also we are given, $d = 1.046$ g/ml

Mol. mass of CH_3COOH $(M) = 60$

Substituting these values in the formula

$$R_m = \left(\frac{n^2-1}{n^2+2}\right)\frac{M}{d}$$

we get

$$12.95 = \left(\frac{n^2-1}{n^2+2}\right)\frac{60}{1.046}$$

or

$$\frac{n^2-1}{n^2+2} = 0.2258$$

$$n^2 - 1 = 0.2258\,n^2 + 0.4516$$

or

$$0.7742\,n^2 = 1.4516$$

$$n^2 = \frac{1.4516}{0.7742} = 1.8753$$

$$\boldsymbol{n = 1.369}$$

Q. 44. Calculate the molar refraction of acetic acid (CH_3COOH) at temperature at which its density is 1.046 g/cm^3. The experimentally observed value of refractive index at this temperature is 1.3715. *(M. Dayanand, 2000)*

Ans. Here we are given

$$d = 1.046 \text{ g/cm}^3$$

$$n = 1.3715$$

Molecular mass of CH_3COOH is $M = 60$

Substituting these value in the formula

$$R_m = \left(\frac{n^2 - 1}{n^2 + 2}\right)\frac{M}{d}$$

$$R_m = \frac{(1.3715)^2 - 1}{(1.3715)^2 + 2} \cdot \frac{60}{1.046}$$

$$= 13.021 \text{ cm}^{-3} \text{ mol}^{-1}$$

Q. 45. A substance having the formula C_3H_6O might be either acetone $\left(\begin{matrix} CH_3 \\ \quad\quad C=O \\ CH_3 \end{matrix}\right)$ or allyl alcohol ($CH_2 = CH—CH_2—OH$). Determine which substance i actually is if its molar refractivity is found to be 16.974 by experiment. Given that refractivities for C =2.418,H = 1.100, O (carbonyl) = 2.211, O(hydroxyl) = 1.525, C = C double bond = 1.733.

Ans. Theoretical molar refractivities of the two compounds are calculated below :

Acetone		*Ally alcohol*	
		3 Carbons	= 7.254
3 Carbons	= 3 × 2.418 = 7.2	6 Hydrogens	= 6.600
6 Hydrogens	= 6 × 1.100 = 6.600	1 Double bond(C = C)	= 1.733
1 Carbonyl oxygen	= 2.211	1 Hydroxyl oxygen	= 1.525
Total	16.065	Total	= 17.112

As the experimental value is close to that calculated value for allyl alcohol, the compound must be allyl alcohol.

MISCELLANEOUS QUESTIONS

Q. 46. Write a note on hydrogen bonding. (*Vidyasagar 2007*)

Ans. See Q. 1 (c)

Q. 47. Viscosity of a liquid will decrease with increase of temperature, Explain. (*Calcutta 2007*)

Ans. See Q. 30

Q. 48. Give Ramsay-Shields equation and its utility. (*Garhwal 2010*)

Ans. This equation is an improvement over Eotvos equation for the calculation of surface tension of a liquid. The commonly used equation (Eotvos equation) is

$$\gamma\left(\frac{M}{D}\right)^{\frac{2}{3}} = k\,(T_c - T)$$

Where M = Molecular mass of the liquid
D = Density of the liquid at temperature T K
T_c = Critical temperature of the liquid
γ = Surface tension of the liquid at temperature T K
k = a constant

William Ramsay and Shields showed an improved version considering that the line normally passes the temperature axis 6 K before the critical point. Thus

$$\gamma\left(\frac{M}{D}\right)^{\frac{2}{3}} = k\,(T_c - 6 - T)$$

3

SOLUTIONS AND DILUTE SOLUTIONS

Q. 1. Define the followi]ng terms :

(*i*) Solution

(*ii*) Solute

(*iii*) Solvent

Ans. (*i*) **Solution :** A solution is defined as a homogeneous mixture of two or more non-reacting components whose relative amounts can be varied upto certain limits. Salt-water and alchohol-water are the examples of solutions.

(*ii*) **Solute :** A solution has two components. The substance which is present relatively in smaller proportion in the solution is called solute. For example, in a sugar solution in water, sugar is the solute.

(*iii*) **Solvent :** The substance present in larger proportion in the solution is called solvent. For example in sugar solution in water, water is the solvent. However in alcohol-water mixture, alcohol can be present as a solute as well as solvent, depending upon whether it is present in smaller or greater amount compared to water.

Q. 2. Describe various methods of expressing the concentration of a solution.

Ans. A number of methods are used to express the concentration of a solution, as given below.

(*i*) **Percentage by weight :** *It is defined as the amount of solute in grams present in 100 grams of the solution.*

$$\text{Thus \% by weight} = \frac{\text{Weight of the solute}}{\text{Weight of the solution}} \times 100$$

(*ii*) **Strength :** *The strength of the solution is defined as the number of grams of the solute dissolved per litre of the solution.*

$$\text{Thus strength} = \frac{\text{Weight of the solute in grams}}{\text{Volume of the solution in litres}}$$

(*iii*) **Molarity :** *It is defined as the number of moles of the solute dissolved per litre of the solution.*

$$\text{Thus molaarity} = \frac{\text{Number of moles of the solute}}{\text{Volume of the solution in litres}}$$

$$= \frac{\text{Strength of the solution in grams per litre}}{\text{Mol. wt. of the solute}}$$

It is denoted by symbol *M*. For example, 0.5 *M* H_2SO_4 means a solution containing 0.5 mol of H_2SO_4 present per litre and is read as 0.5 *molar.*

(*iv*) **Molality:** *It is defined as the number of moles of the solute dissolved in 1000 g of the solvent.*

$$\text{Molality} = \frac{\text{Number of moles of the solute}}{\text{Weight of the solvent in grams}} \times 1000$$

$$= \frac{\text{Number of grams of the solute per 1000 g of the solvent}}{\text{Mol. wt. of the solute}}$$

It is denoted by *m*. For example, 0.1 molal solution means a solution containing 0.1 mol of the solute dissolved per 1000 g of the solvent and is written as 0.1 m.

(*v*) **Normality:** *The normality of a solution is defined as the number of gram equivalents of the solute dissolved per litre of the solution.* Thus

$$\text{Normality} = \frac{\text{Number of gram equivalents of the solute}}{\text{Volume of the solution in litres}}$$

$$= \frac{\text{Strength of the solution in grams per litre}}{\text{Equivalent wt. of the solute}}$$

It is represented by *N*. For example, 1*N* Na_2CO_3 solution means a solution containing one gram equivalent (*i.e.*, 53 g) of Na_2CO_3 dissolved per litre of the solution. It is read as *one normal.*

(*vi*) **Mole Fraction:** *The mole fraction of a component in a solution is defined as the number of moles of that component divided by the total number of moles of all the components.*

Mole fraction of the solute in the solution (x_2)

$$= \frac{\text{No. of moles of the solute}}{\text{No. of moles of solute + No. of moles of solvent}}$$

Mole fraction of solvent in the solution (x_1)

$$= \frac{\text{No. of moles of the solvent}}{\text{No. of moles of solute + No. of moles of solvent}}$$

For example if n_2 moles of the solute are dissolved in n_1 moles of the solvent, we may write

$$x_2 = \frac{n_2}{n_1 + n_2} \quad \text{and} \quad x_1 = \frac{n_1}{n_1 + n_2}$$

Evidently, $x_1 + x_2 = 1$

Q. 3. What is the molarity of a solution of sodium chloride (at wt of Na = 23 and Cl = 35.5) which contains 60 g of sodium chloride in 2000 ml of solution?

Ans. 2000 ml = 2 litre

$$\text{No. of moles of sodium chloride} = \frac{60}{23 + 35.5} = \frac{60}{58.5}$$

$$\text{2 litre of the solution contains} = \frac{60}{58.5} \text{ moles of the solute}$$

$$\text{1 litre of the solution contains} = \frac{60}{58.5} \times \frac{1}{2} = \frac{30}{58.5} = 0.513 \text{ moles}$$

Hence molarity of the solution = 0.513 M.

Q. 4. 0.212 g of Na_2CO_3 with molecular mass 106 is dissolved in 250 ml of the solution. Calculate the molarity of Na_2CO_3 in the solution.

Ans. No. of moles of Na_2CO_3 $= \frac{0.212}{106}$ moles

250 ml of the solution contains $= \frac{0.212}{106}$ moles

1000 ml of the solution contains $= \frac{0.212 \times 1000}{106 \times 250}$ moles

$= 0.008$

Hence molarity of the solution $= 0.008$ M.

Q. 5. 4.0 g of NaOH is contained in a deci-litre of a solution. Calculate the following.

(*i*) Mole fraction of NaOH

(*ii*) Molarity of NaOH

(*iii*) Molality of NaOH

Density of the solution is 1.038 g/ml.

Ans. (*i*) Mole fraction of NaOH

Volume of the solution = 100 ml, Density = 1.038 g/ml

Mass of 100 ml of the solution $= 1.038 \times 100$

$= 103.8$ g

Mass of NaOH $= 4.0$ g

∴ Mass of water $= (103.8 - 4)$ g

$= 99.8$ g

No. of moles of NaOH $= \frac{4}{40} = 0.1$

No. of moles of water $= \frac{99.8}{18} = 5.54$

Mole fraction of NaOH $= \frac{0.1}{0.1 + 5.54}$

$= 0.018$

(*ii*) Molarity of NaOH *(M)*

100 ml of the solution contains = 0.1 moles of NaOH

1000 ml of the solution contains $= \frac{0.1 \times 1000}{100}$

= 1 mole

∴ Molarity of the solution = 1 *M*

(*iii*) Molality of NaOH (*m*)

99.8 g water contains = 0.1 mole of NaOH

1000 g water contains $= \frac{0.1}{99.8} \times 1000$

= 1.002 moles

Hence molality of the solution = 1.002 m

Q. 6. A solution contains 25% water, 25% ethanol and 50% acetic acid by mass. Calculate mole fraction of each component.

Ans. Let the total mass of the solution be 100 g.

(*i*) Mass of water = 25 g

No. of moles of water $= \frac{25}{18} = 1.39$

(*ii*) Mass of ethanol = 25 g

No. of moles of ethanol $= \frac{25}{46} = 0.544$

(*iii*) Mass of acetic acid = 50 g

No. of moles of acetic acid $= \frac{50}{60} = 0.833$

Total No. of moles of the three substances

$= 1.39 + 0.544 + 0.833 = 2.767$

Mole fraction of water $= \frac{1.39}{2.797} = 0.502$

Mole fraction of ethanol $= \frac{0.544}{2.767} = 0.196$

Mole fraction of acetic acid $= \frac{0.833}{2.767} = 0.302$

Q. 7. Find the molarity and molality of a 15% solution of H_2SO_4 (density of H_2SO_4 = 1.10 g/ml and molecular mass of H_2SO_4 = 98).

Ans. Molarity

100 ml of the solution contains = 15 g H_2SO_4

1000 ml of the solution contains $= \frac{15}{100} \times 1000$ g H_2SO_4

= 150 g H_2SO_4

or $= \frac{150}{98}$ moles or 1.53 moles

Hence molarity of the solution = 1.53 *M*

Molality

Mass of 100 ml of solution = 1000 × 1.10 (volume × density)

= 1100 g

Mass of the acid = 150 g

Mass of water = 1100 – 150

= 950 g

950 g water contains = 1.53 moles of H_2SO_4

1000 g water contains $= \frac{1.53}{950} \times 1000 = 1.61$ mole

Hence molality of the solution = 1.61 m

Q. 8. Calculate the mole fraction of water in a mixture of 12 g of water, 108 g of acetic acid and 92 g of ethyl alcohol.

Ans. Mass of water = 12 g

Mol. mass of water = 18 g

No. of moles of H_2O (n_A) $= \frac{12}{18} = 0.66$

Mass of acetic acid = 108 g
Mol. mass of acetic acid = 60 g

No. of moles of acetic acid (n_B) = $\frac{108}{60} = 1.8$

Mass of ethyl alcohol = 92 g
Mol. mass of ethyl alcohol = 46 g

No. of moles of ethyl alcohol (n_c) = $\frac{92}{46} = 2$

Total No. of moles, $n_A + n_B + n_c = 0.66 + 1.8 + 2$

Mole fraction of water $X_A = \frac{n_A}{n_A + n_B + n_C}$

$= \frac{0.66}{0.66 + 1.8 + 2} = 0.148$

Q. 9. 2.82 g of glucose (mol mass = 180) are dissolved in 30 g of water. Calculate the
(*i*) **molality of the solution**
(*ii*) **mole fractions of (*a*) glucose (*b*) water.**

Ans. (*i*) 30 g H_2O contains = 2.82 g of glucose

1000 g H_2O contains = $\frac{2.82}{30} \times 1000$ g of glucose

$= \frac{2.82 \times 1000}{30 \times 180}$ moles of glucose

= 0.52 moles

Hence molality of the solution = 0.52 m

(*ii*) No. of moles of glucose $n_A = \frac{2.82}{180}$

No. of moles of water $n_B = \frac{30}{18}$

Mole fraction of glucose $(X_A) = \frac{2.82/180}{2.82/180 + 30/18} = 0.0093$

As $X_A - X_B = 1$

∴ $X_B = 1 - X_A$

Hence mole fraction of water $X_B = 1 - 0.0093$
$= 0.9907$

Q. 10. What are different types of solutions? Explain with examples.

Ans. There are nine types of solutions depending upon the physical state of the solute and solvent as given in the table below:

Solute	*Solvent*	*Examples*
Gas	Gas	Air, cooking gas.
Liquid	Gas	Moisture in air.
Solid	Gas	Smoke (containing carbon particles in air).
Gas	Solid	Hydrogen absorbed on palladium or platinum or gases absorbed in charcoal.
Liquid	Solid	Mercury in silver (*e.g.* silver amalgam).

Solute	*Solvent*	*Examples*
Solid	Solid	Alloys like brass, bronze etc.
Gas	Liquid	Aerated waters containing CO_2
Liquid	Liquid	Alcohol in water.
Solid	Liquid	Salt or sugar in water.

Q. 11. State and explain Raoult's law. *(Kerala 2000; Lucknow, 2002, Bhopal,2004; G. Nanakdev 2004; Purvanchal 2007, Nagpur 2008)*

Ans. Raoult's law states :

In a solution the vapour pressure of a component at a given temperature is equal to the mole fraction of that component in the solution multiplied by the vapour pressure of that component in the pure state. Mathematically,

$$P_i = x_i \times P^\circ$$

Thus in a *binary solution*

Vapour pressure of the solvent in the solution

= Mole fraction of the solvent in solution × Vapour pressure of the pure solvent ...(*i*)

Now if the solute is non-volatile, it will not contribute to the total vapour pressure of the solution. Thus the vapour pressure of the solution will be the vapour pressure due to solvent in the solution only.

For such solutions,

Vapour pressure of the solution

= Mole fraction of the solvent in solution × Vapour pressure of the pure solvent or

in terms of symbols, we can write

$$p_s = x_1 \times p^o \quad ...(ii)$$

This can be rewritten in the form

$$\frac{p_s}{p^o} = x_1 \quad ...(iii)$$

If the solution contains n_2 moles of the solute dissolved in n_1 moles of the solvent, we have

Mole fraction of the solvent in solution (x_1)

$$= \frac{n_1}{n_1 + n_2}$$

Substituting this value in equation (*iii*), we get

$$\frac{p_s}{p^o} = \frac{n_1}{n_1 + n_2}$$

Subtracting each side from 1, we get

$$1 - \frac{p_s}{p^o} = 1 - \frac{n_1}{n_1 + n_2}$$

or

$$\boxed{\frac{p^o - p_s}{p^o} = \frac{n_2}{n_1 + n_2}} \quad ... (iv)$$

In this expression, $p^o - p_s$ gives the lowering of vapour pressure, $\frac{p^o - p_s}{p^o}$ is called relative lowering of vapour pressure, $\frac{n_2}{n_1 + n_2}$ represents the mole fraction of the solute in the solution. Hence the expression (*v*) may be expressed in words as follows :

The relative lowering of vapour pressure of a solution containing a non-volatile solute is equal to the mole fraction of the solute in the solution.

This is another definition of Raoult's law.

Q. 12. Liquid A (molecular mass 46) and liquid *B* (molecular mass 18) form an ideal solution. At 293 K the vapour pressures of pure *A* and *B* are 44.5 and 17.5 mm of Hg respectively. Calculate (*a*) the vapour pressure of a solution of *A* in *B* containing 0.2 mole fraction of *A*, and (*b*) the composition of the vapour phase.

Ans. (*a*) $P_A^o = 44.5$ mm

$X_A = 0.2$

$P_B^o = 17.5$ mm

$X_B = (1 - 0.2) = 0.8$

According to Raoult's law

$P_A = P_A^o . X_A$

Partial Pressure of *A*, $P_A = P_A^o X_A = 44.5 \times 0.2 = 8.9$ mm.

Partial Pressure of *B*, $P_B = P_B^o X_B$,

$= 175 \times 0.8$

$= 14.0$ mm.

Total pressure $P = P_A + P_B$

$= 8.9 + 14 = 22.9$ mm.

(*b*) In vapour phase

$$\text{Mole fraction of } A = \frac{\text{Partial Presure of A}}{\text{Total pressure}}$$

or

$$X_A = \frac{P_A}{P_A + P_B} = \frac{8.9}{8.9 + 14} = \frac{8.9}{22.9}$$

$$= 0.39$$

$X_B = 1 - 0.39 = 0.61.$

Q. 13 The vapour pressure of two pure liquids A and *B* are 15000 and 30000 Nm^{-2} at 298 K. Calculate the mole fraction of *A* and *B* in the vapour phase when an equimolar solution

of the liquids is made.

Ans. $P_A^o = 15000\ Nm^{-2}$

$P_B^o = 30000\ Nm^{-2}$

In equimolar solution, mole fractions of *A* and *B*, *i.e.*, X_A and X_B are equal. Let

$X_A = X_B = 0.5$

$X_A = 0.5$

$X_{B'} = 1 - 0.5 = 0.5$

Applying Raoult's law of ideal solution.

$P_A = P_A^o X_A$

$= 15000 \times 0.5 = 7500 \text{ Nm}^{-2}$

and $P_B = P_B^o X_B$

$= 30000 \times 0.5 = 15000 \text{ Nm}^{-2}$

Total pressure $P = P_A + P_B$

$= 7500 + 15000 = 22500 \text{ Nm}^{-2}$

In the vapour phase

Mole fraction of $A = \dfrac{\text{Partial pressure A}}{\text{Total pressure}}$

$= \dfrac{7500}{22500} = 0.3333$

Mole fraction of $B = 1 - 0.3333 = 0.6667$

Q. 14. Benzene C_6H_6 (b.p. 353.1 K) and toluene C_7H_8 (b.p. 383.6 K) are two hydrocarbons that form a very nearly ideal solution. At 313 K, the vapour pressure of pure liquids are 160 mm Hg and 60 mm Hg respectively. Assuming an ideal solution behaviour, calculate the partial pressures of benzene and toluene and the total pressure over the following solutions:

(*i*) One made by combining equal number of toluene and benzene molecules.

(*ii*) One made by combining 4 mol of toluene and 1 mol of benzene.

(*iii*) One made by combining equal masses of toluene and benzene.

Ans. (*i*) When the number of molecules of toluene and benzene are equal, that means the number of moles of the two liquids are also equal :

Thus mole fraction of benzene $X_A = \dfrac{1}{1+1} = 0.5$

and mole fraction of toluene $X_B = 1 - 0.5 = 0.5$

According to Raoult's law

$P = P^o . X$

Partial pressure of benzene $P_A = P_A^{\,o} . X_A$

$= 160 \times 0.5 = 80 \text{ mm}$

Partial pressure of toluene $P_B = P_B^{\,o} X_B = 60 \times 0.5 = 30 \text{ mm}$

Total vapour pressure $P_A + P_B = 80 + 30 = 110 \text{ mm}$

(*ii*) Mole fraction of benzene $X_A = \dfrac{1}{1+4} = \dfrac{1}{5} = 0.2$

Mole fraction of toluene $X_B = 1 - 0.2 = 0.8$

Partial pressure of benzene $P_A = P^0_A X_A = 160 \times 0.2 = 32 \text{ mm}$

Partial pressure of toluene $P_B = 60 \times 0.8 = 48 \text{ mm}$

Total vapour pressure $P_A + P_B = 32 + 48 = 80 \text{ mm}$

(*iii*) Here the masses of the two liquids are the same. Let the amount of each be *m* g

Then mole fraction of benzene $X_A = \dfrac{m/78}{m/78 + m/92} = 0.541$

Mole fraction of toluene $X_B = 1 - 0.541 = 0.459$

Partial pressure of benzene $P_A = P_A^{\,o} X_A = 160 \times 0.541$

$= 86.56$ mm of Hg

Partial pressure of toluene $P_B = P_B^o X_B$

$= 60 \times 0.459 = 27.54$ mm of Hg

Total vapour pressure $= 86.56 + 27.54 = 114.1$ mm

Q. 15. The vapour pressures of pure components *A* and *B* are 120 mm and 96 mm Hg. What will be the partial pressures of the components and the total pressure when the solution contains 1 mole of component *A* and 4 mole of component *B* and the solution is ideal? What will be the composition in the vapour phase ?

Ans. $P_A^o = 120$ mm

Here $P_B^o = 96$ mm

$$\text{Mole fraction of } A,\ X_A = \frac{\text{No. of moles of A}}{\text{No. of moles of A + no. of moles of B}}$$

$$= \frac{1}{1+4} = \frac{1}{5} = 0.2$$

Mole fraction of *B*, $X_B = 1 - 0.2 = 0.8$

Partial pressure of component *A*, $P_A = P_A^o \times X_A$

$= 120 \times 0.2 = 24.0$ mm Hg

Partial pressure of *B*, $P_B = P_B^o \times X_B$

$= 96 \times 0.8 = 76.8$ mm

Total pressure $= P_A + P_B = 24 + 76.8$ mm

$= 100.8$ mm

$$\text{Mole fraction of } A \text{ in the vapour phase} = \frac{P_A}{Total\ pressure} = \frac{24}{100.8} = 0.238$$

Mole fraction of *B* in the vapour phase $= 1 - 0.238 = 0.762$

Q. 16. State and explain Henry's law. *(Kurukshetra 2001; Lucknow, 2002; (G. Nanakdev, 2003; Bangalore, 2004; Bhopal 2004)*

Ans. Henry's law states as :

The mass of a gas dissolved in a given volume of the liquid at constant temperature is directly proportional to the pressure of the gas present in equilibrium with the liquid.

Mathematically, $m \propto p$

or $m = kp$

where m = mass of the gas dissolved in a unit volume of the solvent

p = pressure of the gas in equilibrium with the solvent

k = constant called *Henry's law constant,* characteristic of the nature of the gas, the nature of the solvent and the temperature.

Another statement of Henry's law is :

The volume of the gas dissolved, measured at the pressure used, is independent of the pressure.

In fact, the second statement is just a corollary of the first. This can easily be shown as under :

Suppose volume of the liquid taken = V ml

Pressure of the gas = p mm

Mass of the gas dissolved = m grams

Obviously, volume of the gas dissolved $= V$ ml

Now, suppose the pressure is doubled *i.e.*, it becomes $= 2p$

$\therefore$ The mass of the gas dissolved will be $= 2\,m$

But volume of m grams of the gas at pressure p $= V$ ml

$\therefore$ Volume of $2m$ grams of the gas at pressure p $= 2\,V$ ml

or volume of $2m$ grams of the gas at pressure $2p$ $= V$ ml

(by Boyle's law, because temperature is kept constant)

Thus the volume of the gas dissolved at pressure $2p$ is again the same *viz.* V ml

Third statement of Henry's law is as follows :

For a gas in equilibrium with a liquid, the ratio of the concentration of the gas in the solution phase to that in the gaseous phase is constant at constant temperature provided the gas remains in the same molecular state in both the phases.

Q. 17. What are the limitations of Henry's law? *(M. Dayanand 2000; Punjab, 99)*

Ans. Henry's law is applicable only if the following conditions are satisfied :

(*i*) *The pressure should be low and the temperature should be high i.e., the gas should behave like an ideal gas.*

(*ii*) *The gas should not undergo compound formation with the solvent or association or dissociation in the solvent.*

For example, the law is not applicable in case of dissolution of ammonia in water because it undergoes compound formation followed by dissociation.

$$NH_3(g) + H_2O \longrightarrow NH_4OH\ (aq)$$

$$NH_4OH\ (aq) \rightleftharpoons NH_4^+\ (aq) + OH^-\ (aq)$$

Similarly, the law is not applicable to the dissolution of HCl gas in water because it undergoes dissociation after dissolution

$$HCl(g) + aq \longrightarrow HCl\ (aq)$$

$$HCl(aq) \longrightarrow H^+\ (aq) + Cl^-\ (aq)$$

Q. 18. Give the derivation of Duhem-Margules equation.

(M. Dayanand 2002; Kurukshetra 2002)

Ans. Consider a liquid mixture of two components in equilibrium with their vapour. At *constant temperature and pressure,* the condition for an infinitesimal change of composition is given by Gibbs-Duhem equation in the form

$$n_1 d\mu_1 + n_2 d\mu_2 = 0 \qquad \text{... } (i)$$

where n_1 and n_2 are the number of moles of the constituents 1 and 2 respectively in the solution and μ_1 and μ_2 are their chemical potentials.

Dividing eqn. (*i*) by $n_1 + n_2$, we get

$$\frac{n_1}{n_1+n_2} d\mu_1 + \frac{n_2}{n_1+n_2} d\mu_2 = 0$$

or

$$x_1 d\mu_1 + x_2 d\mu_2 = 0 \qquad \text{... } (ii)$$

where x_1 and x_2 represent the mole fractions of the components 1 and 2 in the mixture (solution).

Now, the chemical potential of any component in the mixture depends upon temperature, pressure and composition of the mixture. Hence if the temperature and pressure are kept constant, the chemical potential depends only upon the composition. Then for an infinitesimal change of composition, it is possible to write

$$d\mu_1 = \left(\frac{\partial \mu_1}{\partial x_1}\right)_{T,P} dx_1 \qquad ...(iii)$$

and
$$d\mu_2 = \left(\frac{\partial \mu_2}{\partial x_2}\right)_{T,P} dx_2 \qquad ...(iv)$$

Substituting these values in eqn. (*iv*), we get

$$x_1 \left(\frac{\partial \mu_1}{\partial x_1}\right)_{T,P} dx_1 + x_2 \left(\frac{\partial \mu_2}{\partial x_2}\right)_{T,P} dx_2 = 0 \qquad ...(v)$$

Further we know that the sum of mole fractions is unity *i.e.*

$$x_1 + x_2 = 0$$

Hence
$$dx_1 + dx_2 = 0$$

or
$$dx_1 = -dx_2 \qquad ...(vi)$$

Substituting this value in eqn. (*v*) and simplifying, we get

$$x_1 \left(\frac{\partial \mu_1}{\partial x_1}\right)_{T,P} - x_2 \left(\frac{\partial \mu_2}{\partial x_2}\right)_{T,P} = 0$$

or
$$x_1 \left(\frac{\partial \mu_1}{\partial x_1}\right)_{T,P} = x_2 \left(\frac{\partial \mu_2}{\partial x_2}\right)_{T,P} \qquad ...(vii)$$

The chemical potential of any constituent of a liquid mixture is given by

$$\mu = \mu_o + RT \ln p \qquad ...(viii)$$

where p is the partial pressure of that constituent (assuming that the vapour behave like an ideal gas).

Differentiating eqn. (*viii*) *w.r.t.* x at constant temperature, we get

$$\frac{\partial \mu}{\partial x} = RT \frac{d \ln p}{dx}$$

Thus for the constituents 1 and 2, we can write

$$\frac{\partial \mu_1}{\partial x_1} = RT \frac{d \ln p_1}{dx_2} \qquad ..(ix)$$

and
$$\frac{\partial \mu_2}{\partial x_2} = RT \frac{d \ln p_2}{dx_2} \qquad ...(x)$$

Substituting these values in eqn. (*vii*) and simplifying, we get

$$x_1 \frac{d \ln p_1}{dx_1} = x_2 \frac{d \ln p_2}{dx_2} \qquad ...(xi)$$

or it can be written in the form

$$\frac{d \ln p_1}{d \ln x_1} = \frac{d \ln p_2}{dx_2} \qquad ...(xii)$$

Equations (*xi*) and (*xii*) are two different forms of Duhem-Margules equation.

Q. 19. What is meant by solubility of a solid in a liquid ? What are the factors that affect solubility of a solid ?

Ans. *The solubility of a solid in a liquid at any temperature is defined as the maximum amount of the solid (solute) in grams which can dissolve in 100 g of the liquid (solvent) to form the saturated solution at that particular temperature.*

The following factors affect the solubility of a solid :

(*i*) **Nature of the solute and the solvent:** Solid dissolves in a liquid which is chemically similar to it. This is expressed by saying *"Like dissolves like"*. This statement implies that ionic compounds, dissolve in polar solvents like water and are very little soluble or almost insoluble in non-polar solvents like benzene, ether etc. Similarly, non-polar compounds are soluble in non-polar solvents like benzene, ether, carbon tetrachloride etc. and are very little soluble in water.

(*a*) Common salt (an ionic compound) is more soluble in water than sugar (a covalent compound). Their solubilities in water are 5.3 moles per litre and 3.8 moles per litre respectively.

(*b*) Iodine (a covalent substance) is more soluble in alcohol or carbon tetrachloride (covalent liquids) than in water.

The reason for the behaviour observed above may be explained as follows :

For ionic compounds being soluble in polar solvents, the solubility is on account of the fact that there are strong electrostatic forces of attraction between the ions of the crystal and the polar solvent molecules; the negative ions being attracted by the positive poles of the solvent molecules and the positive ions by the negative poles of the solvent molecules. For example, when water is used as the solvent, the situation may be represented as follows (Fig. 3.1)

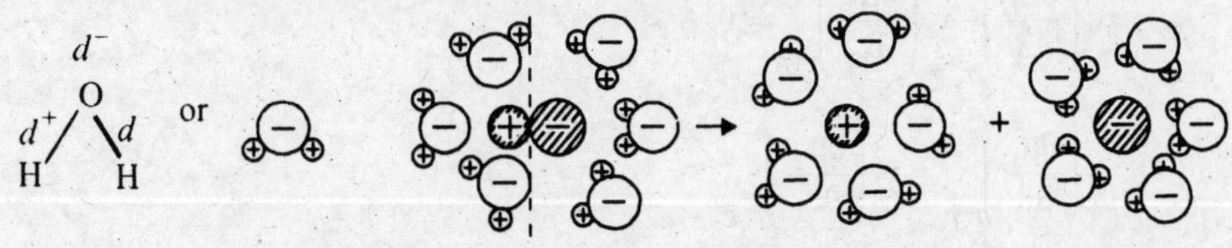

Fig. 3·1. Dissolution of an ionic compound in a polar solvent (H_2O).

Thus the water molecules pull the ions of the crystal apart and the electrostatic forces of attraction existing between the ions of the crystal are cut off. Further, the ions are surrounded by the water molecules which act as a sheath (or envelope) around the ions and prevent the recombination of the ions. The ions thus moving freely in the solution are said to be **hydrated**. It may be mentioned here that whereas energy is required for the splitting of the ionic compound into ions (called *lattice energy*), energy is given out when the ions get hydrated (called hydration energy). A substance dissolves if there is net evolution of energy. Lowering of energy occurs if hydration energy is greater than the lattice energy.

For non-polar compounds being soluble in non-polar solvents, the solubility is due to similar solute-solute, solute-solvent and solvent-solvent interactions.

(*ii*) **Temperature:** Various ionic substances are divided into following categories:

(*a*) *Those whose solubility increases continuously with increase of temperature.* Most of the substances like $NaNO_3$, KNO_3, NaCl, KCl etc. fall into this category. The reason for this behaviour is that in case of all such substances, the process of dissolution is *endothermic i.e.*

$$\text{Solute + Solvent + Heat} \rightleftharpoons \text{Solution}$$

Applying Le Chatelier's principle, as the temperature is increased, equilibrium will shift in a direction in which the heat is absorbed *i.e.*, in the forward direction; consequently more of the solute passes into the solution.

(*b*) *Those whose solubility decreases continuously with increase of temperature.* There are a few substances like cerium sulphate, lithium carbonate, sodium carbonate monohydrate ($Na_2CO_3 . H_2O$) etc. whose solubility decreases with increase of temperature. Obviously it is due to the fact that the process of dissolution of these substances is *exothermic i.e.*, is accompanied by evolution of heat.

Q. 20. What is meant by solubil'ty curve ? Explain the following terms.

(*a*) Continuous solubility curve

(*b*) Discontinuous solubility curve.

Ans. A curve drawn between solubility and temperature is termed Solubility Curve. It gives the effect of temperature on the solubility of a substance. The solubility curves of substances like calcium acetate and calcium chromate show decrease in solubility with increase of temperature while there are others like sodium nitrate and lead nitrate which show a considerable increase of solubility with temperature. The solubility curve of sodium chloride shows very little rise with increase of temperature. Solubility curves of such substances show a continuous change in one direction and thus are called **Continuous Solubility Curves** (Fig. 3.2).

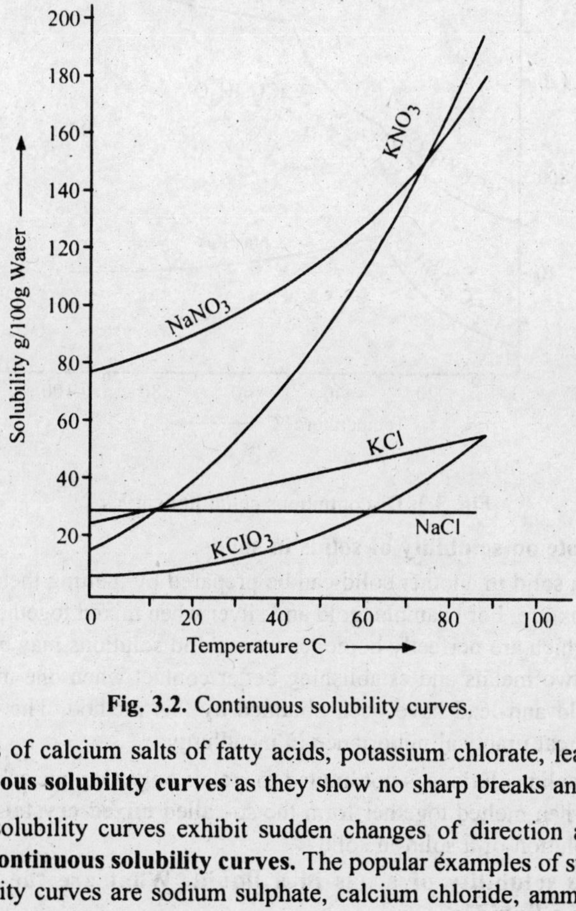

Fig. 3.2. Continuous solubility curves.

Solubility curves of calcium salts of fatty acids, potassium chlorate, lead nitrate and sodium chloride are **Continuous solubility curves** as they show no sharp breaks anywhere.

Sometimes the solubility curves exhibit sudden changes of direction and these curves are, therefore, called **Discontinuous solubility curves.** The popular examples of substances which show discontinuous solubility curves are sodium sulphate, calcium chloride, ammonium nitrate, etc. In fact, at the break a new solid phase appears and another solubility curve of that new phase begins. The break in a solubility curve thus shows a point where the two different curves meet each other (Fig. 3.3).

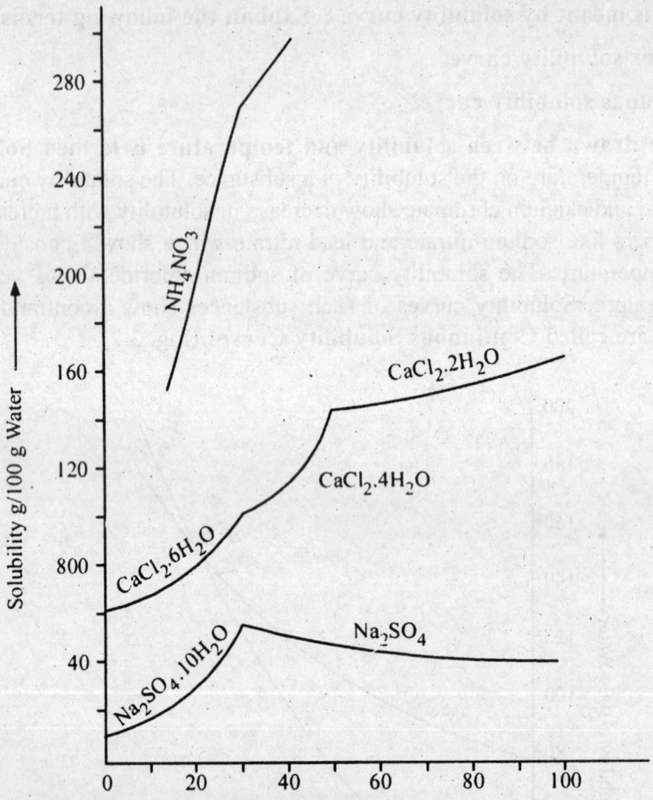

Fig. 3.3. Discontinuous solubility curves.

Q. 21. Write a note on solubility of solids in solids.

Ans. Solution of a solid in another solid can be prepared by melting them together and subsequent cooling of the mixture. For example, gold and silver when mixed together, melted and cooled, yield solid solutions which are perfectly homogeneous. Solid solutions may be obtained by simply pressing together the two metals and establishing better contact when one metal diffuses into the other. Solutions of gold and lead have been obtained by this method. The study of solutions of solids in solids is of great practical importance in metallurgy.

The formation of solid solutions is not limited to metals only. Organic substances like naphthalene and β-naphthol when melted together form the so-called **mixed crystals** on cooling which is another example of solution of a solid in solid.

Q. 22. Define solubility of a gas in a liquid. What are the factors that affect solubility of a gas in a liquid? *(G. Nanakdev; 2003)*

Ans. Solubility of a gas can be defined in two different ways:

(i) The solubility of a gas in a liquid (called the solubility coefficient) is the volume of the gas in ml which will dissolve in 1 ml of the liquid to form a saturated solution, the volume of the gas being measured at the temperature and pressure at which the measurement of the solubility is made.

(*ii*) *The solubility coefficient of a gas in a liquid is also expressed as absorption coefficient which is the volume of the gas in ml which will dissolve in 1 ml of the liquid at the temperature of the experiment under a partial pressure of the gas equal to one atmosphere to form the saturated solution, the volume of the gas being measured at N.T.P.* It is usually represented by α. Thus if v_0 is the volume of the gas dissolved, reduced to S.T.P., V is the volume of the solvent, p is the partial pressure of the gas in atmospheres, then the absorption coefficient α will be given by

$$\alpha = \frac{v_0}{V_{Vp}}$$

The following factors affect the solubility:

(*i*) **Nature of the gas and the solvent:** Gases like hydrogen, oxygen, nitrogen etc. dissolve in water only to a small extent whereas gases like CO_2, HCl, NH_3 etc. are highly soluble. The greater solubility of the latter gases is due to their reaction with the solvent.

Again oxygen, nitrogen and carbon dioxide are much more soluble in ethyl alcohol than in water at the same temperature and pressure while H_2S and NH_3 are more soluble in water than in ethyl alcohol. Evidently, the greater solubility of gas in a solvent is again due to the chemical similarity between the gas and the solvent.

(*ii*) **Effect of temperature:** *The solubility of a gas decreases with increase in temperature.* This is expected because on heating the solution of a gas, some gas is usually expelled out of the solution. The same result also follows in an alternate manner as under :

The dissolution of a gas in a liquid is an exothermic process *i.e.*, it is accompanied by evolution of heat, *i.e.*

$$\text{Gas + Solvent} \rightleftharpoons \text{Solution + Heat}$$

Applying Le Chatelier's principle, it is evident that increase of temperature would shift the equilibrium in the backward direction *i.e.*, the solubility would decrease.

It may be pointed out that though generally the solubility of a gas in a liquid decreases with increase of temperature, there are some exceptions. For example, the solubility of some sparingly soluble gases, such as hydrogen and inert gases, increases slightly with increase of temperature especially in the non-aqueous solvents such as hydrocarbons, alcohols and acetone.

(*iii*) **Effect of pressure:** This is the most important factor influencing the solubility of a gas in a liquid at a particular temperature. It is governed by **Henry's law** according to which the solubility increases with the increase of pressure.

Q. 23. Give a brief account of Henry's Law. (*G. Nanakdev 2004*)

Ans. The relationship between pressure and solubility of a gas in a particular solvent was investigated by William Henry (1803). He gave a generalisation which is known as **Henry's Law.** It may be stated as :

For a gas in contact with a solvent at constant temperature, concentration of the gas that dissolves in the solvent is directly proportional to the pressure of the gas.

Mathematically, Henry's Law may be expressed as

$$C \propto P$$

Or

$$C = kP$$

where P = pressure of the gas; C = concentration of the gas in solution; and k = proportionality constant known as **Henry's Law Constant.** The value of k depends on the nature of the gas and solvent, and the units of P and C used.

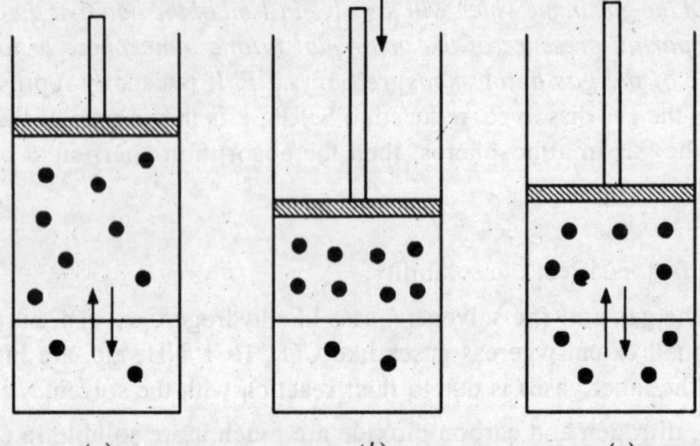

Fig. 3.4. (*a*) Equilibrium exists between the gas molecules and those in solution; (*b*) When pressure is applied, gas volume decreases and concentration in gas phase increases; (*c*) more gas molecules pass into solution to reestablish equilibrium whereby the pressure above the gas is decreased.

Figure 3.4 explains how the concentration of the gas in the solvent increases with rise of pressure.

Limitations of Henry's Law

It applies closely to gases with nearly ideal behaviour under the following conditions :

(*a*) at moderate temperature and pressure.

(*b*) if the solubility of the gas in the solvent is low.

(*c*) the gas does not react with the solvent to form a new species. Thus ammonia (or HCl) which react with water do not obey Henry's Law.

$$NH_3 + H_2O \rightleftharpoons NH_4^+ + OH^-$$

(*d*) the gas does not associate or dissociate in the solvent.

Q. 24. The solubility of pure oxygen in water at 20°C and 1.00 atmosphere pressure is 1.38 × 10^{-3} mole/litre. Calculate the concentration of oxygen at 20°C and partial pressure of 0.20 atmosphere.

Ans. Calculation of k :

$$C = k\,P$$

or
$$k = \frac{C}{P}$$

Substituting values for pure oxygen,

$$k = \frac{1.38 \times 10^{-3} \text{ mole/litre}}{1.00 \text{ atm}} = 1.38 \times 10^{-3} \frac{\text{mol/litre}}{\text{atm}}$$

Calculation of Concentration of O_2 at given pressure

$$C = k\,P$$

$$\text{Concentration of } O_2 = 1.38 \times 10^{-3} \frac{\text{mole / litre}}{\text{atm}} \times 0.20 \text{ atm}$$

$$= 2.76 \times 10^{-4} \text{ mole / litre}$$

Q. 25. What are ideal and non-ideal solutions?

Shivaji, 2000; Lucknow 2000, Kerala 2000, Himachal 2010; Raipur 2010,

Ans. Ideal solutions: A solution of two components is said to be ideal if each component of the solution obeys Raoult's law at all temperatures and concentrations. That is the vapour pressure of each component in the solution at a given temperature is equal to the mole fraction of that component in the solution multiplied by the vapour pressure of that component in the pure state at the same temperature. Thus if two liquids A and B are mixed together, the solution formed by them will be ideal if

$$P_A = x_A \times P_A^o$$

and
$$P_B = x_B \times P_B^o$$

where p_A = vapour pressure of component A in the solution

x_A = mole fraction of the component A in the solution

p_A^o = vapour pressure of the liquid A in the pure state

P_B, x_B, P_B^o are the corresponding terms for the component B.

It is observed that *the solution formed by two liquids A and B is ideal if the interactions (i.e., the intermolecular forces of attraction) among the molecules of A and B in the solution are just the same as among the molecules of pure A or B.*

Further, it is observed that *the solution formed on mixing the two liquids is ideal if*

(*i*) *there is no volume change on mixing i.e.,* $\Delta V_{mix} = 0$

(*ii*) *there is no enthalpy change on mixing i.e.,* $\Delta H_{mix} = 0$

For example, if 50 ml of n-hexane are mixed with 50 ml of n-heptane, the total volume of the solution is found to be 100 ml and the temperature of the solution is found to be same as that of pure n-hexane and n-heptane before mixing showing that no heat is evolved or absorbed on mixing. Such a solution is an ideal solution.

Non-ideal solution: On the other hand, a solution formed by mixing two liquids is said to be non-ideal if it does not obey Raoult's law or the interactions of A and B molecules in the solution are not similar to those of pure A and pure B or $\Delta V_{mix} \neq 0$, and $\Delta H_{mix} \neq 0$.

Q. 26. Draw vapour pressure-composition diagram for ideal solutions. *(Kanpur 2001)*

Ans. In case of ideal solutions, the partial pressure of each component can be calculated using Raoult's law *i.e.*,

$$p_A = x_A \times p_A^o$$

and
$$p_B = x_B \times p_B^o$$

where p_A, p_A^o, p_B and p_B^o are all measured at the same temperature.

Thus $P_{sol.} = x_A p_A^o + x_B p_B^o$

At any particular temperature, the values of $p_{sol.}$ can be determined experimentally for different mole fractions. The values p_A^o and p_B^o for the pure liquids can be measured at the same temperature and hence p_A and p_B can be calculated for any particular mole fraction. For an ideal solution, the straight line plots are obtained for vapour pressure *vs.* composition as shown in Fig. 3.5, where dotted lines represent the plots of the partial pressures and the solid line that of the solution.

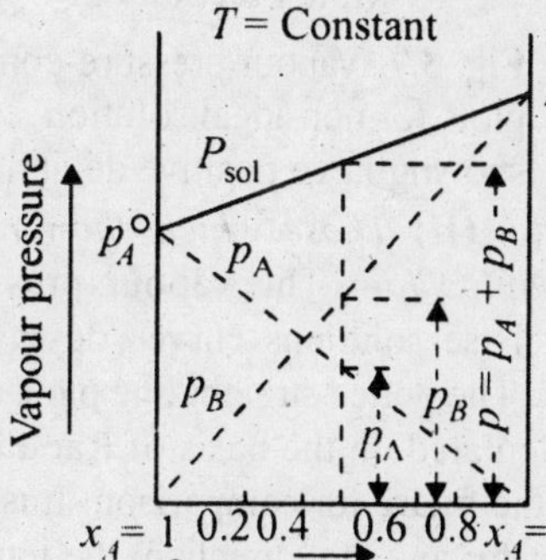

Fig. 3.5. Vapour pressure-composition diagram for an ideal solution.

In these plots, for any mole fraction (say x), the total vapour pressure of the solution is the sum of the partial pressures of its components A and B.

Q. 27. Draw vapour pressure-composition curves for non-ideal solutions.

(Shivaji, 2000; Kanpur, 2001; G. Nanakdev 2003)

Ans. In case of non-ideal solutions, the plots of vapour pressure *vs.* composition (mole fraction) are curved lines instead of straight lines. Non-ideal solutions are divided into the following three types :

Type I: *Those which show small positive deviation from Raoult's law.* The vapour pressure-composition graphs of these solutions curve slightly upwards as shown in Fig. 3.6. The dotted straight line plots are for ideal solution, calculated by using Raoult's law.

Some examples of solutions of this type are benzene + toluene, water + methyl alcohol, carbon tetrachloride + cyclohexane etc.

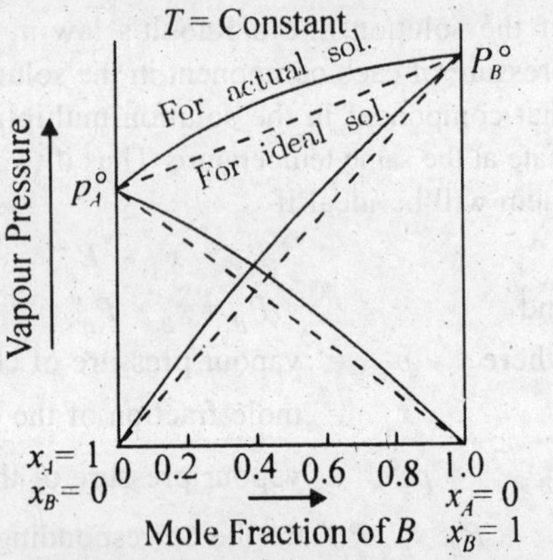

Fig. 3.6. Vapour pressure-composition plots for non-ideal solution showing small positive deviation.

Type II: *Those which show large positive deviation from Raoult's law.* The vapour pressure-composition plots of these solutions curve upwards considerably as shown in Fig. 3.7. The dotted straight line plots, included in the figure for comparison, are for ideal solution (calculated by means of Raoult's law).

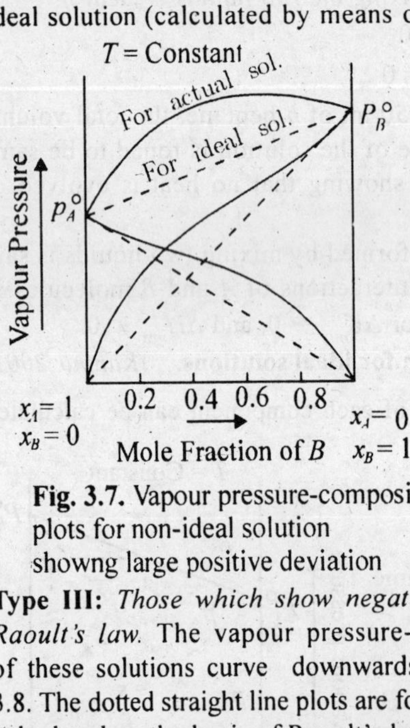

Fig. 3.7. Vapour pressure-composition plots for non-ideal solution showng large positive deviation

It is obvious from Figs. 3.6 and 3.7 that for any mole fraction, the total vapour pressure of the solution is more than that for ideal solution. This means that the tendency for the molecules to escape from the solution is more than from the pure liquids. This indicates that in such solutions, the intermolecular forces of attraction between the molecules of the solution (*A-B* attractions) are weaker than those of either of the pure components (*A-A* attractions or *B-B* attractions). Such a behaviour is associated with an increase in volume and absorption of heat on mixing.

Some examples of liquid pairs showing large positive deviation include water + ethanol, ethanol + chloroform.

Type III: *Those which show negative deviation from Raoult's law.* The vapour pressure-composition plots of these solutions curve downwards as shown in Fig. 3.8. The dotted straight line plots are for the ideal solution, calculated on the basis of Raoult's law and are included in the figure for comparison. It is obvious from the figure that for any mole fraction, the total vapour pressure of the solution is less than that for ideal solution. This means that the tendency for the molecules to escape from the solution is less than that from the pure liquids. This indicates that in such solutions, the intermolecular forces of attraction between the molecules of the solution (*A-B* attractions) are stronger than those of either of the pure

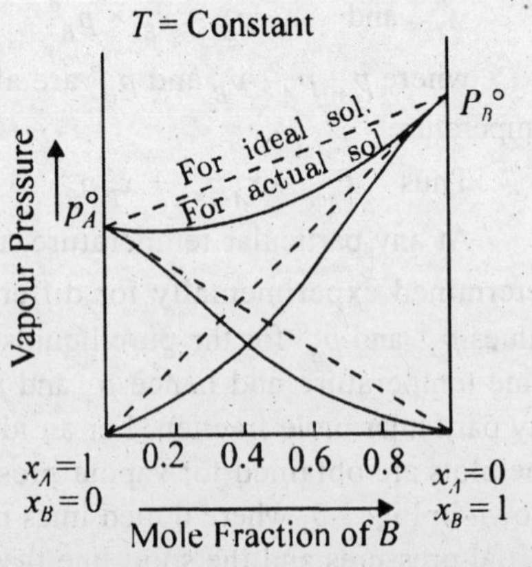

Fig. 3.8. Vapour pressure-composition plots for non-ideal solution showing negative positive deviation.

components (A-A attractions or B-B attractions). It is related to the contraction in volume and evolution of heat in making the solution.

Some examples of the liquid pairs showing negative deviation are water + nitric acid, water + sulphuric acid, acetone + chloroform etc.

Q. 28. Draw vapour pressure-composition diagram showing positive deviation from Raoult's law. Why do positive deviations occur ? *(G. Nanakdev 2003)*

Ans. See Q. 27

Q. 29. State and explain Konowaloff s rule. Derive it from Duhem-Margules equation. *(Himachal 2010)*

Ans. Konowaloff's rule states that in case of ideal solutions or solutions showing small positive or negative deviations, *at any fixed temperature, the vapour phase is always richer in the more volatile component as compared to the solution phase.* In other words, *the mole fraction of the more volatile component is always greater in the vapour phase than in the solution phase.*

The rule is very useful in the separation of binary liquid solutions by distillation.

Derivation of Konowaloff's rule from the Duhem-Margules equation.

According to Duhem-Margules equation,

$$x_1 \frac{d \ln p_1}{dx_1} = x_2 \frac{d \ln p_2}{dx_2} \quad \text{... (i)}$$

It may be rewritten as

$$\frac{x_1}{dx_1}\frac{dp_1}{p_1} = \frac{x_2}{dx_2}\frac{dp_2}{p_2}$$

or

$$\frac{x_1}{p_1}\frac{dp_1}{dx_1} = \frac{x_2}{p_2}\frac{dp_2}{dx_2} \quad \text{... (ii)}$$

But $x_1 + x_2 = 1$ (sum of mole fractions = 1)

$\therefore$ $dx_1 + dx_2 = 0$ or $dx_1 = -dx_2$

$\therefore$ Equation (*ii*) becomes

$$\frac{x_1}{p_1}\frac{dp_1}{dx_1} = -\frac{x_2}{p_2}\frac{dp_2}{dx_1} \quad \text{... (iii)}$$

or

$$\frac{dp_1}{dx_1} = -\frac{x_2 p_1}{x_1 p_2}\frac{dp_2}{dx_1} \quad \text{... (iv)}$$

If P is the total vapour pressure, then

$$P = p_1 + p_2$$

$\therefore$

$$dP = dp_1 + dp_2$$

or

$$\frac{dP}{dx_1} = \frac{dp_1}{dx_1} + \frac{dp_2}{dx_1} \quad \text{... (v)}$$

Substituting the value of dp_1/dx_1 from equation (*iv*), we get

$$\frac{dP}{dx_1} = -\frac{x_2 p_1}{x_1 p_2}\frac{dp_2}{dx_1} + \frac{dp_2}{dx_1}$$

$$= \frac{dp_2}{dx_1}\left(1 - \frac{x_2 p_1}{x_1 p_2}\right) \quad \text{.. (vi)}$$

Now as $dx_1 = -dx_2$

$$\frac{dp_2}{dx_1} = -\frac{dp_2}{dx_2}$$

Thus $\frac{dp_2}{dx_1}$ is a negative quantity. Hence according to equation (*vi*), $\frac{dP}{dx_1}$ can be positive only if

$$\frac{x_2 p_1}{x_1 p_2} > 1$$

or

$$\frac{p_1}{p_2} > \frac{x_1}{x_2}$$

This means that the vapour is richer in component 1 (or *A)* than is the liquid from which it vaporises.

Q. 30. Draw vapour pressure-liquid composition and vapour pressure-vapour composition curves for liquid mixtures.

Ans. At any particular temperature, the composition of the vapour phase is not the same as that of the liquid phase. Hence we have different *vapour pressure-vapour composition curves.* At any particular temperature, these can be obtained by measuring the vapour pressure as well as the composition of the vapour phase corresponding to each composition of the liquid mixture.

For the real solutions, there are following three types of vapour pressure-composition diagrams:

Type I: *Those in which total vapour pressure is intermediate between those of the pure components and do not show any maximum or minimum.*

Type II: *Those which show a maximum in the total vapour pressure curve.*

Type III: *Those which show a minimum in the total vapour pressure curve.*

The vapour pressure-liquid composition curves as well as the vapour pressure-vapour composition curves of all the three types are given in Fig. 3.9, in which liquid *B* is more volatile than liquid *A*

It may be observed that in all the three types, the vapour composition curve lies below the liquid composition curve. This is explained for each type as under :

(*i*) *In type I*, as *B* is more volatile than *A*, corresponding to composition *a* of the liquid phase, vapour phase will be richer in *B i.e.*, corresponding to composition *a′*.

(*ii*) *In type II,* upto the point *C,* behaviour is same as in Type I, at *C,* the liquid phase and the vapour phase have the same composition and after point *C*, the vapour phase is richer in *A* (or less rich in *B*) as seen from points *c* and *c′*.

(*iii*) *In type III,* upto the point *D,* the vapour phase is richer in *A* (or less rich in *B*) than the liquid phase, at the point *D* both have the same composition and after point *D,* the

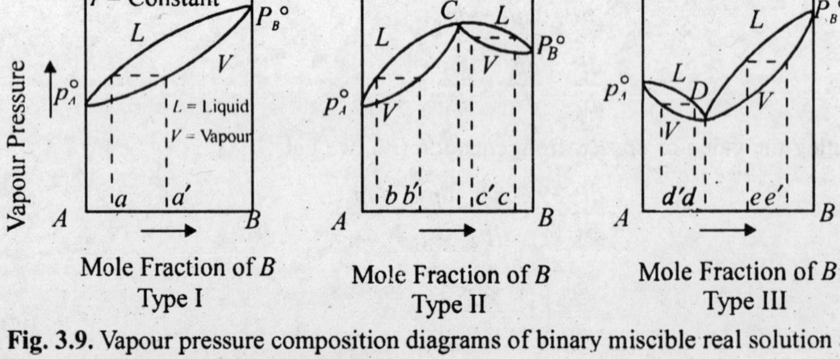

Fig. 3.9. Vapour pressure composition diagrams of binary miscible real solution at constant temperature.

behaviour is similar to Type I. This can be seen from points *d, d′, e* and *e′*.

Q. 31. Draw boiling point-composition diagrams for liquid mixtures.

(Kurukshetra 2001)

Ans. *Boiling point is the temperature at which the total vapour pressure becomes equal to the atmospheric pressure.* Thus a liquid or a solution having low vapour pressure will have a higher boiling point because it will have to be heated more to make the vapour pressure equal to the atmospheric pressure. Thus pure liquid *A* which has less vapour pressure will have a higher boiling point and the pure liquid *B* which has higher vapour pressure will have lower boiling point. In between these two extremes, a solution of any composition having lower vapour pressure will have higher boiling point. These facts help us to construct the boiling point-composition diagrams from the vapour pressure-composition diagrams as shown in Fig. 3.10.

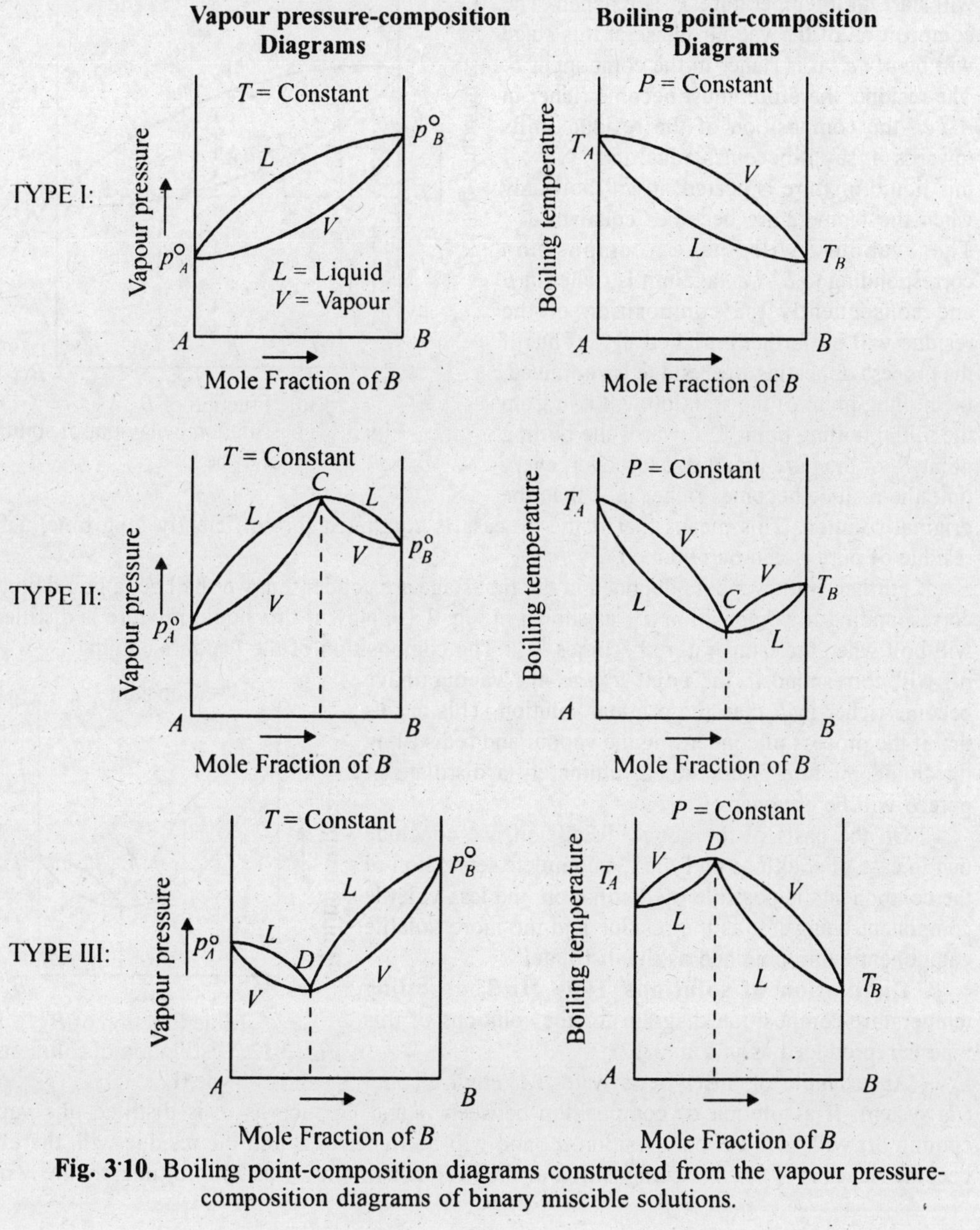

Fig. 3·10. Boiling point-composition diagrams constructed from the vapour pressure-composition diagrams of binary miscible solutions.

The point C in Type II which corresponds to maximum vapour pressure represents the lowest boiling point. Similarly, the point D in type III which represents the lowest vapour pressure corresponds to the highest boiling point. The liquid mixtures having compositions corresponding to points C and D have constant boiling points and are called **azeotropes**.

Q. 32. Discuss the distillation of binary miscible solutions.

(Nagpur 2002; G. Nanakdev 2004)

Ans. Distillation of solutions of Type I-Fractional distillation: The boiling temperature-composition diagram for such a solution involving the liquids A and B in which B is more volatile than A is explained with the help of Fig. 3.11.

Suppose we have a solution corresponding to composition a. No boiling will start until temperature T_a is reached The composition of the vapour phase at this stage will be a' *i.e.,* it is richer in the component B. The residue, therefore, must become richer in A *i.e.,* the composition of the residue shifts towards A, say it becomes equal to b. Now if this liquid mixture is heated, it will boil only when the temperature becomes equal to T_b. The vapour will have composition corresponding to b' *i.e.,* again it is richer in B and consequently the composition of the residue will be further enriched in A. Thus if the process of heating the residue is continued, the boiling point of the solution will rise from the initial boiling point T_a towards the boiling point T_A of the pure liquid A. Moreover, every time the residue becomes richer in A than the original solution. This means that if the process is continued for sufficiently long time, a final residue of pure A can be obtained.

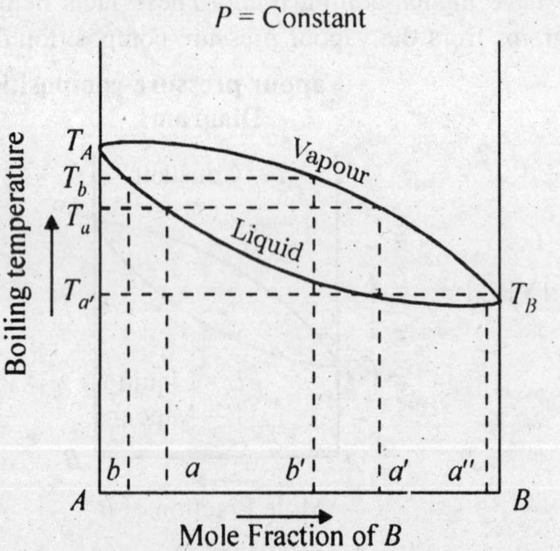

Fig. 3.11. Distillation behaviour of solutions of Type I.

Further, if the vapours obtained in the first stage are condensed, we shall get a liquid mixture corresponding to composition a', as shown in Fig. 3.11. Now if this liquid mixture is distilled, it will boil when the temperature becomes T_a'. The composition of the vapours coming off will correspond to the point a'' *i.e.,* the vapours have become richer in B than the original solution. This means that if the process of condensing the vapour and redistilling the liquid mixture is continued, ultimately a distillate of pure B will be obtained.

On the basis of the above discussion, we conclude that in case of solutions of Type I, a complete separation of the components is possible by distillation, the less volatile component being left as the residue and the more volatile component being obtained as the distillate.

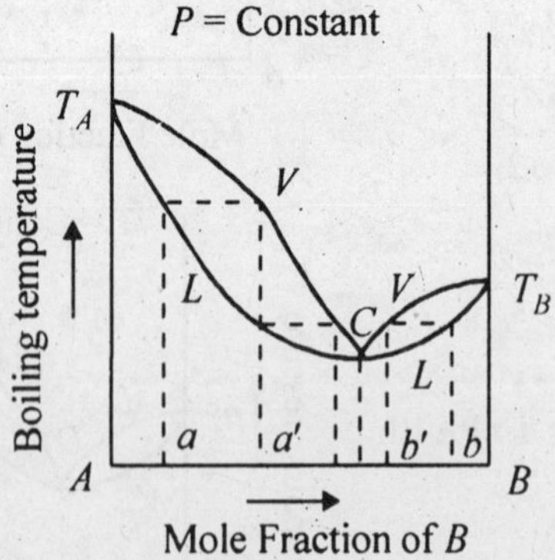

Fig. 3.12. Distillation of solutions of Type II.

Distillation of solutions Type II: The boiling temperature-composition diagram for the solutions of this type is reproduced as in Fig. 3.12.

An example of this type is water (A)-ethyl alcohol (B) system. If a solution of composition between A and C, such as a, is distilled, the vapour coming off will have the composition a' and will be richer in B and the residue will, therefore, become richer in A *i.e.,* the composition of the residue shifts towards A till ultimately the *residue of pure A* is obtained. The liquid obtained on condensing the vapours (corresponding to composition a'), if distilled will give vapour richer in B. If the condensation of vapours and the distillation of the liquid is continued, ultimately the vapours of composition C will be obtained. If these vapours are condensed and the solution distilled, the distillate obtained will have the same composition as the solution. Hence no further separation is possible by distillation.

Thus *"In case of solutions of Type II, a solution of composition between A and C on fractional distillation gives residue of pure A and a final distillate of composition C. No pure B can be recovered."*

Now consider the distillation of a solution having composition between C and B, say corresponding to the point b. The vapour coming off will be richer in A and so the residue will be richer in B. Hence on repeated distillations ultimately a residue of pure B and the final distillate of composition C will be obtained. Thus

"For a solution of composition between C and B, on fractional distillation, a residue of pure B and a final distillate of composition C is obtained. No pure A can be recovered.

Distillation of solutions of Type III: The boiling temperature-composition diagram for solutions of this type, is explained with the help of Fig. 3.13.

The behaviour of solutions of Type III on distillation is analogous to that of solutions of Type II with the exception that the residues tend towards the constant boiling mixture corresponding to composition D whereas distillates tend towards the pure constituents. Further from the Fig. 3.13, it is clear that for a mixture having composition between A and D (say corresponding to a), the vapour coming off are richer in A and hence the residue is richer in B. In other words, the composition of the residue shifts towards D and ultimately becomes equal to that of D, the composition of the vapours shifts towards A and finally a distillate of pure A is obtained. Thus *for a mixture having composition between A and D, distillate of pure A is obtained and the residue has composition corresponding to point D* (*i.e.*, a constant boiling mixture with maximum boiling point). In a similar manner, it can be seen that *for a mixture having composition between D and B, ultimately distillate of pure B and residue of composition corresponding to D is obtained.*

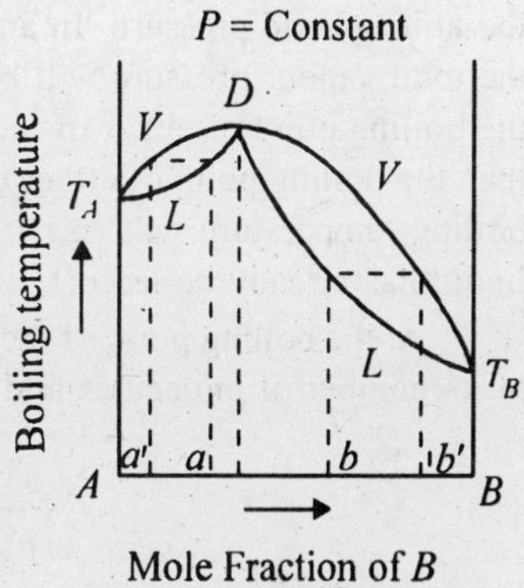

Fig. 3.13. Distillation of solutions of Type III.

Q. 33. Write notes on

(a) Azeotropes *(Lucknow 2000; M. Dayanand 2002; Guwahati 2002; Banglore, 2004, Calcutta 2007, Nagpur 2008)*

(b) Maximum boiling azeotropes

(c) Minimum boiling azeotropes *(Kurukshetra 2002; Assam 2004; G Nanakdev 2004)*

Ans. (*a*) **Azeotropes:** Mixtures of liquids which boil at constant temperature like a pure liquid such that the distillate has the same composition as that of the liquid mixture are called constant boiling mixtures or azeotropic mixtures or simply azeotropes. Evidently, the compo-nents of an azeotrope cannot be separated by fractional distillation. These azeotropes are of two types.

(*b*) **Minimum boiling (point) azeotropes:** These azeotropes are formed by those liquid pairs which show positive deviations from ideal behaviour. Such an azeotrope corresponds to an intermediate composition for which the total vapour pressure is the highest and hence the boiling point is the lowest. Such azeotropes have boiling points lower than either of the pure components. Ethyl alcohol-water is an example of this type of azeotrope.

(*c*) **Maximum boiling (point) azeotropes:** These azeotropes are formed by those liquid pairs which show negative deviations from ideal behaviour. Such an azeotrope corresponds to an intermediate composition for which the total vapour pressure is minimum and hence the boiling point is maximum. Such azeotropes have boiling points higher than either of the pure components. Water HCl is an example of this type of azeotrope.

Q. 34. Discuss the boiling point of a mixture of two immiscible liquids. Derive an expression for the composition of the distillate.

Ans. Immiscible liquids mean those liquids which do not mix with each other at all. Thus if such a mixture of two liquids is heated, each will behave as if the other were not present. Hence the total vapour pressure above the mixture will be same as the sum of the vapour pressures of the two pure constituents *i.e.*,

$$P = p_A^o + p_B^o$$

where P is the total vapour pressure and p_A^o, p_B^o are the vapour pressures of the two pure liquids A and B at the same temperature.

Normal boiling point is the temperature at which the total vapour pressure becomes equal to the atmospheric pressure. In an immiscible liquid pair, since each liquid exerts vapour pressure, the total vapour pressure will become equal to atmospheric pressure at a temperature lower than the boiling point of either of the two liquids. Thus the liquid mixture boils at a temperature lower than the boiling point of either of the two liquids. As soon as one of the liquids is boiled away, the boiling temperature will rise from that of the mixture to either T_A or T_B depending upon which liquid has already boiled off.

At the boiling point of the mixture, the vapour pressure exerted by each liquid is proportional to its number of molecules and hence to the number of moles present in the vapour phase. Thus

$$\frac{p_A^o}{p_B^o} = \frac{n_A}{n_B}$$

where n_A and n_B are the number of moles of A and B respectively present in the vapour phase at the boiling point of the mixture.

Converting moles into weights, we can write

$$n_A = \frac{W_A}{M_A}, \; n_B = \frac{W_B}{M_B}$$

where W_A is the weight of the vapours of A

M_A is the molecular weight of A

W_B is the weight of the vapours of B

M_B is the molecular weight of B

Thus
$$\frac{p_A^o}{p_B^o} = \frac{W_A / M_A}{W_B / M_B}$$

$$= \frac{W_A \cdot M_B}{W_B \cdot M_A}$$

or
$$\boxed{\frac{W_A}{W_B} = \frac{p_A^o \times M_A}{p_B^o \times M_B}}$$

If the vapours are condensed to give the distillate, the ratio of the weights of A and B in the distillate will thus be given by the above expression.

Q. 35. What is steam distillation ? What conditions should be fulfilled by the liquids for carrying out steam distillation ? Give the experimental details of the process.

(Kumaon, 2000; Nagpur, 2002)

Ans. This is method of distillation based upon the principle of boiling of two immiscible liquids and is used in the laboratory or in the industry for the purification of organic liquids.

Applicability of the process: The process is generally applied only under the conditions:

(*i*) The organic liquid to be purified should be immiscible with water

(*ii*) It should have high molecular weight

(*iii*) It should have vapour pressure at about 100°C

(*iv*) Impurities present should be non-volatile

Experimental details of the process: The apparatus used is shown in Fig. 3.14.

The steam produced in the steam generator is passed into the flask containing the impure liquid and kept in a slanting position. This flask is heated from below to prevent the excessive condensation of steam into it. Thus in the space above the liquid in the flask, water vapours as well as the vapours of the liquid are present. The liquid water mixture present in the flask will therefore boil at a temperature lower than the boiling point of either of the two. The vapours on passing through the condenser will give distillate containing the liquid and the water in the ratio calculated using the formula given below :

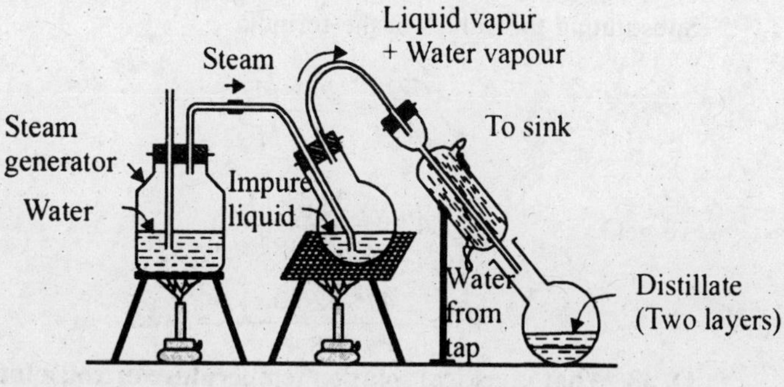

Fig. 3.14. Experimental set-up for steam distillation.

$$\frac{W_l}{W_w} = \frac{p_l^o \cdot M_l}{p_w^o \cdot M_w}$$

The subscripts l and w stand for liquid and water respectively.

Q. 36. A mixture of water and aniline boils at a temperature of 98.5°C at pressure of 760 mm. The vapour pressure of water at this temperature is 717 mm. Find the composition of the distillate.

Ans. Here, $p_w^o = 717$ mm

$p_l^o = 760 - 717 = 43$ mm

Further $M_w = 18, M_l = 93$

$$\therefore \quad \frac{W_l}{W_w} = \frac{p_l^o \cdot M_l}{p_w^o \cdot M_w}$$

$$= \frac{43 \times 93}{717 \times 18} = 0.31$$

i.e. in the distillate, weight of aniline : weight of water = 0.31 : 1

Percentage by weight of aniline $= \frac{0.31}{1+0.31} \times 100 = \frac{0.31}{1.31} \times 100 = \mathbf{23.66}$

Percentage by weight of water $= 100 - 23.66 = \mathbf{76.34}$

Q. 37. An organic liquid was subjected to steam distillation. The liquid in the flask was boiled at 90°C. The external pressure (atmospheric pressure) was found to be 734.4 mm. The vapour pressure of water at 90°C is 526.0 mm. The ratio of the weights of liquid to water in the distillate was found to be 2.47. Calculate the molecular weight of the liquid.

Ans. Here, external pressure = 734.4 mm

Boiling point of the mixture = 90°C

Vapour pressure of water at 90°

i.e., $p_{w'} = 526.0$ mm

∴ Vapour pressure of liquid at 90°

i.e., $p_l^o = 734.4 - 526.0 = 208.4$ mm

Ratio of the weights

$$W_l / W_w = 2.47$$

Molecular weight of water

$$M_w = 18$$

Substituting the values in the formula

$$\frac{W_l}{W_w} = \frac{p_l^o \times M_l}{p_w^o \times M_w}$$

we get $$2.47 = \frac{208.4 \times M_l}{526.0 \times 18}$$

or $$M_l = \frac{2.47 \times 526.0 \times 17}{208.4} = \mathbf{112.2}$$

Q. 38. What is critical solution temperature or consolute temperature? Describe different types of partially miscible liquids. *(Bhopal, 2004; Bangalore 2005 Nagpur 2008; Himachal 2010)*

Ans. The temperature at which the two liquids, which are otherwise partially miscible at ordinary temperature, become completely miscible is called **critical solution temperature** (*CST*). There can be three types of partially miscible systems. Each type is briefly discussed below by taking one example.

Type I. System with an upper critical solution temperature (upper consolute temperature) Phenol-water system. If phenol and water are mixed in roughly equal proportions, two layers are produced, one of a solution of phenol in water and the other of a solution of water in phenol. At any given temperature, the composition of these layers is fixed and can be determined by suitable analytical methods. If the system is warmed, the amount of water in phenol layer increases as well as the amount of phenol in the water layer increases.

Ultimately, at a particular temperature, the composition of both the layers becomes the same and thus the two become completely miscible. Experiments have shown that the minimum temperature at which phenol and water become completely miscible is 66°C and the composition is 34% of phenol by weight. Thus phenol-water system has an upper critical solution temperature of 66°C. The results on being plotted give a parabolic curve as shown in Fig. 3.15.

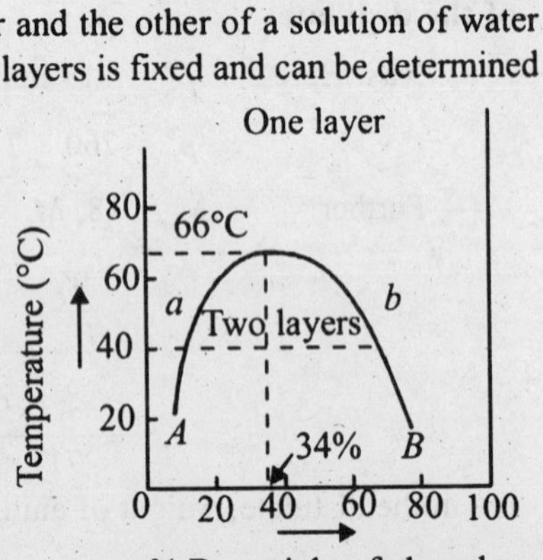

Fig. 3.15. Phenol-water system.

Alternatively, the same curve is obtained if solutions of phenol of different suitable compositions are prepared and the temperature at which the two liquids become completely miscible is determined in each case (from the disappearance of turbidity) and then plotted against the composition.

Type II. System with a lower critical solution temperature (lower consolute temperature): Triethylamine-water system. In this system, it is observed that the mutual solubility of the two liquids increases not with increase of temperature but with decrease of temperature. Thus on cooling such a system, ultimately a stage is reached when the two liquids become completely miscible. This temperature is called the *lower consolute temperature* and is 18.5°C for thriethylamine-water system. This means that below 18.5°C, triethylamine and water are completely miscible in all proportions. The curve obtained by plotting the temperatures against composition is just an inverted form of Type I and is shown in Fig. 3.16.

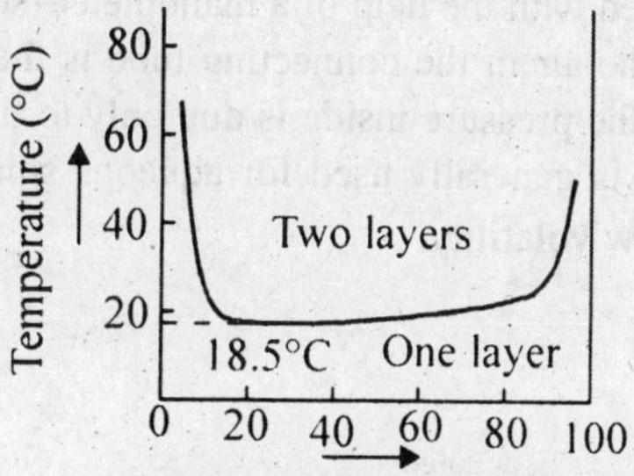

Fig. 3.16. Triethylamine-water system.

Type III. System with an upper as well as a lower critical solution temperature : Nicotine-water system. This is the system in which the mutual solubility of the two liquids increases with increase of temperature as well as with decrease of temperature. Thus in such systems, the two liquids become completely miscible above a particular temperature as well as below a particular temperature. In other words, such systems have an upper critical solution temperature as well as a lower critical solution temperature which are 208°C and 60.8°C respectively for nicotine-water system. The plot obtained between temperature and composition in such cases is a closed curve as shown in Fig. 3.17. As before, for temperature and composition corresponding to any point within the closed curve, there are two layers while outside it, there is one layer only.

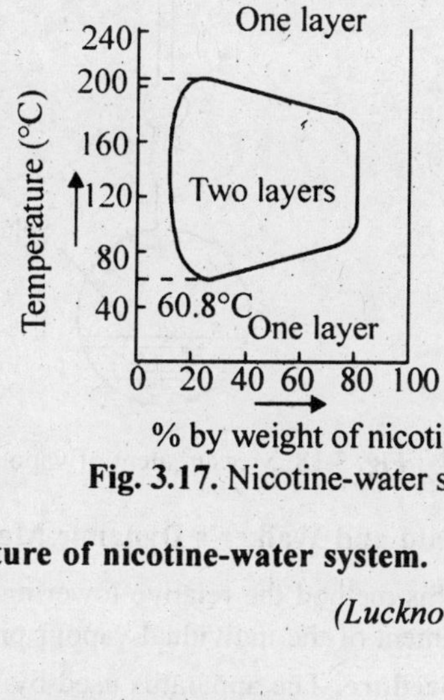

Fig. 3.17. Nicotine-water system.

Q. 39. Briefly explain critical solution temperature of nicotine-water system.

(Lucknow 2000)

Ans. See Q. 38.

COLLIGATIVE PROPERTIES OF DILUTE SOLUTIONS

Q. 40. What are colligative properties? Name four such properties.

(Purvanchal 2002; Nagpur 2004; Calcutta 2007; Vidyasagar 2007)

Ans. *Colligative properties of a dilute solution are those properties which depend entirely upon the number of particles of the solute present in a known volume of a given solvent and not at all upon the nature of the solute.* These properties, are very closely related to each other. The colligative properties of dilute solutions are used to find the molecular weight of the dissolved substance. The various colligative properties are :

(*i*) Lowering of vapour pressure (*ii*) Elevation in boiling point

(*iii*) Depression in freezing point (*iv*) Osmotic pressure

Q. 41. Describe the methods that are employed for the determination of vapour pressure of a liquid or lowering of vapour pressure. *(Guwahati 2006)*

Ans. Two methods that are employed for this purpose are described as under :

1. Manometric Method

The vapour pressure of a liquid or solution can be measured with the help of a manometer (see Fig. 3.18). The bulb *B* is filled with the liquid or solution. The air in the connecting tube is then removed with a vacuum pump. When the stopcock is closed, the pressure inside is due only to the vapour evaporating from the solution or liquid. This method is generally used for aqueous solu- ,ons. The manometric liquid is usually mercury which has low volatility.

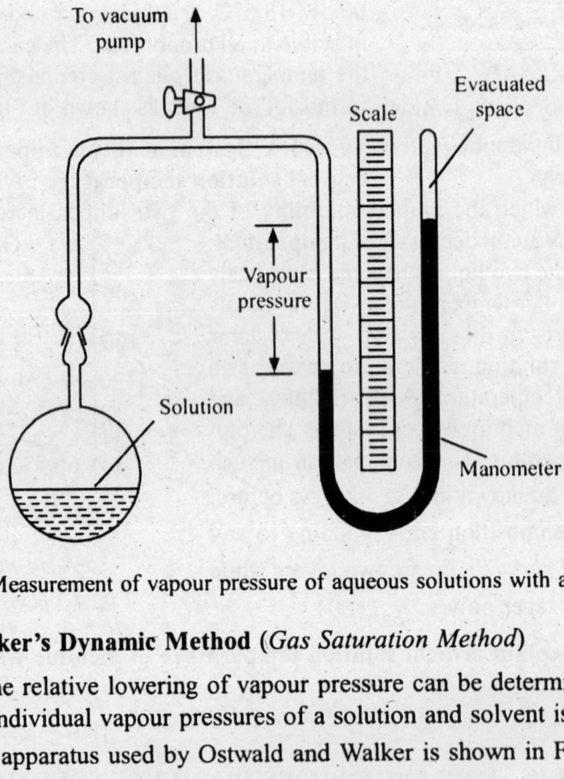

Fig. 3.18. Measurement of vapour pressure of aqueous solutions with a manometer.

2. Ostwald and Walker's Dynamic Method (*Gas Saturation Method*)

In this method the relative lowering of vapour pressure can be determined straightaway. The measurement of the individual vapour pressures of a solution and solvent is thus eliminated.

Procedure. The apparatus used by Ostwald and Walker is shown in Fig. 3.19. It consists of two sets of bulbs :

(*a*) Set *A* containing the solution

(*b*) Set *B* containing the solvent

Each set is weighed separately. A slow stream of dry air is then drawn by suction pump through the two sets of bulbs. At the end of the operation, these sets are reweighed. From the loss of weight in each of the two sets, the lowering of vapour pressure is calculated. The temperature of the air the solution and the solvent must be kept constant throughout.

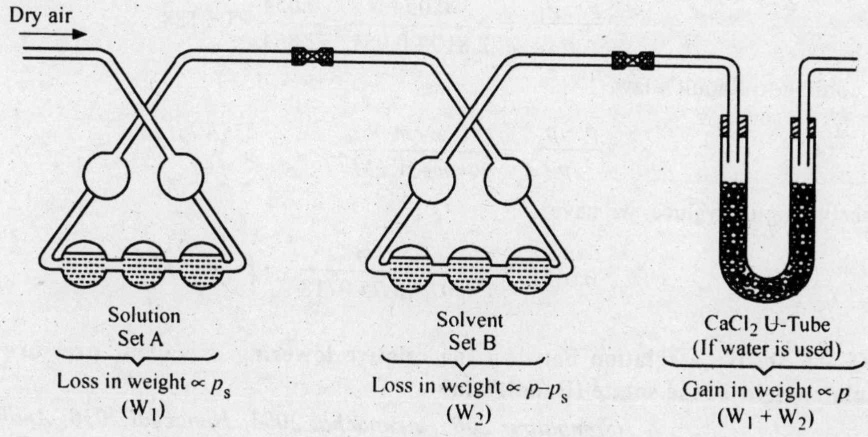

Fig. 3.19. Oswald-Walker method of measuring the relative lowering of vapour pressure.

Calculations. As the air bubbles through set A it is saturated up to the vapour pressure p_s of solution and then up to vapour pressure p of solvent in set B. Thus, the amount of solvent taken up in set A is proportional to p_s and the amount taken up in set B is proportional to $(p - p_s)$.

$$w_1 \propto p_s \quad \ldots(1)$$

$$w_2 \propto p - p_s \quad \ldots(2)$$

Adding (1) and (2), we have

$$w_1 + w_2 \propto p_s + p - p_s$$

$$\propto p \quad \ldots(3)$$

Dividing (2) by (3), we can write

$$\frac{p - p_s}{p} = \frac{w_2}{w_1 + w_2} \quad \ldots(4)$$

Knowing the loss of mass in set B (w_2) and the total loss of mass in the two sets ($w_1 + w_2$), we can find the relative lowering of vapour pressure from equation (4).

If water is the solvent used, a set of calcium chloride tubes is attached to the end of the apparatus to catch the escaping water vapour. Thus, the gain in mass of the $CaCl_2$-tubes is equal to ($w_1 + w_2$), the total loss of mass in sets A and B.

Q. 42. A stream of dry air was passed through a bulb containing a solution of 7.50 g of an aromatic compound in 75.0 g of water and through another globe containing pure water. The loss in mass in the first globe was 2.810 g and in the second globe it was 0.054 g. Calculate the molecular mass of the aromatic compound. (Mol mass of water = 18)

Ans. According to the theory of Ostwald-Walker method,

$$\frac{p - p_s}{p} = \frac{w_2}{w_1 + w_2}$$

In the present case,

w_1, loss of mass of solution $= 2.810$ g

w_2, loss of mass of solvent (water) $= 0.054$ g

Substituting values in the above equation

$$\frac{p-p_s}{p} = \frac{0.054}{2.810+0.054} = \frac{0.054}{2.864} = 0.0188$$

According to Raoult's law

$$\frac{p-p_s}{p} = \frac{w/m}{w/m+W/M}$$

Substituting the values, we have

$$0.0188 = \frac{7.50/m}{7.50/m+75.0/18} \quad \text{or} \quad m = 93.6$$

Q. 43. Derive a relation between the relative lowering of vapour pressure and the molecular weight of the solute (Raoult's law).

(Sambalpur 2003; Arunachal 2003; Himachal 2010; Andhra 2010)

Ans. Suppose vapour pressure of the pure solvent $A = P_A^o$

Let the solute be *B*.

Let the vapour pressure of the solvent in the solution $= P_A$

Since $P_A^o > P_A$

$\therefore$ Lowering of vapour pressure $= P_A^o - P_A$

Relative lowering of vapour pressure $= \dfrac{P_A^o - P_A}{P_A^o}$

According to Raoult's Law, relative lowering of vapour pressure is equal to the mole fraction of the solute in the solution.

Let mole fraction of the solute $B = X_B$.

$$\therefore \quad \frac{P_A^o - P_A}{P_A^o} = X_B \qquad \text{... } (i)$$

Mole fraction of the solute $= \dfrac{n_2}{n_1+n_2}$

where n_2 = No. of moles of the solute

n_1 = No. of the moles of the solvent

$$\therefore \quad X_B = \frac{n_2}{n_1+n_2}$$

Thus relation (*i*) becomes

$$\frac{P_A^o - P_A}{P_A^o} = \frac{n_2}{n_1+n_2} \qquad \text{... } (ii)$$

It is evident from (*ii*) that relative lowering of vapour pressure depends only upon mole fraction or molar concentration of the solute. Therefore, relative lowering of vapour pressure is a colligative property.

Molecular Mass of the Non-volatile Solute

Now, number of moles of the solute $n_2 = \dfrac{w}{m}$

where w = mass of the solute

m = molecular mass of the solute

And number of moles of the solvent $n_1 = \frac{W}{M}$

where W= mass of the solvent.

M= molecular mass of the solvent.

The expression (*ii*) becomes

$$\frac{P_A^o - P_A}{P_A^o} = \frac{\frac{w}{m}}{\frac{w}{m} + \frac{W}{M}} \quad \text{... (iii)}$$

In a dilute solution, n_2 is negligible as compared to n_1. Therefore neglecting n_2 (or w/m) in the denominator, we get from expression (*iii*)

$$\frac{P_A^o - P_A}{P_A^o} = \frac{\frac{w}{m}}{\frac{W}{M}} = \frac{w}{m} \times \frac{M}{W}$$

or

$$\frac{P_A^o - P_A}{P_A^o} = \frac{wM}{mW} \quad \text{...(iv)}$$

Thus by measuring the lowering of vapour pressure of a solution, the molecular mass m of a solute in a given solution of a known concentration can be determined, if other quantities are known.

Q. 44. A solution containing 6.0 gram of benzoic acid in 50 gram of ether ($C_2H_5.O.C_2H_5$) has a vapour pressure equal to 5.466 × $10^4 Nm^{-2}$ at 300 K. Given that vapour pressure of ether at the same temperature is 5.893 × 10^4 Nm^{-2}, calculate the molecular mass of benzoic acid.

Ans. Vapour pressure of ether (Solvent) = P^o = 5.893 × 10^4 Nm^{-2}

Vapour pressure of ether solution = P = 5.466 × 10^4 Nm^{-2}

Molecular mass of solvent ($C_2H_5 - O - C_2H_5$) (M) = 74

Mass of solute (benzoic acid) w = 6 grams

Mass of solvent (ether) W= 50 grams

Let molecular mass of solute (benzoic acid) = m

Substituting the values in the relation

$$\frac{P_A^o - P}{P^o} = \frac{w}{m} \times \frac{M}{W}$$

$$\frac{5.893 \times 10^4 - 5.466 \times 10^4}{5.893 \times 10^4} = \frac{6 \times 74}{m \times 50}$$

or

$$\frac{0.427 \times 10^4}{5.893 \times 10^4} = \frac{6 \times 74}{m \times 50}$$

or

$$\frac{0.427}{5.893} = \frac{6 \times 74}{m \times 50}$$

or

$$m = \frac{6 \times 74 \times 5.893}{50 \times 0.427}$$

$$m = 122.55$$

i.e., the molecular mass of solute (benzoic acid) = **122.55 amu.**

Q. 45. The vapour pressure of water at 293 K is 17.51 mm, lowering of vapour pressure of sugar solution is 0.0614 mm.

Calculate

(*a*) Relative lowering of vapour pressure.

(*b*) Vapour pressure of the solution.

(*c*) Mole fraction of water.

Ans. V.P. of solvent (water) = 17.51

Let V.P. of the solution = P (to be calculated)

∴ Lowering of V.P. = $P^o - P$ = 0.0614 mm

(*a*) ∴ Relative lowering of V.P. = $\dfrac{P^o - P}{P^o} = \dfrac{0.0614}{17.51} = 0.00351$

(*b*) Vapour pressure of the solution $P = P^o - (P^o - P)$

$= 17.51 - (0.0614) = 17.4486$ mm

Now according to Raoult's Law

$$\frac{P^o - P}{P^o} = \text{mole fraction of the solute}$$

$$\frac{P^o - P}{P^o} = \frac{n_1}{n_2 + n_2} = X_2$$

∴ mole fraction of the solute = $\dfrac{P^o - P}{P^o} = \dfrac{0.0614}{17.51} = 0.00351$

(*c*) Hence, mole fraction of the solvent = (1 – 0.00351) = 0.99649

∴ Mole fraction of water = 0.99649

Q. 46. The V.P. of a 5% aqueous solution of non-volatile organic substance at 373 K is 745 mm. Calculate the molecular mass of the solute.

Ans. Weight of non-volatile organic solute, $w = 5$ g

Weight of solvent (water), $W = 95$ g

Molecular mass of solvent (water) $M = 18$

Molecular mass of non-volatile solute$m$ = ?

P^o, the V.P. of the pure solvent (water) at 373 K = 760 mm

V.P. of the solution $P = 745$ mm

Substituting the values in the relation,

$$\frac{P^o - P}{P^o} = \frac{w}{m} \times \frac{M}{W}$$

$$\frac{760 - 745}{760} = \frac{5 \times 18}{m \times 95}$$

or $$m = \frac{5 \times 18 \times 760}{15 \times 95} = 48$$

Q. 47. At 298 K, the vapour pressure of water is 23.75 mm of Hg. Calculate the vapou pressure at the same temperature over 5% aqueous solution of urea (NH_2CONH_2).

Ans. This solution may be considered as a dilute solution and the approximate relation given below may be used

$$\frac{P_A^o - P_A}{P_A^o} = \frac{wM}{mW}$$

In the present case $P_A^o = 23.75$

$w = 5$ g Therefore $W = 100 - 5 = 95$ g

$M = 18$, $m = 60$ (mol. wt of urea)

Substituting these values in the equation above

$$\frac{23.75 - P_A}{23.75} = \frac{5 \times 18}{60 \times 95} \quad \text{or} \quad P_A = 23.375 \text{ mm.}$$

Q. 48. Describe methods to measure boiling point elevation.

Ans. Two methods that are generally employed to measure boiling point elevation are described below :

1. Landsberger-Walker Method

Apparatus. The apparatus used in this method is shown in Fig. 3.20 and consists of : (*i*) An *inner tube* with a hole in its side and graduated in ml; (*ii*) A *boiling flask* which sends solvent vapour into the graduated tube through a bulb with several holes (rose bulb), (*iii*) An *outer tube* which receives hot solvent vapour issuing from the side-hole of the inner tube; (*iv*) A *thermometer* reading to 0.01K, dipping in solvent or solution in the inner tube.

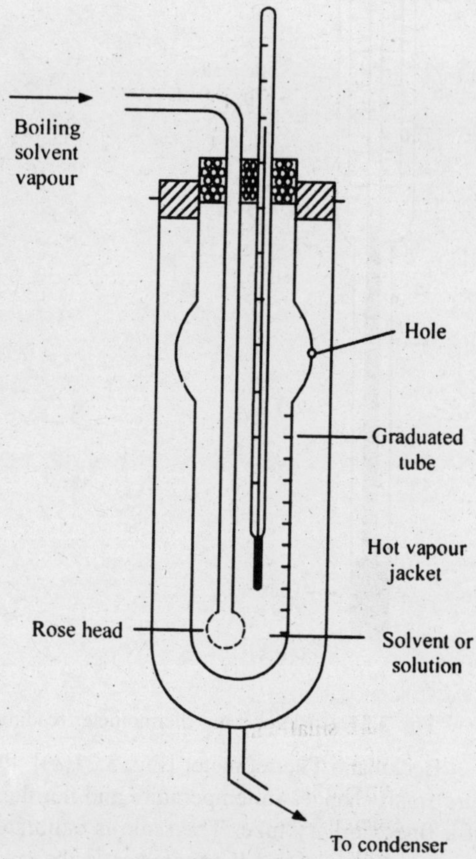

Fig. 3.20. Landsberger-Walker apparatus.

Procedure. Pure solvent is placed in the graduated tube and vapour of the same solvent boiling in a separate flask is passed into it. The vapour causes the solvent in the tube to boil by its latent heat of condensation. When the solvent starts boiling and temperature becomes constant, its boiling point is recorded.

Now the supply of vapour is temporarily cut off and a weighed pellet of the solute is dropped into the solvent in the inner tube. The solvent vapour is again passed through until the boiling point of the solution is reached and this is recorded. The solvent vapour is then cut off, thermometer and rosehead raised out of the solution, and the volume of the solution read.

From a difference in the boiling points of solvent and solution, we can find the molecular weight of the solute by using the expression

$$m = \frac{1000 \times K_b \times w}{\Delta T \times W}$$

where w = weight of solute taken, W = weight of solvent which is given by the volume of solvent (or solution) measured in ml multiplied by the density of the solvent at its boiling point.

2. Cottrell's Method

Apparatus. It consists of : (*i*) a graduated *boiling tube* containing solvent or solution; (*ii*) a reflux condenser which returns the vapourised solvent to the boiling tube; (*iii*) a thermometer reading to 0.01 K, enclosed in a glass hood; (*iv*) A small inverted funnel with a narrow stem which branches into three jets projecting at the thermometer bulb.

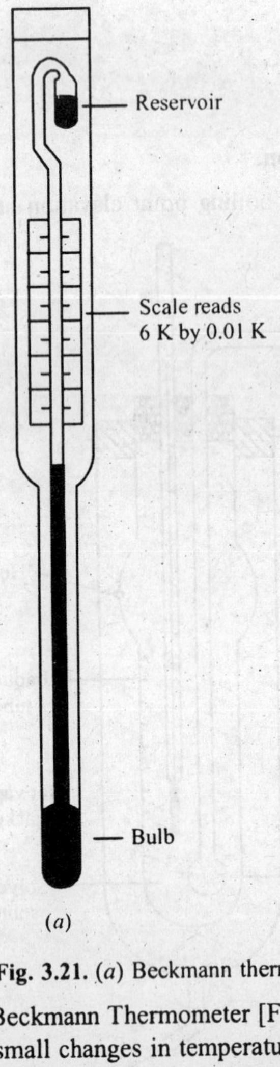

Fig. 3.21. (*a*) Beckmann thermometer reading to 0.01 K.

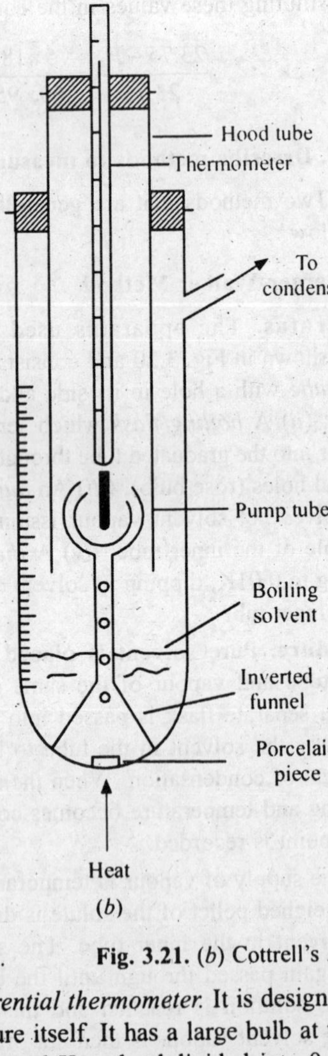

Fig. 3.21. (*b*) Cottrell's Apparatus.

Beckmann Thermometer [Fig. 3.21(*a*)]. It is a *differential thermometer.* It is designed to measure small changes in temperature and not the temperature itself. It has a large bulb at the bottom of a fine capillary tube. The scale is calibrated from 0 to 6 K and subdivided into 0.01 K. The unique feature of this thermometer is the small reservoir of mercury at the top. The amount of mercury in this reservoir can be decreased or increased by tapping the thermometer gently. In this way the thermometer is adjusted so that the level of mercury thread will show up at the middle of the scale when the instrument is placed in the boiling (or freezing) solvent.

Procedure. The apparatus is set up as shown in [Fig. 3.21(*b*)]. Solvent is placed in the boiling tube with a porcelain piece lying in it. It is heated on a small flame. As the solution starts boiling, solvent vapour arising from the porcelain piece pump the boiling liquid into the narrow stem. Thus, a mixture of solvent vapour and boiling liquid is continuously sprayed around the thermometer bulb. The temperature soon becomes constant and the boiling point of the pure solvent is recorded.

Now a weighed amount of the solute is added to the solvent and the boiling point of the solution noted as the temperature becomes steady. Also, the volume of the solution in the boiling tube is noted. The difference of the boiling temperatures of the solvent and solute gives the elevation of boiling point. While calculating the molecular weight of solute, the volume of solution is converted into mass by multiplying with density of solvent at its boiling point.

Q. 49. Derive a relation between elevation in boiling point of the solution and the mol. weight of the solute.* *(Purvanchal 2002; G. Nanakdev 2003; Sambalpur 2003; Osmania 2004; Himachal 2010)*

Ans. The boiling point of a liquid is the temperature at which its vapour pressure becomes equal to the atmospheric pressure. Since, at any temperature, the vapour pressure of a solution of a non-volatile solute is always lower than that of the pure solvent, the boiling point of a solution is always higher than that of the pure solvent.

This fact can be illustrated in Fig. 3.22. The upper curve represents the pressure-temperature relationship of the pure solvent. The lower curve represents the vapour pressure-temperature relationship of the dilute solution of a known concentration. It is evident that the vapour pressure of the solution is less than that of the pure solvent at each temperature. From Fig. 3.22 it is clear that for the pure solvent the vapour pressure becomes equal to the as atmospheric pressure at temperature T_b^o. Similarly the vapour pressure of the solution is equal to the atmospheric pressure at temperature T_b, which is obviously higher than T_b^o. Therefore $T_b - T_b^o$ gives the elevation in the boiling point which is represented as ΔT_b.

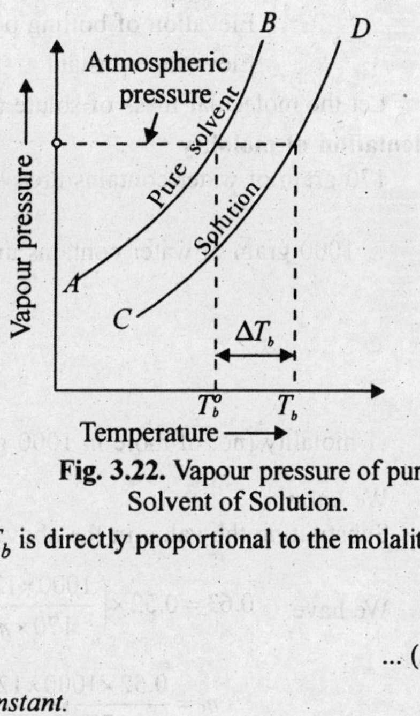

Fig. 3.22. Vapour pressure of pure Solvent of Solution.

or $$T_b - T_b^o = \Delta T_b$$

The elevation in boiling point depends upon concentration of the solute in a solution. In other words ΔT_b is directly proportional to the molality (*m*) of the solution, *i.e.*,

$$\Delta T_b \propto \text{molality}$$

or $$\Delta T_b = K_b \times \text{molality} \quad \text{... } (i)$$

where K_b is called the *molal elevation* (*ebullioscopic*) *constant.*

If $\quad$ molality = 1, $\Delta T_b = K_b$

Thus, *molal elevation constant* may be defined as the elevation in boiling point of a solution containing 1 gram mole of a solute per 1000 gram of the solvent.

Molecular Mass of Non-volatile Solute

By definition,

$$\text{molality} = \frac{1000 \times w}{W \times m}$$

* For thermodynamic derivation, see the chapter thermodynamics II.

∴ eqn. (*i*) becomes

$$\Delta T_b = \frac{1000 \times K_b \times w}{W \times m}$$

or
$$m = \frac{1000 \times K_b \times w}{W \times \Delta T_b} \quad ...(ii)$$

where ΔT_b = Elevation in boiling point.

K_b = molal elevation (Ebullioscopic) constant of the solvent.

m = molecular mass of the solute

w = weight of the solute in grams

W = weight of the solvent in grams.

Q. 50. A solution of 12.5 g of urea in 170 g of water gave boiling point elevation of 0.63 K. Calculate the molar mass of urea. K_b= 0.52 K kg mol^{-1}.

Ans. From the given data

Wt. of the solute, $w = 12.5$ g

Wt of the solvent $W = 170$g

Elevation of boiling point $\Delta T_b = 0.63$ K

Elevation constant $K_b = 0.52$ K kg mol^{-1}

Let the molecular mass of solute (urea) $= m$

Calculation of molality

170 gram of water contains urea = 12.5 gram

∴ 1000 gram of water contains urea $= \frac{1000 \times 12.5}{170}$ g

$= \frac{1000 \times 12.5}{170 \times m}$ mole

∴ molality (no. of mole in 1000 g of solvent) $= \frac{1000 \times 12.5}{170 \times m}$

We know, $\Delta T_b = K_b \times$ molality

Substituting the value in the above relation,

We have, $0.63 = 0.52 \times \left(\frac{1000 \times 12.5}{170 \times m}\right)$

$$m = \frac{0.52 \times 1000 \times 12.5}{170 \times 0.63} = 60.69 \text{ a.m.u.}$$

Q. 51. A solution prepared from 0.3 g of an unknown non-volatile solute in 30.0 g of CCl_4 boils at 350.392 K. Calculate the molecular mass of the solute. The boiling point of CCl_4 and its K_b values are 350.0 K and 5.03 respectively.

Ans. From the given data

Wt. of the solute, $w = 0.3$ g

Wt. of the solvent, $W = 30.0$ g

Elevation of the boiling point = $\Delta T_b = 350.392 - 350.0 = 0.392$ K

Elevation constant $K_b = 5.03$ K kg mol^{-1}

Calculation of molality

30 g of CCl_4 contains = 0.3 g of solute

$\therefore$ 1000 g of CCl_4 contains $= \frac{1000 \times 0.3}{30}$ g

$= \frac{1000 \times 0.3}{30 \times m}$ moles

$\therefore$ Molality (no. of moles in 1000 g of CCl_4) $= \frac{1000 \times 0.3}{30 \times m}$

Substituting the values in the relation

$$\Delta T_b = K_b \times \text{molality}$$

$$0.92 = 5.03 \times \frac{1000 \times 0.3}{30 \times m}$$

$$\therefore \quad m = \frac{5.03 \times 1000 \times 0.3}{30 \times 0.392} = 128.3$$

$\therefore$ Molecular mass = **128.3**

Q. 52. Find the b.p. of a solution containing 0.36 g of glucose ($C_6H_{12}O_6$) dissolved in 100 g of water (K_b = 0.52 K/m).

Ans. Mass of glucose $(w) = 0.36$ g

Mass of water $(W) = 100$ g

Mol. Mass of glucose $(M) = 180$

Molal elevation constant for water $(K_b) = 0.52$

Substituting the values in the relation

$$\Delta T_b = \frac{1000 \times K_b \times w}{W \times m}$$

or

$$\Delta T_b = \frac{1000 \times 0.52 \times 0.36}{100 \times 180}$$

Elevation in b.p. = 0.0104

B.P. of pure water = 373 K

Hence b.p. of the solution = 373 + 0.0104 = 373.0104 K

Q. 53. 10 g of a non-volatile solute when dissolved in 100 g of benzene raises its b.p. by 1°. What is the molecular mass of the solute (K_b for benzene = 2.53 K mol^{-1})?

Ans. In this problem

Mass of the solute $(w) = 10$ g

Mass of the solvent $(W) = 100$ g

Elevation in b.p. $(\Delta T_b) = 1°$, $K_b = 2.53$

Substituting the values in the equation

$$m = \frac{1000 \times K_b \times w}{W \times \Delta T_b} = \frac{1000 \times 2.53 \times 10}{1000 \times 1} = \mathbf{253}$$

Q. 54. A solution containing 0.5126 g of naphthalene (mol mass 128) in 50 g of carbon tetrachloride yields a b.p. elevation of 0.402°C while a solution of 0.6216 g of an unknown solute in the same weight of the solvent gives a b.p. elevation of 0.647°C. Find the molecular mass of the unknown solute.

Ans. 1. Determination of K_b from the first data

$w = 0.5126,\ m = 128,\ W = 50$

$$K_b = \frac{m \times W \times \Delta T_b}{1000 \times w}$$

$$= \frac{128 \times 50 \times 0.402}{1000 \times 0.5126} = \mathbf{5.02}$$

2. Mol. mass of the unknown solute

$w = 0.6216$ g, $W = 50$ g $\Delta T_b = 0.647$, $K_b = 5.02$

The value of K_b remains the same because the solvent is the same. Substituting the values in the equation

$$m = \frac{1000 K_b \times w}{W \times \Delta T_b}$$

$$= \frac{1000 \times 5.02 \times 0.6216}{50 \times 0.647} = \mathbf{96.46}$$

Q. 55. (*a*) Describe Beckman method and (*b*) Rast's Camphor method to determine depression in freezing point of a solvent.

Ans. The two methods are described as under :

1. Beckmann's Method

Apparatus. It consists of : (*i*) A *freezing tube* with a side-arm to contain the solvent or solution, while the solute can be introduced through the side-arm; (*iii*) An outer tube into which is fixed the freezing tube, the space in between providing an air jacket which ensures a slower and more uniform rate of cooling; (*iii*) A *large jar* containing a freezing mixture *e.g.*, ice and salt with a stirrer.

Procedure. About 20 g of the solvent is taken in the freezing point tube and the apparatus set up as shown in Fig. 3.23, so that the bulb of the thermometer is completely immersed in the solvent. Determine the freezing point of the solvent by directly cooling the freezing-point tube in the cooling bath.

The freezing point of the solvent having been accurately determined, the solvent is remelted by removing the tube from the bath, and a weighed amount (0.1 – 0.2 g) of the solute is introduced through the side tube. Now the freezing point of the solution is determined in the same way as that of the solvent. A further quantity of solute may then be added and another reading taken. Knowing the depression of the freezing point, the molecular weight of the solute can be determined by using the expression

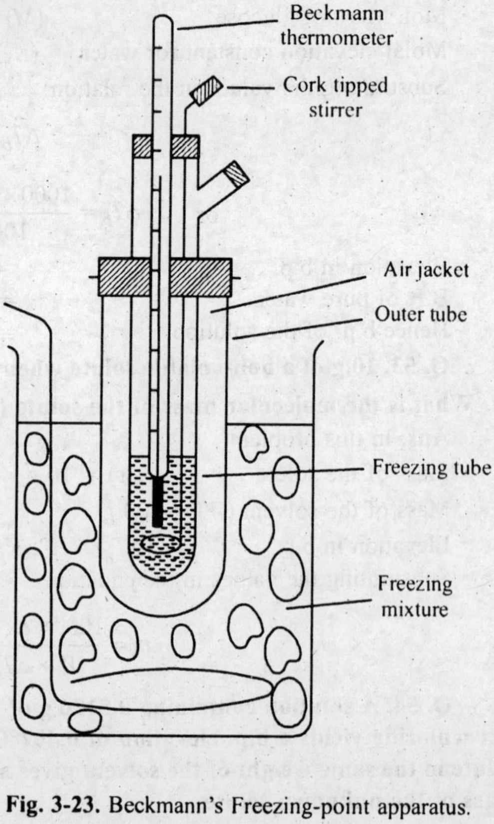

Fig. 3-23. Beckmann's Freezing-point apparatus.

$$m = \frac{1000 \times K_f \times w}{\Delta T \times W}$$

2. Rast's Camphor Method (Cryoscopic method)

This method is used for determination of molecular weights of solutes which are soluble in molten camphor. The freezing point depressions are so large that an ordinary thermometer can also be used.

Pure camphor is powdered and introduced into a capillary tube which is sealed at the upper end. This is tied along a thermometer and heated in a glycerol bath (see Fig. 3.24). The melting point of camphor is recorded. Then a weighed amount of solute and camphor (about 10 times as much) are melted in test-tube with the open end sealed. The solution of solute in camphor is cooled in air. After solidification, the mixture is powdered and introduced into a capillary tube which is sealed. Its melting point is recorded described before. The difference of the melting point of pure camphor and the mixture, gives the depression of freezing point. In modern practice, electrical heating apparatus (Fig. 3.25) is used for a quick determination of melting points of camphor as also the mixture.

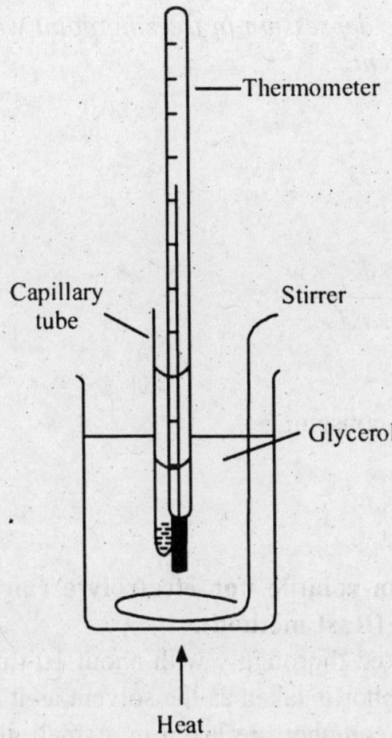

Fig. 3.24. Determination of depression of melting point by capillary method.

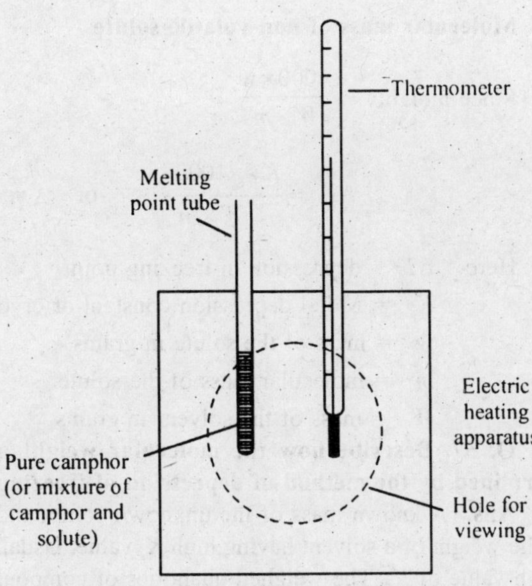

Fig. 3.25. Determination of depression of melting point by electrical apparatus.

Q. 56. Derive a relation between depression in freezing point and molecular weight of the solute*.

(Kerala, 2000; Meerut 2000; Sambalpur 2003; G Nanakdev 2003; Arunachal 2003)

Ans. Freezing point of a substance is the temperature at which solid and liquid states co-exists i.e., the two states have the same vapour pressure. The presence of a non-volatile solute lowers the vapour pressure of the solution. Thus liquid and solid states will have equal vapour pressure at a much lower temperature. Hence there is a depression in the freezing point. In other words, the freezing point of a solution is lower than that of the pure solvent as is clear from the Fig. 3.26.

* For thermodynamic derivation see the chapter thermodynamics II

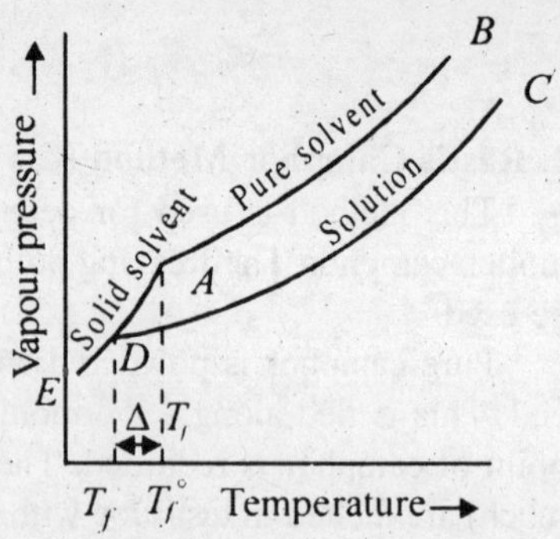

Fig. 3.26. Plot of V.P. versus temp.

The curve AB represents the vapour pressure curve of pure solvent. At point *A*, solvent co-exists with the liquid state. The temperature corresponding to point *A*, gives the freezing point of solvent. Similarly curve *CD* represents the vapour pressure curve of the solution in which point *D* corresponds to two states of the solvent existing simultaneously in the solution. Thus temperature corresponding to point *D* is the freezing point of solution which is clearly lower than the freezing point of the pure solvent. Hence there is a depression in the freezing point given by ΔT_f.

$$\Delta T_f = T_f^o - T_f$$

It has been found that depression in the freezing point is proportional to molal concentration of the solute in the solution *i.e.*, molality as given below.

$$\Delta T_f = K_f \times \text{molality}$$

where K_f is constant known as *molal depression constant* or *cryoscopic constant of the solvent.*

If molality = 1 then $\Delta T_f = K_f$

Hence *molal depression constant may be defined as the depression in freezing point when one mole of the solute is dissolved in 1000 grams of the solvent.*

Molecular mass of non-volatile solute

Since molality $= \dfrac{1000 \times w}{W \times m}$

$$\therefore \quad \Delta T_f = \frac{K_f \times 1000 \times w}{W \times m} \quad \text{or} \quad m = \frac{1000 \times K_f \times w}{W \times \Delta T_f}$$

Here ΔT_f = depression in freezing point

K_f = Molal depression constant or cryoscopic constant

w = mass of the solute in grams

m = molecular mass of the solute

W = mass of the solvent in grams

Q. 57. Describe how the molecular weight of a non-volatile non-electrolyte can be determined by the method of depression of freezing point (Rast method).

Ans. A known mass of the unknown compound is mixed thoroughly with about 10 times the the weight of a solvent having high K_f value. Usually camphor is taken as the solvent as it has a high value of K_f. The weighed quantities of compound and camphor are taken in a small glass tube and placed in a beaker containing hot sulphuric acid. Both the components melt. The tube is removed from the flame and allowed to cool. The mixture of the compound in camphor as solvent solidifies. Its melting point (same thing as freezing point) is noted using a thermometer calibrated to 0·1°. For more accurate measurements, a Beckman thermometer is used. Also the melting point of the camphor alone is noted. Calculations for the molecular weight are done as under.

Let the mass of the compound taken = w g

Let the mass of camphor (solvent) taken = W g

Melting point of camphor = t_1^0C

Melting point of the solution (compound + camphor) = t_2^0C

Depression in melting point (or freezing point) = $t_1 - t_2$

Apply the relation

$$\Delta T_f = \frac{k_f \times 1000 \times w}{W \times m}$$

Substituting for ΔT_f, K_f, w and W; m, the molecular weight can be calculated.

Q. 58. Calculate the freezing point of a solution containing 0.520 g of glucose ($C_6H_{12}O_6$) in 80·2 gram of water. For water K_f = 1·86 K kg mol^{-1}.

Ans. Given

Wt. of solute (glucose), w = 0·520 g

Wt. of solvent water, W = 80·2 g

Molar depression constant K_f = 1·86 K kg mol^{-1}

Molecular wt. of the solute $C_6 11_{12} 0_6$, m = 180

Calculation of molality

80·2 g of water, contains glucose = 0·520 g

$\therefore$ 1000 g of water, contains glucose $= \frac{1000 \times 0.520}{80{\cdot}2}$

$\because$ Molality $= \frac{1000 \times 0.520}{80{\cdot}2 \times 180}$

Applying the relation

$$\Delta T_f = K_f \times \text{molality}$$

$$= 1{\cdot}86 \times \frac{1000 \times 0.520}{80{\cdot}2 \times 180}$$

$\therefore$ Freezing point = 273 – 0·067 ($\because$ Freezing Point of water = 273 K)

= 272·933 K

Q. 59. What is molal depression constant? How is it related to latent heat of fusion? *(Punjab 2003)*

Ans. For definition of molal depression constant. Molal depression constant K_f is also given by the relation $K_f = \frac{RT_f^2}{1000\ L_f}$ where

T_f = Freezing point of solvent

L_f = Molar latent heat of fusion

R = Gas constant

For water, T_f = 273 K, L_f = 336 J g^{-1}

Q. 60. A solution of 1.25 g of a certain non-electrolyte in 20 g of water freezes at 271·95 K. Calculate the molecular mass of the solute. **(K_f= 1·86K kg mol^{-1})**

Ans. Here we are given that:

Wt. of the solute w = 1.25 g

Wt. of the solvent W = 20 g

Molal depression constant K_f = 1·86 K kg mol^{-1}

Depression in freezing point ΔT_f= 273 – 271·95 = 1·05 K

Calculation of molality

20 g of water contains solute $= 1.25$ g

$\therefore$ 1000 g of water contains solute $= \dfrac{1000 \times 1.25}{20}$

Let the molecular mass of the solute = m

$\therefore$ Molality $= \dfrac{1000 \times 1.25}{20 \times m}$

Substituting the values in the relation

$$\Delta T_f = K_f \times m$$

$$1.05 = 1.86 \times \frac{1000 \times 1.25}{20 \times m}$$

$$\therefore \quad m = \frac{1{\cdot}86 \times 1000 \times 1{\cdot}25}{20 \times 1{\cdot}05} = 110{\cdot}71$$

$\therefore$ Molecular mass of the solute (m) = **110·71**

Q. 61. Find (*i*) the boiling point and (*ii*) the freezing point of a solution containing 0·52 g glucose ($C_6H_{12}O_6$) dissolved in 30·2 g of water. For water K_b= 0·52 K kg mol^{-1}; K_f=1.86 K kg mol^{-1}.

Ans. Given

Molecular mass of glucose $m = 180$

Wt. of solute $w = 0{\cdot}52$ g

Wt. of solvent $W = 80{\cdot}2$ g

Calculation of molality

80·2 g of the solvent contains solute $= 0{\cdot}52$ g

$\therefore$ 1000 g of the solvent contains solute $= \dfrac{1000 \times 0{\cdot}52}{80{\cdot}2} g$

$\therefore$ Molality $= \dfrac{1000 \times 0{\cdot}52}{80{\cdot}2 \times 180}$

We know, $\Delta T_b = K_b \times$ molality

$$= 0{\cdot}52 \times \frac{1000 \times 0{\cdot}52}{80{\cdot}2 \times 180} g = 0{\cdot}018\text{K}$$

$\therefore$ Boiling point $= 373 + 0{\cdot}018 = 373{\cdot}018$ K

($\because$ Boiling point of water = 373 K)

$\therefore$ Boiling point of solution = 373.018 K

For freezing point, $\Delta T_f = K_f \times$ molality

$$= 1{\cdot}86 \times \frac{100 \times 0{\cdot}52}{80{\cdot}2 \times 180} = 0{\cdot}067$$

Freezing point of H_2O = 273 K

$\therefore$ Freezing point of solution = 273 – 0·067 = **272·933 K**

Q. 62. The normal freezing point of nitrobenzene $C_6H_5NO_2$ is 278·82 K. A 0·25 molal solution of a certain solute in nitrobenzene causes a freezing point depression of 2 degree.Calculate the value of K_f for nitrobenzene.

Ans. $\Delta T_f = K_f \cdot m$

$$\Delta T_f = 2$$
$$m = 0.25$$

Substituting the values in the first equation

$$2 = K_f \times 0.25$$

or $$K_f = \frac{2}{0.25} = 8 \text{ K kg mol}^{-1}$$

Q. 63. A sample of camphor used in the Rast method of determination molecular masses had a melting point of 176.5°C. The melting point of a solution containing 0.522 g comphor and 0.0386 g of an unknown substance was 158.5°C. Find the molecular mass of the substance. K_f of camphor per kg is 37.7.

Ans.

Applying the expression

$$m = \frac{1000 \times K_f \times w}{\Delta T \times W}$$

to the present case, we have

$$\Delta T = 176.5 - 158.5 = 18$$
$$K_f = 37.7$$
$$w = 0.0386\ g$$
$$W = 0.522\ g$$

Substituting these values

$$m = \frac{1000 \times 37.7 \times 0.0386}{18 \times 0.522} = 154.8$$

Q. 64. Explain the terms
(*i*) Semi-permeable membrane
(*ii*) Osmosis and osmotic pressure.

(Purvanchal 2002; Mysore 2004; Vidyasagar 2007; Andhra 2010)

Ans. (*i*) Semi-permeable membrane: *The membrane which allows the flow of solvent molecules through it but not the solute molecules, is called a semi-permeable membrane.* Examples are parchment, collodion, animal membranes etc. In nature the plant cells are protected by a semi-permeable membrane. Chemically, we can get a semi-permeable membrane of cupric ferrocyanide within the walls of a porous pot.

(*ii*) **Osmosis and osmotic pressure:** A solute tends to dissolve in a solvent as the most predominant randomness factor favours such a tendency. But when a pure solvent is separated from its solution by a semi-permeable membrane, the molecules of the solvent diffuse through the semi-permeable membrane into the solution. This is called **osmosis.** As a result of this transference of solvent to solution, the level of solvent decreases while that of solution increases. It may be demonstrated by taking solvent and solution in two compartments of a box separated by a semi-permeable membrane as shown below in Fig. 3.27.

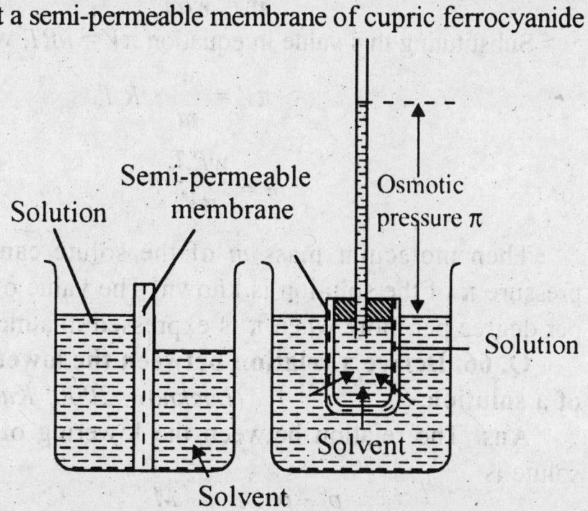

Fig. 3.27. Osmosis and Osmotic pressure.

As more and more of the solvent molecules pass through the membrane into the solution, the concentration of the solution falls gradually. After some time the hydrostatic pressure exerted by the solution column prevents the flow of more of solvent molecules and **osmosis** stops *i. e.*, there is no further rise in the level of solution in the column. Hence, osmotic pressure is defined as follows :

Osmotic pressure *is the equilibrium hydrostatic pressure exerted by the solution column which just prevents the flow of solvent molecules into the solution through a semi- permeable membrane.*

Osmosis *is a phenomenon by which the pure solvent molecules tend to diffuse through a semi-permeable membrane into the solution.*

Q. 65. Derive Van't Hoff relation between the osmotic pressure of a solution and molecular mass of the solute. *(Kerala 2000; Lucknow 2000; Mysore 2004; Calcutta 2007)*

Ans. In dilute solutions the behaviour of solute molecules is similar to that of molecules of a gas. Osmotic pressure (π) of a solution is found to be directly proportional to the molar concentration C of solution and its absolute temperature T. Thus :

$$\pi \propto C$$

and $$\pi \propto T$$

or $$\pi \propto C.T$$

or $$\pi = R.C.T$$

Where R is a constant and its value is found to be same as that of gas constant.

If a solution is prepared by dissolving n moles of the solute in V litres of the solution, the molar concentration

$$C = \frac{n}{V} \text{ moles per litre}$$

$\therefore$ $$\pi = \frac{n}{V} RT$$

or $$\boxed{\pi V = nRT}$$

The above equation is known, as *Van't Hoff equation of dilute solution* and shows that osmotic pressure π is proportional to the molar concentration of the solute in the solution. Hence, it is a colligative property.

Determination of molecular mass from osmotic pressure

If w grams of the solute are dissolved in V litres of the solution and m is the molecular of the solute, then

$$n = w/m$$

Substituting this value in equation $\pi V = nRT$, we get

$$\pi V = \frac{w}{m} \times R.T$$

or $$m = \frac{wR.T}{\pi V}$$

Then molecular mass m of the solute can be calculated from this equation if osmotic pressure π of the solution is known. The value of constant R is taken as 0.0821 litre-atmosphere per degree per mole when π is expressed in atmosphere and T in degree Kelvin.

Q. 66. Derive a relation between the lowering of vapour pressure and osmotic pressure of a solution. *(Lucknow, 2000; Kumaon, 2000; Himachal 2002; Arunachal 2003)*

Ans. The relation between the lowering of vapour pressure and the mole fraction of the solute is

$$\frac{p^o - p}{p^o} = \frac{w}{m} \times \frac{M}{W} \quad \text{(for a dilute solution)} \quad ...(i)$$

or $$\frac{p^o - p}{p^o} = n \times \frac{M}{W} \quad ...(ii)$$

where n is the no. of moles of the solute.

The relation between the osmotic pressure and the no. of moles of the solute is given by

$$\pi = \frac{nRT}{V} \quad ...(iii)$$

From equation (*ii*)

$$n = \frac{p^o - p}{p^o} \times \frac{W}{M} \quad ...(iv)$$

From equation (*iii*)

$$n = \frac{\pi V}{RT} \quad ...(v)$$

Equating R.H.S of equations (*iv*) and (*v*)

$$\frac{p^o - p}{p^o} \cdot \frac{W}{M} = \frac{\pi V}{RT}$$

or $$\frac{p^o - p}{p^o} = \frac{M}{W} \cdot \frac{\pi V}{RT}$$

or $$\frac{p^o - p}{p^o} = \frac{M\pi}{(W/V).RT}$$

or $$\frac{p^o - p}{p^o} = \frac{M\pi}{\rho .RT}$$

where M = Mol. mass of the solvent
ρ = Density of the solution
R = Solution (gas) constant
T = Temperature.

Q. 67. Describe two interesting experiments to demonstrate the phenomenon of Osmosis.

1. The egg experiment

The outer hard shell of two eggs of the same size is removed by dissolving in dilute hydrochloric acid. One of these is placed in distilled water and the other in saturated salt solution. After a few hours, it will be noticed that the egg placed in water swells and the one in salt solution shrinks. In the first case, water diffuses through the skin (a semipermeable membrane) into the egg which swells. In the second case, the concentration of the salt solution being higher than the material, the egg shrinks (Fig. 3.28).

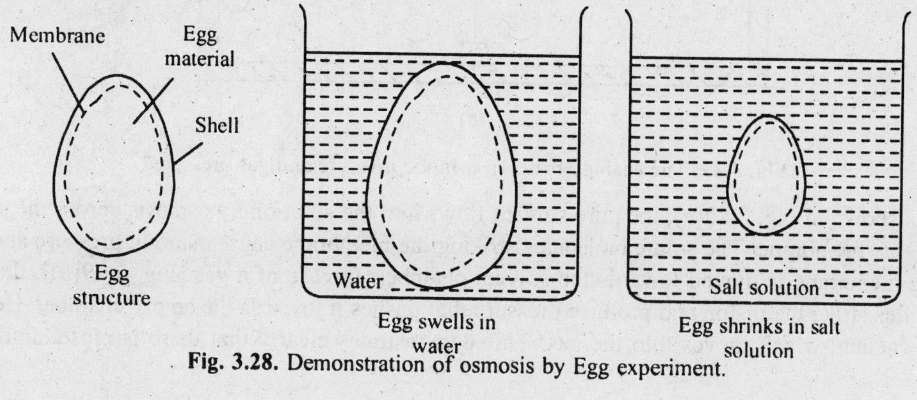

Fig. 3.28. Demonstration of osmosis by Egg experiment.

2. Silica garden

Crystals of many salts *e.g.*, ferrous sulphate, nickel chloride, cobalt nitrate and ferric chloride are placed in a solution of water glass (sodium silicate). The layers of metallic silicates formed on the surface of crystals by double decomposition are semipermeable. The water from outside enters through these membranes which burst and form what we call a *Silica Garden.* It makes an interesting sight.

Q. 68. Explain van't Hoff theory of dilute solution taking the example of Osmosis.

(Arunachal 2003; Vidyasagar 2007)

Ans. van't Hoff noted the striking resemblance between the behaviour of dilute solutions and gases. Dilute solutions obeyed laws analogous to the gas laws. To explain it van't Hoff visualised that gases consist of molecules moving in vacant space (or vacuum), while in solutions the solute particles are moving in the solvent. The exact analogy between solutions and gases is illustrated with Fig. 3.29.

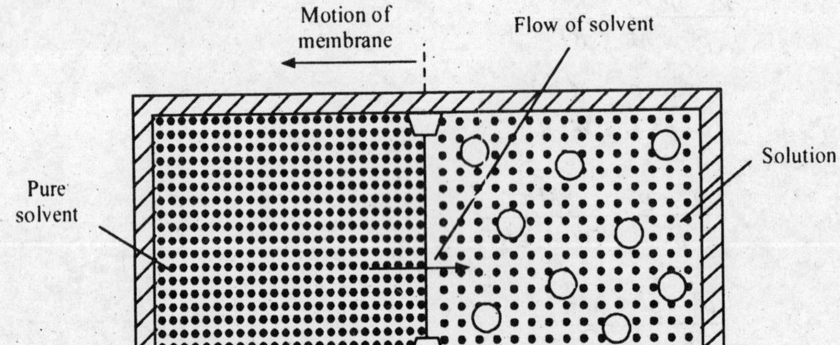

(*a*)

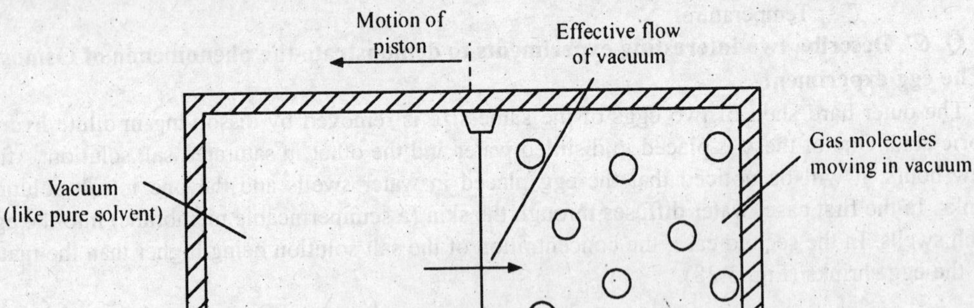

(*b*)

Fig. 3.29. The analogy between osmotic pressure and gas pressure.

As shown in Fig. 3.29(*a*), the pure solvent flows into the solution by osmosis across the semipermeable membrane. The solute molecules striking the membrane cause osmotic pressure and the sliding membrane is moved towards the solvent chamber. In case of a gas [Fig. 3.29(*b*)], the gas molecules strike the piston and produce pressure that pushes it towards the empty chamber. Here it is the vacuum which moves into the gas. This demonstrates clearly that there is close similarity between a gas and a dilute solution.

Thinking on these lines, van't Hoff propounded his theory of dilute solution. The **van't Hoff Theory of Dilute Solutions** states that :

a substance in solution behaves exactly like a gas and the osmotic pressure of a dilute solution is equal to the pressure which the solute would exert if it were a gas at the same temperature occupying the same volume as the solution.

According to the van't Hoff theory of dilute solutions, all laws or relationships obeyed by gases would be applicable to dilute solutions.

From van't Hoff theory it follows that just as 1 mole of a gas occupying 22.4 litres at 0°C exerts 1 atmosphere pressure so 1 mole of any solute dissolved in 22.4 litres would exert 1 atmosphere osmotic pressure.

Q. 69. Describe Berkeley and Hartley method for the determination of osmotic pressure of a solution.

(Purvanchal 2003; Bangalore 2004; Dibrugarh 2004)

Ans. The apparatus used for the determination of osmotic pressure by the above method is shown in Fig. 3.30.

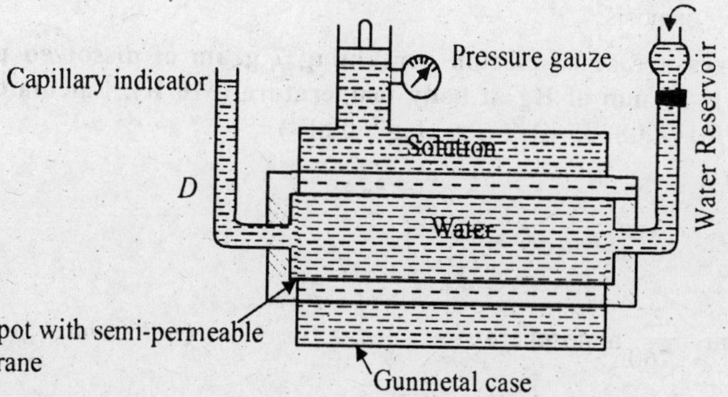

Fig. 3.30. Barkeley and Hartley methods for osmotic pressure determination.

It consists of a porous pot containing potassium ferrocyanide deposited on its walls and fitted into a metallic cylinder. A piston and a pressure gauge is attached to the metallic cylinder which is filled with the solution whose osmotic pressure is to be determined. The porous pot is fitted with a capillary tube and water reservoir on the opposite sides. Water from the porous pot moves towards the solution in the cylinder through the semi-permeable membrane. As a result the level of water tends to fall down. External pressure is applied on the piston to such an extent that the water level in the capillary tube does not change. The magnitude of pressure applied can be read from the pressure gauge. This pressure is equal to the osmotic pressure.

Q. 70. What are isotonic solutions ? *(Bangalore 2004; Andhra 2010)*

Ans. Solutions of equimolecular concentrations at the same absolute temperature have the same osmotic pressure. *Such solutions which have the same osmotic pressure at the same temperature are called isotonic solutions.*

Q. 71. Discuss biological importance of osmosis.

Ans. Osmosis plays a very important role in plants and animals *i. e.,* living organisms.

1. Animal and vegetable cells contain solutions of sugars and salts enclosed in their semi-permeable membranes (cells saps). On placing such a cell in water or in a solution, the osmotic pressure of which is less than that of the cell sap within, water enters the cell. However, on placing the cells in a solution of higher osmotic pressure, water passes out of the cell and the

cell shrinks. This shrinkage of the cell is called plasmolysis which can be illustrated by taking two hen eggs of the same size. Their outer shells are dissolved by placing them in dil HCl, one egg is then placed in distilled water and the other in a saturated solution of NaCl (*i.e.*, of higher Osmotic Pressure). In the first case, water **enters into the egg** through the membrane while in the second case, **water comes out** of the egg through the membrane *i.e.*, the egg **swells** in the first **case and shrinks** in the second case.

Using solutions of varying concentrations and placing plant or animal cells in them, it is possible to find out the concentration when plasmolysis just stops. Such a solution is then said to be **Isotonic** (*i.e.*, having same osmotic pressure) with the cell solution. This process is thus helpful in measuring osmotic pressure of cell saps.

It is found that the red blood corpuscles and a 0.91% sol. of NaCl are isotonic.

2. Plant cells at their roots contain root hair, which are in contact with the soil, since the osmotic pressure inside the cell is higher, water from the soil flows into the cell by osmosis.

3. Damage done to the plant by use of excess of fertilizer is due to higher osmotic pressure of the fertilizer solution than that of the cells and makes the organs stretched and fully expanded.

4. Plant movements such as opening and closing of flowers, opening and closing of leaves etc. are also regulated by osmosis.

Q. 72. Osmotic pressure of a solution containing 7 gram of dissolved protein per 100 cm^3 of a solution is 25 mm of Hg at body temperature (310 K). Calculate the molar mass of the protein. (R = 0.08205 litre atm deg^{-1} mol^{-1})

Ans. We know

$$\pi = \frac{n}{V} \times RT$$

Given $\pi = 25 \text{ mm} = \frac{25}{760}$ atmosphere $\quad$ ($\because$ 760 mm = 1 atmosphere)

R = 0.08205 l atmosphere K^{-1} mole^{-1}

T = 310 K

$V = 100 \text{ cm}^3 = \frac{100}{1000} = \frac{1}{10}$ litres

$\therefore n$, the number of moles of the solute $= \dfrac{\pi \times V}{RT}$ ...(*i*)

Also $n = \dfrac{\text{Wt. of solute}}{\text{Molecular mass of solute}}$ ...(*ii*)

From (*i*) and (*ii*)

$$\frac{\text{Wt. of solute}}{\text{Molecular mass of solute}} = \frac{\pi \times V}{RT}$$

Substituting the values, we have

$$\text{Molecular mass of solute} = \frac{\text{Wt. of solute} \times R \times T}{\pi \times V}$$

$$= \frac{7 \times 0{\cdot}08205 \times 310}{\frac{25}{760} \times \frac{1}{10}}$$

or molecular mass of the solute $= \dfrac{7 \times 0.08205 \times 310 \times 760 \times 10}{25}$

$= 54126{\cdot}7$

$\therefore$ Molecular mass of the solute = **54126·7**

Q. 73. A solution of sucrose (molecular mass 342) is prepared by dissolving 68·4 g of it per litre of the solution. What its osmotic pressure at 300 K ? R= 0·082 litre atm K^{-1} mol^{-1}.

Ans. We know, $\pi = \dfrac{n}{V} \times RT$

Given, molecular mass of the solute = 342

Wt. of the solute = 68·4 g

Volume V of the solution = 1 litre

T= 300 K

Osmotic pressure π= ?

R= 0·082 litre atm K^{-1} mol^{-1}

n, the no. of moles of solute $= \dfrac{\text{wt. of solute}}{\text{molecular mass of solute}}$

or $n = \dfrac{68{\cdot}4}{342}$

Substituting the values in the equation

$$\pi = \frac{n}{V} RT$$

or $\pi = \dfrac{68{\cdot}4}{342} \times \dfrac{0{\cdot}082 \times 300}{1} = \mathbf{4.22\ atm}$

$\therefore$ Osmotic pressure = **4·22 atm**

Q. 74. Calculate the osmotic pressure of a 5% solution of cane sugar at 288 K. R = 0·082 lit atm K^{-1} mol^{-1}

Ans. Molecular mass of the cane sugar ($C_{12}H_{22}O_{11}$) = 342

$\because$ The solution is 5%

$\therefore$ Wt. of sugar per litre = 50 g

$\therefore$ n, the no. of moles $= \dfrac{\text{wt. of solute}}{\text{molecular mass of solute}} = \dfrac{50}{342}$

V = 1 litre

$T = 288K$

Substituting the values in the relation

$$\pi = \frac{n}{V} \cdot R.T.$$

We have $$\pi = \frac{50}{342} \times \frac{0.082 \times 288}{1} = 3.45 \text{ atm}$$

$\therefore$ Osmotic pressure = **3·45 atm**

Q. 75. At 298 K, 100 cm^3 of a solution, containing 3·002 g of an unidentified solute, exhibits an osmotic pressure of 2·55 atmospheres. What would be the molecular mass of the solute?

Ans. $\pi V = nRT$ where n is the no of moles of the solute

Also $\pi V = \frac{w}{M} RT$ where w is the weight and M the molecular mass of the solute

or $$M = \frac{wRT}{\pi V}$$

Substituting the values

$$M = \frac{3.002 \times 0.0821 \times 298}{2.55 \times 0.1} \qquad (V = 100 \text{ cm}^3 = 0{\cdot}1 \text{ litre})$$

$M = 288$ a.m.u.

Q. 76. A solution of sucrose (molar mass = 342 g/mole) is prepared by dissolving 68·4 g in 1000 g of water. What is the

(*i*) Vapour pressure of solution at 293 K

(*ii*) Osmotic pressure at 293 K

(*iii*) Boiling point of the solution

(*iv*) Freezing point of the solution

The vapour pressure of water at 293 K is 17·5 mm, K_b = 0·52, K_f = 1·86.

Ans. (*i*) Mole fraction of sucrose $X_B = \frac{68.4/342}{68{\cdot}4/342 + 1000/18}$

$$= \frac{9}{2509}$$

$$\frac{P_A^o - P_A}{P_A} = X_B$$

Vapour pressure of water P_A^o = 17·5 mm

Substituting the values

$$\frac{17{\cdot}5 - P_A}{17{\cdot}5} = \frac{9}{2509}$$

or $$17{\cdot}5 \times 2509 - 2509\, P_A = 9 \times 17{\cdot}5$$

or $$2509\, P_A = 43750$$

or $$P_A = 17{\cdot}43 \text{ mm}$$

(*ii*) $\pi V = nRT$, $V = 1$ litre, $n = \dfrac{68{\cdot}4}{342}$

$$\pi \times 1 = \frac{68{\cdot}4}{342} \times 0{\cdot}0821 \times 293$$

or $\pi = 4.8$ atmospheres

(*iii*) $\Delta T_b = K_b \times m$ where m is the molality

Molality $(m) = \dfrac{68{\cdot}4}{342}$, $K_b = 0{\cdot}52$

Substituting the values

$$\Delta T_b = 0{\cdot}52 \times \frac{68{\cdot}4}{342} = 0{\cdot}104$$

Boiling point of the solution = 373 + 0.104 = 373·104

(*iv*) $\Delta T_f = K_f \times m$

$$m = \frac{68{\cdot}4}{342},\ K_f = 1.86$$

$$\therefore \quad \Delta T_f = 1{\cdot}86 \times \frac{68{\cdot}4}{342} = 0{\cdot}372$$

∴ Freezing point of the solution = 273 – 0·372 = **272·628 K**

Q. 77. Calculate the osmotic pressure of 20% anhydrous calcium chloride solution at 273 K, assuming that the solution is completely dissociated. (R = 0·082 lit. atm K^{-1} mol^{-1}).

Ans. Molecular mass of $CaCl_2$ = 40 + 71 = 111

Since solution is 20%

∴ Weight of $CaCl_2$ per litre = 200 g

$V = 1$ litre

$T = 273$ K

$$n = \frac{\text{wt. of solute}}{\text{molecular mass of solute}} = \frac{200}{111}$$

Substituting the values in the relation

$$\pi = \frac{n}{V} \times R.T.$$

we have $$\pi = \frac{200}{111} \times \frac{0.082 \times 273}{1} = 40.335 \text{ atm}$$

Since $CaCl_2$ dissolves to give 3 particles for each molecule on complete dissociation ($CaCl_2 \rightarrow Ca^{++} + 2Cl^-$)

∴ Observed osmotic pressure = Normal osmotic pressure × 3

= 40·335 × 3 = 121·00 atm

∴ Observed osmotic pressure = **121·00 atm.**

Q. 78. Give a brief description of the theories of Osmosis.

Ans.

1. The Molecular Sieve theory

According to this theory, the membrane contains lots of fine pores and acts as a sort of 'molecular sieve'. **Smaller solvent molecules can pass through the pores but the larger molecules**

cannot. Solvent molecules flow from a region of higher solute concentration to one of lower concentration across such a membrane (Fig. 3.31). But we observe that some membranes can act as sieves even when the solute molecules are smaller than the solvent molecules. This theory does not provide a satisfactory answer to this.

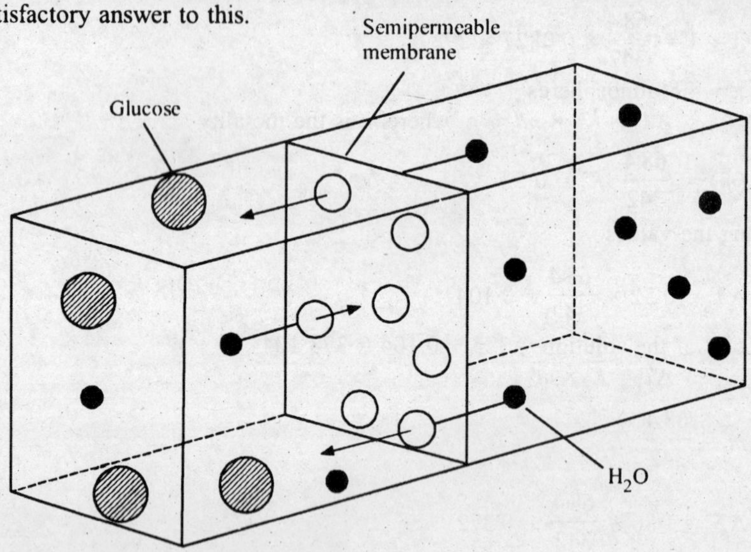

Fig. 3.31. A semipermeable membrane can separate particles on the basis of size. It allows the passage of small water molecules in both directions. But it prevents the passage of glucose molecules which are larger than water molecules.

Recently it has been shown that the pores or capillaries between the protein molecules constituting an animal membrane are lined with polar groups ($-COO^-$, $-NH_3^+$, $-S^{2-}$, etc.). Therefore, the membrane acts not simply as a sieve but also regulates the passage of solute molecules by electrostatic or 'chemical interactions'. In this way even solute molecules smaller than solvent molecules can be held back by the membrane.

2. Membrane Solution theory

Membrane proteins bearing functional groups such as $-COOH$, $-OH$, $-NH_2$, etc., dissolve water molecules by hydrogen bonding or chemical interaction. Thus, membrane dissolves water from the pure water (solvent) forming what may be called '*membrane solution*'. The dissolved water flows into the solution across the membrane to equalise concentrations. In this way water molecules pass through the membrane while solute molecules being insoluble in the membrane do not.

3. Vapour Pressure theory

It suggests that a semipermeable membrane has many fine holes or capillaries. The walls of these capillaries are not wetted by water (solvent) or solution.

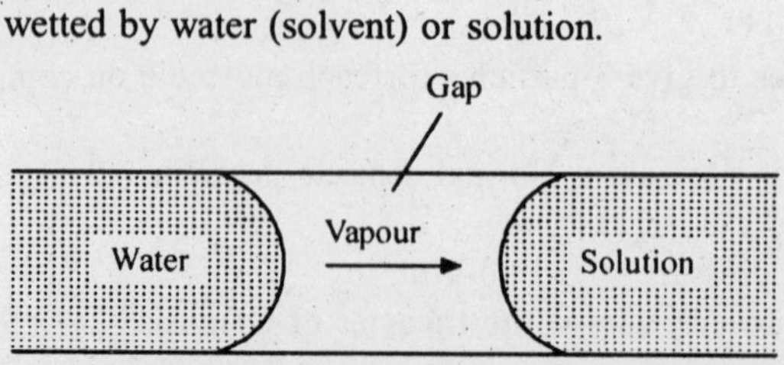

Fig. 3.32. Water vapours diffuse into solution across the gap in a capillary of the membrane.

Thus neither solution nor water can enter the capillaries. Therefore, each capillary will have in it solution at one end and water at the other, separated by a small gap (Fig. 3.32). Since the vapour pressure of a solution is lower than that of the pure solvent, the diffusion of vapour will occur across the gap from water side to solution side. This results in the transfer of water into the solution.

4. Membrane Bombardment theory

This theory suggests that osmosis results from an unequal bombardment pressure caused by solvent molecules on the two sides of the semipermeable membrane. On one side we have only solvent molecules while on the other side there are solute molecules occupying some of the surface area. Thus, there are fewer bombardments per unit area of surface on the solution side than on the solvent side. Hence the solvent molecules will diffuse more slowly through the membrane on the solution side than on the solvent side. The net result causes a flow of the solvent from the pure solvent to the solution across the membrane.

Q. 79. What is meant by reverse osmosis ? What are its applications ? *(Himachal 2010)*

Ans. When a solution is separated from pure water by a semipermeable membrane, osmosis of water occurs from water to solution. This osmosis can be stopped by applying pressure equal to or more than osmotic pressure, on the solution (Fig. 3.33). If pressure greater than osmotic pressure is applied, osmosis is made to proceed in the reverse direction to ordinary osmosis *i.e.*, from solution to water.

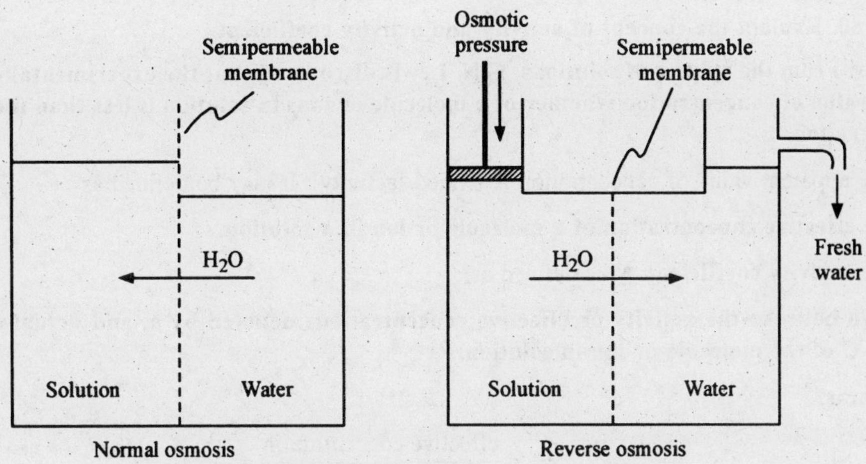

Fig. 3.33. Reverse osmosis versus ordinary osmosis.

The osmosis taking place from solution to pure water by application of pressure greater than osmotic pressure, on the solution, is termed *Reverse Osmosis.*

This technology is used in the commercial production of water purifiers these days (R.O. Water Purifiers).

Desalination of Sea water by Hollow-fibre reverse osmosis

Reverse osmosis is used for the desalination of sea water for getting fresh drinking water. This is done with the help of hollow fibres (nylon or cellulose acetate) whose wall acts as semipermeable membrane. A hollow-fibre reverse osmosis unit is shown in Fig. 3.34.

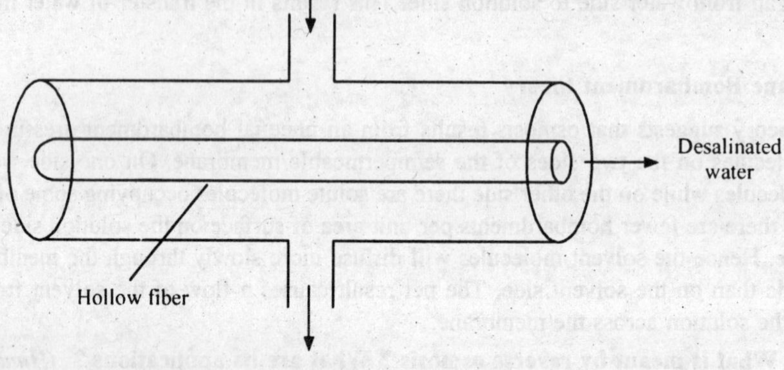

Fig. 3.34. Desalination of sea water by reverse osmosis in a hollow fibre unit.

Water is introduced under pressure around the hollow fibres. The fresh water is obtained from the inside of the fibre. In actual practice, each unit contains more than three million fibres bundled together, each fibre is of about the diameter of a human hair.

Q. 80. Explain the concept of activity and activity coefficient.

Ans. From the studies of solutions, G.N. Lewis discovered that the experimentally determined value of concentration whether of a molecule or ions in solution is less than the actual concentration.

The apparent value of concentration is termed 'activity'. It may be defined as :

the effective concentration of a molecule or ion in a solution.

The **activity coefficient 'γ'** is defined as :

the ratio between the activity or effective concentrations denoted by *a*, and actual concentration C of the molecule or ion in solution.

That is,

$$\gamma = \frac{\text{effective concentration}}{\text{actual concentration}} = \frac{a}{c}$$

or $$a = \gamma c$$

effective concentration

γ can be determined experimentally and is available in the form of Tables. Thus, the value of activity can be calculated by applying the above relation.

Explanation

The effective number of cations and anions in solution becomes less as these tend to form ion-pairs (Fig. 3.35) because of strong electrostatic attractions. Thus, the effective concentration of the ions becomes less than the actual concentration.

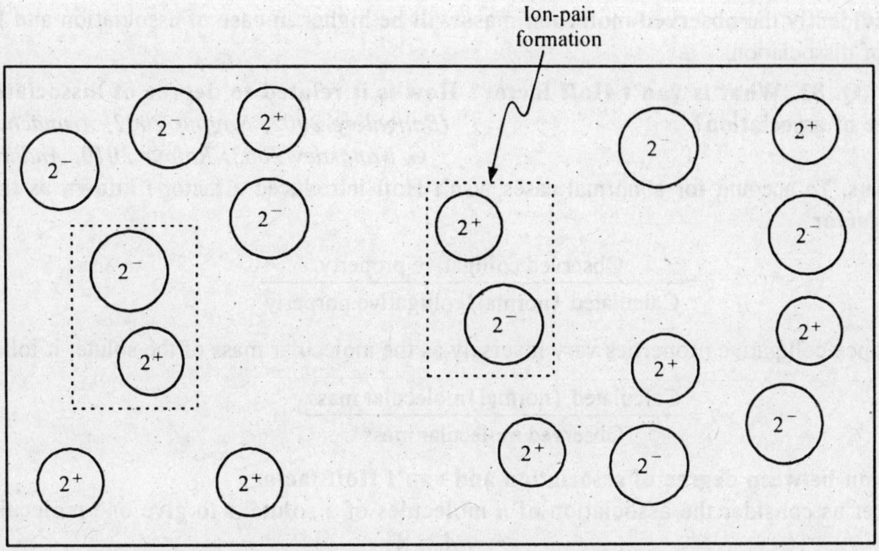

Fig. 3.35. Ion-pair formation of a 2^+ cation and a 2 – anion in a solution reduces the effective concentration of ions compared to actual concentration.

We reach the following conclusions :

(1) The activity coefficient, γ of strong electrolytes is always less than 1.

(2) The value of γ decreases with dilution at the same temperature, and at infinite dilution it approaches 1.

(3) In the mathematical expressions of the various laws of physical chemistry such as Raoult's law, Henry's law, law of Mass action and Ostwald's law, the actual concentration (C) has to be replaced with effective concentration (γC) for accurate experimental work.

Q. 81. What is the cause of abnormal molecular weights of solutes in solutions ?
(Meerut 2000)

Ans. Abnormal molecular mass of solute in solution is due to *association* or *dissociation* of molecules in the solution. These are explained as below.

(*a*) **Association:** If the molecules of a solute undergo association in solution, there will be decrease in the number of species. As a result there will be proportionate decrease in the value of each colligative property. The experimental value of molecular mass of the solute will be higher in such a case. This is because molecular mass is inversely proportional to the value of any colligative property.

For example acetic acid and benzoic acid both associate in benzene. Similarly, chloroacetic acid associates in naphthalene. The molecular mass of solute in such cases is higher than the molecular mass obtained from their molecular formulae. Thus the molecular mass of acetic acid (CH_3COOH) in benzene, as determined from freezing point depression is 118 instead of 60.

(*b*) **Dissociation:** Inorganic acids, bases and salts dissociate or ionise in solution. As a result of this number of effective particles increases and therefore, the value of colligative properties is also increased.

Therefore, the experimental value of molecular mass of the solute will be lower in such a case. This is because molecular mass is inversely proportional to the value of any colligative property as already discussed.

Evidently the observed molecular mass will be higher in case of association and lower in case of dissociation.

Q. 82. What is van't Hoff factor? How is it related to degree of dissociation and degree of association? *(Bangalore 2002; Nagpur 2002; Arunachal 2003; G. Nanakdev 2003; Raipur 2010; Andhra 2010)*

Ans. To account for abnormal cases, van't Hoff introduced a factor *i* **known as the van't Hoff factor**

$$i = \frac{\text{Observed colligative property}}{\text{Calculated (normal) colligative porperty}}$$

Since colligative properties vary inversely as the molecular mass of the solute, it follows that

$$i = \frac{\text{Calculated (normal) molecular mass}}{\text{Observed molecular mass}}$$

Relation between degree of association and van't Hoff factor

Let us consider the association of *n* molecules of a solute *A* to give one molecule of A_n

$$nA = A_n$$

Let the degree of association be α. If we start with 1 mole of *A*, the number of moles that associate is α and the number of moles that remain unchanged is 1 – α.
α moles on association will give α/*n* moles.

$$\text{Total number of moles at equilibrium} = \frac{\alpha}{n} + (1 - \alpha)$$

van't Hoff factor is the ratio of observed colligative property to calculated colligative property. And colligative property is proportional to the number of moles of the solute. Hence

$$\text{van't hoff factor } (i) = \frac{1-\alpha+\dfrac{\alpha}{n}}{1}$$

Relation between degree of dissociation and van't Hoff factor

Let us consider the dissociation of molecule to give *n* molecules or ions in solution

$$A \longrightarrow nB$$

Let us start with 1 mole of *A* and let α be the degree of dissociation.

At equilibrium

No. of moles of undissociated substance = 1 – α

No of moles of dissociated substance = *n*α

(one molecule dissociates to give *n* molecules)

Total number of moles of solute at equilibrium = 1 – α + *n*α.

$$\text{Hence van't Hoff factor } (i) = \frac{1-\alpha+n\alpha}{1}$$

MISCELLANEOUS QUESTIONS

Q. 83. With the help of a colligative property, why is it not possible to determine the correct molecular weight of an electrolyte ?

Ans. See Q. 80.

Q. 84. 0·1 *M* solution of KNO_3 has an osmotic pressure of 4.5 atmosphere at 300 K. Calculate the apparent degree of dissociation of the salt.

Ans. $\pi_{obs} = 4{\cdot}5$ atm $C = 0.1$ moles/litre

$n = 2$ because one KNO_3 molecule dissociates to give two ions, K^+ and NO_3^-

$T = 300$

Substituting the values in the equation

$\pi = C.R.T.$

$\pi = 0.1 \times 0.082 \times 300$

$\pi = 2.6$

$$i = \frac{\text{Observed osmotic pressure}}{\text{Calculated osmotic pressure}}$$

$$= \frac{4.5}{2.5} = 1.83 \qquad ...(1)$$

Also $\quad i = \dfrac{1 - \alpha + n\alpha}{1}$

or $\quad i = 1 + \alpha (n - 1)$ or $\alpha (n - 1) = i - 1$

or $$a = \frac{i-1}{n-1} \qquad ...(2)$$

Substitute the value of i from (1) in (2)

$$\alpha = \frac{1.83-1}{2.-1} \quad \text{or} \quad \alpha = 0.83$$

or $\quad = 83\%$

Q. 85. A 4 percent solution of sucrose $C_{12}H_{22}O_{11}$ is isotonic with 3 per cent solution of an unknown organic substance. Calculate the molecular mass of the unknown substance.

Ans. The two solutions are isotonic, *i.e.* have the same osmotic pressure. Hence, they must have same concentration in moles/litre.

For the solution of sucrose

Concentration = 4 g per 100 ml

= 40 g /litre

$= \dfrac{40}{342}$ moles/litre

For the solution of unknown substance

Let the molecular mass be m

Concentration = 3 g per 100 ml

= 30 g per litre

$= \dfrac{30}{m}$ moles/litre

The two concentrations can be equated

$$\frac{30}{m} = \frac{40}{342} \quad \text{or} \quad m = \frac{30 \times 342}{40} = \mathbf{256.5}$$

Q. 86. Explain giving reasons :

(*a*) *Addition of non-volatile solute lowers the freezing point and elevates the boiling point of the solvent*

(*b*) *Equimolar solutions of sucrose and sodium chloride in water are not isotonic*

(Arunachal 2003)

Ans. (*a*) See Q 49

Ans. (*a*) See Q 82

Q. 87. Account for the following (*i*) Benzoic acid in benzene shows less osmotic pressure than expected (*ii*) The boiling point of 0.1 *m* $BaCl_2$ solution is greater than 0.1 *m* sugar solution (water as solvent). *(G. Nanakdev 1995)*

Ans. See Q. 82

Q. 88. What is moral elevation constant ? How is it related to mole fraction of non-volatile solute ? *(Kalyani 2004)*

Ans. See Q 49

Q. 89. Give the thermodynamic derivation of the expression for boiling point elevation ΔT_b in the from = $\Delta T_b = K_b m$ Where *m* is the solute molality and K_b the ebullioscopic constant. *(G. Nanakdev 2004)*

Ans. See Q 49

Q. 90. Define cryoscopic constant. *(Mumbai 2005)*

Ans. See Q. 56. molal depression constant and Cryoscopic constant represent the same thing.

Q. 91. Tell whether the osmotic pressure of M/10 solution of glucose to be same as that of M/10 solution of sodium chloride. *(Calcutta, 2007)*

Ans. No. colligative property of a solution depends upon the number of moles of the compound or ions present per litre of the solution. M/10 solution of sodium chloride will contain twice the number of particles, as compared to M/10 glucose solution (one sodium chloride molecule gets dissociated into two ions). Consequently, M/10 sodium chloride will show nearly twice the osmotic presusse.

Q. 92. How and why does vapour pressure of a liquid depend upon the temperature ? *(Guwahati 2006)*

Ans. Vapour pressure of a liquid increases with the the increase of temperature. With the increase of temperature, the kinetic energy of the moleculues on the surface increases (K.E. = $\frac{3}{2}$ RT). More and more molecules leave the surface of the liquid and are converted into vapours, thereby raising the vapour pressure.

4

CHEMICAL KINETICS

Q. 1. Explain what is meant by rate of a reaction. How is it expressed ?

Ans. *Rate of a reaction may be defined as the change in the concentration of any one of the reactants or products per unit time.* For example, consider the dissociation of PCl_5 to give PCl_3 and Cl_2.

$$PCl_5 \rightarrow PCl_3 + Cl_2$$

Rate of this reaction

$$= \frac{\text{Decrease in the conc. of } PCl_5}{\text{Time taken}}$$

or $$= \frac{\text{Increase in the conc. } PCl_3 \text{ or } Cl_2}{\text{Time taken}}$$

Since the concentration is usually expressed in moles/litre and the time is taken in seconds or minutes, the units of the rate of reaction are therefore, moles $litre^{-1}$ sec^{-1} or moles $litre^{-1}$ min^{-1}.

Rate of a reaction does not remain constant throughout. Hence we can consider rate of a reaction at a particular instant which is defined as follows.

The rate of a reaction at any time is defined as the change in the concentration of any one of the reactants or products per unit time at that particular instant of time.

Expression of the rate of a reaction

Since the rate of a reaction is changing continuously throughout the course of the reaction, therefore to state the rate of reaction at any particular instant of time, a small interval of time dt may be taken during which the rate of reaction may be assumed to be almost constant. Suppose dx is the small change in concentration of any one of the reactants or products during the small interval of time dt. Then the rate of reaction may be expressed as:

$$\text{Rate of reaction} = \frac{dx}{dt}$$

Further if the rate is expressed in terms of one of the reactants, a negative sign is attached to show that the concentration is decreasing.

For example, consider the reaction

$$A + B \longrightarrow P + Q$$

Then rate of this reaction may be expressed in any one of the following ways :

$$\text{Rate of reaction } \left(\frac{dx}{dt}\right) = -\frac{dC_A}{dt} = -\frac{dC_B}{dt} = +\frac{dC_P}{dt} = +\frac{dC_Q}{dt}$$

where dC_A and dC_B represent small decrease in the concentrations of A and B respectively and dC_p and dC_Q represent small increase in the concentrations of P and Q respectively in the small interval of time dt.

For a more general reaction of the type

$$aA + bB \longrightarrow pP + qQ$$

a unique value of the rate of reaction can be obtained by expressing the rate of reaction as follows:

$$\text{Rate of reaction} = -\frac{1}{a}\frac{d[A]}{dt} = -\frac{1}{b}\frac{d[B]}{dt} = +\frac{1}{p}\frac{d[P]}{dt} = +\frac{1}{q}\frac{d[Q]}{dt}$$

Q. 2. How is the rate of a reaction measured ?

Ans. In order to measure the rate of a reaction, the concentration of one of the reactants or products is observed at different intervals of time. A common practice is to withdraw small amount of the reaction mixture (2 ml or 5 ml) at different intervals of time, cool it down immediately to nearly 0°C to arrest the reaction and then find out the concentration of the reactant or the product by suitable method, usually by titration against a suitable reagent. When some observable property like volume, pressure, optical rotation etc. change with time and can be observed directly at different intervals of time without stopping the reaction, it is preferably utilized. It is important to mention that factors like temperature etc. which affect the rate of the reaction are kept constant during the kinetic study of the reaction.

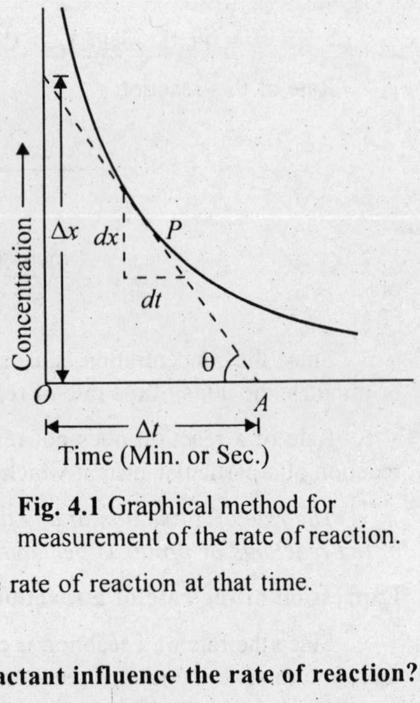

Fig. 4.1 Graphical method for measurement of the rate of reaction.

If concentration of one of the reactants is studied at different intervals of time and these values are plotted against the corresponding times, a graph of the type shown in Fig. 4.1 is obtained. The rate of reaction at any time may be obtained by the following procedure.

A tangent may be drawn on the curve at the point corresponding to the time at which the rate is required (Fig. 4.1). The slope of the tangent will then obviously give the rate of reaction at that time.

$$\text{rate} = \tan\theta = \frac{OB}{OA} = \frac{\Delta x}{\Delta t} = \frac{dx}{dt}$$

Q. 3. How do nature and concentration of the reactant influence the rate of reaction?

Ans. The influence of nature and concentration of the reactant is explained as under:

(*i*) **Nature of reactants:** Consider the oxidation of ferrous ions and oxalate ions by acidified $KMnO_4$ solution. Both reactions are similar as represented below :

$$MnO_4^- + 8H^+ + 5Fe^{+2} \xrightarrow{\text{Fast}} Mn^{+2} + 5Fe^{+3} + 4H_2O$$

$$2MnO_4^- + 16H^+ + 5C_2O_4^{-2} \xrightarrow{\text{Slow}} 2Mn^{+2} + 10CO_2 + 8H_2O$$

At room temperature, the first reaction is very fast whereas the second reaction is quite slow as indicated by the rate of disappearance of pink colour of $KMnO_4$. In fact, in the titrations, the second reaction is so slow that heating is required. Thus in the above case, even if the same concentrations of similar reactants are taken in the two reactions and the temperature is also kept the same, the rates of these two reactions are still different. This is obviously only due to the difference in the nature of the reducing agents *i.e.*, Fe^{+2} ions and $C_2O_4^{-2}$ ions.

The difference in the rates may be ascribed to the fact that a reaction involves breaking of the bonds of the reacting molecules and the formation of new bonds in the product molecules. Energy is required for the breaking of bonds and energy is released when a bond is formed. The net energies involved are different for different reactions and hence the rates are different.

(*ii*) **Concentration of the reactants:** As a reaction proceeds, the reactants are consumed. Therefore, the concentration of the reactants decreases with the passage of time. It is observed experimentally that the rate of reaction also decreases as the time passes. This clearly indicates that the rate of reaction is directly proportional to the concentration of the reactants. The variation of the concentration of a reactant and hence that of the rate of reaction with time may be represented graphically as shown in Fig. 4.2.

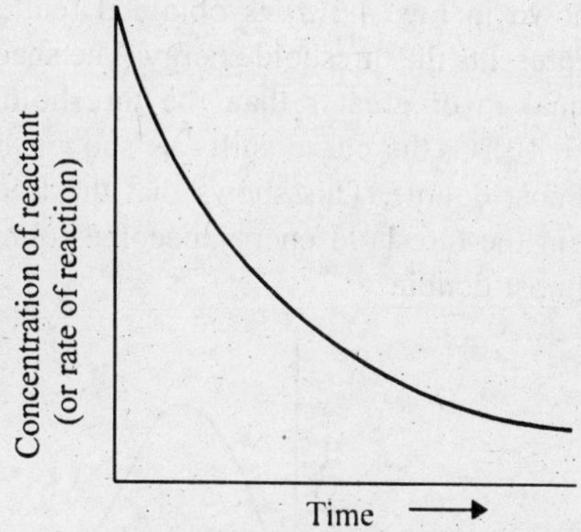

Fig. 4.2. Variation of concentration of a reactant (or the rate of reaction) with time.

The effect of concentration is explained on the basis of collision theory of reaction rates. The rate of a reaction depends upon the number of collisions between the molecules. Higher the concentrations, greater is the number of collisions.

Q. 4. Discuss the effect of temperature on the rate of a reaction.

(Kanpur 2000;Kolhapur, 2000; Lucknow 2000)

Ans. Temperature has a very significant effect on the rate of a reaction. A rise of temperature not only increases the speed of the reaction but in many cases it actually initiates the reaction. In the study of the effect of temperature on the reaction rates, two types of cases are observed:

(*a*) *For most of the reactions, the rate of reaction increases with rise of temperature.* As a rough rule, rate of a reaction becomes nearly double or sometimes even treble for every 10°C rise of temperature. This is generally expressed in the form of temperature coefficient as

$$\text{Temperature coefficient} = \frac{k_{(t+10)}}{k_t} \simeq 2 \text{ to } 3$$

where k_t and $k_{(t+10)}$ are the specific reaction rates at t°C and (t+ 10)°C respectively. It is really a wonderful observation because the number of collisions increase only by about 2% for 10° rise at ordinary temperatures whereas the reaction velocity may increase by 100 or 200 per cent or even more.

(*b*) *A large number of reactions are known which do not take place at room temperature but occur readily at higher temperature e.g.,* reaction between hydrogen and oxygen to form water takes place only when an electric spark is passed. Similarly, it is well known that coal can be kept in oxygen for years without burning but once the ignition is initiated, it goes on till the whole of coal burns away.

The above observations are explained on the basis of **"Collision Theory"**. According to this theory, substances react because their molecules collide with each other. All collisions are, however, not effective collisions. An **effective collision** is that collision which results into a chemical reaction. In order that a collision may be effective, the colliding molecules must possess a certain minimum amount of energy called threshold energy. Thus **threshold energy** *may be defined as the minimum amount of energy which the colliding molecules must possess in order that the collision between them may be an effective collision.*

The two types of cases may be explained as follows :

Explanation of the increase in reaction rate with rise of a temperature: The different molecules of the reactants do not possess the same energy. If the energies of the reacting molecules are plotted against the corresponding fractions, at a particular temperature T, a curve of the type shown in Fig. 4.3(a) is obtained (called *Maxwell's distribution of energies*). In the figure, E_t represents the threshold energy. The shaded area represents the fraction of molecules with energies equal to or greater than the threshold energy E_t. As the temperature is increased from t to $(t + 10)$°C, the curve shifts as shown in Fig. 4·3 (b). The shaded area beyond E_t now becomes almost double. This shows that the fraction of molecules possessing energy equal to or greater than the threshold energy becomes almost double. Hence the rate of the reaction also becomes almost double.

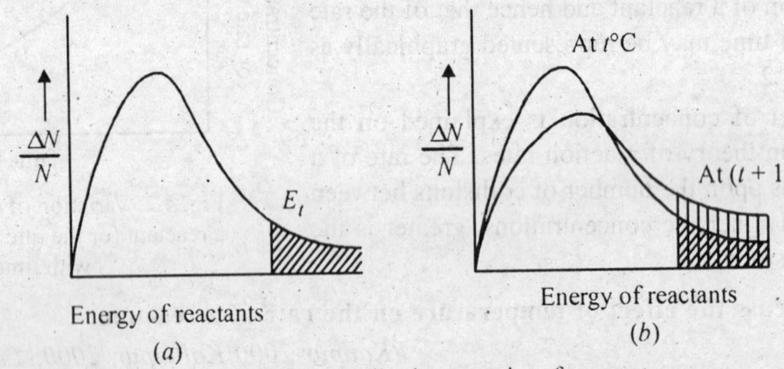

Fig. 4.3 Distribution energies of reactants.

Explanation of the non-occurrence of reactions at room temperature and occurrence at higher temperature: At room temperature, there is no molecule possessing energy greater than the threshold energy and so there is no reaction. When temperature is increased, the molecules of the reactants absorb heat energy and thus a number of molecules start attaining energy equal to or greater than the threshold energy and hence the reaction starts.

The extra amount of energy which the molecules of the reactants have to absorb so that their energy becomes equal to the threshold energy is called the **Activation energy.** In other words,

Activation energy = Threshold energy – Energy actually possessed by the reacting molecules.

Thus in order that the reactants may change over to products, they have to absorb some energy and thus cross a barrier called **Activation Energy Barrier.** The situation may be represented graphically as follows :

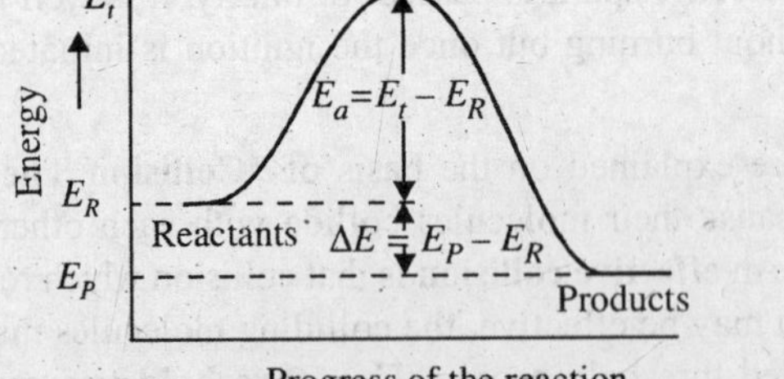

Fig. 4.4. Illustration of the concept of activation energy.

Q. 5. Discuss the effect of catalyst on the rate of a reaction.

Ans. A catalyst is a substance which increases the speed of a reaction without itself undergoing any chemical change. For example, the decomposition of $KClO_3$ is greatly accelerated by the addition of a small quantity of MnO_2 which itself does not undergo any chemical change during the reaction.

The presence of a catalyst provides an alternate path with lower energy barrier, as shown in Fig. 4.5. Thus the energy of activation is lowered and hence a greater number of molecules can cross the barrier and change over to products, thereby increasing rate of the reaction. In activation for the uncatalysed and catalysed reactions respectively E_t and $E_{t(ap)}$ represent total energy of the activated complex in the two paths.

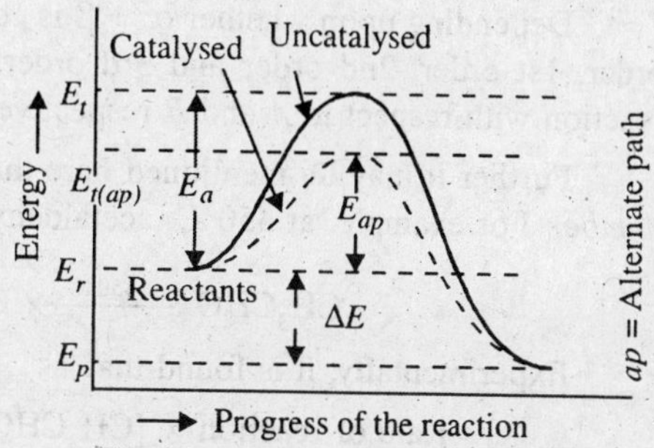

Fig. 4.5. Effect of addition of a catalyst.

Following observations are made from Fig. 4.5.

(*a*) ΔE (= $E_p - E_r$) for the catalysed reaction is the same as ΔE for the uncatalysed reaction.

(*b*) For reversible reactions, the energy of activation for the reverse reaction is lowered to the same extent as for the direct reaction. Hence the presence of the catalyst in such reactions increases the speed of the forward reaction and that of the backward reaction to the same extent. Consequently, the equilibrium is not disturbed.

Q. 6. Write rate law and define rate constant. *(M. Dayanand 2000)*

Ans. Consider the following general reaction

$$aA + bB \longrightarrow \text{Products}$$

Experimentally, it is observed that the rate of this reaction may not depend upon all the *a* concentration terms of *A* and all the *b* concentration terms of *B*. Suppose, experimentally the rate of the reaction is found to depend upon α concentration terms *A* and β concentration terms of *B*. Then we may write

$$\text{Rate} \propto [A]^{\alpha} . [B]^{\beta}$$

$$\text{Rate} = k\,[A]^{\alpha}\;[B]^{\beta}$$

Eq. (*i*) is called rate law.

Here [*A*] and [*B*] are the molar concentrations of *A* and *B* respectively and *k* is a constant called **velocity constant** or **rate constant**. If all concentrations are taken as unity.

i.e., $[A] = [B] = 1$ mol/litre, then

$$\text{Rate} = k.$$

Hence the *rate constant may be defined as the rate of the reaction when the concentration of each reactant is taken as unity.* That is why the rate constant is also called **specific reaction rate.**

Q. 7. What do you mean by rate of reaction and specific reaction rate ? *(Bhopal,2004)*

Ans. See Q. 6.

Q. 8. Define order of a reaction. *(Vidyasagar 2007, Nagpur 2008; Garhwal 2010)*

Ans. According to rate law

$$\text{Rate} \propto [A]^{\alpha}\;[B]^{\beta}$$

The sum of the concentration terms on which the rate of a reaction actually depends is called the order of the reaction. Thus in the above case, order of reaction = α + β.

Hence the order of a reaction may be defined as the sum of the exponents to which the concentration terms in the rate law are raised to express the observed rate of reaction.

Depending upon whether α + β is equal to 0, 1, 2 or 3, the reactions are said to be of zero order, 1st order, 2nd order and 3rd order respectively. α and β are called the orders of the reaction with respect to *A* and *B* respectively.

Further it may be mentioned here that *the order of a reaction may not be always a whole number.* For example, at 450°C, acetaldehyde decomposes as

$$CH_3CHO \xrightarrow{450°C} CH_4 + CO$$

Experimentally, it is found that

$$\text{Rate of reaction} \propto [CH_3CHO]^{1.5}.$$

∴ Order of reaction = 1.5

Q. 9. What is meant by zero order reaction? Describe giving examples.

Ans. *A reaction is said to be of zero order if the rate of the reaction is independent of the concentration of all the reactants.*

For such reactions the rate of the reaction remains constant throughout.

Following reactions are found to be of zero order :

(*i*) **Photochemical reactions:** Many such reactions have been found to be of zero order. For example, the combination between hydrogen and chlorine in presence of light may be studied by enclosing these gases in a tube and inverting the tube in a trough of water (Fig. 4.6). When the reaction takes place between H_2 and Cl_2, HCl is formed which dissolves into water and the level of water in the tube rises. Thus though the quantities of H_2 and Cl_2 decrease, their amounts per unit volume remain the same or the concentrations of H_2 and Cl_2 remain constant.

(*ii*) **Heterogeneous reactions:** Two well known examples of this type are (*a*) the decomposition of hydrogen iodide on the surface of gold and (*b*) the decomposition of ammonia on the surface of tungsten, molybdenum or platinum at high temperatures. The gaseous molecules of the reacting substance get adsorbed on the surface of the catalyst. The surface is thus fully occupied with the molecules of the reactant. Some of these molecules react to form the products but immediately new molecules take up their position and thus the concentration of the reacting molecules remains constant on the surface and hence the rate of reaction remains constant irrespective of the total amount of the gaseous reactants taken initially.

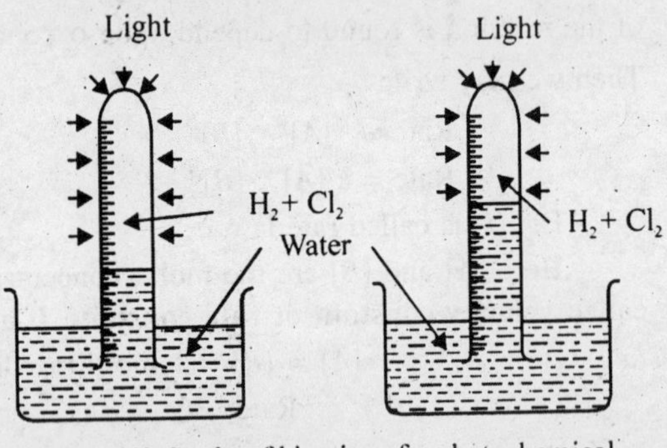

Fig. 4.6. Study of kinetics of a photochemical reaction.

Q. 10. Write mathematical expression for zero order reaction.

(*Kurukshetra 2002, Andhra 2010*)

Ans. If *a* is the initial concentration, the expression for the rate of a reaction of zero order may be written as follows :

$$\frac{dx}{dt} \propto a^o$$

or $$\frac{dx}{dt} = k$$

where k is the rate constant for the zero order reaction.

The above expression may be written as

$$dx = kdt$$

Integrating this equation we get

$$x = kt + I$$

where I is constant of integration

When $t = 0$, $x = 0$; therefore $I = 0$

Hence $$\mathbf{x = kt} \qquad \ldots (i)$$

i.e., the amount of substance reacted ∝ time.

The expression for the half-life period (*i.e.*, the time taken for half the reaction to complete) may be obtained as follows :

When $x = \frac{a}{2}$; $t = t_{1/2}$

Substituting these values in eq. (*i*) above, we get

$$\frac{a}{2} = kt_{1/2}$$

or $$t_{1/2} = \frac{a}{2k}$$

i.e., **the half-life period ∝** initial concentration.

Q. 11. What is a first-order reaction? Derive an expression for rate constant of first order reaction. *(M. Dayanand 2002; Vidyasagar 2007, Nagpur 2008,*

OR

Show that for a first-order reaction

$$k = \frac{2.303}{t} \log \frac{C_0}{C_t}$$

where the symbols have their usual meanings. *(Kurukshetra 200 1. Dayanand, 2000)*

Ans. *A reaction is said to be of the first order if the rate of the reaction depends upon one concentration term only.*

Let us consider a hypothetical case.

$$A \longrightarrow \text{Products}$$

Suppose we start with a moles per litre of the reactant A. After time t, suppose x moles per litre of it have decomposed. Therefore the concentration of A after time $t = (a - x)$ moles per litre. Then according to law of mass action,

Rate of reaction ∝ $(a - x)$

i.e., $$\frac{dx}{dt} \propto (a - x)$$

or $$\frac{dx}{dt} = k(a - x) \qquad \ldots (i)$$

where k is called the *rate constant* or the *specific reaction rate* for the reaction of the first order.

The expression for the rate constant k may be derived as follows :

Equation (*i*) may be rewritten in the form

$$\frac{dx}{a - x} = kdt \qquad \ldots (ii)$$

$$\frac{dx}{a-x} = kdt \qquad ...(ii)$$

Integrating equation (*ii*), we get

$$\int \frac{dx}{a-x} = \int kdt$$

i.e.

$$-\ln(a-x) = kt + I \qquad ...(iii)$$

where I is a constant of integration.

In the beginning, when $t = 0$, $x = 0$ (as no substance has decomposed at the start of the reaction).

Putting these values in equation (*iii*), we get

$$-\ln(a-0) = k \times 0 + I$$

or

$$-\ln a = I \qquad ...(iv)$$

Substituting this value of I in equation (*iii*), we get

$$-\ln(a-x) = kt + (-\ln a)$$

or

$$kt = \ln a - \ln(a-x)$$

$$= \ln \frac{a}{a-x}$$

or

$$k = \frac{1}{t} \ln \frac{a}{a-x} \qquad ...(v)$$

or

$$\mathbf{k = \frac{2.303}{t} \log \frac{a}{a-x}} \qquad ...(vi)$$

Equation (*vi*) is sometimes written in another form which is obtained as follows :

If the initial concentration is C_0 and the concentration after time t is C_t then putting $a = C_0$ and $(a-x) = C_t$, equation (*vi*) becomes

$$\mathbf{k = \frac{2.303}{t} \log \frac{C_0}{C_t}} \qquad ...(viii)$$

Q. 12. Discuss some characteristics of first-order reactions.

(Kerala 2001; M. Dayanand 2002, Nagpur 2008)

Ans. (*i*) *Any reaction of the first order must obey equation* (*vi*) or (*vii*). This may be tested by one of the following methods.

(*a*) *Substitution method.* Starting with a known concentration *a*, the concentration of the reactant $(a-x)$, at different intervals of time may be noted. For every value of *t*, the corresponding value of $(a-x)$ may be substituted in equation (*vi*). If the values of *k* thus obtained are nearly constant (within experimental error), the reaction is of the first order.

(*ii*) *The time taken for any fraction of the reaction to be completed is independent of the initial concentration.* For example, let us test this statement for *half* of the reaction to be completed. Writing the first-order equation.

$$t = \frac{2.303}{k} \log \frac{a}{a-x} \qquad ...(i)$$

When half of the reaction is completed, $x = a/2$. Representing the time taken for half of the reaction to be completed by $t_{1/2}$, above equation becomes

$$t_{1/2} = \frac{2.303}{k} \log \frac{a}{a - a/2}$$

$$= \frac{2.303}{k} \log 2 = \frac{0.693}{k} \qquad ...(ii)$$

Thus a does not appear in this equation so that $t_{1/2}$ is independent of a. Similarly, it can be seen that $t_{1/3}$, $t_{2/3}$ etc. will also be independent of the initial concentration.

The time taken for half of the reaction to be completed *i.e.*, the time in which the concentration of a reactant is reduced to half of its original value is called **half-life period** *of the reaction.*

(*iii*) *The units of k are independent of the units in which the concentrations are expressed.* This is obvious from equation (*i*) because in this equation k depends upon the ratio of two concentrations *viz.* $\frac{a}{a-x}$ and the time t. Thus so long as both a and $(a - x)$ are expressed in the same concentration units, the value of k is not affected. The units of k, however, depend only upon the units of time t.

Q. 13. Discuss decomposition of nitrogen pentoxide as first-order reaction.

Ans. The compound, nitrogen pentoxide, is a volatile solid which decomposes in the gaseous state as well as in the form of its solution in an inert solvent like carbon tetrachloride, chloroform etc. according to the equation

$$N_2O_5 \longrightarrow N_2O_4 + \frac{1}{2}O_2$$
$$N_2O_4 \rightleftharpoons 2NO_2$$

When the reaction is carried out in the solution, N_2O_4 and NO_2 remain in the solution and the volume of oxygen gas collected is noted at different intervals of time. It can be observed that

$$\left.\begin{array}{l}\textit{Volume of oxygen gas collected}\\ \textit{at any time } (V_t)\end{array}\right\} \propto \textit{Amount of } N_2O_5 \textit{ decomposed } (x)$$

i.e., $$x \propto V_t$$

$$\left.\begin{array}{l}\textit{Volume of oxygen gas collected}\\ \textit{at infinite time } (V_\infty)\end{array}\right\} \propto \textit{Amount of } N_2O_5 \textit{ initially taken } (a)$$

i.e., $$a \propto V_\infty$$

Substituting these values in the first order equation *viz.*

$$k = \frac{2.303}{t} log \frac{a}{a-x}$$

we get $$\mathbf{k = \frac{2.303}{t} \log \frac{V_\infty}{V_\infty - V_t}}$$

Q. 14. From the following data for the decomposition of N_2O_5 in carbon tetrachloride solution at 48°C, show that the reaction is of the first order and calculate the rate constant.

Time (in minutes) :	**10**	**15**	**20**	**25**	**∞**
Vol. of O_2 evolved (in ml) :	**6.30**	**8.95**	**11.40**	**13.50**	**34.75**

Ans. If the reaction is of the first order, it must obey the equation

$$k = \frac{2.303}{t} \log \frac{a}{a-x}$$

$$= \frac{2.303}{t} \log \frac{V_\infty}{V_\infty - V_t}$$

In the present case, $V_\infty = 34.75$ ml

The value of k at each time can be calculated as follows :

Time (*min*)	V_t	$V_\infty - V_t$	$k = \frac{2.303}{t}\log\frac{V_\infty}{V_\infty - V_t}$ (min^{-1})
10	6.30	34.75 – 6.30 = 28.45	$k = \frac{2.303}{10}\log\frac{34.75}{28.45} = 0.01997$
15	8.95	34.75 – 8.05 = 25.80	$k = \frac{2.303}{15}\log\frac{34.75}{25.80} = 0.01985$
20	11.40	34.75 – 11.40 = 23.35	$k = \frac{2.303}{20}\log\frac{34.75}{23.35} = 0.01987$

Since the value of k comes out to be nearly constant, given reaction is of the first order. The *average* value of the rate constant = 0.01989 minute^{-1}.

Q. 15. The decomposition of an aqueous solution of ammonium nitrite was studied by placing the apparatus in a thermostat maintained at a particular temperature. The volume of nitrogen gas collected at different intervals of time was as follows :

Time (minutes) :	**10**	**15**	**20**	**25**	**∞**
Vol. of N_2 (ml) :	**6.25**	**9.00**	**11.40**	**13.65**	**35.05**

From the above data prove that the reaction is of the first order.

Ans. Ammonium nitrite decomposes as follows : $NH_4NO_2 \longrightarrow 2H_2O + N_2$. Volume of nitrogen produced is observed after different intervals of time.

Vol of N_2 at any time $\propto x$.

Vol of N_2 after infinite time $\propto a$.

Substitute the values in the equation below :

$$k = \frac{2.303}{t}\log\frac{a}{a-x}$$

$$= \frac{2.303}{t}\log\frac{V_\infty}{V_\infty - V_t}$$

In the present case, V_∞ = 35.05 ml.

The value of k after each time can be calculated as follows:

Time (*min*)	V_t	$V_\infty - V_t$	$k = \frac{2.303}{t}\log\frac{V_\infty}{V_\infty - V_t}$ (min^{-1})
10	6.25	35.05 – 6.25 = 28.80	$k = \frac{2.303}{10}\log\frac{35.05}{28.80} = 0.0196$
15	9.00	35.05 – 9.00 = 26.05	$k = \frac{2.303}{15}\log\frac{35.05}{26.05} = 0.0198$
20	11.40	35.05 – 11.40 = 23.65	$k = \frac{2.303}{20}\log\frac{35.05}{23.65} = 0.0196$

Since the value of k comes out to be nearly constant, the given reaction is of the first order.

Q. 16. Benzene diazonium chloride in aqueous solution decomposes according to the equation

$$C_6H_5N_2Cl \longrightarrow C_6H_5Cl + N_2$$

Starting with an initial concentration of 10 g/litre, the volume of N_2 gas obtained at 50°C at different intervals of time was found to be as under :

t (min) :	**6**	**12**	**18**	**24**	**30**	**∞**
Vol. of N_2 (ml) :	**19.3**	**32.6**	**41.3**	**46.5**	**50.4**	**58.3**

Show that the above reaction follows the first order kinetics. What is the value of the rate constant?

Ans. If the given reaction follows first order kinetics, it must obey the equation

$$k = \frac{2.303}{t} \log \frac{a}{a-x}$$

$$= \frac{2.303}{t} \log \frac{V_\infty}{V_\infty - V_t}$$

In the present case, $V_\infty = 58.3$ ml.

The value of k after each time can be calculated as follows:

Time (*min*)	V_t	$V_\infty - V_t$	$k = \frac{2.303}{t} \log \frac{V_\infty}{V_\infty - V_t}$
6	19.3	58.3 – 19.3 = 39.0	$k = \frac{2.303}{6} \log \frac{58.3}{39.0} = 0.0670 \text{ min}^{-1}$
12	32.6	58.3 – 32.6 = 25.7	$k = \frac{2.303}{12} \log \frac{58.3}{25.7} = 0.0683 \text{ min}^{-1}$
18	41.3	58.3 – 41.3 = 17.0	$k = \frac{2.303}{18} \log \frac{58.3}{17.0} = 0.0685 \text{ min}^{-1}$
24	46.5	58.3 – 46.5 = 11.8	$k = \frac{2.303}{24} \log \frac{58.3}{11.8} = 0.0666 \text{ min}^{-1}$

Since the value of k comes out to be nearly constant, the reaction is of the first order. The mean value of $k = 0.0676 \text{ min}^{-1}$.

Q. 17. Discuss the decomposition of hydrogen peroxide as a first order reaction.

Ans. The decomposition of hydrogen peroxide in aqueous solution takes place according to the equation

$$H_2O_2 \xrightarrow{Pt} H_2O + \frac{1}{2}O_2$$

The kinetics of this reaction may be studied by making use of the fact that H_2O_2 solution can be titrated against $KMnO_4$ solution. Thus by withdrawing equal amounts of the solution (usually 5 ml) at regular intervals of time and titrating against the same $KMnO_4$ solution, the amount of H_2O_2 present can be found everytime. It is obvious that *for the same volume of the reaction solution withdrawn*

Volume of $KMnO_4$ *solution used before the commencement of the reaction i.e., at zero time* (V_0) $\propto$ *Initial concentration of* H_2O_2 (a)

i.e.,

$$V_0 \propto a$$

Volume of $KMnO_4$ *solution used at any instant of time t* (V_t) $\propto$ *Amount of* H_2O_2 *present at that instant i.e.,* ($a - x$)

$$V_t \propto (a - x)$$

Substituting these values in the first order equation, we get

$$k = \frac{2.303}{t} \log \frac{a}{a-x}$$

or $$k=\frac{2.303}{t}\log\frac{V_0}{V_t}$$

The decomposition of hydrogen peroxide, as tested by applying this equation, is found to be of the first order.

Q. 18. From the following data, show that the decomposition of hydrogen peroxide is a reaction of the first order :

t **(min.)**	**0**	**10**	**20**
V **(ml)**	**46.1**	**29.8**	**193**

where *t* is the time in minutes and *V* is the volume of standard $KMnO_4$ solution required for titrating the same volume of the reaction mixture.

Ans. Volume of $KMnO_4$ solution used ∝ amount of H_2O_2 present. Hence if the given reaction is of the first order, it must obey the equation

$$k=\frac{2.303}{t}\log\frac{a}{a-x}$$

$$k=\frac{2.303}{t}\log\frac{V_0}{V_t}$$

In the present case, V_0 = 46.1 ml

The value of k at each instant can be calculated as follows :

t (min)	V_t	$k=\frac{2.303}{t}\log\frac{V_0}{V_t}$
10	29.8	$k=\frac{2.303}{10}\log\frac{46.1}{29.8}=0.0436\ \text{min}^{-1}$
20	19.3	$k=\frac{2.303}{20}\log\frac{46.1}{19.3}=0.0435\ \text{min}^{-1}$

Thus, the value of k comes out to be nearly constant. Hence it is a reaction of the first order.

Q. 19. Discuss conversion of *N*-chloroacetanilide into *p*-chloroacetanilide as a first order reaction.

Ans. The reaction can be represented as under.

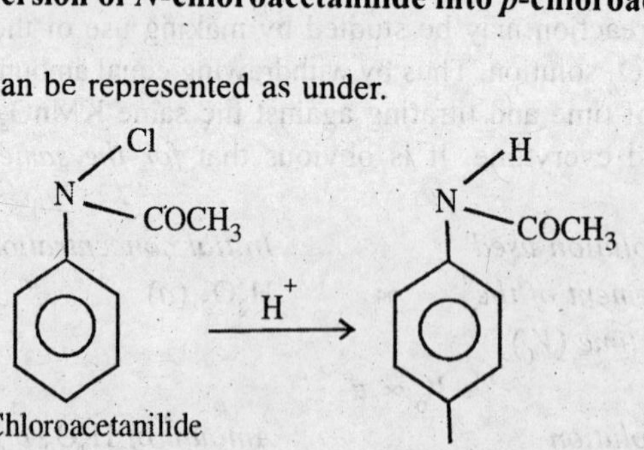

N-Chloroacetanilide

P-Chloroacetanilide

The study of the kinetics of this reaction is based upon the fact that *N*-chloroacetanilide reacts with KI in presence of the acid to give iodine whereas *p*-chloroacetanilide does not. The reactions take place as follows:

$$2KI + H_2SO_4 \longrightarrow K_2SO_4 + 2HI$$

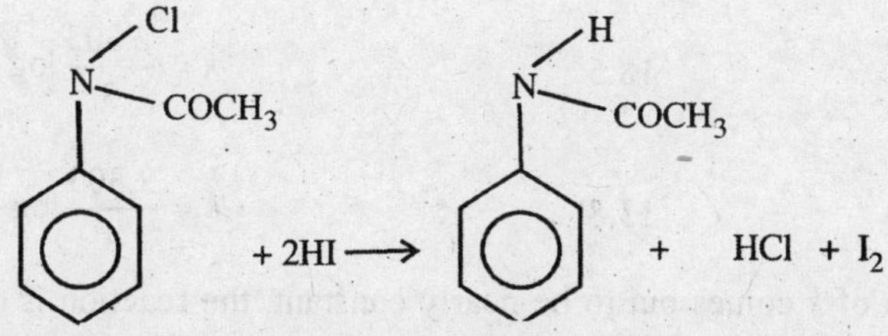

The iodine thus liberated can be found by titration against a standard hypo solution using starch as indicator. For the same volume of the reaction solution withdrawn at different times,

$$\left.\begin{array}{l}\textit{Volume of hypo solution}\\ \textit{used in the beginning}\\ \textit{i.e., at time } t = 0\ (V_0)\end{array}\right\} \propto \left\{\begin{array}{l}\text{Amount of } I_2\\ \textit{liberated at}\\ \textit{that stage}\end{array}\right\} \propto \left\{\begin{array}{l}\textit{Amount of}\\ \textit{N–chloroacctanilide}\\ \textit{present initially } (a)\end{array}\right.$$

or $V_0 \propto a$

$$\left.\begin{array}{l}\textit{Volume of hypo solution}\\ \textit{used at any instant of}\\ \textit{time } t\ (V_t)\end{array}\right\} \propto \left\{\begin{array}{l}\text{Amount of } I_2\\ \textit{liberated at}\\ \textit{that stage}\end{array}\right\} \propto \left\{\begin{array}{l}\text{Amount of } N\text{-chloroacetanilide}\\ \textit{present at that instant of time}\\ (a-x)\end{array}\right.$$

or $V_t \propto (a-x)$

Hence for the above reaction,

$$k = \frac{2.303}{t}\log\frac{a}{a-x}$$

$$k = \frac{2.303}{t}\log\frac{V_0}{V_t}$$

Q. 20. In acidic solution *N*-chloroacetanilide changes into *p*-chloroacetanilide. The former can be estimated volumetrically by withdrawing a definite volume of the reaction mixture, adding KI and titrating the iodine liberated against hypo solution. The following results were obtained in a particular experiment :

Time (hours) :	**0**	**1**	**2**	**3**	**4**	**∞**
Hypo sol. used (ml) :	**49.3**	**35.6**	**25.8**	**18.5**	**13.8**	**0.0**

Show that the above data conform to the first order reaction.

Ans. Volume of hypo solution used ∝ amount of *N*-chloroacetanilide present. Hence if the reaction is of the first order, it must obey the equation

$$k = \frac{2.303}{t}\log\frac{a}{a-x}$$

As $V_0 \propto a$ and $V_t \propto a - x$, we have

$$k = \frac{2.303}{t}\log\frac{V_0}{V_t}$$

In the present case, $V_0 = 49.3$ ml

The value of k at different instants may be calculated as follows:

t *(hrs)*	V_t	$k = \frac{2.303}{t}\log\frac{V_0}{V_t}$
1	35.6	$k = \frac{2.303}{1}\log\frac{49.3}{35.6} = 0.3256\ \text{hr}^{-1}$
2	25.8	$k = \frac{2.303}{2}\log\frac{49.3}{25.8} = 0.3238\ \text{hr}^{-1}$

3	18.5	$k = \frac{2.303}{3}\log\frac{49.3}{18.5} = 0.3267\ \text{hr}^{-1}$
4	13.8	$k = \frac{2.303}{4}\log\frac{49.3}{13.8} = 0.3183\ \text{hr}^{-1}$

Since the value of k comes out to be nearly constant, the reaction is of the first order.

Q. 21. Study the kinetics of hydrolysis of ethyl acetate in acidic medium.

Ans. The reaction can be represented as under :

$$CH_3COOC_2H_5 + H_2O \xrightarrow{H^+} CH_3COOH + C_2H_5OH$$

In this reaction, acetic acid is one of the products the amount of which can be found by titration against standard NaOH solution. But being an acid-catalysed reaction, the acid present originally as a catalyst also reacts with NaOH solution. Hence *for the same volume of reaction mixture withdrawn* at different times.

$$\left.\begin{array}{l}\textit{Volume of NaOH solution}\\ \textit{used in the beginning i.e., at}\\ \textit{zero time } (V_0)\end{array}\right\} \propto \left\{\begin{array}{l}\textit{Amount of acid present only}\\ \textit{as catalyst (as no } CH_3COOH\\ \textit{is produced at } t = 0)\end{array}\right. \quad \ldots (i)$$

$$\left.\begin{array}{l}\textit{Volume of NaOH}\\ \textit{solution used at any}\\ \textit{instant of time } t\ (V_t)\end{array}\right\} \propto \left\{\begin{array}{l}\textit{Amount of acid present}\\ \textit{as catalyst + Amount of}\\ CH_3COOH \textit{ produced}\end{array}\right. \quad \ldots (ii)$$

Combining results (*i*) and (*ii*), we find that

$$\left.\begin{array}{l}\textit{Amount of } CH_3COOH\\ \textit{produced at any instant}\\ \textit{of time}\end{array}\right\} \propto \{(V_t - V_0) \quad \ldots (iii)$$

$$\left.\begin{array}{l}\textit{But amount of } CH_3COOH\\ \textit{produced at any instant of time}\end{array}\right\} \propto \left\{\begin{array}{l}\textit{Amount of } CH_3COOC_2H_5\\ \textit{that has reacted } (x)\end{array}\right.$$

Hence $\quad (V_t - V_0) \propto x \quad \ldots (iv)$

Further,

$$\left.\begin{array}{l}\textit{Volume of NaOH solution used after the}\\ \textit{reaction has taken place for a long time,}\\ \textit{called infinite time } (V_\infty)\end{array}\right\} \propto \left\{\begin{array}{l}\textit{Amount of acid present}\\ \textit{as catalyst + Max. amount}\\ \textit{of } CH_3COOH \textit{ produced}\end{array}\right. \quad \ldots (v)$$

Combining results (*i*) and (*v*), we find that

Max. amount of CH_3COOH *produced* $\propto (V_\infty \quad V_0)$

$$\left.\begin{array}{l}\textit{But Max. amount of}\\ CH_3COOH \textit{ produced}\end{array}\right\} \propto \left\{\begin{array}{l}\textit{Initial concentration of}\\ CH_3COOC_2H_5\ (a)\end{array}\right.$$

Hence $\quad (V_\infty - V_0) \propto a \quad \ldots (vi)$

From equations (*iv*) and (*vi*), we have

$$(a - x) \propto (V_\infty - V_0) - (V_t - V_0)$$

or $\quad (a - x) \propto (V_\infty - V_t) \quad \ldots (vii)$

Substituting the values of a and $(a - x)$ from equations (*vi*) and (*vii*) in the first order equation,

$$k = \frac{2.303}{t}\log\frac{a}{a-x}$$

we have $$k = \frac{2.303}{t} \log \frac{V_\infty - V_0}{V_\infty - V_t}$$

Q. 22. 1.0 ml of ethyl acetate was added to 25 ml of *N*/2 HCl. 2 ml of the mixture were withdrawn from time to time during the progress of the hydrolysis of the ester and titrated against standard NaOH solution. The amount of NaOH required for titration at various intervals is given below :

Time (min)	**0**	**20**	**75**	**119**	**183**	**∞**
NaOH used (ml) :	**20.24**	**21.73**	**25.20**	**27.60**	**30.22**	**43.95**

The value at infinite time was obtained by completing the hydrolysis on boiling. Show that it is a reaction of the first order and find the average value of the velocity constant.

Ans. Amount of NaOH used at $t = 0$ *i.e.*, $(V_0) \propto$ HCl present.

Amount of NaOH used at any time t *i.e.*, $(V_t) \propto$ HCl present + CH_3COOH formed.

CH_3COOH formed at any time $\propto$ ethyl acetate reacted

$$(V_t - V_0) \propto x$$

Amount of NaOH used at time $t = \infty$ *i.e.*, $(V_\infty) \propto$ HCl present + Max CH_3COOH formed.

Max. CH_3COOH formed $\propto$ Initial conc. of ethyl acetate

i.e., $$(V_\infty - V_0) \propto a$$

Hence if the given reaction is of the first order, it must obey the equation

$$k = \frac{2.303}{t} \log \frac{a}{a-x}$$

$$= \frac{2.303}{t} \log \frac{V_\infty - V_0}{V_\infty - V_t}$$

In the present case $V_0 = 20.24$ ml, $V_\infty = 43.95$ ml.

$\therefore \qquad V_\infty - V_0 = 43.95 - 20.24 = 23.71$ ml.

The value of k at different instants can be calculated as follows :

t (min)	V_t	$V_\infty - V_t$	$k = \frac{2.303}{t} \log \frac{V_\infty - V_0}{V_\infty - V_t}$
20	21.73	43.95 – 21.73 = 22.22	$k = \frac{2.303}{20} \log \frac{23.71}{22.22} = 0.00324$
75	25.20	43.92 – 25.20 = 18.75	$k = \frac{2.303}{75} \log \frac{23.71}{18.75} = 0.00313$
119	27.60	43.95 – 27.60 = 16.35	$k = \frac{2.303}{119} \log \frac{23.71}{16.35} = 0.00312$
183	30.22	43.95 – 30.22 = 13.73	$k = \frac{2.303}{183} \log \frac{23.71}{13.73} = 0.00299$

Since the value of k comes out to be nearly constant, it is a reaction of the first order. The average value of $k = 0.00312$ min^{-1}.

Q. 23. Discuss the kinetics of inversion of cane sugar.

Ans. The hydrolysis of sucrose in presence of a mineral acid takes place according to the equation

$$\underset{\text{Sucrose (dextro-rotatory)}}{C_{12}H_{22}O_{11}} + H_2O \xrightarrow{H^+} \underset{\text{Glucose (dextro-rotatory)}}{C_6H_{12}O_6} + \underset{\text{Fructose (laevo-rotatory)}}{C_6H_{12}O_6}$$

(laevo-rotatory)

An important characteristic of this reaction is that sucrose is dextro-rotatory whereas the products, glucose and fructose are dextro-rotatory and laevo-rotatory respectively. Further the laevo-rotation of fructose is more (– 92°) than the dextro rotation of glucose (+ 52.5°) so that the mixture as a whole is laevo-rotatory. Thus on hydrolysis, the dextro-rotatory sucrose gradually changes into the laevo-rotatory mixture. It is for this reason that the reaction is called **inversion of sucrose.**

The kinetics of the above reaction is studied by noting the angle of rotation at different intervals of time with the help of a polarimeter. Suppose

Reading of the polarimeter at zero time $= r_0$

Reading of the polarimeter at any time t $= r_t$

Reading of the polarimeter at infinite time $= r_\infty$

(*i.e.*, after 24 hours or more)

It is evident that the reading at zero time will be positive and it would decrease with the passage of time, pass through zero and ultimately become negative.

Angle of rotation at any instant of time, $(r_0 - r_t)$ $\propto$ *the amount of sucrose hydrolysed* (x)

i.e, $x \propto (r_0 - r_t)$ ...(*i*)

Angle of rotation at infinite $(r_0 - r_\infty) \propto$ *the initial concentration of sucrose* (a)

i.e, $a \propto (r_0 - r_\infty)$ ...(*ii*)

From results (*i*) and (*ii*), we have

$$(a - x) \propto (r_0 - r_\infty) - (r_0 - r_t)$$

i. e., $(a - x) \propto (r_t - r_\infty)$ ...(*iii*)

Substituting the values of a and $(a - x)$ from equations (*ii*) and (*iii*) in the first order equation,

$$k = \frac{2.303}{t}\log\frac{a}{a-x}$$

We have
$$k = \frac{2.303}{t}\log\frac{r_0 - r_\infty}{r_t - r_\infty}$$

Q.24. The inversion of cane sugar was studied in 0.9 *N* HCl at 25°C. The following polarimetric readings were obtained at different intervals of time :

Time (minutes)	**0.0**	**7.18**	**18.00**	**27.05**	**∞**
Reading (degrees)	**+24.09**	**+21.41**	**+17.74**	**+15.00**	**– 10.74**

Show that the inversion of cane sugar is a unimolecular reaction. (*Kurukshetra 2001*)

Ans. If the reaction is of the first order, it must obey the equation

$$k = \frac{2.303}{t}\log\frac{a}{a-x}$$

$$= \frac{2.303}{t}\log\frac{r_0 - r_\infty}{r_t - r_\infty}$$

In the present case, $r_0 = 24.09$, $r_\infty = 10.74$

$\therefore$ $r_0 - r_\infty = + 24.09 - (- 10.74) = 34.83.$

The value of k at different instants may be calculated as under :

$t(min)$	r_t	$r_t - r_\infty$	$k = \frac{2.303}{t} \log \frac{r_0 - r_\infty}{r_t - r_\infty}$
7.18	+ 21.41	21.41 − (− 10.74) = 32.15	$k = \frac{2.303}{7.18} \log \frac{34.83}{32.15} = 0.01116$
18.00	+ 17.74	17.74 − (− 10.74) = 28.48	$k = \frac{2.303}{18.00} \log \frac{34.83}{28.48} = 0.01118$
27.05	+ 15.00	15.00 − (− 10.74) = 25.74	$k = \frac{2.303}{27.05} \log \frac{34.83}{25.74} = 0.01119$

The constant values of k proves that the reaction is of the first order.

Q. 25. Show that in case of a first order reaction, the time required for 99.9% of the reaction to take place is about ten times than that required for half the reaction.

Ans. Time for half-change is worked out as under

$$t_{1/2} = \frac{2.303}{k} \log \frac{a}{a - \frac{a}{2}}$$

$$= \frac{2.303}{k} \log 2 = \frac{2.303}{k}(0.3010)$$

Time for 99.9% change is worked out as under

$$t_{99.9\%} = \frac{2.303}{k} \log \frac{a}{a - 0.999a}$$

$$= \frac{2.303}{k} \log 10^3 = \frac{2.303}{k} \times 3$$

$$\frac{t_{99.9\%}}{t_{1/2}} = \frac{2.303}{k} \times 3 \times \frac{k}{2.303 \times 0.3010}$$

$$= \frac{3}{0.3010} \approx 10$$

Q. 26. A first order reaction is 40% complete in 50 minutes. Calculate the value of the rate constant. In what time will the reaction be 80% complete ?

Ans. (*i*) For the first order reaction

$$k = \frac{2.303}{t} \log \frac{a}{a - x}$$

when $x = \frac{40}{100} a = 0.4\, a,$

$t = 50$ minutes (Given)

$$\therefore \quad k = \frac{2.303}{50} \log \frac{a}{a - 0.4a}$$

$$= \frac{2.303}{50} \log \frac{1}{0.6}$$

or $\quad k = \mathbf{0.010216\ min^{-1}}$.

(*ii*) $\quad t = ?$,

when $\quad x = 0.8\ a$

$$k = 0.010216\ \text{min}^{-1}.\ \text{(Calculated above)}$$

$$\therefore \quad t = \frac{2.303}{k} \log \frac{a}{a - x}$$

$$= \frac{2.303}{0.010216} \log \frac{a}{a - 0.8a}$$

$$= \frac{2.303}{0.010216} \log \frac{1}{0.2}$$

$$= \mathbf{157.58\ min.}$$

Q. 27. The reaction

$$\mathbf{SO_2Cl_2 \xrightarrow{k_1} SO_2 + Cl_2}$$

is a first order reaction with $k_1 = 2.2 \times 10^{-5}$ sec^{-1} at 302°C. What percentage of SO_2Cl_2 will get decomposed in 90 minutes when the reaction is carried out at 302°C ?

Ans. Since the reaction is of the first order (given),

$$k = \frac{2.303}{t} \log \frac{a}{a - x}$$

Here we are given that

$$k = 2.2 \times 10^{-5}\ \text{sec}^{-1}$$

$$t = 90\ \text{minutes} = 90 \times 60 = 5400\ \text{sec}$$

$$\therefore \quad 2.2 \times 10^{-5} = \frac{2.303}{5400} \log \frac{a}{a - x}$$

or $\quad \log \dfrac{a}{a - x} = 0.0516$

or $\quad \dfrac{a}{a - x} = \text{antilog}\ (0.0516) = 1.127$

or $\quad a = 1.127\ a - 1.127x$

or $\quad 0.127\ a = 1.127x$

$$\therefore \quad \frac{x}{a} = \frac{0.127}{1.127} = 0.113$$

$$= \mathbf{11.3\%}$$

Q. 28. At 100°C, the half-life period for the thermal decomposition of N_2O_5 is 4.6 sec and is independent of the initial pressure of N_2O_5. Calculate the specific rate constant at this temperature.

Ans. Since the half-life period is independent of the initial pressure, this shows that the reaction is of the *first order.*

Hence it must obey the equation

$$k = \frac{2.303}{t} \log \frac{a}{a - x}$$

For half life period, $t = t_{1/2}$, $x = a/2$

$$\therefore \quad k = \frac{2.303}{t_{1/2}} \log \frac{a}{a - \frac{a}{2}}$$

$$= \frac{2.303}{t_{1/2}} \log 2$$

$$= \frac{2.303}{4.6}(0.3010) = \mathbf{0.1507\,sec^{-1}}$$

Q. 29. What is meant by a reaction of second order. Derive an expression for rate constant of second order reaction involving one reactant only.

(Bhopal,2004; Madras,2004; Agra,2004; Nankdev. 2004; Purvanchal 2007; Himachal 2010, Andhra 2010)

Ans. *A reaction is said to be of the second order if the rate of reaction depends upon two concentration terms.*

The general equation for a reaction involving one reactant is

$$2A \longrightarrow \text{products}$$

Suppose the initial concentration of $A = a$ moles/litre.

Decrease in the concentration of A at any instant of time = x moles/litre.

$\therefore$ Concentration of A at the instant of time $t = (a - x)$ mole/litre

Applying the law of mass action, the rate of reaction

$$\frac{dx}{dt} \propto [A]^2$$

$$\propto (a - x)^2$$

or $$\frac{dx}{dt} = k\,(a - x)^2 \qquad \ldots(i)$$

where k is the rate constant for the second order reaction.

Equation (i) may be rewritten as

$$\frac{dx}{(a-x)^2} = k\,dt$$

Integrating this equation, we get

$$\int \frac{dx}{(a-x)^2} = \int k\,dt$$

i.e., $$\frac{(a-x)^{-1}}{-1} \times (-1) = kt + I$$

or $$\frac{1}{(a-x)} = kt + I \qquad \ldots(ii)$$

where I is the constant of integration. In the beginning *i. e.,* at time $t = 0$, $x = 0$ (as no substance has reacted at zero time). Putting these values in equation (ii), we get

$$\frac{1}{a} = I$$

Substituting this value of I in equation (ii), we get

$$\frac{1}{a-x} = kt + \frac{1}{a}$$

$$kt = \frac{1}{a-x} - \frac{1}{a} \qquad \ldots(iii)$$

$$= \frac{a-(a-x)}{a(a-x)}$$

$$= \frac{x}{a(a-x)}$$

$$\boldsymbol{k = \frac{1}{t}\frac{x}{a(a-x)}} \qquad ...(iv)$$

If the initial concentration is written as C_0 and concentration at any instant of time t as C_t, then $a = C_0$, and $(a - x) = C_t$. Substituting these values in equation (*iii*), we get

$$k = \frac{1}{t}\left[\frac{1}{C_t} - \frac{1}{C_0}\right] \qquad ...(v)$$

This is another form of the expression for the rate constant of second order reaction.

Q. 30. Derive an expression for rate constant of second order reaction involving two different reactants with different initial concentration. (*Garhwal 2000*)

Ans. General equation for such reaction is

$$A + B \longrightarrow \text{Products}$$

Suppose the initial concentration of $A = a$ moles/litre, the initial concentration of $B = b$ moles/litre, amount of A that reacts in time $t = x$ moles/litre, then the amount of B that would react in the same time would also be x moles/litre

∴ At any instant of time t,

Concentration of $A = (a - x)$ moles/litre

and concentration of $B = (b - x)$ moles/litre

Applying the law of mass action, we get

$$\text{Rate of reaction } \frac{dx}{dt} \propto [A]\,[B]$$

$$\propto (a - x)\,(b - x)$$

or $$\frac{dx}{dt} = k\,(a - x)\,(b - x) \qquad ...(i)$$

where k is the rate constant.

Equation (*i*) may be rewritten as

$$\frac{dx}{(a-x)(b-x)} = kdt \qquad ...(ii)$$

Resolving the left hand side into partial fractions, equation (*ii*) may be rewritten as

$$\frac{1}{(a-b)}\left[\frac{1}{(b-x)} - \frac{1}{(a-x)}\right]dx = kdt$$

Integrating this equation, we get

$$\frac{1}{(a-b)}\left(\int\frac{dx}{(b-x)} - \int\frac{dx}{(a-x)}\right) = \int kdt$$

or $$\frac{1}{(a-b)}[-\ln (b-x) - \{-\ln(a-x)\}] = kt + I$$

or $$\frac{1}{(a-b)}[\ln (a-x) - \ln (b-x)] = kt + I$$

or $$\frac{1}{(a-b)}\left(\ln\frac{a-x}{b-x}\right) = kt + I \qquad ...(iii)$$

Where I is the constant of integration.

But at $t = 0$, $x = 0$, Putting these values in equation (*iii*), we get

$$\frac{1}{(a-b)}\left(\ln\frac{a}{b}\right) = I \qquad ...(iv)$$

Putting this value in equation (*iii*) , we get

$$\frac{1}{(a-b)}\ln\frac{a-x}{b-x} = kt + \frac{1}{(a-b)}\ln\frac{a}{b}$$

or

$$kt = \frac{1}{(a-b)}\ln\frac{a-x}{b-x} - \frac{1}{(a-b)}\ln\frac{a}{b}$$

$$= \frac{1}{(a-b)}\left[\ln\frac{a-x}{b-x} - \ln\frac{a}{b}\right]$$

$$= \frac{1}{(a-b)}\ln\left(\frac{a-x}{b-x}\times\frac{b}{a}\right)$$

or

$$\boxed{k = \frac{2.303}{t(a-b)}\log\frac{b(a-x)}{a(b-x)}} \qquad ...(v)$$

Q. 31. What would be the units of rate constant in the following rate law:

rate = k [A] [B]

if the concentration is expressed in moles per litre and time in seconds. ***(Delhi 2003)***

Ans. For a second order reaction

$$k = \frac{1}{t}\frac{x}{a(a-x)}$$

$$= \frac{1}{\text{time}}\times\frac{\text{Concentration}}{\text{Concentration}\times\text{Concentration}}$$

or

$$k = \frac{1}{\text{time}\times\text{concentration}}$$

Substituting for the units of time and concentration

$$k = \frac{1}{\text{s}\times\text{mol litre}^{-1}} = \text{litre mol}^{-1}\ \text{s}^{-1}$$

Q. 32. Describe some characteristics of second order reaction

Ans. A second order reaction has the following characteristics:

(*i*) *Any reaction of the second order must obey the equation*

$$k = \frac{1}{t}\frac{x}{a(a-x)}$$

If on substituting the values of t, x, a and $(a - x)$, constant values of k are obtained, it means the reaction is of second order.

Graphical method can also be used.

We may use the second order equation in the form

$$kt = \frac{1}{C_t} - \frac{1}{C_0}$$

or

$$\frac{1}{C_t} = kt + \frac{1}{C_0}$$

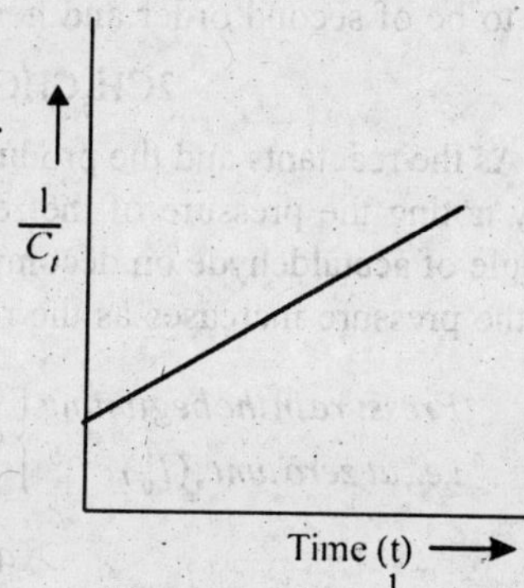

Fig. 4.7. Plot of $\frac{1}{C_t}$ *vs.* t

This equation is like $y = mx + c$ *i.e.*, the equation of a straight line. Hence a plot of $\frac{1}{C_t}$ *vs.* t should be a straight line if the reaction is of the second order (Fig. 4.7). The slope of the line will directly give the value of k.

(*ii*) *The time taken for any fraction of the reaction to be completed is inversely proportional to the initial concentration.* As an example, let us calculate the time taken for half of the reaction to be completed (*i.e.*, half life period). Time taken for half of the reaction to be completed is represented by $t_{1/2}$. Thus when $x = a/2$, $t = t_{1/2}$. Putting these values in equation for second-order reaction, we get

$$t_{1/2} = \frac{1}{k}\frac{a/2}{a(a-a/2)}$$

$$= \frac{1}{k}\cdot\frac{1}{a}$$

i.e,
$$t_{1/2} \propto \frac{1}{a}$$

Similarly, it can be seen that for any fraction of the reaction to be completed, the time taken $\propto \frac{1}{a}$.

(*iii*) *The units of the rate constant k depend not only upon the units of time but also upon the units in which the concentrations are expressed.* This is obvious because there is one extra concentration term in the denominator in the equation for second order reaction.

$$k = \frac{1}{t}\frac{x}{a(a-x)}$$

$$= \frac{1}{\text{Time}}\cdot\frac{\text{Concentration}}{\text{Concentation}\times\text{Concentration}}$$

$$= \frac{1}{\text{Concentation}\times\text{Time}}$$

Q. 33. Discuss the kinetics of thermal decomposition of acetaldehyde. *(Kanpur 2001)*

Ans. The thermal decomposition of acetaldehyde vapour takes place according to the equation

$$CH_3CHO \xrightarrow{791K} CH_4 + CO$$

From this equation, the reaction appears to be of the first order but experimentally it is found to be of second order and hence is preferably written as

$$2CH_3CHO \longrightarrow 2CH_4 + 2CO$$

As the reactants and the products are all gaseous, the progress of the reaction is followed by simply noting the pressure of the reaction mixture at different intervals of time. Further, as one molecule of acetaldehyde on decomposition produces two molecules (one of CH_4 and the other of CO), the pressure increases as the reaction progresses. It is obvious that

$$\left.\begin{matrix}\textit{Pressure in the beginning}\\ \textit{i.e., at zero time, } (P_0)\end{matrix}\right\} \propto \left\{\begin{matrix}\textit{Intital concentration}\\ \textit{of acetaldehyde } (a)\end{matrix}\right.$$

i.e.
$$a \propto P_0 \quad ...(i)$$

At any other instant of time, t, the total pressure of the reaction mixture (P_t) will be given by

$$P_t = P_{CH_3CHO} + P_{CH_4} + P_{CO} \quad ...(ii)$$

If at this instant of time, p is the decrease in pressure of CH_3CHO then p will be increase in the pressure of each of CH_4 and CO so that we have

$$P_{CH_3CHO} = P_0 - p \quad ...(iii)$$

$$P_{CH_4} = p$$

$$P_{CO} = p$$

Substituting these values in equation (ii), we get

$$P_t = (P_0 - p) + p + p$$
$$= P_0 + p$$

or

$$p = P_t - P_0 \quad ...(iv)$$

$$\left.\begin{array}{l}\textit{But decrease in the} \\ \textit{Pressure of} \\ \textit{acetaldelyde } (P)\end{array}\right\} \propto \left\{\begin{array}{l}\textit{Acetaldehyde} \\ \textit{decomposed}(x)\end{array}\right.$$

i.e.

$$x \propto p \quad ...(v)$$

Combining this result with equation (iv), we get

$$x \propto P_t - P_0 \quad ...(vi)$$

From equation (i),

$$a \propto P_0$$

$\therefore$

$$a - x \propto P_0 - (P_t - P_0)$$

or

$$(a - x) \propto 2P_0 - P_t$$

Substituting these values in the second order equation,

$$k = \frac{1}{t}\,\frac{x}{a(a-x)}$$

we have

$$k = \frac{1}{t}\,\frac{P_t - P_0}{P_0(2P_0 - P_t)}$$

Q. 34. The thermal decomposition of acetaldehyde was studied by Hinshelwood and Hutchison at 518°C. Starting with an initial pressure of 363 mm of Hg, the following results were obtained at different intervals of time.

***t* (*sec*)**	**: 42**	**73**	**105**	**190**
Increase in pressure, Δp (mm)	**: 34**	**54**	**74**	**114**

Show that the reaction is of the second order.

Ans. If the reaction is of second order, it must obey the equation.

$$k = \frac{1}{t}\,\frac{P_t - P_0}{P_0(2P_0 - P_t)}$$

Here we are given that

$$P_0 = 363 \text{ mm of Hg}$$

Δp means $P_t - P_0$

The value of k at different intervals may be calculated as follows :

t (sec)	$P_t - P_0$	$2P_0 - P_t = P_0 - (P_t - P_0)$	$k = \frac{1}{t}\frac{P_t - P_0}{P_0(2P_0 - P_t)}$
42	34	363 – 34 = 329	$k = \frac{1}{42} \times \frac{34}{363 \times 329} = 6.79 \times 10^{-6}$
73	54	363 – 54 = 309	$k = \frac{1}{73} \times \frac{54}{363 \times 309} = 6.59 \times 10^{-6}$
105	74	363 – 74 = 289	$k = \frac{1}{105} \times \frac{74}{363 \times 289} = 6.71 \times 10^{-6}$
190	114	363 – 114= 249	$k = \frac{1}{190} \times \frac{114}{363 \times 249} = 6.64 \times 10^{-6}$

The constancy in the value of k shows the reaction to be of second order.

Q. 35. Discuss the kinetics of saponification (alkaline hydrolysis) of esters.

Ans. The reaction of esters with sodium hydroxide solution is called **saponification** of esters. For example, the reaction of ethyl acetate with sodium hydroxide solution takes place as follows:

$$\underset{\text{Ethyl acetate}}{CH_3COOC_2H_5} + NaOH \longrightarrow \underset{\text{Sodium acetate}}{CH_3COONa} + \underset{\text{Ethyl alcohol}}{C_2H_5OH}$$

The reaction involves NaOH as one of the reactants, the concentration of which decreases as the reaction progresses. Starting with known concentration of ethyl acetate and sodium hydroxide (preferably equal concentrations), the concentration of NaOH at any instant of time can be found by withdrawing a definite amount of the reaction mixture and titrating against a standard acid. It is obvious that for the same volume of the reaction mixture withdrawn at different intervals of time,

$$\left.\begin{matrix} \textit{Volume of the standard} \\ \textit{acid used at the beginning} \\ \textit{of the experiment i.e. at} \\ \textit{zero time}, (V_0) \end{matrix}\right\} \propto \left\{\begin{matrix} \textit{Initial concentration} \\ \textit{of NaOH } (a) \end{matrix}\right.$$

i.e. $$a \propto V_0 \qquad ....(i)$$

$$\left(\begin{matrix} \textit{Volume of standard acid} \\ \textit{solution used at any} \\ \textit{instant of time } t, (V_t) \end{matrix}\right) \propto \left\{\begin{matrix} \textit{Concentration of} \\ \textit{NaOH at that} \\ \textit{instant i.e.} (a-x) \end{matrix}\right.$$

$$(a - x) \propto V_t \qquad ...(ii)$$

From equations (*i*) and (*iii*), we get

$$x \propto V_0 - V_t \qquad ...(iii)$$

Substituting the values of a, x and $(a - x)$ from equations (*i*), (*ii*) and (*iii*) in the second order equation,

$$k = \frac{1}{t}\frac{x}{a(a-x)}$$

we have
$$k = \frac{1}{t}\frac{(V_0 - V_t)}{V_0 V_t}$$

Q.36. Starting with equal concentration of ethyl acetate and sodium hydroxide, the saponification of ethyl acetate was carried out. Same volume of the reaction mixture was withdrawn at different intervals of time and titrated with the same acid. The following data were obtained :

Time (min.)	**:**	**0**	**5**	**15**	**25**
Acid used (ml)	**:**	**16**	**10.24**	**6.13**	**4.32**

Show that the reaction is of the second order.

Ans. If the reaction is of the second order, it must obey the equation

$$k = \frac{1}{t}\frac{x}{a(a-x)} = \frac{1}{t}\frac{(V_0 - V_t)}{V_0 \cdot V_t}$$

In the present case, $V_0 = 16$ ml

The value of k at different instants of time can be calculated as follows :

t (min.)	V_t	$(V_0 - V_t)$	$k = \frac{1}{t}\frac{(V_0 - V_t)}{V_0 V_t}$
5	10.24	5.76	$k = \frac{1}{5} \times \frac{5.76}{16 \times 10.24} = 0.00703$
15	6.13	9.87	$k = \frac{1}{15} \times \frac{9.87}{16 \times 6.13} = 0.00671$
25	4.32	11.68	$k = \frac{1}{25} \times \frac{11.68}{16 \times 4.32} = 0.00676$

Thus the value of k is fairly constant. The reaction is of the second order.

Q. 37. A second order reaction in which the initial concentrations of both the reactions are same is 25% complete in 600 sec. How long will it take for the reaction to go to 75% completion?

Ans. For a second order reaction

$$k = \frac{1}{t}\frac{x}{a(a-x)}$$

For the first case, $x = \frac{25}{100} a = 0.25\ a$

$t = 600$ sec.

$$\therefore \quad k = \frac{1}{200} \times \frac{0.25a}{a(a - 0.25a)}$$

$$= \frac{1}{600} \times \frac{0.25}{0.75a} = \frac{1}{1800a}$$

For the second case, $x = \frac{75}{100} a = 0.75a$

t = To be calculated

$$\therefore \quad k = \frac{1}{t} \times \frac{0.75a}{a(a-0.75a)}$$

$$= \frac{1}{t} \times \frac{0.75a}{0.25a} = \frac{3}{ta}$$

But $\quad k = \frac{1}{1800a}$ from the first case

$$\therefore \quad \frac{1}{1800a} = \frac{3}{ta}$$

or $\quad t = 3 \times 1800 = \mathbf{5400\ sec.}$

Q. 38. Decomposition of a gas is of second order. It takes 40 minutes for 40% of a gas to be decomposed when its initial concentration is 4 × 10^{-2} mole/litre. Calculate the specific reaction rate.

Ans. For a second order reaction

$$k = \frac{1}{t} \frac{x}{a(a-x)}$$

Here, we are given that

$$a = 4 \times 10^{-2} \text{ mole/litre}$$
$$t = 40 \text{ minutes}$$
$$x = \frac{40}{100} \times a = 0.4a$$

Substituting the values in the above formula, we get

$$k = \frac{1}{40} \times \frac{0.4a}{a(a-0.4a)}$$

$$= \frac{1}{40} \times \frac{0.4}{0.6a} = \frac{1}{40} \times \frac{0.4}{0.6 \times 4 \times 10^{-2}}$$

$$= \mathbf{0.4167\ litre\ mol^{-1}\ min^{-1}}$$

Q. 39. What is meant by a third order reaction? Derive an expression for the rate constant of a third order reaction.

Ans. *A reaction is said to be of the third order if the rate of reaction depends upon three concentration terms.*

The simplest case of this type can be represented as

$$3A \longrightarrow \text{Products}$$

The expression for the rate constant of such a reaction may be derived as follows :

Suppose a is the initial concentration of A and x is the decrease in the concentration of A in time t. Then the concentration of A after time $t = (a - x)$. Now, according to the law of mass action, the rate of the above reaction at the instant of time t will be given by

$$\frac{dx}{dt} \propto [A]^3$$

i.e. $\quad \frac{dx}{dt} \propto (a-x)^3$

or $\quad \frac{dx}{dt} = k(a-x)^3 \quad \text{...}(i)$

where k is the rate constant for the third order reaction.

Equation (i) may be rewritten as

$$\frac{dx}{(a-x)^3} = kdt$$

Integrating this equation, we get

$$\int \frac{dx}{(a-x)^3} = \int kdt$$

i.e. $$\frac{(a-x)^{-2}}{-2} \times (-1) = kt + I$$

or $$\frac{1}{2(a-x)^2} = kt + I \quad ...(ii)$$

where I is the constant of integration.

Now at $t = 0$, $x = 0$. Putting these values in equation (ii), we get

$$\frac{1}{2a^2} = I \quad ...(iii)$$

Substituting this value in equation (ii), we get

$$\frac{1}{2(a-x)^2} = kt + \frac{1}{2a^2}$$

or $$kt = \frac{1}{2(a-x)^2} - \frac{1}{2a^2}$$

$$= \frac{1}{2}\left[\frac{1}{(a-x)^2} - \frac{1}{a^2}\right]$$

$$= \frac{1}{2}\left[\frac{a^2-(a-x)^2}{a^2(a-x)^2}\right]$$

$$= \frac{1}{2}\left[\frac{a^2-(a^2+x^2-2ax)}{a^2(a-x)^2}\right]$$

$$= \frac{1}{2}\left[\frac{2ax-x^2}{a^2(a-x)^2}\right]$$

$$= \frac{1}{2}\left[\frac{x(2a-x)}{a^2(a-x)^2}\right]$$

or $$\boldsymbol{k = \frac{1}{2t}\left(\frac{x(2a-x)}{a^2(a-x)^2}\right)} \quad ...(iv)$$

Replacing a by C_0 (*i.e.* initial concentration of A) and x by $C_0 - C_t$ (*i.e.* decrease in the concentration of A), equation (iv) may be rewritten as

$$k = \frac{1}{2t}\left[\frac{(C_0-C_t)(2C_0-C_0+C_t)}{C_0^2(C_0-C_0+C_t)^2}\right]$$

$$= \frac{1}{2t}\frac{(C_0 - C_t)(C_0 + C_t)}{C_0^2 C_t^2}$$

$$= \frac{1}{2t}\frac{C_0^2 - C_t^2}{C_0^2 C_t^2}$$

or $$k = \frac{1}{2t}\left[\frac{1}{C_t^2} - \frac{1}{C_0^2}\right] \quad ...(v)$$

Q. 40. Describe some characteristics of third order reactions.

Ans. Third order reactions have the following characteristics.

(*i*) A reaction is of the third order, if it obeys the equation

$$k = \frac{1}{2t}\left[\frac{x(2a-x)}{a^2(a-x)^2}\right] \quad ...(i)$$

Thus on substituting the values of t, x and a, constant values of k must be obtained.

(*ii*) *The time taken for any fraction of the reaction to be completed is inversely proportional to the square of the initial concentration.* For example, the time taken for half of the reaction to be completed (*i.e. half period*) may be calculated as follows :

When $x = \frac{a}{2}$, $t = t_{1/2}$

Putting these values in equation (*i*), we get

$$t_{1/2} = \frac{1}{2k}\left\{\frac{\frac{a}{2}\left(2a - \frac{a}{2}\right)}{a^2\left(a - \frac{a}{2}\right)^2}\right\}$$

or $$t_{1/2} = \frac{1}{2k}\left\{\frac{\frac{a}{2}\cdot\frac{3a}{2}}{a^2\frac{a^2}{4}}\right\}$$

or $$t_{1/2} = \frac{1}{2k}\frac{3}{a^2}$$

i.e. $$t_{1/2} \propto \frac{1}{a^2}$$

(*iii*) *The units of the rate constant k depend upon the units of the concentration.* This may be seen by calculating the dimensions of k with the help of equation for third order.

$$k = \frac{1}{2t}\left[\frac{x(2a-x)}{a^2(a-x)^2}\right]$$

$$= \frac{1}{\text{Time}}\left[\frac{\text{Concentration} \times \text{Concentration}}{(\text{Concentration})^2 \times (\text{Concentration})^2}\right]$$

$$= \frac{1}{\text{Time}}\frac{1}{(\text{Concentration})^2}$$

Thus any change in the unit of concentration will change the value of k.

Q. 41. Give some examples of third order reaction.

Ans. Five reactions in gaseous phase, each involving nitric oxide as one of the reactions are known to be third order reactions.

(*i*) $2NO + Cl_2 \longrightarrow 2NOCl$

(*ii*) $2NO + Br_2 \longrightarrow 2NOBr$

(*iii*) $2NO + O_2 \longrightarrow 2NO_2$

(*iv*) $2NO + H_2 \longrightarrow N_2O + H_2O$

(*v*) $2NO + D_2 \longrightarrow N_2O + D_2O$

Besides, reduction of ferric chloride by stannous chloride is an example of third order reaction in solution.

$$2FeCl_3 + SnCl_2 \longrightarrow 2FeCl_2 + SnCl_4$$

Q. 42. Why are the reactions of higher order rare ?

(M. Dayanand 2000; Kurukshetra 2001; Bhopal, 2004)

Ans. The reason for this may be explained on the basis of **collision theory.** According to this theory, in order that a reaction may take place between certain molecules, these molecules must come together simultaneously to collide with each other. As the probability for three molecules to come together simultaneously and collide are less than those for two molecules, the reactions of third order are fewer than those of second order. Similarly, the chances for four or more molecules to come together simultaneously and collide are still less, hence the reactions of fourth or higher order are very rare.

Q. 43. Formulate relations between half-life period and order of a reaction.

(Punjab 2000; Allahabad 2003; Himachal 2010

Ans. The relations are as under :

(*i*) *For reactions of zero order*

$$t_{1/2} \propto a$$

i.e. half-life period is directly proportional to the initial concentration.

(*ii*) *For reactions of the first order*

$$t_{1/2} = \frac{2.303}{k} \log 2$$

$$= \frac{2.303}{k} (0.3010) = \frac{0.693}{k}$$

i.e. half-life period is independent of the initial concentration.

(*iii*) *For reactions of the second order*

$$t_{1/2} \propto \frac{1}{a}$$

i.e. half-life period is inversely proportional to the initial concentration.

(*iv*) *For reactions of third order*

$$t_{1/2} \propto \frac{1}{a^2}$$

i.e. half-life period is inversely proportional to the square of the initial concentration.

In general, for a reaction of the nth order

$$t_{1/2} \propto \frac{1}{a^{n-1}}$$

Q. 44. What is meant by molecularity of a reaction ? Illustrate with examples.
(Bhopal, 2004; Garhwal, 2010)

Ans. *The number of atoms, ions or molecules that must collide with each other simultaneously so as to result into a chemical reaction is called the molecularity of the reaction.*

In case of **simple reactions,** *the molecularity is simply the sum of the molecules of the different reactants as represented by the balanced chemical equation.* A few examples are given below :

(*i*) *Decomposition of* F_2O_2. *The balanced equation is*

$$F_2O_2 \longrightarrow F_2 + O_2$$

Hence the molecularity of the reaction is 1 and the reaction is called *unimolecular.*

(*ii*) *Dissociation of HI.* The balanced equation is

$$2HI \longrightarrow H_2 + I_2$$

Hence the molecularity is 2 and the reaction is called *bimolecular.*

In case of **complex reactions,** *i.e.* reactions involving three or more atoms, ions or molecules of reactants in the balanced equation, the chances for all the atoms, ions, or molecules of the reactants to come together and collide are very rare. In such cases, the reactions are supposed to take place in a number of steps.

It is obvious that the molecularity of a reaction must always be a whole number whereas the order of a reaction can be fractional also.

Q. 45. What is meant by mechanism of a reaction ? How do we arrive at the mechanism of a reaction. Explain taking examples from first, second and third order reactions.

Ans. *A series of step reactions, proposed to account for the overall reaction is called the mechanism of the reaction.*

The writing of step reactions is usually based upon experimental evidence. *Slowest step (called the rate determining step) must involve the molecules on which the rate of reaction actually depends as observed experimentally.* The following examples will illustrate the point more clearly :

(*i*) *Thermal decomposition of dinitrogen pentoxide. The balanced equation may be written as*

$$2N_2O_5 \longrightarrow 4NO_2 + O_2$$

However, according to the rate law, it is found that

Rate of reaction = $k\,[N_2O_5]$.

Hence the above reaction must be split into steps in such a way that the slowest step should involve only one molecule of N_2O_5. Thus the probable mechanism is as follows:

$$N_2O_5 \xrightarrow{\text{Slow}} NO_2 + NO_3$$

$$N_2O_5 + NO_3 \xrightarrow{\text{Fast}} 3NO_2 + O_2$$

Hence the above reaction is *unimolecular reaction or a reaction of the first order.*

(*ii*) *Bromination of propanone* (*acetone*). The bromination of propanone in dilute aqueous solution is represented by the equation

$$CH_3COCH_3 + Br_2 \longrightarrow CH_3COCH_2Br + HBr$$

Experimentally, it is found that

Rate of reaction $\propto$ $[CH_3COCH_3]$ only (and is independent of the concentration of bromine). Hence the rate determining step (slow step) would involve a molecule of acetone only. The suggested steps are as follows :

(a)
$$\underset{\text{Ketonic from}}{CH_3COCH_3} \xrightarrow{\text{Slow}} \underset{\text{enolic from}}{CH_3-\underset{\underset{OH}{|}}{C}=CH_2}$$

(b)
$$CH_2-\underset{\underset{OH}{|}}{C}=CH_2 + Br_2 \xrightarrow{\text{Fast}} CH_3COCH_2Br + HBr$$

(*iii*) *Combination of* NO_2 *and* F_2 *to form* NO_2F. The reaction may be represented as

$$2NO_2 + F_2 \longrightarrow 2NO_2F$$

Experimentally, it is found that

Rate of reaction = $k\,[NO_2]\,[F_2]$

i.e. the rate law equation involves only one molecule of NO_2 and one molecule of F_2. Hence the probable mechanism is as follows :

$$NO_2 + F_2 \xrightarrow{\text{Slow}} NO_2F + F$$

$$NO_2 + F \xrightarrow{\text{Fast}} NO_2F$$

Hence the above reaction must be a *bimolecular reaction or a reaction of the second order.*

(*iv*) *Oxidation of hydrogen bromide with oxygen.* The oxidation of hydrogen bromide with oxygen in the temperature range 400 – 600°C takes place rapidly as follows :

$$4HBr(g) + O_2(g) \longrightarrow 2H_2O(g) + 2Br_2\,(g)$$

Experimentally, it is found that

Rate of reaction $\propto$ $[HBr]\,[O_2]$

This implies that the slowest step of the mechanism of the above reaction should involve only one molecule of HBr and one molecule of O_2. Hence the probable mechanism is as follows :

$$HBr + O_2 \xrightarrow{\text{Slow}} HOOBr$$

$$HOOBr + HBr \xrightarrow{\text{Fast}} 2HOBr$$

$$HOBr + HBr \xrightarrow{\text{Fast}} H_2O + Br_2] \times 2$$

Overall reaction :

$$4HBr + O_2 \longrightarrow 2H_2O + 2Br_2$$

(*v*) *Reduction of nitric oxide with hydrogen.* The reaction

$$2NO + 2H_2 \longrightarrow N_2 + 2H_2O$$

The rate law equation is found to be

$$\text{Rate}\left(\frac{dx}{dt}\right) = k[NO]^2[H_2]$$

This suggests that the reaction is of the third order and the rate determining step (slow and step) should involve two molecules of NO and one molecule of H_2. Thus the following mechanism has been proposed:

$$2NO + H_2 \xrightarrow{\text{Slow}} N_2 + H_2O_2$$

$$H_2O_2 + H_2 \xrightarrow{\text{Fast}} 2H_2O$$

Q. 46.How does the knowledge of order of reaction and stoichiometry help to know the mechanism of a reaction ? Explain giving examples.

(Kerala 2000;Calcutta 2000; Garhwal 2000)

Ans. See Q. 45.

Q. 47. Bring out the points of difference between order and molecularity of a reaction.

(Lucknow, 2000; Nagpur 2002;Purvanchal 2002; Calcutta, 2007)

Ans. The main points of difference between the order and the molecularity of a reaction are as under :

Order of reaction	*Molecularity of reaction*
1. It is the sum of the concentration terms on which the rate of reaction actually depends or it is the sum of the exponents of the concentrations in the rate law equation.	1. In simple reactions, it is equal to the number of molecules of the reactants as represented in the balanced equation. In complex reactions, it is the number of molecules involved in the slowest step.
2. It need not be a whole number *i.e.* it can be fractional as well as zero.	2. It is always a whole number.
3. Order of even simple reaction *may not* be equal to the number of molecules of the reactants as seen from the balanced equation.	3. For simple reactions,, the molecularity can usually be obtained from the stoichiometry of the equation.
4. It can be determined experimentally only and cannot be calculated.	4. It can be calculated by simply adding the molecules of the slowest step.

Q.48. Order and molecularity of a reaction may not be the same. Explain

(Vidyasagar 2007)

Ans. *Order* of a reaction may be defined as the sum of exponents to which the concentration terms in the rate law are raised to express the observed rate of reaction.

Molecularity, in the case of simple reactions, is the sum of molecules of the reactants as represented in the balanced equation. In complex reactions, it is the number of molecules involved in the slowest step.

Order and molecularity may not be the same.

Consider the hydrolysis of an ester with a mineral acid

$$CH_3COOC_2H_5 + H_2O \xrightarrow{H^+} CH_3COOH + C_2H_5OH$$

This reaction has the molecularity equal to 2, while its order has been found to 1.

Consider the decomposition of acetaldehyde at 450°C

$$CH_3CHO \xrightarrow{450°C} CH_4 + CO$$

It is a unimolecular reaction (molecularity = 1) but experimentally its order has been found to be 1.5.

Q.49. Describe the determination of order of a reaction using integrated rate equations.
(Garhwal 2000; Purvanchal 2003; G.Nanakdev 2004)

Ans. In this method, the initial concentrations of the reactants and the concentration of any one of the reactants or the products at different instants of time are determined experimentally. Then either of the following two methods may be employed :

(*i*) *Substitution method* (*Hit and Trial Method*). The appropriate values of a and x (or C_0 and C_t) are substituted into the integrated forms of the rate equations for the first order, second order and third order one by one. The rate equation which gives the constant value of k gives the order of the reaction.

(*ii*) *Graphical method.* In this method, a suitable function is plotted against time.

(*a*) If the plot of log $(a-x)$ against time is a straight line, the reaction is of the first order.

(*b*) If the plot of $\frac{1}{a-x}$ against time is a straight line, the reaction is of the second order.

Q.50. How will you determine order of a reaction using differential rate equation?
(Punjab 1995; Garhwat 2000; M. Dayanand 2002; Calcutta 2007)

Ans. This method is based upon the fact that the rate of a reaction of the nth order is proportional to the nth power of its concentration, *i.e.*

$$\text{Rate of reaction } \left(\frac{dx}{dt}\right) \propto C^n$$

or $$r = kC^n$$

where k is the rate constant for the reaction and C is the concentration at the instant of time t when the rate is r Hence for two different concentrations,

$$r_1 = kC_1^n$$

$$r_2 = kC_2^n$$

$$\therefore \qquad \frac{r_1}{r_2} = \left(\frac{C_1}{C_2}\right)^n$$

or $$\log\frac{r_1}{r_2} = n\log\frac{C_1}{C_2}$$

or $$\log r_1 - \log r_2 = n\,(\log C_1 - \log C_2)$$

or $$n = \frac{\log r_1 - \log r_2}{\log C_1 - \log C_2}$$

Thus to calculate the value of the order n, we require to know the values of the rates r_1 and r_2 at two different concentrations C_1 and C_2 respectively.

Concentration of the reactant may be plotted against time and the rate at any concentration may be found from the slope of the tangent at that point.

Q.51. For the reaction between gaseous chlorine and nitric oxide

$$2NO + Cl_2 \longrightarrow 2NOCl$$

it is found that doubling the concentration of both reactants increases the rate eight times but doubling the chlorine concentration alone doubles the rate. What is the order of reaction with respect to nitric oxide and chlorine ?

Ans. Suppose the initial concentration of NO is a moles/litre and that of Cl_2 is b moles/litre. Suppose the order of reaction w.r.t. NO is n_1 and w.r.t. Cl_2 is n_2. Then initially the rate of reaction will be given by

$$r = k\,(a)^{n_1}\,(b)^{n_2} \qquad ...(i)$$

In the first case when the concentration of both the reactants is doubled, we will have

$$8r = k\,(2a)^{n_1}\,(2b)^{n_2} \qquad ...(ii)$$

In the second case **when the concentration of only Cl_2 is doubled, we have**

$$2r = k\,(a)^{n_1}\,(2b)^{n_2} \qquad ...(iii)$$

Now to get the values of n_1 and n_2, we have to solve equations (*i*), (*ii*) and (*iii*)

From equations (*i*) and (*iii*), we get

$$2k\,[a]^{n_1}\,[b]^{n_2} = k\,[a]^{n_1}\,[2b]^{n_2}$$

or $$2[b]^{n_2} = [2b]^{n_2}$$

or $$2 = [2]^{n_2}$$

which gives $$n_2 = 1$$

Again, from equations (*ii*) and (*iii*), we get

$$2k\,[2a]^{n_1}\,[2b]^{n_2} = 4k\,[a]^{n_1}\,[2b]^{n_2}$$

or $$[2a]^{n_1} = 4[a]^{n_1}$$

or $$2^{n_1} = 4 \text{ or } n_1 = 2$$

Thus $n_1 = 2$ and $n_2 = 1$. This means that the order of reaction *w.r.t* NO is 2 and *w.r.t.* Cl_2 is 1. The total order of reaction is, therefore, 2 + 1 = 3.

Q.52. Discuss fractional change (half-life period) method to determine order of a reaction.
(Garhwal, 2000; Kurukshetra, 2002; Purvanchal 2003; Madras, 2004; Calcutta 2007)

Ans. In general, we can say that $t_{1/2} \propto \dfrac{1}{a^{n-1}}$

where n is the order of the reaction.

The time taken for half of the reaction to be completed is called half-life period *and is represented by* $t_{1/2}$.

For two different initial concentrations, we may write

$$(t_{1/2})_1 \propto \frac{1}{a_1^{\,n-1}}$$

and $$(t_{1/2})_2 \propto \frac{1}{a_2^{\,n-1}}$$

Dividing one equation by other, we get

$$\frac{(t_{1/2})_1}{(t_{1/2})_2} = \left(\frac{a_2}{a_1}\right)^{n-1}$$

Taking logarithm of both the sides, we get

$$\log\,(t_{1/2})_1 - \log(t_{1/2})_2 = (n-1)\,[\log a_2 - \log a_1].$$

This on simplification gives

$$n = 1 + \frac{\log(t_{1/2})_1 - \log(t_{1/2})_2}{\log a_2 - \log a_1}$$

Thus starting with two different initial concentrations and finding the half life period in each case, the value of n can be calculated.

Q.53. The half-life period of a substance is 50 minutes at a certain concentration. When the concentration is reduced to one half of the initial concentration, the half life period is 25 minutes. Calculate order of the reaction.

Ans. Suppose the initial concentration is a moles/litre.

Then $(t_{1/2})_1 = 50$ minutes, $a_1 = a$

$$(t_{1/2})_2 = 25 \text{ minutes}, a_2 = \frac{a}{2}$$

Substituting the values in the expression

$$\frac{(t_{1/2})_1}{(t_{1/2})_2} = \left(\frac{a_2}{a_1}\right)^{n-1}$$

we get
$$\frac{50}{25} = \left(\frac{a/2}{a}\right)^{n-1}$$

or
$$2 = \left(\frac{1}{2}\right)^{n-1}$$

or
$$\log 2 = (n-1) \log \frac{1}{2}$$
$$= -(n-1) \log 2$$

or
$$1 = -(n-1)$$

or
$$n = 0$$

Hence the reaction is of the *zero order.*

Q.54. What is Ostwald's isolation method for the determination of order of reaction.

(Garhwal 2000)

Ans. This method can be used for finding the order of those reactions which involve two or more reactants. The method is based upon the fact that the order of a reaction does not depend upon the concentration of those reactants which are taken in excess. Turn by turn, all reactants except one are taken in excess. The order thus observed is with respect to the reactant that has been *isolated* from the others. The total or the overall order of the reaction is the sum of the orders found with respect to the different reactants one by one by isolation.

For example, suppose a reaction involves three reactants A, B and C. In the first case B and C are taken in excess (so that A is isolated). The order thus observed is the order with respect to A. Similarly by taking A and C in excess, the order with respect to B can be found and by taking A and B in excess, the order with respect to C can be found. Suppose the orders with respect to A, B and C are found to be a, b and c respectively. Then the overall order of the reaction is $a + b + c$.

Q.55. Write notes on

(*i*) Opposing or Reversible reactions *(Kurukshetra 2000; Awadh 2000; Kerala 2001)*

(*ii*) Consecutive reactions *(Awadh 2000; Lucknow 2000; Allahabad 2003)*

(*iii*) Parallel or side reactions.

(Awadh 2000; Lucknow 2000; Allahabad 2003; Purwanchal 2003)

Ans. Opposing or Reversible reactions: *If in a reaction, the reactants react to form the products and the products also react to give back the original substances under the same conditions, it is called a reversible reaction.* In other words, the reaction takes place not only in the forward direction but also in the backward direction (*i. e.* the forward reaction is opposed by the reverse reaction). Let us consider a general reversible reaction in which both the forward and the backward reactions are of the first order

$$A \underset{k_b}{\overset{k_f}{\rightleftharpoons}} B$$

If the velocity constant for the backward reaction (k_b) is very small as compared to that of the forward reaction (k_f), the former can be neglected in comparison to the latter. However, if the forward and the backward reactions take place at comparable rates, the rate equation has to be modified.

Some examples of the reversible reactions whose kinetics have been studied are as follows:

(*a*) $2HI \rightleftharpoons H_2 + I_2$ (*Both forward and backward reactions are of second order*)

(*b*) $2NO + O_2 \rightleftharpoons 2NO_2$ (*Forward reaction is of third order, backward reaction is of second order*)

(*c*) $CH_3COOH + C_2H_5OH \rightleftharpoons CH_3COOC_2H_5 + H_2O$

(*ii*) **Consecutive reactions:** In case of complex reactions, the order of the reaction is usually not equal to the molecularity of the reaction as observed from the stoichiometry of the equation. Such reactions are, therefore, supposed to take place in a number of steps. The slowest step is the rate determining step. *The various step reactions thus written for the overall reaction are called consecutive reactions.* For example, let us consider the general complex reaction,

$$2A + B \longrightarrow C$$

Suppose by experiment, the rate of reaction $\propto [A]\,[B]$ *i.e.* the rate of reaction depends upon one molecule of *A* and one molecule of *B*. The slow step should, therefore, involve only one molecule of *A* and one molecule of *B*. The step reactions (or the consecutive reactions) may, therefore, written as

(*i*) $$A + B \xrightarrow{\text{Slow } (k_1)} D$$

(*ii*) $$A + D \xrightarrow{\text{Fast } (k_2)} C$$

(*iii*) **Parallel or Side reactions:** *If in a reaction, the reactants react to form two different products by two different paths such that one product is in larger amount than the other, then the reaction giving the larger amount is called main reaction while the other giving the smaller amount of the product is called* **side reaction or parallel reaction** *i.e.*

$$A \xrightarrow{k_1} B\ (90\%) \quad \ldots \quad \textit{Main reaction}$$

$$A \xrightarrow{k_2} C\ (10\%) \quad \ldots \quad \textit{Side reaction}$$

If the side reaction takes place only to a very small extent, it can be neglected in comparison to the main reaction. However, if the two reactions take place in comparable amounts, the rate equation has to be modified.

A few examples of parallel or side reactions are given below :

(*a*) *Nitration of phenol*

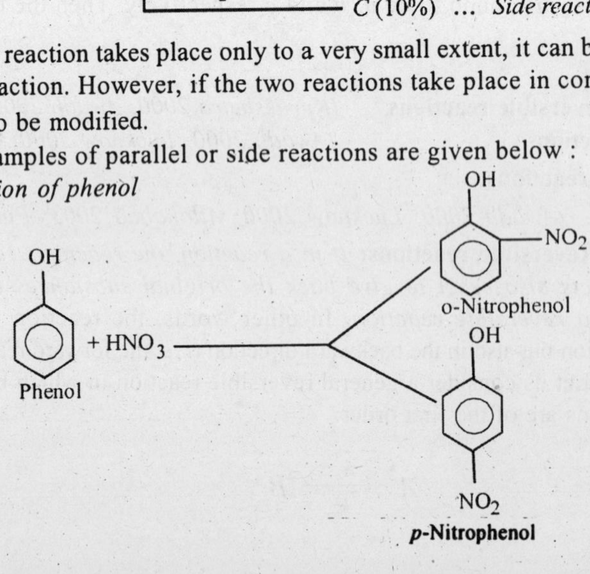

(*b*) *Chlorination of benzene*

$$C_6H_6 + Cl_2 \longrightarrow \underset{\text{Chlorobenzene}}{C_6H_5Cl} + HCl$$

$$C_6H_6 + Cl_2 \longrightarrow \underset{\text{Benzene hexachloride}}{C_6H_6Cl_6}$$

Q.56. What are surface reactions ? What are their characteristics ? *(Himachal 2000)*

Ans. *Gaseous reactions which take place exclusively on the walls of the container are called surface reactions.* The rate of these reactions is affected in a number of ways as follows:

(*i*) *If the surface is sparsely covered* (as in the case at high temperatures) the rate of reaction depends only upon the number of molecules colliding per unit time on the surface *i.e.* on the pressure (p) of the gaseous reactants. We may write

$$\frac{dx}{dt} = kp$$

Hence all such reactions are kinetically the reactions of the first order. Examples of such reactions are the decomposition of phosphine on the surface of glass, silica or tungsten, and decomposition of hydrogen iodide on platinum.

(*ii*) *If the surface is almost completely covered, it* is found that the rate of reaction is independent of pressure *i.e.*

$$\frac{dx}{dt} = \text{constant}$$

Such reactions are then kinetically the reactions of the zero order. Examples of such reactions are decomposition of ammonia on the surface of tungsten and decomposition of hydrogen iodide on the surface of gold.

(*iii*) *If some of the molecules of the reactants are firmly held on the surface of the solid,* the rate of the reaction is obviously retarded.

Q.57. Show that a reaction involving only one gaseous reactant slightly adsorbed on the surface of a solid catalyst follows first order kinetics.

Ans. See Q. 56 under (*i*) when the surface is sparsely covered.

Q.58. What are the chain reactions ? *(Garhwal 2000; M. Dayanand 2002; G.. Nanakdev 2003)*

Ans. A chain reaction takes place through the following steps:

(*i*) *Chain initiating step.* This is the slowest step in which the reactive species are produced.

(*ii*) *Chain propagating step.* The reactive species produced in the first step react immediately with the other molecules of the reactants producing new reactive species and products. The new reactive species on combination with other molecules regenerate the original reactive species. In this way the reaction goes on.

(*iii*) *Chain terminating step.* Towards the end, the reactive species left behind combine with each other and the reaction stops.

The steps (*ii*) and (*iii*) are fast because they involve reactive species.

A reaction proceeding in a series of successive steps initiated by a suitable primary process is called a chain reaction.

Following two examples will illustrate :

(1) Combination between hydrogen and bromine to form hydrogen bromide :

$$H_2 + Br_2 \longrightarrow 2HBr$$

This is an example of a reaction in which the reactive species produced in the chain initiating step are atoms. The various steps may be explained as follows :

(*i*) *Chain initiating step.* In the presence of light, bromine molecules dissociate to form the highly reactive bromine atoms.

$$Br_2 \xrightarrow{h\nu} 2\dot{B}r \text{ (Photochemical reaction)}$$

(*ii*) *Chain propagating step.* The above reaction is immediately followed by the following reactions :

$$\dot{B}r + H_2 \longrightarrow HBr + \dot{H}$$

$$\dot{H} + Br_2 \longrightarrow HBr + \dot{B}r$$

The bromine atom thus produced reacts with more of H_2 and the reaction goes on till whole of hydrogen or bromine present is used up.

(*iii*) *Chain terminating step.* The left off atoms then combine with each other or with the molecule of the product and the reaction stops.

$$\dot{H} + HBr \longrightarrow H_2 + \dot{B}r$$

$$\dot{B}r + \dot{B}r \longrightarrow Br_2$$

(2) Thermal decomposition of acetaldehyde at high temperature :

$$CH_3CHO \longrightarrow CH_4 + CO$$

This is an example of a reaction which proceeds through the free radical mechanism. The measured order of the reaction is 1.5 and the possible mechanism is as follows :

(*i*) *Chain initiating step*

$$CH_3CHO \longrightarrow \underset{\text{Free radicals}}{\dot{C}H_3 + \dot{C}HO}$$

(*ii*) *Chain propagating step*

$$\dot{C}H_3 + CH_3CHO \longrightarrow CH_4 + CH_3\dot{C}O$$

$$\dot{C}HO \longrightarrow CO + \dot{H}$$

$$\dot{H} + CH_3CHO \longrightarrow H_2 + CH_3\dot{C}O$$

(*iii*) *Chain terminating Step*

$$\dot{C}H_3 + \dot{C}H_3 \longrightarrow C_2H_6$$

$$\dot{C}H_3 + CH_3\dot{C}O \longrightarrow CH_3COCH_3$$

$$2CH_3\dot{C}O \longrightarrow CH_3CO.COCH_3$$

Q.59. Write Arrhenius equation for the effect of temperature on rate of reaction.

(M. Dayanand, 2000; Kolhpur 2000; Panjab 2000; Awadh 2000; Awadh 2000; Purvanchal, 2002; Calcutta 2007)

Ans. The effect of temperature on the rate of a reaction was studied by Arrhenius. The equation, called **Arrhenius equation,** is usually written in the form

$$k = Ae^{-E_a/RT} \quad ...(i)$$

where A is a constant called **frequency factor** (because it gives the frequency of collisions of the reacting molecules), E_a is called **energy of activation,** R is gas constant and T is the absolute temperature.

The energy of activation (E_a) is an important quantity as it is characteristic of the reaction. Using the above equation, its value can be calculated as follows :

Taking logarithm of both side of equation (i), we get

$$\ln k = \ln A - \frac{E_a}{RT} \quad ...(ii)$$

If the values of the rate constant at temperatures T_1 and T_2 are k_1 and k_2 respectively, then we have

$$\ln k_1 = \ln A - \frac{E_a}{RT_1} \quad ...(iii)$$

and

$$\ln k_2 = \ln A - \frac{E_a}{RT_2} \quad ...(iv)$$

Subtracting equation (iii) from equation (iv), we get

$$\ln k_2 - \ln k_1 = -\frac{E_a}{RT_2} - \left(-\frac{E_a}{RT_1}\right)$$

$$= \frac{E_a}{RT_1} - \frac{E_a}{RT_2}$$

or

$$\ln \frac{k_2}{k_1} = \frac{E_a}{R}\left(\frac{1}{T_1} - \frac{1}{T_2}\right)$$

$$= \frac{E_a}{R}\left(\frac{T_2 - T_1}{T_1 T_2}\right)$$

or

$$\log \frac{k_2}{k_1} = \frac{E_a}{2.303R}\left(\frac{T_2 - T_1}{T_1 T_2}\right) \quad ...(v)$$

Thus knowing the values of the rate constants k_i and k_2 at two different temperatures T_1 and T_2, the value of E_a can be calculated. Alternatively, knowing the rate constant at any one temperature, its value at another temperature can be calculated provided the value of E_a is known for that reaction.

Equation (ii), may be written as

$$\ln k = -\frac{E_a}{RT} + \ln A$$

$$\log k = -\frac{E_a}{2.303\,RT} + \log A \quad ...(vi)$$

This equation is of the form $y = mx + c$ *i.e.* the equation of a straight line. Thus a plot of log k *vs* $\frac{1}{T}$ is a straight line as shown in Fig. 4.8.

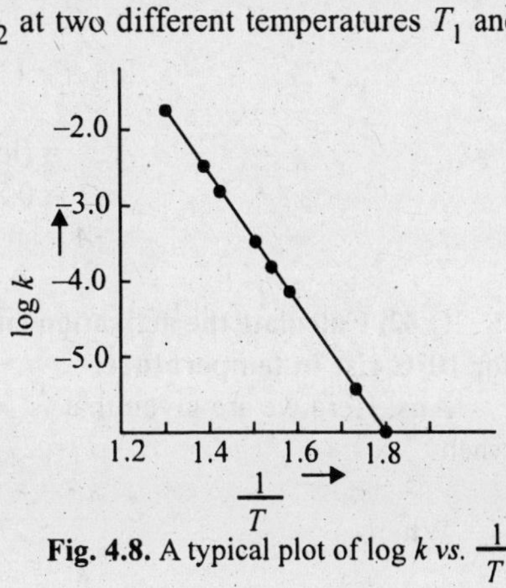

Fig. 4.8. A typical plot of log k *vs.* $\frac{1}{T}$

Further from equation (*vi*), we have

Slope of the line

$$= -\frac{E_a}{2.303\,R}$$

Thus by measuring the slope of the line, the value of E_a can be calculated.

Q. 60. In k versus $1/T$ graph was plotted to calculate the activation energy of a reaction using Arrhening equation for the effect of temperature on reaction rate. The slope of the straight line was found to be – 2.55×10^4. Calculate the activation energy of the reaction.

(G. Nanakdev 2003)

Ans. Slope of the graph straight line is given by :

$$\text{Slope} = -\frac{E_a}{2.303 \times R}$$

or $$-\frac{E_a}{2.303 \times R} = -2.55 \times 10^4$$

Remove negative sign from both sides and put the value of R in joule

$$\frac{E_a}{2.303 \times 8.314} = 2.55 \times 10^4$$

or $$E_a = 2.55 \times 10^4 \times 2.303 \times 8.314$$

or $$E_a = 4.88 \times 10^5 \text{ J.}$$

Q.61. The activation energy of a reaction is 225 kcal mol^{-1} and the value of rate constant at 40°C is 1.8×10^{-5} sec^{-1}. Calculate the frequency factor, A.

Ans. Here, we are given that

$$E_a = 22.5 \text{ kcal mol}^{-1} = 22500 \text{ cal mol}^{-1}$$
$$T = 40° \text{ C} = 40 + 273 = 313 \text{ K}$$
$$k = 1.8 \times 10^{-5} \text{ sec}^{-1}$$

Substituting the values in the equation

$$\log k = -\frac{E_a}{2.303\,RT} + \log A$$

or $$\log A = (1.8 \times 10^{-5}) + \frac{22500}{2.303 \times 1.987 \times 313}$$

$$= (\log 1.8) - 5 + 15.7089$$
$$= 0.2553 - 5 + 15.7089 = 10.9642$$

$\therefore$ $$A = \text{antilog}\,(10.9642)$$
$$= \mathbf{9.208 \times 10^{10}} \textbf{ collisions/sec.}$$

Q.62. Calculate the activation energy of a reaction whose reaction rate at 27°C gets doubled for 10°C rise in temperature. *(Arunachal 2003;Agra 2004)*

Ans. Here we are given that

when $$T_1 = 27 + 273 \text{ K} = 300 \text{ K},$$
$$k = k \text{ (say)}$$

When $$T_2 = 27 + 10°\text{C} = 310 \text{ K}$$
$$k_2 = 2k$$

Substituting these values in the equation

$$\log\frac{k_2}{k_1}=\frac{E_a}{2.303R}\left(\frac{T_2-T_1}{T_1T_2}\right)$$

we get $$\log\frac{2k}{k}=\frac{E_a}{2.303\times8.314}\times\frac{310-300}{300\times310}$$

or $$\log 2=\frac{E_a}{2.303\times8.314}\times\frac{10}{300\times310}$$

or $$\mathbf{E_a = 53598.6\ J\ mol^{-1}\ or\ 53.599\ kJ\ mol^{-1}.}$$

Q.63. Discuss collision theory (Lindemann's theory) of reaction rates.

Lucknow, 2000; Calcutta 2000; M. Dayanand 2002; Guwahati 2002; Purvanchal 2002; Nagpur, 2003; Mumbai 2010)

Ans. This theory is based upon the kinetic theory of gases according to which the molecules of a gas are continuously moving and hence colliding with each other. Since for a collision to take place, at least two molecules must be involved, therefore the simple case to consider is that of *bimolecular* reactions. Hence in the discussion that follows, we shall first discuss the collision theory for bimolecular reactions and then for unimolecular reactions.

(*a*) **Collision theory for bimolecular reactions:** In a bimolecular reaction, in order that two molecules may react, they should not only collide but the collision must be associated with sufficient energy so that the molecules are activated. This means that all collisions are not effective. The number of effective collisions is only a fraction of the total number of collisions. Taking the *specific reaction rate* (*k*) *as equal to the number of effective collisions per cc per second,* if Z is the number of molecules per cc taking part in the collisions and f is the fraction of those molecules that are activated, we can write

$$k = Z \times f \qquad ...(i)$$

Now, according to kinetic theory of gases, the fraction of the molecules in a gas possessing energy greater than a particular value E is given by the equation

$$f=\frac{\Delta N}{N}=e^{-E/RT}$$

where ΔN = number of molecules possessing energy greater than E
and N = total number of molecules.

If E corresponds to E_a, the energy of activation, then the fraction of the activated molecules will be given by

$$f=\frac{\Delta N}{N}=e^{-E_a/RT} \qquad ...(ii)$$

Putting the value of f from equation (*ii*) in equation (*i*), we get

$$\mathbf{k = Ze^{-E_a/RT}} \qquad ...(iii)$$

Modification of the equation of collision theory : A good agreement between the calculated value and the observed value is found for a number of bimolecular reactions in the gaseous phase as well as in the solution. Still there are many reactions for which the calculated value is found to be much higher than the observed value. Hence equation (*iii*) is modified as

$$k = PZe^{-E_a/RT} \qquad ...(iv)$$

where the factor P, introduced in the above equation, is called **probability factor** or **steric factor**. It accounts for the fact that all collisions between the activated molecules may not result into a reaction; only those collisions between the activated molecules result into reaction which collide with proper orientations. The value of P varies from 1 to 10^{-9}. When $P = 1$, the reaction obeys collision theory.

Collision theory for unimolecular reactions (Lindemann hypothesis): A unimolecular reaction, in general, means a reaction of the type

$$A \longrightarrow \text{Products}$$

In such reactions, only one molecule of the reactant takes part in the reaction. Consequently, the question arises : How do molecules in a unimolecular reaction attain their energy of activation ? The answer to this question was first suggested by Lindemann in 1922. He pointed out that the behaviour of unimolecular reactions can be explained on the basis of **reaction.** During this time, the activated molecules may either react or be deactivated to ordinary molecules. Consequently, the rate of reaction will not be proportional to all the molecules activated but only to those which remain active. This concept may be illustrated as follows :

$$\underset{\text{Ordinary Molecules}}{A + A} \underset{k_2}{\overset{k_1}{\rightleftharpoons}} \underset{\text{Activated Molecules}}{A^*} + A$$

$$A^* \xrightarrow{k_3} \text{Products}$$

where k_1 k_2, k_3 are the rate constants of the different processes involved as shown above.

Q. 64. Discuss transition state theory or theory of absolute reaction rates or theory of activated complex formation

(Himachal 2000; Kalhapur, 2000; Gujarat, 2003; Allahabad 2003; Madras, 2004; G.Nanakdev 2004; Nagpur 2008)

Ans. According to this theory, it is believed that before the reacting molecules change into products, they form an intermediate, called activated complex which has higher energy than both the reactants and the products. It is for this reason that the theory is called Transition state theory. The activated complex is in equilibrium with the reacting molecules and the rate of reaction is given by the rate of decomposition of the complex to form the products. Thus we may represent the situation as follows :

$$\underset{\text{Reactants}}{A + B} \rightleftharpoons \underset{\text{Activated complex}}{X^*} \xrightarrow{k} \text{Products}$$

By making use of some fundamental properties of the reacting molecules, it was shown by Eyring that the rate constant (k) for any reaction (irrespective of its molecularity or order) is given by

$$k = \frac{RT}{Nh} K \qquad ...(i)$$

where R = Gas constant

h = Planck's constant

N = Avogadro's number

T = Absolute temperature

and K = Equilibrium constant for the formation of the activated complex from the reacting molecules, that is

$$K = \frac{[X^*]}{[A]\,[B]}$$

Now, from thermodynamics, we know that

$$\Delta G^* = -RT \ln K$$

or

$$\ln K = -\Delta G^* / RT$$

or
$$K = e^{-\Delta G^*/RT} \qquad ...(ii)$$
where ΔG^* is the free energy of activation for the reaction $A + B \rightarrow X^*$

Substituting this value of K in equation (*i*), we get
$$k = \frac{RT}{Nh} e^{-\Delta G^*/RT} \qquad ...(iii)$$

Further substituting for $\Delta G^* = \Delta H^* - T\Delta S^*$, where ΔH^* and ΔS^* are the changes in enthalpy and entropy of activation respectively, we get
$$k = \frac{RT}{Nh} e^{-(\Delta H - T\Delta S^*)/RT}$$
$$= \frac{RT}{Nh} e^{\Delta S^*/R} . e^{-\Delta H^*/RT} \qquad ...(vi)$$

Significance of the equation of the transition state theory: The equation (*iii*) involves two factors *i.e.* = $\frac{RT}{Nh}$ and $e^{-\Delta G^*/RT}$. The first factor is constant at constant temperature. Hence the rate of reaction depends only upon the second factor. This factor involves ΔG^* with a negative sign. This means that at particular temperature, *greater the value of the free energy of activation for a reaction, the slower will be the reaction.*

Comparison with Arrhenius equation: Comparing equation (*iv*) with the Arrhenius equation *viz.* $k = A^{-eE_a/RT}$ we observe that we should have
$$A = \frac{RT}{Nh} e^{-\Delta S^*/R}$$
and
$$E_a = \Delta H^*$$
(*i.e.* enthalpy of activation is equal to the energy of activation).

Comparison with the collision theory equation: Comparing equation (*iv*) *with* the collision theory equation *viz.*
$$k = PZe^{-E_a/RT}$$
we observe that,
$$PZ = \frac{RT}{Nh} e^{\Delta S^*/R}$$

Thus the probability factor P is related to the entropy of activation ΔS^*.

Q.65. What are the advantages of transition state theory over collision theory.

(G.Nanakdev 2004)

Ans. The transition state theory is better than the collision theory in a number of ways. For example,

(*i*) In collision theory, the correction factor P was introduced arbitrarily whereas in the transition state theory, its inclusion has been justified in terms of entropy of activation ΔS^*.

(*ii*) The concept of the formation of an activated complex seems to be more appropriate than assuming that the molecules just collide together and change into products.

Q.66. Discuss kinetics of reactions involving ions in solutions.

Ans. It is believed that the ionic reactions are so rapid that their velocities cannot be determined without special devices. However, this is not always true. There are a number of reactions involving ions only but in which some covalent bonds are broken and some others formed and these take place with a velocity which is measurable at ordinary temperatures.

A well known example is that of *isomeric conversion of ammonium cyanate into urea in aqueous solution.* It involves interaction of NH_4^+ ions and cyanate ions.

$$NH_4^+ + CNO^- \longrightarrow CO(NH_2)_2$$

The rate of this reaction has been measured both by the titration of cyanate and by determining its conductance. The reaction is found to be of the second order. The reaction is in fact reversible but in dilute solution, it goes almost to completion from left to right.

Other examples of relatively slow reactions in which two ions are involved are as follows:

(*a*) Reaction between bromoacetate ions and thiosulphate ions in aqueous solution.

$$CH_2BrCOO^- + S_2O_3^{-2} \longrightarrow CH_2(S_2O_3)COO^- + Br^-$$

(*b*) Reaction between bromoacetate ions and thiocyanate ions in aqueous solution.

$$CH_2BrCOO^- + SCN^- \longrightarrow CH_2(SCN)COO^- + Br^-$$

Q. 67. Discuss primary salt effect in ionic reactions. *(M. Dayanand 2002)*

Ans. Rate constant of a reaction between ions depends upon the ionic strength of the solution and changes with the ionic strength in a manner dependent upon the charges of the reacting ions. This effect is called *primary salt effect* and is explained below.

The reacting ions form an activated complex in equilibrium with the reactants and the rate of reaction is proportional to the concentration of the complex. Thus any ionic reaction between A^{Z_A} and B^{Z_B} can be represented as follows :

$$\underset{\text{Reactants}}{A^{Z_A} + B^{Z_B}} \rightleftharpoons \underset{\text{Activated complex}}{[(A\cdot B)^{Z_A+Z_B}]^*} \xrightarrow{k'} \text{Products} \quad ...(i)$$

where Z_A and Z_B are the charges of the ions A and B respectively while $Z_A + Z_B$ is the charge on the activated complex.

Suppose C_X is the concentration of the complex at any instant of time t, then the rate at this instant will be given by

$$\frac{dx}{dt} = k'C_X \quad ...(ii)$$

Further, as the complex is in equilibrium with the reactants, the equilibrium constant K will be

$$K = \frac{a_X}{a_A \cdot a_B} = \frac{f_X C_X}{(f_A C_A).(f_B C_B)} \quad ...(iii)$$

where the symbols a, C and f represent respectively the activity, concentrations and activity coefficients of the respective species. From eqn. (*iii*) we have

$$C_X = KC_A C_B \frac{f_A f_B}{f_X} \quad ...(iv)$$

Substituting this value in eqn. (*ii*), we get

$$\frac{dx}{dt} = k'KC_A C_B \frac{f_A f_B}{f_X}$$

$$= k_0 \left(\frac{f_A f_B}{f_X}\right) C_A C_B \quad ...(v)$$

where $k_0 = k'K$.

Comparing eqn. (*v*) with the ordinary rate eqn. *viz.*

$$\frac{dx}{dt} = kC_A C_B \qquad ...(vi)$$

we observe that the experimentally determined constant k is not the true reaction constant k_o but involves also the activity coefficients of the reactants and the complex.

In fact, $$k = k_0 \frac{f_A f_B}{f_X} \qquad ...(vii)$$

Since the activity coefficients of the charged reactants and complex depend upon the ionic strength of the solution, the velocity constant k will also depend upon the ionic strength.

If μ represents the ionic strength, it is seen from Debye-Huckel theory that for dilute solutions k varies with the ionic strength as

$$\log \frac{k}{k_0} = 2AZ_A Z_B \sqrt{\mu} \qquad ...(viii)$$

where A is Debye-Huckel constant.

Ionic strength is defined as half the sum of the terms obtained by multiplying the concentration (molality) of each ionic species present in the solution by the square of its valency *i.e.* $$\mu = \frac{1}{2} \Sigma C_i Z_i^2$$

According to eqn. (*viii*), a plot of log (k / k_0) against $\sqrt{\mu}$ should be linear with slope equal to $2AZ_AZ_B$ *i.e.* it is *dependent only on the product $Z_A Z_B$ of the charges of the reacting ions.*

The various reactions involving ions may be divided into the following three categories.

1. *Those for which $Z_AZ_B = 0$.* This is so for a reaction between a non-electrolyte ($Z_A = 0$) and an ion *e.g.* for the reaction

$$CH_2I\ COOH + CNS^- \longrightarrow CH_2(CNS)\ COOH + I^-$$

For such reactions for which primary salt effect is zero, k will not change with ionic strength.

2. *Those for which Z_AZ_B is positive e.g.* for the reaction

$$CH_2Br\ COO^- + S_2O_3^{-2} \longrightarrow CH_2(S_2O_3)\ COO^- + Br^-$$

$$Z_AZ_B = -1 \times (-2) = +2$$

For such reactions in which salt effect is positive, k increases with ionic strength.

3. *Those for which $Z_A Z_B$ is negative e.g.* for the reaction

$$[CO(NH)_3)_5Br^{+2} + OH^- \longrightarrow [CO(NH_3)_5OH]^{+2} + Br^-$$

$$Z_A Z_B = +2 \times (-1) = -2$$

For such reactions in which salt effect is negative, k decreases with increase in ionic strength.

Q.68. Compare the colision theory with transition state theory. *(Mumbai, 2005)*

Ans. See Q. 62 and 63.

MISCELLANEOUS QUESTIONS

Q. 69. The reaction A + B $\longrightarrow$ C + D is a second order reaction. Assuming the initial concentrations of both the reactants to be equal to '*a*', derive the expression for specificre rate constant of the reaction. Find an expression for the half-life of this reaction.

(Vidyasagar 2007, Calcutta 2007)

Ans. Refer to Q 29. Write the general equation as :

$$A + B \longrightarrow \text{Products}$$

Initial conc. of A = *a* moles /litre

Initial conic. of B = *a* moles/ litre

Decrease in conc. of A at any instant = *x* moles/ litre

Decrease in conc. of B at any instant = *x* moles / litre

Applying the law of mass action, the rate of reaction

$$\frac{dx}{dt} \propto [A][B]$$

$$\propto (a-x)(a-x)$$

$$\propto (a-x)^2$$

or
$$\frac{dx}{dt} = k(a-x)^2$$

Rest of the procedure is as given in Q. 29

For half-life of the reaction, refer to Q. 32

5

SURFACE CHEMISTRY (ADSORPTION AND CATALYSIS)

PART-I –ADSORPTION

Q. 1. What do you understand by the term adsorption ? How does it differ from absorption?

(Kumaon, 2000; Panjab 2000; Himachal 2000; Lucknow 2002; G. Nanakdev 2003; Delhi 2004; Awadh 2004)

Ans. Consider a liquid in a vessel. A molecule in the interior of the liquid is being attracted by molecules around it. But all these forces of attraction get cancelled. Thus there is no net force acting on molecules inside the liquid. However, the situation is different about molecules on the surface of the liquid. Such surface molecules are being attracted by molecules below it and inside the liquid. However, there are no molecules above the surface molecules. Hence such molecules on the surface experience a net force of attraction downward.

Because of unbalanced force of attraction existing on the surface, liquids and solids have the property to attract and retain the molecules that come into contact at the surface.

The phenomenon of attracting and retaining the molecules of a substance on the surface of a liquid or solid resulting into higher concentration of the molecules on the surface is called adsorption.

The substance which is adsorbed on the surface is called *adsorbate* and the substance on which adsorption takes place is called *adsorbent.* Adsorption differs from the phenomenon of absorption, in that the former refers to the attraction and retention of the molecules of a substance on the surface only [Fig. 5.1 (*a*)] and the latter involves passing of the substance through the surface into the bulk of the liquid or the solid [Fig. 5.1 (*b*)].

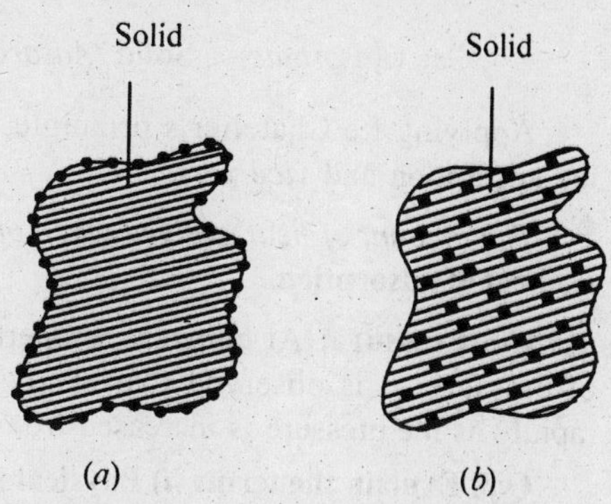

Fig. 5.1. Adsorption vs. absorption

For example, ink is absorbed by the chalk whereas gases are adsorbed by charcoal.

Q.2. Discuss the factors which influence the adsorption of gases by solids.

(Kerala 2000; Lucknow, 2002)

Ans. The following factors influence the adsorption of gases by solids.

(*i*) **Nature and surface area of the adsorbent:** It is observed that the same gas is adsorbed to different extents by different solids at the same temperature. Further, the greater the surface area of the adsorbent, greater is the volume of the gas adsorbed. It is for this reason that substances like charcoal and silica gel are excellent adsorbents because they have highly porous structures and hence large surface areas. The adsorbing power of these substances can be further increased by **activating them** in various ways. For example, wood charcoal can be activated by heating it in vacuum or some inert gas at temperatures between 625 K to 1275 K. As a result of this treatment it appears that hydrocarbons and other gases are removed from charcoal and hence the surface available for adsorption becomes larger.

(*ii*) **Nature of the gas being adsorbed:** Different gases are adsorbed to different extents by the same adsorbent at the same temperature.

It is observed that higher the critical temperature of a gas, greater is the amount of that gas adsorbed. In other words, a gas which is more easily liquefiable or is more soluble in water is more readily adsorbed. This relationship is not surprising as the critical temperature of a gas is related to the intermolecular forces which are an important factor for adsorption too.

(*iii*) **Temperature:** Studying the adsorption of any particular gas by some particular adsorbent, it is observed that the adsorption decreases with increase of temperature and *vice versa*. For example, one gram of charcoal adsorbs about 10 ml of N_2 at 273 K, 20 ml at 244 K and 45 ml at 195 K. The decrease of adsorption with increase of temperature may be explained as follows :

Like any other equilibrium, adsorption is a process involving a true equilibrium. The two opposing processes involved are the *condensation* (*i.e.*, adsorption) of the gas molecules on the surface of the solid and *evaporation* (*i.e.* desorption) of the gas molecules from the surface of the solid into the gaseous phase. Moreover, the process of condensation (or adsorption) is exothermic so that the equilibrium may be represented as

$$\text{Gas } (\textit{Adsorbate}) + \text{Solid } (\textit{Adsorbent}) \underset{\text{Evaporation}}{\overset{\text{Condensation}}{\rightleftharpoons}} \text{Gas adsorbed on solid} + \text{Heat}$$

Applying Le Chatelier's principle, it can be seen that increase of temperature decreases the adsorption and *vice versa*.

The amount of heat evolved when one mol of the gas is adsorbed on the adsorbent is called the **heat of adsorption**.

(*iv*) **Pressure:** At constant temperature, the adsorption of a gas increases with increase of pressure. It is observed that at low temperatures, the adsorption of a gas increases very rapidly as the pressure is increased from very small values.

Q.3. Explain the terms (*i*) Physical adsorption (*ii*) Chemical adsorption.

(Panjab, 2000;Nagpur 2002; Lucknow 2002; Mumbai 2005)

Ans. (*i*) Physical adsorption or van der Waal's adsorption: When a gas is held (adsorbed) on the surface of a solid by van der Waal's forces (which are weak intermolecular forces of attraction) without resulting into the formation of any chemical bond between the adsorbate and the adsorbent, it is called *physical adsorption* or *van der Waal's adsorption*. This type of adsorption involves low heats of adsorption, *i.e.*, about 20-40 kJ per mol. Further, physical adsorption of a gas by a solid is generally reversible. Increase of pressure causes more gas to be adsorbed and the release of pressure frees the adsorbed gas. Similarly decrease of temperature increases adsorption but the gas adsorbed at low temperature can be freed again by heating.

(*ii*) **Chemical adsorption or chemisorption:** *When a gas is held onto the surface of a solid by forces similar to those of a chemical bond, the type of adsorption is called chemical adsorption or chemisorption.* This type of adsorption results into the formation of *surface compound.* That the forces involved are similar to those of a chemical bond is confirmed by the fact that the heats evolved during chemisorption are high (*i.e.* 40 to 400 kJ/mol) which are of the same magnitude as those involved in chemical reactions. Further, as chemisorption is something similar to a chemical change, it is usually irreversible. The efforts to free the adsorbed gas often gives some definite compound instead of the free gas. For example, oxygen adsorbed on tungsten or carbon is liberated as tungsten oxide or as carbon monoxide and dioxide.

Physical adsorption takes place between every gas and a solid *i.e.* is not specific in nature (because it involves van der Waal's force which exists among the molecules of every two substances), whereas chemisorption is specific in nature and occurs only where there is a tendency towards compound formation between the gas and the adsorbent. Further, unlike physical adsorption, chemisorption like the most of chemical changes, increases with increase of temperature. For this reason, a gas may be physically adsorbed at low temperature but chemisorbed at higher temperature.

Q. 4. Distinguish between physical adsorption and chemisorption.

(Kurukshetra 2002; Bhopal 2004; G. Nanakdev 2004)

Ans. The main points of difference between physical adsorption and chemisorption are summed up in the Table below :

Physical Adsorption	*Chemisorption*
1. The forces operating in these cases are weak van der Waal's forces.	1. The forces operating in these cases are similar to those of a chemical bond.
2. No compound formation takes place in these cases.	2. Surface compounds are formed.
3. The heats of adsoption are low *viz.* 20 –40 kJ mol^{-1}.	3. The heats of adsorption are high *viz.* 40– 400 kJ mol^{-1}.
4. The process is reversible *i.e.*, desorption of the gas occurs by increasing the temperature or decreasing the pressure.	4. The process is irreversible. Efforts to free the adsorbed gas give some definite compound.
5. It is not specific in nature *i.e.* all gases are adsorbed on all solids to some extent.	5. It is specific in nature and occurs only when there is some possibility of compound formation between the gas being adsorbed and the solid adsorbent.
6. This type of adsorption decreases with increase of temperature.	6. This type of adsorption increases with increase of temperature.
7. Forms multimolecular layers.	7. Forms unimolecular layer.
8. The extent of adsorption is related to the ease of liquefaction of the gas.	8. There is no such correlation.

Q.5. What are adsorption isotherms ? Show different types of adsorption isotherms with the help of diagrams. *(M. Dayanand, 2000; Dibrugarh 2004; Awadh 2004)*

Ans. Amount of a gas adsorbed by a given amount of the adsorbent depends upon both temperature and pressure. *At constant temperature, there exist some definite relationships between the amount of the gas adsorbed and the equilibrium pressure. Such relationships are known as adsorption isotherms.*

Five different general types of adsorption isotherms have been observed for the adsorption of gases on solids. These are shown in Fig. 5.2 below :

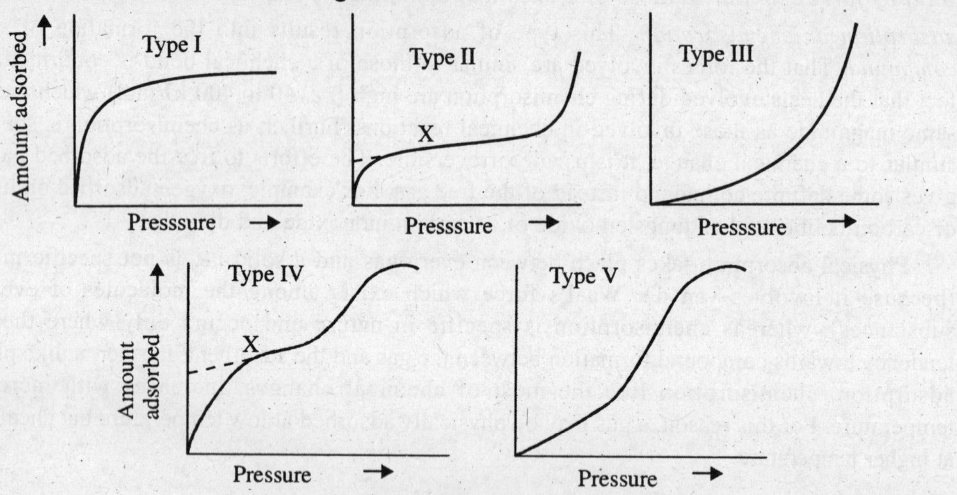

Fig. 5.2. Types of adsorption isotherms.

Q.6. Discuss Freundlich adsorption isotherm of a gas on a solid. How are the constants of this isotherms obtained ? **(*Mumbai, 2004; Bhopal, 2004*)**

Ans. It is an empirical equation which states that at constant temperature

$$\frac{x}{m} = k \times p^{1/n}$$

where x is the mass of the gas adsorbed by m gram of the adsorbent at a pressure P *i.e.*, x/m is the amount of the gas adsorbed per gram of adsorbent. k and n are constants at particular temperature and for a particular adsorbent and gas.

The quantity $1/n$ is generally less than 1 indicating that the amount of the gas adsorbed increases less rapidly than the pressure. Thus the plots of the amount of the gas adsorbed against pressure are of the type shown in Fig. 5.3.

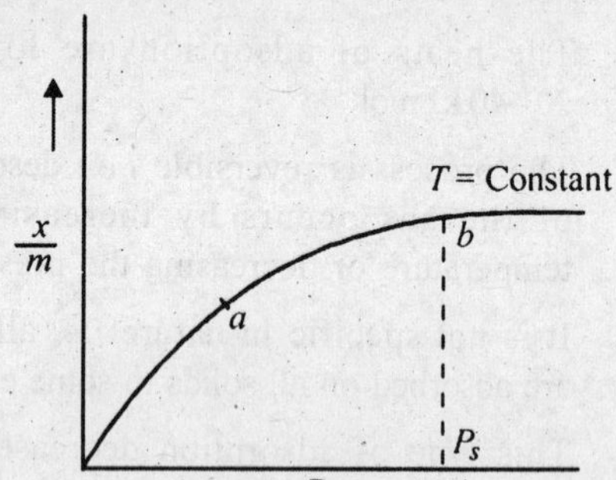

Fig. 5.3. Freundlich adsorption isotherm

At low pressures (say up to the point a) the graph is nearly linear. Hence for low pressures

$$\frac{x}{m} \propto P$$

or

$$\frac{x}{m} \propto P^1 \quad \text{... }(i)$$

For very high pressures (say beyond the point b) the adsorption becomes almost independent of pressure *i.e.*,

$$\frac{x}{m} \propto P^0 \quad \text{...}(ii)$$

The state corresponding to the point b is called the *saturation state* and the corresponding pressure P is called the *saturation pressure.*

For intermediate pressure, obviously x/m will be proportional to P raised to power lying between 0 and 1, *i.e.* fractions. Hence we write

$$\frac{x}{m} \propto P^{1/n} \quad \text{or} \quad \frac{x}{m} = kP^{1/n}$$

To test Freundlich equation further, logarithm of both sides of the equation may be taken, which gives

$$\log \frac{x}{m} = \log k + \frac{1}{n} \log P$$

so that the plot of log x/m against log P should be straight line with an intercept log k on the Y-axis and slope equal to $1/n$. In actual practice, reasonably straight line plot is often obtained (Fig. 5.4).

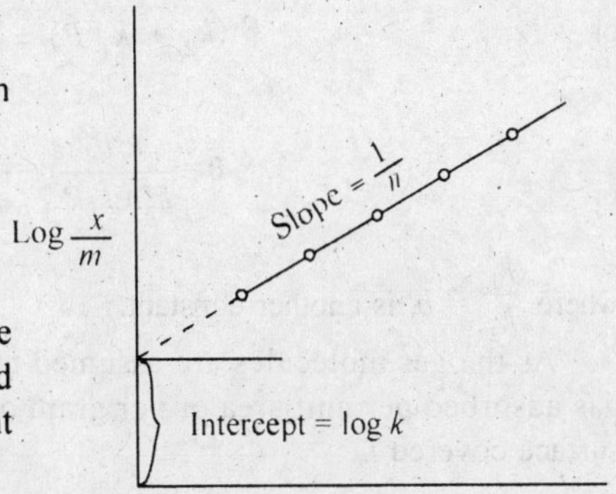

Fig. 5.4. Plot of log x/m vs. log P.

Q.7. Give the main points of Langmuir theory of adsorption.

(Kurukshetra 2002; Nagpur 2002; Guwahati 2002)

Ans. Langmuir put forward a **theory of adsorption,** the main points of which are as follows

(*i*) Adsorption could take place on the surface of a solid only till the whole of the surface was completely covered with a *unimolecular* layer of the gas molecules.

(*ii*) Adsorption consists of two opposing processes, namely *condensation* of the gas molecules on the solid surface and *evaporation* of the gas molecules from the surface back into the gaseous phase.

(*iii*) The rate of condensation depends upon the bare (uncovered) surface available for condensation. Initially as the whole of the surface is bare, the rate of condensation is very high. As the surface is covered more and more, the rate of condensation decreases. On the other hand, the rate of evaporation depends upon the covered surface and hence increases as more and more of the surface is covered. Ultimately a stage is reached when the rate of condensation becomes equal to the rate of evaporation and an equilibrium is said to have been attained.

(*iv*) The rate of condensation also depends upon the pressure of the gas, as according to the kinetic theory, the number of molecules striking per unit area is proportional to the pressure.

Q.8. Deduce Langmuir adsorption isotherm equation

$$\frac{x}{m} = \frac{aP}{1+bP}$$

(Kanpur 2001; Kurukshetra 2001; M. Dayanand 2002; Guwahati 2002; Nagpur 2002; Lucknow 2002; Delhi 2004; Gulberga 2004; G. Nanakdev 2004; Mumbai 2005)

Ans. The relationship was deduced by Langmuir as follows :

Suppose at any instant, the fraction of the surface covered by the adsorbed molecules = θ.

The fraction of the surface bare and available for adsorption = $1 - \theta$.

Suppose the pressure of the gas at that instant = P.

Then according to Langmuir's theory of adsorption,

$$\text{Rate of condensation} \propto (1-\theta) \times P = k_1 (1-\theta) P$$

where k_1 is a constant of proportionality.

$$\text{Rate of evaporation} \propto \theta = k_2 \theta$$

where k_2 is another constant of proportionality.

When the adsorption equilibrium is attained

$$k_1 (1-\theta) P = k_2 \theta$$

or $$\theta (k_2 + k_1 P) = k_1 P \quad \text{...(i)}$$

or $$\theta = \frac{k_1 P}{k_2 + k_1 P} = \frac{\frac{k_1}{k_2} P}{1 + \frac{k_1}{k_2} P} = \frac{bP}{1+bP} \quad \text{...(ii)}$$

where $\frac{k_1}{k_2} = b$, is another constant.

As the gas molecules are assumed to form a layer, one molecule thick on the surface, the gas adsorbed per unit area or per gram of the adsorbent must obviously be proportional to the surface covered *i.e.*,

$$\frac{x}{m} \propto \theta$$

or $$\frac{x}{m} = k_3 \theta \quad \text{...(iii)}$$

where x is the mass of the gas adsorbed on m gram of the adsorbent and k_3 is a constant of proportionality.

$$\therefore \quad \frac{x}{m} = k_3 . \frac{bP}{1+bP}$$

or $$\boxed{\frac{x}{m} = \frac{aP}{1+bP}} \quad \text{...(iv)}$$

where $k_3 b = a$ is another constant.

Equation (*iv*) is known as *Langmuir adsorption isotherm.* The constants a and b depend upon the nature of the gas adsorbate, the nature of the solid adsorbent and the temperature. Their values can be found from the experimental data.

To verify the validity of Langmuir adsorption equation divide both sides of equation (*iv*) by *P.*

$$\frac{x/m}{P} = \frac{a}{1+bP}$$

Take the reciprocal,

$$\frac{P}{x/m} = \frac{1}{a} + \frac{b}{a} P.$$

As a and b are constants, a plot of $\frac{P}{x/m}$ *vs.* P should be a straight line with slope equal to b/a and intercept on the y-axis equal to $1/a$. This is actually found to be so.

Q.9. Show that Freundlich isotherm is a special case of Langmuir isotherm.

(Panjab, 2000; Meerut, 2000; Kumaon, 2000; Gulberga, 2004)

OR

Discuss the behaviour of Langmuir isotherm at very high and very low temperature.

(Kurukshetra 2001; Kanpur 2001)

Ans. When the pressures are very low, adsorption is very little *i.e.* a very small fraction of the surface is covered with molecules. Thus θ is very small so that $1 - \theta = 1$. Equation (*i*), from the previous question is reduced to if, $k_1 P = k_2 \theta$.

But from equation (*iii*) of the previous question

$$\theta = \frac{1}{k_3} \frac{x}{m}$$

$$\therefore \qquad k_1 P = k_2 \frac{1}{k_3} \frac{x}{m}$$

or
$$\frac{x}{m} = \frac{k_1 k_3}{k_2} P$$

i.e.,
$$\frac{x}{m} \propto P \qquad ...(v)$$

When the pressures are very high, there is considerable adsorption *i.e.* the surface is almost completely covered with molecules. Thus $\theta \simeq 1$ so that equation (*i*) becomes

$$k_1 (1-\theta) P = k_2$$

or
$$1-\theta = \frac{k_2}{k_1 P}$$

or
$$\theta = 1 - \frac{k_2}{k_1 P}$$

But from equation (*iii*), as before

$$\theta = \frac{1}{k_3} \frac{x}{m}$$

$$\therefore \qquad \frac{1}{k_3} \frac{x}{m} = 1 - \frac{k_2}{k_1 P}$$

$$\frac{x}{m} = k_3 - \frac{k_3 k_2}{k_1 P}$$

When the pressures are very high, $1/P \simeq 0$ so that we have

$$\frac{x}{m} = k_3$$

or it may be written as
$$\frac{x}{m} = k_3 P^0$$

i.e.
$$\frac{x}{m} \propto P^0 \qquad ...(vi)$$

Thus from equations (*v*) and (*vi*), we have

At very low pressures, $\frac{x}{m} \propto P$

At very high pressures, $\frac{x}{m} \propto P^0$.

Hence for intermediate pressures, we must have

$$\frac{x}{m} \propto P^{1/n}$$

where $1/n$ has a value lying between 0 and 1,

i.e.
$$\frac{x}{m} = kP^{1/n}$$

which is Freundlich isotherm.

Q.10. What is the basic principle of B.E.T. equation. Write B.E.T equation. Explain the meaning of various terms involved. (*M. Dayanand 2002; Dibrugarh 2004; Mumbai 2010*)

Ans. To explain adsorption isotherms of type II and type III (Refer to Q. 5), Brunauer,

Emmett and Teller postulated that the adsorption is not unimolecular but **multimolecular** *i.e.* at low pressure; a single adsorbed layer of molecules is formed but as the pressure is increased many more layers of molecules are attached to the first layer. Based upon this principle they derived an equation, called BET equation. The equation is

$$\frac{P}{v(P^0 - P)} = \frac{1}{v_m^C} + \left[\frac{c-1}{v_m c}\right]\frac{P}{P^0}$$

It can also be written in the form

$$v = \frac{v_m cf}{1-f}\,\frac{1}{1+(c-1)f} \qquad ...(i)$$

where v = volume of the gas (reduced to STP) adsorbed at given pressure P and constant temperature T.

v_m = volume of the gas (reduced to STP) adsorbed when the surface is completely covered by a unimolecular layer.

$f = \frac{P}{P^0}$ where P^0 is the vapour pressure (of the liquefied gas) at the experimental temperature T

c = a constant at the given temperature = $e^{(E_l - E_L)/RT}$

E_l = heat of adsorption of the gas in the first layer

and E_L = heat of liquefaction of the gas.

If $E_l > E_L$ so that c is considerably greater than 1, eq. (*i*) gives a curve of the type II. If $E_l < E_L$, the curve of type III is obtained.

In a special case, when a single unimolecular layer is supposed to be adsorbed on the surface, the BET equation reduces to the same form as Langmuir equation. Then it is able to explain the adsorption isotherm of type I.

To explain adsorption isotherms of type IV and type V, it has been argued that in case of substances showing these types of adsorption, not only multimolecular adsorption takes place but also the condensation of the gas takes place in pores and the capillaries of the adsorbent. Isotherm of type IV is obtained when the condition, as for type II, is that $E_l > E_L$ and isotherm of type V is obtained when the condition, as for type III, is that $E_l < E_L$.

To test the validity of BET equation, it may be rewritten in the form

$$\frac{f}{v(1-f)} = \frac{1+(c-1)f}{v_m c}$$

or

$$\frac{f}{v(1-f)} = \frac{1}{v_m c} + \frac{(c-1)}{v_m c}f \qquad ...(ii)$$

Thus a plot of $f/v\,(1-f)$ *vs.* f should be a straight line with a slope equal to $(c-1)/v_m c$ and the intercept on the y-axis equal to $1/v_m c$.

Q.11.How can BET equation be used for the determination of surface area of an adsorbent?

(Shivaji, 2000; Kurukshetra 2002)

Ans. The plot of $f/v\,(1-f)$ versus f [*i.e.* $P/v\,(P^0 - P)$ *vs.* P/P^0] is a straight line with slope = $(c-1)/v_m c$ and intercept on the y-axis = $1/v_m c$.

From these two quantities the value of v_m *i.e.* the volume of the gas required to completely cover the surface of the adsorbent with a unimolecular layer, can be calculated. As 22400 cc of the gas at STP contains Avogadro's number of molecules, therefore, the number of molecules in the volume v_m can be calculated. Further as the area of cross-section per single gas molecule used for adsorption is known, therefore, the total area occupied by the gas molecules gives the area of the definite weight of the adsorbent taken.

The calculation of surface area of the adsorbent is done as follows :

Volume v_m as obtained from the graph is reduced to N.T.P. conditions by using the general gas equation

$$\frac{P_1V_1}{T_1} = \frac{P_2V_2}{T_2}$$

The surface area of the adsorbent may be calculated directly from the value of v_m as follows :

22400 cc of the gas at STP contain = 6.022×10^{23} molecules

$\therefore$ v_m cc of the gas at STP will contain

$$= \frac{6.022 \times 10^{23} \times v_m}{22400} \text{ molecules}$$

If the area occupied by a single molecule is S, then area occupied by $\frac{6.022 \times 10^{23} \times v_m}{22400}$ molecules *i.e.* the surface area of the adsorbent

$$= \frac{6.022 \times 10^{23}}{22400} \times v_m \times S.$$

Suppose w g of the adsorbent was used in the experiment, then specific surface area *i.e.* area per unit gram of the absorbent is :

$$\boxed{= \frac{6.022 \times 10^{23}}{w \times 22400} \times v_m \times S}$$

It may further be noted that for adsorption isotherms of the types I, II and IV, the volume v_m can be obtained directly from the plots of the adsorption isotherms; for type I directly from the limiting adsorption and for types II and IV by extrapolating the middle flat portion (marked X) to zero pressure.

The gas most commonly used for surface area determination is nitrogen gas at its normal boiling point (–195.8°C). At this temperature, the area occupied by a nitrogen molecule is taken to be 16.2 Å^2 (*i.e.*, 16.2×10^{-20} m^2).

Q.12. The volume of nitrogen gas v_m (measured at S.T.P.) required to cover a sample of silica gel with a unimolecular layer is 129 cm^3 g^{-1} of gel. Calculate the surface area per gram of the gel if each nitrogen molecule occupies 16.2×10^{-20} m^2.

Ans. Here we are given

Volume of N_2 at STP required to cover the gel with unimolecular layer *i.e.*

$$v_m = 129 \text{ cc g}^{-1}$$

Area occupied by each molecule of N_2

i.e. $S = 16.2 \times 20$ m^2

22400 cc of N_2 at STP contains

$= 6.022 \times 10^{23}$ molecules

$\therefore$ 129 cc of N_2 at STP will contain

$$= \frac{6.022 \times 10^{23} \times 129}{22400} \text{ molecules}$$

$$= 3.468 \times 10^{21} \text{ molecules}$$

Area occupied by a single molecule of N_2

$$= 16.2 \times 10^{-20} \text{ m}^2$$

$\therefore$ Area occupied by 3.468×10^{21} molecules of N_2

$$= 16.2 \times 10^{-20} \times 3.468 \times 10^{21}$$

$$= 561.8 \text{ m}^2$$

$\therefore$ Surface area per gram of gel

$$= \mathbf{561.8 \ m^2.}$$

Q.13. What do you understand by positive and negative adsorption ?

(M. Dayanand 2002)

Ans. Positive adsorption: In most of the cases, the solute from the solution is adsorbed on the surface of the solid adsorbent so that the concentration of the solute on the surface of the solid is greater than in the bulk of the solution. This is called **positive adsorption.**

Negative adsorption: In a few cases involving solution of electrolytes, the solvent from the solution is adsorbed by the adsorbent in preference to solute, so that the concentration of the solute in the solution increases *i.e.* becomes greater than the initial concentration. This is called **negative adsorption.** For example, when a dilute solution of KCl is shaken with blood charcoal, it shows negative adsorption but with high concentrations of KCl it shows positive adsorption.

Q.14. Discuss factors that affect adsorption by solids from solutions.

Ans. The adsorption by solids from the solution depends upon following factors :

(*i*) **Nature of the adsorbent:** Some adsorbents are more effective in attracting certain substances to their surfaces than the others.

(*ii*) **Surface area of the adsorbent:** Greater is the surface area of the adsorbent, greater is the adsorption.

(*iii*) **Nature of the solute and the solvent:** Different solutes dissolved in the same solvent are adsorbed to different extents by the same adsorbent. For example, non-electrolytes are adsorbed much more than electrolytes from the aqueous solution by charcoal. Similarly same solute dissolved in different solvents is adsorbed to different extents.

(*iv*) **Effect of temperature:** Increase of temperature results in the decrease of adsorption.

(*v*) **Effect of concentration:** The variation of the extent of adsorption with concentration is given by Freundlich equation which for adsorption from solutions, is written in the form

$$\frac{x}{m} = KC^{1/n}$$

where x is the amount of solute in grams adsorbed by m grams of the adsorbent, C is the equilibrium concentration of the solution in moles/litre. K and n are constants. The above equation is found to represent the adsorption from solutions much better than the adsorption of gases by solids.

To test the validity of the above equation for adsorption from solutions, take logarithm of both the sides. We get

$$\log \frac{x}{m} = \log K + \frac{1}{n} \log C$$

Thus a plot of log x/m vs. log C should be a straight line with slope equal to $1/n$ and intercept on the y-axis equal to log K.

Q.15. Describe the effect of dissolved substances on the surface tension of a solvent.

Ans. Whenever a substance is dissolved in a solvent the surface tension of the solvent changes. This is due to the fact that the concentration of the solute on the surface of the solution is different from the bulk of the solution. Based upon their effect on the surface tension, the various substances are divided into three types :

Type I : *Those substances which increase the surface tension, though the increase is generally not large.* For example, strong electrolytes when dissolved in water exhibit this type of behaviour. This is shown by curve I in Fig. 5.5.

Type II : *Those substances which decrease the surface tension regularly and gradually with increase in the concentration.* For example, non-electrolytes or weak electrolytes when dissolved in water show this type of behaviour. This is in represented by curve II in Fig.5.5.

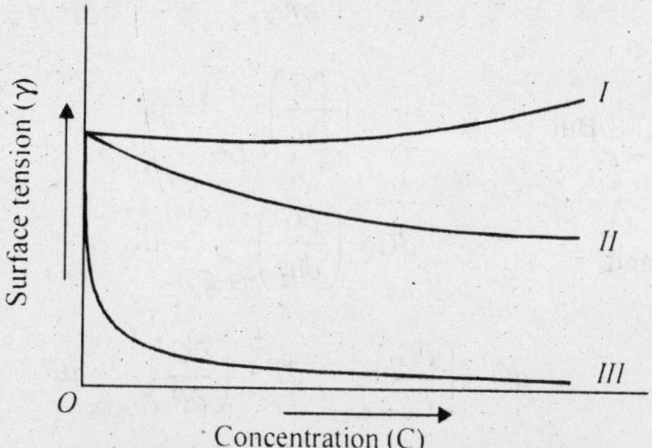

Fig. 5.5. Effect of different types of dissolved substances on the surface tension of the solvent.

Type III: *Those substances which decrease the surface tension considerably even at very low concentration.* For example,this type of behaviour is exhibited by aqueous solutions of soaps. Such substances are called *surface active agents* or *surfactants.* This type of behaviour is represented by curve III in Fig. 5.5.

If a solute brings about a *decrease* of surface tension, its concentration at the surface of the solution is *more* than in the bulk of the solution. This type of behaviour is called *positive adsorption* or the solute is said to exhibit positive surface activity. On the other hand, if a solute brings about an *increase* of surface tension, its concentration at the surface of the solution is found to be *less* than in the remainder of the solution. This type of behaviour is called *negative adsorption* or the solute is said to have negative surface activity.

Q.16. Derive Gibb's adsorption equation. *(Awadh 2000; Allahabad 2002)*

Ans. This equation gives relationship between adsorption and surface tension. Its derivation is based upon the concept of chemical potential which is nothing but *partial molar free energy i.e.* the change in free energy of the system per mol of the component added at constant temperature and pressure, keeping the composition of all other components constant.

Consider a system of two components (*i.e.* a solute and solvent). If n_1 and n_2 are the number of moles of the two components and μ_1 and μ_2 are their respective chemical potentials, then the free energy of the system is given by

$$G = \mu_1 n_1 + \mu_2 n_2 \quad ...(i)$$

However, when the two components are mixed, there is a change in the surface area of the solution and hence there is a change in the surface energy of the system. To account for this, the total free energy of the system should also include the free energy of the surface. If γ is the surface tension *i.e.* surface energy per sq cm and s is the surface area, then

Free energy of the surface = γs

The free energy of the system will be given by

$$G = \gamma s + \mu_1 n_1 + \mu_2 n_2 \quad ...(ii)$$

Differentiating this equation completely, we get

$$dG = \gamma ds + sd\gamma + \mu_1 dn_1 + n_1 d\mu_1 + \mu_2 dn_2 + n_2 d\mu_2 \quad ...(iii)$$

The free energy change of a system can also be obtained in the form of a different expression as follows :

Free energy is a function of temperature, pressure and composition of the various components. It can be written as

$$G = f\,(T, P, n_1, n_2) \quad ...(iv)$$

If there is a small change in the temperature, pressure and the amounts of the two components, the change in the value of G will be given by

$$dG = \left(\frac{\partial G}{\partial T}\right)_{P,n_1,n_2} dT + \left(\frac{\partial G}{\partial P}\right)_{T,n_1,n_2} dP + \left(\frac{\partial G}{\partial n_1}\right)_{T,P,n_2} dn_1 + \left(\frac{\partial G}{\partial n_2}\right)_{T,P,n_1} dn_2 \quad ...(v)$$

But

$$\left(\frac{\partial G}{\partial n_1}\right)_{T,P,n_2} = \mu_1$$

and

$$dG = \left(\frac{\partial G}{\partial n_2}\right)_{T,P,n_1} = \mu_2$$

$$\therefore \quad dG = \left(\frac{\partial G}{\partial T}\right)_{P,n_1,n_2} dT + \left(\frac{\partial G}{\partial P}\right)_{T,n_1,n_2} dP + \mu_1 dn_1 + \mu_2 dn_2 \quad ...(vi)$$

According to total differential equation

$$dG = VdP - SdT \quad ...(vii)$$

If temperature is kept constant, $dT = 0$ eq. (*vii*) reduces to

$$(dG)_T = (VdP)_T$$

or

$$\left(\frac{\partial G}{\partial P}\right)_T = V \quad ...(viii)$$

If pressure is kept constant, $dP = 0$, eq. (*vii*) reduces to

$$(dG)_P = -\ (SdT)_P$$

or

$$\left(\frac{\partial G}{\partial T}\right)_P = -S \quad ...(ix)$$

Substituting the values from equations (*viii*) and (*ix*) in equation (*vi*), we get

$$dG = -\ SdT + VdP + \mu_1 dn_1 + \mu_2 dn_2 \quad ...(x)$$

Total free energy change is obtained by adding the increase in free energy due to increase in the surface area. If ds is the increase in the surface area, then

Free energy change due to increase in surface area = $\gamma.\ ds$

$\therefore$ Total free energy change will be given by

$$dG = SdT + VdP + \gamma ds + \mu_1\ dn_1 + \mu_2 dn_2 \quad ...(xi)$$

Comparing equations (*iii*) and (*xi*), we get

$$sd\gamma + n_1 d\mu_1 + n_2 d\mu_2 = -\ SdT + VdP$$

or

$$SdT - VdP + sd\gamma + n_1 d\mu_1 + n_2 d\mu_2 = 0$$

But at constant temperature and pressure,

$$dT = 0,\ dP = 0$$

$$\therefore \quad sd\gamma + n_1 d\mu_1 + n_2 d\mu_2 = 0 \quad ...(xii)$$

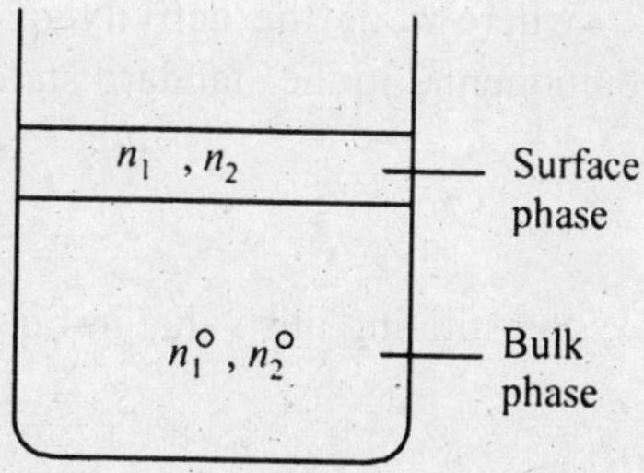

Now imagine that the system under consideration is divided into two parts, one part being that which is under the influence of the surface forces and the other is the remainder of the solution. The former may be called as *surface phase* and the latter the *bulk phase* as shown in Fig.

If n_1^0 and n_2^0 are the number of moles of the two components in the bulk phase corresponding to n_1 and n_2 in the surface phase, then since for the bulk phase, surface considerations are not involved, Eq. (*xii*) for the bulk phase may be written as

$$n_1^0 d\mu_1 + n_2^0 d\mu_2 = 0 \qquad ...(xiii)$$

Multiplying this equation by n_1/n_1^0, we get

$$n_1\, d\mu_1 + \frac{n_1\, n_2^0}{n_1^0} = d\mu_2 = 0 \qquad ...(xiv)$$

Subtracting this equation from Eq. (*xii*), we get

$$sd\gamma + n_2 d\mu_2 - \frac{n_1\, n_2^0}{n_1^0}\, d\mu_2 = 0$$

or
$$sd\gamma + d\mu_2 \left(n_2 - \frac{n_1\, n_2^0}{n_1^0} \right) = 0$$

or
$$sd\gamma = - d\mu_2 \left(n_2 - \frac{n_1\, n_2^0}{n_1^0} \right)$$

or
$$-\frac{d\gamma}{d\mu_2} = \frac{n_2 - n_1 n_2^0 / n_1^0}{s} \qquad ...(xv)$$

n_2 is the number of moles of one component, say the solute, associated with n_1 moles of the solvent in the surface phase.

$n_1 n_2^0 / n_1^0$ is the corresponding number of moles associated with n_1 moles of the solvent in the bulk phase. This is because in the bulk phase,

n_1^0 moles of the solvent are associated with n_2^0 moles of the solute.

$\therefore$ n_1 moles of the solvent are associated with $\frac{n_2^0}{n_1^0} \times n_1$ moles of the solute.

Thus $n_2 - \frac{n_1 n_2^0}{n_1^0}$ is the excess concentration of the surface phase

or $\frac{n_2 - n_1 n_2^0 / n_1^0}{s}$ is the excess concentration of the solute in the surface phase per unit area of the surface.

Denoting this excess concentration of the surface phase by S_2, Eq. (*xv*) can be written as

$$S_2 = -\frac{d\gamma}{d\mu_2} \qquad ...(xvi)$$

Further we know that

$$\mu_2 = \mu_2^0 + RT \text{ In } a_2 \qquad ...(xvii)$$

where a_2 is the activity of the component 2 and μ_2^0 is the chemical potential of the component 2 in the standard state *i.e.* when its activity is unity. Hence at constant temperature

$$d\mu_2 = RTd \ln a_2$$

$$= \frac{RT}{a^2} da_2$$

Substituting this value in Eq. (*xvi*), we get

$$S_2 = -\frac{d\gamma}{RTda_2} \cdot a_2$$

or

$$S_2 = -\frac{a_2}{RT}\frac{d\gamma}{da_2} \qquad ...(xviii)$$

This equation is called Gibbs equation.

Although this equation is usually applied to solutes, it is equally applicable to either of the two components of a binary system. Neglecting the subscripts, Eqn. (*xviii*), therefore, can be written in the form

$$S = -\frac{a}{RT}\frac{d\gamma}{da_2} \qquad ...(xix)$$

For dilute solutions, the activity may be replaced by concentration C so that equation (*xix*) takes the form

$$\boxed{S = -\frac{C}{RT}\frac{d\gamma}{dC}} \qquad ...(xx)$$

Q.17. Discuss applications of Gibbs adsorption equation.

(Kumaon, 2000; Lucknow 2000; Madras 2003)

Ans. 1. In predicting positive or negative adsorption: In the Gibbs equation $d\gamma/dC$ gives the rate of variation of surface tension of the solution with concentration.

If $d\gamma/dC$ is positive, S must be negative *i.e.* the concentration of the solute in the surface is less than in the bulk of the solution. This is a case of negative adsorption.

If $d\gamma/dC$ is negative, S will be positive *i.e.* the concentration of the solute is greater in the surface than in the bulk of the solution. This is a case of positive adsorption. This happens in the case of surface active agents like soaps.

2. In calculation of the excess concentration of a dissolved substance on the surface of a liquid (*called Gibbs adsorption*): Knowing the concentration C of the solute in the solution and knowing the value of $d\gamma/dC$ *i.e.* the rate of variation of surface tension with concentration at temperature T, the excess concentration of the dissolved substance on the surface of the liquid can be obtained.

3. In the calculation of the adsorption of an insoluble gas on the surface of a liquid: For example, the adsorption of saturated hydrocarbons (Component 2) on the surface of water (component 1) can be determined by using Gibbs adsorption equation in a modified form as follows :

According to Gibbs adsorption equation

$$S_2 = -\frac{d\gamma}{d\mu_2} \qquad ...(i)$$

But for the gaseous component 2,

$$\mu_2 = \mu_2^0 + RT \ln p$$

where p is the vapour pressure of component 2 over the liquid

$$\therefore \qquad d\mu_2 = RTd \ln p$$

$$= \frac{RT}{p} dp$$

Substituting this value in Eqn. (*i*), above

$$S_2 = -\frac{d\gamma}{RTdp} \cdot P$$

or $$\boxed{S_2 = -\frac{P}{RT} \cdot \frac{d\gamma}{dp}}$$

Thus knowing the value of $d\gamma / dp$ *i.e* the rate of variation of surface tension with vapour pressure of the gaseous component, the adsorption of the vapour on the surface of the liquid can be calculated.

Q.18. What are adsorption isobars ? Describe adsorption isobars of physical and chemical adsorption.

OR

Explain the effect of temperature on adsorption.

Ans. *The graphs obtained by plotting adsorption* (*x/m*) vs. *temperature, keeping the pressure constant are called adsorption isobars.* Studies show that in case of physical adsorption, the adsorption (*x/m*) decreases regularly with increase of temperature [Fig. 5.6(*a*)]. But in case of chemical adsorption, the adsorption first increases and then begins to decrease with increase of temperature [Fig. 5.6(*b*)]. The increase in the beginning is evidently due to the fact that like any other chemical reaction, the heat supplied acts as the activation energy for the

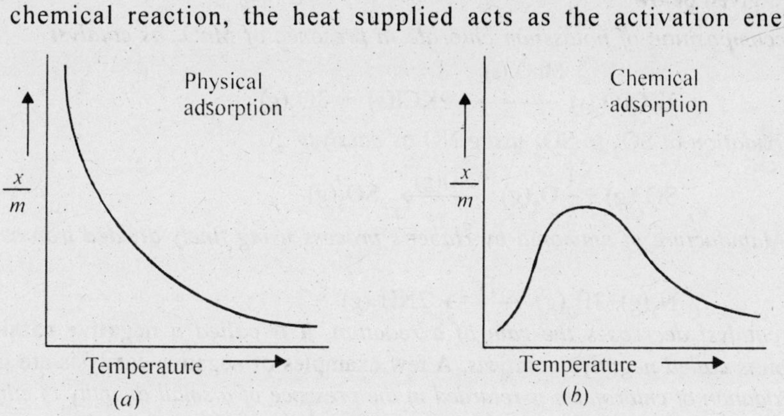

chemisorption.

Fig. 5.6. Adsorption isobars (*a*) Physical adsorption (*b*) Chemisorption.

Q.19. Describe some important applications of adsorption. **(*Awadh 2004*)**

Ans. A few applications are described below :

(*i*) **In gas masks:** Gas masks are devices containing suitable adsorbents so that the poisonous gases present in the atmosphere are preferentially adsorbed and the air for breathing is purified.

(*ii*) **In chromatographic analysis:** The selective adsorption of certain substances from a solution by a particular solid adsorbent has helped to develop technique for the separation of the components of a mixture. This technique is called chromatographic analysis.

(*iii*) **In preserving vacuum:** In Dewar flasks, activated charcoal is placed between the walls of the flask so that any gas which enters into the annular space either due to glass imperfections or diffusion through glass is adsorbed.

(*iv*) **In cleaning of sugar :** Sugar is decolourized by treating sugar solution with charcoal powder. The latter adsorbs the colours present.

(*v*) **In adsorption indicators:** Various dyes, which work on the principle of adsorption, have been introduced as indicators, particularly in precipitation titration. For example, KBr is easily titrated with $AgNO_3$ using eosin as an indicator.

(*vi*) **In paint industry:** The paint should not contain dissolved gases as otherwise the paint does not adhere well to the surface to be painted and thus will have a poor covering power. The dissolved gases are, therefore, removed by suitable adsorbents during manufacture. Further, all surfaces are covered with layers of gaseous liquid or solid films. These have to be removed before the paint is applied. This is done by suitable liquids which adsorb the films. Such liquids are called *wetting agents.* The use of spirit as wetting agent in furniture painting is well known.

(*vii*) **In catalysis:** The action of certain solids as catalysts is best explained in terms of adsorption. The theory is called *adsorption theory.* According to this theory, the gaseous reactants are adsorbed on the surface of the solid catalyst. As a result, the concentration of the reactants increases on the surface and hence the rate of reaction increases.

PART-II—CATALYSIS

Q. 20. What is a catalyst ? Explain positive and negative catalysis with examples.

(Himachal 2010)

Ans. *A catalyst is a substance that changes (usually increases) the rate of a reaction but is recovered chemically unchanged at the end of the reaction and the phenomenon is called catalysis.*

Positive and negative catalysis. *If a catalyst increases the rate of a reaction, it is called a* positive catalyst *and the phenomenon is called positive catalysis.* A few examples of positive catalysis are given below :

(*i*) *Decomposition of potassium chlorate in presence of MnO_2 as catalyst*

$$2KClO_3(s) \xrightarrow{MnO_2(s)} 2KCl(s) + 3O_2(g)$$

(*ii*) *Oxidation of* SO_2 *to* SO_3 *using* NO *as catalyst*

$$SO_2(g) + \frac{1}{2}O_2(g) \xrightarrow{NO(g)} SO_3(g)$$

(*iii*) *Manufacture of ammonia by Haber's process using finely divided iron as catalyst*

$$N_2(g) + 3H_2(g) \xrightarrow{Fe} 2NH_3(g)$$

If a catalyst decreases the rate of a reaction, it is called a **negative catalyst** *and the phenomenon is called negative catalysis.* A few examples of negative catalysis are given below:

(*i*) *Oxidation of chloroform is retarded in the presence of a small quantity of ethyl alcohol*

$$\underset{\text{Chloroform}}{2CHCl_3} + O_2 \xrightarrow{C_2H_5OH} \underset{\text{Phosgene}}{2COCl_2} + 2HCl$$

(*ii*) *Decomposition of* H_2O_2 *is retarded in the presence of traces of acetanilide*

$$2H_2O_2 \xrightarrow{\text{Acetanilide}} 2H_2O + O_2$$

Q. 21. Explain with the help of a diagram working of a catalyst.

(Sambalpur 2003, Delhi 2004)

Ans. A catalyst decreases the activation energy of the reactants. It lowers the energy barrier and thus makes the reaction fast, as shown in Fig. 5·7. However in a reversible reaction, the catalyst decreases the activation energy of the forward reaction as well as that of the backward reaction by the same amount. Thus both the forward and the backward reactions are accelerated

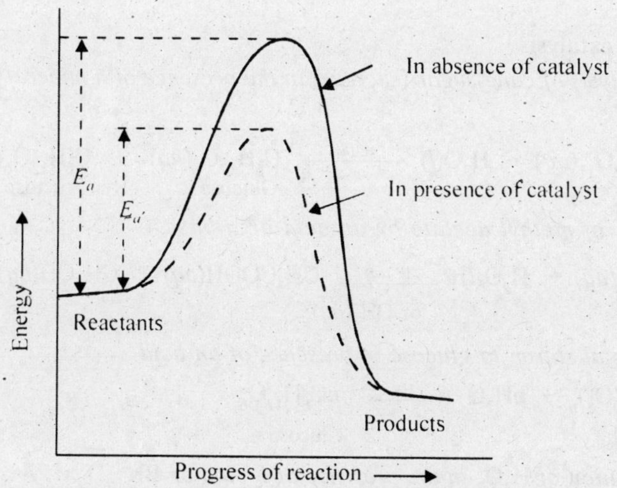

Fig. 57. Activation energy diagram for uncatalysed and catalysed reactions.

to the same extent. As a result, the equilibrium is attained quickly but does not change position.

Q. 22. What are (*i*) Promoters (activators) (*ii*) Inhibitors (catalytic poisons) ?

(M. Dayanand 2002; Nagpur. 2002; Vidyasagar 2007)

Ans. Promoters: Sometimes the presence of some other substances which are not necessarily catalysts increases the efficiency of a catalyst. Such substances thus act as catalyst for the catalyst and are called promoters or activators. Molybdenum acts as a promoter for the iron catalyst in the manufacture of ammonia by Haber's process.

Inhibitors: In some cases, the presence of small quantities of impurities in the reacting substances makes the catalyst inactive. Such substances are called *catalytic poisons* or *inhibitors.* Catalytic poisons are usually also poisonous to organisms. Arsenious oxide and hydrogen cyanide are two of the most powerful catalytic poisons. Arsenious oxide poisons platinized asbestos used as catalyst in the manufacture of H_2SO_4 by contact process.

Q. 23. What is homogenous catalysis ? Explain with examples.

(Kurukshetra 2001; Nagpur, 2003; Purvanchal 2007, Calcutta 2007)

Ans. *When in a reaction the catalyst is present in the same phase as the reactants, it is called homogeneous catalysis.*

Examples of homogeneous catalysis are given below :

A. Gaseous phase

(*a*) *Reaction between* SO_2 *and* O_2 *to form* SO_3 *in presence of nitric oxide* (NO)

$$2SO_2(g) \xrightarrow{NO(g)} 2SO_3(g)$$

(*b*) *Decomposition of acetaldehyde in the presence of iodine vapours*

$$CH_3CHO(g) \xrightarrow{I_2(g)} CH_4(g) + CO(g)$$

(*c*) *Decomposition of vapours of* N_2O *at 1000 K in presence of traces of chlorine gas*

$$2N_2O(g) \xrightarrow[1000\ K]{Cl_2(g)} 2N_2(g) + O_2(g)$$

(*d*) *Decomposition of ozone in presence of* NO or N_2O_5

$$2O_3(g) \xrightarrow{NO\ (g)} 3O_2\ (g)$$

B. Liquid phase

(*a*) **Acid-Base catalysis**

(*i*) *Hydrolysis of cane-sugar (sucrose) in the presence of a mineral acid.* (Inversion of cane-sugar).

$$C_{12}H_{22}O_{11}(aq) + H_2O(l) \xrightarrow{H^+(aq)} \underset{\text{Glucose}}{C_6H_{12}O_6(aq)} + \underset{\text{Fructose}}{C_6H_{12}O_6(aq)}$$

(*ii*)*Hydrolysis of methyl acetate by an acid or a base*

$$CH_3COOCH_3(aq) + H_2O(l) \xrightarrow[\text{or } OH^-(aq)]{H^+(aq)} CH_3COOH(aq) + CH_3OH(aq)$$

(*iii*)*Hydrolysis of starch to glucose in presence of an acid*

$$\underset{\text{Starch}}{(C_6H_{10}O_5)_n} + nH_2O \xrightarrow{H+} \underset{\text{Glucose}}{nC_6H_{12}O_6}$$

(*b*) *Decomposition of* H_2O_2 *in the presence of* I^- *ion or* Br_2

$$2H_2O_2(aq) \xrightarrow[\text{or } Br_2]{I^-(aq)} 2H_2O\,(l) + O_2(g)$$

Q. 24.Explain the mechanism of homogeneous catalytic reactions taking some examples.

Ans. A homogeneous catalytic usually takes place in two steps as follows :

1st step: The catalyst reacts with the reactant to form an intermediate complex

$$A + C \longrightarrow AC$$

2nd step: The complex then decomposes to form the product regenerating the catalyst or reacts with a second reactant to give the product and regenerating the catalyst *i.e.*

$$AC \longrightarrow \text{Product} + C$$

or

$$AC + B \longrightarrow \text{Product} + C$$

This is illustrated with examples given below :

(1)*Mechanism of the reaction* $2SO_2 + O_2 \xrightarrow{NO} 2SO_3$

1st step: $2NO + O_2 \longrightarrow 2NO_2$

2nd step: $2NO_2 + 2SO_2 \longrightarrow 2NO + 2SO_3$

Both the above steps are fast. Hence the overall reaction is also fast whereas in the absence of catalyst the reaction is slow.

(2) *Mechanism of the reaction* $CH_3CHO \xrightarrow{I_2} CH_4 + CO$

1st step: $CH_3CHO + I_2 \longrightarrow CH_3I + HI + CO$

2nd step: $CH_3I + HI \longrightarrow CH_4 + I_2$

(3) *Mechanism of the reaction* $2N_2O \xrightarrow{Cl_2} 2N_2 + O_2$

1st step: $Cl_2 \xrightarrow{1000K} 2Cl$

2nd step: $N_2O + Cl \longrightarrow N_2 + ClO$

3rd step: $2ClO \longrightarrow Cl_2 + O_2$

(4) *Mechanism of the reaction* $2O_3 \xrightarrow{N_2O_5} 3O_2$

1st step: $2N_2O_5 \longrightarrow 2N_2O_4 + O_2$

2nd step: $O_3 + N_2O_4 \longrightarrow O_2 + N_2O_5$

(5) *Mechanism of the reaction* $2H_2O_2 \longrightarrow 2H_2O + O_2$

1st step: $H_2O_2 + I^- \longrightarrow H_2O + IO^-$

2nd step: $H_2O_2 + IO^- \longrightarrow H_2O + O_2 + I^-$

Q. 25. Discuss kinetics of homogeneous catalytic reactions taking some examples.

Ans. The rate equation in the presence of catalyst is found out experimentally. This tells the type and number of the different molecules taking part in the rate determining step (slowest step), which then helps to write the probable mechanism of the reaction. This is illustrated with the help of following examples.

(1) *Kinetics of the decomposition of acetaldehyde in the presence of iodine vapours as catalyst.* The decomposition of a acetaldehyde takes place as

$$CH_3CHO(g) \longrightarrow CH_4(g) + CO(g)$$

The observed rate equation for the this reaction is found to be

$$\text{Rate} = \frac{d[CH_3CHO]}{dt} = k_1 [CH_3CHO]^2$$

which suggests the mechanism as follows :

$$CH_3CHO + CH_3CHO \longrightarrow 2CH_4 + 2CO$$

In the presence of iodine vapours (as catalysts), the rate equation is observed to be

$$\text{Rate} = \frac{d[CH_3CHO]}{dt} = k_1[CH_3CHO][I_2]$$

This suggests the probable mechanism of the catalysed reaction as follows :

$$CH_3CHO + I_2 \xrightarrow{\text{slow}} CH_3I + HI + CO \quad \text{...Rate determining step}$$

$$CH_3I + HI \xrightarrow{\text{fast}} CH_4 + I_2$$

(2) *Kinetics of the dehydration of tert-butyl alcohol in the presence of HBr.* The dehydration of tert-butyl alcohol takes place as

$$\underset{\text{tert-butyl alcohol}}{(CH_3)_3C-OH} \longrightarrow \underset{\text{2-Methyl propene}}{(CH_3)_2C=CH_2} + H_2O$$

In the absence of catalyst, the alcohol does not decompose at any appreciable rate upto about 700 K. On adding a small amount of HBr, rapid decomposition occurs at 500 K. The rate equation for the uncatalysed reaction is

$$\text{Rate} = -\frac{d[t\text{-BuOH}]}{dt} = k_1[t-\text{BuOH}]$$

i.e. it is a first order reaction with very low value of k_1.

In the presence of HBr, the rate equation is observed to be

$$\text{Rate} = -\frac{d[t\text{-BuOH}]}{dt} = k_2[t-\text{BuOH}][\text{HBr}]$$

i.e. it is a second order reaction with high value of k_2.

This suggests that a possible mechanism for the catalysed reaction is

$$(CH_3)_3\,C-OH+HBr \xrightarrow{\text{slow}} (CH_3)_3\,C-Br+H_2O \quad \text{...Rate determining step}$$

$$(CH_3)_3\,C-Br \xrightarrow{\text{fast}} (CH_3)_2\,C=CH_2+HBr$$

Q. 26. Derive an expression for the rate of an acid catalysed reaction.

(*Jamia Millia 2002*)

Ans. *In an acid-catalysed reaction;* the first step is the transfer of a proton (H^+ ion) from the acid HA to the substrate S (one of the reactants)

$$S + HA \underset{k_1}{\overset{k_1}{\rightleftharpoons}} SH^+ + A^- \quad \text{... }(i)$$

This is then followed by one of the two possibilities *i.e.* either proton transfer to the solvent *i.e.*

$$SH^+ + H_2O \xrightarrow{k_2} P + H_3O^+ \qquad (P = \text{Product}) \quad \text{...}(ii)$$

or proton transfer to the base A^- *i.e.*

$$SH^+ + A^- \longrightarrow P + HA \quad \text{...}(iii)$$

Taking reaction (*i*) as the first step and reaction (*ii*) as the second step, then taking the latter as the rate determining step, the reaction rate is given by

$$-\frac{d[S]}{dt} = k_2[SH^+] \quad \text{...}(iv)$$

But the reaction of $[SH^+]$ with time is given by

$$\frac{d[SH^+]}{dt} = \text{Rate of formation of } SH^+ \text{ from S and HA}$$

$$- \text{Rate of reaction of } SH^+ \text{ with } H_2O$$

$$= \{k_1[S][HA] - k_1'[SH^+][A^-]\} - \{k_2\,[SH^+]\}$$

But steady state condition implies that the rate of change of concentration of $[SH^+]$ with time should be zero.

i.e. $$\frac{d[SH^+]}{dt} = 0$$

Thus $k_1\,[S]\,[HA] - k_1'\,[SH^+]\,[A^-] - k_2\,[SH^+] = 0$

or $$[SH^+]\,\{k_1'[A^-] + k_2\} = k_1[S]\,[HA]$$

or $$[SH^+] = \frac{k_1[S][HA]}{k_1'[A^-] + k_2} \quad \text{...}(v)$$

Putting this value in equation (*iv*), we get

$$-\frac{d[S]}{dt} = k_1 k_2 \times \frac{[S][HA]}{k_1'[A^-] + k_2} \quad \text{...}(vi)$$

If $k_2 << k_1'\,[A^-]$, equation (*vi*) becomes

$$\frac{d[S]}{dt} = -\frac{k_1 k_2 [S][HA]}{k_1'[A^-]} \quad \text{...}(vii)$$

Q. 27. What is heterogeneous catalysis ? Give examples of heterogeneous catalytic reactions. *(Lucknow 2000; M. Dayanand 2000; Himachal 2000, Awadh 2000; Nagpur 2003; Purvanchal 2007)*

Ans. *If the catalyst is present in a different phase from that of the reactants, in a reaction, it is called heterogeneous catalysis.* The catalysts are usually solids whereas the reactants are gaseous or liquid. The reactions occur at the surface of the catalyst. That is why this type of catalysis is also called *surface catalysis* or *contact catalysis.* Evidently, greater the surface area of the catalyst, greater is the efficiency of the catalyst. For this reason, finely divided solids are usually used as catalysts.

Examples of heterogeneous catalytic reactions

(*i*) *Manufacture of ammonia from nitrogen and hydrogen using finely divided iron as catalyst*

$$N_2(g) + 3H_2(g) \xrightarrow[\text{divided Fe}]{\text{Finely}} 2NH_3(g)$$

(*ii*) *Formation of* SO_3 *from* SO_2 *and* O_2 *in contact process, using finely divided platinum or vanadium pentoxide as catalyst for manufacture of sulphuric acid*

$$2SO_2\ (g) + O_2\ (g) \xrightarrow[\text{or } V_2O_5]{\text{Pt}} 2SO_3\ (g)$$

(*iii*) *Manufacture of methyl alcohol from* CO *and* H_2 *using* ZnO *and* Cr_2O_3 *as catalysts*

$$CO(g) + 2H_2(g) \xrightarrow{ZnO/Cr_2O_3} CH_3OH(g)$$

Q. 28. Explain (*i*) Heterogeneous catalysis is surface catalysis. (*ii*) Heterogeneous catalysts are specific. *(Purvanchal 2002)*

Ans. (*i*) Consider the bromination of ethylene

$$CH_2 = CH_2(g) + Br_2(g) \longrightarrow BrCH_2-CH_2Br(g)$$

This reaction takes place readily in a glass vessel as 470 K. It was thought to be an ordinary homogeneous catalytic reaction. However it was found that the rate is higher in a smaller reaction vessel or in a vessel packed with glass beads but the rate is reduced when the vessel is coated on the inside with paraffin wax. This shows that the reaction is not occurring in the gas phase but is occurring on the glass surface of the vessel. The glass acts as a catalyst.

(*ii*) Consider the decomposition of formic acid. If the acid vapour is passed through a heated glass tube, the reaction is about 50% dehydration and 50% dehydrogenation *i. e.*

(*a*) $HCOOH \longrightarrow H_2O + CO$

(*b*) $HCOOH \longrightarrow H_2 + CO_2$

If the tube is packed with Al_2O_3, only reaction (*i*) occurs but if it is packed with ZnO, only reaction (*ii*) occurs. This illustrates specificity of a heterogeneous catalyst.

Q. 29. Discuss the mechanism of heterogeneous catalytic reactions with the help of some example *(Awadh 2000; M. Dayanand 2002; Purvanchal 2002)*

Ans. It is believed that a heterogeneous catalytic reaction takes place through the following sequence of steps. (See fig. 5.8).

(*i*) Diffusion of the reactant molecules towards the surface of the catalyst.

(*ii*) Adsorption of the reactant molecules on the surface of the solid.

(*iii*) Chemical reaction of the adsorbed molecules with the surface of the catalyst resulting into stretching and weakening of bonds forming an activated complex with the catalyst which then decomposes or even breaking of the bonds of reactant molecules.

Steps (*ii*) and (*iii*) are jointly called *chemisorption.*

(*iv*) Desorption of the products from the surface.

(*v*) Diffusion of the products away from the surface.

The various steps are illustrated with the following examples .

(*a*) *Hydrogenation of ethylene in presence of finely divided Ni, Pt or Pd*

It is illustrated as under

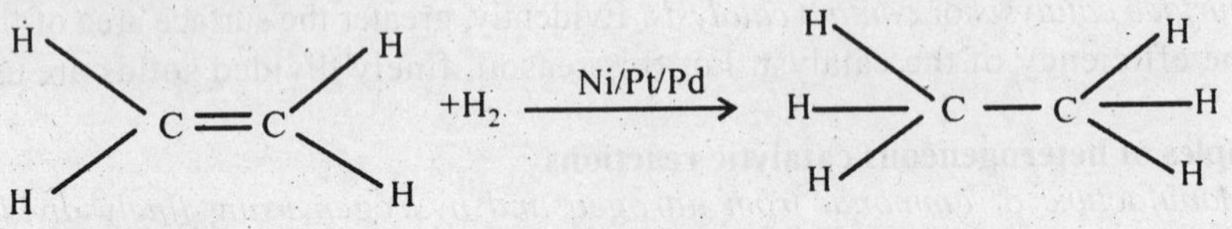

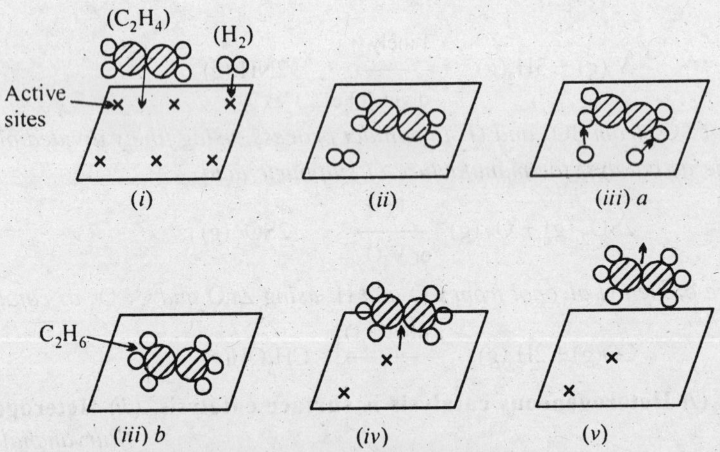

Fig. 5·8. Mechanism for the reaction between ethylene and hydrogen on a catalytic surface.

(*i*) Diffusion of C_2H_4 and H_2 towards surface.

(*ii*) Adsorption of C_2H_4 and H_2 at the active sites.

(*iii*) (*a*) Dissociation of H_2 into H atoms

(*b*) Linking of the H-atom to C_2H_4 to form C_2H_6.

(*iv*) Desorption of C_2H_6 from the surface.

(*v*) Diffusion of C_2H_6 away from the surface.

(*b*) *Decomposition of* HI *in the presence of platinum.*

$$2HI\,(g) \xrightarrow{Pt} H_2(g) + I_2(g)$$

The mechanism of this reaction may be represented as follows :

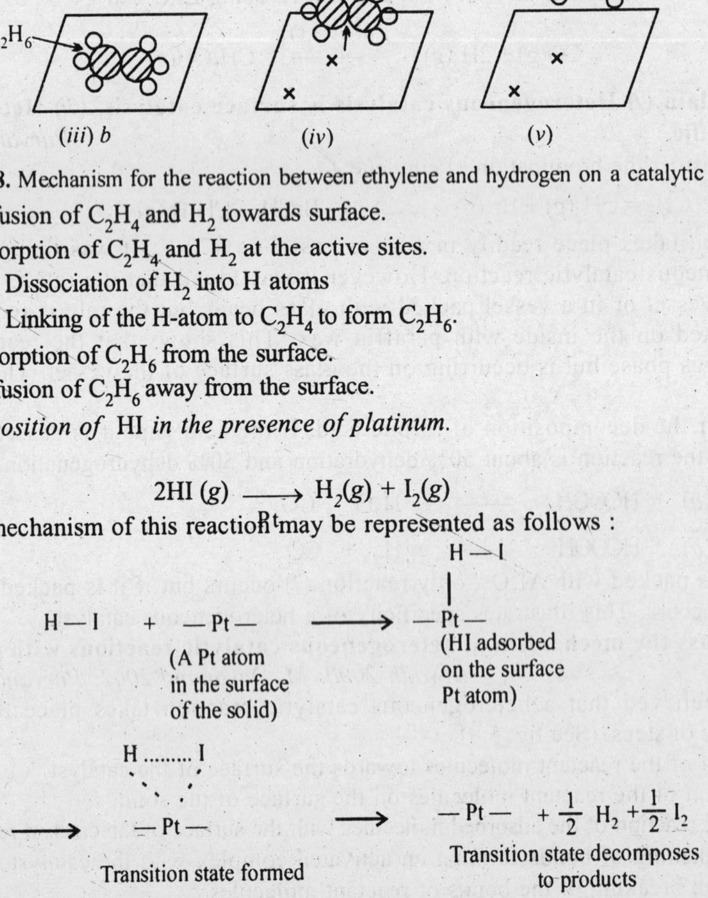

Q. 30. Discuss the kinetics of heterogeneous (surface) catalytic reactions.

(M. Dayanand, 2000)

Ans. A heterogeneous catalytic reaction takes place in a number of consecutive steps. Although any of these steps may be slowest and hence rate determining, yet from the point of view of the study of the kinetics of these reactions, the reaction of the adsorbed molecules is considered to be the rate of determining. Further the rate of reaction is supposed to be proportional to the fraction θ of the surface covered by the reactant molecules.

Thus the two *assumptions* in the study of kinetics of the surface reactions are :

(*i*) The rate determining step of a surface reaction is the reaction of the adsorbed molecules with the surface of the catalyst.

(*ii*)The rate of reactions is proportional to the fraction θ of the surface covered by the reactant molecules.

To understand the kinetics of surface reactions, we consider here the simplest case involving only one *gaseous reactant* decomposing on the surface of a solid catalyst according to the reaction

$$A \longrightarrow \text{Products}$$

Then if θ_A is the fraction of the surface covered by A at any time t when the pressure of the gas is P, then according to assumption (*i*), the rate of reaction would be

$$\text{Rate} = k'\,\theta_A \qquad \ldots(i)$$

where k' is a proportionality constant.

The fraction θ is related to the pressure of the gas according to the Langmuir equation, *viz.*

$$\theta = \frac{bP}{1+bP} \qquad \ldots(ii)$$

where b is a constant.

Substituting this value in equation (*i*), we get

$$\text{Rate} = -\frac{dP}{dt} = k'\frac{bP}{1+bP} \qquad \ldots(iii)$$

Q. 31. Explain the action of catalytic promoters and catalytic poisons (inhibitors).

(M. Dayanand 2000; Kumaon, 2000; Punjabi 2003)

Ans. Promoters *are the substances that enhance the activity of the catalyst.* They are in fact catalysts for the catalysts. They may not act alone as catalyst for the same reaction or may act only as feeble catalysts. A well known example is the manufacture of NH_3 from N_2 and H_2 by Haber's process in which iron is used as catalyst and potassium or vanadium as promoter

$$N_2(g) + 3H_2(g) \xrightarrow[\text{K or V (Promoter)}]{\text{Fe (Catalyst)}} 2NH_3(g)$$

The action of promoters is explained on the basis that the lattice defects or the irregularities on the surface of the catalyst are the active sites for catalysis. The addition of a promoter encourages the formation of lattice defects and the hence the active sites. Consequently, the activity of the catalyst is enhanced.

Inhibitors *are the substances that inhibit the activity of the catalysts.*

A few examples are given below :

(*i*) Oxidation of SO_2 to SO_3 by oxygen in the presence of platinum as catalyst is slowed down if arsenic is present.

(*ii*) Hydrogenation of ethylene in presence of copper is slowed down if CO is present.

The action of inhibitors is explained in different ways as follows :

(*a*)The catalytic poison may have a strong preferential adsorption on the surface of the catalyst. Thus the entire surface of the catalyst may be covered with the catalytic poison and no adsorption of the reactants may take place on the catalytic surface. Example (*ii*) above is a case of this type.

(*b*)The catalytic poison may react chemically with the catalytic surface and form a new surface which may be catalytically inert. For example, the catalytic poison arsenic reacts with platinum to form platinum arsenide at the surface.

The fact that a very small amount of the poison is sufficient to stop the activity of the catalyst completely suggests that the surface of a catalyst is irregular with high catalytic activity only at certain spots on the catalytic surface. As soon as these spots are covered with the poison, the activity of the surfaces reduces to zero.

Q. 32. What is enzyme catalysis ? Describe some characteristics of enzyme catalysis.

(Awadh, 2000; Purvanchal 2002; Nagpur 2003; Punjabi 2003; Calcutta 2007)

Ans. Enzymes are protein molecules present in the living system. Their molecular weights range from 10^4– 10^6 g/mol. They are highly efficient biological catalysts which catalyse the reactions taking place in the living cells and also help in controlling their reaction rates.

The reaction in enzyme catalysis occurs at a specific site on the protein molecule. Thus is called the **active site.** The reactants in an enzyme reaction are referred to **as substrates.**

A few examples of enzyme-catalysed laboratory reactions are given below :

(*i*) $$\text{Starch} \xrightarrow{\text{Diastase}} \text{Maltose} \xrightarrow{\text{Maltase}} \text{Glucose} \xrightarrow{\text{Zymase}} \text{Alcohol} + CO_2$$

(*ii*) $$\text{Cane sugar} \xrightarrow[+H_2O]{\text{Invertase}} \underset{(\text{Glucose} + \text{Fructose})}{\text{Invert sugar}} \xrightarrow{\text{Zymase}} \text{Alcohol} + CO_2$$

The names of the enzymes are written on the arrows.

Enzymes possess certain characteristics which are different from other types of catalysts. A few of these are briefly described below :

(*i*) *Their efficiency is very high i.e.* many of them are so powerful catalysts that they increase the rate of reaction by many orders of magnitude even at quite low concentrations.

(*ii*)*They are highly specific in their catalytic action.* For example, urease even in very low concentrations catalyses the hydrolysis of urea but it has no detectable effect on the hydrolysis rate of substituted ureas such as methyl urea. There are many such examples to quote. However there are some exceptions to this generalization.

(*iii*)*They are highly sensitive to temperature.* The activity of the enzyme is usually found to be maximum at the normal temperature of the organism in which the enzyme is found. Above this temperature, usually the activity is first found to increase temporarily and then drops to zero when the protein is denatured.

(*iv*)*They are highly sensitive to the acidity or basicity of the solution.* It is found that the enzymes show a sharp change in activity if the *p*H of the solution is changed.

Q. 33. Explain the mechanism of enzyme reactions giving Michaelis Menten equation.

(Kurukshetra 2002; Guwahati 2002; Arunachal 2003; Meerut 2004; G. Nanakdev 2004)

Ans. The mechanism of an enzymes reaction as proposed by Michaelis and Menten is as follows :

The first step in the enzyme reaction is the reaction between the enzyme E (at the active site) and the substrate S (*i.e.* reactants of the enzyme reaction) to form the complex ES.

$$E + S \underset{k_1'}{\overset{k_1}{\rightleftharpoons}} ES \qquad ...(i)$$

This reaction is very fast and the equilibrium is rapidly established.

The complex then dissociates forming one or more products P and the enzyme is regenerated

$$ES \underset{k_2'}{\overset{k_2}{\rightleftharpoons}} E+P \qquad ...(ii)$$

This reaction usually goes to completion, so the rate coefficient k_2' is effectively zero.

The steps (*i*) and (*ii*) can be written together as

$$E+S \underset{k_1'}{\overset{k_1}{\rightleftharpoons}} ES \underset{k_2'}{\overset{k_2}{\rightleftharpoons}} E+P \qquad ...(iii)$$

Thus the enzyme-substrate complex ES is an intermediate and the net reaction is S → P. As a result, the rate of appearance of the product must be equal to the rate of disappearance of the substrate S, *i.e.*

$$\frac{d[P]}{dt} = -\frac{d[S]}{dt} \qquad ...(iv)$$

This implies that the concentration of the complex ES, represented by [ES] must be in a *steady state.* In other words, the rate of change of concentration of ES with time must be equal to zero. But rate of change of concentration of ES with time will be given by

$$\frac{d[ES]}{dt} = \text{Rate of formation of ES from E and S}$$

$$- \text{Rate of decomposition of ES into E and P}$$

$$= \{k_1[E][S] - k_1'[ES]\} - \{k_2[ES]\} \qquad ...(v)$$

For steady state condition, $d[ES]/dt = 0$. Hence

$$k_1[E][S] - k_1'[ES] - k_2[ES] = 0$$

or

$$[ES]\{k_1' + k_2\} = k_1[E][S]$$

or

$$[ES] = \frac{k_1[E][S]}{k_1' + k_2} \qquad ...(vi)$$

This expression gives the steady state concentration [ES].

Putting $\frac{k_1' + k_2}{k_1} = K_m$, called *Michaelis constant,* expression (*vi*) can be written as

$$[ES] = \frac{[E][S]}{K_m} \qquad ...(vii)$$

Since [ES] is a steady state, we must have

Rate of disappearance of S = Rate of disappearance of ES.

i.e

$$-\frac{d[S]}{dt} = k_2[ES] \qquad ...(viii)$$

But rate of disappearance of S means the rate of the enzyme catalysed reaction, usually represented by v. Hence

$$v = -\frac{d[S]}{dt} = k_2[\text{ES}] \quad ...(ix)$$

Substituting the value of [ES] from equation (*vii*) in equation (*ix*), we get

$$v = -\frac{d[S]}{dt} = k_2[\text{ES}] = k_2\frac{[\text{E}][\text{S}]}{K_m} \quad ...(x)$$

The equation is, however, not useful in this form since it includes the concentration of free enzyme [E] whereas the experimentally measurable quantity is $[E_0]$, the total concentration of the enzyme *i.e.* free as well as combined in the complex ES. But we have

$$[\text{E}_0] = [\text{E}] + [\text{ES}]$$

so that

$$[\text{E}] = [\text{E}_0] - [\text{ES}]$$

Substituting this value in equation (*vii*), we get

$$[\text{ES}] = \frac{\{[\text{E}_0] - [\text{ES}]\}[\text{S}]}{K_m}$$

or

$$K_m[\text{ES}] = [\text{E}_0][\text{S}] - [\text{ES}][\text{S}]$$

or

$$K_m[\text{ES}] + [\text{ES}][\text{S}] = [\text{E}_0][\text{S}]$$

or

$$[\text{ES}]\{K_m + [\text{S}]\} = [\text{E}_0][\text{S}]$$

or

$$[\text{ES}] = \frac{\{[\text{E}_0][\text{S}]}{K_m + [\text{S}]} \quad ...(xi)$$

Substituting this value in equation (*ix*), we get

$$\boxed{v = -\frac{d[S]}{dt} = k_2[\text{ES}] = \frac{k_2[\text{E}_0][\text{S}]}{K_m + [\text{S}]}} \quad ...(xii)$$

This equation is known as *Michaelis-Menten equation* which contains the measurable quantities, $[E^0]$ and [S].

Equation (*xii*) is usually written in terms of maximum velocity v_m of the enzyme-catalysed reaction which is reached when [S] becomes so large that $[S] >> K_m$ so that K_m can be neglected in comparison to [S]. Then equation (*xii*) takes the form

$$v_m = k_2[E_0] \quad ...(xiii)$$

Hence

$$\frac{v}{v_m} = \frac{[\text{S}]}{K_m + [\text{S}]} \quad ...(xiv)$$

From this equation, we find that

$$\text{when} \quad [\text{S}] = K_m, \frac{v}{\upsilon_m} = \frac{1}{2}$$

i.e.

$$v = \frac{1}{2}v_m.$$

Thus *Michaelis constant is equal to that concentration of the substrate at which the rate of the enzyme-catalysed reaction falls to one-half of its maximum value.* Thus the values of v_m and K_m can be obtained by plotting rate of reaction, v versus the concentration of the substrate S.

MISCELLANEOUS QUESTIONS

Q. 34. What is autocatalysis ? *(Vidyasagar 2007, Calcutta 2007)*

Ans. When one of the products of a reaction acts as a catalyst for that reaction, the phenomenon is known as autocatalysis.

Examples of Autocatalysis

1. Oxidation of oxalic acid. When oxalic acid is oxidised by acidified potassium permanganate, mangamous sulphate or Mn^{2+} ions) produced during the reaction acts as a catalyst for the reaction.

$$2\ KMnO_4 + 5\ H_2C_2O_4 + 3H_2SO_4 \longrightarrow 2MnSO_4 + K_2SO_4 + 8\ H_2O + 10CO_2$$

2. Hydrolysis of an ester. Hydrolysis of ethyl acetate forms acetic acid and ethanol. Acetic acid acts as a catalyst for the reaction

$$CH_3COOC_2H_5 + H_2O \xrightarrow{H+} \underset{\textbf{Catalyst}}{CH_3COOH} + C_2H_5OH$$

Whenever autocatalysis occurs the initial rate of reaction rises as the catalytic product is formed unlike in a normal reaction where the rate of the reaction decreases with the passage of time. The curve shown in Fig. 5.9 shows a maximum when the reaction is complete.

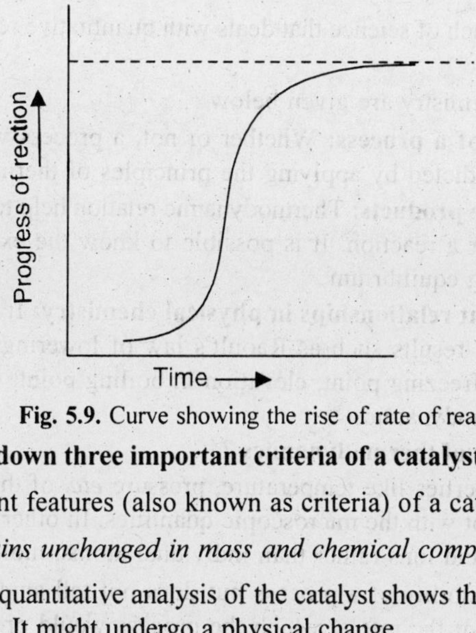

Fig. 5.9. Curve showing the rise of rate of reaction with time.

Q. 35. Write down three important criteria of a catalyst *(Calcutta 2007)*

Ans. Important features (also known as criteria) of a catalyst are given below.

1. *A catalyst remains unchanged in mass and chemical composition at the end of the reaction.*

   Qualitative and quantitative analysis of the catalyst shows that it undergoes no change in mass or chemical nature. It might undergo a physical change.

2. *A small quantity of catalyst is generally sufficient to complete a reaction.*

   It has been observed that in most of the reactions, only a very small amount of the catalyst can produce huge quantities of the products. Only in exceptional cases, like in Friedel Crafts reaction, we require large quantities of the catalyst to obtain good yield.

3. A catalyst is more effective when finely divided powdered than in the coarse state. Finely divided nickel is a better catalyst than lumps of solid nickel.

4. *A catalyst is specific in action*

   A catalyst which works for a reaction need not work for another reaction. The same reactant may yield different products with different catalysts. For example,

$$C_2H_5OH \xrightarrow{Al_2O_3} \underset{\text{Ethene}}{C_2H_4} + H_2O \qquad C_2H_5OH \xrightarrow{Cu} \underset{\text{Ethanal}}{CH_3\ CHO} + H_2$$

6

THERMODYNAMICS-I
(INCLUDING THERMOCHEMISTRY)

Q. 1. What is thermodynamics ? Give important objectives of thermodynamics in chemistry. *(Bhopal, 2004; Gulberga 2004)*

Ans. Thermodynamics is a branch of science that deals with quantitative relationships between heat and other forms of energy.

Its objectives as applied to chemistry are given below :

(*i*) **To predict the feasibility of a process:** Whether or not, a process will occur under the given set of conditions, can be predicted by applying the principles of thermodynamics.

(*ii*) **To estimate the yield of the products:** Thermodynamic relation help to predict the yield of products obtainable in a process or a reaction. It is possible to know the extent upto which the reaction takes place before attaining equilibrium.

(*iii*) **To deduce some important relationships in physical chemistry:** It has been possible to deduce some important and useful results such as Raoult's law of lowering of vapour pressure and expressions for depression in freezing point, elevation in boiling point, distribution law and phase rule from the study of thermodynamics.

Q. 2. What are the limitations of thermodynamics ? *(Gulberga 2004)*

Ans. (*i*) It deals with the properties like temperature, pressure *etc.* of the matter in bulk *i.e.* with macroscopic quantities and not with the microscopic quantities. In other words, it deals with large groups of atoms, molecules and ions rather than individual atoms, molecules or ions.

(*ii*) It helps to predict the feasibility of a process but does not tell anything about the time taken for the process to complete or the rate at which the process would proceed.

(*iii*) It concerns itself only with the initial and final states of a system and is not concerned with the path by which the change is brought about *i.e.* it throws no light on the mechanism of a process.

Q. 3. Explain the following terms : (*i*) System and surroundings (*ii*) Open system, closed system and isolated system. *(M. Dayanand, 2002; Venkateswara 2004; Guwahati 2007)*

Ans. (*i*) **System and surroundings:** *The part of the universe chosen for thermodynamic consideration (i.e. to study the effect of temperature, pressure etc.) is called a system.*

The remaining portion of the universe, excluding the system, is called *surroundings.*

(*ii*) **Open, closed and isolated systems**

(*a*) **Open system:** *A system is said to be an open system if it can exchange both matter and energy with the surroundings.* For example, if some water is kept in an open vessel (Fig. 6.1), exchange of both matter and energy takes place between the system and the surroundings.

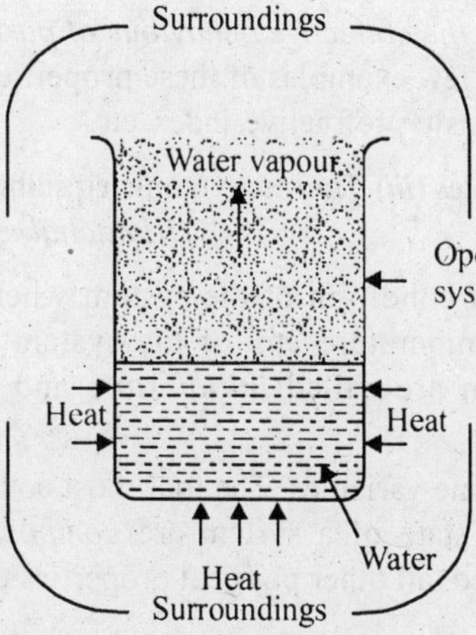

Fig. 6.1. Example of open system: Water kept in an open vessel.

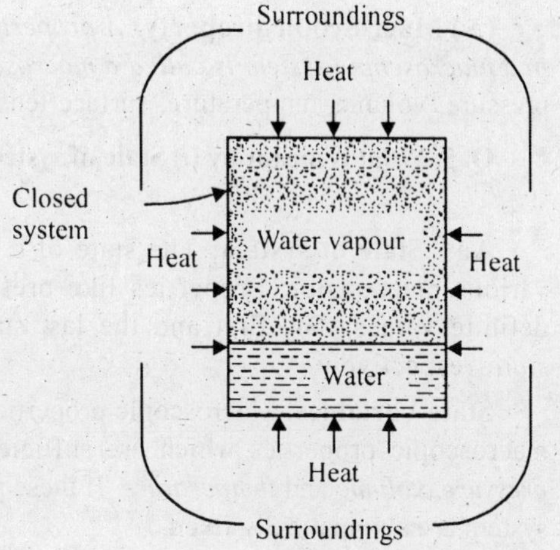

Fig. 6.2. Water placed in an closed vessel.

(*b*) **Closed system:** *If a system can exchange only energy with the surroundings but not matter, it is called a closed system.* For example, if some water is placed in a closed metallic vessel (Fig. 6.2), then as the vessel is closed, no exchange of matter between the system and the surroundings can take place. However, as the vessel has conducting walls, exchange of energy can take place between the system and the surroundings.

(*c*) **Isolated system:** *If a system can exchange neither matter not energy with the surroundings, it is called an isolated system.* For example, if water is placed in a vessel which is closed as well as insulated, no exchange of matter or energy can take place between the system and the surroundings. This constitutes an isolated Fig. 6.3.

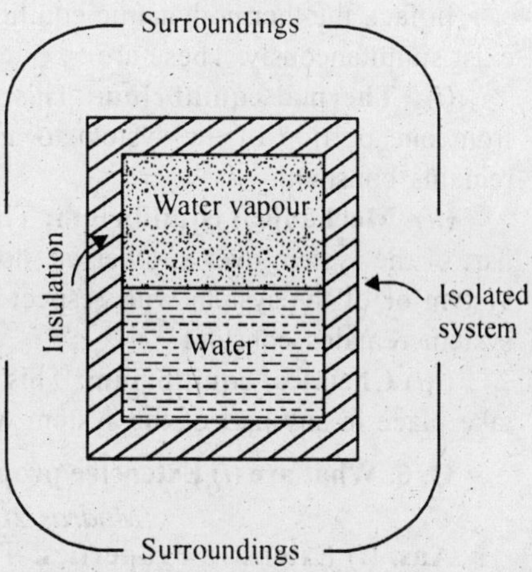

Fig. 6.3. An example of an isolated system.

Q. 4. Explain the following terms

(*i*) Homogeneous system (*Himachal 2000*)

(*ii*) Heterogeneous system (*Himachal 2000*)

(*iii*) Macroscopic system

(*iv*) Macroscopic properties.

Ans. (*i*) **Homogeneous system:** *A system is said to be homogeneous if it is uniform throughout.* For example, a gas or a mixture of gases or a pure liquid or a pure solid or a solution of a solid in a liquid are all homogeneous systems. Such a system is said to have one phase only.

(*ii*) **Heterogeneous system:** *If a system is not uniform throughout, it is said to be heterogeneous.* It then consists of two or more phases which are separated from each other by definite bounding surfaces. A few common examples of the heterogeneous systems are (*a*) a system consisting of a liquid and its vapour, (*b*) two or more immiscible liquids, (*c*) a mixture of two or more solids.

(*iii*) **Macroscopic system:** *Macroscopic system means a system containing a large amount of the substance.* In other words, a macroscopic system is one in which there are large number of particles *i.e.* atoms, ions or molecules.

(*iv*) **Macroscopic property:** *A property associated with the collective behaviour of particles in a macroscopic system is called a macroscopic property.* A few examples of these properties are pressure, volume, temperature, surface tension, viscosity, density, refractive index etc.

Q. 5. What is meant by (*i*) State of system (*ii*) State variables (*iii*) Thermodynamic equilibrium. *(Himachal 2000)*

Ans. State of system: The state of a system implies a condition of the system when the various macroscopic properties like pressure, volume, temperature etc. of the system have definite values. The first and the last state of the system are called *initial state* and *final state* respectively.

State Variables: Macroscopic properties are known as state variables. The four most common macroscopic properties which are sufficient to define the state of a system are *composition, pressure, volume* and *temperature.* If these properties are fixed, all other physical properties of the system are automatically fixed.

Thermodynamic equilibrium: A system is said to be in thermodynamic equilibrium if the macroscopic properties of the system in various phases do not undergo any change with time.

In fact, the thermodynamic equilibrium implies three different types of equilibria which must exist simultaneously. These are :

(*i*) **Thermal equilibrium:** This equilibrium demands that there should be no flow of *heat* from one portion of the system to another. This can happen if the temperature of the system remains constant.

(*ii*) **Mechanical equilibrium:** This equilibrium implies that no work should be done by one part of the system over another *i.e.* no macroscopic movement of matter should occur within the system or of the system with respect to its surroundings. This is possible if the pressure of the system remains constant.

(*iii*) **Chemical equilibrium:** This equilibrium demands that no change in composition should take place in any part of the system with passage of time.

Q. 6. What are (*i*) Extensive properties, (*ii*) Intensive properties? *(Nagpur 2008)*
(Madras 2004; Venkateswara 2004; Guwahati, 2006, Himachal 2010)

Ans. (*i*) **Extensive Properties:** These are the properties which depend upon the quantity of the matter contained in the system. Examples of extensive the properties are *mass, volume, energy, heat capacity* etc. The total value of an extensive property is equal to the sum of the values for the separate parts into which the system may be divided for the sake of convenience.

(*ii*) **Intensive properties:** These are those properties which depend only upon the nature of the substance and are independent of the amount of the substance present in the system. The common examples of these properties are *temperature, pressure, refractive index, viscosity, density, surface tension, specific heat* etc. It is because pressure and temperature are intensive properties independent of the quantity of the matter present in the system that they are frequently used as variables to describe the state of a system.

It is of interest to note that an extensive property may become intensive property by specifying unit amount of the substance concerned. Thus mass and volume are extensive properties but density is intensive property, heat capacity is an extensive property but specific heat is intensive.

Q. 7. Explain the following terms

(*i*) Isothermal process *(Bangalore 2002; Nagpur 2002; Guwahati 2007)*

(*ii*) Adiabatic process *(Bangalore 2002; Guwahati 2007; Nagpur 2008)*

(*iii*) Isochoric process *(M. Dayanand 2002; Punjab 2003; Nagpur 2008)*

(*iv*) Isobaric process.

Ans. Isothermal process: *When a process is carried out in such a manner that the temperature remains constant throughout the process, it is called an isothermal process.* Obviously, when such a process occurs, heat can flow from the system to the surroundings and vice versa in order to keep the temperature of the system constant.

(*ii*) **Adiabatic process:** *When a process is carried out in such a manner that no heat can flow from the system to the surroundings* or *vice versa, i.e., the system is completely insulated from the surroundings, it is called an adiabatic process.*

(*iii*) **Isochoric process:** *It is a process during which the volume of the system is kept constant.*

(*iv*) **Isobaric process:** *It is a process during which the pressure of the system is kept constant.*

Q. 8. What are reversible and irreversible processes?

(Lucknow 2000; Punjab 2003; Calcutta 2007; Bharathiar 2011)

Ans. A reversible process *is a process which is carried out infinitesimally slowly so that all changes occurring in the direct process can be exactly reversed and the system remains almost in a state of equilibrium at all times. In other words, a reversible process may be defined as a process which is conducted in such a manner that at every stage, driving force is only infinitesimally greater than the opposing force and which can be reversed by increasing the opposing force by an infinitesimal amount.*

Strictly speaking, a reversible process is almost impossible as it would require an infinite time for completion. A process can be made very nearly reversible in some cases. For example, if an opposing E.M.F. is applied to a cell, then no current flows if the opposing E.M.F is exactly equal to that of the cell. If the opposing E.M.F. is very slightly (infinitesimally) less than that of the cell, a very small current is given out by the cell. If the opposing E.M.F. is slightly greater than that of the cell, a very small current flows in the opposite direction.

A process which does not meet the above requirements is called an **irreversible process.** In other words, an irreversible process is defined as that process which is not carried out infinitesimally slow (instead, it is carried out rapidly) so that the successive steps of the direct process cannot be retraced and any change in the external conditions disturbs the equilibrium.

It is interesting to note that all naturally occurring processes like the flow of water down a hill or the flow of heat from a hot end to a cold end of an iron bar etc. are thermodynamically irreversible.

Q. 9. Explain *(Kumaon, 2000; Lucknow 2002)*

(i) Internal energy

(ii) Change in internal energy

(iii) Sign conventions in energy.

Ans. (*i*) **Internal energy:** The evolution or absorption of energy in different processes clearly shows that every substance must be associated with some definite amount of energy, the actual value of which depends upon the nature of the substances *(i. e.* arrangement of atoms and electrons within the molecules) and the conditions of temperature, pressure, volume and composition. The energy associated with a substance is called its *internal energy* and is usually denoted by the symbol E or U.

(*ii*) **Change of internal energy:** It is not possible to find the absolute value of internal energy of a substance because it involves certain quantities (like translational, vibrational and rotational kinetic energies) which cannot be measured.

However, it is not required to know the absolute value of internal energy possessed by any substance (or a system). What is required in different processes is simply the *change of internal energy* when the reactants change into products or when system changes from initial state to the final state. This is easily measurable and is represented by ΔE.

Thus if the internal energy of a system in the initial state is E_1 and in the final state, it is E_2 then the change of internal energy (ΔE or ΔU) may be given by

$$\Delta E \text{ (or } \Delta U) = E_2 - E_1$$

Similarly, in a chemical reaction if E_R is the internal energy of the reactants and E_p is the internal energy of the products, then energy change accompanying the process would be

$$\Delta E = E_P - E_R$$

In this book, we shall use the notation E for internal energy and ΔE for change in internal energy.

(*iii*) **Sign conventions in energy:** Obviously, if $E_1 > E_2$ (or $E_R > E_P$), the extra energy possessed by the system in the initial state (or the reactants) would be given out, and ΔE will be negative according to the above equations. Similarly if $E_1 < E_2$ (or $E_R < E_P$), energy will be absorbed in the process and ΔE will be positive.

Hence ΔE *is negative* if energy *is evolved*

and ΔE *is positive* if energy *is absorbed*

In this book, we shall use the notation E for internal energy and ΔE *for internal energy change*

Q. 10. Derive an expression for work of expansion against constant pressure.

Ans. The expression for such a work may be derived as follows :

Consider a gas enclosed in a cylinder fitted with a frictionless piston. (Fig. 6.4).

Area of cross-section of the cylinder = a sq cm

Pressure on the piston (which is slightly less than internal pressure of the gas so that the gas can expand) = P.

Distance through which gas expands = dl cm

As pressure is force per unit area, force (f) acting on the piston will be

$$f = P \times a$$

$\therefore$ Work done by the gas (*i.e.* the system)

$$= \text{Force} \times \text{Distance} = f \times dl$$

$$= P \times a \times dl$$

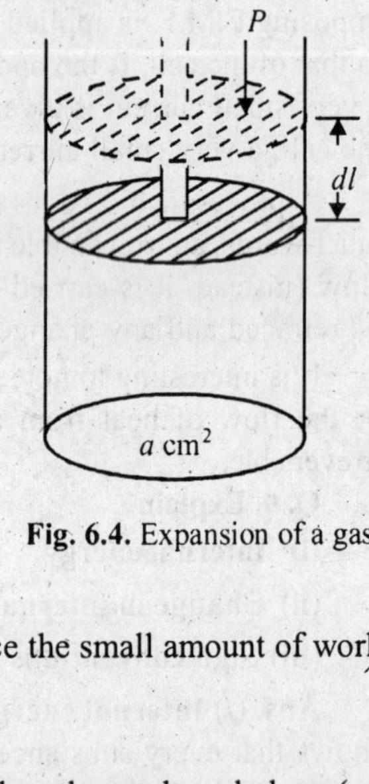

Fig. 6.4. Expansion of a gas.

But $a \times dl = dV$ is a small increase in the volume of the gas. Hence the small amount of work (δw) done by the gas can be written as

$$\delta w = P\,dV$$

If the gas expands from initial volume V_1 to the final volume V_2, then the total work done (w) will be given by

$$w = \int_{V_2}^{V_1} P\,dV$$

If the external pressure P against which the gas expands remains almost constant throughout the process, the above result may be written as

$$w = P\int_{V_2}^{V_1} dV = P(V_2 - V_1)$$

$$= P.\Delta V$$

where $\Delta V = (V_2 - V_1)$ is the total change in the volume of the gas.

If the external pressure (P) is slightly more than the pressure of the gas, the gas will contract *i.e.* the work will be done by the surroundings on the system. However, the same formulae will apply for the work done.

It may be mentioned here that P is the *external* pressure and hence is sometimes written as P_{ext} so that

$$w = P_{ext} \times \Delta V.$$

Note: According to the latest SI conventions of sign, w is taken as positive if work is done on the system (*i.e.* $w_{contraction}$ is positive) and it is taken as negative if work is done by the system (*i.e.* $w_{expansion}$ is negative). Thus $w_{exp} = -P\Delta V$ and $w_{contraction} = +P\Delta V$.

Q. 11. Prove that the work done in a reversible process is the maximum work obtainable.

(G. Nanakdev 2000, Kurukshetra 2001)

Ans. It can be seen that the work done by a system is the maximum work if the process is carried out under reversible conditions. This may be explained as follows ·

We know that work done by the system = $P_{ext} \times \Delta V$. Thus for a given increase in volume (ΔV), the work done can be maximum if P_{ext} is maximum. But in order that the expansion may take place (*i.e.* the work may be done by the system), the external pressure should always be less than the internal pressure of the gas. Thus the maximum value that P_{ext} can have is that it should be *infinitesimally smaller* than the internal pressure of the gas. But these are the conditions of a reversible process. Hence the maximum work is done in a process only if it is carried out under reversible conditions. Hence in reversible processes, w is sometimes replaced by w_{max}.

Q. 12. Define heat. Discuss sign conventions and units of heat.

(Bangalore 2005; Guwahati 2007)

Ans. *Heat is a mode of energy exchanged between the system and the surroundings as a result of the difference of temperature between them.* It is usually represented by the letter q.

Sign of 'q': When heat is given by the system to the surroundings, it is **given a negative** sign. When heat is **absorbed by the system** from the surroundings, it is given a **positive** sign.

Units of 'q'. Heat is usually measured in terms of 'calories'. *A calorie is defined as the quantity of heat required to raise the temperature of one gram of water through 1°C.*

In the S.I. units, heat is expressed in terms of **joules**. The two types of units are related to each other as under:

$$1 \text{ calorie} = 4.184 \text{ joules}$$

Q. 13. State zeroth law of thermodynamics.

(Guwahati 2006; Bharathiar 2011)

Ans. The zeroth law of thermodynamics states that

When two bodies have equality of temperature with a third body, they in turn have equality of temperature with each other.

We are so familiar with this fact that it does not need any further experiments for its support. However, since this fact is not derivable from other laws and since it logically precedes the first and the second law of thermodynamics, it has been called as the zeroth law of thermodynamics.

Q. 14. Give various statements of first law of thermodynamics.

(Nagpur 2002; Calcutta 2007; Bharathiar 2011)

Ans. (*i*) Energy can neither be created nor destroyed, although it may be converted from one form to another.

(*ii*) The total energy of an isolated system remains constant, although it may undergo transformation from one form to the other.

(*iii*) Whenever certain quantity of some form of energy disappears, an exactly equivalent amount of some other form of energy must be produced.

(*iv*) It is impossible to construct a perpetual motion machine *i.e.* a machine which would produce work continuously without consuming energy.

Einstein in 1905 showed that the energy could be created by the destruction of mass. The energy produced (E) by the destruction of mass m is given by the equation

$$E = mc^2 \text{ (called Einstein equation)}$$

where c is the velocity of light. In the light of this idea, the law of conservation of energy has been modified and now it is called *law of conservation of mass and energy.* It may be stated as follows.:

The total mass the energy of an isolated system remains constant in chemical thermodynamics.

Q. 15. Give the mathematical formulation of the first law of thermodynamics.

(Bangalore 2002; Guwahati 2006; Calcutta 2007; Purvanchal 2007; Raipur 2010; Garhwal 2010)

Ans. If heat q is supplied to a system, it may be used up partly to increase the internal energy of the system and partly to produce some mechanical work. If ΔE is the increase in internal energy of the system and w is the work done by the system, then we have

$$q = \Delta E + w \quad \text{... }(i)$$

This equation is the mathematical formulation of the first law of thermodynamics. It may be written in a number of other forms as follows

$$\Delta E = q - w \quad \text{...}(ii)$$

If the changes are very small, it may be written as

$$dE = \delta q - \delta w \quad \text{...}(iii)$$

If the work done is the work of the expansion only and the volume increases by a small amount dV against the external pressure P then we know that $\delta w = P.dV$. Hence the above equation may be written as

$$dE = \delta q - PdV \quad \text{...}(iv)$$

Note: According to the latest SI conventions, as w is negative if work is done by the system, we write eq. (*i*) as

$$q = \Delta E - w$$

or

$$\Delta E = q + w$$

Q. 16. Prove that whereas internal energy is a state function, work and heat are not.

(Purvanchal 2007; Bharathiar 2011)

Ans. Internal energy: The internal energy is a state function *i.e.* when a process occurs, the change in its value depends only upon its value in the initial state and that in the final state and is independent of the path by which the change is brought about. For example, consider a system having a particular pressure, volume and temperature represented by the point A in Fig. 6.5. Suppose the pressure, volume and temperature are changed in such a way that the system is brought to the point B by path I. Now if the system is returned to the point A by path II, then the energy change involved in path I must be equal in magnitude to that involved in path II. If the energy changes involved in paths I and II are not the same, then suppose that the increase in internal energy by path I is greater than the decrease in internal energy along path II. Thus by carrying out the process $A \rightarrow B$ by path I and the reverse process $B \rightarrow A$ by path II, though the original

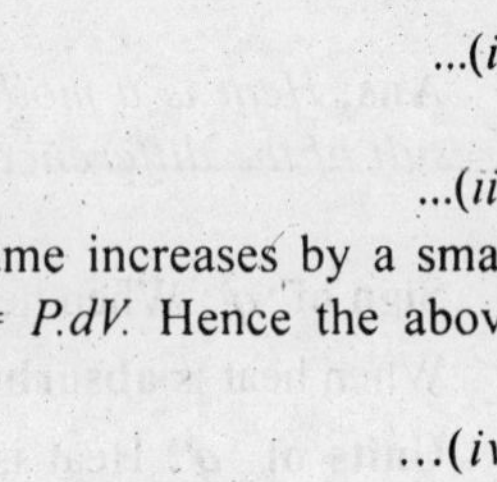

Fig. 6.5. Energy changes in direct and reverse paths.

state has been restored, some energy would have been created. This is contrary to the first law of thermodynamics. Thus it may be concluded that the energy change involved in the process $A \rightarrow B$ must be equal in magnitude to that involved in the reverse process $B \rightarrow A$ by path II. In other words, *the energy change accompanying a process is independent of the path followed and depends only upon the initial and the final states.* Thus if E_A is the internal energy of the system in the initial state and E_B that in the final state, then the change in internal energy accompanying the process would be given by

$$\Delta E = E_B - E_A$$

Work and heat: Unlike internal energy, the work done during a process does not depend merely upon the initial and the final states of the system but it depends upon the path followed. Hence work is not a state function but is a path function. This is explained as under :

Suppose we wish to change the system from state A to state B. This may be done by following different paths as shown in the different figures below (Fig. 6.6). Confining ourselves to the work of expansion only, we know that the work done for a small change in volume dV is PdV where P is the external pressure. Thus the total work done is the sum of $P\,dV$ terms and this is equivalent to the area under the curve in the P–V diagrams. From the figures given below it is obvious that the areas under the curves are different and hence the work done is different when different paths are followed.

That heat is also not a state function follows directly from the mathematical formulation of the first law of thermodynamics *viz.*,

$$q = \Delta E + w$$

Thus the heat q is the sum of two quantities *i.e.* the internal energy change, ΔE and the work

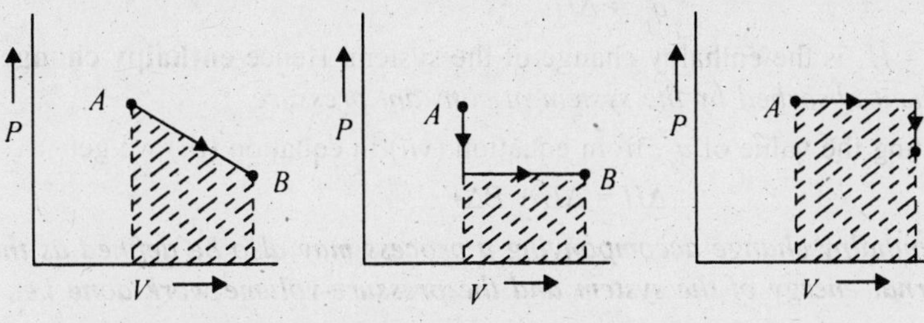

Fig. 6.6. Different paths followed in going from state A to state B.

done, w. Now ΔE is a state function whereas w is not. Now since w depends upon the path followed, therefore, q will also depend upon the path followed. In other words, q is also not a state function but is a path function.

Q. 17. Explain the term enthalpy or heat content.

(Lucknow 2002; Bangalore 2004; Kakatiya 2004; Guwahati 2006)

OR

Deduce a relation between ΔH and ΔE.

(Calcutta 2000; Kurukshetra 2001; G Nanakdev 2002; Guwahati 2007)

Ans. If a process is carried out at constant pressure, the work of expansion is given by

$$w = P\,\Delta V \qquad \ldots (i)$$

where ΔV is the increase in volume and P is the constant pressure.

According to first law of thermodynamics, we know that

$$q = \Delta E + w \qquad \ldots (ii)$$

where q is the heat absorbed by the system, ΔE is the increase in internal energy of the system and w is the work done by the system.

Under condition of constant pressure, putting $w = P\Delta V$ and representing the heat absorbed by q_p, we get

$$q_p = \Delta E + P\Delta V \quad ...(iii)$$

Suppose when the system absorbs q_p calories of heat, its internal energy increases from E_1 to E_2 and the volume increases from V_1 to V_2. Then we have

$$\Delta E = E_2 - E_1 \quad ...(iv)$$

and

$$\Delta V = V_2 - V_1 \quad ...(v)$$

Putting these values in equation (*iii*) above, we get

$$q_p = (E_2 - E_1) + P(V_2 - V_1)$$

$$q_p = (E_2 + PV_2) - (E_1 + PV_1) \quad ...(vi)$$

Now as *E*, *P* and *V* are the functions of state, therefore, the quantity $E + PV$ must also be a state function. The thermodynamic quantity $E + PV$ is called the **heat content or enthalpy** of the system and is represented by the symbol *H i.e.* the enthalpy may be defined mathematically by the equation

$$H = E + PV$$

Thus if H_2 *is* the enthalpy of the system in the final state and H_1 is the value in the initial state, then

$$H_2 = E_2 + PV_2$$

and

$$H_1 = E_1 + PV_1$$

Putting these values in equation (*vi*), we get

$$q_p = H_2 - H_1$$

or

$$q_p = \Delta H \quad ...(vii)$$

where $\Delta H = H_2 - H_1$ is the enthalpy change of the system. Hence **enthalpy change** *of a system is equal to the heat absorbed by the system at constant pressure.*

Further, putting the value of q_p, from equation (*vii*) in equation (*iii*) we get

$$\Delta H = \Delta E + P\Delta V \quad ...(viii)$$

Hence ***the enthalpy change accompanying a process may also be defined as the sum of the increase in internal energy of the system and the pressure-volume work done i.e., the work of expansion.***

$P\Delta V$ can be replaced by ΔnRT, so that eq. (*viii*) can be written as

$$\Delta H = \Delta E + \Delta nRT$$

Q. 18. The heat of formation of methane at 27°C is –19.3 kcal when the measurements are made at constant pressure. What will be the heat of formation at constant volume?

(Burdwan 2004; Delhi 2007)

Ans. Apply the relationship

$$\Delta H = \Delta E + \Delta n \, . \, RT \quad ...(i)$$

ΔH is the heat of reaction at constant pressure whereas ΔE is the heat of reaction at constant volume.

$$\Delta H = -19.3 \text{ kcal}$$

For the reaction

$$C(s) + 2H_2(g) \longrightarrow CH_4(g)$$

No. of moles of the gaseous reactants = 2

No. of moles of the gaseous products = 1

$$\therefore \quad \Delta n = 1-2 = -1$$

$$R = 1.987 \text{ Cal} = 1.987 \times 10^{-3} \text{ kcal}$$

$$T = 27 + 273 = 300$$

Substitute the values in eq. (i)

$$= \Delta E + (-1) \times 1.987 \times 10^{-3} \times 300$$

$$-19.3 = \Delta E - 0.596$$

$$\Delta E = -18.704 \text{ kcal}$$

Q. 19. What is the physical significance of enthalpy ?

Ans. Every substance or system has some definite energy stored in it, called the internal energy. This energy may be of many kinds. The energy stored within the substance or the system that is available for conversion into heat is called the **heat content or enthalpy** of the substance or the system. Absolute value of the heat content or enthalpy of a substance or a system cannot be measured and fortunately this is not required also. In the thermodynamic processes, we are concerned only with the changes in enthalpy (ΔH) which can be easily measured experimentally.

Q. 20. Explain the terms

(*i*) Heat capacity *(G. Nanakdev 2003; Delhi 2003; Madras 2004; Vidyasagar 2007)*

(*ii*) Heat capacity at constant volume *(Awadh 2000, Himachal 2000; M. Dayanand 2002)*

(*iii*) Heat capacity at constant pressure. *(Himachal 2000; Kurukshetra 2001; Arunachal 2003; Delhi 2003)*

Ans. (*i*) **Heat capacity:** *The heat capacity of a system is defined as the amount of heat required to raise the temperature of the system through* 1°C. Thus if q is the amount of heat supplied to a system and as a result the temperature of the system rises from T_1 to T_2, then the heat capacity of the system is given by

$$C = \frac{q}{T_2 - T_1} = \frac{q}{\Delta T} \quad \text{... (i)}$$

However, since the heat capacity varies with temperature, therefore, the value of C has to be considered over a very narrow temperature range.

Thus if δq is the small amount of heat absorbed by a system which raises the temperature of the system by a small amount dT (say from T to $T + dT$), then the heat capacity of the system will be given by

$$C = \frac{\delta q}{dT} \quad \text{...(ii)}$$

Since q is not a state function and depends upon the path followed, therefore, C is also not a state function. Hence to know the value of C, the conditions such as constant volume or constant pressure have to be specified which define the path.

(*ii*) ***Heat capacity at constant volume,*** C_v

According to first law of thermodynamics, we know that

$$\delta q = dE + PdV \quad \text{...(iii)}$$

$$\therefore \quad C = \frac{dE + PdV}{dT} \quad \text{...(iv)}$$

When the volume is kept constant, dV = 0 and, therefore, equation (iv) becomes

$$C_v = \left(\frac{\partial E}{\partial T}\right)_v \quad ...(v)$$

or *for an ideal gas,* this equation may simply be written as

$$C_v = \frac{dE}{dT} \quad ...(vi)$$

C_v is the heat capacity at constant volume.

Thus *the heat capacity at constant volume may be defined as the rate of change of internal energy with temperature at constant volume.*

(*iii*) **Heat capacity at constant pressure, C_p**

When the pressure is kept constant during the absorption of heat, equation (*iv*) becomes

$$C_p = \left(\frac{\partial E}{\delta T}\right)_p + P\left(\frac{\partial V}{\partial T}\right)_p \quad ...(vii)$$

Also, we know that the heat content or enthalpy of a system is given by

$$H = E + PV$$

Differentiating w.r.t. T at constant P, we get

$$\left(\frac{\partial H}{\partial T}\right)_p = \left(\frac{\partial E}{\partial T}\right)_p + P\left(\frac{\partial V}{\partial T}\right)_p \quad ...(viii)$$

Combining equations (*vii*) and (*viii*), we get

$$C_p = \left(\frac{\partial H}{\partial T}\right)_p \quad ...(ix)$$

or for any ideal gas, this equation may simply be put in the form

$$C_p = \frac{dH}{dT} \quad ..(x)$$

C_p is the heat capacity at constant pressure.

Thus *the heat capacity at constant pressure may be defined as the rate of change of enthalpy with temperature at constant pressure.*

Q. 21. Deduce a relation between C_p and C_v.

(Delhi 2002; Nagpur 2008; Himachal 2010)

OR

Explain why C_p is always greater than C_v. *(Meerut 2000; Vidyasagar 2007)*

OR

Derive thermodynamically Cp – Cv = R for one mole of an ideal gas.

(Guwahati 2006; calcutta 2007)

Ans. If the volume of the system is kept constant when the heat is added to a system, then no work is done by the system. Thus the heat absorbed by the system is used up completely to increase the internal energy of the system. Again if the pressure of the system is kept constant when the heat is supplied to the system, then some work of expansion is also done by the system in addition to the increase in internal energy. Thus if at constant pressure, the temperature of the system is to be raised through the same value as at constant volume, then some extra heat is required for doing the work of expansion. Hence $C_p > C_v$.

We have the following equations for C_v and C_p.

$$C_v = \frac{dE}{dT} \quad ...(i)$$

$$C_p = \frac{dH}{dT} \quad ...(ii)$$

The difference between the heat capacities of an ideal gas can be obtained by subtracting equation (*i*) from equation (*ii*). So we have

$$C_p - C_v = \frac{dH}{dT} - \frac{dE}{dT} \quad ...(iii)$$

But $H = E + PV$ (by definition)

and $PV = RT$ (for 1 mole of an ideal gas)

$\therefore$ $H = E + RT$

Differentiating this equation w. r. t T, we get

$$\frac{dH}{dT} = \frac{dE}{dT} + R \quad ...(iv)$$

Or

$$\frac{dH}{dT} - \frac{dE}{dT} = R \quad ...(v)$$

Combining equation (*iii*) and (*v*),

$$\boxed{C_p - C_v = R}$$ for 1 mole of an ideal gas.

Thus C_p is greater then C_v by the gas constant R, *i.e.*, approximately 2 calories or 8.314 joules.

Q. 22. What is Joule-Thomson effect ? Give the experimental set-up of Joule-thomson effect.

(M.Dayanand 2002; Madras 2004
Nagpur 2008; Garhwal 2010; Andhra 2010; Bharathiar 2011

Ans. It was observed by Joule and Thomson that *when a real gas at a certain pressure expands adiabatical through a porous plug or a fine hole into a region of low pressure, it is is accompanied by cooling (except for hydrogen and helium which get warmed up).* This phenomenon is known as **Joule-Thomson effect.**

The experimental set-up of Joule and Thomson is shown in Fig. 6.7.

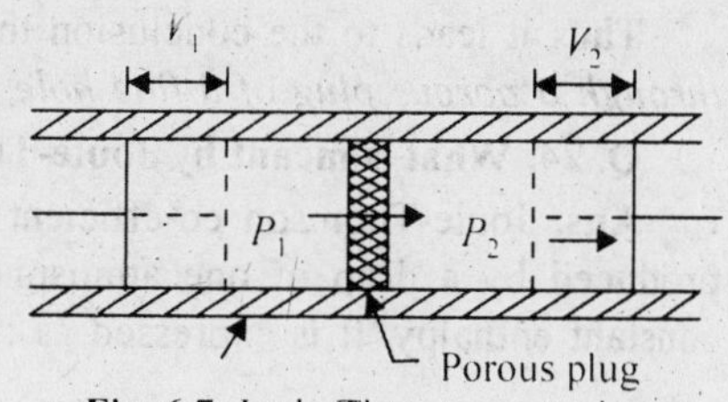

Fig. 6.7. Joule-Thomson experiment (Expansion of real gas through a porous plug).

A tube thoroughly insulated is fitted with a porous plug in the middle and pistons on either side as shown in the figure. The pressures of the gas are kept different on the two sides of the plug, being P_1 on the left side and P_2 on the right side where $P_1 > P_2$. By applying pressure on the piston on the left slowly enough (so as not to change the pressure P_1), a volume of the gas equal to V_1 is forced slowly through the porous plug and then allowed to expand to the pressure P_2 and volume V_2 by moving the piston on the right outward. When the process is being carried out as above, the temperature on the two sides of the porous plug is recorded accurately. The experiment is repeated with a number of gases. It is observed that a fall in temperature of the gas takes place in case of all gases except hydrogen and helium when the experiments are carried out at ordinary temperature *i.e.* at room temperature.

Q. 23. Prove that the enthalpy of the system remains constant in an adiabatic expansion. *(Kurukshetra 2001)*

Ans. Since the process is carried out adiabatically,

$$q = 0$$

$\therefore$ By first law of thermodynamics *viz.* $\Delta E = q - w$

$$\Delta E = -w$$

or $$w = -\Delta E \quad \text{... }(i)$$

Thus the work done, during the expansion of the gas under adiabatic conditions, is at the cost of the internal energy. In other words, when the work of expansion is done adiabatically, the internal energy of the gas decreases and hence the temperature decreases. In fact, a part of the internal energy is used up to overcome the forces of attraction existing among the molecules (van der Waal's forces).

In Joule-Thomson experiment, which is an example of adiabatic process, on the left side of the plug, the work is done *on* the system whereas on the right side of the plug, the work is done *by* the system.

Work done *on* the system on the left side $= P_1 V_1$

Work done *by* the system on the right side $= P_2 V_2$

$\therefore$ Net work done by the system is

$$w = P_2 V_2 - P_1 V_1 \quad \text{...}(ii)$$

Putting the value in equation (*i*), we get

$$\Delta E = -(P_2V_2 - P_1V_1)$$
$$= P_1V_1 - P_2V_2$$

Further putting $\Delta E = E_2 - E_1$, we get

$$E_2 - E_1 = P_1V_1 - P_2V_2$$

or $$E_2 + P_2V_2 = E_1 + P_1V_1$$

or $$H_2 = H_1$$

i.e. $$\Delta H = H_2 - H_1 = 0$$

Thus it leads to the conclusion that *when the expansion of a gas takes place adiabatically through a porous plug of a fine hole, the enthalpy of the system remains constant.*

Q. 24. What is meant by Joule-Thomson coefficient ? *(Kumaon 2000; Lucknow 2002)*

Ans. Joule-Thomson co-efficient μ may be defined as the temperature change in degrees produced by a drop of one atmospheric pressure when the gas expands under conditions of constant enthalpy. It is expressed as :

Thus $$\mu = \left(\frac{\partial T}{\partial P}\right)_H$$

(*i*) For cooling, μ will be positive (because dT and dP both will be negative).

(*ii*) For heating μ will be negative (because dT is positive while dP is negative).

(*iii*) If $\mu = 0$, the gas gets neither heated up not cooled on adiabatic expansion (because $\mu = 0$, only if $dT = 0$ for any value *of* dP).

Q. 25. What is inversion temperature ? *(Nagpur 2002; Madras 2004)*

Ans. Every gas has a definite temperature (at a particular pressure) at which $\mu = 0$. Below this temperature μ is positive and above this temperature μ is negative. *This temperature (at a particular pressure) at which $\mu = 0$ i.e. the gas neither cooled down nor heated up on adiabatic expansion is called the* **inversion temperature.**

Q. 26. Why do helium and hydrogen show heating on adiabatic expansion in Joule-Thomson experiment ?

Ans. The inversion temperatures of most of the gases are very high (much higher than the room temperature). That is why these gases undergo cooling on adiabatic expansion under ordinary conditions of temperature. On the other hand, the inversion temperatures of hydrogen and helium are very low (being much below 0°C). Hence these gases undergo heating on adiabatic expansion under ordinary conditions of temperature. However, if these gases are first cooled to temperatures below their inversion temperatures and then allowed to expand adiabatically, these gases also show cooling effect.

Q. 27. Deduce the relationship $\mu = -\dfrac{1}{C_p}\left(\dfrac{\partial H}{\partial P}\right)_T$

Ans. Enthalpy H is a state function. Taking temperature T and pressure P as the independent variables.

$$H = f(T, P) \qquad \text{... (i)}$$

$$\therefore \qquad dH = \left(\frac{\partial H}{\partial P}\right)_T dP + \left(\frac{\partial H}{\partial T}\right) dT \qquad \text{...(ii)}$$

In Joule-Thomson effect, the enthalpy of the system remains constant *i.e.*

$$dH = 0$$

Putting this value in the above equation, we get

$$\left(\frac{\partial H}{\partial P}\right)_T dP + \left(\frac{\partial H}{\partial T}\right)_p dT = 0 \qquad \text{...(iii)}$$

Dividing this equation throughout by dP and imposing the condition of constant enthalpy, we get

$$\left(\frac{\partial H}{\partial P}\right)_T + \left(\frac{\partial H}{\partial T}\right)_P \left(\frac{\partial T}{\partial P}\right)_H = 0 \qquad \text{...(iv)}$$

But $\left(\dfrac{\partial H}{\partial T}\right)_P = C_p$, heat capacity at constant pressure,

and $\left(\dfrac{\partial T}{\partial P}\right)_H = \mu$, Joule-Thomson coefficient ...(v)

Putting these values in equation (*iv*), we get

$$\left(\frac{\partial H}{\partial P}\right)_T + \mu C_p = 0$$

or

$$\mu = -\frac{1}{C_p}\left(\frac{\partial H}{\partial P}\right)_T \qquad \text{...(vi)}$$

Thus μ can be calculated from $\left(\dfrac{\partial H}{\partial P}\right)_T$ and *vice versa.*

Q. 28. Explain why an ideal gas shows neither heating nor cooling in Joule-Thomson experiment.

OR

Why is joule-Thomson coefficient of an ideal gas zero ? (*Calcutta 2007; Raipur 2010*)

Ans. An ideal gas, undergoes neither cooling nor heating on adiabatic expansion in Joule-Thomson experiment. This is expected as explained below:

The intermolecular forces of attraction (*i.e.* van der Waals forces) in an ideal gas are negligible. Hence no energy is used up in overcoming these forces of attraction when the gas expands adiabatically. Thus the internal energy of the gas does not fall and therefore the temperature also does not fall.

Q. 29. Derive an expression for work done in isothermal reversible expansion of a gas.

(M. Dayanand 2002; Delhi 2003; Calcutta 2007; Vidyasagar 2007, Himachal 2010)

Ans. A general expression for the work of expression in an isothermal reversible process may be derived as follows :

Suppose P is the pressure of the gas within a system undergoing an isothermal expansion by a reversible process. Then in order that the expansion may take place very slowly (*i.e.* reversibly) the external pressure should be $P - dP$, where dP is very small quantity. The small amount of work (δw) done by the system when its volume increases by an infinitesimal amount, dV, is equal to the product of the external pressure and the volume change *i.e.*,

$$\delta w = (P - dP)\, dV \qquad ...(i)$$

Neglecting the term $dP \times dV$ which is very small (being the product of two very small quantities), equation (*i*) becomes

$$\delta w = P\, dV \qquad ...(ii)$$

Hence the total work (*w*) done by the system when its volume increases from V_1 to V_2 will be given by

$$w = \int_{V_1}^{V_2} P dV \qquad ...(iii)$$

Now, for n moles of an ideal gas, the equation of state is

$$PV = nRT$$

so that

$$P = \frac{nRT}{V} \qquad ...(iv)$$

Substituting this value is equation (*iii*), we get

$$w = \int_{V_1}^{V_2} \frac{nRT}{V} dV \qquad ...(v)$$

For an isothermal process, T is constant. Also n and R are constant quantities. Hence equation (*v*) may be written as

$$w = nRT \int_{V_1}^{V_2} \frac{dV}{V}$$

$$= nRT\, In \frac{V_2}{V_1} \qquad ...(vi)$$

or

$$w = 2{\cdot}303\, nRT \log \frac{V_2}{V_1} \qquad ...(vii)$$

A relation can also be obtained in terms of pressure.

As the process is isothermal *i.e.* the temperature remains constant, therefore by Boyle's law

$$P_1 V_1 = P_2 V_2 \qquad ...(viii)$$

where P_1, V_1 are the initial pressure and volume and P_2, V_2 are the final pressure and volume of the system.

Equation (*viii*) can be written as

$$\frac{V_2}{V_1} = \frac{P_1}{P_2}$$

Substituting this value in equation (*vii*), we get

$$\mathbf{w = 2{\cdot}303\ nRT \log \frac{P_1}{P_2}} \qquad ...(ix)$$

Work done in a *reversible process* is the *maximum* work obtainable from a system. Hence w in equations (*vii*) and (*ix*) may be replaced by w_{max}. Expressions (*vii*) and (*ix*) may thus be written as

$$w_{max} = 2{\cdot}303\ nRT \log \frac{V_2}{V_1}$$

$$= 2{\cdot}303\ nRT \log \frac{P_1}{P_2}$$

Expression for *q* (*heat* absorbed): According to first law of thermodynamics

$$q = \Delta E + w$$

But $\Delta E = 0$ (*as explained above*)

$\therefore$ $q = w$

Thus all the heat absorbed is used up completely in doing the work of expression only. Hence the expressions for the heat absorbed in an isothermal reversible expansion of an *ideal gas* are same as for the work done *i.e.*

$$\mathbf{q = w = 2{\cdot}303\ nRT \log \frac{V_2}{V_1} = 2{\cdot}303\ nRT \log \frac{P_1}{P_2}}$$

Q. 30. Deduce the values for ΔE and ΔH in an isothermal expansion process.

Ans. Expression for ΔE: Since the internal energy of an *ideal gas* is function of temperature only, therefore, in an isothermal process as the temperature remains constant, the internal energy of the system (containing an ideal gas) remains constant

i.e. $\Delta E = 0$

Expression for ΔH

$$\Delta H = \Delta (E + PV)$$

$$= \Delta E + \Delta(PV)$$

$$= \Delta E + \Delta(nRT)$$

$$= \Delta E + \Delta nR\ \Delta T$$

$$= 0 + 0 \quad \begin{cases} \Delta E = 0 \text{ as explained above and} \\ \text{and } \Delta T = 0 \text{ as the process is isothermal} \end{cases}$$

i.e. $\mathbf{\Delta H = 0}$

Q. 31. Deduce values of *q*, *w*, ΔE and ΔH in case of isothermal irreversible expansion when (*i*) opposing pressure is zero (*ii*) opposing pressure is constant and less than P_2 but greater than zero.

Ans. The expressions for w, q, ΔE and ΔH for the above expansion may be obtained as follows :

(*i*) **For free expansion:** Here $P_{opp} = 0$

$$\therefore \quad w = \int P_{opp}\, dV = 0$$

Hence
$$q = w = 0$$
$$E = \Delta H = 0$$
(as the gas is *ideal* and expansion is *isothermal*)

(*ii*) **For intermediate expansion:** Here P_{opp} = constant and lies in the range $0 < P_{opp} < P_2$. Hence

$$w = \int_{V_1}^{V_2} P_{opp}\, dV = P_{opp}(V_2 - V_1)$$

$$\therefore \quad q = w = P_{opp}(V_2 - V_1)$$

$$\Delta E = \Delta H = 0$$ (as the gas is *ideal* and expansion is *isothermal*)

It may be noted that the work done in this process will be less than the work done is the reversible process.

Q. 32. Ten moles of an ideal gas at the initial pressure of one atmosphere at 0°C were expanded reversibly under isothermal conditions to a final pressure of 0·1 atmosphere. Calculate the work done by the gas, the change in internal energy and the heat absorbed by the system (R = 8·314 J K^{-1} deg^{-1} mol^{-1}).

Ans. Here we are given that

$$n = 10 \text{ moles}$$
$$P_1 = 1 \text{ atm}$$
$$P_2 = 0{\cdot}1 \text{ atm}$$
$$T = 0°C = 273 \text{ K}$$
$$R = 8.314 \text{ JK}^{-1} \text{ mol}^{-1}$$

$$\therefore \quad w_{max} = 2.303\, nRT \log \frac{P_1}{P_2}$$

$$= 2.303 \times 10 \times 8{\cdot}314 \times 273 \times \log \frac{1}{0.1}$$

$$= \mathbf{52271.7\ J} \qquad (\because \log 10 = 1.0)$$

For an ideal gas during isothermal process

$$\Delta E = 0$$

$\therefore$ By first law of thermodynamics ($q = \Delta E + w$)

$$q = w_{max} = \mathbf{52271.7\ J}$$

Q. 33. One mole of an ideal gas is heated at constant pressure from 0°C to 100°C (*i*) Calculate the work involved (*ii*) If the gas were expanded isothermally and reversibly at 0°C from 1 atmosphere to some other pressure *P*, what must be the final pressure if the isothermal work is equal to the work involved in (*i*) ?

Ans. Work involved when the gas is heated at constant pressure from 0°C to 100°C is obtained as follows :

(*i*) $w = P\Delta V$

$= P(V_2 - V_1)$

$= PV_2 - PV_1$

$= nRT_2 - nRT_1$

$= nR(T_2 - T_1)$

$= 1 \times 8.314\,(373 - 273)$

$= \mathbf{831{\cdot}4\ J}$

(*ii*) Final pressure if the process is carried out isothermally involving the work as in step (*i*) is obtained as under

$$w = 2.303\ nRT \log \frac{P_1}{P_2}$$

This on solving gives

$\boldsymbol{P = 0.6933}$ **atm**

Q. 34. Calculate the pressure-volume work performed by the system during reversible isothermal expansion of two moles of an ideal gas from 2 litres to 10 litres at 20°C.

Ans. Here we are given that

$n = 2$ moles

$V_1 = 2$ litres

$V_2 = 10$ litres

$T = (20 + 273)\ \text{K} = 293\ \text{K}$

Taking $R = 8.314$ Joules degree^{-1} mol^{-1}

$$w_{max} = 2{\cdot}303\ nRT \log \frac{V_2}{V_1}$$

$$= 2.303 \times 2 \times 8.314 \times 293 \times \log \frac{10}{2}$$

$= 11220.2 \times \log 5$

$= 112202 \times 0.6990$

$= \mathbf{7842.9\ joules}$

Q. 35. Calculate the values of *w, q,* ΔE and ΔH in adiabatic expansion of an ideal gas.

(Bangalore 2005)

Ans. An adiabatic process is the one in which no heat enters or leaves the system of any stage. Thus for every infinitesimal change of the process,

$$\delta q = 0$$

Putting this value in the equation of the first law *viz.*

$dE = \delta q - \delta w$, we get

$$dE = -\,\delta w \qquad ...(i)$$

If the work involved is the work of expansion only, then if dV is the small increase in volume and P is the pressure of the gas, then

$$\delta w = PdV \qquad ...(ii)$$

Putting this value in equation (*i*), we get

$$dE = -\,PdV \qquad ...(iii)$$

But for an ideal gas, $C_v = \frac{dE}{dT}$

or $dE = C_v\, dT$

Putting this value in equation (*iii*) we get

$$C_v dT = -PdV \quad ...(v)$$

Further

$$\begin{aligned} \Delta H &= \Delta(E + PV) \\ &= \Delta E + \Delta(PV) \\ &= \Delta E + \Delta(RT) \\ &= \Delta E + R\,\Delta T \\ &= C_v \Delta T + R\,\Delta T \\ &= (C_v + R)\,\Delta T \\ &= (C_v + R)(T_2 - T_1) \\ &= C_p (T_2 - T_1) \end{aligned} \quad ...(vi)$$

To sum up, for any type of adiabatic process (reversible or irreversible)

$$q = 0$$

$$w = -\Delta E$$

$$\Delta E = C_v \Delta T = C_v (T_2 - T_1)$$

$$\Delta H = C_p \Delta T = C_p (T_2 - T_1)$$

Q. 36. Derive the relationship $TV^{\gamma-1}$ = constant, between temperature and volume in reversible adiabatic expansion. *(Kumaon 2000; Kurukshetra 2001; Madras 2004; Delhi 2006)*

Ans. Consider 1 mol of the *ideal gas.* Then according to the gas equation

$$PV = RT$$

or

$$P = \frac{RT}{V} \quad ...(i)$$

Also $C_v\, dT = -PdV$ (Refer to *Q.* 34) ...(*ii*)

Substituting for P from eq. (*i*) in (*ii*)

$$C_v dT = -\frac{RT}{V}\, dV$$

or

$$C_v \frac{dT}{T} = -R\frac{dV}{V} \quad ...(iii)$$

Now suppose the volume of the gas changes from V_1 to V_2 when the temperature changes from T_1 to T_2. Then assuming C_v *to* be independent of temperature, and integrating equation (*iii*) between the limits T_1, T_2 and V_1, V_2, we get

$$C_v \int_{T_1}^{T_2} \frac{dT}{T} = -R\int_{V_1}^{V_2} \frac{dV}{V}$$

$$C_v [\ln T]_{T_1}^{T_2} = -R[\ln V]_{V_1}^{V_2}$$

$$C_v \ln \frac{T_2}{T_1} = -R \ln \frac{V_2}{V_1}$$

$$C_v \ln \frac{T_2}{T_1} = R \ln \frac{V_1}{V_2}$$

Now, for 1 mol of an ideal gas we know that

$$C_p - C_v = R$$

Putting this value of R in equation (*iv*), we get

$$C_v \ln\frac{T_2}{T_1} = (C_p - C_v)\ln\frac{V_1}{V_2}$$

Dividing throughout by C_v we get

$$\ln\frac{T_2}{T_1} = \left(\frac{C_p}{C_v} - 1\right)\ln\frac{V_1}{V_2}$$

Putting $\frac{C_p}{C_v} = \gamma$, the ratio of the two heat capacities, we get

$$\ln\frac{T_2}{T_1} = (\gamma - 1)\ln\frac{V_1}{V_2}$$

or $$\ln\frac{T_2}{T_1} = \ln\left(\frac{V_1}{V_2}\right)^{\gamma-1}$$

Taking antilogarithm of both sides, we get

$$\frac{T_2}{T_1} = \left(\frac{V_1}{V_2}\right)^{\gamma-1} \quad \frac{T_2}{T_1} = \left(\frac{V_1}{V_2}\right)^{\gamma-1}$$

or $$T_1V_1^{\gamma-1} = T_2V_2^{\gamma-1}$$

or $$\boxed{TV^{\gamma-1} = \text{constant}}$$

Q. 37. Derive a relationship between temperature and pressure in adiabatic expansion of an ideal gas. **(*Delhi 2007; Himachal 2010*)**

Ans. Since an ideal gas is under consideration, it follows that

$$P_1V_1 = RT_1$$

and $$P_2V_2 = RT_2$$

i.e. $$V_1 = \frac{RT_1}{P_1}$$

and $$V_2 = \frac{RT_2}{P_2}$$

$$\therefore \quad \frac{V_1}{V_2} = \frac{\frac{RT_2}{P_1}}{\frac{RT_2}{P_2}} = \frac{T_1P_2}{T_2P_1} \qquad \text{...}(i)$$

Also $$\frac{T_2}{T_1} = \left(\frac{V_1}{V_2}\right)^{\gamma-1} \qquad \text{...}(ii)$$

(The relation between T and V in case of adiabatic expansion)

From (*i*) and (*ii*)

$$\frac{T_2}{T_1} = \left(\frac{T_1P_2}{T_2P_1}\right)^{\gamma-1}$$

or $$\frac{T_2}{T_1}\times\left(\frac{T_2}{T_1}\right)^{\gamma-1} = \left(\frac{P_2}{P_1}\right)^{\gamma-1}$$

or $$\left(\frac{T_2}{T_1}\right)^{\gamma} = \left(\frac{P_2}{P_1}\right)^{\gamma-1}$$

or $$\left(\frac{T_2}{T_1}\right)^{\gamma} = \left(\frac{P_1}{P_2}\right)^{1-\gamma}$$

Taking the γth root of both sides, we get ...(*iii*)

$$\frac{T_2}{T_1} = \left(\frac{P_1}{P_2}\right)^{(1-\gamma)/\gamma}$$

or $$T_1P_1^{(1-\gamma)/\gamma} = T_2P_2^{(1-\gamma)/\gamma}$$

or $$TP^{(1-\gamma)/\gamma} = \text{constant}$$

Q. 38. Derive a relationship between pressure and volume in adiabatic expansion of an ideal gas.

(Kurukshetra, 2002; M. Dayanand 2002; Delhi 2003; G. Nanakdev 2003; Calcutta 2007)

OR

Prove that for an adiabatic process PV^{γ} = constant *(Guwahati 2007)*

Ans. Consider the relations between temperature and volume and temperature and pressure

$$\frac{T_2}{T_1} = \left[\frac{V_1}{V_2}\right]^{\gamma-1} \quad ...(i)$$

$$\frac{T_2}{T_1} = \left[\frac{P_1}{P_2}\right]^{(1-\gamma)/\gamma} \quad ...(ii)$$

From (*i*) and (*ii*)

$$\left(\frac{P_1}{P_2}\right)^{(1-\gamma)/\gamma} = \left(\frac{V_1}{V_2}\right)^{\gamma-1}$$

or $$\left(\frac{P_1}{P_2}\right)^{(1-\gamma)/\gamma} = \left(\frac{V_2}{V_1}\right)^{1-\gamma}$$

Taking (1 – γ)th root of both sides, we get

$$\left(\frac{P_1}{P_2}\right)^{1/\gamma} = \frac{V_2}{V_1}$$

Raising both sides to the power γ, we get

$$\frac{P_1}{P_2} = \left(\frac{V_2}{V_1}\right)^{\gamma}$$

or $$P_1V_1^{\gamma} = P_2V_2^{\gamma}$$

$$PV^{\gamma} = \text{constant}$$

Q. 39. Derive an expression for work done in an adiabatic reversible expansion of an ideal gas.

Ans. The expression for the work done in an adiabatic reversible expansion of an ideal gas may be derived as follows :

For adiabatic process, we have

$$PV^{\gamma} = \text{constant}$$

Differentiating this equation, we get

$$P\,(\gamma V^{\gamma-1})\,dV + V^{\gamma}\,dP = 0$$

or $$\gamma P V^{\gamma-1}\,dV + V^{\gamma}\,dP = 0$$

Dividing throughout by $V^{\gamma-1}$, we get

$$\gamma PdV + VdP = 0$$

or $$VdP = -\gamma PdV \qquad ...(i)$$

For n moles of an ideal gas, $PV = nRT$

Differentiating this equation completely, we get

$$PdV + VdP = nRdT \qquad ...(ii)$$

Putting the value of VdP from equation (i) in this equation we get

$$PdV - \gamma PdV = nRdT$$

or $$(1-\gamma)\,PdV = nRdT$$

or $$PdV = \frac{nRdT}{1-\gamma} \qquad ...(iii)$$

Work done in the adiabatic process is given by general equation

$$w = \int_{V_1}^{V_2} PdV \qquad ...(iv)$$

Putting the value of PdV from equation (iii) in this equation, and changing the limits from volumes to temperatures, we get

$$w = \int_{T_1}^{T_2} \frac{nRdT}{1-\gamma}$$

$$= \frac{nR}{1-\gamma}\int_{T_1}^{T_2} dT$$

$$= \frac{nR}{1-\gamma}(T_2 - T_1) \qquad ...(v)$$

Hence $$\boldsymbol{W = \frac{nR(T_2 - T_1)}{1-\gamma}}$$

Expression for work done may be obtained in a different form.

Putting $R = C_p - C_v$ and $\gamma = C_p/C_v$ equation (v) becomes

$$w = \frac{n(C_p - C_v)(T_2 - T_1)}{1 - \frac{C_p}{C_v}}$$

$$= \frac{n(C_p - C_v)(T_2 - T_1)}{\frac{(C_v - C_p)}{C_p}}$$

$$= \frac{n(C_p - C_v)(T_2 - T_1)}{\frac{-(C_p - C_v)}{C_v}}$$

$= - nC_v (T_2 - T_1)$, for n moles of the ideal gas.

$= - C_v (T_2 - T_1)$, for 1 mol of the ideal gas ...(*vi*)

Further for n moles of an ideal gas,

$P_1 V_1 = nRT_1$ ....for the initial state.

and $P_2V_2 = nRT_2$ ....for the final state.

Putting these values in equation (*v*), we get

$$\boldsymbol{w = \frac{P_2V_2 - P_1V_1}{1 - \gamma}} \quad ...(vii)$$

Q. 40. For a certain gas C_p=8.58. Two moles of the gas are expanded adiabatically from an initial temperature of 20°C to a final temperature of – 45.4°C. Calculate the work done in the above case.

Ans. $C_p = 8.58, C_v = 8.58 - 2 = 6.58$

Therefore $r = \frac{C_p}{C_v} = \frac{8{\cdot}58}{6{\cdot}58} = 1.3$

$n = 2,\ R = 8.314 \text{J K}^{-1} \text{mol}^{-1}$

Work done $w = \frac{nR(T_2 - T_1)}{1 - \gamma}$

$$= \frac{2 \times 8.314 \times (-65.4)}{(1 - 1.3)}$$

$$= \frac{2 \times 8.314 \times 65.4}{0.3} \text{J}$$

$= \mathbf{3624.9\ J}$

Q. 41. Thre moles of a perfect gas with C_v = 5.0 cal/mol/degree at 1.0 atmosphere are to be compressed adiabatically and reveribly from a volume of 75 litre at 1·0 atmosphere to a pressure of 100 atmospheres. Predict (a) the final volume of the gas (b) the final temperature of the gas.

Ans. Here we are given that

n = 3 moles

C_v = 5.0 cal/mol/degree

$$P_1 = 1.0 \text{ atm}$$
$$P_2 = 100 \text{ atm}$$
$$V_1 = 75 \text{ litres}$$

For a perfect gas $C_p = C_v + R$

$$= 5{\cdot}0 + 2.0 = 7.0 \text{ cal} \quad (R \cong 2 \text{ cal})$$

$$\therefore \quad \gamma = \frac{C_p}{C_v} = \frac{7.0}{5.0} = 1.4$$

(*a*) *To calculate the final volume*

$$P_1V_1^{\gamma} = P_2V_2^{\gamma}$$

or $$\left(\frac{V_1}{V_2}\right)^{\gamma} = \frac{P_2}{P_1}$$

or $$\left(\frac{75}{V_2}\right)^{1{\cdot}4} = \frac{100}{1.0} = 100,$$

$\therefore$ $$1.4\,[\log 75 - \log V_2] = \log 100$$

or $$1.4\,(1.8751 - \log V_2) = 2$$

or $$1.4 \log V_2 = 0.62514$$

or $$\log V_2 = 0.4465$$

$$V_2 = \text{antilog}\,(0.4465)$$
$$= \mathbf{2{\cdot}796 \text{ litres}}$$

(*b*) *To calculate the final temperature*

For a perfect gas,

$$PV = nRT$$

When
$$P = 100 \text{ atm}$$
$$V = 2.796 \text{ litres (calculated above)}$$
$$n = 3 \text{ moles (given)}$$
$$R = 0{\cdot}0821 \text{ litre atm deg}^{-1} \text{ mol}^{-1}$$

$$\therefore \quad T = \frac{P \times V}{n \times R} = \frac{100 \times 2.796}{3 \times 0.0821} = \mathbf{1135.2\,K}$$

Q. 42. Two moles of H_2 are compressed adiabatically from S.T.P. conditions to occupy a volume of 4.48 litres. Calculate the final temperature (γ for H_2= 1.41).

Ans. Here we are given that $n = 2$

so that
$$\left.\begin{aligned} V_1 &= 2 \times 22.4 = 44.8 \text{ litres} \\ T_1 &= 273 \text{ K} \end{aligned}\right\} \text{at S.T.P.}$$

$$V_2 = 4.48 \text{ litres}$$
$$T_2 = ?$$
$$\gamma = 1.41$$

Putting the values in the formula

$$T_1V_1^{\gamma-1} = T_2V_2^{\gamma-1}$$

or $$\frac{T_2}{T_1} = \left(\frac{V_1}{V_2}\right)^{\gamma-1}$$

We get $$\frac{T_2}{273} = \left(\frac{44.8}{4.48}\right)^{1.41-1} = (10)^{0.41}$$

or $$T_2 = 273 \times (10)^{0.41}$$

$\therefore$ $$\log T_2 = \log 273 + 0.41 \log 10$$
$$= 2.4362 + 0.41$$
$$= 2.8462$$

or $$T_2 = \text{antilog } 2.8462 = \mathbf{701.8\ K}$$

Q. 43. A dry gas at NTP is expanded adiabatically from 1 litre to 5 litre. Calculate the final temperature assuming ideal behaviour ($C_p / C_v = 1.4$) *(Delhi 2007)*

Ans. *Apply the relation* $TV^{\gamma-1}$ = constant

$$T_1V_1^{\gamma-1} = T_2V_2^{\gamma-1} \quad \text{or} \quad \frac{T_1}{T_2} = \left(\frac{V_2}{V_1}\right)^{\gamma-1}$$

$T_1 = 273$ K, $T_2 = ?$ $V_1 = 1$ litre, $V_2 = 5$ litre

Substituting the values in the equation above,

$$\frac{273}{T_2} = 5^{1.4-1} \quad \text{or} \quad \frac{273}{T_2} = 5^{0.4}$$

Taking logarithm of both sides

$$\log 273 - \log T_2 = 0.4 \log 5$$

or $$\log T_2 = \log 273 - 0.4 \log 5$$

or $$\log T_2 = 2.4362 - (0.4 \times 0\ 0.6990)$$
$$= 2.1566$$

or $$\therefore T_2 = 143.4\ \text{K}$$

Q. 44. Compare the magnitude of work in isothermal and adiabatic expansion of an ideal gas. *(G. Nanakdev 2000; M. Dayanand 2002)*

Ans. *For a reversible adiabatic process,* the relationship between pressure and volume is given by

$$PV^{\gamma} = \text{constant}$$

On the other hand *for a reversible isothermal process,* the relationship is simply given by Boyle's law, *viz.*

$$PV = \text{constant}$$

Now since C_p is always greater than C_v,

therefore the ratio $\frac{C_p}{C_v} = \gamma$ is greater than unity.

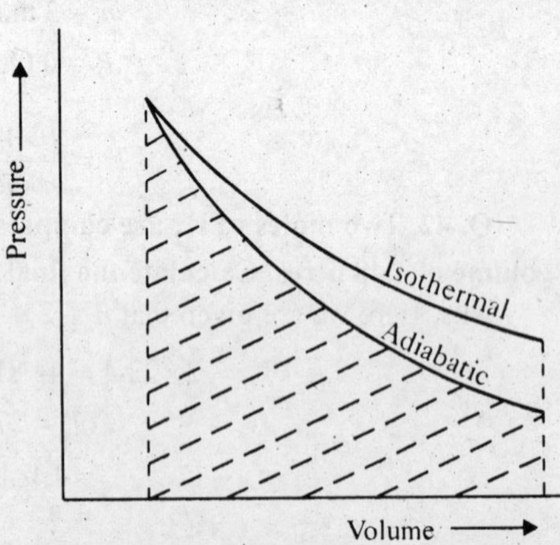

Fig. 6.8. Isothermal and adiabatic curves.

Hence the increase of volume for a given decrease of pressure will be less in an adiabatic than in an isothermal expansion. Thus the plot of pressure against volume *i.e.*, pressure volume curve will be steeper for an adiabatic than for an isothermal process, starting at the same point, as shown is Fig. 6.8.

Further, since the area under the pressure volume curve gives the work of expansion, from Fig. 6.8, it may be observed that for the same increase of volume, the work done in an adiabatic process is less than in an isothermal process.

Q. 45. Calculate w, q, ΔE and ΔH in the expansion of a real gas in isothermal reversible expansion.

Ans. (*i*) **Expression for w,** we know that

$$dw = P_{ext}\, dV$$

For the reversible expansion

$$P_{ext} = P_{int} - dP \simeq P_{int}$$

where P_{int} is the pressure (internal pressure) of the gas

Hence $\quad dw = P_{int}\, dV \quad$ ...(*i*)

As the gas is real, P_{int} can be found from van der Waal's equation *i.e.*

$$\left(P + \frac{an^2}{V^2}\right)(V - nb) = nRT$$

which gives

$$P_{int} = \frac{nRT}{V - nb} - \frac{an^2}{V^2} \quad \text{...}(ii)$$

Substituting this value in eqn (*i*), we get

$$dw = \left(\frac{nRT}{V - nb} - \frac{an^2}{V^2}\right)dV$$

$$\therefore \quad w = \int_{V_1}^{V_2}\left(\frac{nRT}{V - nb} - \frac{an^2}{V^2}\right)dV$$

$$= nRT\left[\ln(V - nb)\right]_{V_1}^{V_2} - an^2\left[-\frac{1}{V}\right]_{V_1}^{V_2}$$

or

$$\boldsymbol{w = nRT \ln\frac{V_2 - nb}{V_1 - nb} + an^2\left(\frac{1}{V_2} - \frac{1}{V_1}\right)} \quad \text{...}(iii)$$

(*ii*) **Expression for ΔE :** According to thermodynamic equation of state

$$\left(\frac{\partial E}{\partial V}\right)_T = T\left(\frac{\partial E}{\partial T}\right)_V - P \quad \text{...}(iv)$$

For a real gas, according to van der Waal's equation

$$P = \frac{nRT}{V - nb} - \frac{an^2}{V^2} \quad \text{...}(v)$$

$$\therefore \quad \left(\frac{\partial P}{\partial T}\right)_V = \frac{nR}{V - nb} \quad \text{...}(vi)$$

Substituting the values from eqns. (*v*) and (*vi*) in eqn. (*iv*) we get

$$\left(\frac{\partial E}{\partial V}\right)_T = \frac{nRT}{V - nb} - \left(\frac{nRT}{V - nb} - \frac{an^2}{V^2}\right) = \frac{an^2}{V^2} \quad \text{...}(vii)$$

For dE, we can write

$$dE = \left(\frac{\partial E}{\partial T}\right)_V dT + \left(\frac{\partial E}{\partial V}\right)_T dV \qquad ...(viii)$$

For the isothermal process, $dT = 0$

$$\therefore \quad dE = \left(\frac{\partial E}{\partial V}\right)_T dV = \frac{an^2}{V^2} dV \qquad \text{[Substituting value from eqn } (vii)\text{]}$$

Hence for a finite process,

$$\Delta E = \int_{V_1}^{V_2} \frac{an^2}{V^2} dV$$

or

$$\Delta \boldsymbol{E} = -\boldsymbol{an^2}\left(\frac{\mathbf{1}}{\boldsymbol{V_2}} - \frac{\mathbf{1}}{\boldsymbol{V_1}}\right) \qquad ...(ix)$$

(*iii*) **Expression for q :** According to first law of thermodynamics

$$\boldsymbol{q = \Delta E - w}$$

Substituting the values of w and ΔE from eqns. (*iii*) and (*ix*), we get

$$q = -an^2\left(\frac{1}{V_2} - \frac{1}{V_1}\right) + nRT \ \ln \frac{V_2 - nb}{V_1 - nb} + an^2\left(\frac{1}{V_2} - \frac{1}{V_1}\right)$$

or

$$\boldsymbol{q = nRT \ln \frac{V_2 - nb}{V_1 - nb}} \qquad ...(x)$$

(*iv*) **Expression for ΔH :** We have

$$\Delta H = \Delta (E + PV)$$
$$= \Delta E + \Delta (PV)$$

Substituting the value of ΔE and P from eqns (*ix*) and (*ii*), we get

$$\Delta H = -an^2\left(\frac{1}{V_2} - \frac{1}{V_1}\right) + \Delta\left[\left(\frac{nRT}{V - nb} - \frac{an^2}{V^2}\right)\right]$$

$$= -an^2\left(\frac{1}{V_2} - \frac{1}{V_1}\right) + nRT\left(\frac{V_2}{V_2 - nb} - \frac{V_1}{V_1 - nb}\right) - an^2\left(\frac{1}{V_2} - \frac{1}{V_1}\right)$$

$$= -2an^2\left(\frac{1}{V_2} - \frac{1}{V_1}\right) + nRT\left[\frac{V_2 - nb + nb}{V_2 - nb} - \frac{V_1 - nb + nb}{V_1 - nb}\right]$$

$$= -2an^2\left(\frac{1}{V_2} - \frac{1}{V_1}\right) + nRT\left[\left(1 + \frac{nb}{V_2 - nb}\right) - \left(1 + \frac{nb}{V_1 - nb}\right)\right]$$

$$\boldsymbol{\Delta H = -2an^2\left(\frac{1}{V_2} - \frac{1}{V_1}\right) + n^2bRT\left(\frac{1}{V_2 - nb} - \frac{1}{V_1 - nb}\right)} \qquad ...(xi)$$

Q. 46. Calculate q, w, ΔE and ΔH in adiabatic reversible expansion of a real gas.

Ans. (*i*) **Value for q:** For the adiabatic process, as already explained,

$$q = 0$$

(*ii*) **Expressions for w and ΔE:** By first law of thermodynamics,

$$\Delta E = q + w$$

As $a = 0$, therefore $\Delta E = w$

The expression for ΔE and hence for w can be obtained as follows :

$$dE = \left(\frac{\partial E}{\partial T}\right)_v dT + \left(\frac{\partial E}{\partial V}\right)_T dV$$

Putting $(\partial E/\partial T)_v = nC_v$ (for n moles) and $\left(\frac{\partial E}{\partial V}\right)_T = \frac{an^2}{V^2}$, we get

$$dE = nC_v dT + \frac{an^2}{V^2} dV \qquad ...(i)$$

Hence for a finite process,

$$\Delta E = \int_{T_1}^{T_2} nC_v dT + \int_{V_1}^{V_2} \frac{an^2}{V^2} dV$$

Taking C_v to be independent of temperature, we have

$$\boldsymbol{\Delta E = nC_v (T_2 - T_1) - an^2 \left(\frac{1}{V_2} - \frac{1}{V_1}\right)} \qquad ...(ii)$$

This expression can be used for calculation of ΔE if T_1, T_2, V_1 and V_2 are known. However if T_2 is not given, it is first calculated as follows :

$$dE = -dw = -PdV$$

or

$$dE = -PdV \qquad ...(iii)$$

For a real gas
$$P = \frac{nRT}{V - nb} - \frac{an^2}{V^2}$$

and
$$dE = nC_v dT + \frac{an^2}{V^2} dV$$
Refer Eqn. (i) above

Substituting these values in eqn. (iii), we get

$$nC_v dT + \frac{an^2}{V^2} dV = -\frac{nRT}{V - nb} dV + \frac{an^2}{V^2} dV$$

$$nC_v dT = -\frac{nRT}{V - nb} dV$$

$$C_v = -\frac{nRT}{V - nb} dV$$

Separating the variables, it can be rewritten as

$$C_v \frac{dT}{T} = -R \frac{dV}{V - nb}$$

Integrating between suitable limits, we get

$$\int_{T_1}^{T_2} C_v \frac{dT}{T} = -R \int_{V_1}^{V_2} \frac{dV}{V - nb}$$

or $$C_v [\ln T]_{T_1}^{T_2} = -R\left[\ln(V-nb)\right]_{V_1}^{V_2}$$

or $$C_v \ln\frac{T_2}{T_1} = -R\ln\frac{V_2-nb}{V_1-nb}$$

or $$\ln\left(\frac{T_2}{T_1}\right)^{C_v} = \ln\left(\frac{V_1-nb}{V_2-nb}\right)^R$$

or $$\left(\frac{T_2}{T_1}\right)^{C_v} = \left(\frac{V_1-nb}{V_2-nb}\right)^R$$

or $$\frac{T_2}{T_1} = \left(\frac{V_1-nb}{V_2-nb}\right)^{R/C_v}$$

or $$\boldsymbol{T_2 = T_1\left(\frac{V_1-nb}{V_2-nb}\right)^{R/c_v}} \qquad ...(iv)$$

Thus knowing V_1, V_2 and T_1, T_2 (temperature after expansion) can be calculated. This may be substituted in eqn. (*ii*) to get ΔE and hence *w*.

(*iii*) Expression for ΔH

$$\Delta H = \Delta(E+PV) = \Delta E + \Delta(PV)$$

$$= \Delta E + \Delta\left[\left(\frac{nRT}{V-nb} - \frac{an^2}{V^2}\right)V\right]$$

$$= \Delta E + nR\left(\frac{V_2T_2}{V_2-nb} - \frac{V_1T_1}{V_1-nb}\right) - an^2\left(\frac{1}{V_2} - \frac{1}{V_1}\right)$$

Substituting the value of ΔE from eqn. (*ii*), we get

$$\boldsymbol{\Delta H = nC_v(T_2-T_1) - 2an^2\left(\frac{1}{V_2}-\frac{1}{V_1}\right) + nR\left(\frac{V_2T_2}{V_2-nb} - \frac{V_1T_1}{V_1-nb}\right)} \qquad ...(v)$$

Q. 47. Derive Kirchoff's equation showing the variation of heat of reaction with temperature

(Delhi 2003; Agra 2004; Calcutta 2007; Nagpur 2008)

OR

Prove that $\left(\frac{\partial \Delta H}{\partial T}\right)_P = \Delta C_P$ *(Kurukshetra 2002; Madras 2004; Delhi 2007; Garhwal 2010. Andhra 2010, Bharathiar 2011)*

Ans. Kirchoff s equation gives the relationship between variation in heat of reaction with temperature. It can be stated as :

The change in the heat of reaction at constant pressure for every degree change of temperature is equal to the change in the heat capacity at constant pressure, accompanying the reaction.

It can be derived as under.

Consider the general reaction

$$A \longrightarrow B$$

as shown in Fig. 6.9.

Suppose the heat of reaction (heat evolved) at temperature $T_1 = \Delta H_1$ and the heat of reaction (heat evolved) at temperature $T_2 = \Delta H_2$.

Now, starting from A at temperature T_1, we can reach B at temperature T_2 by two different paths :

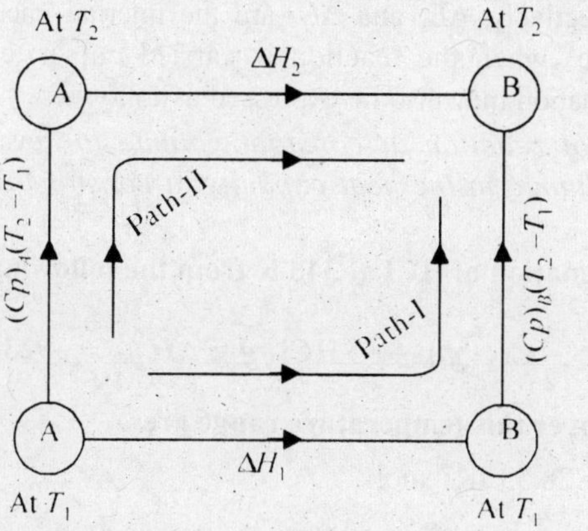

Fig. 6.9. Changing A at T_1 to B at T_2 by two different paths.

Path I : We may allow the reaction to take place at temperature T_l to obtain B at the same temperature and then heat this up to the temperature T_2. Then

Heat **evolved** in the first stage = ΔH_1

Heat **absorbed** in the second stage = $(C_P)_B\,(T_2 - T_1)$

where $(C_P)_B$ is the average molar heat capacity of the products.

$\therefore$ Net heat evolved = $\Delta H_1 - (C_p)_n(T_2 - T_1)$

Path II: We may first heat up the reactants to the temperature T_2 and then allow them to react to give the products at the same temperature. Then if $(C_P)_A$ is the average molar heat capacity of the reactant A,

Heat **absorbed** in the first stage = $(C_p)_A\,(T_2 - T_1)$

Heat **evolved** in the second stage = ΔH_2

$\therefore$ Net heat evolved = $\Delta H_2 - (C_p)_A\,(T_2 - T_1)$.

Nϕw since the initial and the final states are the same in both the above cases, therefore, by law of conservation of energy, we must have

Net heat evolved by path I = Net heat evolved by path II

i.e. $$\Delta H_1 - (C_p)_B\,(T_2 - T_1) = \Delta H_2 - (C_p)_A\,(T_2 - T_1)$$

or $$\frac{\Delta H_2 - \Delta H_1}{T_2 - T_1} = \Delta C_p$$

When the reaction is carried out at constant volume: We have the thermodynamic relation

$$\Delta H = \Delta E + P\Delta V$$

Under conditions of constant volume, $\Delta V = 0$

$\therefore$ $$\Delta H = \Delta E$$

Hence equation derived above takes the form

$$\frac{\Delta E_2 - \Delta E_1}{T_2 - T_1} = \Delta C_v$$

where $\Delta C_v = (C_v)_B - (C_v)_A$ is the change in the heat capacity at constant volume accompanying the reaction, $(C_v)_B$ and $(C_v)_B$ are the mean molar heat capacities at constant volume of the reactants and the products respectively, ΔE_1 and ΔE_2 are the internal energy changes (*i.e.*, heats of reaction at constant volume) when the reaction is carried out at temperatures T_1 and T_2 respectively. Thus Kirchoff's equation may also be expressed as follows :

The change in the heat of reaction at constant volume for every degree change of temperature is equal to the change in the heat capacity at constant volume, accompanying the reaction.

Q. 48. Calculate the heat of formation of HCl at 348 K from the following data :

$$\frac{1}{2}H_2(g) + \frac{1}{2}Cl_2(g) \longrightarrow HCl(g);\ \Delta H^0_{298} = -92300\text{J}$$

The mean heat capacities over this temperature range are

H_2 (g); C_p = 28.53 JK^{-1} mol^{-1}

Cl_2 (g); C_v = 32.26 JK^{-1} mol^{-1}

HCl (g); C_p = 28.49 JK^{-1} mol^{-1}

Ans. The reaction is

$$\frac{1}{2}H_2(g)+\frac{1}{2}Cl_2(g)\longrightarrow HCl\ (g)$$

For this reaction,

$$\Delta C_p = (C_p)_{HCl} - \frac{1}{2}(C_p)_{H_2} - \frac{1}{2}(C_p)_{Cl_2}$$

$$= 28.49 - \frac{1}{2}(28.53) - \frac{1}{2}(32.26)$$

$$= 1.91 \text{ JK}^{-1} \text{ mol}^{-1}$$

Further, we are given that

$\Delta H_1 = -92300$ J

$T_1 = 298$ K

$\Delta H_2 = ?$

According to Kirchoff s equation

$$\frac{\Delta H_2 - \Delta H_1}{T_2 - T_1} = \Delta C_p$$

or $$\Delta H_2 = \Delta H_1 + \Delta C_p (T_2 - T_1)$$

$$= -92300 + (-1{\cdot}91)(348 - 298)$$

$$\mathbf{= -92300 - 95{\cdot}5}$$

$$\mathbf{= -92395.5\ J}$$

THERMOCHEMISTRY

Q. 49. What is thermochemistry? Why are chemical reactions accompanied by evolution or absorption of energy? Give some examples. What are the units of heat and work?

Ans. *The branch of chemistry which deals with heat changes accompanying chemical reactions is callea **thermochemistry.***

A chemical reaction involves rearrangement of atoms, *i.e.*, some bonds within the reactant molecules are broken while some new bonds in molecules of the products are formed. Energy is consumed in the dissociation of bonds while energy is evolved when the bonds are formed. If the total energy evolved is greater than the total energy consumed, the net result is the *release of energy.* On the other hand, if the energy consumed is greater than the energy evolved, the net result is the *absorption of energy.*

All chemical reactions are usually accompanied by energy changes. These energy changes appear in the form of heat, light, work, electricity, etc. These energy changes are of great practical utility both for domestic and industrial purposes. A few reactions alongwith the form in which the accompanying energy changes appear and the use to which they are put are listed below :

(*i*) Burning of coal in air produces *heat.* That is why it is used as a fuel.

$$C(s) + O_2(g) \longrightarrow CO_2(g) + \text{heat}$$

(*ii*) *Digestion and metabolism of carbohydrates* in biological systems produces *heat* which maintains the body temperature.

(*iii*) *Burning of candle* (*wax*) *in air* produces heat and light. It is, therefore, used as a source of light. Wax is a hydrocarbon.

(*iv*) *Reaction that takes place in a galvanic cell* produces electrical energy which may be used to run an electrical motor or to ring a bell, etc.

$$Zn(s) + Cu^{2+}(aq) \longrightarrow Zn^{2+}(aq) + Cu(s)$$

Units of Heat and Work. The heat changes are measured in calories, kilocalories, joules or kilojoules.

1 calorie = 4.184 J 1 kcal = 4.184 kJ

The S.I. unit of heat is joule or kilojoule.

Work is measured in ergs or joules. The S.I. unit of work is joule. 1 Joule = 10^7 ergs

Q. 50. Define internal energy and internal energy change.

Ans. Internal energy. Chemical reactions are usually accompanied by energy changes. Therefore, every substance must have a definite amount of energy stored within itself.

The energy stored within a substance is called its **internal energy.**

It is generally represented by E, its value depends upon temperature, pressure, chemical nature and amount of the substance.

Internal energy is a state function. The internal energy of a definite amount of a particular substance under given conditions of temperature and pressure always has a definite value. For example, a certain amount of CO_2 under given conditions of temperature and pressure will always have the same internal energy, irrespective of the manner by which it has been produced. This is expressed by saying that *internal energy is a state function* (*i.e.*, a property which depends only on the state of the system).

Internal energy change. The absolute or exact value of internal energy possessed by a substance cannot be determined. This is because it is impossible to determine accurately most of the quantities which contribute towards internal energy of the system. But we are not interested in knowing the absolute value of internal energy of a substance. Rather, we are interested in knowing the change in internal energy which accompanies a process and this can be determined. The change in internal energy, ΔE, accompanying a process is given by

$$\Delta E = E_2 - E_1$$

where E_1 is the internal energy of the system in the initial state and E_2 is the internal energy in the final state.

If the heat absorbed at constant volume during the change is represented by q_v, then

$$\boxed{q_v = \Delta E} \quad \text{...(i)}$$

Hence, **Internal energy change** of a system or a reaction is the heat evolved or absorbed at constant volume.

Q. 51. What is meant by enthalpy and enthalpy change? Give sign conventions for heat, energy and work.

Ans. When a process is carried out at constant pressure and temperature, there is a change in volume.

If the volume increases, some mechanical work is done by the system so that the internal energy decreases. Similarly, if the volume decreases, some mechanical work is done on the system so that the internal energy increases. In all such processes occurring at constant temperature when pressure remains constant, the total energy of a system is given by a thermodynamic quantity called **heat content** or **enthalpy** of the system and not by internal energy. It is denoted by H and is defined mathematically by the equation

$$H = E + PV$$

where E is the internal energy, P and V are the pressure and volume of the system respectively.

The **enthalpy** *of a system may be defined as the sum of the internal energy and the product of pressure and volume of the system.*

The *physical significance of enthalpy* may be put as under :

The **enthalpy** *of a substance is the amount of energy stored within the substance, i.e., available for conversion into heat.*

Like internal energy, *the heat content or enthalpy of a substance is also a state function, i.e.,* depends only on the state of the system and is independent of the manner by which the state has been attained.

Enthalpy change. Like internal energy, the absolute value of enthalpy of a substance cannot be determined. However, the change in the enthalpy when a system undergoes a change can be determined experimentally accurately and that serves our purpose.

A relation between enthalpy change and internal energy change exists as follows :

$$\Delta H = \Delta E + P\,\Delta V \qquad \text{...(ii)}$$

where $\Delta E = E_2 - E_1$ represents the change in internal energy

and $\Delta V = V_2 - V_1$ represents the change in volume of the system.

Thus, **enthalpy change** *of a system may be defined as the sum of internal energy change and the pressure-volume work done.*

If q_p is the heat absorbed at constant pressure, then $q_p = \Delta H$ ...(iii)

Enthalpy change *is the heat absorbed or evolved at constant pressure.*

Standard enthalpy change. The enthalpy change of a reaction when all the reactants and the products are in their standard states, *i.e.*, at 298 K and under a pressure of one atmosphere is known as the standard enthalpy change. It is usually represented by $\Delta H°$.

As a chemical reaction involves breaking of bonds of the reactants and the formation of new bonds in the products, some net energy change (called the enthalpy of reaction) takes place.

$$\text{Enthalpy change of a reaction} = \begin{bmatrix}\text{Energy consumed to} \\ \text{break the bonds of} \\ \text{the reactants}\end{bmatrix} - \begin{bmatrix}\text{Energy released in the} \\ \text{formation of the bonds in} \\ \text{the products}\end{bmatrix}$$

If energy consumed > energy released, the net result is the absorption of energy and the reaction is *endothermic*.

If energy released > energy consumed, the net result is the evolution of energy and the reaction is *exothermic*.

Latest sign conventions for q or ΔH, ΔE and w.

(*i*) When heat is absorbed by the system, q is positive.

(*ii*) When heat is given out by the system, q is negative.

(*iii*) When energy is absorbed by the system, ΔE is positive.

(*iv*) When energy is given out by the system, ΔE is negative.

(*v*) When work is done on the system, w is positive.

(*vi*) Work done by the system, w is negative.

Q. 52. Describe giving examples exothermic and endothermic reactions.

Ans. Exothermic Reactions *are those reactions which are accompanied by the evolution of heat.*

The quantity of heat produced is shown alongwith the products with a 'plus' sign. A few examples of exothermic reactions are given below :

$$C(s) + O_2(g) \longrightarrow CO_2(g) + 393.5 \text{ kJ}$$

$$H_2(g) + \frac{1}{2}O_2(g) \longrightarrow H_2O(g) + 248.8 \text{ kJ}$$

$$CH_4(g) + 2O_2(g) \longrightarrow CO_2(g) + 2H_2O(g) + 890.4 \text{ kJ}$$

Endothermic Reactions *are those reactions which are accompanied by absorption of heat.*

Heat absorbed can be written with the products with a 'minus' sign. A few examples of endothermic reactions are given below :

$$N_2(g) + O_2(g) \longrightarrow 2NO(g) - 180.7 \text{ kJ}$$

$$C(s) + H_2O(g) \longrightarrow CO(g) + H_2(g) - 131.4 \text{ kJ}$$

$$C(s) + 2S(s) \longrightarrow CS_2(l) - 92.0 \text{ kJ}$$

The enthalpy change (ΔH) accompanying a reaction is given by

$$\Delta H = \text{Heat content of products} - \text{Heat content of reactants} = H_P - H_R$$

A reaction is exothermic, if the total heat content of the reactants is more than that of the products, *i.e.*, $H_R > H_P$. Thus, ΔH will be negative for an exothermic process.

Thus, the exothermic reactions given above may be written in terms of ΔH as :

$$C(s) + O_2(g) \longrightarrow CO_2(g), \quad \Delta H = -393.5 \text{ kJ}$$

$$H_2(g) + \frac{1}{2}O_2(g)\,O(l) \longrightarrow H_2O(l), \quad \Delta H = -285.8 \text{ kJ}$$

$$CH_4(g) + 2O_2(g) \longrightarrow CO_2(g) + 2H_2O(l), \quad \Delta H = -890.4 \text{ kJ}$$

A reaction is endothermic if the total heat content of reactants is less than that of the products, *i.e.*,

$$H_R < H_P$$

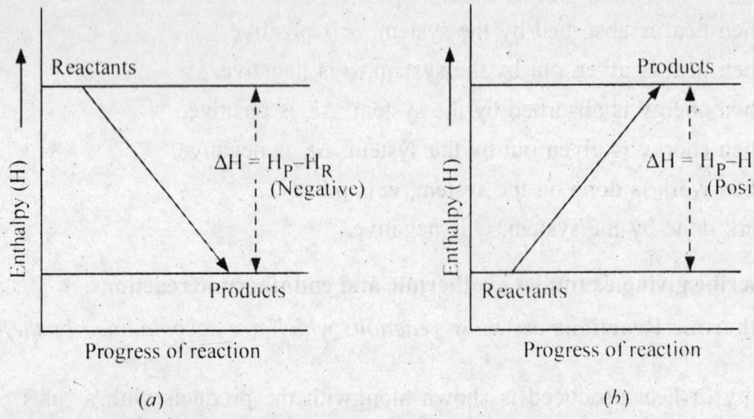

Fig. 6.10. (*a*) Exothermic reaction (*b*) Endothermic reaction.

Thus, ΔH will be positive for an endothermic reaction.

Endothermic reactions given above may be written in terms of ΔH as :

$$N_2\,(g) + O_2\,(g) \longrightarrow 2NO\,(g), \qquad \Delta H = +180.7 \text{ kJ}$$

$$C\,(s) + H_2O\,(g) \longrightarrow CO\,(g) + H_2(g), \quad \Delta H = +131.4 \text{ kJ}$$

$$C\,(s) + 2S\,(s) \longrightarrow CS_2\,(l), \qquad \Delta H = +92.0 \text{ kJ}$$

Exothermic and endothermic reactions may be represented graphically as shown in Fig. 6.10(*a*) and 6.10(*b*).

Q. 53. What is meant by a thermochemical equation?

Ans. *A balanced chemical equation not only indicating the quantities of the different reactants and products but also indicating the amount of heat evolved or absorbed is called a* **thermochemical equation.**

Fractional coefficients are allowed in writing a thermochemical equation. For example, the formation of water is written as

$$H_2(g) + \frac{1}{2}O_2\,(g) \longrightarrow H_2O\,(l), \qquad \Delta H = -\ 285.8 \text{ kJ}$$

Thus, 285.8 kJ of heat is produced when 1 mole of hydrogen reacts with 0.5 mole of oxygen. If the quantities of reactants are doubled, the heat produced will also be doubled. For example, in the above case, we may write

$$2H_2\,(g) + O_2\,(g) \longrightarrow 2\ H_2O\,(l), \qquad \Delta H = -\,571.6 \text{ kJ}$$

SOME POINTS TO REMEMBER ABOUT THERMOCHEMICAL EQUATIONS

(*i*) Unless otherwise mentioned, ΔH values are for the standard state of the substances (*i.e.*, 298 K and 1 atm pressure).

(*ii*) The physical state (*s, l, g*, etc.) of the different substances must be mentioned as the heat evolved or absorbed depends upon the physical state of the substances.

(*iii*) If the coefficients of the substances are multiplied or divided by some number, the value of ΔH is multiplied or divided by the same number.

(*iv*) If the reaction is reversed, the sign of ΔH changes but the magnitude remains the same.

Q. 54. Define heat of reaction. Which factors influence heat of reaction?

Ans. *The amount of heat evolved or absorbed in a chemical reaction when the number of moles of the reactants as represented by the chemical equation have completely reacted, is called the* **heat of reaction.**

Let us consider the following two examples :

$$C(s) + O_2(g) \longrightarrow CO_2(g) \quad \Delta H = -393.5 \text{ kJ}$$

$$C(s) + 2S(s) \longrightarrow CS_2(l), \quad \Delta H = +92.0 \text{ kJ}$$

The first equation indicates that when 1 mole of solid carbon (*i.e.*, 12 g) combines completely with 1 mole of oxygen gas (*i.e.*, 32 g), 393.5 kJ of heat are produced. Similarly, the second equation tells that when 1 mole of solid carbon (*i.e.*, 12 g) reacts completely with 2 moles of sulphur (64 g), 92 kJ of heat are absorbed.

$$\text{Heat of reaction, } \Delta H = \text{Heat content of products} - \text{Heat content of reactants}$$

If the reaction is reversed, the sign of ΔH is also reversed.

Factors Influencing Heat of Reaction : The heat of a reaction depends upon a number of factors as explained below :

(i) Quantities of the reactants involved. If the quantities of the reactants are changed, the heat of reaction changes proportionally.

(ii) Physical state of the reactants and products. Since latent heat is involved in the change of state, therefore, the physical state of the reactants and products affects the heat of reaction. For example, when hydrogen and oxygen gases combine to form liquid water, the heat of reaction is different from the value when they combine to form water vapours.

$$H_2(g) + \frac{1}{2}O_2(g) \longrightarrow H_2O(l), \quad \Delta H = -285.8 \text{ kJ}$$

$$H_2(g) + \frac{1}{2}O_2(g) \longrightarrow H_2O(g), \quad \Delta H = -248.8 \text{ kJ}$$

It is, therefore, essential to write the symbols *s*, *l*, *g* or *aq* to indicate whether a particular substance is solid, liquid, gaseous or an aqueous solution.

(iii) Allotropic modifications. Heat of reaction changes if a different allotropic form is involved in the reaction. For example,

$$C(\text{diamond}) + O_2(g) \longrightarrow CO_2(g), \quad \Delta H = -395.4 \text{ kJ}$$

$$C(\text{graphite}) + O_2(g) \longrightarrow CO_2(g), \quad \Delta H = -393.5 \text{ kJ}$$

(iv) Concentration of solutions. If the solutions are involved in a reaction, their concentrations affect the heat of reaction.

(v) Temperature. The heat of reaction depends upon the temperature at which the reactants and products are taken. Generally, the heat of reaction increases with temperature.

(vi) Conditions of constant pressure or constant volume, i.e., whether the reaction takes place at constant pressure or at constant volume.

If q_p and q_v represent the heats of reaction at constant pressure and volume respectively, then the following equation is applicable.

$$q_p = q_v + \Delta n_g RT$$

where $$\Delta n_g = n_p - n_r$$

n_p and n_r represent the number of moles of gaseous reactants and products respectively. R and T have the usual meanings.

Q. 55. Explain heat (or enthalpy) of combustion giving example. *(Garhwal 2010)*

Ans. *The heat of combustion of a substance is defined as the heat change (usually the heat evolved) when 1 mole of substance is completely burnt or oxidised in oxygen.* For example,

$$C(s) + O_2(g) \longrightarrow CO_2(g), \Delta H = -393.5 \text{ kJ}$$

When one mole of carbon (12 g) burns completely in oxygen, 393.5 kJ of heat are evolved. Thus, heat of combustion of carbon is 393.5 kJ.

The significance of the words *burns completely* must be clearly understood. For example, carbon may be oxidised to carbon monoxide (CO) or carbon dioxide (CO_2). The heats evolved in the two cases are different, viz.,

$$C(s) + \frac{1}{2}O_2(g) \longrightarrow CO(g), \quad \Delta H = -110.5 \text{ kJ}$$

$$C(s) + O_2(g) \longrightarrow CO_2(g), \quad \Delta H = -393.5 \text{ kJ}$$

Complete oxidation means oxidation to CO_2 and not to CO. Hence, the heat of combustion of carbon is 393.5 kJ mol^{-1}.

Q. 56. What is meant by calorific value of foods and fuels? Give calorific value of some common foods and fuels.

Ans. For the working of the human machine, we eat carbohydrates, fats, etc., in the form of food. The carbohydrates are first decomposed in our body by the enzymes to form glucose which is then oxidised by the oxygen that we inhale to produce energy as given by the following equation.

$$C_6H_{12}O_6(s) + 6O_2(g) \longrightarrow 6CO_2(g) + 6H_2O(g) + 2900 \text{ kJ}$$

This oxidation reaction is usually called '*combustion of food*'.

Different foods have different heats of combustion. These are usually expressed in terms of their calorific values.

The calorific value of a fuel or food is the amount of heat in calories or joules produced from the complete combustion of one gram of the fuel or the food.

Thus, according to the above reaction, 1 mole of glucose, *i.e.*, 180 g of glucose produce 2900 kJ of heat.

$$\text{Hence, calorific value of glucose} = \frac{2900}{180} = 16.1 \text{ kJ g}^{-1}$$

A normal person needs about 3000 kcal per day.

Table 6.1 : Calorific Values of Foods per Gram

Food	Calorific value kJ / gram	Food	Calorific value kJ / gram
Ghee	37.65	Fish	5.2
Butter	30.50	Meat	12.0
Cheese	14.56	Egg (yolk)	5.4
Milk (buffalo's)	4.90	Egg (white)	14.0
Milk (cow's)	2.20	Sugar	16.0
Curd	2.51	Rice	16.7
Soyabean	18.07	Potatoes	4.0

Table 6.2 : Calorific Values of some Common Fuels in kJ per kg

Wood	16,000
Charcoal	29,000
Natural gas	46,000
Kerosene	38,000

A normal youngman needs 10,000 kJ of energy per day for normal functioning.

Q. 57. A cylinder of indane gas contains 11.2 kg of butane. A normal family needs 25000 kJ energy per day for cooking. How long will the cylinder of gas last assumping 100% combustion?

If the efficiency of combustion is 85%, how long will the cylinder last? Heat of combustion of butane is 2658 kJ mol^{-1}.

Ans. Mol. mass of butane is 58 g mol^{-1}.

Thus, 58 g butane on complete combustion produces = 2658 kJ

$\therefore$ 11200 g butane on complete combustion produces $= \dfrac{2658 \times 11200}{58}$ kJ = 513268.96 kJ.

$$\text{The cylinder will last } = \frac{513268.96}{25000} = 20.5 \text{ days}$$

If the efficiency of combustion is only 85%, the cylinder will last = 0.85 × 20.5 $\approx$ **17.5 days.**

Q. 58. Explain heat of formation and standard heat of formation.

Ans. Heat of Formation (or Enthalpy of Formation) : *The heat of formation of a substance is defined as the heat change that takes place when 1 mole of the substance is formed from its elements under given conditions of temperature and pressure. It is usually represented by* ΔH_f.

Standard heat of formation *of a substance is defined as the heat change accompanying the formation of 1 mole of the substance in the standard state from its elements, also taken in the standard state (i.e., 298 K, and 1 atmospheric pressure). It is usually represented by* ΔH^o_f.

For example, in the reaction

$$C\,(s) + O_2\,(g) \longrightarrow CO_2\,(g), \quad \Delta H^o_f = -\,393.5 \text{ kJ}$$

When 1 mole of CO_2 (*g*) is formed from its elements, viz., C (*s*) and O_2 (*g*) (all substances being taken in the standard state), 393.5 kJ of heat is produced. Hence, the standard heat of formation of gaseous CO_2 is 393.5 kJ mol^{-1}.

Importance of standard enthalpies of formation. Knowing the standard enthalpies of formation of the different compounds involved in a chemical reaction, the standard enthalpy change of the given reaction can be obtained using the formula.

$$\Delta H^o_{\text{reaction}} = \begin{bmatrix}\text{Sum of the standard heats} \\ \text{of formation of products}\end{bmatrix} - \begin{bmatrix}\text{Sum of the standard heats} \\ \text{of formation of reactants}\end{bmatrix}$$

i.e., $$\Delta H^o_{\text{reaction}} = \Sigma\, \Delta H^o_f \text{ (Products)} - \Sigma\, \Delta H^o_f \text{ (Reactants)}$$

Thus, for a general reaction $\quad aA + bB \longrightarrow cC + dD$

$$\Delta H^o_{\text{reaction}} = [c\, \Delta H^o_f(C) + d\, \Delta H^o_f\,(D)] - [a\, \Delta H^o_f(A) + b\, \Delta H^o_f(B)]$$

Remember ΔH^o_f is taken as zero for the elements in a reaction. Suppose we are interested in finding the heat of combustion of methane. The reaction is represented as :

$$CH_4\,(g) + 2O_2(g) \longrightarrow CO_2\,(g) + 2H_2O\,(l)$$

$$\Delta H^\circ_{reaction} = [\Delta H^\circ_f \text{ for } CO_2 + (2 \times \Delta H^\circ_f \text{ for } H_2O)] - [\Delta H^\circ_f \text{ for } CH_4 + (2 \times \Delta H^\circ_f \text{ for } O_2)]$$

Putting the values of heats of formation in the standard state

$$\Delta H^\circ_{reaction} = [-393.5 + 2 \times (-285.8)] - [-74.9 + 0] = -965.1 + 74.9 = -890.2 \text{ kJ}$$

Q. 59. Enthalpies of formation of CO_2 (*g*) and H_2O (*l*) under standard conditions are 394.65 kJ and 285.84 kJ per mole. If the standard enthalpy of combustion of acetaldehyde (CH_3CHO) be 1167.62 kJ per mole, find its enthalpy of formation.

Ans. From the given data

(*a*) $$C + O_2 \longrightarrow CO_2\,(g) \quad ; \Delta H = -394.65 \text{ kJ}$$

(*b*) $$H_2 + \frac{1}{2}O_2 \longrightarrow H_2O\,(l) \quad ; \Delta H = -285.84 \text{ kJ}$$

(*c*) $$CH_3CHO + \frac{5}{2}O_2 \longrightarrow 2CO_2 + 2H_2O\,(g) \quad ; \Delta H = -1167.62 \text{ kJ}$$

Our aim is to find out the ΔH value for the reaction

$$2C + 2H_2 + \frac{1}{2}O_2 \longrightarrow CH_3CHO$$

Using standard enthalpy values

ΔH°_f for $CO_2 = -394.65$ kJ, ΔH°_f for $H_2O = -285.84$ kJ and ΔH°_f for $O_2 = 0$

($\therefore$ Enthalpy of formation of elements is taken as zero).

Let ΔH°_f for CH_3CHO be x kJ.

ΔH value for any reaction = Enthalpies of the products – Enthalpies of the reactants.

From the equation

$$CH_3CHO + \frac{5}{2}O_2 \longrightarrow 2CO_2\,(g) + 2H_2O \quad ; \quad \Delta H = -1167.62 \text{ kJ}$$

$$\therefore \quad -1167.62 = [2 \times \Delta H^\circ_f\,(CO_2) + 2 \times \Delta H^\circ_f(H_2O)] - [\Delta H^\circ_f(CH_3CHO) + \frac{5}{2} \times \Delta H^\circ_f(O_2)]$$

$$= 2 \times (-394.65) + 2 \times (-285.84) - x + 0 = -789.30 - 571.68 - x$$

$$-1167.62 = -1360.98 - x \quad \text{or} \quad x = -193.36 \text{ kJ}$$

Hence, enthalpy of formation of CH_3CHO is **–193.36 kJ.**

Q. 60. Calculate ΔH° for the reaction : $CO_2(g) + H_2(g) \longrightarrow CO(g) + H_2O(g)$ given that ΔH°_f for $CO_2(g)$, $CO(g)$ and $H_2O(g)$ are – 393.5, – 111.3 and – 241.8 kJ mol^{-1} respectively.

Ans. ΔH° for the reaction $CO_2(g) + H_2(g) \longrightarrow CO(g) + H_2O(g)$ is

$$\Delta H^\circ = [H^\circ_{f\,(products)} - H^\circ_{f\,(reactants)}]$$

$$= [(-111.3) + (-241.8)] - [(-393.5) + 0] = (-111.3 - 241.8) - (-393.5)$$

$$= -353.1 + 393.5$$

ΔH° = **+ 40.4 kJ.** [Enthalpy of every element in standard state is assumed as zero].

Q. 61. Enthalpies of formation of C_2H_5OH (*l*), CO_2, (*g*) and H_2O (*l*) are –277.0, –393.5 and –285.8 kJ mole^{-1} respectively. Calculate enthalpy change for reaction.

$$C_2H_5OH\,(l) + 3O_2\,(g) \longrightarrow 2CO_2(g) + 3H_2O\,(l)$$

Ans. ΔH^o for the reaction $C_2H_5OH\,(l) + 3O_2\,(g) \longrightarrow 2CO_2\,(g) + 3H_2O\,(l)$ is

$$\Delta H^o = \Delta H^o{}_{f\,(\text{products})} - \Delta H^o{}_f\,(\text{reactants})$$

$$= [2\times\Delta H^o{}_f(CO_2)+3\times\Delta H^o{}_f(H_2O)] - [\Delta H^o{}_f(C_2H_5OH) + 3\times\Delta H^o{}_f(O_2)]$$

$$= [2\times(-393.5) + 3\times(-285.8)] - [-277+0] = -787.0 - 857.4 + 277$$

$$= -1644.4 + 277 = \mathbf{-1367.4\ kJ.}$$

Q. 62. What is heat (or enthalpy) of neutralization ? *(Garhwal 2010)*

Ans. *The heat of neutralization of an acid by a base is defined as the heat change when one gram equivalent of the acid is neutralized by a base, the reaction being carried out in dilute aqueous solution.*

The heat of neutralization of a base by an acid is defined in a similar manner.

For example, when one gram equivalent of HCl is neutralized by NaOH or one gram equivalent of NaOH is neutralized by HCl, in dilute and aqueous solution, 57.1 kJ of heat is produced.

$$NaOH + HCl \longrightarrow NaCl + H_2O, \quad \Delta H = -57.1\ kJ$$

Hence, the heat of neutralization of HCl with NaOH or NaOH with HCl is 57.1 kJ.

The heat of neutralization of any strong acid (HCl, HNO_3, H_2SO_4) with a strong base (NaOH, KOH) or vice versa, is always the same, *i.e.*, 57.1 kJ. This is because the strong acids, strong bases and the salts that they form are all completely ionized in dilute aqueous solution. Thus, the reaction between a strong acid and a strong base may be written as

$$Na^+ + OH^- + H^+ + Cl^- \longrightarrow Na^+ + Cl^- + H_2O, \Delta H = -57.1\ kJ$$

Cancelling out common ions on the two sides, we have

$$H^+\,(aq) + OH^-\,(aq) \longrightarrow H_2O\,(l), \Delta H = -57.1\ kJ$$

Thus, neutralization may be regarded as a reaction between the H^+ ions given by the acid with the OH^- ions given by the base. Since strong acids and strong bases ionize completely in dilute aqueous solution, the number of H^+ ions and OH^- ions produced by one gram equivalent of the strong acid and the strong base is always the same. Hence, the heat of neutralization between a strong acid and a strong base is always the same. *i.e.*, –57.1 kJ.

If the acid or the base is weak, the heat of neutralization is usually less than 57.1 kJ. Consider the neutralization of a weak acid like acetic acid with a strong base like NaOH. Acetic acid ionizes to a small extent whereas NaOH ionizes completely as

(*i*) $CH_3COOH \rightleftharpoons CH_3COO^- + H^+$ (*ii*) $NaOH \longrightarrow Na^+ + OH^-$

When H^+ ions given by the acid combine with the OH^- ions given by the base, the equilibrium (*i*) shifts to the right, *i.e.*, more of acetic acid dissociates. A part of the heat produced during the combination of H^+ ions and OH^- ions is used up for the complete dissociation of acetic acid. Hence, the net heat evolved in the above reaction is less than 57.1 kJ.

Compared to a strong acid, a weak acid produces a smaller number of H^+, hence, heat of neutralization is smaller.

Q. 63. Determine the amount of heat released when :

(*a*) 0.75 mole of nitric acid is neutralised by 0.75 mole of sodium hydroxide.

(*b*) 0.75 mole of hydrochloric acid is neutralised by 0.5 mole of potassium hydroxide.

(*c*) 900 ml of 0.5 M hydrochloric acid solution is mixed with 500 ml of 0.4 M sodium hyrdoxide solution. Assuming the specific heat of water as 4.18 J K^{-1} g^{-1}, also calculate the rise in temperature.

(*d*) 200 ml of 0.2 M sulphuric acid is mixed with 300 ml of 0.1 M sodium hydroxide solution. Calculate the rise in temperature.

Ans. (*a*) The reaction is : (0.75 moles) HNO_3 + (0.75 moles) NaOH $\longrightarrow$ $NaNO_3 + H_2O$ When one mole of HNO_3 is neutralised by 1 mole of NaOH, heat released is 57.1 kJ. Therefore, when 0.75 mole of HNO_3 is neutralised by 0.75 mole of NaOH, heat released will be

$$57.1 \times 0.75 = \mathbf{42.8\ kJ.}$$

(*b*) Out of 0.75 mole of HCl, 0.5 mole will react with 0.5 mole of KOH and 0.25 mole of HCl will be left unreacted.

$$\text{Heat released} = 57.1 \times 0.5 = \mathbf{28.5\ kJ.}$$

(*c*) 1000 ml of 1 M HCl = 1 mole

$$\therefore \quad 900 \text{ ml of } 0.5 \text{ M HCl} = \frac{900 \times 0.5}{1000} = 0.45 \text{ mole}$$

$$\text{Similarly,} \quad 500 \text{ ml of } 0.4 \text{ M NaOH} = \frac{500 \times 0.4}{1000} = 0.2 \text{ mole}$$

Out of 0.45 mole of HCl, 0.2 mole of HCl reacts with 0.2 mole of NaOH

$$\therefore \quad \text{Heat evolved} = 57.1 \times 0.2 = 11.4 \text{ kJ.}$$

To calculate the rise in temperature, use the relation

$$\text{Mass} \times \text{Sp. heat} \times \text{Rise in temp.} = \text{Heat evolved}$$

Heat evolved is 11.4 kJ or 11400 J, mass of the mixture is 900 g + 500 g (assuming the densities of the solution as unity) and sp. heat is 4.18 J.

$$\therefore \quad \text{Rise in temp.} = \frac{\text{Heat evolved}}{\text{Mass} \times \text{Sp. heat}} = \frac{11.4 \times 1000}{1400 \times 4.18} = 1.95°$$

(*d*) 1 molecule of H_2SO_4 releases two H^+ ions

$$200 \text{ ml. of } 0.2\text{M } H_2SO_4 = \frac{2 \times 200 \times 0.2}{1000} \text{ moles of } H^+ = 0.08 \text{ moles of } H^+$$

$$300 \text{ ml of } 0.1 \text{ M NaOH} = 0.03 \text{ mole } OH^-$$

Out of 0.08 mole of H^+, 0.03 mole H^+ will react with 0.03 mole OH^-.

$$\therefore \quad \text{Heat evolved} = 57.1 \times 0.03 = 1.7 \text{ kJ.}$$

$$\therefore \quad \text{Rise in temperature} = \frac{1.7 \times 1000}{500 \times 4.18} = \mathbf{0.8°}.$$

Q. 64. Calculate the amount of heat evolved when :

(*i*) 600 cm^3 of 0.1 M hydrochloric acid is mixed with 200 cm^3 of 0.2 M sodium hydroxide solution.

(*ii*) 250 cm^3 of 0.2 M sulphuric acid is mixed with 400 cm^3 of 0.5 M potassium hydroxide solution.

Assuming that the specific heat of water is 4.18 J K^{-1} g^{-1} and ignoring the heat absorbed by the container, thermometer, stirrer, etc., what would be the rise in temperature in each of the above cases?

Ans. (*i*) 600 cm^3 of 0.1M HCl $= \frac{0.1}{1000} \times 600$ mole of HCl

$= 0.06$ mole of HCl $= 0.06$ mole of H^+ ions

200 cm^3 of 0.2M NaOH $= \frac{0.2}{1000} \times 200$ mole of NaOH

$= 0.04$ mole of NaOH $= 0.04$ mole of OH^- ions

Thus, 0.04 mole of H^+ ions will combine with 0.04 mole of OH^- ions to form 0.04 mole of H_2O and 0.02 mole of H^+ ions will remain unreacted.

Heat evolved when 1 mole of H^+ ions combines with 1 mole of OH^- ions = 57.1 kJ.

∴ Heat evolved when 0.04 mole of H^+ ions combines with 0.04 mole of OH^- ions.

$= 57.1 \times 0.04 =$ **2.284 kJ.**

(*ii*) 250 cm^3 of 0.2M H_2SO_4 $= \frac{0.2}{1000} \times 250$ mole of H_2SO_4

$= 0.05$ mole of $H_2SO_4 = 0.10$ mole of H^+ ions

400 cm^3 of 0.5M KOH $= \frac{0.5}{1000} \times 400$ mole of KOH

$= 0.2$ mole of KOH $= 0.2$ mole of OH^- ions

Thus, 0.10 mole of H^+ ions will neutralize 0.10 mole of OH^- ions, (out of 0.2 mole of OH^- ions) to form 0.10 mole of H_2O.

Hence, heat evolved $= 57.1 \times 0.1 =$ **5.71 kJ.**

In case (*i*), heat produced = 2.284 kJ = 2284 J

Total mass of the solution = 600 + 200 = 800 g

Specific heat = 4.18 $JK^{-1}g^{-1}$

$$Q = m \times s \times \Delta t$$

$$\therefore \quad \Delta t = \frac{Q}{m \times s} = \frac{2284}{800 \times 4.18} = \mathbf{0.68^o.}$$

In case (*ii*), heat produced = 5710 J

Total mass of the solution = 250 + 400 = 650 g

$$\therefore \quad \Delta t = \frac{Q}{m \times s} = \frac{4568}{650 \times 4.18} = \mathbf{1.68^o.}$$

Q. 65. Explain the following :

1. Heat of solution

2. Heat of hydration

3. Heat of fusion

4. Heat of vaporisation

5. Heat of sublimation

Ans. 1. Heat of Solution (or Enthalpy of solution) : *The heat of solution of a substance in a particular solvent is defined as the heat change (i.e., the amount of heat evolved or absorbed) when 1 mole of the substance is dissolved in such a large volume of the solvent that further addition of the solvent does not produce any more heat change.*

Thus, the thermochemical equations for the dissolution of KCl and $CuSO_4$ in water may be represented as

$$KCl\,(s) + aq \longrightarrow KCl\,(aq), \quad \Delta H = +18.6 \text{ kJ}$$

$$CuSO_4\,(s) + aq \longrightarrow CuSO_4\,(aq), \quad \Delta H = -66.5 \text{ kJ}$$

Thus, the heat of solution of KCl = +18.6 kJ mol^{-1}

and heat of solution of $CuSO_4$ = – 66.5 kJ mol^{-1}

It is interesting to note that the salts like copper sulphate, calcium chloride, etc., when present in the hydrated state (*i.e.*, $CuSO_4.5H_2O$, $CaCl_2.6H_2O$, etc.) dissolve with the absorption of heat. For example,

$$CuSO_4.5H_2O + aq \longrightarrow CuSO_4\,(aq), \Delta H = +11.7 \text{ kJ}$$

Thus, it can be generalized that the process of dissolution is usually endothermic for :

(*i*) salts which do not form hydrates like NaCl, KCl, KNO_3, etc.

(*ii*) hydrated salts like $CuSO_4.5H_2O$, $CaCl_2.6H_2O$, etc.

2. Heat of Hydration. *The amount of heat change (i.e., the heat evolved or absorbed) when one mole of the anhydrous salt combines with the required number of moles of water so as to change into the hydrated salt, is called the heat of hydration.*

For example, the heat of hydration of copper sulphate is –78.2 kJ mol^{-1}. This is represented as follows :

$$CuSO_4\,(s) + 5H_2O \longrightarrow CuSO_4.5H_2O\,(s), \quad \Delta H = -78.2 \text{ kJ}$$

3. Heat of Fusion. *Heat of fusion is the heat change that takes place when one mole of a solid substance changes into its liquid state at its melting point.*

Heat of fusion (ΔH_{fus}) of ice (m.p. = 273 K) is 6.0 kJ mol^{-1}. It may be represented as follows :

$$\underset{\text{Ice}}{H_2O\,(s)} \longrightarrow \underset{\text{Water}}{H_2O\,(l)}, \quad \Delta H = +6.0 \text{ kJ}$$

4. Heat of Vaporisation. *It is the heat change that takes place when one mole of a liquid changes into its gaseous state at its boiling point.*

For example, the heat of vaporisation ($\Delta H_{vap.}$) of water into its gaseous state at the boiling point (373 K) is 40.7 kJ. It may be represented as follows :

$$\underset{\text{Water}}{H_2O\,(l)} \longrightarrow \underset{\text{Steam}}{H_2O\,(g)}, \quad \Delta H = +40.7 \text{ kJ}$$

5. Heat of Sublimation. *Heat of sublimation of a substance is the heat change that takes place when 1 mole of a solid changes directly into vapour phase at a given temperature below its melting point.*

Sublimation is a process in which a solid on heating changes directly into gaseous state below its melting point. For example, the heat of sublimation of iodine is 62.39 kJ mol^{-1}. It is represented as follows :

$$I_2\,(s) \longrightarrow I_2\,(g), \quad \Delta H = +62.39 \text{ kJ}$$

Most solids, that sublime readily are molecular solids, e.g., iodine and naphthalene, etc.

Q. 66. What is Hess's law of constant heat summation? Explain giving examples.

(Burdwan 2004; Purvanchal 2006; Vidyasagar 2007; Himachal 2010; Bharathiar 2011,

Ans. G.H. Hess, a Russian chemist, in 1840, gave a law about the heats of reactions which states as follows :

The total amount of heat evolved or absorbed in a reaction is the same whether the reaction takes place in one step or in a number of steps. In other words, the total amount of heat change in a reaction depends only upon the nature of the initial reactants and the nature of the final products and not upon the path or the manner by which this change is brought about.

Hess's law may be illustrated with the help of following examples :

(1) Sulphur (Rhombic) burns to form SO_3 to evolve 395.4 kJ of heat.

$$S_R + \frac{3}{2} O_2\ (g) \longrightarrow SO_3\ (g), \quad \Delta H = -395.4 \text{ kJ}$$

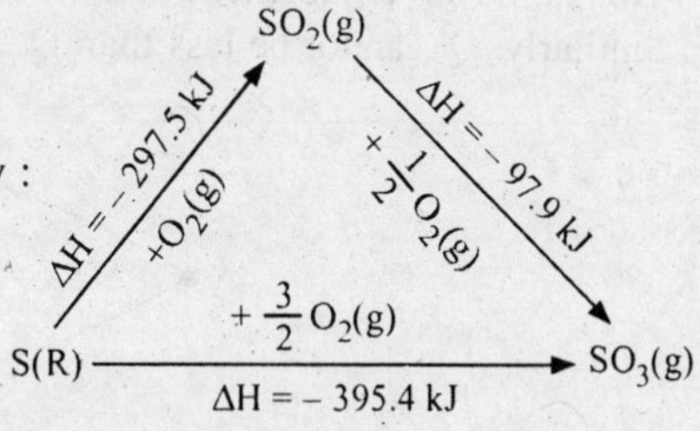

Sulphur may change to SO_3 in two steps as given below :

$$S_R + O_2\ (g) \longrightarrow SO_2\ (g), \quad \Delta H = -297.5 \text{ kJ}$$

$$SO_2 (g) + \frac{1}{2} O_2\ (g) \longrightarrow SO_3\ (g), \quad \Delta H = -97.9 \text{ kJ}$$

(See accompanying figure)

Total heat evolved in the two steps is $-297.5 + (-97.9) = -395.4$ kJ.

This is the same as for the direct reaction in one step.

(2) When carbon (graphite) burns to form carbon dioxide directly in one step, 393.5 kJ mol^{-1} of heat is evolved

$$C\ (s) + O_2\ (g) \longrightarrow CO_2\ (g), \quad \Delta H = -393.5 \text{ kJ}$$

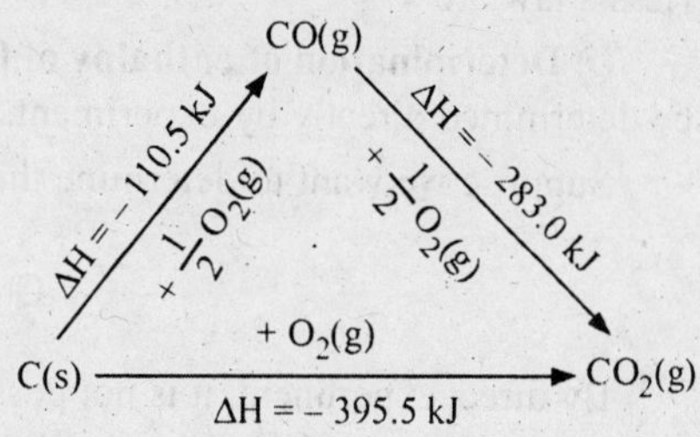

If carbon burns to form carbon monoxide first which then burns to form carbon dioxide, the heats evolved in the two steps are as follows (see accompanying figure).

$$C\ (s) + \frac{1}{2} O_2\ (g) \longrightarrow CO\ (g), \quad \Delta H = -111.5 \text{ kJ}$$

$$CO\ (g) + \frac{1}{2} O_2\ (g) \longrightarrow CO_2\ (g), \quad \Delta H = -282.0 \text{ kJ}$$

Thus, the total heat evolved in the two steps comes out to be $(-111.5) + (-282.0) = -393.5$ kJ, which is the same when the reaction takes place directly in one step.

Proof of Hess's Law. Consider the general reaction : $A \longrightarrow D$

Suppose the heat evolved in this reaction directly is Q joules.

Now suppose the same reaction takes place in three steps as follows :

$$A \xrightarrow{\text{Step 1}} B \xrightarrow{\text{Step 2}} C \xrightarrow{\text{Step 3}} D$$

Suppose the heats evolved in these three steps are q_1, q_2 and q_3 joules respectively (Fig. 6.11).

Thus, the total heat evolved from A to D = $q_1 + q_2 + q_3 = Q'$ joules (say)

According to Hess's law, we must have $Q = Q'$

Suppose Hess's law is not correct. In that case either Q′ < Q or Q′ > Q. Let us suppose Q′ > Q. We shall now complete the cycle, *i.e.*, we shall go from A to D through three steps and return to A in one step and we shall see what heat changes take place.

Heat change from D to A in one step is reverse of heat change from A to D in one step.

In the direct process from A to D, Q joules of heat are evolved. Therefore, in direct process from D to A, Q joules of heat are absorbed.

Now if we move from A to D in steps and return to A directly. Q′ – Q joules of heat is produced. Thus, by repeating the cyclic process a number of times, a large amount of heat can be created. This is, however, against the law of conservation of energy. Hence, Q′ cannot be greater than Q. Similarly. Q′ cannot be less than Q. It has to be equal to Q. Thus, Hess's law is proved.

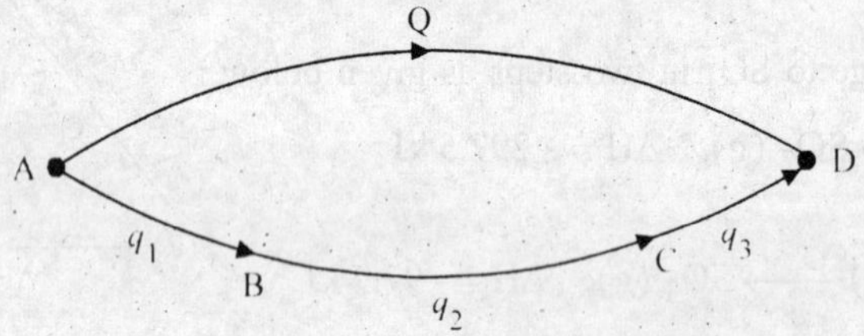

Fig. 6.11. Theoretical proof of Hess's law.

Q. 67. Describe various applications of Hess's Law. *(Allahabad 2002; Vidyasagar 2007)*

Ans. Hess's law is very useful in determining the heats (enthalpies) of many reactions, which cannot otherwise be measured by direct methods. The thermochemical equations are added, subtracted, multiplied or divided just like other algebraic equations. Some important applications of Hess's law are :

***(i)* Determination of enthalpy of formation.** The heat of formation of some compounds can not be determined directly by experiment. This can be calculated by the application of Hess's law.

Suppose we want to determine the heat of formation of CO

$$C + \frac{1}{2}O_2 \longrightarrow CO; \quad \Delta H = ?$$

By direct experiment, it is not possible because in the reaction between carbon and oxygen, we may get a mixture of CO and CO_2. But we require only CO. Therefore, from the experiment it is not possible to determine the heat of formation.

We know experimentally, the heat of combustion of pure carbon and carbon monoxide. It is represented by the following equations :

(a) $$C + O_2 \longrightarrow CO_2; \quad \Delta H = -393.5 \text{ kJ}$$

(b) $$CO + \frac{1}{2}O_2 \longrightarrow CO_2; \quad \Delta H = -283.0 \text{ kJ}$$

Now, applying Hess's law on *(a)* and *(b)*, we get the required equation for the heat of formation of CO. This can be achieved by subtracting *(b)* from *(a)*, *i.e.*, *(a – b)*.

$$C + \frac{1}{2}O_2 \longrightarrow CO; \quad \Delta H = -110.5 \text{ kJ}$$

Therefore, enthalpy of formation of CO is – 110.5 kJ.

***(ii)* Determination of heat (enthalpy) of transition.** Transition of an element from one allotropic form to another is a slow process and requires heating of one allotropic form of the element for a long time at constant temperature. Hence, enthalpies of transition of elements can not be determined experimentally. They are, however, determined indirectly by means of Hess's law. Thus, Hess's law is useful in determining the enthalpies of transition of the allotropic forms such as rhombic sulphur to monoclinic sulphur, yellow phosphorus to red phosphorus, graphite to diamond, etc.

Suppose we want to determine the enthalpy of transition of sulphur (rhombic) into sulphur (monoclinic). Each variety is combustioned and heats of combustion are found out. This is represented by the following equations :

(*a*) $$S(R) + O_2 \longrightarrow SO_2 ; \quad \Delta H = -295.1 \text{ kJ}$$

(*b*) $$S(M) + O_2 \longrightarrow SO_2 ; \quad \Delta H = -296.4 \text{ kJ}$$

Subtracting (*b*) from (*a*), we get

$$S(R) \longrightarrow S(M) ; \quad \Delta H = -295.1 - (-296.4) = +1.3 \text{ kJ}$$

Thus, transition of one gram atom of rhombic sulphur to monoclinic sulphur absorbs 1.3 kJ of heat.

(*iii*) Calculation of enthalpies from given data. Thermochemical equations can be added, subtracted, multiplied or divided by any numerical factor like algebraic equations. Hess's law is, thus, used for the calculation of enthalpy changes for the reactions for which direct experimental measurements are not available.

Q. 68. Calculate the heat of formation of acetic acid if its heat of combustion is 867 kJ mol^{-1}. The heats of formation of CO_2 (g) and H_2O (*l*) are –393.5 kJ mol^{-1} and –285.9 kJ mol^{-1} respectively. *(Agra 2004)*

Ans. The equations given are :

(*a*) $$CH_3COOH + 2O_2 \longrightarrow 2CO_2 + 2H_2O; \Delta H = -867 \text{ kJ}$$

(*b*) $$C + O_2 \longrightarrow CO_2 \quad ; \Delta H = -393.5 \text{ kJ}$$

(*c*) $$H_2 + \frac{1}{2} O_2 \longrightarrow H_2O \quad ; \Delta H = -205.9 \text{ kJ}$$

Our aim is to find out ΔH value for the reaction

$$2C + 2H_2 + O_2 \longrightarrow CH_3COOH ; \quad \Delta H = ?$$

To obtain this equation, multiply equation (*b*) and (*c*) by 2 and add. We get equation (*d*)

$$2C + 2O_2 + 2H_2 + O_2 \longrightarrow 2CO_2 + 2H_2O$$

or

(*d*) $$2C + 2H_2 + 3O_2 \longrightarrow 2CO_2 + 2H_2O ; \Delta H = -1358.8 \text{ kJ}$$

Subtracting (*a*) from (*d*), we get

$$\begin{array}{l} 2C + 2H_2 + 3O_2 \longrightarrow 2CO_2 + 2H_2O ; \Delta H = -1358.8 \text{ kJ} \\ CH_3COOH + 2O_2 \longrightarrow 2CO_2 + 2H_2O ; \Delta H = -867 \text{ kJ} \\ \quad - \qquad\qquad - \qquad\qquad\quad - \qquad - \\ \hline 2C + 2H_2 + O_2 \longrightarrow CH_3COOH \qquad \Delta H = -491.8 \text{ kJ} \end{array}$$

∴ Heat of formation of acetic acid = – **491.8 kJ.**

Q. 69. Find the heat of formation of SO_2 from the following thermochemical equations :

(a) $S(s) + \frac{3}{2} O_2(g) \longrightarrow SO_3(g) + X \text{ kJ}$

(b) $SO_2(s) + O_2(g) \longrightarrow SO_3(g) + Y \text{ kJ}$

Ans. Our aim is to find out ΔH value for the reaction

$$S + O_2 \longrightarrow SO_2 \quad \Delta H = ?$$

This can be obtained by subtracting (*b*) from (*a*), *i.e.*, (*a* – *b*)

(*a*) $$S(s) + \frac{3}{2} O_2(g) \longrightarrow SO_3(g) \quad \Delta H = -X \text{ kJ}$$

(*b*) $$SO_2(g) + \frac{1}{2} O_2(g) \longrightarrow SO_3(g) \quad \Delta H = -Y \text{ kJ}$$

$$\overline{\quad - \quad - \quad\quad - \quad} \quad \Delta H = -X - (-Y) = -X + Y$$

$$S + O_2 \longrightarrow SO_2(g)$$

$$= -(X - Y) \text{ kJ}$$

Q. 70. Calculate enthalpy of formation of anhydrous Al_2Cl_6 from the following data :

(*a*) $2Al(s) + 6HCl(aq) \longrightarrow Al_2Cl_6(aq) + 3H_2$; $\Delta H = 1004.4$ kJ

(*b*) $H_2(g) + Cl_2(g) \longrightarrow 2HCl(g)$; $\Delta H = 754.74$ kJ

(*c*) $HCl(g) + aq \longrightarrow HCl(aq)$; $\Delta H = -73.24$ kJ

(*d*) $Al_2Cl_6(s) + aq \longrightarrow Al_2Cl_6(aq)$ $\Delta H = -642.69$ kJ

Ans. Thermochemical equation for the formation of anhydrous Al_2Cl_6 is

$$2Al + 3Cl_2 \longrightarrow Al_2Cl_6(s); \qquad \Delta H = ?$$

To obtain ΔH value for this equation from the given data we perform as :

$$[a + (3 \times b) + (6 \times c)] - d$$

Rearranging the given equations, we get

$$\left.\begin{array}{l} 2Al(s) + 6HCl(aq) + 3H_2 + 3Cl_2 \\ + 6HCl(g) + aq - Al_2Cl_6(s) - aq \end{array}\right] \longrightarrow \begin{array}{l} Al_2Cl_6(aq) + 3H_2(g) + 6HCl(g) \\ + 6HCl(aq) - Al_2Cl_6(aq) \end{array}$$

$$2Al(s) + 3Cl_2(g) \longrightarrow Al_2Cl_6(s), \Delta H = 1004.4 + 3 \times 754.74 + 6 \times (-73.24)$$

$$-(-642.69)$$

$$= 3471.87 \text{ kJ}$$

Q. 71. Calculate enthalpy of formation of methane. Given :

(a) $C(\text{graphite}) + O_2(g) \longrightarrow CO_2(g)$; $\Delta H = -393.5$ kJ

(b) $2H_2(g) + O_2(g) \longrightarrow 2H_2O(l)$; $\Delta H = -571.8$ kJ

(c) $CH_4(g) + 2O_2(g) \longrightarrow CO_2(g) + 2H_2O(l)$; $\Delta H = -890.3$ kJ

Ans. We want to find out the value of ΔH for the reaction

$$C\ (\text{graphite}) + 2H_2\ (g) \longrightarrow CH_4; \qquad \Delta H = ?$$

This can be obtained by the operation $(a + b) - c$, we get

$$C\ (\text{graphite}) + O_2 + 2H_2 + O_2 \longrightarrow CO_2 + 2H_2O \qquad \text{.... (a + b)}$$

or

$$C\ (\text{graphite}) + 2O_2 + 2H_2 \longrightarrow CO_2 + 2H_2O\ ;\ \Delta H = -965.3\ \text{kJ}$$

$$CH_4 + 2O_2 \longrightarrow CO_2 + 2H_2O\ ;\ \Delta H = -890.3\ \text{kJ}$$

$$- \quad - \qquad\quad - \quad -$$

$$C\ (\text{graphite}) + 2H_2 \longrightarrow CH_4 \quad ;\ \Delta H = -75.0\ \text{kJ} \qquad \text{.... (a + b) – c}$$

Q. 72. From the thermochemical equations :

(a) $\mathbf{N_2\ (g) + 3H_2\ (g) \longrightarrow 2NH_3\ (g)\ \ ;\ \Delta H = -92.40\ kJ}$

(b) $\mathbf{2H_2\ (g) + O_2\ (g) \longrightarrow 2H_2O\ \ ;\ \Delta H = -483.79\ kJ}$

Find ΔH_f for NH_3 (g) and ΔH_f for H_2O (g).

Ans. Heat of formation (ΔH_f) is for one mole

$$\therefore\ \Delta H_f \text{ for } NH_3 = \frac{-92.4}{2} = -46.2\ \text{kJ}$$

$$\therefore\ \Delta H_f \text{ for } H_2O = \frac{-483.79}{2} = \mathbf{-241.895\ kJ.}$$

Q. 73. Calculate the heat of formation of carbon disulphide given that the heats of combustion of carbon disulphide, carbon and sulphur are –1109 kJ, –394.6 kJ and –298.7 kJ respectively.

Ans. The thermochemical equations for the combustion of CS_2, C and S are

(a) $CS_2 + 3O_2 \longrightarrow CO_2 + 2SO_2\ \ ;\ \Delta H = -1109\ \text{kJ}$

(b) $C + O_2 \longrightarrow CO_2\ \ ;\ \Delta H = -394.6\ \text{kJ}$

(c) $S + O_2 \longrightarrow SO_2\ \ ;\ \Delta H = -298.7\ \text{kJ}$

Our aim is to find out the value of ΔH for the following equation,

$$C + 2S \longrightarrow CS_2 \qquad ;\ \Delta H = ?$$

Multiplying (*c*) by 2 and adding to (*b*), we get

$$2S + 2O_2 + C + O_2 \longrightarrow 2SO_2 + CO_2$$

or

(d) $2S + C + 3O_2 \longrightarrow 2SO_2 + CO_2\ ;\ \Delta H = (-298.7) \times 2 + (-394.6)$

Subtracting (*a*) from (*d*), we get

$$2S + C + 3O_2 \longrightarrow 2SO_2 + CO_2\ \ ;\ \Delta H = -992.0\ \text{kJ}$$

$$CS_2 + 3O_2 \longrightarrow CO_2 + 2SO_2\ \ ;\ \Delta H = -1109\ \text{kJ}$$

$$- \quad - \qquad\quad - \quad -$$

$$2S + C \longrightarrow CS_2 \quad ;\ \Delta H = -992 - (-1109) = +117\ \text{kJ}$$

$\therefore$ Heat of formation of CS_2 = + 117 **kJ mol⁻¹**.

Q. 74. The enthalpy of combustion of ethylene (C_2H_4) is – 1410 kJ. What will be calorific value of ethylene per gram?

Ans. The thermochemical equation for the combustion of C_2H_4 is

$$\underset{28 \text{ grams}}{C_2H_4} + O_2 \longrightarrow 2CO_2 + 2H_2O \; ; \Delta H = -1410 \text{ kJ}$$

It is evident from the above equation

$$28 \text{ grams of } C_2H_4 \text{ give heat} = -1410 \text{ kJ}$$

$$\therefore \quad 1 \text{ gram of } C_2H_4 \text{ will give heat } = \frac{-1410}{28} = -50.36 \text{ kJ}$$

$$\therefore \quad \text{Calorific value of ethylene} = \mathbf{-\ 50.36\ kJ.}$$

Q. 75. One gram of an organic liquid A (mol. wt. = 128) liberates 150 joules of heat on solidification. Calculate the enthalpy of fusion of A.

Ans. Molecular weight of liquid A = 128 g

Heat evolved when 1 gram of liquid A solidifies = 150 J

$\therefore$ Heat evolved when 128 grams (1 mole) of liquid A solidify = 150 × 128 = 19200 J

$$\text{Liquid (A)} \longrightarrow \text{Solid (A)} \; ; \Delta H = -19200 \text{ J}$$

Rearranging this equation, we get

$$\text{Solid (A)} \longrightarrow \text{Liquid (A)} \; ; \Delta H = +\ 19200$$

$$\therefore \quad \text{Enthalpy of fusion of A} = \mathbf{+\ 19200\ joules.}$$

Q. 76. Calculate the heat of formation of KCl from the following data :

(*i*) $KOH\,(aq) + HCl\,(aq) \longrightarrow KCl\,(aq) + H_2O\,(l) \quad ; \Delta H = -57.3 \text{ kJ}$

(*ii*) $H_2\,(g) + \frac{1}{2}O_2\,(g) \longrightarrow H_2O\,(l) \quad ; \Delta H = -286.2 \text{ kJ}$

(*iii*) $\frac{1}{2}H_2\,(g) + \frac{1}{2}Cl_2\,(g) + aq \longrightarrow HCl\,(aq) \quad ; \Delta H = -164.4 \text{ kJ}$

(*iv*) $K\,(s) + \frac{1}{2}O_2\,(g) + \frac{1}{2}H_2\,(g) + aq \longrightarrow KOH\,(aq) \quad ; \Delta H = -487.4 \text{ kJ}$

(*v*) $KCl\,(s) + aq \longrightarrow KCl\,(aq) \quad ; \Delta H = +\ 18.4 \text{ kJ}$

Ans. Our aim is to find out ΔH for the following reaction.

(*vi*) $$K(s) + \frac{1}{2}Cl_2\,(g) \longrightarrow KCl\,(s), \Delta H = ?$$

In order to get this thermochemical equation, we follow the following two steps :

Step 1. Adding Eqns. (*iii*) and (*iv*) and subtracting Eqn. (*v*), we have

(*vii*) $K(s) + \frac{1}{2}Cl_2\,(g) + H_2(g) + \frac{1}{2}O_2(g) \longrightarrow KCl\,(s) + HCl\,(aq) + KOH(aq) - KCl\,(aq)$

$$\Delta H = -487.4 + (-164.4) - (18.4) = \mathbf{-670.2\ kJ.}$$

Step 2. To cancel out the terms of this equation which do not appear in the required equation (*vi*), add Eqn. (*i*) to Eqn. (*vii*) and subtract Eqn. (*ii*) from their sum. This gives

$$K(s) + \frac{1}{2}Cl_2(g) \longrightarrow KCl(s); \Delta H = -670.2 + 57.3 - (-286.2) = \mathbf{-441.3\ kJ.}$$

Q. 77. Calculate the heat of hydration of anhydrous copper sulphate ($CuSO_4$) into hydrated copper sulphate ($CuSO_4.5H_2O$). Given that the heats of solutions of anhydrous copper sulphate and hydrated copper sulphate are – 66.5 and +11.7 kJ mol^{-1} respectively.

(*Delhi 2004*)

Ans. Data provided is :

(*i*) $CuSO_4(s) + aq \longrightarrow CuSO_4(aq)$; $\Delta H = -66.5$ kJ

(*ii*) $CuSO_4.5H_2O(s) + aq \longrightarrow CuSO_4(aq)$; $\Delta H = +11.7$ kJ

Our aim is to find out heat of the following reaction.

$$CuSO_4(s) + 5H_2O(l) \longrightarrow CuSO_4.5H_2O(s) \quad ; \Delta H = ?$$

Equation (*i*) can be written in two steps as :

(*iii*) $CuSO_4(s) + 5H_2O(l) \longrightarrow CuSO_4.5H_2O(s)$; $\Delta H = q_1$ kJ

(*iv*) $CuSO_4.5H_2O(s) + aq \longrightarrow CuSO_4(aq)$; $\Delta H = q_2$ kJ

(*v*) $q_1 + q_2 = -66.5$ kJ (According to Hess's law)

Further, equations (*ii*) and (*iv*) are same. $\therefore\ q_2 = +11.7$ kJ

Putting this value in eq. (*v*) above, we get

$$q_1 + 11.7 = -66.5 \text{ or } q_1 = -66.5 - 11.7 \text{ kJ} = -78.2 \text{ kJ}$$

Thus, equation (*iii*) may be written as

$$CuSO_4(s) + 5H_2O(l) \longrightarrow CuSO_4.5H_2O(s) ; \Delta H = -78.2 \text{ kJ}$$

This is what we aimed at. Hence, the required value of the heat of hydration is $\Delta H = -78.2$ kJ.

Q. 78. Calculate the heat of formation of carbon monoxide (CO) from the following data :

(*i*) $C(s) + O_2(g) \longrightarrow CO_2(g)$; $\Delta H = -393$ kJ

(*ii*) $CO(g) + \frac{1}{2}O_2(g) \longrightarrow CO_2(g)$; $\Delta H = -282$ kJ

Ans. Our aim is to find out ΔH for the following reaction :

(*i*) $C(s) + \frac{1}{2}O_2(g) \longrightarrow CO(g)$; $\Delta H = ?$

Subtracting equation (*ii*) from (*i*), we get

$$C(s) + \frac{1}{2}O_2(g) - CO(g) \longrightarrow 0; \Delta H = -393.3 - (-282.8) = -110.5 \text{ kJ}$$

or $$C(s) + \frac{1}{2}O_2(g) \longrightarrow CO(g); \Delta H = -110.5 \text{ kJ}$$

∴ Heat of formation of CO, ΔH_f = **110.5 kJ.**

Q. 79. Calculate the heat change accompanying the transformation of C (graphite) to C (diamond). Given that the heats of combustion of graphite and diamond are 393.5 and 395.4 kJ mol^{-1} respectively.

Ans. Data provided is :

(*i*) C (graphite) + O_2 (*g*) ⟶ CO_2 (*g*) ; $\Delta H = -393.5$ kJ

(*ii*) C (diamond) + O_2 (*g*) ⟶ CO_2 (*g*) ; $\Delta H = -395.4$ kJ

Our aim is to find out ΔH for the following reaction

C (graphite) ⟶ C (diamond) ; ΔH = ?

Subtracting equation (*ii*) from (*i*), we get

C (graphite) – C (diamond) ⟶ 0 ; $\Delta H = -393.5 - (-395.4) = +1.9$ kJ

or C (graphite) ⟶ C (diamond) ; ΔH = **+ 1.9 kJ.**

Q. 80. The heat of combustion of ethyl alcohol (C_2H_5OH) is 1380.7 kJ mol^{-1}. If the heats of formation of CO_2 and H_2O are 394.5 and 286.6 kJ mol^{-1} respectively, calculate the heat of formation of ethyl alcohol.

Ans. Data provided is :

(*i*) $C_2H_5OH + 3O_2 \longrightarrow 2CO_2 + 3H_2O$; $\Delta H = -1380.7$ kJ

(*ii*) $C + O_2 \longrightarrow CO_2$; $\Delta H = -394.5$ kJ

(*iii*) $H_2 + \frac{1}{2}O_2 \longrightarrow H_2O$; $\Delta H = -286.6$ kJ

Our aim is to determine ΔH for the following reaction

$$2C + 3H_2 + \frac{1}{2}O_2 \longrightarrow C_2H_5OH$$

In order to get this thermochemical equation multiply Eqn. (*ii*) by 2 and Eqn. (*iii*) by 3 and substract Eqn. (*i*) from their sum, *i.e.*, operating : 2 × Eqn. (*ii*) + 3 × Eqn. (*iii*) – Eqn. (*i*), we get

$$2C + 3H_2 + \frac{1}{2}O_2 \longrightarrow C_2H_5OH ; \Delta H = 2(-394.5) + 3(-286.6) - (-1380.7) = -268.1 \text{ kJ}$$

Thus, the heat of formation of ethyl alcohol = ΔH_f = **–268.1 kJ mol^{-1}.**

Q.81. Calculate the standard enthalpy of formation of n-butane given that the standard enthalpies of combustion of n-butane (g), C (graphite) and H_2(g) are – 2878.5 kJ mol^{-1}, 393.5 kJ mol^{-1} and –285.3 kJ mol^{-1} respectively. ***(Delhi 2003)***

Ans. We aim at the equation

$4C + 5H_2 \longrightarrow C_4H_{10}$ ΔH = ?

The given data can be expressed in the form of the following equations.

(*i*) $C_4H_{10} + 6\frac{1}{2}O_2 \rightarrow 4CO_2 + 5H_2O$ **ΔH = – 2878.5 kJ**

(*ii*) $4C + 4O_2 \longrightarrow 4CO_2$ $\Delta H = 4 \times (-393.5)$ kJ

(*iii*) $5H_2 + \frac{5}{2}O_2 \longrightarrow 5H_2O$ $\Delta H = 5 \times (-285.3)$ kJ

Add (*ii*) and (*iii*) and subtract (*i*)

$$(-1574.0) + (-1426.5) - (-2878.5)$$
$$= -122.0 \text{ kJ}$$

Q. 82. Heats of solution (ΔH) for $BaCl_2.2H_2O$ and $BaCl_2$ are 8.8 and –20.6 kJ respectively. Calculate the heat of hydration of $BaCl_2$ to $BaCl_2.2H_2O$.

Ans. We are given

(*i*) $BaCl_2.2H_2O\ (s) + aq \longrightarrow BaCl_2\ (aq) \quad ; \Delta H = 8.8 \text{ kJ}$

(*ii*) $BaCl_2\ (s) + aq \longrightarrow BaCl_2\ (aq) \quad ; \Delta H = -20.6 \text{ kJ}$

We aim at :

$$BaCl_2\ (s) + 2H_2O \longrightarrow BaCl_2.2H_2O\ (s)\ ;\ \Delta H = ? \qquad ...(iii)$$

Equation (*ii*) may be written in two steps as

$$BaCl_2\ (s) + 2H_2O \longrightarrow BaCl_2.2H_2O\ (s)\ ;\ \Delta H = \Delta H_1 \text{ (say)} \qquad ...(iv)$$

$$BaCl_2.2H_2O\ (s) + aq \longrightarrow BaCl_2\ (aq)\ ;\ \Delta H = \Delta H_2 \text{ (say)} \qquad ...(v)$$

Then according to Hess's law

$$\Delta H_1 + \Delta H_2 = -20.6 \text{ kJ}$$

But $\quad \Delta H_2 = 8.8 \text{ kJ} \qquad$ [∵ Equation (*i*) = Equation (*v*)]

∴ $\quad \Delta H_1 = -20.6 - \mathbf{8.8} = -29.4 \text{ kJ}$

But Equation (*iii*) = Equation (*iv*)

Hence, the heat of hydration of $\quad BaCl_2 = \mathbf{-29.4\ kJ.}$

Q. 83. Given the following thermochemical equations :

(*i*) $S \text{ (rhombic)} + O_2\ (g) \longrightarrow SO_2\ (g)\ ; \qquad \Delta H = -297.5 \text{ kJ}$

(*ii*) $S \text{ (monoclinic)} + O_2\ (g) \longrightarrow SO_2\ (g)\ ; \qquad \Delta H = -300.0 \text{ kJ}$

Calculate ΔH for the transformation of one gram atom of rhombic sulphur into monoclinic sulphur.

Ans. We aim at :

$$S \text{ (rhombic)} \longrightarrow S \text{ (monoclinic)};\ \Delta H = ?$$

Equation (*i*) – Equation (*ii*) gives

$$S \text{ (rhombic)} - S \text{ (monoclinic)} \longrightarrow 0;\ \Delta H = 297.5 - (-300.0) = 2.5 \text{ kJ}$$

or $\quad S \text{ (rhombic)} \longrightarrow S \text{ (monoclinic)};\ \Delta H = \mathbf{+2.5\ kJ.}$

Thus, for the transformation of one gram atom of rhombic sulphur into monoclinic sulphur, 2.5 kJ of heat is absorbed.

Q. 84. Calculate the enthalpy of formation of $N_2O_5(g)$:

$$N_2(g) + \frac{5}{2}O_2(g) \longrightarrow N_2O_5(g)$$

The following information is given :

(i) $N_2(g) + O_2(g) \longrightarrow 2NO(g) \qquad \Delta H° = +180 \text{ kJ}$

(ii) $4NO_2(g) + O_2(g) \longrightarrow 2N_2O_5(g) \qquad \Delta H° = -120.6 \text{ kJ}$

(iii) $2NO(g) + O_2(g) \longrightarrow 2NO_2(g) \qquad \Delta H° = -114.0 \text{ kJ}$ *(Delhi 2006)*

Ans. Perform the operation (*i*) $+ \frac{1}{2}$ (*iii*) to get the desired equation as

$$N_2(g) + \frac{5}{2}O_2(g) \longrightarrow N_2O_5(g)$$

Applying Hess's law, perform the same operation on enthalpies to obtain the enthalpy of formation of N_2O_5 (g)

$$\Delta H_f = [180 - \frac{1}{2}(120.6) - 114.0] \text{ kJ}$$
$$= 5.7 \text{ kJ}$$

Q. 85. What is meant by bond energy? How does bond energy help to determine the heat of a reaction ? *(Nagpur 2008; Himachal 2010; Bharathiar 2011)*

Ans. *The bond energy of a particular bond is defined as the average amount of energy released when one mole of bonds are formed from isolated gaseous atoms or the amount of energy required when 1 mole of bonds are broken so as to get the separated gaseous atoms.*

For diatomic molecules (like H_2, HCl, etc.), the bond energy is equal to the dissociation energy of the molecules but for a *polyatomic molecule* like CH_4, the bond dissociation energies of the four C — H bonds are different. Hence, an average value is taken.

Heat of reaction can be calculated from bond energy data. The following formula can be used directly in these calculations.

$$\Delta H_{reaction} = \Sigma \text{ Bond energies of reactants} - \Sigma \text{ Bond energies of products}$$

Q. 86. Calculate the enthalpy change for the reaction

$$H_2(g) + Br_2(g) \longrightarrow 2HBr(g)$$

Given that the bond energies of H — H, Br — Br and H — Br are 435, 192 and 364 kJ mol^{-1} respectively.

Ans. First method. *By calculating the total energy absorbed and released.*

Energy absorbed in the dissociation of 1 mole of H – H bonds = 435 kJ

Energy absorbed in the dissociation of 1 mole of Br – Br bonds = 192 kJ

Total energy absorbed = 435 + 192 = 627 kJ

Energy released in the formation of 1 mole of H – Br bonds = 364 kJ

∴ Energy released in the formation of 2 moles of H – Br bonds = 2 × 364 = 728 kJ

Energy released is greater than energy absorbed.

Hence, net result is the release of energy.

Energy released = 728 kJ – 627 kJ = 101 kJ

Thus, for the given reaction, ΔH = **–101 kJ.**

Second method. *Using the formula directly.*

$$\Delta H_{reaction} = \Sigma \text{ Bond energies of reactants} - \Sigma \text{ Bond energies of products}$$

$$= [\text{Bond energy (H—H)} + \text{Bond energy (Br—Br)}] - [2 \times \text{Bond energy (H—Br)}]$$

$$= 435 + 192 - 2 \times 364 = 627 - 128 = \mathbf{-101\ kJ.}$$

Hess's law can also be applied to solve the problem.

Q. 87. Calculate the bond energy of C — H bond, given the heat of combustion of methane (CH_4), graphite and hydrogen are 891 kJ, 394 kJ and 286 kJ respectively while the heat of sublimation of graphite is 717 kJ and the heat of dissociation of hydrogen molecule is 436 kJ.

Ans. The data given is :

(*i*) $CH_4(g) + 2O_2(g) \longrightarrow CO_2(g) + 2H_2O(g)$; ΔH = –891 kJ

(*ii*) $C(s) + O_2(g) \longrightarrow CO_2(g)$; ΔH = –394 kJ

(*iii*) $H_2(g) + \frac{1}{2}O_2(g) \longrightarrow H_2O(g)$; ΔH = –286 kJ

(*iv*) $C(s) \longrightarrow C(g)$; ΔH = +717 kJ

(*v*) $H_2(g) \longrightarrow 2H(g)$; ΔH = +436 kJ

We want ΔH for the following reaction

$$\frac{1}{4}[CH_4(g) \longrightarrow C(g) + 4H(g)] \quad ; \quad \Delta H = ? \qquad \ldots (vi)$$

The problem is solved in two steps.

First we calculate ΔH for $CH_4(g) \longrightarrow C(g) + 4H(g)$

and then divide the result by 4.

Operating Eqn. (*i*) + Eqn. (*iv*) + 2 × Eqn. (*v*), we get

$$CH_4\,(g) + 2O_2\,(g) + C\,(s) + 2H_2\,(g) \longrightarrow CO_2\,(g) + 2H_2O\,(g) + C\,(g) + 2H\,(g) \quad ...(vii)$$

To cancel out the terms not needed now operate Eqn. (*vii*) – 2 × Eqn. (*iii*) – Eqn. (*ii*).

$$CH_4\,(g) \longrightarrow C\,(g) + 4H\,(g);$$

$$\Delta H = -891 + 717 + 2 \times 436 - 2 \times (-286) - (-394) = +1664 \text{ kJ.}$$

This is the energy required for the dissociation of 4 moles of C—H bonds.

Bond dissociation energy for C—H bond $= \frac{1664}{4}$ kJ = **416 kJ mol^{-1}.**

Q. 88. If a man takes a diet which gives him energy equal to 9500 kJ per day and he expends energy in all forms to a total of 12000 kJ per day, what is the change in internal energy per day? If the energy lost was stored as sucrose (1632 kJ per 100 g), how many days should it take to lose 1 kg? Ignore water loss.

Ans. Loss of energy per day = 12000 – 9500 = 2500 kJ

For a loss of 1632 kJ of energy, sucrose ($C_{12}H_{22}O_{11}$) lost = 100 g (given)

∴ For a loss of 2500 kJ of energy, sucrose lost $= \frac{100}{1632} \times 2500 \text{ g} = 153.2 \text{ g}$

Thus, loss of 153.2 g in weight takes place in 1 day.

∴ Time taken to lose 1 kg or 1000 g glucose $= \frac{1}{153.2} \times 1000 =$ **6.5 days.**

Q. 89. From the thermochemical equation

$$C_6H_6\,(l) + 7\frac{1}{2}\,O_2\,(g) \longrightarrow 3H_2O\,(l) + 6CO_2\,(g); \qquad \Delta H = -3264.64 \text{ kJ mol}^{-1}$$

Calculate the energy evolved when 39 g of C_6H_6 are burnt in an open container.

Ans. 1 mole of benzene = 78 g

When 78 g of C_6H_6 burn, heat evolved = 3264.64 kJ

∴ When 39 g of C_6H_6 burn, heat evolved $= \frac{3264.64}{78} \times 39 =$ **1632.32 kJ.**

Q. 90. The heat evolved in the combustion of methane is given by the following equation :

$$CH_4\,(g) + 2O_2\,(g) \longrightarrow CO_2\,(g) + 2H_2O\,(l)\,;\ \Delta H = -890.3 \text{ kJ}$$

(*a*) How many grams of methane would be required to produce 445.15 kJ of heat on combustion?

(*b*) How many grams of carbon dioxide would be formed when 445.15 kJ of heat are evolved?

(*c*) What volume of oxygen at STP would be used in the combustion process (*a*) or (*b*)?

Ans. (*a*) From the given equation,

Heat produced from 1 mole of CH_4, *i.e.*, 16 g of CH_4 = 890.3 kJ

Heat produced from 8 g of CH_4 = **445.15 kJ.**

(*b*) From the given equation,

When 890.3 kJ of heat are evolved, CO_2 formed = 44 g

∴ When 445.15 kJ of heat are evolved, CO_2, formed = **22 g.**

(*c*) O_2 used in the production of 890.3 kJ of heat = 44.8 litres at STP.

Hence, O_2 used in the production of 445.15 kJ of heat = **22.4 litres at STP.**

Q. 91. The heat evolved in the combustion of glucose is shown in the following equation :

$$C_6H_{12}O_6 + 6O_2\,(g) \longrightarrow 6CO_2\,(g) + 6H_2O\,(g);\ \Delta H = -2840 \text{ kJ}$$

What is the energy requirement for production of 0.36 g of glucose by the reverse reaction?

Ans. Data provided is

$$C_6H_{12}O_6 + 6O_2\,(g) \longrightarrow 6CO_2\,(g) + 6H_2O\,(g);\ \Delta H = -2840 \text{ kJ}$$

Writing the reaction in reverse direction

$$6CO_2\,(g) + 6H_2O\,(g) \longrightarrow C_6H_{12}O_6\,(s) + 6O_2\,(g);\ \Delta H = +\,2840\ \text{kJ}$$

For production of 1 mole of $C_6H_{12}O_6$ (180 g) heat required (absorbed) = 2840 kJ

$\therefore$ For production of 0.36 g of glucose, heat absorbed $= \dfrac{2840}{180} \times 0.36 = \mathbf{5.68\ kJ.}$

Q. 92. The thermochemical equation for solid and liquid rocket fuel are given below :

$$2Al\,(s) + 1\frac{1}{2}O_2\,(g) \longrightarrow Al_2O_3\,(s)\ ;\ \Delta H = -1667.8\ \text{kJ}$$

$$H_2\,(g) + \frac{1}{2}O_2\,(g) \longrightarrow H_2O\,(l)\ \ ;\ \Delta H = -285.9\ \text{kJ}$$

(*a*) If equal masses of aluminium and hydrogen are used, which is better rocket fuel?

(*b*) Determine ΔH for the reaction $Al_2O_3\,(s) \longrightarrow 2Al\,(s) + 1\frac{1}{2}O_2\,(g)$

Ans. 2 moles of Al = 54 g

54 g of Al on combustion give heat = 1667.8 kJ

$\therefore$ 1 g of Al on combustion gives heat $= \dfrac{1667.8}{54} = 30.9$ kJ

1 mole of H_2 = 2 g

2 g of H_2 on combustion give = 285.9 kJ of heat

$\therefore$ 1 g of H_2 on combustion will give $= \dfrac{285.9}{2} = 142.95$ kJ of heat

Thus, H_2 is a better rocket fuel.

(*b*) Writing the reverse of the first reaction, we have

$$Al_2O_3\,(s) \longrightarrow 2Al\,(s) + 1\frac{1}{2}O_2\,(g);\ H = +\,1667.8\ \text{kJ}$$

Thus, for the reaction given in part (*b*) of the problem,

$$\Delta H = +\,1667.8\ \text{kJ.}$$

MISCELLANEOUS QUESTIONS

Q. 93. Heat of neutralisation of a strong acid with strong base is always constant. Explain. (*Calcutta 2007*)

Ans. See Q. 61

Q. 94. Explain integral heat of solution and differential heat of solution giving examples. (*Bharathiar 2011*)

Ans. There are two heats of solution as described below :

Integral heat of solution. It is defined as the enthalpy change when 1 mole of the solute is dissolved in n moles of the solvent.

Differential heat of solution. It is defined as the enthalpy change when one mole of the solute is dissolved in an infinite amount of the solution such that the concentration does not change by adding one mole of the solute.

Mathematically,

$$\text{Differential heat of solution} = \frac{d(m\,\Delta\,H)}{dm}$$

Heat change is plotted as a function of number of moles of the solute. Differential heat is obtained by finding the slope at any concentration.

Thus the differential heat of solution depends upon the concentration of the solution.

7

THERMODYNAMICS-II

Q. 1. Give various statements of second *law of thermodynamics.* (*Himachal 2000;*
(*Himachal 2000; Bhopal 2004;Indore 2004;*
Purvanchal 2007; Nagpur 2008; Himachal 2010)

Ans. Various statements of second law of thermodynamics are given below:

1. *All spontaneous processes like the flow of heat from hot end to cold end, diffusion of gas from high pressure to low pressure or the flow of water down a hill etc. are thermodynamically irreversible.*

Spontaneous process means a process which can take place without the help of any external agency. All *natural processes,* some of which are mentioned in the definition above, are spontaneous processes.

Further, the first law states that when heat is converted into work, the work obtained is equivalent to the heat absorbed. However it has been seen from experience that the heat absorbed cannot be completely converted into work without leaving some change in the system or the surroundings. Hence the second law is also stated as follows:

2. *The complete conversion of heat into work is impossible without leaving some effects elsewhere.*

3. *It is impossible to construct a machine, functioning in cycles, which can convert heat completely into the equivalent amount of work without producing changes elsewhere.*

4. *Without the use of an external agency, heat cannot by itself pass from a colder to a hotter body.*

Q. 2. What is meant by efficiency of an engine? Explain the terms source and sink.
(*Himachal 2000*)

Ans. *The fraction of the heat absorbed by a machine that it can transform into work is called the* **efficiency** *of the machine.* Thus if Q is the heat absorbed and W is the work done, then the efficiency of the machine is given by

$$\eta = \frac{W}{Q}$$

The machine used for the conversion of heat into work is called **Heat Engine.** In order to bring about this conversion, the engine absorbs heat from a heat reservoir at a higher temperature, called the **source**, converts a part of it into work and returns the remainder to the heat reservoir at a lower temperature, called the sink.

Q. 3. What is Carnot cycle? Calculate the efficiency of a Carnot engine working between two temperatures T_1 and T_2. (*Madurai 2004; Purvanchal 2007; Calcutta 2007;*
Vidyasagar 2007, Himachal 2010, Andhra 2010)

Ans. *A Carnot cycle is a process in which a system returns to its original state after a number of successive changes. A process conducted in this manner is called a cyclic process.* The carnot cycle consists of four different operations which can be shown on pressure-volume diagram as shown in Fig. 7.1.

A Carnot engine consists of a cylinder containing 1 mole of an ideal gas as the working substance and fitted with weightless, frictionless piston so shat all the operations in the cycle are carried out reversibly. For this reason, it is also called *reversible heat engine.* The cylinder is supposed to be insulated on all sides except at the bottom so that heat can flow to or from the system only through the bottom. Further, it is supposed that there are two heat reservoirs, one at a higher temperature T_2 (called the *source*) and the other at a lower temperature T_1 (called the *sink*). If some operation is carried out by placing the cylinder in the source or the sink, it can exchange heat with it and hence the temperature remains constant so that the process is *isothermal.* On the other hand, if during an operation, the cylinder is placed on an insulating material, no heat exchange can take place between the system and the surroundings and hence the process is *adiabatic.*

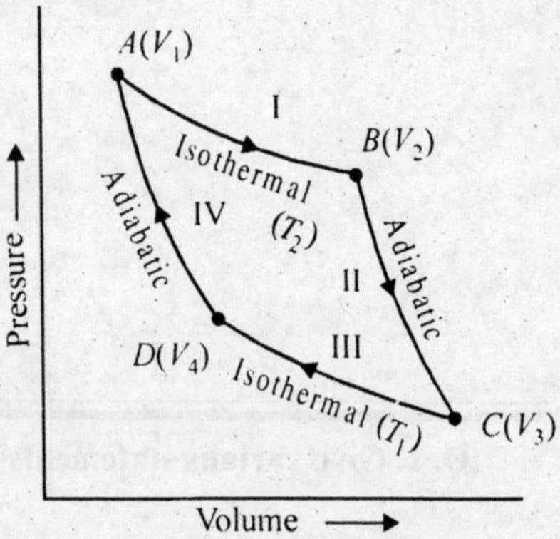

Fig. 7.1. Carnot cycle.

Four different operations are carried out as shown in Fig. 7.2 and are described below:

I. Isothermal expansion: The cylinder containing one mole of the ideal gas, occupying a volume V_1, is placed in contact with the source (*i.e.* heat reservoir at temperature T_2). The gas absorbs heat q_2 from the source and expands *isothermally* and *reversibly* till its volume

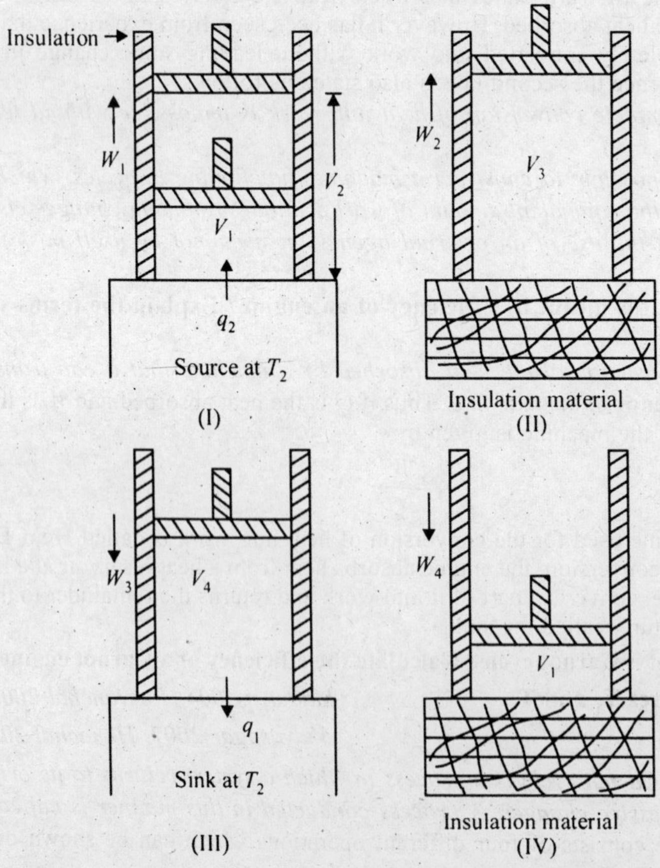

Fig. 7.2. Different operations of the Carnot cycle.

has increased to V_2. Since the gas is ideal, the work done w_1 by one mole of the ideal gas will be given by

$$w_1 = RT_2 \ln \frac{V_2}{V_1} \qquad ...(i)$$

Further for an ideal gas, the work done is equal to the heat absorbed, q_2 so that

$$w_1 = q_2 = RT_2 \ln \frac{V_2}{V_1} \qquad ...(ii)$$

Here, both w_1 (work done by the system) and q_2 are positive (old conventions)*

II. Adiabatic expansion: The cylinder is now removed from the source and placed in perfectly insulating material so that the gas now expands adiabatically and reversibly. Work is done in the expansion but since no heat enters or leaves the system, the temperature must fall. The reversible adiabatic expansion is continued till the temperature has fallen to T_1 which is the temperature of the sink. Suppose the corresponding volume becomes V_3. The path is shown by the adiabatic curve BC in Fig. 7·1. The work done w_2 is be given by

$$w_2 = C_v\,(T_2 - T_1) \qquad ...(iii)$$

where C_v is the heat capacity of the ideal gas. Here, again w_2 *is positive* by old conventions.

III. Isothermal compression: The cylinder is now removed from the insulating material and placed in contact with the sink (*i.e.* the heat reservoir at the lower temperature T_1). the gas is compressed isothermally and reversibly till the volume decreases from V_3 to V_4. The process is represented by the path CD in Fig. 7·1. The work done on the system is taken as *negative* by (old conventions) hence will be given by

$$-w_3 = RT_1 \ln \frac{V_4}{V_3} \qquad ...(iv)$$

During the operation, an amount of heat q_1, exactly equal to w_3, will be returned to the sink at T_1. Both q_1 and w_3 will be *negative* so that

$$-q_1 = -w_3 = RT_1 \ln \frac{V_4}{V_3} \qquad ...(v)$$

IV. Adiabatic compression: The cylinder is now removed from the sink and placed again on the insulating material. The gas is then compressed adiabatically and reversibly along DA till the initial state A is regained. The temperature of the gas rises from T_1 to T_2. The work done on the system is *negative* by the old conventions of sign and will be given by

$$w_4 = C_v\,(T_1 - T_2)$$
$$= -C_v\,(T_2 - T_1) \qquad ...(vi)$$

As a result of these four operations, the system has returned to its original state so that a reversible cycle has been completed. The *net work* (w) done by the system will be given by

$$w = w_1 + w_2 + (-w_3) + (-w_4) \qquad ...(vii)$$

$$= RT_2 \ln \frac{V_2}{V_1} + C_v(T_2 - T_1) + RT_1 \ln \frac{V_4}{V_3} - C_v(T_2 - T_1)$$

$$= RT_2 \ln \frac{V_2}{V_1} + RT_1 \ln \frac{V_4}{V_3} \qquad ...(viii)$$

* For new sign conventions, refer to chapter 6.

Since V_1 and V_4 lie on one adiabatic curve and V_3 and V_2 lie on another, applying volume-temperature relationships, we have

$$\left(\frac{V_4}{V_3}\right)^{\gamma-1} = \frac{T_2}{T_1} \qquad ...(ix)$$

and

$$\left(\frac{V_3}{V_2}\right)^{\gamma-1} = \frac{T_2}{T_1} \qquad ...(x)$$

Comparing equations (*ix*) and (*x*), we get

$$\left(\frac{V_4}{V_3}\right)^{\gamma-1} = \left(\frac{V_3}{V_2}\right)^{\gamma-1}$$

or

$$\frac{V_4}{V_1} = \frac{V_3}{V_2}$$

or

$$\frac{V_4}{V_3} = \frac{V_1}{V_2} \qquad ...(xi)$$

Putting this value in equation (*viii*), we get

$$w = RT_2 \ln\frac{V_2}{V_1} + RT_1 \ln\frac{V_1}{V_2}$$

$$= RT_2 \ln\frac{V_2}{V_1} - RT_1 \ln\frac{V_2}{V_1}$$

$$= R(T_2 - T_1) \ln\frac{V_2}{V_1} \qquad ...(xii)$$

Dividing equation (*xii*) by equation (*ii*), we get

$$\frac{w}{q_2} = \frac{T_2 - T_1}{T_2} \qquad ...(xiii)$$

which gives the *efficiency* of the Carnot cycle or engine.

From equation (*xiii*), it is clear that the efficiency of the reversible heat engine depends only upon the temperatures of the source and the sink and is independent of the nature of the working substance.

Since the quantity $\frac{T_2 - T_1}{T_2}$, which represents efficiency is always less than unity, the efficiency of the heat engine is thus always less than unity.

Q. 4. State and explain Carnot theorem. How can the efficiency of a heat engine be increased. (*M. Dayanand 2000; G. Nanakdev 2000; Calcutta 2000; Bhopal 2004; Indore 2004*)

Ans. Carnot theorem in stated as follows:

The efficiency of a reversible heat engine depends only upon the temperatures of the source and the sink and is independent of the nature of the working substance. In other words, all reversible heat engines working between the same two temperatures have the same efficiency.

Hence the efficiency of a heat engine may be given by any one of the following expressions:

$$\eta = \frac{w}{q_2} = \frac{q_2 - q_1}{q_2} = \frac{T_2 - T_1}{T_2}$$

Obviously, the efficiency of the engine can be increased by widening the difference in the temperatures of the source and sink.

Q.5. Calculate the amount of heat supplied to Carnot's cycle working between 368 K and 288 K if the maximum work obtained is 895 joules.

Ans. Here we are given that

$$T_1 = 368 \text{ K}$$
$$T_2 = 288 \text{ K}$$
$$w = 895 \text{ Joules}$$

Putting these values in the formula,

$$\eta = \frac{w}{q_2} = \frac{T_2 - T_1}{T_2}$$

we get

$$\frac{895}{q_2} = \frac{368 - 288}{368}$$

$$q_2 = \frac{895 \times 368}{80} = \mathbf{4117 Joules}$$

Q. 6. Calculate the maximum efficiency of a steam engine operating between 110° and 25°C. What would be the efficiency of the engine if the boiler temperature is raised to 140°C, the temperature of the sink remaining the same.

Ans. In the first case, we are given that

$$T_2 = 110°C = 110 + 273K = 383 \text{ K}$$
$$T_1 = 25°C = 25 + 273 \text{ K} = 298 \text{ K}$$

$$\therefore \quad \eta = \frac{T_2 - T_1}{T_2} = \frac{383 - 298}{383} = 0.222 = \mathbf{22.2\%}$$

In the second case, we are given that

$$T_2 = 140°C = 140 + 273 \text{ K} = 413 \text{ K}$$
$$T_1 = 25°C = 25 + 273 \text{ K} = 298 \text{ K}$$

$$\therefore \quad \eta = \frac{T_2 - T_1}{T_2} = \frac{413 - 298}{413} = 0.278 = \mathbf{27.8\%}$$

Q. 7. Compare the thermodynamic efficiencies to be expected:

(*a*) when an engine is allowed to operate between 1000 K and 300 K;

(*b*) when an engine is allowed to operate between 1000 K and 600 K and then the waste heat is passed on to another engine which operates between 600 K and 300 K.

Ans. *In case* (*a*)

$$\eta = \frac{1000 - 300}{10000} = 0.70 = \mathbf{70\%}$$

In case (*b*)

(*i*) $$\eta = \frac{1000 - 600}{1000} = 0.40 = \mathbf{40\%}$$

(*ii*) $$\eta = \frac{600 - 300}{600} = 0.50 = 50\% \text{ of the rejected } 60\% \text{ of } (i) = 30\%$$

$\therefore$ Total $\eta = 40 + 30$

$= 70$

Q. 8. How does Carnot cycle lead to the definition of entropy? (*Madurai 2004*)

Ans. For a reversible Camot cycle working between temperatures T_2 and T_1,

$$\frac{q_2 - q_1}{q_2} = \frac{T_2 - T_1}{T_2} \quad ...(i)$$

where q_2 is the heat absorbed isothermally and reversibly at temperature T_2 and q_1 *is* the heat lost isothermally and reversibly at temperature T_1.

The above equation may be simplified as

$$1 - \frac{q_1}{q_2} = 1 - \frac{T_1}{T_2}$$

or $$\frac{q_2}{T_2} = \frac{q_1}{T_1} \quad ...(ii)$$

or $$\frac{q}{T} = \text{constant}$$

Thus heat absorbed or lost isothermally and reversibly divided by the temperature at which the heat is absorbed or lost is a constant quantity for a particular system.

Further, in equation (*ii*), q_2 is the heat absorbed at temperature T_2 and q_1 is the heat lost at temperature T_1. If q_1 were the heat absorbed at temperature T_1, equation (*ii*) would be rewritten as

$$\frac{q_2}{T_2} = -\frac{q_1}{T_1}$$

or $$\frac{q_2}{T_2} + \frac{q_1}{T_1} = 0$$

ie. $$\sum \frac{q}{T} = 0. \quad ...(iii)$$

Now consider any reversible cycle *ABA*. It may be regarded as being made up of a number of Carnot cycles drawn across the diagram as shown in Fig. 7.3. Starting from A and going through all the cycles successively from *A* to *B* it can be shown that all paths inside the closed curve *ABA* cancel each other leaving only the zig-zag outer path. This zig-zag path is almost the same as the path *ABA* of the reversible cycle. By making each Carnot cycle smaller and increasing their number, it can be made to match more closely to *ABA*. The reversible cycle can be regarded as being made up of an infinite number of Carnot cycles. For each of these cycles, we have

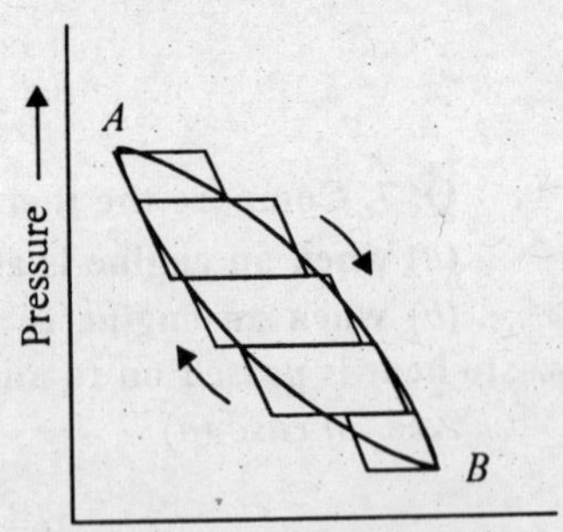

Fig. 7.3. Carnot cycles drawn across the reversible cycle.

$$\sum \frac{q}{T} = 0.$$

Hence for an infinite number of Camot cycles,

$$\sum \frac{\delta q}{T} = 0. \quad (iv)$$

where δq *is* a very small quantity of the heat absorbed at temperature T during the small isothermal change of the reversible process.

Further, the above term may be divided into two parts,one for the path A to B and the other for the back path B to A. Thus

$$\sum \frac{\delta q}{T} = \sum_{A \to B} \frac{\delta q}{T} + \sum_{B \to A} \frac{\delta q}{T} = 0$$

or we may write

$$\sum_{A \to B} \frac{\delta q}{T} = -\sum_{B \to A} \frac{\delta q}{T}$$

or

$$\left(\frac{q}{T}\right)_{A \to B} = -\left(\frac{q}{T}\right)_{B \to A} \quad ...(v)$$

This implies that the total value of the quantity $\frac{q}{T}$ for the path *A to B* is equal to the total value of the quantity $\frac{q}{T}$ for the reverse path *B* to *A*.

Thus *quantity* $\frac{q}{T}$ *is a state function, i.e. its value depends only upon the initial state (A) and the final state (B) and is independent of the path.* Obviously $\frac{q}{T}$ is the change in the value of a quantity or a function having some definite values at *A* and *B* depending upon the values of pressure, volume and temperature. This quantity or the function is called entropy and is usually represented by the symbol *S*. Thus if S_A is the value of the entropy at *A* and S_B is the value at *B*, then we must have

$$\frac{q}{T} = S_B - S_A = \Delta S \quad ...(vi)$$

where ΔS represents the total change in entropy in going from the initial state to the final state *B*.

Hence **entropy** is a state function, the change in the value of which from the initial state A to the final state is equal to the quantity *q/T i.e.* the total heat absorbed reversibly and isothermally in going from the initial state to the final state divided by the absolute temperature at which the heat is absorbed. From equation (*vi*), the **entropy change** *may be defined directly as the quantity of heat absorbed isothermally and reversibly divided by the absolute temperature (T) at which the heat is absorbed.*

Since the derivation of the above result is based upon Carnot cycle in which the heat is absorbed reversibly, therefore the expression (*vi*) for the entropy change is valid only when the heat is absorbed reversibly. Thus it is better to use the symbol q_{rev} in place of *q*. The expression (*vi*) may, therefore, be written as

$$\Delta S = \frac{q_{rev}}{T} \quad ...(vii)$$

Further, as entropy is a state function, a *small change* in its value can be represented by *dS*. Hence we may write

$$dS = \frac{q_{rev}}{T}$$

Q. 9. Explain the term entropy. Show that it is a state function.

(*Himachal 2000; Calcutta 2007; Andhra 2010*)

Ans. See Q. 8.

Q. 10. Prove that in a reversible process net entropy change for the system and surroundings is zero.

Ans. If q_{rev} is the heat absorbed by the system reversibly, then the heat lost by the surroundings will also be q_{rev}. If the process takes place isothermally at the absolute temperature T, then

(*i*) *Entropy change of the system* is given by

$$\Delta S_{\text{system}} = \frac{q_{rev}}{T} \quad \ldots(i)$$

(*ii*) *Entropy change of the surroundings* is given by

$$\Delta S_{\text{surroundings}} = -\frac{q_{rev}}{T} \quad \ldots(ii)$$

Hence the *total entropy change* for the combined system and the surroundings will be

$$\Delta S_{system} + \Delta S_{\text{surroundings}} = \frac{q_{rev}}{T} - \frac{q_{rev}}{T} = 0 \quad \ldots(iii)$$

Thus we may conclude that *in a reversible process, the net entropy change for the combined system and the surroundings is zero.*

Q. 11. Prove that in an irreversible process, net entropy change for the system and surroundings is greater than zero. (*G. Nanakdev 2003*)

OR

Prove that all spontaneous processes are accompanied by an increase of entropy. (*Delhi 2006*)

Ans. Suppose that the total heat lost by the surroundings is q_{irrev}. This heat is upon absorbed by the system but the *entropy change* of the system *does not depend the heat actually absorbed but it depends upon the heat absorbed reversibly i.e.,* on the quantity q_{rev}. Thus if the heat is absorbed isothermally by the system at the absolute temperature T, the *entropy change of the system* will be

$$\Delta S_{system} = \frac{q_{rev}}{T} \quad \ldots(i)$$

Further suppose that the loss of heat (q_{irrev}) by the surroundings takes place infinitesimally slowly because the surroundings are much bigger in size and magnitude compared to the system (*i.e.* reversibly) and isothermally at the temperature T. Then the *entropy change of the surroundings* is given by

$$\Delta S_{\text{surroundings}} = -\frac{q_{irrev}}{T} \quad \ldots(ii)$$

The *total entropy change for the combined system and the surroundings will, therefore, be*

$$\Delta S_{system} + \Delta S_{\text{surroundings}} = \frac{q_{rev}}{T} - \frac{q_{irrev}}{T} \quad \ldots(iii)$$

Now, we know that the work done in a reversible process is the maximum work *i.e.*

$$w_{\text{rev}} > w_{irrev} \quad \ldots(iv)$$

Further, as the internal energy (E) is state function, the value of AE is same whether the process is carried out reversibly of irreversibly. Hence

$$\Delta E = q_{\text{rev}} - w_{\text{rev}} = q_{\text{irrev}} - w_{\text{irrev}} \quad \ldots(v)$$

Combining results (*iv*) and (*v*), we conclude that

$$q_{\text{rev}} > q_{\text{irrev}} \quad \ldots(vi)$$

$$\therefore \quad \frac{q_{rev}}{T} > \frac{q_{ir\,rev}}{T} \qquad \text{...(vii)}$$

or
$$\frac{q_{rev}}{T} - \frac{q_{ir\,rev}}{T} > 0 \qquad \text{...(viii)}$$

From (*iii*) and (*viii*), we have

$$\Delta S_{\text{system}} + \Delta S_{\text{surroundings}} > 0 \qquad \text{...(ix)}$$

Thus it may be concluded that *in an irreversible process, the entropy change for the combined system and the surroundings is greater than zero i.e., an irreversible process is accompanied by a net increase of entropy.*

Since all spontaneous processes are thermodynamically irreversible it may be stated that *All spontaneous processes are accompanied by a net increase of entropy.*

Q. 12. Justify the statement "Entropy of the universe is increasing".

(*Gulberga 2004; Purvanchal 2007*)

Ans. All processes occurring in nature are thermodynamically irreversible and these are accompanied by increase of entropy. Hence it may be concluded that

The entropy of the universe is continuously increasing.

Thus the main ideas of the *first and the second law of thermodynamics* may be summed up as:

The energy of the universe is constant whereas the entropy of the universe is continuously increasing and tends to a maximum value.

Q. 13.Derive expressions for entropy change for an ideal gas under different conditions.

(*Nagpur 2002; Purvanchal 2002*)

Ans. Consider one mole of an ideal gas enclosed in a cylinder fitted with a frictionless piston. If a small amount of heat δq_{rev} is supplied to the system reversibly and isothermally at the temperature *T*, then the entropy change accompanying the process is given by

$$dS = \frac{\delta q_{rev}}{T} \qquad \text{...(i)}$$

According to first law of thermodynamics, for an infinitesimal process,

$$\delta q = dE + \delta w \qquad \text{...(ii)}$$

If the process is carried out *reversibly,* equation (*ii*) may be written as

$$\delta q_{rev} = dE + \delta w \qquad \text{...(iii)}$$

Further, if the *work is restricted to the work of expansion,* ...(*iv*)

$$\delta w = PdV$$

where dV is the small increase in volume and P is the pressure of the system.

Putting the value of δw from equation (*iv*) in equation (*iii*), we get

$$\delta q_{re} = dE + PdV \qquad \text{...(v)}$$

Substituting this value in equation (*i*), we get

$$dS = \frac{dE + PdV}{T}$$

or it may be written as

$$TdS = dE + PdV \qquad \text{...(vi)}$$

Now, for 1 mole of an ideal gas, we know that

$$C_v = \frac{dE}{dT}$$

i.e, $$dE = C_v\, dT \qquad ...(vii)$$

where C_v is the molar heat capacity at constant volume and

$$PV = RT$$

i.e. $$P = \frac{RT}{V} \qquad ...(viii)$$

where V is the volume of the system at temperature T and pressure P and R is the gas constant.

Putting the values of dE and P from equation (*vii*) and (*viii*) in equation (*vi*), we get

$$TdS = C_V\, dT + \frac{RT}{V} dV$$

or $$dS = C_V \frac{dT}{T} + R\frac{dV}{V} \qquad ...(ix)$$

If the volume changes from V_1 to V_2 when the temperature changes from T_1 to T_2, then the entropy change accompanying the complete process is given by

$$\int_{S_1}^{S_2} dS = \int_{T_1}^{T_2} C_V \frac{dT}{T} + \int_{V_1}^{V_2} R\frac{dV}{V} \qquad ...(x)$$

Assuming that C_v remains constant in the temperature range T_1 to T_2, equation (x) may be written as

$$\int_{S_1}^{S_2} dS = C_V \int_{T}^{T_2} \frac{dT}{T} + R\int_{V_1}^{V_2} \frac{dV}{V}$$

$$\boldsymbol{\Delta S = C_V \ \ln \frac{T_2}{T_1} + R \ \ln \frac{V_2}{V_1}} \qquad ...(xi)$$

This is an expression for the calculation of entropy change of 1 mole of an ideal gas accompanying a process when temperature changes from T_1 to T_2 and the volume changes from V_1 to V_2.

Further for 1 mole of an ideal gas, we may write

$$P_1V_1 = RT_1 \text{ for the initial state} \qquad ...(xii)$$

and $$P_2V_2 = RT_2 \text{ for the final state} \qquad ...(xiii)$$

Dividing equation (*xiii*) by equation (*xii*), we get

$$\frac{P_2V_2}{P_1V_1} = \frac{T_2}{T_1}$$

or $$\frac{V_2}{V_1} = \frac{T_2P_1}{T_1P_2} \qquad ...(xiv)$$

Putting this value in equation (*xi*), we get

$$\Delta S = C_V \ \ln \frac{T_2}{T_1} + R \ \ln \frac{T_2P_1}{T_1P_2} \qquad ...(xv)$$

Further, we know that

$$C_P - C_v = R \quad \text{or} \quad C_v = C_P - R$$

Putting this value in equation (*xv*), we get

$$\Delta S = (C_P - R)\ln\frac{T_2}{T_1} + R\ln\frac{T_2 P_1}{T_1 P_2}$$

$$= C_P \ln\frac{T_2}{T_1} - R\ln\frac{T_2}{T_1} + R\ln\frac{T_2}{T_1} + R\ln\frac{P_1}{P_2}$$

or
$$\Delta S = C_P \ln\frac{T_2}{T_1} + R\ln\frac{P_1}{P_2} \quad ...(xvi)$$

This is an expression for entropy change of 1 mole of an ideal gas accompanying a process when temperature changes from T_1 to T_2 and the pressure changes from P_1 to P_2.

(*i*) *If temperature is kept constant* (*i.e.* for **isothermal process**)

$$T_1 = T_2$$

Equations (*xi*) and (*xvi*) are reduced to

$$\Delta S = R\ln\frac{V_2}{V_1} = R\ln\frac{P_1}{P_2} \quad ...(xvii)$$

(*ii*) *If pressure is kept constant* (*i.e. for* **isobaric process**)

$$P_1 = P_2$$

Equation (*xvi*) is reduced to

$$\Delta S = C_P \ln\frac{P_1}{P_2} \quad ...(xviii)$$

(*iii*) *If volume is kept constant* (*i.e.,* for isochroic process)

$$V_1 = V_2$$

Equation (*xi*) is reduced to

$$\Delta S = C_v \ln\frac{T_2}{T_1} \quad ...(xix)$$

Q.14. Obtain an expression for entropy change of a gas when the tempetature changes from T_1 to T_2 and volume from V_1 to V_2. (*G.Nanakdev 2002; Kurukshetra 2002*)

Ans. See Q. 13.

Q. 15. Derive an expression for entropy change accompanying isothermal expansion of ar ideal gas. (*Nagpur 2008*

Ans. See Q. 13.

Q.16. Derive an expression for entropy change of an ideal gas when the temperature changes from T_1 to T_2 and pressure from P_1 to P_2.

Ans. See Q. 13.

Q. 17.Give expressions for entropy of melting and entropy of vaporisation.

Ans. Entropy of melting

$$\Delta S_m = \frac{\Delta H_m}{T_m}$$

where ΔH_m is the latent heat of melting and T_m is the melting temperature.

Entropy of vaporisation

$$\Delta S_v = \frac{\Delta H_v}{T_b}$$

where ΔH_v is the latent heat of vaporisation and T_b is the boiling temperature.

Processes like melting and evaporation involve *absorption of heat* which is taken as *positive* by conventions of sign, therefore ΔS for these processes is positive *i.e.* these processes are accompanied by *increase of entropy.* On the other hand, the reverse processes like freezing and condensation involve *evolution of heat which* is taken as *negative,* therefore for these processes ΔS is negative *i.e.* these processes are accompanied by *decrease of entropy.*

Q.18. Calculate the increase in entropy in the evaporation of one mole of water at 3.73 K. Latent heat of vaporization of water is 2.26 kJ/g (molecular mass of water = 18)

Ans. Here, we are given that

$$T_1 = 373 \text{ K}$$
$$\Delta H_v = 2260 \text{ Jg}^{-1}$$
$$= 2260 \times 18 \text{ Jmol}^{-1}$$
$$= 40680 \text{ J mol}^{-1}$$
$$\Delta S = \frac{\Delta H_v}{T_b} = \frac{40680}{373} = \mathbf{109\,JK^{-1}mol^{-1}}$$

Q. 19. Calculate the entropy change for the fusion of 1 mole of a solid which melts at 300 K. The latent heat of fusion is 2·51 kJ mol⁻¹.

Ans. Here, we are given that

$$\Delta H_m = 2510 \text{ J mol}^{-1}$$
$$T_m = 300 \text{ K}$$
$$\Delta S = \frac{\Delta H_m}{T_m}$$
$$= \frac{2510}{300} = \mathbf{8.37 JK^{-1}mol^{-1}}$$

Q. 20. Calculate the temperature at which the reaction $Ag_2O(s) \longrightarrow 2Ag\,(s) + \frac{1}{2}O_2\,(g)$ at 1 atmospheric pressure will be in equilibrium. The values of ΔH and ΔS for the reaction are 30.58 kJ and 66.11 JK⁻¹ respectively and these values do not change much with temperature.

Ans. For a reaction in equilibrium

$$\Delta S = \frac{\Delta H}{T}$$

Here we are given that

$$\Delta H = 30.58 \text{ kJ} = 30580 \text{ J}$$
$$\Delta S = 66.11 \text{ JK}^{-1}$$
$$66.11 = \frac{30580}{T}$$

or $$T = \mathbf{462.6\ K}$$

Q. 21. Calculate the entropy change involved in the isothermal reversible expansion of 5 moles of an ideal gas from a volume of 10 litres to a volume of 100 litres at 300 K.

Ans. $\Delta S = C_v \ln \frac{T_2}{T_1} + R \ln \frac{V_2}{V_1}$ for 1 mole of an ideal gas

∴ **For isothermal process $T_1 = T_2$ so that**

$$\Delta S = R \ln \frac{V_2}{V_1} \quad \text{for 1 mole of an ideal gas}$$

or $$\Delta S = nR \ln \frac{V_2}{V_1} \quad \text{for } n \text{ mole of an ideal gas}$$

or $$\Delta S = 2.303\, nR \log \frac{V_2}{V_1}$$

Here we are given that

$n = 5$

$V_1 = 10$ litres

$V_2 = 100$ litres

$R = 8.314$ JK^{-1} mol^{-1}

$$\therefore \quad \Delta S = 2.303 \times 5 \times 8.314 \times \log \frac{100}{10}$$

$$= \mathbf{95.75\ JK^{-1}\ mol^{-1}}$$

Q. 22. 100 g of ice at 0°C was added to an insulated beaker containing 150 g of water at 100°C. Calculate

***(i)* The final temperature of the system and**

***(ii)* The entropy change of the process.**

Heat of fusion of ice is 80 cal/g and heat capacity of water is 1.0 cal/g/degree (Ignore the heat capacity of the beaker).

Ans. (*i*) Suppose the final temperature after mixing

$= t$°C

Heat lost by water = $150 \times (100 - t)$ cal

Heat gained by ice = $100 \times 80 + 100 \times (t - 0)$

$= 800 + 100\, t$ cal

Heat gained = Heat lost

$800 + 100t = 150\,(100 - t)$

or $\quad$ **t = 56.8°C**

(*ii*) the process involves the following steps :

(*a*) 100 g of ice melts at 0°C to give 100 g of liquid water.

(*b*) 100 g of liquid water gets heated up from 0°C to 56.8°C.

(*c*) 150 g of water gets cooled from 100°C to 56.8°C.

Making use of the fact that when temperature remains constant (*i.e.* for isothermal process), $\Delta S = \frac{q_{rev}}{T}$. When temperature varies (but volume or pressure remains constant), $\Delta S = C \ln \frac{T_2}{T_1}$, where C is C_v or C_p.

Entropy change in step (*a*)

$$= \frac{(100\text{g}) \times (80\text{cal/g})}{273} = \frac{8000}{273} = 29.30\,\text{cal deg}^{-1}$$

Entropy change in step (*b*)

$$= (100\text{ g})\ (1\text{ cal/g}) \times 2.303 \times \log \frac{56.6+273}{0+273}$$

$$= 230.3 \log \frac{329.6}{273}$$

$$= 18.84\text{ cal deg}^{-1}$$

Entropy change in step (*c*)

$$= (150\text{ g}) \times (1\text{ cal/g}) \times 2.303 \log \frac{56.6+273}{100+273}$$

$$= 345.45 \log \frac{329.6}{373} = -18.56\,\text{cal deg}^{-1}$$

$\therefore$ Net entropy change in the process

$$= 29.30 + 18.84 - 18.55$$

$$= \mathbf{29.55\ cal\ deg^{-1}}$$

Q. 23. Explain the physical significance of entropy.

(Himachal, 2000; Garhwal 2000; Calcutta 2007)

Ans. Various processes which are accompanied by a net increase of entropy are associated with an increased randomness of distribution. For example :

(*i*) **The melting of a solid** is accompanied by a net increase of entropy and at the same time we know that the molecules, atoms or ions in a solid have fixed positions and they become free to move about in the molten state. This is expressed by saying that there is no *disorder* among the molecules, atoms or ions in the solid but the disorder sets in when the solid changes into liquid. In other words, it is stated that the *randomness* has increased.

(*ii*)**The vaporization of a liquid** is accompanied by a net increase of entropy and at the same time, the disorder or the randomness also increases because the molecules of vapour are more random than the molecules in the liquid.

Thus

Entropy is a measure of randomness or disorder of the system.

This concept may be further understood with the help of the following analogy.

In a college when all classes are being held, all the students are sitting in their respective class rooms seriously and the disorder is minimum. As soon as the bell goes, the students of different classes move out to go to other rooms and thus get mixed up. In other words, the disorder or the randomness increases.

Q. 24. Discuss the term work function (Helmholtz free energy).

(Kolhapur 2000; Bhopal 2004; Guwahati 2007, Nagpur 2008)

Ans. This is denoted by *A* and is defined by the equation

$$A = E - TS \quad ...(i)$$

where *E* is the internal energy of the system, *T* is the temperature and *S* is the entropy.

Since, *E, T* and *S* are the functions of the state of the system only (and do not depend upon its previous history), therefore *A* also must be a state function.

In order to understand the *physical significance* of the work function, consider an isothermal change taking place at temperature *T*. then

$$A_1 = E_1 - TS_1 \quad \textit{for the initial state} \quad ...(ii)$$

and

$$A_2 = E_2 - TS_2 \quad \textit{for the final state} \quad ...(iii)$$

where A_1, E_1 and S_1 are respectively the values of work function, internal energy and entropy of the system in the initial state and A_2, E_2 and S_2 are the corresponding values in the final state.

$\therefore$ Change in the function A accompanying the process is given by

$$A_2 - A_1 = (E_2 - TS_2) - (E_1 - TS_1)$$
$$= (E_2 - E_1) - T(S_2 - S_1)$$

or
$$\Delta A = \Delta E - T\Delta S \quad ...(iv)$$

According to the definition of entropy,

$$S = \frac{q_{rev}}{T} \quad ...(v)$$

Further, according to the first law of thermodynamics as applied to an isothermal reversible process, we have

$$\Delta E = q_{rev} - w_{max} \quad ...(vi)$$

Putting the values of ΔS and ΔE from equation (v) and (vi) in equation (iv), we get

$$\Delta A = (q_{rev} - w_{max}) - T.\frac{q_{rev}}{T}$$

or
$$-\Delta A = w_{max} \quad ...(vii)$$

Thus *for a process occurring at constant temperature, the decrease in the work function A is equal to the maximum work done by the system*. In fact it is for this reason that this thermodynamic quantity has been termed as work function.

Q. 25. Discuss the term Gibb's free energy. What is its physical significance?

(*Meerut 2000; Awadh 2000; Kumaon 2000; Purvanchal 2002; Bhopal 2004; Calcutta 2007*)

Ans. This is usually denoted by G and is defined by the equation

$$G = H - TS \quad (i)$$

where H, T and S are the heat content, temperature and entropy of the system respectively.

Since H, T and S are the function of the state only, therefore G is also function of the state of the system only. Hence for the isothermal process occurring at temperature T, we can write

$$G_1 = H_1 - TS_1. \text{ for the initial state} \quad ...(ii)$$

$$G_2 = H_2 - TS_2 \text{ for the final state} \quad ...(iii)$$

or
$$G_2 - G_1 = (H_2 - TS_2) - (H_1 - TS_1)$$
$$= (H_2 - H_1) - T(S_2 - S_1)$$

or
$$\Delta G = \Delta H - T\Delta S \quad ...(iv)$$

where ΔG = Change in Gibb's free energy of the system

ΔH = Enthalpy change of the system

and ΔS = Entropy change of the system

The physical significance of Gibb's free energy may be achieved as follows:

At constant temperature (T),

$$\Delta S = \frac{q_{rev}}{T}$$

i.e.
$$T\Delta S = q_{rev} \quad ...(v)$$

At constant pressure (P), $\Delta H = \Delta E + P\,\Delta V$ ...(*vi*)

Substituting the values of $T\Delta S$ and ΔH from equation (*v*) and (*vi*) in equation (*iv*), we get

$$\Delta G = (\Delta E + P\Delta V) - q_{rev}$$
$$= (\Delta E - q_{rev}) + P\,\Delta V \quad ...(vii)$$

Now, according to first law of thermodynamics

$$q_{rev} = \Delta E + w_{max}$$

or
$$\Delta E - q_{rev} = -w_{max}$$

Substituting this value in equation (*vii*), we get

$$\Delta G = -w_{max} + P\Delta V$$
$$-\Delta G = w_{max} - P\Delta V$$

But $P\Delta V$ is the *work of expansion* done by the system corresponding to the increase in volume ΔV. Hence ($w_{max} - P\Delta V$) stands for the maximum work other than the work of expansion. This is called the *maximum useful work* available from the process. Hence we conclude that *for a process occurring at constant temperature and constant pressure, the decrease in Gibb's free energy is equal to the* **maximum useful work** *obtainable from the process i.e. the total work minus the pressure volume work (or work of expansion).*

Q. 26. Differentiate between Gibb's free-energy and Helmholtz free energy.

(*Calcutta 2000*)

Ans. Gibb's free energy is represented by the symbol *G* and Helmholtz free energy is represented by the symbol *A*. The following relations hold good for the two quantities

$$A = E - TS$$
$$G = H - TS$$

$-\Delta A$ gives the maximum work obtainable from a system whereas $-\Delta G$ gives the maximum useful work (other than that of expansion) obtainable from the system.

Q. 27. Show that Gibb's free energy change at constant pressure and temperature is a measure of maximum useful work obtainable from a system (*Vidyasagar 2007*)

Ans. See Q. 25.

Q. 28. How does work function vary with temperature and volume?

Ans. The work function is given by the equation

$$A = E - TS \quad ...(i)$$

Complete differentiation of this equation gives

$$dA = dE - TdS - SdT \quad ...(ii)$$

By definition of entropy change,

$$dS = \frac{\delta q_{rev}}{T} \quad ...(iii)$$

From first law of thermodynamics,

$$\delta q_{rev} = dE - \delta w \quad ...(iv)$$

and if the work is restricted to the work of expansion only,

$$\delta w = -PdV \quad ...(v)$$

Putting this value in equation (*iv*), we have

$$\delta q_{rev} = dE + PdV \quad ...(vi)$$

Putting this value in equation (*iii*), we get

$$dS = \frac{dE + PdV}{T}$$

or $$TdS = dE + PdV \quad \text{... (vii)}$$

Putting this value in equation (*ii*), we obtain

$$dA = dE - dE - PdV - SdT$$
$$= - PdV - SdT \quad \text{...(viii)}$$

(*a*) *If temperature is kept constant, dT*= 0. *Equation* (*viii*) takes the form

$$(dA)_T = -(PdV)_T \quad \text{...(ix)}$$

or $$\left(\frac{\partial A}{\partial V}\right)_T = -P \quad \text{...(x)}$$

(*b*) *If volume is kept constant, dV*= 0. *Equation* (*viii*) becomes

$$(dA)_V = -(SdT)_V$$

$$\left(\frac{\partial A}{\partial T}\right)_V = -S \quad \text{... (xi)}$$

Q. 29. How does free energy vary with temperature and pressure?

(*M. Dayanand 2000 ; Delhi 2003*)

Ans. The free energy is given by the equation

$$G = H - TS \quad \text{...(i)}$$

But, $$H = E + PV \quad \text{...(ii)}$$

Putting this value in equation (*i*), we get

$$G = E + PV - TS \quad \text{...(iii)}$$

Complete differentiation of this equation gives

$$dG = dE + PdV + VdP - TdS - SdT \quad \text{...(iv)}$$

But $$dS = \frac{\delta q_{\text{rev}}}{T} = \frac{dE + PdV}{T}$$

or $$TdS = dE + PdV \quad \text{...(v)}$$

Substituting this value in equation (*iv*), we get

$$dG = dE + PdV + VdP - (dE + PdV) - SdT = VdP - SdT$$

i.e. $$\mathbf{dG = VdP - SdT} \quad \text{...(vi)}$$

This is an important expression which gives the change in free energy with change in pressure and change in temperature in a reversible process. Equation (*vi*) is called **total differential equation.**

(*a*) *If temperature is kept constant, dT* = 0. Equation (*vi*) takes the form

$$(dG)_T = (VdP)_T \quad \text{...(vii)}$$

or $$\left(\frac{\partial G}{\partial P}\right)_T = V \quad \text{...(viii)}$$

(*b*) *If pressure is kept constant, dP* = 0. Equation (*vi*) becomes

$$(dG)_p = -(SdT)_p$$

or $$\left(\frac{\partial G}{\partial T}\right)_P = -S \quad \text{... (ix)}$$

Q. 30. Derive the thermodynamics relation

$$\left(\frac{\partial G}{\partial T}\right)_P = -S$$ *(M. Dayanand, 2000)*

Ans. See Q. 29.

Q. 31. Show that $\left[\frac{\partial G}{\partial P}\right]_T = V$

Ans. See Q. 29.

Q. 32. Derive an expression for change in free energy for a process under isothermal condition. *(Bangalore 2002; Delhi 2003)*

Ans. The change in free energy for a complete process under isothermal conditions can be calculated as follows:

For an infinitesimal change, under isothermal conditions, we have the equation

$$(dG)_T = VdP \qquad ...(i)$$

For a complete process under isothermal conditions, when the pressure changes from P_1 to P_2, we have

$$(\Delta G)_T = \int_{P_1}^{P_2} VdP \qquad ...(ii)$$

Putting $PV = nRT$, *i.e.*, $V = \frac{nRT}{P}$ for n moles of an ideal gas, we get

$$(\Delta G)_T = \int_{P_1}^{P_2} \frac{nRT}{P} dP$$

$$= nRT \int_{P_1}^{P_2} \frac{dP}{P}$$

$$= nRT \ln \frac{P_2}{P_1} \qquad ...(iii)$$

Further for an isothermal process,

$$P_1V_1 = P_2V_2 \text{ or } \frac{P_2}{P_1} = \frac{V_1}{V_2}$$

Substituting this value in equation (*iii*), we get

$$\boxed{(\Delta G)_T = nRT \ln \frac{P_2}{P_1} = nRT \ln \frac{V_1}{V_2}} \qquad ...(iv)$$

Q. 33. Calculate the free energy change accompanying the compression of 1 mole of CO_2 at 57°C from 5 atm to 50 atm. Assume that CO_2 behaves like an ideal gas.

Ans. Here it is given that

$n = 1$ mole

$T = 57 + 273$ K $= 330$ K

$P_1 = 5$ atm

$P_2 = 50$ atm

Taking $R = 1.987$ cal deg^{-1} mol^{-1}, we get

$$(\Delta G)_T = 2.303 nRT \log \frac{P_2}{P_1}$$

$$= 2.303 \times 1 \times 1.987 \times 303 \times \log \frac{50}{5}$$

$$= \mathbf{1510.1\ cal}$$

Q. 34. Calculate the free energy change which occurs when one mole of an ideal gas expands reversibly and isothermally at 300 K from the initial volume of 5 litres to 50 litres.

Ans. Here it is given that $n = 1$ mole

$T = 300$ K

$V_1 = 5$ litres

$V_2 = 50$ litres

Taking $R = 8.314$ JK^{-1} mol^{-1}

$$(\Delta G)_T = 2.303\ nRT \log \frac{V_1}{V_2}$$

$$= 2.303 \times 1 \times 8.314 \times 300 \times \log \frac{5}{50}$$

$$= \mathbf{-5744.1\ jouls}$$

Q. 35. Derive Maxwell relations as given below:

(*i*) $\left[\frac{\partial T}{\partial V}\right]_S = -\left[\frac{\partial P}{\partial S}\right]_V$ (*ii*) $\left[\frac{\partial T}{\partial P}\right]_S = \left[\frac{\partial V}{\partial S}\right]_P$

(*iii*) $\left[\frac{\partial S}{\partial V}\right]_T = \left[\frac{\partial P}{\partial T}\right]_V$ (*iv*) $-\left[\frac{\partial S}{\partial P}\right]_T = \left[\frac{\partial V}{\partial T}\right]_P$

(Kumaon2000;Calcutta 2000; Coimbatore 2000; Delhi 2006)

Ans. According to first law of thermodynamics,

$$dE = dq - dw$$

or $$dE = dq - PdV \quad \text{...}(i)$$

Also $$dS = \frac{dq}{T}$$

or $$dq = TdS \quad \text{...}(ii)$$

Substituting this value in Eqn. (*i*), we get

$$dE = TdS - PdV \quad \text{...}(iii)$$

Also we know that

$$H = E + PV$$

$$A = E - TS$$

$$G = H - TS = E + PV - TS$$

Differentiating these equations, we get

$$dH = dE + PdV + VdP$$

$$dA = dE - TdS - SdT$$

$$dG = dE + PdV + VdP - TdS - SdT$$

Substituting the result of Eqn. (*iii*) in these equations,

$$dH = TdS + VdP \quad \text{...}(iv)$$

$$dA = -SdT - PdV \quad ...(v)$$

$$dG = -SdT + VdP \quad ...(vi)$$

By applying Euler's theorem for exact differentials to equations (*iii*) to (*vi*), the results obtained are called Maxwell relations. Let us, first understand Euler Theorem.

If z is a function of x and y, and dz is an exact differential, then we have

$$z = f(x, y)$$

$$dz = \left(\frac{\partial z}{\partial x}\right)_y dx + \left(\frac{\partial z}{dy}\right)_x dy \quad ...(vii)$$

This can be written as

$$dz = M(x, y)\, dx + N(x, y)\, dy$$

or simply as

$$dz = Mdx + Ndy \quad ...(viii)$$

where M and N are functions of the variables x and y.

Comparing equations (*vii*) and (*viii*), we have

$$M = \left(\frac{\partial z}{\partial x}\right)_y$$

$$N = \left(\frac{\partial z}{dy}\right)_x$$

As the mixed second derivatives are equal *i.e.*

$$\left[\frac{\partial}{\partial y}\left(\frac{\partial z}{\partial x}\right)_y\right]_x = \left[\frac{\partial}{\partial x}\left(\frac{\partial z}{\partial y}\right)_x\right]_y$$

$$\therefore \quad \left(\frac{\partial M}{\partial y}\right)_x = \left(\frac{\partial N}{\partial x}\right)_y \quad ...(ix)$$

This is Euler's condition for exact differentials. Now as dE, dH, dA and dG in equations (*iii*) to (*vi*) are exact differentials, the mixed second derivatives of the coefficients of the two terms on the right must be equal. The results obtained corresponding to these four equations are

$$\left(\frac{\partial T}{\partial V}\right)_S = -\left(\frac{\partial P}{\partial S}\right)_V \quad ...(x)$$

$$\left(\frac{\partial T}{\partial P}\right)_S = \left(\frac{\partial V}{\partial S}\right)_P \quad ...(xi)$$

$$\left(\frac{\partial S}{\partial V}\right)_T = \left(\frac{\partial P}{\partial T}\right)_V \quad ...(xii)$$

$$-\left(\frac{\partial S}{\partial P}\right)_T = \left(\frac{\partial V}{\partial T}\right)_P \quad ...(xiii)$$

Q. 36. Establish the following thermodynamic relations

$$\left(\frac{\partial E}{\partial S}\right)_V = T \qquad \left(\frac{\partial E}{\partial V}\right)_S = -P$$

$$\left(\frac{\partial H}{\partial S}\right)_P = T \qquad \left(\frac{\partial H}{\partial P}\right)_S = V$$

$$\left(\frac{\partial A}{\partial T}\right)_V = -S \quad \left(\frac{\partial A}{\partial V}\right)_T = -P$$

$$\left(\frac{\partial G}{\partial T}\right)_P = -S \quad \left(\frac{\partial G}{\partial P}\right)_T = V$$

Ans. Rewriting the four fundamental equations of thermodynamics

$$dE = TdS - PdV \quad \text{... (i)}$$
$$dH = T\,dS + VdP \quad \text{...(ii)}$$
$$dA = -SdT - PdV \quad \text{...(iii)}$$
$$dG = -SdT + VdP \quad \text{...(iv)}$$

If volume is kept constant, $dV = 0$, Eqn. (*i*) takes the form

$$\left(\frac{\partial E}{\partial S}\right)_V = T \quad \text{...(v)}$$

If entropy is kept constant, $dS = 0$, Eqn. (*i*) becomes

$$\left(\frac{\partial E}{\partial V}\right)_S = -P \quad \text{...(vi)}$$

Proceeding exactly in the same manner, different thermodynamic relations obtained from equations (*ii*) to (*iv*) are

$$\left(\frac{\partial H}{\partial S}\right)_P = T \quad \left(\frac{\partial H}{\partial P}\right)_S = V \quad \text{...(vii)}$$

$$\left(\frac{\partial A}{\partial T}\right)_V = -S \quad \left(\frac{\partial A}{\partial V}\right)_T = -P \quad \text{...(viii)}$$

$$\left(\frac{\partial G}{\partial T}\right)_P = -S \quad \left(\frac{\partial G}{\partial P}\right)_P = V \quad \text{...(ix)}$$

Q. 37. Establish the criteria for feasibility or spontaneity of a process.

(Lucknow 2000; G. Nanakdev 2000; Delhi 2003)

Ans. (*i*) If $\Delta S_{system} + \Delta S_{surroundings} > 0$; the process is irreversible, *i.e.*, it is spontaneous or feasible.

(*ii*) If $\Delta S_{system} + \Delta S_{surroundings} = 0$; the process is reversible *i.e.*, the system is in equilibrium.

The above two results may be combined and put in the form

$$\Delta S_{system} + \Delta S_{surroundings} \geq 0 \quad \text{...(i)}$$

where the sign *greater than* refers to an irreversible process (*i.e.* spontaneous process) and e sign *equal to* for a reversible process.

For an infinitesimal change, the above criterion may be written as

$$dS_{system} + dS_{surroundings} \geq 0 \quad \text{...(ii)}$$

where the sign, > stands for irreversible process and the sign, = for the reversible process.

If the surroundings lose heat δq reversibly and isothermally at the temperature T (which may be absorbed by the system reversibly or irreversibly), then

$$dS_{surroundings} = -\frac{\delta q_{rev}}{T} \quad \text{...(iii)}$$

Further according to first law of thermodynamics,

$$\delta q_{rev} = dE + pdV \quad \text{...(iv)}$$

Substituting this value in equation (*iii*), we get

$$dS_{\text{surroundings}} = -\frac{dE + PdV}{T} \quad ...(v)$$

Substituting this value in equation (*ii*), we get

$$dS_{\text{system}} - \frac{dE + PdV}{T} \geq 0$$

or $$TdS - dE - PdV \geq 0$$

or $$\mathbf{TdS \geq dE + PdV} \quad ...(vi)$$

where dS stands for the entropy change of the system and the subscript *system* has been omitted.

Equation (*vi*) is the basic equation which leads to a number of criteria for predicting the feasibility of a process. In this equation, the sign '=' stands for the reversible process and the sign '>' for the irreversible process. From equation (*vi*), the various criteria may be deduced as under.

(*i*) *In terms of entropy change of the system.* If the internal energy and the volume of the system are kept constant,

$$dE = 0 \text{ and } dV = 0$$

Then from equation (*vi*), we have

$$(TdS)_{E,\,V} \geq 0$$

or $$\mathbf{(dS)_{E,\,V} \geq 0} \quad ...(vii)$$

where the sign '=' is for the reversible process and the sign '>' for the irreversible process. The subscripts E and V indicate the constancy of these properties.

(*ii*) *In terms of internal energy change of the system.* If the entropy and volume of the system are kept constant,

$$dS = 0,\ dV = 0$$

The equation (*vi*) becomes

$$0 \geq (dE)_{S,\,V} \quad ...(viii)$$

or $$\mathbf{(dE)_{S,V} \leq 0} \quad ...(viii)$$

where the sign '=' stands for the reversible process and the sign '<' for the irreversible process. This implies that *if a process under constant entropy and volume is accompanied by a decrease of internal energy, the process is irreversible; if no change of internal energy takes place, the process is reversible.*

(*iii*) *In terms of enthalpy change of the system.*

$$H = E + PV$$

Complete differentiation of this equation gives

$$dH = dE + PdV + VdP$$

or $$dE + PdV = dH - VdP$$

Putting this value in equation (*vi*), we get

$$TdS \geq dH - VdP$$

If entropy and pressure of the system are kept constant,

$$dS = 0 \text{ and } dP = 0, \text{ so that}$$

$$0 \geq (dH)_{S,P}$$

or $$\mathbf{(dH)_{S,P} \leq 0} \quad ...(ix)$$

where the sign '=' refers to the reversible process and the sign '<' refers to the irreversible process.

(*iv*) *In terms of change in work function of the system.*

$$A = E - TS$$

Differentiating this equation completely, we get

$$dA = dE - TdS - SdT$$

or $$TdS = dE - SdT - dA$$

Putting this value in equation (*vi*), we get

$$dE - SdT - dA \geq dE + PdV$$

or $$-SdT - dA \geq PdV$$

or $$SdT + dA \leq -PdV$$

If temperature and volume of the system are kept constant, $dT = 0$, $dV = 0$, so that we have

$$(dA)_{T,V} \leq 0 \quad ...(x)$$

whereas usual sign '=' is for the reversible process and the sign '<' for the irreversible process.

(*v*) *In terms of free energy change of the system.*

$$G = H - TS$$

Further $$H = E - PV$$

$$\therefore \quad G = E + PV - TS$$

Complete differentiation of this equation gives

$$dG = dE + PdV + VdP - TdS - SdT$$

or $$dE + PdV = dG - VdP + TdS + SdT$$

Putting this value in equation (*vi*), we get

$$TdS \geq dG - VdP + TdS + SdT$$

or $$0 \geq dG - VdP + SdT$$

or $$dG - VdP + SdT \leq 0$$

If pressure and temperature of the system are kept constant, $dP = 0$, $dT = 0$, so that we obtain

$$(dG)_{P,T} \leq 0 \quad ...(xi)$$

where, as before, the sign '=' refers to the reversile process and the sign '<' refers to the irreversible process.

The criterion in terms of free energy change is most important because most of the processes take place at constant temperature and pressure.

(*i*) If $(dG)_{P,T} < 0$, the process is irreversible, *i.e.*, it is feasible.

(*ii*) If $(dG)_{P,T} = 0$, the process is reversible, *i.e.*, the system is in equilibrium.

(*iii*) If $(dG)_{P,T} > 0$, the process does not occur, *i.e.*, it is not feasible.

Q. 38. Show that $\Delta A \leq 0$ is a criterion for spontanecity and equilibrium.

Ans. See Q. 37.

Q. 39. Derive conditions under which ΔG and ΔA can be used as criteria for spontaneity of a process.

Ans. See Q. 37.

Q. 40. Derive Gibb's—Helmholtz equation. *(Kurukshetra 2002; Nagpur 2002; Bhopal 2004; Delhi 2006 ; Raipur 2010)*

Ans. The Gibbs free energy (G) is given by the equation,

$$G = H - TS \quad ...(i)$$

Hence for an *isothermal* process, we can write

$$G_1 = H_1 - TS_1 \quad \textit{for the initial state} \quad ...(ii)$$

and $$G_2 = H_2 - TS_2 \quad \textit{for the final state} \quad ...(iii)$$

Subtracting equation (*ii*) from equation (*iii*), we get

$$G_2 - G_1 = (H_2 - TS_2) - (H_1 - TS_1) = (H_2 - H_1) - T(S_2 - S_1)$$

or $$\Delta G = \Delta H - T\Delta S \quad ...(iv)$$

where ΔG is the change in free energy of the system

ΔH is the change in enthalpy of the system

and ΔS is the entropy change of the system.

Taking up equation (*i*) again and putting $H = E + PV$, we get

$$G = E + PV - TS \quad ...(v)$$

Differentiating this equation completely, we get

$$dG = dE + PdV + VdP - TdS - SdT \quad ...(vi)$$

But $$dS = \frac{\delta q_{rev}}{T} = \frac{dE + PdV}{T}$$

or $$TdS = dE + PdV$$

Substituting this value in equation (*vi*), we get

$$dG = VdP - SdT$$

If pressure is kept constant, $dP = 0$, so that we have

$$(dG)_P = -(SdT)_P$$

or $$\left(\frac{\partial G_1}{\partial T}\right)_P = -S \quad ...(vii)$$

$\therefore$ For the initial state, $$\left(\frac{\partial G_1}{\partial T}\right)_P = -S_1 \quad ...(viii)$$

For the final state, $$\left(\frac{\partial G_2}{\partial T}\right)_P = -S_2 \quad ...(ix)$$

Subtracting equation (*viii*) from (*ix*), we get

$$\left(\frac{\partial G_2}{\partial T}\right)_P - \left(\frac{\partial G_1}{\partial T}\right)_P = -S_2 - (-S_1)$$

$$= -S_2 + S_1 = -(S_2 - S_1)$$

$$= -\Delta S$$

or it may be written

$$\left[\frac{\partial (G_2 - G_1)}{\partial T}\right]_P = -\Delta S$$

$$\left[\frac{\partial (\Delta G)}{\partial T}\right]_P = -\Delta S \quad ...(x)$$

Substituting this value of $-\Delta S$ in equation (*iv*), we get

$$\Delta G = \Delta H + T\left[\frac{\partial (\Delta G)}{\partial T}\right]_P \quad ...(xi)$$

This is the most common form of Gibbs-Helmholtz equation.

Q. 41. Derive the following relationship

$$\left[\frac{\partial (\Delta G / T)}{\partial T}\right]_P = -\frac{\Delta H}{T^2}$$ **, where the symbols have their usual meanings.**

Ans. Writing the Gibbs-Helmholtz equation in its common form

$$\Delta G = \Delta H + T\left[\frac{\partial(\Delta G)}{\partial T}\right]_P$$

Dividing the above equation by T^2

$$\frac{\Delta G}{T^2} = \frac{\Delta H}{T^2} + \frac{1}{T}\left[\frac{\partial(\Delta G)}{\partial T}\right]_P$$

or
$$\frac{1}{T}\left[\frac{\partial(\Delta G)}{\partial T}\right]_P = \frac{\Delta G}{T^2} - \frac{\Delta H}{T^2}$$

or
$$\left[\frac{\partial(\Delta G/T)}{\partial T}\right]_P = -\frac{\Delta H}{T^2}$$

Q. 42. Give the applications of Gibbs-Helmholtz equation. ***(Calcutta,1999)***

Ans. Gibbs-Helmholtz equation has the following applications.

1. Knowing the free energy change at any two different temperatures, the enthalpy change accompanying the process can be calculated.

2. When a redox reaction (*e.g.* $Zn + CuSO_4 \longrightarrow ZnSO_4 + Cu$) is allowed to take place in a single beaker, heat is evolved. Earlier it was believed that when the same reaction is allowed to take place in an electrochemical cell, the electrical energy produced was equivalent to the heat evolved. However, later on it was found that in a number of cases, the electrical energy produced is not equivalent to the heat evolved. This was explained by Gibbs and Helmholtz by stating that the *electrical energy produced was equal to the decrease in the free energy* ($-\Delta G$) *accompanying the cell reaction.* They pointed out that if n is the number of electrons given out by one of the electrodes (*e.g.* for $Zn \longrightarrow Zn^{+2} + 2e$, $n = 2$), or taken up by the other electrode (*e.g.* for $Cu^{+2} + 2e \longrightarrow Cu$, $n = 2$), then the quantity of electricity produced = nF faradays, where F = 1 Fraday = 96,500 coloumbs. If E is the E.M.F. of the cell, then

Electrical energy produced by the cell = nFE

$\therefore$ Decrease in free energy, $-\Delta G = nFE$

Substituting this value in the Gibbs-Helmholtz equation

$$\Delta G = \Delta H + T\left[\frac{\partial(\Delta G)}{\partial T}\right]_P$$

we get
$$-nFE = \Delta H + T\left[\frac{\partial(-nFE)}{\partial T}\right]_P$$

or
$$\boldsymbol{nFE = -\Delta H + TnF\left(\frac{\partial E}{\partial T}\right)_P}$$

Thus knowing the temperature coefficient of the E.M.F. of the cell *i. e.* $(\partial E/\partial T)_p$, ΔH can be calculated if E is known or *vice versa.* Evidently, the electrical energy produced (nFE) is equal to the heat evolved ($-\Delta H$) only when $(\partial E/\partial T)_p = 0$.

Q. 43. Derive Clausius-Clapeyron equation. What are its applications ?

(Delhi 2003; Jamia Millia 2003; Mumbai, 2004; Raipur 2010)

When a system is in equilibrium, the free energy change accompanying the process is zero *i.e.*

$$\Delta G = 0, \text{ for a system in equilibrium} \qquad ...(i)$$

Further, the variation of free energy with temperature and pressure is given by the total differential equation, *viz.*

$$dG = VdP - SdT \qquad ...(ii)$$

Consider the system consisting of two phase I and II in equilibrium at temperature T and pressure P. Under these conditions, suppose that free energy of phase I is G_1 and that of phase II is G_2. If the temperature of the system is changed from T to $T + dT$ and pressure from P to $P + dP$, suppose the corresponding change in free energy of phase I is dG_1 and that of phase II is dG_2 (Fig. 7.4). Then according to equation (*ii*)

$$dG_1 = V_1dP - S_1dT \qquad ...(iii)$$

and

$$dG_2 = V_2dP - S_2dT \qquad ...(iv)$$

where V_1 and S_1 are the molar volume and entropy of phase I and V_2 and S_2 are the corresponding values for phase II.

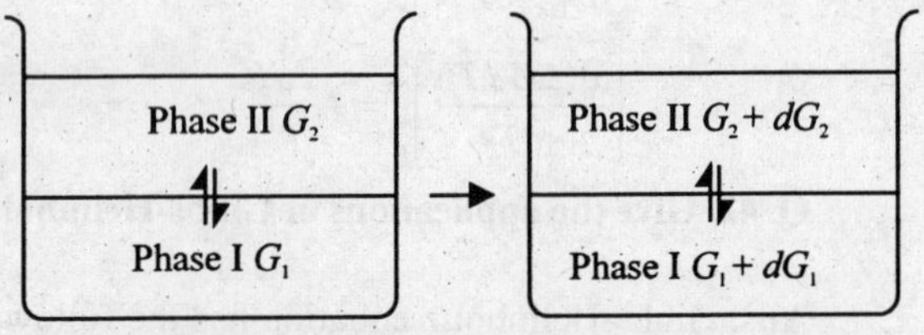

Fig. 7.4. Phase equilibria at T and P and at $T + dT$ and $P + dP$

But when phase I is in equilibrium with phase II,

$$\Delta G = 0$$

i.e. $G_2 - G_1 = 0$

where G_1 is the molar free energy of phase I and G_2 that of phase II at temperature T and pressure P.

Similarly, for the equilibrium at temperature $T + dT$ and pressure $P + dP$, we must have

$$(G_2 + dG_2) - (G_1 + dG_1) = 0$$

But $G_2 - G_1 = 0$ (as already stated above)

$\therefore$ $dG_2 - dG_1 = 0$

or

$$dG_2 = dG_1 \qquad ...(v)$$

Substituting the values of dG_1 and dG_2 from equations (*iii*) and (*iv*), we get

$$V_2dP - S_2dT = V_1dP - S_1dT$$

or $(V_2 - V_1)\, dP = (S_2 - S_1)\, dT$

or $\Delta V.dP = \Delta S.\, dT$

or

$$\frac{dP}{dT} = \frac{\Delta S}{\Delta V} \qquad ... (vi)$$

where $\Delta V = V_2 - V_1$ is the molar volume change

and $\Delta S = S_2 - S_1$ is the molar entropy change

where phase I changes into phase II.

If this change occurs at the temperature T (*i.e.* melting point, boiling point etc.) and ΔH is latent heat of transformation, then

$$\Delta s = \frac{\Delta H}{T}$$

Putting this value in equation (*vi*), we get

$$\boxed{\frac{dP}{dT} = \frac{\Delta H}{T\Delta V}} \qquad ...(viii)$$

This equation was first derived by Clapeyron (1934) and is called **Clapeyron equation.**

Applications

It can be applied to various physical equilibria like melting, vaporisation and sublimation.

For the melting equilibrium *i.e.*

$$\text{Solid} \rightleftharpoons \text{liquid}$$

$\Delta H = \Delta H_m$, latent heat of melting and $T = T_m$, the melting point

Equation (*vii*) may then be written as

$$\frac{dP}{dT} = \frac{\Delta H_m}{T_m \Delta V} \quad ...(viii)$$

Further $\Delta V = V_l - V_s$

where V_s = molar volume of the solid

and V_l = molar volume of the liquid (melt)

Hence equation (*viii*) may be written as

$$\frac{dP}{dT} = \frac{\Delta H_m}{T_m(V_l - V_s)}$$

or

$$\frac{dT}{dP} = \frac{T_m(V_l - V_S)}{\Delta H_m} \quad ...(ix)$$

This equation gives the variation of melting point with pressure.

For the vaporisation equilibrium *i.e.*

$$\text{Liquid} \rightleftharpoons \text{Vapour}$$

The Clapeyron equation, *i.e.* equation (*viii*) takes the form

$$\frac{dP}{dT} = \frac{\Delta H_v}{T_b(V_v - V_l)} \quad ...(x)$$

where ΔH_v = latent heat of vaporisation

T_b = boiling point of the liquid

V_l = molar volume of the liquid

and V_v = molar volume of the vapour.

Volume of the liquid (V_l) is very small as compared to the volume of the vapour (V_v), V_l can be neglected in comparison to V_v. Hence equation (*x*) may be written as (omitting the subscript in T_b).

$$\frac{dP}{dT} = \frac{\Delta H_v}{TV_v} \quad ...(xi)$$

Further assuming that the vapour behave like an ideal gas, we will have

$$PV_v = RT$$

or

$$V_v = \frac{RT}{P} \quad ...(xii)$$

Substituting this value in equation (*xii*), we get

$$\frac{dP}{dT} = \frac{\Delta H_v}{RT^2}.P$$

or

$$\frac{1}{P}.\frac{dP}{dT} = \frac{\Delta H_v}{RT^2}$$

or

$$\frac{d \ln P}{dT} = \frac{\Delta H_v}{RT^2} \quad ...(xiii)$$

This equation is called **Clausius-Clapeyron equation** as it was derived by Clausius during a detailed study of Clapeyron equation.

The equation (*xiii*) is very commonly written in another form called **integrated form of Clausius-Clapeyron equation.** This may be derived as follows:

Equation (*xiii*) may be rewritten as

$$d \ln P = \frac{\Delta H_v}{RT^2} dT \qquad \text{...}(xiv)$$

If the temperature changes from T_1 to T_2 when the vapour pressure changes from P_1 to P_2 then integrating equation (*xiv*) between the appropriate limits, we get

$$\int_{P_1}^{P_2} d \ln P = \int_{T_1}^{T_2} \frac{\Delta H_v}{RT^2} dT$$

or

$$\ln \frac{P_2}{P_1} = \frac{\Delta H_v}{R} \int_{T_1}^{T_2} \frac{1}{T^2} dT$$

$$= \frac{\Delta H_v}{R}\left(\frac{1}{T_1} - \frac{1}{T_2}\right)$$

Changing the natural logarithm to the logarithm to base 10, we get

$$2.303 \log \frac{P_2}{P_1} = \frac{\Delta H_v}{R}\left(\frac{T_2 - T_1}{T_1 T_2}\right)$$

or

$$\log \frac{P_2}{P_1} = \frac{\Delta H_v}{2.303 R}\left(\frac{T_2 - T_1}{T_1 T_2}\right) \qquad \text{...}(xv)$$

This is called the *integrated form of Clausius-Clapeyron equation,* and has a number of applications.

(*i*) If the vapour pressure of a liquid is known at one particular temperature, the value at any other temperature can be calculated, if the latent heat of vaporization of the liquid is known.

(*ii*) If the vapour pressures of a liquid at two different temperatures are known, the latent heat of vaporization of the liquid can be calculated.

(*iii*) If the boiling point of a liquid is known at some particular pressure, the boiling point at any other pressure can be calculated, provided the latent heat of vaporization of the liquid is known.

Q. 44. The normal boiling point of water is 100°C. Its vapour pressure at 80°C is 0.4672 atmosphere. Calculate the enthalpy of vaporisation per mole of water.

Ans. Here, we have

$T_1 = 100°C = 100 + 273 = 373$ K

$P_1 = 1$ atm

$T_2 = 80°C = 80 + 273 = 353$ K

$P_2 = 0.4672$ atm

ΔH_v = to be calculated

Substituting the values in the equation,

$$\log \frac{P_2}{P_1} = \frac{\Delta H_v}{2.303R}\left(\frac{T_2 - T_1}{T_1 T_2}\right)$$

we get $$\log\frac{0.4672}{1} = \frac{\Delta H_v}{2.303\times 8.314}\left(\frac{353-373}{353\times 373}\right)$$

or $$-0.3305 = \frac{\Delta H_v}{2.303\times 8.314}\times\left[\frac{-20}{353\times 373}\right]$$

or ΔH_v **= 41.66 kJ mol^{-1}**

Q. 45. Water boils at 373 K at one atmospheric pressure. At what temperature will it boil when atmospheric pressure becomes 528 mm of Hg at some space station.

Latent heat of H_2O = 2.28 kJ g^{-1}

R = 8.314 JK^{-1} mol^{-1}

Ans. We are given that

$$T_1 = 373 \text{ K}$$
$$P_1 = 1 \text{ atm} = 760 \text{ mm}$$
$$P_2 = 528 \text{ mm},\ T_2 = \text{To be calculated}$$
$$\Delta H_v = 2.28 \text{ kJ/g}$$
$$= 2.28 \times 18 \text{ kJ mol}^{-1}$$
$$= 41.04 \text{ kJ mol}^{-1}$$
$$= 41040 \text{ J mol}^{-1}$$
$$R = 8.314 \text{ JK}^{-1} \text{ mol}^{-1}$$

Substituting the values in the equation,

$$\log\frac{P_2}{P_1} = \frac{\Delta H_v}{2.303R}\left(\frac{T_2 - T_1}{T_2\ T_1}\right)$$

we get $$\log\frac{521}{760} = \frac{41040}{2.303\times 8.314}\times\left(\frac{T_2 - 373}{373T_2}\right)$$

or $$-0.1582 = 2143.4\times\frac{T_2 - 373}{373T_2}$$

or $$-59.0\,T_2 = 2143\,T_2 - 799488.2$$

or $$2202.4\,T_2 = 799488.2$$

$\therefore$ $$T_2 = 363 \text{ K}$$

Q. 46. Water boils at 100°C at a pressure of 1 atm. Calculate the vapour pressure of water at 90°C. The heat of vaporisation of water is 9.80 kcal mol^{-1}.

Ans. Here, we have

$$T_1 = 100 + 273 = 373 \text{ K}$$
$$P_1 = 1 \text{ atm}$$
$$T_2 = 90 + 273 = 363 \text{ K}$$
$$P_2 = \text{To be calculated}$$
$$\Delta H_v = 9.80 \text{ kcalmol}^{-1} = 9800 \text{ cal mol}^{-1}$$
$$R = 1.987 \text{ cal deg}^{-1} \text{ mol}^{-1}$$

Substituting these values in the equation

$$\log\frac{P_2}{P_1} = \frac{\Delta H_v}{2.303R}\left(\frac{T_2 - T_1}{T_1\ T_2}\right)$$

we get $$\log \frac{P_2}{P_1} = \frac{9800}{2.303 \times 1.987} \times \frac{363 - 373}{363 \times 373}$$

or $$\log P_2 = -0{\cdot}1582 = \bar{1}.8418$$

$$\therefore \quad P_2 = \mathbf{0{\cdot}6947\ atm}$$

Q. 47. Derive from thermodynamic consideration a relationship between elevation in boiling point and molecular weight of a non-volatile solute. (*Kerala 2000; M.Dayanand 2000;* ***Lucknow 2000; Mumbai 2010***)

Ans. Fig. 7·5 gives the vapour pressure-temperature curves for the solvent and solution. The horizontal dotted line drawn corresponding the external pressure cuts the solvent and the solution curves at the points *E* and *F* which correspond to temperatures T_0 and T_s respectively. Then by definition, T_0 is the boiling point of the pure solvent and T_s that of the solution. Obviously $T_s > T_0$. The increase $(T_s - T_0)$ is called the elevation in boiling point and is usually represented by ΔT_b. Thus

$$\Delta T_b = T_s - T_0 \qquad ...(i)$$

The relationship between the elevation in boiling point and the concentration of the solution can be obtained by applying Clausius-Clapeyron equation and Raoult's law to the different conditions shown in Fig. 7.5.

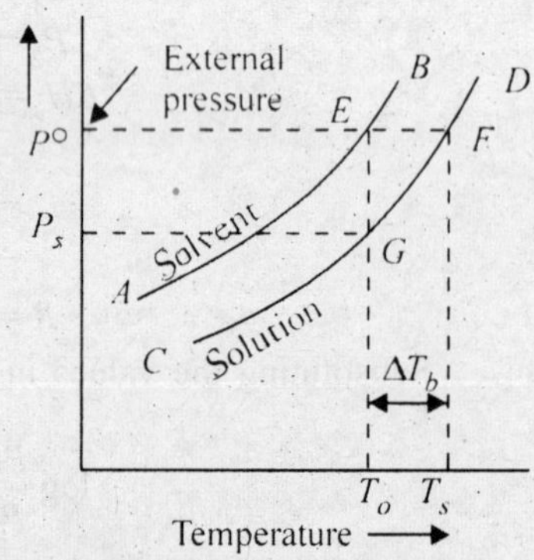

Fig. 7.5. Relationship between elevation in boiling point and lowering of vapour pressure.

Clausius-Clapeyron equation which is applicable to a phase equilibrium, in the integrated form is given by

$$\ln \frac{p_2}{p_1} = \frac{\Delta H}{R}\left(\frac{1}{T_1} - \frac{1}{T_2}\right)$$

where p_1 is the vapour pressure at temperature T_1, p_2 is the vapour pressure at temperature T_2 and ΔH is the molar latent heat of transition.

Referring to Fig. 7.5 points *G* and *F* lie on the same solution curve. Suppose the vapour pressure of the pure solvent and the solution at temperature T_0 are p_0 and p_s respectively. Then from Fig. 7·5, we have

(*a*) Corresponding to point *G*, temperature = T_0, vapour pressure = p_s

(*b*) Corresponding to point *F*, temperature = T_s, vapour pressure = p^o.

Hence for the equilibrium, solution $\rightleftharpoons$ vapour, according to Clausius-Clapeyron equation, we have (for the points *G* and *F*)

$$\ln \frac{p^o}{p_s} = \frac{\Delta H_v}{R}\left(\frac{1}{T_0} - \frac{1}{T_s}\right) \qquad ...(ii)$$

where ΔH_v is the latent heat of vaporisation of one mole of the solvent from the solution. When the solution is dilute, ΔH_v is nearly equal to the latent heat of evaporation of the pure solvent.

Equation (*ii*) can be written as

$$\ln \frac{p^o}{p_s} = \frac{\Delta H_v}{R}\left(\frac{T_s - T_0}{T_0 T_s}\right)$$

or $$\ln \frac{p^o}{p_s} = \frac{\Delta H_v}{R} \frac{\Delta T_v}{T_b T_s} \qquad [\because T_s - T_0 = \Delta T_b] \qquad ...(iii)$$

Further, when the solution is dilute T_s is nearly equal to T_0. Hence equation (*iii*) can written as

$$\text{In}\frac{p^o}{p_s}=\frac{\Delta H_v}{R}\frac{\Delta T_b}{T_0{}^2} \quad ...(iv)$$

According to Raoult's law, we have

$$\frac{p^o-p_s}{p^o}=x_2 \quad ...(v)$$

where x_2 is the mole fraction of the solute in the solution.

Equation (*v*) can be written as

$$1-\frac{p_s}{p^o}=x_2 \text{ or } \frac{p_s}{p^o}=1-x_2$$

$$\therefore \quad \text{In}\frac{p_s}{p^o}=\text{In}(1-x_2)$$

or

$$\text{In}\frac{p_s}{p_o}=-\text{In}(1-x_2) \quad ...(vi)$$

Since x_2 is very small (less than 1), the expansion of ln $(1 - x_2)$ as an infinite series is given by

$$\ln(1-x_2)=-x_2-\frac{1}{2}x_2^2-\frac{1}{3}x_2^3 ....$$

Neglecting x_2^2, x_2^3 etc. (as x_2 is much smaller than 1), we have

$$\ln(1-x_2)\simeq -x_2 \quad ...(vii)$$

or

$$-\ln(1-x_2)\simeq x_2$$

From eq. (*vi*) and (*vii*) $\quad \text{In}\frac{p^o}{p^s}=x_2 \quad ...(viii)$

Substituting this value in equation (*iv*)

$$x_2=\frac{\Delta H_v}{R}\frac{\Delta T_b}{T_0^2} \quad ...(ix)$$

or

$$\Delta T_b=\frac{RT_0^2x_2}{\Delta H_v} \quad ...(x)$$

However the common practice in the studies on elevation in boiling point is not to express the concentrations in terms of mole fractions but in terms of *moles of the solute per 1000 g of the solvent i.e.* in terms of *molality, m*. Hence equation (*x*) is further modified as follows :

If n_2 moles of the solute are dissolved in n_1 moles of the solvent, the mole fraction of the solute (x_2) in the solution will be given by

$$x_2=\frac{n_2}{n_1+n_2}$$

As the solution is supposed to be dilute, $n_2 \ll n_1$ so that in the denominator n_2 can be neglected in comparison to n_1. Hence for dilute solution, above equation can be written as

$$x_2=\frac{n_2}{n_1}$$

Further, if w_2 g of the solute is dissolved in w_1 g of the solvent and M_2 and M_1 are the molecular weights of the solute and solvent respectively, then

$$n_2 = \frac{w_2}{M_1} \text{ and } n_1 = \frac{w_1}{M_1}$$

Hence
$$x_2 = \frac{n_2}{n_1} = \frac{w_2 / M_2}{w_1 / M_1} = \frac{w_2 \times M_1}{w_1 \times M_2}$$

Putting this value of x_2 in equation (*x*), we get

$$\Delta T_b = \frac{RT_0^2}{\Delta H_v} \frac{w_2 M_1}{w_1 M_2}$$

$$= \frac{RT_0^2}{\Delta H_v / M_1} \cdot \frac{w_2}{w_1 M_2} \quad ...(xi)$$

Putting $\frac{\Delta H_v}{M_2} = l_v$, latent heat of vaporisation per gram of the solvent, equation (*xi*) becomes

$$\Delta T_b = \frac{RT_0^2}{l_v} \frac{w_2}{w_1 M_2} \quad ...(xii)$$

Let us say molality of the solution is *m*. Thus *m* moles of the solute are dissolved in 1000 g of the solvent, we can write

$$\frac{w_2}{M_2} = m \text{ and } w_1 = 1000$$

Putting these values in equation (*xii*), we get

$$\Delta T_b = \frac{RT_0^2}{l_v} \frac{m}{1000} \quad ...(xiii)$$

or
$$\Delta T_b = \frac{RT_0^2}{1000 l_v} m \quad ...(xiv)$$

For a given solvent, the quantity $\frac{RT_0^2}{1000 l_v}$ is a constant quantity because l_v, T_0 and R are constant. It is represented by K_b and is called **molal elevation constant or ebullioscopic constant.**

i.e.
$$\boxed{K_b = \frac{RT_0^2}{1000 l_v}} \quad ...(xv)$$

Hence equation (*xiv*) can be written as

$$\boldsymbol{\Delta T_b = K_b \cdot m} \quad ...(xvi)$$

If $m = 1, \Delta T_b = K_b$

Thus molal elevation constant may be defined as the elevation in boiling point when the molality of the solution is unity.

Q. 47. Derive from thermodynamic consideration a relationship between the freezing point and molecular weight of a non-volatile solute. (*Himachal 2010, Raipur 2010*)

Ans. *Freezing point may be defined as the temperature at which the liquid and the solid states of a substance have the same vapour pressure.*

The depression in the freezing point of solution can be explained again on the basis of lowering of vapour pressure of a solution. If vapour pressures are plotted against temperatures, the curve AB is obtained for the liquid solvent as shown in Fig. 7·6. At the point B, the liquid solvent starts solidifying and hence the liquid solvent and the solid solvent are in equilibrium with each other. The temperature corresponding to the point B is thus the freezing point of the pure solvent (T_0). When whole of the liquid solvent has been solidified, there is a sharp change in the vapour pressure-temperature curve beyond the point B, as shown by the curve BC for the solid solvent. Vapour pressure temperature curve for the solution lies below the vapour pressure curve of the liquid solvent, as represented by the curve DE in Fig. 7·6. As the cooling of the solution is continued, at the point E, the separation of the solid solvent starts. Thus the point E represents the freezing point of the solution (T_s). From Fig. 7·6, it is evident that $T_s < T_0$. The decrease ($T_0 - T_s$) is called the depression in freezing point and is usually represented by ΔT_f. Thus

$$\Delta T_f = T_0 - T_s \qquad ...(i)$$

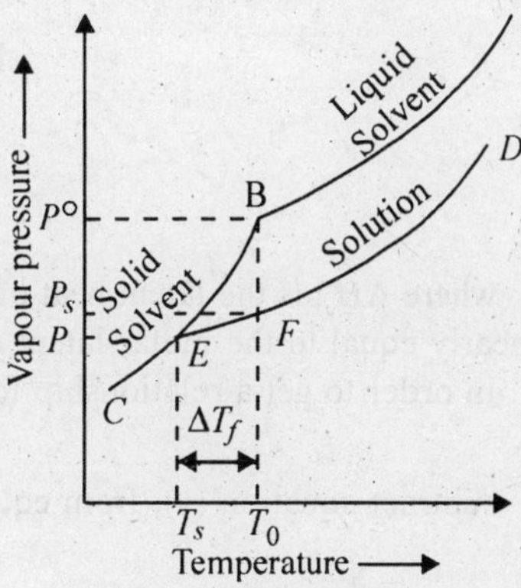

Fig. 7.6. Relationship between depression in freezing point and lowering of vapour pressure.

The relationship between the depression in freezing point and the concentration of the solution can be derived by applying Clausius-Clapeyron equation and Raoult's law to the curves shown in Fig. 7·6.

Suppose the vapour pressures corresponding to the point B, E and F are $p°$, p and p_s respectively.

The equilibrium existing along the curve BC is

Solid $\rightleftharpoons$ Vapour

The curve BC is, therefore, the sublimation curve of the solid solvent. For the points B and E lying on this curve, we have

(*a*) Corresponding to point B, temperature = T_0, vapour pressure = $p°$

(*b*) Corresponding to point E, temperature = T_s, vapour pressure = p.

Applying Clausius-Clapeyron equation corresponding to points B and E, we have

$$\ln \frac{p^o}{p} = \frac{\Delta H_s}{R}\left(\frac{1}{T_s} - \frac{1}{T_0}\right) \qquad ...(ii)$$

where ΔH_s is the molar latent heat of sublimation of the solid solvent.

Equation (*ii*) can be written as

$$\ln \frac{p^o}{p} = \frac{\Delta H_s}{R} \cdot \frac{T_0 - T_s}{T_s T_0}$$

$$= \frac{\Delta H_s}{R} \cdot \frac{\Delta T_f}{T_0 T_s} \qquad ...(iii)$$

As the solution is dilute $T_s \simeq T_0$ and therefore equation (*iii*) becomes

$$\ln \frac{p^o}{p} = \frac{\Delta H_s}{R} \cdot \frac{\Delta T_f}{T_0^2} \qquad ...(iv)$$

The equilibrium existing along the curve DE is

Solution $\rightleftharpoons$ Vapour

The curve DE is thus the vaporisation curve of the solution. For the points E and F lying on this curve, we have

(*a*) Corresponding to point *E*, temperature = T_s,vapour pressure = $p°$
(*b*) Corresponding to point *F*, temperature = T_0, vapour pressure = p_s
Applying Clausius-Clapeyron equation to these conditions, we have

$$\ln \frac{p_s}{p} = \frac{\Delta H_v}{R}\left(\frac{1}{T_s} - \frac{1}{T_0}\right) = \frac{\Delta H_v}{R} \cdot \frac{T_0 - T_s}{T_s \cdot T_0}$$

$$= \frac{\Delta H_v}{R} \frac{\Delta T_f}{T_0^2} \text{ (Taking } T_s \simeq T_0) \quad ...(v)$$

where ΔH_v; is the latent heat of vaporisation of one mole of the solvent from the solution and is nearly equal to the molar latent of vaporisation of the pure solvent for a dilute solution.

In order to get a relationship for the equilibrium

$$\text{Solid} \rightleftharpoons \text{Liquid}$$

subtract equation (*v*), from equation (*iv*), we get

$$\ln \frac{p^o}{p} - \ln \frac{p_s}{p} = \frac{\Delta H_s}{R} \frac{\Delta T_f}{T_0^2} - \frac{\Delta H_v}{R} \frac{\Delta T_f}{T_0^2}$$

or
$$(\ln p^o - \ln p) - (\ln p_s - \ln p) = \frac{\Delta T_f}{RT_0^2} (\Delta H_s - \Delta H_v) \text{ .}$$

$$\ln p^o - \ln p_s = \frac{\Delta T_f}{RT_0^2} \Delta H_f$$

$$\ln \frac{p^o}{p_s} = \frac{\Delta T_f}{RT_0^2} \Delta H_f \quad ...(vi)$$

where $\Delta H_s - \Delta H_v = \Delta H_f$ is the molar latent heat of fusion of the solid solvent.

The left hand term of equation (*vi*) can be expressed in terms of mole fraction by applying Raoult's law,

$$\frac{p^o - p_s}{p^o} = x_2 \quad ...(vii)$$

where x_2 is the mole fraction of the solute in the solution.

Equation (*vii*) can be written as

$$1 - \frac{p_s}{p^o} = x_2$$

or
$$\frac{p_s}{p^o} = 1 - x_2$$

or
$$\ln \frac{p_s}{p^o} = \ln(1 - x_2)$$

or
$$\ln \frac{p_s}{p^o} = -\ln(1 - x_2) \quad ...(viii)$$

The expansion of $\ln (1 - x_2)$ as an infinite series is given by

$$\ln (1 - x_2) = -x_2 - \frac{1}{2} x_2^2 - \frac{1}{3} x_2^3 ......$$

Neglecting x_2^2, x_2^3 etc. (because x_2 is very small, higher powers will be still smaller), we have

$$\ln(1 - x_2) = -x_2$$

Hence equation (*viii*) becomes

$$\ln \frac{p^o}{p_s} = x_2$$

Substituting this value in equation (*vi*), we get

$$x_2 = \frac{\Delta T_f}{RT_0^2} . \Delta H_f$$

or

$$\Delta T_f = \frac{RT_0^2}{\Delta H_f} . x_2 \qquad ...(ix)$$

This equation gives the relationship between depression in freezing point and the mole fraction x_2 of the solute in the solution.

To obtain the equation in terms of molality instead of mole fraction x_2, we have

$$x_2 = \frac{n_2}{n_1 + n_2}$$

where n_2 = number of moles of the solute in the solution

and n_1 = number of moles of the solvent in the solution.

For dilute solution $n_2 \ll n_1$ so that on neglecting n_2 in comparison to n_1, we have

$$x_2 = \frac{n_2}{n_1} = \frac{w_2 / M_2}{w_1 / M_1} = \frac{w_2 M_1}{w_1 M_2} \qquad ...(x)$$

where w_2 = weight of the solute dissolved

M_2 = molecular weight of the solute

w_1 = weight of the solvent

M_1 = molecular weight of the solvent.

Putting the value of x_2 from equation (*x*) in equation (*ix*), we get

$$\Delta T_f = \frac{RT_0^2}{\Delta H_f} \frac{w_2 M_1}{w_1 M_2} \qquad ...(xi)$$

$$= \frac{RT_0^2}{\Delta H_f / M_1} . \frac{w_2}{w_1 M_2}$$

$$= \frac{RT_0^2}{l_f} . \frac{w_2}{w_1 M_2} \qquad ...(xii)$$

where $l_f = \dfrac{\Delta H_f}{M_1}$ is the latent heat of fusion *per gram* for the solid solvent.

If m is the molality of the solution *i.e.* m moles of the solute are dissolved in 1000 g of the solvent, then

$$\frac{w_2}{M_2} = m \text{ and } w_1 = 1000\text{g}$$

Putting these values in equation (*xii*), we get

$$\Delta T_f = \frac{RT_0^2}{l_f} \frac{m}{1000}$$

$$= \frac{RT_0^2}{l_f} \cdot m \quad \text{...(xiii)}$$

Since T_0 and l_f for a given solvent are constant, the quantity $\frac{RT_0^2}{1000 l_f}$ in equation (*xiii*) must be a constant quantity. It is usually represented by K_f and is called **molal depression constant or cryoscopic constant.** Thus

$$K_f = \frac{RT_0^2}{1000 l_f} \quad \text{...(xiv)}$$

Substituting this value in equation (*xiii*), we get

$$\boxed{\Delta T_f = K_f . m} \quad \text{...(xv)}$$

If $m = 1$, $\Delta T_f = K_f$

Thus *molal depression constant may be defined as the depression in freezing point which takes place when the molality of the solution is unity.*

THIRD LAW OF THERMODYNAMICS

Q. 49. Describe Nernst heat theorem. ***(Himachal 2010, Raipur 2010)***

Ans. This theorem gives the variation of enthalpy change (ΔH) and free energy change (ΔG) of a system with decrease of temperature.

According to Gibbs-Helmholtz equation,

$$\Delta G = \Delta H + T \left(\frac{\partial \Delta G}{\partial T}\right)_p$$

From this equation, it is evident that at the absolute zero *i.e.* when $T = 0$, $\Delta G = \Delta H$. However Nernst observed that as the temperature is lowered towards absolute zero, the value of $\partial(\Delta G)/\partial T$ decreases and then approaches zero asymptotically. This means that ΔG and ∂H *are not only equal to the absolute zero but the values approach each other asymptotically in the vicinity of this temperature.* This result is known as Nernst heat theorem. Mathematically, it may be expressed as

$$\lim_{T \to 0} \frac{d(\Delta G)}{dT} = \lim_{T \to 0} \frac{d(\Delta H)}{dT} = 0 \quad \text{...(i)}$$

Graphically, the result may be represented by Fig. 7·7.

In Fig. 7.7, ΔG has been shown as greater than ΔH at temperature away from absolute zero. However the reverse is also possible because $\partial(\Delta G)/\partial T$ can be both positive or negative.

Since gases do not exist at the absolute zero, this means that the heat theorem is not applicable to gases. Similarly, it has been found to be inapplicable to liquids also. Nernst heat theorem is applicable to solids only.

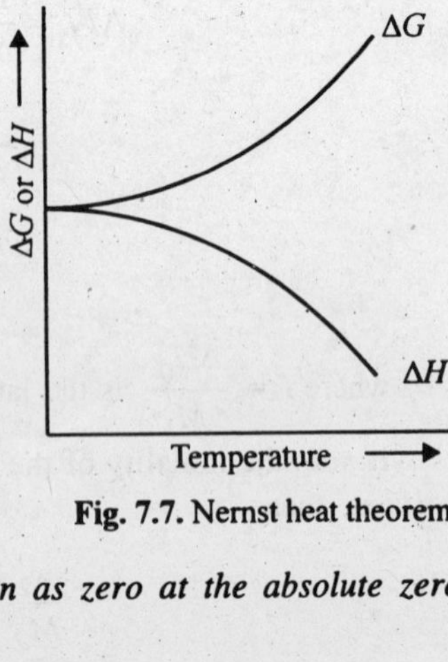

Fig. 7.7. Nernst heat theorem

Q. 50. State third law of thermodynamics. Explain its importance. *(G. Nanakdev 2000)*

Ans. Third law of thermodynamics states:

The entropy of all perfectly crystalline solids may be taken as zero at the absolute zero temperature.

The law is also stated as follows :

Every substance has finite positive entropy but at the absolute zero of temperature, the entropy may become zero and in fact it does become zero in case of perfectly crystalline solids.

Importance of the Third Law

Calculation of absolute entropies: It helps in the calculation of the absolute entropies of chemical compounds at any desired temperature. This is illustrated as under :

The infinitesimal entropy change is given by

$$dS = \frac{dq}{T} \quad ...(i)$$

But $$C_p = \frac{dq}{dT} \text{ so that } dq = C_p\, dT$$

Substituting this value in equation (*i*), we have

$$dS = \frac{C_p dT}{T}$$

Hence the entropy change of a substance when its temperature changes from absolute zero to the temperature (T) can be obtained from the equation

$$\int_{S=S_0}^{S=S} dS = \int_{T=0}^{T=T} C_p \frac{dT}{T}$$

or $$S - S_0 = \int_0^T C_p \frac{dT}{T} \quad ...(ii)$$

where S_0 is the entropy of the substance at absolute zero and S is the entropy of the substance at temperature T.

According to third law of thermodynamics, $S_0 = 0$.

$$\therefore \quad S = \int_0^T C_p \frac{dT}{T}$$

$$= \int_0^T C_p d \ln T \quad ...(iii)$$

Thus the entropy (S) of the substance at temperature T can be calculated from the measurements of heat capacities (C_p) at a number of temperatures between 0 K to T K. The integral in equation (*iii*) can be valuated by plotting C_p *vs* $\ln T$ *i.e.* 2·303 log T and then measuring the area under the curve between T = 0 to T = T as shown in Fig. 7.8.

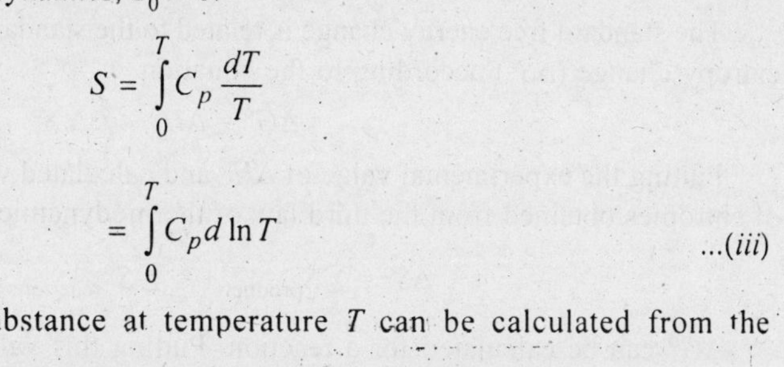

Fig. 7.8. Plot of heat capacities (C_p) *vs* T in the first case and *vs* log T in the second case.

If C_p is supposed to remain almost constant in the temperature range 0 to T °K, equation (*iii*) can be simplified to

$$S = C_p \ln T = 2{\cdot}303\, C_p \log T$$

This equation helps to calculate the value of S directly from the values of C_p and T.

For liquid and gaseous substances, the total absolute entropy of the substance at the given temperature will be the sum of all the entropy changes which the substance has to undergo in order to reach that particular state starting from the crystalline solid at absolute zero. Thus if a substance is gaseous at 1 atm pressure and 25°C, the entropy of the gas at 25°C will be the sum of the following entropies involved at different stages :

(*i*) the entropy of heating the crystalline solid from $T = 0$ to $T = T_m$ the melting point;

(*ii*) the entropy of melting, ΔH_m is latent heat of melting;

(*iii*) the entropy of heating the liquid from T_m to T_b, the boiling point;

(*iv*) the entropy of vaporization, $\dfrac{\Delta H_v}{T_b}$ where ΔH_v is latent heat of vaporisation;

(*v*) the entropy of heating the gas from T_b to 25°C (*i.e.* 298 K).

Thus the complete expression for the calculation of entropy at temperature T may be written as

$$S = \underbrace{\int_0^{T_m} C_{p(s)} d\ln T}_{\text{I}} + \underbrace{\frac{\Delta H_m}{T_m}}_{\text{II}} + \underbrace{\int_{T_m}^{T_b} C_p(l)\, d\ln T}_{\text{III}} + \underbrace{\frac{\Delta H_v}{T_b}}_{\text{IV}} + \underbrace{\int_{T_b}^{T} C_{P(g)} d\ln T}_{\text{V}}$$

Q. 51. How can you test the validity of third law of thermodynamics?

Ans. (1) By calculation of equilibrium constants: The equilibrium constant (K) is related to standard free energy change ($\Delta G°$) according to the equation

$$-\Delta G° = RT \ln K \qquad ...(i)$$

The standard free energy change is related to the standard enthalpy change ($\Delta H°$) and standard entropy change ($\Delta S°$) according to the equation

$$\Delta G° = \Delta H° - T\Delta S° \qquad ...(ii)$$

Putting the experimental value of $\Delta H°$ and calculated value of $\Delta S°$, using the absolute values of entropies obtained from the third law of thermodynamics, and applying the relation

$$\Delta S° = \Sigma S°_{\text{products}} - \Sigma S°_{\text{reactions}}$$

$\Delta G°$ can be calculated for a reaction. Putting this value of $\Delta G°$ in Eq. (*i*), the equilibrium constant K can be calculated. If the value of equilibrium constant as calculated above comes out to be in close agreement with the experimental value, validity of the law in proved.

(2) By comparing with entropies calculated from statistical mechanics: The absolute entropies calculated on the basis of the third law are found to be in complete agreement with those calculated using statistical mechanics. This proves the validity of third law.

Q. 52. What is residual entropy ? (*Delhi, 2003*)

Ans. Entropy is a state function and is given by the relation

$$\Delta S = \frac{q}{T}$$

The entropy of a system as it approaches absolute zero is called residual entropy. Of course, a perfectly crystalline solid at absolute zero has zero residual entropy.

8

THERMODYNAMICS-III

Q. 1. What do you understand by partial molar quantities? Write the general expression for partial molar quantity of a component *i* in a mixture.

(G. Nanakdev 2003; Mumbai 2010)

Ans. Let an extensive property such as volume, free energy, entropy, energy content etc be represented by X. Suppose there are n constituents in the system having $n_1, n_2, n_3 ...$ moles of individual components. Then the property X is a function of temperature, pressure as well as of the amounts of different constituents.

Thus $$X = f(T, P, n_1, n_2, n_3 ...) \qquad ...(i)$$

If there is a small change in the temperature, pressure and the amounts of the constituents, then change in property X is given by

$$dX = \left(\frac{\partial X}{\partial T}\right)_{P, n_1, n_2...} dT + \left(\frac{\partial X}{\partial P}\right)_{T, n_1, n_2...} dP + \left(\frac{\partial X}{\partial n_1}\right)_{T, P, n_2, n_3...} dn_1 + \left(\frac{\partial X}{\partial n_2}\right)_{T, P, n_1, n_3...} dn_2 + ... \qquad ...(ii)$$

The first term on R.H.S. gives the change in the value of X with temperature when pressure and composition are kept constant. The second term on R.H.S. gives the change in the value of X with pressure when temperature and composition are kept constant. The remaining quantities give the change in the value of X with a change in the amount of a constituent, when temperature, pressure and the amounts of other constituents are kept constant.

If the temperature and pressure of the system are kept constant, then

$$dT = 0 \text{ and } dP = 0$$

so that equation (*ii*) becomes

$$(dX)_{T,P} = \left(\frac{\partial X}{\partial n_1}\right)_{T, P, n_2, n_3...} dn_1 + \left(\frac{\partial X}{\partial n_2}\right)_{T, P, n_1, n_3...} dn_2 + ... \qquad ...(iii)$$

First quantity on right hand side is partial molar quantity of component 1, the second quantity for the component 2 and so on.

These are represented by putting a bar over the symbol of that particular property *i.e.* $\overline{X_1}, \overline{X_2}$ for the 1st, 2nd component etc. respectively. Thus

$$\left(\frac{\partial X}{\partial n_1}\right)_{T,P,n_2,n_3...} = \overline{X}_1, \left(\frac{\partial X}{\partial n_2}\right)_{T,P,n_1,n_3...} = \overline{X}_2 \quad \text{etc.}$$

In general, for any component i

$$\left(\frac{\partial X}{\partial n_1}\right)_{T,P,n_1,n_2...} = \overline{X}_i \qquad ...(iv)$$

Q. 2. What is chemical potential or partial molar free energy ? Derive an expression for it in terms of other quantities. It is an extensive or intensive property?

(Gujarat 2003; Madras 2004; Indore 2004; Nagpur 2008; Himachal 2010)

Ans. Consider the *extensive* property, free energy. Let it be represented by G. Suppose that the system consists of n constituents (in a solution or in a heterogeneous system), the amounts of which present in the system are n_1, n_2, n_3.... moles. Then the property G is a function not only of temperature and pressure but of the amounts of the different constituents as well, so that we can write

$$G = f(T, P, n_1, n_2, n_3 .....) \qquad ...(i)$$

Now, if there is a small change in the temperature, pressure and the amounts of the constituents, then the change in the property F is given by

$$dG = \left(\frac{\partial G}{\partial t}\right)_{P,n_1,n_2...} dT + \left(\frac{\partial G}{\partial P}\right)_{T,n_1,n_2...} dP$$

$$+ \left(\frac{\partial G}{\partial n_1}\right)_{T,P,n_2,n_3...} dn_1 + \left(\frac{\partial G}{\partial n_2}\right)_{T,P,n_1,n_3...} dn_2 + ... \qquad ...(ii)$$

On the right hand side, the first quantity gives the change in the value of G with temperature when pressure and composition are kept constant; the second quantity gives the change in the value of G with pressure when temperature and composition are kept constant; the remaining quantities give the change in the value of G with a change in the amount of a constituent, when temperature, pressure and the amounts of other constituents are kept constant.

If the temperature and pressure of the system are kept constant, then

$$dT = 0 \text{ and } dP = 0$$

so that equation (*ii*) becomes

$$(dG)_{T,P} = \left(\frac{\partial G}{\partial n_1}\right)_{T,P,n_2,n_3....} dn_1 + \left(\frac{\partial G}{\partial n_2}\right)_{T,P,n_1,n_3....} dn_2 + ... \qquad ...(iii)$$

Each derivative on the right hand side is called partial molar property and is represented by putting a bar over **the** symbol of that particular property *i.e.* $\overline{G_1}$, $\overline{G_2}$ for the 1st, 2nd component etc. respectively. Thus,

$$\left(\frac{\partial G}{\partial n_1}\right)_{T,P,n_2,n_3....} = \overline{G}_1; \left(\frac{\partial G}{\partial n_2}\right)_{T,P,n_1,n_3....} = \overline{G}_2 \text{ etc.}$$

In general, for any compone]nt i

$$\left(\frac{\partial G}{\partial n_i}\right)_{T,P,n_1,n_2....} = \overline{G}_i \qquad ...(iv)$$

This quantity is called *partial molar free energy or chemical potential* and is usually represented by the symbol μ. Thus

$$\mu_i = \overline{G}_1 = \left(\frac{\partial G}{\partial n_i}\right)_{T,P,n_1,n_2....} \quad ...(v)$$

If $dn_i = 1$ mole, $\mu_i = (dG)_{T,P,n_1,n_2....}$

Hence chemical potential may be defined as follows:

The chemical potential of a constituent in a mixture is the increase in the free energy which takes place at constant temperature and pressure when 1 mol of that constituent is added to the system, keeping the amounts of all other constituents constant i.e. when 1 mol of the constituents is added to such a large quantity of the system that its composition remains almost unchanged.

Eq. (*iii*) can be written as:

$$(dG)_{T,P} = \mu_1 dn_1 + \mu_2 dn_2 + ... \quad ...(vi)$$

For a system of definite composition represented by the number of moles n_1, n_2, n_3 etc., equation (*vi*) on integration gives

$$G_{T,P,N} = n_1\mu_1 + n_2\mu_2 + ...$$

where the subscript N stands for constant composition.

On the right hand side, the first term gives contribution of the first constituent to the total free energy of the system, the second term gives the contribution of the second constituent and so on. Obviously, μ_1, μ_2 etc. give the *contribution per mole* to the total free energy. Hence chemical potential may also be defined as under :

The chemical potential of a constituent in a mixture is its contribution per mol to the total free energy of the system of a constant composition at constant temperature and pressure. It may be noted that whereas free energy is an **extensive property,** the chemical potential is an **intensive property** because it refers to one mole of the substance.

Q. 3. Define the terms fugacity, activity, activity coefficient.

(G. Nanakdev 2003; Punjab 2003; Himachal 2010)

Ans. *Fugacity may be defined as a substitute for pressure to explain the behaviour of real gas and activity may be defined as the substitute for the concentration to explain the behaviour of a non-ideal solution.*

Thus *in a mixture of real gases,* the chemical potential μ_i of any constituent i is given by

$$\mu_i = \mu_i^0 + RT \ln f_i \quad ...(i)$$

where f_i is the fugacity of the constituent i in the mixture.

Similarly, *in a non-ideal solution,* the chemical potential of any component i is given by

$$\mu_i = \mu_i^0 + RT \ln a_i \quad ...(ii)$$

where a_i is the activity of component i in the solution.

Thus to explain the behaviour of real gases, the pressure has to be corrected by multiplying with a suitable factor to get the fugacity *i.e.*

$$f = \gamma \times P \quad ...(iii)$$

γ is called the activity coefficient of the gas.

From equation (*iii*), we have

$$\gamma = \frac{f}{p} \quad ...(iv)$$

Hence *activity coefficient of a gas may be defined as the ratio of the fugacity of the gas to the pressure of the gas in the same state.*

Evidently, for the ideal gas, $\gamma = 1$. For real gas, γ is always less than 1 and greater the departure in values from unity, the greater is the non-ideal behaviour.

Similarly, in a non-ideal solution, the concentration has to be corrected to give the activity as follows :

$$a = \gamma \times C \quad \text{...}(v)$$

γ represents the activity coefficient of that component in the solution whose concentration is C.

From equation (v), we have

$$\gamma = \frac{a}{C} \quad \text{...}(vi)$$

Hence *activity coefficient of any component in the solution may be defined as the ratio of activity o that component in the solution to the concentration of the same component in the solution.*

For an ideal solution (*i.e.* for very dilute solution), $\gamma = 1$ (so that $a = C$).

For other solutions, greater is the departure in the value of γ from 1, more non-ideal is the solution.

Q. 4. Calculate the fugacity of a gas at 450 atmosphere pressure and 300 K with α equal to -7.5×10^{-4} dm^{3}mol^{-1}.

Ans. For an ideal gas

$$PV = RT$$

or

$$V = \frac{RT}{P} \quad \text{...}(i)$$

where P, V and T are the pressure, volume and temperature of an ideal gas. R is the gas constant.

But the equation has to be modified for a real gas.

For a real gas

$$V = \frac{RT}{P} - \alpha \quad \text{...}(ii)$$

where α is the correction factor accounting for the deviation of the gas from ideal behaviour.

A relation that has been established between the fugacity and the pressure is as follows:

$$\ln \frac{f}{P} = -\frac{\alpha P}{RT} \quad \text{...}(iii)$$

In the present case

$$\alpha = -7.5 \times 10^{-4} \text{ dm}^3 \text{ mol}^{-1}$$

$$P = 450 \text{ atmosphere}$$

$$T = 300 \text{ K}$$

and

$$R = 0.082$$

$$\ln \frac{f}{P} = -\frac{\alpha P}{RT}$$

or

$$\log \frac{f}{P} = -\frac{\alpha P}{2.303RT}$$

or

$$\log f = \log P - \frac{\alpha P}{2.303RT} \quad \text{...}(iv)$$

Substitute the values in equation (iv)

$$\log f = \log 450 + \frac{7.5\times10^{-4}\times450}{2.303\times0.082\times300}$$

or $$\log f = 2.6592$$

Taking anti-logarithm of above

$$f = 456.2 \text{ atm}$$

Q. 5. Derive Gibbs-Duhem equation. *(Mumbai 2010; Nagpur 2008)*

Ans. Free energy (G), being an extensive property, depends not only upon the temperature and pressure of the system but also upon the composition of the system. If the system consists of a number of constituents, the amounts of which are n_1 n_2 n_3 ... moles respectively, then we can write

$$G = f(T, P, n_1, n_2, n_3, \ldots) \qquad \ldots(i)$$

For small changes in temperature, pressure and the quantity of different constituents, the small change in free energy can be obtained by the partial differentiation of equation (i). This gives

$$(dG) = \left(\frac{\partial G}{\partial T}\right)_{P,n_1,n_2\ldots} dT + \left(\frac{\partial G}{\partial n_2}\right)_{T,n_1,n_2\ldots} dP$$

$$+\left(\frac{\partial G}{\partial n_1}\right)_{T,P,n_2,n_3\ldots} dn_1 + \left(\frac{\partial G}{\partial n_2}\right)_{T,P,n_1,n_3\ldots} dn_2 + \ldots \qquad \ldots(ii)$$

If temperature and pressure are kept constant

$$dT = 0, dP = 0$$

so that equation (ii) becomes

$$(dG)_{T,P} = \left(\frac{\partial G}{\partial n_1}\right)_{T,P,n_2,n_3\ldots} dn_1 + \left(\frac{\partial G}{\partial n_2}\right)_{T,P,n_1,n_3\ldots} dn_2 + \ldots \qquad \ldots(iii)$$

Putting $$\left(\frac{\partial G}{\partial n_1}\right)_{T,P,n_2,n_3\ldots} = \mu_1,$$

$$\left(\frac{\partial G}{\partial n_2}\right)_{T,P,n_1,n_3\ldots} \text{ and so on}$$

we get $$(dG)T, p = \mu_1 dn_1 + \mu_2 dn_2 + \ldots \qquad \ldots(iv)$$

For a system of definite composition, represented by the number of moles n_1, n_2, n_3 etc., equation (iv) on integration gives

$$G_{T,P,N} = n_1\mu_1 + n_2\mu_2 \ldots$$

Differentiating this equation under conditions of constant temperature and pressure but varying composition, we get

$$(dG)_{T,P} = (n_1 d\mu_1 + \mu_1 dn_1) + (n_2 d\mu_2 + \mu_2 dn_2) + \ldots$$

$$= (\mu_1 dn_1 + \mu_2 dn_2 + \ldots) + (n_1 d\mu_1 + n_2 d\mu_2 + \ldots) \qquad \ldots(v)$$

Comparing equations (iv) and (v), we get

$$n_1 d\mu_1 + n_2 d\mu_2 + \ldots = 0$$

$$\boxed{\sum n_i d\mu_i = 0}$$

This equation which is applicable to a system under constant temperature and pressure is called **Gibbs-Duhem equation.**

Q. 6. Derive expressions for variation of chemical potential
(*i*) With temperature, (*ii*) With pressure.

(G. Nanakdev 2003; Calcutta 2000; Madras 2004)

In order to study the variation of chemical potential with temperature and pressure, let us first derive expressions for the *variation of* free energy *with temperature and pressure* to be used therein.

Ans. Free energy is a function of temperature, pressure and composition of the system *i.e.*

$$G = f(T, P, n_1, n_2, n_3 \ldots)$$

$$\therefore \quad (dG) = \left(\frac{\partial G}{\partial T}\right)_{P,N} dT + \left(\frac{\partial G}{\partial P}\right)_{T,N} dP + \left(\frac{\partial G}{\partial n_1}\right)_{T,P,n_2,n_3 \ldots} dn_1 + \left(\frac{\partial G}{\partial n_2}\right)_{T,P,n_1,n_3 \ldots} dn_2 + \ldots \quad \ldots(i)$$

If the system is a **closed** one, there is no change in the number of moles of the various constituents present *i.e.* dn_1, dn_2, dn_3 etc. will be all zero, so that the above equation reduces to

$$(dG) = \left(\frac{\partial G}{\partial T}\right)_{P,N} dT + \left(\frac{\partial G}{\partial P}\right)_{T,N} dP \quad \ldots(ii)$$

For a closed system, we also know that the *total differential equation* is

$$dG = VdP - SdT$$
$$= -SdT + VdP \quad \ldots(iii)$$

where V is the volume and S is the entropy of the system.

Comparing equations (*ii*) and (*iii*), we get

$$\left(\frac{\partial G}{\partial T}\right)_{P,N} = -S \quad \ldots(iv)$$

$$\left(\frac{\partial G}{\partial P}\right)_{T,N} = V \quad \ldots(v)$$

These two equations give the variation of free energy with temperature and pressure, for a closed system.

Variation of chemical potential with temperature: By definition, chemical potential of any constituent of a system is given by

$$\mu_i = \left(\frac{\partial G}{\partial n_i}\right)_{T,P,n_1,n_2 \ldots}$$

Differentiating this equation with respect to temperature, we get

$$\left(\frac{\partial \mu_i}{\partial T}\right)_{P,N} = \frac{\partial^2 G}{\partial n_i \partial T} \quad \ldots(vi)$$

Differentiating equation (*iv*) with respect to n_i, we get

$$\frac{\partial^2 G}{\partial T \partial n_i} = -\left(\frac{\partial S}{\partial n_i}\right)_{T,P,n_1,n_2 \ldots} = -\overline{S}_i \quad \ldots(vii)$$

where $\left(\frac{\partial S}{\partial n_i}\right)_{T,P,n_1,n_2 \ldots} = -\overline{S}_i$ is the partial molar entropy of the *i*th constituent.

Combining equations (*vi*) and (*vii*), we get

$$\left(\frac{\partial \mu_i}{\partial T}\right)_{P,N} = -\bar{S}_i \qquad ...(viii)$$

Thus *the rate of variation of chemical potential, of any constituent i of a system, with temperature at constant pressure and composition is equal to the partial molar entropy of that constituent.*

Variation of chemical potential with pressure: Chemical potential of a component i is given by

$$\mu_i = \frac{\partial \mu_i}{\partial n_i}_{T,P,n_1,n_2....}$$

Differentiating this equation with respect to pressure, we get

$$\left(\frac{\partial \mu_i}{\partial P}\right)_{T,N} = \frac{\partial^2 G}{\partial n_i \partial P} \qquad ...(ix)$$

Differentiating equation (*v*) with respect to n_i we get

$$\frac{\partial^2 G}{\partial P \partial n_i} = \left(\frac{\partial V}{\partial n_i}\right)_{T,P,n_1,n_2....} \qquad ...(x)$$

where $\left(\frac{\partial V}{\partial n_i}\right)_{T,P,n_1,n_2....} = \bar{V}_i$, represents the partial molar volume of the ith constituent.

Combining equations (*ix*) and (*x*), we get

$$\left(\frac{\partial \mu_i}{\partial P}\right)_{T,N} = \bar{V}_i \qquad ...(xi)$$

Hence *the rate of variation of chemical potential of any constituent i of the system, with pressure is equal to the partial molar volume of that constituent.*

Q. 7. For a system of ideal gases, prove the relation

$$\mu_i = \mu_i^0 + RT \, In \, p_i$$

Ans. For ideal gas, we know that

$$PV = nRT$$

As the system consists of n_1, n_2, n_3 etc. moles of the different constituents,

$$n = n_1 + n_2 + \ ... \ n_i + ... \qquad ...(ii)$$

Substituting this value in equation (*i*), we get

$$PV = (n_1 + n_2 + \ ...n_i + ...)RT \qquad ...(iii)$$

or

$$V = (n_1 + n_2 + \ ...n_i + ...) \ \frac{RT}{P} \qquad ...(iv)$$

Differentiating this equation with respect to n_i, keeping all other n's constant as well as temperature and pressure constant, we get

$$\left(\frac{\partial V}{\partial n_i}\right)_{T,P,n_1,n_2....} = \frac{RT}{P} \qquad ...(v)$$

or

$$\bar{V}_i = \frac{RT}{P} \qquad ...(vi)$$

We have another relation showing change of chemical potential with pressure

$$\left[\frac{\partial \mu_i}{\partial P}\right]_{T,N} = \bar{V}_i \qquad ...(vii)$$

From eq. (*vi*) and (*vii*)

$$\boxed{\left(\frac{\partial \mu_i}{\partial P}\right)_{T,N} = \frac{RT}{P}} \qquad ...(viii)$$

This equation can be rewritten as

$$d\mu_i = \frac{RT}{P}dP = RTd \ln P \qquad ...(ix)$$

In the mixture of gases, if p_i is the partial pressure of the *i*th constituent, we have

$$p_iV = n_iRT \qquad ...(x)$$

where n_i is the number of moles of the *i*th constituent.

For the complete mixture of gases containing *n* moles, for ideal behaviour, we know that

$$PV = nRT \qquad ...(xi)$$

Dividing equation (*x*) by equation (*xi*), we get

$$\frac{p_i}{P} = \frac{n_i}{n}$$

$$p_i = \frac{n_i}{n}P = K.P$$

where $K\left(=\frac{n_i}{n}\right)$, is a constant.

$\therefore$ $\qquad \ln p_i = \ln K + \ln P$

Hence $\qquad d \ln p_i = d \ln P \qquad [\because d \ln K = 0]$

Substituting this value in equation (*ix*), we get

$$d\mu_i = RTd\, ln\, p_i$$

Integrating this equation, we get

$$\boxed{\mu_i = \mu_i^0 + RT \ln p_i} \qquad ...(xii)$$

where μ_i^0 is the integration constant, the value of which depends upon the *nature of the gas* and also on the *temperature.*

From equation (*xii*), it is obvious that when p_i = 1 atm, $\mu_i = \mu_i^0$. Thus μ_i^0 is the chemical potential of the *i*th constituent when its partial pressure is unity in the mixture of ideal gases.

Q. 8. Derive the criteria for phase equilibrium for multicomponent system.

Ans. Let us consider the simplest case consisting of only two phases *A* and *B*. Suppose a number of components ($C_1, C_2, ...C_c$) are distributed between them (Fig. 8.1). Further suppose that at constant temperature and pressure, a small amount δn_1 of component 1 is transferred from phase *A* to phase *B*. If $(\mu_1)_A$ and $(\mu_1)_B$ represent the chemical potentials of component 1 in phases *A* and *B* respectively, we have

$$\left.\begin{array}{l}\text{Decrease in the free energy of component 1}\\ \text{in phase } A\end{array}\right\} = (\mu_1)_A\delta n_1$$

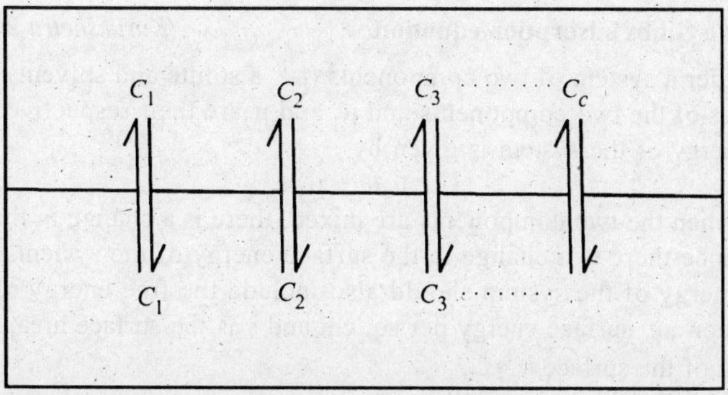

Fig. 8.1. Distribution of a number of components between two phases.

$$\left.\begin{array}{l}\text{Increase in the free energy of component 1}\\ \text{in phase } B\end{array}\right\} = (\mu_1)_B \delta n_1$$

Since the system is a closed one, $(\Delta, G)_{T,P} = 0$

$$\therefore \qquad -(\mu_1)_A \delta n_1 + (\mu_1)_B \delta n_1 = 0$$

(minus sign has been used to represent the decrease)

$$\text{or} \qquad (\mu_1)_A = (\mu_1)_B$$

It means *for a system consisting of two phases in equilibrium,the chemical potential of any given component is same in both the phases.*

Similarly, if a system consists of three phases, *A*, *B* and C in equilibrium, the above criterion will apply to the equilibrium between phases *A* and *B* as well as to the equilibrium between phases *B* and C so that we will have

$$(\mu_1)_A = (\mu_1)_B = (\mu_1)_C$$

In general for a system consisting of a number of phases P_1, P_2, P_2 etc in equilibrium (*i.e.*, a multiphase equilibria) for any component *i*,

$$(\mu_i)_{P_1} = (\mu_i)_{P_2} = (\mu_i)_{P_3} = ...$$

This result may be expressed as :

For a multiphase equilibria containing a number of components distributed among them, the chemical potential of any component is the same in all the phases.

Q. 9. Derive the criteria for two-phase equilibrium for one-component system.

Ans. For a system consisting of one pure substance only which may exist in two different phases in equilibrium, *e.g.*, liquid water $\rightleftharpoons$ water vapour or ice $\rightleftharpoons$ liquid water or sulphur (monoclinic) $\rightleftharpoons$ sulphur (rhombic) etc., the criterion is obtained in a simple manner in terms of *free energy.* In such a case, if a certain amount of the substance is transferred from one phase to the other, the molar free energy of one phase decreases while that of the other phase increases by an equal amount. Hence the net result is that there is no **change** in free energy, *i.e.*

$$\Delta G = 0$$

This implies that if at equilibrium, G_1 is the molar free energy of phase I and G_2 that of the phase II, we will have

$$G_2 - G_1 = 0 \quad \text{or} \quad G_1 = G_2$$

Thus, *whenever two phases of the same single substance (one component system) are in equilibrium, at a given temperature and pressure, the molar free energy is the same for each phase.*

Q. 10. Derive Gibbs adsorption equation. *(Kurukshetra 2002; Guwahati 2002)*

Ans. Consider a system of two components (*i.e.* a solute and solvent). If n_1 and n_2 are the number of moles of the two components and μ_1 and $\mu_2$are their respective chemical potentials, then the free energy of the system is given by

$$G = \mu_1 n_1 + \mu_2 n_2 \quad ...(i)$$

However, when the two components are mixed, there is a change in the surface area of the solution and hence there is a change in the surface energy of the system. To account for this, the total free energy of the system should also include the free energy of the surface. If γ is the surface tension *i.e.* surface energy per sq. cm and s is the surface area, then

Free energy of the surface = γs

The free energy of the system will be given by

$$G = \gamma s + \mu_1 n_1 + \mu_2 n_2 \quad ...(ii)$$

Differentiating this equation completely, we get

$$dG = \gamma ds + s d\gamma + \mu_1 dn_1 + n_1 d\mu_1 + \mu_2 dn_2 + n_2 d\mu_2 \quad ...(iii)$$

The free energy change of a system can also be obtained in the form of a different expression as follows :

Free energy is a function of temperature, pressure and composition of the various components, *i.e.*

$$G = f(T, P, n_1, n_2) \quad ...(iv)$$

If there is a small change in the temperature, pressure and the amounts of the two components, the change in the value of G will be given by

$$dG = \left(\frac{\partial G}{\partial T}\right)_{P,n_1,n_2} dT + \left(\frac{\partial G}{\partial P}\right)_{T,n_1,n_2} dP + \left(\frac{\partial G}{\partial n_1}\right)_{T,P,n_2} dn_1 + \left(\frac{\partial G}{\partial n_2}\right)_{T,P,n_1} dn_2 \quad ...(v)$$

But $\left(\frac{\partial G}{\partial n_1}\right)_{T,P,n_2} = \mu_1$ and $\left(\frac{\partial G}{\partial n_2}\right)_{T,P,n_1} = s\mu_2$

$$\therefore \quad dG = \left(\frac{\partial G}{\partial T}\right)_{P,n_1,n_2} dT + \left(\frac{\partial G}{\partial P}\right)_{T,n_1,n_2} dP + \mu_1 dn_1 + \mu_2 dn_2 \quad ...(vi)$$

We have the total differential equation

$$dG = VdP - SdT \quad ...(vii)$$

If temperature is kept constant, $dT = 0$ so that

$$(dG)_T = (VdP)_T \text{ or } \left(\frac{\partial G}{\partial P}\right)_T = V \quad ...(viii)$$

If pressure is kept constant, $dP = 0$, so that

$$(dG)_P = -(SdT)_P \text{ or } \left(\frac{\partial G}{\partial T}\right)_P = -S \quad ...(ix)$$

Substituting the values from equations (*viii*) and (*ix*) in equation (*vi*), we get

$$dG = -SdT + VdP + \mu_1 dn_1 + \mu_2 dn_2 \quad ...(x)$$

To get the total free energy change, the increase in free energy due to increase in the surface area should be added. If ds is the increase in the surface area and γ the surface tension, then

Free energy change due to increase in surface area = $\gamma . ds$

$\therefore$ Total free energy change will be given by

$$dG = - SdT + VdP + \gamma ds + \mu_1 dn_1 + \mu_2 dn_2 \qquad ...(xi)$$

Comparing equations (*iii*) and (*xi*), we get

$$sd\gamma + n_1 d\mu_1 + n_2 d\mu_2 = - SdT + VdP$$

or $$SdT - VdP + sd\gamma + n_1 d\mu_1 + n_2 d\mu_2 = 0$$

But at constant temperature and pressure,

$$dT = 0, dP = 0$$

$\therefore$ $$sd\gamma + n_1 d\mu_1 + n_2 d\mu_2 = 0 \qquad ...(xii)$$

Now imagine that the system is divided into two parts, the surface phase and the bulk phase as illustrated in Fig. 8.2.

If n_1^0 and n_2^0 are the number of moles of the two components in the bulk phase corresponding to n_1 and n_2 in the surface phase, then since for the bulk phase, surface considerations are not involved, eq. (*xii*) for the bulk phase may be written as

$$n_1^0 d\mu_1 + n_2^0 d\mu_2 = 0 \qquad ...(xiii)$$

Multiplying this equation by n_1 / n_1^0, we get

$$n_1 d\mu_1 + \frac{n_1 n_2^0}{n_1^0} d\mu_2 = 0 \qquad ...(xiv)$$

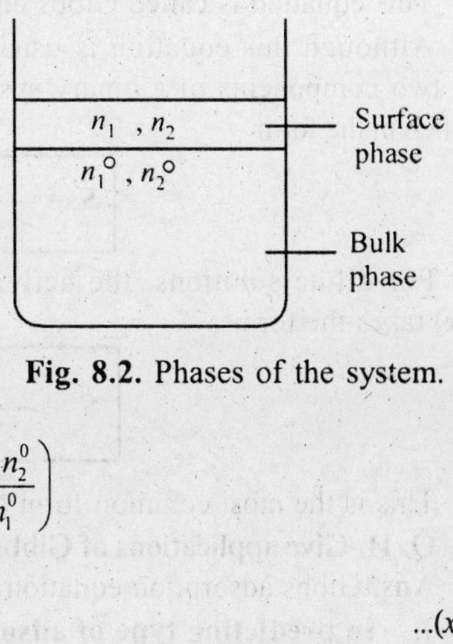

Fig. 8.2. Phases of the system.

Subtracting this equation from eq. (*xii*), we get

$$sd\gamma + n_2 d\mu_2 - \frac{n_1 n_2^0}{n_1^0} d\mu_2 = 0$$

or $$sd\gamma + d\mu_2 \left(n_2 - \frac{n_1 n_2^0}{n_1^0} \right) = 0 \text{ or } sd\gamma = - d\mu_2 \left(n_2 - \frac{n_1 n_2^0}{n_1^0} \right)$$

or $$-\frac{d\gamma}{d\mu_2} = \frac{n_2 - n_1 n_2^0 / n_1^0}{s} \qquad ...(xv)$$

n_2 is the number of moles of one component, say the solute, associated with n_1 moles of the solvent in the surface phase.

$n_1 n_2^0 / n_1^0$ is the corresponding number of moles associated with n_1 moles of the solvent the bulk phase. This is because in the bulk phase n_1^0 moles of the solvent are associated with n_2^0 moles of the solute.

$\therefore$ n_1 moles of the solvent will be associated with $\frac{n_2^0}{n_1^0} \times n_1$ moles of the solute.

Thus $\left(n_2 - \frac{n_1 n_2^0}{n_1^0} \right)$ is the excess concentration of the surface phase or $\left(\frac{n_2 - n_1 n_2^0 / n_1}{s} \right)$ is the excess concentration of the solute in the surface phase per unit area of the surface.

Representing this excess concentration of the surface phase by S_2, eq. (*xv*) can be written as

$$S_2 = -\frac{d\gamma}{d\mu_2} \qquad ...(xvi)$$

Also we have a standard result

$$\mu_2 = \mu_2^0 + RT \ln a_2 \qquad ...(xvii)$$

where a_2 is the activity of the component 2 and μ_2^0 is the chemical potential of the component 2 in the standard state *i.e.* when its activity is unity. Hence at constant temperature

$$d\mu_2 = RTd \ln a_2 = \frac{RT}{a_2} da_2$$

Putting this value in eq. (*xvi*), we get

$$S_2 = \frac{d\gamma}{RTda_2} . a_2$$

$$S_2 = \frac{a_2}{RT} \frac{d\gamma}{da_2} \qquad ...(xviii)$$

This equation is called Gibbs equation.

Although this equation is usually applied to solutes, it is equally applicable to either of the two components of a binary system. Removing the subscripts, eq. (*xviii*), therefore, can be written in the form

$$\boxed{\boldsymbol{S = - \frac{a}{RT} \frac{d\gamma}{da}}} \qquad ...(xix)$$

For dilute solutions, the activity may be replaced by concentration C so that equation (*xix*) takes the form

$$\boxed{\boldsymbol{S = - \frac{C}{RT} \frac{d\gamma}{dC}}} \qquad ...(xx)$$

This is the most common form of Gibbs adsorption equation.

Q. 11. Give applications of Gibbs adsorption equation.

Ans. Gibbs adsorption equation has the following important applications:

1. **In predicting type of adsorption:** If $d\gamma/dC$ is positive *i.e.* when the surface tension of the solution increases with concentration, S must be negative *i.e.* the concentration of the solute in the surface is less than in the bulk of the solution. This is a case of negative adsorption. This is so in the case of many electrolytes.

If $d\gamma/ dC$ is negative *i.e.* when the surface tension of the solution decreases with concentration, S will be positive *i.e.* the concentration of the solute is greater in the surface than in the bulk of the solution. This is a case of positive adsorption. This happens in the case of surface active agents like soaps as already discussed.

2. **In calculation of the excess concentration of a dissolved substance on the surface of a liquid:** Knowing the concentration C of the solute in the solution and knowing the value of $d\gamma/dC$, *i.e.* the rate of variation of surface tension with concentration at temperature T, the excess concentration of the dissolved substance on the surface of the liquid can be calculated.

3. **In the calculation of the adsorption of an insoluble gas (vapour) on the surface of a liquid:** For example, the adsorption of saturated hydrocarbons (component 2) on the surface of water (component 1) can be determined by using Gibbs adsorption equation in a slightly modified form as follows:

We have the equation for the excess concentration of the solute in the surface phase

$$S_2 = - \frac{d\gamma}{d\mu_2} \qquad ...(i)$$

But for the gaseous component 2,

$$\mu_2 = \mu_2^0 + RT \ln p$$

where p is the vapour pressure of component 2 over the liquid

$$\therefore \qquad d\mu_2 = RTd \ln p = \frac{RT}{p} dp$$

Substituting this value in Eq. (*i*), we get

$$S_2 = -\frac{d\gamma}{RTdp} . p$$

or

$$\boxed{S_2 = -\frac{p}{RT} . \frac{d\gamma}{dp}}$$

Thus knowing the value of $d\gamma/dp$ *i.e.* the rate of variation of surface tension with vapour pressure of the gaseous component, the adsorption of the vapour on the surface of the liquid can be calculated.

Q. 12. Explain positive and negative adsorption with the help of Gibbs adsorption equation.

Ans. See Q. 11.

9

POLYMERS OR MACROMOLECULES

Q. 1. What are polymers or macromolecules?

(*Kanpur, 2001; Awadh, 2000; Purvanchal 2002*)

Ans. Polymers or macromolecules are long chain high molecular weight compounds obtained by the combination of a large number of simple molecules. The simple molecules taking part in the formation of polymers are called *monomers* and this process of polymer formation is called *polymerisation.*

The product formed is called a dimer, trimer or tetramer according as two, three or four molecules combine to give the product. If a large number of molecules (say n molecules) combine together, the product is called *polymer.* Thus n molecules of ethylene combine to produce polyethylene.

$$\underset{\text{Ethylene}}{n\,CH_2 = CH_2} \longrightarrow \underset{\text{Polyethylene}}{[-CH_2-CH_2-]_n}$$

Q. 2. Explain the following terms :

(*a*) Degree of polymerisation

(*b*) High polymers and oligomers (*M. Dayanand 2000*)

(*c*) Homopolymers and copolymers. (*M. Dayanand, 2000; Himachal 2000; Kerala 2000*)

Ans. (*a*) The number of repeat units in the polymer chain is called *degree of polymerisation.* The product of degree of polymerisation (n) and the molecular mass of the monomeric unit M_M is equal to the molecular mass of the polymer.

Thus Mol. mass of the polymer $= n \times M_M$

(*b*) *High polymers and oligomers.* Polymers with high degree of polymerisation are called high polymers while those with low degree of polymerisation are called oligomers.

(*c*) The polymer formed from only one kind of monomers is called a *homopolymer* while a polymer made from two or more different kinds of monomers is called a *copolymer* or *mixed polymer.* For example, polythene is an example of homopolymer whereas Buna-S rubber is an example of copolymer formed from the monomers 1, 3-butadiene $CH_2= CH–CH = CH_2$)

and stryrene 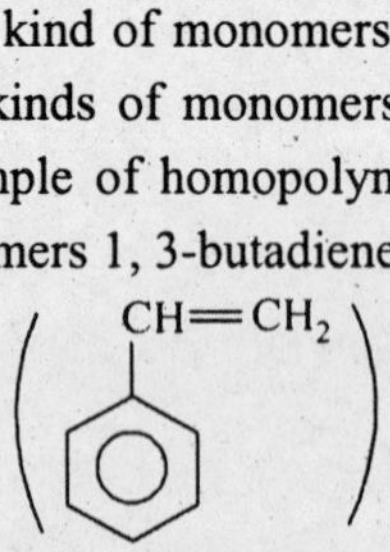

Q. 3. Differentiate between natural and synthetic polymers. (*Kanpur 2001*)

Ans. Natural polymers: Polymers obtained as such from a natural source are called natural polymers. A few examples of the naturally occurring polymers include proteins, polysaccharides, gums, natural rubber, silk and wool. Proteins, silk and wool are the polymers of amino acids, natural rubber is a polymer of isoprene, polysaccharides are polymers of glucose, fructose etc.

(*ii*) **Synthetic polymers:** Polymers prepared by synthesis from simple substances are called synthetic polymers. A few examples of the polymers prepared synthetically include synthetic rubbers, polyethylene, polystyrene, nylon etc.

Q. 4. Differentiate between linear, branched and cross-linked polymers. (*Kurukshetra 2002*)

Ans. Linear polymers: These are those polymers whose macromolecules contain the monomeric units linked together to form long straight chains. Representing the monomeric units by X, a linear polymer may be represented as follows :

$$...—X—X—X—X—X—X—...$$

Branched polymers: These are those polymers in which the monomeric units are linked to form long chains (called the *main chain*) with side branches (called the *side chains*), the number and length of which may vary widely from polymer to polymer. Branched polymers may be

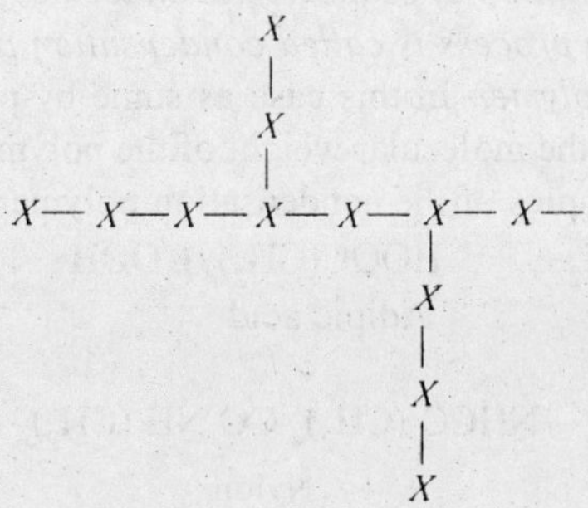

represented as under.

Cross-linked or three dimensional polymers: When the monomers are linked together to form a three dimensional network, the links involved are called cross-links and the polymer formed

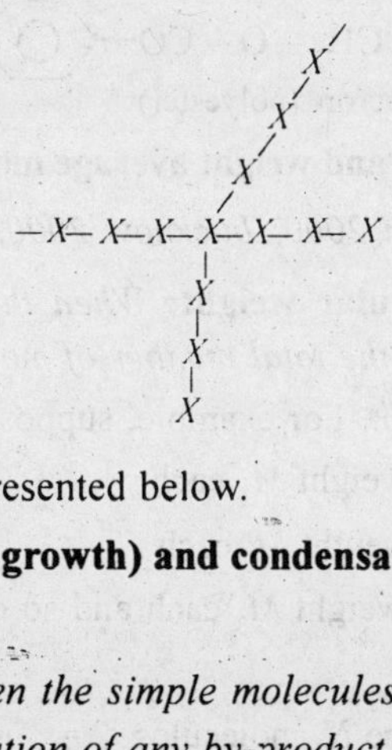

is called cross-linked polymer as represented below.

Q. 5. What are addition (chain growth) and condensation (step growth) polymers ? Give examples of each. (*Himachal 2000*)

Ans. Addition polymers: *When the simple molecules (monomers) are linked together to form long chains without the elimination of any by-product molecules, the products formed are called addition polymers and the process involved is called addition polymerization.*

As no by-product molecule is eliminated during the process, the molecular formula and hence the molecular weight of the polymer formed is an integral multiple of that of the monomer. A few examples of the addition polymerization are given below :

1 ...+ $CH_2 = CH_2 + CH_2 = CH_2 +$ .... $\longrightarrow$ .... $CH_2 - CH_2 - CH_2 - CH_2$ ...
Ethene — Polyethene

2 ....+ $CH_2 = CH(Cl) + CH_2 = CH(Cl) +$ .... $\longrightarrow$ .... $CH_2 - CH(Cl) - CH_2 - CH(Cl)$ ...
Vinyl chloride — Polyvinyl Chloride (PVC)

3.

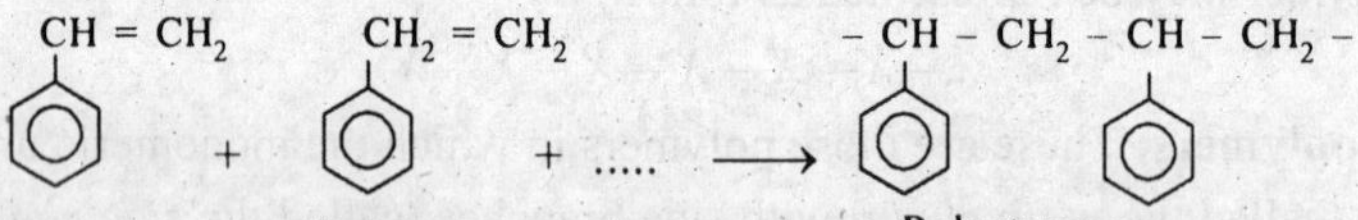

Styrene — Polystyrene

Condensation polymers: *When the monomers contain some active functional groups which can react together with the elimination of some simple molecules (usually water) and result into the formation of long chains, the process is called condensation polymerization and the product formed is called condensation polymer.* In this case as some by-product molecule is eliminated, the molecular formula and hence the molecular weight of the polymer is not an integral multiple of that of the monomer. A few examples of the condensation polymerization are given below :

1. $H_2N(CH_2)_6NH_2$ + $HOOC(CH_2)_4COOH$ + $H_2N-(CH_2)_6-NH_2$
Hexamethylene diamine — Adipic acid — Hexamethylene diamine
$\downarrow$
$-NH-(CH_2)_6-NHCO(CH_2)_4CONH(CH_2)_6-NH-$
Nylon

2.

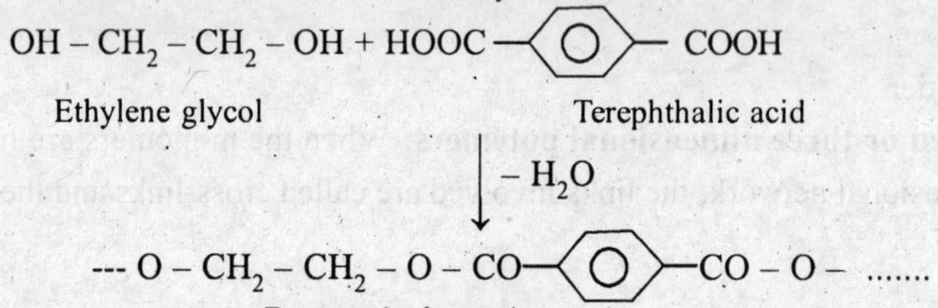

Dacron (polyester)

Q. 6. Explain number average and weight average molecular weights of polymers.

(Awadh 2000; Lucknow 2000; Nagpur 2002; G. Nanakdev 2003)

Ans. Number-average molecular weight: *When the total molecular weight of all the molecules of a sample is divided by the total number of molecules, the result obtained is called the number average molecular weight.* For example, suppose in a particular sample

N_1 molecules have molecular weight M_1 each

N_2 molecules have molecular weight M_2 each

N_3 molecules have molecular weight M_3 each and so on.

Then we have

Total molecular weight of all the N_1 molecules $= N_1M_1$

Total molecular weight of all the N_2 molecules $= N_2M_2$

Total molecular weight of all the N_3 molecules $= N_3M_3$ and so on

$\therefore$ Total molecular weight of all the molecules

$$= N_1M_1 + N_2M_2 + N_3M_3 + ...$$
$$= \Sigma N_i M_i$$

Total number of all the molecules

$$= N_1 + N_2 + N_3 + ...$$
$$= \Sigma N_i$$

Hence the number-average molecular weight will be given by

$$\overline{M_n} = \frac{N_1M_1 + N_2M_2 + N_3M_3 + ...}{N_1 + N_2 + N_3 + ...}$$

or
$$\overline{M_n} = \frac{\Sigma N_i M_i}{\Sigma N_i}$$

Weight-average molecular weight: *When the total molecular weight of group of molecules having particular molecular weights are multiplied with their respective molecular weight, the products are added and the sum is divided by the total weight of the all the molecules, the result obtained is called the weight-average molecular weight. Supposing N_1, N_2, N_3 etc. molecules have molecular weights M_1, M_2, M_3 etc.* respectively.

Total molecular weight of N_1 molecules $= N_1 M_1$

Total molecular weight of N_2 molecules $= N_2 M_2$

Total molecular weight of N_3 molecules $= N_3 M_3$ and so on.

The products with their respective molecular weights will be $N_1M_1 \times M_1$, $N_2M_2 \times M_2$, $N_3 M_3 \times M_3$, etc. *i.e.* $N_1M_1^2$, $N_2 M_2^2$, $N_3 M_3^2$, etc.

Sum of the products $= N_1 M_1^2 + N_2 M_2^2 + N_3 M_3^2 + ...$

$= \Sigma N_i M_i^2$

Total weight of all the molecules $= N_1 M_1 + N_2 M_2 + N_3M_3 + ...$

$= \Sigma N_i M_i$,

Hence weight-average molecular weight is given by

$$\overline{M_w} = \frac{N_1M_1^2 + N_2M_2^2 + N_3M_3^3 + ...}{N_1M_1 + N_2M_2 + N_3M_3 + ...}$$

or
$$\overline{M_w} = \frac{\Sigma N_i M_i^2}{\Sigma N_i M_i}$$

Condition under which $\overline{M_n} = \overline{M_w}$:

$$\overline{M_n} = \frac{N_1M_1 + N_2M_2 + ...}{N_1 + N_2 + ...}$$

$$\overline{M_w} = \frac{N_1M_1^2 + N_2M_2^2 + ...}{N_1M_1 + N_2M_2 + ...}$$

If $\overline{M_n} = \overline{M_w}$, then

$$\frac{N_1M_1 + N_2M_2 + ...}{N_1 + N_2 + ...} = \frac{N_1M_1^2 + N_2M_2^2 + ...}{N_1M_1 + N_2M_2 + ...}$$

The two sides can be equal only if

$$M_1 = M_2 = ...$$

i.e. all molecules should have the same molecular weight. In that case, average has no significance.

Q. 7. In a particular sample of a polymer, 100 molecules have molecular weight 10^3 each, 200 molecules have molecular weight 10^4 each and 200 molecules have molecular weight 10^5 each. Calculate the number-average and weight-average molecular weight.

Ans. $$\overline{M}_n = \frac{100\times10^3+200\times10^4+200\times10^5}{100+200+200}$$

$$= \frac{10^5(1+20+200)}{500} = 4.4\times10^4 = \mathbf{44,000}$$

$$\overline{M}_w = \frac{100\times(10^3)^2+200\times(10^4)^2+200\times(10^5)^2}{100\times10^3+200\times10^4+200\times10^5}$$

$$= \frac{10^8(1+200+20000)}{10^5(1+20+200)}$$

$$= \frac{10^3\times20201}{221} = \mathbf{91000}$$

Q. 8. A suspension contains equal masses of particles with molecular weights 20000 and 40000. Calculate $\overline{M}_n$ and $\overline{M}_w$.

Ans. Since the suspension contains equal masses of particles with molecular weights 20000 and 40000, we have

$$N_1 M_1 = N_2 M_2$$

Here $M_1 = 20000$ and $M_2 = 40000$

Substitute the values in the equation above

$$N_1 \times 20000 = N_2 \times 40000$$

or $$\frac{N_1}{N_2} = 2 \text{ or } N_1 = 2 N_2$$

$$\overline{M}_n = \frac{N_1M_1+N_2M_2}{N_1+N_2}$$

Substitute 2 N_2 for N_1 and putting the values of M_1 and M_2, we have

$$\overline{M}_n = \frac{(2N_2\times20000)+(N_2\times40000)}{2N_2+N_2}$$

$$= \frac{N_2(40000+40000)}{N_2(2+1)} = \mathbf{26667}$$

$$\overline{M}_w = \frac{N_1M_1^2+N_2M_2^2}{N_1M_1+N_2M_2}$$

Substitute $2N_2$ for N_1 and put the values of M_1 and M_2

$$\overline{M}_w = \frac{2N_2\times20000^2+N_2\,40000^2}{2N_2\times20000+N_2\times40000}$$

$$= \frac{N_2(2\times20000^2+40000^2)}{N_2(2\times20000+40000)}$$

$$= \frac{2\times2^2\times10^8+4^2\times10^8}{2\times2\times10^4+4\times10^4} = \frac{8\times10^8+16\times10^8}{4\times10^4+4\times10^4} = \frac{24\times10^8}{8\times10^4} = 30000$$

Q. 9. A polymer consists of 40% by weight of macromolecules of molecular weight 10000 and 60% by weight of macromolecules of molecular weight 75000. Calculate the number-average and weight-average molecular weight of the polymer.

Ans. $M_1 = 10000$, $M_2 = 75000$

Let N_1 and N_2 be the number of molecules of the two types

$$\frac{N_1M_1}{N_2M_2} = \frac{4}{6} \text{ or } \frac{N_1M_1}{N_2M_2} = \frac{2}{3} \text{ or } N_2M_2 = 1.5N_1M_1$$

Also $$\frac{N_1\times10000}{N_2\times75000} = \frac{2}{3} \text{ or } N_2 = \frac{N_1\times10000}{75000}\times\frac{3}{2} = 0.2N_1$$

$$\overline{M}_n = \frac{N_1M_1+N_2M_2}{N_1+N_2} \qquad \text{Substitute for } N_2M_2$$

$$= \frac{N_1M_1+1.5N_1M_1}{N_1+0.2N_1}$$

$$= \frac{2.5N_1M_1}{1.2N_1} = \frac{2.5}{1.2}\times10000 = \mathbf{20833}$$

$$\overline{M}_w = \frac{N_1M_1^2+N_2M_2^2}{N_1M_1+N_2M_2} = \frac{N_1M_1.M_1+N_2M_2.M_2}{N_1M_1+N_2M_2}$$

$$= \frac{(N_1M_1\times10000)+(1.5N_1M_1\times75000)}{N_1M_1+1.5N_1M_1}$$

$$= \frac{N_1M_1(10000+112500)}{N_1M_1(1+1.5)}$$

$$= \frac{122500}{2.5} = \mathbf{49000}$$

Q. 10. Calculate $\overline{M}_w$ (weight-average molecular weight) for a system containing equal number of particles with molecular weights 10000 and 20000.

Ans. In this case $N_1 = N_2$

$$\overline{M}_w = \frac{N_1M_1^2+N_2M_2^2}{N_1M_1+N_2M_2}$$

Substitute N_1 for N_2

$$\frac{N_1M_1^2+N_1M_2^2}{N_1M_1+N_1M_2} = \frac{N_1(M_1^2+M_2^2)}{N_1(M_1+M_2)}$$

$$= \frac{M_1^2+M_2^2}{M_1+M_2}$$

Substitute for M_1 and M_2

$$\overline{M}_w = \frac{10000^2+20000^2}{10000+20000} = \frac{10^8+(4\times10^8)}{10^4+(2\times10^4)} = \frac{5\times10^8}{3\times10^4} = 16667$$

Q. 11. Describe osmotic pressure method for the determination of molecular weight of a polymer. *(Guwahati 2002; G. Nanakdev 2003; Purvanchal 2003; Madras 2011)*

Ans. Osmosis is the phenomenon of the passage of solvent molecules into the solution through a semipermeable membrane.

The hydrostatic pressure of the liquid column set up as a result of osmosis is called the *osmotic pressure* or osmotic pressure is the external pressure that has to be applied on the solution to prevent the entry of the solvent into the solution through the semipermeable membrane.

The solution of a polymer in a suitable solvent can be subjected to the phenomenon of osmosis because the size of the polymer molecules in the solution is large enough not to pass through the semipermeable membrane while the solvent molecules can easily pass through. If the solution were *dilute,* the osmotic pressure is given by the van't Hoff equation *viz.*

$$PV = nRT$$

or

$$P = \frac{n}{V}RT \qquad ...(i)$$

where P = osmotic pressure
n = number of moles of the polymer
V = volume of the polymer solution
T = absolute temperature at which measurements are made
R = solution constant or the gas constant.

Substituting, $n = \frac{w}{M}$ eq. (*i*) reduces to

$$P = \frac{w}{M} \times \frac{1}{V}RT = \frac{w}{V}\frac{RT}{M} = C\frac{RT}{M} \qquad ...(ii)$$

where $C = \frac{w}{V}$ is the concentration of the solution in grams/litre.

Equation (*ii*) may be rewritten as

$$\frac{P}{C} = \frac{RT}{M} \qquad ...(iii)$$

According to this equation, $\frac{P}{C}$ should be constant at constant temperature *i.e* a plot of $\frac{P}{C}$ *vs* C should be a straight line.

However this result is found to hold good only for dilute solutions of low molecular weight polymers. For high polymer solutions, it is observed that the osmotic pressure varies with concentration according to the following relationship

$$\frac{P}{C} = \frac{RT}{M} + AC + BC^2 + ... \qquad ...(iv)$$

where A, B etc. are constant. The quantity $\frac{P}{C}$ is called *reduced osmotic pressure.*

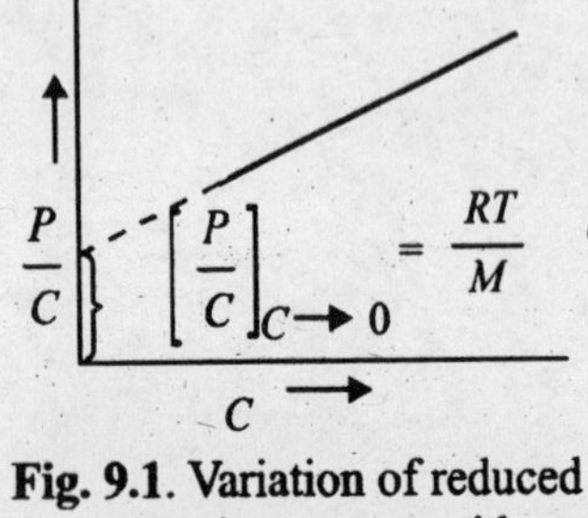

Fig. 9.1. Variation of reduced osmotic pressure with concentration.

From equation (*iv*), it is evident that

$$\frac{P}{C} = \frac{RT}{M} \text{ if } C = 0$$

i.e.,

$$\text{limit} \left(\frac{P}{C}\right)_{C \to 0} = \frac{RT}{M}$$

Thus if $\frac{P}{C}$ values are plotted against C and the graph is extrapolated to $C = 0$ then the intercept on the Y-axis is equal to $\frac{RT}{M}$ (Fig. 9·1).

From this the value of M can be calculated as :

$$M = \frac{RT}{\left(\frac{P}{C}\right)_{C\to 0}}$$

As osmotic pressure is a colligative property *i.e.*, it depends upon the number of particles present, therefore the molecular weight obtained by this method is the *number-average molecular weight.*

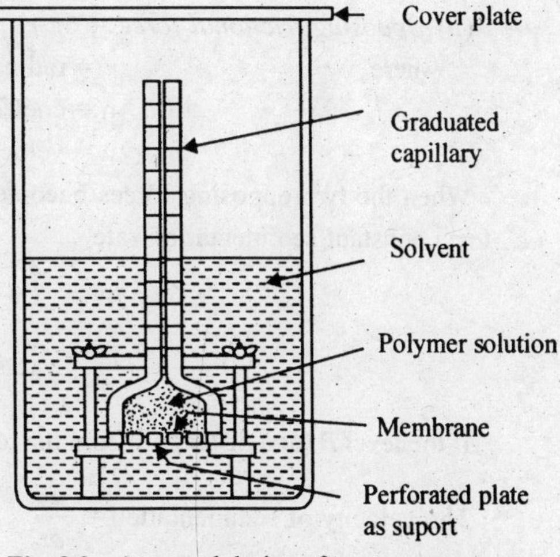

Fig. 9.2 . A general design of an osmometer.

Thus to determine the molecular weight of a polymer, it is required to measure the osmotic pressure of solutions of several concentrations and extrapolate the line to zero concentration.

The instruments used for determining the molecular weight by this method are called *osmometers.* A simple design of an osmometer is shown in Fig. 9.2. Two different techniques are used :

(*i*) **Static method:** The solvent is allowed to enter the polymer solution through the semipermeable membrane till level in the capillary tube stops rising.

(*ii*) ***Dynamic*** **method:** In this method, the *rate of penetration* of the solvent into the solution through the semipermeable membrane is noted under a known applied pressure.

Dynamic method is preferred because it gives quick results.

Q. 12. Describe sedimentation or ultracentrifugation method for the determination of molecular weight of polymers.

(Nagpur 2002)

Ans. Sedimentation is the process of settling down of the particles in a solution. Polymers because of large size of their molecules start settling down due to the force of gravity, though very very slowly. Based upon this idea, a method has been developed for the determination of molecular weights of polymers.

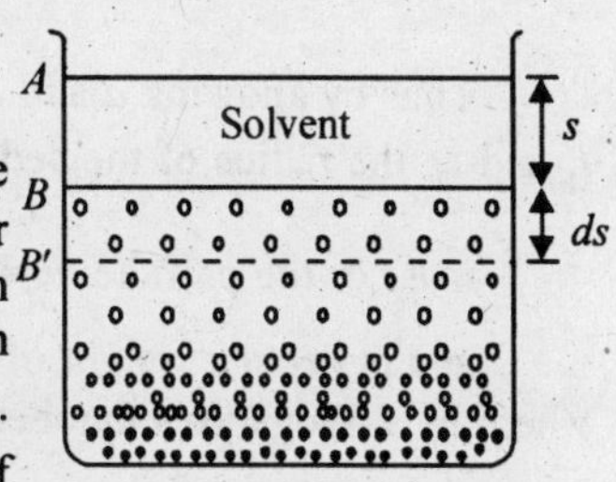

Fig. 9.3. Sedimentation under gravity.

This method is based upon the principle that the rate of sedimentation depends upon the size of the particles. Thus by measuring the rate of sedimentation, the size of the particles can be calculated from which the mass of the particles and then the molecular weight can be calculated. The rate of sedimentation can be followed from the fact that if the solution is kept undisturbed for quite some time, the solvent forms a separate layer at the top (Fig. 9.3). As the time passes, the level B of the solution moves downwards which can be located either directly or by optical method. The mathematical expression for the calculation of the result is obtained as follows :

For any sedimenting particle, there are two forces acting on it. These are :

(*i*) *The gravitational force acting downwards*

$$= \frac{4}{3}\pi r^3 (d-d')g$$

where r = radius of the particle

d = density of the particles

d' = density of the solvent (medium)

g = acceleration due to gravity

(*ii*) *Opposing frictional force* = $6\pi\, r\, \eta v$ (*Stokes' law*)

where r = radius of the particle, as mentioned above

η = coefficient of viscosity of the medium

v = velocity of sedimentation of the particle

When the two opposing forces become equal, the particle falls with a constant velocity v, *i.e.*, for a constant sedimentation rate

$$6\pi\, r\, \eta v = \frac{4}{3}\pi r^3 (d-d')g$$

or,

$$v = \frac{2r^2 g(d-d')}{9\eta} \quad \text{....}(i)$$

If the level B (at a distance x from the top) travels a distance dx in a small interval of time dt,

$$\text{The velocity of sedimentation} = \frac{dx}{dt}$$

$$\therefore \quad \frac{dx}{dt} = \frac{2r^2 g(d-d')}{9\eta} \quad \text{....}(ii)$$

or

$$dx = \frac{2r^2 g(d-d')}{9\eta}\, dt \quad \text{....}(iii)$$

Integrating this equation between the limits x_1 at t_1 and x_2 at t_2, we get

$$\int_{x_1}^{x_2} dx = \frac{2r^2 g(d-d')}{9\eta} \int_{t_1}^{t_2} dt$$

i.e.,

$$x_2 - x_1 = \frac{2r^2 g(d-d')\,(t_2-t_1)}{9\eta} \quad \text{....}(iv)$$

Thus by knowing d and d', and by measuring the distances x_1 and x_2 at two different times t_1 and t_2, the radius of the sedimenting particles can be calculated. Then we have

mass of the particle, $m = \frac{4}{3}\pi r^3 d$

and molecular weight of the polymer = Nm

where N is Avogadro's number.

A limitation of this method is that sedimentation under gravity is very very slow. This difficulty was solved by performing the experiment in an ultracentrifuge. Thus gravity is substituted by the centrifugal force of rotation. The solution is taken in a cell placed in a specially designed rotor. The force exerted on the particles in this machine may be as high as one million times the force of gravity or even more. Thus the time of sedimentation is reduced considerably. The rate of sedimentation can be observed by optical methods without stopping the ultracentrifuge. Thus using this technique, the acceleration due to gravity g in expression (*ii*), is replaced by the acceleration due to the centrifugal field. The latter at a distance x from the axis of rotation is given by $\omega^2 x$ where ω is the angular velocity of rotation. Thus equation (*ii*) is modified to

$$\frac{dx}{dt} = \frac{2r^2\omega^2 x(d-d')}{9\eta} \quad \text{or} \quad \frac{1}{x}\,dx = \frac{2r^2\omega^2 x(d-d')}{9\eta}\,dt$$

Integrating between the limits x_1 at t_1 and x_2 at t_2, we get

$$\ln\frac{x_2}{x_1}=\frac{2r^2\omega^2(d-d')}{9\eta}(t_2-t_1) \quad ...(v)$$

Using this equation, the radius r of the particle and hence the mass m and then the molecular weight can be calculated, as before.

The molecular weight obtained by the sedimentation method is called sedimentation average molecular weight ($\overline{M}_g$). It is nearly equal to the weight average molecular weight ($\overline{M}_w$).

Q. 13. Discuss viscosity method for the determination of molecular weight of a polymer.
(*Himachal 2000; Kurukshetra 2002; G. Nanakdev 2003; Gujarat 2003*)

Ans. This method requires the measurement of viscosities of polymer solutions of different concentrations which can be done easily with the help of Ostwald viscometer. However, this method is not an absolute method because it requires calibration using certain other substances of known molecular weights. The theory of the method is discussed below :

If η is the viscosity of a solution and η_0 that of pure solvent at the same temperature, the ratio η/η_0 is called the **relative viscosity** of the solution and is usually represented by η_r *i.e.*,

$$\frac{\eta}{\eta_0} = \eta_r \quad ...(i)$$

Another quantity **specific viscosity** η_{sp} is given by

$$\eta_{sp} = \frac{\eta-\eta_0}{\eta_0} \quad ...(ii)$$

Further, it is observed that the ratio η_{sp}/C where C is the concentration of the solution, generally expressed in grams per 100 ml of the solution, varies linearly with concentration. The extrapolation of η_{sp}/C value to zero concentration gives what is known as **intrinsic viscosity**, represented usually by $[\eta]$ *i.e.*,

$$[\eta] = \left(\frac{\eta_{sp}}{C}\right)_{C\to 0} \quad ...(iii)$$

The intrinsic viscosity is found to be related to the molecular weight M by the expression

$$[\eta] = k M^a \quad ...(iv)$$

where k and a are constants for the given polymer series in a particular solvent. Thus if k and a are known for a particular polymer series in a particular solvent, the value of M can be determined by finding $[\eta]$ for the polymer under investigation.

$[\eta]$ for the polymer under investigation can be found by plotting η_{sp}/C *vs.* C and extrapolating the result to zero concentration as shown in Fig. 9.4 for polyvinyl chloride in cyclohexane at 25°C.

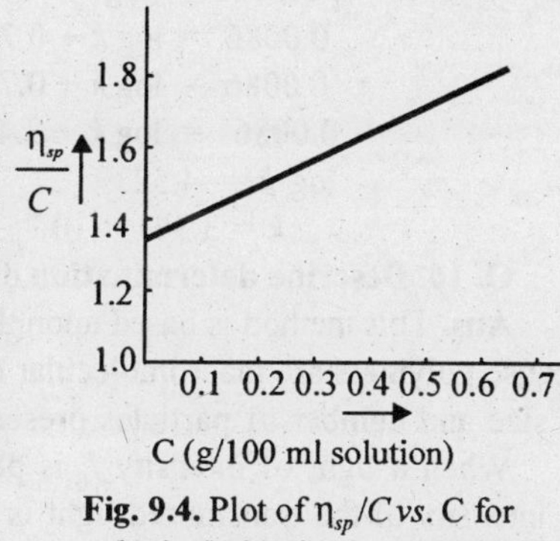

Fig. 9.4. Plot of η_{sp}/C *vs.* C for polyvinyl chloride in cyclohexane at 25°C.

To determine k and a, a number of samples of the same polymer series are taken whose molecular weights have been determined by other methods like osmometry or light scattering. Writing equation (*iv*) in the form

$$\log \eta = \log k + a \log M$$

a plot of $\log \eta$ *vs.* $\log M$ will be a straight line with slope a and intercept $\log k$. Thus the values of k and a can be obtained for the given polymer series in a particular solvent.

The molecular weight obtained by this method is called *viscosity-average molecular weight* ($\overline{M}_v$). If $a = 1$, it is equal to the weight-average molecular weight. However as a usually lies between 0.5 to 1, the viscosity- average molecular weight ($\overline{M}_v$) usually lies between number-average molecular weight ($\overline{M}_n$), and weight-average molecular weight ($\overline{M}_w$).

Q.14.Define the term intrinsic viscosity. How is it related to molecular weight of a polymer? When does the viscosity average molecular weight become equal to weight average molecular weight? (*M. Dayanand 2000*)

Ans. See Q. 13.

Q. 15. Write intrinsic viscosity-molecular weight relationship. Determine the values of two constants appearing in the above relation when fractions of a polymer of molecular weights 34000, 61000 and 130000 dissolved in an organic solvent gave the intrinsic viscosities as 1.02, 1.60 and 2.75 respectively at 25°C.

Ans. Intrinsic viscosity—molecular weight relation is

$$[\eta] = k\,M^a$$

Substitute the values in the equation above

$$1.02 = k\,34000^a \quad ...(i)$$

$$1.60 = k\,61000^a \quad ...(ii)$$

Divide eq. (*ii*) by (*i*),

$$\frac{1.60}{1.02} = \left(\frac{61000}{34000}\right)^a \text{ or } \frac{1.60}{1.02} = \left(\frac{61}{34}\right)^a$$

or

$$1.568 = 1.794^a$$

Taking logarithm of both sides

$$a \log 1.794 = \log 1.568 \text{ or } a = \frac{\log 1.568}{\log 1.794} = \frac{0.1953}{0.2539} = 0.769$$

Substitute the value of a in (*i*)

$$1.02 = k\,34000^{0.769}$$

Taking logarithm of both sides

$$\log 1.02 = \log k + 0.769 \log 34000$$

$$0.0086 = \log k + 0.769 \log (3.4 \times 10^4)$$

$$0.0086 = \log k + 0.769\,[0.5315 + 4]$$

$$0.0086 = \log k + 0.769 \times 4.5315$$

$$0.0086 = \log k + 3.4847$$

or

$$\log k = \bar{4}.5239$$

or

$$k = 3.341 \times 10^{-4}$$

Q. 16. Describe determination of molecular weight by light scattering method.

Ans. This method is based upon the phenomenon of 'scattering of light' by colloidal solutions because polymers are macromolecular colloids. The amount of light scattered depends directly on the size and number of particles present in the solution.

When a light of intensity I_0 is passed through the solution contained in a cell of length l, the intensity of the transmitted light is reduced to I due to scattering. The relationship between I and I_0 is given by

$$\tau = \frac{1}{l}\log\frac{I_0}{I} \quad ...(i)$$

where τ is called the turbidity of the scattering medium.

The turbidity τ is related to the apparent molecular weight of the polymer in the solution according to the following equation, given by Debye

$$\frac{HC}{\tau} = \frac{1}{M'} + AC + BC^2 + ... \qquad ...(ii)$$

where C = concentration of the solution in g/cm^3

A, B etc. are constants and M' is the apparent molecular weight of the polymer

$$H = \frac{32\pi^2\eta_0^2(d\eta/dc)^2}{25N\lambda_0^4} \qquad ...(iii)$$

where η_0 = refractive index of the solvent

$d\eta/dc$ = variation of refractive index of the solution with concentration

λ_0 = wavelength in the vacuum of the light used

N = Avogadro's number

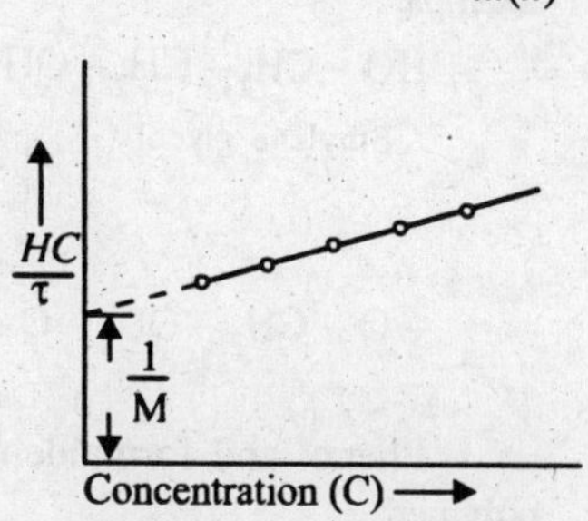

Fig. 9.5. Plot of HC/τ Vs C

Eq. (*ii*) leads to the following result :

$$\text{limit}\left(\frac{HC}{\tau}\right)_{C\to 0} = \frac{1}{M'} \qquad ....(iv)$$

Thus to get M' we should plot $\frac{HC}{\tau}$ versus C and then extrapolate the plot to zero concentration. Intercept gives $1/M'$ from which M' can be calculated.

To get the value of $\frac{HC}{\tau}$ corresponding to different concentrations C, solutions of experimental polymers in a particular suitable solvent are prepared. In each case, the turbidity τ is measured and the quantity H is calculated. To measure turbidity τ [using eq. (*i*)], the intensity of the light scattered at 90° to the direction of the incident beam is measured with simple photometer. To get the value of H, the quantities involved in eq. (*iii*) have to the measured. The refraction index n_0 of the solvent is measured with a refractometer, *dn/dc* is measured with special differential refractometers.

Further the apparent molecular weight M' is related to a the correct molecular weight M by the relation

$$M = M' \alpha \beta \qquad ...(v)$$

where α and β are corrections depending upon the size and the nature of the solute particles. α is called *disymmetry correction* and β is called *depolarization correction.* The correction factor α is obtained by measuring the intensities of scattered light at angles of 45° and 135° to the primary beam and β is obtained from the ratio of the horizontal to the vertical components of the light scattered by the solute at 90° to the direction of the incident beam.

It may be pointed out that the molecular weight obtained by this method is the *weight-average molecular weight,* M_w.

Q.17. What are different methods for determination of molecular weight of macromolecules? Describe any one method shortly. ***(Kanpur 2001)***

Ans. See Q.11, 12 and 13

Q.18. What are Macromolecules ? What difficulties do we feel in determining their molecular weight ? ***(Lucknow 2000; Kanpur 2001; Kerala 2000)***

Ans. See Q.1.

Q.19. Give the classification of polymers. ***(Awadh 2000)***

Ans. See Q. 2, 3, 4 and 5.

Q.20. What is step polymerisation? Under what conditions the linear and cross linked polymers are obtained? Give suitable examples. ***(Mumbai 2010)***

Ans. When the monomers contain some active functional groups which can react together with the elimination of some simple molecules like H_2O, HCl, etc and result into the formation of long chains, we obtain step growth polymers and this process is called step growth polymerisation.

Example

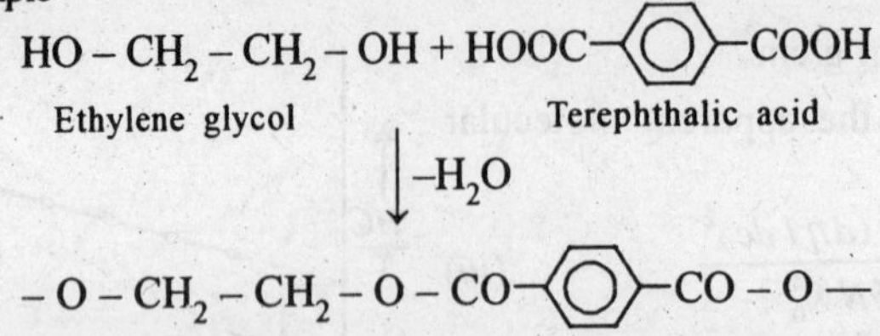

Dacron (polyester)

Phenol and formaldehyde form bakelite as given below. It is an example of cross linked polymer

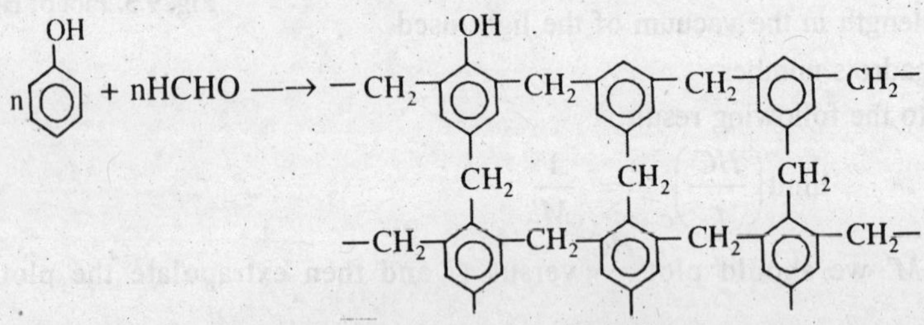

Bakelite

Cross-linked polymers are generally obtained where there is a possibility of chemical reaction at more than one point in the molecule. In the absence of such an opportunity, linear polymers are formed.

Q. 21. Equal number of molecules with $M_1 = 10000$ and $M_2 = 100000$ are mixed. Calculate the number average and weight average molecular weight of the polymer. Also calculate the polydispersity of the polymer sample. ***(Mumbai 2010)***

Ans. Number average molecular weight $\overline{M}_n = \dfrac{\Sigma N_i M_i}{\Sigma N_i}$

$$= \frac{N \times 10000 + N \times 100000}{N + N} = \frac{N \times 110000}{2N} = 55000$$

Weight average molecular weight $\overline{M}_w = \dfrac{\Sigma N_1 M_1^2}{\Sigma N_1 M_1}$

$$= \frac{N \times (10000)^2 + N \times (100000)^2}{(N \times 10000) + (N \times 100000)} = \frac{10000^2 + 100000^2}{110000}$$

$$= \frac{10^8 + 10^{10}}{11 \times 10^4} = \frac{10^8(1+10^2)}{11 \times 10^4} = \frac{10^4 \times 101}{11} = 0.919 \times 10^4$$

$$\text{Polydispersity index} = \frac{\text{Number average molecular weight}}{\text{Weight average molecular weight}}$$

$$= = \frac{55 \times 10^3}{0.919 \times 10^4} = \frac{5.5 \times 10^4}{0.919 \times 10^4} = 5.98$$

Polydispersity index gives the width of molecular weight distribution.

10

ELECTROCHEMISTRY

Q.1. Give the postulates of Arrhenius theory of ionisation.

Ans. The main points of Arrhenius theory are given below :

(*i*) When dissolved in water, an electrolyte splits or dissociates into positively and negatively charged particles called *ions*. This process is called electrolytic *dissociation or ionisation*.

$$NaCl \longrightarrow Na^+ + Cl^-$$

$$Mg\,Cl \longrightarrow Mg^{2+} + 2\,Cl^-$$

$$Al_2(SO_4)_3 \longrightarrow 2\,Al^{3+} + 3SO_4^{2-}$$

(*ii*) On passing electricity, the positive ions migrate towards the cathode and are therefore called *cations*. Similarly, negative ions called the *anions* move towards the positive electrode *viz.* anode.

(*iii*) The solution of the electrolyte is electrically neutral as a whole. That means total charge carried by the positive ions is equal to the charge carried by negative ions.

(*iv*) The ions obtained by dissociation of the electrolyte combine back to give the undissociated molecules. Ultimately an equilibrium is reached between the ions and the undissociated molecules.

$$Na\,Cl \rightleftharpoons Na^+ + Cl^-$$

$$Mg\,Cl_2 \rightleftharpoons Mg^{2+} + 2\,Cl^-$$

The fraction of the total number of molecules of the electrolyte that dissociates into ions is called the **degree of ionization or dissociation.** It is usually denoted by α, *i.e.*

$$\text{Degree of dissociation}\,(\alpha) = \frac{\text{No. of molecules dissociated}}{\text{Total No. of molecules taken}}$$

(*v*) The degree of dissociation increases with dilution and becomes maximum at the infinite dilution.

(*vi*) The conductance of an electrolytic solution is due to the presence of the ions, produced by the dissociation of the electrolyte. Greater the dissociation, the higher is the conductance.

(*vii*) The properties of an electrolyte are the properties of its ions. For example, the properties of an acid are due to the H^+ ions furnished by it in the aqueous solution. Similarly, the properties of a base are due to the OH^- ions produced in the aqueous solution.

(*viii*) The ions produced by the dissociation of an electrolyte behave like molecules towards colligative properties. For example, each NaC1 molecule dissociates to give two ions *viz.* Na^+ and Cl^-. Hence the observed value of any colligative property is double the expected value.

Q.2. What are the factors that govern degree of dissociation of an electrolyte ?

Ans. The degree of dissociation of an electrolyte in the solution depends upon the following factors :

(*i*) **Nature of the solute:** Some electrolytes like mineral acids, alkalies and most of the salts are almost completely ionized in water. These are called *strong electrolytes.* On the other hand, some electrolytes like organic acids and bases and some inorganic acids and bases ionize only to a small extent. These are called *weak electrolytes.*

(*ii*) **Nature of the solvent:** Every solvent has some definite value of the *dielectric constant* which is defined as its power to weaken the forces of attraction present between the ions of the electrolyte. Water has the highest dielectric constant (≈ 80 compared with vacuum taken as unity). Thus the dissociation of an electrolyte is maximum in water.

(*iii*) **Concentration:** Degree of dissociation of an electrolyte increases with dilution and becomes maximum at infinite dilution *i.e.*, zero concentration.

(*iv*) **Temperature:** As the temperature is increased, more of the molecules split into ions *i.e.*, degree of dissociation increases.

(*v*) **Presence of common ions:** The degree of dissociation of an electrolyte decreases when another electrolyte is present which furnishes a common ion. This effect is called *common ion effect.* For example, dissociation of NH_4OH decreases in the presence of NH_4C1.

Q.3. Give evidence in support of Arrhenius theory of ionisation. *(Lucknow2000)*

Ans. The following evidence lends support to the theory.

(*i*) **Electrolytic conductance:** Arrhenius theory says that the conductance of the solution of an electrolyte is due to the presence of the ions, produced by the electrolyte on dissociation. Strong electrolytes possess high degree of dissociation and hence show greater conductance whereas weak electrolytes dissociate to a small extent and hence show low conductance.

(*ii*) **Variation of conductance with dilution:** It is observed that the specific conductance decreases whereas conductance as well as equivalent conductance increases with dilution. The explanation is evidently based on postulates of the Arrheniuus theory. The equivalent conduc-tance and the dissociation of the electrolyte vary in such a similar manner with dilution that the degree of dissociation (α) of an electrolyte at any concentration can be expressed in terms of equivalent conductances as follows :

$$\alpha = \wedge_c / \wedge_0$$

where $\wedge_c$ is the equivalent conductance at the given concentration and $\wedge_0$ is the equivalent conductance at infinite dilution.

(*iii*) **Electrolysis:** According to Arrhenius theory, an electrolyte when dissolved in water dissociates into ions. Thus the function of the electricity is only to direct the ions towards the electrodes.

(*iv*) **Colligative properties:** Anomalous results are obtained while measuring colligative properties like osmotic pressure, elevation in boiling points etc. of the strong electrolytes. This is because every ion produced from the electrolyte acts as an independent particle as far as colligative properties are concerned.

(*v*) **Ionic nature of reactions:** According to Arrhenius theory, the properties of an electrolyte are the properties of its ions. Thus the reaction between NaCl solution and $AgNO_3$ solution forming a precipitate of AgCl may be explained as follows :

NaCl in aqueous solution ionizes as

$$NaCl \longrightarrow Na^+ + Cl^-$$

$AgNO_3$also ionises in aqueous solution as

$$AgNO_3 \longrightarrow Ag^+ + NO_3^-$$

When the above two the solutions are mixed, we have

$$Na^+ + Cl^- + Ag^+ + NO_3^- \longrightarrow AgCl + Na^+ + NO_3^-$$

The reaction between any two ionic compounds in aqueous solution takes place in a similar manner.

(*vi*) **Heat of neutralization:** *The heat of neutralization is defined as the amount of heat evolved when one gram equivalent of the acid is neutralized by the base or one gram equivalent of the base is neutralized by the acid.* The heat of neutralization of a strong acid with a strong base is always the same *viz.* 13.7 kcal (or 57.3 kJ). This is explained by the Arrhenius theory as follows :

Suppose the strong acid is HCl and the strong base is NaOH. As both are completely dissociated in the aqueous solution, the reaction between them may be written as

$$Na^+ + OH^- + H^+ + Cl^- \longrightarrow Na^+ Cl^- + H_2O$$

Cancelling out the common ions on both sides, the above reaction reduces to

$$H^+ + OH^- \longrightarrow H_2O$$

Thus neutralization may be defined simply as the process in which H^+ *ions given by acid combine with the* OH^- *ions given by the base to form unionized water molecules.*

As one gram equivalent of any strong acid in aqueous solution gives the same number of H^+ ions and one gram equivalent of any strong base gives the same number of OH^- ions, the neutralization involves the combination between the same number of H^+ ions and OH^- ions. Hence the heat of neutralization of a strong acid with a strong base is always the same.

(*vii*) **Salt hydrolysis:** *It is defined as the process in which the salt reacts with water to give back the acid and base.*

$$\text{Salt} + \text{Water} \longrightarrow \text{Acid} + \text{Base}$$

Depending upon the relative strengths of the acid and the base produced, the resulting solution is acidic, basic or neutral. These observations are also explained on the basis of Arrhenius theory.

Q.4. Discuss the limitations of Arrhenius theory. (*Nagpur 2008*)

Ans. (*i*) Arrhenius suggested that *ionization* of the electrolyte takes place when dissolved in water. However, X-rays studies have shown that many such electrolytes are already in the ionized state (*e.g.* Na^+Cl^-).

(*ii*) As an electrolyte can conduct electricity even in the molten state, it shows that the dissociation of the electrolyte takes place even in the molten state. As such the function of water could not be explained by Arrhenius theory.

(*iii*) The most serious objection against the Arrhenius theory *was its failure when applied to strong electrolytes:*

(*a*) Ostwald's dilution law which is derived on the basis of Arrhenius theory, using the concept of ionic equilibrium, fails for the strong electrolytes, *i.e.*, a constant value of the dissociation constant K is not obtained.

(*b*) The degree of dissociation (α) of strong electrolytes calculated from conductance measurements ($\alpha = \wedge_c / \wedge_0$) was found to be different from that calculated from colligative properties.

(*c*) Strong electrolytes are almost completely dissociated even at moderate concentrations, *i.e.*, α = 1 or 100% at all concentrations. Hence Arrhenius' view that conductance increases with dilution because the dissociation increases is irrelevant in case of strong electrolytes. The increase in conductance with dilution must be theretore, due to some other factors

Q.5. Give the main points of difference between metallic and electrolytic conductors.

Ans. The main points of difference between the two types of conductors are given in Table 10.1.

Table 10.1. Metallic and Electrolytic Conductors

Metallic Conductor	Electrolytic Conductor
1. In metallic conductors, the electricity is conducted by the flow of electrons from negative potential to a positive (or less negative) potential.	**1.** Here, the electric current is carried by the charged ions moving towards the oppositely charged electrodes.
2. No chemical change takes place when a metallic (or electronic) conductor is conducting electricity. No new products are formed.	**2.** Conduction of electricity by electrolytic conductors is accompanied by some definite chemical process with the evolution of some gases or deposition of some metals at the electrodes.
3. In metallic conduction, temperature has the effect of decreasing the conductance, *i.e.*, conductance of a metallic conductor decreases with the increase of temperature. Metallic conductance becomes pronounced at low temperature. This is because at low temperature, the kernels of the atoms have the least vibrations and offer the least resistance to the flow of electrons.	**3.** In electrolytic conductors, the conductance increases with the rise of temperature. This is due to increase in the mobility of the ions and increase in the degree of ionisation of the electrolytes.

Q. 6. What is meant by electrical resistance and electrical conductance ?

(*G. Nanakdev 2000; Himachal 2000*)

Ans. Resistance. The law that gives exact relationship between the resistance, the current and voltage is called *Ohm's Law.* It states :

If to the ends of a conductor a voltage 'E' is applied anmd a current 'I' flows through it, then the resistance 'R' of the conductor is given by E / I.

Current is generally measured in amperes, whereas voltage is measured in volts. If one ampere current flows through a conductor when a voltage of one volt is applied to it, the resistance of the conductor is taken as 1 ohm (written Ω).

Mathematically, the law can be put as

$$R = \frac{E}{I}$$

All electrolytic solutions like metallic conductors obey Ohm's law.

Conductance. The reciprocal of the electrical resistance is called the conductance. It is usually represented by C. Thus,

$$C = \frac{1}{R}$$

It is more common practice to speak of the conductance of a solution rather than its resistance.

Units. The units of conductance are siemen (S) or ohm^{-1} or mho ($1\ S = 1\Omega^{-1}$)

Q.7. Explain the term specific conductance.

(Bhopal 2004; Vidyasagar 2007; Calcutta 2007; Raipur 2010)

Ans. It is known that resistance (R) of a conductor is

(*i*) directly proportional to its length (l) and

(*ii*) inversely proportional to its area of cross-section (a)

i.e.,

$$R \propto \frac{l}{a}$$

or

$$R = \rho\frac{l}{a} \quad \text{...}(i)$$

where ρ is a constant of proportionality, called Specific Resistance or Resistivity. Its value depends upon the material of the conductor.

"The reciprocal of resistivity is known *specific conductivity* or *simply conductivity*". It is denoted by K. Thus, if K is the specific conductivity and C is the conductance of the solution, then

$$R = \frac{1}{C} \text{ and } \rho = \frac{1}{K}$$

Substituting the values of R and ρ in eq (*i*)

$$\frac{1}{C} = \frac{1}{K} \times \frac{l}{a}$$

$$K = C \times \frac{l}{a}$$

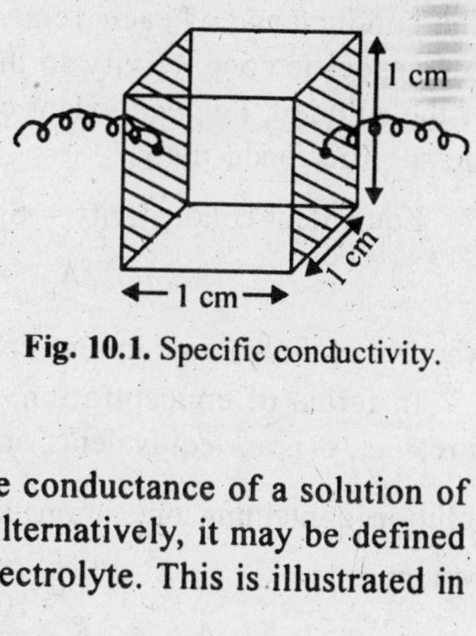

Fig. 10.1. Specific conductivity.

Now if $l = 1$ cm and $a = 1$ sq cm, then

$$K = C$$

Hence, specific conductivity of a solution is defined as the conductance of a solution of 1 cm length and having 1 sq cm as the area of cross-section. Alternatively, it may be defined as conductance of one centimetre cube of the solution of the electrolyte. This is illustrated in Fig. 10.1.

If the volume of the solution is V cm^3, the specific conductivity of such a solution at this dilution V is written as K_V.

Units. Resistivity or specific resistance (ρ)

$$= R\frac{a}{l} = \text{ohms}\frac{(\text{cm})^2}{\text{cm}} = \text{ohm cm}$$

$$\text{Specific conductivity } (K) = \frac{1}{\rho} = \frac{1}{\text{ohm cm}}$$

$$= \text{ohm}^{-1}\ \text{cm}^{-1} \text{ or } \Omega^{-1}\ \text{cm}^{-1} \text{ or } S\ \text{cm}^{-1}.$$

Q.8. What is meant by equivalent conductance and molar conductance ?

(Kalyani 2003; Delhi 2004; Bangalore 2004; Madras 2004; Bhopal 2004; G. Nanakdev 2003; Vidyasagar 2007; Calcutta 2007)

Ans. Equivalent conductivity of a solution at a dilution V is defined as the conductance of all the ions produced from one gram equivalent of the electrolyte dissolved in $V cm^3$ of the solution when the distance between the electrodes is one cm and the area of the electrodes is so large that whole of the solution is contained between them. It is represented by λ_{eq} (lambda).

Relationship between equivalent conductivity and specific conductivity. The equivalent conductivity of a solution is calculated from the specific conductivity. The relationship between equivalent conductivity and specific conductivity may be obtained as follows :

Consider a rectangular vessel with its two opposite walls one cm apart and made of some metal sheets so that they act as the electrodes.

Case I. Suppose $1 cm^3$ of the solution containing 1 g eq. of the electrolyte is taken in the vessel (Fig. 10.1). The conductance of this solution will be its specific conductivity. Further $1\ cm^3$ of the solution taken contains one gram equivalent of the electrolyte, the conductance of the solution will be its equivalent conductivity (by definition). Thus, when $1 cm^3$ of the solution containing one gram equivalent of the electrolyte is considered, the equivalent conductivity is equal to its specific conductivity.

Case II. Suppose $4\ cm^3$ of the solution containing one gram equivalent of the electrolyte is taken. The conductance of the solution will be still equal to its equivalent conductivity at this dilution but now there will be four cubes each of volume $1 cm^3$ as shown in Fig. 10.2. The conductance of each $1 cm^3$ of the solution is equal to its specific conductivity so that the total conductance of the solution, *i.e.*, equivalent conductivity is four times the specific conductivity.

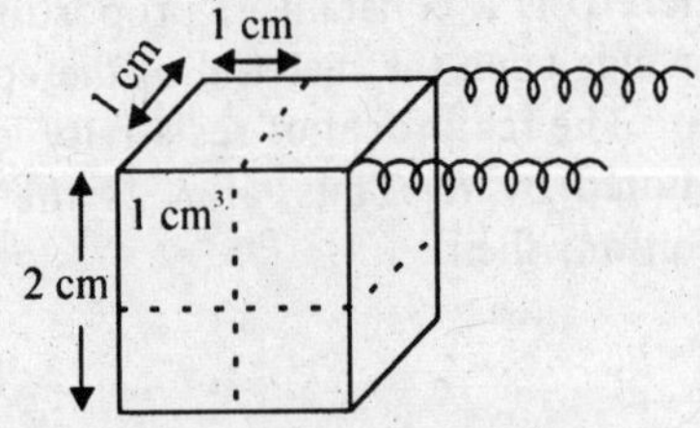

Fig. 10.2. Relationship between equivalent conductivity and specific conductivity.

Equivalent conductivity = Specific conductivity × V

$$\Lambda_{eq} = K_V \times V$$

where V is the volume in cm^3 containing one gram equivalent of the electrolyte.

In terms of concentration, if the solution has a concentration of c gram equivalent per litre, *i.e.*, c gram equivalents are present in $1000\ cm^3$ of the solution, then the volume of the solution containing one gram equivalent will be $\frac{1000}{c}$. Hence, the above expression may be written as

$$\Lambda_{eq} = K \times \frac{1000}{c_{eq}} = K \times \frac{1000}{\text{Normality}}$$

Units of equivalent conductivity

$$\Lambda_{eq} = K \times V$$

$$= \text{ohms}^{-1}\ \text{cm}^{-1} \times \frac{\text{cm}^3}{\text{Gram Eq.}}$$

$$= \text{ohm}^{-1}\ \text{cm}^2\ (\text{g eq})^{-1} \text{ or } S\ \text{cm}^2\ \text{eq}^{-1} \text{ or } \Omega^{-1}\ \text{cm}^2\ \text{eq}^{-1}.$$

Molar conductivity. The molar conductivity of a solution is defined as under :

The molar conductivity of a solution at a dilution V is the conductance of all the ions produced from one mole of the electrolyte dissolved in $V\ cm^3$ of the solution when the electrodes are one cm apart and the area of the electrodes is so large that the whole of the solution is contained between them. It is usually represented by Λ_m.

Relationship between molar conductivity and specific conductivity. Molar conductivity is related to the specific conductivity as follows.

Molar conductivity = Specific conductivity × Volume in cm^3 containing one mole of the electrolyte.

or
$$\Lambda_m = K \times V$$
$$= K \times \frac{1000}{C} = K \times \frac{1000}{\text{Molarity}}$$

where K is the specific conductivity and V is the volume of the solution containing one mole of the electrolyte and c is the molar concentration.

Units of Λ_m will be $ohm^{-1}\ cm^2\ mol^{-1}$ (or $S\ cm^2\ mol^{-1}$) or $\Omega^{-1}\ cm^2\ mol^{-1}$.

Q.9. Describe the measurement of (*i*) electrolytic conductance (*ii*) specific conductivity (*iii*) equivalent conductivity (*iv*) molar conductivity (conductance).

(*G. Nanakdev 2000; Indore 2001*)

Ans. (*i*) *Electrolyte conductance.* Conductance is the reciprocal of resistance. Hence, the conductance can be obtained by the measurement of the resistance and the latter can be found by Wheatstone bridge method shown in Fig. 10.3. It consists of four resistances R_1, R_2, R_3 and R_4 arranged in a manner as shown in figure. R_1 is the variable resistance and R_2 is the unknown resistance. When null point is obtained, *i.e.*, there is no deflection in the galvanometer G, we have,

$$\frac{R_1}{R_2} = \frac{R_3}{R_4}$$

Knowing the values of R_1, R_3 and R_4, R_2 can be calculated.

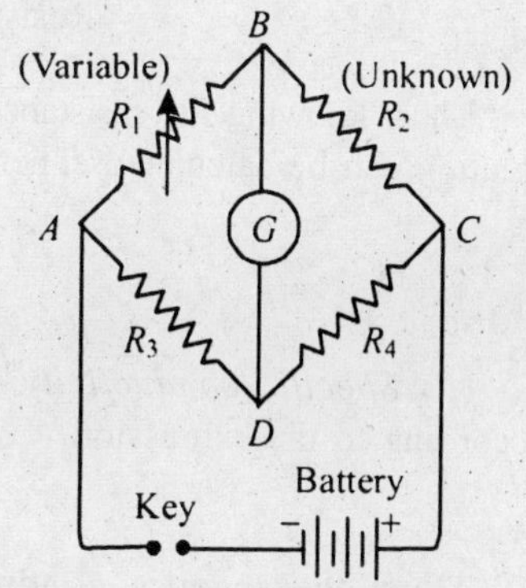

Fig. 10.3. Principle of wheatstone bridge method.

In finding the resistance of the solution of an electrolyte, a special type of cell has to be used such that the solution is present between the two electrodes. The cell thus used is called *conductivity cell.* It is made up of pyrex glass and two platinum electrodes at a fixed distance apart.

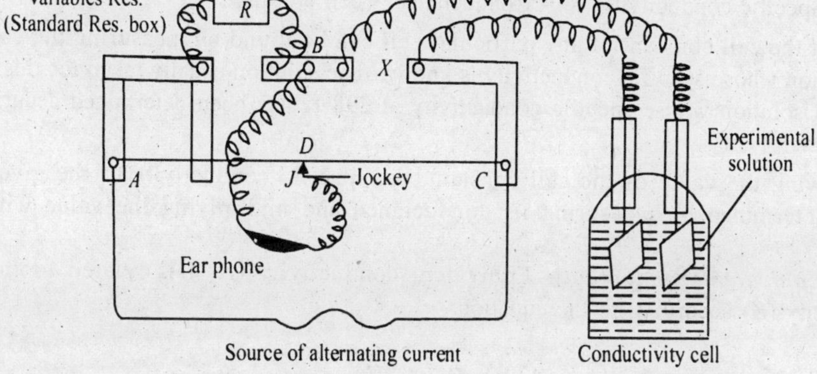

Fig. 10.4. Apparaturs for the measurement of electrolytic conductance

Various designs of conductivity cells are available. If direct current is used, it causes electrolysis of the solution and this results in the change in the resistance of the solution. These effects are called *polarization effects.* To overcome this problem, an alternating current of frequency 50-200 cycles per second is used. Also we have to use a head phone instead of a galvanometer to detect the null point.

The complete assembly for the measurement of the electrolytic conductance is shown in Fig. 10.4.

A suitable value of the resistance R is introduced from the standard resistance box such that when the sliding contact, *i.e.*, the jockey J is moved along the streched wire, the sound in the earphone is reduced to minimum at a point somewhere in the middle of the wire AC, say at the point D. Then if X is resistance of the electrolytic solution, by Wheatstone bridge principle,

$$\frac{\text{Resistance } R}{\text{Resistance } X} = \frac{\text{Resistance of wire } AD}{\text{Resistance of wire } CD} = \frac{\text{Length } AD}{\text{Length } CD}$$

$$\text{Resistance } X = \text{Resistance } R \times \frac{\text{Length } CD}{\text{Length } AD}$$

Thus, knowing the resistance R and the balance point D, the resistance X of the electrolytic solution can be calculated. Then,

$$\text{Conductance, } C = \frac{1}{X}$$

(ii) Specific conductivity. Specific conductivity (K) is related to the conductance (C) according to the expression

$$K = C \times \frac{l}{a}$$

Thus, the specific conductivity of a solution can be determined by measuring its conductance and the distance (l) between the electrodes and the area of cross-section (a) of each of the electrodes. However, for a particular cell, l/a is constant and this constant is called *cell constant.* The value of cell constant is generally written on the cell. It can also be determined experimentally.

$$\frac{l}{a} = \text{cell constant}$$

∴ Specific conductivity (K)= Conductance × cell constant.

Thus, the cell constant of any particular cell can be found by measuring the conductance of a solution whose specific conductivity is known. The solution usually taken for this purpose is $N/50$ KCl solution whose specific conductivity at 298 K has been determined accurately to be 0.002765 $\text{ohm}^{-1}\ \text{cm}^{-1}$

Knowing the value of the cell constant, the specific conductivity of the given solution can be determined by measuring its conductance and multiplying the value with the cell constant.

(iii) Equivalent conductivity. Equivalent conductivity (Λ_{eq}) is related to the specific conductivity (K) according to the equation.

$$\Lambda_{eq} = K \times \frac{1000}{c_{eq}}$$

where c_{eq} the concentration of the solution in gram equivalents per litre (*ie.*, the normality of the solution). Thus, knowing the normality of the solution and the specific conductivity K, Λ_{eq} can be calculated.

(*iv*) *Molar conductivity.* The molar conductivity is related to the specific conductivity according to the relation

$$\Lambda_m = K \times \frac{1000}{c}$$

where c is the molar concentration. Thus, knowing molar concentration c and the specific conductivity K, Λ_m can be calculated.

NUMERICAL PROBLEMS ON SPECIFIC, EQUIVALENT AND MOLAR CONDUCTIVITIES

Formulas to be used :

(*i*) $$R = \rho\frac{l}{a} \quad \text{or} \quad \rho = R\frac{a}{l}$$

(*ii*) $$K = \frac{1}{\rho} = \frac{1}{R} \times \frac{l}{a} = C \times \frac{l}{a}$$

= observed conductance × cell constant.

(*iii*) $$\text{Cell constant} = \frac{l}{a} = \frac{\text{Specific conductivity}}{\text{Observed conductance}}$$

= Specific conductivity × Observed resistance

(*iv*) $$\Lambda_{eq} = K \times \frac{1000}{c}$$

where c = concentration of the solution in gram equivalent/litre, *e.*, normality of the solution.

(*v*) $$\Lambda_m = K \times \frac{1000}{c}$$

where c = Molar concentration of the solution. Sometimes the term conductivity is used in place of specific conductivity. It can be understood from the units attached to it. If the units attached to conductivity are ohm^{-1} cm^{-1}, it is obvious that it is specific conductivity.

Q.10. The resistance of 0.01 N NaCl solution at 298 K is 200 ohm. Cell constant of the conductivity cell is unity. Calculate the equivalent conductance.

Ans. *Calculation of specific conductance*

$$\text{Specific conductance, } K = \frac{1}{R} \times \text{cell constant}$$

$$= \frac{1}{200\,\text{ohm}} \times 1\,\text{cm}^{-1}$$

$$= 5 \times 10^{-3}\ \text{ohm}^{-1}\ \text{cm}^{-1}$$

Calculation of equivalent conductance

$$\Lambda_m = \frac{K \times 1000}{\text{Normality}}$$

$$= \frac{5 \times 10^{-3} \times 1000}{0.01}$$

$$= 5 \times 10^{2}\ \text{ohm}^{-1}\ \text{cm}^{2}\ \text{eq}^{-1}$$

Q.11. The conductivity of a solution containing 1 gram of anhydrous $BaCl_2$ in 200 cm^3 of water has been found to be 0.0058 mho cm^{-1}. What are the molar conductivity and equivalent conductivity of the solution (At. wt. of Ba = 137 and Cl = 35.5).

Ans. Conductivity (K) = 0.0058 mho cm^{-1}

Mol. wt of $BaCl_2$ = 137 + 2 × 35.5 = 208

As 1 gram of $BaCl_2$ is present in 200 cm^3 of the solution, therefore,

$$\text{Molar concentration, } (c) = \frac{1}{208} \times \frac{1}{200} \times 1000 \text{ mol L}^{-1}$$

$$= 0.0240 \text{ mol L}^{-1}$$

$$\text{Molar conductivity, } \quad \Lambda_m = K \times \frac{1000}{c} = 0.0058 \times \frac{1000}{0.0240}$$

$$= 241.67 \text{ ohm}^{-1} \text{ cm}^2 \text{ mol}^{-1}$$

$$\text{In case of } BaCl_2 \text{ equivalent weight } \frac{\text{Mol. wt.}}{2} = \frac{208}{2} = 104$$

∴ Concentration of the solution in gram equivalent per litre

$$c_{eq} = \frac{1}{104} \times \frac{1}{200} \times 1000 = 0.0480$$

$$\therefore \text{ Equivalent conductivity, } \Lambda_{eq} = K \times \frac{1000}{c_{eq}}$$

$$= 0.0058 \times \frac{1000}{0.480}$$

$$= 120.83 \text{ ohm}^{-1} \text{ cm}^2 \text{ eq}^{-1}$$

Q.12. The resistance of a conductivity cell when filled with 0.02M KCl solution is 164 ohm at 298 K. However, when filled with 0.05 M $AgNO_3$ solution, its resistance is found to be 78.5 ohm. If specific conductivity of 0.02 M KCl is 2.768 × 10^{-3} ohm^{-1} cm^{-1}, calculate :

(i) The conductivity of 0.05 M Ag NO_3.

(ii) The molar conductivity of $AgNO_3$ solution.

Ans. *Calculation of cell constant*

Specific conductivity of 0.02 M KCl = 2.768 × 10^{-3} ohm^{-1} cm^{-1}

Resistance = 164 ohm

$$\text{Specific conductivity} = \frac{1}{R} \times \text{cell constant.}$$

$$\text{Cell constant} = R \times \text{specific conductivity}$$

$$= 164 \times 2.768 \times 10^{-3}$$

$$= 0.4539 \text{ cm}^{-1}$$

$$\text{Conductivity} = \frac{1}{R} \times \text{cell constant}$$

$$= \frac{1}{78.5} \times 0.4539 = 0.00578 \text{ ohm}^{-1}\text{cm}^{-1}$$

Calculation of Λ_m of 0.05M $AgNO_3$

$$\Lambda_m = \frac{1000 \times K}{\text{Molarity}}$$

$$\Lambda_m = \frac{1000 \times 0.0058}{0.05}$$

$$= 115.6 \text{ ohm}^{-1} \text{ cm}^2 \text{ mol}^{-1}$$

Q.13. What is meant by cell constant ? How is it determined ?

(Kerala 2000; Lucknow 2000; Bangalore 2002; Nagpur 2003; Delhi 2004)

Ans. Cell constant = $\frac{l}{a}$

where l is the distance between the electrodes of the cell and a is the surface area of the electrode. Thus cell constant may be obtained by measuring l and a. Alternatively, this may be obtained as follows with the help of the following equation

$$R = \rho \frac{l}{a} \quad \text{or} \quad \frac{1}{C} = \frac{1}{K} \cdot \frac{l}{a}$$

or $$\frac{l}{a} = \frac{K}{C} \quad \text{or} \quad \frac{l}{a} = \frac{\text{Specific conductance}}{\text{Observed conductance}}$$

Q.14. 0.5 normal solution of a salt placed between two platinum electrodes 3.0 cm apart and of area of cross-section 6.0 sq cm has a resistance of 25 ohms. Calculate the equivalent conductivity of solution.

Ans. *1st step — Calculation of specific conductivity.*

Here l = 2.0 cm, a = 4.0 sq cm, R = 25 ohms.

$$\text{Conductance, C} = \frac{1}{R} = \frac{1}{25} \text{ ohm}^{-1}$$

$$\text{Cell constant} = \frac{l}{a} = \frac{3}{6} = \frac{1}{2} \text{ cm}^{-1}$$

Specific conductivity (K) = Observed conductance × Cell constant

$$= \frac{1}{25} \times \frac{1}{2} = 0.02 \text{ ohm}^{-1}\text{cm}^{-1}$$

2nd step — Calculation of equivalent conductivity.

$$\Lambda_{eq} = K \times \frac{1000}{c_{eq}}$$

$$c = 0.5 \text{ N}$$

$$K = 0.02 \text{ ohm}^{-1} \text{ cm}^{-1} \text{ (calculated above)}$$

$$\Lambda_{eq} = \frac{1000 \times 0.02}{0.5} = 40 \text{ ohm}^{-1}\text{cm}^2\text{eq}^{-1}$$

Q. 15. What is the effect of dilution on specific conductance and equivalent conductance

(Himachal 2010

Ans. The specific conductance decreases with dilution. This is due to the fact that th number of current carrying particles *i.e.* ions per ml of the solution become less. Of course on dilution, the degree of dissociation also increases but the no. of ions per ml is decreased tc such an extent that even complete dissociation does not produce as many ions per ml as were present before dilution.

The equivalent conductance on the other hand increases with dilution and finally attains a maximum value at infinite dilution. This is due to the fact that this value is product of specific conductance and the volume V of the solution containing 1 gm equivalent of the electrolyte. Decreasing value of specific conductance is more than compensated by increasing value of V and therefore the value of $\wedge_V$ increases with dilution.

Q.16. The resistance of a decinormal solution of a salt occupying a volume between two platinum electrodes 1·8 cm apart and 5·4 sq. cm in area was found to be 32 ohm. Calculate the equivalent conductance of the solution.

Ans. $R = \frac{1}{K} \cdot \frac{l}{a}$ or $K = \frac{1}{R} \cdot \frac{l}{a}$

Substituting the values

Specific conductance $K = \frac{1}{32} \times \frac{1.8}{5.4} = 0.0104 \text{ ohm}^{-1} \text{ cm}^{-1}$

Eq. conductance = Sp. conductance × Vol in ml containing 1gm eq.of electrolyte

= 0.0104 × 10000

= **104 ohm^{-1} cm^{2} eq^{-1}**

Q. 17. What is Ostwald dilution law? *(Kurukshetra 2002; H. S. Gaur 2002; Vidyasagar 2007; Raipur 2010; Andhra 2010,*

Ans. According to Arrhenius theory of ionization, when an electrolyte is dissolved in water, there is an equilibrium between the ions and the undissociated molecules. For example, for a binary electrolyte AB, we have

$$AB \rightleftharpoons A^+ B^-$$

Ostwald pointed out the law of chemical equilibrium can be applied to this *ionic equilibrium.*

Suppose the initial concentration of the binary electrolyte AB in the solution is C moles/litre and α is the degree of dissociation, then we will have

	AB	$\rightleftharpoons$ A^+	+ B^-
Initial concentration :	**C moles/litre**	**0**	**0**
Concentrations at equilibrium :	**C (1– α)**	**Cα**	**Cα moles/litre**

Applying the law of chemical equilibrium, we get

$$K = \frac{[A^+][B^-]}{[AB]} = \frac{C\alpha . C\alpha}{C(1-\alpha)}$$

or

$$\boxed{K = \frac{C\alpha^2}{1-\alpha}} \quad \textit{(Exact Formula)} \qquad ...(i)$$

This is Ostwald dilution law.

The equilibrium constant K for the equilibrium $AB \rightleftharpoons A^+ + B^-$ is called the dissociation constant or ionization constant of the electrolyte AB and may be represented by K_d (or K_i). Like any other equilibrium constant, K_d is constant only at constant temperature. The units of K_d will be same as those of the concentration C.

If the degree of dissociation α is very small, as in the case of *weak electrolytes at ordinary concentrations,* α can be neglected in comparison to 1 in the denominator of expression (*i*). Then the approximate formula becomes

$$K_d = C\alpha^2 \text{ or } \alpha^2 = \frac{K_d}{C}$$

or $$\boxed{\alpha = \sqrt{\frac{K_d}{C}}} \qquad \textit{(Approx. Formula)} \qquad ...(ii)$$

Equations (*i*) and (*ii*) are the two expressions of Ostwald's dilution law. Expression (*ii*) can be expressed in words as follows :

For a weak binary electrolyte, with a small degree of dissociation, the degree of dissociation is inversely proportional to the square root of the initial molar concentration.

Q. 18. How will you test the validity of Ostwald dilution law ? *(Mumbai 2005)*

Ans. To test the validity of Ostwald's dilution law, *i.e.*, to verify the law experimentally, the degree of dissociation α is determined from conductance measurements (using the formula $\alpha = \wedge_c / \wedge_0$) at different concentrations for a number of electrolytes. For each electrolyte, the data are fitted in equation (*i*) to see whether the value of K_d comes out to be constant or not. It is observed that :

(*i*) The value of K_d comes out to be constant only for weak electrolytes and that too when the concentration of the solution is not too high *i.e.*, for dilute solutions only.

(*ii*) The value of K_d does not come out to be constant for strong electrolytes.

Thus it may be concluded that *Ostwald's dilution law is valid only for weak electrolytes in dilute solutions* (that is why it is called dilution law) *and it does not hold good for strong electrolytes.*

Q.19. What is the important of Ostwald's dilution law ?

Ans. Ostwald's dilution law is applicable to weak electrolytes only. It is very useful in the calculation of the dissociation constants (K_d) of the weak acids and weak bases, by determining the degree of dissociation (α) from conductance measurements ($\wedge_c / \wedge_0$) at any concentration C. Having determined the value of K_d, which is constant for a particular weak acid or weak base, at a particular temperature, the degree of dissociation (α) of that weak acid or weak base can be calculated at any concentration (C).

Q. 20. A base has a dissociation constant equal to 1.8 × 10⁻⁵ at 298 K. Calculate its degree of dissociation at a concentration of 0.1 N at the same temperature.

Ans. Suppose the base is BOH and α is its degree of dissociation. Then we have

	BOH $\longrightarrow$	B^+ +	OH^-
Initial concentration	0.1 M	0	0
Concentration at equilibrium	$0.1\,(1-\alpha)$	0.1α	0.1α

$$\therefore \quad K_d = \frac{[B^+][OH^-]}{[BOH]}$$

or $$1.8 \times 10^{-5} = \frac{0.1\alpha \times 0.1\alpha}{0.1\,(1-\alpha)}$$

$$= \frac{0.1\alpha^2}{1-\alpha} = 0.1\,\alpha^2 \quad \text{[Neglecting } \alpha \text{ compared to 1 in the denominator]}$$

or $$\alpha^2 = \frac{1.8 \times 10^{-5}}{0.1} = 1.8 \times 10^{-4}$$

$$\therefore \quad \alpha = \sqrt{1.8 \times 10^{-4}} = 1.34 \times 10^{-2}$$

$$= \mathbf{0.0134}$$

Q. 21. A solution of ammonia gives the following values of equivalent conductivities at different dilutions :

V (in litres)	**8**	**16**	**32**
$\wedge_v$	**3.20**	**4.45**	**6.32 ohm^{-1} cm^{2}**

If equivalent conductivity in infinite dilution of NH_4OH is 237 ohm^{-1} cm^2, show that the above given data establish the truth of Ostwald's dilution law.

Ans. The ionisation constant K_d for weak electrolytes is given by the expression

$$K_d = \frac{C\alpha^2}{1-\alpha} \quad \text{or} \quad K_d = \frac{\alpha^2}{(1-\alpha)V}$$

where C is the concentration of the solute in mol/litre and V is the volume of the solution in litre containing 1 mole of the solute.

Putting the values of α and V in the above expression, we have

S. No.	V (litres)	$\alpha = \frac{\wedge_V}{\wedge_\infty}$	$K = \frac{\alpha^2}{(1-\alpha)V}$
1	8	$\frac{3.2}{237} = 0.0135$	$\frac{(0.0135)^2}{(1-0.0135)8} = 2.31\times10^{-5}$
2	16	$\frac{4.45}{237} = 0.0188$	$\frac{(0.0188)^2}{(1-0.0188)16} = 2.26\times10^{-5}$
3	32	$\frac{6.32}{237} = 0.0265$	$\frac{(0.0265)^2}{(1-0.0265)32} = 2.25\times10^{-5}$

Thus we see that the value of K comes out to be almost constant. Hence the truth of the law is established.

Q.22. Discuss the variation of conductivity with concentration of strong electrolytes.

OR What is Debye - Huckel - Onsager equation ? *(Nagpur 2003; Raipur 2010,*

Ans. The following equation governs the change of conductance of a strong electrolytes with concentration.

$$\Lambda_m^c = \Lambda_m^0 - (A + B\Lambda_m^0)\sqrt{C}$$

where, Λ_m^c = molar conductance of the solution at the concentration C.

Λ_m^0 = molar concentration at infinite dilution.

C = concentration of the solution. A and B are constants for a particular solvent at a particular temperature.

The solution is known as Debye-Huckel-Onsager equation. A plot of Λ_m^c against $\sqrt{C}$ is a straight line as shown in Fig. 10.5. Taking the case of KC1, which is a strong electrolyte, we find that there is not much variation in conductance with dilution. This is because of the fact that a strong electrolyte is 100% ionised even in a concentrated solution. In a concentrated solution, there are interionic attractions which decrease the mobility of the ions. With dilution, these attractive forces are weakened resulting in greater mobility of the ions. Thus, there is slight increase in the conductance of the solution of a strong electrolyte on dilution.

In order to obtain the value of Λ_m^0 for a strong electrolyte, the plot which becomes linear at low concentration can be extrapolated to *y*-axis.

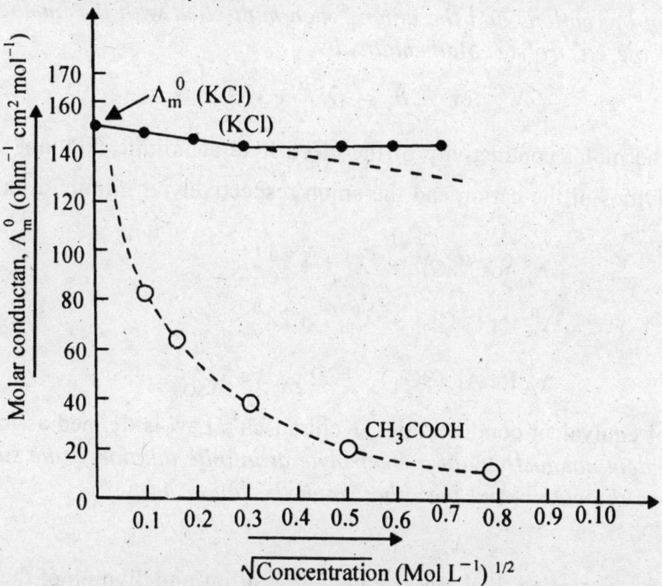

Fig. 10.5. Variation of molar conductivity of a strong electrolyte (KCl) and weak electrolyte (CH_3COOH) with concentration.

Q. 23. State and explain Kohlrausch Law. *(G. Nanakdev 2000; Meerut 2000; Kerala 2000; Awadh 2000; Kumaon 2000; Bangalore 2002; Calcutta 2007)*

Ans. Kohlrausch studied the molar conductivities at infinite dilution, Λ_m^0 for a number of pairs of strong electrolytes each having a common cation or anion. Then, he calculated the differences of these Λ_m^0 values of each pair. A few of these results are given in Table below :

Table

Λ_m^0 values of some pairs of electrolytes (at 298 K) having common ions (in ohm^{-1}cm^2)

Electrolyte	*Λ_m^0 at 298 K*	*Difference*	*Electrolyte*	*Λ_m^0 at 298 K*	Difference
KCl NaCl	149.86 126.45	23.41	KBr KCl	151.92 149.86	2.06
KBr NaBr	151.92 128.51	23.41	NaBr NaCl	128.51 126.45	2.06
KNO_3 $NaNO_3$	144.96 121.55	23.41	LiBr LiCl	117.09 115.03	2.06

We find that difference between Λ_m^0 values for each pair of sodium and potassium salts having a common anion is same, irrespective of the anion. Similarly, the difference in the Λ_m^0 values for each pair of salts having the same different anions and a common cation is same, irrespective of the cation. In fact, column 3 in the Table gives the Λ_m^0 values for K^+ ion minus the Λ_m^0 values for Na^+ ion. Similarly, the last column gives Λ_m^0 value for Br^- ion minus Λ_m^0 value for Cl^- ion.

Based on this, Kohlrausch concluded that *each ion makes a definite contribution to the total molar conductivity of an electrolyte at infinite dilution, irrespective of the nature of the other ion of the electrolyte. This individual contribution of an ion towards the total molar conductivity is called molar ionic conductivity.* Kohlrausch in 1876 put forward a generalization, known as `*Kohlrausch Law*'. It states as under :

The molar conductivity of an electrolyte at infinite dilution is the sum of the ionic conductivities of the cations and the anions each multiplied with the number of ions present in one formula of the electrolyte. Mathematically,

$$\Lambda_m^0 \text{ for } A_xB_y = x\lambda_{A^+}^0 + y\lambda_{B^-}^0$$

where Λ_m^0 is the molar conductivity of the electrolyte at infinite dilution, $\lambda_{A^+}^0$ and $\lambda_{B^-}^0$ are the molar conductivities of the cation and the anion respectively at infinite dilution.

Thus,

$$\Lambda_m^0 \text{ for NaCl} = \lambda_{Na^+}^0 + \lambda_{Cl^-}^0$$

$$\Lambda_m^0 \text{ for } BaCl_2 = \lambda_{Ba^{2+}}^0 + 2\lambda_{Cl^-}^0$$

$$\Lambda_m^0 \text{ for } Al_2(SO_4)_3 = 2\lambda_{Al^{3+}}^0 + 3\lambda_{SO_4^{2-}}^0$$

In terms of equivalent conductivities, Kohlrausch's Law is defined as follow .

The equivalent conductivity of a electrolyte at infinite dilution is the sum of two values one depending upon the cation and the other upon the anion, i.e.,

$$\Lambda_{eq}^0 = \lambda_c^0 + \lambda_a^0$$

where λ_c^0 and λ_a^0 are called the *ionic conductivities* at infinite dilution of the cation and the anion respectively.

Q.24. Discuss the applications of Kohlrausch Law. *(Jammu 2002; Madurai 2004; Mumbai 2005; Nagpur 2008; Raipur 2010)*

Ans. Applications of Kohlrausch Law are discussed as under :

1. *Calculation of molar conductivity at infinite dilution* (Λ^0) *for weak electrolytes.* Molar conductivity of a weak electrolyte at infinite dilution cannot be determined experimentally. This can be calculated using Kohlrausch's Law.

Consider the example of acetic acid [CH_3COOH] as the weak electrolyte. By Kohlarausch's Law.

$$\Lambda^0(CH_3COOH) = \lambda_{CH_3COO^-}^0 + \lambda_{H^+}^0$$

This equation can be arrived at by knowing the molar conductivities at infinite dilution for the strong electrolytes KCl, CH_3COOK and HCl. As per Kohlrausch's Law,

$$\Lambda^0(KCl) = \lambda_{K^+}^0 + \lambda_{Cl^-}^0$$

$$\Lambda^0(CH_3COOK) = \lambda_{CH_3COO^-}^0 + \lambda_{K^+}^0$$

$$\Lambda^0(HCl) = \lambda_{H^+}^0 + \lambda_{Cl^-}^0$$

Hence, we have

$$\lambda_{CH_3COO^-}^0 + \lambda_{H^+}^0 = (\lambda_{CH_3COO^-}^0 + \lambda_{K^+}^0) + (\lambda_{H^+}^0 + \lambda_{Cl^-}^0) - (\lambda_{K^+}^0 + \lambda_{Cl^-}^0)$$

i.e., $$\Lambda^0(CH_3COOH) = \Lambda^0(CH_3COOK) + \Lambda^0(HCl) - \Lambda^0(KCl)$$

2. *Calculation of the degree of dissociation.* For a weak electrolyte, if Λ_m^0 is the molar conductivity of a solution at any concentration c and Λ_m^0 the molar conductivity at infinite dilution (*i.e.*, zero concentration); then we have

$$\text{Degree of dissociation, } (\alpha) = \frac{\Lambda_m^c}{\Lambda_m^0}$$

The value of Λ_m^0 for the weak electrolytes can be calculated, using Kohlrausch's Law.

3. *Calculation of solubility of a sparingly soluble salt.* Salts such as AgC1, $BaSO_4$, $PbSO_4$ etc., which dissolve to a very small extent in water are called sparingly soluble salts. As they dissolve very little, their solutions are considered as infinitely dilute. Further as their solutions are saturated, their concentration is equal to their solubility. Thus, by determining the specific coductivity (K) and the molar conductivity (Λ_m) of such a solution, we can obtain solubility as follows :

$$\Lambda_m^0 = K \times \frac{1000}{\text{Molarity}} = K \times \frac{1000}{\text{Solubility}}$$

or

$$\text{Solubility} = \frac{K \times 1000}{\Lambda_m^0}$$

Λ_m^0 can be obtained by using Kohlrausch Law.

Q. 25. If the molar conductivities at infinite dilution of NaCl, HCl and CH_3COONa are 126.4, 426.1 and 91.0 $ohm^{-1}cm^2\ mol^{-1}$ respectively, what will be that of acetic acid ?

(*Delhi 2003*)

Ans. According to Kohlrausch Law.

$$\Lambda^0 \text{ for } CH_3COOH = \lambda^0_{CH_3COO^-} + \lambda^0_{H^+}$$

$$\Lambda^0 \text{ for NaCl} = \lambda^0_{Na^+} + \lambda^0_{Cl^-} = 126.4 \text{ ohm}^{-1}\text{cm}^2 \quad ...(i)$$

$$\Lambda^0 \text{ for HCl} = \lambda^0_{H^+} + \lambda^0_{Cl^-} = 426.1 \text{ ohm}^{-1}\text{cm}^2 \quad ...(ii)$$

$$\Lambda^0 \text{ for } CH_3COONa = \lambda^0_{CH_3COO^-} + \lambda^0_{Na^+} = 91.0 \text{ ohm}^{-1}\text{cm}^2 \quad ...(iii)$$

Adding equations (*ii*) and (*iii*) and sbtracting (*i*), we get,

$$\lambda^0_{H^+} + \lambda^0_{Cl} = \lambda^0_{CH_3COO^-} + \lambda^0_{Na^+} - \lambda^0_{Na^+} - \lambda^0_{Cl} = 426.1 + 91.0 - 126.4 \text{ ohm}^{-1} \text{cm}^2$$

$$\lambda^0_{CH_3COO^-} + \lambda^0_{H^+} = 390.7 \text{ ohm}^{-1} \text{cm}^2 \text{ mol}^{-1}$$

$$\Lambda^0 \text{ for } CH_3COOH = 390.7 \text{ ohm}^{-1} \text{cm}^2 \text{ mol}^{-1}$$

Q.26. At 291 K, the molar conductivities, at infinite dilution of NH_4Cl, NaOH and NaCl are 129.8, 217.4 and 108.9 ohm^{-1} cm^2 respectively. If the molar conductivity of a centinormal solution of NH_4OH is 9.33 ohm^{-1} cm^2, what is the percentage dissociation of NH_4OH at this dilution ?

Ans. We are given that

$$\Lambda^0 \text{ for } NH_4C1 = 129.8 \text{ ohm}^{-1} \text{cm}^2$$

$$\Lambda^0 \text{ for NaOH} = 217.4 \text{ ohm}^{-1} \text{cm}^2$$

$$\Lambda^0 \text{ for NaC1} = 108.9 \text{ ohm}^{-1} \text{cm}^2$$

By Kohlrausch law

$$\Lambda^0 \text{ for } NH_4OH = \lambda^0_{NH_4^+} + \lambda^0_{OH^-}$$
$$= \Lambda^0 (NH_4C1) + \Lambda^0 (NaOH) - \Lambda^0 (NaC1)$$

Substituting the values

$$\Lambda^0(NH_4Cl) = 129.8 + 217.4 - 108.9$$
$$= 238.3 \text{ ohm}^{-1} \text{ cm}^2$$
$$\Lambda_c = 9.33 \text{ ohm}^{-1} \text{ cm}^2 \text{ (Given)}$$

$$\text{Degreee of dissociation } (\alpha) = \frac{\Lambda_c}{\Lambda^0} = \frac{9.33}{238.3} = 0.0392$$

Percentage dissociation = 0.0392 × 100 = 3.92

Q. 27. From the following molar conductivities at infinite dilution:

$$\Lambda_m^0 \text{ for } Ba(OH)_2 = 457.6 \text{ ohm}^{-1} \text{ cm}^2 \text{ mol}^{-1}$$

$$\Lambda_m^0 \text{ for } BaCl_2 = 240.6 \text{ ohm}^{-1} \text{ cm}^2 \text{ mol}^{-1}$$

$$\Lambda_m^0 \text{ for } NH_4Cl = 129.8 \text{ ohm}^{-1} \text{ cm}^2 \text{ mol}^{-1}$$

calculate Λ_m^0 for NH_4OH.

Ans. Λ_m^0 for $[Ba(OH)_2] = \lambda^0_{Ba^{2+}} + 2\lambda^0_{OH^-} = 457.6 \text{ ohm}^{-1} \text{ cm}^2 \text{ mol}^{-1}$ ... (*i*)

Λ_m^0 for $(BaCl_2) = \lambda^0_{Ba^{2+}} + 2\lambda^0_{Cl^-} = 240.6 \text{ ohm}^{-1} \text{ cm}^2 \text{ mol}^{-1}$ ... (*ii*)

Λ_m^0 for $NH_4Cl = \lambda^0_{NH_4^+} + \lambda^0_{Cl^-} = 129.8 \text{ ohm}^{-1} \text{ cm}^2 \text{ mol}^{-1}$ ... (*iii*)

In order to find Λ_m^0 (NH_4Cl), we need to find the value of $\lambda^0_{NH_4^+} + \lambda^0_{OH^-}$

Multiplying (*iii*) by 2, add to (*i*) and substract eq. (*ii*) from the sum

$$2\lambda_{NH_4^+} + \lambda^0_{Cl^-} + \lambda^0_{Ba^{2+}} + 2\lambda^0_{OH^-} - \lambda^0_{Ba^{2+}} - 2\lambda^0_{Cl^-}$$
$$= (2 \times 129.8) + 457.6 - 240.6$$

$$2\left(\lambda^{+0}_{NH_4^+} + 2\lambda^0_{OH^-}\right) = 476.6$$

$$\lambda^0_{NH_4^+} + \lambda^0_{OH^-} = 238.6 \text{ ohm}^{-1} \text{ cm}^2 \text{ mol}^{-1}$$

Q. 28. The specific conductance of a 0.01 N solution of acetic acid at 25°C is 1.63×10^{-4} mhos cm^{-1} and the ionic conductance at infinite dilution but at the same temperature for H^+ and CH_3COO^- ions are 349.8 and 40.9 mhos cm^2 equiv^{-1} respectively. Calculate the dissociation constant of acetic acid at 25°C.

Also calculate the H^+ ion concentration in the solution.

Ans. Here we are given that

$$K = 1.63 \times 10^{-4} \text{ mhos cm}^{-1}$$

$$\lambda^0_{H^+} = 349.8 \text{ mhos cm}^2 \text{ eqiv}^{-1}$$

$$\lambda^0_{CH_3COO^-} = 40.9 \text{ mhos cm}^2 \text{ equiv}^{-1}$$

Calculation of $\wedge_0$ for CH_3COOH

$\wedge_0$ for $CH_3COOH = 349.8 + 40.9$ (By Kohlrausch's law)

$= 390.7 \text{ mhos cm}^2 \text{ equiv}^{-1}$

Calculation of $\wedge_c$ for CH_3COOH

As the solution is 0.01 N, volume of solution in cm^3 containing one gram equivalent of acetic acid $= \dfrac{1000}{0.01} = 10^5 \text{ cm}^3$

$$\therefore \quad \wedge_c = \text{Sp. conductance} \times \text{Vol. in cm}^3 \text{ containing one gm equiv. of electrolyte.}$$

$$\wedge_c = 1.63 \times 10^{-4} \times 10^5 = 16.3 \text{ mhos cm}^2 \text{ equiv}^{-1}$$

Calculation of α

$$\therefore \quad \alpha = \frac{\wedge_c}{\wedge_0} = \frac{16.3}{390.7} = 0.0417$$

Calculation of K_d

	CH_3COOH	$\rightleftharpoons$	CH_3COO^-	+	H^+
Initial concentration	0.01		0		0
Concentration at equilibrium	$0.01(1-\alpha)$		$0.01\,\alpha$		$0.01\,\alpha$

$$K_d = \frac{[CH_3COO^-][H^+]}{[CH_3COOH]} = \frac{0.01\alpha \times 0.01\alpha}{0.01(1-\alpha)}$$

$$= \frac{0.01\alpha^2}{1-\alpha}$$

$$= 0.01\,\alpha^2 \qquad [\text{Taking } 1 - \alpha = 1, \text{ as } \alpha \text{ is negligible compared to } 1]$$

$$= 0.01 \times (0.0417)^2$$

$$= \mathbf{1.74 \times 10^{-5}}$$

Calculation of H^+

$$[H^+] = 0.01\,\alpha = 0.01 \times 0.0417$$

$$= \mathbf{4.17 \times 10^{-4} \text{ g ions/litre}}$$

Q. 29. At 18°C equivalent conductivities of 0.01 N solutions of hydrochloric acid, NaCl, CH_3COONa and CH_3COOH are 369.3, 101.7, 71.7 and 14.3 ohm^{-1} cm^2 respectively. Calculate the hydrogen ion concentration in 0.01 N acetic acid at 18°C.

Ans. Equivalent conductance at infinite dilution $\wedge_0$ of CH_3COOH can be obtained using Kohlrausch law

$$\wedge_0 (CH_3COOH) = 369.3 + 71.7 - 101.7$$

$$= 339.3$$

$$\text{Degree of dissociation of } CH_3COOH = \frac{\wedge_c}{\wedge_0}$$

$$= \frac{14.3}{339.3} = 0.042$$

$\therefore$ Hydrogen ion concentration in $CH_3COOH = 0.042 \times 0.01$

$$= \mathbf{4.2 \times 10^{-4} \text{ g ion/litre}}$$

Q. 30. Calculate the hydrogen ion concentration of a solution of formic acid containing 0.092 g of acid per litre. The dissociation constant at 25°C is 2.14×10^{-4}.

Ans. Concentration of formic acid (HCOOH) in mole/litre $= \dfrac{0.092}{46} = 0.002$

Using Ostwald dilution law,

$$K = C\alpha^2$$

or $$2.14 \times 10^{-4} = 0.002\,\alpha^2$$

or $$\alpha^2 = \frac{2.14\times10^{-4}}{2\times10^{-3}}$$

or $$\alpha^2 = 0.107$$

or $$\alpha = 0.327$$

Hence $$[H^+] = 0.002 \times 0.327 \text{ g ion/litre}$$

$$= 2 \times 10^{-3} \times 3.27 \times 10^{-1}$$

$$= \mathbf{6.54 \times 10^{-4}}$$

Q. 31. Discuss Debye-Huckel theory of strong electrolytes.

(Purvanchal 2002; Guwahati 2002; G. Nanakdev 2003; Delhi 2003; Nagpur 2003)

Ans. The increase of equivalent conductance with dilution in case of weak electrolytes can be easily explained on the basis of Arrhenius theory which suggests that the conductance increases because the dissociation of the weak electrolyte increases with dilution. However, in case of strong electrolytes, as they are almost completely dissociated even at moderate concentration, the increase in equivalent conductance with dilution must be due to some other factors.

Debye-Huckel theory explains the increase in conductance of strong electrolytes on dilution based upon the following two effects:

1. **Relaxation effect or Asymmetry effect:** In the solution, each ion is surrounded by an **ionic atmosphere** of opposite charge. So long as no electric field is applied, the ionic atmosphere remains symmetrical around the central ion [Fig. 10.6 (*a*)]. However, when a current is passed through the solution, the central ion moves towards the oppositely charged electrode. As it is moving out of the ionic atmosphere, it has to rebuild an ionic atmosphere of opposite charge around it and the old ionic atmosphere dies out. However, the destruction of the old ionic atmosphere and the formation of the new ionic atmosphere do not take place at the same time. **There is some time lag called *time of relaxation* between the destruction of the old and the formation of new ionic atmosphere. During this time, the old ionic atmosphere pulls the moving ion backward and hence retards its motion [Fig. 10.6 (*b*)]. Hence this effect is called *relaxatio***

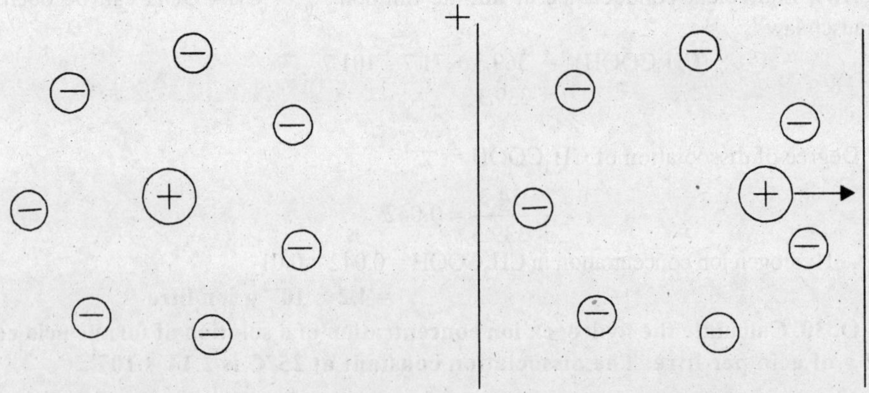

(*a*) Symmetrical ionic atmosphere around a positive ion.

(*b*) Ionic atomsphere becoming asymmetric when central ion moves.

Fig. 10.6. Relaxation or asymmetry effect.

effect. Alternatively, it may be argued that as the central ion moves, the symmetry of the ionic atmosphere is lost; more ions of the ionic atmosphere are left behind than are present in the front [Fig. 10.6 (*b*)]. The excess ions of the ionic atmosphere present behind the moving ion drag the ion backward and retard its motion. Thus the effect arises because of the asymmetry of the ionic atmosphere of the moving ion and hence is also called *asymmetry effect.*

2. **Electrophoretic effect:** When EMF is applied, the central ion moves in one direction and the oppositely charged ionic atmosphere moves in the opposite direction. As this ionic atmosphere moves, the solvent molecules associated with it also move. Thus the flow of the ionic atmosphere and that of the solvent molecules attached to it takes place in a direction opposite to that of the movement of the central ion. In other words, the central ion is moving against the stream. Hence motion of the ions is retarded. This effect is called *electrophoretic effect* (Fig. 10.7).

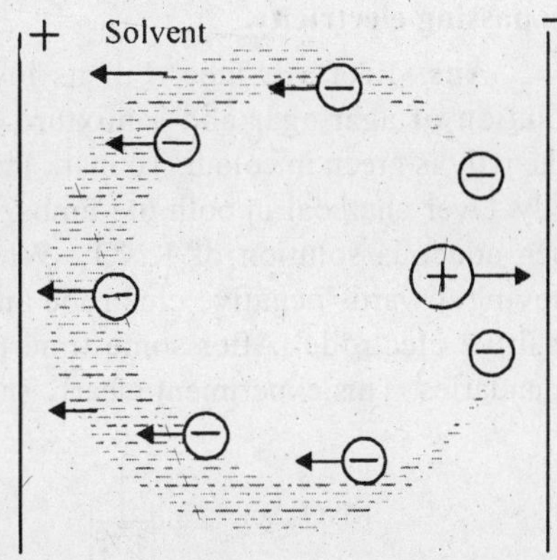

Fig. 10.7. Movement of a central ion in a direction opposite to that of the ionic atmosphere and the associated solvent molecules.

Besides the above two effects, the third retarding force is the normal frictional resistance offered by the medium which depends on the viscosity of the medium, its dielectric constant etc.

Based upon the above ideas, Debye and Huckel (1923) derived a mathematical expression for the variation of equivalent conductance with concentration. This equation was further improved by Onsagar (1926-27) and now the equation is know as **Debye-Huckel-Onsagar equation** or simply **Onsagar equation.** For a uni-univalent electrolyte, it is written in the form

$$\wedge_c = \wedge_0 - \left[\frac{82.4}{(DT)^{1/2}\eta} + \frac{8.20\times10^5}{(DT)^{3/2}}\wedge_0\right]\sqrt{c} \qquad \ldots (i)$$

where $\wedge_c$ = Equivalent conductance at concentration c

$\wedge_0$ = Equivalent conductance at infinite dilution

D = Dielectric constant of the medium

η = Viscosity of the medium

T = Temperature of the solution in degrees absolute

c = Concentration of the solution in moles/litre.

As D and η are constants for a particular solvent, therefore, at constant temperature, the above equation can be written in the form

$$\wedge_c = \wedge_0 - (A + B\wedge_0)\sqrt{c} \qquad \ldots (ii)$$

where A and B are constants for a particular solvent at a particular temperature.

Q. 32. How does Debye-Huckel theory explain the behaviour of strong electrolytes.

(G. Nanakdev 2000)

Ans. Strong electrolytes are completely ionised even in concentrated solution. But their ionic mobility is small. This is explained by Debye-Huckel theory in terms of asymmetry effect and electrophoretic effect. Both these effects oppose the movement of ions of the strong electrolyte and retard its speed thereby bringing down ionic mobility.

Increase in ionic conductance of strong electrolytes on dilution can be explained by saying that asymmetry and electrophoretic effects will be weakened when the solution is made dilute.

There will be smaller number of ions in the oppositely charge ionic sphere per unit volume of the solution. Also for the similar reasons, electrophoretic effect will also be weakened. This will result in higher conductance on dilution.

Q. 33. Explain asymmetric effect and electrophoretic effect found in strong electrolytes.
(Kurukshetra 2002; Bangalore 2004; Nagpur 2008)

Ans. See Q. 31.

Q. 34. Describe an experiment to demonstrate migration of ions towards the electrodes on passing electricity.

Ans. Take a U-tube. Fill its lower portion by $Cu_2Cr_2O_7$ jelly obtained by mixing a 5% solution of agar-agar and a mixture of $CuSO_4$ and $K_2Cr_2O_7$; and allow it to set (Fig. 10.8). The jelly is green in colour. To mark the position of jelly, some charcoal is scattered over the green jelly. Over charcoal in both the limbs, is added KNO_3 solution in agar-agar, *i.e.*, KNO_3 jelly and then aqueous solution of KNO_3. When current is passed, blue coloured Cu^{2+} ions are seen moving towards negative electrode and yellow coloured $Cr_2O_7^{2-}$ ions are seen moving towards positive electrode. After some time the two types of ions are seen moving with well-defined boundaries. This experiment clearly shows the migration of ions.

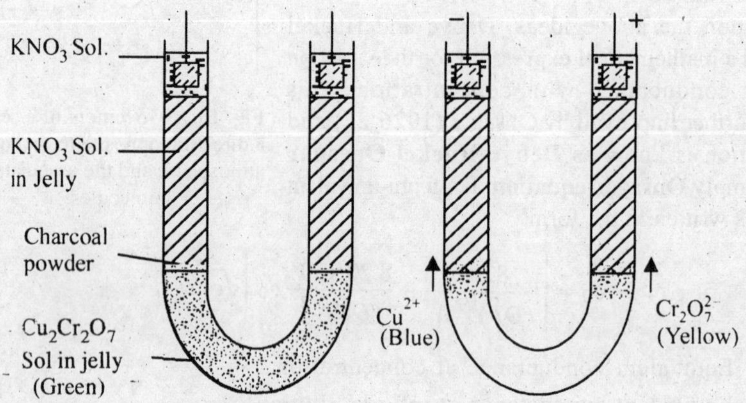

Fig. 10.8. Demonstration of the migration of ions.

Q. 35. With the help of a theoretical device (Hittorf's Device) show that fall in concentration around an electrode is proportional to the speed of the anion leaving that electrode.

Ans. Consider a cell containing the solution and provided with the anode *A* and the cathode *C* (Fig. 10.9). Let the solution lying between the electrodes *A* and *C* be divided into three compartments (the anodic, the middle and the cathodic) with the help of the imaginary porous partitions *a* and *c* shown by vertical dotted lines. Before electrolysis, suppose there are 13 pairs of ions, 4 being present in each of the anodic and the cathodic compartments and 5 in the middle (as shown in position I). The following different cases are considered :

When the electrodes are not attacked

Case I. When only anions (–) move: Let two anions leave cathodic compartment and enter anodic compartment. We get the position II. Although the cations have not moved still we see that the number of ions liberated is the same (*viz.* 2) at both the electrodes. But we see that concentration in anodic chamber has not altered while that in cathodic chamber has fallen by two molecules.

Similarly we can show that if cations are capable of movement and two of them enter cathodic compartment from anodic compartment then there will be a fall in concentration of two molecules in the anodic chamber. Thus we see that the fall of concentration of chamber is proportional to the speed of the ion leaving that chamber.

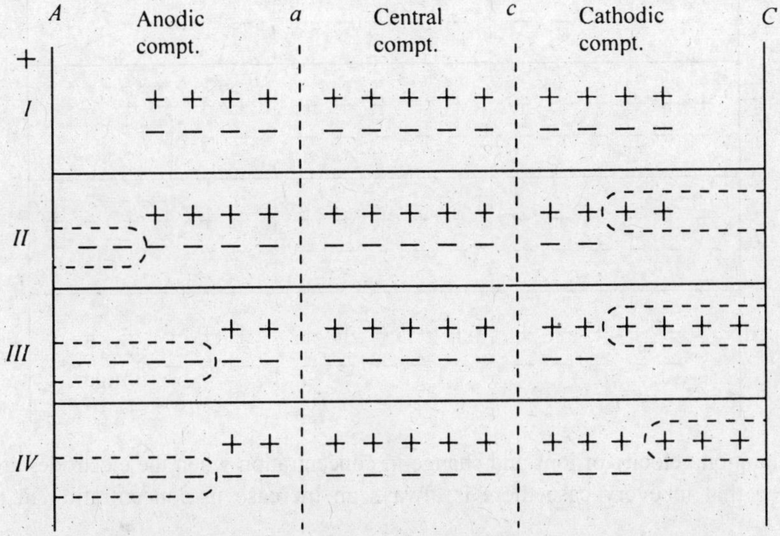

Fig. 10.9. Migration velocity of ions and change in concentration when the electrodes are not attacked.

Case II. When cations (+) and anions (–) move at the same rate: Let us assume that two cations move from anodic compartment to the cathodic compartment and two anions move from the cathodic compartment to the anodic compartment. Then we get position III. In this case there is fall in concentrations of 2 molecules in each compartment. But we find that the number of ions liberated at both the electrodes is the same, *i. e.*, 4.

Case III. When cations (+) move at double the speed of the anions (–): Suppose that two cations (+) from anodic compartment leave and enter the cathodic compartment. Since the speed of anions is half the speed of cations, therefore, only one anion enters the anodic compartment. The position is shown in IV. We also observe that the falls in concentration in anodic and cathodic compartments are 2 and 1 molecules respectively. The number of the ions discharged is again the same, *viz.*, 3 on each electrode.

From the above observation we draw the following conclusions :

(*i*) *Fall in concentration around any electrode ∝ the speed of the ion moving away from it.*

It means: Fall in concentration around cathode ∝ speed of the anion (u_a) and fall in concentration around anode ∝ speed of the cation (u_c).

i.e.,
$$\frac{\text{Fall in conc. around anode}}{\text{Fall in conc. around cathode}} = \frac{\text{Speed of cation}}{\text{Speed of anion}}$$

(*ii*) *Irrespective of the speeds of the ions, the number of ions liberated on both the electrodes is equal.*

When the electrodes are attackable

In a silver voltameter (containing $AgNO_3$ solution and having silver electrodes) each NO_3^- ion that reaches the anodic compartment reacts with Ag of the electrode forming $AgNO_3$ and instead of decrease, there is an increase of concentration in the anodic compartment. In a silver voltameter, the above three cases are shown again in Fig. 10.10 below :

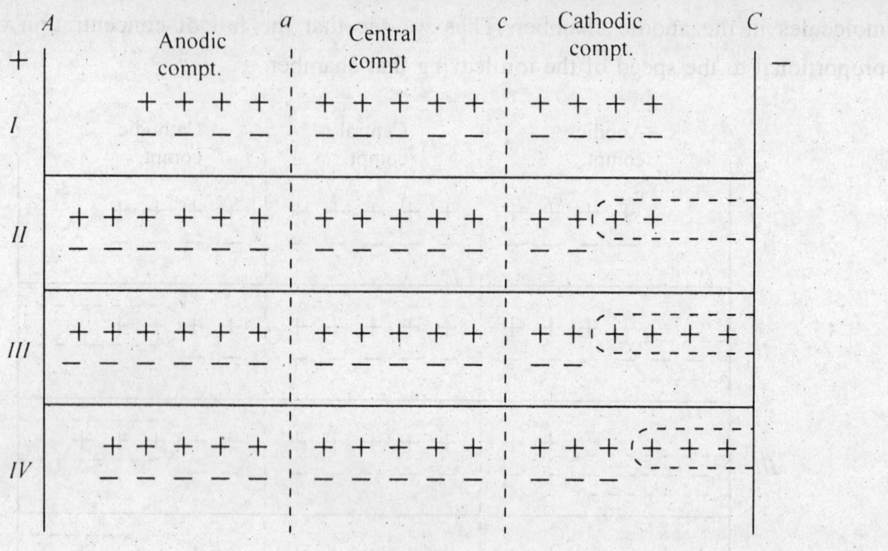

Fig. 10.10. Migration velocity of ions and change in concentration when the electrodes are attacked.

It is seen that in every case there is always an increase in concentration in the anodic compartment.

This increase in concentration takes place even when the Ag^+ ions migrate from the anodic compartment to the cathodic compartment (cases III and IV). In order to find the increase in concentration that would have taken place, if no Ag^+ ions had migrated from the anodic compartment to the cathodic compartment, a silver or a copper voltameter or coulometer is placed in series. Suppose the number of gram equivalents of Ag or Cu deposited in this voltameter is x. Obviously, x gram equivalents would have been the increase in the concentration of anodic compartment if no Ag^+ ions had migrated from this compartment. However the actual increase is less due to migration of Ag^+ ions. Suppose the actual increase in the concentration of Ag in the anodic compartment is y gram equivalents. Then

$$\left.\begin{array}{l}\textit{Fall in conc. in the anodic compartment}\\ \textit{due to migration of } Ag^+ \textit{ ions}\end{array}\right\} = (x-y) \textit{ gram equivalents.}$$

Further from Fig. 10.10, it may be seen that

Fall in conc. around cathode = Increase in conc. around anode = y gram equivalents

Thus the speed ratio in this case will be given by

$$\frac{\text{Speed of } Ag^+ \text{ ions}}{\text{Speed of } NO_3^- \text{ ions}} = \frac{x-y}{y}$$

Q. 36. What is meant by transport number of an ion? Show that sum of transpor umbers of cation and anion is unity. *(Delhi 2004; Nagpur 2008; Himachal 2010; Andhra 2010)*

OR

Write the expression for transport number of ions in terms of its velocity.

(Vidyasagar 2007)

Ans. *The fraction of the total current carried by an ion is called its* **transport number** or **transference number** or **Hittorf's number**.

Greater the speed of the ion, greater the amount of current it will carry with it and hence greater will be its transport number.

If u_a and u_c are the speeds of the anion and the cation respectively.

Then Current carried by the anion $\propto u_a$

$= k.u_a$

Current carried by the cation $\propto u_c = k.u_c$

Since current is carried by ions.

$\therefore$ Total current carried = current carried by anion + current carried by cation

$= k.u_a + k.u_c = k(u_a + u_c)$

$\therefore$ Transport number of anion, $n_a = \dfrac{ku_a}{k(u_a + u_c)}$

or $$\boldsymbol{n_a = \frac{u_a}{u_a + u_c}}$$

Similarly, transport number of the cation will be

$$\boldsymbol{n_c = \frac{u_c}{u_a + u_c}}$$

It may be seen that sum of transport numbers of cation and anion is unity.

or $$\boldsymbol{n_a + n_c = 1}$$

Q. 37. What is meant by the term transport number ? Is it constant for an ion?

(Punjab 2002)

Ans. For the definition of transport number, see Q. 36.

Transport number for an ion is not constant. It depends upon the second ion which is present in the salt.

Q. 38. Describe Hittorf's method for the determination of transport number.

(Punjab 2002; Arunachal 2002; Madurai 2004; Bhopal; 2004; Nagpur 2008)

Ans. Principle: The method is based upon the principle that the fall in concentration around an electrode is proportional to the speed of the ion moving away from it,

i.e., $$\frac{\text{Fall in conc. around cathode}}{\text{Fall in conc. around anode}} = \frac{\text{Speed of anion}}{\text{Speed of cation}} = \frac{u_a}{u_c}$$

Adding 1 to both sides, we get

$$1 + \frac{\text{Fall in conc. around cathode}}{\text{Fall in conc. around anode}} = \frac{u_a}{u_c} + 1$$

or $$\frac{\text{Fall in conc. around anode + Fall in conc. around cathode}}{\text{Fall in conc. around anode}} = \frac{u_a + u_c}{u_c} \text{ or}$$

$$\frac{\text{Total fall in conc. around both the electrodes}}{\text{Fall in conc. around anode}} = \frac{u_a + u_c}{u_c}$$

or $$\frac{\text{Fall in conc. around anode}}{\text{Total fall in conc. around both the electrodes}} = \frac{u_c}{u_a + u_c} = n_c$$

If the concentrations are expressed in terms of gram equivalents, the above expression may be written as

$$n_c = \frac{\text{Number of gram equivalents lost from the anodic compartment}}{\text{Number of gram equivalents lost from both the compartments}}$$

It may be seen that the number of gram equivalents lost from both the compartments is equal to the number of gram equivalents discharged on each electrode. This, in turn, can be found by placing a silver or copper voltameter in series, as according to Faraday's second law, when the same quantity of electricity flows through the silver voltameter, the same number of gram equivalents of Ag or Cu will be deposited. Thus the above formula reduces to

$$n_c = \frac{\text{Number of gram equivalents lost from the anodic compartment}}{\text{Number of gram equivalents deposited in the voltameter}}$$

Thus the determination of the transport number of the cation (n_c) involves the determination of the number of gram equivalents lost from the anodic compartment and the number of gram equivalents of Ag or Cu deposited in the voltameter placed in series with the apparatus.

To determine the transport number of the anion (n_a), there is no need to find the loss in concentration of the cathodic compartment because n_a can be calculated directly from the equation

$$n_a = 1 - n_c$$

Apparatus: It consists of U-tube joined on its two sides to two vertical bulbed glass tubes (Fig. 10.11). All the three *i.e.*, the U-tube and the two vertical tubes have got the stop-cocks at their bottom. The U-tube is also provided with two stop-cocks, one on either side at the top so as to cut off the various compartments from each other when required. In the two vertical tubes are placed the anode and the cathode separately. These are of platinum, if it is desired that these should not be attacked by the ions. Alternatively, the anode and the cathode of silver or copper are taken depending upon whether transport number of Ag^+ ions or Cu^{2+} ions is to be determined, in which case the anode is attacked by the anions. The apparatus is connected to a silver or copper voltameter in series with the apparatus together with a battery, a rheostat, a key and a milli-ammeter, as shown in Fig. 10.11.

Procedure: (*i*) The apparatus is filled with the solution of the electrolyte, containing ions whose transport number is to be determined.

(*ii*) A silver or copper voltameter is placed in series with the apparatus and the various connections are made as shown in Fig. 10.11.

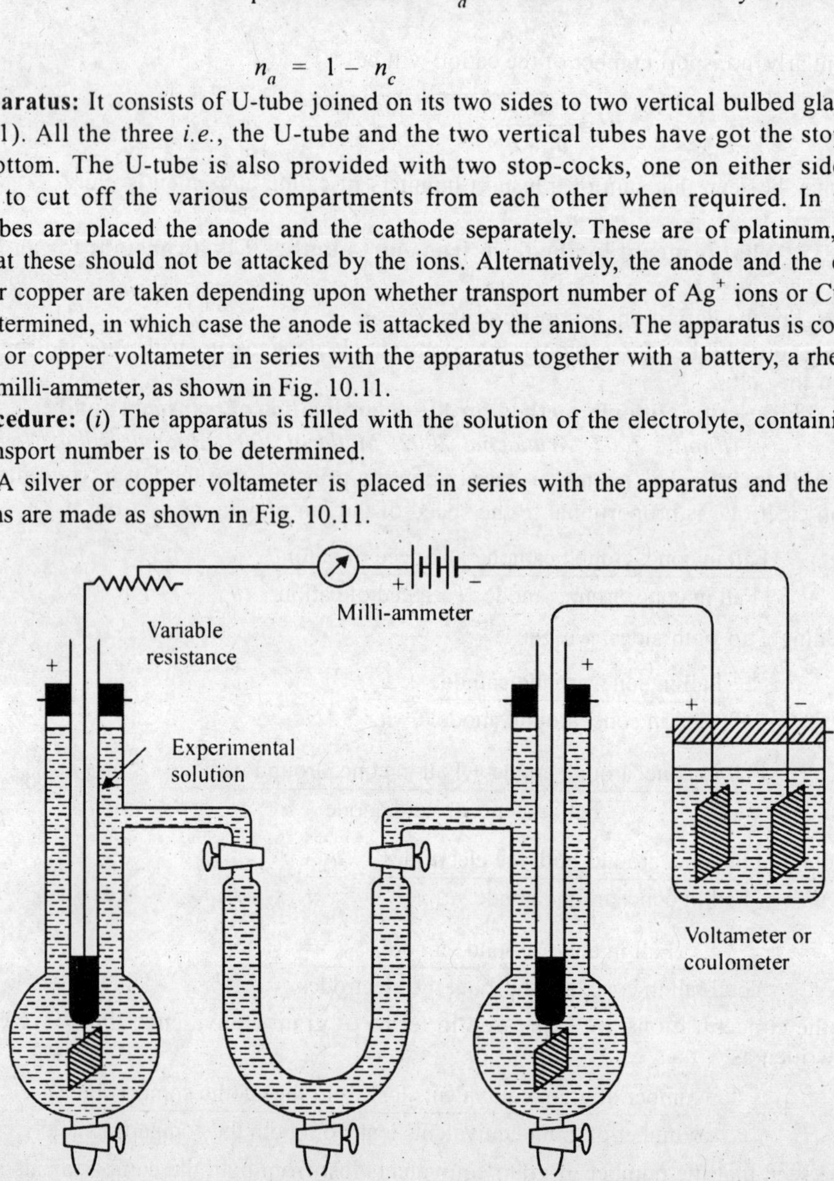

Fig. 10.11. Apparatus for the determination of transport number by Hittorf 's method.

(*iii*) Before electrolysis, the concentration of the electrolyte is found by titrating a known weight of the solution against a suitable reagent.

(*iv*) Now a current of 0.01 to 0.02 ampere is passed for 2-3 hours. A low current is used so that the change in concentration around the electrodes is not too high as otherwise the diffusion may take place.

(*v*) After electrolysis, the solution from the anodic compartment is taken out, weighed and then titrated as before to know the amount of the electrolyte present.

(*vi*) The weight of the silver deposited in the silver voltameter (or that of copper in case copper voltameter is used) is also determined.

It is assumed that there is no change in conc. of the middle compartment during the experiment.

Calculations: The method of calculation is illustrated by taking the example of $AgNO_3$ as the electrolyte and using silver voltameter in series.

When the electrodes are not attacked (*i.e., when platinum electrodes are used*).

After electrolysis :

Suppose, weight of the anodic solution taken out = w_1g

Weight of $AgNO_3$ found to be present (by titration) = w_2g

$\therefore$ Weight of water present = $(w_1 - w_2)$ g

Thus $(w_1 - w_2)$ g of water contains w_2 g of $AgNO_3$ after electrolysis.

$$w_2 \text{ g of } AgNO_3 \doteq \frac{w_2}{170} \text{ g eq. of } AgNO_3 = \frac{w_2}{170} \text{ g of Ag} = x \text{ g eq. (say)}$$

Hence we can say that after electrolysis $(w_1 - w_2)$ g of water contains x *g* eq. of Ag.

Before electrolysis :

Suppose w_3 g of solution contains w_4 g of $AgNO_3$

$\therefore$ Weight of water present = $(w_3 - w_4)$ g

Thus $(w_3 - w_4)$ g of water contains w_4 g of $AgNO_3$ before electrolysis.

$\therefore$ $(w_1 - w_2)$ g of water contains

$$= \frac{w_4}{(w_3 - w_4)} \times (w_1 - w_2) \text{ g of } AgNO_3$$

$$= \frac{w_4}{(w_3 - w_4)} \times \frac{(w_1 - w_2)}{170} \text{ g eq. of } AgNO_3$$

$$= \frac{w_4}{(w_3 - w_4)} \times \frac{w_1 - w_2}{170} \text{ g eq. of Ag}$$

$$= y \text{ g eq. (say)}$$

When the electrodes are not attacked, $x < y$

$\therefore$ Fall in the concentration of the anodic compartment

$$= (y - x) \text{ g eq. of Ag}$$

Suppose the weight of silver deposited in the silver voltameter

$$= W \text{ g} = \frac{W}{108} \text{ g eq. of Ag} = z \text{ g eq. (say)}$$

$\therefore$ Transport number of Ag^+ ion (n_{Ag^+})

$$= \frac{\text{Fall in concentration around anode in g eq.}}{\text{Amount of Ag deposited in voltameter in g eq.}}$$

$$= \frac{y-z}{z}$$

Transport number of NO_3^- ions $(n_{NO_3^-}) = 1 - n_{Ag^+}$

When the electrodes are attacked (*i.e. when silver electrodes are used*).

In this case, as the NO_3^- ; ions reaching the anode attack the silver anode, there is increase in the concentration of Ag^+ ions in the anodic solution (instead of decrease) *i.e.*, in this case we will have

$$x > y$$

∴ Increase in the concentration of the anodic solution

$$= (x - y) \text{ g eq. of Ag}$$

This increase in the concentration takes place in spite of the migration of Ag^+ ions from the anodic compartment to the cathodic compartment. If no Ag^+ ions had migrated from the anodic compartment, the increase in the concentration of Ag^+ ions in the anodic compartment would have been higher and equal to z g eq. of Ag (*i.e.*, the same as the amount deposited in the silver voltameter). Hence

Fall in concentration in the anodic solution due to migration of Ag^+ ions $= z - (x - y)$ g eq. of Ag.

$$\therefore \text{ Transport number of } Ag^+ \text{ ions } = \frac{z-(x-y)}{z}$$

Transport number of NO_3^- ions $= 1 - n_{Ag^+}$

In case a copper voltameter is used, the number of gram equivalents of copper deposited in the voltameter is obtained by dividing the weight of copper deposited (Wg) by its equivalent weight *viz.* 31.75.

Q. 39. During the electrolysis of a solution of potassium chloride between platinum electrodes, 0.0137 g of the chloride was lost from the anodic compartment and 0.0857 g of silver was deposited in a silver coulometer connected in series with the cell. Determine the transport number of K^+ and Cl^- ions.

Ans. No. of g eq. of the Cl^- ions lost from the anodic compartment

$$= \frac{0.0137}{35.5} = 0.0003859. \qquad \{\because \text{Eq. wt. of Cl} = 35.5\}$$

No of g eq. of Ag deposited in the coulometer

$$= \frac{0.0857}{108} = 0.0007935 \qquad \{\because \text{Eq. wt. of Ag} = 108\}$$

∵ Transport number of the cation (*i.e.* K^+) will be given by

$$n_{K^+} = \frac{\text{No. of g eq. lost from the anodic compartment}}{\text{No. of g eq. deposited in the coulometer}}$$

$$= \frac{0.0003859}{0.0007935} = \mathbf{0.486}$$

$$\therefore \quad n_{Cl^-} = 1 - n_{K^+}$$

$$= 1 - 0.486 = \mathbf{0.514}$$

Q. 40. A solution containing 10.09 g of $CuSO_4$ in 189.9 g of water was electrolysed. After electrolysis, 275.4 g of the solution around the anode was found to contain 15.4 g of $CuSO_4$ and 3.096 g of silver was deposited in a coulometer placed in series. Calculate the Hittorf's numbers of Cu^{2+} and SO_4^{2-} ions.

Ans. *After electrolysis:*

Weight of the anode solution = 275.4 g

Weight of $CuSO_4$ present = 15·4 g

Weight of water present = 27.54 – 15.4 g = 260 g

Before electrolysis

Weight of water = 189.9 g

Weight of $CuSO_4$ = 10.09 g

i.e. 189.9 g of water are associated with $CuSO_4$ = 10.09 g

∴ 260 g of water is associated with $CuSO_4$

$$= \frac{10.09}{189.9} \times 260$$

$$= 13.815 \text{ g}$$

Increase in the concentration of $CuSO_4$ around anode

$$= 15.4 - 13.815 = 1.585 \text{ g}$$

$$= \frac{1.585}{79.75} \text{ g eq.}$$

$$= 0.01987 \text{ g eq.}$$

$$\left\{\because \text{Eq. wt. of } CuSO_4 = \frac{63.5+32+64}{2} = 79.75\right\}$$

As there has been an increase in concentration around anode, it is a case of attack of electrode.

Silver deposited in coulometer = 3.096 g

$$= \frac{3.06}{108} \text{ g eq.} \qquad \{\because \text{Eq. wt. of Ag} = 108\}$$

$$= 0.02867 \text{ g eq}$$

Thus if no Cu^{2+} ions had migrated from the anodic compartment to the cathodic compartment, the increase around anode would have been = 0.02867 g eq. Actual increase =0.01987 g eq. Hence *fall in conc. around anode* due to migration of Cu^{2+} ions.

$$= 0.02867 - 0.01987 = 0.0088 \text{ g eq.}$$

$$\therefore \quad n_{Cu^{2+}} = \frac{\text{No. of g eq. of ion lost from anodic compartment}}{\text{No. of g eq. deposited in the coulometer}}$$

$$= \frac{0.0088}{0.02867} = \mathbf{0.307}$$

$$\therefore \quad n_{SO_4^{2-}} = 1 - 0.307$$

$$= \mathbf{0.693}$$

Q. 41. A solution of silver nitrate was electrolysed between silver electrodes. Before electrolysis 10 g of the solution contained 0.01788 g of silver nitrate. After the experiment 20.09 g of the anodic solution contained 0.06277 g of silver nitrate and 0.00947 g of copper

was deposited in the copper coulometer placed in series. Calculate the transport numbers of silver and nitrate ions. The atomic weights of Ag and Cu are 108 and 63.6 respectively.

Ans. *After electrolysis*:

Weight of the anodic solution = 20.09 g

Weight of $AgNO_3$ present = 0.06277 g

$\therefore$ Weight of water present = 20.09 – 0.06277 = 20.02723 g

Before electrolysis :

Weight of solution = 10 g

Weight of $AgNO_3$ present = 0.01788 g

$\therefore$ Weight of water present = 10 – 0.1788 = 9.98212 g

Thus 9.98212 g of water is associated with $AgNO_3$ = 0.01788 g

20.02723 g of water is associated with $AgNO_3$

$$= \frac{0.01788}{9.98212} \times 20.02723 = 0.03587 \text{ g}$$

Increase in conc. of $AgNO_3$ around anode = 0.06277 – 0.03587

= 0.0269 g

$$= \frac{0.0269}{170} \text{ g eq.} \qquad \{\because \text{ Eq. wt of } AgNO_3 = 170\}$$

= 0.0001582 g eq.

Copper deposited in the coulometer = 0.00947 g

$$= \frac{0.00947}{31.8} \text{ g eq.} \qquad \{\because \text{ Eq. wt. of Cu} = \frac{63.6}{2} = 31.8$$

= 0.0002977 g eq.

Thus if no Ag^+ ions had migrated from the anodic compartment to the cathodic compartment, the increase in the conc. of the anodic compartment would have been = 0.0002977 g eq.

Actual increase in conc. = 0.0001582g eq. Hence fall in concentration around anode due to migration of Ag^+ ions from the anodic compartment

= **0.0002977 – 0.0001582 g eq.**

= 0.0001395 g eq.

$$\therefore \quad n_{Ag^+} = \frac{\text{No. of g eq. lost from the anodic compartment}}{\text{No. of eq. deposited in the coulometer}}$$

$$= \frac{0.0001395}{0.0002977} = \mathbf{0.469}$$

$$\therefore \quad n_{NO_3^-} = 1 - n_{Ag^+} = 1 - 0.469 = \mathbf{0.531}$$

Q. 42. A solution of $AgNO_3$ containing 0.5066 g of the salt in 60 g of solution was electrolysed between silver electrodes. After electrolysis 50 g of the anode liquid was found to contain 0.5186 g of $AgNO_3$. A current of one ampere was passed for 1 minute and 58 sec. What are the transport numbers of Ag^+ and NO_3^- ions ?

Ans. Here quantity of electricity passed = Current × time

(amp.) (sec)

= 1 × 118 = 118 Coulomb

Now 96500 Coulomb of electricity deposits 1 gm eq. of Ag

$$\therefore \text{ 118 Coulomb of electricity deposits Ag} = \frac{118}{96500} \text{ g eq.}$$

After electrolysis
Weight of anode solution = 50 g
Weight of $AgNO_3$ = 0.5186 g
Weight of water = 49.4814 g
Before electrolysis
Weight of anode solution = 60 g
Weight of $AgNO_3$ = 0.5066 g
Weight of water = 59.4934 g
59.4934 g water contain = 0.5066 g $AgNO_3$ (Before electrolysis)

$\therefore$ 49.4814 g water contain $= \dfrac{0.5066}{59.4934} \times 49.4814 \text{ g } AgNO_3$

= 0.4213 g $AgNO_3$

Increase in conc. of $AgNO_3$ = 0.0973 g $AgNO_3$

$= \dfrac{0.0973}{170}$ g equivalent

= 0.00057 g equivalent

Silver deposited in the coulometer $= \dfrac{118}{96500}$ g eq.

= 0.00122

Hence decrease in the cone of $AgNO_3$ due to migration of cations from the anodic compartment

= 0.00122 – 0.00057
= 0.00065 g eq.

$\therefore$ Transport number of $Ag^+ = \dfrac{0.00065}{0.00122} = \mathbf{0.533}$

Transport no. of $NO_3^- = 1 - 0.533 = \mathbf{0.467}$

Q. 43. Describe moving boundary method for the determination of transport numbers. *(Bangalore, 2004; Madras, 2004; Purvanchal, 2007; Himachal 2010)*

Ans. This method is based upon the fact that the transport numbers of the cation and the anion are given by

$$n_c = \frac{u_c}{u_c + u_a}$$

and

$$n_a = \frac{u_a}{u_c + u_a}$$

Thus by determining the speeds of the cation and the anion (u_c and u_a), the transport numbers can be calculated.

The apparatus used for transport number determination is shown in Fig. 10.12. It consists of a long tube mounted vertically and fitted with the platinum electrodes on the two sides. The salt *BA* under investigation is dissolved in water and then placed between the solutions of two other salts *B'A* and *BA'* one having a common cation and the other having a common anion. For example, if *BA* is KCl, *B'A* may be LiCl and *BA'* may be CH_3COOK. The three solutions are arranged in order of increasing density downwards to avoid mixing

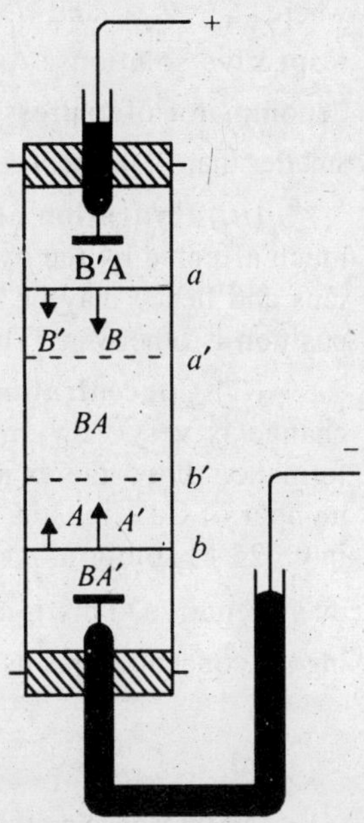

Fig. 10.12. Moving boundary method for determination of transport number.

solutions. The original boundaries are *a* and *b*. Even if the solutions are colourless, the boundaries can be observed because of the differences in the refractive indices of the solution. In order that the boundaries may remain sharp throughout the experiment, *B'* and *A'* must have smaller velocities than *B* and *A* respectively so that *B'* does not overtake *B* and *A'* does not overtake *A*. When the current is passed, the boundaries move to *a'* and *b'*. Thus

Speed of the cation (B^+), $u_c \propto$ length *aa'*

Speed of anion (A^-), $u_a \propto$ length *bb'*

Hence $$n_c = \frac{aa'}{aa'+bb'}$$

and $$n_a = \frac{bb'}{aa'+bb'}$$

Q. 44. What are the factors that influence transport number of ions ?

(M. Dayanand 2002; Kalyani 2003)

Ans. Following factors govern the transport number

(*i*) **Nature of the ion:** Smaller the size of the ion, greater is its speed and hence higher is its transport number. For example, the transport number of H^+ ion in 0.1 N HCl is greater than that of Na^+ ion in 0.1 N NaCl or the K^+ ion in 0.1 N KCl etc.

(*ii*) **Nature of the other ion present:** For example, the transport number of Cl^- ion (n_{Cl}) in HCl is different from that of Cl^- ion in NaCl. This may be explained as follows :

$$n_{Cl^-} \text{ in HCl} = \frac{u_{Cl^-}}{u_{H^+} + u_{Cl^-}} \quad \text{... (i)}$$

$$n_{Cl^-} \text{ in NaCl} = \frac{u_{Cl^-}}{u_{Na^+} + u_{Cl^-}} \quad \text{... (ii)}$$

where u_{H^+}, u_{Na^+} and u_{Cl^-} represent the speeds of H^+ ion, Na^+ and Cl^- ions respectively in their respective solutions. As Na^+ ion is larger in size than H^+ ion $u_{H^+} > u_{Na^+}$. As a result, the denominator of expression (*i*) is greater than that of expression (*ii*) and hence n_{Cl^-} in HCl is smaller than n_{Cl^-} in NaCl.

(*iii*) **Hydration of the ions:** The speed of an ion and hence its transport number is very much affected by the hydration of the ions. For example, Li^+ ions are smaller is size than Na^+ ions and hence may be expected to have higher transport number than Na^+ ions but the actual position is otherwise. This is because Li^+ ions are more hydrated than Na^+ ions.

(*iv*) **Concentration:** The transport number of an ion changes with concentration but the change is very very small. However, in case the increase in concentration results into the formation of a complex ion, the transport number changes significantly. For example, the transport number of Cd^{2+} ions in cadmium iodide solution falls from 0.444 in 0.005 M solution to 0.003 in 0.025 M solution. At still higher concentrations, the transport number appears to be zero or even negative. This is explained on the basis of the formation of the complex CdI_4^{2-} ions at higher concentrations as follows :

$$CdI_2 \longrightarrow Cd^{2+}(aq) + 2I^-(aq)$$

$$Cd^{2+}(aq) + 4I^-(aq) \longrightarrow CdI_4^{2-}(aq)$$

As the concentration is increased, more and more of CdI_4^{2-} ions are formed, which migrate to the anodic compartment. A stage comes when as many CdI_4^{2-} ions migrate into the anodic compartment as the number of Cd^{2+} ions migrating from it to the cathodic compartment. Thus there is no change in the concentration of Cd^{2+} ions in the anodic compartment and so the Cd^{2+} ions appear to have zero transport number.

(*v*) **Temperature:** As the temperature is increased, the ions tend to approach a transport number value of 0·5 *i.e.*, at this temperature, equal fraction of the current is carried by the cations and the anions. In case of ions having transport number less than 0.5 at ordinary temperature, the transport number increases with increase of temperature. In case of ions having transport number greater than 0·5 at ordinary temperature, the transport number decreases with increase of temperature.

Q. 45. Derive a relationship between the ionic conductance and transport number.

(Lucknow 2000; Himachal 2000; Bangalore, 2004)

Ans. Ionic conductance (λ) $\propto$ ionic mobility (u)

i.e. $\lambda_c \propto u_c$ or $\lambda_c = ku_c$ ...(*i*)

and $\lambda_a \propto u_a$ or $\lambda_a = ku_a$ ...(*ii*)

where λ_c and λ_a are the ionic conductances and u_c and u_a are the ionic mobilities of the cation and the anion respectively and k is a constant.

Dividing equation (*i*) by equation (*ii*), we get

$$\frac{\lambda_c}{\lambda_a} = \frac{u_c}{u_a}$$

Adding 1 to both sides, we get

$$\frac{\lambda_c}{\lambda_a} + 1 = \frac{u_c}{u_a} + 1 \quad \text{or} \quad \frac{\lambda_c + \lambda_a}{\lambda_a} = \frac{u_c + u_a}{u_a}$$

Taking reciprocal of both sides, we get

$$\frac{\lambda_a}{\lambda_c + \lambda_a} = \frac{u_a}{u_c + u_a}$$

But $\dfrac{u_a}{u_c + u_a} = n_a$, transport number of the anion

and $\lambda_c + \lambda_a = \wedge_0$, eq. conductance at infinite dilution. *(By Kohlrausch's law)*

$$\therefore \quad \frac{\lambda_a}{\wedge_0} = n_a \quad \text{or} \quad \lambda_a = n_a \times \wedge_0$$

Similarly $\lambda_c = n_c \times \wedge_0$

Q. 46. Define ionic mobility and ionic conductance and how are they related.

(Calcutta 2000; Lucknow 2000; Rajasthan 2003; Mumbai 2005 Vidyasagar 2007)

Ans. Ionic mobility is the absolute velocity of an ion. It is the distance travelled by an ion in cm/sec under a potential gradient of 1 V/cm. Contribution by an ion towards total equivalent conductance of an electrolyte at infinite dilution is called ionic conductance. The relationship between the two is as follows :

$$\text{Ionic mobility} = \frac{\text{Ionic conductance}}{96500}$$

Q. 47. The transport number of silver ion is 0.466 and that of nitrate ion is 0.534 at 18°C. The molar conductivity of a very dilute solution of $AgNO_3$ is 115.8 ohm^{-1} cm^2 at the same temperature. Calculate (*a*) the ionic conductance of silver and nitrate ions (*b*) the ionic mobilities of the two ions.

Ans. (*a*) Ionic conductance = Transport number of the ion $\times \wedge_0$ for strong electrolyte containing that ion

i.e. $\lambda_a = n_a \times \wedge_0$

or $\lambda_{NO_3^-} = 0.533 \times 115.8 = 61.837\ ohm^{-1}\ cm^2$

and $\lambda_c = n_c \times \wedge_0$

or $\lambda_{Ag^+} = 0.466 \times 115.8$

$= 53.962\ ohm^{-1}\ cm^2$

(*b*) Ionic mobility (*u*) $= \dfrac{\text{Ionic conductance}(\lambda)}{96,500}$

$\therefore$ Ionic mobility of $Ag^+ = \dfrac{53.962}{96,500} = \mathbf{0.000559\ cm / sec}$

and Ionic mobility of $NO_3^- = \dfrac{61.837}{96,500} = \mathbf{0.000641\ cm / sec}$

Q. 48. The equivalent conductance of a very dilute solution of $NaNO_3$ at 18°C is 105.2 ohm^{-1} cm^2. If the ionic conductance of NO_3^- ion in the solution is 61.7 ohm^{-1} cm^2, calculate the transport number of Na^+ ions in the solution.

Ans. Here we are given

$\wedge_0$ for $NaNO_3 = 105.2\ ohm^{-1}\ cm^2$

$\lambda_{NO_3^-} = 61.7\ ohm^{-1}\ cm^2$

Putting the values in the equation

$$\lambda^0_{NO_3^-} = n_{NO_3^-} \times \wedge^0_{NaNO_3}$$

$$61.7 = n_{NO_3^-} \times 105.2$$

$$n_{NO_3^-} = \frac{61.7}{105.2} = 0.586$$

$\therefore$ Transport No. of $Na^+ = 1 - 0.516 = \mathbf{0.414}$

Q. 49. At 18°C, the equivalent conductivities at infinite dilution of NH_4Cl, NaOH and NaCl are 129.8, 217.4 and 108.9 ohm^{-1} cm^2 respectively. If the equivalent conductivity of centi-normal solution of NH_4OH is 9.33 ohm^{-1} cm^2 what is the percentage dissociation of NH_4OH at this dilution ?

Ans. Here we are given :

$\wedge_0$ for $NH_4Cl = 129.8\ ohm^{-1}\ cm^2$

$\wedge_0$ for $NaOH = 217.4\ ohm^{-1}\ cm^2$

$\wedge_0$ for $NaCl = 108.9\ ohm^{-1}\ cm^2$

By Kohlrausch's law

$$\wedge_0 \text{ for } NH_4OH = \lambda_{NH_4^+} + \lambda_{OH^-}$$

$$= \wedge_0 (NH_4Cl) + \wedge_0 (NaOH) - \wedge_0 (NaCl)$$

$$= 129.8 + 217.4 - 108.9$$

$$= 238.3\ ohm^{-1}\ cm^2$$

Eq. conductance of centinormal NH_4OH ($\wedge$) = 9.33 ohm^{-1} cm^2 (Given)

$\therefore$ Degree of dissociation $(\alpha) = \frac{\wedge}{\wedge_0} = \frac{9.33}{238.3} = 0.0392$

or Percentage dissociation = 0.0392 × 100 = **3.92%**

Q. 50. If ionic conductance for H^+ and OH^- ions are 394.8 and 198.5 $ohm^{-1}\ cm^2$ respectively at 25°C and if the specific conductance of water at this temperature is $5.54 \times 10^{-8}\ ohm^{-1}\ cm^{-1}$, calculate the ionic product of water.

Ans.

$$\lambda_{H^+} = 394.8\ ohm^{-1}\ cm^2$$

$$\lambda_{OH^-} = 198.5\ ohm^{-1}\ cm^2$$

$$\therefore \quad \wedge_0 (H_2O) = \lambda_{H^+} + \lambda_{OH^-} \qquad \textit{(By Kohlrausch's law)}$$

$$= 349.8 + 198.5\ ohm^{-1}\ cm^2$$

$$= 548.3\ ohm^{-1}\ cm^2.$$

Specific conductance of pure water at 25°C is found to be

$$K = 5.54 \times 10^{-8}\ ohm^{-1}\ cm^{-1}.$$

As water dissociates to a very small extent to give H^+ ions and OH^- ions, water may be considered to be a very dilute solution of H^+ ions and OH^- ions in water so that we have

Eq. conductance of water = Specific conductance of water × Vol. in cm^3 containing 1g eq. of ions

$\therefore$ Volume in cm^3 containing 1 g eq. of ions

$$= \frac{\text{Eq.conductance of water}}{\text{Sp.conductance of water}} = \frac{548.3}{5.54 \times 10^{-8}} = 9.897 \times 10^9$$

$9.897 \times 10^9\ cm^3$ of water contains 1g eq. of H^+ ions or OH^- ions.

$\therefore$ 1 cm^3 of water contains H^+ ions or OH^- ions

$$= \frac{1}{9.897 \times 10^9} = 1.01 \times 10^{-10} \text{g eq.}$$

or 1000 cm^3 of water contains H^+ ions or OH^- ions

$$= 1.01 \times 10^{-10} \times 1000 = 1.01 \times 10^{-7} \text{g eq.}$$

i.e. $[H^+]$ or $[OH^-] = 1.01 \times 10^{-7}$ g eq. per litre

$$= 1.01 \times 10^{-7} \text{ moles/litre.}$$

As pure water dissociates to give equal number of H^+ ions and OH^- ions so that we have

$$[H^+] = [OH^-] = 1.01 \times 10^{-7} \text{ M at 25°C}$$

Hence

$$K_w = [H^+][OH^-]$$

$$= (1.01 \times 10^{-7}) \times (1.01 \times 10^{-7})$$

$$= 1.02 \times 10^{-14} \text{ at 25° C.}$$

Q. 51. How will you determine the solubility of a sparingly soluble salt by conductance measurement ? *(Lucknow 2001; Kurukshetra 2002)*

Ans. Salts such as AgCl, $BaSO_4$, $PbSO_4$ etc. dissolve in water to a very small extent. That is why these salts are called *sparingly soluble salts.* The solubility of such salts can be calculated by application of Kohlraush's law and conductance measurements as follows :

Specific conductance of the solution is determined. Suppose it comes out to be *K*. If *S* is the solubility of the salt in *gram equivalents per litre* then the volume of the solution in cm^3 containing 1 g eq will be 1000/S.

Eq. conductance of the solution = Specific conductance $\times \frac{1000}{S}$

i.e., $$\wedge = K \times \frac{1000}{S}$$

Since the salt dissolves very little, the solution is very dilute so that $\wedge$ may be considered as equal to $\wedge_0$, *i.e.*, equivalent conductance at infinite dilution. Hence we have

$$\wedge_0 = K \times \frac{1000}{S}$$

So that $$\boxed{S = \frac{1000 \times K}{\wedge_0}}$$

The value of $\wedge_0$ can be found by the application of Kohlrausch's law *e.g.* for AgC1, $\wedge_0 = \lambda_{Ag^+} + \lambda_{Cl^-}$

This will give the solubility in gram-equivalent per litre.

The solubility in grams/litre can be obtained by multiplying S with the equivalent weight of the substance.

Q. 52. Specific conductivity of a saturated solution of AgCl at 25°C was found to be 3.41×10^{-5} ohm^{-2} cm^{-1}. The specific conductivity for water used to make up the solution was 1.60×10^{-6} ohm^{-1} cm^{-1}. Determine the solubility of AgCl in water. Ionic conductances of Ag^+ and Cl^- at 25°C are 60.3 and 78.0 ohm^{-1} cm^2 respectively.

Ans. It is given that

$$K \text{ for AgC1 solution} = 3.41 \times 10^{-6} \text{ ohm}^{-1} \text{ cm}^{-1}$$
$$K \text{ for } H_2O = 1.60 \times 10^{-6} \text{ ohm}^{-1} \text{ cm}^{-1}$$
$$\lambda_{Ag^+} = 60.3 \text{ ohm}^{-1} \text{ cm}^2$$
$$\lambda_{Cl^-} = 78.0 \text{ ohm}^{-1} \text{ cm}^2.$$

By Kohlrausch's law, we have

$$\wedge_0 \text{ for AgCl} = \lambda_{Ag^+} + \lambda_{Cl^-}$$
$$= 60.3 + 78.0 = 138.3 \text{ ohm}^{-1} \text{ cm}^2$$
$$K \text{ for AgCl} = K \text{ for solution} - \text{K for } H_2O$$
$$= (3.41 - 1.60) \times 10^{-6} = 1.81 \times 10^{-6} \text{ ohm}^{-1} \text{ cm}^{-1}$$

Substituting the values directly in the formula

$$S = \frac{1000 \times K}{\wedge_0}, \text{ we get}$$
$$S = \frac{1000 \times 1.81 \times 10^{-6}}{138.3} \text{ g eq / litre}$$
$$= \frac{1000 \times 1.81 \times 10^{-6}}{138.3} \times 143.5 \text{ g / litre}$$
$$= \mathbf{0.001878 \text{ g/litre}}$$

Q. 53. Discuss the curve obtained in conductometric titration of a strong acid with a strong base.

(Nagpur 2002; Purvanchal 2007; Calcutta 2007; Andhra 2010; Himachal 2010)

Ans. Suppose a solution of HCl is to be titrated against NaOH solution. For this purpose, the acid solution is taken in a beaker and the NaOH solution in the burette. The conductance of the acid solution is noted initially as well as after successive additions of small amount of NaOH solution. Evidently, the conductance of the acid solution in the beginning is very high because it contains highly mobile H^+ ions ($HC1 \rightarrow H^+ + Cl^-$). As NaOH solution is added to the

HCl solution, the H^+ ions are replaced by the slow moving Na^+ ions and hence the conductance of the solution keeps on falling till the end point is reached, (*i.e.* all the H^+ ions are replaced by Na^+ ions).

$$\underbrace{H^+ + Cl^-}_{} + Na^+ + OH^- \longrightarrow Na^+ + Cl^- + H_2O$$

After the end point, further addition of NaOH solution brings in fast moving OH^- ions and hence the conductance of the solution again starts increasing. If conductance values are plotted against the volume of the alkali added, a curve of type *ABC* is obtained as shown Fig. 10.13. Point of intersection (*i.e.*, point *B*) corresponds to the end point.

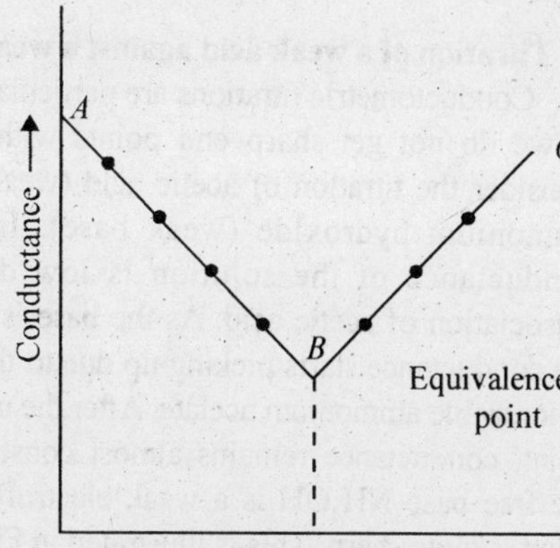

Fig. 10.13. Conductometric titration of a strong acid with a strong base.

Q. 54. Discuss the curve obtained by conductometric titration of a weak acid with a strong base. (Say CH_3COOH vs. NaOH) (*Arunachal 2002; Madras, 2004; Calcutta, 2007; Nagpur 2008*)

Ans. Suppose a weak acid like acetic acid is titrated against a strong base like NaOH. The conductance of the acid initially is very low because of low ionization of the acid. However as NaOH solution is added more and more to the acid solution, the salt produced *viz.* CH_3COONa is highly ionized and hence the conductance keeps on increasing along the line *AB* as shown in Fig. 10.14. When the whole of acetic acid has been neutralized, further addition of NaOH solution causes the conductance to increase sharply along the line *BC* because the NaOH added introduces the fast moving OH^- ions. Intersection of the two lines *AB* and *BC,* at the point *B* gives the equivalent point.

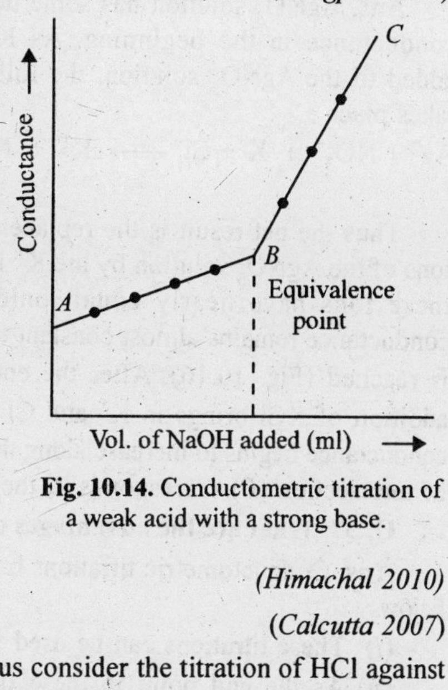

Fig. 10.14. Conductometric titration of a weak acid with a strong base.

Q. 55. Discuss the curve obtained by conductometric titration of

(a) a strong acid with a weak base. (*Himachal 2010*)

(b) a weak acid with a weak base (*Calcutta 2007*)

Ans. (a) Titration of a strong acid with weak base Let us consider the titration of HCl against NH_4OH solution. The conductance of HCl solution is initially high because of the presence of the fast moving H^+ ions. As NH_4OH solution is added, the fast moving H^+ ions are replaced by the slower NH_4^+ ions and hence the conductance falls along the line *AB*.

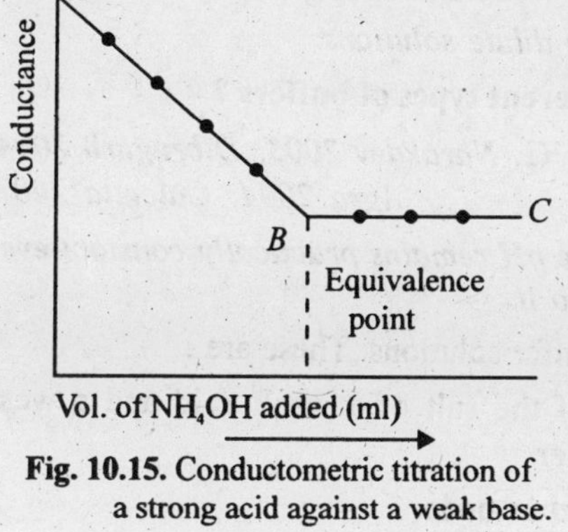

Fig. 10.15. Conductometric titration of a strong acid against a weak base.

$$\underbrace{H^+ + Cl^-}_{} + NH_4OH \longrightarrow NH_4^+ + Cl^- + H_2O$$

When the end point is passed, further addition of NH_4OH does not cause much change in the conductance because NH_4OH is a weakly ionized substance. The line *BC* is thus almost horizontal (Fig. 10.15).

(b) Titration of a weak acid against a weak base

Conductometric titrations are particularly valuable as we do not get sharp end points with indicators. Consider the titration of acetic acid (weak acid) with ammonium hydroxide (weak base). Initially, the conductance of the solution is low due to poor dissociation of acetic acid. As the base is added to it, the conductance starts picking up due to the formation of ionizable ammonium acetate. After the neutialization point, conductance remains almost constant, because the free base NH_4OH is a weak electrolyte. The end point is quite sharp. This is illustrated in Fig. 10.15 (b).

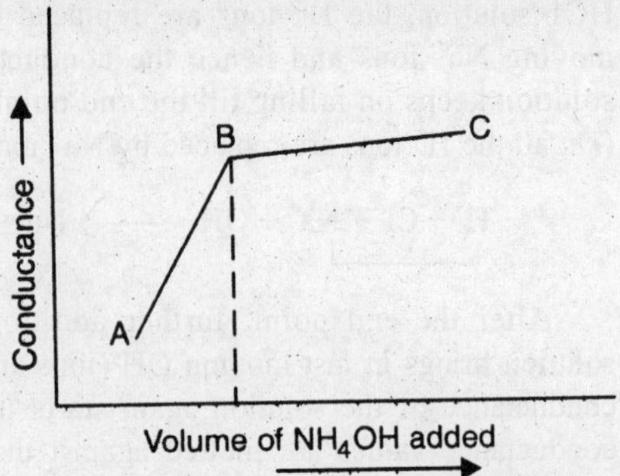

Fig. 10.15 (b) Conductometric titration of a weak acid with a weak base

Q. 56. Discuss the shape of the curve in conductometric titration of silver nitrate against KCl (precipitation reaction).

(Bangalore 2004; Madras, 2004)

Ans. $AgNO_3$ solution has some definite value of conductance in the beginning. As KCl solution is added to the $AgNO_3$ solution, the following reaction takes place :

$$Ag^+ + NO_3^- + K^+ + Cl^- \longrightarrow K^+ + NO_3^- + \underset{(ppt)}{AgCl}$$

Thus the net result is the replacement of the Ag^+ ions of the $AgNO_3$ solution by the K^+ ions. Since both these ions have nearly equal ionic mobility, the conductance remains almost constant till the end point is reached (Fig. 10.16). After the end point, further addition of KCl brings in K^+ and Cl^- ions. Thus the conductance begins to increase along the line *BC*. Point of intersection (*B*) corresponds to the equivalence point.

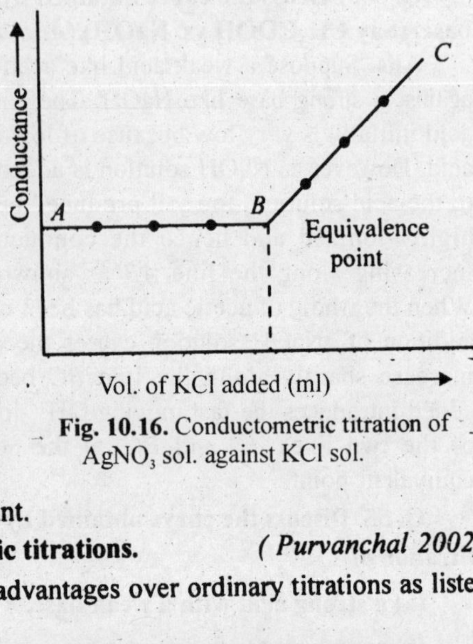

Fig. 10.16. Conductometric titration of $AgNO_3$ sol. against KCl sol.

Q. 57. What are the advantages of conductometric titrations. *(Purvanchal 2002)*

Ans. Conductometric titrations have a number of advantages over ordinary titrations as listed below.

(*i*) These titrations can be used for *coloured solutions* where ordinary indicators fail.

(*ii*) As the end point in these titrations is the intersection of two lines, *no extra care* is needed near the end point.

(*iii*) These can be used for the titration of *mixture of weak and strong acids.*

(*iv*) These can be used for the titration of even *very dilute solutions.*

Q. 58. What is a buffer solution ? What are the different types of buffers ?

(Awadh 2001; Purvanchal 2002; G. Narakdev 2003; Dibrugarh 2004; Agra 2004; Calcutta 2007)

Ans. *A buffer solution is defined as a solution whose pH remains practically constant even when small amounts of the acid or the base are added to it.*

Types of buffer solutions. There are two types of buffer solutions. These are :

(*i*) *Solutions of single substances:* The solution of the salt of a weak acid and a weak base *e.g.* ammonium acetate (CH_3COONH_4) acts as a buffer.

(*ii*) *Solutions of mixtures:* These are further classified as under :

(*a*) *Acidic buffer:* It is the solution of a mixture of equimolar quantities of a weak acid and a salt of this weak acid with a strong base (*e.g.* $CH_3COOH + CH_3COONa$).

(*b*) *Basic buffer:* It is solution of a mixture of equimolar quantities of a weak base and a salt of this weak base with a strong acid (*e.g.* $NH_4OH + NH_4Cl$).

Q. 59. Explain the buffer action of ammonium acetate. *(Meerut 2000)*

Ans. Ammonium acetate, is almost completely dissociated in the aqueous solution as follows :

$$CH_3COONH_4 \rightleftharpoons CH_3COO^- + NH_4^+$$

Thus in the solution, there are plenty CH_3COO^- ions and NH_4^+ ions.

When a small amount of an acid is added to the above solution the H^+ ions given by the acid combine with the CH_3COO^- ions to form CH_3COOH molecules which are almost unionised.

$$\mathbf{CH_3COO^- + H^+ \longrightarrow CH_3COOH}$$

Thus the H^+ ion concentration of the solution does not change practically and hence the *p*H of the solution remains almost constant. Similarly, when a small amount of a base is added to the above solution, the OH^- ions given by the base combine with the NH_4^+ ions to form NH_4OH molecules which are almost unionised.

$$NH_4^+ + OH^- \longrightarrow NH_4OH$$

Thus the OH^- ion concentration and hence the H^+ ion concentration or the *p*H of the solution remains almost constant.

Q. 60. Explain the buffer action of a mixture of acetic acid and sodium acetate (Acidic buffer). *(G. Nanakdev 2003)*

Ans. Acetic acid dissociates to a small extent whereas sodium acetate is almost completely dissociated in the aqueous solution as follows :

$$CH_3COOH \rightleftharpoons CH_3COO^- + H^+ \quad ...(i)$$

$$CH_3COONa \longrightarrow CH_3COO^- + Na^+ \quad ...(ii)$$

By *common ion effect,* the ionization of CH_3COOH is further suppressed. Thus in the solution, there are excess of acetate (CH_3COO^-) ions and a small amount of H^+ ions.

When a small amount of an acid is added to the above mixture solution, the H^+ ions given by the acid combine away with the CH_3COO^- ions given by the acid to form CH_3COOH which is almost unionised.

$$CH_3COO^- + H^+ \longrightarrow CH_3COOH$$

Thus the H^+ ion concentration and hence the *p*H of the solution remains almost constant.

Similarly, when a small amount of a base is added, the OH^- ions given by the base combine with the H^+ ions already present to form H_2O which is almost unionised.

$$\underset{\text{[From }(i)]}{H^+} + \underset{\text{[Given by the base }(ii)]}{OH^-} \longrightarrow H_2O$$

As the H^+ ions are consumed, the equilibrium (*i*) shifts towards right (according to Le Chatelier's principle). Thus more of CH_3COOH dissociates to make up the loss of H^+ ions. Hence the H^+ ion concentration or the pH of the solution does not change.

Q. 61. Explain the buffer action of a mixture of ammonium hydroxide and ammonium chloride (Basic buffer). *(Kalyani 2003)*

Ans. NH_4OH dissociates to a small extent whereas NH_4Cl dissociates completely in the aqueous solution as follows :

$$NH_4OH \rightleftharpoons NH_4^+ + OH^- \quad ...(i)$$

$$NH_4Cl \longrightarrow NH_4^+ + Cl^- \quad ...(ii)$$

By common ion effect, the ionization of NH_4OH is further suppressed. Thus in the solution, there are excess of NH_4^+ ions and a small amount of OH^- ions.

When a small amount of a base is added, the OH^- ions given by it immediately combine away to form the NH_4OH which is weakly ionised.

$$NH_4^+ + OH^- \longrightarrow NH_4OH$$

Thus the H^+ ion concentration or the pH of the solution remains unaffected.

Similarly, when a small amount of an acid is added, the H^+ ions given by it combine with the OH^- ions already produced by NH_4OH in equilibrium (*i*).

$$\underset{\text{[Given by the acid]}}{H^+} + \underset{\text{[Given by (i)]}}{OH^-} \longrightarrow H_2O$$

As the OH ions are consumed away the equilibrium (*i*) shifts in the forward direction. Thus more of NH_4OH dissociates to produce OH^- ions which make up the loss of OH^- ions. Hence the OH^- ion concentration and therefore the H^+ ion concentration or the pH of the solution remains almost constant.

Q. 62. Discuss the importance of buffer solutions.

Ans. Importance of buffer solutions is discussed as under :

(1) In biological processes: The *p*H of our blood is maintained constant (at about 7.4) inspite of various acid and base producing reactions going on in our body. In the absence of its buffer nature, we can not eat variety of foods.

(2) In industrial processes: The use of buffers is an important part of many industrial processes, ***e.g.,***

(*i*) in electroplating;

(*ii*) in the manufacture of leather, photographic materials and dyes ;

(*iii*) in bacteriological research, culture media are generally buffered to maintain the pH required for the growth of the bacteria being studied ;

(*iv*) in analytical chemistry.

Q. 63. Derive Henderson equation for the pH of a buffer mixture.

(Purvanchal 2002; Punjab 2003; Mumbai 2005; Delhi 2007; Raipur 2010)

Ans. (*a*) **For acidic buffer mixture:** Suppose it is required to calculate the pH of an *acidic buffer mixture,* obtained by mixing a weak acid HA and its salt with a strong base, BA. The former being weak dissociates to a small extent whereas the latter being strong electrolyte dissociates almost completely in the solution *i.e.*, we have

$$HA \rightleftharpoons H^+ + A^-$$

$$BA \longrightarrow B^+ + A^-$$

Applying the law of chemical equilibirum to the dissociation equilibrium of the weak acid, we get

$$K_a = \frac{[H^+][A^-]}{[HA]} \quad ...(i)$$

where K_a is the dissociation constant of the acid

$$\therefore \quad [H^+] = K_a \frac{[HA]}{[A^-]} \quad ...(ii)$$

Let the concentration of the acid and the salt taken originally in the buffer solution be c_1 and c_2 moles/litre respectively.

As HA dissociates to a small extent which is further suppressed by the common ion effect (as the salt BA produces the common A^- ions), the A^- ions in the mixture are mainly those produced from the salt BA. Further as the salt BA dissociates completely, we have

$$[A^-] = [BA] = c_2 \text{ moles/litre} \quad ...(iii)$$

Also as HA dissociates to a very small extent, the concentration of HA at equilibrium will be almost same as the original concentration *i.e.*

$$[HA] = c_1 \text{ moles/litre} \quad ...(iv)$$

Substituting the values of $[A^-]$ and [HA] from equations (*iii*) and (*iv*) in equation (*ii*), we get

$$\boxed{[H^+] = K_a \frac{c_1}{c_2}}$$

or
$$[H^+] = K_a \frac{[Acid]}{[Salt]} \quad ...(v)$$

Taking logarithm of both the sides, we get

$$\log[H^+] = \log K_a + \log \frac{[Acid]}{[Salt]}$$

or
$$-\log[H^+] = -\log K_a - \log \frac{[Acid]}{[Salt]}$$

or
$$pH = pK_a + \log \frac{[Salt]}{[Acid]} \quad ...(vi)$$

This equation is known as *Henderson equation.* Using this equation, the *p*H of an acidic buffer mixture can be calculated from the known concentrations of the acid and salt provided the dissociation constant of the acid is known.

(*b*) For basic buffer mixture: Proceeding exactly in a similar manner as above, the expression for the *p*OH of a *basic buffer mixture* (BOH + BA) can be obtained. Analogous to the expression (*vi*), we get

$$pOH = pK_b + \log \frac{[Salt]}{[Base]} \quad ...(vii)$$

where K_b is the dissociation constant of the base.

The *p*H of the basic buffer solution can be obtained from the following expression,

$$pH + pOH = 14$$

Q. 64. Determine the *p*H of a solution obtained by mixing equal volumes of 0.10 N ammonium nitrate and 0.02 N ammonium hydroxide, K_b for NH_4OH being 1.8×10^{-5}.

Ans. *Before mixing :*

$$[NH_4NO_3] = 0.10 \text{ N} = 0.10 \text{ M}$$

$$[NH_4OH] = 0.02 \text{ N} = 0.02 \text{ M}$$

After mixing :

As the volume of the solution becomes two-fold

$$[NH_4NO_3] = \frac{0.10}{2} = 0.05M$$

$$[NH_4OH] = \frac{0.02}{2} = 0.01M$$

Hence
$$pOH = pK_b + \log\frac{[\text{Salt}]}{[\text{Base}]}$$

$$= -\log(1.8 \times 10^{-5}) + \log\frac{0.05}{0.01}$$

$$= 4.7497 + 0.6990$$

$$= 5.4487$$

$\therefore$
$$pH = 14 - pOH$$

$$= 14 - 5.4487 = 8.5513 \simeq 8.55$$

Q. 65. What will be the *p*H value of a solution obtained by mixing 5 g of acetic acid and 7.5 g of sodium acetate and making the volume to 500 ml (K_{CH_3COOH} = 1.8 × 10^{-5} at 25°C).

Ans. Concentration of acetic acid

$$= 5 \text{ g in } 500 \text{ ml}$$

$$= 10 \text{ g/litre}$$

$$= \frac{10}{60} \text{M} \qquad \{\because \text{Mol. wt. of } CH_3COOH = 60\}$$

i.e.
$$[CH_3COOH] = \frac{1}{6} \text{ M}$$

Concentration of sodium acetate

$$= 7.5 \text{ g in } 500 \text{ ml}$$

$$= 15 \text{ g/litre}$$

i.e.,
$$[CH_3OO\,Na] = \frac{15}{82} \text{ M} \qquad \{\because \text{Mol. wt. of } CH_3COONa = 82\}$$

$$pK_a = -\log K_a$$

$$= -\log(1.8 \times 10^{-5})$$

$$= 4.7447$$

$\therefore$
$$pH = pK_a + \log\frac{[\text{Salt}]}{[\text{Acid}]}$$

$$= 4.7447 + \log\frac{15/82}{1/6}$$

$$= 4.7447 + \log 1.097$$

$$= 4.7447 + 0.0402$$

$$= 4.7849 \simeq 4.78$$

Q. 66. Calculate the *p*H of a mixture containing 0.01 M acetic acid and 0.03 M sodium acetate solutions (pK_a = 4.8).

Ans. *p*H of an acid-buffer is given by

$$pH = pK_a + \log\frac{[\text{Salt}]}{[\text{Acid}]}$$

Substituting the values

$$pH = 4.8 + \log\frac{0.03}{0.01}$$

or $$pH = 4.8 + \log 3$$

or $$pH = 4.8 + 0.4771$$

$$= 5.2771$$

Q. 67. Discuss the neutralisation of a strong acid with a strong base.

Ans. The strong acids and strong bases are completely dissociated in the aqueous solutions. Thus the neutralization reaction taking place is simply

$$H^+ + OH^- \rightleftharpoons H_2O$$

Heat of neutralization: Obviously, the neutralization reaction, as written above, does not depend upon the nature of the acid and the base. Now if the solution containing equivalent amounts of the strong acid and the strong base are taken, they are dissociated completely and produce the same quantity of H^+ ions and OH^- ions respectively. Thus on mixing the two solutions, the same amount of neutralization taken place and hence same amount of heat *i.e.*, 13.7 kcal/mole or 57.3 kJ mol^{-1} is produced always.

***p*H of the solution:** In the present case, excess of H^+ ions in the acid solution is equal to the excess of OH^- ions present in the solution of the base. On mixing, the excess of H^+ ions and OH^- ions combine with each other so that the resulting solution will have

$$[H^+] = [OH^-] = 10^{-7} M$$

$\therefore$ *p*H of the resulting solution will be = 7 *i.e.* the resulting solution will be neutral.

Q. 68. Discuss the neutralisation of a weak acid with a strong base.

Ans. Consider the reaction between CH_3COOH (weak acid) and NaOH (strong base). In this case, acetic acid is dissociated to a small extent and NaOH is dissociated completely in the aqueous solution as follows :

$$CH_3COOH \rightleftharpoons CH_3COO^- + H^+ \quad ...(i)$$

$$NaOH \longrightarrow Na^+ + OH^- \quad ...(ii)$$

The heat of neutralization and the *p*H of the resulting solution will be affected as explained below:

Heat of neutralization: The small amount of H^+ ions produced from the dissociation of an acetic acid combine with the OH^- ions produced from NaOH. Thus as the H^+ ions are removed, the equilibrium (*i*) shifts towards right (according to Le Chatelier's principle). In other words, more of CH_3COOH dissociates producing more of H^+ ions which again combine with the OH^- ions. The process continues till all the H^+ ions produced combine with the OH^-. In this case, *some heat is used up for the dissociation of acetic acid.* Hence the net reaction may be represented as

$$CH_3COOH \rightleftharpoons CH_3COO^- + H^+ \quad \Delta H = 0.5 \text{ kcal/mole } (2.1 \text{ kJ mol}^{-1})$$

$$H^+ + OH^- \longrightarrow H_2O \quad \Delta H = -13.7 \text{ kcal/mole } (57.3 \text{ kJ mol}^{-1})$$

$$CH_3COOH + OH^- \rightleftharpoons CH_3COO^- + H_2O, \quad \Delta H = -13.2 \text{ kcal/mole } (55.2 \text{ kJ mol}^{-1})$$

Thus the net heat evolved is less than in the first case.

***p*H of the solution:** As represented above, the final reaction does not proceed to completion. Instead, it reaches a state of equilibrium. The backward reaction produces some acetic acid (CH_3COOH) molecules and OH^- ions. Thus the final solution contains some undissociated acetic acid molecules and an equal number of OH^- ions. Hence the solution obtained is alkaline *i.e.* has *p*H > 7.

Q. 69. Discuss the neutralization of a strong acid with a weak base. *(Lucknow 2000)*

Ans. Let us consider the reaction between HCl (strong acid) and NH_4OH (weak base). Here, NH_4OH is not completely dissociated, whereas HCl is completely dissociated. Thus we have

$$NH_4OH \rightleftharpoons NH_4^+ + OH^-$$

$$HCl \longrightarrow H^+ + Cl^-$$

Heat of neutralization: The reaction taking place may be represented as

$$NH_4OH \rightleftharpoons NH_4^+ + OH^- \qquad \Delta H = +1.4 + \text{kcal/mole } (5.9 \text{ kJ mole}^{-1})$$

$$H^+ + OH^- \longrightarrow H_2O \qquad \Delta H = -13.7 \text{ kcal/mole } (57.3 \text{ kJ mol}^{-1})$$

On adding the two equations, we get

$$NH_4OH + H^+ \rightleftharpoons NH_4^+ + H_2O \qquad \Delta H = -12.3 \text{ kcal/mole } (51.4 \text{ kJ mol}^{-1})$$

In this case, 1.4 kcal (5.9 kJ) of heat is used up for the dissociation of one mole of NH_4OH. Hence the net heat evolved is only 12.3 kcal/mole (51.4 kJ mol^{-1}).

As represented by the final equilibrium equation above, the final solution contains some undissociated NH_4OH molecules and an equal number of H^+ ions. Hence the final solution is acidic *i.e.* has pH < 7.

Q. 70. Discuss the neutralization of a weak acid with a weak base.

Ans. Consider the reaction between CH_3COOH (weak acid) and NH_4OH (weak base). In this case, both dissociate to a small extent in the aqueous solution. Thus we have

$$CH_3COOH \rightleftharpoons CH_3COO^- + H^+ \qquad ...(i)$$

$$NH_4OH \rightleftharpoons NH_4^+ + OH^- \qquad ...(ii)$$

Heat of neutralization: The small amount of H^+ ions and OH^- ions thus produced combine with each other as follows :

$$H^+ + OH^- \longrightarrow H_2O \qquad ...(iii)$$

As the H^+ ions and the OH^- ions are being removed from equilibria (*i*) and (*ii*) respectively, these equilibria shift towards right producing more of H^+ ions and OH^- ions which again combine with one another and thus the process goes on. In this case, some heat is required for the dissociation of CH_3COOH as well as NH_4OH. Hence the net heat evolved is much less.

***p*H of the solution:** Although both the acid and the base involved dissociate to a small extent, yet depending upon their degrees of dissociation, one of the two may dissociate more than the other. Thus they may produce unequal number of H^+ ions and OH^- ions. Depending upon which of the two ions are in excess, the resulting solution may be alkaline or acidic.

Q. 71. What is salt hydrolysis ? *(Purvanchal 2002; Delhi 2004; Purvanchal 2007, Calcutta 2007)*

Ans. *It is defined as the process in which a salt reacts with water to give back the acid and the base.*

$$\text{Salt} + \text{Water} \longrightarrow \text{Acid} + \text{Base}$$

or

$$\underset{\text{Salt}}{BA} + \underset{\text{Water}}{H_2O} \longrightarrow \underset{\text{Acid}}{HA} + \underset{\text{Base}}{BOH}$$

All salts are strong electrolytes and thus ionize completely in the aqueous solution. If the acid (HA) produced is strong and the base (BOH) produced is weak, we can write the above equation as

$$B^+ + A^- + H_2O \longrightarrow H^+ + A^- + BOH$$

or $$B^+ + H_2O \longrightarrow H^+ + BOH$$

Thus, in this case the cation reacts with water to give an acidic solution. This is called **cationic hydrolysis.**

Again, if the acid produced is weak and the base produced is strong, we can write

$$B^+ + A^- + H_2O \longrightarrow HA + B^+ + OH^-$$

or $$A^- + H_2O \longrightarrow HA + OH$$

Here, the anion reacts with water to give the basic solution. This is called **anionic hydrolysis.**

Hence *salt hydrolysis may be defined as the reaction of the cation or the anion of the salt with water to produce acidic or basic solution.*

Q. 72. Discuss the salt hydrolysis in the following cases:

(*i*) Salts of strong acid and strong base

(*ii*) Salts of weak acid and strong base

(*iii*) Salts of strong acid and weak base

(*iv*) Salts of weak acid and weak bases. (*Purvanchal 2007*)

Ans. Hydrolysis of salts in the four types of salts can be represented as under

$$NaCl + aq \longrightarrow Na^+ + Cl^-$$

$$H_2O \rightleftharpoons H^+ + OH^-$$

or the overall process will be

$$NaCl + H_2O \rightleftharpoons Na^+ + Cl^- + H^+ + OH^-$$

i.e., there is only dissociation of sodium chloride and no hydrolysis

(*ii*) $$CH_3COONa + H_2O \rightleftharpoons CH_3COOH + NaOH$$

or $$CH_3COO^- + Na^+ + H_2O \rightleftharpoons CH_3OOOH + Na^+ + OH^-$$

or the overall reaction will be

$$CH_3COO^- + H_2O \rightleftharpoons CH_3COOH + OH^-$$

i.e.,it is *anionic hydrolysis*

(*iii*) $$NH_4Cl + H_2O \rightleftharpoons NH_4OH + HCl$$

or $$NH_4^+ + Cl^- + H_2O \rightleftharpoons NH_4OH + H^+ + Cl^-$$

or the overall reaction will be

$$NH_4^+ + H_2O \rightleftharpoons NH_4OH + H^+$$

i.e., it is *cationic hydrolysis*

(*iv*) $$CH_3COONH_4 + H_2O \rightleftharpoons CH_3COOH + NH_4OH$$

or the overall reaction will be

$$CH_3COO^- + NH_4^+ + H_2O \rightleftharpoons CH_3COOH + NH_4OH$$

i.e., it involves both *anionic as well as cationic hydrolysis.*

Q. 73. Explain the following terms:

(*i*) Hydrolysis constant

(*ii*) Degree of hydrolysis. (*M. Dayanand 2000; Lucknow 2000;Delhi 2004*)

Ans. Hydrolysis constant: Hydrolysis of a salt (BA) may be represented as

$$BA + H_2O \rightleftharpoons HA + BOH$$

Applying the law of chemical equilibrium, we get

$$\frac{[HA][BOH]}{[BA][H_2O]} = K, \text{ the equilibrium constant}$$

Since water is present in very large excess in the aqueous solution, its concentration $[H_2O]$ may be regarded as constant so that we have

$$\frac{[HA][BOH]}{[BA]} = K[H_2O] = K_h$$

where K_h is called the hydrolysis constant.

Since [HA] and [BOH] represent the molar concentrations of the free acid and the free base produced and [BA] represents the molar concentration of the unhydrolysed salt, the expression for the hydrolysis constant may be written in a more general way as :

$$\boxed{K_h = \frac{[\text{Free Acid}][\text{Free Base}]}{[\text{Unhydrolysed salt}]}}$$

(*ii*) **Degree of hydrolysis:** *The degree of hydrolysis of a salt is defined as the fraction (or percentage) of the total salt which is hydrolysed.* For example, if in a solution of aniline hydrochloride, 90% of the salt is hydrolysed into aniline and hydrochloric acid, the degree of hydrolysis is 0.90 or may be expressed as percentage *viz.* 90%. It is usually represented by the symbol *h*.

Q. 74. Derive expressions for the hydrolysis constant, degree of hydrolysis and *p*H of solution of a salt of weak acid and strong base. *(Lucknow 2000; Calcutta 2007; Delhi 2007)*

Ans. (*a*)**Hydrolysis constant:** Representing the salt by BA, the hydrolysis may be represented as follows :

$$\underset{\text{Salt}}{BA} + H_2O \rightleftharpoons \underset{\text{Strong}}{BOH} + HA$$

$$\downarrow \qquad\qquad \downarrow$$

or $$B^+ + A^- + H_2O \qquad B^+ + OH^- + HA$$

or $$A^- + H_2O \qquad OH^- + HA$$

i.e., it is a case of *anion hydrolysis*.

The hydrolysis constant K_h for the above reaction will be given by

$$K_h = \frac{[OH^-][HA]}{[A^-]} \qquad ...(i)$$

For the weak acid HA, the dissociation equilibrium is

$$HA \rightleftharpoons H^+ + A^-$$

∴ The dissociation constant K_a of the acid HA will be given by

$$K_a = \frac{[H^+][A^-]}{[HA]} \qquad ...(ii)$$

Further the ionic product of the water K_w is given by

$$K_w = [H^+][OH^-] \qquad ...(iii)$$

Multiplying equation (*i*) with (*ii*) and dividing by (*iii*), we get

$$\frac{K_h K_a}{K_w} = \frac{[OH^-][HA]}{[A^-]} \times \frac{[H^+][A^-]}{[HA]} \times \frac{1}{[H^+][OH^-]} = 1$$

or $$K_h = \frac{K_w}{K_a} \quad ...(iv)$$

(*b*) **Degree of hydrolysis:** Suppose the original concentration of the salt in the solution is *c* moles/litre and *h* is its degree of hydrolysis at this concentration. Then we have

	$A^- + H_2O \rightleftharpoons$	OH^-	+	HA
Original molar conc.	c	0		0
Molar conc. at equilibrium	$c(1-h)$	ch		ch

The hydrolysis constant (K_h), will, therefore, be given by

$$K_h = \frac{[OH^-][HA]}{[A^-]}$$

$$= \frac{ch.ch}{c(1-h)} = \frac{ch^2}{1-h}$$

If *h* is very small as compared to 1,we can take $1 - h \simeq 1$ so that the above expression becomes

$$ch^2 \simeq K_h$$

or $$h^2 \simeq \frac{K_h}{c}$$

or $$h \simeq \sqrt{\frac{K_h}{c}} \quad ...(v)$$

i.e $$h \propto \sqrt{\frac{1}{c}}$$

Thus *the degree of hydrolysis of a salt of a weak acid and a strong base is inversely proportional to the square-root of its molar concentration.*

If *V* is the volume of the solution in litres containing 1 mole of the solute, we have $c = \frac{1}{V}$ so that the above expression can also be written as

$$h \simeq \sqrt{K_h.V}$$

Substituting the value of K_h from equation (*iv*) in eq. (*v*) we get

$$h \simeq \sqrt{\frac{K_w}{K_a.C}} \quad ...(vi)$$

(*c*) ***p*H:** In the present, we have

	$A^- + H_2O \rightleftharpoons$	OH^-	+ HA
Original concentration :	c	0	0
Conc. at equilibrium :	$c(1-h)$	ch	ch

i.e., $$[OH^-] = ch$$

$\therefore$ $$[H^+] = \frac{K_w}{[OH^-]} = \frac{K_w}{ch}$$

Substituting the value of *h* from equation (*vi*), we get

$$[H^+] = \frac{K_w}{c}\sqrt{\frac{K_a.c}{K_w}}$$

or
$$[H^+] = \sqrt{\frac{K_w.K_a}{c}}$$

$\therefore$
$$pH = -\log[H^+] = -\log\sqrt{\frac{K_w K_a}{c}}$$
$$= -\log\left(\frac{K_w.K_a}{c}\right)^{1/2}$$

or
$$\mathbf{pH = -\frac{1}{2}[\log K_w + \log K_a - \log c]}$$

Q. 75. Derive the relation

$$\mathbf{pH = -\frac{1}{2}[\log K_h + \log K_a - \log C]}$$

in case of hydrolysis of salt of weak acid and strong base.

Ans. See Q. 74.

Q. 76.Derive expressions for the hydrolysis constant, degree of hydrolysis and *p*H of solutions of salts of strong acid and weak base.

(Himachal 2000; Kumaon; 2002; Purvanchal 2002; Delhi 2006)

Ans. Hydrolysis constant: For the salt BA, the hydrolysis may be represented as

$$\underset{\text{Salt}}{BA} + H_2O \rightleftharpoons \underset{\text{Weak}}{BOH} + \underset{\text{Strong}}{HA}$$

or
$$B^+ + A^- + H_2O \rightleftharpoons BOH + H^+ + A^-$$

or
$$B^+ + H_2O \rightleftharpoons BOH + H^+$$

i.e., it is a case of *cation hydrolysis.*

The hydrolysis constant K_h will be given by

$$K_h = \frac{[BOH][H^+]}{[B^+]} \quad ...(i)$$

For the weak base BOH, the dissociation equilibrium is

$$BOH \rightleftharpoons B^+ + OH^-$$

$\therefore$ The dissociation constant K_b of the weak base BOH will be given by

$$K_b = \frac{[B^+][OH^-]}{[BOH]} \quad ...(ii)$$

The ionic product of water, K_w, is given by

$$K_w = [H^+][OH^-] \quad ...(iii)$$

Multiplying equation (*i*) with (*ii*) and dividing by (*iii*), we get

$$\frac{K_h K_b}{K_w} = 1$$

or $$K_h = \frac{K_w}{K_b} \qquad ...(iv)$$

(*b*) **Degree of hydrolysis:** Proceeding as in the earlier case, we get the same relationship

$$h = \sqrt{\frac{K_h}{c}}$$

Substituting the value of K_h from equation (*iv*), we get

$$h = \sqrt{\frac{K_w}{K_b.c}} \qquad ...(v)$$

(*c*) ***p*H:** In this case, we have

$$B^+ + H_2O \rightleftharpoons BOH + H^+$$

Original concentration :	c	0	0
Conc. at equilibrium :	$c(1-h)$	ch	ch

i.e $[H^+] = ch$

Substituting the value of *h* from equation (*v*), we get

$$[H^+] = c.\sqrt{\frac{K_w}{K_b.c}} = \sqrt{\frac{K_w.c}{K_b}} = \left(\frac{K_w.c}{K_b}\right)^{1/2}$$

$$\therefore \qquad pH = -\log[H^+] = -\log\left(\frac{K_w.c}{K_b}\right)^{1/2}$$

or $$pH = -\frac{1}{2}[\log K_w - \log K_b + \log c] \qquad ...(vi)$$

Q. 77. Derive expressions for hydrolysis constant, degree of hydrolysis and *p*H of solutions of salts of weak acid and weak base.

(G.Nanakdev 2000; Punjab 2003; Bangalore 2004; Delhi 2004; Purvanchal 2007)

Ans. (*a*) **Hydrolysis constant:** Representing the salt by BA, we have

$$\underset{\text{Salt}}{BA} + H_2O \rightleftharpoons \underset{\text{Weak}}{BOH} + \underset{\text{Weak}}{HA}$$

$$\downarrow$$

$$B^+ + A^- + H_2O \rightleftharpoons BOH + HA$$

i.e., it involves both *anionic hydrolysis as well as cationic hydrolysis.* Writing the various equations, we have

$$K_h = \frac{[BOH][HA]}{[B^+][A^-]} \qquad ...(i)$$

For the weak acid, $HA \rightleftharpoons H^+ + A^-$, so that

$$K_a = \frac{[H^+][A^-]}{[HA]} \qquad ...(ii)$$

For the weak base, $BOH \rightleftharpoons B^+ + OH^-$, so that

$$K_b = \frac{[B^+][OH^-]}{[BOH]} \qquad ...(iii)$$

Also $$K_w = [H^+][OH^-] \qquad ...(iv)$$

Multiplying equation (*i*), (*ii*) and (*iii*) and dividing by equation (*iv*), we get

$$\frac{K_h . K_a . K_b}{K_w} = 1$$

or
$$K_h = \frac{K_w}{K_a \, K_b} \qquad ...(v)$$

(*b*) **Degree of hydrolysis:** In this case, we have

	B^+ + A^- + H_2O	$\rightleftharpoons$	BOH + HA
Original concentration	c c		0 0
Conc. at equilibrium	$c(1-h)$ $c(1-h)$		ch ch

$$K_h = \frac{[\mathrm{BOH}][\mathrm{HA}]}{[\mathrm{H}^+][\mathrm{A}^-]} = \frac{ch.ch}{c(1-h).c(1-h)} = \frac{h^2}{(1-h)^2} \qquad ...(vi)$$

Here, the relationship between K_h and h does not involve c. Thus the *degree of hydrolysis of such a salt is independent of the concentration of the solution.*

If h is very small in comparison to 1,we can take $1 - h \simeq 1$ so that equation (*vi*) becomes

$$K_h \simeq h^2$$

or
$$h \cong \sqrt{K_h}$$

Substitute the value of K_h from eqn.(*v*)

$$h \simeq \sqrt{\frac{K_w}{K_a \cdot K_b}} \qquad ...(vii)$$

(*c*) *p***H**: For the weak acid HA,

$$\mathrm{HA} \rightleftharpoons \mathrm{H}^+ + \mathrm{A}^-$$

$$\therefore \quad K_a = \frac{[\mathrm{H}^+][\mathrm{A}^-]}{[\mathrm{HA}]}$$

or
$$[\mathrm{H}^+] = K_a \frac{[\mathrm{HA}]}{[\mathrm{A}^-]}$$

$$= K_a \frac{ch}{c(1-h)}$$

$$= K_a \frac{h}{1-h}$$

But from equation (*vi*), $\frac{h}{1-h} = \sqrt{K_h}$ so that we have

$$[\mathrm{H}^+] = K_a \sqrt{K_h}$$

$$= K_a \sqrt{\frac{K_w}{K_a \cdot K_b}}$$

$$= \sqrt{\frac{K_a \, K_w}{K_b}}$$

$$pH = -\log \mathrm{H}^+ = -\log \sqrt{\frac{K_a \, K_w}{K_b}}$$

$$= -\log \left(\frac{K_a \, K_w}{K_b} \right)^{1/2}$$

or $$pH = -\frac{1}{2}[\log K_a + \log K_w - \log K_b] \quad ...(viii)$$

Q. 78. Show that for a salt like ammonium acetate the degree of hydrolysis is independent of concentration.

Ans. See Q. 77.

Q. 79. What would be the pH of a 0.1 molar sodium acetate solution when the dissociation constant for acetic acid is 1.8×10^{-5} ?

Ans. The pH of the solution of a salt depends upon the products of hydrolysis. In the present case, sodium acetate is hydrolysed as follows :

$$CH_3COONa + H_2O \rightleftharpoons CH_3COOH + NaOH$$

or $$CH_3COO^- + Na^+ + H_2O \rightleftharpoons CH_3COOH + Na^+ + OH^-$$

or $$CH_3COO^- + H_2O \rightleftharpoons CH_3COOH + OH^-$$

If c is the initial concentration of the salt and h is the degree of hydrolysis, we have

	CH_3COO^- + H_2O ⇌	CH_3COOH +	OH^-
Before hydrolysis:	c	0	0
After hydrolysis :	$c(1-h)$	ch	ch

Thus $$[OH^-] = ch = c\sqrt{\frac{K_h}{c}}$$

or $$[OH^-] = \sqrt{K_h \cdot c} = \sqrt{\frac{K_w}{K_a}\cdot c}$$

$$\therefore \quad [OH^-] = \sqrt{\frac{10^{-14}}{1.8\times10^{-5}}\times 0.1} = \sqrt{55.55\times10^{-12}} = 7.45\times10^{-6}$$

$$\therefore \quad [H^+] = \frac{K_w}{[OH^-]} = \frac{10^{-14}}{7.45\times10^{-6}}$$

$$= 1.342 \times 10^{-9}$$

$$\therefore \quad pH = -\log[H^+] = -\log(1.342\times10^{-9})$$

$$= \mathbf{8.8723}$$

Alternatively, using the formula directly, we have

$$pH = -\frac{1}{2}[\log K_w + \log K_a - \log c]$$

$$= -\frac{1}{2}[\log 10^{-14} + \log(1.8\times10^{-5}) - \log 10^{-1}]$$

$$= -\frac{1}{2}[-14 - 5 + 0.2553 + 1]$$

$$= \mathbf{8.8723}$$

Q. 80. Ionization constant of acetic acid and ionic product of water at 25°C are 1.75×10^{-5} and 1×10^{-14} respectively. Calculate the hydrolysis constant of sodium acetate and its degree of hydrolysis in 0.1 molar solution at 25°C. *(Kumaon 2000)*

Ans. Since sodium acetate is a salt of weak acid and strong base, therefore

$$K_h = \frac{K_w}{K_a}$$

Here we are given :

$$K_w = 10^{-14} \text{ and } K_a = 1.75 \times 10^{-5}$$

$$\therefore \quad K_h = \frac{10^{-14}}{1.75\times10^{-5}} = \mathbf{5.7\times10^{-10}}$$

Further, for a salt of weak acid and strong base,

$$h = \sqrt{\frac{K_h}{c}}$$

where c is the molar concentration *i.e.* number of moles/litre

$$\therefore \quad h = \sqrt{\frac{5.7\times10^{-10}}{0.1}} = 7.55\times10^{-5}$$

Q. 81. At 25°C, the ionisation constant of anilinium hydroxide is 4.6 × 10^{-10}. Taking ionic product of water as 1 × 10^{-14}, calculate (*a*) the hydrolysis constant of anilinium chloride (*b*) the degree of hydrolysis and (*c*) pH value in 0.02 molar solution of the salt.

Ans. (*a*) *Calculation of hydrolysis constant.* As anilinium chloride $C_6H_5NH_3Cl$ is salt of weak base (aniline) and strong acid (HCl) hydrolysis constant is given by

$$K_h = \frac{K_w}{K_b}$$

Here we are given

$$K_w = 10^{-14} \text{ and } K_b = 4.6\times 10^{-10}$$

$$\therefore \quad K_h = \frac{10^{-14}}{4.6\times10^{-10}} = \mathbf{2.17\times10^{-5}}$$

(*b*)*Calculation of degree of hydrolysis.* For such a salt

$$h = \sqrt{\frac{K_h}{c}} = \sqrt{\frac{2.17\times10^{-5}}{0.02}} = \mathbf{3.29\times10^{-2}}$$

(*c*)*Calculation of pH.* For such a salt

$$pH = -\frac{1}{2}[\log K_w - \log K_b + \log c]$$

$$= -\frac{1}{2}[\log10^{-14} - \log(4.6\times10^{-10}) + \log 2\times10^{-2}]$$

$$= -\frac{1}{2}[-14 - (0.6628 - 10) + (0.3010 - 2)]$$

$$= \mathbf{3.18}$$

Q. 82. If the dissociation constant of HCN be 7.2 × 10^{-10} and that of ammonia be 1.75 × 10^{-5}, find out the hydrolysis constant for ammonium cyanide and also its degree of hydrolysis.

Ans. For a salt of weak acid and weak base

$$K_h = \frac{K_w}{K_a \cdot K_b}$$

In the present case $K_a = 7.2\times10^{-10}$

$K_b = 1.75\times10^{-5}$ and $K_w = 1\times10^{-14}$

Hence $$K_h = \frac{1\times10^{-14}}{7.2\times10^{-10}\times1.75\times10^{-5}}$$

or $$K_h = \frac{10}{7.2 \times 1.75} = 0.7937$$

Degree of hydrolysis $$h = \sqrt{\frac{K_w}{K_a \cdot K_b}}$$

or $$h = \sqrt{K_h}$$

or $$h = \sqrt{0.7937} = 0.8909$$

Q. 83. What is electrolysis ? How can we explain this phenomenon on the basis of ionization theory ?

Ans. *Electrolysis* may be defined as a process of decomposition of an electrolyte by passage of electricity through its aqueous solution or molten state.

The apparatus used to carry out electrolysis is called electrolytic cell (Fig. 10.17). It consists of a glass vessel in which the electrolyte in form of an aqueous solution or in the molten state is taken. Two metal rods are dipped into it and are connected to a source of electricity, *i.e.*, battery. These rods are called electrodes; the one connected to the negative pole of the battery is called *cathode* and the other connected to the positive pole of the battery is called *anode.*

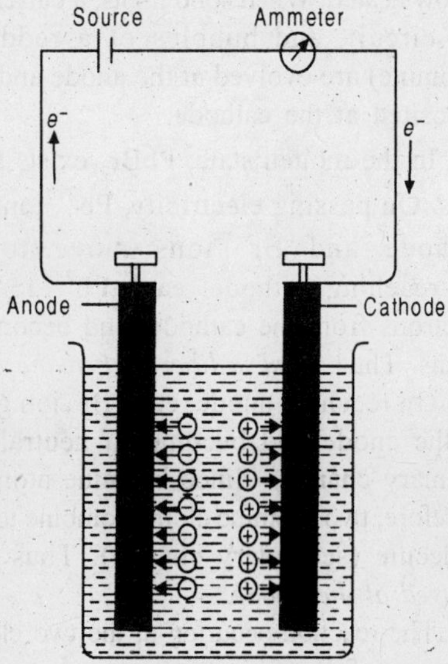

Fig. 10.17. Electrolytic cell.

The process of electrolysis can be easily explained on the basis of *ionization theory.* According to this theory, whenever an electrolyte (acid, base or salt) is dissolved in water or is taken in the molten state, the electrolyte dissociates to produce positively and negatively charged ions. The positively charged ions are called *cations* while the negatively charged ions are called *anions.* On passing electric current, these ions move towards the oppositely charged electrodes. Thus, *cations move towards the cathode while anions move towards the anode.* On reaching their respective electrodes, ions lose their charges and become neutral species. Cations accept electrons from the cathode to become neutral species and the anions give electrons to the anode to become neutral species. Thus, *oxidation occurs at the anode while reduction takes place at the cathode. The conversion of ions into neutral species at their respective electrodes is called the* **primary change.** The product formed as a result of primary change may be collected as such or it may undergo a **secondary change** to give the final products.

When the electrolysis of an electrolyte is carried out in the molten state, the products of electrolysis are actually the substances obtained from the ions of the electrolyte. But the situation is not so simple when the electrolysis of an electrolyte is carried out in the aqueous solution. This is due to reason that in the aqueous solution, besides the ions of the electrolyte which are present in large amounts, there are also some H^+ and OH^- ions furnished by the slight ionization of water. Consequently, when electric current is passed, both the cations of the electrolyte and H^+ ions from water would move towards the cathode, while the anions of the electrolyte and OH^- ions from water would move towards the anode. In other words, at each electrode, there are two different ions which can be discharged. The probability of the ion actually getting discharged depends upon :

(*i*) *their relative discharge potentials or their relative positions in the electrochemical series* (*Refer to electrochemical cells in the following sections*).
(*ii*) *the material of the electrode used.*

Usually ions with lower discharge potentials or which occur low in the electrochemical series are discharged in preference to those which either have high discharge potentials or occur high in the electrochemical series.

Q. 84. Discuss the electrolysis of molten lead bromide.

Ans. Solid lead bromide is taken in a silica vessel and two graphite electrodes are introduced into it. These electrodes are connected to a D.C. battery source through an ammeter as shown in Fig. 10.18. The vessel is now heated. When solid melts, a current flows through the circuit, and bubbles of a reddish brown gas (bromine) are evolved at the anode and metallic lead is deposited at the cathode.

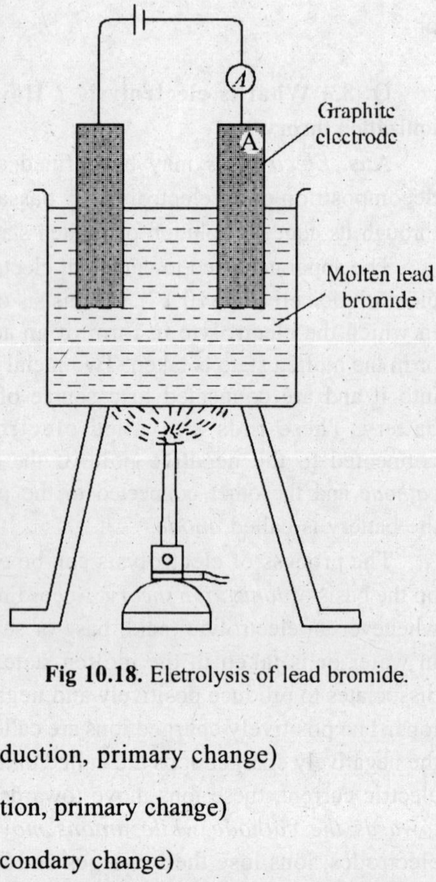

Fig 10.18. Eletrolysis of lead bromide.

In the molten state, $PbBr_2$ exists as Pb^{2+} and Br^- ions. On passing electricity, Pb^{2+} ions move towards cathode and Br^- ions move towards anode. On reaching cathode, each Pb^{2+} ion picks up two electrons from the cathode, and becomes neutral lead atoms. Thus, *lead is liberated at the cathode.*

On reaching anode, each Br^- ion gives an electron to the anode and becomes a neutral bromine atom (primary change). Since bromine atoms are unstable, therefore, two bromine atoms combine to form a bromine molecule (secondary change). Thus *bromine gas is evolved at the anode.*

The reaction occuring at the two electrodes may be written as follows :

At cathode : $Pb^{2+}\ (l) + 2e^- \longrightarrow Pb\ (l)$ (reduction, primary change)

At anode : $Br^-\ (l) - e^- \longrightarrow \cdot Br$ (oxidation, primary change)

$\cdot Br + \cdot Br \longrightarrow Br_2\ (g)$ (secondary change)

Overall reaction :

$$Pb^{2+}\ (l) + 2Br^- \xrightarrow{\text{electrolysis}} Pb\ (l) + Br_2\ (g)$$

Q. 85. Discuss the electrolysis of molten sodium chloride.

Ans. Experimental setup for this process is the same as for electrolysis of lead bromide except that sodium chloride is taken in place of lead bromide. Molten sodium chloride contains Na^+ and Cl^- ions.

$$NaCl\ (l) \rightleftharpoons Na^+(l) + Cl^-\ (l)$$

On passing electricity, Na^+ ions move towards cathode while Cl^- ions move towards anode.

On reaching the anode, each Cl^- ions gives an electron to the anode and becomes a neutral chlorine atom. Since chlorine atoms are unstable, these two atoms of chlorine combine to form a chlorine molecule. Thus, *Cl_2 is evolved at the anode.*

The reactions occuring at the two electrodes may be written as follows :

At cathode : $Na^+ + e^- \longrightarrow Na$ (reduction, primary change)

At anode :
$Cl^- - e^- \longrightarrow \cdot Cl$ (oxidation, primary change)
$\cdot Cl + \cdot Cl \longrightarrow Cl_2 (g)$ (secondary change)

Overall reaction :

$$2\,Na^+ (l) + 2\,Cl^- \xrightarrow{\text{electrolysis}} 2\,Na (l) + Cl_2(g)$$

Q. 86. Discuss the electrolysis of water.

Ans. Water is taken in the electrolytic cell as shown in Fig. 10.19. Two graphite electrodes are introduced into it. These electrodes are connected via an ammeter to a battery in series. It is observed that ammeter does not show any deflection. This is due to the reason that water is a *bad conductor of electricity.* On adding a few drops of dil H_2SO_4 to water, the ammeter shows a deflection and electrolysis begins. As a result of electrolysis, oxygen is produced at the anode and hydrogen at the cathode. Further, the volume of hydrogen evolved is found to be double the volume of oxygen.

Water is only weakly ionized but the presence of an acid increases the degree of ionization of water. Thus, in solution, we have three ions, *viz.*, H^+, OH^- and SO_4^{2-}

$$H_2O (l) \rightleftharpoons H^+ (aq) + OH^- (aq)$$

$$H_2SO_4 (aq) \longrightarrow 2\,H^+ (aq) + SO_4^{2-} (aq)$$

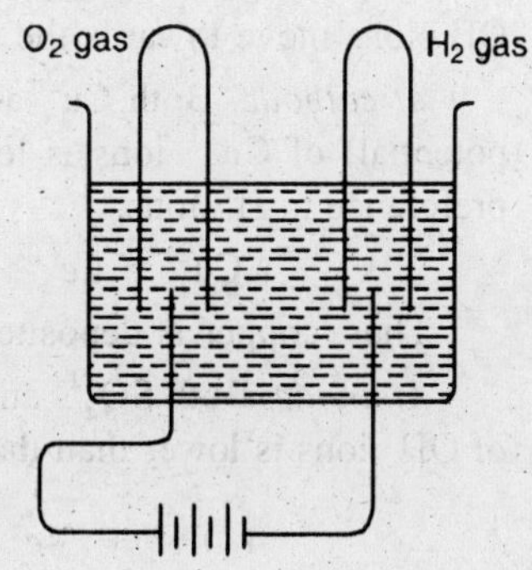

Fig. 10.19 Electrolysis of water.

When electric current is passed through acidulated water, OH^- and SO_4^{2-} ions, move towards anode. Since the discharged potential of OH^- ions is much lower than that of SO_4^{2-} ions, OH^- ions are discharged at anode while SO_4^{2-} ions remain in the solution, H^+ ions are discharged at the cathode. The reactions, occuring at the two electrodes may be written as :

At cathode : $H^+ (aq) + e^- \longrightarrow H\bullet$ (reduction, primary change)

$H\bullet + \bullet H \longrightarrow H_2 (g)$ (secondary change)

At anode : $OH^- (aq) - e^- \longrightarrow OH$ (oxidation, primary change)

$4\,(OH) \longrightarrow 2\,H_2O + O_2$ (secondary change)

Overall reaction :

$4\,H^+ (aq) + 4\,OH^- (aq) \longrightarrow 2\,H_2 (aq) + 2\,H_2O (l) + O_2(g)$

Q. 87. Discuss the electrolysis of an aqueous solution of sodium chloride.

Ans. Sodium chloride and water ionize as under :

$NaCl(aq) \longrightarrow Na^+ (aq) + Cl^- (aq)$ (almost completely ionized)

$H_2O (l) \rightleftharpoons H^+ (aq) + OH^-(aq)$ (only slightly ionized)

On passing electricity, Na^+ *(aq)* and H^+ *(aq)* move towards the cathode while Cl^- *(aq)* and OH^- *(aq)* ions move towards the anode.

At cathode. Both Na^+ *(aq)* and H^+ *(aq)* are present near the cathode. Since the discharge potential of H^+ ions is lower than that of Na^+ ions, H^+ ions are discharged in preference to Na^+ ions.

$$H^+ + e^- \longrightarrow H \quad \text{(reduction, primary change)}$$

$$H\cdot + \cdot H \longrightarrow H_2\,(g) \quad \text{(secondary change)}$$

Thus, H_2 gas is evolved at the cathode while Na^+ ions remain in the solution.

At anode. Both Cl^- and OH^- ions are present near the anode. Since the discharge potential of Cl^- ions is lower than that of OH^- ions, Cl^- ions are discharged in preference to OH^- ions.

$$Cl^- - e^- \longrightarrow \bullet Cl \quad \text{(oxidation, primary change)}$$

$$\bullet Cl + \bullet Cl \longrightarrow Cl_2\,(g) \quad \text{(secondary change)}$$

Thus, Cl_2 gas is evolved at the anode while OH^- ions remain in the solution.

Q. 88. Discuss electrolysis of aqueous copper sulphate solution using inert electrodes.

Ans. Copper sulphate and water ionize as under :

$$CuSO_4\,(aq) \longrightarrow Cu^{2+}\,(aq) \quad \text{(almost completely ionised)}$$
$$H_2O\,(l) \rightleftharpoons H^+\,(aq) + OH^-\,(aq) \quad \text{(only slightly ionized)}$$

On passing electricity, Cu^{2+} *(aq)* and H^+ *(aq)* move towards the cathode while SO_4^{2-} ions and OH^- ions move towards the anode.

At cathode. Both Cu^{2+} and H^+ are present near the cathode. Discharge potential (oxidation potential) of Cu^{2+} ions is lower than that of H^+ ions. Therefore, Cu^{2+} ions are discharged in preference to H^+ ions,

$$Cu^{2+} + 2e^- \longrightarrow Cu \quad \text{(primary change)}$$

Thus, copper is deposited on the platinum electrode.

At anode. Both SO_4^{2-} and OH^- ions are present near the anode. Since the discharge potential of OH^- ions is lower than that of SO_4^{2-} ions, OH^- are discharged in preference to SO_4^{2-} ions.

$$4\,OH^- + 4e^- \longrightarrow 4\,OH \quad \text{(oxidation, primary change)}$$

$$4\,OH^- \longrightarrow 2\,H_2O\,(l) + O_2\,(g) \quad \text{(secondary change)}$$

Q. 89. Describe electrolysis of an aqueous solution of copper sulphate using copper electrodes.

Ans. Copper sulphate and water ionize as under :

$$CuSO_4\,(aq) \longrightarrow Cu^{2+}\,(aq) + SO_4^{2-}\,(aq) \quad \text{(almost completely ionized)}$$

On passing electricity, Cu^{2+}*(aq)* and H^+ *(aq)* move towards the cathode while SO_4^{2-} ions and OH^- ions move towards the anode.

At cathode. Both Cu^{2+} *(aq)* and H^+ ions are present near the cathode. Discharge potential (oxidation potential) of Cu^{2+} ions is lower than that of H^+ ions. Therefore, Cu^{2+} ions are discharged in preference to H^+ ions

$$Cu^{2+} + 2e^- \longrightarrow Cu \quad \text{(Primary change)}$$

At anode. Unlike electrolysis of $CuSO_4$ using platinum electrodes, no ions are liberated here. Instead, anode itself undergoes oxidation (*i.e.*, loses electrons) to form Cu^{2+} ions which go into the solution. This is due to the reason that Cu is more easily oxidised than that OH^- and SO_4^{2-} ions.

$$Cu\,(s) \longrightarrow Cu^{2+}\,(aq) + 2e^- \quad \text{(oxidation, primary change)}$$

In general, if more than one substance are present in the electrolytic cell, the substance, liberated at the cathode is one which has highest reduction potential and that liberated at the anode is one which has the highest oxidation potential.

Q. 90. Describe electrolysis of aqueous solution of copper chloride using platinum (inert) electrodes.

Ans. $CuCl_2 \longrightarrow Cu^{2+} + 2Cl^-$

$H_2O \longrightarrow H^+ + OH^-$

At anode : $Cl^- \longrightarrow Cl + e^-$

$\bullet Cl + \bullet Cl \longrightarrow Cl_2$

At anode : $OH^- \longrightarrow OH + e^-$

$OH^- + OH \longrightarrow H_2O + \frac{1}{2}O_2$

At cathode : $H^+ + e \longrightarrow H$

$H + H \longrightarrow H_2$

At cathode : $Cu^{2+} + 2e \longrightarrow Cu$

When several ions are completing for transfer of electrons, the actual product depends upon the energy required to discharge the ion called the *discharge potential* or *deposition potential.* In the above case, out of the OH^- and Cl^-, the latter has the lower discharge potential, hence it is preferentially discharged as chlorine gas at the anode.

Out of H^+ and Cu^{++} near the cathode, the latter has the lower discharge potential, hence it is deposited as copper metal on the cathode.

We expect similar results if molten copper chloride is subjected to electrolysis.

Q. 91. Discuss electrolysis of copper chloride using copper (reacting) electrodes.

Ans. In this case, copper from the anode material loses electrons and goes into the solution as Cu^{2+}. It is also accompanied by the release of electrons. It happens because it requires greater energy for the chloride ions to get liberated at the anode and release the electrons. However, as usual, deposition of copper metal takes place at the cathode.

At anode. $Cu \longrightarrow Cu^{2+} + 2e$

At cathode. $Cu^{2+} + 2e \longrightarrow Cu$

Q. 92. Discuss electrolysis of an aqueous solution of sulphuric acid.

Ans. Sulphuric acid is ionized as under :

$$H_2SO_4 \longrightarrow 2H^+ + SO_4^{2-}$$

$$H_2O \rightleftharpoons H^+ + OH^-$$

Here hydrogen gas is liberated at the cathode. SO_4^{2-} ions have a greater deposition potential than OH^- ions. Hence, OH^- ions are discharged to give oxygen gas :

At cathode. $H^+ + e \longrightarrow H$

$H + H \longrightarrow H_2$

At anode. $OH^- \longrightarrow OH + e^-$

$OH + OH \longrightarrow OH + e^-$

$OH + OH \longrightarrow H_2O + O_2$

Q. 93. Predict the products of electrolysis of each of the following :

(*i*) An aqueous solution of $AgNO_3$ with silver electrodes.

(*ii*) An aqueous solution of $AgNO_3$ with inert electrodes.

(*iii*) A dilute solution of H_2SO_4.

Ans. (*i*) $AgNO_3\ (aq) \rightleftharpoons Ag^+\ (aq) + NO_3^-\ (aq)$

$$H_2O \rightleftharpoons H^+\ (aq) + OH^-\ (aq)$$

(*a*) *At cathode.* $Ag^+\ (aq) + e^- \longrightarrow Ag\ (s)$

Ag gets deposited due to lower discharge potential or higher reduction potential as compared to H^+.

At anode. NO_3^- : ions move towards the silver anode and react with it to dissolve some $AgNO_3$ into the solution.

$$NO_3^- + Ag \longrightarrow AgNO_3 + e^-$$

(*ii*) $AgNO_3 \longrightarrow Ag^+\ (aq) + NO_3^-\ (aq)$

$$H_2O \longrightarrow H^+\ (aq) + OH^-\ (aq)$$

At cathode. Ag will get deposited because of higher reduction potential than hydrogen.

$$Ag^+\ (aq) + e^- \longrightarrow Ag\ (s)$$

At anode. $H_2O\ (l) \longrightarrow \frac{1}{2} O_2 + 2H^+\ (aq) + 2e^-$

(*iii*) $H_2SO_4 \longrightarrow 2H^+\ (aq) + SO_4^{2-}\ (aq)$

$$H_2O\ (l) \longrightarrow H^+\ (aq) + OH^-\ (aq)$$

At cathode. Hydrogen gas is evolved at the cathode.

$$2H^+ + 2e^- \longrightarrow H_2\ (g)$$

At anode. OH^- will get oxidised in preference to SO_4^{2-}

$$H_2O\ (l) \longrightarrow \frac{1}{2}O_2 + 2H^+\ (aq) + 2e^-$$

Q. 94. Discuss various applications of electrolysis.

Ans. Various applications of electrolysis are discussed as under :

1. *In electro-refining of metals.* Here electrolysis is used to finally purify metals upto 100%. Here, we take a tank made of glass or stone. The impure metal block is made the anode and connected to positive electricity. A thin wire or sheet made of the same metal in pure form is made the cathode. An electrolytic solution containing some salt of the same metal is taken in the tank. Electricity is passed. Pure metal is deposited on the cathode. The anions in the solution move towards the anode, dissolve the metal out into the solution. Thus, metal is progressively transferred from the anode to the cathode. The impurities present in the anode settle down as *anode mud* and are removed, see Fig.10.20.

Similar reactions, as given above under the heading of electroplating take place.

2. *Electroplating.* It is the process of (cathode) depositing a metal over another metal. This is done, in order to protect a metal which is easily rusted or corroded, with the help of a metal which is resistant to oxidation or corrosion. Another utility is to deposit a thin layer of gold or silver an objects made of iron or copper.

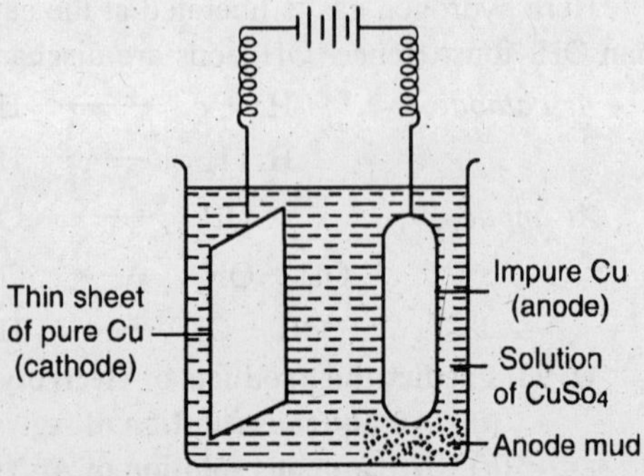

Fig. 10.20 Electro-refining of copper.

The process of electroplating is carried out in a electrolytic tank made of glass or stone. The articles to be electroplated on suspended from the cathode and connected to negative pole of the battery as shown in Fig. 10.21.

The anode consists of a metal which is to be deposited. A suitable solution of a salt containing the same metal is taken in the tank. Low temperature is maintained during the maintained for smooth deposition of the metal on the objects to be electroplated. The electrolyte chosen should be a good conductor of electricity, have high solubility in water, be cheap and stable towards oxidation- reduction and hydrolysis.

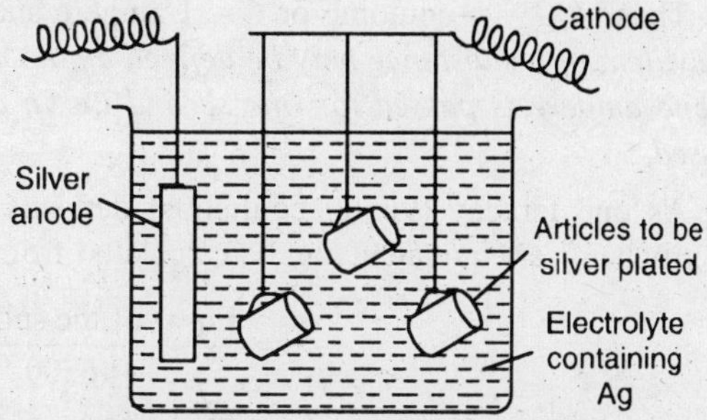

Fig. 10.22 Electroplating Process.

For copper plating, a mixture of H_2SO_4 and $CuSO_4 \cdot 5H_2O$ is taken whereas for silverplating, potassium argento-cyanide, $K[Ag(CN)_2]$ is taken. $K[Au(CN)_2]$ is taken if goldplating is to be done and nickel ammonium sulphate $NiSO_4 (NH_4)_2 SO_4 \cdot 6H_2O$ is used for depositing nickel.

On passing electricity, the metal ion in the solution moves towards the cathode and is deposited on the object. Let us say, we are depositing Cu on the object. The following reactions take place :

$$CuSO_4 \rightleftharpoons Cu^{2+} + SO_4^{2-} \text{ (Ionisation)}$$

$$Cu^{2+} + 2e \longrightarrow Cu \text{ (At cathode)}$$

$$SO_4^{2-} + Cu \longrightarrow CuSO_4 + 2e \text{ (At anode)}$$

Sulphate ions move towards the copper anode and react with the electrode material. Copper sulphate formed comes into the solution and maintains the concentration of the solution.

3. *Electrotyping.* It is a process by which we can reproduce the works of arts such as wood carving or stone sculpture work.

An impression of the original work is taken on a plate coated with wax. Graphite powder is sprinkled on the grooves thus obtained. This plate is made the cathode. Copper is deposited on the graphite powder. After some time an extra replica of the original work is obtained.

4. *Determination of Equivalent Weights.* When the same amount of electricity is passed through different electrolytes, the amounts of substances deposited are proportional to their equivalent weights. This can help us to determine the eq. wt. of an unknown metal.

5. *Chemical Industries.* The principle of electrolysis finds application in a number of chemical process. For example, sodium hydroxide is prepared by the electrolysis of a solution of sodium chloride. Hydrogen and oxygen gases are obtained by the electrolysis of water containing some acid or alkali.

Q. 95. Discuss Faraday's Laws of electrolysis.

Ans. 1. *Faraday's First Law of Electrolysis.* *It states that "the mass of any substance deposited or liberated at any electrode is directly proportional to the quantity of electricity passed". Thus, if W g of the substance is deposited on passing Q coulombs of electricity, then*

$$W \propto Q$$

or $$W = ZQ$$

where Z is a constant of proportionality and is called *electrochemical equivalent* of the substance deposited. If a current of C amperes is passed for t seconds, then

$$Q = C \times t$$

so that $$W = Z \times C \times t$$

Thus, if $Q = 1$ coulomb or $C = 1$ ampere and $t = 1$ second, $W = Z$. Hence, *electrochemical equivalent of a substance may be defined as the mass of the substance deposited when a current of one ampere is passed for one second i.e., a quantity of electricity equal to one coulomb is passed.*

As one faraday (96500 coulumbs) deposits one gram equivalent of the substance, hence electrochemical equivalent can be calculated from the equivalent weight, *i.e.*,

$$Z = \frac{\text{Eq.wt.of the substance}}{96500}$$

2. *Faraday's Second Law of Electrolysis.* *It states that "when the same quantity of electricity is passed through solutions of different electrolytes connected in series, the weights of the substances produced at the electrolydes are directly proportional to their equivalent weights. For example, for $CuSO_4$ solution and $AgNO_3$ solution connected in series, if the same quantity of electricity is passed then*

$$\frac{\text{Weight of Cu deposited}}{\text{Weight of Ag deposited}} = \frac{\text{Eq.weight of Cu}}{\text{Eq.weight of Ag}}$$

Q. 96. What is the utility of Faraday's Laws ?

Ans. 1. Using Faraday's First Law, it is possible to calculate the amount deposited of a substance on passing a certain amount of electricity for a certain time if the electrochemical equivalent is known to us. Alternatively if the amount deposited is determined, it is possible to calculate the electrochemical equivalent of a substance. Knowing both the amount deposited and the electrochemical equivalent, we can proceed to calculate, the quantity of electricity passed.

2. The quantity of electricity passed through a circuit (where these is no deposition), can be determined using Faraday's Second Law. The circuit is connected in series with a silver voltameter or coulometer having a platinum anode (Fig. 10.22).

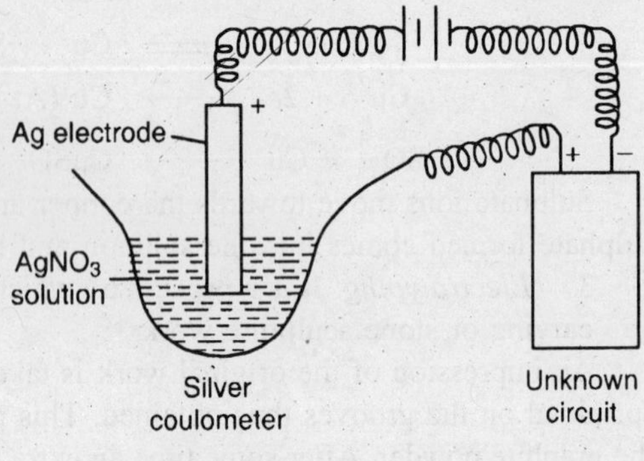

Fig. 10.22 Use of Silver Coulometer.

A solution of silver nitrate is taken in the coulometer. It is connected in series to the unknown circuit. Silver is deposited on the silver cathode. After a certain time, the process of passing electricity is stopped, the silver coulometer (cathode) is washed and weighted. Knowing the initial weight of the coulometer, we can obtain the weight of silver deposited. The quantity of electricity is calculated by using the relation $W = Z, Q$. Since the circuit is connected in series with the silver coulometer, same quantity of electricity passes through the circuit also.

Q. 97. How many coulombs are required for the following reductions :

(*i*) 1 mol of Al^{3+} to Al

(*ii*) 1 mol of Cu^{2+} to Cu

(*iii*) 1 mol of MnO^-_4 to Mn^{2+}

Ans. (*i*) $Al^{3+} + 3e^- \longrightarrow Al$

Charge required = 3 faraday

= 3 × 96500 coulumbs

= 289500 coulumbs

(*ii*) $Cu^{2+} + 2e^- \longrightarrow Cu$

Charge required = 2 Faraday

= 2 × 96500 coulombs

= 193000 coulombs

(*iii*) $MnO_4^- + 8H^+ + 5e^- \longrightarrow Mn^{2+} + 4H_2O$

Charge required = 5 faraday

= 5 × 96500 coulombs

= 482500 coulombs

Q. 98. How many coulombs are required to produce ?

(*i*) 20 g of calcium from molten $CaCl_2$.

(*ii*) 50 g of aluminium from molten Al_2O_3.

Ans. (*i*) $Ca^{2+} + 2e^- \longrightarrow Ca\ (s)$

1 mol or 40 g

40 g calcium require = 2 × 96500 coulombs

$$20 \text{ g calcium require} = \frac{2 \times 96500}{27} \times 20$$

= 96500 coulombs

(*ii*) $Al^{3+} + 3e^- \longrightarrow Al\ (s)$

1 mol or 27g

27 g aluminium require = 3 × 96500 coulombs

$$50 \text{ g aluminium require} = \frac{2 \times 96500}{27} \times 50$$

= 536111 (approx.) coulombs.

Q. 99. How many coulombs are required for the following oxidation ?

(*i*) 1 mol of H_2O to O_2

(*ii*) 1 mol of FeO to Fe_2O_3.

Ans. (*i*) $2H_2O \longrightarrow O_2 + 2e^-$

$$H_2O \longrightarrow \frac{1}{2}O_2 + 2H^+ + 2e^- \quad \text{(For 1 mol of } H_2O)$$

Hence, charge = 2 faraday

= 2 × 96500 coulombs

= 193000 coulombs

(*ii*) $Fe^{2+} \longrightarrow Fe^{3+} + e^-$

Hence, charge = 1 faraday

= 96500 coulombs

Q. 100. Three electrolytic cells A, B and C containing electrolytes zinc sulphate, silver nitrate and copper sulphate respectively were connected in series. A steady current of 1.50 ampere was passed through them until 1.45 g of Ag were deposited at the cathode of cell B. How long did the current flow ? What weight of copper and of zinc were deposited ?

Ans. $Ag^+ + e^- \longrightarrow Ag$ (1 mol or 108 g)

Thus, 108 g Ag is deposited by = 96500 coulombs

$$1.45 \text{ g Ag is deposited by} = \frac{96500}{108} \times 1.45 \text{ coulombs}$$

$$= 1295.6 \text{ coulombs}$$

$$\text{Charge} = \text{Current} \times \text{Time in seconds}$$

$$\text{Time} = \frac{\text{Charge}}{\text{Current}}$$

$$\text{Time} = 863.73 \text{ seconds}$$

(*ii*) $Zn^{2+} + 2e^- \longrightarrow Zn\,(s)$ (1 mol or 65 g)

2 × 96500 coulombs deposit = 65 g Zn

$$1295.6 \text{ coulombs deposit} = \frac{65 \times 1295.6}{2 \times 96500}$$

$$= 0.436 \text{ g Zn}$$

Again $Cu^{2+} + 2e \longrightarrow Cu\,(s)$ (1 mol or 63.5 g)

2 × 96500 coulombs deposit = 63.5 g

$$1295.6 \text{ coulombs deposit} = \frac{63.5 \times 1295.6}{2 \times 96500}$$

$$= 0.426 \text{ g Copper}$$

Q. 101. A solution of Ni $(NO_3)_2$ is electrolysed between platinum electrodes using current of 5.0 ampere for 30 minutes. What weight of Ni will be produced at the cathode ?

Ans. The reaction is

$$Ni^{2+} + 2e^- \longrightarrow Ni \text{ (1 mol or 58.7 g)}$$

Thus, 2 faraday or 2 × 96500 coulombs produce 58.7 g Ni.

Total charge produced Q = Current × Time (seconds)

= 5 × 30 × 60

= 9000 coulombs

2 × 96500 coulombs of charge produce = 58.7 g Ni

$$9000 \text{ coulombs of charge produce} = \frac{58.7 \times 9000}{2 \times 96500} \text{ g Ni}$$

$$= 2.737 \text{ g Ni}$$

Q. 102. A current of 4.5 ampere was passed through silver nitrate solution in a silver voltameter for 20 minutes. It deposited 6.0 g of silver on the cathode. What is the electrochemical equivalent of silver ?

Ans. I = 4.5 ampere

t = 20 minutes or 1200 seconds

Applying the relation

$$W = Z \times I \times t$$

Substituting the values

$$6 = Z \times 4.5 \times 1200$$

$$Z = \frac{6}{4.45 \times 1200} = 0.00111$$

Q. 103. A current of 3 ampere is passed for 20 minutes through a voltameter containing a solution of sodium chloride. What is the amount of chlorine produced at the anode ?

Ans. $I = 3$ ampere

$t = 20$ minutes or 1200 seconds

$Q = Z \times t$

$3 = 3 \times 1200$

$= 3600$ coulombs

We know that

96500 coulombs of electricity deposit = 1 g eq. of chlorine

= 35.5 g of chlorine

Hence, 3,600 coulombs of electricity deposit

$$= \frac{35.5}{96500} \times 3600 g$$

$$= 1.32 \text{ g}$$

Q. 104. A current of 1 ampere is passed through a solution of copper sulphate for 90 seconds using platinum electrodes at 298 K and 740 mm pressure. Calculate the amount of copper deposited at the cathode and the volume of oxygen evolved.

Ans. $I = 1$ ampere

$t = 90$ seconds

$Q = I \times t$

$= 1 \times 90$

$= 90$ coulombs

96500 coulombs deposit = 31.75 g copper

$$90 \text{ coulombs deposit} = \frac{31.75 \times 90}{96500} \text{ g copper}$$

$$= 0.0296 \text{ g copper}$$

Again 96500 coulomb evolve = 35.5 g chlorine

$$90 \text{ coulomb evolve} = \frac{35.5 \times 90}{96500} \text{ g chlorine}$$

$$= 0.031 \text{ g}$$

We know that

71 g chlorine occupy = 22400 ml at N.T.P.

$$0.0331 \text{ g chlorine occupy} = \frac{22400}{71} \times 0.0331 \text{ ml}$$

$$= 10.44 \text{ ml}$$

We require the volume at 298 K and 740 mm pressure. Applying the gas equation and substituting the values

$$\frac{P_1V_1}{T_1} = \frac{P_2V_2}{T_2}$$

$P_1 = 740$ mm		$P_2 = 760$ mm
$T_1 = 298$ K		$T_2 = 273$ K
$V_1 = ?$		$V_2 = 10.44$ ml

$$\frac{V_1 \times 740}{298} = \frac{760 \times 10.44}{273}$$

$$V_1 = \frac{760 \times 10.44}{273} \times \frac{298}{740} = 11.7\text{ml}$$

Q. 105. An electric current was passed through two voltameters connected in series and containing solutions of copper sulphate and silver nitrate using inert electrodes. 0.215 g of copper was deposited in the copper sulphate voltameter. Calculate the amount of silver deposited in the other voltameter. Eq. weights of copper and silver are 31.8 and 108 respectively.

Ans. Applying Faraday's Second Law

$$\frac{\text{Wt. of silver deposited}}{\text{Wt. of copper deposited}} = \frac{\text{Wt. wt. of silver}}{\text{Eq. wt. of copper}}$$

Substituting the values in the above relation.

$$\frac{\text{Wt. of silver deposited}}{0.215} = \frac{108}{31.8}$$

$$\text{Wt. of silver deposited} = \frac{108}{31.8} \times 0.215 \text{ g} = 0.73 \text{ g}$$

Q. 106. A current of 0.5 ampere was passed through a voltameter for 30 minutes. If resulted in the deposition of 4.0 g of the metal on the cathode. Calculate the equivalent weight of the metal.

Ans. $I = 0.5$ ampere

$t = 30$ minutes or 1800 sec.

$$Q = I \times t$$

$$= 0.5 \times 1800 = 900 \text{ coulombs}$$

900 coulombs of electricity deposit = 0.4 g metal

$$96500 \text{ coulombs of electricity deposit} = \frac{0.4}{900} \times 96500 = 42.9$$

Hence, eq. wt. of the metal = 42.9

Q. 107. A current of 5 amperes flowing for 30 minutes deposits 3.048 g of Zn at the cathode. Calculate the equivalent weight of Zn.

Ans. Current = 5 ampere, Time = 30 minutes

$= 30 \times 60$ seconds = 1800 seconds

$$Q = I \times t$$

$$Q = 5 \times 1800 = 9000 \text{ coulombs}$$

9000 coulombs deposit = 3.048g

$$96500 \text{ coulombs deposit} = \frac{3.048}{9000} \times 96500$$

$$= 32.7 \text{ g}$$

Q. 108. How many grams of chlorine can be produced by the electrolysis of molten NaCl by a current of 20 amperes for 3 minutes.

Ans. Current = 20 ampere, Time = 3 minutes = 180 seconds

Quantity of electricity, $Q = I \times t$

$$= 20 \times 180 = 3600 \text{ coulombs}$$

96500 coulombs of electricity deposit = 35.5 g chlorine

$$3600 \text{ coulombs of electricity deposit} = \frac{35.5 \times 3600}{96500} = 1.33\text{g}$$

MISCELLANEOUS QUESTIONS

Q. 109. Which indicators woruld be suitable in the titration of .

(i) HCl *vs.* NaOH

(ii) CH_3COOH *vs.* NaOH

(iii) NH_4 OH *vs* HCl

(*Delhi 2006*)

Ans. A suitable indicator for a given titration is one which has as narrow a pH range as possible lying entirely on the upright (vertical) part of the titration curve. Every indicator has its working range of pH.

(i) HCl vs NaOH. The vertical portion of this titration curve extends from pH 3.2 to pH 10

Both methyl orange (3.1 – 4.4 pH) and phenolphthalein (8.3 – 10 pH) are suitable for this titration.

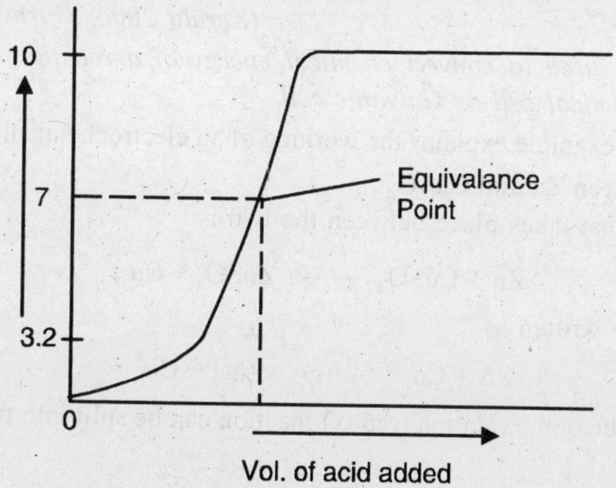

(ii) CH_3COOH vs NaOH. The vertical portion of this titration curve extends from pH 6 to pH 11.

So phenolphthalein (pH 8.3 – 10.0) is suitable

(iii) NH_4OH vs HCl. The vertical portion lies between pH 3 to 8. So both methyl orange (3.1 – 4.4) and methyl red (4.4 – 6.0) are suitable.

Q. 110. What is buffer index? How is it different from buffer capacity? (*Himachal 2010*)

Ans. Buffer index may be defined as the differential ratio of the increase in the amount of strong acid or strong base added, to the variation in pH.

Thus, Mathematically it may be written as

$$\text{Buffer index} = \frac{\text{Amount of strong acid or base added}}{\text{change in pH}}$$

The buffer capacity is numerically equal to the minimum concentration of strong acid or strong base which causes the variation of buffer's pH by one unit.

On the face of it, the two quantities appear to be equivalent. However, there is a difference in the field of application of the two notions. Buffer capacity is used in quantitative chemical analysis and the buffer index is used in studying biological systems.

11

ELECTROCHEMICAL CELLS

Q. 1. What is an electrochemical cell or Galvanic cell? Explain its working.

(Kerala 2000; Garhwal 2000; Punjabi 2003)

Ans. *A device used to convert chemical energy of a reaction into electrical energy is called an electrochemical cell or Galvanic cell.*

The following example explains the working of an electrochemical cell.

Reaction between Zn and $CuSO_4$

The reaction that takes place between the two is

$$Zn + CuSO_4 \longrightarrow ZnSO_4 + Cu\downarrow$$

It may also be written as

$$Zn + Cu^{2+} \longrightarrow Zn^{2+} + Cu$$

The above reduction oxidation (redox) reaction can be split into two half-reactions as under

$$Zn \longrightarrow Zn^{2+} + 2e^- \qquad \textit{(Oxidation half reaction)}$$

$$Cu^{2+} + 2e^- \longrightarrow Cu \qquad \textit{(Reduction half reaction)}$$

The first reaction is called *oxidation half reaction* whereas the second reaction is called *reduction half reaction.* The addition of the two half reactions gives the *overall reaction.*

Thus if a redox reaction is allowed to take place in such a way that oxidation half reaction takes place in one beaker and the reduction half reaction in the other breaker, the electrons given out by the former will be taken up by the latter and a current will flow.

A zinc rod is placed in zinc sulphate solution, taken in a beaker. A copper rod is placed in copper sulphate solution taken in another beaker. The two rods are connected by a wire and the two solutions are connected by a *salt-bridge* (Fig. 11.1). A salt-bridge is a U-shaped tube containing concentrated solution of an electrolyte like

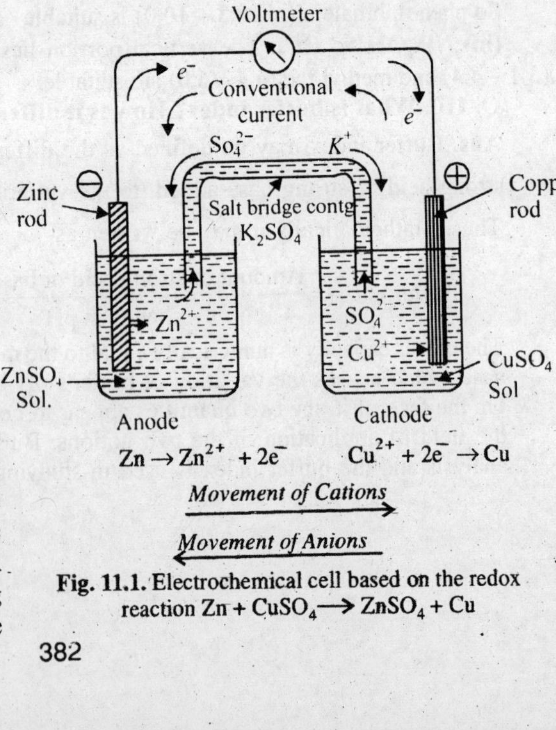

Fig. 11.1. Electrochemical cell based on the redox reaction $Zn + CuSO_4 \longrightarrow ZnSO_4 + Cu$

KCl or K_2SO_4. The function of the salt-bridge is to allow the movement of ions from one solution to the other, without mixing of the two solutions. Thus, whereas electrons flow in the outer circuit in the wire, the inner circuit is completed by the flow of ions from one solution to the other, through the salt-bridge.

Some important features of the above cell are as follows:

(*i*) The zinc electrode at which *oxidation* takes place is called the *anode*. The copper electrode at which the *reduction takes place* is called the *cathode*.

(*ii*) Since electrons are produced at the zinc electrode, it is designated as the *negative pole*. The other electrode *i.e.*, the copper electrode acts as the *positive pole*.

(*iii*) The electrons flow from the negative pole to the positive pole in the external circuit. However conventionally this current is said to flow in the opposite direction.

(*iv*) The oxidation of zinc into zinc ions produces excess of Zn^{2+} ions in the left beaker. Similarly, the reduction of copper ions to copper atoms leaves the excess SO_4^{2-} ions in the solution in the right beaker. To maintain electrical neutrality of the solutions in the two beakers, the cations and anions move through the salt bridge. This helps to complete the inner circuit.

(*v*) As copper from copper sulphate solution is deposited on the copper electrode and sulphate ions migrate to the other side, the concentration of copper sulphate solution decreases as the cell operates. Consequently the current falls with the passage of time.

(*vi*) Evidently, the weight of copper rod will increase while that of zinc rod will decrease as the cell works.

Q. 2. Draw a labelled diagram and explain the working of Ag – Cu cell.

Ans. The cell reactions that take place are represented as under.

$$Cu + 2AgNO_3 \longrightarrow Cu(NO_3)_2 \quad + \quad 2Ag$$

or it may be written as

$$Cu + 2Ag^+ \longrightarrow Cu^{2+} + 2Ag$$

The two half reactions are

$$Cu \longrightarrow Cu^{2+} + 2e^-$$

(*Oxidation half reaction*)

$$2Ag^+ + 2e^- \longrightarrow Ag$$

(*Reduction half reaction*)

Hence the cell may be set up as shown in Fig. 11.2. Potassium nitrate may be taken as electrolyte in the salt bridge.

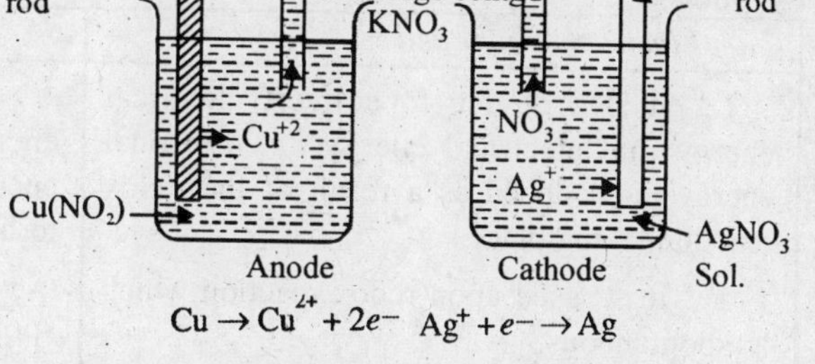

Fig. 11.2. Electrochemical cell based on the reaction $Cu + 2AgNO_3 \rightarrow Cu(NO_3)_2 + 2Ag$

Here oxidation takes place at the copper electrode, therefore it acts as *anode* or *negative pole*. Reduction takes place at the silver electrode, it acts as *cathode* or *positive pole*. The conventional flow of current is from silver to copper.

Q. 3. Explain why KCl is used as electrolyte in salt bridge. ***(Mumbai 2010)***

Ans. KCl is specifically chosen as an electrolyte in salt bridge because the transport numbers of K^+ and Cl^- are almost equal.

Q. 4. What is an electrolytic cell? ***(Bombay 1994)***

Ans. If the aqueous solution of an electrolyte or the molten electrolyte is taken in a vessel and two metal rods are immersed into it and then connected to a battery as shown in Fig. 11.3, the cations (+ve ions) move towards the cathode and anions (–ve ions) towards the anode.

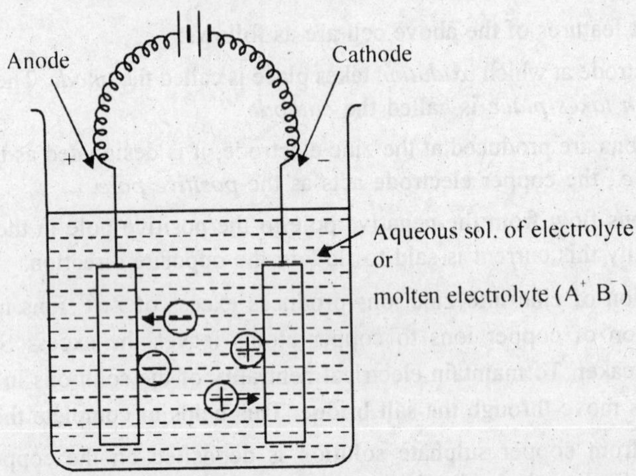

Fig. 11.3. Electrolytic cell.

The anions on reaching the anode, lose electrons. The cations on reaching the cathode gain electrons. Thus

At Anode : $B^- \longrightarrow B + e$ *(Oxidation)*

At Cathode: $A^+ + e^- \longrightarrow A$ *(Reduction)*

Thus a redox reaction takes place on supplying electrical energy resulting into the decomposition of the substance.

An arrangement in which electrical energy is supplied to bring about the redox reaction which was otherwise non-spontaneous i.e. to bring about the decomposition of the electrolyte is called an electrolytic cell.

Q. 5. Give points of difference between an electrochemical cell and an electrolytic cell. ***(G.Nanakdev 2004)***

Ans. The main points of difference between an electrochemical cell and electrolytic cell are as follows.

Electrochemical cell	*Electrolytic cell*
1. It is a device to convert chemical energy into electrical energy *i.e.* electrical energy is produced as a result of the redox reaction taking place.	1. It is a device to convert electrical energy into chemical energy *i.e.* electrical energy is supplied to the electrolytic solution to bring about the redox reaction.
2. It is based upon redox reaction which is spontaneous.	2. The redox reaction is non-spontaneous.
3. The two electrodes may be set up in solutions of the same or different electrolytes.	3. The two electrodes are set up in the same electrolytic solution.
4. The electrode on which oxidation takes place is called the anode and the electrode on which reduction takes place is called the cathode.	4. The electrode which is connected to the negative terminal of the battery is called cathode, the cations migrate to it which gain electrons and hence reduction takes place here. The other electrode is called the anode.
5. A salt bridge or a porous pot is used in this cell.	5. No salt bridge is used in this case.

Q. 6. Give the schematic representation of an electrochemical cell.

(M. Dayanand 2000)

Ans. An electrochemical cell is represented as under for the Daniel cell

$$Zn \mid Zn^{2+} (C_1) \parallel Cu^{2+} (C_2) \mid Cu$$

By convention, the electrode on which oxidation takes place is written on the left hand side and the other electrode on which reduction takes place is written on the right hand side. The electrode on the left hand side is written by writing the symbol of the metal (or the gas) first followed by the symbol of the ion with its concentration in brackets. The electrode on the right hand side is written by first writing the ion alongwith its concentration in brackets followed by the symbol of the metal (or the gas). Single vertical lines represent the phase boundaries of the electrodes and the double line represents the salt bridge. C_1 and C_2 represent the concentration of Zn^{2+} ions and Cu^{2+} ions respectively.

In a similar manner, Cu-$AgNO_3$ cell may be represented as

$$Cu \mid Cu^{2+} (C_1) \parallel Ag^+ (C_2) \mid Ag$$

Q. 7. What is meant by electrode potential and standard electrode potential.

(Purvanchal 2007; Vidyasagar 2007; Calcutta 2007, Andhra 2010)

Ans. *The tendency of an electrode to lose or gain electrons when it is in contact with its own ions in the solution is called its* **electrode potential.**

The electrode potential is further termed as **oxidation potential** if oxidation takes place at the electrode (with respect to normal hydrogen electrode) and is called reduction potential if reduction takes place at the electrode (with respect to normal hydrogen electrode). Further, if in the half cell, the metal rod is suspended in a solution of unit activity and the temperature is kept at 298 K the electrode potential is called **standard electrode potential,** represented usually by $E°$. If a gas is involved, the standard conditions chosen are 1 atmosphere pressure.

Q. 8. What is meant by EMF of a cell? How is it measured?

(Garhwal 2000; Madurai 2004)

Ans. *The force which causes the flow of electrons from one electrode to the other and thus results in the flow of current is called electromotive force, or EMF, and measured in volts.* This force arises because of the different potentials possessed by the two electrodes. Thus *electromotive force may also be defined as the difference of potential which causes the current to flow from an electrode at higher potential to an electrode at a lower potential.* If all the reactants and products are present in their standard state at 25°C, the EMF is called **standard EMF.**

Potentiometric method which is based upon *compensation principle,* is used for the accurate measurement of EMF. In this method, the unknown EMF is opposed by another known EMF until the two are equal. The general principle involved may be explained with the help of Fig. 11.4 given below:

AB is a stretched wire of uniform thickness, *S* is a standard cell whose EMF is known. It is connected to the end *A* and *B*. *X* is a cell whose EMF is to be determined. One end of the cell *X* is connected to the end *A* and the other to the sliding contact *J* through the galvanometer *G* is such a manner that it sends EMF in a direction opposite to that of the cell *S*. The sliding contact is moved along the wire *AB* till at a particular point, say *C*, no deflection is observed in the galvanometer. If E_s represents

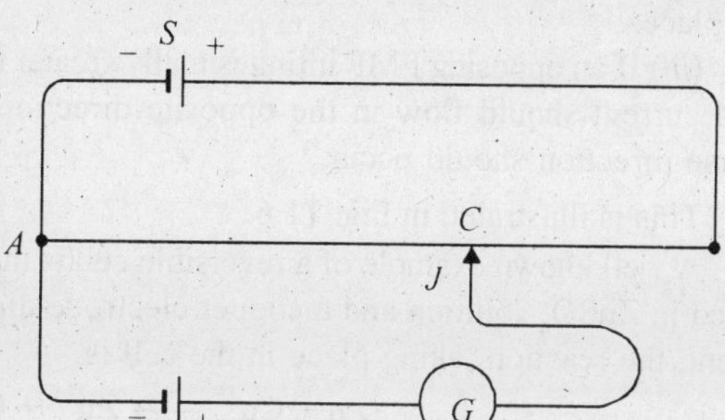

Fig. 11.4. Compensation method for measuring EMF.

the EMF of the standard cell S and E_x that of the experimental cell X, then

$$\frac{E_x}{E_s} = \frac{\text{length } AC}{\text{length } AB}$$

Or
$$E_x = \frac{\text{length } AC}{\text{length } AB} \times E_s$$

Thus knowing E_s, E_x can be calculated.

Q. 9. Write a note on Weston standard cell. *(Bangalore 2004)*

Ans. It consists of an H-shaped glass vessel (Fig. 11.5). One of the limbs contains mercury, covered with a paste of mercurous sulphate and mercury, crystals of solid $CdSO_4.8/3H_2O$ and saturated solution of cadmium sulphate. This limb acts as the anode. The other limb which acts as the cathode contains cadmium amalgam (containing 12-14% of Cd by weight), covered with crystals of solid $CdSO_4.8/3H_2O$ and then saturated $CdSO_4$ solution. Short platinum wires are sealed into the bottoms of the limbs to make contact with the active materials. The limbs are finally closed with corks and sealing wax. The purpose of placing solid crystals of $CdSO_4.8/3H_2O$ is to keep the $CdSO_4$ solution saturated at all temperatures.

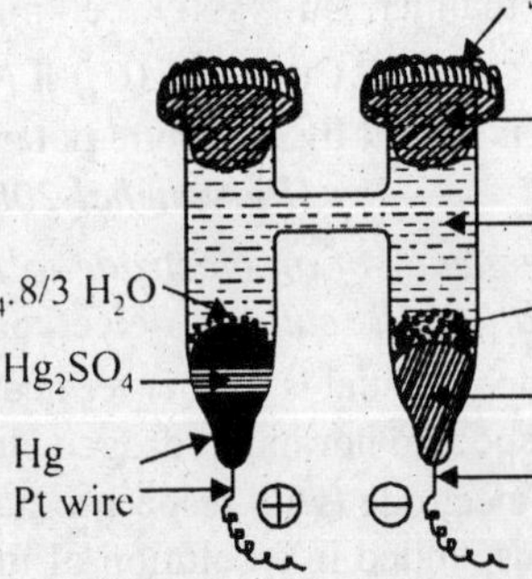

Fig. 11.5. Weston standard cell.

This cell has a constant reproducible value of e.m.f. which does not change with time. Also the temperature coefficient of e.m.f. is very low.

The EMF of the above cell is 1.0185 volt at 15°C and 1.0181 volt at 25°C.

The reaction taking place in the cell is as follows:

$$Cd + Hg_2SO_4 + aq \longrightarrow CdSO_4.8/3\ H_2O + 2Hg$$

Q. 10. What is a reversible cell? How do you measure its e.m.f.? **(Nagpaur 2008)**

Ans. A cell is said to be reversible if it satisfies the following three conditions:

(*i*) If an opposing EMF equal to that of the cell is applied to the cell, no current should flow in the circuit and hence no chemical reaction should take place.

(*ii*) If an opposing EMF slightly (infinitesimally) smaller than that of the cell is applied, an infinitesimally small current should flow from the cell and an infinitesimally small reaction should take place.

(*iii*) If an opposing EMF infinitesimally greater than that of the cell is applied, an infinitesimally small current should flow in the opposite direction and an infinitesimally small reaction in the reverse direction should occur.

This is illustrated in Fig. 11.6.

A well known example of a reversible cell is that of Daniell cell, consisting of a zinc electrode dipped in $ZnSO_4$ solution and a copper electrode dipped in $CuSO_4$ solution. When it is producing current, the reaction taking place in the cell is

$$Zn + Cu^{2+} \longrightarrow Zn^{2+} + Cu$$

The EMF of a reversible cell can be measured potentiometrically.

In the potentiometric methods for the determination of EMF, as no current flows at the time

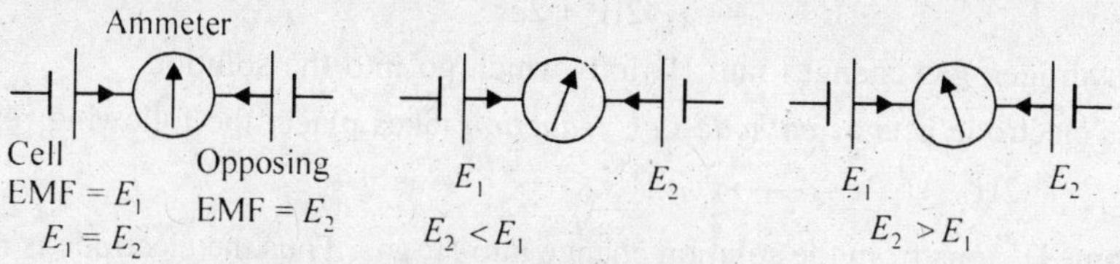

Fig. 11.6. Concept of reversible cell.

of measurement, the conditions of reversibility are satisfied.

Q. 11. Give an example of irreversible cell. Explain as to why it is irreversible.

(Garhwal 2000)

Ans. Consider a cell consisting of a zinc rod and a silver rod dipped in dilute sulphuric acid. When the two rods are connected with a wire, the following reaction takes place :

$$Zn + H_2SO_4 \longrightarrow ZnSO_4 + H_2$$

i.e., zinc is oxidized to Zn^{2+} ions and H^+ ions are reduced to H_2 gas. In other words, zinc dissolves at one electrode and H_2 gas is evolved at the other.

Now if the cell is connected to an external source of EMF whose EMF is slightly greater than that of the cell and is opposing the cell, silver dissolves at one electrode and H_2 gas is evolved at the other

$$2Ag + H_2SO_4 \longrightarrow Ag_2SO_4 + H_2$$

i.e. now oxidation takes place at the silver electrode and reduction at the other. Thus condition of reversibility is not satisfied *i.e.* the chemical change is not reversed (*i.e.* $Zn + 2H^+ \longrightarrow Zn^{2+} + H_2$ does not become $Zn^{2+} + H_2 \longrightarrow Zn + 2H^+$). Instead, a new chemical reaction takes place. Hence the above cell is not reversible but is irreversible.

Q. 12. Describe the following (*i*) Metal-metal ion electrode (*ii*) Hydrogen electrode (*iii*) Calomel electrode.

(Lucknow 2002; Nagpur 2003, Dibrugarh 2004; Arunachal 2003; Calcutta 2007; Raipur 2010)

OR

What are different types of reversible electrodes ? *(Purvanchal 2007)*

Ans. (*i*) **Metal-metal ion electrodes:** Such an electrode consists of a metal dipped in the solution of its ions *e.g.* zinc rod dipped in zinc sulphate solution, copper rod dipped in copper sulphate solution etc.

Taking the example of copper-copper ion electrode, when *oxidation* takes place at the electrode, the reaction may be represented as

$$Cu \longrightarrow Cu^{2+} + 2e^-$$

On the other hand, when *reduction* takes place at the electrode, the reaction may be represented as

$$Cu^{2+} + 2e^- \longrightarrow Cu$$

Thus the electrode is reversible with respect to copper ions.

(*ii*) **Hydrogen electrode:** The design of the hydrogen electrode is shown in Fig. 11.7. In the standard or normal hydrogen electrode, hydrogen gas at 1 atmospheric pressure is passed into 1 M HCl acid in which the foil of platinized platinum remains immersed.

If in a cell, this electrode acts as the *anode i.e., oxidation* takes place, the following reaction takes place:

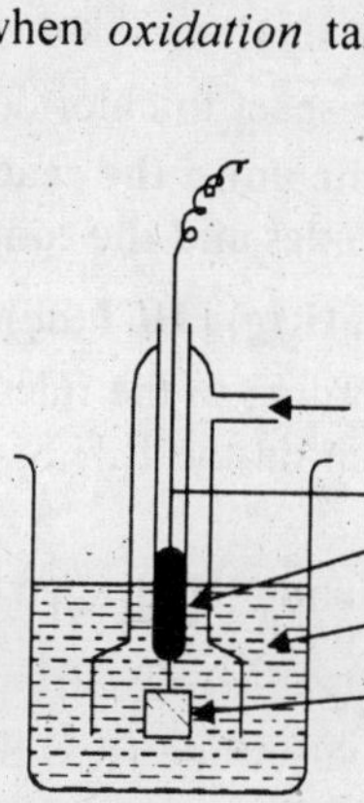

Fig. 11.7. Hydrogen electrode.

$$H_2 \longrightarrow 2H^+ + 2e^-$$

i.e., some hydrogen gas changes into H^+ ions which go into the solution.

If this electrode acts as *cathode i.e., reduction* takes place, the following reaction occurs:

$$2H^+ + 2e^- \longrightarrow H_2$$

i.e. some H^+ ions from the solution change into H_2 gas. Thus the electrode is reversible with respect of H^+ ions. This electrode is usually represented as

$$Pt, H_2(g) \mid H^+ \text{ (conc.} = c)$$

(*iii*) **Hg — Hg_2Cl_2 – KCl electrode (Calomel electrode):** It is prepared as follows:

Pure mercury is placed at the bottom of a tube *T* (Fig. 11.8). It is covered with a paste of mercurous chloride (calomel) and then filled with either *decinormal* or *normal* or *saturated* solution of KCl from the side tube *A* such that the side tube *B* is also filled with the KCl solution. A tube *C* having a piece of platinum wire sealed at its end is fixed into the tube *T* such that the platinum wire remains dipped into the mercury. A little mercury is placed in the tube *C* and a copper wire put into it to make an electrical contact.

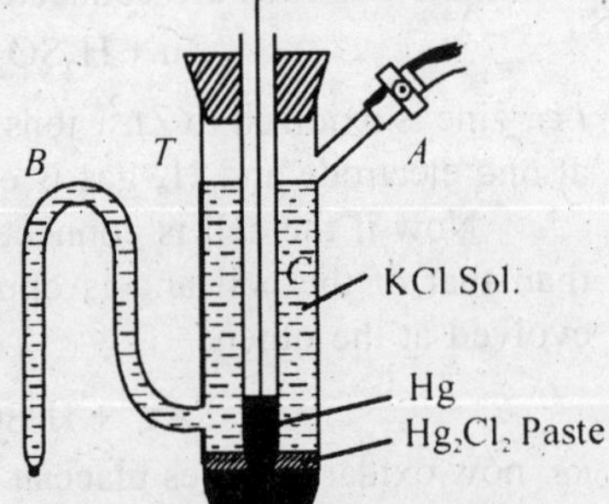

Fig. 11.8. Calomel electrode.

A more convenient assembly of the calomel electrode used in the laboratory is shown in Fig. 11.9.

Oxidation can occur at this electrode through the following reactions:

$$2Hg(l) \longrightarrow Hg_2^{2+} + 2e^-$$

$$Hg_2^{2+} + 2Cl^- \longrightarrow Hg_2Cl_2(s)$$

Overall reaction $\quad 2Hg(l) + 2Cl^- \longrightarrow Hg_2Cl_2(s) + 2e^-$

Reduction can occur at this electrode as follows:

$$Hg_2Cl_2(s) \longrightarrow Hg_2^{2+} + 2Cl^-$$

$$Hg_2^{2+} + 2e^- \longrightarrow 2Hg\ (l)$$

Overall reaction $\quad Hg_2Cl_2(s) + 2e^- \longrightarrow 2Hg(l) + 2Cl$

This overall reaction is just the reverse of the overall reaction when oxidation takes place. Hence the calomel is reversible with respect to chloride ions.

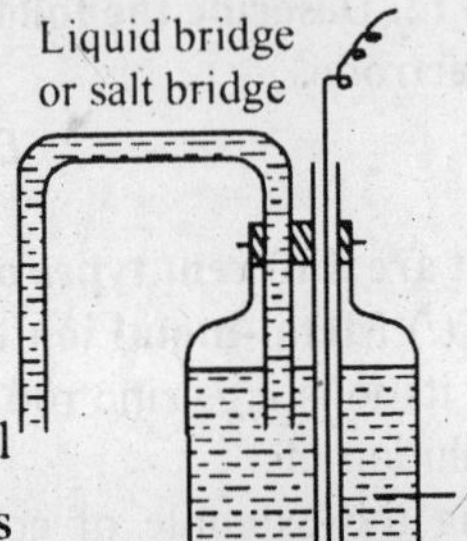

Fig. 11.9. Another form of calomel electrode.

Q. 13. Write down the reactions taking place separately at the two electrodes and the complete cell reaction

Pt, H_2(*g*) | HCl (*aq*) || KCl (*aq*), AgCl(*s*) | Ag

Ans. According to the rules of representation of a cell, anode is written on the L.H.S. and cathode is written on the R.H.S. Hence in this case, hydrogen electrode is anode *i.e.* oxidation takes place here.

Anode reaction (Hydrogen electrode)

$$H_2(g) \longrightarrow 2H^+\ (aq) + 2e^-$$

The other electrode (R.H.S is metal-metal insoluble salt-salt anion electrode. In this case reduction takes place here.

Cathode reaction

$$\begin{array}{l} \text{Ag Cl}(s) \longrightarrow Ag^+ + Cl^- \\ Ag^+ + e \longrightarrow Ag(s) \\ \hline AgCl(s) + e^- \longrightarrow Ag(s) + Cl^- \end{array}$$

The net reaction is written as under

$$\begin{array}{l} H_2(g) \longrightarrow 2H^+ + 2e \\ 2\,\text{Ag Cl}(s) + 2e \longrightarrow 2Ag(s) + 2Cl^- \\ \hline H_2(g) + 2\text{Ag Cl}(s) \longrightarrow 2H^+ + 2Ag\,(s) + 2Cl \end{array}$$

Q. 14. Devise a cell in which the overall reaction is

$$\mathbf{Zn + Hg_2SO_4(s) \longrightarrow ZnSO_4 + 2Hg(\mathit{l})}$$

and write the separate electrode reactions.

Ans. The cell can be represented as under

$$Zn \mid ZnSO_4(ag) \parallel K_2SO_4(aq), Hg_2SO_4(s) \mid Hg$$

Thus the anode (L.H.S.) consists of zinc electrode and the cathode (R.H.S.) consists of calomel electrode. The electrode reactions are as under

Anode $\qquad Zn(s) \longrightarrow Zn^{2+}\ (aq) + 2e$

Cathode $\qquad Hg_2SO_4(s) \longrightarrow Hg_2^{\ 2+} + SO_4^{\ 2-}$

$$\begin{array}{l} Hg_2^{\ 2+} + 2e \longrightarrow 2Hg\ (l) \\ \hline Hg_2SO_4(s) + 2e \longrightarrow 2Hg(l) + SO_4^{\ -2} \end{array}$$

The overall cell reaction can be obtained by adding the two electrode reactions.

Q. 15. Give a relationship between electrical energy and chemical energy.

(*Panjab 2000; M. Dayanand 2000; Bombay 2004*)

Ans. *The electrical energy produced in a cell is defined as the product of the quantity of electricity that passes through the cell and the EMF of the cell i.e.,*

$$\text{Electrical energy} = \text{Quantity of electricity} \times \text{EMF}$$

For every one mol of electrons transferred in the cell reaction (*i.e.*, for every one gram equivalent of the ions deposited), the quantity of electricity that flows through the cell is one Faraday (1F = 96500 Coulombs). Thus if n moles of electrons are transferred in any cell reaction (*e.g.* for the reaction $Zn + Cu^{2+} \rightarrow Zn^{2+} + Cu$, $n = 2$), the quantity of electricity passing through the cell = nF Faradays. If E is the EMF of the cell, then

$$\text{Electrical energy} = nFE$$

It was observed that for a large number of reactions the electrical energy produced was not equal to the chemical energy. Hence it was suggested by Gibbs (1875) and Helmholtz (1882) that the electrical energy produced in a reversible cell is equal to the decrease in free energy accompanying the cell reaction. Thus if ΔG represents the change in free energy, we must have

$$-\Delta G = nFE \qquad \text{... }(i)$$

Further, according to the Gibbs-Helmholtz equation, we have

$$\Delta G = \Delta H + T\left(\frac{\partial(\Delta G)}{\partial T}\right)_P \qquad \text{.. }(ii)$$

Substituting the value of ΔG from equation (*i*) in equation (*ii*), we get

$$-nEF = \Delta H + T\left(\frac{\partial(-nFE)}{\partial T}\right)$$

$$nEF = -\Delta H + TnF\left(\frac{\partial E}{\partial T}\right)_P \quad \text{... (iii)}$$

or $$E = -\frac{\Delta H}{nF} + T\left(\frac{\partial E}{\partial T}\right)_P \quad \text{... (iv)}$$

From equation (*iii*), it is evident that whether electrical energy (nFE) produced is equal to less than or greater than the enthalpy change (ΔH), it depends upon the sign of $\partial E/\partial T$ *i.e.*, the *temperature coefficient* of the EMF. Thus

(*i*) If $\partial E/\partial T$ is extremely small (almost equal to zero), as is the case in Daniell cell, the electrical energy produced is equal to the enthalpy change.

(*ii*) If $\partial E/\partial T$ is positive, the electrical energy produced is greater than the heat of reaction at constant pressure, the additional energy being absorbed from the surroundings.

(*iii*) If $\partial E/\partial T$ is negative, the electrical energy produced is less than the heat of reaction at constant pressure, the additional energy being absorbed by the surroundings.

Q. 16. Derive expressions for ΔG, ΔH and ΔS in terms of EMF of the cell and the temperature coefficient of the EMF. *(Mumbai 2005; Dibrugarh 2004)*

Ans. ΔG is related to EMF according to the expression

$$\Delta G = -nFE \quad \text{... (i)}$$

Thus knowing the EMF, E, *ΔG can be calculated.*

Equation (*iii*) of Q. 15. can be rearranged as

$$\Delta H = nF\left[T\left(\frac{\partial E}{\partial T}\right)_P - E\right] \quad \text{... (ii)}$$

Thus knowing the EMF, E and the temperature coefficient of the EMF, $\left(\frac{\partial E}{\partial T}\right)_P$, the heat of reaction, ΔH, can be calculated.

Having known the values of ΔG and ΔH, the entropy change (ΔS) accompanying the reaction can be calculated using the relationship

$$\Delta G = \Delta H - T\Delta S \quad \text{...(iii)}$$

Q. 17. Explain how Gibbs-Helmholtz equation can be utilized for calculating enthalpy change of reaction occurring in a reversible cell.

Ans. See Q. 15 and 16.

Q. 18. The EMF of the cell

$$\mathbf{Cd / CdCl_2.2\frac{1}{2}H_2O \text{ (saturated)} \parallel AgCl(s) \mid Ag}$$

is found to be 0.6753 volt at 25°C and 0.6915 volt at 0°C. Calculate ΔG, ΔH and ΔS of the cell reaction at 25°C.

Ans. The cell is

$$Cd + 2AgCl + aq \longrightarrow CdCl_2.2\frac{1}{2}H_2O \text{ (sat.)} + 2Ag$$

or $$Cd + 2Ag^+ \longrightarrow Cd^{2+} + 2Ag$$

$\therefore$ $$n = 2$$

(*i*) *Calculation of free energy change* (ΔG)

$$\begin{aligned}\Delta G &= -nFE\\ &= -2\times 96500\times 0.6753\\ &= -130333 \text{ joules}\\ &= -130.33 \text{ kJ mol}^{-1}\end{aligned}$$

(*ii*) *Calculation of enthalpy change* (ΔH)

At atmospheric pressure, the temperature coefficient of EMF in the range 0° – 25°C will be

$$\left(\frac{\partial E}{\partial T}\right)_P = \frac{E_2 - E_1}{T_2 - T_1} = \frac{0.6915 - 0.6753}{0 - 25}$$

$$= -\frac{0.0162}{25} = -0.00065 \text{ volt per degree}$$

Now, using the relationship

$$E = -\frac{\Delta H}{nF} + T\left(\frac{\partial E}{\partial T}\right)_P$$

$$\Delta H = -nFE + TnF\left(\frac{\partial E}{\partial T}\right)_P$$

Using the data at 25°C, we get

$$\begin{aligned}\Delta H &= -2\times 96500\times 0.6753 + 298\times 2\times 96500\times(-0.00065)\\ &= -167717 \text{ joules}\\ &= \mathbf{-167.72\ kJ\ mol^{-1}}\end{aligned}$$

(*iii*) *Calculation of entropy change* (ΔS)

We know that

$$\Delta G = \Delta H - T\Delta S$$

$$\therefore \quad \Delta S = \frac{\Delta H - \Delta G}{T}$$

$$= \frac{-167717-(-130333)}{298} = \mathbf{-125.4\ JK^{-1}\ mol^{-1}}$$

The negative signs with ΔG, ΔH and ΔS indicate the *decrease* in these quantities.

Q. 19. How will you determine the standard electrode potential of (*i*) Zinc electrode (*ii*) Copper electrode? ***(Panjab 2000)***

Ans. (*i*) Determination of standard electrode potential of Zn/Zn^{2+} electrode: A cell comprising

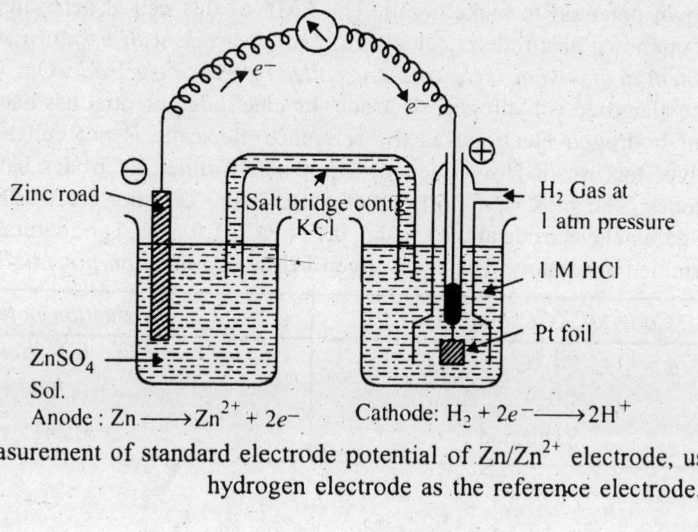

Fig. 11.10. Measurement of standard electrode potential of Zn/Zn^{2+} electrode, using standard hydrogen electrode as the reference electrode.

of zinc electrode immersed in 1M $ZnSO_4$ solution and a standard hydrogen electrode is set up as shown in Fig. 11.10.

Standard electrode potential of hydrogen electrode is zero.

EMF of the above cell comes out to be 0.76 volt. Hence standard electrode potential of Zn/Zn^{2+} electrode = 0.76 volt. Further the direction of flow of current indicates that oxidation takes place at the zinc electrode. Hence 0.76 volt is the **standard oxidation potential** of the Zn/Zn^{2+} electrode.

(*ii*) **Determination of standard electrode potential of Cu/Cu^{2+} electrode:** A cell comprising of copper electrode immersed in 1 M $CuSO_4$ solution and a standard hydrogen electrode is set up as shown in Fig. 11.11.

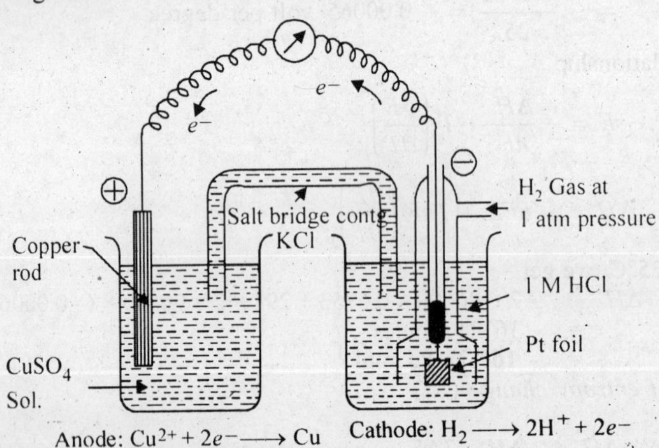

Fig. 11.11. Measurement of standard electrode potential of Cu/Cu^{2+} electrode using hydrogen electrode as teh reference electrode.

Standard electrode potential of hydrogen electrode is zero.

EMF of the cell comes out to be 0.34 volt. Further the direction of flow of current indicates that reduction takes place at the copper electrode. Hence 0.34 volt is the **standard reduction potential** of Cu/Cu^{2+} electrode.

Q. 20. What is meant by reference electrode?

(Gujarat 2003; G. Nanakdev 2003;Banglore,2004; Vidyasagar 2007)

Ans. The potential of an unknown electrode can be obtained only by combining it with an electrode of known potential to make a cell. The EMF of this cell is determined and then the potential of the unknown electrode is calculated. *The electrode with a known potential used to determine the potential of unknown electrodes is called reference electrode.* One of the electrodes used as reference electrode is hydrogen electrode. Its electrode potential has been taken as zero.

But use of hydrogen electrode as the reference electrode is not convenient obviously because it involves the use of H_2 gas. Hence a number of other electrodes have been used as reference electrodes. The most common out of these is the **calomel electrode**. The electrode potentials of the calomel electrode at 25°C using 0.1 N KCl, 1.0 N KCl and saturated KCl solution have been determined accurately. These are given below as *reduction potentials.*

Conc. of KCl solution	*Reduction potential*
0.1 N	+ 0.3338 volt
1.0 N	+ 0.2800 volt
Saturated	+ 0.2415 volt

To determine the electrode potential of any electrode, a cell using this as one of the electrodes and calomel electrode as the second electrode is set up. The EMF of the cell is measured and then the following expression is used to calculate the required electrode potential.

EMF of the cell = Reduction potential of cathode – Reduction potential of anode.

Q. 21. Illustrate how can a calomel electrode be used as a reference electrode to determine the potential of

(*i*) Zinc electrode (*ii*) Copper electrode

Ans. (*i*) **Determination of the standard electrode potential of Zn/Zn^{2+} electrode:** A cell comprising of zinc electrode immersed in 1 M $ZnSO_4$ solution and the calomel electrode (say, using 1N KCl solution) is set up as shown in Fig. 11.12.

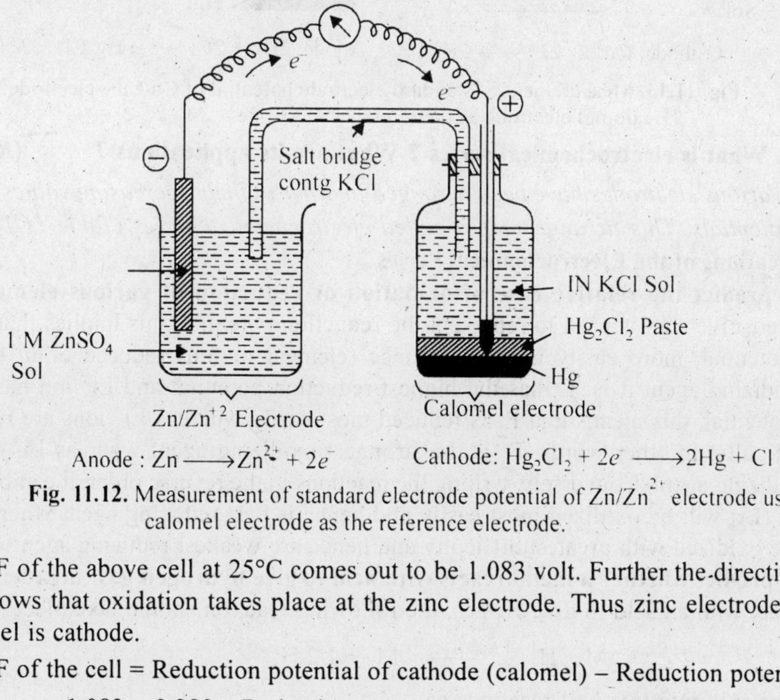

Fig. 11.12. Measurement of standard electrode potential of Zn/Zn^{2+} electrode using calomel electrode as the reference electrode.

EMF of the above cell at 25°C comes out to be 1.083 volt. Further the direction of flow of current shows that oxidation takes place at the zinc electrode. Thus zinc electrode is the anode and calomel is cathode.

EMF of the cell = Reduction potential of cathode (calomel) – Reduction potential of anode

1.083 = 0.280 – Reduction potential of anode (Zn electrode)

or Reduction potential of anode (Zn electrode)

= 0.280 – 1.083

= – 0.803 volt

(*ii*) **Determination of the standard electrode potential of Cu/Cu^{2+} electrode:** A cell comprising of copper electrode immersed in 1 M $CuSO_4$ solution and the calomel electrode (say using 1N KCl solution) is set up as shown in Fig. 11.13.

EMF of the above cell comes out to be 0.018 volt. Further the direction of flow of current shows that reduction takes place at the copper electrode. Thus copper electrode is the cathode and calomel is the anode.

EMF of the cell = Reduction potential of cathode (copper) – Reduction potential of anode (calomel)

or 0.018 = Reduction potential of copper cathode – 0.280

or Reduction potential of copper cathode = 0.018 + 0.280

= 0.298 volt

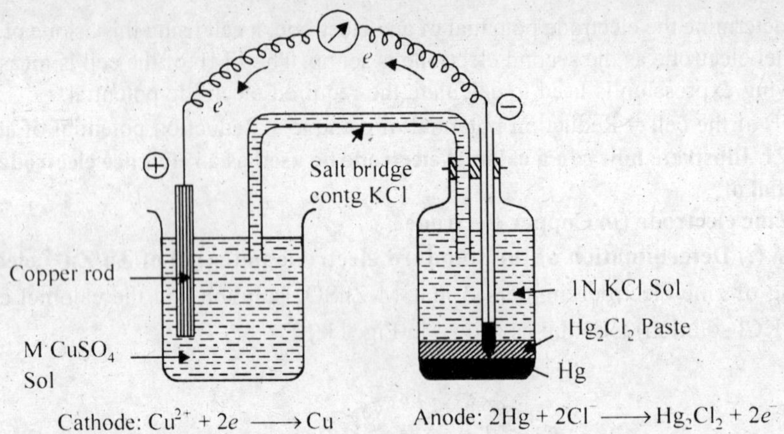

Fig. 11.13. Measurement of standard electrode potential of Cu/Cu^{2+} electrode using calomel electrode as the reference electrode.

Q. 22. What is electrochemical series ? What are its applications ? ***(Raipur 2010)***

Ans. *Various electrodes have been arranged in order of their decreasing values of standard reduction potentials. This arrangement is called electrochemical series* (*Table 11.1*).

Applications of the Electrochemical Series

1. To predict the relative ease of oxidation or reduction of various elements: As by convention positive sign is used to represent the reduction potential, this implies that greater the reduction potential, more easily is the substance (element or ion) reduced or in other words stronger oxidizing agent it is. F_2 has the highest reduction potential and Li^+ ion has the lowest reduction potential, this means that F_2 is reduced most easily whereas Li^+ ions are reduced with greatest difficulty. In other words, F_2 is the strongest oxidizing agent whereas Li^+ ions are the weakest oxidizing agent. Conversely, writing the reactions in the reverse order, it can be postulated that lithium (Li) will be oxidized most easily and hence is best reducing agent whereas fluoride ions (F^-) are oxidized with greatest difficulty and hence are weakest reducing agent.

2. To predict whether a metal reacts with acid to give hydrogen gas: In order that a metal (M) may react with an acid to give H_2 gas, the following reaction should take place:

$$M + H^+ \longrightarrow M^+ + 1/2\, H_2$$

which can be split into two half reactions as

$$M \longrightarrow M^+ + e^- \quad \text{(Oxidation half reaction)}$$

$$H^+ + e^- \longrightarrow 1/2\, H_2 \quad \text{(Reduction half reaction)}$$

i.e. the metal should have the tendency to lose electrons *i.e.* undergoes oxidation with respect to hydrogen. In other words, the metal would have a negative reduction potential. Thus all metals lying below hydrogen (having negative reduction potential) in the electrochemical series (Table 11.1) react with the acid to give hydrogen gas. Further evidently, lower the reduction potential, higher is the reactivity.

3. To calculate the standard EMF of any electrochemical cell: Electrochemical cell is based on a redox reaction which can be split into two half-reactions *viz.*,

(*i*) Oxidation half reaction

(*ii*) Reduction half reaction

In the representation of a cell, the electrode on which oxidation takes place is written on the left hand side and the electrode on which reduction takes place is written on the right hand side

$$\text{Standard EMF of the cell} = \begin{bmatrix}\text{Standard reduction} \\ \text{potential of the R.H.S.} \\ \text{electrode}\end{bmatrix} - \begin{bmatrix}\text{Standard reduction} \\ \text{potential of the L.H.S.} \\ \text{electrode}\end{bmatrix}$$

Table 11.1. Standard reduction potentials at 25°C (Electrochemical series)

Reduction half-reaction	*Standard reduction potential E° (in volts)*
$F_2 + 2e^- \longrightarrow 2F^-$	2.87
$Co^{3+} + e^- \longrightarrow Co^{2+}$	1.82
$Au^{3+} + 3e^- \longrightarrow Au$	1.50
$Cl_2 + 2e^- \longrightarrow 2Cl^-$	1.36
$Br_2 + 2e^- \longrightarrow 2Br^-$	1.09
$Ag^+ + e^- \longrightarrow Ag$	0.80
$Fe^{3+} + e^- \longrightarrow Fe^{2+}$	0.77
$I_2 + 2e^- \longrightarrow 2I^-$	0.53
$Cu^{2+} + 2e^- \longrightarrow Cu$	0.34
$AgCl + e^- \longrightarrow Ag + Cl^-$	0.22
$Cu^{2+} + e^- \longrightarrow Cu^+$	0.15
$Sn^{4+} + 2e^- \longrightarrow Sn^{2+}$	0.13
$2H^+ + 2e^- \longrightarrow H_2$	0.00
$Pb^{2+} + 2e^- \longrightarrow Pb$	–0.13
$Sn^{2+} + 2e^- \longrightarrow Sn$	–0.14
$Ni^{2+} + 2e^- \longrightarrow Ni$	–0.25
$Co^{2+} + 2e^- \longrightarrow Co$	–0.28
$PbSO_4 + 2e^- \longrightarrow Pb + SO_4^{2-}$	–0.31
$Cd^{2+} + 2e^- \longrightarrow Cd$	–0.40
$Fe^{2+} + 2e^- \longrightarrow Fe$	–0.44
$Cr^{3+} + 3e^- \longrightarrow Cr$	–0.74
$Zn^{2+} + 2e^- \longrightarrow Zn$	–0.76
$Mn^{2+} + 2e^- \longrightarrow Mn$	–1.18
$Al^{3+} + 3e^- \longrightarrow Al$	–1.66
$Mg^{2+} + 2e^- \longrightarrow Mg$	–2.37
$Na^+ + e^- \longrightarrow Na$	–2.71
$Ca^{2+} + 2e^- \longrightarrow Ca$	–2.87
$Ba^{2+} + 2e^- \longrightarrow Ba$	–2.90
$K^+ + 2e^- \longrightarrow K$	–2.93
$Li^+ + e^- \longrightarrow Li$	–3.05

4. To predict spontaneity of a reaction: To find out whether a reaction is feasible or not, EMF of the cell based upon the given redox reaction is calculated. If the EMF of the cell comes out to be positive, the reaction is feasible. If it comes out to be negative, the reaction is not feasible.

Q. 23. A cell is prepared by dipping a copper rod in 1 M $CuSO_4$ solution and a nickel rod in 1 M $NiSO_4$ solution. The standard reduction potentials of copper electrode and nickel electrode are 0.34 volt and – 0.25 volt respectively.

(*a*) What will be cell reaction ?

(*b*) What will be the standard EMF of the cell ?

(*c*) Which electrode will be positive ?

(*d*) How will the cell be represented ?

Ans. (*a*) The cell reaction can be

either $Ni + CuSO_4 \longrightarrow NiSO_4 + Cu$

i.e. $Ni + Cu^{2+} \longrightarrow Ni^{2+} + Cu$

or $Cu + NiSO_4 \longrightarrow CuSO_4 + Ni$

i.e. $Cu + Ni^{2+} \longrightarrow Cu^{2+} + Ni$

The correct reaction will be the one which gives positive EMF.

If the first reaction is the cell reaction, we will have

$Ni \longrightarrow Ni^{2+} + 2e^-$ Std. oxdn. potential = – (– 0.25) volt

$Cu^{2+} + 2e \longrightarrow Cu$ Std. redn. potential = + 0.34 volt

Overall reaction

$Ni + Cu^{2+} \longrightarrow Ni^{2+} + Cu$ Std. EMF = + 0.59 volt

Thus EMF is positive for this reaction. Hence this is the correct cell reaction. (If the other reaction is tried, EMF will come out to be – 0.59 volts)

(*b*) Standard EMF of the cell as calculated above = 0.59 volt.

(*c*) Since oxidation takes place at the *nickel electrode,* therefore it acts as *anode* or since electrons are produced at this electrode it acts as a *negative pole.* Obviously, copper electrode will act as *cathode* or a *positive pole.*

(*d*) By convention, the cell will be represented as

$$Ni \mid NiSO_4 \ (1M) \parallel CuSO_4(1M) \mid Cu$$

Q. 24. Calculate the standard EMF of a cell which involves the following cell reaction

$$Zn + 2Ag^+ \longrightarrow Zn^{2+} + 2Ag$$

Given that $E^0_{Zn, Zn^{2+}}$ = 0.76 volt and E^0_{Ag, Ag^+} = – 0.80 volt.

Ans. As oxidation takes place at the zinc electrode and reduction at the silver electrode, by convention, the cell may be represented as

$$Zn \mid Zn^{2+} \parallel Ag^+ \mid Ag$$

Standard EMF of the cell (E^0_{cell}) = [Standard reduction potential of the R.H.S. electrode] – [Standard reduction potential of the L.H.S. electrode]

As standard oxidation potentials of zinc electrode and silver electrode are given to be 0.76 volt and – 0.80 volt, their standard reduction potentials will be – 0.76 volt and + 0.80 volt respectively. Hence

$$E^0_{cell} = 0.80 - (-0.76) = \mathbf{1.56}$$

Q. 25. Can a solution of 1 M copper sulphate be stored in a vessel made of nickel metal ?

Given that $E^0_{Ni, Ni^{2+}} = +0.25$ volt and $E^0_{Cu, Cu^{2+}} = -0.34$ volt.

Ans. In this problem, we want to see whether the following reaction takes place or not

$$Ni + CuSO_4 \longrightarrow NiSO_4 + Cu$$

i.e.

$$Ni + Cu^{2+} \longrightarrow Ni^{2+} + Cu$$

By convention, the cell may be represented as

$$Ni \mid Ni^{2+} \parallel Cu^{2+} \mid Cu$$

Oxidation potentials are given as

$$E^0_{Ni, Ni^{2+}} = 0.25 \text{ volt and } E^0_{Cu,Cu^{2+}} = -0.34 \text{ volt}$$

Hence the reduction potentials will be

$$E^0_{Ni^{2+}, Ni} = -E^0_{Ni, Ni^{2+}} = -0.25 \text{ volt}$$

and

$$E^0_{Cu^{2+},Cu} = -E^0_{Cu, Cu^{2+}} = +0.34 \text{ volt}$$

$$\therefore \quad E^0_{Cell} = E^0_{R.H.S.electrode} - E^0_{L.H.S.electrode}$$

$$= +0.34 - (-0.25) = 0.59 \text{ volt}$$

Thus EMF comes out to be positive. This implies that $CuSO_4$ reacts with nickel. Hence $CuSO_4$ cannot be stored in nickel vessel.

Q. 26. Predict whether zinc and silver react with 1 N sulphuric acid to give out hydrogen gas or not. Given that the standard reduction potentials of zinc and silver are – 0.76 volt and 0.80 volt respectively.

Ans. (*a*) *To predict reaction of zinc with sulphuric acid:* If it reacts, the following reaction should take place

$$Zn + H_2SO_4 \longrightarrow ZnSO_4 + H_2$$

i.e.

$$Zn + 2H^+ \longrightarrow Zn^{2+} + H_2$$

By convention, the cell will be represented as

$$Zn \mid Zn^{2+} \parallel H^+ \mid H_2$$

$\therefore$ Standard EMF of the cell (E^0_{Cell})

$$\begin{bmatrix}\text{Standard reduction potential} \\ \text{of the R.H.S. electrode}\end{bmatrix} - \begin{bmatrix}\text{Standard reduction potential} \\ \text{of the L.H.S. electrode}\end{bmatrix}$$

$$= 0 - [-0.76] = +0.76 \text{ volt}$$

Thus the EMF of the cell comes out to be positive. Hence the reaction takes place.

(*b*) *To predict the reaction of silver with sulphuric acid:* If it reacts, the following reaction should take place.

$$2Ag + H_2SO_4 \longrightarrow Ag_2SO_4 + H_2$$

i.e.

$$2Ag + 2H^+ \longrightarrow 2Ag^+ + H_2$$

By convention, the cell may be represented as

$$Ag \mid Ag^+ \parallel H^+ \mid H_2$$

$$E^0_{cell} = E^0_{H^+, H_2} - E^0_{Ag^+, Ag}$$

$$= 0 - 0.80 \text{ volt} = -0.80 \text{ volt.}$$

Thus EMF of the supposed cell comes out to be negative. Hence this reaction does not take place.

Q. 27. What is meant by activity and activity coefficient ? *(M. Dayanand 1995)*

Ans. *The effective concentration of an ion or the electrolyte in a solution is called its activity.*

The activity of an electrolyte in a solution is usually represented by *a*. It is related to the actual concentration, expressed as *molality m* by the equation

$$a = \gamma \times m$$

where γ is called the *activity coefficient i.e.*

$$\gamma = \frac{a}{m}$$

Hence *activity coefficient may be defined as the ratio of the activity to the actual concentration (molality) of the solution.*

As $a < m$ for the real solutions, γ is less than 1. For ideal solutions (*i.e.* for very dilute solution), $a = m$ and hence $\gamma = 1$.

It is observed that as the concentration is increased, the activity coefficient first decreases below unity, passes through a minimum and then begins to increase and then may attain values much greater than one.

Activity of an electrolyte is, in fact, the product of the activities of its ions (for uni-univalent electrolyte)

i.e. $$a = a_+ \, a_- \qquad ...(iii)$$

Since the activities of the individual ions cannot be determined, a term called **mean ionic activity** has been introduced. The mean ionic activity ($a_\pm$) for a uni-univalent electrolyte is defined by the expression

$$a_\pm = \sqrt{a} = \sqrt{a_+ . a_-} \qquad ...(iv)$$

In a similar manner, the term **mean activity coefficient** ($\gamma_\pm$) has been introduced. For a uni-univalent electrolyte, it is defined by the expression

$$\gamma_\pm = \sqrt{\gamma_+ . \gamma_-} \qquad ...(v)$$

Further, putting

$$a_+ = \gamma_+ m_+$$

and $$a_- = \gamma_- m_-$$

equation (*iv*) becomes

$$a_\pm = \sqrt{a} = \sqrt{\gamma_+ m_+ \times \gamma_- m_-}$$

For a uni univalent electrolyte,

$$m_+ = m_- = m \text{ and } \gamma_+ = \gamma_- = \gamma_\pm$$

$$\therefore \quad \sqrt{a} = \sqrt{m^2 \gamma_\pm^2}$$

or $$a = m^2 \gamma_\pm^2 \qquad ... (vi)$$

Q. 28. Derive Nernst equation for measuring EMF of a cell.

(Lucknow 2003; Madras 2004; Bangalore 2004; Vidyasagar2007)

Ans. Suppose the cell reaction involved in any reversible cell is represented by the general equation.

$$aA + bB + ... \longrightarrow xX + yY + ...$$

The free energy change of this reaction is given by the equation

$$\Delta G = \Delta G^0 + RT \ln \frac{a_X^x . a_Y^y ...}{a_A^a . a_B^b ...} \qquad ... (i)$$

where ΔG^0 represents the standard free change and a_A, a_B, a_X and a_Y represent the activities of *A, B, X* and *Y* respectively in any state, other than the standard state at the state of equilibrium.

If the cell reaction involves transference of *n* moles of the electrons, this corresponds to the flow of *nF* faradays of electricity. If *E* is the EMF of the cell, then electrical energy produced in the cell = *nFE.* Since electrical energy produced is equal to the decrease in free energy of the cell reaction, we have

$$-\Delta G = nFE \qquad ...(ii)$$

Similarly, for the standard state, we will have

$$-\Delta G^0 = nFE^0 \qquad ...(iii)$$

where $E°$ is the standard EMF of the cell.

Substituting the values of ΔG and ΔG^0 from equations (*ii*) and (*iii*) in equation (*i*), we get

$$-nFE = -nFE^0 + RT \ln \frac{a_X^x . a_Y^y ...}{a_A^a . a_B^b ...}$$

or

$$E = E^0 - \frac{RT}{nF} \ln \frac{a_X^x . a_Y^y ...}{a_A^a . a_B^b ...} \qquad ...(iv)$$

Thus knowing the cell reaction (so that *n* is known), and the standard EMF of the cell, the EMF of the cell for known activities of the various reactants and products can be calculated. Equation (*iv*) is called **Nernst equation** for the calculation of EMF of cell.

When the concentrations are not very high, replacing activites by concentrations, equation (*iv*) becomes

$$E = E^0 - \frac{RT}{nF} \ln \frac{[X]^x [Y]^y ...}{[A]^a . [B]^b ...} \qquad ...(v)$$

Substituting for $R = 8.314$ and $F = 96500$, at 25°C *i.e.* $T = 298$ K, Eq. (*v*)

changes

$$E = E^0 - \frac{0.0591}{n} \ln \frac{[X]^x [Y]^y ...}{[A]^a . [B]^b ...}$$

It must be borne in mind that the activities or the concentrations of the pure liquids or pure solids are taken as unity.

Taking a particular case of Daniell cell, the reactions involved are

$$Zn \longrightarrow Zn^{2+} + 2e^-$$

$$Cu^{2+} + 2e^- \longrightarrow Cu$$

$n = 2$. The overall reaction is

$$Zn + Cu^{2+} \longrightarrow Zn^{2+} + Cu$$

Hence we will have

$$E = E^0 - \frac{RT}{2F} \ln \frac{a_{Zn^{2+}} . a_{Cu}}{a_{Zn} . a_{Cu^{2+}}} \qquad ...(vi)$$

Putting the activities of solids equal to unity *i.e.*, $a_{Cu} = 1$ and $a_{Zn} = 1$, equation (*vi*) reduces to

$$E = E^0 - \frac{RT}{2F} \ln \frac{a_{Zn^{2+}}}{a_{Cu^{2+}}} \qquad ...(vii)$$

Using concentrations in place of activities, equation (*vii*) becomes

$$E = E^0 - \frac{RT}{2F} \ln \frac{[Zn^{2+}]}{[Cu^{2+}]} \quad ...(viii)$$

Substituting the values of R = 8.314 joule deg^{-1} mol^{-1} and F = 96500 coulombs and substituting In by log, at 25°C, eq. (*viii*) changes to

$$E = E^0 - 0.0295 \log \frac{[Zn^{2+}]}{[Cu^{2+}]} \quad ...(ix)$$

Important: If the potential is given for the reaction

$$M \longrightarrow M^{n+} + ne$$

It shows that it is oxidation potential. Similarly, if the value of potential is given as $E_{M/M^{n+}}$ it again expresses oxidation potential. Reduction potential may be obtained by changing the sign.

Q. 29. Calculate the EMF of a zinc-silver cell at 30°C when activity of Zn^{2+} ions is 0.5 and the activity of Ag^+ ions is 10.

Standard reduction potentials at 30°C are :

(*i*) Ag^+, Ag electrode = + 0.799 volt

(*ii*) Zn^{2+}, Zn electrode = – 0.760 volt

Ans. Since reduction potential of silver (0.799 volt) is greater than that of zinc (– 0.760 volt), this means that in the cell reaction, reduction takes place at the silver electrode and oxidation at the zinc electrode. Hence the cell reaction will be

$$Zn + 2Ag^+ \longrightarrow Zn^{2+} + 2Ag$$

Here $n = 2$.

The cell may be represented, by convention, as

$$Zn \mid Zn^{2+}\ (a = 0.5) \parallel Ag^+\ (a = 10) \mid Ag$$

∴ Standard EMF of the cell, $E^0 = E^0_{Ag^+, Ag} - E^0_{Zn^{2+}, Zn}$

$$= +\,0.799 - (-\,0.760) \text{ volt} = 1.559 \text{ volt}$$

By Nernst equation, the EMF of the given cell is

$$E^0 = E^0 - \frac{RT}{F} \ln \frac{a_{Zn^{2+}}}{a^2_{Ag^+}}$$

$$= 1.599 - \frac{8.314 \times (30+273) \times 2.303}{2 \times 96500} \log \frac{0.5}{(10)^2}$$

$$= 1.559 - 0.03006 \log (5 \times 10^{-3})$$

$$= 1.559 + 0.069 = \mathbf{1.628 \text{ volt}}$$

Q. 30. Represent schematically the cell made up of the following half cell reactions:

$$\mathbf{Mg \longrightarrow Mg^{2+}\ (0.01\ \mathit{M}) + 2e^-,\ E^0 = +\ 2.34\ volts}$$

$$\mathbf{Sn^{2+}\ (0.1\ \mathit{M}) + 2e^- \longrightarrow Sn,\ \ E^0 = -\ 0.136\ volt}$$

Calculate the EMF of the above cell at 25°C.

Ans. Oxidation potential of Mg-electrode is +2.34 volt. Its reduction potential will therefore be – 2.34 volt. The potential of tin electrode given is reduction potential.

By convention, the cell will be represented as

$$Mg \mid Mg^{2+}\ (0.01\ M) \parallel Sn^{2+}\ (0.1\ M) \mid Sn$$

Standard EMF of the cell (E^0)

$$= -0.136 - (-2.34)$$
$$= 2.204 \text{ volts.}$$

The overall reaction is

$$Mg + Sn^{2+} (0.1\ M) \longrightarrow Mg^{2+} (0.01\ M) + Sn$$

Here $n = 2$.

By Nernst equation, the EMF of the cell at 25°C is given by

$$E = E^0 - \frac{0.0591}{2} \log \frac{[Mg^{2+}]}{[Sn^{2+}]} \qquad \text{[as [Mg] = [Sn] = 1]}$$
$$= 2.204 - \frac{0.0591}{2} \log \frac{0.01}{0.1}$$
$$= 2.204 - 0.02955 \times (-1)$$
$$= 2.23355 \text{ volt} \simeq \mathbf{2.234 \text{ volt}}$$

Q. 31. Derive Nernst equation giving the effect of concentration and temperature of surrounding electrolyte on the electrode potential.

(Kanpur 2001; Nagpur 2002; Purvanchal 2007; Raipur 2010; Himachal 2010)

Ans. Take the example of zinc electrode in which zinc metal is in equilibrium with zinc ions. The electrode reaction, written as reduction reaction will be

$$Zn^{2+} + 2e^- \longrightarrow Zn \ (n = 2)$$

Applying Nernst equation, the electrode potential of the zinc electrode is given by

$$E_{Zn^{2+},Zn} = E^0_{Zn^{2+},Zn} - \frac{RT}{2F} \ln \frac{a_{Zn}}{a_{Zn^{2+}}} \qquad ...(i)$$
$$= E^0_{Zn^{2+},Zn} + \frac{RT}{2F} \ln \frac{a_{Zn^{2+}}}{a_{Zn}}$$

where $E^0_{Zn^{2+},Zn}$ represents the standard electrode potential of the zinc electrode. In the symbols, $E_{Zn^{2+}}$ *Zn* and $E^0_{Zn^{2+},Zn}$ the ions have been written first because these represent *reduction* potentials.

Putting $a_{Zn} = 1$ (because the activity of a pure solid is unity), equation (*i*) becomes

$$E_{Zn^{2+},Zn} = E^0_{Zn^{2+},Zn} + \frac{RT}{2F} \ln a_{Zn^{2+}}$$

Further, assuming that the solution is dilute, the activity of Zn^{+2} ions may be replaced by molar concentration so that we have

$$E_{Zn^{2+},Zn} = E^0_{Zn^{2+},Zn} + \frac{RT}{2F} \ln [Zn^{2+}]$$

In a similar manner, the expression for the silver electrode may be derived. The electrode reaction, written as reduction reaction, is

$$Ag^+ + e^- \longrightarrow Ag \ (n = 1)$$

Applying Nernst equation and simplifying the result obtained, we shall get

$$E_{Ag^+,Ag} = E^0_{Ag^+,Ag} + \frac{RT}{F} \ln [Ag^+]$$

In general, for any metal electrode, consisting of the metal (*M*) in equilibrium with its ions, (M^{n+}), the electrode reaction, written as reduction reaction, is

$$M^{n+} + ne^- \longrightarrow M$$

The electrode potential will then be given by the expression

$$E_{M^{n+},M} = E^0_{M^{n+},M} + \frac{RT}{nF} \ln [M^{n+}]$$

Putting $R = 8{\cdot}314$ joules, $F = 96500$ coulombs, $\ln[M^{n+}] = 2.303 \log [M^{n+}]$ and $T = 25 + 273$K (assuming that the measurements are carried out at 25°C), the above equation may be simplified to the form

$$E_{M^{n+},M} = E^0_{M^{n+},M} + \frac{0.0591}{n} \log [M^{n+}]$$

This is **Nernst equation** for the calculation of electrode potential of metal-metal ion electrode.

Q. 32. Derive Nernst equation for the potential of hydrogen-electrode.

Ans. Taking the case of hydrogen electrode, consisting of H_2 gas in equilibrium with H^+ ions, the electrode reaction, written as reduction reaction, is

$$H^+ + e^- \longrightarrow \frac{1}{2}H_2 (n = 1)$$

Applying Nemst equation, the electrode potential of the hydrogen electrode is given by

$$E_{H^+,H_2} = E^0_{H^+,H_2} - \frac{RT}{F} \ln \frac{(a_{H_2})^{1/2}}{a_{H^+}}$$

If H_2 gas is at 1 atmospheric pressure, $a_{H_2} = 1$

$$\therefore \quad E_{H^+,H_2} = E^0_{H^+,H_2} - \frac{RT}{F} \ln \frac{1}{a_{H^+}}$$

$$= E^0_{H^+,H_2} + \frac{RT}{F} \ln a_{H^+}$$

Replacing the activity of H^+ *ions* by its molar concentration, we have

$$E_{H^+,H_2} = E^0_{H^+,H_2} + \frac{RT}{F} \ln[H^+]$$

Since the standard electrode potential of hydrogen electrode is taken as zero

$$E^0_{H^+,H_2} = 0$$

Hence $$E_{H^+,H_2} = \frac{RT}{F} \ln [H^+] = 0.0591 \log[H^+] \text{ at } 25°C$$

Q. 33. Derive Nernst equation showing the effect of electrolyte concentration on the potential of

(*i*) Calomel electrode

(*ii*) Oxidation-reduction electrode. *(Kalyani 2003)*

Ans. (*i*) **Calomel electrode:** The electrode reaction, written as reduction reaction, is

$$Hg_2Cl_2(s) + 2e^- \longrightarrow 2Hg(l) + 2Cl^- \ (n = 2)$$

By Nernst equation, the electrode potential of this electrode is given by

$$E_{cal} = E^0_{cal} - \frac{RT}{2F} \ln \frac{a^2_{Cl^-} \times a^2_{Hg}}{a_{Hg_2Cl_2}}$$

As Hg is a pure liquid and Hg_2Cl_2 is a pure solid, therefore $a_{Hg} = 1$ and $a_{Hg_2Cl_2} = 1$. Hence we have

$$E_{cal} = E^o{}_{cal} - \frac{RT}{2F} \ln a^2_{Cl^-}$$

$$= E^o{}_{cal} - \frac{RT}{F} \ln a_{Cl^-}$$

or in terms of molar concentration in place of activity, we have

$$E_{cal} = E^o{}_{cal} - \frac{RT}{F} \ln [Cl^-]$$

(*ii*) **Oxidation-reduction electrodes:** Consider the electrode prepared by putting a platinum wire in a solution of a mixture of a ferrous salt and a ferric salt. The electrode reaction, expressed as reduction reaction, is

$$Fe^{3+} + e^- \longrightarrow Fe^{2+} \quad (n = 1)$$

Applying Nemst equation, the electrode potential of this electrode will be given by

$$E_{Fe^{3+}, Fe^{2+}} = E^0_{Fe^{3+}, Fe^{2+}} - \frac{RT}{2F} \ln \frac{a_{Fe^{2+}}}{a_{Fe^{3+}}}$$

or using molar concentrations in place of activities, we have

$$E_{Fe^{3+}, Fe^{2+}} = E^0_{Fe^{3+}, Fe^{2+}} - \frac{RT}{F} \ln \frac{[Fe^{2+}]}{[Fe^{3+}]}$$

or $$\boldsymbol{E_{Fe^{3+}, Fe^{2+}} = E^0_{Fe^{3+}, Fe^{2+}} - \frac{RT}{F} \ln \frac{[Fe^{3+}]}{[Fe^{2+}]}}$$

The electrode reaction can be written in a general form as

Oxidized state + n electrons = Reduced state

Applying Nerst equation, the expression for the electrode potential will be

$$E = E^0 - \frac{RT}{nF} \ln \frac{[\text{Reduced state}]}{[\text{Oxidized state}]}$$

or $$\boldsymbol{E = E^0 + \frac{RT}{nF} \ln \frac{[\text{Oxidized state}]}{[\text{Reduced state}]}}$$

Here E represents the electrode potential for the given concentrations of the ions involved and $E°$ represents the standard electrode potential.

Q. 34. Calculate the electrode potentials (reduction potentials) of each of the following single electrodes at 25°C.

(*i*) Sn | Sn^{++} (*a* = 0.01) $\quad$ $\boldsymbol{E^0_{Sn^{++}, Sn} = -0.14V}$

(*ii*) Ag | AgCl(*s*), Cl^- (*a* = 0.0001) $\quad$ $\boldsymbol{E^0_{AgCl, Cl^-} = 0.22V}$

Ans. (*i*) The electrode reaction is

$$Sn^{++} + 2e^- \longrightarrow Sn \ (n = 2)$$

$$\therefore \quad E_{Sn^{2+}, Sn} = E^0_{Sn^{2+}, Sn} + \frac{RT}{2F} \ln a_{Sn^{2+}}$$

Substituting the values

$$E_{Sn^{2+}, Sn} = -0.14 + \frac{0.0591}{2} \log 0.01$$

$$= -0.14 - 0.0591$$

$$= -0.1991 \text{ V}$$

(*ii*) The electrode reaction is

$$AgCl(s) + e^- \longrightarrow Ag(s) + Cl^- \; (n = 1)$$

$$\therefore \quad E_{AgCl,Cl^-} = E^0_{AgCl,Cl^-} - \frac{RT}{F} \ln a_{Cl^-}$$

Substituting the values

$$E_{AgCl,\,Cl^-} = 0.22 - 0.0591 \log 0.0001$$
$$= 0.22 + 0.2364$$
$$= 0.4564 \text{ V}$$

Q. 35. Calculate the electrode potential of a copper wire dipped in 0.1 molar copper sulphate solution at 25°C. At this temperature, the standard electrode potential of copper is 0.34 volt. (F = 96500 coulombs, R = 8.314 joules/degree/mol). Assume $CuSO_4$ to be completely ionized and take the activity of copper ions equal to the molar concentration.

Ans. The electrode reaction as reduction reaction may be written as

$$Cu^{2+} + 2e^- \longrightarrow Cu$$

Here $n = 2$

By Nernst equation, the electrode potential is given by

$$E = E^0 + \frac{RT}{nF} \ln \frac{[Cu^{2+}]}{[Cu]}$$

The concentration of solid is taken as unity, thus [Cu] = 1

$$E = E^0 + \frac{2.303RT}{nF} \log [Cu^{2+}]$$

$$= 0.34 + \frac{2.303 \times 8.314 \times 298}{2 \times 96500} \log (0.1)$$

$$= 0.34 + (-0.0296) = \mathbf{0.3104 \text{ volt}}$$

Q. 36. A zinc electrode is placed in 0.1 M solution of zinc sulphate at 25°C. If the degree of dissociation of salt at this concentration is found to be 0.95, calculate the electrode potential of the electrode at 25°C. Given that $E^0_{Zn^{2+},Zn} = -0.76$ volt.

Ans. As $\alpha = 0.95$ *i.e.*, the salt is 95% dissociated,

$$[Zn^{+2}] = \frac{95}{100} \times 0.1 = 0.095$$

$$E = E^0 + \frac{0.0591}{2} \log 0.095$$

$$= -0.76 + \frac{0.0591}{2} \times (-1.022)$$

$$= -0.76 - 0.03$$

$$= -\ \mathbf{0.79 \text{ volt}}$$

Q. 37. What are concentration cells ? Give examples.

(Bangalore 2002; Banaras 2004; Nagpur 2008)

Ans. Cells in which the EMF produced is only due to the difference in the concentrations of the electrodes or that of the solutions of the electrolytes with which they are in contact are called **concentration cells.** When the difference is in the concentration of the electrodes, the cells are called **electrode concentration cells.** When the difference is in the concentration of the electrolytes, the cells are called **electrolyte concentration cells.**

Example of electrode concentration cell:

$$H_2(P_{H_2} = P_1) | H^+ | H_2(P_{H_2} = P_2)$$

In the above cell both electrodes are hydrogen electrodes immersed in the same solution of H^+ ions but the H_2 gas passed in the two electrodes is at different pressures (Fig. 11.14).

In the above cell, there is no transference of electrolyte and hence this is called concentration cell without transference.

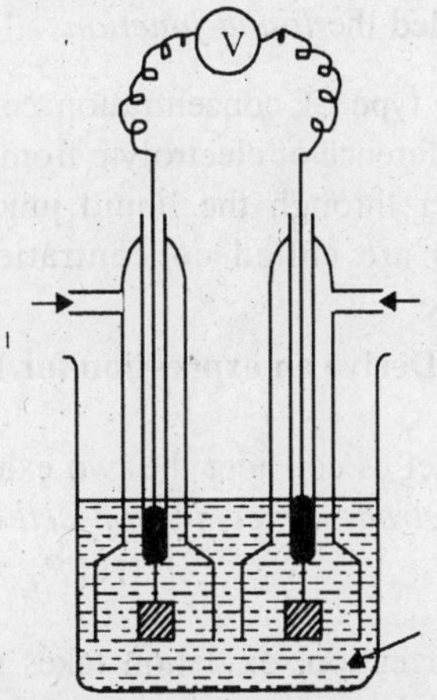

Fig. 11.14. Electrode concentration cell consisting of hydrogen electrodes.

Examples of electrolyte concentration cells :

(*A*) *In which the two electrodes are same, having different concentrations of the electrolytes and connected by a salt bridge, e.g.*

(*i*) H_2 (1 atm.) | H^+ (a_1) || H^+ (a_2) | H_2 | (1 atm.)
i.e. both the electrodes are hydrogen electrodes and in each electrode the H_2 gas at one atmospheric pressure is passed but the activity of the H^+ ions is different in the two electrodes. Further the two electrodes are connected by a salt bridge.

(*ii*) Zn | Zn^{2+} (a_1) || Zn^{2+} (a_2) | Zn

i.e., both the electrodes are zinc electrodes but in one electrode, the zinc metal is in contact with Zn^{+2} ions having activity a_1, and in the other electrode, the zinc metal is in contact with Zn^{2+} ions having activity a_2. Again the two electrodes are connected by a salt bridge (Fig. 11.15).

Again this is a concentration cell without transference.

(*B*) *In which the two electrodes are same having different concentrations of the electrolytes but the two solutions are directly in contact with each other.* This can be done by putting the two solutions in a tube very carefully so that there is no mixing of the two solutions. (Fig. 11.16)

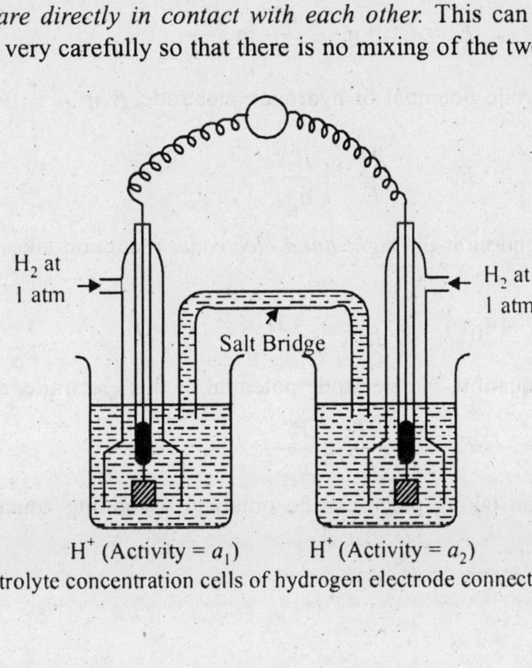

Fig. 11.15. Electrolyte concentration cells of hydrogen electrode connected by a salt bridge.

In such a case, the cells are represented as

(*i*) H_2 (1 atm.) 1 HCl (a_1) : HCl(a_2) | H_2 (1 atm)

(*ii*) Zn | $ZnSO_4$ (a_1) : $ZnSO_4$ (a_2) | Zn

i.e., the double line is replaced by a dotted line.

The junction at which the two solutions meet is called the *liquid junction.*

In this type of concentration cell, there is a direct transference of electrolyte from one solution to the other through the liquid junction. Hence these cells are called concentration cell with transference.

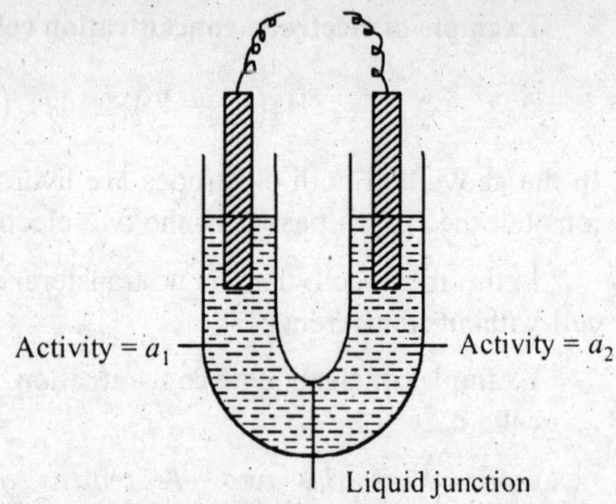

Fig. 11.16. Electrolyte concentration with a liquid junction.

Q. 38. Derive an expression for EMF of electrode concentration cell without transference.
(*Kolhapur 2000; Panjab 2000; G. Nankdev 2003*)

Ans. Let us consider the two examples of electrode concentration cells.

(*i*) *Electrode concentration cell consisting of gas electrodes e.g.,*

$$H_2\left(P_{H_2} = P_1\right) | H^+\left(a_{H^+}\right) | H_2\left(P_{H_2} = P_2\right)$$

By convention, oxidation takes place in the *left had electrode i.e.,* the reaction in the left hand electrode is

$$\frac{1}{2}H_2\,(P_1) \longrightarrow H^+\left(a_{H^+}\right) + e^- \qquad ...(i)$$

To apply Nernst equation, this reaction may be written as reduction reaction *i.e.,*

$$H^+(a_{H^+}) + e^- \longrightarrow \frac{1}{2}H_2(P_1)(n=1) \qquad ...(ii)$$

∴ Electrode potential (reduction potential) of this electrode according to Nernst equation will be

$$E_1 = E^0{}_{H^+.H_2} - \frac{RT}{F}\ln\frac{P_1^{1/2}}{a_{H^+}}$$

But standard electrode potential of hydrogen electrode, $E^0{}_{H^+.H_2} = 0$

$$\therefore \qquad E_1 = -\frac{RT}{F}\ln\frac{P_1^{1/2}}{a_{H^+}} \qquad ...(iii)$$

Similarly by convention at the *right hand electrode,* reduction takes place *i.e* the reaction is

$$H^+\left(a_{H^+}\right) + e^- \longrightarrow \frac{1}{2}H_2\,(P_2)\,(n=1)$$

Applying Nemst equation, the electrode potential of this electrode, as before, will be

$$E_2 = -\frac{RT}{F}\ln\frac{P_2^{1/2}}{a_{H^+}} \qquad ...(v)$$

The overall reaction taking place can be obtained by adding equations (*i*) and (*iv*) which gives

$$\frac{1}{2}H_2(P_1) \longrightarrow \frac{1}{2}H_2(P_2) \quad ...(vi)$$

The EMF (E) of the cell is given by the relation

$$E = \begin{pmatrix}\text{Electrode pot.of the}\\ \text{right hand electrode}\end{pmatrix} - \begin{pmatrix}\text{Electrode pot. of the}\\ \text{left hand electrode}\end{pmatrix}$$

$$= E_2 - E_1$$

$$= -\frac{RT}{F}\ln\frac{P_2^{1/2}}{a_{H^+}} + \frac{RT}{F}\ln\frac{P_1^{1/2}}{a_{H^+}}$$

$$= -\frac{RT}{F}\left(\ln P_2^{1/2} - \ln a_{H^+}\right) + \frac{RT}{F}\left(\ln P_1^{1/2} - \ln a_{H^+}\right)$$

$$= -\frac{RT}{F}\ln P_2^{1/2} + \frac{RT}{F}\ln P_1^{1/2}$$

$$= -\frac{RT}{2F}\ln P_2 + \frac{RT}{2F}\ln P_1$$

$$= \frac{RT}{2F}\left[\ln P_1 - \ln P_2\right]$$

$$= \frac{RT}{2F}\ln\frac{P_1}{P_2} \quad ...(vii)$$

From equation (*vii*), we come to know that there is no net chemical reaction. There is no transference of electrolyte. Further equation (*vii*) shows that the EMF of the cell depends only upon the pressures P_1 and P_2 and does not depend upon the activity of the H^+ ions.

(*ii*) *Electrode concentration cell consisting of amalgam electrodes e.g.*

$$Zn\,(Hg)(a_{Zn} = a_1) |\, Zn^{2+}\,(a_{Zn^{2+}})\,|\, Zn\,(Hg)(a_{Zn} = a_2)$$

For left hand electrode

The reaction is

$$Zn\,(Hg)\,(a_1) \longrightarrow Zn^{2+}\,(a_{Zn^{2+}}) + 2e^- \quad ...(ix)$$

Writing as reduction half reaction, we have

$$Zn^{2+}\,(a_{Zn^{2+}}) + 2e^- \longrightarrow Zn\,(Hg)(a_1)\ (n = 2).$$

By Nernst equation, electrode potential will be

$$E_1 = E^0_{Zn^{2+},Zn} - \frac{RT}{2F}\ln\frac{a_1}{a_{Zn^{2+}}} \quad ...(x)$$

For right hand electrode

The reaction is

$$Zn^{2+}\left(a_{Zn^{2+}}\right) + 2e^- \longrightarrow Zn(Hg)(a_2) \quad ...(xi)$$

By Nernst equation, electrode potential will be

$$E_2 = E^0_{Zn^{2+},Zn} - \frac{RT}{2F}\ln\frac{a_2}{a_{Zn^{2+}}} \quad ...(xii)$$

Overall reaction is obtained by adding equations (*ix*) and (*xi*). This gives

$$Zn(Hg)(a_1) \longrightarrow Zn\,(Hg)\,(a_2) \quad ...(xiii)$$

EMF of the cell is obtained by the expression

$$E = E_2 - E_1$$

$$= -\frac{RT}{2F}\ln\frac{a_2}{a_{Zn^{2+}}} + \frac{RT}{2F}\ln\frac{a_1}{a_{Zn^{2+}}}$$

$$= \frac{RT}{2F}\ln\frac{a_1}{a_2} \quad ...(xiv)$$

From equation (*xiii*), it is obvious that the EMF of the cell originates due to the transfer of zinc from amalgam where its activity is a_1 to the amalgam where its activity is a_2. There is no transference of the electrolyte (namely Zn^{2+} ions). Further equation (*xiv*) shows that the EMF of the cell depends only upon the activities of zinc in the two amalgams and not at all on the activity of the Zn^{2+} ions in the solution. Further, in the final expression for EMF of the cell, E^0 does not appear. This is true for all the concentration cells. We may, therefore, conclude that for concentration cells, $E^0 = 0$. Hence for a concentration cell in which each electrode reaction involves transference of n electrons, the EMF will be given by the general expression

$$E_{cell} = \frac{RT}{nF}\ln\frac{a_1}{a_2} \quad ...(xv)$$

Q. 39. Derive an expression for EMF of electrolyte concentration cells without transference
(Panjab 2000; G. Nanakdev 2003; Nagpur 2008)

Ans. Let us take the example of hydrogen concentration cell having different concentration of electrolytes.

(*i*) H_2 (1 atm) | H^+ (a_1) || H^+ (a_2) | H_2 (1 atm)

On the left hand electrode, by convention, oxidation takes place *i.e.* the electrode reaction is

$$\frac{1}{2}H_2\ (1\,\text{atm}) \longrightarrow H^+(a_1) + e^- \quad ...(i)$$

To apply Nemst equation, this reaction may be written as reduction half reaction as

$$H^+(a_1) + e^- \longrightarrow \frac{1}{2}H_2\ (1\text{ atm})\ (n = 1)$$

∴ Electrode potential of the left hand electrode will be

$$E_1 \ll E^0_{H,H_2} + \frac{RT}{F}\ln\frac{1}{a_1} \quad [\because a_{H_2} = 1]$$

$$= E^0_{H^+,H_2} - \frac{RT}{F}\ln a_1$$

$$= \frac{RT}{F}\ln a_1 \quad \left[\because E^0_{H^+,H_2} = 0\right] \quad ...(ii)$$

On the *right hand electrode,* by convention, reduction takes place *i.e.* the electrode reaction is

$$H^+(a_2) + e^- \longrightarrow \frac{1}{2}H_2\ (1\text{ atm}) \quad ...(iii)$$

By Nemst equation, as before, the EMF of this electrode will be

$$E_2 = \frac{RT}{F}\ln a_2 \quad ...(iv)$$

The *overall reaction* is obtained by adding the reactions (*i*) and (*iii*). This gives

$$H^+(a_2) \longrightarrow H^+(a_1) \quad ...(v)$$

The EMF of the cell is given by the expression

$$E = E_2 - E_1$$

$$= \frac{RT}{F} \ln a_2 - \frac{RT}{F} \ln a_1$$

$$= \frac{RT}{F} \ln \frac{a_2}{a_1} \quad \text{... (vi)}$$

From equation (*v*), it is obvious that for 1 faraday of electricity to be produced, there is a transfer of 1 mol of H^+ ions from solution having activity a_2 to the solution having activity a_1. The transfer takes place through the salt bridge. Secondly, from equation (*vi*), it is evident that for EMF to be positive, a_2 must be greater than a_1.

(*ii*) For the zinc concentration cell

$Zn \mid Zn^{2+} (a_1) \parallel Zn^{2+} (a_2) \mid Zn$

Proceeding exactly in the same way as above, it can be seen that the overall reaction is

$$Zn^{2+} (a_2) \longrightarrow Zn^{2+} (a_1) \quad \text{...(vii)}$$

and the EMF of the cell is

$$E = \frac{RT}{2F} \ln \frac{a_2}{a_1} \text{ (as } n=2) \quad \text{...(viii)}$$

In general, for any such cell, involving *n* faradays of electricity, the EMF will be

$$E = \frac{RT}{nF} \ln \frac{a_2}{a_1} \quad \text{...(ix)}$$

Further, replacing the activities of the ions by the products of their molalities and activity coefficients *i.e.*, putting $a_1 = m_1 \gamma_1$ and $a_2 = m_2\gamma_2$, we get

$$E = \frac{RT}{nF} \ln \frac{m_2 \gamma_2}{m_1 \gamma_1} \quad \text{...(x)}$$

(*iii*) *Cells obtained by combination of two chemical cells* :

Consider the following chemical cell

H_2 (1 atm) | HC1(a_1), AgCl(s)|Ag

i.e., the hydrogen electrode and the silver-silver chloride electrode are dipped in the same solution of HC1 (of activity = a_1)

For this cell, the reaction is

$$\frac{1}{2} H_2 (1 \text{ atm}) + AgCl(s) \longrightarrow Ag(s) + HCl (a_1) \quad \text{...(xi)}$$

By Nernst equation the EMF of this cell will be

$$E_1 = E^0 - \frac{RT}{F} \ln a_1 \quad \text{...(xii)}$$

$n = 1$ and take the activities of solids Ag and AgCI and that of H_2 at 1 atm as unity.

Again suppose we have the same cell but with a different activity of HC1 acid (say a_2) *i.e.*, we have

H_2 (1 atm.) | HC1(a_2) , AgCl(s) | Ag

The cell reaction will be

$$\frac{1}{2} H_2 (1 \text{ atm.}) + AgCl(s) \longrightarrow Ag(s) + HCl(a_2) \quad \text{...(xiii)}$$

The EMF of this cell by Nernst equation will be

$$E_2 = E^0 - \frac{RT}{F} \ln a_2 \qquad ...(xiv)$$

If the above two cells are connected so as to oppose each other, we have the cell

H_2 (1 atm.) HCl (a_1), AgCl(*s*) Ag – Ag | AgCl(*s*), HCl (a_2) | H_2 (1 atm.)

The *overall reaction* of this cell (a combination of two cells) will be the *difference* between equations (*xi*) and (*xiii*). We have

$$\frac{1}{2}H_2 \text{ (1 atm)} + AgCl(s) \;-\; \frac{1}{2}H_2 \text{ (1 atm)} \;-\; AgCl(s)$$

$$= Ag(s) + HCl\ (a_1) - Ag(s) - HCl\ (a_2)$$

or $$HCl\ (a_2) = HCl\ (a_1) \qquad ...(xv)$$

Similarly, the EMF (*E*) of combination will be the *difference* between equations (*xii*) and (*xiv*) *i.e.*,

$$E = E_1 - E_2 = \left(E^0 - \frac{RT}{F}\ln a_1\right) - \left(E^0 - \frac{RT}{F}\ln a_2\right)$$

$$= \frac{RT}{F}\ln\frac{a_2}{a_1} \qquad ...(xvi)$$

From equation (*xv*), it is obvious that for one faraday of electricity passing through the cell, the overall change taking place is simply the transfer of 1 mol of HCl acid from the solution where its activity is a_2 to the solution where the activity is a_1. Hence it constitutes an *electrolyte concentration cell.* However, there is no direct transference of the electrolyte, HCl from one solution to the other. Hence more correctly speaking, it is an *electrolyte concentration cell without transference.*

The cells of the above type are useful *in the determination of the activity coefficient of the electrolytes* involved. For the purpose, in the cell combination, the concentration of the electrolyte in one of the cells is kept constant and the concentration of the electrolyte in the second cell is varied. For each concentration, the EMF of the cell is measured. Putting $a_1 = m_1^2\gamma_1^2$ and $a_2 = m_2^2\gamma_2^2$ in equation (*xvi*), we get

$$E = \frac{RT}{F}\ln\frac{m_2^2\gamma_2^2}{m_1^2\gamma_1^2} = \frac{2RT}{F}\ln\frac{m_2\gamma_2}{m_1\gamma_1}$$

$$= \frac{2RT}{F}\ln\frac{m_2}{m_1} + \frac{2RT}{F}\ln\frac{\gamma_2}{\gamma_1} \qquad ...(xvii)$$

Knowing γ_2 corresponding to m_1 (which has been kept constant) the values of γ_2 can be calculated for different values of m_2.

Q. 40. Derive an expression for the EMF of an electrolyte concentration cell with transference.

(Meerut 2000; Panjab 2000; Nagpur 2003; Madras 2004; Mumbai2004; Banaras 2004)

Ans. Consider the cell (Fig. 11.17)

$$H_2 \text{ (1 atm)} \mid HCl(a_1) : HCl\ (a_2) \mid H_2 \text{ (1 atm)}$$

Processes taking place at the electrodes: By convention, oxidation takes place on the left hand electrode *i.e.*, the electrode reaction on the left hand electrode is

$$\frac{1}{2}H_2 \text{ (1 atm)} \longrightarrow H^+ (a_1) + e$$

Reduction takes place on the right hand electrode *i.e.*, the electrode reaction on the right hand electrode is

$$H^+ (a_2) + e^- \longrightarrow \frac{1}{2} H_2 \text{ (1 atm)}$$

The sum of the two electrode reactions gives

$$H^+ (a_2) \longrightarrow H^+ (a_1) \qquad ...(i)$$

Processes taking place at the liquid junction: Electrons flow in the external circuit from left to right. However as H^+ ions are produced in the left electrode and Cl^- ions are left in the right electrode, the inner circuit is completed by the transference of Cl^- ions from right to left and the transference of H^+ ions from left to right across the liquid junction.

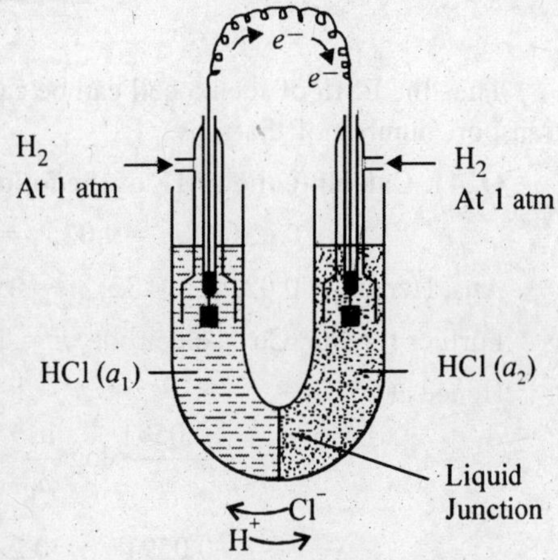

Fig. 11.17. Concentration cell with transference.

If the transport number of Cl^- ions is represented by t_- , then for one faraday of electricity passing transference through the cell, t_- gram equivalents of the Cl^- ions will be transferred from right to left (*i.e.* from solution of activity a_2 to the solution of activity a_1) *i.e.*

$$t_- Cl^- (a_2) \longrightarrow t_- Cl^- (a_1) \qquad ...(ii)$$

Further, transport number of H^+ ions = $1 - t_-$. Therefore during the same time, $(1 - t_-)$ gram equivalents of H^+ ions will be transferred from left to right, *i.e.*, (from solution of activity a_1 to the solution of activity a_2) *i.e.*

$$(1 - t_-) H^+ (a_1) \longrightarrow (1 - t_-) H^+ (a_2) \qquad ...(iii)$$

Overall process: To calculate the *net transfer* of the material, equations (*i*), (*ii*) and (*iii*) must be added

$$H^+ (a_2) + t_- Cl^- (a_2) + (1 - t_-) H^+ (a_1)$$
$$= H^+ (a_1) + t_- Cl^- (a_1) + (1 - t_-) H^+ (a_2)$$

or $$t_- H^+ (a_2) + t_- Cl^- (a_2) = t_- H^+ (a_1) + t_- Cl^- (a_1)$$

or $$t_- HCl(a_2) = t_- HCl (a_1) \qquad ...(iv)$$

This equation communicates that for one faraday of electricity passing through the cell, t_- gram equivalents of HCl from the solution of activity a_2 are transferred to the HCl solution of activity a_1. There for the passage of one faraday of electricity through the cell, one *full gram equivalent* of HCl is transferred.

Applying Nernst equation to the process (*iv*), the EMF of the cell is given by

$$E = -\frac{RT}{F} \ln \frac{a_1^{t_-}}{a_2^{t_-}}$$

$$= \frac{t_- RT}{F} \ln \frac{a_2}{a_1} \qquad ...(v)$$

Here again, for EMF to be positive, a_2 must be greater than a_1.

Replacing the activities of the electrolytes by the molalities and activity coefficients of the ions, we have

$$E = \frac{t_- RT}{F} \ln \frac{m_2^2 \gamma_2^2}{m_1^2 \gamma_1^2}$$

$$= \frac{2t_- RT}{F} \ln \frac{m_2 \gamma_2}{m_1 \gamma_1} \qquad ...(vi)$$

Thus the EMF of such a cell can be calculated if we know molalities, activity coefficients and transport number of the ions.

Q. 41. Calculate the EMF of the following concentration cell at 25°C

Cu | $CuSO_4$ (*m* = 0.02, γ = 0.32) || $CuSO_4$ (*m* = 0.2, γ = 0.11) | Cu

Ans. Here $m_1 = 0.02$, $\gamma_1 = 0.32$, $m_2 = 0.2$, $\gamma_2 = 0.11$

Further for Cu | Cu^{2+} electrode, $n = 2$.

Hence at 25°C

$$E = \frac{0.0591}{n} \log \frac{a_2}{a_1} = \frac{0.0591}{n} \log \frac{m_2 \gamma_2}{m_1 \gamma_1}$$

$$= \frac{0.0591}{n} \log \frac{0.2 \times 0.11}{0.02 \times 0.32}$$

$$= 0.02955 \log 3.4375$$

$$= 0.02955 \times 0.5363 = \mathbf{0.0158\ volt}$$

Q. 42. What is the EMF of a concentration cell consisting of zinc electrodes, one immersed in a solution of 0.05 molar concentration and the other in a solution of 0.5 molar concentration of its ions at 25°C ? The two solutions are connected by a salt bridge.

Ans. Here we are given $C_2 = 0.5$ M, $C_1 = 0.05$ M.

Further, as the cell involves Zn, Zn^{2+} ions, $n = 2$.

Hence at 25°C,

$$E = \frac{0.0591}{n} \log \frac{C_2}{C_1}$$

$$= \frac{0.0591}{2} \log \frac{0.5}{0.05} = \mathbf{0.02955\ volt}$$

Q. 43. What is liquid junction potential ? How can it be minimised ?

(Gujarat 2003; G.Nanakdev 2003; Madurai 2004; Raipur 2010

Ans. *The potential set up at the junction of the two solutions because of the difference in the speeds of the ions moving across the boundary is called liquid junction potential.*

The EMF of a cell should thus not only be the sum of the oxidation potential (E_1) of the electrode where oxidation occurs and reduction potential (E_2) of the electrode where reduction occurs but should also include the junction potential (E_1). In other words, the EMF of the cell should be

$$E = E_1 + E_2 + E_j$$

However, the type of junction employed most commonly is the salt bridge containing an electrolyte such as KC1. It is believed that the K^+ ions and Cl^- ions move almost with equal speed and thus the junction potential is almost reduced to zero. In such a case, the EMF of the cell can be simple taken as $E_1 + E_2$.

Q. 44. Derive an expression for liquid junction potential. *(Allahabad 1994; Gujarat 2003)*

Ans. Consider the following concentration cell with transference

$$H_2(1\ atm) \mid HCl(a_1) \ \vdots\ HCl(a_2) \mid H_2\ (1\ atm)$$

Sum of the two electrode reactions gives

$$H^{+}(a_2) \longrightarrow H^{+}(a_1)$$

Hence the sum of the two electrode potentials *i.e.*, $E_1 + E_2$ is

$$E_1 + E_2 = -\frac{RT}{F}\ln\frac{(a_{H^+})_1}{(a_{H^+})_2}$$

$$= \frac{RT}{F}\ln\frac{(a_{H^+})_2}{(a_{H^+})_1}$$

$$= \frac{RT}{F}\ln\frac{(m_{H^+}\gamma_{H^+})_2}{(m_{H^+}\gamma_{H^+})_1} \qquad ...(i)$$

EMF of the cell considered above is given by

$$E = \frac{2t_{-}RT}{F}\ln\frac{m_2\gamma_2}{m_1\gamma_1} \qquad ...(ii)$$

Subtracting eqn. (*i*) from eqn. (*ii*), we get

$$E_j = E - (E_1 + E_2)$$

$$= \frac{2t_{-}RT}{F}\ln\frac{m_2\gamma_2}{m_1\gamma_1} - \frac{RT}{F}\ln\frac{(m_{H^+}\gamma_{H^+})_2}{(m_{H^+}\gamma_{H^+})_1} \qquad ...(iii)$$

But $(m_{H^+})_2 = m_2$ and $(m_{H^+})_1 = m_1$

Further if the activity coefficient of an ion is taken to be equal to the mean activity coefficient of the electrolyte,

$(\gamma_{H^+})_2 = \gamma_2$ and $(\gamma_{H^+})_1 = \gamma_2$. Hence eqn. (*iii*) becomes

$$E_j = \frac{2t_{-}RT}{F}\ln\frac{m_2\gamma_2}{m_1\gamma_1} - \frac{RT}{F}\ln\frac{m_2\gamma_2}{m_1\gamma_1}$$

$$= (2t_{-} - 1)\frac{RT}{F}\ln\frac{m_2\gamma_2}{m_2\gamma_1} \qquad ...(iv)$$

Also, since $t_{+} + t_{-} = 1$

$\therefore$ $2t_{-} - 1 = t_{-} + (t_{-} - 1) = t_{-} + (-t_{+})$

$= t_{-} - t_{+}$

Hence eqn. (*iv*) may be written in the form

$$\boldsymbol{E_j = (t_{-} - t_{+})\frac{RT}{F}\ln\frac{m_2 g_2}{m_1 g_1}}$$

Thus we find that the liquid junction potential depends not only on the activities of the two solutions constituting the junction but also depends upon the difference between the two transport numbers of the electrolyte. When $t_{-} = t_{+}$, then $E_j = 0$. In case of an electrolyte for which $t_{-} > t_{+}$, E_j will be positive and thus E will be greater than $E_1 + E_2$ by E_j. In case of an electrolyte for which $t_{-} < t_{+}$, E_j will be negative and hence E will be less than $E_1 + E_2$ by E_j.

Q. 45. Describe determination of activity and activity coefficient from EMF measurement.

Ans. To know the activity of a solution corresponding to the given molality, we need to measure the activity coefficient of the electrolyte in the given solution. To understand the method for this determination, consider the example of the *determination of activity coefficient of hydrochloric* acid. For this purpose, consider the following chemical cell without transference :

$$H_2\text{ (1 atm)} \mid HCl\text{ (activity} = a), AgCl(s) \mid Ag$$

i.e. it consists of hydrogen and silver-silver chloride electrodes in hydrochloric acid as the electrolyte. As the two electrodes are reversible with respect to the ions of the electrolyte, they may be immersed directly in the acid to give a cell with no liquid junction. The cell reaction will be

$$\frac{1}{2}H_2(1\text{atm}) + AgCl(s) \longrightarrow Ag(s) + H^+(a_{H^+}) + Cl^-(a_{Cl^-})$$

and the EMF of the cell is given by

$$E_{cell} = E^0_{Ag|AgCl} - \frac{RT}{F}\ln a_{H^+}a_{Cl^-}$$

$$= E^0_{Ag|AgCl} - \frac{RT}{F}\ln a \qquad ...(i)$$

where a is the activity of HCl as a whole.

Now as the activity (a) of HCl at any molality m is related to the mean activity coefficiency $\gamma_\pm$ by the expression $a = m^2\gamma_\pm^2$, substituting this value in eqn. (i), we get

$$E_{cell} = E^0_{Ag|AgCl} - \frac{RT}{F}\ln m^2\gamma_\pm^2$$

$$E_{cell} = E^0_{Ag|AgCl} - \frac{2RT}{F}\ln m - \frac{2RT}{F}\ln\gamma_\pm$$

$$\mathbf{E_{cell} + \frac{2RT}{F}\ln m = E^0_{Ag|AgCl} - \frac{2RT}{F}\ln\gamma_\pm} \qquad ...(ii)$$

All the quantities on the left hand side of this equation are known experimentally. Hence to calculate $\gamma_\pm$, the value of $E^0_{Ag|AgCl}$ is required. To determine $E^0_{Ag|AgCl}$ the quantity $E_{cell} + \frac{2RT}{F}\ln m$ is plotted against m and the result is extrapolated to $m = 0$. When $m = 0$, $\gamma_\pm = 1$, therefore from eqn. (ii) $E^0_{Ag|AgCl} = E_{cell} + \frac{2RT}{F}\ln m$ *i.e.* equal to the value of the ordinate.

Q. 46. Calculate the mean ionic activity coefficient of 0.1 molar hydrochloric acid at 25°C given that the EMF of the cell

$$\mathbf{H_2\ (1\ atm) \mid HCl\ (a), AgCl(s) \mid Ag}$$

is 0.3524 V at 25°C and that the standard electrode potential of Ag – AgCl is 0.2224 V at 25°C.

Ans. For the given cell

$$E_{cell} + \frac{2RT}{F}\ln m = E^0_{Ag|AgCl} - \frac{2RT}{F}\ln\gamma_2$$

or

$$E_{cell} + \frac{2\times2.203RT}{F}\log m = E^0_{Ag|AgCl} - \frac{2\times2.203RT}{F}\log\gamma_\pm$$

Putting $T = 298$ K, $R = 8.314$ JK^{-1} mol^{-1} and $F = 96500$ C, this equation becomes

$$E_{cell} + 0.1183\log m = E^0_{Ag|AgCl} - 0.1183\log\gamma_\pm$$

Substituting $E_{cell} = 0.3524$ V, $E^0_{Ag|AgCl} = 0.2224$ V, and $m = 0.1$ mol kg^{-1}

$$0.3524 + 0.1183 \log 0.1 = 0.2224 - 0.1183 \log \gamma_{\pm}$$

or $$\log\gamma_{\pm} = \frac{-0.3524 + 0.2224 + 0.1183}{0.1183} = -0.0989$$

or $$\boldsymbol{\gamma_{\pm} = 0.796}$$

Q. 47. How do you calculate standard free energy change and equilibrium constant of a given reaction from EMF measurement. ***(Delhi 2003)***

Ans. Electrical energy produced in a cell is equal to the decrease in free energy accompanying the cell reaction *i.e.*

$$-\Delta G = nFE \qquad \text{... (i)}$$

where ΔG is the change in free energy accompanying the reaction, n is the number of the moles of electrons transferred in the cell reaction, F is one faraday (= 96500 columns), E is the EMF of the cell.

If the reactants and products are taken in the standard state, we can write

$$-\Delta G^\circ = nFE^\circ \qquad \text{... (ii)}$$

Thus knowing the value of E°, the value of ΔG° can be calculated.

Further, for a reversible reaction in equilibrium, ΔG° is related to the equilibrium constant K according to the equation

$$\Delta G^\circ = -RT \text{ In } K = -2.303\, RT \log K \qquad \text{... (iii)}$$

Thus knowing ΔG°, K can be calculated at the given temperature T.

Q. 48. Calculate the free energy change of the following cell at 25°C.

$$\mathbf{Sn|Sn^{2+}\ (a = 0.6)\ ||\ Pb^{2+}\ (a = 0.3)|\ Pb}$$

Standard EMF of the cell is 0.014 volt.

Ans. The cell reaction is

$$Sn + Pb^{2+} \longrightarrow Sn^{2+} + Pb$$

Here $n = 2$

By Nernst equation, the EMF of this cell at 25°C is given by

$$E = E^0 - \frac{0.0591}{n} \ln \frac{a_{Sn^{2+}}}{a_{Pb^{2+}}}$$

$$= E^0 - \frac{0.0591}{n} \log \frac{a_{Sn^{2+}}}{a_{Pb^{2+}}}$$

$$= 0.014 - \frac{0.0591}{n} \log \frac{0.6}{0.3} = 0.0051 \text{ volt}$$

$$\therefore \qquad \Delta G = -nFE = -2 \times 96500 \times 0.0051 \text{ joules} = -\mathbf{984.3\ joules}$$

Q. 49. How will you determine the following from EMF measurement

(*i*) Transport number of the ions

(*ii*) Valency of the ion.

Ans. (*i*) **Determination of the transport number of the ions:** For a cell with transference *e.g*

$$H_2(1 \text{ atm}) \mid HCl\ (a_1) : HCl\ (a_2) \mid H_2\ (1 \text{ atm})$$

the EMF is given by

$$E = \frac{t_- RT}{F} \ln \frac{a_2}{a_1} \quad ...(i)$$

By measuring the EMF of the cell and knowing the activities of the two solutions, the transport number of the anion (t_-) can be calculated.

(*ii*) **In the determination of the valency of the ions:** This determination is based upon the fact that the EMF of a concentration cell without transference is given by

$$E = \frac{RT}{nF} \ln \frac{a_2}{a_1} \simeq \frac{RT}{nF} \ln \frac{c_2}{c_1} \quad ...(ii)$$

In this expression n is the valency of the ion involved. Thus knowing all other quantities, n can be calculated.

For example, the valency of mercurous ions, which remained doubtful for a very long time, was determined by this method, by setting up the following cell :

$$\text{Hg} \left| \frac{N}{20} \begin{array}{l}\text{Mercurous nitrate}\\ \text{solution}\end{array} \right\| \frac{N}{2} \begin{array}{l}\text{Mercurous nitrate}\\ \text{solution}\end{array} \left| \text{Hg} \right.$$

Here $\frac{c_2}{c_1} = \frac{1/2}{1/20} = 10$

Putting the values of the various constants in equation (*ii*) at 25°C, we get

$$E = \frac{0.0591}{n} \log \frac{c_2}{c_1}$$

Putting $\frac{c_2}{c_1} = 10$ *i.e.*, $\log \frac{c_2}{c_1} = \log 10 = 1$, we get

$$E = \frac{0.0591}{n}$$

The EMF of the above cell was found to be 0.0295 volt.

Hence $n = \frac{0.0591}{E} = \frac{0.0591}{0.0295} = 2$

Thus the valency of mercurous ion is 2 and it should be represented as Hg_2^{2+} and its salts such as nitrate by $Hg_2(NO_3)_2$.

Q. 50. How will you determine solubility and solubility product of sparingly soluble salts from EMF measurement? ***(Lucknow 2000; Awadh 2000; Dibrugarh 2004)***

Ans. The EMF of a concentration cell is given by

$$E = \frac{RT}{nF} \ln \frac{c_2}{c_1}$$

Thus knowing the EMF of the cell and the concentration of ions in one of the solutions that in the other can be calculated. If the second solution is that of the sparingly soluble salt, its ionic concentration is equal to its solubility assuming that the substance is completely dissociated.

Suppose it is required to find the solubility of the sparingly soluble salt, AgC1.The following cell is set up and its EMF is measured :

$$\text{Ag} \left| \frac{N}{100} \begin{array}{l}AgNO_3\\ \text{solution}\end{array} \right\| \begin{array}{c}\text{Saturated}\\ NH_4NO_3\end{array} \left\| \frac{N}{10} \begin{array}{l}\text{KCl solution saturated}\\ \text{with AgCl}\end{array} \right| \text{Ag}$$

This cell is prepared by dipping a silver electrode in N/100 $AgNO_3$ solution taken in one beaker and dipping another silver electrode in N/10 KCl solution to which 1 or 2 drops of $AgNO_3$ solution are added, forming AgC1 which is enough to make the solution saturated with AgCl. The two beakers are then connected by a salt bridge containing saturated NH_4NO_3 solution.

Calculations. The cell, as set up above, is a concentration cell having higher concentration of Ag^+ ions in the left hand electrode and low concentration of Ag^+ ions in the right hand electrode (produced by the dissociation of AgC1). By convention, writing the electrode with higher concentration of ions on the right hand side, the above cell may be written as

$$\text{Ag}\left|\begin{array}{l}\frac{\text{N}}{10}\ \text{KCl solution saturated}\\ \quad\ \text{with AgCl}\\ \quad\ [\text{Ag}^+]=c\,(\text{say})\end{array}\right\|\begin{array}{l}\text{Saturated}\\ \text{NH}_4\text{NO}_3\end{array}\left\|\begin{array}{l}\frac{\text{N}}{100}\ \text{AgNO}_3\ \text{solution}\\ \quad [\text{Ag}^+]=0.01\end{array}\right|\text{Ag}$$

EMF of this cell at 25°C is given by

$$E = 0.0591 \log \frac{0.01}{c}$$

From this equation, the value of c can be calculated.

Thus in the solution of AgCl in $\frac{N}{10}$ KC1,

$$[Ag^+] = c \text{ gram equivalents/litre.}$$

As AgC1 dissolves very little, the Cl^- ions produced form AgC1 are negligible as compared to hat produced from KCl. Thus in the solution of AgCI in $\frac{N}{10}$ KCl,

$$[Cl^-] = [KCl] = 0.1$$

The solubility product (K_s) of AgC1 will, therefore, be given by

$$K_s = [Ag^+][Cl^-]$$
$$= c \times 0.1$$

If the solubility of AgC1 is s moles/litre, then

$[Ag^+] = [Cl^-] = s$ gram equivalent/litre

$$K_s = [Ag^+][Cl^-] = s \times s = s^2$$

or

$$s = \sqrt{K_s}$$

Knowing the value of K_s, the value of s can be calculated.

Q. 51. How the solubility of AgBr can be determined by EMF measurement ?

Ans. See Q. 50. In the R.H.S. electrode take KBr in place of KCl.

Q. 52. Consider the cell :

$$\textbf{Ag} \mid \textbf{0.005 M AgNO}_3 \parallel \underset{\textbf{salt bridge}}{\textbf{KNO}_3\ \textbf{(satd)}} \parallel \textbf{AgCl (satd)} \mid \textbf{Ag}$$

If the EMF of this cell at 25°C is 0.1585 volt, calculate the solubility of AgCl at this temperature.

Ans. As the solubility of AgCl in water is very small, the concentration of Ag^+ ions in AgCl solution will be very small as compared to that of Ag^+ ion in 0.005 $AgNO_3$ solution. On writing the electrode with higher concentration of Ag^+ ions on the right hand side (by convention), we have

$$Ag \mid AgCl(satd) \parallel KNO_3\ (satd) \parallel 0.005M\ AgNO_3 \mid Ag$$

Let C be the concentration of Ag^+ ions in the left hand electrode of the above concentration cell. Then we have

$$0.1585 = \frac{0.0591}{1}\log\frac{0.005}{C}$$

which gives $\log C = \log 0.0005 - 2.6819$

$$= -3.301 - 2.6819$$

$$= -5.9829$$

$$= \bar{6}.0171$$

or $C = 1.040 \times 10^{-6}$ g eq/litre

$$= 1.040 \times 10^{-6} \times 143.5 \text{ g /litre}$$

$$= \mathbf{1.49 \times 10^{-4}} \textbf{ g/litre}$$

Q. 53. Write detailed notes on

(a) Dipping calomel electrode

(b) Glass electrode *(Nagpur 2008)*

(c) Quinhydrone electrode *(Mumbai 2010)*

Ans. (a) Dipping calomel electrode

Calomel electrode is now commercially available as a self-contained unit. It is used by simply dipping the tip of the electrode in the solution of another electrode whose emf is to be determined. This type of electrode is often referred as the dipping calomel electrode (Fig. 11.18). This saves us from the cumbersome method of preparing the calomel.

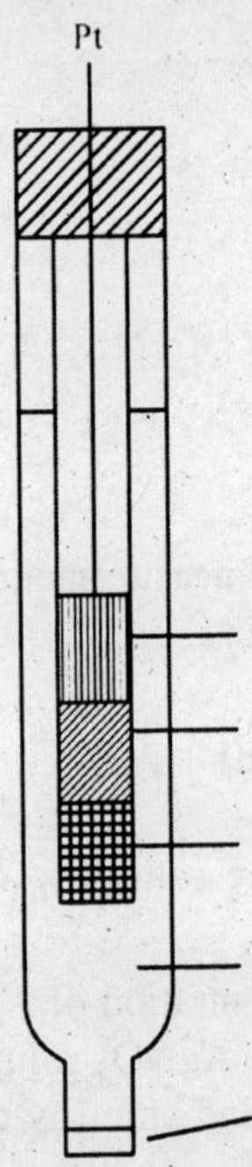

Fig. 11.18. A dipping calomel electrode

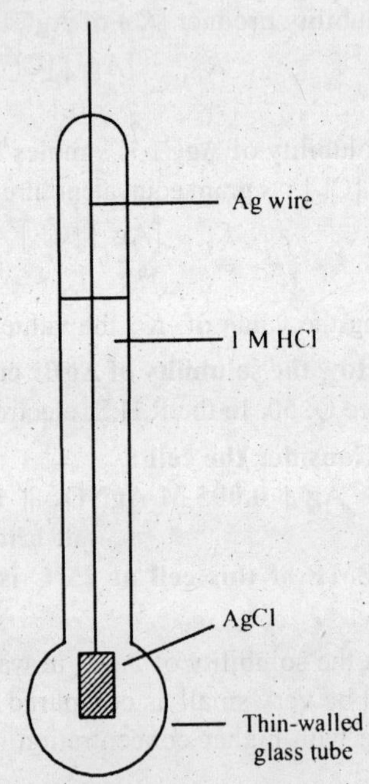

Fig. 11.19. Glass-electrode

The mercury and calomel (Hg_2Cl_2) are held in a narrow central tube by a porous cotton wool plug. A platinum wire dips into the mercury. The KCl saturated solution contained in the electrode makes contact with the other electrode solution through a sintered glass disc. This disc functions as a 'salt bridge' between the two electrode solutions. Thus, a galvanic cell is set up and its emf is measured with the help of a voltmeter.

(*b*) Glass Electrode

A commonly used secondary standard electrode is glass electrode. Its emf is determined by coupling with a standard calomel electrode (SCE). The glass electrode provides one of the easiest methods for measuring the pH of a given solution.

A glass electrode (Fig. 11.19) consists of a glass tube having a thin-walled bulb at the lower end. The bulb contains a 1M HCl solution. A silver wire coated with silver chloride is sealed into the glass tube at its lower end. The lower end of this silver wire dips into the hydrochloric acid, forming silver-silver chloride electrode. The glass electrode may be represented as

$$\text{Ag, AgCl} \mid \text{1 M HCl} \mid \text{H}^+ \text{ (TEST SOLUTION)}$$

When placed in a solution, the potential of the glass electrode depends on the H^+ ion concentration of the solution. The potential develops across the glass membrane as a result of a concentration difference of H^+ ions on the two sides of the membrane. This happens much in the same way as the emf of a concentration cell develops.

The potential of a glass electrode can be determined against a standard calomel electrode (SCE).

Quinhydrone Electrode

It is a widely used secondary standard electrode. It involves the redox reaction between quinone (Q) and hydroquinone (QH_2),

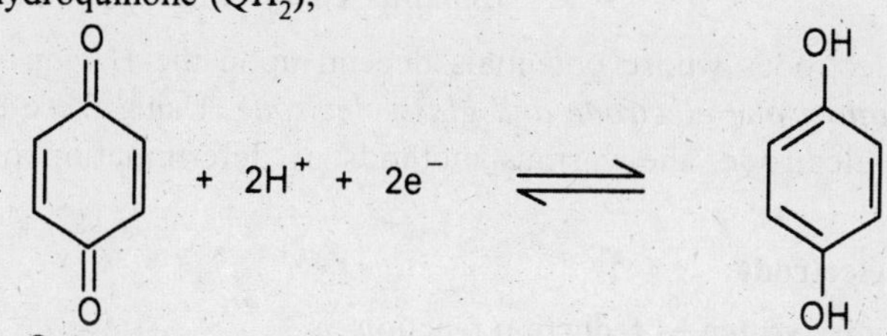

Quinone Hydroquinone

or, simply

$$Q + 2H^+ + 2e^- \rightleftharpoons QH_2$$

The hydroquinone half-cell consists of a platinum strip immersed in a saturated solution of quinhydrone at a definite H^+ ion concentration (Fig. 11.20). Quinhydrone is a molecular

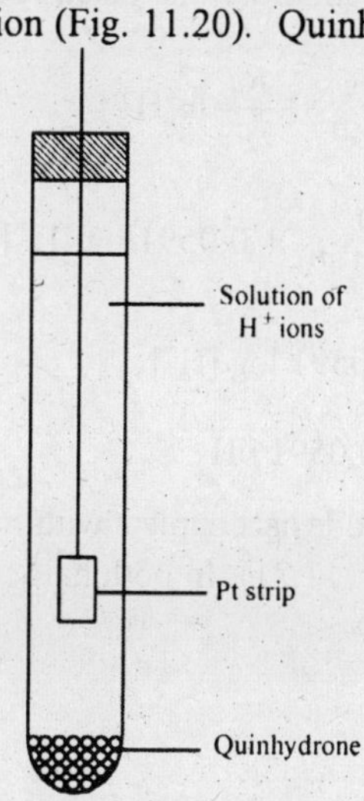

Fig. 11.20. The Quinhydrone Electrode

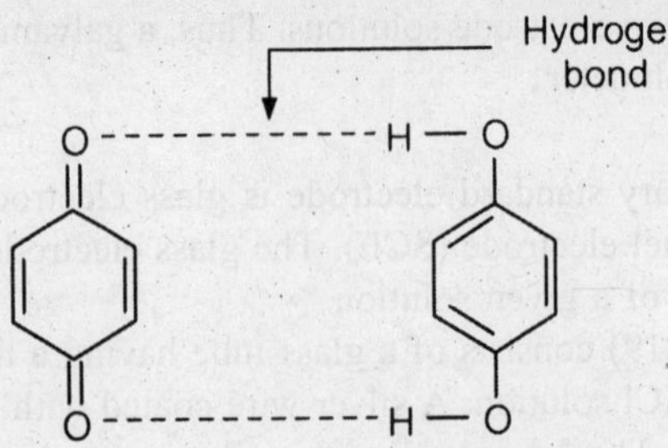

compound which gives equimolar amounts of quinone and hydroquinone in solution. Quinhydrone may be represented as

Pt | QH_2, Q, H^+

The potential developed is measured against a hydrogen electrode or calomel electrode. The emf with respect to a standard hydrogen electrode is 0.2875 V at 25°C.

Q. 54. Describe determination of *p*H or H^+ concentration of a solution by EMF measurement.

(Kanpur 2001; Purvanchal 2002; Lucknow 2002; G. Nanakdev 2003; Banglore 2004; Mumbai 2004)

OR

Describe potentiometric method to determine the *p*H of a solution.

(Banaras 2004; G.Nanakdev 2004; Nagpur 2003)

Ans. The various electrodes whose potentials depend upon the H^+ ion concentration are *hydrogen electrode, quinhydrone electrode and glass electrode.* Using these electrodes against the reference calomel electrode, the various methods of determination of *p*H are briefly described below :

1. Using hydrogen electrode:

The electrode reaction, written as reduction reaction, is

$$H^+ + e^- \longrightarrow \frac{1}{2} H_2 \text{ (1 atm.)}$$

By Nernst equation, the potential of this electrode is given by

$$E_{H^+,H_2} = E^0_{H^+,H_2} + \frac{RT}{F} \ln[H^+]$$

$$= E^0_{H^+,H_2} + 0.0591 \log [H^+] \text{ at } 25°C$$

$$= 0.0591 \log [H^+] \qquad [\text{as } E^0_{H^+,H_2} = 0]$$

$$= -0.0591\, pH \qquad [\text{as } -\log [H^+] = pH]$$

A standard hydrogen electrode is coupled with another hydrogen electrode which contains the solution of unknown pH (Fig. 11.21). In both half cells, hydrogen gas is used at 1 atm pressure and 25°C. The e.m.f. of the cell

Pt | H_2(1 atm) | H^+ (1M) || H^+ (unknown) | H_2 (1 atm) | Pt.
is recorded from the experiment.

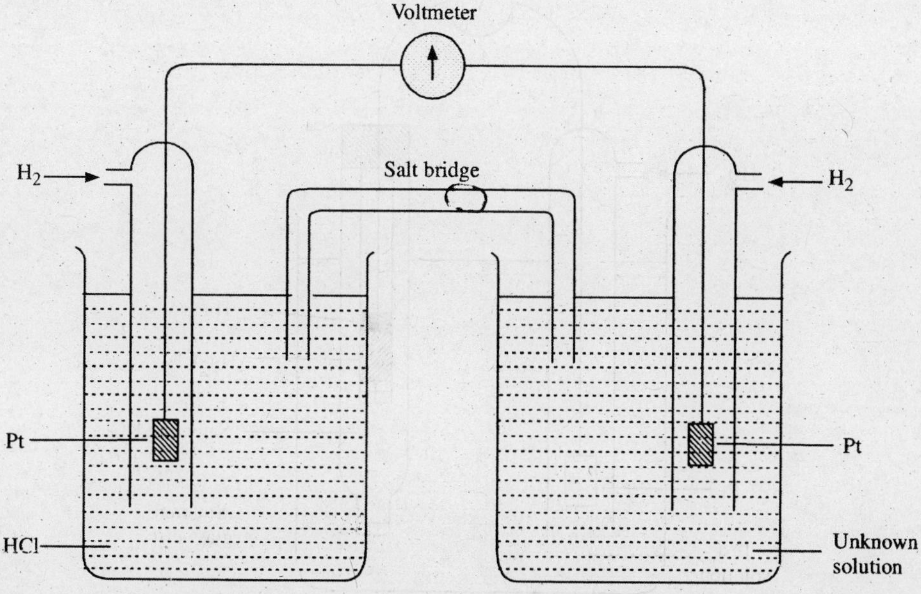

Fig. 11.21. Determination of pH with hydrogen electrode.

The emf of the cell, E_{cell} is given by

$$E_{cell} = E_{R.H.S} - E_{L.H.S}$$

The left hand electrode is standard hydrogen electrode and its e.m.f. is zero. Thus,

$$E_{cell} = E_{R,H.S} = -0.0591 \text{ pH}$$

$$\text{pH} = -\frac{E_{cell}}{0.0591}$$

2. Using Calomel Electrode

When the hydrogen electrode is combined with the calomel electrode, oxidation takes place at the hydrogen electrode and reduction at the calomel electrode. Hence by convention, the cell may be written as

Pt, H_2 (1 atm) | H^+ (unknown conc.) symbol || KCl (saturated) lHg_2Cl_2(s)|Hg

EMF of this cell is given by

$$E = \begin{bmatrix}\text{Electrode potential of} \\ \text{R.H.S. electrode}\end{bmatrix} - \begin{bmatrix}\text{Electrode potential of} \\ \text{the L.H.S. electrode}\end{bmatrix}$$

$$= 0.2415 - (-0.0591\, pH)$$

$$= 0.2415 + 0.0591\, pH$$

or
$$pH = \frac{E - 0.2415}{0.0591}$$

Thus measuring the EMF (E) of the cell potentiometrically, the pH of the solution can be (Fig. 11.22) calculated.

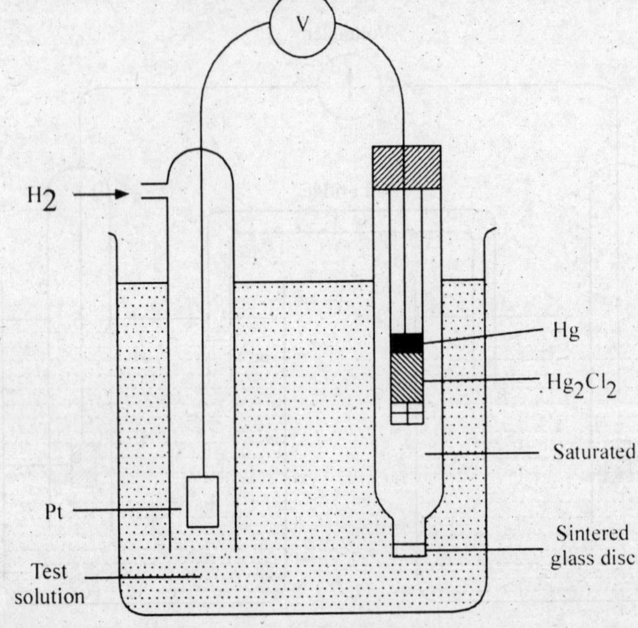

Fig. 11.22. A calomel electrode coupled with unknown hydrogen electrode.

3. Using the quinhydrone electrode:

Quinhydrone is a compound which contains equimolar amounts of quinone ($C_6H_4O_2$) and hydroquinone ($C_6H_6O_2$). The working of the quinhydrone electrode is based upon the fact that quinone is reduced to hydroquinone ions

i.e.,

$$\underset{\text{Quinone}}{C_6H_4O_2} + 2e^- \longrightarrow \underset{\text{Hydroquinone ion}}{C_6H_4O_2^{\,2-}}$$

and the hydroquinone ions combine reversibly with the H^+ ions to form hydroquinone *i.e.,*

$$C_6H_4O_2^{\,2-} + 2H^+ \rightleftharpoons C_6H_6O_2$$

so that the overall reaction is

$$\underset{\text{Quinone}}{C_6H_4O_2} + 2H^+ + 2e^- \rightleftharpoons \underset{\text{Hydroquinone}}{C_6H_6O_2}$$

If quinone is represented by Q, hydroquinone can be represented by QH_2. So the above reaction may be written as

$$Q + 2H^+ + 2e^- \rightleftharpoons QH_2$$

Thus if a platinum wire is dipped into a solution containing H^+ ions and a pinch of a quinhydrone (*i.e.* $Q + QH_2$), it can act as a reversible electrode. The reaction as written above is reduction reaction. Applying Nernst equation, the electrode potential of the above electrode is given by

$$E_Q = E_Q^0 - \frac{RT}{2F} \ln \frac{[QH_2]}{[Q][H^+]^2} \qquad [n = 2 \text{ for the above reaction}]$$

Since quinhydrone is an equimolar mixture of Q and QH_2, taking $[Q] = [QH_2]$ *i.e.,* putting $\frac{[QH_2]}{[Q]} = 1$, in the above equation, we get

$$E_Q = E_Q^0 - \frac{RT}{2F} \ln \frac{1}{[H^+]^2}$$

$$= \quad E_Q^0 + \frac{RT}{2F}\ln[H^+]$$

$$= \quad E_Q^0 + \frac{2.303RT}{F}\log[H^+]$$

$$= \quad E_Q^0 + 0.0591 \log [H^+] \text{ at } 25° \text{ C}$$

$$= \quad E_Q^0 - 0.0591\, pH$$

The standard electrode potential (E_Q^0) of the quinhydrone electrode has been found to be 0.6996 volt. Hence we may write

$$E_Q = 0.6996 - 0.0591\, pH$$

When the quinhydrone is combined with calomel electrode, oxidation takes place at the calomel electrode and reduction at the quinhydrone electrode. Hence by convention, the cell may be written as

$$Hg \mid Hg_2Cl_2(s) \mid KCl \text{ (saturated)} \parallel H^+ \text{ (unknown conc.)} \mid Q \mid QH_2 \mid Pt$$

EMF of this cell is given by

$$E = \begin{bmatrix}\text{Electrode potential of} \\ \text{R.H.S. electrode}\end{bmatrix} - \begin{bmatrix}\text{Electrode potential of} \\ \text{the L.H.S. electrode}\end{bmatrix}$$

$$= (0.6996 - 0.0591\, pH) - 0.2415$$

$$= 0.4581 - 0.0591\, pH$$

or

$$pH = \frac{0.4581 - E}{0.0591}$$

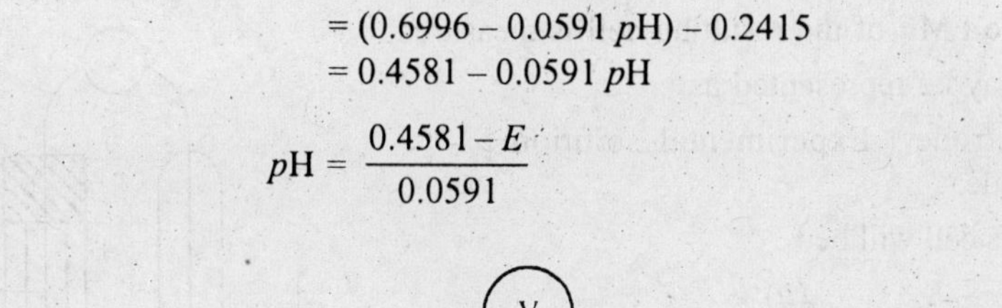

Fig. 11.23. Quinhydrone electrode coupled with standard calomel electrode.

After, measuring the EMF (E) of the cell potentiometrically, the pH of the solution can be calculated.

4. Using glass electrode: If two solutions of different pH are present on the two sides of the glass surface, a potential is established across the glass membrane whose magnitude depends upon the difference in the pH of the two solutions. For a particular variety of a glass, if on one side of the glass surface, a solution of fixed pH value is placed, the potential developed across the glass membrane will depend only upon the pH of the other solution, placed in contact with the other side of the glass surface. The assembly thus obtained is called a glass electrode. (Fig. 11.24).

A tube of a special variety of glass (usually soda glass) is taken which has low melting point and high electrical conductivity. At the end of this tube is blown a glass bulb having a very thin wall. This is filled with a solution of constant pH. To make the electrical contact with this solution, a platinum wire is dipped into this solution, the other end of which is attached to a terminal.

The electrode potential of the glass electrode, as prepared above, is given by the equation

$$E_G = E_G^0 + 0.0591 \log [H^+] \text{ at } 25°C$$

$$= E_G^0 - 0.0591\,pH \quad ...(i)$$

where E_G^0 is a constant depending upon the nature of the glass and the pH of the solution taken inside the glass bulb. To determine the value of E_G^0 for a particular glass electrode, the value of E_G is determined for a solution of known pH value.

To determine the value of E_G the glass electrode is combined with the calomel electrode as usual and the EMF of the resulting cell is measured.

The cell may be represented as

Glass electrode | Experimental solution | Calomel electrode.

EMF of this cell will be

$$E = E_C - E_G \quad ...(ii)$$

where E_C is the electrode potential of the calomel electrode whose value depends upon the concentration of KCl solution used.

From equation (*ii*),

$$E_G = E_C - E$$

Combining this result with equation (*i*), we get

$$E_C - E = E_G^0 - 0.0591\,pH$$

or $$pH = \frac{E_G^0 - E_C + E}{0.0591} \quad ...(iii)$$

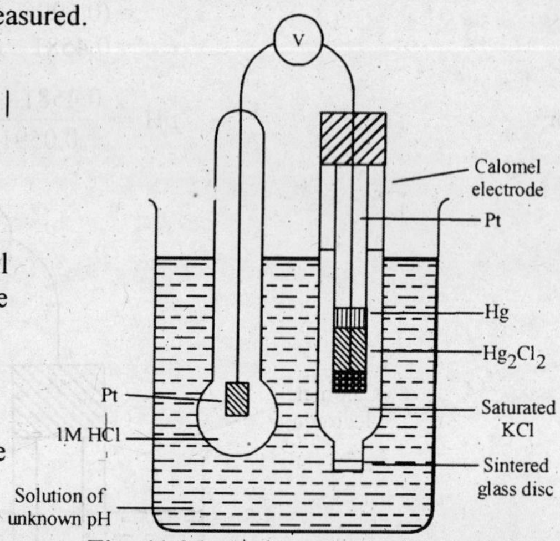

Fig. 11.24. A glass electrode coupled with standard calomel electrode.

Q. 55. Explain the working of a glass electrode in determining the pH of a solution.
(Bangalore 2004; Madras, 2004;Bombay, 2004)

Ans. See Q. 53.

Q. 56. What are the advantages of using glass electrodes ?

Ans. The advantages of the glass electrodes over the other electrodes are as mentioned below :

(*i*) It is easy to use.
(*ii*) It is not poisoned easily.
(*iii*) It is not affected by oxidizing or reducing agents.
(*iv*) It can be used for very small quantities of the solutions.

At present, the glass electrode is the universal pH electrode.

Q. 57. Consider the cell

$$\text{Pt} \mid H_2\text{(g, 1 atm)} \mid H^+ \text{(}x\text{ molar)} \parallel \text{KCl (0.1 M)} \mid Hg_2Cl_2 \mid \text{Hg}$$

If the EMF of this cell is 0.50 volt at 25°C, what would be the pH of the *x* molar acid solution ? (Electrode potential of the calomel electrode is 0.281 volt at 25°C)

Ans. EMF of the given cell is given by

$$E_{\text{cell}} = \begin{bmatrix}\text{Electrode potential} \\ \text{or R.H.S. electrode}\end{bmatrix} - \begin{bmatrix}\text{Electrode potential} \\ \text{or L.H.S. electrode}\end{bmatrix}$$

$$= E_{\text{calomel}} - E_{H^+,H_2}$$

$$\therefore \quad E_{H^+,H_2} = E_{\text{calomel}} - E_{\text{cell}}$$

But by Nernst equation

$$E_{H^+,H_2} = 0.0591 - \log [H^+]$$

$$= -0.0591\, pH \qquad [\text{as } pH = -\log [H^+]]$$

$$\therefore \quad -0.0591\, pH = E_{\text{calomel}} - E_{\text{cell}}$$

or

$$pH = \frac{E_{\text{Cell}} - E_{\text{calomel}}}{0.0591}$$

$$= \frac{0.50 - 0.281}{0.0591} = \mathbf{3.71}$$

Q. 58. Discuss the applications of EMF measurement in potentiometric titrations.

(Lucknow 2000; Awadh 2000 ; Kurukshetra 2002; Purvanchal 2007)

Ans. The principle of these titrations is as under :

In all types of titrations like acid-base titrations, redox titrations, precipitation titrations etc. one of the solutions is taken in the titration flask and the other solution is added to it from the burette. As a result, the concentration of some ion in the titration flask keeps on decreasing. If an electrode reversible with respect to this ion is set up in the titration flask, its electrode potential will keep on changing. If the electrode thus set up in the titration flask (or beaker) is combined with a calomel electrode, the EMF of the cell will keep on changing and show a sudden change in value when the reaction is complete.

Each type of titration is discussed hereunder :

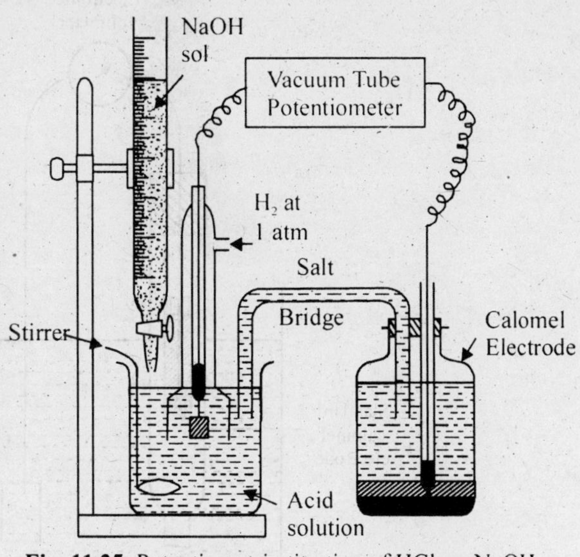

Fig. 11.25. Potentiometric titration of HCl *vs.* NaOH.

(*i*) **Acid Base titrations:** Suppose a solution of HCl acid is to be titrated against NaOH solution. The acid solution is taken in a beaker and a hydrogen electrode (quinhydrone or glass or hydrogen gas) is set up into it. This electrode is then coupled with the calomel electrode. A burette containing the NaOH solution is fitted over the beaker (Fig. 11.25). NaOH solution is added to the beaker first in the larger amounts (say 1 ml or 0.5 ml at a time) and then in very small amounts (say 0.1 ml or even less) near the end point. EMF of the cell is noted after each addition. The EMF is then plotted against the volume of NaOH added when a curve of the type shown in Fig. 11.26 is obtained. From this curve the volume of NaOH solution used corresponding to the equivalence point can be found.

However, when the solutions are very dilute, or weak acids or bases are involved, the steepness of the curve is less marked and it is less marked and it is difficult to judge the end point. In such a case, we plot the slope of the curve $\Delta E / \Delta V$ against the volume of alkali used. The maximum of **the curve indicates the end point (Fig. 11.27).**

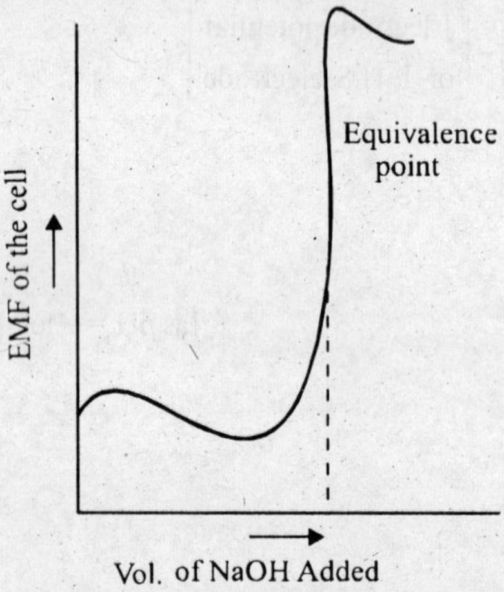

Fig. 11.26 Plot of EMF of the cell vs. vol. of the NaOH solution added.

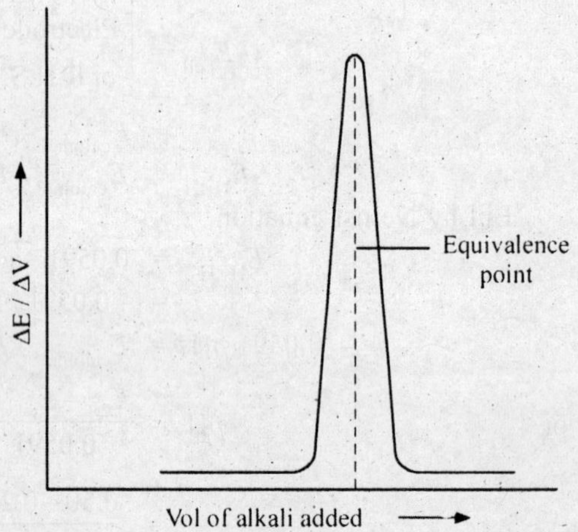

Fig. 11.27. Titration curve of a weak acid and weak base.

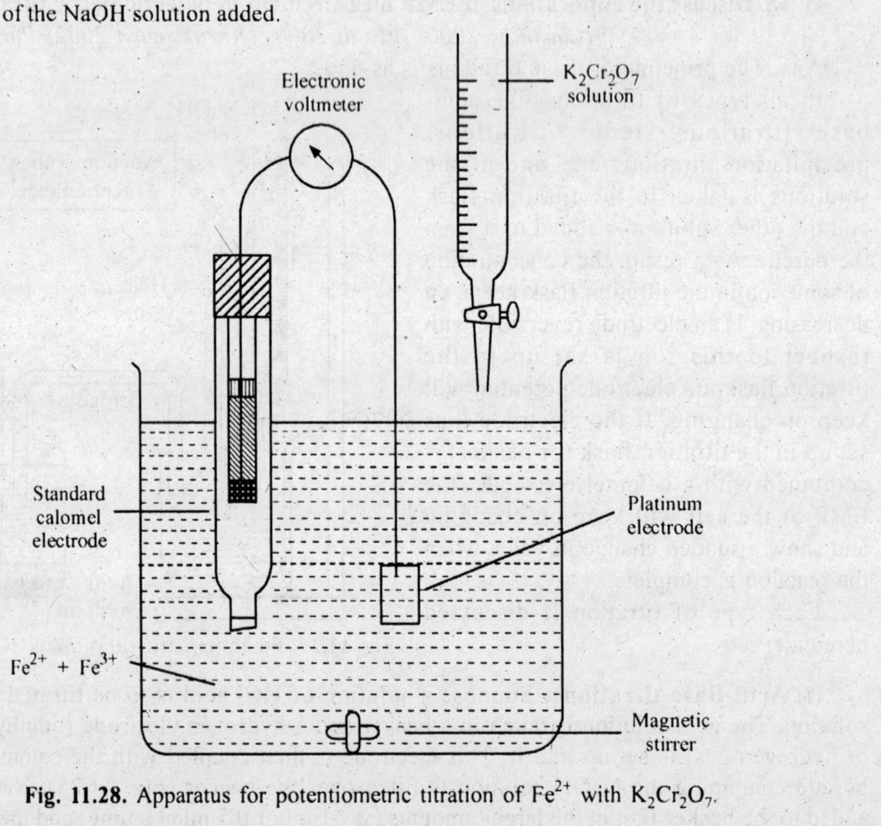

Fig. 11.28. Apparatus for potentiometric titration of Fe^{2+} with $K_2Cr_2O_7$.

(*ii*) **Redox titrations:** Suppose a solution of ferrous sulphate is to be titrated against $K_2Cr_2O_7$ solution. The ferrous sulphate solution along with dilute H_2SO_4 is taken into a beaker and a platinum wire is inserted into it. The electrode thus formed is combined with the calomel electrode. (Fig. 11.28). A burette containing $K_2Cr_2O_7$ solution is fixed over the beaker and after every addition of $K_2Cr_2O_7$ solution, the EMF of the cell is noted. The end point is then determined in a manner similar to that of acid-base titrations (Fig. 11.29).

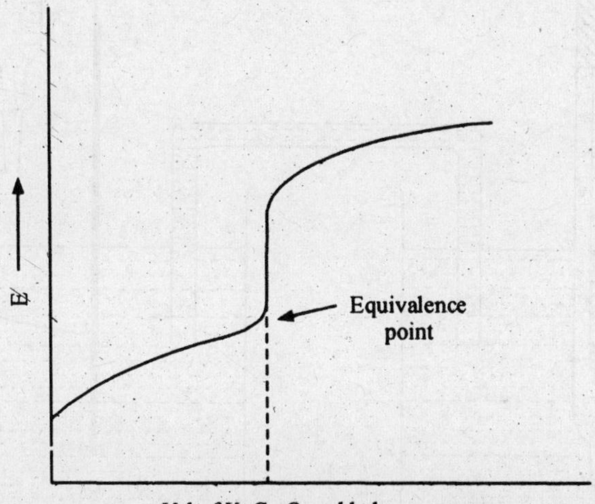

Fig. 11.29. Potentiometric titration curve of Fe^{2+} ions with $K_2Cr_2O_7$

Similarly Fe^{2+} can be estimated potentiometrically using a standard solution of $KMnO_4$. We shall take $KMnO_4$ solution in place of $K_2Cr_2O_7$ solution. Rest of the procedure remains the same. The equivalence point will be determined in cases of Fe^{2+} — $K_2Cr_2O_7$ titration.

(*iii*) **Precipitation titrations:** Suppose a solution of $AgNO_3$ is to titrated against NaCl solution. The silver nitrate solution is taken in a beaker and a silver wire is inserted into it. The electrode thus formed is then combined with the calomel electrode. A burette containing the NaCl solution is fixed over the beaker and after every addition of NaCl solution, the EMF of the cell is noted. The end point is then determined as in the case of acid-base titrations. The obvious reason for getting a similar curve is that the electrode potential of the silver electrode depends upon the concentration of Ag^+ ions in the solution. The addition of NaCl solution results into the precipitation of Ag^+ ions to form AgCl (which is sparingly soluble). The concentration of Ag^+ ions in the solution thus keeps on falling with the addition of NaCl solution till at the end point, all the Ag^+ ions are precipitated. At this stage, a very small amount of Ag^+ ions is still present which are obtained due to the ionization of the AgCl, that has gone into the solution. Further addition of NaCl solution has no effect on the Ag^+ ion concentration except that it suppresses the ionization of AgCl due to common ion effect and as a result the Ag^+ ion concentration decreases very slightly.

$$AgCl \rightleftharpoons Ag^+ + Cl^-$$

$$NaCl \longrightarrow Na^+ + Cl^-$$

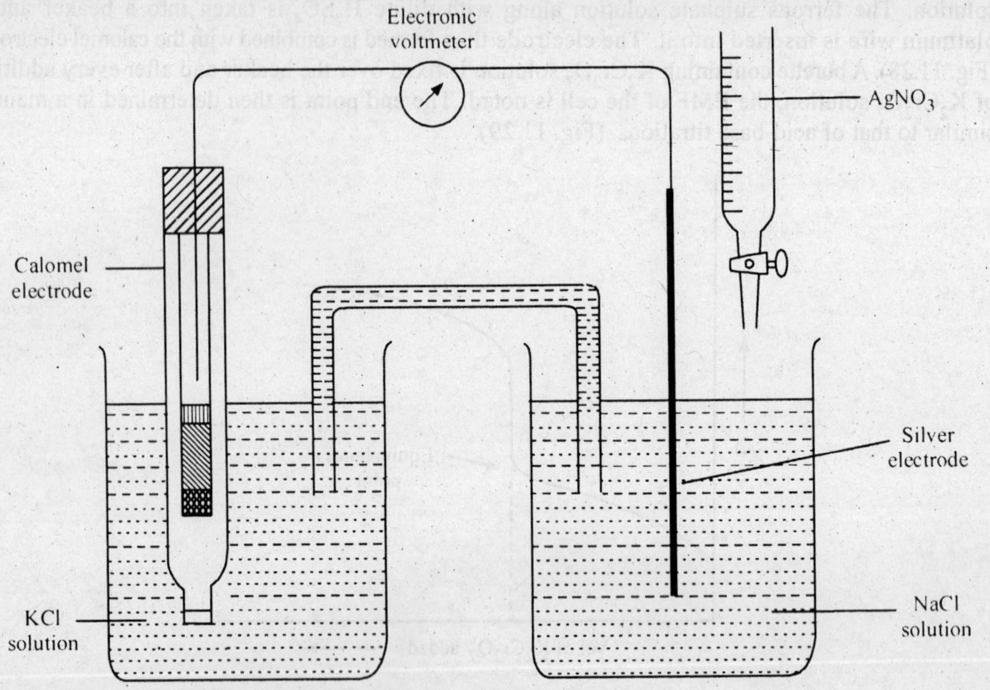

Fig. 11.30. Potentiometric titration of sodium chloride against silver nitrate solution.

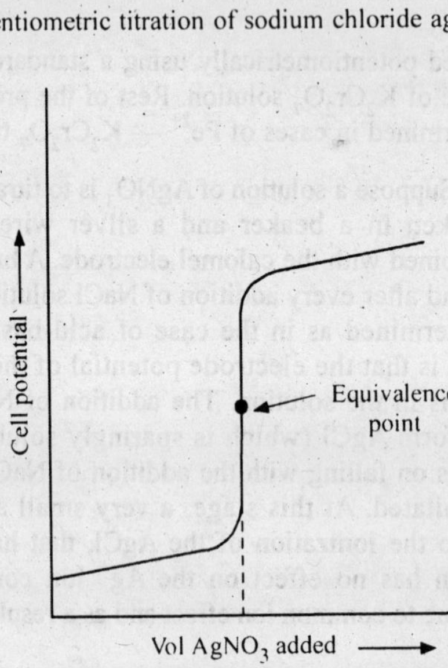

Fig. 11.31. Potentiometric titration curve for NaCl and $AgNO_3$.

Q. 59. Mention some advantages of potentiometric titrations. (*Awadh 2000; Bangalore 2004*)

Ans. (*i*)These titration are very useful in case of coloured solutions where internal indicators fail.

(*ii*) For acid-base titration, no prior knowledge of the *p*H range in which the indicator changes colour is required.

(*iii*)These titrations can be used for the titrations of a mixture of two acids (having different degrees of ionization), with the same NaOH solution. In such a case, two different points of inflexion are obtained. In a similar manner, a mixture of chloride, bromide and iodide can be estimated by titration with $AgNO_3$ solution.

Q. 60. Discuss the theory of and apparatus used in polarography. How is it used in the analysis of ions ? *(Kerala 2000)*

Ans. It is a technique of analysing qualitatively as well as quantitatively a mixture of metal ions present in a very dilute solution (10^{-3} to 10^{-4}M) solution. It was developed by Heyrovsky in 1922.

When the electrolysis of the solution of an electrolyte is carried out the products of electrolysis on reaching the electrodes set up an EMF opposite to the applied EMF. This phenomenon is called **polarization.** For example, in the electrolysis of dilute sulphuric acid using platinum electrodes, as a small EMF is applied, the electrolysis starts but soon the electrodes are covered with the bubbles of the gas which form gas electrodes and produce EMF in the reverse direction. As a result, the electrolysis almost stops. Similarly in the electrolysis of copper sulphate solution using copper electrodes, in the beginning a very small EMF in enough to start the electrolysis but soon the concentration of copper sulphate in the neighbourhood of the anode and cathode becomes different. As a result, a concentration cell is set up with EMF opposite to the applied EMF. Thus in the process of electrolysis, in order that the electrolysis may take place continuously, some minimum EMF must be applied. *The minimum EMF that must be applied to bring about the electrolysis of the solution of a particular substance continuously is called its* **decomposition potential** *or* **discharge potential.**

Thus, if to the solution of an electrolyte, staring from zero, an increasing EMF is applied, a very small current flows initially till the applied EMF becomes equal to the discharge potential of the ion. Further increase in the applied EMF by a slight amount increases the rate of discharge of ions appreciably and so does the current and the rate of diffusion. However, there is a limit to the rate of diffusion, depending upon the concentration of ions in the solution. As soon as this limit is reached, further increase in the applied EMF does not cause any further increase in the current. The current flowing at this stage is called the **limiting current i_L.** The small current that flows through the cell before the decomposition potential is reached is called **residual current** (i_r). The difference between the limiting current and the residual current is called **diffusion current i_d.** The way in which the current changes with increasing applied EMF is shown in Fig. 11.32.

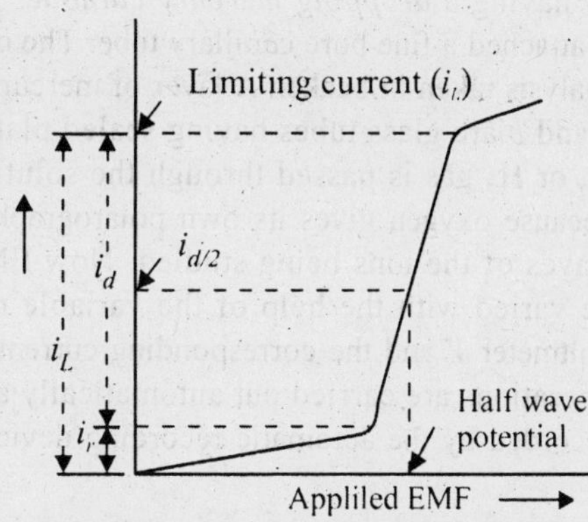

Fig. 11.32. Variation of current with applied EMF in the process of electrolysis.

The half wave potential is characteristic of a particular ion. Thus if a number of metal ions are present in the solution, every ion will give its characteristic wave, having a definite value of half-wave potential. From the half-wave potentials the ions present in the solution can be identified. The results obtained for a mixture are shown in Fig. 11.33. Such a plot of EMF *vs.* current is called a **polarogram.** The small oscillations in the current are due to the growing size of mercury drops, acting as the cathode.

On comparing the heights of the waves with those of the standard solutions, the approximate amount of the various ions can also be determined.

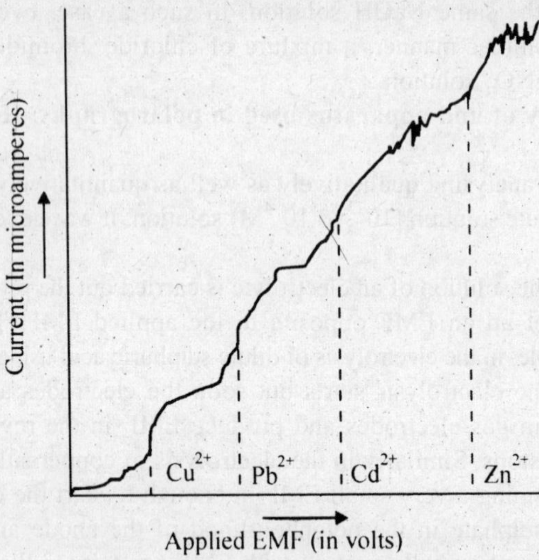

Fig. 11.33. Plot of applied EMF vs. current (Polarogram) for a mixture.

Apparatus: The apparatus used is called a **polorograph** (Fig. 11.34). As the metal ions are discharged on the cathode, therefore for analysing a solution containing a number of metal ions, fresh cathode surface should be available for the discharge of each kind of ions. This is achieved by having a *dropping mercury cathode.* There is a mercury reservoir to the bottom of which is attached a fine bore capillary tube. The other end of the capillary is dipped into the solution for analysis taken in beaker. A layer of mercury taken at the bottom of the beaker acts as the anode. *a* and *b* are glass tubes having sealed platinum wires to make electrical contact. To start with, N_2 or H_2 gas is passed through the solution to remove the dissolved oxygen. This is essential because oxygen gives its own polarographic wave which may interfere with the polarographic waves of the ions being studied. Now EMF is applied with the help of the cell *C*, which can be varied with the help of the variable resistance *R*. The applied EMF can be read from the voltmeter *V* and the corresponding current from the ammeter *A*. In the instruments, the various operations are carried out automatically and the polarograms (graphs of current *vs.* voltage) are recorded by the automatic recording device.

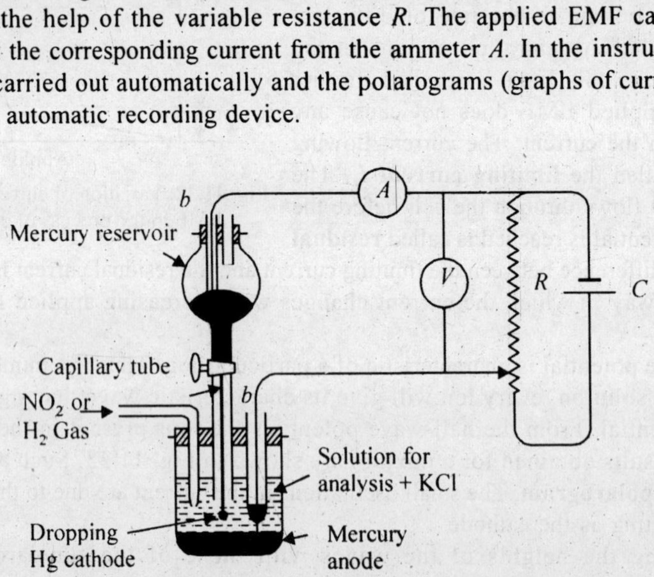

Fig. 11.34. Polarograph dropping mercury cathode method for obtaining a polarogram.

As the surface area of the dropping mercury cathode is very small, the resistance of the cell is very high. Hence to increase the electrical conductivity of the solution, KCl is added. It is called *a supporting electrolyte.* The K^+ ions are reduced at the mercury cathode only at high voltage, therefore they do not interfere in the polarographic analysis of other ions.

Advantages of Polarographic Analysis

(*i*) The greatest advantage of the polarographic analysis is that with this technique, ions even in microconcentration can be analysed.

(*ii*)*By* comparing the heights of the waves, the relative amounts of the ions in the mixture can be estimated.

(*iii*) A number of metal ions present in the solution can be analysed simultaneously in a very short time.

Q. 61. Explain the terms

(*i*) Half-wave potential

(*ii*) Limiting diffusion current. *(Kurukshetra 2002)*

Ans. See Q. 59.

Q.62. Mention the type of electrodes. Write the representation of (*i*) Negative electrode (*ii*) Positive electrode (*iii*) Complete cell. *(Kolhapur 2000)*

Ans. See Q. 1 and 2.

Q.63. Following table gives *p*H values obtained for the titration of 10.0 cm^3 HCl vs NaOH.

vol of NaOH added (cm^3) :	0.0	1.0	2.0	3.0	4.0	5.0	6.0	7.0	8.0	9.0	10.0	11.0	12.0	13.0
*p*H :	2.0	2.3	2.5	2.8	3.0	3.3	3.5	3.8	4.0	4.3	7.0	8.7	9.0	9.4

Plot the graph of *p*H as a function of vol, of NaOH added and determine the equivalence point for the titration using graphical method.

Ans. The graph between *p*H and volume of NaOH added is plotted as under : (Fig. 11.35)

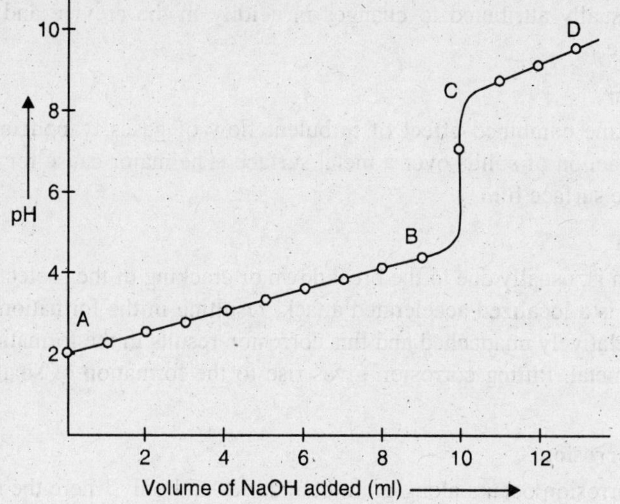

Fig. 11.35 Graph between pH and volume of alkali added

As the equivalence point is reached, the curve gets vertical The equivalent point corresponds to a volume of 10 ml

Q. 64. What is meant by corrosion ? Discuss different types of corrosion. (*Purvanchal 2007*)

Ans. The process of slow eating away of the metal due to attack of the atmospheric gases on the surface of the metal resulting into the formation of compounds such as oxides, sulphides, carbonates, sulphates, etc. is called corrosion.

ates, sulphates, etc. is called corrosion.

Rusting of iron is the most common example of rusting. Rust is hydrated ferric oxide, $Fe_2O_3 . x H_2O$. Other examples include tarnishing of silver, green coating on copper, bronze, etc.

TYPES OF CORROSION

Galvanic Corrosion

This is the most common form of corrosion and can be dry or wet, chemical or electrochemical. When two different metals (*e.g.*, zinc and copper) are electrically connected and exposed to an electrolyte, the metal higher in electrochemical series undergoes corrosion. Such type of corrosion is called ***galvanic corrosion.*** In this corrosion, a uniform decrease in the volume of a metal takes place as a result of chemical action and soluble corrosion products are formed.

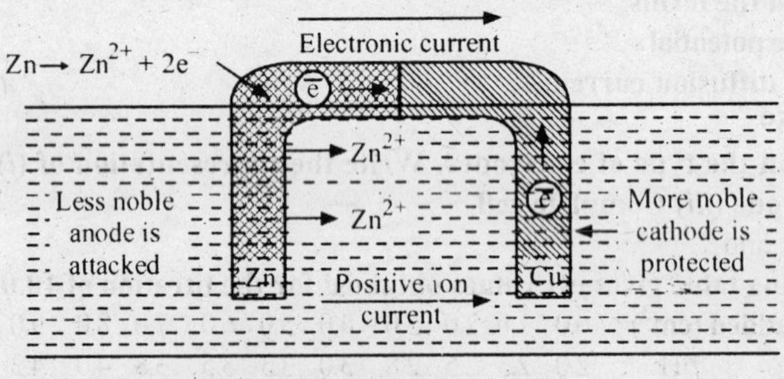

Fig. 11.36. Volatile oxide layer.

2. Crevice Corrosion

It is a local corrosion and is usually caused by dirt deposits, corrosion products, crack in paint coatings etc. It is usually attributed to changes in acidity in the crevice and lack of O_2 in the species in the crevice.

3. Erosion Corrosion

It is caused by the combined effect of turbulent flow of gases, vapour and liquids and the mechanical rubbing action of solids over a metal surface. The major cause for its corrosion is the removal of protective surface film.

4. Pitting Corrosion

Pitting corrosion is, usually due to the breakdown or cracking of the protective film on a metal at specific points. It is a localized accelerated attack, resulting in the formation of cavities around which the metal is relatively unattached and this corrosion results in the formation of pinholes, pits and cavities in the metal. Pitting corrosion gives rise to the formation of small anodic and large cathodic areas.

5. Intergranular Corrosion

Intergranular corrosion occurs along grain boundaries and only where the material, especially sensitive to corrosive attack exists, and corrosive liquid possesses a selective character of attacking only at the grain boundaries, but leaving the grain interiors untouched or only slightly attacked. This type of corrosion is due to the fact that the grain boundaries contain material, which shows electrode potential more anodic than that of the grain centre in the particular corroding medium. This may be due to the precipitation of certain compounds at the grain boundaries.

6. Stress-Corrosion Cracking

Stress corrosion is due to the combined effect of static tensile stresses and the corrosive environment stresses on a metal that cause cracking result from residual cold working and quench under stress become more anodic and tend to increase the rate of corrosion. For stress corrosion to occur presence of tensile stress and a specific corrosive environment are necessary. The corrosive agents are highly specific and selective.

7. Exfoliation and Selective Leaching

Exfoliation and selective leaching is a sub-surface corrosion which starts on a clean surface and spreads below it. It differs from pitting in that the attack has a rather laminated appearance and whole layers of material are corroded. The attack is usually recognized from a flaky or blistered surface. This type of corrosion is known to take place in aluminium alloys and can be avoided by heat treatment and proper alloying.

Selective leaching (parting) is the removal of one of the elements in an alloy. De-zincification from copper-zinc alloys is the most common example. This type of corrosion is highly undesirable as it yields a porous metal with poor mechanical properties. This can be remedied by the use of non-susceptible alloys.

This type of corrosion involves in a localized electrochemical corrosion, occurring along narrow paths, forming anodic areas with respect to the more cathodic areas at the metal surface.

8. Soil Corrosion

Corrosion by soils is very important in case of water mains, electric cables and other underground structures, which are embedded in the soil. The various factors that are responsible for soil corrosion include the following :

(*i*) Acidity of the soil (*ii*) Moisture content
(*iii*) Content of electrolytes (*iv*) Micro-organisms present
(*v*) Content of organic matter (*vi*) Physical properties of the soil.

Soil corrosion is purely electrochemical in character. The texture of a soil is determined by the percentage of particles of various sizes.

9. Waterline Corrosion

Waterline corrosion results from differential aeration leading to the formation of oxygen concentration cells. It has generally been observed that maximum corrosion takes place in a steel tank containing water along a line just beneath the level of water, because access of oxygen is much less there (Fig. 11.37). The area above the waterline is highly oxygenated and hence acts as cathodic area.

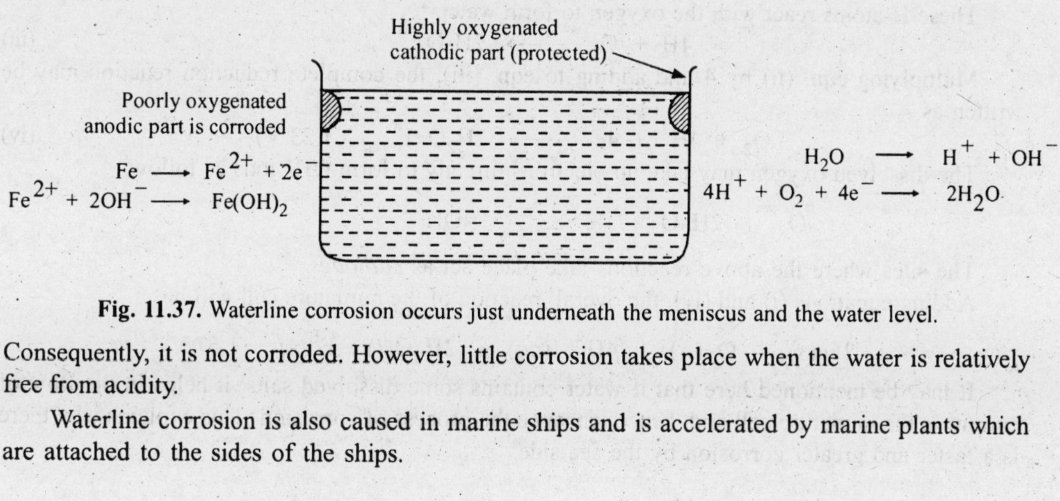

Fig. 11.37. Waterline corrosion occurs just underneath the meniscus and the water level.

Consequently, it is not corroded. However, little corrosion takes place when the water is relatively free from acidity.

Waterline corrosion is also caused in marine ships and is accelerated by marine plants which are attached to the sides of the ships.

10. Micro-Biological Corrosion

Corrosion caused by the metabolic activity of various micro-organisms, is called *microbiological corrosion.* The micro-organisms can develop in an environment with or without oxygen and are classed as aerobic or anaerobic.

Bacteria, fungi, algae, diatom etc. are capable of forming micro-biological film on the surface of iron and generally known as film-forming micro-organism. Such film maintains concentration gradients of dissolved salts, gases, acids etc. on the surface of iron and as a result, local biological concentration cells are developed on the surface leading to micro-biological corrosion.

Q. 65. Describe electrochemical theory of corrosion.

Ans. Electrochemical theory of corrosion. The theory of corrosion can be explained by taking the example of rusting of iron. This theory explains the formation of rust on the basis of the formation of electrochemical cells on the surface of the metal.

The following steps are involved in the formation of rust :

(*i*) The water vapours on the surface of the metal dissolve CO_2 and O_2 from the air. Thus, the surface of metal is covered with the solution of CO_2 in water, *i.e.*, carbonic acid (H_2CO_3)

$$H_2O + CO_2 \longrightarrow H_2CO_3$$

This acts as an electrolytic solution of the cell. The carbonic acid and water ionise to a small extent as :

$$H_2CO_3 \rightleftharpoons 2H^+ + CO_3^{2-}$$

$$H_2O \rightleftharpoons H^+ + OH^-$$

(*ii*) In contact with the dissolved CO_2 and oxygen iron undergoes oxidation as follows :

$$Fe \longrightarrow Fe^{2+} + 2e^- \quad (E^0_{oxid.} = 0.44 \text{ V}) \qquad \text{...(i)}$$

Thus, the sites where the above reaction takes place act as *anodes.* As a result of the above reaction, iron is converted into ferrous (Fe^{2+}) ions.

(*iii*) The electrons lost by iron are taken up by the H^+ ions produced by the dissociation of H_2CO_3 and H_2O present on the surface of the metal. Thus, H^+ ions are converted into H-atoms.

$$H^+ + e^- \longrightarrow H \qquad \text{...(ii)}$$

These H-atoms react with the oxygen to form water.

$$4H + O_2 \longrightarrow 2H_2O \qquad \text{...(iii)}$$

Multiplying eqn. (ii) by 4 and adding to eqn. (iii), the complete reduction reaction may be written as

$$O_2 + 4H^+ + 4e^- \longrightarrow 2H_2O \ (E^0_{red} = 1.23 \text{ V}) \qquad \text{...(iv)}$$

The dissolved oxygen may take up electrons directly to form OH^- ions as follows :

$$O_2 + 2H_2O + 4e^- \longrightarrow 4OH^-$$

The sites where the above reactions take place act as *cathodes.*

Adding equations (i) and (iv), the overall reaction of the miniature cell will be

$$2Fe(s) + O_2(g) + 4H^+ \ (aq) + 2H_2O(l) : E^0_{cell} = 1.67 \text{ V}$$

It may be mentioned here that if water contains some dissolved salts, it helps in the flow of current in the miniature cell and, hence, enhance the process of corrosion. This explains why there is a faster and greater corrosion by the sea-side.

(iv) The ferrous ions formed react with the dissolved oxygen or oxygen from the air to form ferric oxide

$$4Fe^{2+} + O_2 + 4H_2O \longrightarrow 2Fe_2O_3 + 8H^+(aq)$$

Ferric oxide then undergoes hydration to form rust.

$$Fe_2O_3 + xH_2O \longrightarrow \underset{\text{Hydrated ferric oxide (Rust)}}{Fe_2O_3xH_2O}$$

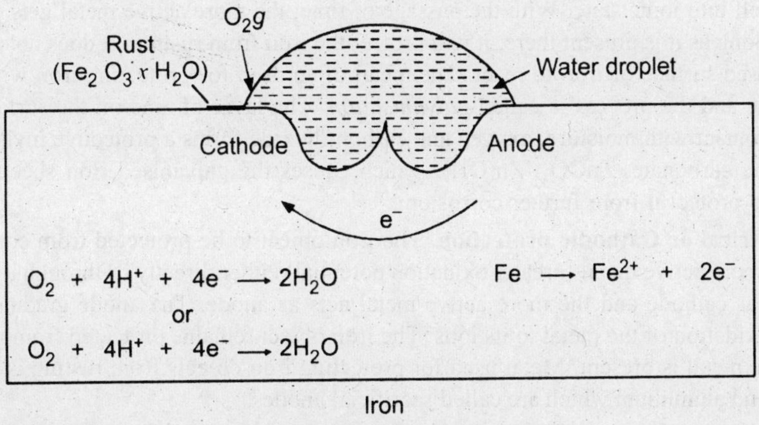

Fig. 11.38. Corrosion of iron.

It may be noted that the rust does not stick to the surface. It peels off exposing fresh iron surface for further rusting. It, thus, reduces the life of an article.

The rusting phenomenon may be represented diagrammatically as shown in Fig. 11.38

Factors responsible for corrosion.

(*i*) *Reactivity of the metal.* More active metals having high oxidation potential values are readily corroded.

(*ii*) *Presence of air and moisture.* Air and moisture accelerate corrosion phenomenon. Presence of gases like SO_2 and CO_2 in air enhance the process of corrosion. Iron does not rust when placed in vacuum.

(*iii*) *Presence of impurities.* Presence of impurities in metals enhances the chances of corrosion. Pure metals do not corrode.

(*iv*) *Presence of electrolytes.* Electrolytes increase the rate of corrosion. For example, iron rusts faster in saline water than in pure water.

(*v*) *Strains in metals.* Corrosion takes place rapidly at bends, scratches and cuts in the metal.

Q. 66. Describe different methods of prevention of corrosion.

Ans. Some of the methods for prevention of corrosion are explained below :

1. By external coating.

(*i*) By applying a film of oil or greases on the surface of the iron tools and machinery, the rusting of iron can be prevented. It keeps the iron surface protected from moisture, oxygen and carbon dioxide.

(*ii*) The metal surface is coated with paint which prevents the action of air, moisture, etc., till the paint layer develops cracks.

(*iii*) The iron surface is coated with phosphate to give a tough adherent insoluble film which does not allow air and moisture to react with iron surface.

(*iv*) The iron surface is coated with non-corroding metals such as nickel, chromium, aluminium, etc., by electroplating with tin, zinc, etc., or by dipping the iron article in the molten metal. This again cuts off the supply of oxygen and water to iron surface.

2. Sacrificial Protection. Sacrificial protection implies covering the iron surface with a layer of metal which is *more active* (electropositive) than iron. It, thus, prevents the iron from losing electrons and getting oxidised. The more active metal loses electrons in preference to iron and converts itself into ionic state. With the passage of time, the more active metal gets consumed into ions but so long as it is present there, it will protect the iron from rusting. It does not allow even the nearly exposed surface of iron to react. The metal often used for protecting iron with more active metal is zinc and the process is called *galvanisation.* The layer of zinc on the surface of iron, on coming in contact with moisture, oxygen and carbon-dioxide forms a protective invisible thin layer of basic zinc carbonate, $ZnCO_3$. $Zn(OH)_2$ which causes the galvanised iron sheets to lose their lustre and to protect it from further corrosion.

3. Electrical or Cathodic protection. The iron object to be protected from corrosion is connected to a more active metal (greater oxidation potential) either directly or through a wire. The iron object acts as cathode and the more active metal acts as anode. The anode gradually disappears due to the oxidation of the metal to its ions. The iron object remains protected from rusting as long as the active metal is present. Metal used for protecting iron objects from rusting is out of magnesium, zinc and aluminium which are called sacrificial anodes.

Cathode protection is often employed in the protection of iron pipes buried in the moist soil, canals, storage tanks, etc. Pieces of magnesium are buried along the pipeline and connected to it by the wire as illustrated in Fig. 11.39. Electrode reactions are given alongside.

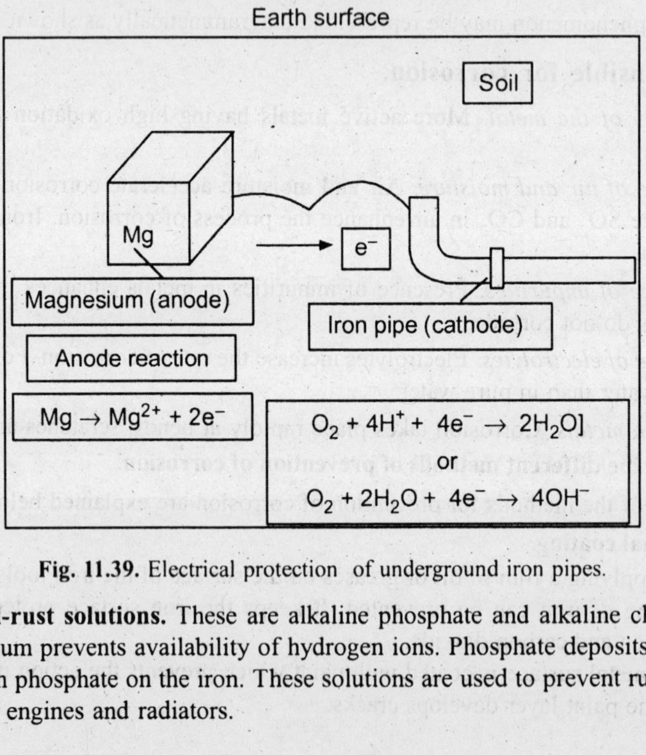

Fig. 11.39. Electrical protection of underground iron pipes.

4. Using anti-rust solutions. These are alkaline phosphate and alkaline chromate solutions. The alkaline medium prevents availability of hydrogen ions. Phosphate deposits an insoluble protective film or iron phosphate on the iron. These solutions are used to prevent rusting of iron parts of the automobile engines and radiators.

12

PHOTOCHEMISTRY

Q. 1. Define photochemistry. What are photophysical and photochemical reactions?

(Awadh, 2000; Purvanchal 2002; Gujarat 2003)

Ans. *Photochemistry is defined as that branch of chemistry which deals with the process involving emission or absorption of radiation.*

Photophysical processes: These processes take place in the presence of light but do not involve any chemical reaction. These processes take place due to absorption of light by the substances followed by the emission of the absorbed light. If there is instantaneous emission of the absorbed light, the process is called *fluorescence.*

If a high amount of energy is absorbed, the electrons may leave the atoms completely. Such a process is called *photoelectric effect.* Photophysical processes include fluorescence, phosphorescence and photoelectric effect.

Photochemical reactions: These are the reactions that take place by absorption of radiations of suitable wavelength. In such processes, the absorbed light energy is first stored in the molecule and then it is further used to bring about the reaction. Examples of photochemical reactions are :

(*i*) *Reaction between hydrogen and chlorine*

$$H_2(g) + Cl_2(g) \xrightarrow{\text{Light}} 2HCl(g)$$

(*ii*) *Photosynthesis of carbohydrates in plants taking place in presence of chlorophyll, the green colouring matter, present in the leaves.*

$$\underset{\text{From air}}{\underline{6CO_2 + 6H_2O}} \xrightarrow[\text{(chlorophyll)}]{\text{Light}} \underset{\text{Glucose}}{C_6H_{12}O_2} + 6O_2$$

Q. 2. Demonstrate an experiment to explain a photochemical reaction.

Ans. A mixture of hydrogen and chlorine remains unchanged with lapse of time. But when exposed to light, the reaction occurs with a loud explosion.

$$H_2 + Cl_2 \xrightarrow{\text{Dark}} \text{No reaction}$$

$$H_2 + Cl_2 \xrightarrow{\text{Light}} 2HCl$$

A bottle is filled with equimolar amounts of hydrogen and chlorine (Fig. 12.1). It is tightly stoppered with a handball. When the lamp is turned on, a beam of light falls on the mixture through the bottom of the bottle. The reaction occurs with an explosion. The ball is expelled with high velocity so that it strikes the opposite wall of the room.

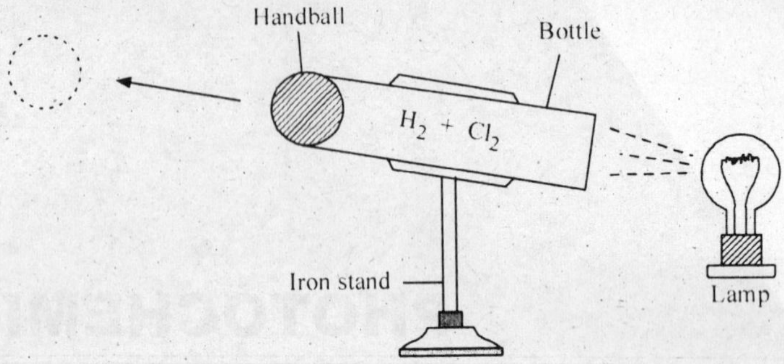

Fig. 12.1. The 'HCl-cannon' experiment.

Q. 3. Bring out the points of difference between Thermochemical reactions and photochemical reactions. *(Garhwal 2000, Raipur 2010)*

Ans.

	Thermochemical reactions		*Photochemical reactions*
(*i*)	These reactions involve absorption or of heat.	(*i*)	These reactions involve absorption of evolution light.
(*ii*)	They can take place even in the dark.	(*ii*)	The presence of light is the primary requisite for the reaction to take place.
(*iii*)	The free energy change (ΔG) of a thermochemical reaction is always negative.	(*iii*)	The free energy change (ΔG) of a photochemical reaction may not be negative. A few examples of photochemical reactions for which ΔG is positive and still they are spontaneous are synthesis of carbohydrates in plants and decomposition of HC1 into H_2 and $C1_2$.
(*iv*)	Temperature has significant effect on the rate of a thermochemical reaction.	(*iv*)	Temperature has very little effect on the rate of photochemical reaction. Instead, the intensity of light has a marked effect on the rate of a photochemical reaction.

Q. 4. Discuss Lambert's law of transmission of light. *(Bombay 2004; Punjab, 2000; Awadh 2000; Garhwal, 2000; Meerut 2000; Nagpur 2002; Guwahati 2002; Calcutta 2007)*

Ans. This law put forward by Lambert states as under :

When a monochromatic light is passed through a pure homogeneous medium, the decrease in the intensity of light with thickness of the absorbing medium at any point X is proportional to the intensity of the incident light. Mathematically, the law can be put as

$$-\frac{dI}{dx} \propto I$$

or

$$-\frac{dI}{dx} = kI \qquad \text{...}(i)$$

where dI is the small decrease in intensity of the light on passing through a small thickness dx, I is the intensity of the incident light just before entering the thickness dx and k is the constant of proportionality called the *absorption coefficient.* It depends upon the nature of the absorbing medium.

The intensity I at any point X (at a distance x from the start of the medium), can be found in terms of the original intensity I_0 as follows:

Equation (*i*) can be rewritten as

$$\frac{dI}{I} = -k\,dx \qquad \text{...}(ii)$$

When $x = 0, I = I_0$.

Integrating equation (*ii*) between the limits $x = 0$ to x and $I = I_0$ to I, we get

$$\int_{I_0}^{I} \frac{dI}{I} = \int_{x=0}^{x=x} -k\,dx$$

or
$$\ln \frac{I}{I_0} = -kx \qquad \text{...}(iii)$$

or
$$\frac{I}{I_0} = e^{-kx}$$

or
$$\boldsymbol{I = I_0 e^{-kx}} \qquad \text{...}(iv)$$

This equation expresses how the original intensity I_0 is reduced to intensity I after passing through a thickness x of the medium.

Equation (*iii*) can be written as

$$2.303 \log \frac{I}{I_0} = -kx$$

or
$$\log \frac{I}{I_0} = -\frac{k}{2.303} x$$

or
$$\log \frac{I}{I_0} = -k'x \qquad \text{...}(v)$$

or
$$\frac{I}{I_0} = 10^{-k'x}$$

or
$$\boldsymbol{I = I_0 10^{-k'x}} \qquad \text{...}(vi)$$

where $k' = \dfrac{k}{2.303}$ *is* called **extinction coefficient** of the substance (*i.e.* of the absorbing medium)

This quantity is now called **absorption coefficient or adsorptivity** of the substance.

$-\log I/I_0$ is called **absorbance** of the medium

The physical significance of the extinction coefficient or absorptivity follows more clearly from equation (*v*), which can be rewritten as

$$k' = -\frac{1}{x} \log \frac{I}{I_0} = \frac{1}{x} \log \frac{I_0}{I}.$$

If
$$\log \frac{I_0}{I} = 1, \; k' = \frac{1}{x}$$

But
$$\log \frac{I_0}{I} = 1, \text{ means } \frac{I_0}{I} = 10 \;\; i.e. \; I = \frac{1}{10} I_0.$$

Hence extinction coefficient or absorptivity may be defined as follows :

The extinction coefficient or absorptivity is the reciprocal of the thickness (expressed in cm) at which the intensity of the light falls to one-tenth of its original value.

Q. 5. Discuss Beer's law. What is the significance of molar extinction coefficient ?

(Arunachal 2003; Punjab 2003; Calcutta 2007; Himachal 2010)

Ans. Beer's law states as :

When a monochromatic light is passed through a solution, the decrease in the intensity of light with the thickness of the solution is directly proportional not only to the intensity of the incident light but also to the concentration c of the solution. Mathematically, we have

$$-\frac{dI}{dx} \propto I \times c \qquad ...(i)$$

or

$$-\frac{dI}{dx} = \varepsilon\, Ic (\varepsilon \text{ is pronounced as epsilon})$$

where ε is a constant of proportionality and is called **molar absorption coefficient.** Its value depends upon the nature of the absorbing solute and the wavelength of the light used.

Equation (i) can be rewritten as

$$\frac{dI}{I} = -\varepsilon c\, dx.$$

Integrating this equation between the limits $x = 0$ to x and $I = I_0$ to I, we get

$$\int_{I_0}^{I} \frac{dI}{I} = \int_{x=0}^{x=x} -\varepsilon c\, dx$$

or

$$\ln \frac{I}{I_0} = -\varepsilon cx \qquad ...(ii)$$

or

$$\frac{I}{I_0} = e^{-\varepsilon cx} \qquad ...(iii)$$

or

$$\boldsymbol{I = I_0 e^{-\varepsilon cx}} \qquad ...(iv)$$

This equation expresses how the intensity of a monochromatic light falls from I_0 to I on passing through a thickness x of a solution of concentration c.

Equation (ii) can be expressed in another form.

$$2.303 \log \frac{I}{I_0} = -\varepsilon cx$$

or

$$\log \frac{I}{I_0} = -\frac{\varepsilon}{2.303} cx$$

or

$$\log \frac{I}{I_0} = -\varepsilon' cx \qquad ...(v)$$

or

$$\log \frac{I}{I_0} = 10^{-\varepsilon' cx}$$

or

$$\boldsymbol{I = I_0 \times 10^{-\varepsilon' cx}} \qquad ...(vi)$$

$\varepsilon' = \dfrac{\varepsilon}{2.303}$ was earlier called **molar extinction coefficient** of the absorbing solution. Now this quantity is called molar absorption coefficient or molar absorptivity of the absorbing solution.

The physical significance of molar extinction coefficient or **molar absorptivity** follows from equation (*v*), which can be rewritten as

$$\varepsilon' = \frac{1}{cx} \log \frac{I}{I_0}$$

If $$c = 1M \text{ and } \log \frac{I}{I_0} = 1$$

then $$\varepsilon' = \frac{1}{x}.$$

$$\log \frac{I}{I_0} = 1 \text{ means } \frac{I}{I_0} = 10 \text{ i.e. } I = \frac{1}{10} I_0$$

Hence *molar extinction coefficient or molar absorptivity may be defined as the reciprocal of that thickness of the solution layer of 1 molar concentration which reduces the intensity of the light passing through it to one-tenth of its original value.*

Q. 6. How do you apply Lambert-Beer law to determine unknown concentration in solution? *(Purvanchal 2003)*

Ans. See Q.4 and 5.

Q. 7. Calculate the transmittance, absorbance and absorption co-efficient of a solution which absorbs 90% of a certain wavelength of light beam passed through a 1 cm cell containing 0.25 M solution.

Ans. It is given that

$$I_{abs} = 0.90\ I_0;\ x = 1 \text{ cm};\ c = 0.25 \text{ M}$$

$$\therefore \quad I_t = I_0 - I_{abs} = I_0 - 0.90\ I_0 = 0.10\ I_0$$

$\therefore$ Transmittance $$T = \frac{I_t}{I_0} = \mathbf{0.10}$$

Absorbance $$A = -\log \frac{I}{I_0} = \log \frac{I_0}{I} = \log \frac{1}{0.1} = 1$$

Applying Beer's law, we have

$$\log = \frac{I_t}{I_0} = -\varepsilon' cx$$

i.e. $$\log 0.10 = -\varepsilon' \times 0.25 \times 1$$

$$-1 = -\varepsilon' \times 0.25$$

or $$\varepsilon' = 0.4 \text{ L mol}^{-1}\text{cm}^{-1}$$

Q. 8. A certain substance in a cell of length *l* absorbs 10% of the incident light. What fraction of the incident light will be absorbed in a cell which is five times long ?

Ans. *In the first case,* $I_{abs} = 0.10\ I_0$ so that $I_t = 0.90\ I_0$

or $$\frac{I_t}{I_0} = 0.90$$

$$\log = \frac{I_t}{I_0} = -k'l \text{ i.e. } \log 0.90 = -k'l$$

or $$k' = \frac{0.0458}{l}$$

In the second case, $x = 5l$ so that

$$\log = \frac{I_t}{I_0} = -k' \times 5l = -\frac{0.0458}{l} \times 5l = -0.2290$$

or $$\log \frac{I_t}{I_0} = \bar{1}.7710$$

i.e. $$\frac{I_t}{I_0} = 0.5902 \text{ or } I_t = 0.5902 I_0$$

$\therefore$ $$I_{abs} = I_0 - I_t = I_0 - 0.5902 I_0 = 0.4098 I_0$$

$$= 40.98 \% \text{ of } I_0$$

Q. 9. When an incident beam of wavelength 3000 Å was allowed to pass through 2 mm thick pyrex glass, the intensity of the radiation was reduced to one-tenth of its initial value. What part of the same radiation will be absorbed by 1 mm thick glass ?

Ans. *In the first case,* it is given that

$$\frac{I_t}{I_0} = \frac{1}{10} \qquad \text{(It denotes transmitted light)}$$

$$x = 2 \text{ mm} = 0.2 \text{cm}$$

Applying Lambert's law, we have

$$\log \frac{I_t}{I_0} = -k'x$$

i.e. $$\log \frac{1}{10} = -k'(0.2)$$

or $$k' = \frac{1}{0.2} = 5$$

or *In the second case,* it is given that

$$x = 1 \text{ mm} = 0.1 \text{ cm}$$

As the medium is the same, k' will have the same value *viz.* 5

Applying Lambert's law, we have

$$\log \frac{I_t}{I_0} = -k'x = -5 \times 0.1 = -0.5$$

$\therefore$ $$\frac{I_t}{I_0} = \text{antilog}(-0.5)$$

$$= \text{antilog}(\bar{1}.5) = 0.3162$$

i.e., $$I_t = 0.3162 I_0$$

$$I_{abs} = I_0 - I_t = I_0 - 0.3162 I_0 = 0.6838 I_0$$

Q. 10. State and explain Grotthus law (first law of photochemistry).

(Punjab 2000; Kanpur 2001; Andhra 2010)

Ans. *When light falls on a body, a part of it is reflected a part of it is transmitted and the rest of it is absorbed It is only the absorbed light which is effective in bringing about a chemical reaction.*

The above law, however, does not imply that the absorbed light must always result into chemical reaction. The absorbed light may simply bring about phenomena such as fluorescence, phosphorescence etc. Similarly, the absorbed light energy may be simply converted into thermal energy *e.g.* in case of potassium permanganate solution, the light energy is absorbed strongly but no chemical effect is produced. Further in some cases it is observed that light energy may not be absorbed by the reacting substance directly but may be absorbed by some other substance present along with the reacting substance. The energy thus absorbed it then passed onto the reacting substance which then starts reacting. This process is called *photosensitization.* A well known example is that of photosynthesis of carbohydrates in plants where chlorophyll acts as a photosensitizer.

Q. 11. State and explain Stark-Einstein law of photochemical equivalence (Second law of photochemistry). *(Purvanchal 2002; Gujarat 2003; Bhopal 2004 Meerut. 2004 : Andhra 2010,*

Ans. The law states:

Every atom or molecule that takes part in a photochemical reaction absorbs one quantum of the radiation to which the substance is exposed.

If v is the frequency of the absorbed radiation, then the energy absorbed by each reacting atom or molecule is one quantum *i.e.* hv where h is Planck's constant. The energy absorbed by one mole of the reacting molecules will then be given by

$$E = Nhv \qquad ...(i)$$

where N is Avogadro's number

Putting $v = \frac{c}{\lambda}$,we can write $E = Nh\frac{c}{\lambda}$ ...(*ii*)

where c is the velocity of light and λ is the wavelength of the absorbed radiation.

Substituting $N = 6.022 \times 10^{23}$ mol^{-1}

$h = 6.626 \times 10^{-27}$ erg sec and $c = 3.0 \times 10^{10}$ cm /sec

we get
$$E = \frac{(6.022 \times 10^{23})(6.626 \times 10^{-27})(3.0 \times 10^{10})}{\lambda}$$

$$= \frac{119.7 \times 10^6}{\lambda} \text{ ergs per mole or } \frac{119.7 \times 10^6}{4.184 \times 10^7 \lambda} \text{ cal per mole}$$

or
$$E = \frac{2.86}{\lambda} \text{ cal per mole} \qquad \text{(in CGS units)} \qquad ...(iii)$$

In this equation, λ is expressed in cm. However as λ is usually expressed in Angstrom units (Å) and we know that 1Å = 10^{-8} cm, therefore equation (*iii*) can be written as

$$E = \frac{2.86}{\lambda} \times 10^8 \text{ cal per mole} \qquad ...(iv)$$

where λ is in Angstrom units

Further putting 10^3 cal = 1 kcal, equation (*iv*) can be written as

$$\mathbf{E = \frac{2.86 \times 10^5}{\lambda} \text{ kcal per mole}} \qquad ...(v)$$

This equation gives *the energy possessed by one mole of photons and is called* **one Einstein.** Thus the *energy per Einstein is inversely proportional to the wavelength of the radiation.* Shorter the wavelength of a radiation, greater is the energy per Einstein.

The unit of energy used most frequently in photochemistry is electron volt (eV).

$$1 \text{ eV} = 23.06 \text{ kcal/mole.}$$

To obtain the equation in **SI units,** we substitute

$N = 6.022 \times 10^{23}$ mol^{-1}, $h = 6.626 \times 10^{-34}$ J s, $c = 3.0 \times 10^{8}$ ms^{-1} we get

$$E = \frac{\mathbf{0.1197}}{\lambda} \textbf{ Jmol}^{-1} = \frac{\mathbf{11.97 \times 10^{-5}}}{\lambda} \textbf{ kJ mol}^{-1}$$

In this equation λ is in metre.

Q. 12. If the value of an Einstein is 72 kcal, calculate the wavelength of the light.

Ans. Here $E = 72$ kcal

From the equation, $E = \dfrac{2.86 \times 10^5}{\lambda}$ kcal/mole

$$\lambda = \frac{2.86 \times 10^5}{E} \text{Å} = \frac{2.86 \times 10^5}{72} = \mathbf{3972.2Å}$$

Q. 13.Calculate the value of Einstein of energy for radiation of wavelength 4000 Å.

Ans. $E = \dfrac{2.86 \times 10^5}{\lambda}$ kcal /mole (λ in Å units)

$$= \frac{2.86 \times 10^5}{4000} \text{ kcal /mole}$$

$$= \mathbf{71.5 \text{ kcal/mole}}$$

Q. 14. Calculate the value of an Einstein of energy in electron volts for radiation of frequency 3×10^{13} s^{-1}.

Ans. Here we are given that

$$\nu = 3 \times 10^{13} \text{ s}^{-1}$$

$$\therefore \quad \lambda = \frac{c}{\nu} = \frac{3 \times 10^{10} \text{ cm s}^{-1}}{3 \times 10^{13} \text{ s}^{-1}}$$

$$= 10^{-3} \text{ cm} = 10^5 \text{Å}$$

$$\therefore \quad E = \frac{\mathbf{2.86 \times 10^5}}{\lambda} \quad \textbf{kcal/mole} \ (\lambda \text{ in Å units})$$

Q. 15. Calculate the energy of one photon of light of wavelength 2450 Å. Will it be able to dissociate a bond in diatomic molecule which absorbs this photon and has a bond energy equal to 95 kcal per mole.

Planck's constant (h) = 6.626×10^{-27} erg sec molecule^{-1}

Velocity of light (c) = 3×10^{10} cm sec^{-1}

Avogadro's number (N) = 6.02×10^{23}.

Ans. Energy of one photon $= h\nu = h\dfrac{c}{\lambda}$

$$= \frac{(6.026 \times 10^{-27}) \times (3 \times 10^{10})}{2450 \times 10^{-8}} \text{ ergs}$$

$$= 8.11 \times 10^{-12} \text{ ergs}$$

$$= \frac{8.11 \times 10^{-12}}{4.184 \times 10^{7}} \text{ cal} = 1.938 \times 10^{-19} \text{ cal}$$

Energy required for dissociation of one bond

$$= \frac{95000}{6.02 \times 10^{23}} \text{ cal}$$

$$= \mathbf{1.578 \times 10^{-19}} \textbf{ cal}$$

Thus energy of photon is greater than the energy required to break the bond. Hence the bond will be dissociated.

Q. 16. What is meant by quantum efficiency or quantum yield ? *(Nagpur 2008)*

(Gujarat 2003; Arunachal 2003; Calcutta 2007; Andhra 2010)

Ans. To relate the number of quanta absorbed with the number of reacting molecules, a term called **quantum efficiency** or **quantum yield** (ϕ) has been introduced. It is expressed as

$$\phi = \frac{\text{Number of molecules reacting in a given time}}{\text{Number of quanta of light absorbed in the same time}}$$

$$= \frac{\text{Number of moles reacting in a given time}}{\text{Number of Einsteins of light absorbed in the same time}}$$

Hence quantum efficiency may be defined as the number of moles reacting per Einstein of the light absorbed.

If Stark-Einstein's law is strictly obeyed in the form already stated, ϕ should be equal to unity. However, studies have shown that the law is applicable only to the primary processes. Hence it is sometimes preferred to state it as follows :

Each molecules that gets activated to initiate the reaction absorbs one quantum of light.

Q. 17. Describe how to determine the quantum yield of a photochemical reaction. *(Raipur 2010)*

Ans. To determine the quantum yield of any photochemical reaction, we have to measure

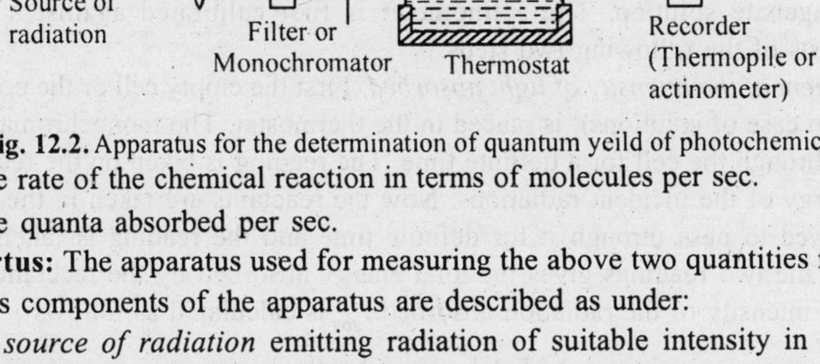

Fig. 12.2. Apparatus for the determination of quantum yeild of photochemical reaction.

(*i*) the rate of the chemical reaction in terms of molecules per sec.

(*ii*) the quanta absorbed per sec.

Appartus: The apparatus used for measuring the above two quantities is shown in Fig.12.2 Various components of the apparatus are described as under:

(*i*) *A source of radiation* emitting radiation of suitable intensity in the desired spectral range. The commonly used sources are filament lamps, carbon and metal arcs and various gas discharge tubes.

(*ii*) *Lens* of suitable focal length.

(*iii*) *Monochromator or filter* which cuts off all radiations except the radiations of the desired wavelength.

(*iv*) *Slit.*

(*v*) *A cell placed in a thermostat* and containing the reaction mixture. The cell is made of glass or quartz and has optically plane windows for the entrance and exit of light. Glass is used only if the wavelength of the light used lies in the visible range. For radiations with wavelength below 3500 Å the cell completely made of quartz is used.

(*vi*) *Recorder* is usually a *thermopile* or an *actinometer* which is used to measure intensity of the radiation. The principle of a thermopile and that of an actinometer are briefly explained below :

A *thermopile* is a set of thermocouples. Rods of two different metals (*e.g.* Ag and Bi) are joined alternately as shown in Fig. 12.3. One set of junctions is soldered to metal strip blackened with platinum or lamp black and the other set of junctions is protected completely from radiation by placing the system in a box. The radiations falling on the blackened metal strip are almost completely absorbed by it. As a result, this set of junctions becomes hot. The temperature difference between the hot end and the cold end produces a current which can be measured by connecting a milliammeter to the thermopile. By calibrating the thermopile with radiations of known intensity, the intensity of the desired radiation can be measured.

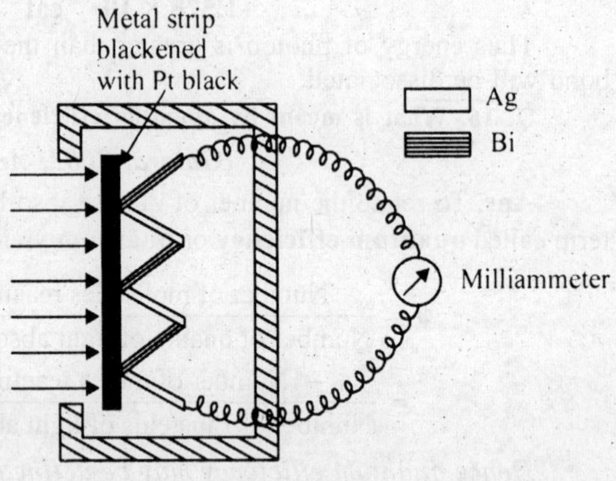

Fig. 12.3 A thermopile.

A *chemical actinometer* is a device in which solutions sensitive to light are used. The working of this device is based upon the fact that a definite amount of the radiation absorbed brings about a definite amount of chemical reaction. The most common actinometer is the uranyl oxalate actinometer which consists of 0.05 molar oxalic acid and 0.01 molar uranyl sulphate (UO_2SO_4) solutions in water. When exposed to light, the following reaction takes place:

$$\begin{array}{c}\text{COOH}\\ |\\ \text{COOH}\end{array}(aq) \xrightarrow[(UO_2SO_4)]{h\nu} CO_2(g) + CO(g) + H_2O(l)$$

Uranyl sulphate acts as photosensitizer *i.e.* absorbs the light and then passes it on to oxalic acid. The extent of reaction is measured by titrating the oxalic acid solution with potassium permanaganate solution. The actinometer is first calibrated against a thermopile.

Procedure It consists of the following two steps.

(*i*) *Measurement of the intensity of light absorbed.* First the empty cell or the cell filled with the solvent alone (in case of solutions), is placed in the thermostat. The monochromatic radiation is allowed to pass through the cell for a definite time. The reading is taken on the recorder. From this we get the energy of the incident radiations. Now the reactants are taken in the cell and the radiations are allowed to pass through it for definite time and the reading is taken again. The difference between the two readings gives the total energy absorbed by the reactants in a given time. From this the intensity of the radiation absorbed I_{abs} is calculated as follows :

$$I_{abs} = \frac{\text{Total energy absorbed}}{\text{Volume of the reaction mixture} \times \text{Time in sec}}$$

(*ii*) *Measurement of the rate of reaction.* The rate of the reaction can be studied by noting the change in some physical property. Alternatively, samples of the reaction mixtures are removed periodically from the cell and analysed.

Knowing the rate of reaction and the intensity of light absorbed, the quantum yield of the reaction can be obtained.

Q.18. For the photochemical reaction $A \rightarrow B$, 1.0×10^{-5} moles of *B* were formed on absorption of 6.0×10^{7} ergs at 3600 Å. Calculate the quantum efficiency.

$N = 6.02 \times 10^{23}$, $h = 6.0 \times 10^{-27}$ erg sec, $c = 3 \times 10^{10}$ cm /sec.

Ans. Wavelength of the light absorbed

$$= 3600\text{Å} = 3600 \times 10^{-8}\text{cm}$$

$\therefore$ Energy of one quantum

$$= h\nu = h\frac{c}{\lambda} = \frac{(6.0\times10^{-27})(3\times10^{10})}{3600\times10^{-8}}\text{ergs}$$

$$= 5.0\times10^{-12} \text{ ergs}$$

Total energy absorbed

$$= 6.0 \times 10^{7} \text{ ergs (Given)}$$

$\therefore$ Number of quanta absorbed

$$= \frac{6.0\times 10^{7}}{5.0\times10^{-12}} = 1.2\times10^{19}$$

Number of *moles* reacting $= 1.0 \times 10^{-5}$

Number of *molecules* reacting

$$= 1.0 \times 10^{-5} \times 6.02 \times 10^{23} = 6.02 \times 10^{18}$$

$$\text{Quantum efficiency } (\phi) = \frac{\text{Number of molecules reacting}}{\text{Number of quanta absorbed}}$$

$$= \frac{6.02\times10^{18}}{1.2\times10^{19}} \cong \mathbf{0.5}$$

Q.19. A certain system absorbs 3×10^{18} quanta of light per second. On irradiation for 20 minutes 0.003 mole of the reactant was found to have reacted. Calculate the quantum yield for the process (Avogadro's number = 6.02×10^{23})

Ans. Number of quanta absorbed per second

$$= 3 \times 10^{18}$$

Number of *moles* reacting in 20 minutes

$$= 0.003$$

$\therefore$ Number of *molecules* reacting in 20 minutes

$$= 0.003 \times 6.022 \times 10^{23}$$

Number of *molecules* reacting per second

$$= \frac{0.0303\times6.022\times10^{13}}{20\times60} = 1.506 \times 10^{18}$$

$$\therefore \text{ Quantum yield } (\phi) = \frac{\text{Number of molecules reacting per second}}{\text{Number of quanta absorbed per second}}$$

$$= \frac{1.506\times10^{18}}{3\times10^{18}} = \mathbf{0.5}$$

Q.20. In the photobromination of cinnamic acid using blue light of 4358 Å at 30.6°C an intensity of 14,000 ergs per second produced a decrease of 0.075 millimole of Br_2 during an exposure of 1105 seconds. The solution absorbed 80.1% of the light passing through it. Calculate the quantum yield for the photoreaction of bromine.

Ans. Intensity of incident light

$$= 14{,}000 \text{ ergs per sec.}$$

Energy of one quantum

$$= \frac{hc}{\lambda} = \frac{(6.62\times10^{-27}) \times (3 \times 10^{10})}{4358 \times 10^{-8}} \text{erg}$$

$$= 4.56 \times 10^{-12} \text{ erg}$$

No. of quanta/sec $= \dfrac{14{,}000}{4.56 \times 10^{-12}}$

$$= 3.07 \times 10^{-15}$$

As the light absorbed is 80.1%

Number of quanta absorbed/sec $= 3.07\times10^{15} \times \dfrac{80.1}{100} = 2.46 \times 10^{15}$

Amount of substance reacting $=$ 0.075 millimole $=$ 0.000075 moles

Number of molecules reacting/sec $= \dfrac{0.000075\times6.023\times10^{23}}{1105}$

$$= 4.09 \times10^{16}$$

Quantum yield $(\phi) = \dfrac{4.09\times 10^{16}}{2.46\times10^{15}}$ 16.6 molecules/quantum

Q.21. A uranyl oxalate actinometer is irradiated for 15 minutes with light of 4350 Å. At the end of this time, it is found that oxalic acid equivalent to 12.0 ml of 0.001 molar $KMnO_4$ solution has been decomposed by the light. At this wavelength, the quantum efficiency of the actinometer is 0.58. Find the average intensity of the light used in (*a*) ergs per sec and (*b*) in quanta per sec.

Ans. Oxalic acid decomposed in 15 minutes

$$= 12.0 \text{ ml of } 0.001 \text{ molar} = \frac{0.001}{1000}\times12 \text{ moles}$$

$$= \frac{0.001}{1000} \times 12 \times 6.022 \times10^{23} \text{ molecules}$$

$$= 7.23 \times 10^{18} \text{ molecules}$$

∴ Number of molecules decomposed per sec

$$= \frac{7.23\times10^{18}}{15 \times60} = 8 \times 10^{15}$$

∴ Number of quanta absorbed per sec

$$= \frac{\text{Number of molecules reacting/sec}}{\text{Quantum efficiency}}$$

$$= \frac{8 \times 10^{15}}{0.58} = \mathbf{1.38 \times 10^{16}}$$

Further, wavelength of the light absorbed $= 4350\text{ Å} = 4350 \times 10^{-8}$ cm

$\therefore$ Energy of one quantum $= h\nu = \dfrac{hc}{\lambda}$

$$= \frac{(6.62\times10^{-27})\times(3\times10^{10})}{4350\times10^{-8}}$$

$$= 4.57 \times 10^{-12} \text{ ergs}$$

$\therefore$ Energy absorbed per second

$$= (1.38 \times 10^{16}) \times (4.57 \times 10^{-12}) \text{ ergs}$$

$$= \mathbf{6.31 \times 10^4 \text{ ergs}}$$

Q.22. What are the reasons for very high and very low quantum yields in photochemical reactions?

(Bangalore 2004 ; Himachal 2010)

Ans. The reason for very high or very low quantum yield is found in the fact that a photochemical reaction takes place in two steps, which are :

(*i*) **Primary process:** This involves the activation of certain molecules by absorption of light (one quantum per molecule) *i.e.*

$$AB + h\nu \longrightarrow \underset{\text{Activated molecule}}{AB^*}$$

or the dissociation of certain molecules to produce active atoms or free radical *i.e.*

$$AB + h\nu \longrightarrow A + \underset{\text{Active or excited atom}}{B^*}$$

(*ii*) **Secondary process:** This involves the reaction of the activated molecules or the products of the primary process with other molecules or the deactivation of the activated molecules. Actually a number of reactions are involved in the secondary process.

Some of these are endothermic reactions and thus take place slowly thereby affecting the overall reaction. Some of these are exothermic reactions and the heat evolved decomposes the products again affecting overall reaction. The effect of these factors brings about sudden rise or fall of quantum yield.

Q.23. What are primary and secondary processes ?

(Bangalore 2002; Sambalpur 2003)

Ans. See Q. 22.

Q.24. Discuss the photochemical decomposition of hydrogen iodide.

(Awadh 2000; Purvanchal 2003)

Ans. The photolysis of HI to give H_2 and I_2 has been studied in the wavelength range 2070 –2820Å and the quantum yield of this reaction has been found to be approximately 2. Mechanism for the photolysis of HI has been proposed as under.

(*a*) *Primary process:*

$$HI + h\nu \longrightarrow H + I^*$$

i.e. a molecules of HI absorbs one photon of light and dissociates to give a normal hydrogen atom and an excited iodine atom.

(*b*) *Secondary processes:* The products of the primary process then take part in the secondary processes as follows :

(*i*) $H + HI \longrightarrow H_2 + I$

(*ii*) $I^* + HI \xrightarrow{\text{Endo}} H + I_2$

(*iii*) $$H + H \longrightarrow H_2$$

(*iv*) $$I + I \longrightarrow I_2$$

(*v*) $$H + I \xrightarrow{\text{Exo}} HI$$

All these secondary processes are not important. The reactions (*iii*) and (*v*) are highly exothermic and the heat produced causes the dissociation of the products of these reactions. Reaction (*ii*) is endothermic and hence takes place slowly. Thus the only secondary processes of importance are the reaction (*i*) and (*iv*). Rewriting the reaction of the primary process and the two important reactions *viz.* (*i*) and (*iv*) of the secondary processes, we have

$$HI + h\nu \longrightarrow H + I^{*} \} \text{ Reaction of the primary process}$$

$$\left.\begin{array}{l} H + HI \longrightarrow H_2 + I \\ I + I \longrightarrow I_2 \end{array}\right\} \text{The only Important reactions of the secondary processes}$$

Adding these three equations, we get the overall reaction

$$2HI + h\nu \longrightarrow H_2 + I_2$$

Thus for every one quantum of light absorbed, two molecules of HI are decomposed. Hence the quantum yield value of 2 is explained.

Q. 25. Discuss the photolysis (photochemical decomposition) of HBr.

Ans. The quantum yield of this reaction is found to be 2. Mechanism of this reaction is suggested to be similar to that of photolysis of HI. See Q. 24. Use HBr in place of HI, Br_2 in place of I_2 and Br in place of I.

Q. 26. Discuss the photosynthesis of HBr from hydrogen and bromine. (*Mumbai 2005*)

Ans. The reaction may be represented as

$$H_2 + Br_2 \xrightarrow{h\nu} 2HBr$$

The quantum efficiency of this reaction is very low *i.e.* about 0.01 at ordinary temperatures. This is explained by proposing the following mechanism for the above reaction

(*a*) *Primary process :*

$$Br_2 + h\nu \longrightarrow 2Br$$

(*b*) *Secondary processes :*

(*i*) $$Br + H_2 \xrightarrow{\text{Endo}} HBr + H$$

(*ii*) $$H + Br \longrightarrow HBr + Br$$

(*iii*) $$H + HBr \longrightarrow H_2 + Br$$

(*iv*) $$Br + Br \longrightarrow Br_2$$

The H-atom produced in reaction (*i*) attacks Br_2 according to reaction (*ii*) producing HBr and Br atom. The Br-atom thus produced attacks more of H_2 according to reaction (*i*) producing H-atom again. As a result, reactions (*i*) and (*ii*) should repeat over and again *i.e.* a chain reaction should be set up and the quantum yield of the reaction should be very high. But actually the quantum yield is very low. This is explained as follows :

Reaction (*i*) is endothermic and hence takes place slowly or almost does not tend to take place. The reactions (*ii*) and (*iii*) depend upon the formation of H atoms in reaction (*i*); consequently, reactions (*ii*) and (*iii*) are very slow. So the only reaction of importance in the secondary processes is the reaction (*iv*). This is just the reversal of the primary process. Thus the Br atom formed in the primary process do not combine to form HBr; instead, they combine together to give back Br_2. As a result, the quantum yield of the reaction is very low at ordinary temperatures.

The quantum yield of the above reaction increases with increase of temperature. This is obviously because of the fact that the reaction (*i*) which is endothermic and requires high energy of activation becomes faster at high temperatures because the required energy of activation becomes available at higher temperatures. This reaction produces more H-atoms thereby increasing the rate of reactions (*ii*) and (*iii*) also.

Q.27. Discuss the photosynthesis of HCl from hydrogen and chlorine.

(Bangalore 2004)

Ans. This reaction whose quantum yield is very high *i.e.* 10^4 to 10^6 may be represented as

$$H_2 + Cl_2 \xrightarrow{hv} 2HCl$$

The high quantum yied of this reaction is explained by the chain mechanism. The different steps involved are as follows:

(*a*) *Primary process:*

$$Cl_2 + hv \longrightarrow 2\dot{C}l \qquad \textit{chain initiating step}$$

i.e. a chlorine molecule absorbs one quantum of light and dissociates to give Cl atoms.

(*b*) *Secondary processes :*

(*i*) $\dot{C}l + H_2 \xrightarrow{Exo} HCl + \dot{H}$

(*ii*) $\dot{H} + Cl_2 \longrightarrow HCl + Cl$

} *Chain propagating steps*

(*iii*) $\dot{C}l + \dot{C}l \longrightarrow Cl_2$ *Chain terminating step*

The H-atoms produced in reaction (*i*) attacks Cl_2 according to reaction (*ii*) producing HC1 and Cl atom. The Cl atom thus produced attacks more of H_2 according to reaction (*i*) producing H-atom again. In this way, the reactions (*i*) and (*ii*) repeat over and again till almost whole of H_2 and Cl_2 have reacted to form HC1. The reaction stops when the Cl atoms left off combine with each other to form $C1_2$, according to reaction (*iii*). *This reaction takes place on the walls of the reaction vessel.*

In this reaction, the reaction (*i*) of the secondary processes which immediately follows the primary process is *exothermic* and therefore, takes place very easily and consequently the chain reaction is set up very easily. In the photosynthesis of HBr, the reaction (*i*) of the secondary process is *endothermic* and thus has very little tendency to take place.

The quantum yield of the above reaction decreases if the reaction vessel contains small traces of oxygen. This explained on the basis that oxygen brings about chain termination by the following reactions :

$$\dot{H} + O_2 \longrightarrow HO_2$$

$$\dot{C}l + O_2 + HCl \longrightarrow HO_2 + Cl_2$$

Q. 28. Give one example of photochemical reaction and show primary and secondary processes occurring in it.

Ans. See Q. 26 or 27.

Q. 29. Describe photolysis of ammonia.

Ans. This is an example of a photochemical reaction whose quantum yield is very low *i.e.* about 0·25 at ordinary temperatures and pressures. The probable mechanism is as follows:

(*a*) *Primary process* :

$$NH_3 + h\nu \longrightarrow NH_2 + H$$

The presence of H-atoms has been established.

(*b*) *Secondary processes:* The ultimate products of the photolysis of ammonia have been found to be nitrogen, hydrogen and hydrazine. Hence the following probable secondary reactions have been proposed.

(*i*) $NH_2 + NH_2 \longrightarrow N_2 + 2H_2$

(*ii*) $NH_2 + NH_2 \longrightarrow N_2H_4$

(*iii*) $H + H \longrightarrow H_2$

(*iv*) $NH_2 + H \longrightarrow NH_3$

Since the quantum yield of the reaction is very low, the reaction (*i*) to (*iii*) take place only to a very small extent. The important secondary reaction is the reaction (*iv*) only which involves the reversal of the primary process leading to decline in the value of (ϕ).

Q. 30. Describe photolysis of acetone. *(Garhwal 2000; Calcutta 2000)*

Ans. The reaction may be represented as

$$\underset{\text{Acetone}}{CH_3COCH_2} \xrightarrow{h\nu} \underset{\text{Ethane}}{C_2H_6} + \underset{\substack{\text{Carbon}\\\text{monoxide}}}{CO}$$

The main products of the photolysis of acetone are ethane and carbon monoxide but small amounts of methane (CH_4) and appreciable amounts of diacetyl, $(CH_3CO)_2$ are also found to be present. The quantum yield of the reaction is found to be nearly unity. All these observations suggest the following probable mechanism for the photolysis of acetone.

(*a*) *Primary process* :

$$\underset{\text{Acetone}}{CH_3COCH_3} \xrightarrow{h\nu} \underset{\substack{\text{Methyl}\\\text{free radical}}}{\dot{C}H_3} + \underset{\substack{\text{Acetyl}\\\text{free radical}}}{CH_3\dot{C}O}$$

(*b*) *Secondary processes* : A number of secondary reactions are possible. A few of these are as under :

(*i*) $CH_3\dot{C}O \longrightarrow \dot{C}H_3 + CO$

(*ii*) $\dot{C}H_3 + CH_3\dot{C}O \longrightarrow C_2H_6 + CO$

(*iii*) $CH_3\dot{C}O + CH_3\dot{C}O \longrightarrow (CH_3CO)_2$

Thus no chain reaction is set up. The free radicals formed in the primary process simply react with each other. Hence the quantum efficiency of the reaction is unity.

Q. 31. Write notes on the following

(*i*) Chemiluminescence *(Kurukshetra 2002; Guwahati 2002;Arunachal 2003; Mumbai 2005;)*

(*ii*) Fluorescence *(Arunachal 2003; Andhra 2010; Himachal, 2010)*

(*iii*) Phosphorescence *(Gujarat 2003; Bangalore 2004; Madras, 2004; Raipur 2010; Andhra 2010; Himachal 2010)*

Ans. (*i*) **Chemiluminescence:** When a photochemical reaction takes place, light is absorbed. However, there are certain reactions in which light is produced. *The emission of light in chemical reactions at ordinary temperatures is called chemiluminescence.* Thus chemiluminescence is just the reverse of a photochemical reaction.

A few examples of chemiluminescence are as under :

(*a*) The light emitted by glow-worms is due to the oxidation of the protein, *luciferin, present in the glow-worms.*

(*b*) The oxidation of yellow phosphorus in oxygen or air to give P_2O_5 at ordinary temperatures (–10 to 40°C) is accompanied by the emission of visible greenish-white luminiscence.

In the above cases, a part or whole of the energy emitted during the reaction, is used up for the excitation of electrons. When they jump back to the inner orbits, the emission of light takes place.

(*ii*) **Fluorescence:** There are certain substances which when exposed to light or certain radiations absorb the energy and then immediately or instantaneously start re-emitting the energy. Such substances are called fluorescent substances and the phenomenon is called *fluorescence.* Obviously the absorption of energy results into the excitation of the electrons followed immediately by the jumping back of the excited electrons to the lower levels. As a result, the absorbed energy is emitted back.

Fluorescence starts as soon as the substance is exposed to light and the fluorescence stops as soon as the light is cut off.

A few examples of the substances showing the phenomenon of fluorescence are :

(*a*) fluorite, CaF_2

(*b*) certain organic dyes such as eosin, fluorescein etc. ;

(*c*) certain inorganic compounds such as uranyl sulphate, UO_2SO_4.

It is interesting to observe that in the phenomenon of fluorescence, the wavelength of the emitted light is usually greater than that of the absorbed light. This is explained on the basis that the energy absorbed raises the electron to a sufficiently higher level but the return of the excited electron to the original level takes place in steps, through the intermediate levels. The energy thus produced in every jump is smaller and hence the wavelength of the light emitted is larger than that of the absorbed light.

(*iii*) **Phosphorescence:** There are certain substances which continue to glow for some time even after the external light is cut off. Such substances are called *phosphors or phosphorescent substances* and the phenomenon is called *phosphorescence.* Thus in a way phosphorescene is a slow fluorescence. It is found mostly in solids, because the molecules have least freedom of motion. The excited electrons thus keep on jumping back slowly for quite sometime.

The examples of the substances exhibiting phosphorescence include *zinc sulphide* and *sulphides* of the *alkaline earth metals.* It has been found that fluorescent substances become phosphorescent if fixed by suitable methods, *e.g.* by fusion with other substances etc. For example, many dyes which show fluorescence when dissolved in fused boric acid or glycerol and the cooled to rigid mass become phosphorescent.

Chemiluminescence, fluorescence and phosphorescence together are called *luminescence.*

Q.32. Give Jablonski diagram and explain fluorescence and phosphorescence in terms of singlet and triplet states.

OR

Explain the mechanism of fluorescence and phosphorescence showing various ground and excited states of a molecule, using Jablonski diagram.

Ans. When radiations are incident on certain substances, some of the electrons may absorb energy and jump to higher levels. The electrons may retrurn to the ground state or may bring about a chemical change.

The electronic spins are expressed in terms of **spin multiplicity** which is equal to 2S + 1, where S is the total electron spin. Taking $+\frac{1}{2}$ and $-\frac{1}{2}$ as spin values of electrons spinning clockwise and anti-clockwise, we can say that if a molecule contains all the paired electrons, their spins will neutralise and S will be equal to zero. Substituting the value of S = 0 in the spin multiplicity relation

Spin multiplicity = 2S + 1 = 1

The molecule having spin multiplicity equal to 1 is said to be in **singlet ground state**

After irradiation, again suppose, the electrons in the excited state have opposite spin, the spins will neutralise so that S = 0 and spin multiplicity will be 2S + 1 i.e 1. This is **singlet excited state** of the molecules.

If after excitation of electrons, the electrons occupy different orbitals with similar spins, $S = \frac{1}{2} + \frac{1}{2} = 1$ and spin multiplicity will be 2S + 1 = 3. The molecule is said to be in **triplet excited state.**

Ground singlet state is denoted by S_0 and successive singlet excited state are denoted by S_1, S_2, S_3, etc.

Corresponding triplet excited states are denoted by T_1, T_2, T_3, etc. Excitation of electrons in the molecules can be represented as under.

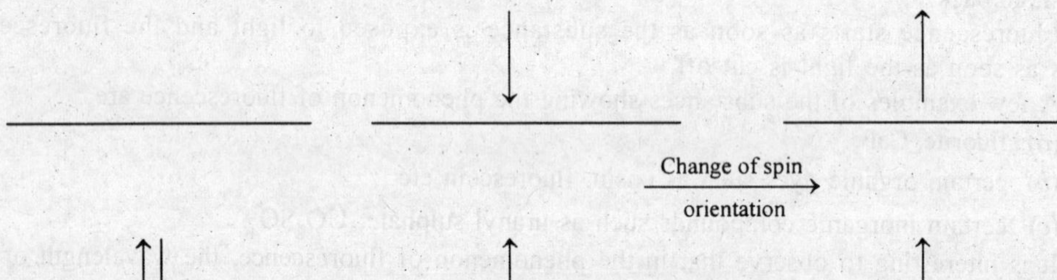

singlet ground state (pair of electrons with opposite spin in one orbital) | singlet excited state (pair of electrons with opposite spin but in different orbital) | Triplet excited state (pair of electrons with parallel spin in different orbital)

Any singlet excited state possesses higher energy than the corresponding triplet excited state.

Thus $E_{S_1} > E_{T_1}$, $E_{S_2} > E_{T_2}$ and so on.

Excited molecules return to the original (ground) state by emitting energy in a number of ways as shown in the Fig 12.4. This is known as Jablonski diagram.

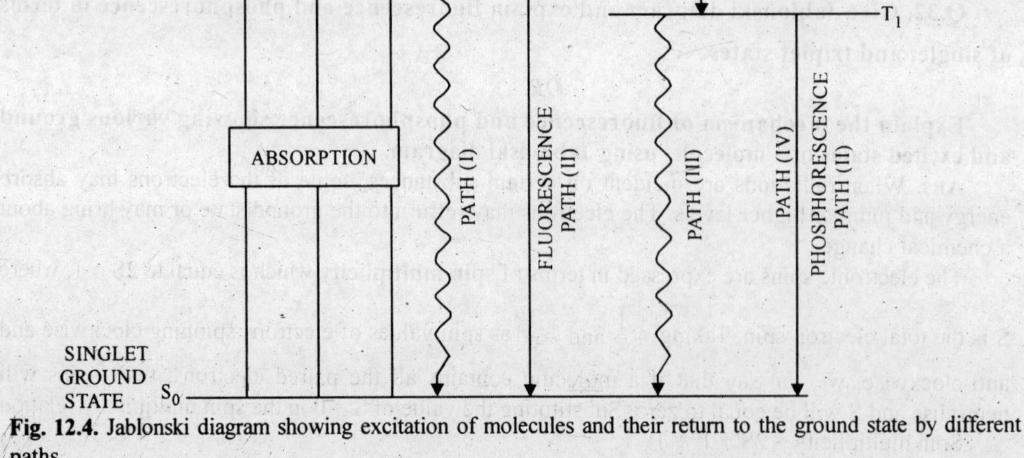

Fig. 12.4. Jablonski diagram showing excitation of molecules and their return to the ground state by different paths.

In the first step of reverting to ground state, transition from higher excited singlet states (S_2, S_3, etc) to lowest excited singlet state state S_1 occurs. This is called internal conversion (IC). Loss of energy takes place only in the form of heat, no radiation are emitted. Therefore this step is called **non-radiative** or **radiationless** transition. From the first excited singlet state, the molecules return to the ground state through one of the following paths.

Path I If the molecule emits rest of the energy also in the form of heat, the whole process becomes non-radiative. This is shown by a wavy line in the figure.

Path II If the molecule loses energy in the form of radiation to reach the ground state, it is radiative transition. This transition represents **fluorescence** and is shown by a straight line in the figure.

Path III It may sometimes so happens that some energy is lost in moving from first excited singlet state to first excited triplet state i.e. from S_1 to T_1 in the form of heat. This is called **ntersystem cossing** (ISC) and is a non-radiative step. Further transition from first excited triplet state o the ground state occurs again in the form of loss of heat. Thus path III is completely non-radiative

Path IV. After reaching first excited triplet state from first excited singlet state, further transition to the ground state may occur in the form of emission of radiation, thus path IV is a radiative step. It represents the phenomenon of phosphorescence.

Triplet states have much longer life time than singlet state. Therefore phosphorescence continues even after the absorbed radiations is removed.

Q.33. Write a note on photosensitization. *(Himachal, 2000; Dayanand 2000; Garhwal 2000; Kurukshetra 2002; Bangalore 2004)*

Ans. There are many substances which do not react directly when exposed to light. However, if another substance is added, the photochemical reaction starts. The substance thus added itself does not undergo any chemical change.

It merely absorbs the light energy and then passes it on to one of the reactants. *Such a substance which when added to a reaction mixture helps to start the photochemical reaction but itself does not undergo any chemical change is called a photosensitizer and the process is called photosensitization.* Thus a photosensitizer simply acts as a *carrier of energy.*

At the first instance it appears that a photosensitizer is similar to a catalyst. However, this is not so as it acts by different mechanism.

Examples: A few well known examples of the photosensitized reactions are briefly described as under :

1. *Decomposition of oxalic acid in presence of uranyl sulphate.* This reaction forms the basis of the actinometer used to measure the intensity of radiation. The coloured uranyl ions absorb the light and then pass it on to the colourless oxalic acid which then undergoes decomposition; the extent of decomposition depending upon the light energy absorbed.

$$UO_2^{2+} + h\nu \longrightarrow UO_2^{2+*}$$

$$UO_2^{2+*} + \begin{matrix} COOH \\ | \\ COOH \end{matrix} \longrightarrow UO_2^{2+} + CO_2 + CO + H_2O$$

The uranyl ions thus act as photosensitizer.

2. *Dissociation of hydrogen molecules in presence of mercury vapour.* Hydrogen molecules do not dissociate when exposed to ultraviolet light. However when hydrogen gas is mixed with mercury vapour and then exposed to the ultraviolet light, hydrogen molecules dissociate to give hydrogen atoms through the following reactions :

$$Hg + h\nu \longrightarrow Hg^*$$

$$Hg^* + H_2 \longrightarrow Hg + 2H$$

where Hg^* represents the activated mercury atom. Thus in the above reaction mercury acts a photosensitizer.

The H-atom being highly reactive can easily reduce metallic oxides, nitrous oxide, carbon monoxide etc.

3. *Photosynthesis of carbohydrates in plants.* Carbon dioxide and water vapour present in the air do not absorb the visible light emitted by the sun. However, the green colouring matter, namely *chlorophyll* present in the plants can absorb the visible light. After absorption, it passes on the energy to the carbon dioxide and water molecules which then combine to form carbohydrates alongwith the evolution of oxygen.

$$\underbrace{CO_2 + H_2O}_{\text{Form air}} + \underbrace{h\nu}_{\substack{\text{From}\\\text{Sunlight}}} \xrightarrow{\text{Chlorophyll}} \frac{1}{6}\underbrace{(C_6H_{12}O_6)}_{\text{Glucose}} + O_2$$

Thus chlorophyll acts as a photosensitizer in the above reaction.

Q.34. What are photo-inhibitors ? How do they act ?

Ans. Presence of certain substances considerably reduces the quantum yield of some photochemical reactions. For example, in the photosynthesis of HC1, the presence of traces of oxygen reduces the quantum yield of this reaction. Substances like nitric oxide, sulphur dioxide, propylene etc. also show similar effects. *Such substances which reduce the quantum yield of photochemical reactions are called photo-inhibitors.* It is believed that such substances react with the chain propagating atoms or radicals resulting into the chain termination. The chain termination in the photosynthesis of HCl by the presence of traces of oxygen takes place as follows:

$$\dot{H} + O_2 \rightarrow HO_2$$

$$\dot{Cl} + O_2 + HCl \rightarrow HO_2 + Cl_2$$

Q.35. What is meant by photochemical equilibrium ? Give two examples. Show that for the reaction: *(Shivaji 2000; Kolhapur 2000)*

$$A \underset{\text{Dark}}{\overset{\text{Light}}{\rightleftharpoons}} B$$

the concentration of the products at equilibrium is proportional to the intensity of light.

Ans. Suppose a substance A changes into a substance B by the absorption of light. If the reverse reaction can also occur either as a *photochemical reaction* or *as thermal reaction (dark reaction), the* situation may be represented as

$$A \underset{\text{Light}}{\overset{\text{Light}}{\rightleftharpoons}} B$$

or

$$A \underset{\text{Thermal}}{\overset{\text{Light}}{\rightleftharpoons}} B$$

In either case, a stage may reach when the rate of forward reaction may become equal to the rate of backward reaction. Now further absorption of light produces no chemical change. The reaction is then said to have attained a *photochemical equilibrium.*

Examples: (*i*) *Photochemical decomposition of nitrogen dioxide*

$$2NO_2 \underset{\text{Thermal}}{\overset{\text{Light}}{\rightleftharpoons}} 2NO + O_2$$

(*ii*) *Photochemical decomposition of sulphur trioxide*

$$2SO_3 \underset{\text{Light}}{\overset{\text{Light}}{\rightleftharpoons}} 2SO_2 + O_2$$

Equilibrium constant: The equilibrium constant of a photochemical equilibrium is different from that of the ordinary chemical equilibrium. This is due to the fact that the rate of a photochemical reaction does not depend upon the concentration of the reactants but it depends upon the intensity of the light absorbed. Thus for the reaction

$$A \underset{\text{Thermal}}{\overset{\text{Light}}{\rightleftharpoons}} B$$

Rate of forward reaction $\propto I_{abs} = k_1 I_{abs}$

Rate of backward reaction $\propto [B] = k_2$ [B]

When the reaction is in equilibrium,

Rate of forward reaction = Rate of backward reaction

i.e. $$k_1 I_{abs} = k_2 [B]$$

or $$\frac{k_2}{k_1} = \frac{I_{abs}}{[B]} \quad ...(i)$$

or $$K = \frac{I_{abs}}{[B]} \quad ...(ii)$$

where $K = \dfrac{k_2}{k_1}$ is the **photochemical equilibrium constant.**

From eqn (*ii*), it is clear that photochemical equilibrium constant depends upon the intensity of the light absorbed.

Further eqn (*i*) may be written as

$$[B] = \frac{k_2}{k_1} I_{abs}$$

It follows that the concentration of the products formed at equilibrium is directly proportional to the intensity of the light absorbed.

Q 36. Write a short note on mechanism of photochemical reactions and flash photolysis. *(Kanpur 2001)*

or

What are various steps involved in photochemical reaction? *(M. Dayanand 2000)*

Ans. See Q. 24 and 30

13

QUANTUM MECHANICS

Q. 1. What is classical mechanics ? What are its limitations ? *(Nagpur 2008)*

Ans. *The branch of science based on Newton's laws of motion and Maxwell's electromagnetic wave theory to explain phenomenon related to motion and energy is known as classical mechanics or Newtonian mechanics.* According to classical mechanics, it should be possible to determine simultaneously the position and velocity of a moving particle. But Heisenberg uncertainty principle contradicts it. Classical mechanics assumes that energy is emitted or absorbed in a continuous manner but Planck's quantum dictates that emission or absorption of energy takes place in an uncontinous manner in the form of packets of energy called *quanta.*

Limitations of classical mechanics

Classical mechanics does not provide satisfactory explanation for the following phenomena:

(*i*) Blackbody radiation

(*ii*) Photoelectric effect

(*iii*) Atomic and molecular spectra

(*iv*) Heat capacities of solids.

Q. 2. Trace the origin of quantum mechanics. *(Himachal 2010)*

Ans. The classical point of view is adequate for objects of appreciable size but is quite unsatisfactory for describing the behaviour of particles of atomic dimensions. It has been necessary therefore to devise a new mechanics for the treatment of electrons and atomic nuclei. In this new mechanics, which takes into account the de-Broglie's dual nature of matter and Heisenberg's uncertainty principle, the exact position of the moving object such as the orbit of an electron around the nucleus of an atom, is replaced by a function which determines the probability of the object being in the particular position. These probability functions satisfy differential equations which are analogous to those representing the variation of the amplitude of a wave. This new atomic mechanics has been referred to as the **wave mechanics**. This new approach to the study of small particles provides a satisfactory basis for many of the quantum postulates and the term quantum mechanics is generally used for this new approach.

Q. 3. Define black body and black body radiations. *(Calcutta 2000; Madras 2004)*

Ans. If the radiant energy is allowed to fall on the blackened metallic surface or carbon black, it is found that the energy is almost completely absorbed. *A body which completely absorbs the radiant energy falling on it is called a perfectly black body.* The absorption is found to be more perfect if we take a hollow sphere blackened on the inside and having a small hole for the entry of the radiation. This is so because any radiation that enters through the hole is reflected over and again by the walls of the sphere till finally it is completely absorbed.

The most commonly used black body is the one shown in Fig. 13.1.

It is a hollow double walled metallic sphere having a conical projection P opposite to the hole H and is coated on the inside with lamp black. The projection helps to avoid any direct reflection.

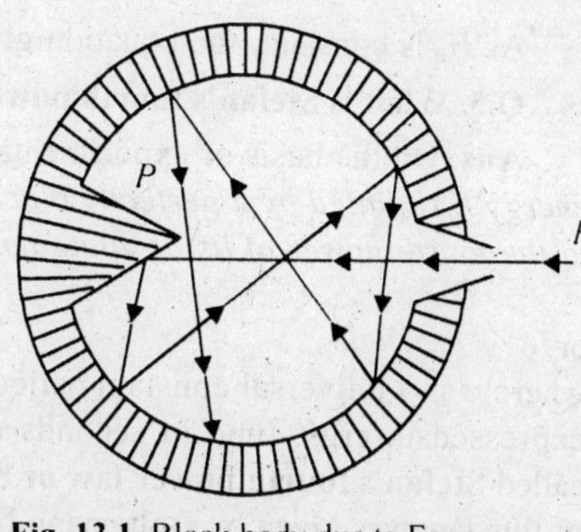

Fig. 13.1. Black body due to Ferny.

A black body is not only a perfect absorber of the radiant energy, but also a perfect radiator. In fact, of all the bodies, the black body radiates the maximum amount of energy for the given temperature. The radiations thus emitted are called *black body radiation.*

Q. 4. State and explain Kirchoff's law of black body radiations.

Ans. The *quantitative* relationship between the energy absorbed and the energy emitted by a body was put forward by Kirchoff in 1859. It is called Kirchoff s law and it states as follows :

At any temperature, the ratio of the emissive power (or emittance) of a body to the absorptive power (or absorbance or absorptivity) is constant, independent of the nature of the surface and equal to the emissive power (emittance) of a perfectly black body.

The term *emissive power* or *emittance* means the energy emitted by the surface per unit time per unit area whereas the term *absorptive power* or *absorbance* or *absorptivity* is the fraction of the incident energy absorbed by the surface per unit time per unit area.

Thus if E_s and A_s represent the emissive power and absorptive power of a substance and E_B represents the emissive power of a perfectly black body, then according to Kirchoff's law,

$$\frac{E_S}{A_S} = E_B$$

Derivation of Kirchoff's law : Kirchoff's law can be derived in a simple manner as follows:

Suppose Q is tne amount of radiation incident per unit area per second on a body. If A, is the absorptive power of the surface, then

the amount of radiation absorbed by the body per unit area per second = $A_s \times Q$.

If E_S represents the emissive power of the surface, then

the amount of radiation emitted by the surface per unit area per second = E_S

When the body is in *thermal equilibrium* with the enclosure,

Amount of radiation emitted by the surface per unit area per second = Amount of radiation absorbed by the body per unit area per second.

Thus
$$E_s = A_s \times Q \qquad ...(i)$$

For a perfectly black body, $A_s = A_B = 1$ and E_s may be replaced by E_B where A_B and E_B represent the absorptive power and the emissive power of a perfectly black body. Thus we have

$$E_B = Q \qquad ...(ii)$$

From equations (*i*) and (*ii*)

$$E_s = A_s \times E_B$$

or

$$\boxed{\frac{E_S}{A_S} = E_B} \qquad ...(iii)$$

As E_B is constant, the equation gives the expression for the Kirchoff s law.

Q.5. What is Stefan's fourth power law ?

Ans. On the basis of experimental studies, Stefan in 1879 showed that the *total amount of energy E radiated by a perfectly black body per unit area per unit time is directly proportional to the fourth power of its absolute temperature T, i.e.*

$$E \propto T^4$$

or

$$E = \sigma T^4$$

where σ is a universal constant called Stefan's constant. Its value is 5.6697×10^{-5} if energy is expressed in ergs, time in seconds and area in square centimeters. The above statement is called **Stefan's fourth power law or Stefan-Boltzmann fourth power law.** The theoretical proof of this law was given by Boltzmann by applying the laws of thermodynamics.

Q.6. What is a bolometer ?

Ans. Bolometer is an extremely sensitive instrument for measuring heat radiations. It consists of two very thin blackened platinum gratings, forming two arms of a Wheatstone bridge circuit. Radiant heat falling on one of gratings raises its electrical resistance, thus causing a deflection of the needle of a galvanometer in the circuit.

Q.7. Describe spectral distribution of black body radiation. *(M. Dayanand 2000; Calcutta 2000)*

Ans. The energy emitted by a black body at any temperature does not consist of a single frequency and is also not uniformly distributed along the spectrum. Lummer and Pringsheim in 1899 studied experimentally the distribution of energy amongst different wavelengths of the black body spectrum as described below.

The black body radiations were obtained by heating a carbon tube electrically. After passing through the slit S_1 these were made parallel with the help of a concave reflector mirror M_1 and then allowed to fall on a fluorspar prism to get the spectrum. The radiation after refraction through the prism were focussed by another concave mirror M_2 and slit S_2 on to a linear bolometer as shown in Fig. 13.2.

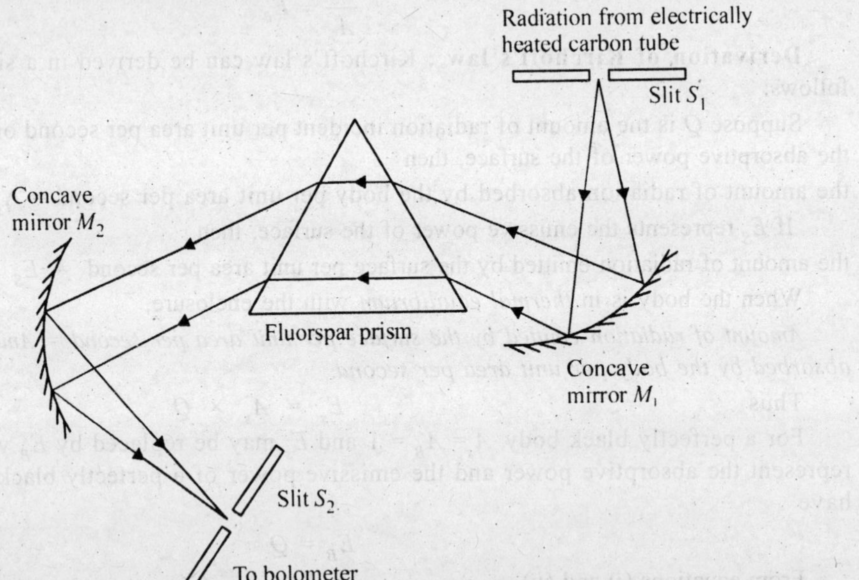

Fig. 13.2. Lummer and Pringsheim experiment for the study of spectral distribution of black body radiation.

By gradually rotating the prism, the radiations corresponding to different wavelengths are focussed on to the bolometer one by one and their intensities measured. By keeping the carbon tube (black body) at different temperatures a number of observations were made between 621 K to 1646 K. The curves were then plotted between the monochromatic emittance E_λ along the ordinate and the wavelengths along the abscissae for each of the temperatures. These curves are shown in Fig. 13.3. Following observations are made in respect of these curves.

(*i*) At a particular temperature, the distribution of energy is not uniform among the various wavelengths of the radiation emitted by the black body.

(*ii*) For each temperature,there is a wavelength (λ_m) at which the energy radiated is maximum (= E_m).

(*iii*) With increase in temperature, the maximum shifts higher but towards lower wavelengths. This means that with increase in temperature E_m increases but the corresponding λ_m decreases.

(*iv*) Higher the temperature, more pronounced is the maximum.

(*v*) The area under the curve for a particular temperature gives the total energy emitted by the black body per unit area per second for the complete spectrum *i.e.* corresponding to all the wavelengths at that particular temperature (from $\lambda = 0$ to ∞)

(*vi*) The area under the curve increases with increase of temperature and it is found that the area is proportional to the fourth power of the absolute temperature. As the area under the curve represents the total energy emitted by the black body per unit area per second, hence $E_B \propto T^4$. This is **Stefan Boltzmann's law.**

(*vii*) From the curves, it is also found that

$$\lambda_{m_1} T_1 = \lambda_{m_2} T_2 = \lambda_{m_3} T_3 \ldots\ldots$$

i.e. $$\lambda_m T = \text{constant}$$

This implies that $$\lambda_m \propto \frac{1}{T}$$

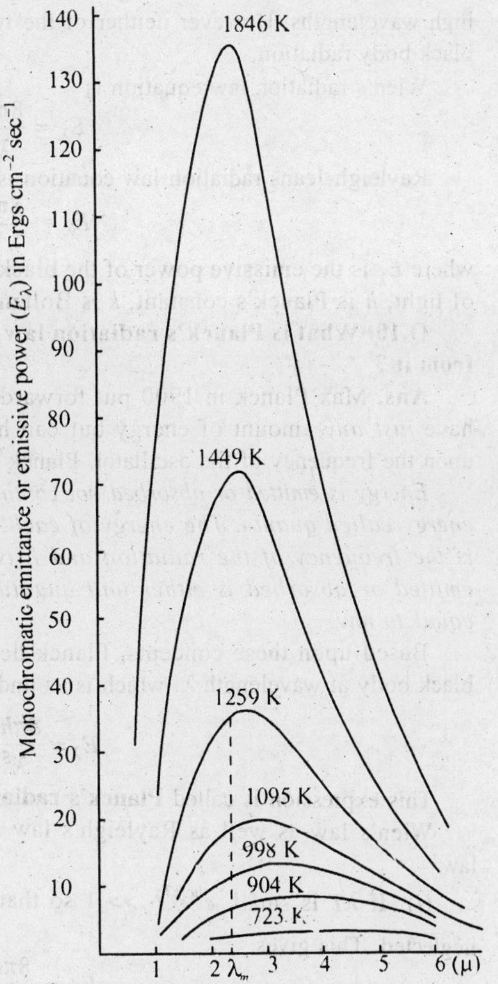

Fig. 13.3. Distribution of energy in the spectrum of a black body.

Q.8. What is Wien's displacement law ?

Ans. ***The wavelength λ_m for which the emittance of a black body is maximum is inversely proportional to its absolute temperature.*** This is called Wien's displacement law. Mathematically, the law may be represented as :

$$\lambda_m \propto \frac{1}{T}$$

where λ_m = Wavelength corresponding to maximum radiation of energy

T = Temperature on Kelvin scale.

This law is able to explain why the colour of the visible light radiation changes from red to yellow as the temperature of the hot body is increased.

Q.9. How does classical mechanics fail to explain the spectral distribution of black body radiation ? *(Himachal 2010)*

Ans. According to Maxwell electromagnetic wave theory of classical mechanics, the source of radiant energy are the vibrating particles called oscillators and these oscillators could radiate or absorb any amount of energy. The frequency of the energy (or the wave) emitted is equal to the frequency of the oscillator. Based on these classical concepts, Wien in 1896 and Lord Rayleigh and Jeans in 1900 explained the distribution of black body radiation using the classical concepts of continuous emission of radiation. Wien's equation was able to explain the distribution only at low wavelengths whereas Rayleigh-Jeans equation was able to do so only at high wavelengths. However neither of the two was able to explain the complete distribution of black body radiation.

Wien's radiation law equation is

$$E_\lambda = \frac{8\pi hc}{\lambda^5} e^{-hc/kT\lambda}$$

Rayleigh-Jeans radiation law equation is

$$E_\lambda = \frac{8\pi}{\lambda^4} kT$$

where E_λ is the emissive power of the black body corresponding to wavelength λ, c is velocity of light, h is Planck's constant, k is Boltzmann's constant and T is the absolute temperature.

Q.10. What is Planck's radiation law ? How do Wien's law and Rayleigh's law follow from it ? *(Raipur 2010)*

Ans. Max Planck in 1900 put forward a theory that the oscillator of black body cannot have *just any* amount of energy but can have only a *discrete* amount of energy depending upon the frequency of the oscillator. Planck summed up his result as under :

Energy is emitted or absorbed not continuously but discontinuously in the form of packets of energy called quanta. The energy of each quantum is given by the relation, E= hv, where v is the frequency of the radiation and h is called Planck's constant. Thus the total energy emitted or absorbed is either unit quantum i.e. hv or a whole number multiple of hv i.e. equal to nhv.

Based upon these concepts, Planck deduced expression for the energy E_λ radiated by a black body at wavelength λ, which is as under :

$$E_\lambda = \frac{8\pi hc}{\lambda^5} \frac{1}{e^{hc/kT\lambda} - 1} \quad ...(i)$$

This expression is called **Planck's radiation law.** Different symbols have usual meanings.

Wien's law as well as Rayleigh's law are simply particular cases of Planck's radiation law.

(*i*) If λT is small, $e^{hc/kT\lambda} >> 1$ so that 1 in the denominator in above equation can be neglected. This gives

$$E_\lambda = \frac{8\pi hc}{\lambda^5} e^{-hc/kT\lambda}$$

which is Wien's law.

(*ii*) If λT is large, $e^{hc/kT\lambda}$ may be expanded by the exponential theorem

$$e^x = 1 + x + x^2 + \cdots$$

we get

$$e^{hc/kT\lambda} = 1 + \frac{hc}{kT\lambda} + \cdots \quad \text{(Higher term may be neglected)}$$

So that eqn. (*i*) reduces to the form

$$E_\lambda = \frac{8\pi hc}{\lambda^5} \times \frac{kT\lambda}{hc} = \frac{8\pi}{\lambda^4} . kT$$

which is Rayleigh-Jeans law.

Q.11. Derive Planck's radiation law.

Ans. Planck considered the black body radiations to consist of linear oscillators of molecular dimensions and that the energy of a linear oscillator can assume only the discrete values

$$0, h\nu, 2h\nu, 3h\nu.....nh\nu$$

If N_0, N_1, N_2...... are the number of oscillators per unit volume possessing energies 0, $h\nu$, $2h\nu$ respectively, then the total number of oscillator N per unit volume will be

$$N = N_0 + N_1 + N_2 + ... \qquad ...(i)$$

Number of oscillators, N_r having energy E_r is given by (Maxwell's formula)

$$N_r = N_0 e^{-E_r/kT} \qquad ....(ii)$$

Putting these values in eqn. (*i*), we get

$$\begin{aligned} N_r &= N_0 + N_0 e^{-h\nu/kT} + N_0 e^{-2h\nu/kT} + \cdots\cdots + N_0 e^{-nh\nu/kT} \\ &= N_0(1 + e^{-h\nu/kT} + e^{-2h\nu/kT} + \cdots) \\ &= \frac{N_0}{1-e^{-h\nu/kT}} \quad [\because (1-e^{-x})^{-1} = 1 + e^{-x} + e^{-2x} + \cdots] \qquad ...(iii) \end{aligned}$$

The total energy of N oscillators will be

$$\begin{aligned} E &= N_1 . h\nu + N_2 . 2h\nu + N_3 . 3h\nu + ... \\ &= h\nu (N_1 + 2N_2 + 3N_3 + ...) \\ &= h\nu (N_0 e^{-h\nu/kT} + 2N_0 e^{-2h\nu/kT} + 3N_0 e^{-3h\nu/kT} + ...) \\ &= N_0 h\nu (e^{-h\nu/kT} + 2e^{-2h\nu/kT} + 3e^{-3h\nu/kT} + ...) \\ &= N_0 h\nu e^{-h\nu/kT} (1 + 2e^{-h\nu/kT} + 3e^{-2h\nu/kT} + ...) \\ &= N_0 h\nu e^{-h\nu/kT} (1 + 2x + 3x^2 + ...) \qquad (\text{where } x = e^{-h\nu/kT}) \\ &= N_0 h\nu e^{-h\nu/kT} (1 - x)^{-2} \\ &= \frac{N_0 h\nu e^{-h\nu/kT}}{(1-x)^2} \\ &= N_0 h\nu \frac{e^{-h\nu/kT}}{(1 - e^{-h\nu/kT})^2} \qquad ...(iv) \end{aligned}$$

Hence the average energy per oscillator is

$$\bar{E} = \frac{E}{N} = N_0 h\nu \frac{e^{-h\nu/kT}}{(1-e^{-h\nu/kT})^2} \times \frac{1-e^{-h\nu/kT}}{N_0}$$

$$\frac{h\nu e^{-h\nu/kT}}{1-e^{-h\nu/kT}} = \frac{h\nu}{e^{h\nu/kT} - 1} \qquad \left(\begin{array}{l}\text{On dividing numerator and} \\ \text{denominator by } e^{-h\nu/kT}\end{array}\right) \qquad ...(v)$$

Thus we see that the average energy of the oscillator is not kT (as given by classical theory) but equal to $(e^{h\nu/kT} - 1)$ according to Planck's quantum theory.

Further it can be deduced that the number of oscillators per unit volume having frequency in the range of ν and $\nu + d\nu$ is equal to

$$\frac{8\pi\nu^2 d\nu}{c^3} \qquad ...(vi)$$

Hence the average energy per unit volume (*i.e.* energy density) inside the enclosure is obtained by multiplying (v) with (vi) *i.e.*, it is given by

$$E_v dv = \frac{8\pi v^2 dv}{c^3} \times \frac{hv}{e^{hv/kT} - 1} \quad ...(vii)$$

Putting $v = \frac{c}{\lambda}$ and $dv = -\frac{c}{\lambda^2} d\lambda$, the average energy per unit volume in the enclosure for the wavelengths between λ and $\lambda + d\lambda$ will be

$$E_v d\lambda = \frac{8\pi hc}{\lambda^5} \times \frac{1}{e^{hc/\lambda kT} - 1} d\lambda \quad ...(viii)$$

or the energy radiated by the black body corresponding to wavelength λ is

$$\boxed{E_\lambda = \frac{8\pi hc}{\lambda^5} \frac{1}{e^{hc/\lambda kT} - 1}} \quad ...(ix)$$

which is Planck's radiation law or Planck's distribution law.

Q. 12. Derive Schrodinger wave equation using de Broglie equation.

(M. Dayanand 2002; Guwahati 2006; Himachal 2010; Mumbai 2010)

Ans. Consider the simplest type of wave motion like that of the vibration of a stretched string travelling along the x-axis with a velocity u (Fig. 13·4). If ω is the amplitude of the wave at any point whose co-ordinate is x, at any time t, then the equation for such a wave motion is

$$\frac{\partial^2 \omega}{\partial x^2} = \frac{1}{v^2} \frac{\partial^2 \omega}{\partial t^2} \quad ...(i)$$

This differential equation indicates that the amplitude ω of the wave at any time travelling with a particular velocity depends upon the displacement x and the time t. In other words, ω is a function of x and t. Hence we may write

$$\omega = f(x) f'(t) \quad ...(ii)$$

where $f(x)$ is a function of the co-ordinate x only and $f'(t)$ is a function of the time t only.

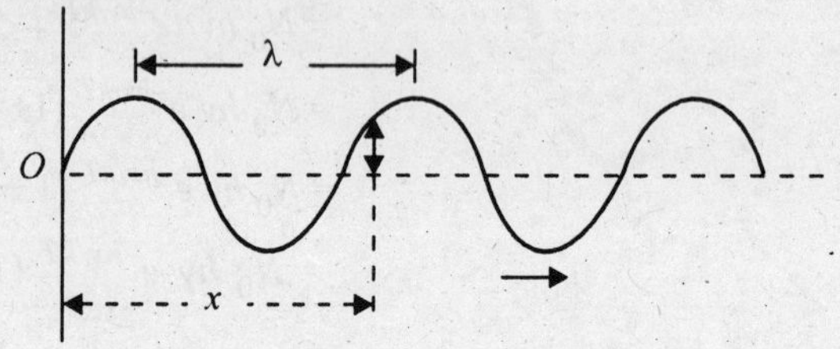

Fig. 13.4. Vibration in a stretched string.

But for the stationary waves, we have

$$f'(t) = A \sin 2\pi vt \quad ...(iii)$$

where v is the frequency of vibration and A is a constant, equal to the maximum amplitude of the wave.

Substituting the value of $f'(t)$ from eqn. (iii), in eqn. (ii), we get

$$\omega = f(x) A \sin 2\pi vt \quad ...(iv)$$

Differentiating this equation twice w.r.t. t, we get

$$\frac{\partial^2 \omega}{\partial t^2} = -f(x) 4\pi^2 v^2 (A \sin 2\pi vt)$$

$$= -4\pi^2 v^2 f(x) f'(t) \quad ...(v)$$

Differentiating eqn. (ii) twice w.r.t x, we get

$$\frac{\partial^2 \omega}{\partial x^2} = \frac{\partial^2 f(x)}{\partial x^2} f'(t) \quad ...(vi)$$

Substituting the values of $\partial^2\omega/\partial t^2$ and $\partial^2\omega/\partial x^2$ from equations (v) and (vi) into the equation (i), we get

$$\frac{\partial^2 f(x)}{\partial x^2} f'(t) = -\frac{1}{v^2} 4\pi^2 \nu^2 f(x) f'(t)$$

or
$$\frac{\partial^2 f(x)}{\partial x^2} = -\frac{4\pi^2 \nu^2}{v^2} f(x) \quad ...(vii)$$

This equation has become **time-independent** and hence represents the variation of the amplitude function $f(x)$ with x only.

Further, the velocity v is related to the frequency ν and the wavelength λ of the wave by the expansion

$$v = \nu\lambda \quad ...(viii)$$

Substituting this value in equation (*vii*), we get

$$\frac{\partial^2 f(x)}{\partial x^2} = -\frac{4\pi^2}{\lambda^2} f(x) \quad ...(ix)$$

This equation has been derived for the wave motion in one direction only. For extending this equation to the wave motion in three dimensions, the amplitude function $f(x)$ for one co-ordinate may be replaced by $\psi(x, y, z)$ or simply written as ψ for the sake of brevity which represents the *amplitude function* for three co-ordinates. Thus the equation for wave motion in three dimensions becomes

$$\frac{\partial^2 \psi}{\partial x^2} + \frac{\partial^2 \psi}{\partial y^2} + \frac{\partial^2 \psi}{\partial z^2} = -\frac{4\pi^2}{\lambda^2} \psi \quad ...(x)$$

This equation is applicable to all particles like electrons, nuclei, atoms, molecules, photons etc. Further applying de Broglie equation, according to which $\lambda = h/mv$ where m is the mass of the particle, v is its velocity and h is Planck's constant, eqn. (x) becomes

$$\frac{\partial^2 \psi}{\partial x^2} + \frac{\partial^2 \psi}{\partial y^2} + \frac{\partial^2 \psi}{\partial z^2} = -\frac{4\pi^2 m^2 v^2}{h^2} \psi \quad ...(xi)$$

Further we know that

Total energy (E) of the particle = Kinetic energy + Potential energy

i.e.,
$$E = \frac{1}{2} mv^2 + V$$

or
$$\frac{1}{2} mv^2 = E - V \quad ...(xii)$$

where V represents the potential energy of the particle.

Substituting the value of mv^2 from eqn. (*xii*) in eqn. (*xi*), we get

$$\frac{\partial^2 \psi}{\partial x^2} + \frac{\partial^2 \psi}{\partial y^2} + \frac{\partial^2 \psi}{\partial z^2} + \frac{8\pi^2 m (E - V)}{h^2} \psi = 0 \quad ...(xiii)$$

This equation which represents the wave motion of the particle in three dimensions is called *Schrodinger wave equation.*

Q.13. What do you understand by eigen operator, eigen values and eigen functions (wave functions) ? Explain giving examples. (*Raipur 2010*)

Ans. For a stationary (or standing) wave in a stretched string, the amplitude function $f(x)$ can have significance only for certain definite values of λ. These functions will be those which must satisfy the following conditions :

(*i*) $f(x)$ must be equal to zero at each end of the string, as the string is fixed at these points and the amplitude of vibration is zero.

(*ii*) $f(x)$ must be single valued and finite *i.e.*, at every point on the vibrating string, the amplitude has a definite value at any given instant.

Similarly, Schrodinger wave equation, being a second order differential equation, can have many solutions for the amplitude function ψ but only those values of ψ are acceptable and have significance which correspond to some definite values of the total energy E. Such values of the total energy E are called **eigen values.** These values are also called *proper values or characteristic values.* The corresponding values of the function ψ are called **eigen functions** or **wave functions.** In the case of a stretched string, the eigen functions of the Schrodinger equation will be those which must satisfy the following conditions :

(*i*) ψ must be single values and finite *i.e.*, for each value of the variable x, y, z there is only one definite value of the function ψ.

(*ii*) ψ must be continuous *i.e.*, there must not exist any sudden changes in ψ as its variables are changed.

(*iii*) ψ must become zero at infinity.

We define a class of functions as all those functions which obey certain specified conditions. It may happen that in a given class *there are functions which when operated on by an operator,* α *are merely multiplied by some constant a, or in symbols*

$$\alpha f(x) = a f(x) \qquad ...(1)$$

Those members of the class which obey such a relation are known as characteristic or eigen functions of the operator α. The various possible values of a are called the characteristic values or eigen values of the operator.

For example, when function $\Psi = \sin kx$ on which the operator $\delta = \frac{d}{dx}$ operates, we get

$$\frac{d\psi}{dx} = \frac{d}{dx}(\sin kx) = k \cos kx$$

Since $\cos kx \neq$, ψ, $\frac{d}{dx}$ is not an eigen.operator for the function $\psi = \sin kx$. Now let us consider $\frac{d^2}{dx^2}$ as the operator for the same functions $\psi = \sin kx$, then upon operating, we get

$$\frac{d^2\Psi}{dx^2} = \frac{d}{dx}\left[\frac{d}{dx}\sin kx\right]$$

$$= \frac{d}{dx}[k \cos kx]$$

$$= -k^2 \sin kx = -k^2 \psi.$$

Here $\frac{d^2}{dx^2}$ is an eigen operator for $\psi = \sin kx$. Thus $\psi = \sin kx$ is an eigen function for operator, $\frac{d^2}{dx^2}$ having eigen value $-k^2$.

Let us take another example when function $\psi = e^{imx}$, operators are $\frac{d}{dx}$ and $\frac{d^2}{dx^2}$.

Then $\frac{d}{dx}(\psi) = ime^{imx} = im\psi$

Here $\frac{d}{dx}$ is an eigen operator, *im* is the eigen value for the eigen function $\psi = e^{imx}$.

Also, $\frac{d^2\psi}{dx^2} = im\ im\ e^{imx} = -\ m^2\ \psi.$

Here, $\frac{d^2}{dx^2}$ is an eigen operator with eigen value $-\ m^2$ for eigen function $\psi = e^{imx}$. As is clear different eigen operators have different eigen values for the same function.

Q. 14. Evaluate the expression for the operator $\left(x\frac{d}{dx}\right)^2$.

Ans. Consider any function $\psi(x)$. When the given operator operates on it, we have

$$\left(\frac{d}{dx}\right)^2 \psi = \left(x\frac{d}{dx}\right)\left(x\frac{d}{dx}\right)\psi$$

$$= x\frac{d}{dx}\left(x\frac{d\psi}{dx}\right)$$

$$= x\left[x\frac{d^2\psi}{dx^2} + \frac{d\psi}{dx}\right]$$

$$= x^2\frac{d^2\psi}{dx^2} + x\frac{d\psi}{dx}$$

Hence,
$$\left(x\frac{d}{dx}\right)^2 = x^2\frac{d^2}{dx^2} + x\frac{d}{dx}.$$

Q. 15. Show that the function cos *ax* cos *by* cos *cz* is an eigen function of the operator

$$\nabla^2 = \left[\frac{\partial^2}{\partial x^2} + \frac{\partial^2}{\partial y^2} + \frac{\partial^2}{\partial z^2}\right].$$

Ans. Here $\psi = \cos ax \cos by \cos cz$

$$\nabla^2\psi = \left[\frac{\partial^2}{\partial x^2} + \frac{\partial^2}{\partial y^2} + \frac{\partial^2}{\partial z^2}\right][\cos ax \cos by \cos cz]$$

or $\nabla^2\psi = \cos by \cos cz \frac{\partial^2}{\partial x^2}(\cos ax) + \cos ax \cos cz \frac{\partial^2}{\partial x^2}(\cos by) + \cos ax \cos by \frac{\partial^2}{\partial x^2}$ $(\cos cz)$. Now, $\frac{\partial}{\partial x}(\cos ax) = -\ a \sin ax$ or $\frac{\partial^2}{\partial x^2}(\cos ax) = \frac{\partial}{\partial x}(-a \sin ax) = -\ a^2 \cos ax.$

Similarly, $\frac{\partial^2}{\partial y^2}(\cos by) = -\ b^2 \cos by$ and $\frac{\partial^2}{\partial z^2}(\cos cz) = -\ c^2 \cos cz.$

Thus, $\nabla^2\psi = \nabla^2[\cos ax \cos by \cos cz]$

$$= -\ a^2 \cos ax \cos by \cos cz - b^2 \cos ax \cos by \cos cz - c^2 \cos ax \cos by \cos cz.$$

$$= -\ (a^2 + b^2 + c^2)\ \cos ax \cos by \cos cz.$$

$$= -\ (a^2 + b^2 + c^2)\ \psi$$

Hence, the function cos ax cos by cos cz is an eigen function of operator ∇^2 and the eigen value is $-(a^2 + b^2 + c^2)$.

Q. 16. Show that the function $A\, e^{-ax}$ is an eigen function of the operator $\frac{d^2}{dx^2}$. What is the eigen value?

Ans. Here $\psi = A\, e^{-ax}$, $\frac{d\psi}{dx} = -A\, a\, e^{-ax}$, $\frac{d^2\psi}{dx^2} = +A\, a^2\, e^{-ax} = a^2\, e^{-ax} = a^2\, A\, e^{-ax} = a^2\, \psi$.

Thus, the function $A\, e^{-ax}$ is an eigen function of the operator $\frac{d^2}{dx^2}$ with eigen value a^2.

Q. 17. Under what conditions is the function e^{-aq^2} is an eigen function of the operator $\frac{d^2}{dq^2} - Kq^2$, where K is constant? What is the eigen value under these conditions?

Ans. Here $\psi = e^{-aq^2}$ and operator is $\frac{d^2}{dq^2} - Kq^2$

Then $\left[\frac{d^2}{dq^2} - Kq^2\right] e^{-aq^2} = \frac{d^2}{dq^2}(e^{-aq^2}) - Kq^2\,(e^{-aq^2})$.

Now $\frac{d^2}{dq^2}(e^{-aq^2}) = -2aqe^{-aq^2}$

$$\frac{d^2}{dq^2}(e^{-aq^2}) = \frac{d}{dq}(-2aqe^{-aq^2})$$

$$= -2aqe^{-aq^2} - 2aq\,(-2aq)\,e^{-aq^2}$$

$$= -2aqe^{-aq^2} + 4a^2q^2e^{-aq^2}$$

Thus $\left[\frac{d^2}{dq^2} - Kq^2\right] e^{-aq^2} = -2ae^{-aq^2} + 4a^2q^2e^{-aq^2} - Kq^2\,e^{-aq^2}$

$$= (-2a + 4a^2q^2 - Kq^2)e^{-aq^2}$$

For the function e^{-eq^2} to be an eigen function with respect to the given operator $(-2\,a + 4a^2\,q^2 - Kq^2)$ must be constant. This is possible only when the sum of the terms containing q is equal to zero *i.e.,*

$$4a^2\,q^2 - Kq^2 = 0$$

$$(4a^2\,q^2 - K)q^2 = 0$$

Thus, $4a^2 - K = 0$ (since $q \neq 0$)

$$K = 4\,a^2$$

Also, the eigen value is $-\,2a$.

Q.18. Explain Born's interpretation and significance of wave function. *(Madras 2004)*

Ans. If the particle under consideration is an electron, the wave function ψ represents the amplitude of the electron wave at any instant of time when the co-ordinates of the electron are x, y, z. However, it was suggested by Max Born that, just as in the case of light or sound, the square of the amplitude of the wave at any point gives the intensity of the light or sound at that point, the square of the amplitude of the electron wave *i.e.*, ψ^2 at any point gives the intensity of the electron wave at that point. However, keeping in view the Heisenberg's uncertainty principle, the intensity of electron at any point may be interpreted as the probability of finding the particle electron at that point (Fig. 13.5). Thus ψ^2 *at any point gives the probability of finding, the electron at that point.* That is why the wave function, ψ^2 is also called **probability amplitude.** But the probability is directly related to the density (concentration) of the electron cloud. Thus ψ^2 gives the electron density at any given point. Since the region around the nucleus which represents the electron density at different points is called an orbital, that is why the wave function for an electron in an atom is called *orbital wave function* or simply atomic orbital.

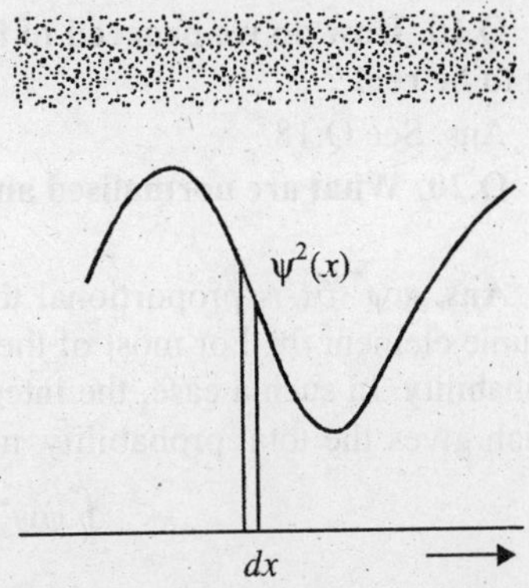

Fig. 13.5. Born interpretation of wave function in one dimensional system. Probability of finding the particle at points along x-axis. $\psi^2(x)$ is represented by the density of shading in the upper half of the diagram.

Further, it may be noted that ψ may not always come out to be real. In some cases it may come out to be imaginary *i.e.*, may include the imaginary quantity $i(=\sqrt{-1})$. Then ψ^2 may be real or imaginary depending upon the nature of the expression for ψ. However, since the probability of finding a material particle at any point in space must always be real, the value taken is $\psi\psi^*$ in place of ψ^2 where, ψ^* is called the *complex conjugate* of ψ and is obtained by replacing i by $-i$. This product (*i.e.*, $\psi\psi^*$) is always real irrespective of whether ψ is real or imaginary. For example, suppose $\psi = a + ib$. Then the complex conjugate will be $\psi^* = a - ib$. The product $\psi\psi^* = (a + ib)\ (a - ib) = a^2 + b^2$ which is real. The product $\psi\psi^*$ is usually represented by $|\psi|^2$ where, $|\psi|$ represents *modulus* or *absolute* value of ψ. $|\psi|^2$ is called the **square modulus** of ψ. If ψ is real, then $\psi = \psi^*$ and the product will automatically be equal to ψ^2.

Born interpretation of the wave function may be summed up as follows :

(*i*) *For a one-dimensional system,* if the amplitude of the wave function of a particle is ψ at some point x, then the probability of finding the particle between x and $x + dx$ is proportional to $\psi(x)\psi^*(x)\ dx$ or in short it is written as $\psi\psi^*\ dx$.

(*ii*) *For a particle free to move in three-dimensions,* (*e.g.* an electron near the nucleus of an atom), if ψ is the amplitude of the wave function at some point at distance r, then the probability of finding the particle in the small volume element $dx\ dy\ dz$ at that point whose co-ordinates are x, y, z is proportional to $\psi\ (x, y, z)\ \psi^*\ (x, y, z)\ dx\ dy\ dz$ or in short it is written as $\psi\psi^*\ d\tau$ where $d\tau = dx\ dy\ dz$ represents the small volume element of the space at some distance r (Fig. 13.6). For this reason the product $\psi\psi^*$ is called the *probability distribution function.*

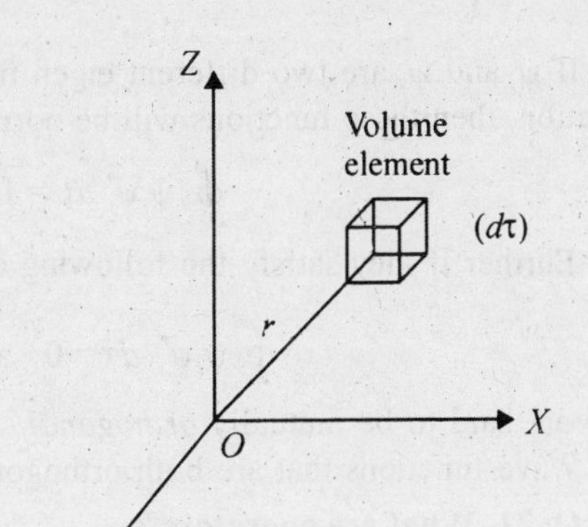

Fig. 13.6. Born interpretation of the wave function in three dimensional space.

Q.19. How do we proceed to find out probability of finding an electron if ψ does not come out to be real ?

Ans. See Q.18.

Q.20. What are normalised and orthogonal wavefunctions ?

(*Himachal 2010*)

Ans. $\psi\psi^* \partial\tau$ is proportional to the probability of finding the particular system in the small volume element $\partial\tau$. For most of the purposes it is taken as equal to rather than proportional to the probability. In such a case, the integration of $\psi\psi^* \partial\tau$ over the whole of the configuration space which gives the total probability, must be equal to unity *i.e.*,

$$\oint \psi\psi^* d\tau = 1 \qquad ...(i)$$

where $\oint$ represents the integration over the whole space.

A wave function which satisfies the above equation is said to be normalised.

In some cases, the result of integration is found to be equal to some constant, say N. Then we have

$$\oint \psi\psi^* d\tau = N \qquad ...(ii)$$

To make the result equal to unity, we have

$$\frac{1}{N}\oint \psi\psi^* d\tau = 1 \qquad ...(iii)$$

or $$\oint \left(\frac{1}{N^{1/2}}\psi\right)\left(\frac{1}{N^{1/2}}\psi^*\right)d\tau = 1 \qquad ...(iv)$$

The factor $1/N^{1/2}$ thus introduced is called the *normalisation constant* and the function $\left(\frac{1}{N^{1/2}}\psi\right)$ is called the *normalised function.*

If ψ_i and ψ_j are two different eigen functions obtained as satisfactory solutions of the wave equation, then these functions will be normalized if

$$\oint \psi_i\psi_i^* d\tau = 1 \quad \text{and} \quad \oint \psi_j\psi_j^* d\tau = 1 \qquad ...(v)$$

Further if they satisfy the following conditions

$$\oint \psi_i\psi_j^* d\tau = 0 \quad \text{and} \quad \oint \psi_i^*\psi_j\, d\tau = 0 \qquad ...(vi)$$

they are said to be mutually *orthogonal.*

Wave functions that are both orthogonal and normalized are called *orthonormal.*

Q. 21. What are operators ? **(*Bombay 2004*)**

Ans. *An operator is a mathematical instruction or procedure to be carried out on a function so as to get another function.*

It is written in the form

(Operator) . (function) = (Another function)

The function on which the operation is carried out is called an *operand.* The left hand side of the above equation does not mean that the function is multiplied with the operator. Evidently, an operator written alone has no significance.

A few examples are given below :

(*i*) $\frac{d}{dx}(x^3) = 3x^2$. Here $\frac{d}{dx}$ which stands for differentiation w.r.t. x is the operator, x^3 is the operand and $3x^2$ is the result of the operation.

(*ii*) $\int x^3\, dx = x^4/4 + C$. Here $\int (\ )\, dx$ which stands for integration w.r.t x is the operator, x^3 is the operand and $x^4/4 + C$ is the result of the operation.

In case the symbol used for the operator is not self-explanatory, a suitable letter or some symbol for the operator is used with the symbol (^) over it.

Q.22. Explain (*i*) addition and subtraction of operators (*ii*) multiplication of operators.

Ans. (1) **Addition and subtraction of operators:** If $\hat{A}$ and $\hat{B}$ are two different operators and f is the operand, then

$$(\hat{A}+\hat{B})f = \hat{A}f+\hat{B}f \quad ...(i)$$

$$(\hat{A}-\hat{B})f = \hat{A}f-\hat{B}f \quad ...(ii)$$

(2) **Multiplication of operators:** If $\hat{A}$ and $\hat{B}$ are two different operators and f is the operand, then the expression $\hat{A}\hat{B}f$ implies that first f is operated by the operator $\hat{B}$ to get the result, say f', and then f' is operated by the operator $\hat{A}$ to get the final result say f'', *i.e.*, $\hat{A}\hat{B}f$ implies that $\hat{B}f = f'$ and then $\hat{A}f' = f''$ so that we have

$$\hat{A}\hat{B}f = f'' \quad ...(iii)$$

Thus the order of using the operators is from right to left as written in the given expression.

If the same operation is to be done a number of times in succession, it is shown by a power of operator. For example,

$$\hat{A}\hat{A}f \text{ is written as } \hat{A}^2 f$$

i.e.,

$$\hat{A}\hat{A}f = \hat{A}^2 f \quad ...(iv)$$

Q. 23. Explain with the help of example that operators generally don't commute.

Ans. It may be noted that usually $\hat{A}\hat{B}f \neq \hat{B}\hat{A}f$

This is expressed in words by saying that operators usually donot commute.

This may be illustrated by the following example :

Suppose $\hat{A} = x, \hat{B} = \frac{d}{dx}$ and $f = f(x) = x^2$. Then

$$\hat{A}\hat{B}f = x\frac{d}{dx} = (x^2) = x(2x) = 2x^2$$

$$\hat{B}\hat{A}f = \frac{d}{dx}\, x\,(x^2)\, \frac{d}{dx} = x^3 = 3x^2$$

If, however, it is found that

$$\hat{A}\hat{B}f = \hat{B}\hat{A}f$$

i.e the change of order of the operators gives the same result, the operators are said to *commute.*

Q. 24. If ψ_1 and ψ_2 are the proper functions of an atom, show that any linear combination of these wave functions will also be a proper wave function of the Schrodinger equation.

Ans. Consideration for proper or eigen wave functions for a system of constant energy (E) is given by

$$\hat{H}\psi = E\psi$$

where $\hat{H}$ is Hamiltonian operator.

Thus, $\hat{H}\psi_1 = E\psi_1$, $\hat{H}\psi_2 = E\psi_2$.

Linear combination of ψ_1 and ψ_2 may be represented by ψ as

$$\psi = c_1\psi_1 + c_2\psi_2$$

Where c_1 and c_2 are constants.

$$\begin{aligned}\therefore \quad \hat{H}\psi &= \hat{H}[c_1\psi_2 + c_2\psi_2] \\ &= c_1\hat{H}\psi_1 + c_2\hat{H}\psi_2 \\ &= c_1 E\psi_1 + c_2 E\psi_2 \\ &= E(c_1\psi_1 + c_2\psi_2) \\ &= E\psi\end{aligned}$$

Hence, ψ is also proper function of the Schrodinger equation.

Q. 25. Find the commutor $\left[x, \frac{d}{dx}\right]$.

Ans. By definition of commuter,

$$\left[x, \frac{d}{dx}\right] = \left[x\frac{d}{dx} - \frac{d}{dx}x\right].$$

Let us operate this on some arbitrary function $\psi(x)$.

$$\begin{aligned}\left[x\frac{d}{dx} - \frac{d}{dx}x\right]\psi &= x\frac{d\psi}{dx} - \frac{d}{dx}(x\psi) \\ &= x\frac{d\psi}{dx} - x\frac{d\psi}{dx} - \psi\frac{dx}{dx} \\ &= (-1)\psi\end{aligned}$$

Thus $$\left[x, \frac{d}{dx}\right] = -1$$

Q. 26. What of the following wave functions are acceptable in quantum mechanics over the range when x goes from 0 to 2 .

(i) sin x (ii) tan x (iii) cos x + sin x

Ans.

(*i*) sin x is continuous, single valued and finite in the given range of x. It is acceptable. (sin x varies between –1 and 1 in the given range).

(*ii*) tan x is infinite for $x = \frac{\pi}{2}$. It is not acceptable.

(*iii*) cos x+ sin x is continuous, single valued and finite in the given range. It is acceptable (cos x + sin x) varies between – 1 to + 1 in the given range).

Q. 27. If two operators $\hat{A}$ and $\hat{B}$ commute then they have the same set of eigen functions." Justify it. (*Mumbai 2010*)

Ans. Suppose the operator $\hat{A}$ has eigen function ψ, then

$$\hat{A}\psi = a\psi$$

where "a" is the eigen value

since $\hat{A}$ and $\hat{B}$ commute

$$\hat{A}\hat{B} = \hat{B}\hat{A}\psi = \hat{B}a\psi = a\hat{B}\psi$$

This shows that $\hat{B}\psi$ is an eigen function of the operator $\hat{A}$ with eigen value a.

This is possible only if ($\hat{B}\psi$) is a multiple of ψ *i.e.*

$$\hat{B}\psi = b\psi$$

where "b" is constant.

In other words, "ψ" is also an eigen function of $\hat{B}$.

Q.28. Explain the terms (*i*) Linear operator (*ii*) Laplacian operator (*iii*) Hamiltonian operator. (*Bombay2004*)

Ans. Linear operators: An operator A is said to be linear if for the two functions f and g.

$$\hat{A}(f+g) = \hat{A}f + \hat{A}g \qquad \text{... } (i)$$

i.e., the operator on the sum of two functions gives the same result as the sum of two results obtained by carrying out the same operation on the two functions separately. For example, d/dx, d^2/dx^2 etc. are linear operators whereas 'taking the square', 'taking the square root' etc. are non-linear.

(*ii*) **Laplacian operator:** This is a very common operator used in quantum mechanics. It is represented by ∇^2 (pronounced as *del squared*) and is defined as

$$\nabla^2 = \frac{\partial^2}{\partial x^2} + \frac{\partial^2}{\partial y^2} + \frac{\partial^2}{\partial z^2} \qquad \text{...}(ii)$$

Thus Schrodinger wave equation *viz.*,

$$\frac{\partial^2 \psi}{\partial x^2} + \frac{\partial^2 \psi}{\partial y^2} + \frac{\partial^2 \psi}{\partial z^2} + \frac{8\pi^2 m\,(E-V)}{h^2}\psi = 0$$

may be written in terms of Laplacian operator as

$$\nabla^2\psi + \frac{8\pi^2 m\,(E-V)}{h^2}\psi = 0 \qquad \text{...}(iii)$$

(*iii*) **Hamiltonian operator:** Schrodinger wave equation as written above may be rewritten in the form

$$\nabla^2\psi = -\frac{8\pi^2 m\,(E-V)}{h^2}\psi$$

or

$$\nabla^2\psi = -\frac{8\pi^2 m}{h^2}(E\psi - V\psi)$$

or

$$-\frac{h^2}{8\pi^2 m}\nabla^2\psi + V\psi = E\psi$$

or
$$\left(-\frac{h^2}{8p^2 m}\nabla^2 + V\right)\psi = E\psi \qquad ...(iv)$$

This equation implies that the operation $\left(-\frac{h^2}{8p^2 m}\nabla^2 + V\right)$ carried on the function ψ, is equal to the total energy multiplied with the function ψ. The operator $\left(-\frac{h^2}{8\pi^2 m}\nabla^2 + V\right)$ is called Hamiltonian operator and is represented by $\hat{H}$. Thus

$$\hat{H} = -\frac{h^2}{8p^2 m}\nabla^2 + V \qquad ...(v)$$

Eqn. (*iv*) may, therefore, be written as

$$\hat{H}\psi = E\psi \qquad ...(vi)$$

Q.29. Write the Schrodinger wave equation and explain the meaning of various terms. Also write the equation in terms of Laplacian operator.

Ans. See Q.28.

Q. 30. What are the postulates of quantum mechanics ?

(M. Dayanand 2002; Nagpur 2008)

Ans. The postulates of quantum mechanics are given as under :

(1) For every time-independent state of system, a function ψ of the coordinates can be written which is single valued, continuous and finite throughout the configuration space. This function describes completely the state of the system.

(2) To each observable quantity in classical mechanics, like position, velocity, momentum, energy etc., there corresponds a certain mathematical operator in quantum mechanics, the nature of which depends upon the classical expression for the observable quantity. For example,

(*i*) The operator corresponding to a position co-ordinate is multiplied by the value of that co-ordinate *i.e.,* operator for a position coordinate x is the multiplier x.

(*ii*) The operation representing the momentum (p) in the direction of any co-ordinate q is the differential operator

$$\frac{h}{2\pi i}\cdot\frac{\partial}{\partial q} \quad \text{or} \quad -\frac{ih}{2\pi}\frac{\partial}{\partial q}$$

where h is Planck's constant and $i = \sqrt{-1}$

(3) If ψ is a well-behaved function for the given state of a system and $\hat{A}$ is a suitable operator for the observable quantity or property, then the operation on ψ by the operator $\hat{A}$ gives ψ multiplied by a constant value (say, a) of the observable property *i. e.,*

$$\hat{A}\psi = a\psi \qquad ...(i)$$

The given state is called the eigen state of the system, ψ is called the eigen function and a is called the eigen value.

(4) *If a number of measurements are made over the configuration space, then the average value of the quantity (represented by $\bar{a}$) is given by*

$$\bar{a} = \frac{\oint y^* \hat{A} y \, dt}{\oint y^* y \, dt} \qquad ...(ii)$$

where $\oint$ represents integration over the whole of the configuration space.

Q. 31. Derive Schrodinger wave equation based on the postulates of quantum mechanics. *(M. Dayanand 2002; Raipur 2010)*

Ans. Using the postulates of quantum mechanics, Schrodinger wave equation can be derived in a simple manner as follows :

Consider a single particle *e.g.* an electron of mass m moving with a velocity v. Its total energy E is the sum of its kinetic energy (T) and potential energy (V), *i.e.*,

$$E = T + V \qquad ...(i)$$

Put $$T = \frac{1}{2}mv^2 = \frac{p^2}{2m} \qquad ...(ii)$$

where p represents the total momentum of the particle.

Further we know that

$$p^2 = p_x^2 + p_y^2 + p_z^2$$

where p_x, p_y and p_z are the momenta of the electron along the three mutually perpendicular directions along the three axes

$$T = \frac{p_x^2 + p_y^2 + p_z^2}{2m} \qquad ...(iv)$$

Substitute this value in equation (i), we get

$$E = \frac{p_x^2 + p_y^2 + p_z^2}{2m} + V \qquad ...(v)$$

But according to the second postulate of quantum mechanics, the operators for the momenta p_x,p_y and p_z are

$$p_x = \frac{h}{2\pi i}\frac{\partial}{\partial x}$$

$$p_y = \frac{h}{2\pi i}\frac{\partial}{\partial y}$$

$$p_z = \frac{h}{2\pi i}\frac{\partial}{\partial z}$$

V is a function of position co-ordinates and hence the operator for V is V itself.

The operator for the energy E is the Hamiltonian operator, H.

Hence equation (v) takes the form (taking into consideration the algebraic rules of the operators)

$$\hat{H} = \frac{1}{2m}\left[\left(\frac{h}{2\pi i}\frac{\partial}{\partial x}\right)^2 + \left(\frac{h}{2\pi i}\frac{\partial}{\partial y}\right)^2 + \left(\frac{h}{2\pi i}\frac{\partial}{\partial z}\right)^2\right] + V$$

$$= -\frac{h^2}{8\pi^2 m}\left(\frac{\partial^2}{\partial x^2} + \frac{\partial^2}{\partial y^2} + \frac{\partial^2}{\partial z^2}\right) + V$$

$$= -\frac{h^2}{8\pi^2 m}\nabla^2 + V \qquad ...(vii)$$

where $\nabla^2 = \frac{\partial^2}{\partial x^2} + \frac{\partial^2}{\partial y^2} + \frac{\partial^2}{\partial z^2}$ represents the Laplacian operator.

Further according to the third postulate, we have

$$\hat{H}\psi = E\psi$$

$$\hat{H}\psi - E\psi = 0 \qquad ...(viii)$$

where E is the eigen value of the energy for the given state of the system.

Substituting the value of $\hat{H}$ from equation (vii) into equation ($viii$) we get

$$\left(-\frac{h^2}{8\pi^2 m}\nabla^2 + V\right)\psi - E\psi = 0 \qquad ...(ix)$$

which can be rearranged and written in the form

$$\nabla^2\psi + \frac{8\pi^2 m}{h^2}(E-V)\psi = 0 \qquad ...(x)$$

which is the required Schrodinger wave equation in one of the common forms.

Q. 32. What is the concept of particle in a one dimensional box ? What is Schrodinger wave equation in such a case ? How can this equation be solved for ψ and E ?

(Calcutta 2000; M. Dayanand 2002; Nagpur 2008)

Ans. The motion of a particle in a one dimensional box is like the flow of an electron in a wire but still it is called a "particle in a box" and a general, case can be considered in which we may assume that a single particle *e.g.*, a gas molecule of mass *m* is restricted to move in a region of space from $x = 0$ to $x = a$ and that its potential energy within the box is constant and taken as equal to zero for the sake of convenience. (Fig. 13.7). In order that the particle may remain within the box, it is essential to assume that the potential energy on or outside the walls is very high ($= \infty$) so that as soon as the particle reaches the walls, it is reflected back into the box.

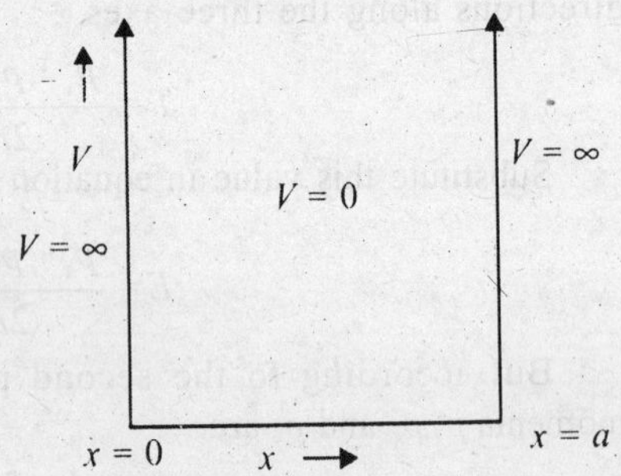

Fig. 13.7. A particle in a one dimensional box with P.E. = 0 inside the box and P.E. = ∞ on the walls of the box and outside the box.

Now, the Schrodinger wave equation for one dimension is

$$\frac{d^2\psi}{dx^2} + \frac{8\pi^2 m}{h^2}(E-V)\,\psi = 0 \qquad ...(i)$$

where ψ has been taken as the function of x co-ordinate only.

As outside the box, $V = \infty$, therefore for outside the box, equation (*i*) becomes

$$\frac{d^2\psi}{dx^2} + \frac{8\pi^2 m}{h^2}(E-\infty)\,\psi = 0 \qquad ...(ii)$$

Neglecting E in comparison to ∞, equation (*ii*) reduces to

$$\frac{d^2\psi}{dx^2} - \infty\,\psi = 0 \qquad \left(\frac{8\pi^2 m}{h^2} \times \text{infinity} = \text{infinity}\right)$$

$$\frac{d^2\psi}{dx^2} = \infty\,\psi$$

$$\psi = \frac{1}{\infty}\frac{d^2\psi}{dx^2} = \text{zero} \qquad ...(iii)$$

This proves that outside the box, $\psi = 0$ which implies that the particle cannot go outside the box.

For the particle within the box, $V = 0$, therefore the Schrodinger wave equation (*i*) takes the form

$$\frac{d^2\psi}{dx^2} + \frac{8\pi^2 m}{h^2} E\psi = 0 \quad ...(iv)$$

For the given state of the system, the energy E is constant which is one of the postulates of the quantum mechanics. Therefore we put

$$\frac{8\pi^2 m}{h^2} E = k^2 \quad ...(v)$$

where k^2 is a constant, independent of x.

Equation (*iv*), then becomes

$$\frac{d^2\psi}{dx^2} + k^2\psi = 0 \quad ...(vi)$$

A general solution of this differential equation is given by

$$\psi = A \sin kx + B \cos kx \quad ...(vii)$$

where A and B are constants.

Depending upon the value of A, B and k, ψ can have many values. But all the values are not acceptable. Only those values of ψ are acceptable which satisfy the boundary conditions, *viz.*,

$$\psi = 0 \text{ at } x = 0 \text{ and } x = a$$

Putting $\psi = 0$ when $x = 0$, equation (*vii*) becomes

$$0 = B \sin 0 + B \cos 0$$

$$= 0 + B \qquad (\because \sin 0 = 0 \text{ and } \cos 0 = 1)$$

i.e., $\quad B = 0$

Thus when $x = 0$, equation (*vii*) becomes (by putting $B = 0$),

$$\psi = A \sin kx \quad ...(viii)$$

Putting $\psi = 0$ when $x = a$, equation (*viii*) becomes

$$0 = A \sin ka$$

i.e., $\quad \sin ka = 0 \quad ...(ix)$

This equation holds good only when the value of ka are integral multiplies of π *i.e.*,

$$ka = n\pi \quad ...(x)$$

where n is an integer *i.e.*, $n = 0, 1, 2, 3.....$ However the value $n = 0$ may be excluded because it makes $k = 0$ and hence $\psi = 0$ for any value of a between 0 and a *i.e.*, within the box. This is not true because the particle is always assumed to be present within the box,

From equation (*x*)

$$k = \frac{n\pi}{a} \quad ...(xi)$$

Substituting this value in equation (*viii*), we get

$$\boldsymbol{\psi = A \sin\left(\frac{n\pi}{a}x\right)} \quad ...(xii)$$

This gives the expression for the eigen function ψ.

The expression for the eigen value of the energy may be obtained as follows :

From equation (*v*)

$$E = \frac{k^2h^2}{8\pi^2 m} \quad ...(xiii)$$

Substituting the value of k from equation (*xi*), we get

$$E = \frac{(n\pi/a)^2 h^2}{8\pi^2 m}$$

i.e.

$$\boxed{E = \frac{n^2 h^2}{8ma^2}} \qquad ...(xiv)$$

Equations (*xii*) and (*xiv*) are the solutions of the Schrodinger wave equation for a particle in one dimensional box.

Q. 33. How can eigen function for a particle in a one dimensional box be normalized? Write expression for the normalized wave function.

Ans. The process of *normalisation of the wave function* can be carried out as follows :

$$\int_0^a \psi\psi^* d\tau = 1$$

Substituting the value of $\psi(=\psi^*)$ from equation (*xii*) in Q. 32, we get

$$A^2 \int_0^a \sin^2 \frac{n\pi}{a} x\, dx = 1$$

or

$$A^2 \cdot \frac{a}{2} = 1 \qquad \left[\text{as} \int_0^a \sin^2\left(\frac{n\pi}{a}x\right) dx = \frac{a}{2}\right]$$

$$A = \sqrt{\frac{2}{a}}$$

Hence the normalised wave function (which will also be a solution of the Schrodinger equation) is

$$\boldsymbol{\psi = \sqrt{\frac{2}{a}} \sin\left(\frac{n\pi}{a} x\right)}$$

Q. 34. Show that if there are any two eigen functions ψ_m and ψ_n corresponding to two different values of n then ψ_m and ψ_n will be orthogonal to each other. *(Calcutta 2000)*

Ans.

$$\int_0^a \psi_m \psi_n dx = \frac{2}{a} \int_0^a \sin\left(\frac{m\pi x}{a}\right) \sin\left(\frac{n\pi x}{a}\right) dx$$

$$= \frac{2}{a} \int_0^a \frac{1}{2}\left[\cos(m-n)\frac{\pi x}{a} - \cos(m+n)\frac{\pi x}{a}\right] dx$$

$$[\because 2 \sin A \sin B = \cos(A-B) - \cos(A+B)]$$

$$= \frac{1}{a}\left[\frac{\sin(m-n)\frac{\pi x}{a}}{(m-n)\frac{\pi}{a}} - \frac{\sin(m+n)\frac{\pi x}{a}}{(m+n)\frac{\pi}{a}}\right]_0^a$$

$$= \frac{1}{\pi}\left[\frac{\sin(m-n)\frac{\pi x}{a}}{(m-n)} - \frac{\sin(m+n)\frac{\pi x}{a}}{(m+n)}\right]_0^a$$

$$= \frac{1}{\pi}\left[\frac{\sin(m-n)\pi}{(m-n)} - \frac{\sin(m+n)\pi}{(m+n)}\right] = 0$$

[∵ m and n are integers and m# n, therefore $(m - n)$ and $(m + n)$ are also integers and $\sin N\pi = 0$, where N is an integer]

Q. 35. Write expression for energy for a particle in one-dimensional box. Explain (*i*) quantization of energy (*ii*) existence of zero point energy. (*Himachal 2010; Mumbai 2010*)

Ans. Expression for energy for a particle in one dimensional box is

$$E = \frac{n^2 h^2}{8ma^2} \qquad ...(i)$$

where m is the mass of the particle and a is the length of the box ($x = 0$ to $x = a$)

(*i*) **Quantization of energy:** Since n can have only integral values equal to 1, 2, 3 etc., therefore from equation (*i*), it follows that the energy E associated with the motion of a particle in a box can have only discrete values *i.e.*, the energy is quantized.

The integer n is called the *quantum number* of the particle.

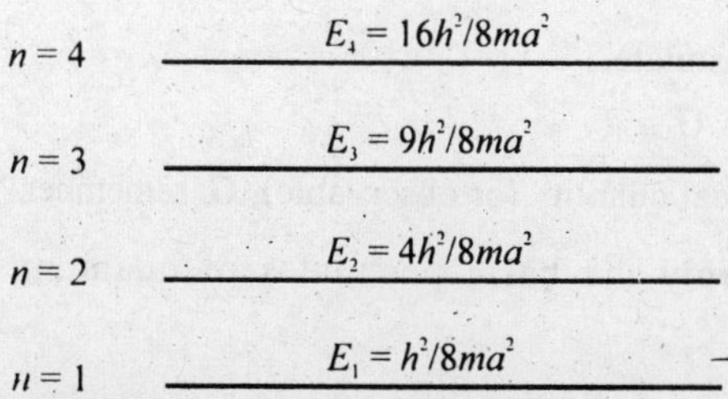

Fig. 13.8. The discrete energy levels of a particle of mass m confined in a box of length *a*.

Further putting $n = 1, 2, 3....$ etc., the discrete energy levels obtained for the particle of mass m confined in the box of length a are shown in Fig. 13.8. It is important to note that as the quantum number n increases, the separation between them increases. It may also be noted that energy levels also depends upon the box length a. As a increases, *i.e.*, the space available to a particle increases, energy quanta become smaller and energy levels move closer together. If the box length becomes very large, quantization disappears and there is a smooth transition from quantum behaviour to classical behaviour. (Continuous change).

(*ii*) **Existence of zero point energy :** Minimum value of $n = 1$ ($n = 0$ is ruled out, as discussed earlier). Substituting this value in eq. (*i*), we get

$$E = \frac{h^2}{8ma^2} \qquad ...(ii)$$

This implies that the minimum energy possessed by the particle is not zero but has a definite value. This is called *zero point energy.*

Q. 36. Explain that energy of a free particle is not quantised.

Ans. If the walls of the box are removed, so that the particle is free to move in a field whose potential energy may be supposed to be zero, then the boundary conditions are no longer applicable and so the constants *A, B* and k can have any value. The energy of the particle is then given by equation

$$E = \frac{k^2 h^2}{8\pi^2 m}$$

As the constant k can have any value, the energy of the free particle is not quantised.

Q. 37. What do you understand by expectation value of an observable quantity in quantum mechanics ? Write expressions for expectation values of

(*i*) **position, x**

(ii) any other observable quantity, *G*.

Ans. Any physical property of a particle moving in one dimension which does not depend upon time can be calculated from the time-independent wave function $\psi(x)$ of the system. In these calculations, the probability of finding the particle in any region of the one-dimensional space is calculated using Born's postulate, according to which the probability $P(x)dx = \psi(x)\,\psi^*(x)\,dx$. For example, the wave function helps us to calculate observables like position, momentum etc. However, if we carry out an experiment to measure the position x of the particle, $\psi(x)$ may not give the result of that particular experiment. But if we repeat the experiment a number of times, the value of the position x calculated from $\psi(x)$ is the same as the mean value of the different experiments. *This mean (or average) value, denoted by* $< x >$, *is called the expectation value of the observable x.*

(*i*) *Expectation value of position x*

According to one of the postulates of quantum mechanics, as operator for the position co-ordinate x is the multiplier x, therefore the expectation value $< x >$ will be given by

$$< x > = \int \psi^* x \psi \, dx = \int x\, \psi^* \psi \, dx = \int x\, P(x)\, dx \qquad ...(i)$$

where $P(x)\,dx$ is the probability calculated by Born's postulate.

(*ii*) *Expectation value for any observable quantity, G*

To calculate the expectation value of any other physical quantity (or observable), G, remember that the operator for G is represented by $\hat{G}$ and apply the basic postulates of quantum mechanics. The expectation value $<G>$ is given by

$$<G> = \int_{-\infty}^{\infty} \psi^* \hat{G} \psi dx \qquad ...(ii)$$

which would be same as the mean value of the results of the repeated experiments for the measurement of G in the state $\psi(x)$.

It may be noted that $\hat{G}$ is not a simple factor like x. It is an operator for the quantity G. Hence $\int \psi^* \hat{G} \psi \, dx$ cannot be rearranged to give $\int \hat{G} \psi^* \psi \, dx$. Thus $\hat{G}$ must be allowed to operate on ψ. For example, according to one of the postulates of quantum mechanics, for the momentum, p_x, the operator p_x is $-\dfrac{ih}{2\pi}\dfrac{d}{dx}$ *i.e.* take derivative with respect to x and multiply the result by $-ih/2\pi$.

Q. 38. Derive expressions for expectation values of x, x^2, p_x and p_x^2 for a particle in one dimensional box of length *a*.

Ans. For a particle in one dimensional box, we have

$$\psi_n = \sqrt{\frac{2}{a}} \sin\left(\frac{n\pi x}{a}\right)$$

As ψ_n is a normalized wave function, we have

(*i*) $$<x> = \int_0^a \psi x \psi dx = \int_0^a x\psi^2 dx$$

$$= \int_0^a x . \frac{2}{a} \sin^2\left(\frac{n\pi x}{a}\right) dx$$

$$= \frac{2}{a}\int_0^a x.\sin^2\left(\frac{n\pi x}{a}\right)dx$$

$$= \frac{2}{a}\int_0^a x.\left[\frac{1-\cos\left(\frac{2n\pi x}{a}\right)}{2}\right]dx$$

$$= \frac{2}{a}\left(\frac{a^2}{4}\right) = \frac{a}{2} \quad ...(i)$$

(*ii*) $$<x^2> = \int_0^a \psi x^2 dx. = \int_0^a x^2 \psi^2 dx = \frac{2}{a}\int_0^a x^2 \sin^2\left(\frac{n\pi x}{a}\right)dx$$

$$= \frac{2}{a}\int_0^a x^2\left[\frac{1-\cos\left(\frac{2n\pi x}{a}\right)}{2}\right]dx$$

$$= \frac{2}{a}\left[\frac{a^3}{6} - \frac{a^3}{4n^2\pi^2}\right] = \frac{1}{a}\left[\frac{a^3}{3} - \frac{a^3}{2n^2\pi^2}\right]$$

$$= \left(\frac{a^2}{3} - \frac{a^2}{2n^2\pi^2}\right) \quad ...(ii)$$

(*iii*) $$< p_x >_n = \int_0^a \psi\left(\frac{-ih}{2\pi}\frac{d}{dx}\right)\psi dx$$

$$= \int_0^a\left[\left(\frac{2}{a}\right)^{1/2} \sin\frac{n\pi x}{a}\right]\left(\frac{-ih}{2\pi}\frac{d}{dx}\right)\left[\left(\frac{2}{a}\right)^{1/2} \sin\frac{n\pi x}{a}\right]dx$$

$$= -\frac{ih}{2\pi}\left(\frac{2}{a}\right)\int_0^a \sin\left(\frac{n\pi x}{a}\right)\left(\frac{n\pi}{a}\right)\cos\left(\frac{n\pi x}{a}\right)dx$$

$$= -\frac{ih}{2\pi}\left(\frac{2}{a}\right)\frac{n\pi}{a}\int_0^a \sin\left(\frac{n\pi x}{a}\right)\cos\left(\frac{n\pi x}{a}\right)dx$$

= 0 (on integration by parts) ...(*iii*)

This result is justified, because the average momentum has to be zero as the particle in a box cannot continue to travel in one direction only. Also note that the limits used are 0 and a in place of $-\infty$ to $+\infty$ because the wave function is known to be zero outside the box.

(*iv*) $$<p_x^2>_n = \int_0^a\left[\left(\frac{2}{a}\right)^{1/2} \sin\frac{n\pi x}{a}\right]\left(\frac{-ih}{2\pi}\frac{d}{dx}\right)^2\left[\left(\frac{2}{a}\right)^{1/2} \sin\frac{n\pi x}{a}\right]dx^*$$

* The operator for p_x^2 is $(\hat{p}_x)(\hat{p}_x)$

$$\hat{p}_x^2 = (\hat{p}_x)(\hat{p}_x) = \left(-\frac{ih}{2\pi}\frac{d}{dx}\right)\left(-\frac{ih}{2\pi}\frac{d}{dx}\right) = -\frac{h^2}{4\pi^2}\frac{d^2}{dx^2}$$

$$= -\frac{h^2}{4\pi^2}\left(\frac{2}{a}\right)\int_0^a \sin\left(\frac{n\pi x}{a}\right)\frac{d^2}{dx^2}\left(\sin\frac{n\pi x}{a}\right)dx$$

$$= -\frac{h^2}{4\pi^2}\left(\frac{2}{a}\right)\int_0^a \sin\left(\frac{n\pi x}{a}\right)\left[(-)\left(\frac{n\pi}{a}\right)^2 \sin\left(\frac{n\pi x}{a}\right)\right]dx^*$$

$$= \left(\frac{h^2}{4\pi^2}\right)\left(\frac{2}{a}\right)\left(\frac{n\pi}{a}\right)^2 \int_0^a \sin^2\left(\frac{n\pi x}{a}\right)dx$$

$$= \frac{n^2h^2}{2a^3}\int_0^a \sin^2\left(\frac{n\pi x}{a}\right)dx$$

$$= \frac{n^2h^2}{2a^3}\int_0^a \frac{1-\cos\frac{(2n\pi x)}{a}}{2}dx \qquad [\because 2\sin^2 A = 1-\cos 2A]$$

$$= \frac{n^2h^2}{4a^3}\left[\frac{x-\sin\frac{(2n\pi x)}{a}}{\frac{2n\pi x}{a}}\right]_0^a$$

$$= \frac{n^2h^2}{4a^3}\cdot a = \frac{n^2h^2}{4a^2} = 2mE \qquad \left[\because E = \frac{n^2h^2}{8ma^2}\right]$$

Q. 39. Derive expressions for standard deviation in x and p_x. Derive Heisenberg uncertainty principle from this. ***(Kolhapur 2000)***

Ans. The standard deviation in the position x for the particle in an eigen state of a one dimensional box is given by

$$\Delta x = (<x^2> - <x>^2)^{1/2} \qquad ...(i)$$

Substituting the values derived in Q. 38, we get

$$\Delta x = \left[\left(\frac{a^2}{3} - \frac{a^2}{2n^2\pi^2}\right) - \left[\frac{a}{2}\right]^2\right]^{1/2} = \left[\left(\frac{a^2}{12} - \frac{a^2}{2n^2\pi^2}\right)\right]^{1/2}$$

$$= a\left(\frac{1}{12} - \frac{1}{2n^2\pi^2}\right)^{1/2} \qquad ...(ii)$$

The standard deviation in the momentum, p_x for the particle is given by

$$\Delta p_x = (<p^2> - <p>^2)^{1/2} \qquad ...(iii)$$

Substituting the values derived in Q. 38, we get

$$\Delta p_x = (<p^2>)^{1/2} = \left(\frac{n^2h^2}{4a^2}\right)^{1/2} = \frac{nh}{2a} \qquad ...(iv)$$

---

* $\int \sin x dx = -\cos x + C, \int \cos x dx = \sin x + C.$

$\frac{d}{dx}(\sin x) = \cos x, \frac{d}{dx}(\cos x) = -\sin x.$ That is why (–) sign appears.

$$\therefore \qquad \Delta x.\Delta p = a\left(\frac{1}{12}-\frac{1}{2n^2\pi^2}\right)^{1/2}.\frac{nh}{2a}$$

$$= \frac{nh}{2}\left(\frac{1}{12}-\frac{1}{2n^2\pi^2}\right)^{1/2}$$

$$= \frac{nh}{2}\times\frac{1}{2n\pi}\left(\frac{4n^2\pi^2}{12}-\frac{4n^2\pi^2}{2n^2\pi^2}\right)^{1/2}$$

$$= \frac{h}{4\pi}\left(\frac{n^2\pi^2}{3}-2\right)^{1/2} \qquad ...(v)$$

The quantity on the right hand side is $> \frac{h}{4\pi}$ because $(n^2 \pi^2/3) > 2$ always. Hence

$$\Delta x\Delta p > \frac{h}{4\pi} \qquad ...(vi)$$

which is Heisenberg's uncertainty principle.

Q. 40. Calculate the probability of finding the particle between 0.49 *a* and 0.51 *a* for the states ψ_1 and ψ_2.

Ans. Probability $= \int_{0.49a}^{0.51a} \psi^2 dx$

$$= \int_{0.49a}^{0.51a} \frac{2}{a}\sin^2\left(\frac{n\pi x}{a}\right)dx$$

$$= \frac{2}{a}\int_{0.49a}^{0.51a} \sin^2\left(\frac{n\pi x}{a}\right)dx$$

$$= \frac{2}{a}\int_{0.49a}^{0.51a} \frac{1-\cos\left(\frac{2n\pi x}{a}\right)}{a}dx$$

$$= \frac{1}{a}\int_{0.49a}^{0.51a}\left[1-\cos\left(\frac{2n\pi x}{a}\right)\right]dx$$

$$= \frac{1}{a}\left[x-\frac{\sin\left(\frac{2n\pi x}{a}\right)}{\frac{2n\pi}{a}}\right]_{0.49a}^{0.51a}$$

$$= \frac{1}{n\pi}\left[\frac{n\pi x}{a}-\frac{1}{2}\sin\frac{2n\pi x}{a}\right]_{0.49a}^{0.51a}$$

Putting $n = 1$ and reading the value of sines of the angles from the tables,
Probability = **0.0399**

Similarly, for $n = 2$, Probability = **0.0001**

Q. 41. What is the ground state energy for an electron which is confined to a potential well (one-dimensional box) having a width of 0.2 nm ?

Ans. The energy for a particle in a one-dimensional box is given by

$$E = \frac{n^2h^2}{8ma^2}$$

For the ground state, $n = 1$

For the electron, $m = 9.11 \times 10^{-31}$ kg

We are given $a = 0.2 \text{ nm} = 0.2 \times 10^{-9}$ m

Also we know that $h = 6.626 \times 10^{-34}$ Js

$$\therefore \quad E = \frac{(1)^2 (6.626 \times 10^{-34})^2}{8 \times 9.11 \times 10^{-31} \times (0.2 \times 10^{-9})^2}$$

$$= 1.506 \times 10^{-18} \text{J}$$

$\therefore$ For 1 mole of electrons, the energy will be

$$E = \frac{(1.506 \times 10^{-18}) \times (6.022 \times 10^{23})}{10^3} \text{kJ mol}^{-1}$$

$$= \mathbf{907 \text{ kJ mol}^{-1}}$$

Q. 42. Derive expressions for normalized eigen function and total energy of a particle in a three-dimensional box.

Ans. Consider a single particle *e.g.*, a gas molecule of mass m restricted to move within a rectangular box of dimensions a, b and c along the X-axis, Y-axis and Z-axis respectively (Fig 13.9). As before, it is assumed that the potential energy of the particle within the box is

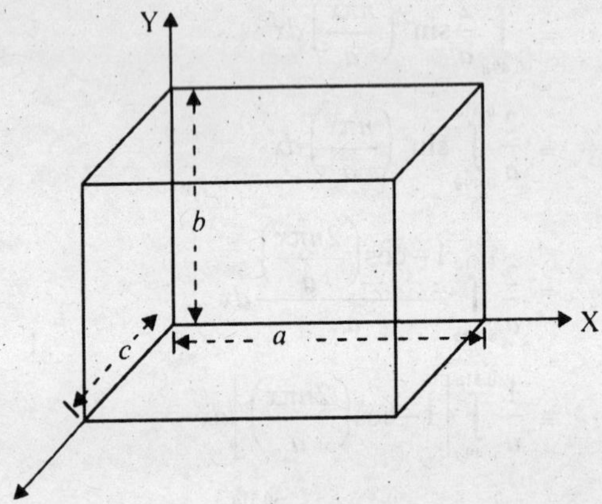

Fig. 13.9. Particle in a three dimensional box.

constant (taken as equal to zero) and on the walls or outside, it suddenly rises to a very large value (= ∞).

Now, the Schrodinger wave equation for three dimensions is

$$\frac{\partial^2 \psi}{\partial x^2} + \frac{\partial^2 \psi}{\partial y^2} + \frac{\partial^2 \psi}{\partial z^2} + \frac{8\pi^2 m}{h^2}(E-V)\psi = 0 \qquad ...(i)$$

where E represents the total energy of the particle, V its potential energy and ψ is a function of the three co-ordinates x, y and z.

For the particle within the box, $V = 0$. Hence equation (*i*) takes the form

$$\frac{\partial^2 \psi}{\partial x^2} + \frac{\partial^2 \psi}{\partial y^2} + \frac{\partial^2 \psi}{\partial z^2} + \frac{8\pi^2 m}{h^2} E\psi = 0 \qquad ...(ii)$$

To solve this differential equation it is convenient to separate the variables. For this purpose, ψ, which is a function of x, y and z is taken as equal to the product of three functions $X(x)$, $Y(y)$ and $Z(z)$, each in one variable, *i.e.*,

$$\phi = X(x).\ Y(y).\ Z(z) \qquad ...(iii)$$

or for the sake of simplicity, it may be written as

$$\phi = XYZ \qquad ...(iv)$$

Substituting this value of ϕ in equation (*ii*), we get

$$\frac{\partial^2(XYZ)}{\partial x^2}+\frac{\partial^2(XYZ)}{\partial y^2}+\frac{\partial^2(XYZ)}{\partial z^2}+\frac{8\pi^2 m}{h^2}EXYZ=0$$

or

$$YZ\frac{\partial^2 X}{\partial x^2}+XZ\frac{\partial^2 Y}{\partial y^2}+XY\frac{\partial^2 Z}{\partial z^2}+\frac{8\pi^2 m}{h^2}EXYZ=0$$

Dividing throughout by XYZ, we get

$$\frac{1}{X}\frac{\partial^2 X}{\partial x^2}+\frac{1}{Y}\frac{\partial^2 Y}{\partial y^2}+\frac{1}{Z}\frac{\partial^2 Z}{\partial z^2}+\frac{8\pi^2 m}{h^2}E=0$$

$$\frac{1}{X}\frac{\partial^2 X}{\partial x^2}+\frac{1}{Y}\frac{\partial^2 Y}{\partial y^2}+\frac{1}{Z}\frac{\partial^2 Z}{\partial z^2}=-\frac{8\pi^2 m}{h^2}E \qquad ...(v)$$

Since the total energy, E of the particle is constant, therefore R.H.S.of equation (*v*) is constant and we may put $\frac{8\pi^2 mE}{h^2}=k^2$. So we may write equation (*v*) as

$$\frac{1}{X}\frac{\partial^2 X}{\partial x^2}+\frac{1}{Y}\frac{\partial^2 Y}{\partial y^2}+\frac{1}{Z}\frac{\partial^2 Z}{\partial z^2}=-k^2 \qquad ...(vi)$$

Thus the sum of the three terms is constant. Now suppose we change the value of x, keeping y and z constant, the sum of three terms should be again equal to the same constant, *viz.* k^2. This indicates that the first term on L.H.S. of equation (*vi*) is not only independent of y and z but is also independent of x. Likewise the second and the third terms must also be independent of x and z. In other words, each term on the L.H.S. of equation (*vi*) must be constant and the sum of the three constants should be equal to k^2. Hence we may write

$$\frac{1}{X}\frac{\partial^2 X}{\partial x^2}=-k_x^2 \qquad ...(vii)$$

$$\frac{1}{Y}\frac{\partial^2 Y}{\partial y^2}=-k_y^2 \qquad ...(viii)$$

$$\frac{1}{Z}\frac{\partial^2 Z}{\partial z^2}=-k_z^2 \qquad ...(ix)$$

where k_x^2 and k_y^2 and k_z^2 are constants such that

$$k_x^2+k_y^2+k_z^2=k^2 \qquad ...(x)$$

and the values of these constants are as follows :

$$k_x^2=\frac{8\pi^2 mE_x}{h^2} \qquad ...(xi)$$

$$k_y^2=\frac{8\pi^2 mE_y}{h^2} \qquad ...(xii)$$

$$k_z^2=\frac{8\pi^2 mE_z}{h^2} \qquad ...(xiii)$$

where E_x, E_y and E_z represent the components of the energy in the directions parallel to X-axis, Y-axis and Z-axis respectively such that

$$E = E_x + E_y + E_z \quad ...(xiv)$$

Equations (*vii*) to (*ix*) may be rewritten as

$$\frac{\partial^2 X}{\partial x^2} + k_x^2 X = 0, \frac{\partial^2 Y}{\partial y^2} + k_y^2 Y = 0 \text{ and } \frac{\partial^2 Z}{\partial z^2} + k_z^2 Z = 0$$

Each of these equations is of the same form as for particle in one-dimensional box.

The solutions of these equations are :

$$X = \sqrt{\frac{2}{a}} \sin \frac{n_x \pi}{a} x, Y = \sqrt{\frac{2}{b}} \sin \frac{n_y \pi}{b} y$$

and
$$Z = \sqrt{\frac{2}{c}} \sin \frac{n_z \pi}{c} z \quad ...(xv)$$

and
$$E_x = \frac{n_x^2 h^2}{8ma^2}, E_y = \frac{n_y^2 h^2}{8mb^2} \text{ and } E_z = \frac{n_z^2 h^2}{8mc^2} \quad ...(xvi)$$

where n_x, n_y and n_z having integral values represent three quantum numbers in directions parallel to the three axes.

The normalised wave function for a particle in three dimensional box is

$$\psi = XYZ$$

$$= \left(\sqrt{\frac{2}{a}} \sin \frac{n_x \pi}{a} x\right)\left(\sqrt{\frac{2}{b}} \sin \frac{n_y \pi}{b} y\right)\left(\sqrt{\frac{2}{c}} \sin \frac{n_z \pi}{c} z\right)$$

$$= \sqrt{\frac{8}{abc}} \sin\left(\frac{n_x \pi}{a} x\right) \sin\left(\frac{n_y \pi}{b} y\right) \sin\left(\frac{n_z \pi}{c} z\right) \quad ...(xvii)$$

The total energy (E) will be given by

$$E = E_x + E_y + E_z$$

$$= \frac{n_x^2 h^2}{8ma^2} + \frac{n_y^2 h^2}{8mb^2} + \frac{n_z^2 h^2}{8mc^2}$$

or
$$\boxed{E = \frac{h^2}{8m}\left(\frac{n_x^2}{a^2} + \frac{n_y^2}{b^2} + \frac{n_z^2}{c^2}\right)} \quad ...(xviii)$$

Thus three quantum numbers follow from the solution of the Schrodinger wave equation in three dimensions.

Q. 43. How does the study of particle in three dimensional box lead to the concept of degeneracy. ***(Nagpur 2008)***

Ans. When the sides of the three dimensional box are equal

$$a = b = c$$

The equation for the energy of a particle in three dimensional box

$$E = \frac{h^2}{8m}\left(\frac{n_x^2}{a^2} + \frac{n_y^2}{b^2} + \frac{n_z^2}{c^2}\right) \text{ reduces to}$$

$$E = \frac{h^2}{8ma^2}\left(n_x^2 + n_y^2 + n_z^2\right) \quad ...(i)$$

$$E = \frac{h^2}{8ma^2}\left(n_x^2 + n_y^2 + n_z^2\right) \quad ...(i)$$

The different combinations of the values of quantum number n_x, n_y and n_z may give the same value of the energy E. For example, the following six combinations of the values 1, 2 and 3 give the same value of E.

n_x	1	1	2	2	3	3
n_y	2	3	1	3	1	2
n_z	3	2	3	1	2	1

Such an energy level having different states of the system but having the same energy is said to be **degenerate** and the number of independent wave functions associated with the given energy level is called its **degeneracy.**

The degeneracies of a few energy levels with the different quantum numbers are given below:

$n_x\, n_y\, n_z$	111	211	221	311	222	321	322
Degeneracy	1	3	3	3	1	6	3

As an example, the three states corresponding to the quantum numbers 211 will have the quantum numbers, 211, 121 and 112. All these states will have energy $= \frac{h^2}{8ma^2}(4 + 1 + 1)$

$\frac{3h^2}{4ma^2}$ [obtained by putting the values of n_x, n_y and n_z in eqn. (*i*)]

This is illusrated in Fig. 13.10

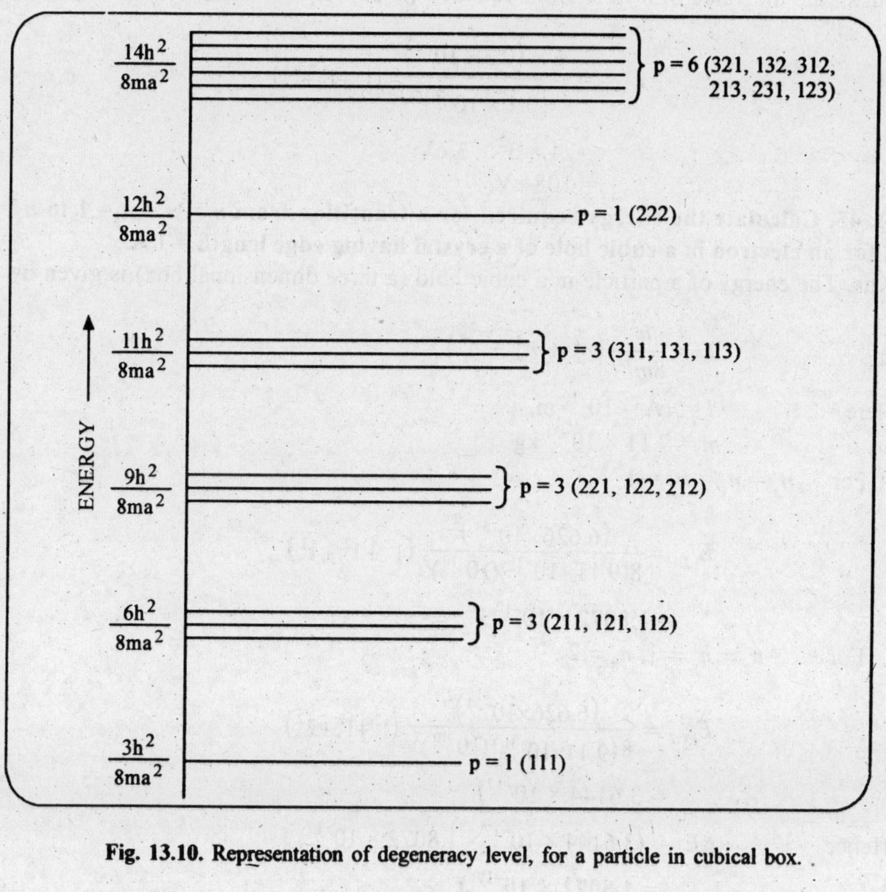

Fig. 13.10. Representation of degeneracy level, for a particle in cubical box.

It is important to mention here that degeneracies arise in quantum mechanics when there is some element of symmetry. If the symmetry is broken *e.g.* by taking a box whose sides have different lengths, the degeneracy vanishes.

Q. 44. Calculate the expected ground state energy of a hydrogen atom electron assumed to be present in a three dimensional cubical box of length 0.1 nm if the ground state energy of the electron in a one dimensional box of length 0.3 nm is 4 eV.

Ans. For one-dimensional box, for ground state $n = 1$ so that

$$E_1 = \frac{n^2h^2}{8ma^2} = \frac{(1)^2(h)^2}{8m(0.3\times10^{-9})^2} = 4eV$$

or, $$\frac{h^2}{8m} = 4\times(0.3\times10^{-9})^2 \quad ...(i)$$

For three dimensional box, for ground state,

$$n_x = n_y = n_z = 1 \text{ so that}$$

$$E_{1,1,1} = \frac{h^2}{8ma^2}\left(n_x^2 + n_y^2 + n_z^2\right)$$

or $$E_{1,1,1} = \frac{h^2}{8m(0.1\times10^{-9})^2}(1^2+1^2+1^2) \quad ...(ii)$$

Substitute the value of $h^2/8\ m$ from equation (*i*) into eq. (*ii*)

$$E_{1,1,1} = \frac{4\times(0.3\times10^{-9})^2}{(0.1\times10^{-9})^2}(1^2+1^2+1^2)$$

$$= 4 \times 3^2 \times 3 \text{ eV}$$

$$= 108 \text{ eV}$$

Q. 45. Calculate the energy required for a transition from $n_x = n_y = n_z = 1$ to $n_x = n_y = 1$, $n_z = 2$ for an electron in a cubic hole of a crystal having edge length = 1 Å.

Ans. The energy of a particle in a cubic hold (a three dimensional box) is given by

$$E = \frac{h^2}{8ml^2}\left(n_x^2 + n_y^2 + n_z^2\right)$$

Here $l = 1\text{Å} = 10^{-10}$ m,

$m = 9.11 \times 10^{-31}$ kg

∴ For $n_x = n_y = n_z = 1$

$$E_{1,1,1} = \frac{(6.626\times10^{-34})^2}{8(9.11\times10^{-31})(10^{-10})^2}(1^2+1^2+1^2)$$

$$= 1.8072 \times 10^{-17} \text{ J}$$

For $n_x = n_y = 1, n_z = 2$

$$E_{1,1,2} = \frac{(6.626\times10^{-34})^2}{8(9.11\times10^{-31})(10^{-10})^2}(1^2+1^2+2^2)$$

$$= 3.6144 \times 10^{-17}\text{J}.$$

Hence $$\Delta E = (3.6144 \times 10^{-17} - 1.8072 \times 10^{-17}) \text{ J}$$

$$= \mathbf{1.8072 \times 10^{-17}\ J}$$

Q. 46. An electron is confined to an infinite one dimensional box of width 4Å. Calculate its energy (in eV) in the fourth energy level.

Ans. For one dimensional box,

$$E = \frac{n^2h^2}{8mL^2}$$

Here, $n = 4, m = 9{\cdot}1 \times 10^{-31}$ kg

$L = 4 \times 10^{-10}\ m,\ h = 6{\cdot}626 \times 10^{-34}$ Js

$$E = \frac{(4)^2 \times (6{\cdot}26 \times 10^{-31}\ \text{Js})^2}{8\times(9{\cdot}1\times10^{-31}\ \text{kg})\times(4\times10^{-10}\ m)^2}$$

$$= 6{\cdot}03 \times 10^{-18}\ \text{J} \qquad (1\ \text{J} = 1\ \text{kg}\ m^2\ s^{-1})$$

$$= \frac{6{\cdot}03 \times 10^{-18}}{1{\cdot}6 \times 10^{-19}}\ \text{eV} \qquad (1\ \text{eV} = 1{\cdot}6 \times 10^{-19}\ \text{J})$$

$$= 37{\cdot}68\ \text{eV}.$$

Q. 47. An electron is confined to a molecule of length 1nm. What is its minimum energy? What is the probability of finding it is the region of the molecule lying between $x = 0$ and $x = 0.2$ nm?

Ans. The system of an electron confined to length of 1 nm is just like a particle in one dimensional box; so energy is

$$E = \frac{n^2h^2}{8mL^2} \text{ and wave function } \psi = \sqrt{\frac{2}{L}} \sin \frac{n\pi}{L} x.$$

For minimum energy $n = 1$, we get

$$E = \frac{h^2}{8mL^2} \text{ and } \psi = \sqrt{\frac{2}{L}} \sin \frac{\pi}{L} x.$$

Now

$$E = \frac{h^2}{8mL^2} = \frac{(6{\cdot}626 \times 10^{-34}\ \text{Js})^2}{8\times(9{\cdot}1 \times 10^{-31}\ \text{kg}) \times (1 \times 10^{-9}\ m)^2}$$

$$= 6{\cdot}025 \times 10^{-20}\ \text{J}.$$

The probability of finding the electron in the given region is:—

$$= \int_0^{0{\cdot}2\,nm} \psi\psi^*\, dx = \int_0^{0{\cdot}2\,nm} \psi^2\, dx$$

$$= \frac{2}{L} \int_0^{0{\cdot}2\,nm} \sin^2 \frac{\pi}{L} x\, dx$$

$$= \frac{2}{L} \int_0^{0{\cdot}2\,nm} \frac{1 - \cos \frac{2\pi}{L} x}{2}\, dx$$

$$= \frac{2}{L}\int_0^{0.2\,nm}\frac{1}{2}dx - \int_0^{0.2\,nm}\frac{1}{2}\cos\frac{2\pi}{L}x\,dx$$

$$= \frac{2}{L}\left[\left|\frac{x}{2}\right|_0^{0.2\,nm} - \frac{1}{2}\left|\frac{\sin\frac{2\pi}{L}}{\frac{2\pi}{L}}\right|_0^{0.2\,nm}\right]$$

$$= \frac{2}{1nm}\left[\frac{0.2\,nm}{2} - \frac{1}{2}\times\frac{1nm}{2\pi}\sin\frac{2\pi\times 0.2\,nm}{1\,nm}\right]$$

$$= \left[0.2 - \frac{1}{2\pi}\sin 0.4\,\pi\right]$$

$$= \left[0.2 - \frac{1}{2\pi}\times 0.9511\right]$$

$$= [0.2 - 0.1514]$$

$$= 0.0486.$$

Q. 48. What will happen if the walls of the one dimensional box are suddenly removed?

Ans. When the walls of a one-dimensional box are suddenly removed, the particle in it becomes free to move without any restriction on the value of potential energy.

Accordingly, it can possess any value of energy i.e., the energy of the particle is no longer quantized rather it becomes continuous.

14

STATISTICAL MECHANICS

Q. 1. What is statistical mechanics? What are the two main points of difference between classical statistical mechanics and quantum statistical mechanics?

Ans. Statistical mechanics is that branch of science which helps us to determine the macroscopic properties of the system from its microscopic properties by statistical methods and using the laws of mechanics.

The main points of difference between the two statistical mechanics are:

(*i*) According to classical statistical mechanics, individual atoms and molecules have definite positions and momenta, whereas according to quantum statistical mechanics, molecules have only quantized energies.

(*ii*) Classical statistical mechanics treats different particles as distinguishable, whereas quantum statistical mechanics does not distinguish between the particles.

Q. 2. Compare the important features of Maxwell-Boltzmann, Bose-Einstein and Fermi-Dirac statistics. **(*Kurukshetra 2002*)**

Ans. The important features of the three different statistics may be summed up as follows:

M — B Statistics	*B—E Statistics*	*F—D Statistics*
1. The laws of *classical mechanics* are applied according to which individual atoms or molecules have definite positions and momenta.	1. The laws of *quantum mechanics* are applied according to which individual atoms or molecules have only quantized values of energy.	1. In this statistics also, laws of *quantum mechanics* are applied.
2. It treats the different particlesas *distinguishable.*	2. It treats the different particlesas *indistinguishable.*	2. Like B—E statistics, it also treats the particles as *indistinguishable.*
3. According to this, *any number* of particles may occupy the same *energy level.*	3. Here also, any number of particles may occupy the same energy level.	3. In this case, *only one particle* is supposed to occupy a *particular energy level.*

M — B Statistics	*B — E Statistics*	*F — D Statistics*
4. The particles obeying M—B statistics are called *maxwellslet* or *boltzmannous.*	**4. It is applied to those particles which have functions like photons or atoms and molecules having *even* no. of particles in the nucleus. Such particles are called** *bosons.*	4. It is applied to those par ticles which have a symmetric wave functions like electrons, protons or atoms and molecules having *odd* no. of particles in the nucleus. Such particles are called *fermions.*

Q. 3. What do you understand by statistical method? Explain with the help of suitable examples.

Ans. *The method of averaging the behaviour of a large number of individuals is called a statistical method.* In calculating the statistical average, it is not necessary to know the behaviour of any one individual. This is illustrated below with the help of following examples:

(*i*) The price index is calculated on the basis of the statistical average of rise in prices of a large number of essential commodities without laying emphasis on one particular commodity.

(*ii*) The death rate and the birth rate of any country are a statistical average of a large number of individuals without going into the exact details of the persons born or died in different places of a country.

(*iii*) Kinetic energy of a gas at a particular temperature is a statistical average of the kinetic energies of the individual molecules.

Q. 4. What do you understand by state of a system in (*i*) Thermodynamics (*ii*) Statistical mechanics?

Ans. In **thermodynamics,** the state of a system is completely defined by specifying the various macroscopic properties of the system such as temperature, pressure, volume and composition. If any of these properties is changed, the system is said to be in a different state. These macroscopic properties are, therefore, called *state variables.*

If these properties are fixed, all other physical properties of the system are automatically fixed. Further, if the system is homogeneous and consists of a single substance, the composition is fixed (because it is 100%), and hence the state of the system depends only upon pressure, volume and temperature. In actual practice, it is not necessary to specify even these three thermodynamic properties because experiments have shown that these three properties of a simple homogeneous system of definite mass are related to one another. Thus out of P, V and T, if any two are fixed, the third is automatically fixed. Hence we conclude that the *state of a simple homogeneous system may be completely defined by specifying only two of the properties out of P, V and T.*

In **statistical mechanics,** the state of a system is completely described by specifying the state of the individual atoms or molecules. In classical statistics, the state of an individual particle is described by specifying its position coordinates (x, y, z) and the components of its momentum (p_x, p_y, p_z). In quantum statistics, the state of the individual particles is described by specifying the particular quantum state (energy level) to which the particle belongs.

Q. 5. What is Stirling formula for a large number N of molecules?

(M.Dayanand,2002)

Or

Deduce the relation $\ln x! = x \ln x - x$ for large values of x.

Ans. In the study of statistical mechanics, we usually consider a system consisting of very large number of particles. If this number is represented by N, the calculation of $\ln N!$ becomes very laborious. Hence a simple approximation is used for the value of $\ln N!$ which is called **Stirling approximation**. This is obtained as follows:

We know that

$$N! = N \times (N-1)(N-2).......3 \times 2 \times 1$$

or

$$N! = 1 \times 2 \times ..... \quad (N-2) \times (N-1) \times N \quad ...(i)$$

$$\therefore \quad \ln N! = \ln 1 + \ln 2 + ...+ \ln (N-2) + \ln (N-1) + \ln N$$

$$= \sum_{m1}^{N} \ln m \quad ...(ii)$$

In this summation, except for the first few terms whose values are small, as m increases and attains large values, the increase in the value of $\ln m$ with increase in the value of m by unity is very small. Hence in the above summation, $\ln m$ can be approximately treated as continuous so that it is given by the area under the curve from $m = 1$ to $m = N$ obtained by plotting $\ln m$ *vs* m. This in turn is equal to the integration of $\ln x\, dx$ between the limits $x = 1$ to $x = N$. Hence eqn. (*ii*) can be written as

$$\ln N! = \int_{x=1}^{x=N} \ln x\, dx$$

The integral on the R.H.S. can be solved by integration by parts. Thus

$$\ln N! = [x \ln x]_1^N - \int_1^N x.\frac{1}{x} dx$$

$$= N \ln N - N$$

Hence, for large values of N

$$\boxed{\ln N! = N \ln N - N}$$

This is called Stirling's formula for large values of N.

Q. 6. What is an ensemble as used in statistical mechanics?
Describe the following:

(*i*) microcanononical ensemble

(*ii*) canonical ensemble

(*iii*) grand canonical ensemble. *(M. Dayanand 2000)*

Ans. *A collection of a large number of systems which are identical with the system under consideration in a number of aspects such as total volume, total number of molecules, etc. is called an ensemble of system.* They may, however, differ in some thermodynamic property such as energy etc.

Depending upon the thermodynamic variables kept constant, the ensembles are of different kinds. The most common are as follows:

(*i*) **Microcanonical ensemble.** If the original system is an isolated one having volume V, total number of molecules N and energy E, then we can have an ensemble of n such systems, each member (system) having the same value of N, V and E. This ensemble is shown schematically in Fig. 14.1. It can be set up by imagining rigid impermeable walls separating the system so that neither energy nor material particles can flow from one system to the other. Thus each system in this ensemble is like the isolated system in the thermodynamic sense. *Such an ensemble of systems in which each member has the same values of N, V and E is called a microcanonical ensemble.* In the microcanonical ensemble, shown in Fig. 14.1, there are 36 members ($n = 36$). However, in statistical mechanics, we consider a very large value for n and in a number of calculations, we take the limit $n \to \infty$.

Rigid Impermeable Adiabatic Walls

N,V,E	N,V,E	N,V,E	N,V,E	N,V,E	N,V,E
N,V,E	N,V,E	N,V,E	N,V,E	N,V,E	N,V,E
N,V, E	N,V,E	N,V,E	N,V,E	N,V,E	N,V,E
N,V,E	N,V,E	N,V,E	N,V,E	N,V,E	N,V,E
N,V,E	N,V,E	N,V,E	N,V,E	N,V,E	N,V,E
N,V,E	N,V,E	N,V,E	N,V,E	N,V,E	N,V,E

Fig. 14.1. Schematic representation of a microcanonical ensemble of n systems each with the same value of N, V and E.

(*ii*) **Canonical ensemble.** *If all the members (systems) of an ensemble have the same value for N, V and T, then it is called a cononical ensemble.* It can be set up by imagining rigid but conducting walls separating the different systems through which energy can pass but not the material particles. Thus each system in this ensemble is like a closed system in the thermodynamic sense. Such an ensemble is shown schematically in Fig. 14.2 (*a*). Due to the conducting walls, when the equilibrium is attained, each member of the ensemble has the same temperature T but may not have the same energy E. Hence diagrammatically, a canonical ensemble may also be represented as shown in Fig. 14.2 (*b*).

Canonical ensemble is the most commonly used ensemble in statistical mechanics.

(*iii*) **Grand canonical ensemble.** In this ensemble, each system is an open system. Hence the matter can flow between the systems and the composition of each one may fluctuate. Consequently, the number of molecules in different systems is not same. However V, T and μ (chemical potential) for each component is same for each member of the ensemble.

Rigid conducting walls

N, V, T	N, V, T	N, V, T	N, V, T	N, V, T	N, V, T
N, V, T	N, V, T	N, V, T	N, V, T	N, V, T	N, V, T
N, V, T	N, V, T	N, V, T	N, V, T	N, V, T	N, V, T
N, V, T	N, V, T	N, V, T	N, V, T	N, V, T	N, V, T
N, V, T	N, V, T	N, V, T	N, V, T	N, V, T	N, V, T
N, V, T	N, V, T	N, V, T	N, V, T	N, V T	N, V, T

(*a*)

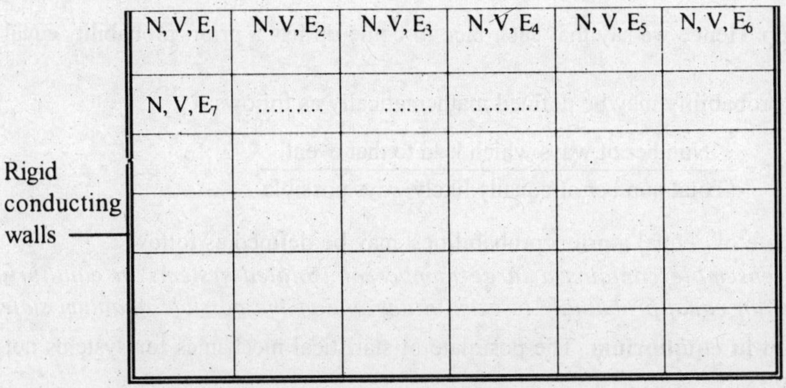

(*b*')

Fig. 14.2. Two different schematic representations of a canonical ensemble.

Q. 7. Give the postulates of statistical mechanics. **(Nagpur 2008)**

Or

Explain the postulate of 'equal a priori probabilities'.

Ans. Statistical mechanics deals with both, the systems in equilibrium as well as the systems not in equilibrium.

For systems in equilibrium. To understand the postulate of statistical mechanics on which the study of systems in equilibrium is based, consider the canonical ensemble as represented in Fig. 14.2 (*b*).

After the equilibrium has been attained so that each system in the ensemble has the same values of *N, V* and *T,* suppose each system is insulated and hence isolated. Now different systems possess different energies (or some of them may possess the same energy, called the degenerate systems). These energies represented by E_1, E_2, E_3, ... etc. correspond to the eigen values of the energy for the molecules in the *N*-molecule system. In other words, they correspond to certain specific quantum states. After the equilibrium has been attained, suppose

n_1 systems possess energy E_1,

n_2 systems possess energy E_2,

n_3 systems possess energy E_3,

and so on.

This is called *'distribution'* of systems over different possible quantum states.

Now, as we are considering an isolated ensemble having total energy *E* and total number of systems equal to *n,* the above distribution must satisfy the following two conditions:

$$\Sigma n_i = n \qquad ...(i)$$

and

$$\Sigma n_i E_i = E \qquad ...(ii)$$

Many distributions N_1, N_2, N_3 ..... etc are possible which satisfy equations (*i*) and (*ii*). Each of these distributions has an equal probability. This postulate (or principle) is called the **"postulate of equal a priori probabilities"**. The word "a priori" means something which exists in our mind prior to its verification by actual observations. This concept may be illustrated with the following examples:

(*i*) *Tossing of a coin.* Without actually tossing a coin, we know that there is equal probability for the head to come up or the tail to come up. Hence we say that a priori probability for head up or tail up is equal and each is equal to 1/2.

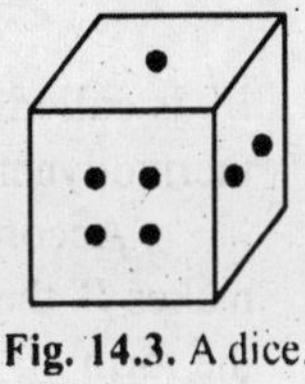

Fig. 14.3. A dice.

(*ii*) *Throwing of a dice.* When a six-faced dice is thrown, each has an equal probability to come up. Hence we say that each face to come up has a priori probability equal to 1/6.

In general, the probability may be defined mathematically as follows:

$$\text{Probability of an event} = \frac{\text{Number of ways which lead to that event}}{\text{Total number of equally likely ways possible}}$$

Thus the postulate of 'equal a priori probabilities' may be defined as follows:

If an isolated ensemble contains a large number of isolated systems in equilibrium then each system has an equal probability to exist in any one of the possible quantum states.

For systems not in equilibrium. The postulate of statistical mechanics for systems not in equilibrium is as follows:

If an isolated system in an ensemble is not in equilibrium it is not found with equal probability in the different possible quantum states; it then tends to change with time until it ultimately attains the equilibrium where it has an equal possibility to exist in any one of the possible quantum states.

There is no proof of these postulates. However, the values of the thermodynamics properties calculated on the basis of these postulates agree well with the experimental measurements. It gives enough proof of the validity of these postulates.

Q. 8. Derive the Maxwell-Boltzmann distribution law.

(M. Dayanand 2000)

Ans. Consider a *system* of *N identical but distinguishable* particles such as gas molecules at temperature *T.* Suppose the total volume of the system is *V* and total energy is *E.* Then, all the molecules do not have the same energy. Suppose n_1 molecules are present in the level with energy ε_1, n_2 in the level with energy ε_2, n_3 in the level with energy ε_3 and so on. However, whatever may be the type of distribution, the following two conditions must be satisfied:

$$\Sigma\, n_i = N \text{ (total number of molecules)} \quad \text{... (i)}$$

and

$$\Sigma n_i \varepsilon_i = E \text{ (total energy of the system)} \quad \text{...(ii)}$$

where n_i is the number of molecules in the *i*th level with energy ε_i.

Many distributions are possible which satisfy the above two conditions. Now let us first see what is the total number of such possible distributions. The situation is similar to putting *N* numbered balls into *n* boxes such that n_1 are in the first box, n_2 in the second box and so on. Mathematically, the number of possible arrangements is given by

$$G = \frac{N!}{n_1!\, n_2! \ldots . n_n!}$$

When applied to the system of *N* molecules, the number of possible distributions is usually represented by *W* and is given by

$$W = \frac{N!}{n_1!\, n_2! \ldots . n_n!} \quad \text{... (iii)}$$

W is called the **thermodynamic probability** of the system, because it is found that most of the thermodynamic properties are intimately related to the value of *W.*

According to principles of statistical mechanics, the most probable distribution is one that makes *W* the maximum. This condition for maximum probability may be written mathematically in the form

$$d \ln W = 0 \quad \text{...(iv)}$$

$d \ln W$ stands for differential of $\ln W$ which is equal to $\frac{1}{W}$ and for maximum value of W,

$\frac{1}{W} \to 0$.

Taking logarithm of both sides of equation (*iii*), we get

$$\ln W = \ln N! - [\ln n_1! + \ln n_2! + ... + \ln n_n!]$$
$$= \ln N! - \Sigma \ln n_i! \quad ...(v)$$

Applying Stirling's formula, according to which

$$\ln N! = N \ln N - N$$

and $\ln n_i! = n_i \ln n_i - n_i$

eqn. (*v*) takes the form

$$\ln W = N \ln N - N - \Sigma(n_i \ln n_i - n_i)$$
$$= N \ln N - N - \Sigma n_i \ln n_i + \Sigma n_i$$
$$= N \ln N - N - \Sigma n_i \ln n_i + N \quad (\because \Sigma n_i = N)$$
$$= N \ln N - \Sigma n_i \ln n_i \quad ...(vi)$$

Differentiating both sides of this equation, we get

$$d \ln W = -\Sigma d(n_i \ln n_i) \quad [d(N \ln N) = 0, \text{ because } N \text{ is constant}]$$

$$= -\Sigma\left[n_i \times \frac{1}{n_i} + (\ln n_i \times 1)\right] dn_i$$

($n_i \ln n_i$ has been differentiated as a product of two functions)

$$= -\Sigma(\ln n_i + 1)\, dn_i \quad ...(vii)$$

But $d \ln W = 0$ (according to eqn. (*iv*))

∴ Eqn. (*vii*) gives

$$\Sigma(\ln n_i + 1)\, dn_i = 0 \quad ...(viii)$$

For the system under consideration, we have the following relations:

$$\Sigma n_i = N = \text{constant}$$

and $\Sigma n_i \varepsilon_i = E = \text{constant}$

Differentiation of these equations gives

$$\Sigma dn_i = 0 \quad ...(ix)$$

and $\Sigma \varepsilon_i dn_i = 0 \quad ...(x)$

Equations (*viii*), (*ix*) and (*x*) represent the conditions which must be satisfied simultaneously for the most probable distribution.

Multiplying equations (*ix*) and (*x*) by the undetermined constants α' and β respectively and adding to eqn. (*viii*), we get

$$\Sigma(\ln n_i + 1 + \alpha' + \beta\varepsilon_i)\, dn_i = 0 \quad ...(xi)$$

This is called method of *Langrange undetermined multipliers* according to which the constraints (restrictions/conditions) are multiplied with undetermined multipliers and added to differential of the function that we want to maximise.

Putting $\alpha' + 1 = \alpha$, another constant, eqn. (*xi*) is simplified to the form

$$\Sigma(\ln n_i + \alpha + \beta\varepsilon_i)\, dn_i = 0 \quad ...(xii)$$

Now within the restrictions imposed on the system for constant N and E values, the variations in n_i are independent of each other and need not be zero, *i.e.*, $dn_i \neq 0$. Consequently, equation (*xii*) will be satisfied only if

$$\ln n_i + \alpha + \beta\varepsilon_i = 0$$

or $$\ln n_i = -(\alpha + \beta\varepsilon_i) \qquad ...(xii\ (a))$$

or $$n_i = e^{-(\alpha+\beta\varepsilon_i)} \qquad ...(xii\ (b))$$

or $$\boldsymbol{n_i = \frac{1}{e^{\alpha+\beta\varepsilon_i}}} \qquad ...(xiii)$$

This expression is called **Maxwell-Boltzmann distribution law**. It gives the most probable distribution of molecules among the various possible energy states at equilibrium for a system of constant total energy.

Eqn. (*xiii*) can be written in the form.

$$n_i = e^{-\alpha} \cdot e^{-\beta\varepsilon_i} \qquad ...(xiv)$$

The value of $e^{-\alpha}$ can be found from eqn. (*i*) as follows:

$$N = \Sigma n_i = \Sigma e^{-\alpha} \cdot e^{-\beta\varepsilon_i}$$

$$= e^{-\alpha} . \Sigma e^{-\beta\varepsilon_i}$$

$$\therefore \qquad e^{-\alpha} = \frac{N}{\Sigma e^{-\beta\varepsilon_i}} \qquad ...(xv)$$

Substituting this value in eqn. (*xiv*), we get

$$n_i = \frac{Ne^{-\beta\varepsilon_i}}{\Sigma e^{-\beta\varepsilon_i}} \qquad ...(xvi)$$

Further by applying eqn. (*ii*) to the energy of an ideal gas, it can be shown that

$$\beta = \frac{1}{kT} \qquad ...(xvii)$$

where k is Boltzmann constant.

(Refer to *Q*. 30. in this chapter)

Substituting this value of β in eqn. (*xvi*), we get

$$n_i = \frac{Ne^{-\varepsilon_i/kT}}{\Sigma e^{-\varepsilon_i/kT}}$$

or which can be written as

$$\boldsymbol{\frac{n_i}{N} = \frac{e^{-\varepsilon_i/kT}}{\sum e^{-\varepsilon_i/kT}}} \qquad ...(xviii)$$

This expression is called **Boltzmann distribution law.** This is another form of equation which gives the most probable distribution *i.e.*, at temperature T it gives the fraction of the total number of molecules which in the most probable state will possess the energy ε_i.

In the derivation of the above expression, it has been assumed that each energy level is non-degenerate *i.e.*, consists of a single level. However, in many cases it may not be so. Corresponding to each eigen value of the energy there may be a number of eigen functions (called the eigen states). *The number of eigenstates corresponding to a particular energy* ε_i *is called* **degeneracy** or **statistical weight**. It is usually represented by g_i, (for the level with energy ε_i). Thus the eqn. (*xviii*) is modified to a more general form as follows:

$$\frac{n_i}{N} = \frac{g_i e^{-\varepsilon_i/kT}}{\Sigma g_i e^{-\varepsilon_i/kT}} \qquad \text{.... (xix)}$$

Q. 9. Derive Maxwell-Boltzmann distribution law in the form $n_i = g_i \exp(-\alpha - \beta\varepsilon_i)$.

Ans. If degeneracy is taken into consideration then the number of possible distributions (thermodynamic probability *W*) of *N* **distinguishable** particles in the different energy levels such that the level with energy ε_i has g_i states (any of which may be occupied by a particle with energy ε_i) and n_i is the number of particles which have energy ε_i, is given by

$$W = N! \frac{g_0^{n_0} g_1^{n_1} \ldots g_n^{n_n}}{n_0 \mid n_1 \ldots n_n !} = N! \prod \frac{g_i^{n_i}}{n_i !} \qquad \text{... (i)}$$

From eqn. (*i*), In $W = \ln N! + \Sigma(\ln g_i^{n_i} - \ln n_i !)$

$= N \ln N - N + \Sigma\,(n_i \ln g_i - n_i \ln n_i + n_i)$

(applying Stirling's formula)

$$= N \ln N + \Sigma n_i \ln \frac{g_i}{n_i} \qquad (\because \Sigma n_i = N) \qquad \text{...(ii)}$$

Derivation. Differentiating both sides of eqn. (*ii*), we get (keeping in mind that *N* is constant)

$$d \ln W = \Sigma d \left(n_i \ln \frac{g_i}{n_i} \right)$$

$$= \Sigma\, d\,(n_i \ln g_i - n_i \ln n_i)$$

$$= \Sigma \ln g_i\, dn_i - \Sigma\, n_i\, d \ln n_i - \Sigma \ln n_i\, dn_i$$

$$= \Sigma \ln g_i\, dn_i - \Sigma\, dn_i - \Sigma \ln n_i\, dn_i \qquad \text{...(iii)}$$

But $d \ln W = 0$

$\therefore$ Eqn. (*iii*) gives

$$\Sigma \ln g_i\, dn_i - \Sigma dn_i - \Sigma \ln n_i\, dn_i = 0 \qquad \text{...(iv)}$$

Further for the system under consideration

$$\Sigma\, n_i = N = \text{constant}$$

$$\Sigma\, n_i\, \varepsilon_i = E = \text{constant}$$

therefore, differentiation of these equations give

$$\Sigma\, dn_i = 0 \qquad \text{....(v)}$$

$$\Sigma \varepsilon_i dn_i = 0 \qquad \text{... (vi)}$$

Substituting the value from eqn. (*v*) into eqn. (*iv*), we get

$$\Sigma \ln g_i\, dn_i - \Sigma \ln n_i\, dn_i = 0 \qquad \text{... (vii)}$$

Now applying the method of Langrange undetermined multipliers *i.e.*, multiplying the constraints (*v*) and (*vi*) by undetermined constants α and β and subtracting from eqn. (*vii*), we get

$$\Sigma\,(\ln g_i - \ln n_i - \alpha - \beta\varepsilon_i)\, dn_i = 0 \qquad \text{...(viii)}$$

or

$$\Sigma \left(\ln \frac{g_i}{n_i} - \alpha - \beta\varepsilon_i \right) dn_i = 0$$

Now within the restrictions imposed on the system for constant *N* and *E* value, $dn_i \neq 0$. Hence

$$\ln \frac{g_i}{n_i} - \alpha - \beta\varepsilon_i = 0$$

or $$\ln \frac{g_i}{n_i} = \alpha + \beta\varepsilon_i$$

$$\ln \frac{g_i}{n_i} = -(\alpha + \beta\varepsilon_i)$$

or $$\frac{n_i}{g_i} = e^{-(\alpha + \beta\varepsilon_i)}$$

$$n_i = g_i\, e^{-(\alpha+\beta\varepsilon_i)}$$

or $$n_i = \frac{g_i}{e^{\alpha+\beta\varepsilon_i}}$$

Q. 10. What is statistical thermodynamics?

Ans. *The branch of science dealing with the calculation of thermodynamic properties of the systems using the methods of statistical mechanics is called statistical thermodynamics.* In other words, statistical thermodynamics deals with the calculation of the macroscopic properties of the systems such as pressure, entropy, internal energy, Gibbs free energy *etc.* from the microscopic properties *i.e.*, properties of the individual molecules such as their position and momenta (in classical statistics) and their quantum states (in quantum statistics).

Q. 11. What do you understand by thermodynamics probability? Derive the relationship between entropy and thermodynamic probability ($S = k \ln W$).

Ans. Thermodynamic probability is the number of microstates corresponding to the given macroscopic state.

From thermodynamics, we know that when a system with constant volume and energy is in equilibrium, the entropy is maximum. According to statistical mechanics, for such a system in equilibrium, the probability is also maximum. Hence it was suggested by Boltzmann that a relationship between entropy and probability must exist.

The relationship between entropy and probability can be obtained by considering the mixing of two systems and remembering that the total entropy (S) is the sum of entropies (S_1 and S_2) of the individual systems and the total probability (W) is the product of the probabilities (W_1 and W_2) of the individual systems.

Thus, we have

$$S = f(W) \quad \text{... (i)}$$

$$S = S_1 + S_2 \quad \text{... (ii)}$$

$$W = W_1 \times W_2 \quad \text{... (iii)}$$

Combining these equations, we get

$$f(W) = S_1 + S_2$$

or $$f(W_1 \times W_2) = f(W_1) + f(W_2) \quad \text{... (iv)}$$

This relationship is true only if the function f stands for the logarithm, ln. Hence the relationship between entropy (S) and the thermodynamic probability (W) must be logarithmic. Hence we may write

$$S = k_1 \ln W + k_2 \quad \text{...(v)}$$

where k_1 and k_2 are constants.

The constant k_2 is found to be zero as discussed below:

Applying eqn. (v) to a perfect crystal at 0 K, $S = 0$ (third law of thermodynamics). Further for a perfect crystal at $0K$, there is only one arrangement possible. Hence $W = 1$. Substituting these values in equation (v), we get $k_2 = 0$. So equation (v) becomes

$$S = k_1 \ln W \qquad ...(vi)$$

The constant k_1 is found to be equal to the Boltzmann constant, k (*viz.* gas constant per molecule, R/N). Hence equation (vi) becomes

$$S = k \ln W \qquad ...(vii)$$

Q. 12. Explain : Thermodynamic probability (W) is a measure of disorder of a system.

(M. Dayanand 2002)

Ans. Consider a system existing in a macrostate having only one microstate *i.e.*, $W = 1$. This means that the particles of the system can be arranged in only one way and in no other way. Thus in this state we have *full and definite* information about the particles of the system. In such a case, the system is said to be in a *perfect order.* For example a system at 0 K should be in a state of perfect order.

Let us now consider a macrostate of the system for which $W = 2$. Since, in this case, the system can exist in any of the two equally probable microstates, we cannot say with certainty in which of the two states the system exists at any instant of time. Thus there is a certain amount of loss of information about the system. In this case, we can only say that the system can exist in either of the two microstates with a probability = 1/2. *This statement of uncertainty or the loss of information about the state of a system is known as disorder.*

Thus increase in the value of W means increase in the amount of disorder or in other words, decrease in the amount of information as to which microstate the system occupies. Hence *W can be taken as measure of the disorder in a system and is sometimes called the disorder number of a given macrostate.*

Q. 13. What is meant by partition function? *(Punjabi 1996; G. Nanakdev 1996)*

Ans. According to Boltzmann distribution law, the fraction of molecules which in the most probable state at temperature T possesses the energy ε_i is given by

$$\frac{n_i}{N} = \frac{g_i e^{-\varepsilon_i/kT}}{\Sigma g_i e^{-\varepsilon_i/kT}} \qquad ...(i)$$

where n_i is the number of molecules having energy ε_i at temperature T, N is the total number of molecules and g_i is the degeneracy (or statistical weight) of the energy level ε_i.

The denominator of the above equation which gives the sum of the terms $g_i e^{-\varepsilon_i/kT}$ for all the energy levels is called the partition function. It is usually represented by Q. Thus

$$\textbf{Partition function} \quad Q = \Sigma g_i e^{-\varepsilon_i/kT} \qquad ...(ii)$$

The partition function is so called because it indicates how the particles, are distributed among the various energy states.

Q. 14. Derive expression for the following thermodynamic functions in terms of partition function.

(*i*) Energy | **(*ii*) Enthalpy**
(*iii*) Heat capacity C_V | **(*iv*) Heat capacity C_p**
(*v*) Entropy | **(*vi*) Helmholtz free energy**
(*vii*) Gibb's free energy.

Ans. The expressions for the thermodynamic functions are usually obtained in terms of the partition function Q and its derivatives

$$(\delta \ln Q / \delta T)_V \text{ and } (\delta^2 \ln Q/\delta T^2)_V$$

The expressions for these derivatives are found to be as follows:

$$\left(\frac{\partial \ln Q}{\partial T}\right)_V = \frac{1}{T}\left(\frac{Q'}{Q}\right) \qquad \text{... (i)}$$

and
$$\left(\frac{\partial^2 \ln Q}{\partial T^2}\right)_V = \frac{1}{T^2}\left[\left(\frac{Q''}{Q}\right) - \left(\frac{Q'}{Q}\right)^2 - 2\left(\frac{Q'}{Q}\right)\right] \qquad \text{... (ii)}$$

where
$$Q' = \Sigma g_i \left(\frac{\varepsilon_i}{kT}\right) e^{-\varepsilon_i / kT} \qquad \text{... (iii)}$$

and
$$Q'' = \Sigma g_i \left(\frac{\varepsilon_i}{kT}\right)^2 e^{-\varepsilon_i / kT} \qquad \text{... (iv)}$$

(1) **Expression for Energy.** If the molecules are non-interacting *i.e.*, the gas is ideal, the total energy E of the system is given by

$$E = \Sigma\, n_i\, \varepsilon_i \qquad \text{...(v)}$$

where n_i is the number of particles present in the level of energy ε_i relative to the ground state.

From equations (*i*) and (*ii*), Q 13 we have

$$\frac{n_i}{N} = \frac{g_i e^{-\varepsilon_i / kT}}{Q}$$

i.e.,
$$n_i = \frac{N}{Q} g_i e^{-\varepsilon_i / kT} \qquad \text{... (vi)}$$

Substituting this value in equation (*v*), we get

$$E = \frac{N}{Q} \Sigma \frac{\varepsilon_i}{kT} \varepsilon_i g_i e^{-\varepsilon_i / kT} \qquad \text{... (vii)}$$

which can be rearranged as

$$E = \frac{NkT}{Q} \Sigma \frac{\varepsilon_i}{kT} g_i e^{-\varepsilon_i / kT} \qquad \text{... (viii)}$$

If one mole of the gas is considered, the total number of molecules N will be equal to Avogadro's number so that $Nk = R$, the molar gas constant. The eqn. (*viii*) becomes

$$E = \frac{RT}{Q} \Sigma g_i \left(\frac{\varepsilon_i}{kT}\right) e^{-\varepsilon_i / kT} \qquad \text{... (ix)}$$

But
$$\Sigma g_i \left(\frac{\varepsilon_i}{kT}\right) e^{-\varepsilon_i / kT} = Q'$$
[according to eqn. (*iii*) above]

∴ Eqn. (*ix*) can be written as

$$\boldsymbol{E = RT\left(\frac{Q'}{Q}\right)} \qquad \text{... (x)}$$

Further substituting the value of Q'/Q from eqn. (*i*) in equation (*x*), we get

$$\boldsymbol{E = RT^2\left(\frac{\partial \ln Q}{\partial T}\right)_V} \qquad \text{... (xi)}$$

(2) **Expression for heat content/Enthalpy.** The heat content H is given by an expression similar to the above one as

$$H = RT^2\left(\frac{\partial \ln Q}{\partial T}\right)_P \qquad \text{...}(xii)$$

Alternatively the heat content H is given by

$$H = E + PV$$

Substituting the value of E from eqn. (xi) and $PV = RT$ for 1 mol of the gas, we get

$$H = RT^2\left(\frac{\partial \ln Q}{\partial T}\right)_V + RT \qquad \text{...}(xiii)$$

(3) **Expression for Heat capacity at constant volume (C_v).**

By definition $$C_v = \left(\frac{\partial E}{\partial T}\right)_V$$

Using the value of E given by equation (xi), we get

$$C_v = \frac{\partial}{\partial T}\left\{RT^2\left(\frac{\partial \ln Q}{\partial T}\right)\right\}_V \qquad \text{...}(xiv)$$

It can be written as

$$C_v = \frac{R}{T^2}\left\{\frac{\partial \ln Q}{\partial (1/T)^2}\right\}_V \qquad \text{...}(xv)$$

(4) **Expression for Heat Capacity at Constant Pressure (C_P).**

By definition $$C_P = \left(\frac{\partial H}{\partial T}\right)_P$$

Substituting the value of H given by equation (xii), we get

$$C_P = \frac{\partial}{\partial T}\left\{RT^2\left(\frac{\partial \ln Q}{\partial T}\right)\right\}_P \qquad \text{...}(xvi)$$

(5) **Expression for the Entropy.** From thermodynamics, we get

$$dS = \frac{C_v}{T}\,dT$$

Integrating both sides between suitable limits, we get

$$\int_{S=S_0}^{S=S} dS = \int_0^T \frac{C_v}{T}\,dT$$

or $$S - S_0 = \int_0^T \frac{C_v}{T}\,dT \qquad \text{...}(xvii)$$

where S is the molar entropy of the gas as temperature T and S_0 is the hypothetical value at absolute zero. Substituting the value C_v from equation (xiv) in equation $(xvii)$, we get

$$S - S_0 = \int_0^T \frac{1}{T}\frac{\partial}{T}\left(RT^2\left(\frac{\partial \ln Q}{\partial T}\right)\right)_V dT$$

Carrying out the integration by parts, we get

$$S - S_0 = \frac{1}{T} \times RT^2 \left(\frac{\partial \ln Q}{\partial T}\right)_V + \int_0^T \frac{RT^2}{T^2}\left(\frac{\partial \ln Q}{\partial T}\right)_V dT$$

or
$$\boldsymbol{S - S_0 = RT\left(\frac{\partial \ln Q}{\partial T}\right)_V + R\ln Q - R\ln Q_0} \qquad \text{...(xviii)}$$

where Q_0 is the value of Q at $T = 0$.

(6) Expression for Helmholtz Free Energy. The Helmholtz free energy (A) is given by

$$A = E - TS$$

Substituting the value of S from equation (*xviii*) after incorporating eqn. (*xi*) in it, we get

$$A = E - T\left(R\ln Q + \frac{E}{T}\right)$$

$$= - RT \ln Q$$

i.e.,
$$\boldsymbol{A = - RT \ln Q} \qquad \text{...(xix)}$$

(7) Expression for Gibbs Free Energy. Gibbs free energy (G) is given by

$$G = H - TS = E + PV - TS$$
$$= (E - TS) + PV = A + PV$$

Substituting the value of A from equation (*xix*), we get

$$\boldsymbol{G = - RT \ln Q + PV} \qquad \text{...(xx)}$$

Q. 15. Prove that complete partition function for a system is the product of translational, rotational, vibrational and electronic partition functions.

Ans. The total energy of system containing an ideal gas or any other non-interacting molecules must be the sum of translational, rotational, vibrational and electronic energies *i.e.*,

$$E = E_t + E_r + E_v + E_e \qquad \text{... (i)}$$

Using the expression for energy given by the following equation *viz.*,

$$E = RT^2\left(\frac{\partial \ln Q}{\partial T}\right)_V \qquad \text{...(ii)}$$

we can write

$$E_t = RT^2\left(\frac{\partial \ln Q_t}{\partial T}\right)_V \qquad \text{..(iii)}$$

$$E_r = RT^2\left(\frac{\partial \ln Q_r}{\partial T}\right)_V \qquad \text{.. (iv)}$$

$$E_v = RT^2\left(\frac{\partial \ln Q_v}{\partial T}\right)_V \qquad \text{...(v)}$$

$$E_e = RT^2\left(\frac{\partial \ln Q_e}{\partial T}\right)_V \qquad \text{... (vi)}$$

where Q_t, Q_r, Q_v and Q_e represent the partition functions for translational, rotational, vibrational and electronic energies.

Substituting these values in equation (*i*), we get

$$E = RT^2\left[\left(\frac{\partial \ln Q_t}{\partial T}\right)_V + \left(\frac{\partial \ln Q_r}{\partial T}\right)_V + \left(\frac{\partial \ln Q_v}{\partial T}\right)_V + \left(\frac{\partial \ln Q_e}{\partial T}\right)_V\right]$$

$$= RT^2\left[\frac{\partial \ln(Q_t Q_r Q_v Q_e)}{\partial T}\right] \quad ...(vii)$$

Comparing this equation with equation (*ii*), we find that

$$Q = Q_t Q_r Q_v Q_e \quad ...(viii)$$

Q. 16. Derive expression for translational partition function *viz.*,

$$Q_t = \frac{(2\pi m kT)^{3/2}}{h^3}.V$$

Ans. Partition function of a system is given by

$$Q = \Sigma g_i e^{-\varepsilon_i / kT} \quad ... (i)$$

For translational energy, $g_i = 1$. Representing the translational energy of a molecule by ε_i, the translational partition function, Q_t will be given by

$$Q_t = \Sigma e^{-\varepsilon_i / kT} \quad ...(ii)$$

Further we know that the translational energy of a molecule moving in a rectangular box of dimensions *a*, *b*, and *c* is given by

$$\varepsilon_t = \frac{h^2}{8\ m}\left(\frac{n_x^2}{a^2} + \frac{n_y^2}{b^2} + \frac{n_z^2}{c^2}\right)$$

Substituting this value in eqn. (*ii*), we get

$$Q_t = \Sigma \exp\left[-\frac{h^2}{8\ mkT}\left(\frac{n_x^2}{a^2}\right)\right] \exp\left[-\frac{h^2}{8\ mkT}\left(\frac{n_y^2}{b^2}\right)\right]$$

$$\exp\left[-\frac{h^2}{8\ mkT}\left(\frac{n_z^2}{c^2}\right)\right] \quad ... (iii)$$

Let us consider the motion in *x*-direction only. The partition Q_x for this motion will be

$$Q_x = \sum_{n_x=0}^{n_x=\infty} \exp\left[-\frac{h^2}{8\ mkT}\left(\frac{n_x^2}{a^2}\right)\right] \quad ...(iv)$$

Since the energy levels are very closely spaced, the summation in the equation may be replaced by integration *i.e.*, we have

$$Q_x = \int_0^\infty \exp\left[-\frac{h^2}{8\ mkT}\left(\frac{n_x^2}{a^2}\right)\right] dn_x$$

$$= \frac{(2\pi mkT)^{1/2}}{h} a \quad ... (v)$$

Similarly, for the motion in the *y* and *z*-directions, we will have

$$Q_y = \frac{(2\pi mkT)^{1/2}}{h} b$$

and

$$Q_z = \frac{(2\pi mkT)^{1/2}}{h} c$$

Hence from equation (*iii*)

$$Q_t = Q_x . Q_y . Q_z = \frac{(2\pi mkT)^{3/2}}{h^3}(abc)$$

or $$Q_t = \frac{(2\pi mkT)^{3/2}}{h^3}V \quad ...(vi)$$

where V is the volume of the container occupied by the particle.

Q. 17. Calculate translational partition function for helium molecule confined in a vessel 10^2 cm^3 volume at 300 K. Given that $k = 1.38 \times 10^{-23}$ J K^{-1}, $h = 6.625 \times 10^{-34}$ J s. Mass of He atom $m = 4$ amu, $N = 6.023 \times 10^{23}$.

Ans. Here $$m = \frac{4}{6.023 \times 10^{23}} \times \frac{1}{1000} \text{ kg} = \frac{4}{6.023 \times 10^{26}} \text{ kg}$$

$$k = 1.38 \times 10^{-23} \text{ JK}^{-1} = 1.38 \times 10^{-23} \text{ kg m}^2 s^{-2} \text{ K}^{-1}$$

$$h = 6.625 \times 10^{-34} \text{ Js} = 6.625 \times 10^{-34} \text{ kg m}^2 s^{-1}$$

$$V = 10^2 \text{ cm}^3 = \frac{10^2}{100 \times 100 \times 100} m^2 = 10^{-4} m^3$$

The translational partition function for He molecule is given by $Qt \text{ (He)} = \left[\frac{2\pi mkT}{h^2}\right]^{3/2}$

Putting various values in the above expression

$$Qt(\text{He}) = \left[\frac{2\pi \times \left(\frac{4}{6.023 \times 10^{26}} \text{ kg}\right)(1.38 \times 10^{-22} \text{ gm}^2 \text{ s}^{-2} \text{ K}^{-1}) \times 300\text{K}}{(6.625 \times 10^{-34} \text{ kgm}^2 \text{ s}^{-1})^2}\right]^{3/2} \times 10^{-4} \text{ m}^3$$

$\therefore$ $Qt \text{ (He)} = \mathbf{7.808 \times 10^{26}}$

Q. 18. Calculate the translational partition function of NO molecule at 300 K in a volume 1000 m^3. Assuming the gas to behave ideally.

Ans. Here V = 1000 cm^3 = 10^3 cm^3, T = 300 K, m = Mass of NO molecule $= \frac{30}{6.023 \times 10^{23}}$ g

$$h = 6.625 \times 10^{-27} \text{ erg s} = 6.625 \times 10^{-27} \text{ g cm}^2 \text{ cm}^2 \text{ s}^{-1}$$

$$k = 1.38 \times 10^{-16} \text{ erg K}^{-1} = 1.38 \times 10^{-16} \text{ g cm}^2 \text{ s}^{-2} \text{ K}^{-1}$$

$$[\because 1 \text{ erg} = 1 \text{ g cm}^2 \text{ s}^{-2}]$$

The translational partition function of NO molecule is given by the expression

$$Q_t = \left[\frac{2\pi\, mkT}{h^2}\right]^{3/2} .V \quad ...(i)$$

Substituting various values in equation (i), we get

$$Q_t = \left[\frac{2 \times \pi \times \left(\frac{30\text{g}}{6.023 \times 10^{23}}\right)(1.38 \times 10^{-16} \text{ gcm}^2 \text{ s}^{-2}\text{K}^{-1}) \times 300\text{K}}{(6.625 \times 10^{-27} \text{ Kgm}^2 \text{ s}^{-1})^2}\right] \times 10^3 \text{ cm}^3$$

$$= \mathbf{1.6038 \times 10^{29}}$$

Q. 19. Calculate the ratio of Q_t of helium molecule (He) to that of hydrogen (H_2) at the same temperature and volume.

Ans. For helium molecule (He), $m = 4$ a.m.u.

$$Q_t\,(\text{He}) = \left(\frac{2\pi \times 4 \times kT}{h^2}\right)^{3/2} .V \qquad ...(i)$$

For hydrogen molecule *molecule* (H_2), $m = 2$ a.m.u.

$$Q_t\,(H_2) = \left(\frac{2\pi \times 2 \times kT}{h^2}\right)^{3/2} .V \qquad ...(ii)$$

Dividing (*i*) by (*ii*)

$$\frac{Q_t\,(\text{He})}{Q_t\,(H_2)} = \frac{\left(\frac{2\pi \times 4 \times kT}{h^2}\right)^{3/2} .V}{\left(\frac{2\pi \times 2 \times kT}{h^2}\right)^{3/2} .V} = 2^{3/2} = 2\sqrt{2} = 2(1.414)$$

or

$$\frac{Q_t\,(\text{He})}{Q_t\,(H_2)} = \mathbf{2.828}$$

In other words, Q_t (He) is 2.828 times larger than Q_t (H_2).

Q. 20. Derive expression for rotational partition function *viz.*,

$$Q_r = \frac{8\pi^2\, IkT}{\sigma h^2}$$

Ans. The rotational energy of a molecule, ε_r depends upon the rotational quantum number, *J*. Further, for rotation

$$g_i = (2J + 1)$$

Hence the general partition function for rotation is given by

$$Q_r = \Sigma(2J+1)e^{-\varepsilon_r/kT} \qquad ...(i)$$

The simplest model for a diatomic molecule is the rigid rotator, the energy for which is given by

$$\varepsilon_r = J(J+1)\frac{h^2}{8\pi^2 I} \qquad ...(ii)$$

where *J* is the rotational quantum number and *I* is the moment of inertia of the rotator.

Substituting this value of ε_r in equation (*i*), we get

$$Q_r = \Sigma(2J+1)e^{-aJ(J+1)} \qquad ...(iii)$$

where

$$a = \frac{h^2}{8\pi^2 IkT} \qquad ...(iv)$$

Assuming that the energy levels are very closely spaced, the summation in equation (*iii*) may be replaced by integration so that we have

$$Q_r = \int_0^\infty (2J+1)e^{-aJ(J+1)} \qquad ...(v)$$

Solution of this equation gives

$$Q_r = \frac{1}{a}$$

Substituting the value of a from equation (iv), we get

$$\boldsymbol{Q_r = \frac{8\pi^2 IkT}{h^2}} \quad \text{... (vi)}$$

Taking the symmetry effect of the molecule into consideration, the above expression is modified to

$$\boldsymbol{Q_r = \frac{8\pi^2 IkT}{\sigma h^2}} \quad \text{...(vii)}$$

where σ is called *symmetry number.* Its value is 1 for unsymmetrical molecules like HCl, CO etc. and it is 2 for symmetrical molecules like O_2, Cl_2 etc.

Q. 21. Derive expression for vibrational partition function *viz.,*

$$\boldsymbol{Q_v = \frac{1}{1 - e^{-h\nu_0/kT}}}$$

Ans. Considering the diatomic molecule as a simple harmonic oscillator, its vibrational energy is given by

$$e'_v = (v + 1/2)\, h\nu_0 \quad \text{... (i)}$$

where v is the vibrational quantum number and ν_0 is the fundamental vibrational frequency.

Equation (i) can be rewritten as

$$e_v = \frac{1}{2} h\nu_0 + vh\nu_0$$

$$= \varepsilon_0 + \varepsilon_v$$

where $\varepsilon_0 = \frac{1}{2} h\nu_0$ is the zero point energy of the oscillator and

$\varepsilon_v = vh\nu_0$ is the energy of the oscillator relative to the ground state.

Further for vibration $g_i = 1$. Substituting these values of g_i and ε_v the vibrational partition function becomes

$$Q_v = \sum_{v=0}^{\infty} e^{-vh\nu_0/kT} \quad \text{... (ii)}$$

Putting $h\nu_0/kT = x$, this equation becomes

$$Q_v = \sum_{v=0}^{\infty} e^{-vx} \quad \text{...(iii)}$$

$$= 1 + e^{-x} + e^{-2x} + \ldots$$

$$= (1 - e^{-x})^{-1}$$

$$= \frac{1}{1 - e^{-x}}$$

or

$$\boldsymbol{Q_v = \frac{1}{1 - e^{-h\nu_0/kT}}} \quad \text{...(iv)}$$

Q. 22. Derive an expression for electronic partition function.

Ans. The electronic partition function of an atom or a molecule is given by

$$Q_e = \Sigma g_i e^{-\varepsilon_e / kT} \qquad \text{... }(i)$$

where ε_e is the electronic energy relative to the ground state.

The above summation is carried over the energy levels instead of over the states. Hence we write

$$Q_e = \sum_i g_i e^{-\varepsilon_i / kT} \qquad \text{... }(ii)$$

Thus if the degeneracy of the ground state level is g_0, (for which $\varepsilon_i = 0$) and that of the first excited state is g_1 (for which we can write $\varepsilon_i = \varepsilon_1$) and so on, the above expression can be written as

$$\boldsymbol{Q_e = g_0 + g_i e^{-\varepsilon_1 / kT} + ...} \qquad \text{... }(iii)$$

Q. 23. Derive expressions for translational contribution to energy, enthalpy, heat capacity, entropy, Helmholtz free energy and Gibb's free energy.

Ans. The total value of any thermodynamic property is the sum of translational, rotational, vibrational and electronic contributions.

Let us consider a systems of perfect gases.

(*i*) **Translational energy.** Translational energy is given by

$$E_r = RT^2 \left(\frac{\partial \ln Q_t}{\partial T} \right)_V \qquad \text{...}(i)$$

Let us now find the value of $(\partial \ln Q_t / \partial T)_V$

The expression for translational partition function is

$$Q_t = \frac{(2\pi mkT)^{3/2}}{h^3} V$$

Taking logarithm of both sides, we get

$$\ln Q_t = \ln \frac{(2\pi mkT)^{3/2}}{h^3} V$$

$$= \ln \frac{(2\pi mk)^{3/2} V}{h^3} + \ln T^{3/2}$$

$$= \frac{(2\pi mk)^{3/2} V}{h^3} + \frac{3}{2} \ln T \qquad \text{...}(ii)$$

Differentiating w.r.t T at constant V, we get

$$\left(\frac{\partial \ln Q_t}{\partial T} \right)_V = \frac{3}{2} \left(\frac{1}{T} \right) \qquad \text{...}(iii)$$

Substituting this value in equation (*i*), we get

$$E_t = RT^2 \times \frac{3}{2} \left(\frac{1}{T} \right)$$

or

$$\boldsymbol{E_t = \frac{3}{2} RT} \qquad \text{...}(iv)$$

(*ii*) **Translational enthalpy.** Enthalpy is given by

$$H = E + PV$$

For 1 mole of an ideal gas, $PV = RT$ so that

$$H = E + RT$$

For translational enthalpy, we can write

$$H_t = E_t + RT$$

Substituting the value of E_t from equation (*iv*), we get

$$H_t = \frac{3}{2}RT + RT$$

i.e.,

$$\boldsymbol{H_t = \frac{5}{2}RT} \qquad ...(v)$$

(*iii*) **Translational Heat Capacity at Constant Volume.** By definition

$$C_{v(t)} = \left(\frac{\partial E_t}{\partial T}\right)_v = \left[\frac{\partial\left(\frac{3}{2}RT\right)}{\partial T}\right]_v$$

i.e.,

$$\boldsymbol{C_{v(t)} = \frac{3}{2}R} \qquad ...(vi)$$

(*iv*) **Translational Heat Capacity at Constant Pressure.** By definition

$$C_{p(t)} = \left(\frac{\partial H_t}{\partial T}\right)_p = \left[\frac{\partial\left(\frac{5}{2}RT\right)}{\partial T}\right]_p$$

i.e.,

$$\boldsymbol{C_{p(t)} = \frac{5}{2}R} \qquad ...(vii)$$

(*v*)**Translational entropy.** The expression for entropy is

$$S - S_0 = RT\left(\frac{\partial \ln Q}{\partial T}\right)_v + R\ln Q - R\ln Q_0 \qquad \text{(See Q. 14)} \qquad ...(viii)$$

where S_0 is the entropy and Q_0 is the value of Q at $T = 0$.

For translational motion, it is found that

$$S_0 = R\ln Q_0 - R\ln N + R$$

Substituting this value in equation (*viii*), we get

$$S_t = RT\left(\frac{\partial \ln Q_t}{\partial T}\right)_v + R\ln\frac{Q_t}{N} + R \qquad ...(ix)$$

$$= R\left[T\left(\frac{\partial \ln Q_t}{\partial T}\right) + \ln Q_t - \ln N + 1\right]$$

Substituting the values of $\ln Q_t$ and $\left(\frac{\partial \ln Q_t}{\partial T}\right)_v$ from equations (*ii*) and (*iii*), we get

$$S_t = R\left[T\cdot\frac{3}{2}\left(\frac{1}{T}\right) + \ln\frac{(2\pi mk)^{3/2}}{h^3}V + \frac{3}{2}\ln T - \ln N + 1\right]$$

$$= R\left[\frac{3}{2} + \ln\left(\frac{(2\pi mk)}{h^2}\right)^{3/2}V + \frac{3}{2}\ln T - \ln N + 1\right]$$

$$= R\left[\frac{5}{2} + \ln\left(\frac{(2\pi mk)}{h^2}\right)^{3/2} V + \frac{3}{2}\ln T - \ln N\right]$$

$$= R\left[\frac{5}{2} + \ln\left(\frac{(2\pi mkT)}{h^2}\right)^{3/2} + \ln\frac{V}{N}\right]$$

or
$$\boldsymbol{S_t = R\left[\frac{5}{2} + \ln\left\{\left(\frac{(2\pi mkT)}{h^2}\right)^{3/2}\frac{V}{N}\right\}\right]} \quad ...(x)$$

This equation can be expressed in another form by putting $V = RT/P$ (as we are considering one mole of an ideal gas) so that

$$\frac{V}{N} = \frac{RT}{NP} = \frac{kT}{P}$$

where $R / N = k$ is Boltzmann's constant. Hence we have

or
$$\boldsymbol{S_t = R\left[\frac{5}{2} + \ln\left\{\left(\frac{(2\pi mkT)}{h^2}\right)^{3/2}\frac{kT}{P}\right\}\right]} \quad ...(xi)$$

Equations (x) and (xi), are known as Sackur-Tetrode equations.

The above equations are sometimes written in a number of convenient forms which may be obtained from the above equations as follows:

Putting mass of the molecule, $m = M / N$ (*i.e.*, Molecular mass of the gas/Avogadro's number and expanding, eqn. (x) can be written as

$$S_t = \left[\frac{5}{2} + \frac{3}{2}(\ln 2\pi + \ln M - \ln N + \ln k + \ln T - \ln h^2 + \ln V - \ln N)\right]$$

$$= R\left(\frac{5}{2} + \frac{3}{2}\ln 2\pi - \frac{5}{2}\ln N + \frac{3}{2}\ln k - 3\ln h\right) + \left(\frac{3}{2}\ln M + \frac{3}{2}\ln T + \ln V\right)$$

or
$$\boldsymbol{S_t = C_1 + R\left(\frac{3}{2}\ln M + \frac{3}{2}\ln T + \ln V\right)} \quad ...(xii)$$

where
$$C_1 = R\left(\frac{5}{2} + \frac{3}{2}\ln 2\pi - \frac{5}{2}\ln N + \frac{3}{2}\ln k - 3\ln h\right) = \text{constant}$$

$$= -11.073 \text{ cal degree}^{-1} = -46.329 \text{ JK}^{-1} \text{ mol}^{-1}$$

Similarly, eqn. (xi) on expanding takes up the form

$$\boldsymbol{S_t = C_2 + R\left(\frac{3}{2}\ln M + \frac{5}{2}\ln T - \ln P\right)} \quad ...(xiii)$$

where $C_2 = C_1 + R\ln R$ = constant (Remember $R = Nk$

$$= -2.315 \text{ cal degree}^{-1} = -9.686 \text{ JK}^{-1} \text{ mol}^{-1}.$$

Further if standard state conditions are taken, the value obtained is for standard entropy ($S°$). Under these conditions, putting $P = 101.32$ kPa and $T = 298.15$ K, the above equation is further simplified to the form

$$\boldsymbol{S_t^0 = 195 + \frac{3}{2}R\ln M \ (\text{JK}^{-1}\text{mol}^{-1})} \quad ...(xiv)$$

(vi) **Translational contribution to Helmholtz free energy.** By definition

$$A_t = E_t - TS_t$$

Substituting the values of E_t and S_t from equations (iv) and (xi), we get

$$A_t = \frac{3}{2}RT - RT\left[\frac{5}{2} + \ln\left\{\left(\frac{(2\pi mkT)}{h^2}\right)^{3/2}\frac{kT}{P}\right\}\right]$$

or
$$A_t = -RT\left[1 + \ln\left\{\left(\frac{(2\pi mkT)}{h^2}\right)^{3/2}\frac{kT}{P}\right\}\right] \qquad ...(xv)$$

(vii) **Translational contribution to Gibb's free energy.** By definition

$$G_t = H_t - TS_t.$$

Substituting the values of H_t and S_t from equation (v) and (xi), we get

$$G_t = \frac{5}{2}RT - RT\left[\frac{5}{2} + \ln\left\{\left(\frac{(2\pi mkT)}{h^2}\right)^{3/2}\frac{kT}{P}\right\}\right]$$

or
$$\boldsymbol{G_t = -RT\ln\left\{\left(\frac{(2\pi mkT)}{h^2}\right)^{3/2}\frac{kT}{P}\right\}} \qquad ...(xvi)$$

Q. 24. Determine the translational contribution to the kinetic energy of O_2 gas at 300 K.

Ans. Translational contribution to the kinetic energy of 1 mole of O_2 gas is given by

$$E_t = \frac{3}{2}RT$$

$$\mathbf{R = 1.987\ cal\,degree^{-1}\ mol^{-1}\ and\ T = 300\ K}$$

Equation can be expressed as

$$S^0 = \text{Enthalpy function} + \text{Free energy function}$$

To find the translational enthalpy function

$$H^0 = E_t + PV = E_t + RT \qquad \text{(for 1 mole of an ideal gas)}$$

$\therefore$
$$H^0 = \frac{3}{2}RT + RT = \frac{5}{2}RT$$

$$H_0^0 = \frac{5}{2} \times R \times 0 = 0$$

$\therefore$ Translational enthalpy function of an ideal gas

$$H_t = \left(\frac{H^0 - H_0^0}{T}\right)$$

or
$$H_t = \frac{H^0 - H_0^0}{T} = \frac{\frac{5}{2}RT - 0}{T} = \frac{5}{2}R$$

$$H_t = \frac{5}{2}R = \frac{5}{2} \times 1.987\ \text{cal mol}^{-1}$$

$\therefore$
$$H_t = 4.9675\ \text{cal mol}^{-1}$$

But $$H_t = \frac{5}{2} R = C_p \qquad ...(iv)$$

∴ Heat capacity C_p of an ideal gas = Translational enthalpy function.

To calculate translational free energy function we proceed as follows :

$$\text{Translational free energy function} = -\left(\frac{G^0 - H_0^0}{T}\right)$$

$$G_t = S_t^0 - \text{enthalpy function of gas}$$

Using the equation [Refer to Q. 23, eq. xiii and substituting the value of c_2 as –2.315 cal degree *M* stands for molecular mass.

$$S_t = 2.303R\left[\log \frac{M^{3/2}.T^{5/2}}{P} - 0.5058\right], \text{ we get}$$

$$G_t = 2.303 \times 1.987\left[\frac{3}{2}\log M + \frac{5}{2}\log T - \log P - 0.5058\right] - 4.9675$$

$$G_t = 4.576\left[\frac{3}{2}\log M + \frac{5}{2}\log T - \log P - 0.5058\right] - 4.9675$$

when P = 1, G_t beccomes G_t^0

$$\boxed{G_t^0 = 6.86 \log M + 11.44 \log T - 7.28}$$

Q. 25. Derive equation of state of an ideal gas from partition function.

Ans. The realtion between pressure 1 and molecular partition function Q for 1 mole of an ideal gas

$$P = RT\left(\frac{\partial \ln Q}{\partial V}\right)_T \qquad ...(i)$$

Only Q_t is volume dependent partition function. The other partition such as rotation, vibrational and electronic are independent of volume.

∴ Equation (*i*) can be written as

$$P = RT\left(\frac{\partial \ln Q_t}{\partial V}\right)_T \qquad ...(ii)$$

But $$Q_t = \left(\frac{2\pi m kT}{h^3}\right)^{3/2} - V$$

Putting in (*ii*) the value of Q_t

$$P = RT\left[\frac{\partial}{\partial V}\left[\ln \frac{(2\pi m kT)^{3/2}.V}{h^3}\right]\right]_T$$

$$= RT\left[\frac{\partial}{\partial V}\left[\ln \frac{(2\pi m kT)^{3/2}}{h^3} + \ln V\right]\right]_T$$

$$P = RT\left[0 + \frac{1}{V}\right] = \frac{RT}{V}$$

or $$\boxed{PV = RT} \qquad ...(iii)$$

which is an ideal gas equation for 1 more of an ideal

For n moles of an ideal gas, the equation of state become $PV = nRT$

Q. 26. Calculate the translational contribution to H°, S° and G° for oxygen gas at 298 K. Boltzmann constant $k = 1.381 \times 10^{-23}$ and $h = 6.626 \times 10^{-34}$.

Ans. Mass of the oxygen molecule is

$$m = \frac{\text{Mol. mass of } O_2}{\text{Avogadro's No.}} = \frac{32}{6.022 \times 10^{23}} \text{g} = 5.314 \times 10^{-23} \text{g}$$

$$= 5.314 \times 10^{-26} \text{ kg}$$

$$T = 298 \text{ K}$$

$$P = 10^5 \text{ Nm}^{-2}.$$

(*i*) $H_t^0 = \frac{5}{2} RT = \frac{5}{2} (8.314 \text{ JK}^{-1} \text{ mol}^{-1})(298 \text{ K})$

$= \mathbf{6.194 \text{ kJ mol}^{-1}}.$

(*ii*) $$S_t^0 = R\left[\frac{5}{2} + \ln\left\{\left(\frac{(2\pi mkT)}{h^2}\right)^{3/2} \frac{kT}{P}\right\}\right]$$

$$= R\left[\frac{5}{2} + \ln\left(\frac{(2\pi mkT)}{h^2}\right)^{3/2} + \ln\frac{kT}{P}\right]$$

$$= R\left[\frac{5}{2} + \frac{3}{2}\ln\left(\frac{(2\pi mkT)}{h^2}\right) + \ln\frac{kT}{P}\right]$$

$$= R\left[\frac{5}{2} + \frac{3}{2} \times 2.303 \log\frac{2\pi mkT}{h^2} + 2.303 \log\frac{kT}{P}\right]$$

$$= 8.314\left[\frac{5}{2} + \frac{3 \times 2.303}{2} \log\frac{2 \times 3.143 \times 5.314 \times 10^{-26} \times 1.381 \times 10^{-23} \times 298}{(6.626 \times 10^{-34})^2} + 2.303 \log\frac{1.381 \times 10^{-23} \times 298}{10^5}\right]$$

$$= 8.314\left[\frac{5}{2} + 3.4545 \log(3.131 \times 10^{21}) + 2.303 \log(4.115 \times 10^{-26})\right]$$

$$= 8.314\left[\frac{5}{2} + 3.4545 \times (21.4956) + 2.303 \times (-25.3857)\right]$$

$= 8.314\, [2.5 + 74.2565 - 58.4633]$

$= 8.314 \times 18.2932$

$= \mathbf{152.09 \text{ JK}^{-1} \text{ mol}^{-1}}.$

(*iii*) $G_t^0 = H_t^0 - TS_t^0$

$= 6.194 - 298 \times 0.15209$

$= \mathbf{-39.129 \text{ kJ mol}^{-1}}$

Q. 22. Derive expressions for rotational contribution to the following thermodynamic functions:

(*i*) **Energy**
(*ii*) **Enthalpy**
(*iii*) C_v
(*iv*) C_p
(*v*) **Entropy**
(*vi*) **Helmholtz free energy**
(*vii*) **Gibb's free energy.**

Ans. (*i*) **Rotational energy.** The rotational energy is given by the equation

$$E_r = RT^2\left(\frac{\partial \ln Q_r}{\partial T}\right)_V \qquad ...(i)$$

The rotational partition function (Q_r) is given by the equation

$$Q_r = \frac{8\pi^2 IkT}{\sigma h^2}$$

Taking logarithm of both sides

$$\ln Q_r = \ln\left(\frac{8\pi^2 Ik}{\sigma h^2}\right) + \ln T \qquad ...(ii)$$

$$\therefore \qquad \left(\frac{\partial \ln Q_r}{\partial T}\right)_V = \frac{1}{T} \qquad ...(iii)$$

Substituting this value in equation (*i*), we get

$$E_r = RT^2 \times \frac{1}{T}$$

or

$$\boxed{E_r = RT} \qquad ...(iv)$$

(*ii*) **Rotational enthalpy.** Equation for enthalpy is

$$H_r = RT^2\left(\frac{\partial \ln Q_r}{\partial T}\right)_P$$

But as derived above

$$\left(\frac{\partial \ln Q_r}{\partial T}\right)_P = \frac{1}{T}$$

$$\boxed{H_r = RT} \qquad ...(v)$$

(*iii*) **Rotational Heat Capacity at constant volume.** By definition

$$C_{v(r)} = \left(\frac{\partial E_r}{\partial T}\right)_V = \left(\frac{\partial (RT)}{\partial T}\right)_V$$

i.e.,

$$\boxed{C_{v(r)} = R} \qquad ...(vi)$$

(*iv*) **Rotational Heat Capacity at constant pressure.** By definition

$$C_{P(r)} = \left(\frac{\partial H_r}{\partial T}\right)_P = \left(\frac{\partial (RT)}{\partial T}\right)_P$$

i.e.,

$$\boxed{C_{P(r)} = R} \qquad ...(vii)$$

(*v*) **Rotational entropy.** Expression for entropy is

$$S - S_0 = RT\left(\frac{\partial \ln Q}{\partial T}\right)_v + R \ln Q - R \ln Q_0 \quad ...(viii)$$

where S_0 is the entropy and Q_0 is the value of Q at $T = 0$.

For rotational (as well as vibrational) motion, it is found that

$$S_0 = R \ln Q_0$$

Hence eqn. (*viii*) becomes

$$S_r = RT\left(\frac{\partial \ln Q_r}{\partial T}\right)_v + R \ln Q_r \quad ...(ix)$$

But
$$Q_r = \frac{8\pi^2 IkT}{\sigma h^2}$$

so that
$$\ln Q_r = \ln \frac{8\pi^2 IkT}{\sigma h^2}$$

$$= \ln \frac{8\pi^2 Ik}{\sigma h^2} + \ln T$$

and
$$\left(\frac{\partial \ln Q_r}{\partial T}\right)_v = \frac{1}{T}$$

Substituting these values in equation (*ix*), we get

$$S_r = RT \cdot \frac{1}{T} + R \ln \frac{8\pi^2 IkT}{\sigma h^2}$$

or
$$S_r = R\left[1 + \ln \frac{8\pi^2 IkT}{\sigma h^2}\right] \quad ...(x)$$

or
$$S_r = R \ln \frac{eT}{\sigma} \frac{8\pi^2 Ik}{h^2} \quad (\because 1 = \ln e)$$

or
$$\boxed{S_r = R \ln \frac{eT}{\sigma\alpha}} \quad ...(xi)$$

where $\alpha = h^2/8\pi^2 Tk$

(*vi*) **Rotational contribution to Helmoltz free energy .** By definition.

$$A_r = E_r - TS_r \quad ...(xii)$$

Substituting the values of E_r and S_r, A_r can be calculated.

(*vii*) **Rotational contribution to Gibb's free energy.** By definition

$$G_r = H_r - TS_r$$

Sbustituting the values of H_r and S_r, we get

$$G = RT - RT \ln \frac{eT}{\sigma\alpha}$$

$$= RT\left(1 - \ln \frac{eT}{\sigma\alpha}\right)$$

$$= -RT\left(\ln\frac{eT}{\sigma\alpha} - 1\right)$$

$$= -RT\left(\ln\frac{eT}{\sigma\alpha} - \ln e\right)$$

$$\boxed{G_r = -RT\ln\frac{eT}{\sigma\alpha}}$$

Q. 28. Find the rotational partition function for Cl_2 molecule at 25°C and 1 atmospheric pressure.

Given (*i*) Internuclear distance Cl — Cl = 1.988Å.

(*ii*) Reduced mass μ_{Cl_2} = 17.4894 a.m.u.

Ans. Step I. [*To calculate moment of inertia I*]

$$\mu_{Cl_2} = 17.4894 \text{ a.m.u.} = \frac{17.4894}{6.023\times10^{23}} g$$

$$r_{Cl_2} = 1.988 \text{ Å} = 1.988\times10^{-8} \text{ cm}$$

$$I = \mu r^2 = \left(\frac{17.4894}{6.023\times10^{23}} g\right)\times(1.988\times10^{-8}\text{ cm})^2$$

$$I = 1.1476\times10^{-38} \text{ g cm}^2$$

Step II. [*To calculate Q_r*]

$$Q_r = \frac{8\pi^2 I kT}{\sigma h^2}$$

$$\sigma = 2,\ T = 298,\ k = 1.38\times10^{-16} \text{ g cm}^2\text{ s}^{-2}\text{ K}^{-1}$$

$$h = 6.625\times10^{-27}\text{ g cm}^2\text{ s}^{-1}$$

$$Q_r = \frac{8\pi^2(1.1476\times10^{-38} g\text{ cm}^2)(1.38\times10^{-16} g\text{ cm}^2\text{ s}^{-2}\text{ K}^{-1})\,298\text{ K}}{2\times(6.625\times10^{-27} g\text{ cm}^2\text{s}^{-1})^2}$$

$$Q_r = \mathbf{424.84}$$

Q. 29. The rotational constant (B) of gaseous HCl, determined from microwave spectroscopy is 10.59 cm^{-1}, calculate the rotational partition function of HCl at (*i*) 100 K and (*ii*) 500 K.

Ans. Rotational constant B = 10.59 cm^{-1}

To convert this value to Joules, we multiply by *hc*

$$\therefore\quad B = (10.59\text{ cm}^{-1})(6.625\times10^{-34}\text{ Js})(3\times10^{10}\text{ cm s}^{-1})$$

$$\sigma = 1$$

(*i*) $$\therefore Q_r = \frac{kT}{\sigma B} = \frac{1.38\times10^{-23}\text{ JK}^{-1}\times100\text{K}}{1\times10.59\text{ cm}^{-1}\times6.625\times10^{-34}\text{ Js}\times3\times10^{10}\text{ cm s}^{-1}}$$

Q_r = **6.56** at 100 K

(*ii*) $$Q_r = \frac{1.38\times10^{-23}\text{ JK}^{-1}\times500\text{K}}{1\times10.59\text{ cm}^{-1}\times6.625\times10^{-34}\text{ Js}\times3\times10^{10}\text{ cm s}^{-1}}$$

Q. 30. Calculate the rotational contributions to H^o, S^o and G^o for oxygen gas at 298 K. The moment of inertia for O_2 (g), $I = 1.937 \times 10^{-46}$ kg m^2.

Ans. (*i*) $H_r^0 = RT = (8.314 \text{ JK}^{-1} \text{ mol}^{-1}) \times (298\text{K})$

$= \mathbf{2.477 \text{ kJ mol}^{-1}}$

(*ii*)
$$S_r^0 = R \ln \frac{eT}{\sigma \alpha}$$

$$\alpha = \frac{h^2}{8\pi^2 Ik}$$

$$= \frac{(6.626\times10^{-34} \text{ Js})^2}{8\times(3.143)^2\times(1.937\times10^{-46} \text{ kg m}^2)\times(1.384\times10^{-23} \text{ JK}^{-1})} = 2.507\text{K}$$

$e = 2.7183$

$\sigma = 2$ for O_2

$$\therefore \quad S_r^0 = 8.314\times2.303 \log \frac{2.718\times298}{2\times2.507}$$

$= 8.314 \times 2.303 \log 161.5$

$= \mathbf{42.28 \text{ kJ mol}^{-1}}$.

(*iii*)
$$G_r^0 = H_r^0 - TS_r^0$$

$= 2.477 - 298 \times 0.04228$

$= \mathbf{-10.12 \text{ kJ mol}^{-1}}$.

Q. 31. Calculate S_r for O_2 in a mole of O_2 at 298 K. Given I for O_2 is 1.937×10^{-46} kg m^2.

Ans. Rotational entropy is given by the equation

$$S_r = R\left(1 + \ln \frac{8\pi^2 I kT}{\sigma h^2}\right)$$

$\sigma = 2$ (for O_2); $I = 1.937 \times 10^{-46}$ kg m^2

$T = 298$ K; $k = 1.38 \times 10^{-23}$ kg m^2 s^{-2} K^{-1}

$h = 6.625 \times 10^{-34}$ kg m^2 s^{-1}

Substituting in (*i*) various values, we have

$$S_r = R\left[1 + \ln\left(\frac{8 \times \pi^2 \times 1.937 \times 10^{-46} \times 1.38 \times 10^{-23} \times 298}{2 \times (6.625 \times 10^{-34})^2}\right)\right]$$

$= 8.314 [1+4.2726] = 8.314\times5.2726 \text{ JK}^{-1} \text{ mol}^{-1}$ $\quad [1 EU = 1 JK^{-1}]$

$\therefore \quad S_r = 43.84$ EU mol^{-1}

Q. 32. Derive expressions for vibrational contribution to the following thermodynamic functions:

(*i*) Energy **(*ii*) Enthalpy**

(*iii*) C_V **(*iv*) C_p**

(*v*) Entropy **(*vi*) Helmholtz free energy**

(*vii*) Gibb's free energy.

Ans. (*i*) Vibrational energy. The vibrational energy is given by the following equation

$$E_v = RT^2\left(\frac{\partial \ln Q_v}{\partial T}\right)_V \quad ...(i)$$

The vibrational partition function is given by

$$Q_v = \frac{1}{1-e^{-h\nu_0/kT}}$$

where ν_0, is the fundamental frequency of vibration. ...(*ii*)

Putting $\frac{h\nu_0}{k} = \beta$, eqn. (*ii*) becomes

$$Q_v = \frac{1}{1-e^{-\beta/T}} \quad ...(iii)$$

Substituting this value in equation (*i*) and solving, we get

$$\boxed{E_v = RT\frac{\beta/T}{e^{\beta/T}-1}} \quad ...(iv)$$

where $\beta = \frac{h\nu_0}{k} = \frac{hc\bar{\nu}}{k}$ in which $\bar{\nu}$ is the frequency in wave number *i.e.* cm^{-1}.

(*ii*) **Vibrational Enthalpy.** It is found to be same *viz.*

$$\boxed{H_v = RT\frac{\beta/T}{e^{\beta/T}-1}} \quad ...(v)$$

(*iii*) **Vibrational Heat Capacity at constant volume.** By definition

$$C_{v(v)} = \left(\frac{\partial E_v}{\partial T}\right)_V$$

Using the value of E_v derived in equation (*iv*), we get

$$C_{v(v)} = R\left(\frac{\beta}{T}\right)^2 \frac{e^{\beta/T}}{(e^{\beta/T}-1)^2} \quad ...(vi)$$

(*iv*) **Vibrational Heat Capacity at constant pressure.** It is found to be same as $C_{v(v)}$ *viz.*

$$\boxed{C_{p(v)} = R\left(\frac{\beta}{T}\right)^2 \frac{e^{\beta/T}}{(e^{\beta/T}-1)^2}} \quad ...(vii)$$

(*v*) **Vibrational entropy.** Vibrational entropy is given by

$$S_v = RT\left(\frac{\partial \ln Q_v}{\partial T}\right)_v + R\ln Q_v \quad ...(viii)$$

Substituting $Q_v = (1-e^{-\beta/T})^{-1}$ and solving, we get

$$\boxed{S_v = R\left[\frac{\beta/T}{(e^{\beta/T}-1)} - \ln(1-e^{-\beta/T})\right]} \quad ..(ix)$$

(*vi*) **Vibrational contribution to Helmholtz free energy.** By definition

$$A_v = E_v - T S_v$$

Substituting the values of E_v and S_v, A_v can be calculated.

(*vii*) **Vibrational contribution to Gibb's free energy.** By definition

$$G_v = H_v - T S_v$$

Substituting the values of H_v and S_v and simplifying, we get

$$\boxed{G_v = RT \ln (1 - e^{-\beta/T})}$$

Q. 33. Fundamental vibrational frequency of O_2 at 27°C is s^{-1}, calculate Q_v for O_2. Assuming that O_2 molecule behaves as a simple harmonic oscillator.

Ans. The vibrational partition function is given by

$$Q_v = \frac{1}{1 - e^{-h\nu/kT}}$$

Here $\nu = 4.74 \times 10^{-13} \text{ s}^{-1}, h = 6.625 \times 10^{-27}$ ergs

$$k = 1.38 \times 10^{-16} \text{ erg K}^{-1}, \ T = 300 \text{ K}$$

$$\therefore \quad \frac{h\nu}{kT} = \frac{(6.625 \times 10^{-27}) \times (4.74 \times 10^{-10})}{(1.38 \times 10^{-16}) \times (300)}$$

or $$\frac{h\nu}{kT} = 7.585$$

Now $$Q_v = \frac{1}{1 - e^{-7.585}} = \frac{1}{1 - 0.000508}$$

$$= \frac{1}{0.99492} = 1.0005$$

Q. 34. For CO_2 the characteristic vibration temperature θ_v is 3084 K at 300 K. Calculate θ_v for CO_2.

Ans. Q_v in terms of θ_v is given by

$$Q_v = \frac{1}{1 - e^{-\theta_v/T}}$$

$$\theta_v = 3084\ k \ \& \ T = 300 \text{ K}$$

$$\therefore \quad Q_v = \frac{1}{1 - e^{-3084/300}} = \frac{1}{1 - 0.0000343} = \frac{1}{0.9999657}$$

$$Q_v = \mathbf{1.0000343}$$

Q. 35. Calculate the vibrational contribution to *H*°, *S*° and *G*° for oxygen gas at 298 K. The vibrational frequency is 1580.25 cm^{-1}.

Ans. (*i*) $$H_v^0 = RT \frac{\beta/T}{e^{\beta/T} - 1}$$

$$\beta = \frac{hc\bar{\nu}}{k} = \frac{(6.626 \times 10^{-34} \text{ Js}) (3.0 \times 10^{8} \text{ ms}^{-1}) (1.5802 \times 10^{5} \text{ m}^{-1})}{1.381 \times 10^{-23} \text{ JK}^{-1}}$$

$$= 2274.5 \text{ K}$$

$$\therefore \quad H_v^0 = (8.314\,\text{JK}^{-1}\text{mol}^{-1})\,(298\text{K})\left(-\frac{2274.5/298}{e^{2274.5/298}-1}\right)$$

$$= 8.314 \times 298 \times \frac{7.63}{e^{7.63}-1}$$

$$= 8.314 \times 298 \times \frac{7.63}{2041}$$

[To solve $e^{7.63}$, Put $x = e^{7.6}$ $\therefore$ $\ln x = 7.63$]

or $2.303 \log x = 7.63$ or $\log x = \dfrac{7.63}{2.303} = 3.31$ $\therefore$ $x = 2.042 \times 10^3 = 2042$]

$$= \mathbf{9.2J\ mol^{-1}}$$

(ii) $$S_v^0 = R\left[\frac{\beta/T}{(e^{\beta/T}-1)} - \ln(1-e^{-\beta/T})\right]$$

$$= 8.314\left[\left(\frac{7.63}{e^{7.63}-1}\right) - 2.303\log(1-e^{-7.63})\right]$$

$$= 8.314\left[\frac{7.63}{2041} - 2.303\log(1)\right]$$

$$= \mathbf{0.031JK^{-1}mol^{-1}}$$

(iii) $$G_r^0 = H_r^0 - TS_r^0 = 9.2 - 298 \times (0.031)$$

$$= \mathbf{-0.038\ Jmol^{-1}}$$

Q. 36. Calculate standard vibrational entropy of a mole of HCl gas, given x = 4.1 cm K.

Ans. We know that

$$S_v = \frac{Rx}{e^x - 1} - R\ln(1-e^{-x})$$

$$x = 4.1$$

$$e^x = e^{4.1} = 60.34$$

$$e^{-x} = e^{-4.1} = 0.01657,\ R = 1.987 \text{ cal deg}^{-1} \text{ mole}^{-1}$$

$$S_v = \frac{1.987 \times 4.1}{60.34 - 1} - 1.987 \ln(1 - 0.01657)$$

$$= \frac{8.1465}{59.34} - 1.987 \ln 0.9834$$

$$= 0.1374 - 1.987\,(-0.0167)$$

$$= 0.1373 + 0.03326$$

$$\therefore \quad S_v = \mathbf{0.17055\ eu\ mol^{-1}}$$

Q. 37. What is the electronic contribution to entropy?

Ans. The electronic partition function is given by the equation

$$Q_e = g_0 + g_1 e^{-\varepsilon_1/kT} + \ldots \qquad \ldots(i)$$

The electronic contribution to entropy is then given by

$$S_e^0 = R \ln Q_e$$
$$= R \ln (g_0 + g_1 e^{-\varepsilon_1/kT} + ...)$$

If the energies of the excited states are all large with respect to kT, only the degeneracy g_0 of the ground state has to be taken into account so that we have

$$\boxed{S_e^0 = R \ln g_0}$$

Q. 38. Calculate the electronic contribution to the entropy S^o for O_2 (g) at 298 K. Given that the electronic ground state is triplet and the energy of the first excited electronic state is so high that it has not to be considered.

Ans. Here $g_0 = 3$

$$\therefore \quad S_e^0 = 8.314 \ln 3$$
$$= 8.314 \times 2.303 \log 3$$
$$= \mathbf{9.135\ JK^{-1}\ mol^{-1}}.$$

Q. 39 The energy of the first three energy levels of F-atom determined spectroscopically are:

Term Symbol	Frequency in ω (cm^{-1})
$^2P_{3/2}$	0
$^2P_{1/2}$	404.0
$^2P_{5/2}$	102406.5

Calculate the electronic partition function of F-atom at 1500 K.

Ans. The degeneracies of the above electronic levels are calculated as follows :

For $\quad {}^2P_{3/2}.J = \frac{3}{2}$

$$= (2J+1) = \left(2 \times \frac{3}{2} + 1\right) = 4$$

For $\quad {}^2P_{1/2}.J = \frac{1}{2}$

$$g_e(1) = \left(2 \times \frac{5}{2} + 1\right) = 2$$

For $\quad {}^2P_{5/2}.J = \frac{5}{2}$

$$g_e(2) = \left(2 \times \frac{5}{2} + 1\right) = 6$$

The electronic partition function of F-atom is given by

$$Q_e = g_e(0) + g_e(1)e^{-hc\omega_1/kT} + g_e(2)e^{-hc\omega_2/kT} + ...$$

$$\omega_1 = 404\ cm^{-1}$$

$$\omega_2 = 102,406.5\ cm^{-1}$$

$$\frac{hc\omega_1}{kT} = \frac{1.439 \times 404}{1500} = 0.3876$$

$$\frac{hc\omega_2}{kT} = \frac{1.439 \times 102,406.5}{1500} = 98.2419$$

Now $$Q_e = g_e(0)+g_e(1)e^{-0.386} + g_e(2)\, e^{-98.2419}$$

$$= 4 + 2 \times 0.67868 + 6 \times 2.15816 \times 10^{-43}$$

$$= 4+1.35736+0$$

$$Q_e = 5.35736$$

Q. 40. For Tellurium vapours, there is low lying electronic state 0.94 eV above the ground state. Calculate Q_e at 10,000 K. If degeneracies of the ground is 2 and of the first excited state is 4.

Ans. $$\varepsilon_e(0) = 0\,;\ \varepsilon_e(1) = 0.94\ eV$$

As $$1\ eV = 1.602 \times 10^{-19}\ J$$

$\therefore$ $$0.94\ eV = 1.602 \times 10^{-19} \times 0.94 = 1.50588 \times 10^{-19}\ J$$

$$\varepsilon_e(1) = 1.50588 \times 10^{-10}\ J$$

$$g_e(0) = 2\,;\ g_e(1) = 4$$

$$\frac{\varepsilon_e(1)}{kT} = \frac{1.50588 \times 10^{-19}\ J}{1.38 \times 10^{-23}\ JK^{-1} \times 10{,}000\ K} = 1.0912$$

The electronic partition function Q_e is given by

$$Q_e = g_{e(0)} + g_{e(1)} e^{-\varepsilon_e(1)/kT}$$

$$= 2+4 \times e^{-1.0912}$$

$$= 2+4\,(0.3858)$$

$$Q_e = 2+1.5432 = 3.5432$$

$\therefore$ $$Q_e = 3.5432$$

Q. 41. In Si-atom at 25°C, the ground and two excited states have frequencies (ω) in terms cm^{-1} are respectively 10,77.15 and 223.31 cm^{-1}. The corresponding degeneracies are 1,3 and 5. Determine Q_e for Si-atom.

Ans. $$g_e(0) = 1,\ g_e(1) = 3\,;\ g_e(2) = 5$$

$$\varepsilon_e(0) = 10\,;\ \omega_1 = 77.15\,;\ \omega = 223.31$$

$$\frac{\varepsilon_e(1)}{kT} = \frac{hc\omega_1}{kT} = \frac{1.439 \times 77.15}{298} = 0.36725$$

$$\frac{\varepsilon_e(2)}{kT} = \frac{hc\omega_2}{kT} = \frac{1.439 \times 223.31}{298} = 1.07883$$

The electronic partition function Q_e is given by

$$Q_e = g_e(0) + g_e(1)\, e^{-0.3725} + g_e(2)^{-1.07883}$$

$$= 1+3\, e^{-0.3725} + 5e^{-1.07883}$$

$$Q_e = 1+2.067+1.69995$$

$\therefore$ $$Q_e = 4.76695$$

Q. 42. Write expressions for Bose-Einstein and Fermi-Dirac distribution laws. How does Maxwell-Boltzmann's law follow from these?

Ans. According to Bose-Einstein distribution law, the most probable distribution of N particles among the various levels is given by

$$n_i = \frac{g_i}{e^{\alpha+\beta\varepsilon_i} - 1} \qquad ...(i)$$

Here n_i is the number of molecules in the ith level with energy ε_i, g_i represents the number of eigenfunctions corresponding to the energy ε_i and is called the degeneracy or statistical weight. α and β are undetermined constants.

According to **Fermi-Dirac** distribution law, the most probable distribution is given by

$$n_i = \frac{g_i}{e^{\alpha+\beta\varepsilon_i} + 1} \qquad ...(ii)$$

It may be noted that in the eqns. (i) and (ii) if n_i's are small which can be so when the denominator is large *i.e.*, $e^{\alpha+\beta\varepsilon_i}$ is large as compared to 1, then 1 can be neglected in comparison to the exponential term, then these equations reduce to

$$n_i = \frac{g_i}{e^{\alpha + \beta\varepsilon_i}}$$

which is the same as Maxwell-Boltzmann's law.

Q. 43. Prove that the constant k_1 in the expression $S = k_1 \ln W$ is Boltzmann's constant, K.

Ans. Thermodynamic probability, W is given by the relation

$$W = \frac{N!}{n_1!n_2!....n_n!}$$

This gives the total number of arrangements (distributions) possible for N particles among n energy levels (taking the particles to the distinguishable).

However, as each energy level may correspond to a number of eigen states, it is essential to introduce the degeneracy or statistical weight corresponding to each energy level. Also as identical particles are indistinguishable, this factor should also be taken into consideration. With these two modifications, we have

$$W = \frac{g_0^{n_0}\, g_1^{n_1}\, g_2^{n_2} ....g_n^{n_n}}{n_0!n_1!n_2!n_n!} = \Pi \frac{g_i^{n_i}}{n_i!}$$

where n_0 is the number of particles in ground state energy level ε_0 with degeneracy g_0.

n_1 is the number of particles in the energy level ε_1 with degeneracy g_1 and so on.

$$\ln W = (n_0 \ln g_0 + n_1 \ln g_1 + ... n_n \ln g_0) - (\ln n_0! + \ln n_1! + ... \ln n_n!)$$

$$= \Sigma\, n_i \ln g_i - [(n_0 \ln n_0 - n_0) + (n_1 \ln n_1 - n_1) + ... (n_n \ln n_n - n_n)]$$

(by Stirling formula *viz.*, $\ln n_i = n_i \ln n_i - n_i$)

or

$$\ln W = \Sigma n_i \ln g_i - \Sigma\, n_i \ln n_i + (n_0 + n_1 + \;..... n_n)$$

$$= \Sigma n_i \ln g_i - \Sigma\, n_i \ln n_i + N \qquad ...(i)$$

But according to Boltzmann's distribution law

$$\frac{n_i}{N} = \frac{g_i e^{-\varepsilon_i/kT}}{\Sigma g_i e^{-\varepsilon_i/kT}}$$

Applying to it translational energy, we can write

$$\frac{n_i}{N} = \frac{g_i e^{-\varepsilon_i/kT}}{Q_t} \qquad \text{(where } Q_t = \text{translational partition function)}$$

or

$$n_i = \frac{N g_i e^{-\varepsilon_i/kT}}{Q_t}$$

Substituting this value in the second term of eqn. (i), we get

$$\ln W = \Sigma n_i \ln g_i - \Sigma n_i \ln\left[\left(\frac{N}{Q_t}\right) g_i e^{-\varepsilon_i/kT}\right] + N$$

$$= \Sigma n_i \ln g_i - \Sigma n_i \ln\left(\frac{N}{Q_t}\right) - \Sigma n_i \ln g_i + \Sigma \frac{n_i \varepsilon_i}{kT} + N$$

$$= -N \ln\left(\frac{N}{Q_t}\right) + \frac{1}{kT}\Sigma n_i \varepsilon_i + N$$

$$= N \ln\left(\frac{Q_t}{N}\right) + \frac{1}{kT}\Sigma n_i \varepsilon_i + N$$

putting $\Sigma n_i \varepsilon_i = E_t = RT^2\left(\frac{\partial \ln Q}{\partial T}\right)_V$, we get

$$\ln W = N \ln\left(\frac{Q_t}{N}\right) + \frac{RT}{k}\left(\frac{\partial \ln Q}{\partial T}\right)_V + N$$

Multiplying both sides by Boltzmann's constant, k, we get

$$k \ln W = Nk \ln\left(\frac{Q_t}{N}\right) + RT\left(\frac{\partial \ln Q}{\partial T}\right)_V + Nk$$

$$= R \ln\left(\frac{Q_t}{N}\right) + RT\left(\frac{\partial \ln Q}{\partial T}\right)_V + R \quad ...(ii)$$

But for the translation entropy,

$$S_t = R \ln\left(\frac{Q_t}{N}\right) + RT\left(\frac{\partial \ln Q}{\partial T}\right)_V + R \quad ...(iii)$$

Comparing equations (*ii*) and (*iii*), we have

$S_t = k \ln W$ (*i.e.*, $k_1 = k$)

or in general $S = k \ln W$

Q. 44. Derive the value of β used in statistical mechanics.

Ans. From Maxwell-Boltzmann distribution law,

$$n_i = \frac{g_i}{e^{\alpha+\beta\varepsilon_i}} = g_i e^{-\alpha-\beta\varepsilon_i} = g_i e^{-\alpha}.e^{-\beta\varepsilon_i} \quad ...(i)$$

$$\therefore \quad \Sigma n_i = \Sigma g_i e^{-\alpha} e^{-\beta\varepsilon_i}$$

or

$$N = e^{-\alpha} \Sigma g_i e^{-\beta\varepsilon_i} \qquad (\because \Sigma n_i = N) \quad ...(ii)$$

$$\therefore \quad e^{-\alpha} = \frac{N}{\Sigma g_i e^{-\beta\varepsilon_i}} = \frac{N}{Q} \quad ...(iii)$$

where $Q = \Sigma g_i e^{-\beta\varepsilon_i}$ is the molecular partition function.

Substituting the value of $e^{-\alpha}$ from eqn. (*iii*) into eqn. (*i*), we get

$$n_i = \frac{N g_i e^{-\beta\varepsilon_i}}{Q} \quad ...(iv)$$

We know that the number of possible distributions (thermodynamic probability, W) of N distinguishable particles is given by

$$W = N! \Pi \frac{g_i^{n_i}}{n_i!} \quad ...(v)$$

Taking logarithm of both the sides, we get

$$\ln W = \ln N! + \Sigma (n_i \ln g_i - \ln n_i!)$$

$$= N \ln N - N + \Sigma (n_i \ln g_i - n_i \ln n_i + n_i) \text{ (by applying Stirling's formula)}$$

$$= N \ln N + \Sigma n_i \ln g_i - \Sigma n_i \ln n_i \quad ...(vi)$$

Taking logarithm of both sides of eqn. (*iv*), we get

$$\ln n_i = \ln N + \ln g_i - \ln Q - \beta\varepsilon_i \quad ...(vii)$$

Substituting this value in eqn. (*vi*), we get

$$\ln W = N \ln N + \Sigma n_i \ln g_i - \Sigma n_i (\ln N + \ln g_i - \ln Q - \beta\varepsilon_i)$$

$$= N \ln N + \Sigma n_i \ln g_i - \Sigma n_i \ln N - \Sigma n_i \ln g_i + \Sigma n_i \ln Q + \beta \Sigma n_i \varepsilon_i$$

$$= N \ln N + \Sigma n_i \ln g_i - N \ln N - \Sigma n_i \ln g_i + N \ln Q + \beta E \quad (\text{where } E = \Sigma n_i \varepsilon_i)$$

$$= N \ln Q + \beta E \quad ...(viii)$$

Substituting this value in the Boltzmann's equation viz.,

$S = k \ln W$, we get

$$S = Nk \ln Q + k\beta E \quad ...(ix)$$

From first law of thermodynamics

$$q = dE + PdV \quad ...(x)$$

From second law of thermodynamics

$$dS = \frac{q}{T} \text{ i.e., } q = TdS \quad ...(xi)$$

Combining equations (*x*) and (*xi*), we get

$$dE = TdS - PdV \quad ...(xii)$$

At constant volume, $dV = 0$ so that

$$(dE)_V = (TdS)_V$$

or

$$\left(\frac{\partial S}{\partial E}\right)_V = \frac{1}{T} \quad ...(xiii)$$

Differentiating (*ix*) with respect to E at constant volume, we get

$$\left(\frac{\partial S}{\partial E}\right)_V = \frac{Nk}{Q}\left(\frac{\partial Q}{\partial E}\right)_V + k\beta + kE\left(\frac{\partial \beta}{\partial E}\right)_V$$

$$= \frac{Nk}{Q}\left(\frac{dQ}{d\beta}\right)\left(\frac{\partial \beta}{\partial E}\right)_V + k\beta + kE\left(\frac{\partial \beta}{\partial E}\right)_V \quad ...(xiv)$$

Now, as already mentioned, the molecular partition function is

$$Q = \Sigma g_i e^{-\beta\varepsilon_i}$$

$$\therefore \quad = g_1 e^{-\beta\varepsilon_1} + g_2 e^{-\beta\varepsilon_2} + ...$$

Differentiating w.r.t. β, we get

$$\frac{dQ}{d\beta} = g_1 e^{-\beta\varepsilon_1}(-\varepsilon_1) + g_2 e^{-\beta\varepsilon_2}(-\varepsilon_2) + ...$$

Taking $\varepsilon_1 = \varepsilon_2 .... = \bar{\varepsilon} = E/N$, we get

$$\frac{dQ}{d\beta} = -\frac{E}{N}(g_1 e^{-\beta\varepsilon_1} + g_2 e^{-\beta\varepsilon_2} + ...)$$

$$= -\frac{E}{N}\sum g_i e^{-\beta\varepsilon_i}$$

$$= -\frac{E}{N}Q \quad ...(xv)$$

Substituting this value in (*xiv*), the first and the last term cancel out. Hence we get

$$\left(\frac{\partial S}{\partial E}\right)_V = k\beta \quad ...(xvi)$$

Comparing with equation (*xiii*), we get

$$k\beta = \frac{1}{T}$$

or

$$\beta = \frac{1}{kT} \quad ...(xvii)$$

Q. 45. Why standard third law entropies are less than statistical entropies for CO and N_2O, whereas it is reverse for $(CH_3)_2$ Cd and CH_3 CCl_3 ?

Ans. Statistical and third laws standard entropies of the four substances are given as under in Table 14.1:

Table 141. Comparison of standard entropies calculated from statistical mechanics and third law

Substance	$S°(JK^{-1}\ mol^{-1})$	
	Statistical	*Third law*
CO	197.9	193.4
N_2O	219.9	215.2
$CH_2 - Cd - CH_3$	290.5	302.9
$CH_3 - CCl_3$	290.5	302.9

CO and N_2O have third law entropies lower than the statistical values.

The explanation for this is that at T= 0, the crystals of these substances are not perfect.

Hence the assumption that $S^0 = 0$ at 0 K is not correct. The explanation for higher third law entropies in case of $(CH_3)_2Cd$ and CH_3CCl_3 is that the rotational contribution towards entropy has been taken for the molecule as a whole whereas actually in such molecules, groups of atoms can rotate as a unit about other group to which they are joined by single bonds. *e.g.*, CH_3 around the bond linking it to Cd or CCl_3 around the bond linking it to CH_3.

Q. 46. What is residual entropy? How is it calculated?

Ans. *The entropy which the crystal of a substance has at T = 0, is called the residual entropy.* It can be calculated by applying Boltzmann's formula *viz.*, $S = k \ln W$ as follows :

Suppose a sample consists of N molecules and each molecule can have two orientations that are equally probable. Then the same energy can be achieved in 2^N different ways *i.e.*, $W = 2^N$. Hence

$$S = k \ln 2^N = kN \ln 2$$

If 1 mol of the sample is taken, kN =R. Then

$$S = R \ln 2 = 2.303\, R \log 2 = 5.85\ JK^{-1}\ mol^{-1}$$

Thus residual molar entropy = 5.85 $JK^{-1}\ mol^{-1}$

In general if a molecule can have p possible orientations with about the same energy then residual molar entropy will be

$$S_R = R \ln p$$

For example, $FClO_3$ molecule can have four possible orientations with approximately same nergy, hence its residual entropy will be $=R \ln 4 = 11.5\ JK^{-1}\ mol^{-}$

15

DISTRIBUTION LAW

Q. 1. State and explain Nernst distribution law.

(*Indore 2001; Kumaun 2002; Garhwal 2000; M. Dayanand 2002; Meerut 2002; Bhopal 2004; Bangalore 2004; Mumbai 2005; Purvanchal 2007*)

Ans. Nernst distribution law states as follows:

A solute distributes itself between two immiscible solvents in such a way that the ratio of the concentrations of the solute in the two solvents is constant at constant temperature, independent of any other molecular species present, provided that the solute has the same molecular state in both the solvents.

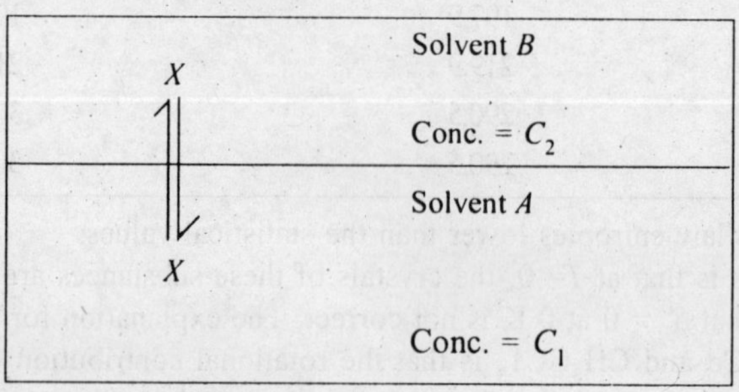

Fig. 15.1. Distribution of solute X between immiscible solvents A and B.

Thus if a solute X distributes itself between the two immiscible solvents A and B (Fig. 15.1), Then we have

$$\frac{\text{Concentration of } X \text{ in solvent } A}{\text{Concentration of } X \text{ in solvent } B} = \frac{C_1}{C_2} = K \text{ (constant)}$$

provided X remains in the same molecular species in both the solvents. The constant K is called *Distribution coefficient* or *Partition coefficient.*

Q. 2. What are the limitations of distribution law ?

(*Lucknow 2000; M. Dayanand 2000;Garhwal 2000; Kumaon 2000; Bangalore 2005*)

Ans. The distribution law is valid only if the following conditions are fulfilled:

(*i*) The temperature must remain constant throughout the experiment.

(*ii*) The two liquids (solvents) should be mutually immiscible.

(*iii*) The amount of the solute added should be small *i.e.*, the solution should be dilute. This is essential because in some cases, the solute added affects the mutual solubility of the solvents.

(*iv*) The solute must have the same molecular state in both the solvents *i.e.*, it should not undergo association or dissociation in any of the solvents. Thus the law is not valid for the distribution of benzoic acid between water and benzene, because it has been observed that in benzene, benzoic acid undergoes association to form dimers *i.e.*, it exists as $(C_6H_5COOH)_2$. Similarly, the law is also not valid for the distribution of an electrolyte between water and a nonpolar solvent, because the electrolytes undergo dissociation in water.

Q. 3. Succinic acid was shaken with a mixture of water and ether. The concentrations of the acid in the two layers are as follows (per 10 cm^3 of solution). Find out the partition coefficient.

In water layer :	**0.0244**	**0.071**	**0.121**
In ether layer :	**0.0046**	**0.013**	**0.022**

If succinic acid has normal molecular weight of 118 in water, find its mol. wt. in ether.

Ans. Partition coefficient of succinic acid between water and ether is given by the ratio

$$K = \frac{\text{conc. in water } (C_1)}{\text{conc. in ether } (C_2)}$$

Thus we have :

From 1st set : $K = \dfrac{0.0244}{0.0046} = 5.30$

From 2nd set : $K = \dfrac{0.071}{0.013} = 5.47$

From 3rd set : $K = \dfrac{0.121}{0.022} = 5.50$

$\therefore$ Mean value of Partition Coeffi., $K = \dfrac{5.30 + 5.47 + 5.50}{3} = 5.42$

Further, since the value of $K = C_1/C_2$ is almost constant, therefore, the succinic acid must have the same molecular weight in both the solvents.

As mol. wt. of succinic acid in water = 118 (Given), therefore, mol. wt. of succinic acid in ether = **118**.

Q. 4. When varying amounts of iodine were shaken with CCl_4 – water mixture, the following concentrations of iodine (in g/100 cm^3) were obtained:

CCl_4 layer	**5.1**	**10.2**	**15.2**	**20.3**
Water layer	**0.06**	**0.119**	**0.178**	**0.236**

Show that these results illustrate the distribution law.

Ans. Here we have :

Conc. in CCl_4 layer (C_1)	*Conc. in H_2O layer* (C_2)	$C_1/C_2 = K$
5.1	0.06	5.1/0.06 = 85.0
10.2	0.119	10.2/0.119 = 85.7
15.2	0.178	15.2/0.178 = 85.4
20.3	0.236	20.3/0.236 = 86.0

Since the value of K comes out to be almost constant, hence the given data illustrate the distribution law.

Q. 5. An aqueous solution of succinic acid at 15°C containing 0.07 g in 10 cm^3 is in equilibrium with an ethereal solution containing 0.013 g in 10 cm^3 Succinic acid has its normal molecular weight in both the solvents. What will be the concentration of an ethereal solution which is in equilibrium with an aqueous solution containing 0.024 g in 10 cm^3 ?

Ans. It is given that

Conc. of succinic acid in water

$$(C_1) = 0.07 \text{ g in } 10 \text{ cm}^3$$

$$= \frac{0.07}{10} \text{ g / cm}^3$$

Conc. of succinic acid in ether

$$(C_2) = 0.013 \text{ g in } 10 \text{ cm}^3$$

$$= \frac{0.013}{10} \text{ g / cm}^3$$

As succinic acid has the same molecular weight in both the solvents, therefore, distribution law is applicable, *i.e.*,

$$K = \frac{C_1}{C_2}$$

$$= \frac{0.07/10}{0.013/10} = 5.38$$

In the second experiment, it is given that

Conc. of succinic acid in water

$$(C_1) = 0.024 \text{ g in } 10 \text{ cm}^3$$

$$= \frac{0.024}{10} \text{ g / cm}^3$$

We want to calculate conc. of succinic acid in ether, *i.e.*, C_2.

According to the distribution law

$$\frac{C_1}{C_2} = K$$

Substituting the various values, we get

$$\frac{0.024/10}{C_2} = 5.38$$

$$C_2 = \frac{0.024}{10} \times \frac{1}{5.38} = 0.00045$$

Thus, Conc. of succinic acid in ether layer

$$= \mathbf{0.00045 \text{ g / cm}^3}$$

Q. 6. If the distribution coefficient of benzoic acid between water and benzene is 0.304 at 20°C, calculate the number of moles of benzoic acid which may be extracted from 100 cm^3 of 0.2 molar aqueous solution by 10 cm^3 of benzene.

Ans. 1000 cm^3 of aqueous solution contains benzoic acid = 0.2 mole

$\therefore$ 100 cm^3 of aqueous solution contains benzoic acid = 0.02 mole

Suppose the amount of benzoic acid extracted into benzene = x mole

$\therefore$ Amount of benzoic acid left in aqueous solution = $(0.02 - x)$ mole

$\therefore$ Conc. of benzoic acid in aqueous layer $= \frac{(0.02 - x)}{100}$ mole /cm^3

and conc. of benzoic acid in benzene layer $= \frac{x}{10}$ mole/cm^3

Hence $\dfrac{(0.02-x)/100}{x/10} = 0.34$ which gives $x = 0.005$.

Q. 7. Derive distribution law from thermodynamic considerations.

(Arunachal 2003; Baroda 2004; Himachal 2010)

Ans. The thermodynamic derivation of the distribution law is based upon the principle that *if there are two phases in equilibrium (in this case two immiscible solvents containing the same solute dissolved in them), the chemical potential of a substance must be same in both the phases.*

From thermodynamics, we know that the chemical potential (μ) of a substance is a solution given by

$$\mu = \mu^\circ + RT \ln a$$

where μ° is the standard chemical potential and 'a' is activity of the substance (solute) in the solution.

Thus for the solute in liquid A, we have

$$\mu_A = \mu_A{}^\circ + RT \ln a_A \quad \text{.... } (i)$$

Similarly for the solute in liquid B we have

$$\mu_B = \mu_B{}^\circ + RT \ln a_B \quad \text{.... } (ii)$$

Since the liquids A and B are in equilibrium,

$$\mu_A = \mu_B \quad \text{.... } (iii)$$

$$\therefore \quad \mu_A{}^\circ + RT \ln a_A = \mu_B{}^\circ + RT \ln a_B$$

or
$$RT \ln a_A - RT \ln a_B = \mu_B{}^\circ - \mu_A{}^\circ$$

or
$$\ln \frac{a_A}{a_B} = \frac{\mu_B{}^\circ - \mu_A{}^\circ}{RT} \quad \text{... } (iv)$$

Further at a given temperature, $\mu_A{}^\circ$ and $\mu_B{}^\circ$ are constant for given substance in the particular solvents. Hence at constant temperature, we have from equation (*iv*)

$$\ln \frac{a_A}{a_B} = \text{constant} \quad \text{... } (v)$$

and therefore,
$$\frac{a_A}{a_B} = \text{constant} \quad \text{... } (vi)$$

This is the exact expression for the distribution law. However, if the solutions are dilute, the activities are equal to the concentrations so that the expression (*vi*) is modified to

$$\frac{C_A}{C_B} = \text{constant}$$

which is the original form of the distribution law.

Q. 8. How is distribution law modified when the solute undergoes association in one of the solvents. *M. Dayanand 2002; Baroda 2004; Himachal 2010)*

Ans. Suppose the solute, X remains in the normal molecular state in the phase I but n molecules of it associate to form complex molecules X_n in phase II. Then the following two equilibria will exist simultaneously, as shown in the Fig. 15.2.

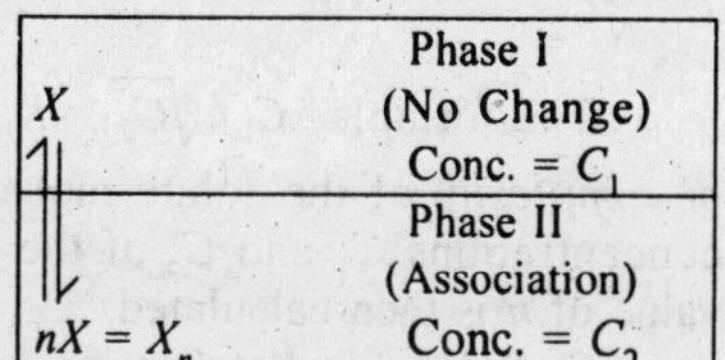

Fig. 15.2. Association in Phase II.

(*i*) $\underset{\text{Phase I}}{X} \rightleftharpoons \underset{\text{Phase II}}{X}$

(*ii*) $\underset{C_2(1-\alpha)}{nX} \rightleftharpoons \underset{\frac{1}{n}C_2\alpha}{X_n}$ in the same phase II.

Suppose the total conc. of the solute X in phase $I = C_1$ and the total conc. of solute X in phase $II - C_2$

Also suppose α is the degree of association of the solute in phase II.

Then conc. of the associated molecules in phase II, *i.e.* $[X_n] = \frac{1}{n} C_2 \alpha$

and conc. of the unassociated (normal) molecules in phase $II = (1 - \alpha)\, C_2$.

Applying the distribution law to equilibrium (*i*), we get

$$\frac{\text{Conc. of solute } X \text{ in phase I}}{\text{Conc. of normal moles of } X \text{ in phase II}} = \text{constant}$$

We have to take into account concentration of identical molecular species in the two phases.

i.e., $$\frac{C_1}{(1-\alpha)C_2} = \text{constant} \qquad \text{... (i)}$$

Again applying the law of chemical equilibrium to the equilibrium (*ii*), we get

$$\frac{\text{Conc. of associated molecules in phase II}}{(\text{Conc. of normal molecules in phase II})^n} = \text{constant}$$

i.e., $$\frac{\frac{1}{n}C_2\alpha}{[(1-\alpha)C_2]^n} = \text{constant}$$

or $$\frac{C_2\alpha}{[(1-\alpha)C_2]^n} = \text{constant} \qquad (\because n \text{ is constant})$$

Taking the *n*th root, we get

$$\frac{\sqrt[n]{C_2\alpha}}{(1-\alpha)C_2} = \text{constant} \qquad \text{...(ii)}$$

Dividing eqn. (*i*) by eqn. (*ii*), we get

$$\frac{C_1}{\sqrt[n]{C_2.\alpha}} = \text{constant} \quad \text{.... (Exact formula)}$$

When the molecules in the phase II are almost completely associated, $\alpha = 1$ and, therefore,

$$\frac{C_1}{\sqrt[n]{C_2}} = \text{constant} \quad \text{.....(Approx formula)}$$

Thus in such cases when the solute undergoes association in one of the solvents, the constant value for the partition coefficient is obtained not for the ratio C_1/C_2 but for the ratio $C_1/\sqrt[n]{C_2}$.

The formula, $C_1/\sqrt[n]{C_2} = K$, is commonly used for finding the value of *n*, *i.e.*, *the extent of complexity* of the solute molecules in one of the solvents. This is done by finding out the concentrations C_1 and C_2 of the solute in the two solvents in a number of experiments. The value of *n* is then calculated.

Q. 9. How is distribution law modified when the solute undergoes dissociation in one of the solvents? ***(Indore 2001; Bangalore 2005; Himachal 2010)***

Ans. Suppose the solute X dissolves in the phase I without any change in the molecular state but undergoes dissociation into *Y*, *Z* etc., in the phase II. Then the following two equilibria will be set up as shown in Fig. 15.3.

(*i*) $\underset{\text{Phase I}}{X} \rightleftharpoons \underset{\text{Phase II}}{X}$

(*ii*) $X \rightleftharpoons Y + Z$ in the same phase II

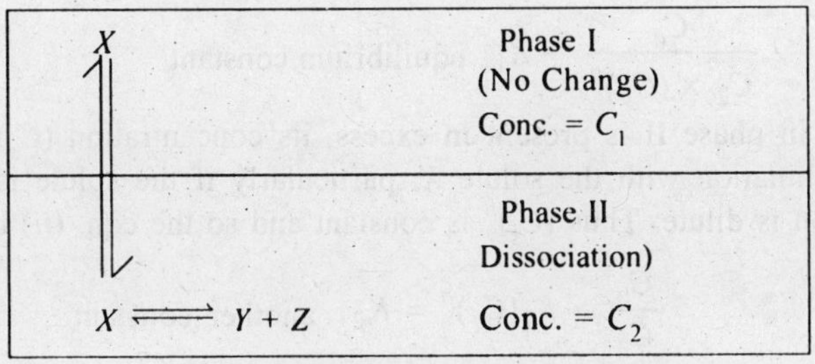

Fig. 15.3. Dissociation in Phase II.

Suppose total conc.of solute *X* in phase I $= C_1$

and total conc. of solute *X* in phase II $= C_2$.

Also, suppose the degree of dissociation of the solute in phase II is α.

Then, conc. of undissociated (simple) molecules in phase II

$$= (1 - \alpha)\, C_2$$

Applying the distribution law to equilibrium (*i*), we get

$$\frac{C_1}{(1-\alpha)C_2} = K \text{ (Partition coefficient)}$$

Note. It is obvious that if the solute undergoes dissociation in phase I (with degree of dissociation $= \alpha$) but remains normal in phase II, then the formula obtained will be

$$\frac{C_1(1-\alpha)}{C_2} = K$$

From the above expression, the degree of dissociation, α, can be calculated.

Q. 10. Derive the modified form of distribution law when the solute combines with one of the solvents.

Ans. Suppose the solute *X* remains in the normal molecular state in phase I but one molecule of it combines with *n* molecules of the solvent *S* to form the compound molecules, *X.nS* in phase II. Then, the following two equilibria will be established as shown in Fig. 15.4.

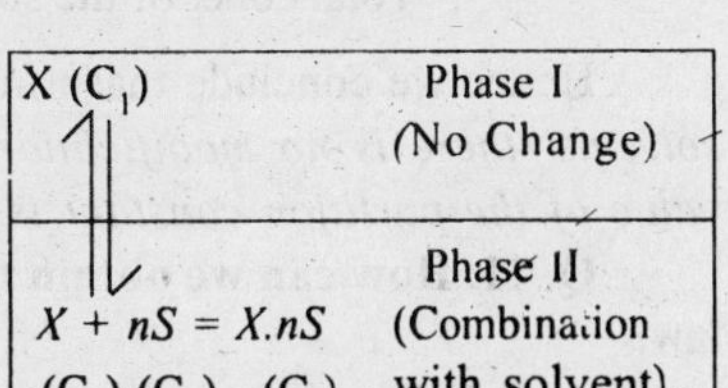

Fig. 15.4. Combination with solvent in Phase II.

(*i*) $\underset{\text{Phase I}}{X} \rightleftharpoons \underset{\text{Phase II}}{X}$

(*ii*) $X + nS \rightleftharpoons X.nS$

in the same phase II.

Suppose, conc. of solute *X* (normal molecules) in phase $= C_1$

conc. of the normal molecules in phase II $= C_2$

conc. of the solvent molecules in phase II $= C_s$

and conc. of the compound molecules in phase II $= C_c$

Applying the distribution law to equilibrium (*i*), we get

$$\frac{C_1}{C_2} = K, \text{ the distribution coefficient} \quad ...(i)$$

Again, applying the law of chemical equilibrium to equilibrium (*ii*), we get

$$\frac{C_c}{C_2 \times (C_s)^n} = K_1 \text{ equilibrium constant} \quad ...(ii)$$

Since the solvent in phase II is present in excess, its concentration (C_s) remains almost constant inspite of combination with the solute *X*, particularly if the solute is added in small quantity, *i.e.*, the solution is dilute. Thus $(C_s)^n$ is constant and so the eqn. (*ii*) becomes

$$\frac{C_c}{C_2} = K_1(C_s)^n = K_2, \text{ another constant}$$

Adding 1 to both the sides, we get

$$1 + \frac{C_c}{C_2} = 1 + K_2$$

or

$$\frac{C_2 + C_c}{C_2} = 1 + K_2 \quad ...(iii)$$

Dividing eqn. (*i*) by eqn. (*iii*), we get

$$\frac{C_1}{C_2} \times \frac{C_2}{C_2 + C_c} = \frac{K}{1 + K_2}$$

or

$$\frac{C_1}{C_2 + C_c} = \frac{K}{1 + K_2} = K_3, \text{ another constant.}$$

But $C_2 + C_c$ is the total conc. of the normal as well as the solvated molecules of the solute *X* in phase II. Thus the above equation, in fact, implies that

$$\frac{\text{Total conc. of the solute } X \text{ in phase I}}{\text{Total conc. of the solute } X \text{ in phase II}} = \text{constant}$$

Hence we conclude that *when the solute enters into chemical combination with one of the solvents, there is no modification of the distribution law equation except that the numerical value of the partition constant is changed.*

Q. 11. How can we obtain the solubility of a solute in a given solvent using distribution law? **(*M. Dayanand 2002*)**

Ans. When the two immiscible solvents are in contact with each other and the solute is added in such a large excess that saturated solutions are formed, then the concentrations of the solutions at this stage will be equal to their solubilities. Assuming that the distribution law is applicable for such concentration, we will have

$$\frac{C_1}{C_2} = \frac{S_1}{S_2} = K$$

where conc. $C_1 = S_1$ is the solubility of the solute in solvent 1.

and conc. $C_2 = S_2$ is the solubility of the solute in solvent 2.

Thus knowing the distribution coefficient *K* and the solubility of the solute in one of the solvents, that in the other can be obtained.

Q. 12. At 15°C an aqueous solution of oxalic acid containing 5.0 g of oxalic acid per 100 cm^3 of water in equilibrium with an ethereal solution containing 0.50 g per 100 cm^3. The

solubility of oxalic acid in water at 15°C is 10 g per 100 cm³. Calculate the solubility in water.

Ans. Here, we are given :

Conc. of oxalic acid in water (C_1) = 5.0 g in 100 cm³
= 50.0 g/litre

Conc. of oxalic acid in ether (C_2) = 0.50 g in 100 cm³
= 5.0 g/litre

∴ Distribution coefficient of oxalic acid between water and ether will be

$$K = \frac{C_1}{C_2} = \frac{50}{5.0} = 10.0$$

When the solutions are saturated, the ratio of concentrations is equal to the ratio of solubilities

i.e.,
$$K = \frac{\text{Solubility in water } (S_1)}{\text{Solubility in ether } (S_2)}$$

But $S_1 = 10 \text{ g}/100 \text{ cm}^3$ (given)

∴
$$10 = \frac{10}{S_2}$$

or
$$S_2 = 1.0 \text{ g}$$

i.e Solubility of oxalic acid in ether = **1.0 g/100 cm³.**

Q. 13. How can we determine the extent of association or dissociation of a solute in a solvent using distribution law? *(M. Dayanand 2002; Panjab 2003)*

Ans. When a solute undergoes association in one of the solvents ($nX \rightleftharpoons X_n$) in which its total concentration is C_2 and has the normal molecular state in the second solvent in which its concentration is C_1, then distribution law formula is modified to

$$\frac{C_1}{\sqrt[n]{C_2}} = K, \text{ the distribution coefficient.}$$

From this formula, the value of n can be calculated by either of the two methods given below.

(1) **Hit and Trial method.** The value of C_1 and C_2 obtained in different experiments are put in the equation

$$\frac{C_1}{\sqrt[n]{C_2}} = K.$$

Then putting n = 2, 3, 4 etc., one by one, the values of K are calculated in each case. The number which gives constant value for K is the correct value for n.

(2) **Logarithmic method.** The formula for finding n is

$$\frac{C_1}{\sqrt[n]{C_2}} = K.$$

or
$$\frac{C_1}{C_2^{1/n}} = K$$

or
$$C_1 = (C_2)^{1/n}.K$$

Taking logarithm of both the sides, we get

$$\log C_1 = \frac{1}{n}\log C_2 + \log K$$

For two sets of values: C_1, C_2 and C_1', C_2', we have

$$\log\ C_1 = \frac{1}{n}\log C_2 + \log K \qquad \text{... }(i)$$

and

$$\log C_1' = \frac{1}{n}\log C_2' + \log K \qquad \text{...}(ii)$$

Subtracting eqn. (*ii*) from eqn. (*i*), we get

$$\log C_1 - \log C_1' = \frac{1}{n}(\log C_2 - \log C_2')$$

or

$$n = \frac{\log C_2 - \log C_2'}{\log C_1 - \log C_1'}$$

Hence n can be calculated.

Similarly, when a solute undergoes dissociation in one of the solvents, ($X \rightleftharpoons Y + Z$) in which its total concentration is C_2 and has the same normal molecular state in the second solvent, in which its concentration is C_1, then we have

$$\frac{C_1}{C_2(1-a)} = K, \text{ the distribution coefficient.}$$

Q. 14. In the distribution of benzoic acid between water and benzene, the following results were obtained :

C_1 (in water)	**1.50**	**1.95**	**2.97**
C_2 (in benzene)	**24.20**	**41.2**	**97.0**

Assuming that benzoic acid exists as single molecules in water, show that it exists as double molecules in benzene.

Ans. If the benzoic acid exists as single molecules in water and as double molecules in benzene, then the ratio $C_1/\sqrt{C_2}$ should be constant. This can be seen as follows :

C_1 (in water)	C_2 (in benzene)	$\frac{C_1}{\sqrt{C_2}} = K$
1. 5	24.20	$\frac{1.5}{\sqrt{24.20}} = 0.305$
1.95	41.2	$\frac{1.95}{\sqrt{41.2}} = 0.304$
2.97	97.0	$\frac{2.97}{\sqrt{97.0}} = 0.302$

As the ratio $C_1/\sqrt{C_2}$ comes out to be almost constant, this proves that benzoic acid exists as double molecules in benzene.

Q. 15. In the distribution of a solute between water (C_1) and chloroform (C_2), the following results were obtained :

C_1	**C_2**
0.163	**0.761**
0.436	**5.43**

What information do you gather regarding the molecular state of the solute in chloroform?

Ans. If the solute exists as simple molecules in water as well as in chloroform then the ratio $C_1/ C_2 = K$ should be constant. Now, from the given data:

For 1st set : $$\frac{C_1}{C_2} = \frac{0.163}{0.761} = 0.212$$

For 2nd set : $$\frac{C_1}{C_2} = \frac{0.436}{5.43} = 0.080$$

Thus we observe that the ratio C_1/C_2 is not constant and hence the distribution law in its simple from is not applicable. In other words, in the distribution of the solute between water and chloroform, the molecular state of the solute is not the same.

Suppose the solute is associated as double molecules in chloroform, then the ratio $C_1/\sqrt{C_2}$ should be constant.

For 1st set : $$K = \frac{C_1}{\sqrt{C_2}} = \frac{0.163}{\sqrt{0.761}} = 0.1869$$

For 2nd set : $$K = \frac{C_1}{\sqrt{C_2}} = \frac{0.436}{\sqrt{5.43}} = 0.1871$$

Thus as the ratio $C_1/\sqrt{C_2}$ comes out to be constant, this proves that the solute exists as *double* molecules in chloroform.

Q. 16. Succinic acid shows normal molecular weight in water and is associated in benzene. The following data was obtained on the distribution of succinic acid between water and benzene.

C_1 in water (g / litre)	**1.50**	**1.95**	**2.89**
C_2 in benzene (g / litre)	**24.20**	**41.20**	**96.50**

Calculate the molecular weight of succinic acid in benzene.

Ans. *By hit and trial method.* Suppose succinic acid exists as double molecules in benzene, then the ratio $C_1/\sqrt{C_2}$ should be constant.

This can be seen as follows:

C_1 (in water)	C_2 (in benzene)	$\frac{C_1}{\sqrt{C_2}} = K$
1.50	24.20	$\frac{1.50}{\sqrt{24.20}} = 0.305$
1.95	41.20	$\frac{1.95}{\sqrt{41.20}} = 0.304$
2.89	96.50	$\frac{2.89}{\sqrt{96.50}} = 0.303$

As $C_1/\sqrt{C_2} = K$ comes out to be almost constant, this proves that $n = 2$.

(*ii*) *Logarithmic method.* According to this method, the value of n is given by

$$n = \frac{\log C_2 - \log C_2'}{\log C_1 - \log C_1'}$$

From the given data, we have :

S. No.	C_1	$\log C_1$	C_2	$\log C_2$
(1)	1.50	0.176	24.2	1.384
(2)	1.95	0.290	41.2	1.616
(3)	2.89	0.461	96.3	1.985

From (1) and (2),

$$n = \frac{1.616 - 1.384}{0.290 - 0.176} = 2.04$$

From (2) and (3),

$$n = \frac{1.985 - 1.616}{0.461 - 0.290} = 2.16$$

From (3) and (1),

$$n = \frac{1.985 - 1.384}{0.461 - 0.176} = 2.11$$

Thus exact value of $n = 2$

i.e. **succinic acid exists as** $\begin{bmatrix} CH_2COOH \\ | \\ CH_2COOH \end{bmatrix}_2$ **in benzene. Hence mol. wt. of succinic acid in benzene $= 2 \times 118 = 236$.**

Q. 17. Describe how distribution can be used to study chemical equilibria involving formation of complex ions. *(Kumaon 2000)*

Or

Determine the equilibrium constant of the following reaction using distribution law

$$KI + I_2 \rightleftharpoons KI_3$$

(M. Dayanand 2002)

Ans. Consider the following reaction

$$KI + I_2 \rightleftharpoons KI_3 \quad \text{or} \quad I^- + I_2 \rightleftharpoons I_3^-$$

Distribution law helps us to determine the equilibrium constant for such equilibrium reactions and to establish the formula of complex ion on the basis of the constancy in the value of the equilibrium constant. The study of this equilibrium involves the following steps:

(*i*) A solution of iodine in CS_2 is shaken with water to find out the distribution coefficient of iodine between water and CS_2.

(*ii*) A solution of iodine in CS_2, of known concentration, is shaken with KI solution, also of known concentration (See Fig. 15.5)

(*iii*) After the distribution of iodine is complete between the two layers, the concentration of iodine left in the CS_2 layer is determined (by titration against standard hypo solution).

Calculations are done in the following manner.

Suppose that

Initial conc. of iodine in CS_2 = a moles/litre

Initial conc. of KI solution = b moles/litre

Conc. of I_2 left in CS_2 layer after distribution = c moles/litre

Distribution coefficient of I_2 between water and CS_2 = K

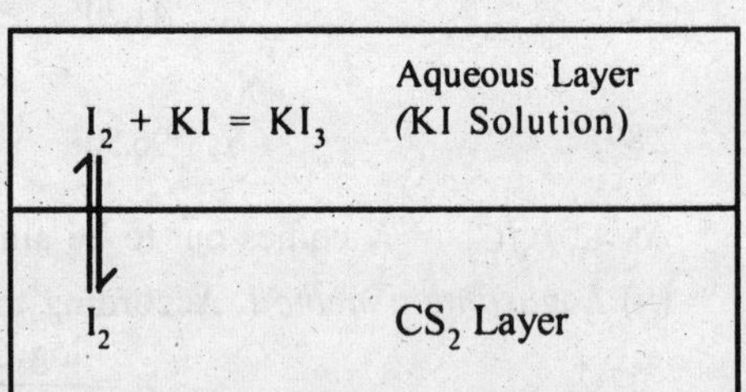

Fig. 15.5 Study of the formation of the complex KI_3.

As iodine reacts with KI, the total concentration of iodine (free + combined) in the aqueous layer is much higher (than if no KI were present). Obviously,

Total conc. of I_2 (free + combined) in aqueous layer

$= \text{Initial conc. of } I_2 \text{ in } CS_2 - \text{conc. of } I_2 \text{ left in } CS_2 \text{ layer}$

$= a - c$ moles/litre

Conc. of free I_2 in aqueous layer can be calculated by applying distribution law *i.e.*,

$$\frac{\text{Conc. of free } I_2 \text{ in aqueous layer}}{\text{Conc. of } I_2 \text{ in } CS_2 \text{ layer}} = K, \text{ distribution coeff.}$$

But conc. of I_2 in CS_2 layer $= c$ moles/litre

$\therefore$ Conc. of *free* I_2 in aqueous layer $= Kc$ mole/litre

$\therefore$ Amount of I_2 that has *combined* with KI

$= (a - c) - Kc$ moles/litre

$= d$ moles/litre (say)

Now, according to the reaction

$$KI + I_2 \rightleftharpoons KI_3$$

No. of moles of I_2 combined = No. of moles of KI combined

= No. of moles of KI_3 formed

$\therefore$ Conc. of KI_3 in the aqueous layer $= d$ moles/litre

Conc. of KI in the aqueous layer $= (b - d)$ moles/litre

Conc. of I_2 in the aqueous layer $= Kc$ moles/litre

Substituting these values in the equilibrium law expression *viz.*,

$$K' = \frac{[KI_3]}{[KI][I_2]}$$

the equilibrium constant K' can be calculated.

Starting with different initial concentrations of iodine and KI solutions, the value of K' is determined in each case. A fair constancy in the value of K' confirms the formula of the complex (KI_3).

Q. 18. The partition coefficient of iodine between CS_2 and H_2O is 410. A solution of KI containing 8 g of the salt per litre was taken with CS_2 until equilibrium was attained. It was found that at equilibrium, the aqueous layer contained 2.15 g of iodine per litre and the CS_2 layer 35.42 g iodine per litre. Calculate the equilibrium constant for the reaction.

$$KI + I_2 \rightleftharpoons KI_3$$

Given that the concentration of iodine in the aqueous layer is the sum of free and combined iodine.

Ans. The equilibrium constant for the given reaction will be

$$K = \frac{[KI_3]}{[KI][I_2]} \qquad \text{...}(i)$$

In CS_2 layer, the whole iodine is present as free iodine. Concentration of *free iodine* in the aqueous layer can be calculated as follows:

$$\frac{\text{Conc. of free iodine in } CS_2 \text{ layer } (C_1)}{\text{Conc. of iodine in aqueous layer } (C_2)} = \text{K (Partition coeff.)}$$

$$\therefore \qquad \frac{35.42}{C_2} = 410$$

or $$C_2 = \frac{35.42}{410} \text{ g /litre}$$

$$= \frac{35.42}{410 \times 254} \text{ moles/litre}$$

$$= 0.0003401 \text{ moles/litre}$$

or $$[I_2] = 0.0003401 \text{ M}$$

Total iodine in the aqueous layer

$$= 2.15 \text{ g/litre}$$

$$= \frac{2.15}{254} \text{ moles/litre}$$

$$= 0.008465 \text{ moles/litre}$$

∴ Combined iodine in the aqueous layer

$$= 0.008465 - 0.0003401$$

$$= 0.008125 \text{ moles/litre}$$

But $$[KI_3] = [\text{combined } I_2]$$

$$[KI_3] = 0.008125 \text{ M}$$

Molar conc. of $$KI = \frac{8}{166} \text{ moles/litre}$$

$$= 0.04819 \text{ moles/litre}$$

KI used in forming KI_3 = 0.008125 moles/litre

∴ Cone of free KI *i.e.*,

$$[KI] = 0.04819 - 0.008125$$

$$= 0.04007 \text{ moles/litre}$$

Substituting the values in eqn. (*i*), we get

$$K = \frac{[KI_3]}{[KI][I_2]} = \frac{0.008125}{0.04007 \times 0.0003401}$$

$$= \mathbf{596.2}$$

Q. 19. Justify the statement : In the process of extraction of a solute from a solution, the extracting liquid should not be used in one lot but in a number of smaller lots.

(M. Dayanand 2000)

Ans. Suppose X g of the substance are present in 1 litre of the aqueous solution. Also suppose 1 litre of ether is available as the extracting solvent and that the distribution coefficient of the substance between ether and water is 2. Now if whole of the ether is used in one lot and x g be the amount of the substance extracted into ether, then we will have

$$\frac{\text{Conc. in ether}}{\text{Conc. in } H_2O} = \frac{x/1000}{(X-x)/1000} = 2$$

or $$x = \frac{2}{3} X = 0.67 X$$

Thus only 67% of the total amount will be extracted in this case.

Now, suppose the whole of the ether is used in two instalments, each of 500 cm^3. After shaking with first instalment of 500 cm^3, suppose x_1 g of the substance passes into the ether layer. Then clearly,

$$\frac{x_1/500}{(X-x_1)/1000} = 2$$

or $$x_1 = \frac{1}{2}X = 0.5\ X$$

Thus 50% of the total amount is extracted out in this operation. Therefore, the amount of the substance left unextracted = $X - \frac{1}{2}X = \frac{1}{2}X$. Now one litre of the aqueous solution containing $\frac{1}{2}X$ g of the substance is shaken with the remaining 500 cm^3 of ether. Suppose the amount extracted into ether by this operation is x_2 g. Then,

$$\frac{x_2/500}{\left(\frac{X}{2}-x_2\right)/1000} = 2$$

or $$x_2 = \frac{1}{4}X = 0.25\ X$$

i.e. 25% of the total substance passes into the ether layer in this operation. Therefore, total amount of the substance extracted = 50 + 25 = 75%.

Thus when 1 litre of ether is used in two equal instalments, the percentage extraction increases from 67 to 75. Similarly, it can be seen that 1 litre of ether is used in four equal instalments, the extraction can be increased to 93.75%.

Q. 20. Derive a general formula for the amount of substance left unextracted and the volume of extracting liquid used each time. **(*M. Dayanand 2002*)**

Ans. Suppose W g of the substance is present in V ml of the solution made in solvent A. Also suppose that v ml of the extracting solvent B is used each time and K is the distribution coefficient of the substance between the solvents A and B.

Now if w_1 g is the amount of the substance left *unextracted* in solvent A after the first operation, then

$$\frac{\text{Conc. in } A}{\text{Conc. in } B} = K,\ i.e., \frac{w_1/V}{(W-w_1)v} = K$$

or $$K = \frac{(W-w_1)}{v} = \frac{w_1}{V}$$

or $$KV(W-w_1) = w_1 v$$

or $$w_1(v+KV) = KVW$$

or $$w_1 = \frac{KV}{(v+KV)}W \qquad \text{... (i)}$$

Again, after the second operation, if w_2 g be the amount of the substance left unextracted in solvent A so that the amount extracted is ($w_1 - w_2$) g, then

$$\frac{w_2/V}{(w_1-w_2)v} = K$$

or $$w_2 = \left(\frac{KV}{v+KV}\right)w_1 \qquad \text{... (ii)}$$

Putting the value of w_1 from eqn. (*i*), we get

$$w_2 = \left(\frac{KV}{v+KV}\right)\left(\frac{KV}{v+KV}\right)W$$

$$= \left(\frac{KV}{v+KV}\right)^2 W$$

In general, after nth operation, the amount of the substance left unextracted will be

$$w_n = \left(\frac{KV}{v+KV}\right)^n W \quad ...(iii)$$

In order that the extraction may be most efficient, clearly w_n should have the smallest possible value. This can be so only if the factor $\left(\frac{KV}{v+KV}\right)^n$ is as small as possible. Now in this factor, the denominator, $v + KV$ is greater than the numerator, KV. Therefore, $KV/(v + KV)$ is less than 1. Hence the above factor will be smallest possible when n has the largest possible value. So it follows that the *multi-step extraction is more economical than the single-step extraction.*

Note for the Students. For solving numerical problems, it may be mentioned here that K is the distribution coefficient of the substance between the solvent A already present and the extracting solvent B. If K is given for the distribution of the substance between solvents B and A, the reciprocal of this value should be used in the above equation.

Q. 21. How is the principle of extraction used in Parke's process for desilverisation of lead?

Ans. Silver is more soluble in zinc than in lead. Zinc is immiscible with molten lead. Thus when zinc is added to argentiferous lead, a large quantity of silver goes into the molten zinc layer leaving little silver in the molten lead layer, because the distribution coefficient of silver between zinc and lead is 300 at 880°C. Zinc containing silver forms the upper layer. Thus by repeating the process, almost total quantity of silver can be extracted from lead.

Q. 22. 0.83 g succinic acid was shaken up with 100 ml each of water and ether. The water layer was found to contain 0.70 g of succinic acid, the rest having passed into the ether layer. Calculate the quantity of succinic acid which can be extracted from 1000 ml of ether solution containing 1 g of the acid, using 100 ml of water in one lot and in two equal fractions.

Ans. Here total amount of succinic acid taken

$= 0.83$ g

Amount present in 100 ml of water layer

$= 0.70$ g

$\therefore$ Amount gone into 100 ml ether layer

$= 0.83 - 0.70 = 0.13$ g

As water is the extracting solvent, therefore, we shall take

$$K = \frac{\text{Conc. in ether}}{\text{Conc. in water}} = \frac{0.13/100}{0.70/100} = 0.19.$$

(*a*) *When 100 ml of water is used in one lot :*

$v = 100$ ml, $n = 1$

Also $W = 1$ g, $V = 1000$ ml, $K = 0.19$

Using the general equation for extraction

$\therefore$ Amount left unextracted will be

$$w_n = \left(\frac{KV}{v+KV}\right)^n \times W = \left(\frac{0.19\times1000}{100+0.19\times1000}\right)^1 \times 1 = \frac{190}{290} = 0.655 \text{ g}$$

$\therefore$ Amount extracted 1 – 0.655 = **0.345 g.**

(*b*) *When 100 ml of water is used in two equal fractions:*

v = 50 ml, n = 2

Also W = 1 g, V= 1000 ml, K= 0.19

$\therefore$ Amount left unextracted will be

$$w_n = \left(\frac{KV}{v+KV}\right)^2 \times W = \left(\frac{0.19\times1000}{50+0.19\times1000}\right)^2 \times 1 = \left(\frac{190}{240}\right)^2 = 0.627 \text{ g}$$

$\therefore$ Amount extracted = 1 – 0.627 = **0.373 g.**

Q. 23. The distribution coefficient of a substance between chloroform and water is 5. Show that the amount of substance extracted from 100 ml of an aqueous solution is more by using 50ml chloroform at a time in two instalments rather than using 100 ml of in one lot.

Ans. Here, it is given that :

$$K = \frac{\text{Conc. in Chloroform } (C_1)}{\text{Conc. in Water } (C_2)} = \frac{5}{1}$$

Suppose the total amount of the substance to be extracted is W g, present in 100 ml of aqueous solution.

(*a*) *When 100 ml of chloroform is used in one lot:*

Suppose the amount of substance extracted into chloroform

= w g

$\therefore$ Amount of the substance left unextracted in water

= $(W - w)$ g

i.e., conc. of the substance in chloroform = $w/100$ g per ml and Conc. of the substance in water = $(W - w)/100$ g per ml.

$\therefore$ By distribution law

$$\frac{w/100}{(W-w)/100} = 5$$

or $$w = 5\,(W - w)$$

or $$w = \frac{5}{6}W = 0.833\,W \text{ g}$$

i.e., the amount extracted will be 0.833 W g

(*b*) *When 50 ml of chloroform is used at a time:*

(*i*) *In the first extraction,* suppose the amount of the substance extracted into chloroform

= w_1 g

$\therefore$ Amount of the substance left unextracted in water

= $(W - w_1)$ g

i.e., conc. of the substance in chloroform layer = $\frac{w_1}{50}$ g /ml

and conc. of the substance in water layer = $\frac{(W - w_1)}{100}$ g/ml

$\therefore$ By distribution law

$$\frac{w_1/50}{(W-w_1)/100} = 5$$

or $$\frac{w_1}{50} \times \frac{100}{W-w_1} = 5$$

or $$2w_1 = 5\,(W - w_1)$$

or $$w_1 = \frac{5}{7}\,W = 0.714\,W\text{ g}$$

∴ Amount of the substance left unextracted in 100 ml of the aqueous solution

$$= W - 0.714\,W = 0.286\,W\text{g}$$

(*ii*) *In the second extraction, suppose the amount of the substance extracted into chloroform* $= w_2$ g.

Then amount of substance left unextracted in water

$$= (0.286\,W - w_2)\text{ g}$$

i.e., conc. of substance in chloroform $= w_2/50$ g per ml and conc. of substance in water $= (0.286\,W - w_2)/100$ g per ml

Applying distribution law, we get

$$\frac{w_2/50}{(0.286W-w_2)/100} = 5$$

or $$\frac{w_2}{50} \times \frac{100}{0.286W-w_2} = 5$$

or $$2w_2 = 5\,(0.286\,W - w_2)$$

or $$w_2 = 0.204\,W\text{ g}$$

∴ Total amount of the substance extracted

$$= 0.714\,W + 0.204\,W = 0.918\,W$$

Since 0.918 Wg is more than 0.833 W g, the required result is proved.

Q. 24. At 25°C the distribution ratio of H_2S between water (C_1) and benzene (C_2) is 0.17.

(*i*) What is the minimum volume of C_6H_6 necessary at 25°C to extract in one step 90% of H_2S present in one litre of 0.1 molar aqueous solution?

(*ii*) What total volume of benzene will be required if the 90% extraction were to be completed in three steps of equal volumes of benzene?

(*iii*) Using 100 ml of benzene every time, how many extractions will be required to complete 90% extraction?

Ans. Suppose volume of benzene required = V litre

As extraction is 90%

∴ Amount left unextracted = 10%

$$\frac{w_n}{W} = \frac{10}{100} = 0.1$$

Now
$$n = 1$$
$$V = 1\text{ litre}$$
$$K = 0.17$$

Putting these values in the formula

$$\frac{w_n}{W} = \left(\frac{KV}{v+KV}\right)^n$$

$$0.1 = \left(\frac{0.17\times1}{v+0.17\times1}\right)^1$$

which gives $v = 1.53$ litres

(*ii*) Suppose the total volume of benzene required = $3v$ litre

Then clearly $v = v$ litre, $n = 3$

Further, as the extraction = 90%

$\therefore$ Amount left unextracted = 10%

i.e.
$$\frac{w_n}{W} = \frac{10}{100} = 0.1$$

Also $V = 1$ litre and $K = 0.17$

Putting these values in the formula

$$\frac{w_n}{W} = \left(\frac{KV}{v+KV}\right)^n \text{, we get}$$

$$0.1 = \left(\frac{0.17 \times 1}{v + 0.17 \times 1}\right)^3$$

or
$$(v + 0.17)^3 = \frac{(0.17)^3}{0.1}$$

or
$$(v + 0.17)^3 = 10 \times (0.17)^3$$

Taking cube roots of both the sides, we get

$$v + 0.17 = (10)^{1/3} \times 0.17$$

Taking logarithm of both the sides, we get

$$\log(v + 0.17) = \frac{1}{3}\log 10 + \log 0.17$$

$$= \frac{1}{3} + \bar{1}.2304$$

$$= 0.3333 + \bar{1}.2304$$

$$= \bar{1}.5637$$

$$v + 0.17 = \text{antilog}\,(\bar{1}.5637) = 0.3662$$

$$\therefore \quad v = 0.3662 - 0.17 = 0.1962 \text{ litres}$$

Total volume required $= 3v = 3 \times 0.1962$ litres

= 0.5886 litres.

(*iii*) Here, $\frac{w_n}{W} = \frac{1}{10} = 0.1$ (as before)

$v = 100$ ml$= 0.1$ litre, $V = 1$ litre, $K = 0.17$, $n = ?$

Substituting these values in the formula

$$\frac{w_n}{W} = \left(\frac{KV}{v+KV}\right)^n \text{ we get}$$

$$0.1 = \left(\frac{0.17'1}{0.1 + 0.17 \times 1}\right)^n$$

$$0.1 = \left(\frac{0.17}{0.27}\right)^n$$

Taking logarithm of both the sides, we get

$$n \log \frac{0.17}{0.27} = \log 0.1$$

or $\quad n(\log 17 - \log 27) = \log 1/10$

or $\quad n(0.2304 - 0.4314) = -1$

or $\quad n(-0.2010) = -1$

or $$n = \frac{1}{0.201} = 5$$

Thus 5 extractions are needed to accomplish the work.

Q. 25. How will you determine the degree of hydrolysis of aniline hydrochloride applying the distribution law? *(M. Dayanand 2000)*

Ans. The following steps are involved:

(*i*) **Determination of the hydrolysis constant.** Aniline hydrochloride is hydrolysed as

$$C_6H_5NH_3Cl + H_2O \rightleftharpoons C_6H_5\overset{+}{N}H_3\overset{-}{O}H + HCl$$
$$\downarrow$$
$$C_6H_5NH_2 + H_2O$$

The hydrolysis constant will, therefore, be given by

$$K_h = \frac{[\text{Free aniline}][\text{Free HCl acid}]}{[\text{Unhydrolysed salt}]}$$

Thus to get the value of K_h, we need to find the molar concentrations of free aniline, free HCl acid and the unhydrolysed salt in an aqueous solution of aniline hydrochloride. This is done by shaking aqueous solution of aniline hydrochloride of known concentration with benzene. As a result, the free aniline distributes itself between the two solvents *i.e.*, water and benzene, whereas HCl acid and the unhydrolysed salt remain in the aqueous layer only as shown in Fig. 15.6.

(*a*) *Calculation of concentration of free aniline in the aqueous layer.* For this purpose, the concentration of aniline present in the benzene layer is found by taking a known volume of this layer, evaporating off the benzene and finding weight of aniline left. Having known the concentration of aniline in the benzene layer, that in the aqueous layer can be calculated if the distribution coefficient of aniline between water and benzene is known which can be found by a separate experiment.

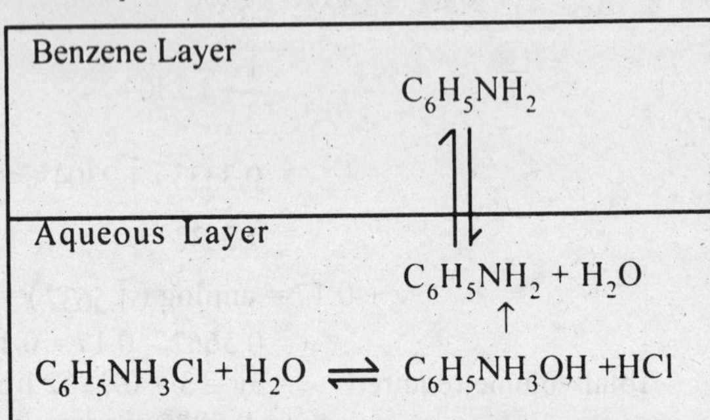

Fig. 15.6. Shaking an aqeueous solution of aniline hydrochloride with benzene.

(*b*) *Calculation of concentration of free HCl produced* From the hydrolysis equation,

Number of moles of HCl produced = Total number of moles of aniline produced.

But total amount of aniline produced = Amount of aniline present in the benzene layer + Amount of aniline present in the aqueous layer.

Each of these amounts has already been calculated in step (*a*) above. Hence the total number of moles of aniline produced and the free HCl produced can be calculated. Knowing the volume of the aqueous layer the concentration of free HCl in the aqueous layer can be calculated.

(*c*) *Calculation of the concentration of the unhydrolysed salt.* From the hydrolysis equation, it is clear that the number of moles of the salt hydrolysed is equal to total number of moles of the aniline produced or free HCl produced, which has already been calculated in step (*b*) above. Hence knowing the initial number of moles of the salt taken, the number of moles of the salt left unhydrolysed can be calculated. Further knowing the volume of the aqueous layer, the concentration of the unhydrolysed salt can be calculated.

(*ii*) **Calculation of the degree of hydrolysis.** Knowing the hydrolysis constant K_h, the degree of hydrolysis of a solution of aniline hydrochloride having concentration c moles/litre can be calculated directly from the expression

$$h = \sqrt{\frac{K_h}{c}}$$

This was a case of salt of weak base and strong acid. When other salts are being considered *viz.*, salts of strong base and weak acid or weak base and weak acid, the relevant equation between hydrolysis constant and degree of hydrolysis may be used.

Q. 26. In an experiment to determine the hydrolysis constant of aniline hydrochloride at 20°C, 15 g of the salt were shaken with 1000 cm³ of water and 100 cm³ of benzene until equilibrium was established. 50 cm³ of benzene layer was then found to contain 0.2360 g of aniline. The partition coefficient of aniline between benzene and water at 20°C is found to be 11.3. Calculate the hydrolysis constant and the degree of hydrolysis of aniline hydrochloride.

Ans. Aniline hydrochloride is hydrolysed as

$$C_6H_5NH_3Cl + H_2O \rightleftharpoons C_6H_5\,NH_3\,OH + HCl$$
$$\downarrow$$
$$C_6H_5NH_2 + H_2O$$

Hence $$K_h = \frac{[\text{Free aniline}][\text{Free HCl acid}]}{[\text{Unhydrolysed salt}]}$$

Thus to find K_h, we need find the molar concentrations of free aniline, free HCl and hydrolysed salt in the aqueous solution.

(*i*) *To calculate the concentration of free aniline in the aqueous layer i.e., [Free Aniline]*

It is given that 50 cm³ of benzene layer contains 0.2360 g of aniline *i.e.*, concentration of aniline present in the benzene layer

$$= 0.2360 \text{ g of aniline in 50 cm}^3 \text{ of benzene}$$

$$= \frac{0.2360}{93} \times \frac{1000}{50} \text{ moles/litre}$$

($\because$ Mol mass of aniline, $C_6H_5NH_2 = 93$)

$$= 0.05075 \text{ moles/litre}$$

As partition coefficient of aniline between benzene and water is given to be 11.3, this means

$$\frac{\text{Conc. in the benzene layer}}{\text{Conc. in the aqueous layer}} = 11.3$$

But conc in benzene layer = 0.05075 moles/litre

$$\therefore \text{ Conc. in the aqueous layer} = \frac{0.05075}{11.3} \text{ moles/litre}$$

$$= 0.00449 \text{ moles/litre}$$

$\therefore$ Amount of aniline present in 1000 cm³ of the aqueous layer

= 0.00449 moles

As the volume of the aqueous layer is 1000 cm³, therefore,

[Free Aniline] = 0.00449 moles/litre.

(*ii*) *To calculate the concentration of free HCl acid produced i.e., [Free HCl]*

From the hydrolysis equation

No. of moles of HCl produced = No. of moles of aniline produced

But total amount of aniline produced

= Amount of aniline present in the benzene layer

+ Amount of aniline present in the aqueous layer

Conc. of aniline in the benzene layer

= 0.05075 moles/litre

∴ Amount of aniline present in 100 cm^3 of benzene layer

= 0.005075 moles.

Also conc. of aniline in aqueous layer

= 0.00449 moles/litre

∴ Amount of aniline present in 1000 cm^3 of aqueous layer

= 0.00449 moles

∴ Total amount of aniline produced

= 0.005075 + 0.00449

= 0.009565 moles

Thus no of moles of HCl produced

= 0.009565 moles

As volume of the aqueous layer is 1000 cm^3.

[Free HCl] = 0.009565 moles/litre

(iii) To calculate the concentration of the unhydrolysed salt i.e., [unhydrolysed salt]

From the hydrolysis equation

No. of moles of aniline hydrochloride hydrolysed

= No. of moles of aniline or HCl produced

= 0.009565 moles

As aniline hydrochloride taken was 15 g and molecular mass of aniline hydrochloride ($C_6H_5NH_3Cl$) is 129.5, Therefore,

No. of moles of aniline hydrochloride taken

$$= \frac{15}{129.5} \text{ moles} = 0.11583 \text{ moles}$$

∴ No. of moles of salt left unhydrolysed

= 0.11583 – 0.009565 = 0.10626 moles

As the volume of the aqueous layer is 1000 cm^3.

[Unhydrolysed salt] = 0.10626 moles/litre

Hence $$K_h = \frac{[\text{Free Aniline}]\,[\text{Free HCl acid}]}{[\text{Unhydrolysed salt}]}$$

$$= \frac{0.0049 \times 0.009565}{0.10626}$$

$$= \mathbf{4.04 \times 10^{-4}}$$

As aniline is a salt of weak base and strong acid, the degree of hydrolysis will be

$$h = \sqrt{\frac{K_h}{c}}$$

where c is the original concentration of the salt.

But c = 0.11583 moles/litre (calculated above)

$$h = \sqrt{\frac{4.04\times10^{-4}}{0.11583}} = \mathbf{5.9\times10^{-2}}$$

Q. 27. Mention some applications of distribution law.

Ans. Some applications of distribution law are given below:

(*i*) To determine the solubility of a solute in a solvent.

(*ii*) To determine the extent of association or dissociation of a solute in a solvent.

(*iii*) To study chemical equilibrium of reactions involving complex ions.

(*iv*) In the process of extraction.

(*v*) In the Parke's process of desilverisation of lead.

(*vi*) To determine hydrolysis constant and degree of hydrolysis.

Q. 28. Describe in detail the separation of a mixture of small amounts of organic using liquid-liquid chromatography.

Ans. This is a versatile technique for separation of a mixture of small amounts of organic substances. The separation is carried in a glass tube packed with an inert solid material (silica) soaked in a solvent 1. The mixture solution is dropped at the column top. Then another immiscible solvent 2 is allowed to flow down the column (Fig. 15.7). A film of solvent 1 held by the silica forms the stationary liquid phase, while the running solvent 2 the mobile liquid phase. Each component of the mixture distributes itself between the stationary liquid phase and the mobile liquid phase.

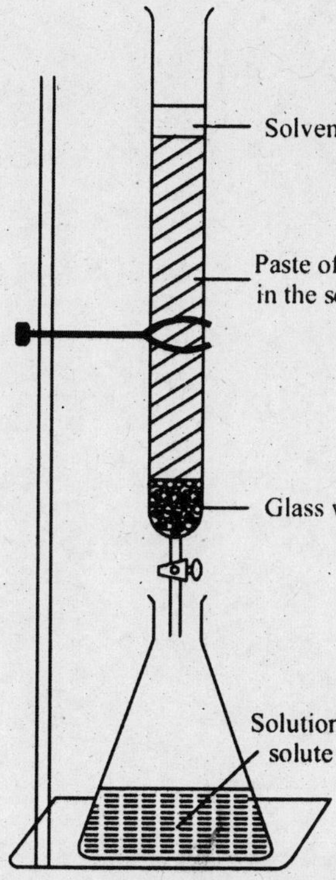

Fig. 15.7. Set-up of the apparatus for liquid-liquid chromatography

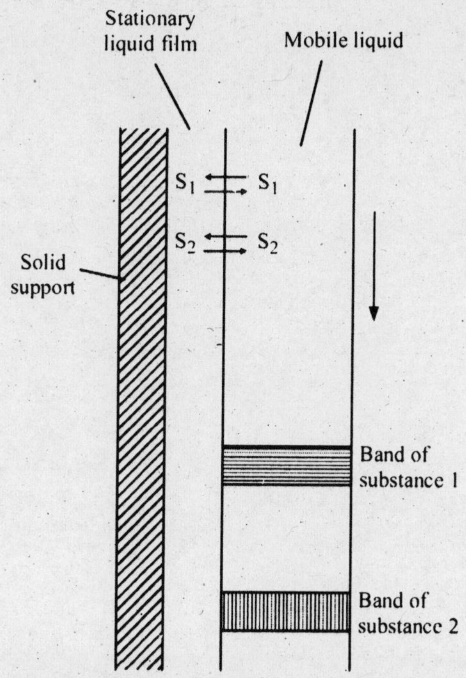

Fig. 15.8. Separation of substances 1 and 2 (S_1 and S_2) in the form of bands

The liquid film on the solid support acts as the **stationary** state while the second liquid which is immiscible with the first liquid forms the mobile liquid phase. The solute which is more soluble in solvent 2 comes down in the column first (Distribution coefficient of substance 2 in the liquids 2 and 1 is more than unity).

Thus, $$\frac{C_s}{C_m} = K_D$$

where C_s = concentration in stationary liquid phase and C_m = concentration in the mobile liquid phase. K_D = Distribution coefficient.

Different components in the mixture have different distribution coefficients. They pass into the mobile phase in order of their distribution coefficients. **The component (say S_2) with higher distribution coefficient is extracted first and is the first to move down the column. The location of the components is easy if these are coloured.** Otherwise they are visualised with the help of a locating agent. The portion of the solvent (or band) containing each component is received in a separate vessel and the solute recovered from it by distillation.

Q. 29. A solute distributes in the ratio of 10 : 1 between ether and water. If 5 g of the solute is added to a liquid mixture containing 100 cm³ each of ether and water, what is the mass of solute dissolved in aqueous layer? ***(Bangalore 2005)***

Ans. Let the mass of the solute in aqueous layer be = x g

Then, the mass of the solute in ether layer = $(5 - x)$ g

conc. of solute in ether layer = $(5 - x)$ g per 100 cm^3

conc. of solute in aqueous layer = 5 g per 100 cm^3

According to Distribution Law

$$\frac{5-x}{x} = \frac{10}{1} \text{ or } 5 - x = 10x \text{ or } x = 0.45 \text{ g}$$

16

PHASE RULE

Q.1. State Gibb's phase rule and explain the various terms involved.

(M. Dayanand 2002; Guwahati 2007; Nagpur 2008)

Ans. Gibb's phase rule cannot be expressed in words. It is described by a mathematical equation as under:

For a heterogeneous system in equilibrium

$$F = C - P + 2$$

where F = No. of degrees of freedom

C = No. of components

P = No. of phases

Q.2. Explain the following terms:

***(i)* Phase** *(Banaras 2004; Purvanchal 2007; Himachal 2010)*

***(ii)* Component** *(Purvanchal 2007; Himachal 2010)*

***(iii)* Degree of freedom** *(Kumaon 2000; Banaras 2004; Purvanchal 2007; Himachal 2010)*

Ans. (1) Phase: A phase may be defined as any part of a system which is

(*a*) *homogeneous and separated from other parts of the system by a distinct boundary,*

(*b*) *physically and chemically different from other parts of the same system, and* (*c*) *mechanically separable from other parts of the system.*

Examples illustrating phases are given below:

System	*Phases*
(*i*) Ice and water.	2 phases: 1 solid and 1 liquid.
(*ii*) Water and water vapour.	2 phases: 1 liquid and 1 gaseous.
(*iii*) Ice, water and water vapour.	3 phases: 1 solid, 1 liquid and 1 gaseous.
(*iv*) Two immiscible liquids, e.g., CS_2/H_2O, CCl_4/H_2O, etc.	2 phases: both liquids.
(*v*) Two miscible liquids, *i.g.*, water/alcohol	1 phase: only liquid.
(*vi*) $CaCO_3$, CaO and CO_2	3 phases: 2 solids and 1 gaseous.
(*vii*) Monoclinic and rhombic sulphur.	2 phases: both solids.

Miscible gases and miscible liquids constitute a *single* phase.

Solid + liquid, solid + gas, liquid + gas, two solids and two immiscible liquids constitute *two* phases.

Different solids are regarded as different phases, howsoever intimately they may be mixed.

(2) **Components:** *The number of components of a system at equilibrium is defined as the smallest number of independently variable constituents by means of which the composition of each phase present can be expressed either directly or in the form of a chemical equation, negative and zero signs being permitted.*

Examples : (*i*) Water system consists of the quilibrium

$$\underset{\text{(Solid)}}{\text{Ice}} \rightleftharpoons \underset{\text{(Liquid)}}{\text{Water}} \rightleftharpoons \underset{\text{(Gaseous)}}{\text{Vapour}}$$

This system consists of one component only, *i.e.*, *it is a one component system* because the composition of each of the three phases present can be directly expressed as H_2O.

(*ii*) **Sulphur system** consisting of four phases—Monoclinic (*s*), Rhombic (*s*), Liquid (*l*) and Vapours (*g*) is also *one component system* because the composition of each phases can be expressed in terms of one constituent—sulphur.

(*iii*) Decomposition of $CaCO_3$ by heat can be expressed according to the equilibrium

$$\underset{\text{(Solid)}}{CaCO_3} \rightleftharpoons \underset{\text{(Solid)}}{CaO} + \underset{\text{(Gas)}}{CO_2}$$

Thus there are three different constituents forming three different phases. But the composition of each phase can be expressed in terms of any two of the constituents. For example,

(*a*) If the constituents chosen are CaO and CO_2, then

Phase		*Composition*
$CaCO_3$	$\longrightarrow$	$CaO + CO_2$
CaO	$\longrightarrow$	$CaO + 0\,CO_2$
CO_2	$\longrightarrow$	$0\,CaO + CO_2$

(*b*) If the constituents chosen are $CaCO_3$ and CO_2, then

Phase		*Composition*
$CaCO_3$	$\longrightarrow$	$CaCO_3 + 0\,CO_2$
CaO	$\longrightarrow$	$CaCO_3 - CO_2$
CO_2	$\longrightarrow$	$0\,CaCO_3 + CO_2$

Similarly composition of the three phases can be expressed in terms of $CaCO_3$ and CaO. Thus the *dissociation of $CaCO_3$ by heat* is a *two component system.*

(*iv*) Sodium sulphate–water system. Such a system at equilibrium contains a number of phases like $Na_2SO_4.10H_2O$, $Na_2SO_4.7H_2O$, Na_2SO_4 solution, ice, water and vapour, but the composition of each phase can be expressed, by chemical equations, in terms of only anhydrous sodium sulphate (Na_2SO_4) and water (H_2O) as follows:

Phase		*Composition*
$Na_2SO_4.10H_2O$	$\longrightarrow$	$Na_2SO_4 + 10H_2O$
$Na_2SO_4.7H_2O$	$\longrightarrow$	$Na_2SO_4 + 7H_2O$
Na_2SO_4	$\longrightarrow$	$Na_2SO_4 + 0H_2O$
Na_2SO_4 Solution	$\longrightarrow$	$x\,Na_2SO_4 + y\,H_2O$
Ice	$\longrightarrow$	$0\,Na_2SO_4 + H_2O$
Water	$\longrightarrow$	$0Na_2SO_4 + H_2O$
Vapour	$\longrightarrow$	$0Na_2SO_4 + H_2O$

Thus the *smallest number* of independently variable constituents by which the composition of the phases present at equilibrium can be expressed is two —Na_2SO_4 and H_2O. Hence it is also a *two component system.*

(3) **Degrees of freedom or Variance.** *Degrees of freedom or variance of a system is the smallest number of variable factors, i.e., temperature, pressure and concentration of the components which must be arbitrarily fixed in order that the conditions of the system may be completely defined.*

Examples : (*i*) **Ice-water-vapour system.** In this system, there are three phases of one component (H_2O). These three phases can co-exist in equilibrium only at one particular temperature and under one particular pressure. Any variation of these factors will lead to the ultimate disappearance of one or more of the phases and, therefore, will destroy the system. Hence this system has no degree of freedom or in other words, it is *invariant.* The same result follows from Phase Rule *i.e.*,

$$F = C - P + 2 = 1 - 3 + 2 = 0$$

(*ii*) **Water-vapour system.** This system consists of two phases of one component. As the vapour pressure of water is definite at a definite temperature, independent of the amount of water present, it follows that if the temperature is fixed, the vapour pressure is also fixed and conversely, fixing the pressure fixes the temperature. Thus the system has only one degree of freedom or in other words, it is *univariant.*

(*iii*) **For a gas.** The temperature, pressure and concentration (volume) are related according to the equation $PV=RT$. Thus in order that the system may be completely defined, any two factors out of P, V and T must be fixed. Hence it follows that a gas has two degrees of freedom or it is *blvariant.*

Q.3. What the number of components, phases and degrees of freedom in the following equilibrium system :

$$CaCO_3\,(s) \rightleftharpoons CaO\,(s) + CO_2\,(g)$$ *(Kerala 2001; Meerut 2002; Arunaachal; 2002; Guwahati 2007)*

Ans. No. of components = 2

No. of phases = 3 (2 solid + 1 gas)

Degree of freedom $F = C - P + 2$ or $F = 2 - 3 + 2$

$F = 1$

Q. 4. Explain the following:

(*i*) System in equilibrium (*ii*) True equilibrium (*iii*) Metastable equilibrium.

(M. Dayanand 2002; Mumbai 2010; Himachal 2010

Ans. (*i*) **Equilibrium.** *A system is said to be in equilibrium if the properties like temperature, composition etc. of the various phases do not undergo any change with time.*

(*ii*) *A system is said to be in a state of true equilibrium if the same state can be achieved by approach from either direction.* For example, the equilibrium between ice and water at 1 atm. pressure and 0°C is a true equilibrium, because it can be attained by melting of ice or by freezing of water.

(*iii*) *If the state of a system can be attained only from one direction and that too by very careful change of conditions, it is called a state of* **metastable equilibrium.** For example, water at –2°C can be obtained only by very careful cooling of liquid water but not by melting of ice. Hence water at –2°C is said to be in a state of metastable equilibrium.

Q.5. Deduce thermodynamically phase rule equation *viz.*, F = C – P + 2

(Nagpur 2003; G. Nanakdev 2004; Madras 2004)

Ans. Consider a heterogeneous system consisting of C components (C_1, C_2, C_3 ...... C_c) distributed among P phases (P_1, P_2, P_3 ..... P_p) as shown in Fig.16.1. Degree of freedom of a system is equal to the number of independent variables which must be fixed arbitrarily to define the system completely. Clearly, the number of these variables is equal to the total number of variables of the system minus the number of variables which are defined by the system itself because of its being in equilibrium. Thus

Degrees of freedom = Total number of variables
– Number of variables defined by the system because of its being in equilibrium.

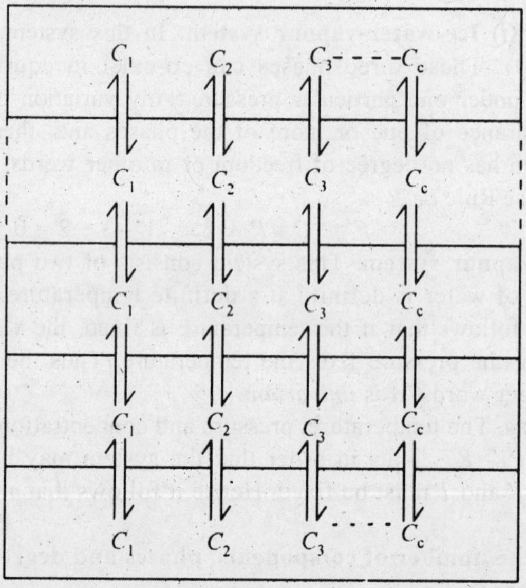

Fig. 16.1. Distribution of *C* components among *P* phases.

The total number of variables can be calculated as follows:

Suppose a phase contains two components then if the molar concentration of one component is known, that of the other can be calculated by difference. Thus the phase containing two components can be defined completely by one concentration term. Similarly, if a phase contains three components, then to define it completely, two concentration terms should be known (because then the concentration of the third can be found by difference). Hence in general, if a phase contains C components it can be defined completely by $(C-1)$ concentration terms or variables. Thus to define the system completely, *i.e.*, all the *P* phases, the total number of concentration terms or variables required will be $P(C-1)$. Besides these concentration variables, the temperature and pressure of the system, which are the same for all the phases, are variables. It, therefore, follows that

Total number of variables $= P(C-1)+2$.

The number of variables defined by the system itself, because of its being in equilibrium, can be calculated as follows:

When the system is in equilibrium the *chemical potential* of a component (μ) is the same in all the *P* phases. Thus if a system consists of three phases, say *A*, *B* and C, then for any components *i*

$$(\mu_i)_A = (\mu_i)_B = (\mu_i)_C$$

These are two equations *viz.*,

$$(i)\ (\mu_i)_A = (\mu_i)_B \quad \text{and} \quad (ii)\ (\mu_i)_B = (\mu_i)_C$$

Thus for a system of three phases, two equations are known for each component, Similarly for a system of four phases, three equations are known for each component. In general, for a system of *P* phases, $(P-1)$ equations are known for each component. Hence for a system of *P* phases and C components, the number of equations known will be $C(P-1)$. This is clearly the number of variables defined by the system itself. Thus, we have

Number of variables defined by the system itself = $C(P-1)$

Hence *the degree of freedom F* of the system is given by:

$$F = [P(C-1)+2] - [C(P-1)]$$
$$= C - P + 2$$

Q. 6. Discuss different techniques for the determination of transition point.

Ans. Transition temperature of a change can be obtained by observing the change in physical property such as colour, density, solubility, etc. Various methods that are employed to determine the transition temperature are discussed as under :

(1) Colour change

If a little mercuric iodide is placed in a melting point tube attached to a thermometer and heated in a conc. H_2SO_4 bath, it is possible to record temperature at which the red mercuric iodide change to the yellow form.

(2) Density change

As rhombic sulphur changes to monoclinic sulphur, there is a decrease in density and, therefore, an increase in volume. The change in volume is employed to measure the transition temperature by using an apparatus known as **dilatometer** shown in Fig. 16.2.

Some powdered rhombic sulphur is placed in the glass bulb and liquid paraffin (an inert liquid) is introduced above the sulphur [Fig. 16.2(*a*)]. The apparatus is then immersed in a hot water-bath, the temperature of which is raised by heating. The scale reading and the temperature are recorded every minute. A plot of liquid level in the capillary against temperature is drawn [Fig. 16.2(*b*)]. On cooling of the dilatometer, reverse changes take place but due to thermal lag the curve takes a different route. The transition temperature is taken as the mean of the respective temperatures where expansion starts (T_1) and contraction begins (T_2).

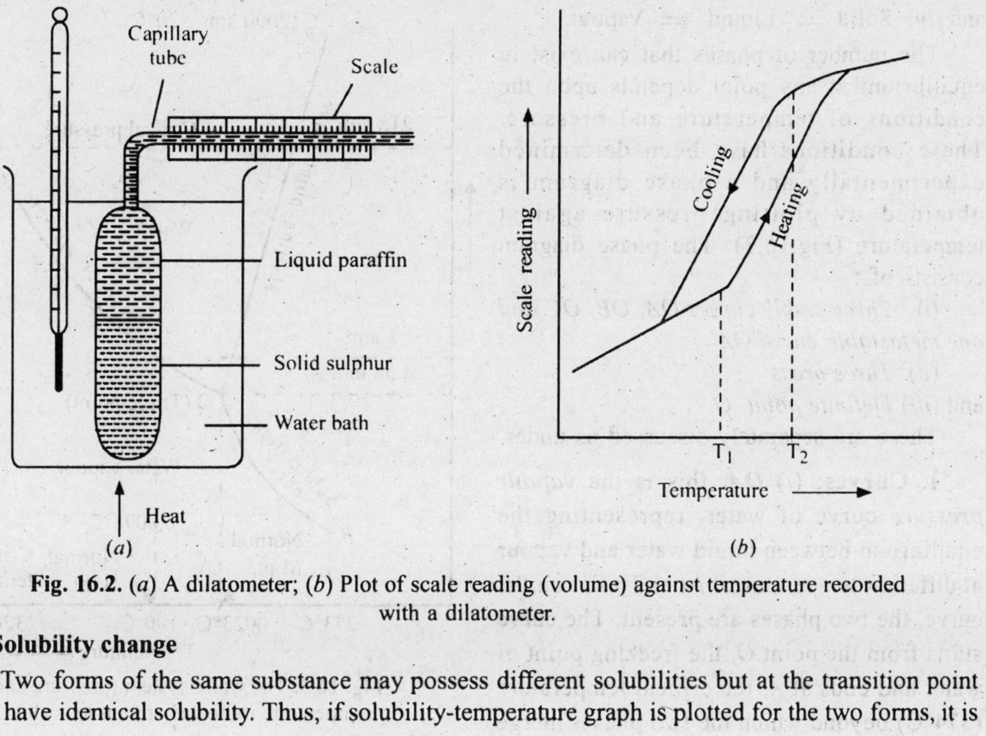

Fig. 16.2. (*a*) A dilatometer; (*b*) Plot of scale reading (volume) against temperature recorded with a dilatometer.

(3) Solubility change

Two forms of the same substance may possess different solubilities but at the transition point they have identical solubility. Thus, if solubility-temperature graph is plotted for the two forms, it is found to consist of two parts with a sharp break. While one part represents the solubility curve for one form, the second part represents that for the other. At the meeting point of the two curves, the solubility of the two forms is the same. This is how the transition temperature is located.

For example, for the system sodium sulphate–water, the solubility curves of Na_2SO_4 (rhombic) and Na_2SO_4 $10H_2O$ meet at 32.2°C. Thus, 32.2°C is the transition temperature where $Na_2SO_4.10H_2O$ changes to Na_2SO_4.

(4) Cooling curve Method

There is often an evolution or absorption of heat when one form changes into the other. Suppose form *A* is converted into form *B* on heating. Now let *B* be allowed to cool and a curve obtained by plotting the temperature against the time. **The otherwise steady curve has a distinct break at a temperature which corresponds to the transition point.**

This method is found to be suitable for determination of transition point for different hydrates of a salt or for a hydrate and anhydrous salt. The method can also be employed for different allotropes of a metal.

Q.7. Give the labelled phase diagram of water system and discuss the importance of various points, lines and areas. ***(Kerala 2000; Madras 2004; G. Nanakdev 2004)***

Ans. The system consists of three phases.

$$\underset{(s)}{\text{Ice}} \rightleftharpoons \underset{(l)}{\text{Water}} + \underset{(g)}{\text{Vapour}}$$

These three single phases may occur in four possible combinations in equilibrium as

(*i*) Liquid $\rightleftharpoons$ Vapour

(*ii*) Solid $\rightleftharpoons$ Vapour

(*iii*) Solid $\rightleftharpoons$ Liquid

and (*iv*) Solid $\rightleftharpoons$ Liquid $\rightleftharpoons$ Vapour.

The number of phases that can exist in equilibrium at any point depends upon the conditions of temperature and pressure. These conditions have been determined experimentally and a phase diagram is obtained by plotting pressure against temperature (Fig.16.3). The phase diagram consists of :

(*i*) *Three stable curves OA, OB, OC and one metastable curve OA′*

(*ii*) *Three areas*

and (*iii*) *Definite point, O.*

These are separately discussed as under

I. Curves: (*i*) ***OA.*** this is the *vapour pressure* curve of water, representing the equilibrium between liquid water and vapour at different temperatures. At any point on the curve, the two phases are present. The curve starts from the point *O,* the freezing point of water and ends at *A,* the critical temperature (374°C) beyond which the two phases merge

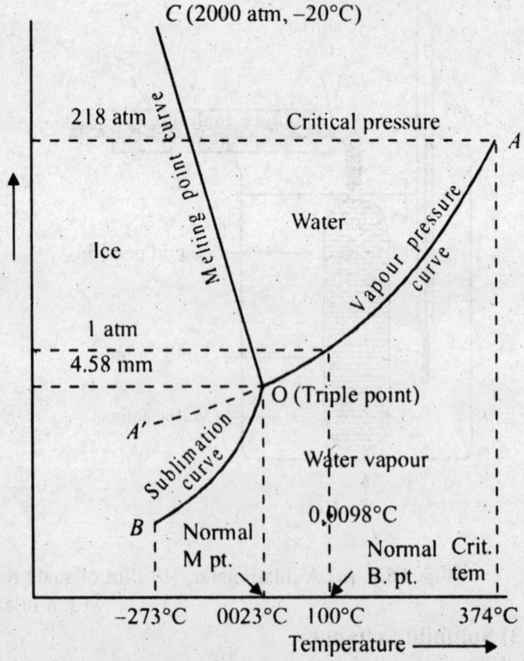

Fig. 16.3. The water system (pressure and temperature indicated are not to scale).

into each other. Consider any point on the curve. Now for a fixed value of pressure, temperature has also a fixed value. Hence if by keeping the pressure constant, an attempt is made to raise the temperature then the liquid phase will disappear and if the temperature is lowered then the vapour phase will disappear. Therefore, to define the system completely at any time, it is necessary to mention either temperature or pressure. In other words, pressure being fixed, it is not necessary to mention temperature; it is automatically fixed. Thus water-vapour system is univariant, that is, it has one degree of freedom. It also follows from the phase rule equation,

$$F = C - P + 2$$

$$F = 1 - 2 + 2 = 1$$

(*ii*) ***OA'***. The dotted curve *OA'*, a conti-nuation of curve *OA* represents the *vapour pressure curve of super-cooled water.* It is a metastable curve. On slight disturbance, the supercooled phase at once changes to solid ice and curve merges into *OB*.

(*iii*) ***OB***. It is the *sublimation curve* or the vapour pressure curve of ice. It gives various values of temperature and pressure at which ice and vapours can exist together. It is obtained by studying the effect of pressure on the freezing point of water. The curve starts from the point *O*, the freezing point of water and ends at *B* (–273°C). As temperature decreases, the vapour-pressure of ice tends to become insignificant. Again we see that for a particular temperature, there can be one and only one pressure value and *vice versa*. In other words, ice-vapour system is univariant, *i.e.*, has one degree of freedom.

(*iv*) ***OC***. It is *melting point curve* and represents the equilibrium between ice and liquid water at various pressures. In other words, it shows the effect of pressure on the melting point of ice. It should be noted that the line *OC* is inclined towards the pressure axis. The slope of the curve *OC* clearly indicates that melting point of ice is lowered by increase of pressure. There is, however, no limit for the curve *OC*. It goes up to a point corresponding to 2000 atmospheres and –20°C. At this point, one-type of ice changes into another solid modification but the solid-liquid equilibrium still remains.

At any point on the curve *OC*, two phases (ice and liquid water) are present in equilibrium. Hence the system is univariant, *i.e.*, degree of freedom is one.

II. Areas: The areas, *i.e.*, the regions bounded by two lines give the conditions of temperature and pressure under which the single phase *i.e.*, ice or liquid water or vapour is capable of stable existence. Area below *AOB* has water vapour, area *BOC* has ice and area *COA* has liquid water. These areas are bivariant as to locate any point in the area, temperature as well as pressure must be fixed, thus indicating two degrees of freedom. This also follows from phase rule equation.

$$F = C - P + 2$$

$$F = 1 - 1 + 2 = 2$$

III. Triple point: It is a point where the three curves *OA*, *OB*, *OC* meet. At this point the three phases–ice, water and vapour are in equilibrium. It should be noted that it is only a point indicating that there is only one set of conditions under which it is possible for the three phases to exist together. The point *O* is a self-defined point corresponding to definite temperature of 0.0098°C and a definite pressure of 4.58 mm. If either pressure or temperature is changed, then one of the three phases will disappear. Since on changing either pressure or temperature one of the phases disappear, therefore, the system at the point *O* has no degree of freedom, *i.e.*, it is invariant. This also follows from phase rule equation.

$$F = C - P + 2$$

$$\therefore \quad F = 1 - 3 + 2 = 0$$

Q.8. Explain why the fusion curve of ice has a negative slope whereas the sublimation curve has a positive slope in the phase diagram of water.

Ans. This is because the freezing point of water is depressed by the increase of pressure. herefore, the fusion curve is inclined towards the pressure axis *i.e.*, it has a negative slope. But the sublımation curve has a positive slope *i.e.*, it is inclined away from the pressure axis. This is because the vapour pressure of water increases with the rise of temperature.

Q. 9. Give the labelled phase diagram of carbon dioxide system and explain the importance of various points, lines and areas. ***(Mumbai 2010)***

Ans. Phase diagram of carbon dioxide system resembles that of water system with some differences.

The phase diagram for carbon dioxide system is shown in Fig. 16.4. It has three distinct areas in which carbon dioxide can exist either as *solid, liquid* or *gas.*

Areas

Area on the R.H.S. of the curve ABD consists of vapour phase. Area above the curve CBD consists of the liquid phase while the area on the L.H.S. of the curve ABC consists of solid phase.

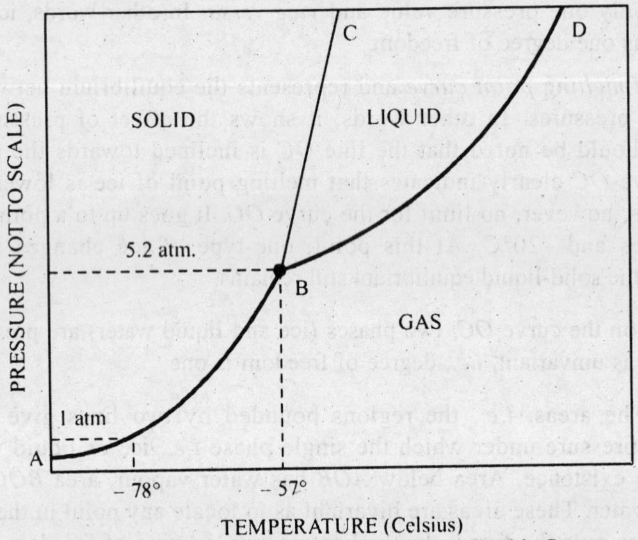

Fig. 16.4. The Phase Diagram for the Carbon dioxide System.

Curves

AB is the *sublimation curve* along which solid carbon dioxide is in equilibrium with the gas, *BD* is the *vaporisation curve* along which liquid carbon dioxide is in equilibrium with the gas while *BC* is the *fusion curve* along which solid and liquid carbon dioxide are in equilibrium with each other.

Point

B is the *triple point* at which all the three phases of carbon dioxide coexist in equilibrium with one another. The temperature of the system at this point is –57°C while the pressure is 5·2 atm. A slight variation in temperature or pressure at this point may result in the disappearance of one of the two phases. For example, a slight *increase* in temperature will result in the disappearance of the solid phase and the equilibrium will shift along the curve *BD* while a slight *decrease* in temperature will result in the disappearance of the liquid phase and the equilibrium will shift along the curve *BA*.

Keeping the temperature constant, if the pressure is increased, the gaseous phase will disappear and the equilibrium will shift along the curve *BC*.

The phase diagram of carbon dioxide resembles that of water in showing three distinct areas for solid, liquid and gaseous phases. But it differs from the latter in several respects. In the first place, the fusion curve *slopes away from the pressure axis.* This indicates that increase of pressure raises the melting point of solid carbon dioxide.

The second difference is that solid carbon dioxide can exist in equilibrium with its liquid only at a very high pressure equal to 5·2 atm. This is unlike the water system in which ice and water can exist in equilibrium even at a very low pressure equal to 4·58 mm of Hg (Refer to phase diagram of water).

The third difference is that the vapour pressure of solid carbon dioxide even at extremely low temperature is very high and many times higher than that of ice.

A significant feature of carbon dioxide system is that even at atmospheric pressure, carbon dioxide gas can be directly solidified without the appearance of liquid phase merely on cooling to **–78°C (Fig. 16.4). It is for this reason that solid carbon dioxide is commonly known as dry ice.**

Q. 10. Write a note on polymorphism.

Ans. *The existence of a given substance in more than one crystalline form with different physical properties is known as* **polymorphism.** This phenomenon occurs in a number of elements and compounds. When it occurs in elements, it is referred to as **allotropy.**

Each polymorphic form constitutes a separate phase. The temperature at which one form changes into another, at a given pressure, is known as the **transition temperature.** For example, rhombic sulphur when heated under a pressure of one atm. changes into monoclinic sulphur at 95·6°C.

Similarly, monoclinic sulphur on cooling at a pressure of one atmosphere changes into rhombic sulphur at 95.6°C. Thus, 95.6°C is the transition temperature at which the two forms of sulphur (rhombic and monoclinic) change reversibly into each other.

Q.11. Draw the labelled phase diagram of sulphur system and discuss its salient features.

(Guwahati, 2002; Bombay 2004; Purvanchal 2007; Guwahati 2007; Nagpur 2008; Himachal 2010)

Ans. It is a one component system comprising **four phases,** *viz.*, (*i*) *Rhombic sulphur,* (*ii*) ***Monoclinic sulphur,* (*iii*) *Sulphur liquid* and (*iv*) *Sulphur vapour.* This is a typical example of** one component system exhibiting solid-solid transformation.

Sulphur exists in two well known crystalline forms, namely rhombic sulphur and monoclinic sulphur. The two forms are in equilibrium with each other at 95.5°C, which is a **transition temperature.** It is defined as the *"temperature at which one stable form of a solid is transformed into another stable solid form of it, without any change in temperature, pressure on the system being one atmosphere."* Rhombic sulphur is stable below 95.5°C, whereas monoclinic is stable above this temperature.

According to phase rule equation, all the four phases cannot co-exist in equilibrium. This is because on substituting $P = 4$ and $C = 1$ in the phase rule equation, we have

$$F = C - P + 2$$
$$= 1 - 4 + 2 = -1$$

i.e., we get $F = -1$, which is meaningless. Hence for a system consisting of single component, it is not possible to have four phases together in equilibrium. However, it is possible to have three or fewer number of phases to exist together.

The phase diagram of the sulphur system as obtained experimentally by plotting pressure vs. temperature is shown in Fig. 16.5. The diagram consists of *six stable curves BO, OA, AE, OC, AC* and *CF* and *four metastable curves OD, OG, AD* and *CD*. The curves divide the diagram into *four areas, i.e.*, to the left of *BOCF*, below *BOAE*, to the right of *FCAE* and the one enclosed by *OCA*.

I. Areas : The areas are bivariant because to locate any point in the areas both temperature and pressure are to be fixed up. Each area represents the conditions for the stable existence of a single phase. In areas thus there are two degrees of freedom, which also follows from the phase rule equation $F = C - P + 2$; putting the values $C = P = 1$, we get $F = 1 - 1 + 2 = 2$.

In the area to the left of *BOCF only* S_R (Rhombic sulphur) exists while in the area below *BOAE*, only S_V (sulphur vapour) exists; to the right of *FCAE*, S_L (liquid sulphur) is present and in the region enclosed by *OCA*, S_M (monoclinic sulphur) exists.

II. Curves: (*i*) ***OB***. It is *sublimation curve of rhombic sulphur*. The two phases present in equilibrium along this curve are S_R and S_V. The curve starts from the point *O*, the transition temperature of S_R to S_M and end at *B* (50°C) below which vapour-pressure of S_R cannot be measured. If heating is done rapidly, then it is possible to heat rhombic sulphur above the transition point along *OD*. *OD is the metastable curve* and on slight disturbance it merges into curve *OA*, the vapour pressure curve of monoclinic sulphur.

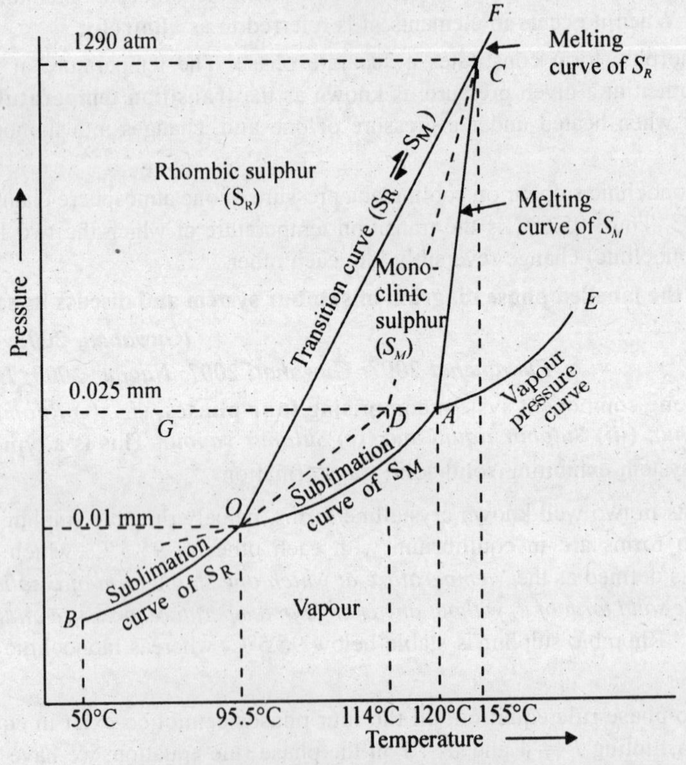

Fig. 16.5. Phase diagram sulphur system.

(*ii*) ***OA***. It is the *sublimation curve of monoclinic sulphur*. The two phases present in equilibrium along this curve are S_M and S_V. The curve starts from the point *A*, melting point of S_M (120°) and ends at *O*, the transition temperature of S_R (95.5°C).

Except at the point *O*, where three phases S_R, S_M and S_V are in equilibrium, at any other point all along the curve only two phases S_V and S_M are in equilibrium. For a fixed value of temperature, pressure is automatically fixed. The degree of freedom, therefore, is one.

If the temperature of monoclinic sulphur is lowered, then normally it would change into rhombic sulphur at 95.5°C, but it is possible to prevent this conversion by carefully cooling it. Then we will go along the curve *OG* which is a *metastable vapour pressure curve of monoclinic sulphur.*

(*iii*) ***AE.*** *It is the vapour pressure curve of liquid sulphur.* The two phases present in equilibrium along this curve are liquid sulphur, S_L; and vapour sulphur, S_V. The curve starts at *A*, the melting point of S_M, and ends at *E*, the critical point beyond which out of the two phases S_L and S_V only S_V (vapour sulphur phase) can exist. Since for any point on the curve, for a fixed value of pressure, the temperature has a fixed value, the system also has one degree of freedom.

(*iv*) ***OC.*** *It is the transition curve* and represents the equilibrium between rhombic and monoclinic sulphur and is obtained from the effect of pressure on the transition temperature of rhombic sulphur. The slope of the curve *OC* is away from the vertical pressure axis, indicating that the transition point is raised by the rise of pressure. The curve ends at the point C, after which monoclinic sulphur phase disappears.

(*v*) ***AC.*** *It is the melting point curve of monoclinic sulphur.* It represents the equilibrium between monoclinic sulphur and liquid sulphur and is obtained from the effect of pressure on the melting point of monoclinic sulphur. The curve is inclined away from the pressure axis showing that the melting point is raised by the rise of pressure. The curve ends at C beyond which monoclinic sulphur ceases to exit in a stable condition. The system has degree of freedom equal to one.

(*vi*) ***CD.*** this curve represents the equilibrium between rhombic and liquid sulphur and is the *metastable melting point curve of rhombic sulphur.* Since the curves *OC* and *AC* meet at C, for which the conditions are 155°C and 1290 atmospheres, it follows that if liquid sulphur were cooled at a pressure higher than 1290 atmospheres, then rhombic sulphur would crystallise at once, no monoclinic sulphur being formed. The curve passes through the point C and beyond this point it becomes a stable melting point curve *CF* of rhombic sulphur.

III. Triple points: There are three stable triple points *viz.*, *O*, *A* and *C* in the phase diagram. At the point *O*, three phases in equilibrium are sulphur rhombic, monoclinic and vapour. If one of the variables is changed, one of the phases disappears. Similarly *A* is also a triple point between monoclinic, vapour and liquid sulphur. This point is again invariant. Further, lines *OC* and *AC* intersect at *C* to yield a three phase equilibrium between rhombic, monoclinic and liquid sulphur. Since on changing one of the variables, either pressure or temperature, one of the phases disappears, therefore, the system at the triple point has no degree of freedom. This also follows from phase rule equation:

$$F = C - P + 2$$
$$F = 1 - 3 + 2 = 0$$

D is a metastable triple point; the phases in equilibrium being S_R, S_L and S_V.

Q.12. Why a three dimensional model is needed for two-component systems? How is this difficulty overcome ?

Ans. *A two-component system can be defined as one in which it is possible to express the composition of all the phases in terms of two substances.* A two-component system may have as many as three degrees of freedom. This is because when $P = 1$, then

$$F = C - P + 2 = 2 - 1 + 2 = 3.$$

Thus the system will have, besides temperature and pressure variables, *one concentration variable for each phase present at the time of equilibrium.* Naturally, a three dimensional model is required in which pressure, temperature and concentration would be plotted along the three axes at right angles to one another (Fig. 16.6). However, in actual practice, a simple plain diagram with two variables only is considered, the third variable is considered as constant. In this way, use is made of the following three types of diagrams (Fig. 16.7).

(*a*) Pressure-temperature diagrams
(*b*) Composition-pressure diagrams
(*c*) Temperature-composition diagram

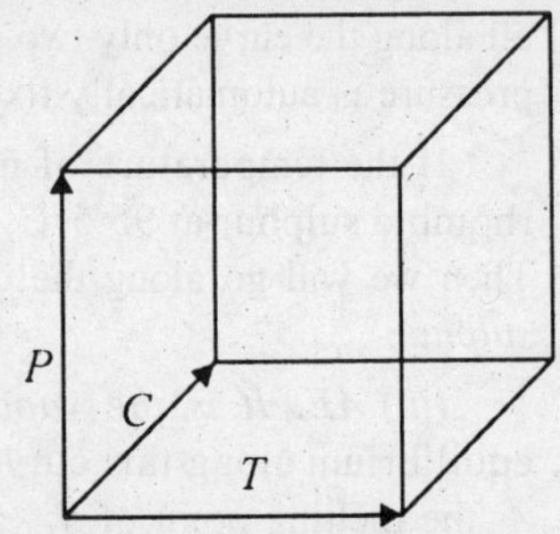

Fig. 16.6. Space model for Two-component System.

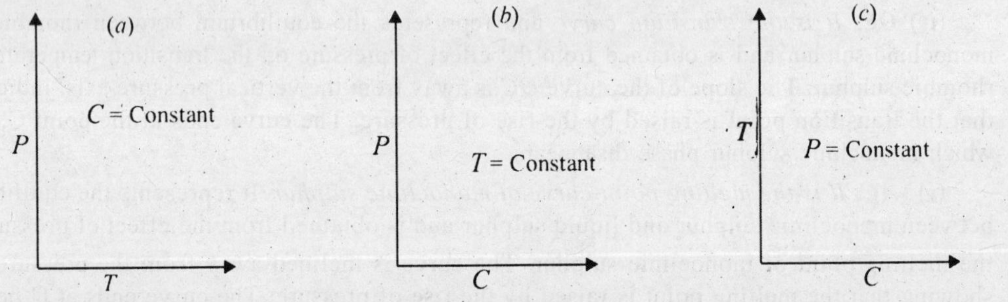

Fig. 16.7. Two dimensional diagrams.

But in most of the cases it is usually convenient to keep pressure constant at the atmospheric value and thus a $T - C$ diagram is often used. A diagram drawn for a constant value of temperature is called *isothermal,* while one obtained by keeping composition constant, is called *isoplethal* diagram.

Q.13. What is a condensed system? Write the reduced phase rule equation.

(Gujarat 2003; Arunachal 2003; G. Nanakdev 2004; Mumbai 2010)

Ans. In case of systems consisting of only solid and liquid phases, small changes of pressures have very little effect on the systems. Such systems are called *condensed systems.* Now as the degree of freedom in such a case is reduced by one as pressure is taken as fixed at one atmosphere pressure, therefore, we can write a *'reduced' phase rule equation* as

$$F' = C - P + 1$$

F' in the above equation represents the number of degrees of freedom which the system can possess in addition to the pressure.

Q. 14. Why is it justified to use reduced phase rule equation for solids and liquids systems? ***(M. Dayanand 2000)***

Ans. See Q. 13.

Q. 15. Give examples of two-component systems in which

(*i*) the components do not react with each other

(*ii*) the components react to form a component with congruent melting point

(*iii*) The components react to form a compound with incongruent melting point.

(Lucknow 2000)

Ans. (*i*) Those in which the two components do not react with each other but simply mix into each other in the molten state or in the solution. Examples of such systems are lead-silver system and KI – H_2O system.

(*ii*) *Those in which the components react to form a compound with congruent melting point.* These are those systems in which the two pure components react to form a compound which is stable up to its melting point. At the melting point, it melts to give a liquid of the same composition as that of the solid compound. The compound formed is then said to have a *congruent melting point.* For example, in $FeCl_3 - H_2O$ system, each hydrate formed is stable up to the melting point and possesses a congruent melting point.

(*iii*) *Those in which the components react to form a compound with incongruent melting point.* These are those two-component systems in which the compound formed does not remain stable up to the melting point. When such a compound is heated, it starts decomposing before the melting point to give a new solid phase and a solution with composition different from that of the solid phase. When this happens, the compound is said to undergo a *transition* or *peritectic reaction* and is said to have *incongruent melting point.* In general, a peritectic reaction may be represented by the equation

$$C_2 \rightleftharpoons C_1 + \text{solution (or melt)},$$

when C_2 is the compound formed as a result of reaction between the two pure components and C_1 is the new compound (or a pure constituent) formed as a result of decomposition of compound C_2 below its melting point. An example of this type of system is $Na_2SO_4 - H_2O$ system in which the hydrates formed undergo peritectic reaction.

Q. 16. What is meant by a peritectic reaction? **(*M.Dayanand 2000*)**

Ans. See Q. 15 (*iii*).

Q. 17. Discuss the features of phase diagram of two components system (solid) miscible in liquid state.

Ans. If a number of mixtures of two components *A* and *B* of different composition are prepared and their freezing points determined, carrying out the process in an open vessel in each case so that the pressure remains equal to the atmospheric pressure and then, these freezing points are plotted against the compositions of the liquid mixtures, a phase diagram of the type shown in Fig. 16.8 is obtained.

The diagram consists of the curves *AC, BC,* the areas enclosed by *ADC, BCE,* areas above *ACB,* below *DCE* and the point *C*. Each of these is described below.

I. Curves *AC* and *BC*. The point *A* and *B* repre sent the freezing points of pure *A* and pure *B* respectively. The addition of *B* to *A* lowers the freezing point along the curve *AC*. Similarly the addition of *A* to *B* lowers the freezing point along the curve *BC*. In other words, *AC* represents the freezing point curve of *A* to which different amounts of *B* are being added. Similarly, *BC* represents the freezing point curve of *B* to which different amounts of *A* are being added. Further it is evident that if liquid mixture rich in *A* (*i.e.*, for any composition between *B* and *C*) is cooled, solid *B* will separate out. Thus the curves *AC* and *BC* represent the conditions of temperature at which the liquid mixture is in equilibrium with solid *A* and solid *B* respectively.

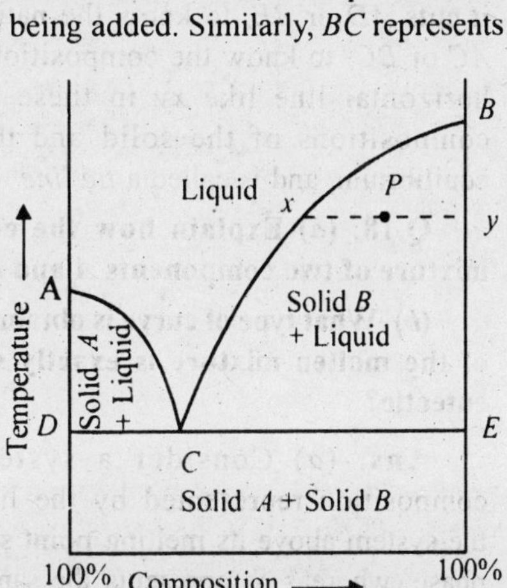

Fig. 16.8. General shape of the phase diagram of two-component system miscible in the liquid phase.

Ignoring the vapour phase, all along the curve *AC* or *BC* there are two phases in equilibrium (solid *A* + liquid along the curve *AC* and solid *B* + liquid along the curve *BC*), *i.e.*, $P = 2$.

As $C = 2 \therefore \quad F = C - P + 1$ (reduced equation)

$$= 2 - 2 + 1 = 1$$

Thus the condensed phase has one degree of freedom. This is otherwise also evident because either the temperature or the composition of the liquid phase alone is sufficient to define the system completely.

II. Point *C*. *C* is the point where the curves *AC* and *BC* meet. At this point, three phases are in equilibrium *viz.*, solid *A*, solid *B* and the liquid mixture, *i.e.*, $P = 3$.

Hence $F = C - P - 1 = 2 - 3 + 1 = 0$,

i.e., condensed system becomes invariant at the point C. In other words, there is only one temperature, at the atmospheric pressure, where the liquid phase can be in equilibrium with both the solids. The point C corresponds to a temperature below which any liquid mixture of *A* and *B* will freeze as a whole. The point *C* is, therefore, called **eutectic point.** The composition corresponding to C is called *eutectic composition.*

III. Areas: *Area above the curve ACB.* In this area only liquid phase is present as the temperature is above the melting point of any mixture, *i.e.*, $P = 1$. Hence

$$F' = C - P + 1 \quad \text{(reduced phase rule equation)}$$

$$= 2 - 1 + 1 = 2$$

So there are two variables, *i.e.*, the areas are bivariant. This means that to define the system completely, both temperature and composition need to be specified. This is otherwise also evident because to define any point within the area, both the coordinates are required.

Area below DCE. As *DCE* is passing through the eutectic point *C*, so in the area below *DCE*, only the two solid phases *viz.*, pure *A* and pure *B* are present, *i.e.*, $P = 2$. Hence $F = C - P + 1 = 2 - 2 + 1 = 1$, *i.e.*, the condensed system is univariant. This is so because the composition of each phase is fixed (as each of them is a pure component) *i.e.*, composition is not variable. The only variable is the temperature.

Areas enclosed by ADC and BEC. Each of the these areas has two phases, being solid *A* + liquid in the area *ADC* and solid *B* + liquid in the area *BEC*. Thus for each of these areas,

$$F' = C - P + 1 = 2 - 2 + 1 = 1$$

i.e., the condensed system is univariant.

The composition of the phases in equilibrium at any point within these areas can be obtained by drawing a horizontal line passing through that particular point and finding where it cuts *AD* or *BE*, to know the nature of the solid and *AC* or *BC*, to know the composition of the liquid. Any horizontal line like *xy* in these areas connects the compositions of the solid and the liquid phases in equilibrium, and is called a *tie line.*

Q.18. (*a*) Explain how the cooling of a molten mixture of two components *A* and *B* proceeds.

(*b*) What type of curve is obtained if the composition of the molten mixture is exactly same as that of the eutectic?

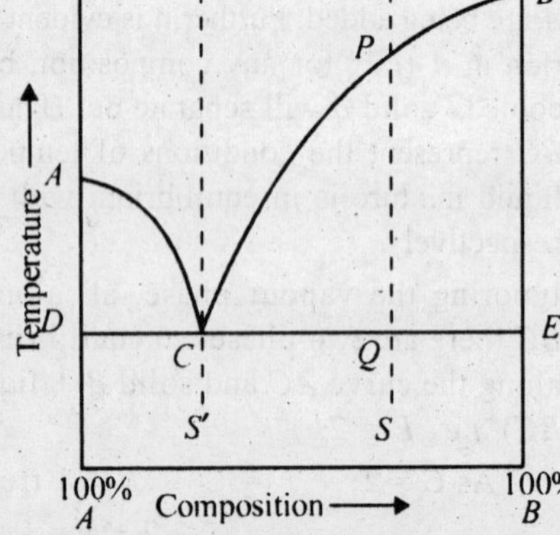

Fig. 16.9. Cooling of liquid mixtures.

Ans. (*a*) Consider a system (Fig. 16.9) of composition represented by the line *LS*, *L* represents the system above its melting point so that it is the liquid phase, whereas *S* represents the same system completely in the solidified state. If the liquid *L* is cooled, no solid will separate out till the point *P* on the freezing point curve BC is reached. At this temperature the solid B will begin to separate out. The solidification thus taking place results in the evolution of heat (called *heat of solidification).* As a result, the rate of cooling will be slowed down upto the point Q. During cooling from P to Q the composition of liquid phase will change from P to C along the curve PC. On reaching the eutectic point C, the second solid *viz.*, solid A will also begin to separate out and the temperature will remain constant till all the 100% liquid has been solidified. This is because according to phase rule, there is only one solid phase in a condensed system. Thus eutectic is point indicated by complete arrest in the rate of cooling. The temperature will further fall say from Q to S only after the whole of the system has been solidified.

Starting from a liquid mixture (molten mixture) corresponding to the temperature and composition represented by the point L, if cooling is allowed to take place and temperature is plotted versus time, a curve of the type shown in Fig. 16.10 is obtained.

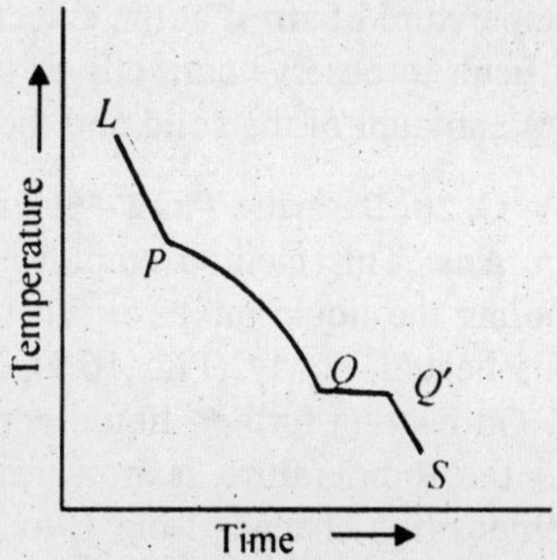

Fig. 16.10. Cooling curve.

From L to P, the liquid cools rapidly. At P solid begins to separate out, heat is evolved, the rate of cooling becomes slow. Solid B keeps on separating out and the composition of the *liquid phase keeps on changing along the curve* PQ. After reaching the point Q which corresponds to the eutectic point, solid A also starts separating out and the temperature remains constant till whole of the liquid solidifies at the point Q'. Now further cooling results in the fall of temperature from Q' to S.

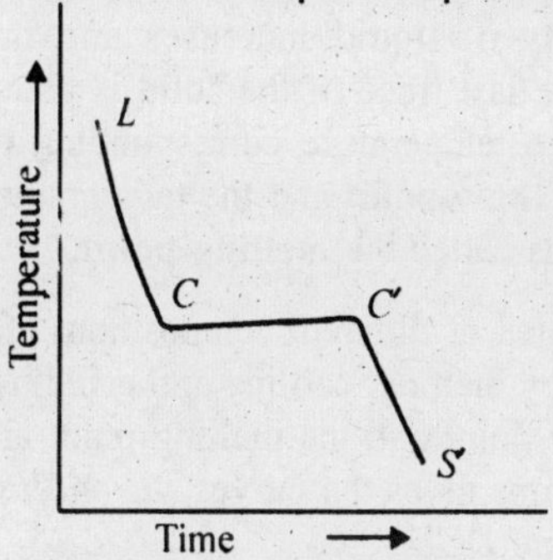

Fig. 16.11. Cooling curve of eutectic mixture.

(*b*) If the composition of the liquid mixture is exactly same as that of the eutectic, as shown by point L' in Fig. 16.9, then on cooling, the solid will separate only when the eutectic point is reached at C. the system will now solidify completely. the cooling curve will show only one break as shown in Fig. 16.11. Solid separates at the point C and is complete at the point C'. Further cooling results in the fall of temperature of the solid along $C'S'$.

Q. 19. Describe experimental determination of phase diagrams of two component systems by the method of cooling curves (thermal analysis method). What are its limitations?

(Madras 2004)

Ans. This method is based on the principle illustrated in Fig. 16.9. A number of mixtures of the two components of different compositions are made. Each mixture is melted to form a homogeneous liquid and then allowed to cool. In this way, a cooling curve of the type shown in Fig. 16.10 is obtained for each of the mixtures. the point P corresponds to the freezing point and point Q corresponds to the eutectic point for the composition taken. In this way, the freezing point and the eutectic points for mixtures of different compositions are obtained. Plotting the freezing points versus composition, the curves AC and BC are obtained, whereas plotting the eutectic temperatures versus composition, the line DCE is obtained. To complete the diagram the freezing points of pure A and pure B are also determined by the cooling curve method as a pure liquid solidifies at a constant temperature.

Limitations of the Cooling curve Method

(*i*) In case of some liquid mixtures, the separation of the solid does not occur readily and supercooling is observed. Thus no halt or break is observed at *P*. **Instead, cooling continues beyond *P* along *PP'* and** then the temperature rises suddenly and then follows the path *RQ*. The point *P* is then located by extrapolation of the curve *QR* as shown in Fig. 16.12.

(*ii*) Even if there is no supercooling, the temperature of arrest at the eutectic point is sometimes difficult to observe correctly as the mixture contains a large amount of the solid and the stirring is difficult.

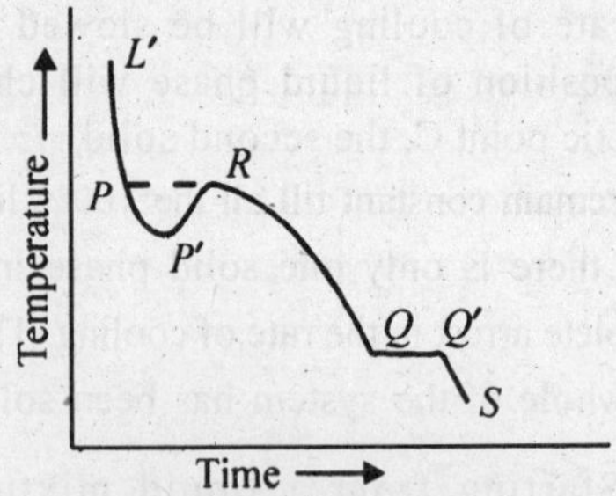

Fig. 16.12. Cooling curve with super cooling.

Q. 20. Describe Thaw-Melt method to determine phase diagrams. (*M. Dayanand 2002*)

Ans. This method employs heating the solid mixture of a known composition instead of cooling the liquid mixture. When a mixture of solids *A* and *B* corresponding to the composition *S* is heated slowly (Fig. 16.9), no change in composition takes place upto the eutectic point *Q*. On heating further, liquid corresponding to the eutectic composition *C* begins to be formed. As the temperature is now raised from *Q* to *P*, the quantity of liquid increases and the composition changes from *C* to *P* (Refer to Fig. 16.9). At *P* the last trace of the solid is about to disappear and above *P* the system is completely liquid. The temperature corresponding to point *Q* where the solid mixture just *starts* melting is called the **thaw point** and the temperature corresponding to point *P* where the last trace of the solid melts is called the **melting point.**

Thus the method requires preparing a number of solid mixtures of different compositions. To ensure thorough mixing of the solids, the mixtures are prepared by melting, cooling and grinding. Each mixture is packed into a wide capillary tube and the thaw point and the melting point are determined as usual. The plot of melting point versus compositions gives the curves *AC* and *BC* whereas the plots of thaw point versus composition gives the line *DCE*.

STUDY OF SOME TWO-COMPONENT SYSTEMS

Q.21. Discuss salient features of lead-silver system.

(*M. Dayanand 2002; Guwahati 2002; Arunachal 2003; Madras 2004; Mumbai 2005; Nagpur 2008*)

Ans. The various phases possible are (*i*) solid lead (*ii*) solid silver (*iii*) solution of lead and silver in the molten state and (*iv*) vapour.

Since small changes of pressure have negligible effect on the equilibrium existing between lead and silver in the solid state or in the molten state, we can use the *reduced phase rule equation*, $F' = C - P + 1$. According to this equation, when $F' = 0$, *P* will be 3. Thus the maximum number of phase (other than vapour) that can coexist would be three.

The phase diagram of lead-silver system is shown in the Fig. 16.13. The diagram is drawn taking temperature and composition along the axes. the equilibrium diagram consists of two curves *OA* and *OB*, intersecting at point *O*.

Various curves and points are described as under:

I. Curves: (*i*) **Curve *OA*.** It represents the *freezing point curve of silver* to which successive small quantities of lead are added. The point *A* indicates the melting point of pure silver (961°C). It is observed that the melting point of silver is lowered gradually by the addition of lead into it. All along this curve the added lead goes into the solution and the separation of silver occurs when the point *O* is reached. At this point no more lead goes into the solution, as the solution becomes saturated with respect to lead and hence the melting point of silver does not fall any longer. Thus the point *O* represents the lowest possible temperature (303°C) in the system that corresponds to fixed composition, *viz.*, 2.4% Ag and 97.6% Pb. The point *O* is called the **Eutectic point.**

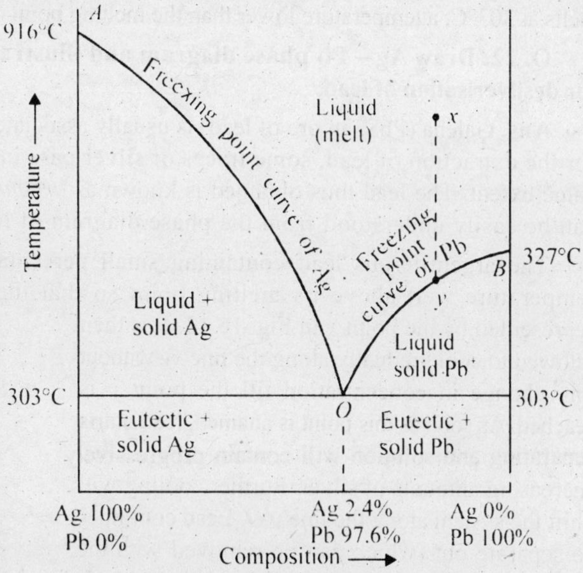

Fig. 16.13. Lead – Silver system.

All along the curve *AO,* two phases—solid silver and liquid (melt) are present in equilibrium neglecting the vapour phase. Therefore, the system is univariant as shown below:

$$F' = C - P + 1 = 2 - 2 + 1 = 1$$

(*ii*) **Curve *BO*.** It represents the *freezing point curve of lead* to which successive small amounts of silver are added. The point *B* represents the melting point of pure lead (327°C). It is seen that the addition of silver to lead lowers the melting point of lead. All along this curve, the silver which is added goes into the solution while the separation of lead takes place. This state of affairs continues till the point *O* is reached. At this point, no more silver goes into the solution and the solution is said to have become saturated with respect to silver. Hence the melting point of lead does not fall any longer.

All along this curve, two phases — solid lead and liquid (melt) are present in equilibrium. Therefore, as before $F' = C - P + 1 = 2 - 2 + 1 = 1$, *i.e.*, the system is univariant.

II. Eutectic point, *O*. This is the point where two curves *AO* and *BO* meet. At this point solid lead and solid silver are in equilibrium with the fused mass, *i.e.*, three phases are present. Therefore, there are no degrees of freedom, as according to *reduced phase rule equation,*

$$F' = C - P + 1$$

$$= 2 - 3 + 1 = 0$$

The point *O* represents the lowest possible temperature (303°C) below which a liquid phase cannot exist and above which the liquid phase cannot be enriched in either component by freezing out the other component. *Such a liquid mixture of two components—Ag and Pb, which has the lowest freezing point compared to all other liquid mixtures is called* **eutectic mixture.** This mixture freezes out completely at a constant temperature and on freezing gives the two components as solid phases in the same proportion in which they are present in the liquid phase. The temperature corresponding to the eutectic point *O* is called the **eutectic temperature.** This is always lower than the melting points of the components constituting the eutectic mixture.

For example, in the above system, we see that the alloy containing 2.6% silver and 97.4% lead melts at 303°C, a temperature lower than the melting point of Ag (961°C) and that of Pb (327°C).

Q. 22. Draw Ag – Pb phase diagram and illustrate the principle of Pattinson's process for desilverisation of lead.

Ans. Galena (PbS, an ore of lead) is usually associated with silver. When galena is worked out for the extraction of lead, some traces of silver pass into it, because silver is soluble in lead to some extent. The lead thus obtained is known as *argentiferous lead.* The desilverisation of lead can be easily understood from the phase diagram of lead-silver system, shown in Fig. 16.13.

The argentiferous lead containing small percentage of silver is first of all heated to a temperature well above its melting point so that the system consists only of liquid phase represented by the point *x* in Fig. 16.13. It is then allowed to cool gradually along the line *xy* without any change in concentration till the point *y* is reached. As soon as this point is attained, lead starts **separating and solution will contain progressively** increasing amounts of silver. Further cooling will shift the system along the line *y O.* Lead continues to separate out (which can be removed with the help of ladles/perforated spoons) and the melt continues to be richer and richer in silver, till the point *O* is reached where an alloy containing about 2.4% of silver is obtained.

The above principle is employed in Pattinson's process which is used for increasing the relative proportions of silver in the alloy. The relative increase in the amount of solid lead that separates out and in the proportion of silver in the melt on cooling is shown in Fig. 16.14.

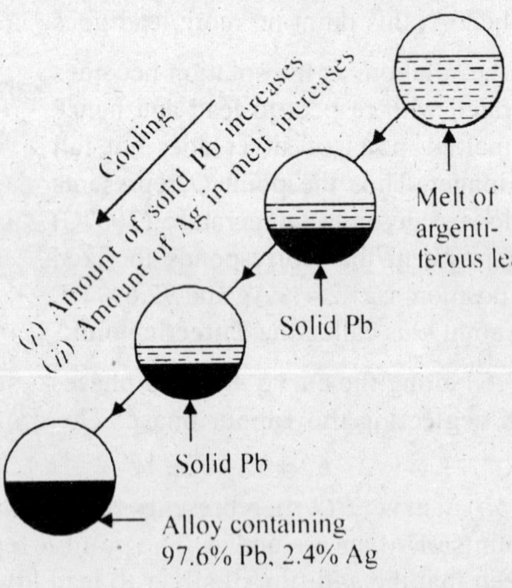

Fig. 16.14 Cooling the melt of argentiferous lead.

Q. 23. Draw a labelled diagram of zinc - cadmium system and discuss the significance of various points, lines and areas.

Ans. The phase diagram is shown in Fig. 16.15. It consists of :

Curve *AO*; the Freezing point curve of zinc. *A* represents the freezing point (or melting point) of zinc (419°C). The curve *AO* shows that the melting point of zinc is lowered by the addition of cadmium. The phases in equilibrium along *AO* are solid zinc and liquid solution of zinc and cadmium. Applying the reduced phase rule equation and substituting the values,

$$F' = C - P + 1 = 2 - 2 + 1 = 1$$

Thus, the system solid zinc/solution is *monovariant i.e.* its degree of freedom is one.

Curve *BO*; the Freezing point curve of cadmium. The point *B* represents the melting point of cadmium (321°C). The curve *BO* represents the fall of melting point of cadmium on addition of zinc. Along this curve, solid cadmium is in equilibrium with the liquid solution of zinc and cadmium.

Applying the reduced phase rule equation to the equilibrium cadmium/solution and substituting the values,

$$F' = C - P + 1 = 2 - 2 + 1 = 1$$

Thus, the degree of freedom is one and the equilibrium is *monovariant.*

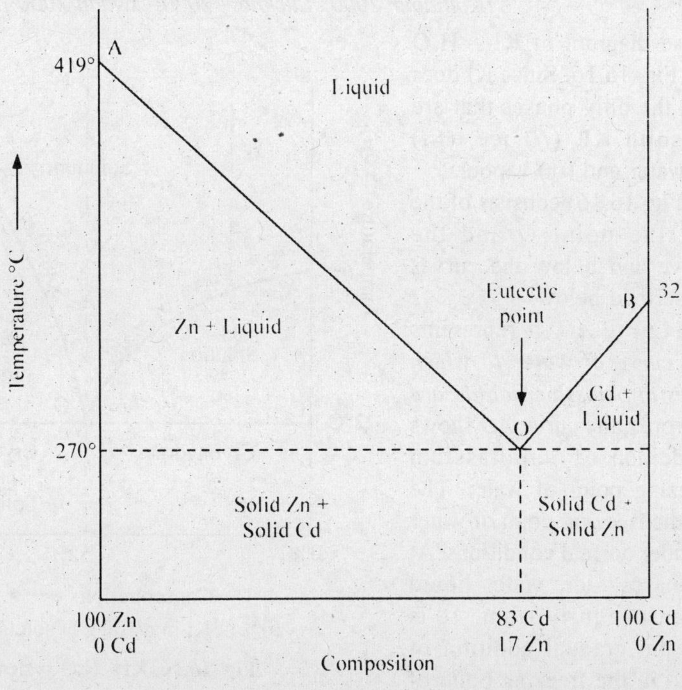

Fig. 16.15. Phase diagram of Zn–Cd system.

The Eutectic point *O.* The curves *AO* and *BO* meet at *O*. Here, solid zinc, solid cadmium and solution are in equilibrium. The number of phases is three. Substituting values in the reduced phase rule equation,

$$F' = C - P + 1 = 2 - 3 + 1 = 0$$

Thus, the system at point *O* is *nonvariant.* The point *O* is called the **eutectic point.** Both the temperature and composition being fixed, the system has no degree of freedom. The eutectic temperature is 270° and the eutectic composition is 83% cadmium and 17% zinc.

The Area above *AO.* In this area, both the components zinc and cadmium are present as liquid solution of varying composition. The solution of the two metals being homogeneous, constitutes one phase only. Thus, the system represented by the area above *AO* is *bivariant i.e.,* it has degree of freedom as 2.

$$F' = C - P + 1 = 2 - 1 + 1 + 1 = 2$$

The Area above *BO.* In this area, zinc and cadmium exist as liquid solution. The composition of the solution is indicated on the composition axis. The liquid solution, regardless of its composition, represents one phase only. Thus, the system in the area above *BO* is *bivariant i.e.* degree of freedom is 2.

$$F' = C - P + 1 = 2 - 1 + 1 = 2$$

The effect of cooling. If the solution in the area above *AO* is cooled, zinc separates as the curve itself is reached. This continues till the point *O* when the solid eutectic mixture (83% Cd + 17% Zn) separates. Similarly, the solution above *BO* on cooling allows the separation of cadmium. This continues till the point *O* is reached at 270°C and eutectic mixture separates. As is clear from figure, the area below the curve *AO* represents zinc and solution; the area below *BO* shows cadmium and solution. The area below the eutectic temperature represents the system solid zinc/solid cadmium.

Q. 24. Draw a labelled phase diagram of KI–H_2O system and describe the main features.

(Kolhapur 2000; Lucknow 2000; Awadh 2000; Meerut; 2002)

Ans. The phase diagram of KI – H_2O system is shown in Fig. 16.16. Since KI does not form hydrates, the only phases that are possible are (*i*) solid KI, (*ii*) ice, (*iii*) solution of KI in water and (*iv*) vapour.

The diagram (Fig. 16.16) consists of the curves *AO, BO,* the point *O* and the various areas above and below the curves *AO and BO,* as described below:

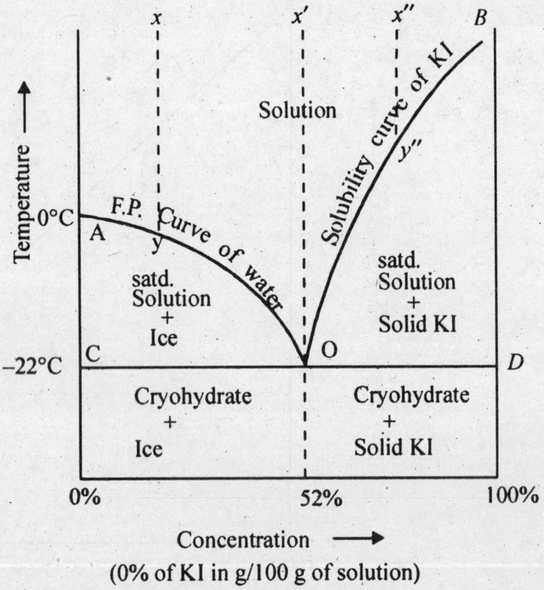

Fig. 16.16. KI – H_2O system.

I. Curves: (*i*) Curve *AO*. It represents *the freezing point curve of water to which successive quantities of potassium iodide are added* or in other words, the curve *AO* shows the effect of the addition of potassium iodide on the freezing point of water. The point *A* represents the freezing point of water (or m. pt. of ice) under normal conditions. At this point three phases—ice, water liquid and vapours are in equilibrium. It is observed that by the gradual addition of potassium iodide to it, the freezing point of water is lowered and curve *AO* slants downwards. This, however, does not go indefinitely, for, as soon as the solution is saturated with respect to potassium iodide, the lowest temperature is reached at the point *O;* then further addition of KI will not produce further lowering in the freezing point of water.

The point *O* corresponding to a concentration of 52% potassium iodide represents the lowest temperature that can be attained in this system. This eutectic temperature in this case is called **Cryohydric point** and the eutectic itself is called a **Cryohydrate.**

All along the curve *OA,* two phases namely solution and ice are in equilibrium (neglecting the vapour phase). Therefore, the system is univariant as

$$F' = C - P + 1$$
$$= 2 - 2 + 1$$
$$= 1$$

(*ii*) **Curve *OB*.** It is the *solubility curve of potassium iodide in water at different temperatures.* It represents the effect of temperature on the solubility of solid potassium iodide. It is observed that solubility of potassium iodide increases with the rise of temperature and is maximum at the point *B*, the boiling point of the saturated solution. However, the curve comes to an end before it can touch the 100% potassium iodide axis. Conversely, if the saturated solution of potassium iodide in water is cooled, potassium iodide separates out and the solution becomes dilute and the curve *BO* is followed till the point *O* is reached. Here the solution freezes as a whole with a fixed composition.

All along the curve *OB,* there are two phases — solid KI and solution in equilibrium, and being a two-component system, the number of degree of freedom according to reduced phase rule equation would be

$$F' = C - P + 1 = 2 - 2 + 1 = 1$$

II. Cryohydric point, *O*. It is the point where the two curves *AO* and *BO* intersect. Here, three phases—ice, solid salt and solution are in equilibrium. If the vapour phase is not neglected, then the fourth phase is also present. Point *O* is, therefore, called a *quadruple point.* If we apply the phase rule, we find that the system is invariant, *i.e.*,

$$F = C - P + 2 = 2 - 4 + 2 = 0$$

Therefore, any change in temperature or composition will cause one of the phases to disappear. The point *O* corresponds to a definite temperature (– 22°C) and composition (52% KI and 48% Ice).

Q. 25. Discuss with the help of KI – H_2O phase diagram the effect of cooling KI solution.

Ans. Refer to Fig. 16.16.

The solution phase exists in the area above *AOB*. Let the solution represented by the point *x* be allowed to cool gradually along the line *xy* without any change in composition till the point *y* is reached. As soon as this point is reached, ice starts separating, and on further cooling, solution will continue becoming more and more concentrated till the cryohydric point *O* is reached where solid KI also makes its appearance. Similarly if a solution of composition x'' lying on the right hand of point *O* is cooled then KI will begin to separate as soon as point y'' is reached. The composition will, therefore, change with temperature along y'' *O* and more and more of KI will continue to separate out until point *O* is reached. Ultimately, as before, the whole of the mixture will freeze to give eutectie mixture.

Now consider a solution of composition represented by the point x' lying vertically above the cryohyric point *O*. When such a solution is allowed to cool gradually along the line $x'O$ without any change in composition, the solution will solidify as a whole, *i.e.*, on reaching the point *O*, both ice and KI start separating out simultaneously.

Hence from the above discussion we conclude that all solutions on cooling show no further change in temperature at the eutectic point *O*. Moreover, on cooling a solution of eutectic composition, the solution freezes at the eutectic temperature without any change in composition.

Q. 26. Explain with the help of phase rule, the use of salt in freezing mixture.

Ans. See Q. 25.

Q. 27. What are the essential features of a freezing mixture? ***(Himachal 2010)***

Ans. (*i*) *It should have a low cryohydric temperature.* This condition is fulfilled by having a salt which is highly soluble.

(*ii*) *The heat of solution of the salt should be high.* This is generally true of salts, solubility of which rapidly increases on increasing the temperature.

(*iii*) *The materials should be cheap.*

Q. 28. Discuss magnesium–zinc system and discuss the salient features.

(Andhra 2010, Raipur 2010)

Ans. It is a typical two-component system which involves the formation of an intermetallic compound $MgZn_2$. It has four phases : solid magnesium, solid zinc, solid $MgZn_2$, and the liquid solution of Mg and Zn.

The complete phase diagram of the system magnesium–zinc is shown in Fig. 16.17. It appears to be made of two simple eutectic diagrams. The one toward the left represents the eutectic system Mg–$MgZn_2$, while the one to the right is the system Zn–$MgZn_2$.

The curves AC, *CDE*, and *BE*. *AC* is the freezing point curve of magnesium; *BE* is the freezing point curve of zinc; and *CDE* is that of the compound $MgZn_2$.

The curve *AC* shows that the melting point of magnesium is (651°C). It is lowered on the addition of zinc. This continues until the point *C* is reached. Here a new phase, solid $MgZn_2$ appears.

The curve *CD* shows the increase of concentration of zinc in the melt with the rise of temperature. At the maximum point *D*, the composition of the melt and the solid compound become the same *i.e.*, $MgZn_2$. The point *D*, therefore, represents the melting point of $MgZn_2$ (575°C). The curve *DE* now shows the lowering of the melting point with the addition of zinc until the lowest point is attained. Here solid zinc appears.

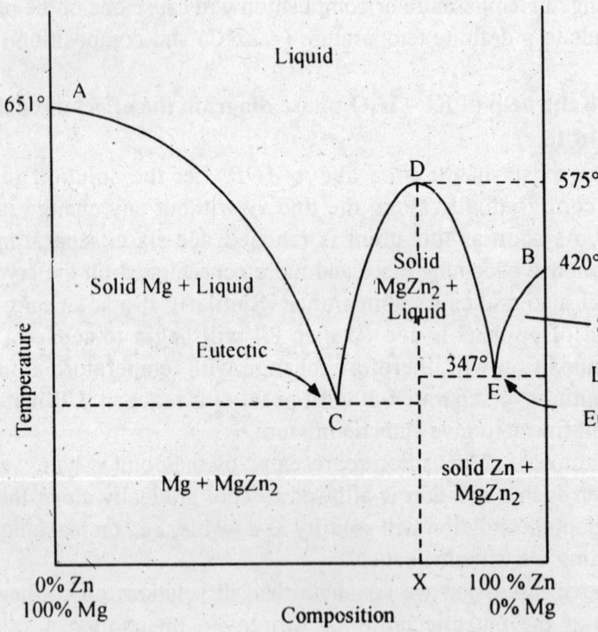

Fig. 16.17. Phase diagram of Magnesium-Zinc system.

The curve *BE* exhibits that the melting point of zinc (420°C) falls with the addition of magnesium until the point *E* is reached.

Along the freezing point curves *AC, CDE* and *BE,* there are two phases in equilibrium *viz.*, one solid phase (Mg, Zn, or $MgZn_2$) and the other liquid phase. Applying the reduced phase rule equation, and substituting values, we have

$$F' = C - P + 1 = 2 - 2 + 1 = 1$$

This shows that the systems Mg/liquid, Zn/liquid, and $MgZn_2$/liquid are all *monovariant* having degree of freedom equal to one.

Eutectic points *C* and *E*. There are two eutectic points in the phase diagram. The systems at the points *C* and *E* have two components and three phases in equilibrium.

Point	Phases Present
C	Solid Mg, solid $MgZn_2$, Liquid
E	Solid Zn, solid $MgZn_2$, Liquid

These systems are, therefore, *nonvariant.*

$$F' = C - P + 1 = 2 - 3 + 1 = 0$$

Congruent Melting point. The composition of the compound $MgZn_2$ and the melt at *D* is identical. The corresponding temperature is the *congruent melting point* of the compound. Here the system has two phases *viz.*, the solid compound and the melt. Both these can be represented by one component ($MgZn_2$). Therefore the system at D is *nonvariant,*

$$F' = C - P + 1 = 1 - 2 + 1 = 0$$

The Areas. The area above the curves *AC, CDE,* and *BE* represents the solution of magnesium and zinc (the melt). The single phase system at any point in this area is *bivariant.*

The phases present in the other regions of the phase diagram have been labelled.

Q. 29. Draw the labelled phase diagram of $FeCl_3 - H_2O$ system. Describe the important features. **(*M. Dayanand 2002*)**

Ans. This is an example of a two-component system in which the two components, namely ferric chloride and water react to form four stable hydrates, represented by the formulae $Fe_2Cl_6.12H_2O$, $Fe_2Cl_6.7H_2O$, $Fe_2.Cl_6.5H_2O$ and $Fe_2\ Cl_6.4H_2O$. The double formula is used for ferric chloride in order to avoid writing of fractional number of water molecules of crystallization. Each of these hydrates possesses a congruent melting point *i.e.*, is stable up to the melting point and melts at the melting point to give liquid of the same composition as that of the solid compound or in other words, at this temperature, the saturated solution of the hydrate has the same composition as the solid phase. Besides the above four solid phases, there are two more solid phases, namely anhydrous Fe_2Cl_6 and ice, one liquid phase *viz.*, solution and one gaseous phase *viz.*, vapour. Thus in all there are *eight* possible phases. The different phases existing in equilibrium under different conditions of temperature and composition are shown in the phase diagram given in Fig. 16.18. Corresponding to the congruent melting point of each hydrate, a maximum is obtained in curve, as shown in the figure.

The various curves and points contained in the above diagram are explained as follows:

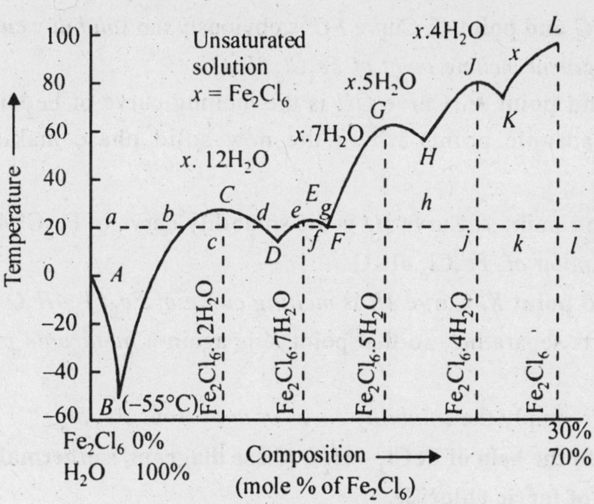

Note: The points corresponding to capital letters *A, B, C,* .... lie on the curve while those corresponding to small letters *a, b, c,* ..... lie on the dotted horizontal line.

Fig. 16.18. $FeCl_3 - H_2O$ system.

(*i*) **Curve *AB* and Point *B*.** Point *A* represents the freezing point of water. Curve *AB* represents the *freezing curve of water to which increasing amounts of $FeCl_3$ have been added.* When point *B* is reached, the solution is saturated with the hydrate $Fe_2Cl_6.12H_2O$ which then starts separating alongwith the ice. thus *B* is a *cryohydric point* where the phases in equilibrium are ice, solid $Fe_2Cl_6.12H_2O$, saturated solution and vapour *i.e.*, there are 4 phases in equilibrium. Hence point *B* is *quadruple point.* The degree of freedom is $F = C - P + 2 = 2 - 4 + 2 = 0$ *i.e.*, the system is *invariant.* Point *B* is a *eutectic point.*

(*ii*) **Curve *BC* and point *C*.** As more of $FeCl_3$ is added, the ice phase disappears and now there are three phases in equilibrium *viz.*, solid $Fe_2Cl_6.12H_2O$, saturated solution and vapour.
Hence $F = C - P + 2 = 2 - 3 + 2 = 1$ *i.e.*, the system becomes *univari ant.* The equilibrium is represented by curve *BC*. Thus the curve *BC* represents the *solubility curve of $Fe_2Cl_2.12H_2O$.* At the point C (corresponding to 37°C) the saturated solution has the same composition as that of the solid $Fe_2Cl_6.12H_2O$. The point C thus represents a *congruent melting point.*

(*iii*) **Curve *CD* and point *D*.** On further addition of $FeCl_3$, the melting point of $Fe_2Cl_6.12H_2O$ decreases and the curve *CD* is followed. Thus the curve *CD* represents the *melting curve of* $Fe_2Cl_6.12H_2O$. At the point *D*, a new solid phase *viz.*, $Fe_2Cl_6.7H_2O$ starts separating out. Hence at the point *D*, again there are four phases in equilibrium namely $Fe_2Cl_6.12H_2O$, $Fe_2Cl_6.7H_2O$, saturated solution and vapour *i.e.*, point *D* is a *quadruple point* and is *invariant* Point *D* is in fact a *eutectic point* or a *transition point* representing the transition.

$$Fe_2Cl_6.12H_2O \rightleftharpoons Fe_2Cl_6.7H_2O$$

(*iv*) **Curve *DE* and point *E*.** Like the curve *BC*, the curve *DE* is the solubility curve of $Fe_2Cl_6.7H_2O$ and point *E* is the congruent melting point of $Fe_2Cl_6.7H_2O$.

(*v*) **Curve *EF* and point *F*.** Like the curve *CD*, the curve *EF* is the melting curve of $Fe_2Cl_6.7H_2O$. At the point *F*, a new solid *viz.*, $Fe_2Cl_6.5H_2O$ starts appearing. The point *F* is thus a *quadruple point*, the four phases in equilibrium being $Fe_2Cl_6.7H_2O$, $FeCl_6.5H_2O$, saturated solution and vapour. Point *F* is another *transition point*, representing the transition.

$$Fe_2Cl_6.7H_2O \rightleftharpoons Fe_2Cl_6.5H_2O.$$

(*vi*) **Curve *FG* and point *G*.** Curve *FG* is obviously the *solubility curve of $Fe_2Cl_6.5H_2O$ and point G is the congruent melting point of $Fe_2Cl_6.5H_2O$.*

(*vii*) **Curve *GH* and point *H*.** Curve *GH* is the melting curve of $Fe_2Cl_6.5H_2O$ and *H* is a *transition point* (quadruple point) where the new solid phase making appearance is $Fe_2Cl_6.4H_2O$.

(*viii*) **Curve *HJ* and point *J*.** Curve *HJ* is the solubility curve of $Fe_2Cl_64H_2O$ and point *J* is the *congruent melting point of* $Fe_2Cl_6.4H_2O$

(*ix*) **Curve *JK* and point *K*.** Curve *JK* is *melting curve of $Fe_2Cl_6.4H_2O$* and at point *K*, the anhydrous Fe_2Cl_6 starts separating, so that point *K* is again a *quadruple point* as well as the *transition point.*

(*x*) **Curve *KL*.** It is simply the *solubility curve of anhydrous Fe_2Cl_6.*

Q. 30. Explain with the help of $FeCl_3 - H_2O$ phase diagram, isothermal evaporation of an unsaturated solution of ferric chloride. ***(Kolhapur 2000)***

Ans. Using the phase diagram of $FeCl_3 - H_2O$ system Fig.16.18, Q. 29 the behaviour on evaporation of an unsaturated ferric chloride solution at constant temperature can be explained. Suppose an unsaturated solution, at temperature between the temperatures of points *E* and *F* (say at 31°C) is evacuated at constant temperature. Suppose such a solution is represented initially by the point *a*. As the solution is heated, water evaporates off and the horizontal path *al* is followed. The various changes occurring on the way are as follows:

(*i*) The concentration of the solution increases till the point *b*, where solid $Fe_2Cl_6.12H_2O$ begins to separate out.

(*ii*) On heating beyond the point *b*, more and more of water evaporates off and more and more of the hydrate $Fe_2Cl_6.12H_2O$ separates out till at the point *c* there is only solid $Fe_2Cl_6.12H_2O$ and no water at all.

(*iii*) On heating beyond the point *c*, the water comes out of the hydrate $Fe_2Cl_6.7H_2O$. The latter keeps on dissolving into the former till a solution of composition corresponding to point *d* is obtained.

(*iv*) Continued evaporation leads to the point *e* where solid $Fe_2Cl_6.7H_2O$ starts separating out, then to the point *f* where there is only solid $Fe_2Cl_6.7H_2O$ which then begins to lose water and form solution till a solution having composition corresponding to point *g* is obtained.

(*v*) At *h*, solid $Fe_2Cl_6.5H_2O$ starts separating out till the point *j*, where there is only solid $Fe_2Cl_6.5H_2O$.

(*vi*) Beyond the point *j*, no solution is formed because the point *j* lies at a temperature below that of the unsaturated solution. Evaporation beyond the point *j* results into the formation of next lower hydrate *viz.*, $Fe_2Cl_6.4H_2O$. Thus all along the path *jk* there are $Fe_2Cl_6.5H_2O$ and $Fe_2Cl_6.4H_2O$ in equilibrium. At the point *k*, there is only $Fe_2Cl_6.4H_2O$.

(*vii*) Heating beyond the point *k* leads to the gradual formation of anhydrous salt which will ultimately be the only residue.

Q. 31. Draw the labelled phase diagram of $Na_2SO_4 - H_2O$ system and describe the main features.

Ans. Various phases present in this system are (*i*) solid $Na_2SO_4.10H_2O$ (*ii*) solid $Na_2SO_4.7H_2O$ (*iii*) solid Na_2SO_4 (rhombic) (*iv*) solid Na_2SO_4 (monoclinic) (*v*) ice (*vi*) solution in water and (*vii*) water vapour. The equilibrium diagram obtained between composition and temperature is shown in Fig. 16.19.

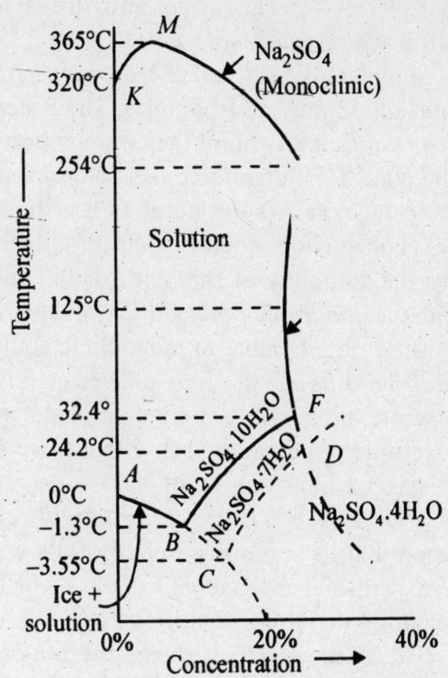

Fig. 16.19. $Na_2SO_4 - H_2O$ system

This is an example of a two-component system with incongruent melting point.

The diagram consists of the curves *AB*, *BC*, *BF*, *CD*, *FD*, *FG* and *GH*; triple point *A* and quadruple points *B*, *C*, *D*, *F* and *G*. However, in the discussion of these curves and points given below, the vapour phase has not been neglected.

(*i*) **Curve *AB* and point *B*.** It is the *freezing point curve* or the curve showing the composition of solutions in equilibrium with ice at different temperatures. At the point *A*, ice is in equilibrium with water and vapour at 0°C. *A* is, therefore, a triple point. With the addition of anhydrous sodium sulphate, the temperature falls as shown by the freezing point curve *AB*. All along the curve *AB*, we have three phases namely ice, solution and vapour in equilibrium. Hence curve *AB* constitutes a univariant system.

$$F = C - P + 2$$

$$= 2 - 3 + 2 = 1.$$

At the point *B* (– 1.3°C) the fourth phase, namely sodium sulphate decahydrate, also appears, so that the system is now invariant,

$$F = C - P + 2 = 2 - 4 + 2 = 0.$$

The point B is, therefore, a stable quadruple point.

(*ii*) **Curve *BC* and point *C*.** Under certain conditions, with care, however, it is possible to travel along the curve BC in which case decahydrate fails to appear at B. At the point C sodium sulphate heptahydrate separates out. The curve BC is, therefore, a *metastable curve* and the point C is a *metastable quadruple point* ; the four phases coexisting in metastable equilibrium at this point C are ice, heptahydrate, solution and vapour.

(*iii*) **Curve *CD* and point *D*.** If now the temperature is raised and more of anhydrous sodium sulphate is added, ice phase disappears as we travel along the curve CD. The curve CD represents the *metastable solubility curve* of sodium sulphate heptahydrate. All along this curve, three phases *viz.*, heptahydrate, solution and vapour are present in equilibrium. Hence the curve CD represents a univariant system,

$$F = C - P + 2$$
$$= 2 - 3 + 2 = 1.$$

At the *point D*, the fourth phase (anhydrous sodium sulphate) makes its appearance. *Point D*, therefore, represents a *second metastable quadruple point.* By lowering the temperature, it is theoretically possible to reach point E, at which sodium sulphate tetrahydrate would separate out. However, in practice it has not been obtained.

(*iv*) **Curve *BF* and point *F*.** It is *stable solubility curve of sodium sulphate decahydrate.* All along this curve three phases namely sodium sulphate decahydrate, solution and vapour co-exist. This, therefore, represents a stable univariant system, however, only upto the point F (32.4°C). At the point F, rhombic form of the salt appears as one of the solid phases. Hence at this point all the four phases—decahydrate, anhydrous salt, solution and vapour are in equilibrium. Point F is thus a *stable quadruple point.*

Point F is also called *transition point* because at this point decahydrate changes into the solid anhydrous salt (rhombic). The water lost by the decahydrate is taken up by the solution and as a result, at this point the composition of the solid phase and liquid phase becomes different. The point F is, therefore, also said to represent incongruent melting point.

(*v*) **Curve *FG* and point *G*.** It is the *solubility curve for anhydrous rhombic sodium sulphate.* This curve is obtained by warming decahydrate. From the shape of the curve, it is quite clear that the solubility of rhombic sodium sulphate decreases upto 125°C, after which it again rises upto the *quadruple point G* (254°), the pressure at this point being 27.5 atmosphere. At this point, transition of rhombic to monoclinic sodium sulphate takes place. So point G is another *stable quadruple point,* the four phases coexisting in equilibrium at this point are rhombic sodium sulphate, monoclinic sodium sulphate, solution and vapour. This point differs from the previous quadruple points in the sense that there are no water molecules present in the two solid phases. It is also a transition point.

(*vi*) **Curve *GH*.** It is the *solubility curve of monoclinic sodium sulphate.* It is clear from the graph that its solubility steadily falls with the rise of temperature and finally the curve ends at H, the critical temperature (365°C) at which the solution and vapour phase merge into each other, the solubility thereby reducing to zero.

(*vii*) **Curve *HK*.** It shows the *solubility of anhydrous sodium sulphate in water vapour,* the two curves GH and HK being continuous.

Q. 32. Explain the following

(*i*) Eutectic point in Pb – Ag system,

(*ii*) Importance of tie – line in univariant system. (*Nagpur 2002*)

Ans. See Q. 21.

Q. 33. Is it possible to have a quadruple point in a phase diagram for one-component system.

Ans. Quadruple point is a point where four different phases meet. Thus $P = 4$, substituting the values in the phase rule equation

$$F = C - P + 2$$

or $$F = 1 - 4 + 2$$

or $$F = -1$$

This is absurd. Therefore, a one-component system cannot have a quadruple point in the phase diagram.

Q. 34. Calculate the degree of freedom for the following:

(*i*) A mixture of nitrogen and oxygen gases contained in a vessel,

(*ii*) Rhombic sulphur in equilibrium with monoclinic sulphur.

Ans. (*i*) **Mixture of nitrogen and oxygen**

No. of phases = 1 (Gases form a single phase)

No. of component = 2

Degree of freedom is given by

$$F = C - P + 2$$
$$= 2 - 1 + 2$$
$$= 3$$

(*ii*) **Rhombic sulphur in equilibrium with monoclinic sulphur.**

No. of phases = 2

No. of components = 2

Degree of freedom is obtained as follows:

$$F = C - P + 2$$
$$= 2 - 2 + 2$$
$$= 2$$

Q. 35. Calculate the number of phases, components and degree of freedom in the following systems:

(*i*) N_2O_4 (g) $\rightleftharpoons$ $2NO_2$ (g)

(*ii*) Solid carbon in equilibrium with gaseous CO, CO_2 and O_2 at 100°C.

Ans. (*i*) N_2O_4 (g) $\rightleftharpoons$ $2NO_2$ (g)

No. of phases = 1 (Gases mix uniformly to give one phase)

No. of components = 1

Degree of freedom $F = C - P + 2$

$$= 1 - 1 + 2$$

or $$F = 2$$

(*ii*) Solid carbon with gaseous CO, CO_2 and O_2

No. of phases = 2 (1 for solid carbon and 1 for gases)

No. of components = 2

Degree of freedom $F = C - P + 2$

$$= 2 - 2 + 2$$
$$= 2$$

Q. 36. Explain (*i*) Transition point (*ii*) Eutectic point (*iii*)Thaw Point

(*M. Dayanand 2000; Nagpur 2002*)

Ans. See Q. 11, 20 and 21

Q.37. Briefly describe a two component system with labelled diagram which involved the compound formation with incongruent melting point. (*M.Dayanand 2000*)

Ans. See Q. 21.

17

SOLID STATE

Q. 1. What are:

(*i*) Crystalline solids

(*ii*) Amorphous solids?

Ans. The two types of solids are described as under:

(*i*) **Crystalline solids:** If various ions, atoms and molecules constituting the solid are arranged in definite geometric pattern within the solid, it is called a crystalline solid. All solid elements and compounds exist in this form.

(*ii*) **Amorphous solids:** A solid is known as amorphous solid if the constituent particles of the substance are not arranged in any regular fashion. Glass, pitch and polymers of high molecular mass appear in this category. Amorphous solids are not accepted as *true solids.* They are regarded as supercooled liquids of high viscosity.

Q. 2. List the points of difference between crystalline and amorphous solids.

(Kerala 2000; Patna 2004; Madras 2010)

Ans. Points of difference between the two solids are given as under:

(*i*) **Arrangement of particles:** The constituent atoms or ions are arranged in a definite pattern in a crystalline solids whereas there is no regular pattern of arrangement of particles in the amorphous solids.

(*ii*) **Melting Point:** Crystalline solids melt sharply at their melting points whereas amorphous solids melt gradually over a range of temperature.

(*iii*) **Isotropy and anisotropy:** Properties like refractive index and electrical conductivity are identical in all directions in the case of amorphous substances. This property is also shown by gases and liquids, and is called *isotropy* and substances showing this property are called *isotropic.* On the other hand, in the case of crystalline substan- ces, the properties mentioned above have different values in different directions. This type of behaviour is called *anisotropy* and the substances exhibiting this type of behaviour are called *anisotropic.* Thus whereas amorphous substances are isotropic in nature, the crystalline substances are anisotropic. The anisotropy exhibited by crystalline substances is obviously due to the reason that we come across different particles in different directions (Fig. 17.1.).

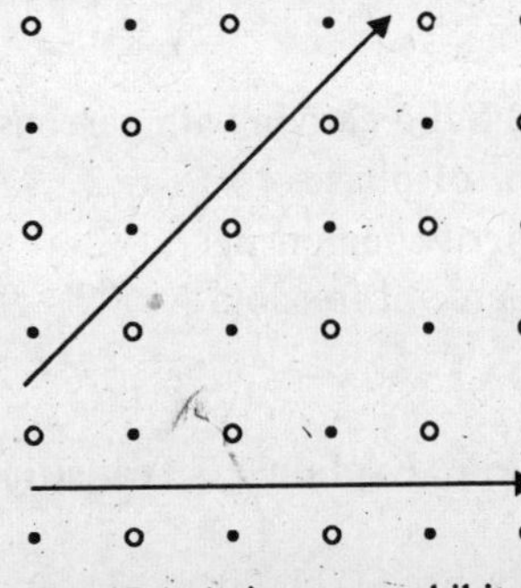

Fig. 17.1. Anisotropy exhibited by crystalline substances.

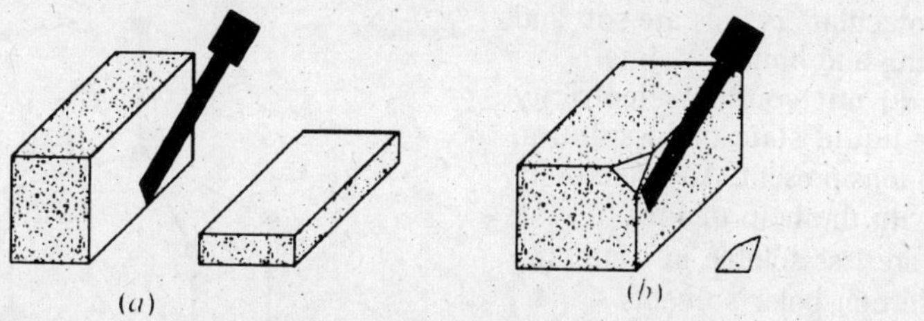

Fig. 17.2. Cutting of solids.

(*iv*) **Regular and irregular cut:** A crystalline solid on being cut with a sharp-edged knife gives a clean cleavage, whereas an amorphous solid breaks irregularly as shown in Fig. 17.2.

(*a*) Acrystalline solid gives a clean cut. (*b*) An amorphous solid gives an irregular cut.

Q. 3. Describe the following: ***(Garhwal 2000)***

(*i*) Ionic crystals

(*ii*) Molecular crystals

(*iii*) Network or covalent crystals

(*iv*) Metallic crystals.

Ans. (*i*) **Ionic Crystals:** In these crystals the constituent particles occupying the lattice points are the *positive and negative ions.* These ions are held together by the *strong electrostatic forces of attraction.* Some of the important characteristics of these crystals are as follows:

(*i*) Because of strong electrostatic forces of attraction among the ions, they have high melting and boiling points.

(*ii*) They are good conductors of electricity in the molten state and in the solution.

(*iii*) They are soluble in polar solvents but insoluble in non-polar solvents.

(*iv*) Because of strong electrostatic forces of attraction, the ions are closely packed and hence the ionic crystals are hard. However, they are brittle because on applying force, their arrangement pattern is disturbed. Sodium chloride crystal is a common example of this type of solid.

(*ii*) **Molecular Crystals:** The constituent particles occupying the lattice points, in this case are the *molecules.* The intermolecular forces holding the molecules together **in the crystal lattice are the *weak van der Waals forces.* In** the case of polar molecules, (*e.g.* HCl, H_2O, NH_3 etc.) these van der Waals forces are the ***dipole-dipole attractions*** as shown in Fig. 17.3 (*a*). In case of non-polar **molecules (*e.g*., H_2, Cl_2, CH_4 etc.) the van der Waals** forces are the *London dispersion forces.*

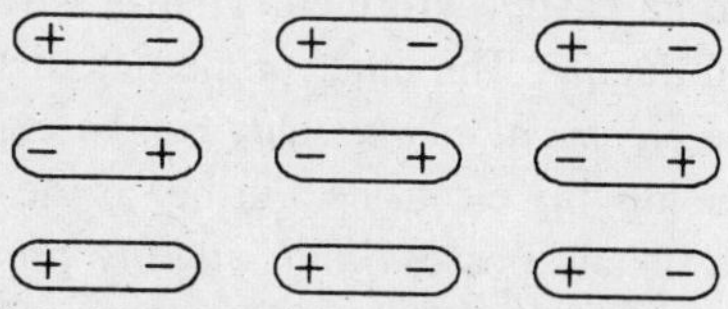

Fig. 17.3. (*a*) Dipole-dipole attraction in molecular crystals.

These forces are supposed to arise due to momentary dipole produced as a result of distortion of the electron cloud of one molecule which produces an induced dipole in the other molecule as shown in Fig. 17.3 (*b*).

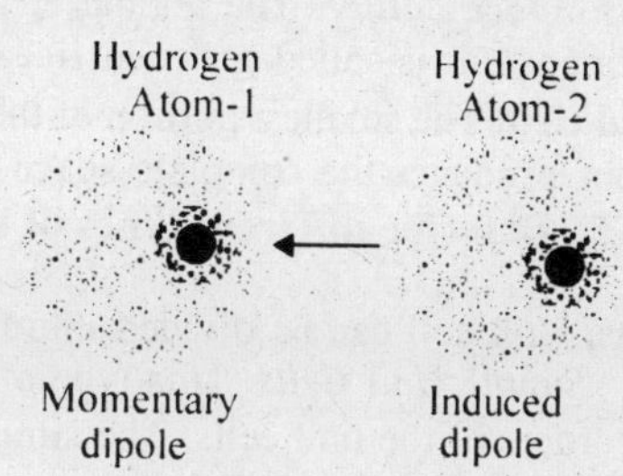

Fig. 17.3. (*b*) London dispersion forces in the molecular of hydrogen.

The main characteristics of the molecular crystals are as follows:

(*a*) As the van der Waal's forces are much weaker than the electrostatic forces existing in ionic crystals, the molecular crystals are soft and have low melting and boiling points.

(*b*) They do not conduct electricity in the solid or liquid state or in solution as there are no ions present. And electricity is conducted with the help of ions.

(*c*) They are less soluble in water and more soluble in non-polar solvents.

(*d*) As dipole-dipole attractions are stronger than London forces, therefore, polar molecular crystals generally have higher melting points and boiling points than crystals of non-polar molecules of *comparable molecular size and shape.*

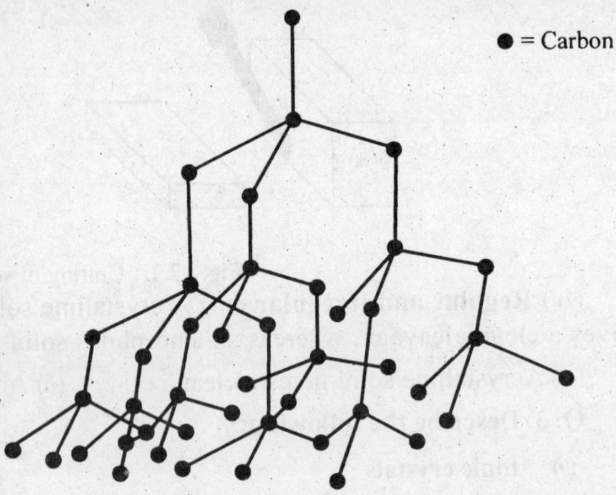

Fig. 17.4. Structure of diamond.

(*iii*) **Network or Covalent Crystals:** In these crystals, the lattice points are occupied by atoms which are linked together by a network of covalent bonds to form a giant molecule. A common example of the crystals of this type is that of diamond in which the carbon atoms are linked together by covalent bonds to give a three dimensional structure as shown in Fig. 17.4.

Substances of this type show high melting points, high boiling points and are extremely hard because of the large number of covalent bonds that have to be broken to destroy the crystal structure. Further, they are non-conductors of electricity.

(*iv*) **Metallic crystals:** These crystals consist of positively charged metals ions (kernels) occupying the lattice points which are held together by the metallic bond. The metallic bond arises due to the presence of *mobile electrons.* These mobile electrons undergo simultaneous attraction by a number of positive ions and hence the ions are held together (Fig. 17.5).

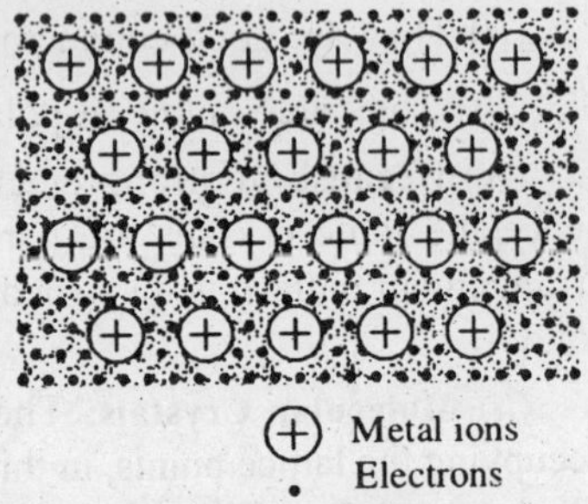

Fig. 17.5. Metallic crystal.

The positions of the positive ions can be changed without destroying the crystal because of the uniform charge distribution provided by the freely moving electrons. Thus metallic crystals can be easily deformed. That is why the metals are *malleable* and *ductile.* The other properties of metals like *lustre, thermal* and *electrical conductivity* can be explained on the basis of the free moving electrons. Further as the positive ions are closely packed in the crystal lattice, most of the metals possess *high melting points* and *high densities.*

Q. 4. What is meant by Space Lattice and Unit Cell ?

(Garhwal 2000; Purvanchal 2003; Mumbai 2004; Nagpur 2008)

Ans. *Space Lattice.* The regular arrangement of the constituents of a crystalline substance in three dimensions is called **space lattice or crystal lattice**.

Unit Cell. The smallest portion of the arrangement of particle which when repeated in different directions produces the complete space lattice, is called **unit cell.**

Q. 5. Describe different types of unit cells (crystal systems). **(Also see Q. 46)**

(Garhwal 2000)

Ans. Unit cell can be divided into four types.

(*i*) *Simple Unit Cells.* This type of unit cell is produced when the particles are present only at the corners of the unit cells. The simple unit cell is also called **primitive unit cell.**

(*ii*) *Face-Centred Unit Cell.* If the particles are located at the centre of each face in addition to the corners, **face-centred unit cell** is produced.

(*iii*) *End-Face Centred Unit Cell.* If the particles are located at the centre of end faces, in addition to the corners, **end face centred unit cell** is produced.

(*iv*) *Body-Centred Unit Cell.* If the particles are present at the centres of the cell in addition to the corners, **body-centred unit cell** is produced.

Different unit cells are shown in Fig. 17.6.

Q. 6. What is meant by Co-ordination Number ?

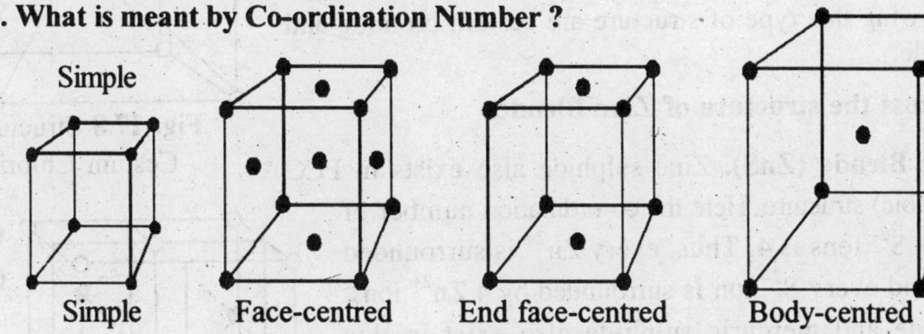

Fig. 17.6. Types of unit cells

Ans. In solid state chemistry, we represent constituent particles by spheres.

The number of spheres which are in direct contact with a particular sphere is called co-ordination number of that sphere or (constituent particle).

In case of an ionic solid, the co-ordination number of a particular ion is the number of oppositely charged ion surrounding that particular ion.

Q. 7. Describe the structure of Sodium Chloride (NaCl).

(Purvanchal 2003, Garhwal 2010; Raipur 2010)

Ans. NaCl has the structure as shown in Fig. 17.7.

(*i*) It has face-centred cubic (FCC) lattice.

Na^+ are present at the corners and face-centres while Cl^- are present at body-centre and edge centres. It can also be vice-versa.

(*ii*) Every sodium ion is surrounded by six Cl^- ions, one above, one below, one on the left, one on the right, one in front and one at the back.

Similarly, every chloride is surrounded by six sodium ions. We express it by saying that sodium ion as well as chloride ion have co-ordination number equal to 6. Co-ordination number is the no. of atoms or ions immediately around a particular atom or ion. This structure has co-ordination number 6 : 6.

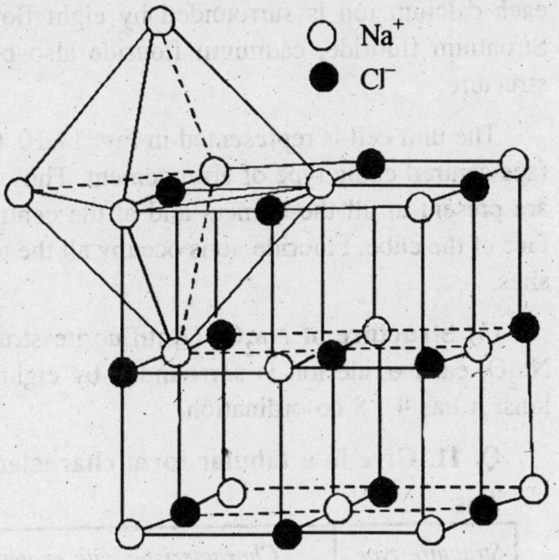

Fig. 17.7 Structure of Sodium Chloride.

(*iii*) If we look carefully at the structure of sodium chloride, we shall find that there are 4 units of sodium and chloride ions each in a unit cell. Sodium bromide, sodium iodide, silver chloride and magnesium oxide also possess sodium chloride type structure, *i.e.*, FCC lattice.

Q. 8. Describe the structure of cesium chloride.

Ans. Cesium chloride has the structure as shown in Fig. 17.8.

This structure is called simple **cubic** structure. Is is commonly misunderstood as body centred cubic structure. It may be understood that the ion in the centre of the cube is different from the ions at the corners, therefore it should not be called body centred cubic structure. The eight corners of the cube are occupied by Cl^-. **Cesium** occupies the centre of body of the cube. Similarly, we shall find that cesium occupies the corners of a cube and chloride is at the centre of that cube. It can be seen from the figure that number of chloride ions surrounding a particular cesium ion is 8. Similarly, the number of cesium ions surrounding a particular chloride ion would be 8. Hence, the co-ordination number of chloride as well as cesium ion is 8. The number of units of Cs and Cl per unit is one. Other substances showing this type of structure are cesium bromide and thalium chloride.

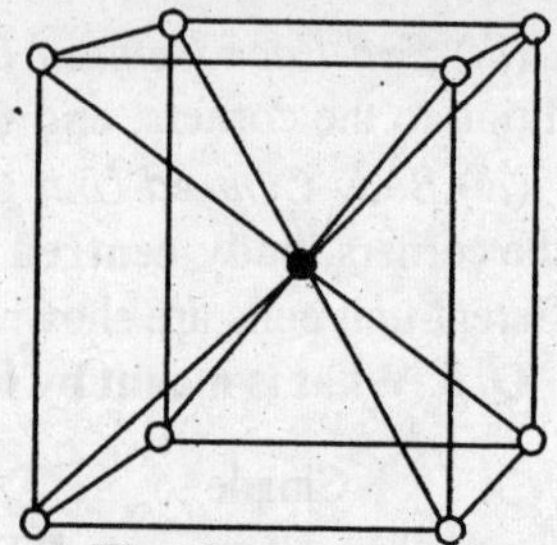

Fig. 17.8 Structure of Cesium Chloride.

Q. 9. Discuss the structure of Zinc Blende.

Ans. Zinc Blende (ZnS). Zinc sulphide also exists in FCC (face-centred cubic) structure. Here the co-ordination number of Zn^{2+} as well as S^{2-} ions is 4. Thus, every Zn^{2+} is surrounded by 4 S^{2-} ions and every S^{2-} ion is surrounded by 4 Zn^{2+} ions. Copper chloride and mercuric sulphide also exist in this structure.

A unit cell representation of zinc blende type structure is shown in Fig. 17.9. The sulphide ions are present at the corners and at the centre of each face of the cube. Zn^{2+} ions occupy alternate tetrahedral sites.

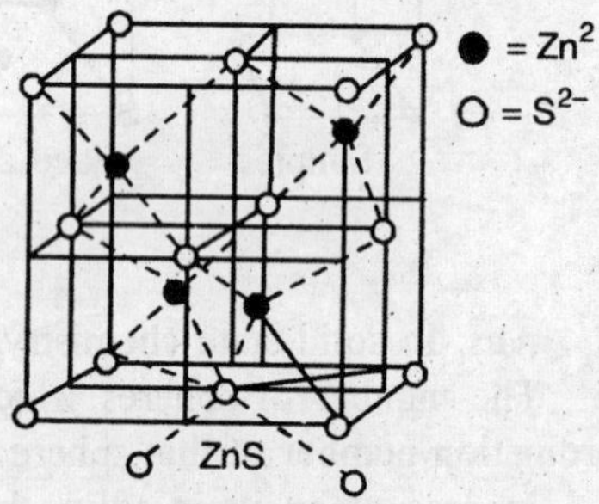

Fig. 17.9 Structure of Zinc Blende.

Q. 10. Give the structure of (*a*) Calcium Flouride (CaF_2) (*b*) Na_2O.

Ans. (*a*) **Structure of CaF_2.** In calcium flouride, each flouride ion is surrounded by four calcium ions and each calcium ion is surrounded by eight flouride ion. Strontium fluoride, cadmium flouride also posses this structure.

The unit cell is represented in Fig. 17.10. Ca^{2+} form face-centred cubic type of arrangement. Thus, Ca^{2+} ions are present at all the corners and at the centre of each face of the cube. Fluoride ions occupy all the tetrahedral sites.

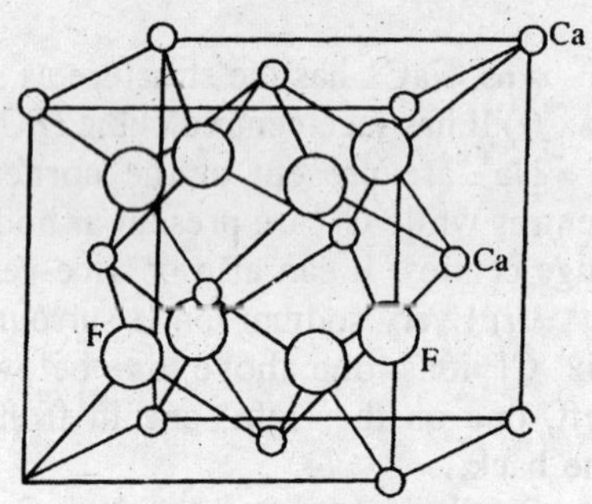

Fig. 17.10 Structure of CaF_2. (Flourite Structure)

(*b*) **Structure of Na_2O.** (Antifluorite structure). In Na_2O, each oxide ion is surrounded by eight sodium ions and each sodium ion by four oxide ions. It has 4 : 8 co-ordination.

Q. 11. Give in a tabular form characteristics of different type of structures.

Ans.

Structure type	*Characteristic with examples*	*Co-ordination number*
1. NaCl	It is FCC structure. The two ions inter-penetrate each other in sodium halides : AgCl, MgO.	$Na^+ = 6$ $Cl^- = 6$
2. CsCl	Cs^+ at the body centre of a cube of Cl^- ions and vice-versa : CsCl, CsBr.	$Cs^+ = 8$ $Cl^- = 8$
3. ZnS	FCC arrangement of both ions ZnS, CuCl, HgS.	$Zn^{2+} = 4$ $S^{2-} = 4$

Q. 12. What is the effect of temperature and pressure on crystal lattice ?

Ans. It is noticed that substances undergo changes in their structure when temperature and pressure are changed. As a general rule, high pressure favours greater co-ordination number and high temperature favours lower co-ordination number. For example, when sodium chloride is subjected to high pressure, it transforms into body-centred cubic structure of cesium chloride. Thus, the co- ordination number is increased from 6 : 6 to 8 : 8. On the other hand, if we heat CsC1 to a high temperature of 760 K, it converts to a structure of sodium chloride.

$$\underset{(6:6 \text{ co-ordination})}{\textbf{NaCl structure}} \underset{\text{Heat}}{\overset{\text{Pressure}}{\rightleftharpoons}} \underset{(8:8 \text{co-ordination})}{\textbf{CsCl structure}}$$

Q. 13. Discuss packing of particles in a crystal.

Ans. Considering the particles as spheres of equal size, we can visualize the formation of a solid as shown in Fig. 17.11.

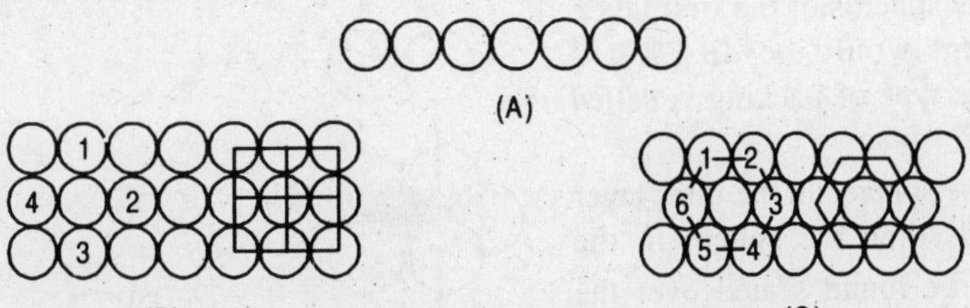

Fig. 17.11. Packing of spheres for efficient use of space. (A) Edge formation (B) Square close packing (C) Hexagonal close packing.

One-Dimensional Packing. We can arrange the spheres in row, see Fig. 17.11 (A), such that they touch one another. This is to utilize the space economically and to account for the high density of the solids.

Two-Dimensional Packing. Then we can place another row of spheres by the side of the first row. This can be done in two ways. One way is depicted in Fig. 17.11 (B). Second method is shown in Fig. 17.11 (C). The first type [Fig. 17.11 (B)] is called **square close packing** and second type [Fig. 17.11 (C)] is called **Hexagonal close packing.**

In square close packing, a sphere is surrounded by four spheres in the plane whereas in hexagonal close packing, a sphere is surrounded by six spheres in the plane. Thus, the empty space in square close packing is more as compared to the hexagonal close packing. Or in other words, hexagonal close packing is more dense than the square close packing.

Three-Dimensional Packing. Now let us form a second layer of spheres on the first layer in hexagonal close packing. (See Fig. 17.12)

Spheres in the first layer are named *A*.There are voids in between them. These voids are *C* voids and voids are indicated in Fig. 17.12 (A). We could place the spheres of the second layer on the first layer in such

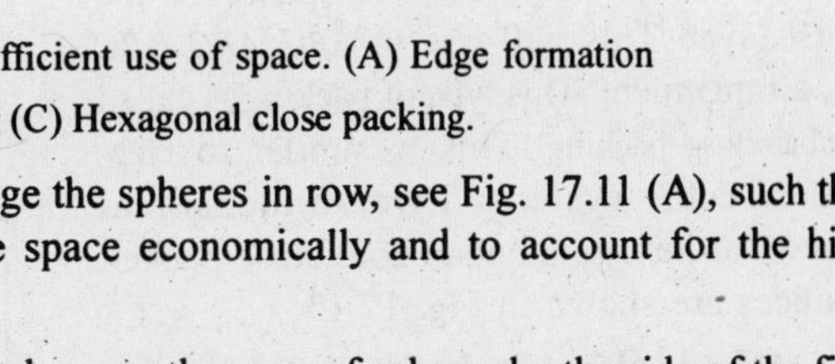

(A)

Fig. 17.12 (A) Hexagonal closed packed layer.

a way that every sphere in the second layer lies vertically over the sphere in the first layer. But this would not to be an efficient packing as this would leave lots of gaps between them.

So let us place the spheres in the second layer over the voids *B* in the first layer as shown in Fig. 17.12 (B). It will then not be possible to place spheres over the voids *C* because of space limitation. Now in the second layer of sphere there are *B′* voids and *C′* voids. **In fact, *C′* voids of the second layer lie over the C voids of the first layer.**

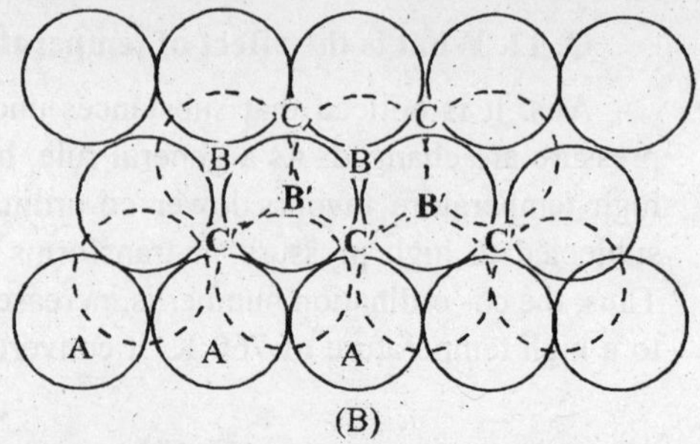

Fig. 17.12 (B) Formation of the second layer.

Let us now place a third layer of spheres over the second layer. There are two possibilities. Either we place spheres on the *B′* voids or on *C′* voids. If we place the spheres on the *B′* voids in third layer, these spheres in fact would be seated over the spheres of the first layer. This arrangement would be *ABABAB* . arrangement. This type of packing is called **Hexagonal Close Packing.**

If we place the spheres in the third layer over the *C′* voids then the sphere in the fourth layer will be found seated over the Cubic close packing Hexagonal spheres in the first layer. This will rise to *ABC, ABC, ABC* ... arrangement. This type of packing is called cubic-close packing. This is similar to face-centre cubic system. Three-dimensional views of hexagonal and cubic close packed lattices are shown in Fig. 17.13.

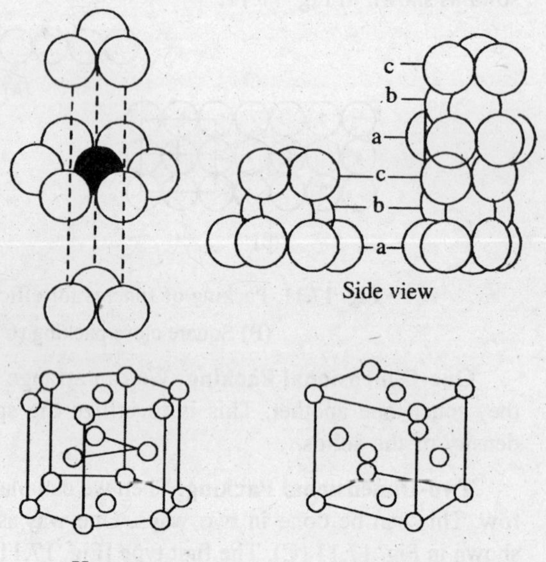

Fig. 17.13 Three-dimensional view of hexagonal and cubic close packed lattices.

Q. 14. What is body-centred cubic (BCC) arrangement ?

Ans. This is deviation from the close packing. If the spheres of the first layer are loosened up. We obtain body-centred cubic structure. Each sphere in the second layer is in contact with four spheres of the first layer (see Fig. 17.14).

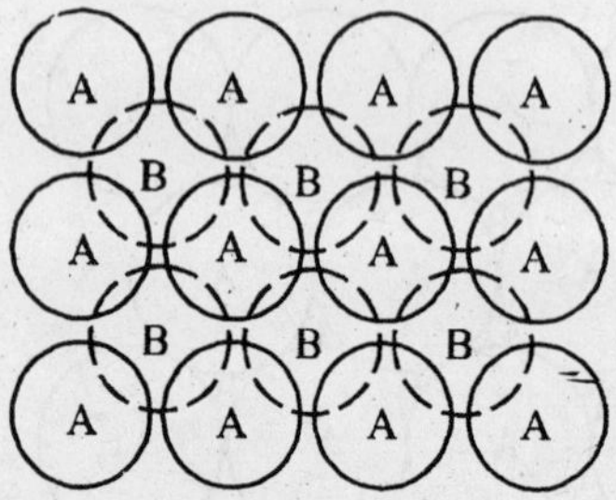

Fig. 17.14. Body-centred cubic packing.

The spheres in the second layer have the same packing as those in the first layer. The spheres in the third layer, having the same packing will be laying exactly over the corresponding spheres in the first layer. Thus, every sphere will be in contact with 8 spheres, four of the layer above and four of the layer below it. The co-ordination number is thus 8.

Alkali metals like Li, Na, K, etc. crystallise in BCC structure. The structure is less closely packed than the CCP and HCP structures and the empty space is about 32%.

Q. 15. Explain tetrahedral and octahedral voids ?

Ans. A void or empty space in a lattice which is surrounded by four spheres is called a tetrahedral void. For example, the void *B* in Fig. 17.12 are tetrahedral voids because they are surrounded by four spheres three in first layer and the fourth in the second layer.

A void which is surrounded by six spheres, is called the octaheral void. The voids *C* in Fig. 17.12 are octaheral voids because, they are surrounded by six spheres, three in the first layer and three in the second layer. These voids are shown in Fig. 17.15

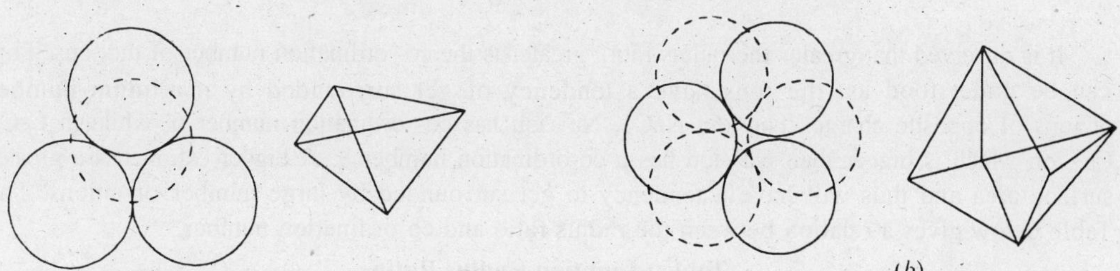

Fig. 17.15. Two types of voids (*a*) Tetrahedral (*b*) Octaheral.

Q. 16. What is the co-ordination number in HCP and CCP ? ***(Kerala 2000)***

Ans. Every shpere in either HCP or CCP has 12 spheres directly touching it, 6 in its own layer, 3 in layer below it and 3 in the layer above it. Thus, the co-ordination number of a sphere in HCP or CCP is 12. In other types of crystal, co-ordination number of 4, 6, 8 is quite common.

Q. 17. Discuss some general observations that have been made in the study of crystals.

Ans. *Some observations are* :

(*i*) Noble gases are most of metals crystallise in closed packed structures.

(*ii*) A large number of ionic compounds crystallise in structures which are related to close packed structures. Here, one of the ions forms the close packed structures and the other ion goes into voids.

(*iii*) If in a compound of the general formula *AB,* the *B* ions form the close packed structures, then *A* ions will occupy all the octahedral voids or half the tetrahedral voids.

(*iv*) In a compound of the type A_2B, all the tetrahedral sites will be occupied by *A* ions.

(*v*) In NaCl, Cl^- forms the CCP structure and Na^+ occupy the octahedral voids. In ZnS, S^{2-} forms the CCP structures with Zn^{2+} occupying alternate tetrahedral voids. In CaF_2, it is the Ca^{2+} which form the CCP lattice and F^- occupy all the tetrahedral voids.

(*vi*) In the packing of oxides of ion, *viz.*, FeO and Fe_2O_3, FeO has the NaCl structure. Thus, Fe^{2+} should occupy the octahedral voids. But this oxide has stootimetry with a formula $Fe_{0.95}O$. In converting two thirds of Fe_2^+ in FeO into Fe^{3+}, we get FeO. Fe_2O_3 or Fe_3O_4. In Fe_2O_4 all the Fe^{2+} are in octahedral voids and Fe^{3+} are equally distributed between octahedral and tetrahedral voids. $MgFe_2O_4$ has also a similar structure.

(*vii*) $MgAl_2O_4$ and $ZnFe_2O_4$ (ferrites) form spinal structures, Mg^{2+} in the former and Zn^{2+} in the latter (divalent ions) are in tetrahedral holes and Al^{3+} and Fe^{3+} are in the octahedral voids. These spinal type compounds are important magnetic materials used in memory loops in computers.

Q. 18. What is the radius ratio ? How does co-ordinate number vary with the radius ratio ?

Ans. In a solid, the co-ordination number of an ion, *i.e.*, the number of ions surrounding a particular ion depend upon the radius ratio. In ionic solids, radius ratio is defined *as the ratio of the radius of the cation to the radius of anion.*

$$\text{Radius ratio} = \frac{\text{Radius of cation}\,(r^{+})}{\text{Radius of anion}\,(r^{-})}$$

It is observed that greater the radius ratio, greater is the co- ordination number of the ions. This can be understood as: the ions nave a tendency of get surrounded by maximum number of ions of opposite charge. Thus, in NaCl, Na^+ ion has co- ordination number 6, while in CsCl, Cs^+ ion which is bigger than Na^+ ion has a co-ordination number 8. A bigger cation has a greater surface area and thus will have a tendency to get surrounded by large number of anions. The Table below gives a relation between the radius ratio and co-ordination number.

Table : Limiting Radius Ratio

Radius Ratio	*Co-ordination No.*	*Arrangement*
1.0 – 0.732	8	Cubic
0.732 – 0.414	6	Octahedral
0.414 – 0.225	4	Tetrahedral
0.225 – 0.155	3	Triangular

We may visualize the above results as follows :

If the anions are touching one another and also the cation together, it leads to stable arrangement. The stable arrangement for triangular structure will be as shown in Fig. 17.16 (*a*). If the cation becomes bigger in size, the anion will no longer be touching one another, leading to instability [Fig. 17.16 (*b*)]. To counter this they will transform into tetrahedral arrangement [Fig. 17.16 (*c*)], so that anions continue to touch one-another and also the cation in contact with all the three anions. Let us start from the octahedral arrangement. If the cation becomes smaller (radius ratio decrease), the arrangement changes to tetrahedral in order to maintain close contact between the anions and between the anions and the cations.

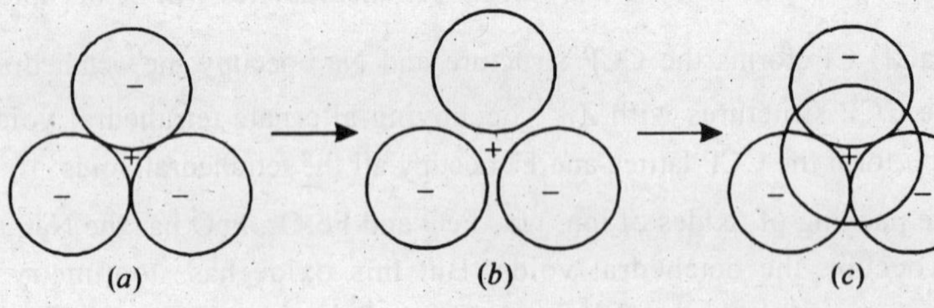

Fig. 17.16. The stable planar triangular arrangement (*a*) changing to tetrahedral arrangement (*c*) as r^{+}/r^{-} increases.

On the other hand if the cation becomes bigger (radius ratio increases), the structure changes to BCC (co-ordination no. 8) again to maintain close contact between the anions and between anion and cation. This is illustrated in Fig. 17.17.

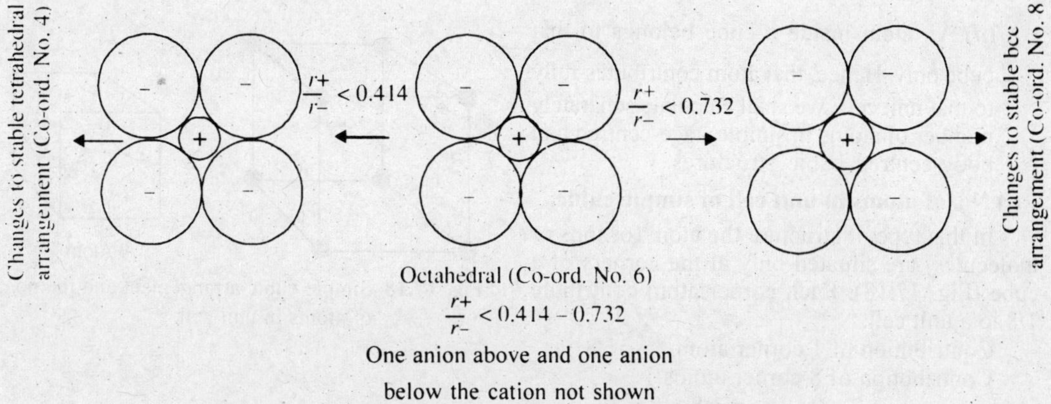

Fig. 17.17. Effect of radius ratio on co-ordination number.

PROBLEMS BASED ON RADIUS RATIO

Q. 19. The radius of the cation in a solid measures 33 pm and the anion measures 70 pm. Predict the co-ordination no. of cation and structure of solid.

Ans. Radius of cation (r^+) = 30 pm

Radius of anion (r^-) = 70 pm.

$$\text{Radius ratio } (r^+/r^-) = \frac{30}{70} = 0.471 \text{ pm}$$

As the value lies in the range 0.732 – 0.414, the co- ordination no. of the cation is 6 and octahedral arrangement is predicted.

Q. 20. A solid AB has the NaCl structure. If the radius of cation A is 100 pm, what is the radius of the anion B ?

Ans. Since the solid has NaCl structure the co-ordination number of *A* and *B* each is 6. This is possible when the radius ratio is in the range 0.414 – 0.732 thus

$$\frac{\text{Radius of cation}}{\text{Radius of anion}} = 0.414 \text{ to } 0.732$$

or
$$\text{Radius of anion} = \frac{\text{Radius of cation}}{0.414 \text{ to } 0.732}$$

$$= 241.6 \text{ to } 133.0 \text{ pm}$$

$$= 133.6 \text{ to } 241.6 \text{ pm}$$

Q. 21. How do you calculate the number of atoms in a unit cell ? Calculate the same for simple cubic, FCC and BCC structures? (*Guwahati 2006, Calcutta 2007*)

Ans. The following points are kept in mind while calculating the number of atoms in unit cell.

(*a*) Each corner atom in a cube is shared between eight unit cells. Therefore, a corner atom contributes 1/8 to each unit cell.

(*b*) An atom on the face of unit cell is shared by the two unit cells. Thus, a face atom contributes 1/2 to each unit cell.

(*c*) An atom on the edge of a cube is shared by four unit cells. Hence, an edge atom contribute 1/4 to each unit cell.

(*d*) An atom inside a cube belongs to that cube only. Hence, that atom contributes fully to that unit cell. We shall calculate separately number of atoms in simple, face-centred and body-centred cubic structures.

(*i*) **No. of atoms in unit cell of simple cubic.**

In this type of structure the atom (or ions or molecules) are situated only at the corners of a cube (Fig. 17.18). Each corner atom contribute 1/8 to a unit cell.

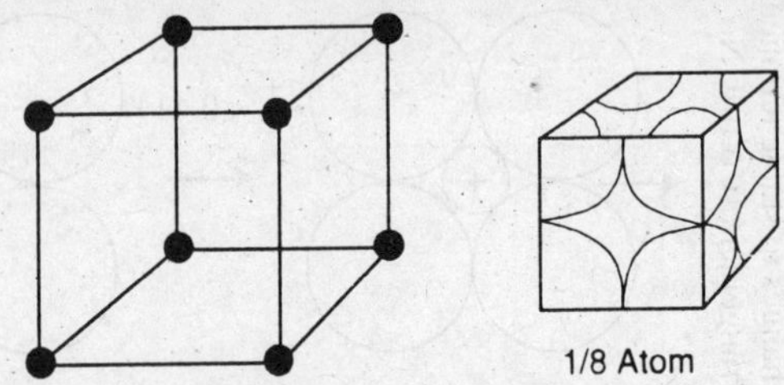

Fig. 17.18 Simple cubic arrangement and the no. of atoms in unit cell.

Contribution of 1 corner atom = 1/8

Contribution of 8 corner atoms

$= 1/8 \times 8 = 1$

Hence, the no. of atoms per unit cell in a simple cubic structure is 1.

(*ii*) **No. of atoms in a unit cell of FCC.**

In a face-centred cubic, there are 8 atoms at the 8 corners of the cubic and 6 atoms at the centre of 6 faces of the cube. (Fig. 17.19)

Contribution of 1 corner atom = 1/8

Contribution of 8 corner atoms

$= 1/8 \times 8 = 1$

Contribution of 1 face atom = 1/2

Contribution of 6 face atoms = $6 \times 1/2 = 3$

$\therefore$ Total no. of atoms in unit cell of FCC = $1 + 3 = 4$

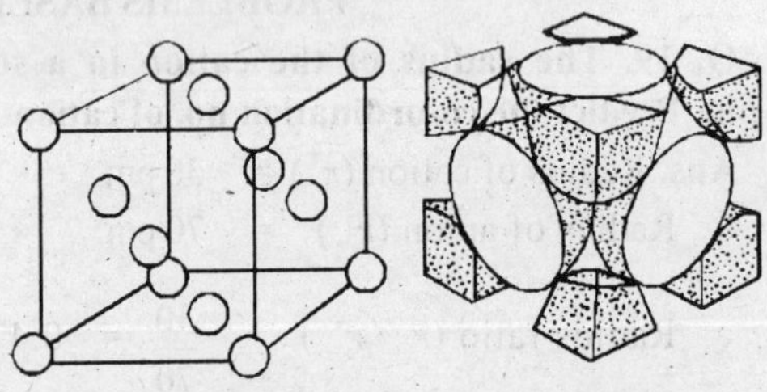

Fig. 17.19 Face-centred cubic arrangement.

(*iii*) **No. of atoms in a unit cell of BCC.**

In Body centred cubic structure, there are 8 atoms at the centre of the body of the cube (Fig. 17.20).

Contribution of 1 corner atom = 1/8

Contribution of 8 corner atoms = $1/8 \times 8 = 1$

Contribution of 1 central atom = 1

Thus, total no. of atoms in a unit cell of body centred cubic = $1 + 1 = 2$.

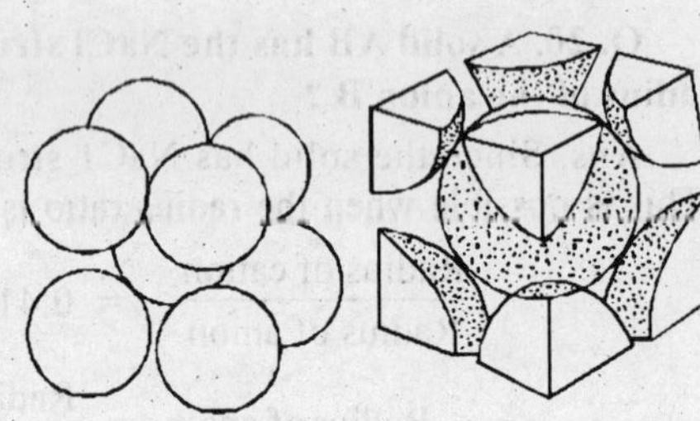

Fig. 17.20. Body-centred cubic arrangement.

Q. 22. A compound formed by elements *X* and *Y* crystallizes in the cubic structures. *X* atoms are situated at the corners and *Y* atoms are the centre of faces. What is the formula of compound ?

Ans. There are 8 atoms of *X* at 8 corners and 6 atoms of Y at the centre of six faces of the face-centred cubic.

Contribution of 1 corner atom = 1/8

Contribution of 8 corner atoms = $1/8 \times 8 = 1$

Contribution of 1 face-centred atom = 1/2

Contribution of 6 face-centred atoms = $6 \times 1/2 = 3$

Thus, there is 1 atom of *X* and 3 atoms of *Y*/unit cell.

Hence, the formula of the compound is XY_3.

Q. 23. A cubic solid is made up of two elements *P* and *Q*. Atoms *Q* are present at the corner of cube and atoms *P* at the body centre. What is the formula of the compound ? What are the co-ordination number of *P* and *Q* ?

Ans. No. of atoms of *Q* (at the corner) in unit cell = $1/8 \times 8 = 1$

No. of atoms of P (at body-centre) in unit cell = 1

Hence, the formula of the compound is PQ.

Co-ordination of no. of P and Q each = 8.

Q. 24. Write the relation between atomic radius (r) and the edge length (a) in crystals.

Ans. The following relations exist between d and a for various structures.

Structure	*Relation*
Simple	$d = a$
Body-centred	$d = \frac{\sqrt{3}}{2} a = 0.866\ a$
Face-centred	$d = \frac{a}{\sqrt{2}} = 0.707\ a$

Relation between atomic radius (r) which is equal to $d/2$ for pure elements and the edge (a).

The following relations exist between the two quantities for various structures.

Structure	*Relation*
Simple	$r = \frac{a}{2}$
Body-centred	$r = \frac{\sqrt{3}}{4} a = 0.433\ a$
Face-centred	$r = \frac{a}{2\sqrt{2}} = 0.3535a$

Q. 25. CsCl has bcc arrangement and its unit cell edge length is 400 pm. Calculate the interionic distance in CsCl.

Ans. Fig. 17.21 shows the arrangement of CsCl. Open circles represent Cl^- and solid circle Cs^+. Half of the distance CE is the interionic distance

$$AE = \sqrt{a^2 + a^2} = \sqrt{2}\, a$$

$$CE = \sqrt{2a^2 + a^2} = \sqrt{3}\ a$$

$$= \sqrt{3} \times 400$$

$$\text{Interionic distance} = \frac{1}{2} \times \sqrt{3} \times 400 = 346.4 \text{ pm.}$$

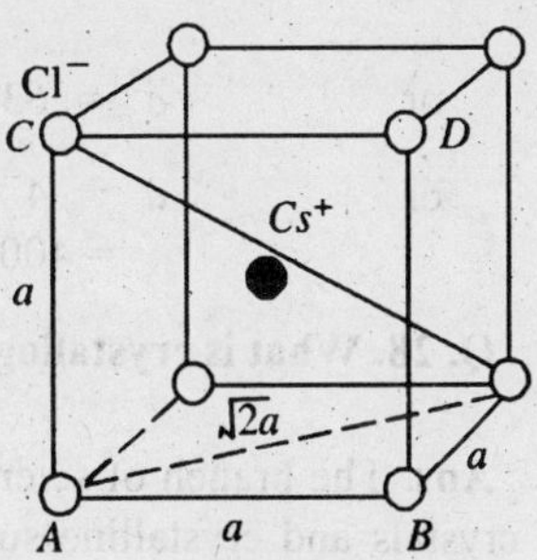

Fig. 17.21

Q. 26. Derive the relation between the density and edge length of a cube.

Ans. If we know the structure of the solid, then it is possible to calculate the density provided we know the cell parameters, *i.e.*, length of edge of the cube.

Let the length of edge of the cube = a pm.

$$= 9 \times 10^{-10} \text{cm.}$$

The volume of the cube = $(\text{Edge})^3$

$$= (a \times 10^{-10})^3 = a^3 \times 10^{-30} \text{ cm}^3$$

$$\text{Density of the unit cell} = \frac{\text{Mass of the unit cell}}{\text{Volume of the unit cell}} \quad ...(1)$$

But mass of unit cell = No. of atoms in the unit cell × Mass of each atom

If we represent the number of atoms in the unit cell by Z and mass of each atom m, then

$$\text{Mass of unit cell} = Z \times m \qquad ...(2)$$

Also, $$\text{Mass of each atom, } m = \frac{\text{Atomic Mass}}{\text{Avogadro's no.}} \qquad ...(3)$$

Substituting the value of m from eqn. (3), in eqn. (2)

$$\text{Mass of unit cell} = Z \times \frac{\text{Atomic Mass}}{6.023 \times 10^{23}} \qquad ...(4)$$

Substituting the value of mass of unit cell from eqn. (4) in eqn. (1)

$$\text{Density of the unit cell of the solid} = \frac{Z \times \text{Atomic Mass}}{6.023 \times 10^{23} \times a^3 \times 10^{-30}} \text{g/cm}^3$$

Q. 27. An element (atomic mass = 60 a.m.u.) having face-centred cubic unit cell has a density of 6. 23 gcm^{-3} Calculate the edge length of the unit cell. (*Guwahati 2007*)

Ans. Use the relation between density and edge length (a) :

$$\text{Density} = \frac{Z \times \text{atomic mass}}{6.023 \times 10^{23} \times a^3 \times 10^{-3}} \text{g/cm}^3 \quad (a \text{ has the units of picometre pm})$$

or $$a^3 = \frac{Z \times \text{atomic mass}}{6.023 \times 10^{23} \times 10^{-30} \times \text{density}} \text{cm}^3$$

For a face-centred cubic, $Z = 4$ Substitute the values in the above equation

$$a^3 = \frac{4 \times 60}{6.023 \times 10^{-7} \times \text{density}}$$

or $$a^3 = \frac{4 \times 60}{6.023 \times 10^{-7} \times 6.23}$$

or $$a^3 = 63.96 \times 10^6$$

or $$a = 4 \times 10^2 \text{ pm}$$

$$= 400 \text{ pm} \quad \text{or} \quad 400 \times 10^{-10} \text{ cm} \quad \text{or} \quad 4 \times 10^{-8} \text{ cm or } 4 \text{ Å}$$

Q. 28. What is crystallography? Name three fundamental laws of crystallography. . (*Purvanchal 2002*)

Ans. The branch of science which deals with the study of geometry, properties and structure of crystals and crystalline substances is known as *Crystallography*. Three fundamental laws of crystallography are:

(1) The law of constancy of interfacial angles
(2) The law of rationality of indices
(3) The law of symmetry.

Q. 29. Explain the following terms used in crystallography:

(*i*) Face (*ii*) Edge (*iii*) Form (*iv*) Zone

(*v*) Crystal habit(*vi*) Interfacial angle.

Ans. These terms are described as under:

Face: The plane surfaces of a crystal are called the faces of the crystal.

Edge: The line along which the two adjacent faces of a crystal intersect is called the edge.

Form: A set of similar faces of a crystal is called a form.

Zone: Faces are found to occur in sets, which meet in parallel edges or would meet in parallel edges if the planes of the faces were extended. Such a set of faces is known as zone.

Crystal habit: The external shape of the crystal is called crystal habit

Interfacial angle: The angle between the adjacent faces of a crystal is called interfacial angle.

Q. 30. Explain the law of constancy of interfacial angles (*M. Dayanand 2002; Agra 2004, Nagpur 2008*)

Ans. The crystals of a substance are obtained by cooling the liquid (or the melt) or the solution of that substance. The size of the crystals depends upon the rate of cooling. If cooling is carried out slowly, crystals of large size are obtained because the particles (*viz.*, ions, atoms or molecules) get sufficient time to arrange themselves in proper positions. Similarly, it is found that the shape of the crystals of a substance also depends upon the conditions under which the crystallization takes place *i.e.*, the absence or presence of any other substance *etc.* For example, shapes of the crystals of a particular substance are shown in Fig. 17.22.

However, inspite of the different sizes and shapes of the crystal of the same substance, it is observed that the *angles between the corresponding faces called the interfacial angles of the crystals of a particular substance are always the same.* This is called the **law of constancy of interfacial angles.** The instrument used for the measurement of interfacial angles is called a **goniometer.**

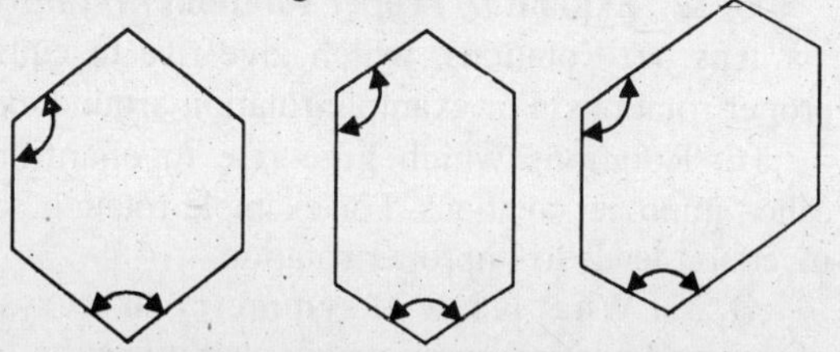

Fig. 17.22. Shapes of crystals (crystal habits) of a particular substance.

Q. 31. Explain the following elements of symmetry:

i)	**Centre of symmetry**	*(Nagpur 2008; Andhra 2010)*
ii)	**Plane of symmetry**	*(Nagpur 2008; Andhra 2010)*
(iii)	**Rotation axis of symmetry**	*(Himachal 2010; Mumbai 2010)*
(iv)	**Rotation-reflection axis of symmetry**	*(Nagpur 2008)*
v)	**Rotation-inversion axis of symmetry.**	*(Andhra 2010; Himachal 2010)*

Or

Write a note on symmetry elements of a crystal. ***(M. Dayanand 2000; Delhi 2003)***

Ans. (*i*) Centre of symmetry: *It is defined as an imaginary point within the crystal such that any line passing through this point intersects the opposite faces of the crystal at equal distances.* The centre of symmetry is also called the **centre of inversion** because if the crystal or the molecule is inverted through the centre of symmetry, it gives results equivalent to the original and hence indistinguishable from the original one. For example, the centre of inversion

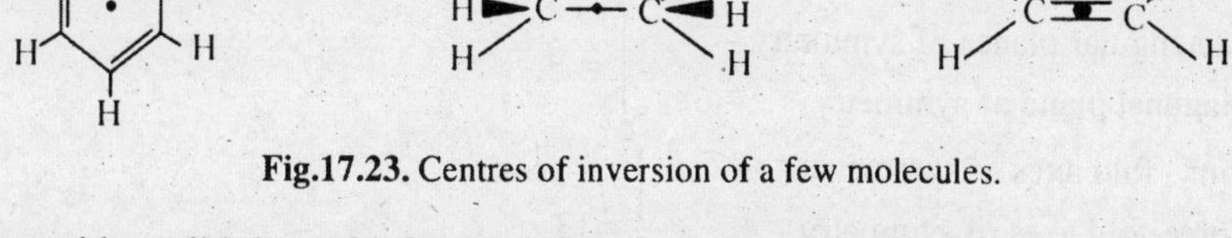

Fig.17.23. Centres of inversion of a few molecules.

represented by solid dots of a few molecules are shown in Fig. 17.23.

(*ii*) Plane of symmetry: *It is defined as an imaginary plane passing through, the crystal such that it divides the crystal into two parts in such a way that one part is the mirror image of the other.*

(*iii*) Rotation axis of symmetry: *The rotation axis of symmetry or simply called the axis of symmetry is defined as the imaginary line passing through the crystal such that when the crystal is rotated about this line, exactly similar appearance occurs more than once in one complete* *revolution, i.e., in a rotation through 360°.* If similar appearance occurs twice in one complete revolution (*i.e.*, after every 180°), the axis is called *two-fold axis* of symmetry or a **diad axis**. Similarly the other possibilities found for similar appearance in one complete revolution are three, four times or six times (*i.e.*, after rotation through 120°, 90°, or 60°). The corresponding axes are called *three fold* (**triad**), *four-fold* (**tetrad**) and *six fold* (**hexad**) axes of symmetry.

(*iv*) Rotation-reflection axis of symmetry : When the rotation of a molecule or a crystal about an *n*-fold axis of proper rotation is followed by reflection through a plane perpendicular to the axis, the axis is called an *n*-fold rotation-reflection axis. The new configuration obtained is not congruent but is enantiomorphous.

(*v*) Rotation-inversion axis of symmetry: When the rotation of a molecule or a crystal is followed by inversion through the centre of symmetry the axis is called an *n*-fold rotation-inversion axis. The new configuration obtained is not congruent but is enantiomorphous, as in the above case.

Q. 32. Explain *(i)* Proper rotation, *(ii)* Improper rotation. *(M. Dayanand 2000)*

Ans. (*i*) Rotations, which give rise to equivalent and congruent configurations are called proper rotations. For example rotation around axis of symmetry is *proper rotation.*

(*ii*) Rotations which give rise to enantiomorphous but incongruent configurations are called improper rotations. For example rotation around axis of symmetry followed by reflection or inversion leads to improper rotation.

Q. 33. What is law of symmetry in a crystal? Show with the help of diagrams various planes and axes of symmetry in a simple cube. *(Vidyasagar 2003; Banglore 2005)*

Ans. *The crystals of a particular substance always possess the same elements of symmetry.* This is called **law of symmetry**. A cubic crystal always possesses the following 23 different elements of symmetry:

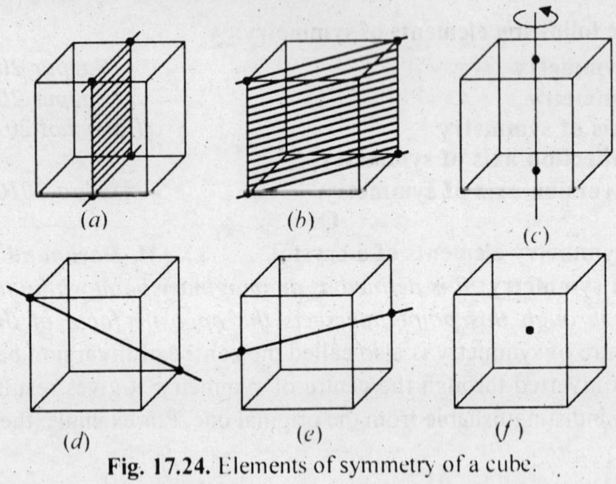

Fig. 17.24. Elements of symmetry of a cube.

(*a*) Rectangular planes of symmetry	– 3	} = 9
(*b*) Diagonal plane of symmetry	– 6	
(*c*) Four - fold axes of symmetry	– 3	} =13
(*d*) Three-fold axes of symmetry	– 4	
(*e*) Two-fold axes of symmetry	– 6	
(*f*) Centre of symmetry (located at the centre of gravity of the cube)		= 1

An example of each type of symmetry possessed by a cube is shown in Fig. 17.24.

Q. 34. What is meant by crystallographic axes and axial ratio ?

Ans. Crystallographic Axes: In order to describe the relative directions and orientations of the faces and also the planes present within the crystal, three suitable axes are chosen which meet at a point, say O, (Fig. 17.25). These axes can be selected in a number of ways depending upon the symmetries of the crystal. The best choice is usually the lines, coinciding with or parallel to the edges between the principal faces. For example, in a cubic crystal, the best choice of the axes is the three lines mutually at right angles to each other. The three lines thus selected meeting at a point are called *crystallographic axes.*

The angles between the axes opposite to the intercepts a, b and c are represented by α, β and γ respectively. These angles are the same as the inter axial ratio. facial angles if the axes chosen are the lines coinciding with or parallel to the edges of the principal faces of the crystal.

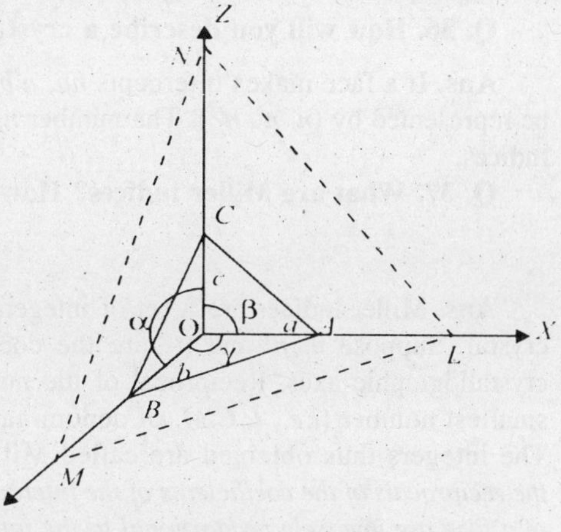

Fig. 17.25. Crystallographic axes and axial ratio.

Axial Ratio: To explain axial ratio, consider a plane called **unit plane** which meets the crystallographic axes. For example, suppose the three axes are represented by OX, OY and OZ and the unit plane ABC cuts these axes at A, B and C respectively, making the intercepts $OA = a$, $OB = b$, $OC = c$. The ratio $a : b : c$ is called the **axial ratio.** Thus *the axial ratio may by defined as the ratio of the intercepts made by the unit plane on the crystallographic axes.*

Q. 35. Explain the law of rational indices.

(M. Dayanand 2002; Delhi 2002; Purvanchal 2007; Mumbai 2010

Ans. If the intercepts of any a face of the crystal on the three axes are measured, the following generalization is observed which is called law of rational indices:

The intercepts made by any face of the crystal on the crystallographic axes are either:

(*i*) *same as those of the unit plane, or*

(*ii*) *simple whole number multiples of those of the unit plane, or*

(*iii*) *one or two intercepts may be infinity, if the face is parallel to one or two axes, i.e., the face does not cut one or two axes.*

For example, in Fig. 17.25 (Q. 34), the intercepts made by the face LMN are $2a$, $2b$ and $3c$ which are simple whole number multiples of those of the unit plane. Again consider the case of cube. Suppose the origin O of the crystallographic axes lies at the centre of symmetry and the axes OX, OY, OZ are parallel to the edges as shown in Fig. 17.26. Suppose the axis OX cuts the face $ABCD$ at the point L. Similarly, suppose the axes OY and OZ cut the faces at M and N. If LMN is taken as the unit plane, the intercepts made by the unit plane on the three axes are OL, OM and ON. The face $ABCD$ cuts the X-axis at L *i.e.*, makes the same intercept as that made by the unit plane but it does not cut the Y-axis and Z-axis at all, because these two axes are parallel to this face. In other words, the intercepts made by the face $ABCD$ on the Y-axis and Z-axis are *infinity.*

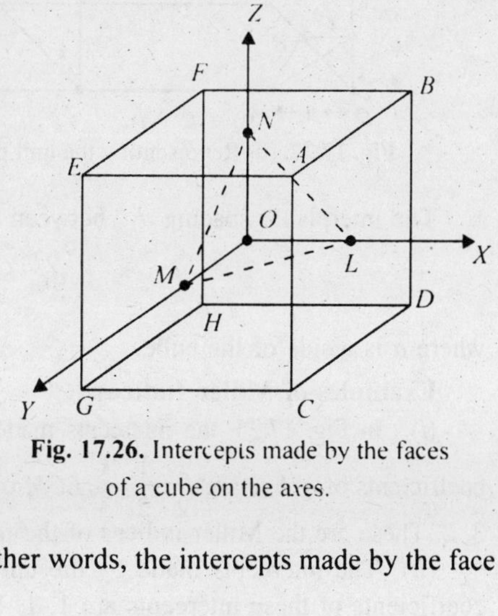

Fig. 17.26. Intercepts made by the faces of a cube on the axes.

In general if a face PQR makes intercepts p, q and r on the three axes and the intercepts made by the unit plane are a, b, c then we have $p = na$, $q = n'b$ and $r = n''c$ where n, n' and n'' are simple whole numbers or some out of these may be *infinity*.

Q. 36. How will you describe a crystal plane in terms of Weiss indices ?

Ans. If a face makes intercepts na, $n'b$ and $n''c$ on the crystallographic axes, the face may be represented by (n, n', n''). The number n, n', n'' thus used to represent a face are called **Weiss indices.**

Q. 37. What are Miller indices? How are they used to describe a plane?

(Bangalore 2004; Patna 2004; Calcutta 2007; Purvanchal 2007; Nagpur 2008; Madras 2011)

Ans. Miller indices are a set of integers (hkl) which are used to describe a given plane in a crystal. Suppose n, n' and n'' are the coefficients of the intercepts of a given plane on the crystallographic axes. Reciprocal of the numbers n, n' and n'' are taken and multiplied by the smallest number (*i.e., L.C.M.* of denominators) so as to express all the reciprocals as integers. The integers thus obtained are called Miller indices. Thus *Miller indices may be defined as the reciprocals of the coefficients of the intercepts, expressed as integers.* Obviously, *the Miller indices of a face are inversely proportional to the intercepts of that face on the crystallographic axes.*
Alternatively, the intercepts made by any crystal plane such as DEF, in Fig. 17.27, are usually expressed in terms of the intercepts of the unit plane (a, b, c) as follows:

$$OD = \frac{a}{h},\ OE = \frac{b}{k} \text{ and } OF = \frac{c}{l}$$

Then the Miller indices of the plane are represented as (hkl)
The unit plane is represented as shown in Fig. 17.27 (a) and any other plane of the crystal may be represented as shown in Fig. 17.27 (b).

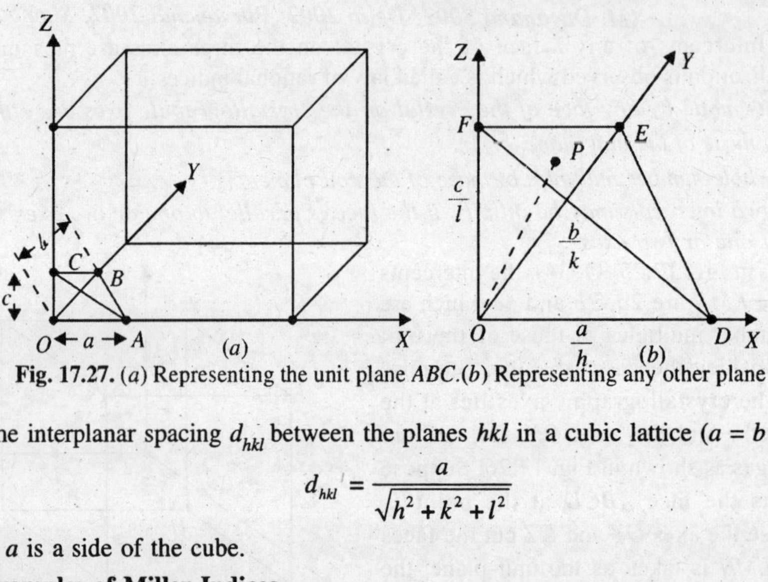

Fig. 17.27. (a) Representing the unit plane ABC.(b) Representing any other plane DEF.

The interplanar spacing d_{hkl} between the planes hkl in a cubic lattice ($a = b = c$) is given by

$$d_{hkl} = \frac{a}{\sqrt{h^2 + k^2 + l^2}}$$

where a is a side of the cube.

Examples of Miller Indices

(*i*) In Fig. 17.25, the intercepts made by the face LMN are $2a$, $2b$, $3c$. The reciprocals of the coefficients of a, b, c are $\frac{1}{2}, \frac{1}{2}, \frac{1}{3}$, *LCM* of 2, 2, 3 = 6. Multiplying the reciprocals by 6, we have 3, 3, 2. These are the Miller indices of the face LMN in Fig. 17.25. This face is represented as (332).

(*ii*) The intercepts made by the unit ABC [Fig. 17.25 or 17.27 (a)] plane are a, b, c. The coefficients of these intercepts are 1, 1, 1. Hence the unit plane is always represented as (111).

(*iii*) If some face does not cut some axis at all, the intercept on this axis is taken as infinity. For example, suppose a face makes intercepts $2a$ and $3b$ on the X-axis and Y-axis respectively and does not cut the Z-axis at all, the intercepts will be $2a$, $3b$ and ∞ (or ∞x). The reciprocals of the coefficients are 1/2, 1/3, 0. *LCM* of 2, 3 = 6. Multiplying by 6, the Miller indices will be 3, 2, 0, *i.e.*, the face will be represented as (320). Similarly, if a face makes intercepts a on the X-axis and does not cut the Y-axis and Z-axis at all (see Fig. 17.26, face *ABCD*), its Miller indices will be 1, 0, 0, *i.e.*, it will be represented as (100).

(*iv*) If some intercept made by a face is negative *i.e.*, the face is cut by the axis when extended in the opposite direction, it is represented by putting a bar over it. For example in Fig. 17.26, the intercept made by the face *EFGH* on the X-axis is $-a$ and this face does not cut the Y-axis and Z-axis at all. Hence this face is represented as $(\bar{1}00)$

(*v*) **Miller Indices of the Faces of a Cube:** In Fig. 17.26, in which the origin of the crystallographic axes is the centre of symmetry of the cube, following the method as explained in (*iii*) and (*iv*) above, the various faces of the cube in terms of Miller indices may be represented as (100), (010), (001), $(\bar{1}00)$, $(0\bar{1}0)$ and $(00\bar{1})$.

Q. 38. A crystal plane has intercepts on the three axes of crystal in the ratio of $\frac{3}{2}$: 2 : 1. What are the Miller indices of the plane?

Ans. The reciprocal of the ratio of intercepts are

$$\frac{2}{3} : \frac{1}{2} : 1$$

Multiply by the L.C.M of 3 and 2 *.i.e.*, 6

$$4 : 3 : 6$$

The Miller indices of the plane are 436.

Q. 39. Write Miller indices for any four faces of a Cubic unit Cell.

Refer to Fig. 17.26, Q. 35.

If O is taken as the origin of crystallographic axes, then Miller indices of :

Plane *ABCD* are 100

Plane *ACGE* are 010

Plane *ABFE* are 001

Plane *EFGH* are $\bar{1}00$.

Q. 40. The Weiss indices of a planes are $\frac{1}{2}\frac{1}{2}\frac{1}{2}$ What are the Miller indices?

Ans. Miller indices are obtained by taking the reciprocal of Weiss indices and then by multiplying, if required, with a suitable number to obtain integral numbers.

Millar indices will be

$$\frac{1}{1/2}\ \frac{1}{1/2}\ \frac{1}{1/2} \quad \text{or } 222$$

Q. 41. Determine the interplanar spacing between the (221) planes of a cubic lattice of length 4·5 Å (450 pm).

Ans. In a cubic lattice, the interplanar spacing d_{221} is given by

$$d_{221} = \frac{a}{(h^2+k^2+l^2)^{1/2}} \text{Å}$$

$$= \frac{4.5}{(4+4+1)^{1/2}} \text{Å}$$

$$= 1.5 \text{ Å} = 150 \text{ pm}$$

Q.42. Find the miller Indices of the plane that intersect the crystallographic axes at the distance a, 2b, ∞ c *(Calcutta 2000)*

Ans. Miller indices of the plane are : $\frac{1}{1}\ \frac{1}{2}\ \frac{1}{\infty}$ or 210

Q. 43. Differentiate between isomorphism and polymorphism.

Ans. Certain crystalline substances which have similar chemical composition, (for example, **sodium phosphate and sodium arsenate** ***i.e.*****,** Na_3PO_4 **and** Na_3AsO_4**) are found to possess the same** crystal shape. This phenomenon is known as **isomorphism** and the substances are said to be **isomorphous.**

There are certain substances which exist in more than one crystalline form which are formed under different conditions of temperature and pressure. This phenomenon is known as **polymorphism.** For example, carbon exists in two crystalline forms called graphite and diamond; **calcium carbonate exists in two crystalline forms called calcite and aragonite. The polymorphism** exhibited by the elements is commonly called **allotropy.**

Q. 44. Name different crystallographic systems. Give the axial characteristics, symmetry elements and examples of each. *(M. Dayanand 2000, Madras 2011)*

Ans. Various crystallographic systems alongwith the required information are given in the tabular form as shown below in Table 17.1:

Table 17.1. Related Information Regarding Crystallographic Systems

System	*Axial characteristics*	*Maximum Symmetry*	*Examples*
1. Cubic	$\alpha = \beta = \gamma = 90°$ $a = b = c$	Plane = 9 Axes = 13	NaC1, KC1, Alums, Diamond
2. Tetragonal	$\alpha = \beta = \gamma = 90°$ $a = b \neq c$	Planes = 5 Axes = 5	Sn, SnO_2, TiO_2
3. Orthorhombic or Rhombic	$\alpha = \beta = \gamma = 90°$ $a \neq b \neq c$	Planes = 3 Axes = 3	KNO_3, K_2SO_4, $BaSO_4$, Pb,CO_3 Rhombic sulphur
4. Monoclinic	$\alpha = \gamma = 90° \neq \beta$ $a \neq b \neq c$	Planes = 1 Axes = 1	$Na_2SO_4 . 10H_2O$ $CaSO_4 . 2H_2O$ Monoclinic sulphur
5. Triclinic	$\alpha \neq \beta \neq \gamma \neq 90°$ $a \neq b \neq c$	Planes = Nil Axes = Nil	$CuSO_4 . 5H_2O$ $K_2Cr_2O_2$
6. Rhombohedral or Trigonal	$\alpha = \beta = \gamma \neq 90°$ $a = b = c$	Planes = 7 Axes = 7	$NaNO_3$, ICl, As, Sb **Bi**
7. Hexagonal,	$\alpha = \beta = 90°$ $\gamma = 120°$ $a = b \neq c$	Planes = 7 Axes = 7	HgS, Ice Graphite, Mg, Zn, Cd

Q. 45. Identify the crystal system to which some solids having the following dimensions for their unit cell belong. Give examples of the solids.

(*i*) $a \neq b \neq c \qquad \alpha \neq \beta \neq \gamma \neq 90°$

(*ii*) $a \neq b \neq c \qquad \alpha = \beta = \gamma = 90°$ *(M. Dayanand 2002)*

Ans. (*i*) Solids having these characteristics are *triclinic.* Examples of this crystal system are $Cu\ SO_4 . 5H_2O$ and $K_2Cr_2O_7$.

(*ii*) Solids having these characteristics are *orthorhombic or rhombic.* Examples of this category of solids are KNO_3 and K_2SO_4.

Q. 46. What is meant by space lattice and unit cell of a crystal?

(Kumaon 2000;Nagpur 2003; Kerala 2004; Calcutta 2007)

Ans. It is found that the particles constituting the crystalline substance are arranged in a regular fashion within the crystal in the three dimensional space. For example, the arrangement of particles (represented by black circles) for a cubic crystal is shown in Fig. 17.28.

The regular arrangement of ions, atoms, or molecules constituting the crystal in the three dimensional space within the crystal is called the **space lattice.**

If a big-sized crystal is broken more and more, ultimately a stage is reached when we get the smallest possible crystal. If it is broken further, it will break to give the constituent particles, *i.e.*, ions, atoms or molecules. For example, if a big cubic crystal is broken, the smallest crystal obtained will be a small cube, as represented by thick lines in Fig. 17.28. This small cubic crystal has all the elements of symmetry as possessed by the big cubic crystal. Moreover it is evident that the complete lattice has been obtained by the repetition of this smallest unit in different directions.

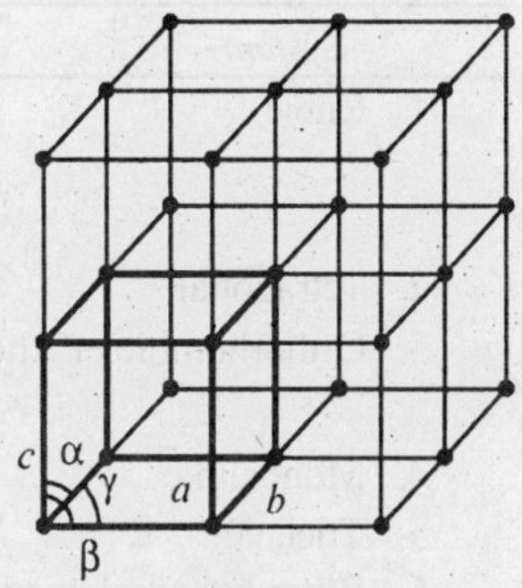

Fig. 17.28. Space lattice and unit cell.

The smallest portion of the complete space lattice, which when repeated over and again in different directions produces the complete space lattice, is called the **unit cell.**

Q. 47 Explain: We do not come across 5-fold, 7-fold, 8-fold or higher fold axis of rotation in crystals.

Ans. We have only 2-fold, 3-fold, 4-fold and 6-fold axes of rotation and we do not have 5-fold, 7-fold, 8-fold or higher fold axes of rotation. This is due to the fact that the most essential requirement of a lattice is that it should be built up by the repetition of unit cells, the faces of each unit cell touching the faces of the surrounding unit cells without leaving any gap in-between. Such a close packing is possible if an object has 2-fold, 3-fold, 4-fold or 6-fold axis of symmetry but is not possible in case an object has 5-fold, 7-fold, 8-fold or higher-fold axis of symmetry. This is illustrated along side for objects having 6-fold axis of symmetry and 5-fold axis of symmetry. (Fig. 17.29). We find that a gap is created in a solid with 5-fold axis of symmetry.

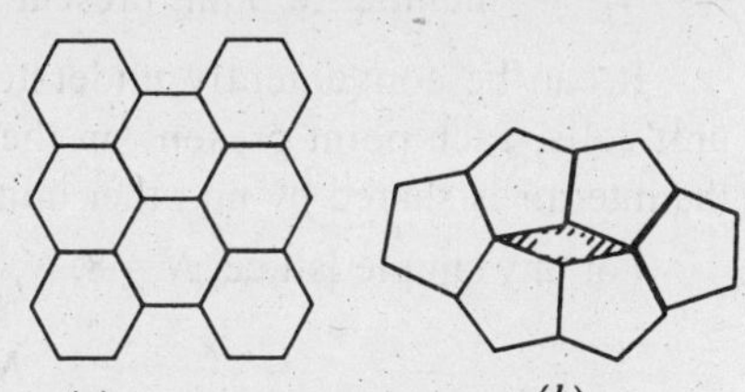

(*a*) (*b*)

Fig. 17.29.(*a*) the close packing of hexagonal objects. (*b*) a gap created in the packing of pentagonal objects.

Q. 48.What are Bravais lattices? How do you calculate the number of atoms per unit cell?

(Delhi, 2003; Nagpur, 2003; Vidyasagar 2003; Burdwan 2004; Madras 2011,

Ans. The detailed study of the internal structure by Bravais, has shown that the points (particles) may be present not only at the corners of unit cells but also at the centre of the faces or within the body of the unit cells. This gives rise to different types of lattices corresponding to the seven crystallographic systems. These 14 types of lattices are called **Bravais lattices.** The description of these lattices is given in Table 17.2. and the shapes are shown in Fig. 17.30. The terms *simple, face-centresd* (*end-centred* and *body-centred* are explained as under:

(*i*) **Simple:** When the points are present only at the corners of the unit cell.

(*ii*) **Face-centred:** When in addition to the points at the corners, there are points at the centre of the each face of the unit cell.

(*iii*) **End-centred:** When there are points present at the centre of the end faces of the unit cell, in addition to points at the corners.

(*iv*) **Body-centred:** When in addition to the points at the corners, there is one point present within the body of each unit cell.

Table 17.2. Bravais lattices corresponding to different crystal systems.

System	*Number of Bravais lattices*	*Description of the lattices*
1. Cubic	3	Simple, Face-centred, Body-centred.
2. Tetragonal	2	Simple, Body-centred
3. Orthorhombic or Rhombic	4	Simple, Face-centred End-centred, Body-centred.
4. Monoclinic	2	Simple, End-centred
5. Triclinic	1	Simple
6. Rhombohedral or trigonal	1	Simple
7. Hexagonal	1	Simple

The number of points per unit cell (N) is given by the expression

$$N = \frac{N_c}{8} + \frac{N_f}{2} + N_i$$

where

N_c = number of points present on the corners of the unit cell

N_f = number of points present on the faces of the unit cell

N_i = number of point present in the interior of the unit cell.

It can be conveniently understood that each point present on the corner is shared by 8 unit cells, each point present on the face is shared by two unit cells and the point present in the interior is shared by no other unit cell.

For any simple latt.ce, $N = 8$, $N_f = 0$, $N_i = 0$, so that

$$\mathbf{N = \frac{8}{8} + 0 + 0 = 1}$$

For a face-centred cubic lattice, $N_c = 8$, $N_f = 6$, $N_i = 0$ so that

$$N = \frac{8}{8} + \frac{6}{2} + 0 = 4$$

For body-centred orthorhombic lattice, $N_c = 8$, $N_f = 0$, $N_i = 1$ so that

$$N = \frac{8}{8} + 0 + 1 = 2.$$

For end-centred monoclinic lattice, $N_c = 8$, $N_f = 2$, $N_i = 0$ so that

$$N = \frac{8}{8} + \frac{2}{2} + 0 = 2.$$

Q. 49. Explain the terms:

(*i*) Point groups

(*ii*) Space groups. (*Meerut 2000*)

Ans. Theoretically, there are 32 different ways of arranging the elements of symmetry about a point. These different arrangements are called *points groups.* However, in a complete space lattice,

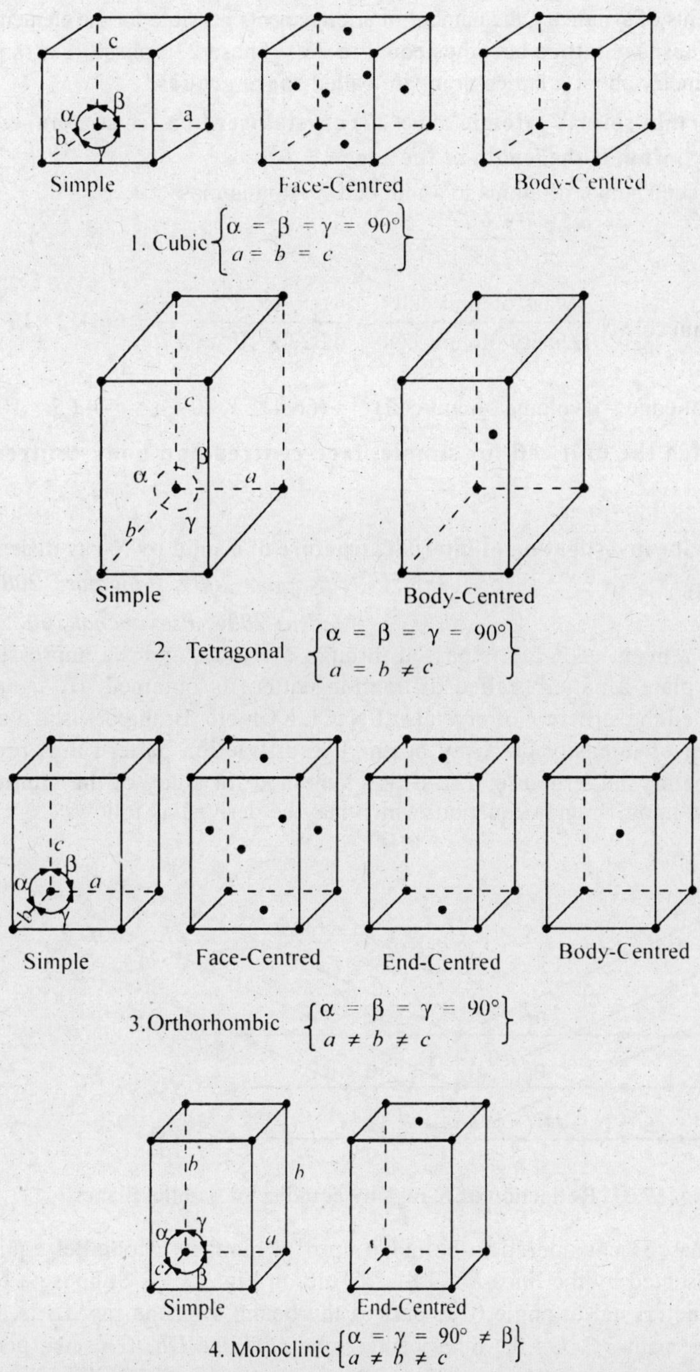

Fig. 17.30 Bravais lattices corresponding to different crystal systems

because of the regular repetition of the pattern in the three dimensional space, two additional symmetry elements called **glide planes** and **screw axes** become possible. Because of these additional elements of symmetry, the number of arrangements possible for the elements of symmetry about a point of a space lattice becomes equal to 230. These 230 possible arrangements of the elements of symmetry about a lattice point are called **space groups.**

Q. 50. A certain solid *X* (atomic mass 27) crystallises in a fcc structure. If the density of *X* is 2.7 *g* cm^{-3}, what is the length of the edge ?

Mass of unit cell = No. of atoms in a unit cell × Atomic mass

$$= \frac{4 \times 27}{6.023 \times 10^{23}}$$

$$\text{Valume of unit cell} = \frac{\text{Mass of unit cell}}{\text{Density of unit cell}} = \frac{4 \times 27}{6.023 \times 10^{23} \times 2.7} = 66.412 \times 10^{-24}\ \text{cm}^3$$

$$\text{Length of edge} = (\text{volume of unit cell})^{\frac{1}{3}} = (66.412 \times 10^{-24})^{\frac{1}{3}} = 4.05 \times 10^{-8}\ \text{cm} = 4.05\text{Å}$$

Q. 51. Sketch the unit cell for simple, face-centred and body-centred cubic space lattices.

Ans. See Q. 48, Fig. 17.30.

Q. 52. Describe investigation of internal structure of a solid by X-ray diffraction (Bragg's method). *(Nagpur, 2003; Bangalore, 2004; Jiwaji 2004, Madras 2004; Purvanchal 2007; Mumbai 2010*

Ans. When a beam of X-ray is passed through a crystal, and the impression is taken on a photographic plate a pattern called diffraction pattern is obtained. By using this pattern, Bragg investigated the structure of crystals of NaCl, KCl, etc. Bragg devised a spectromter for the measurement of intensity of X-ray beam. The diffraction pattern thus produced can be used to study the crystal structure. The Bragg's method for study of the internal structure of crystals is based upon Bragg's equation, which may be derived as follows·

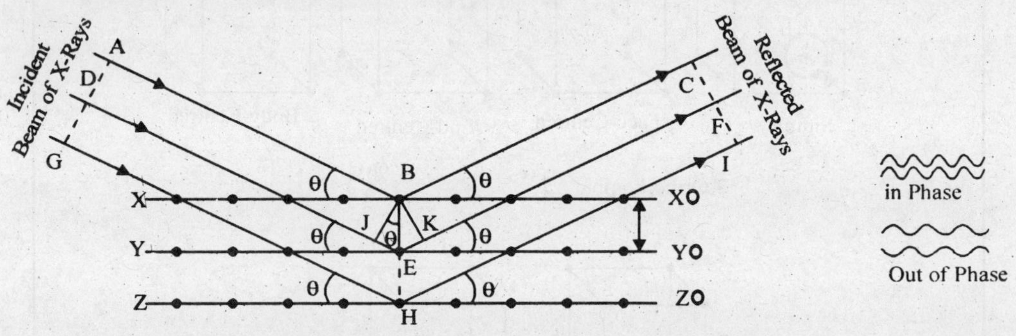

Fig. 17.31. Reflection of X-rays by equidistant parallel planes.

A crystal may be considered to be made up of a number of parallel equidistant atomic planes, as represented by the lines *XX'*, *YY'*, *ZZ"* etc. in Fig. 17.31. Suppose a beam of X-rays is incident on the crystal at angle θ. A part of the beam, *e.g.*, the ray *AB* is reflected at the point *B* along the path *BC*. On the other hand, some rays like *DE*, *GH*, etc., penetrate into the crystal and are reflected by the atoms present in the atomic planes *YY'*, *ZZ'*, etc., as shown below :

Evidently, as compared to the ray *AB*, a ray like *DE* has to travel a longer distance, equal to *JEK* in order to emerge out of the crystal where *BJ* and *BK* are perpendiculars drawn on the lines DE and EF respectively. The reflected beams like BC, EF, etc. then undergo interference with each other. If these reflected rays are in phase (*i.e.*, in the reflected rays, crests fall over the crests and troughs over the troughs), they reinforce each other and the intensity of the reflected rays is maximum. On the other hand, if the reflected rays are out of phase (*i.e.*, crests fall over the troughs), the intensity of the reflected beam is very low. If a photographic plate is placed to receive the reflected rays, *diffraction pattern* is obtained.

It is obvious that in order that the reflected rays BC and EF may be in phase, the extra distance JEK traversed by the ray DE should be an integral multiple of the wavelength λ of the X-rays, *i.e.*,

Distance $JEK = n\lambda$ ...(*i*)

where n is an integer, *i.e.* 1, 2, 3, 4, etc. If d is the distance between the successive atomic planes, it is obvious from Fig. 17.31 that

$$JE = EK = d \sin \theta$$

so that $$JE K = 2d \sin \theta \quad ...(ii)$$

Putting this value in equation (*i*), we get

$$\boxed{2d \sin \theta = n\lambda} \quad ...(iii)$$

This equation is called Bragg's equation. Using monochromatic X-rays, λ is constant. Also for a particular crystal when a particular face is facing the X-rays, d is constant. Thus by gradually increasing the value of θ, a number of positions are observed corresponding to n= 1, 2, 3, 4 etc., where the reflected beam will have maximum intensity. For other values of θ lying in between, the reflected beam will have intensity less than the maximum. Thus a diffraction pattern will be obtained in which the various maxima correspond to n= 1, 2, 3 etc. These are respectively called diffraction maxima of the first order, second order, third order etc.

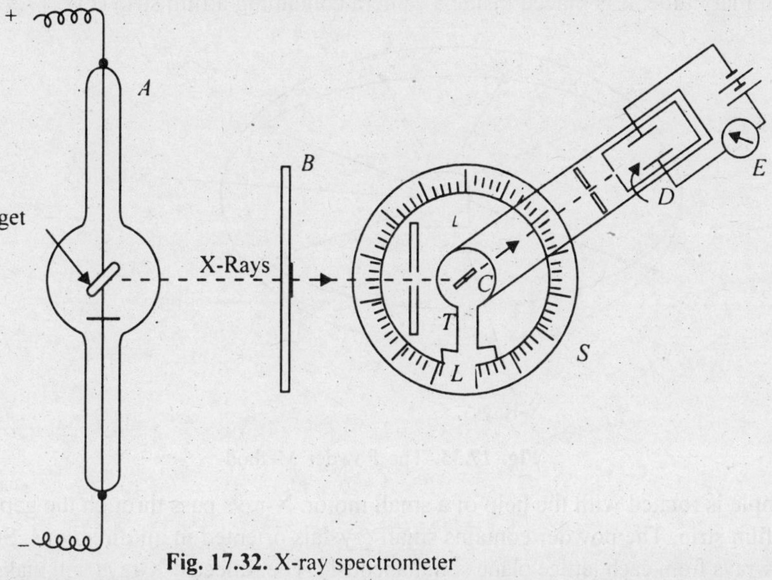

Fig. 17.32. X-ray spectrometer

Measuring the angle θ at which the first maxima occurs (*i.e.*, first maximum intensity is observed) so that n = 1, and knowing the wavelength λ of X-rays used, the value of d can be calculated, using equation (*iii*).

Apparatus: The apparatus used by Bragg for studying the internal structure of crystals is called Bragg's X-ray spectometer and is shown in Fig. 17.32.

X-ray are generated in the *X*-ray tube *A*. These are made monochromatic *i.e.*, of single wavelength by passing through the absorbing screens *B*. Then they are passed through the slits *L*, to obtain the rays in a fine beam. This beam is then allowed to fall on some particular face of the crystal *C* which is mounted on a turn-table *T*. The turn-table can be moved over a circular scale *S*. The reflected beam is passed into the ionization chamber *D* which is connected to an electrometer *E*. The ionization chamber can also be moved on the circular scale *S* independent of the turn table *T*. The chamber contains some easily ionizable gas such as sulphur dioxide. Greater the intensity of the reflected beam, greater is the ionization produced in the gas and larger is the reading on the electrometer. When the crystal face is rotated by rotating the turn-table, the direction of the reflected beam changes. To receive the reflected beam into the ionization chamber every time, the ionization chamber is rotated through double the angle θ through which the crystal face is rotated. From these plots, the angles corresponding to the first, second, third maxima etc., are noted and the value of *d* calculated using Bragg's equation.

Calculation of *d*. For the face (100) of NaCl crystal, when $n = 1$, $\theta = 5.9°$ so that $\sin\theta = 0.103$ Substituting these values in the Bragg equation, we get

$$2d\sin\theta = n\lambda$$

$$d = \frac{n\lambda}{2\sin\theta} = \frac{\lambda}{2\times 0.103} = 4.85\lambda$$

Thus knowing the wavelength of the X-ray used, distance '*d*' between the planes parallel to the face on which the X-rays are incident can be calculated.

Q. 53. Describe the powder method for determination of crystal structure of a solid.

(Jamia Millia 2003; Raipur 2010)

Ans. When a single undistorted crystal of a solid is not available, we can use the powder method for crystal structural determination. In this method, the powdered crystalline substance is taken in a capillary tube. It is placed inside a camera containing a film strip (Fig. 17.33).

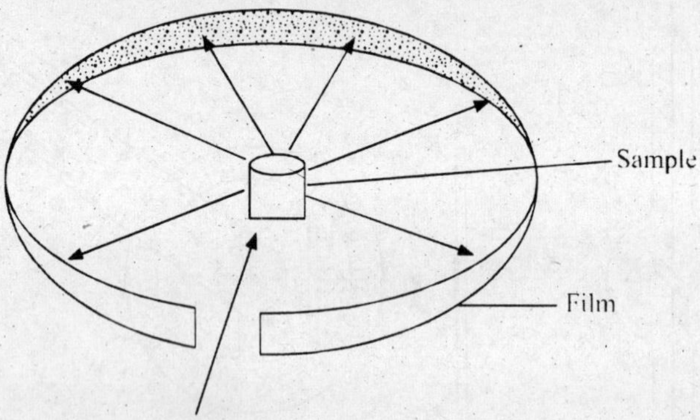

Fig. 17.33. The Powder Method

The sample is rotated with the help of a small motor. X-rays pass through the gap between the ends of the film strip. The powder contains small crystals oriented in all directions. Some of these will reflect X-rays from each lattice plane simultaneously. The reflected X-rays will make an angle 2θ with the original direction. Lines of constant θ are obtained on the film. From the design of the camera, θ can be calculated for different crystal planes.

Fig. 17.34 illustrates the reflection of X-rays from the crystal planes

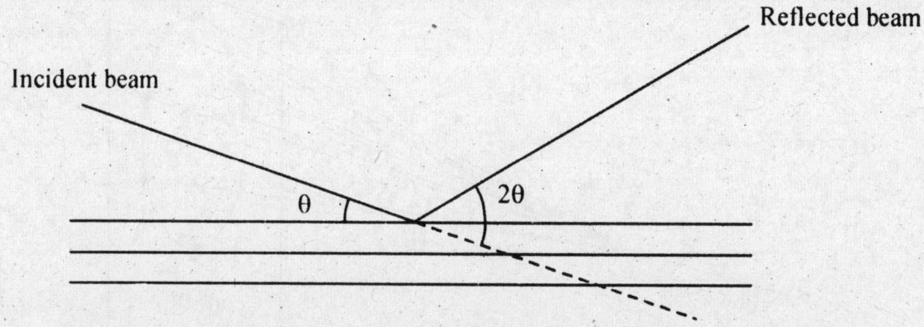

Fig. 17.34. Reflection of X-rays from crystal planes (powder method)

To calculate d, the interplanar distance, the Bragg's equation

$$n\lambda = 2d \sin \theta$$

as discussed in Q. 52 is used.

Q. 54. Describe Laue photographic method for the determination of structure of solids.

(*Nagpur 2008*)

Ans. The Laue method of examining the diffraction of X-rays by a crystal differs from other methods, in that a homogeneous beam of X-rays is used. The beam falls upon a small crystal and the diffracted beams are recorded on a photographic plate placed at a distance of a few centimeters from the crystal. The rays are limited to a narrow pencil by the slit system A, the crystal C is set on a holder which permits the adjustment of its orientation, and the photographic plate is placed at S (Fig. 17.35). A beam with a continuous range of wavelengths is conveniently obtained by using a bulb with a tungsten anticathode and a potential of about 65,000 volts.

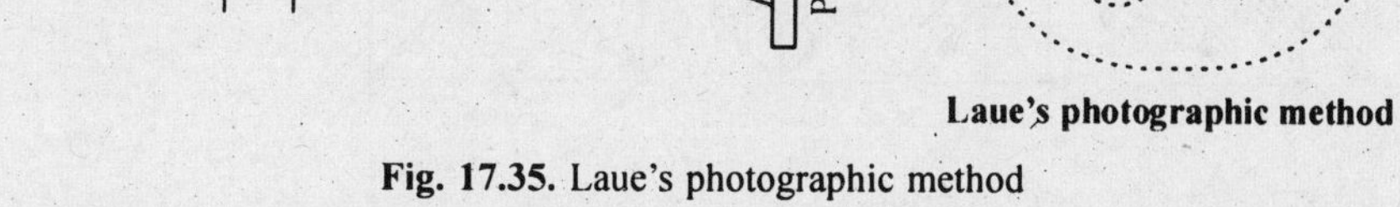

Fig. 17.35. Laue's photographic method

A feature of the Laue photograph is the arrangement of the spots on ellipses which have one end of their major axis at the central spot. The same feature is shown more or less distinctly by all Laue photographs, the spots occurring where the ellipses intersect. All the spots which lie on any one ellipse are due to reflections by planes parallel to a common zone axis (Fig. 17.36). Incident rays fall upon a crystal at C, which has a zone axis parallel to ZC', and we consider rays which are reflected by a plane of the corresponding zone. Suppose a reflecting plane is to be rotated about ZC'; the reflected beam will trace out a curve EE' on the plate. If the incident rays make an angle θ with the zone axis, the reflected rays will always make the same angle with it, so that they lie on a cone of semi-vertical angle, θ around the zone axis. The ellipses in the Laue photograph are the intersections of such cones with the photographic plate. Each spot is at an intersection of ellipses, since a plane parallel to two zone axes is a crystal plane and can reflect the X-ray beam.

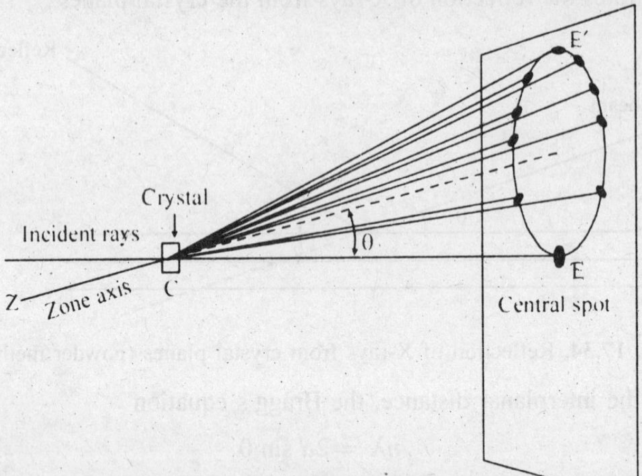

Fig. 17.36. Reflection of *X*-rays

Q.55. How can Bragg's equation be used to determine the structure of sodium chloride ?

Ans. There are three types of cubic lattices possible, namely, simple, face-centred and body-centred. Possible planes passing through the atoms are (100), (110) and (111), as shown in Fig. 17.37. If a is the distance between the 100 planes of the simple cubic lattice, the distances

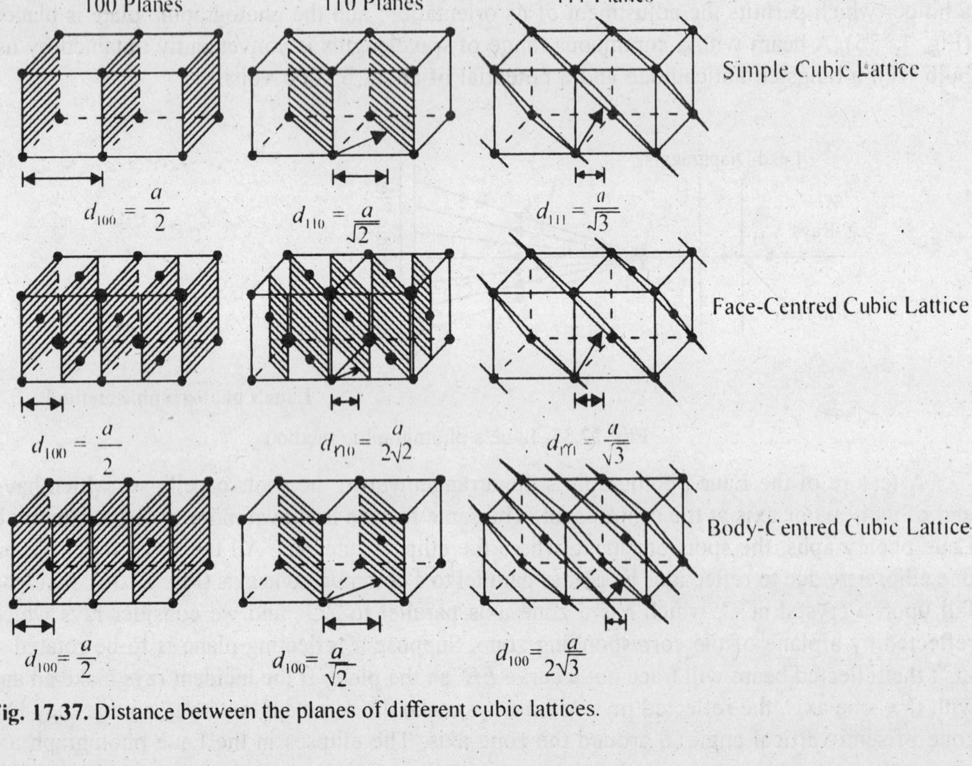

Fig. 17.37. Distance between the planes of different cubic lattices.

between the planes of the planes of the three types of cubic lattices can be calculated theoretically and are given below each figure.

Thus the ratios $d_{100} : d_{110} : d_{111}$ for simple cubic, face-centred cubic and body-centred cubic lattices will be

Simple cube: $d_{100} : d_{110} : d_{111} = a : \frac{a}{\sqrt{2}} : \frac{a}{\sqrt{3}}$

$= 1 : 0.707 : 0.577$

Face-centred cubic: $d_{100} : d_{110} : d_{111} = \frac{a}{2} : \frac{a}{2\sqrt{2}} : \frac{a}{\sqrt{3}}$

$= 1 : 0.707 : 1.154$

Body-centred cubic : $d_{100} : d_{110} : d_{111} = \frac{a}{2} : \frac{a}{\sqrt{2}} : \frac{a}{2\sqrt{3}}$

$= 1 : 1.414 : 0.577$

In order to find out the type of the cubic lattice in the given crystal, the angles at which the first, second, or third order reflections, that take place from different planes are studied. The values of d_{100}, d_{110}, and d_{111} are calculated using Bragg's equation, *viz.*, $d = \frac{n\lambda}{2\sin\theta}$ where *n* is the order of reflection. The ratio $d_{100} : d_{110} : d_{111}$ is then calculated and compared with the ratios given above for different types of lattices.

Carrying out the study as above, it is found that NaC1 (as well as KC1) has face-centred cubic lattice as shown in Fig. 17.38.

The basis of this structure is Na^+ at (000) and Cl^- at $\left(\frac{1}{2}00\right)$

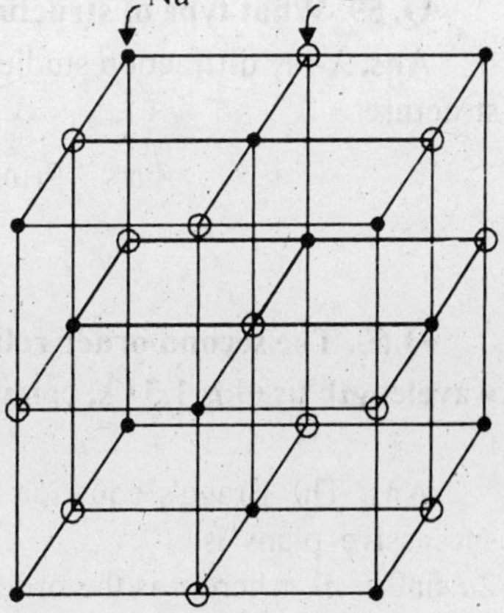

Fig. 17.38. Structure of NaCl (*fcc*).

Q. 56. Describe the structure of sodium chloride. *(G. Nanakdev 2000)*

Ans. The structure of sodium chloride is as shown in Fig. 17.38, See Q. 55.

Q. 57. The first order reflections from the 100, 110 and 111 planes of a given cubic crystal (NaCl crystal) were found to occur at angles 5.9°, 8.4° and 5.2° respectively. Determine the type of cubic lattice to which the crystal belongs. *(Delhi 2003)*

Ans. For first order reflections, $n = 1$ so that we have

$$d = \frac{n\lambda}{2\sin\theta} = \frac{\lambda}{2\sin\theta}$$

$$\therefore \quad d_{100} = \frac{\lambda}{2\sin 5.9^\circ} = \frac{\lambda}{2\times 0.103} = \mathbf{4.845\lambda}$$

$$d_{110} = \frac{\lambda}{2\sin 8.4^\circ} = \frac{\lambda}{2\times 0.146} = 3.425\lambda$$

$$d_{111} = \frac{\lambda}{2\sin 5.2^\circ} = \frac{\lambda}{2\times 0.094} = 5.319\lambda$$

$$\therefore \quad d_{100} : d_{110} : d_{111} = 4.845 : 3.425 : 5.319 = 1 : 0.706 : 1.096$$

This ratio is very close to that of the face-centred cubic. Hence the given crystal is face-centred cubic.

Q. 58. Answer the following:

(a) $a \neq b \neq c$, $\alpha = \gamma = 90°$, $\beta \neq 90°$, identify the crystal system. *(G. Nanakdev 2000)*

(b) The interplanar distance d_{hkl} ratio is 1 : 0.707: 0.577, identify the cubic lattice.

(c) The Weiss indices of a plane are 2, 2, ∞. What are its Miller indices ?

Ans. (a) The crystal system is monoclinic.

(b) It is a simple cubic system

(c) The Millar indices may be obtained as follows :

Take the reciprocal of Weiss indices

$$\frac{1}{2}\ \frac{1}{2}\ \frac{1}{\infty} \text{ or } \frac{1}{2}\ \frac{1}{2}\ 0$$

The miller indices are 110.

Q. 59. What type of structure is exhibited by KCl ?

Ans. *X*-ray diffraction studies reveal that KCl possesses simple cubic structure. In simple cubic structure

$$d_{100} : d_{110} : d_{111} = a : \frac{a}{\sqrt{2}} : \frac{a}{\sqrt{3}}$$

$$= 1 : 0.707 : 0.577$$

Q.60. The second order reflection of x-rays from (100) plans of NaCl occurs at 29.3°.If the wavelength used is 1.54Å, calculate the distance between two successive (100) plans in NaCl. *(Mumbai 2005)*

Ans. The Bragg's equation relating the angle, wavelength of light and the distance between successive plans is

$2d \sin\theta = n\lambda$ where n is the order of reflection

Substituting the values,we have

$$2d \sin 29.3° = 2 \times 1.54 \text{ or } d = \frac{1.54}{\sin 29.3} = 3.17\text{Å}$$

Q.61. Calculate the interplant distance and inter planer ratio d_{100}:d_{110}: d_{111} for simple and face -centred crystals. *(Mumbai 2005)*

Ans. Sec Q.55.

LIQUID CRYSTALS

Q. 62. What are liquid crystals? What are their uses?

(M. Dayanand 2000; Kerala 2000; Calcutta 2007)

Ans. A large number of organic compounds with long chain molecules such as:

Cholesteryl acetate ($CH_3COOC_{27}H_{45}$),

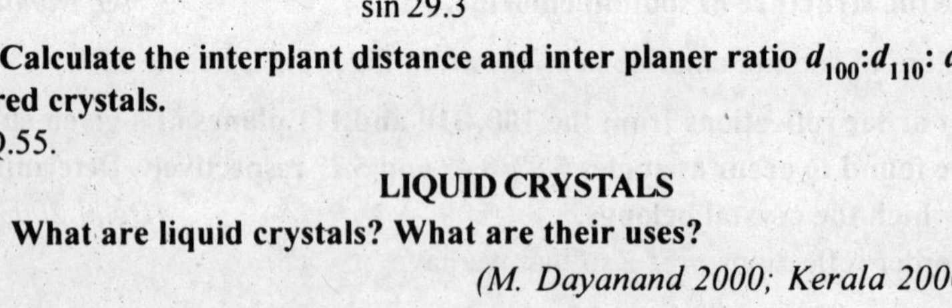

Cholesteryl benzoate ($C_6H_5COOC_{27}H_{45}$), and ammonium oleate ($C_{17}H_{33}COONH_4$) on heating melt sharply to give a milky (turbid) liquid which on further heating becomes clear again at a definite temperature.

On cooling the clear liquid (melt), the reverse changes take place exactly at the same temperatures. The turbid liquids thus obtained are found to be *anisotropic,* a property possessed by crystalline substances. Hence these turbid liquids are called **liquid crystals.** However, it is found that the molecules of a liquid crystal do not have the arrangement like that of a crystal lattice. Moreover, they behave like true liquids, *e.g.*, they possess surface tension and have the property to flow. Hence in place of 'liquid crystals' certain other names have been proposed such as *anisotropic liquids or para crystals.* The best name for the substances in this state is **mesomorphic state.** However, in spite of all this the name 'liquid crystals' is still frequently used.

Uses of Liquid Crystals

(*i*) Liquid crystals are used for digital display of numbers and letters in watches, calculators and other electronic devices.

(*ii*) Thin films of cholesteric liquid crystals are used to determine temperature and in the rings to study a person's mood.

Q. 63. What are different kinds of liquid crystals?

(*M. Dayanand, 2000; Kerala, 2000; Nagpur, 2008*

Ans. Liquid crystals are of the following three types:

(*i*) **Smectic Liquid Crystals:** The smectic liquid crystals have limited mobility. They flow in layers as if different sheets are gliding over one another. The distribution of velocity in different layers is different from that found in a true liquid. The concept of viscosity is, therefore, not applicable to the liquid crystals of this type. When spread over a clean glass surfaces, they form a series of *terraces or strata,* as shown in Fig. 17.39. The edges of these terraces often show a series of fine lines even in ordinary light, though more so in polarised light.

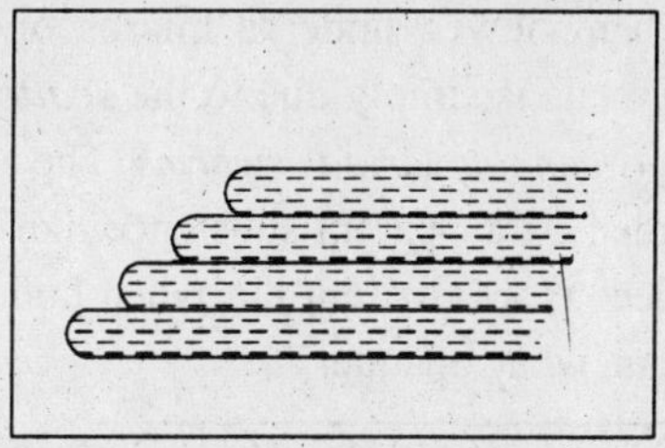

Fig. 17.39. Layer structure of smectic liquid crystals.

Smectic liquid crystals also give X-ray diffraction patterns like solid crystals but it is in one direction only. Thus, smectic phase has a complex internal structure.

When viewed in polarized light, smectic phases appear to have fan-like appearance. They are always uniaxial and are not affected by a magnetic field.

Ethyl-*p*-azoxy benzoate, ethyl-*p*-azoxy cinnamate are some examples of this class of liquid crystals.

(*ii*) **Nematic Liquid Crystals:** Nematic type of liquid crystals show near normal flow behaviour of liquids. They flow more readily than the smectic liquid crystals. Their flow is also Newtonian and the concept of viscosity is applicable to their flow. Their viscosity, however, is rather low as compared to that of liquids. Thus, nematic phases have liquid-like character and yet being *turbid and anisotropic,* they are unlike true liquids. *This type is, therefore, regarded less truly anisotropic than the smectic type.*

In polarised light, substances in nematic phase appear to have *thread-like structures*. They are uniaxial like smectic substances, but unlike the latter, they are affected by a strong magnetic field. Thus, when viewed in the direction of the lines of force of a magnetic field, the turbid liquid appears quite clear. If the magnetic field is cut off, the liquid appears turbid again. *p*-Azoxy anisole and *p*-Azoxy phenetole are some examples of this type.

(*iii*) **Cholesteric Liquid Crystals:** Certain optically active compounds show many characteristics of nematic behaviour but at the same time they show strong colour effects in polarised light. This indicates that they have a *layer structure* like smectic liquid crystals. But they differ from smectic liquid crystals in having much thicker layers which are of the order of 500 to 5000 molecules thick. Such liquid crystals having some nematic and some smectic characters are called **cholesteric liquid crystals.** The first substance in which this type of mesomorphism was detected was cholesteryl benzoate. Its transition point is 145° and melting point 178°C. *l*-cholesteryl formate is another example of this type of liquid crystals.

Q. 64. Discuss Swarm theory of liquid crystals.

Ans. The best explanation for turbidity is furnished on the basis of the 'Swarm Theory' proposed by Bose. According to this theory, since mesomorphic behaviour is shown by substances having long chains, their free rotation is restricted and they tend to arrange themselves in parallel formations. In this way, a number of groups or 'swarms' are obtained. The molecules in any one swarm are parallel to one another but may not be parallel to those present in another swarm (Fig. 17.38).

Each swarm consisting of several molecules, arranged parallel to one another, is taken as a small crystal. Thus, a turbid liquid is viewed as consisting of a series of small 'crystals' distributed at random in the body of the liquid. Lehmann actually found the each particular swarm appears as a *transparent liquid crystal* when viewed under a microscope. The turbidity of the liquid crystals is simply due to the *scattering of light at the boundaries between the various swarms.* The effect is similar to that observed when a transparent substance like glass is finely powdered. Each piece is transparent but when huddled together the mass appears somewhat opaque.

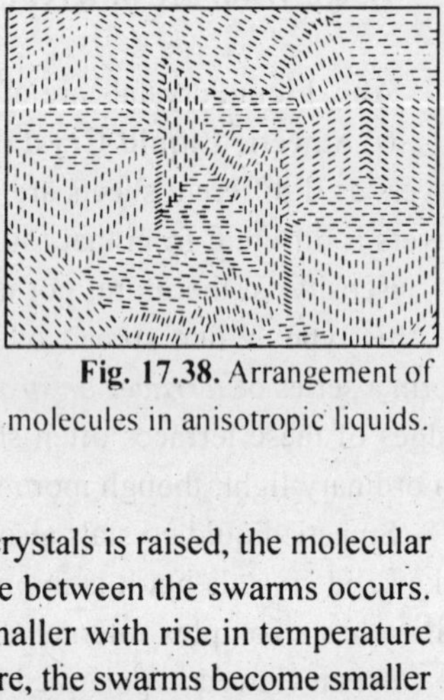

Fig. 17.38. Arrangement of molecules in anisotropic liquids.

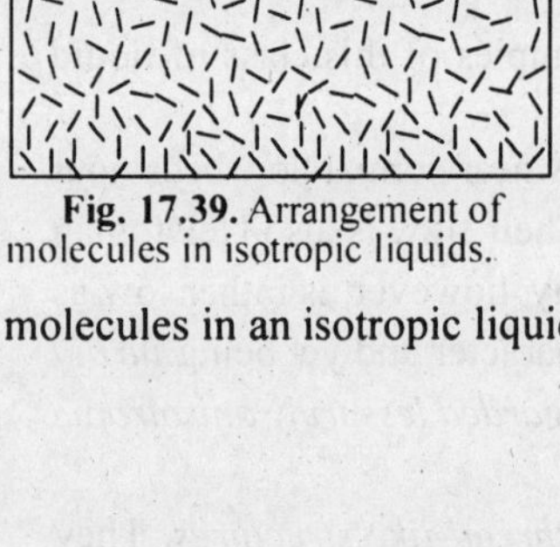

Fig. 17.39. Arrangement of molecules in isotropic liquids.

As the temperature of the liquid crystals is raised, the molecular motion increases and an interchange between the swarms occurs. The swarms become smaller and smaller with rise in temperature and ultimately at a certain temperature, the swarms become smaller than half the wavelength of light. They are then too small to scatter light. The turbidity now disappears. The liquid becomes clear. The double refraction and other anisotropic properties due to orientation of molecules within the swarms disappear. The liquid becomes singly refracting and isotropic. This temperature is the melting point of the substance. The arrangement of molecules in an isotropic liquid is depicted in Fig. 17.39.

18

PHYSICAL PROPERTIES AND MOLECULAR STRUCTURE (INCLUDING SPECTROSCOPY)

Q. 1. Name various physical properties that help in establishing the structure of a compound.

Ans. Various physical properties that support the determination of molecular structure are:

(*i*) Molar refraction

(*ii*) Parachor (Surface Tension)

(*iii*) Optical rotation

(*iv*) Dipole moment

(*v*) Magnetic susceptibility.

Q. 2. What is meant by (*i*) Additive property (*ii*) Constitutive property.

Ans. (*i*) An *additive property* is defined as the property, the aggregate value of which is the sum of values of its constituents.

(*ii*) A *constitutive property* is defined as the property which depends upon the constitution or structure of the molecule like arrangement of atoms within the molecule, multiple bonds and rings etc.

OPTICAL ROTATION

Q. 3. What is plane polarized light?

Ans. If a ray of light travels in any direction, it has vibrations in all directions at right angles to the path of propagation (Fig. 18.1). If this light is passed through Nicol prism, then

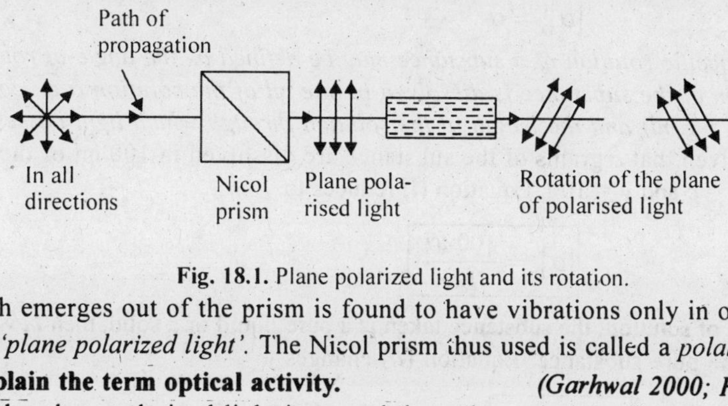

Fig. 18.1. Plane polarized light and its rotation.

the light which emerges out of the prism is found to have vibrations only in one plane. This light is called *'plane polarized light'*. The Nicol prism thus used is called a *polarizer*.

Q. 4. Explain the term optical activity. *(Garhwal 2000; Himachal 2010)*

Ans. If the plane polarized light is passed through a solution of sugar, it is found that the plane of the polarized light gets rotated when it comes out of the solution. *The substances like* *quartz, sugar etc., which rotate the plane of the polarized light are called optically active substances and the property of a substance to rotate the plane of the polarized light is called optical activity.* The substances which rotate the plane of the polarized light to the right are called dextero-rotatory and those which rotate the plane of the polarized light to the left are called laevo-rotatory.

Q. 5. Explain the term specific rotation. *(Himachal 2010*

Ans. The angle (in degree) through which the plane of the polarized light is rotated on passing through the *solution* of an optically active substance depends upon the following factors:

(*i*) Nature of the optically active substance.

(*ii*) Concentration of the solution (in g/ml).

(*iii*) Length of the solution *i.e.*, the length of the tube through which the light passes (in decimeters).

(*iv*) Wavelength of the light used.

(*v*) Temperature of the solution.

Thus if m grams of the substance are dissolved in v ml of the solution. (so that the concentration of the solution is m/v g/ml), l is the length of the tube (in decimeters) and α is the angle of rotation (in degrees), then it is found that

$\alpha \propto$ (concentration in g/ml) × (length of the tube in dm)

i.e.,
$$\alpha \propto \frac{m}{v} \times l$$

or
$$\alpha = [\alpha]_\lambda^t \, \frac{m \times l}{v} \qquad ...(i)$$

where $[\alpha]_\lambda^t$ is a constant. It depends upon the nature of the substance, the wavelength of light used (λ) and the temperature of the solution (t°C). If D line of the sodium light is used and the temperature of the solution is 25°C, then it is written as $[\alpha]_D^{25°}$. This constant is called **'specific rotation'** of the substance. Thus from equation (*i*)

$$[\alpha]_\lambda^t = \frac{v \times \alpha}{l \times m} \qquad ...(ii)$$

Now, if m= 1 g, v= 1 ml and l = 1 dm, then

$$[\alpha]_\lambda^t = \alpha$$

Hence *specific rotation of a substance may be defined as the angle of rotation produced when one gram of the substance is dissolved in one ml of the solution (i.e., concentration of the solution is 1 g/ml) and the length of the solution through which light passes is 1 dm.*

If it is given that c grams of the substance are dissolved in 100 ml of the solution, then we may put $m = c$ and v = 100. Equation (*i*) reduces to

$$\boxed{[\alpha]_\lambda^t = \frac{100 \times \alpha}{l \times c}} \qquad ...(iii)$$

If instead of solution, the substance taken is a pure liquid or a solid, then m/v is replaced by density d of the pure substance. Equation (*ii*) changes to

$$\boxed{[\alpha]_\lambda^t = \frac{\alpha}{l \times d}}\text{, for pure liquids or solids} \qquad ...(iv)$$

Q. 6. Describe the measurement of optical activity of a substance using polarimeter.

Ans. The instrument used for the measurement of the amount of rotation caused by an optically active substance to the plane of the polarized light is called a *polarimeter.* It consists of two Nicol prisms, one called the *polarizer* and the other called the *analyser.* Both of these are fitted on the opposite ends of the same metal tube. However, the polarizer is fixed but the analyser can be rotated about the axis of the tube. In between the polarizer and the analyser, there is space for keeping the glass tube containing the solution. The glass tube is usually 5 cm, 10 cm, 20 cm or 25 cm in length and has glass windows on both ends. A circular scale is fixed on the metal tube near the analyser, as shown in Fig. 18.2.

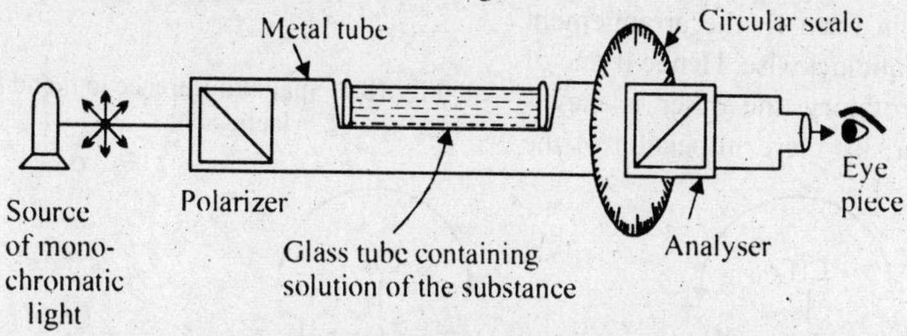

Fig. 18.2. Working of a polarimeter.

In the procedure, first of all monochromatic light from a suitable source (usually sodium vapour lamp) is passed through the instrument, without the liquid or the solution. The light emerging out of the analyser is viewed through the eye-piece. The analyser is then rotated slowly till the field of view is completely dark. This gives the initial reading of the instrument. Now the tube is fitted with the liquid or the solution and the shutter is closed. On seeing through the eye-piece, the field of view is no longer found to be dark. The analyser is rotated slowly till the field of view is completely dark again. This gives the final reading. The difference between the *final reading* and the *initial reading* gives the angle of rotation 'α'. If it comes out to be positive the substance is dextero-rotatory and if negative, it is laevo-rotatory.

Q. 7. What is the cause of optical activity in a compound? Explain with examples.

(Meerut 2000)

Ans. Optical activity is a *constitutive* property *i.e.*, it depends upon the arrangement of atoms within the molecule. Organic compounds which are found to be optically active contain at least one carbon atom to which four different atoms or groups are attached. Such a carbon atom is called an *asymmetric carbon* atom and is usually shown by putting an asterisk over it. A very simple example of such a compound is lactic acid, as represented below:

$$\begin{array}{c} CH_3 \\ | \\ H—{}^{*}C—OH \\ | \\ COOH \end{array}$$

Such a compound exists in two forms, one of which rotates the plane of the polarized light to the right and is called dextero-rotatory (*d*-form or '+' form) and the other which rotates the plane of the polarized light to the left and is called laevo-rotatory (*l*-form or '–' form). The two forms have the same constitution but differ in the arrangement of atoms or groups in space. Thus the two forms are called *optical isomers.*

Le Bel and van't Hoff pointed out the differences in the two forms by suggesting that in case of such molecules, the asymmetric carbon atom is present in the centre of a regular tetrahedron and the four atoms or groups attached to it are present at the corners of the tetrahedron. This gives rise to two different spatial arrangement which are mirror images of each other, as shown in Fig. 18.3.

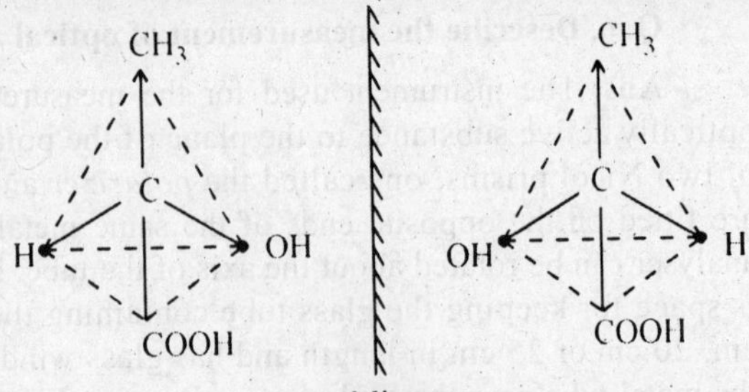

(*i*) (*ii*)

Fig. 18.3. Spatial difference in *d* and *l* forms of lactic acid.

The two configuration are different in the sense that they are not superimposable over each other. The above two forms may be represented in a simplified way as follows:

In form I, the arrangement from H to OH is clockwise and in form II, the arrangement from H to OH is anticlockwise. Hence if one of them is dextero-rotatory, the other is laevo-rotatory. However, by mere inspection of the

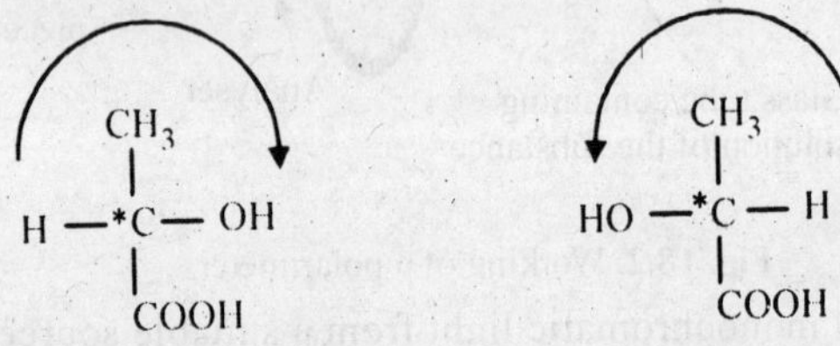

(I) (II)

configurations, it is not possible to predict as to which form is dextero or laevo. The two forms are also said to be *enantiomorphs.*

Q. 8. Explain the terms (*i*) External compensation (*ii*) Internal compensation.

Ans. (*i*) External compensation: If equimolar quantities of *d*-and *l*-forms are mixed together, the mixture obtained is optically inactive. This is obviously due to the fact that the rotation caused by one form to the right is compensated by the equal rotation caused by the second form to the left. This process is called *external compensation* and the substances thus obtained are called *racemic mixtures* (or *dl*-form or ± form). This form can be separated into *d* and *l* forms by suitable methods.

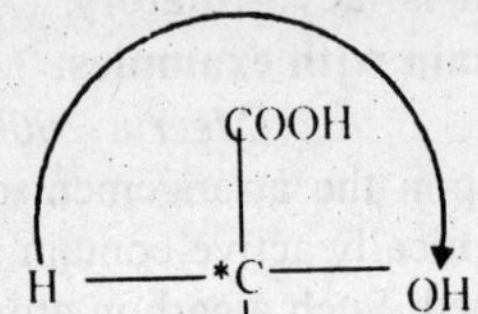

Fig. 18.4. Meso form of tartaric acid.

(*ii*) **Internal Compensation:** In case of compounds containing two asymmetric carbon atoms (*e.g.*, tartaric acid, Fig. 18.4), another form is also found to exist. In this form, the rotation caused by half of the molecule is compensated by the other half within the molecule. Thus the molecule is optically inactive. This is a case of *internal compensation.*

This form which is optically inactive due to *internal compensation* is called *meso form.*

Q. 9. The value of $[\alpha]_D^{20}$ for lactose is 55.4°. What is the concentration in grams per litre of a solution of lactose which gives a rotation of 7.24° in a 10 cm cell at 20° with sodium *D* light?

Ans. Here, we are given

$$[a]_D^{20} = 55.4°$$

$$\alpha = 7.2°$$

$$l = 10 \text{ cm} = 1 \text{ dm}$$

Putting the values in the formula

$$[\alpha]_\lambda^t = \frac{100 \times \alpha}{l \times c}$$

or $$c = \frac{100\times\alpha}{[\alpha]_\lambda^t \times l} = \frac{100 \times 7.24}{55.4 \times 1}$$

$$= 13.07 \text{ g /100 ml} = \mathbf{130.7 \text{ g/litre}}$$

Q. 10. A solution of a certain optically active substance in water containing 1.56 g in 100 ml rotated polarised light by 4.91° in a polarimeter which had a cell 20 cm long. The *D* line of sodium was used as a light source. Calculate the specific rotation.

Ans. Here, we are given

$$c = 1.56 \text{ g/100 ml}$$

$$\alpha = 4.91°$$

$$l = 20 \text{ cm} = 2 \text{ dm}$$

$$\therefore \quad [\alpha]_\lambda^t = \frac{100 \times \alpha}{l \times c} = \frac{100 \times 4.91}{2 \times 1.56} = \mathbf{157.37°}$$

Q. 11. When α-D glucose $\left([\alpha]_D^{20} = +112.2°\right)$ is dissolved in water the optical rotation decreases as β-D glucose is formed until at equilibrium $[\alpha]_D^{20}$ = + 52.7°. As expected, when β-D glucose $\left([\alpha]_D^{20} = +18.7°\right)$ is dissolved in water, the optical rotation increases until $[\alpha]_D^{20}$ = + 52.7° is obtained. Calculate the percentage of β-form in the equilibrium mixture.

Ans. Suppose x is the fraction of D-glucose in β form. Then fraction of α-form = $1 - x$

$$\therefore \quad x(+18.7) + (1-x)(+112.2) = 52.7$$

or $$x = 0.636 = 63.6\%\ \beta$$

$$1 - x = 0.364 = 36.4\%\ \alpha$$

DIPOLE MOMENT

Q. 12. Explain the term dipole moment. *(Lucknow 2000; Panjab 2000; Garhwal 2000; Indore 2001; M.Dayanand 2002)*

Ans. The extent of polarity of a molecule is expressed in terms of dipole moment which is defined as follows:

The dipole moment of a polar diatomic molecule is the product of the charge on each end of the molecule and the average distance between the centres of their nuclei i.e., the bond length.

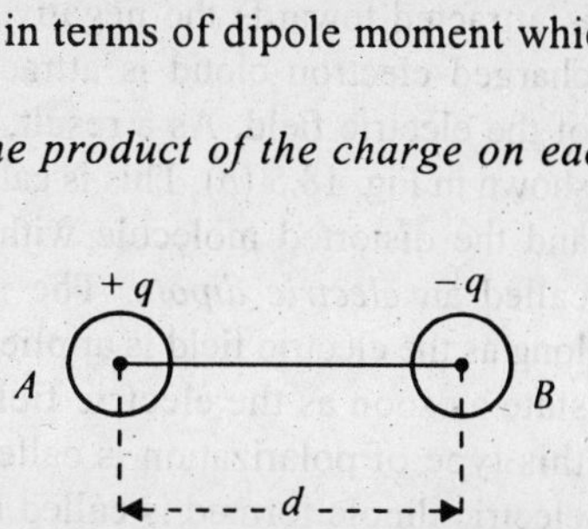

In general, for a polar diatomic molecule AB in which atom B is more electronegative than A, if $+q$ is the charge on the end A and $-q$ on the end B of the molecule and d is the average distance between the centres of their nuclei, then the dipole moment of the molecule AB is given by

$$\mu = q \times d$$

Obviously if the molecule is non-polar, $q = 0$ and hence $\mu = 0$.

The charge present on the ends of the molecule is usually of the order of 10^{-10} esu and the distance between the atoms (*i.e.*, bond length) is usually of the order of 10^{-8} cm (*i.e.* Å units), therefore, μ must be of the order of $10^{-10} \times 10^{-8} = 10^{-18}$ esu cm. This value is called 1 Debye. Thus

$$1 \text{ Debye} = 10^{-18} \text{ esu cm.}$$

The dipole moment is a vector quantity *i.e.*, it has both magnitude as well as direction. It is, therefore, represented by an arrow pointing from the +ve end to the –ve end of the molecule *i.e.*,

$$+ \longrightarrow -$$

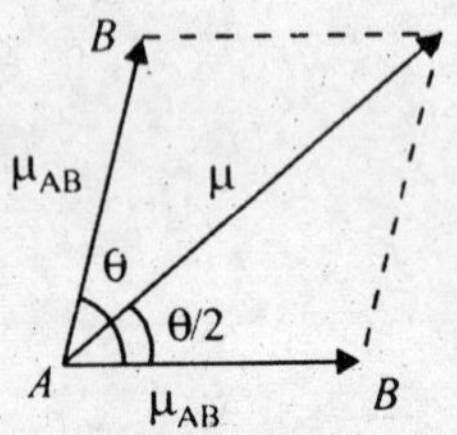

In a polyatomic molecule every two bonded atoms have their dipole moment. This is called the **bond moment.** The *net dipole moment of the polyatomic molecule is the resultant of the different bond moments.*

The resultant will depend upon the directions of the bonds. Thus if a molecule, AB_2 contains two $A-B$ bonds at an angle of θ and μ_{AB} is the value of each bond moment, then the dipole moment of the molecule will be

$$\mu = 2\mu_{AB} \cos \frac{\theta}{2}$$

For a more general molecule ABC with two bonds AB and AC having bond moments μ_1 and μ_2 and inclined at an angle θ, the dipole moment of the molecule (by parallelogram law) will be

$$\mu = \sqrt{\mu_1^2 + \mu_2^2 + 2\mu_1\mu_2\cos\theta}$$

As μ is known from experiments, therefore, knowing the value of θ, the method is usually used for the calculation of bond moments. Alternatively, knowing the value of μ and μ_{AB}, the method can be used for the calculation of θ and hence for predicting the shapes of molecules.

Q. 13. Describe the measurement of dipole moment of a substance using Clausius-Mussott equation. *(Nagpur, 2002; Purvanchal 2003; Raipur 2010; Himachal 2010)*

Ans. Every molecule is made up of positively charged nuclei and negatively charged electron cloud as shown in Fig. 18.5 (*a*). The centre of positive charge coincides with the centre of negative charge. Hence when such a molecule is introduced between the two plates of an electric field, the positively charged nuclei and hence the centre of positive charge is attracted towards the negative plate, whereas the negatively charged electron cloud is attracted towards the positive plate of the electric field. As a result, the molecule gets distorted, as shown in Fig. 18.5 (*b*). This is called *polarization* of the molecule and the distorted molecule with positive and negative ends is called an *electric dipole.* The molecule remains polarized so long as the electric field is applied but goes back to the original state as soon as the electric field is switched off. That is why this type of polarization is called *induced polarization* and the electric dipole formed is called *induced dipole.*

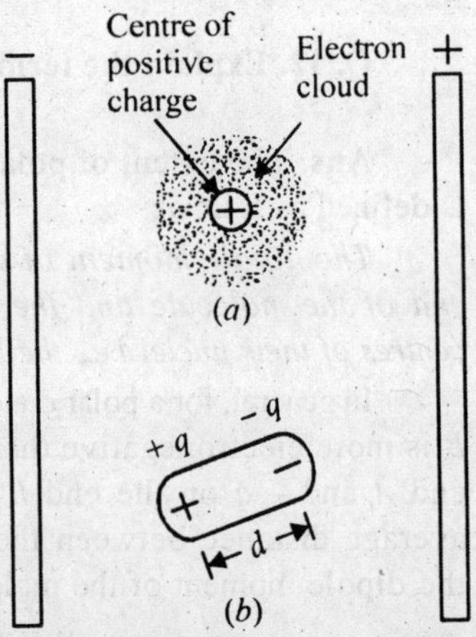

Fig. 18.5. Polarization of molecule an electric field (*a*) original state (*b*) polarized state.

Since even after polarization, the molecule is neutral as a whole, this means that the positive charge on one end must be equal to the negative charge on the other. Suppose this charge is q. If d is the distance of separation between the charges, then dipole moment of the induced dipole μ_i will be given by

$$\mu_i = q \times d \qquad ...(i)$$

Evidently, the value of μ_i depend upon the nature of the molecule and the strength of the electric field, say X. Thus

$$\mu_i = \alpha \times X \qquad ...(ii)$$

where α is a constant called the *polarizability* of the given molecule.

α is found to be related to the dielectric constant, D, of the medium present between the plates *i.e.*, **the molecules being studied. According to the equation given by Clausius and Mossott**

$$\left(\frac{D-1}{D+2}\right)\frac{M}{\rho} = \frac{4}{3}\pi N\alpha \qquad ...(iii)$$

where M is molecular mass of the molecules

ρ is the density of the molecules

N is Avogadro's number.

As all quantities on the right hand side of eqn. (*iii*) are constant (α being constant for a substance) independent of temperature, therefore, the quantity on the left hand side of eqn. (*iii*) must also be constant, independent of temperature and depending only upon the nature of the molecules. This quantity is called *induced molar polarization* and is represented by P_i. Thus

$$P_i = \left(\frac{D-1}{D+2}\right)\frac{M}{\rho} \qquad ...(iv)$$

Induced molar polarization is defined as the amount of polarization produced in 1 mole of the substance when placed between the plates of an electric field of unit strength.

Using eqn. (*iv*), the values of molar polarization have been determined for a number of substances. It is found that whereas molar polarization of a number of substances such as O_2, CO_2, N_2 and CH_4 are constant and independent of temperature, the polarization values for some other substances such as HCl, $CHCl_3$, $C_6H_5NO_2$ and CH_3Cl do not come out to be constant, but decrease with increase of temperature.

In case of molecules such as HCl, $CHCl_3$ etc., their centres of positive and negative charges do not coincide with each other. Thus they possess some dipole moment even in the

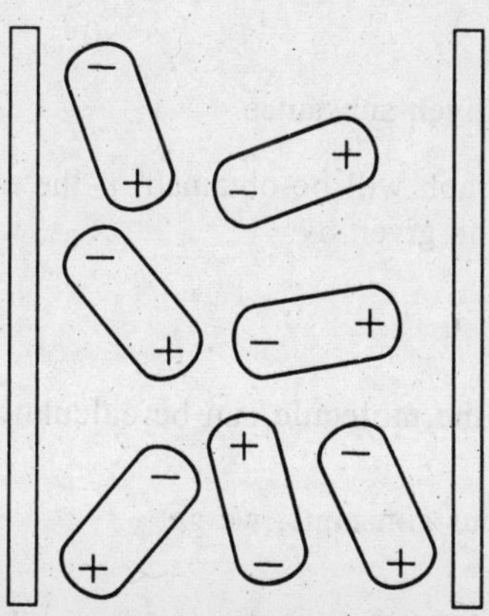

Fig. 18.6. Permanent dipoles oriented in a random manner in the absence of an electric field.

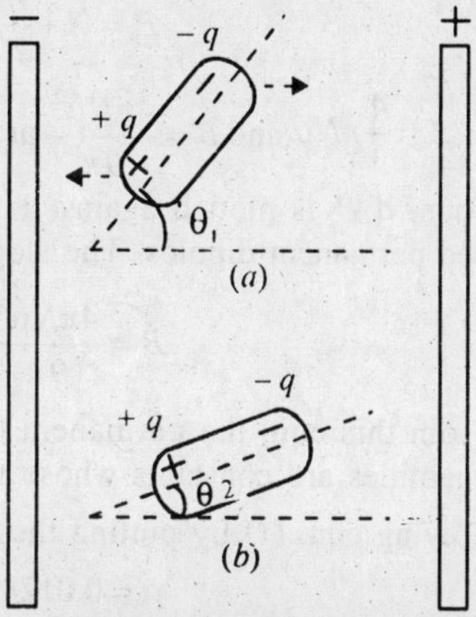

Fig. 18.7. (*a*) Original state of a permanent dipole: (*b*) Final state after polarization and orientation.

absence of electric field. The dipole moment thus possessed by the molecules is called *permanent dipole moment.* In the absence of electric field these permanent dipoles are oriented in a random manner as shown in Fig. 18.6. However in the presence of an electric field, two disturbing effects take place. These are:

(*i*) The electric field will tend to rotate and orient these dipoles in the direction of the field.

(*ii*) The electric field will tend to polarize the molecules.

If the molecules were stationary, the dipoles would have been oriented perfectly parallel to the electric field *i.e.*, at right angles to the plates of the electric field. However, due to thermal agitation, they take up position in between that of the original state and the perfectly parallel state, as shown at (*b*) in Fig. 18.7 for any one permanent dipole.

Thus in case of molecules possessing permanent dipoles, the right-hand side of eqn **(*iv*) will give the total polarization (P_t) which will be the sum of the induced molar polarization (P_i) and the *orientation molar polarization* (P_o)**

i.e.
$$P_t = \left(\frac{D-1}{D+2}\right)\frac{M}{\rho} = P_i + P_o \quad ...(v)$$

And, the induced polarization is given by [Refer eq. (*iii*)]

$$P_i = \frac{4}{3}\pi N\alpha \quad ...(vi)$$

The orientation molar polarization is given by the equation put forward by Debye, *viz.*,

$$P_o = \frac{4}{3}\pi N\left(\frac{\mu^2}{3kT}\right) \quad ...(vii)$$

where μ is the *permanent dipole moment* of the molecule, k is Boltzmann's constant (=R/N) and T is the temperature.

Substituting the values of P_i and P_o from equations (*vi*) and (*vii*) in eqn. (*v*), we get

$$P_t = \left(\frac{D-1}{D+2}\right)\frac{M}{\rho} = \frac{4}{3}\pi N\alpha + \frac{4}{3}\pi N\left(\frac{\mu^2}{3kT}\right) \quad ...(viii)$$

This equation can be written in the form

$$P_i = A + \frac{B}{T} \quad ...(ix)$$

where $A = \frac{4}{3}\pi N\alpha$ and $B = \frac{4\pi N\mu^2}{9k}$ are constant for the given substance.

Thus, if P_t is plotted against $1/T$, a straight line graph will be obtained, if the molecules possessed permanent dipoles. The slope of the plot will be given by

$$B = \frac{4\pi N\mu^2}{9k} \quad ...(x)$$

From this eqn. the permanent dipole moment of the molecule can be calculated, as all other quantities are constants whose values are known.

Solving eqn. (*x*) by putting the values of the various constants, we get

$$\mu = 0.0128\sqrt{B} \times 10^{-18}$$

Thus the experiment involves the determination of the dielectric constant D and the density ρ at different temperatures T. The values of P_t are then calculated and plotted against $1/T$. If P_t is found to vary linearly with $1/T$, the substance possesses a permanent dipole moment and the value of the dipole moment (μ) is calculated from the slope of the line. On the other hand, if P_t is found to be independent of temperature, the substance does not possess any permanent dipole moment (*i.e.*, it is non-polar).

Q. 14. Discuss applications of dipole moment measurement in the study of structure of compounds. (*Lucknow 2000; Panjab 2000; Awadh 2000; Indore 2001; Nagpur 2002; M. Danayand 2002)*

Ans. A few important applications of dipole moment are briefly described below:

1. In comparing the relative polarities of molecules: The dipole moments of HF, HCl, HBr, HI are 2.0, 1.03, 0.78 and 0.38 Debye respectively. This shows that the polar character of the molecules decreases as we move from HF to HI.

2. In calculating the percentage ionic character of a polar covalent bond: For example, the dipole moment of HCl is 1.03 D. If the bond were 100% ionic, each end would acquire full one unit charge *i.e.* 4.8×10^{-10} esu. The bond distance in HCl is taken to be 1.275 Å *i.e.*, 1.275×10^{-8} cm. Hence if HCl were 100% ionic, the dipole moment moment would have been

$$= (48 \times 10^{-10}) \times (1.275 \times 10^{-8}) \text{ esu cm.}$$
$$= 6.11 \times 10^{-18} \text{ esu cm} = 6.11 \text{ D.}$$

Actual value of dipole moment of HCl = 1.03 D.

$\therefore$ percentage age ionic character H–Cl bond $= \dfrac{1.03}{6.11} \times 100 = 17\%$ approx.

3. In predicting the shapes of molecules: The fact that the dipole moment is a vector quantity and that a particular bond has almost a constant value of bond moment is very useful in predicting the shapes of molecules. A few cases are discussed below:

(*a*) **Triatomic molecules:** In triatomic molecules of the type AB_2, there are two possibilities.

(*i*) *The dipole moment of the molecule may be zero as* in case of CO_2 and CS_2. This shows that in these molecules, one bond moment is equal and opposite to the other. Hence the molecules must be linear, as shown below:

$$\delta^- \leftarrow \delta^+ \rightarrow \delta^- \qquad\qquad \delta^- \leftarrow \delta^+ \rightarrow \delta^-$$
$$O = C = O \qquad\qquad S = C = S$$
$$\mu = 0 \qquad\qquad \mu = 0$$

(*ii*) The molecule may have a definite value of dipole moment (*e.g.*, μ for H_2O = 1.84 D, μ for SO_2= 1.63 D). Obviously, in these cases, the possibility of linear shape is ruled out. The two bonds must have a definite angle between them which can be calculated, if we know the bond moments. For example, the bond moment for H—O is 1.6 D, the bond angle is calculated to be nearly 105°. Thus H_2O may be represented as shown in the figure. Such molecules are called *bent molecules* and the shape is called *V-shape.*

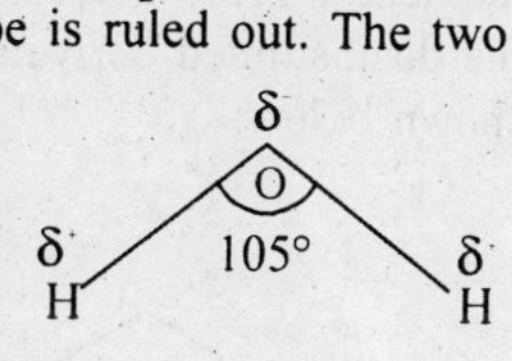

(*b*) **Tetratomic molecules:** In tetratomic molecules of the type AB_3, the two simple possibilities are :

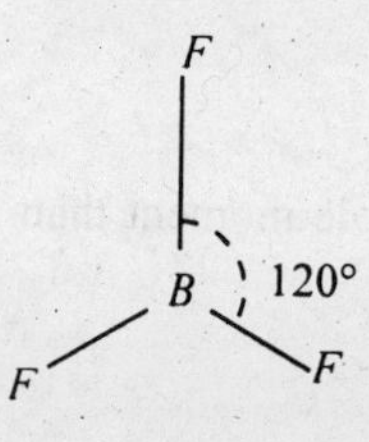

(*i*) *The dipole moment of the molecule may be zero as in* BF_3. In such a case, all the four atoms must be in one plane and the angles between the bonds must be equal. Such a shape is called **planar trigonal.**

(*ii*) *The molecule may have some definite value of dipole.* (*e.g.*, NH_3, PF_3, PCl_3). This indicates that one atom lies out of plane of rest of the atoms. For example, the molecule of NH_3 may be represented as shown in the figure.

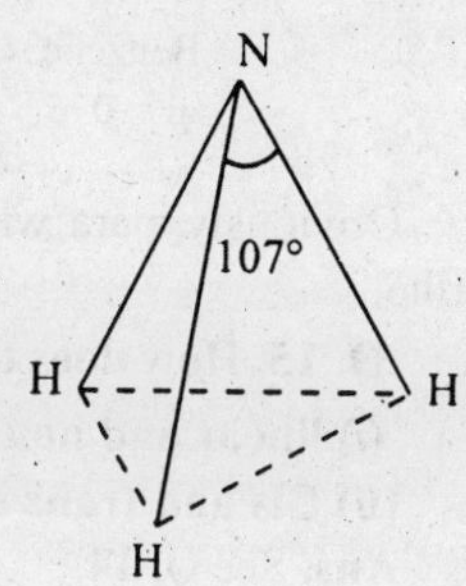

Such a shape is called *pyramidal shape.*

(*c*) **Pentatomic molecules:** The pentatomic molecules of the type AB_4 which we commonly come across are CH_4, CCl_4, $SiCl_4$, $SnCl_4$ etc. The dipole moments of all these molecules is found to be zero. This is possible only if these molecules have symmetrical or regular geometry so that the different bond moments cancel out one another. The experimental facts confirm that these molecule have **tetrahedral** shape as represented below:

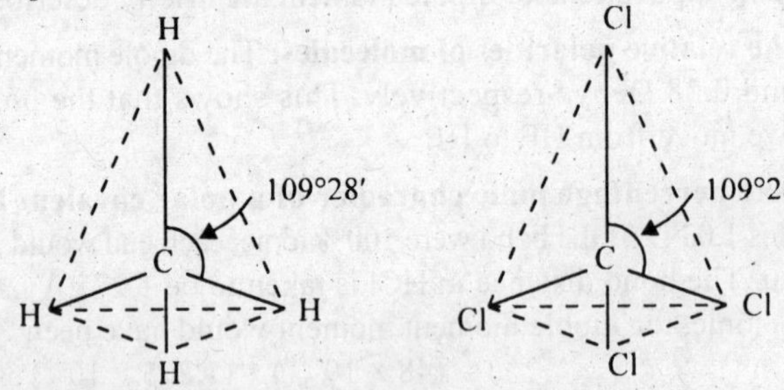

4. **In differentiating between the cis- and trans-isomers:** The *cis*- and *trans*- isomers are represented as:

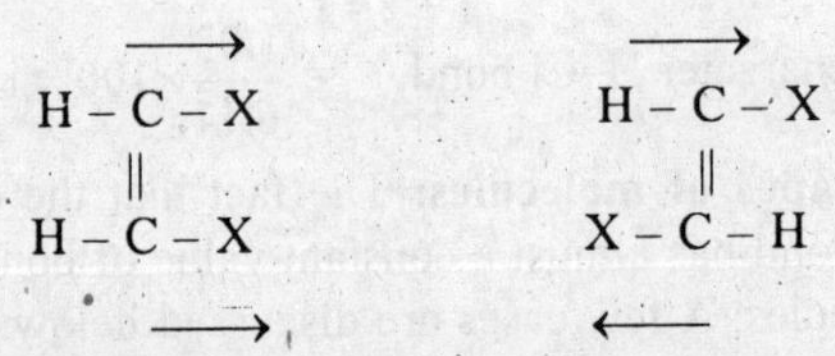

cis- *trans-*

The bond moments will cancel each other in *trans* isomer. Hence *trans* isomer will show zero value for dipole moment. On the other hand, the *cis* isomer will show some definite value of dipole moment.

For example, in case of dichloroethylene (putting X = Cl), cis-form has been found to have a dipole moment of 1.9 D, while the trans- form is found to have zero dipole moment.

5. **In differentiating between *o, m* and *p* isomers:** As benzene is a regular hexagon its dipole moment is zero, therefore, substituting two atoms or groups in the ortho, meta or para positions, the disubstituted derivative will have different dipole moment for the *o, m* and *p* positions, as shown below:

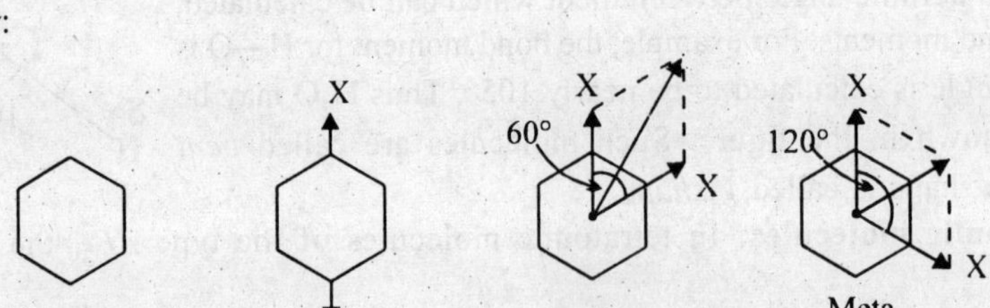

Benzene $\mu = 0$ Para $\mu = 0$ Ortho $\mu \neq 0$ Meta $\mu \neq 0$

Obviously, para will have zero dipole moment, and meta will have less dipole moment than ortho.

Q. 15. How does the knowledge of dipole moment help in distinguishing

(*i*) linear and nonlinear molecules

(*ii*) Cis and trans isomers

Ans. See Q. 14.

Q. 16. How can you establish the planar nature of benzene using dipole moment measurement?

Ans. Dipole moment of benzene molecule is zero *i.e.*, $\mu = 0$

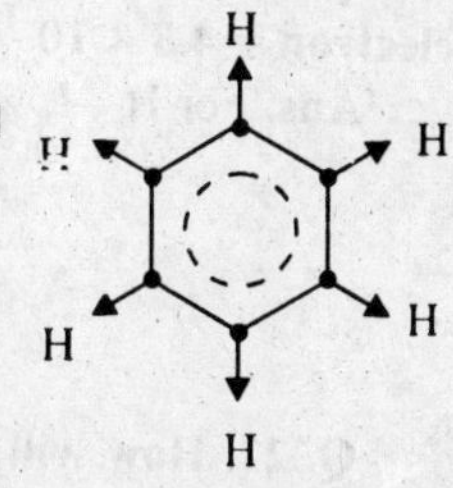

This is possible only if benzene has planar structure as represented in the figure. In this structure, dipole moments of six C—H bonds cancel one another and the net dipole moment of the benzene molecule is zero.

Had the structure not been planar, molecule would have shown a net dipole moment which is not the case.

Q. 17. The molecule of CS_2 is linear and the molecule of NH_3 is nonlinear. Explain on the basis of their dipole moments.

Ans. See Q. 14,

Q. 18. Explain how p-chlorobenzene has zero dipole moment while p-dihydroxy benzene has a definite value.

Ans. The structure of *p*-dichlorobenzene and *p*-dhydroxybenzene are given below

It is clear from the structures of the two compounds that the dipole moments of two C—Cl

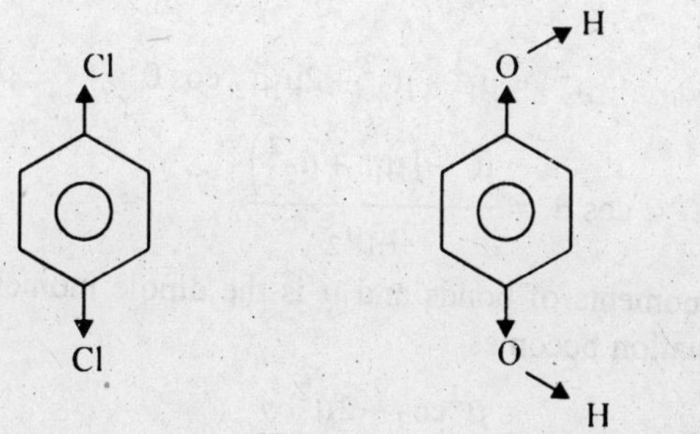

p-dichlorobenzene *p*-dihydroxybenzene

bonds will cancel each other, so that the net dipole moment of the molecule will be zero.

This cancellation of dipole moments of two C—O—H groups in *p*-dihydroxybenzene is not possible. Hence it shows a definite value of dipole moment.

Q. 19. The bond length of H-I bond is 1.60 Å and its dipole moment is 0.38 D. Calculate the percentage ionic character of H–I bond.

Ans. Assuming H–I to be 100% ionic bond, each end would acquire full one unit charge *i.e.*, $q = 4.8 \times 10^{-10}$ esu

$$\mu = q \times d$$

$$= (4.8 \times 10^{-10} \text{ esu}) (1.60 \times 10^{-8} \text{ cm})$$

$$= 7.68 \times 10^{18} \text{ esu cm}$$

$$= 7.68 \text{ D}$$

Observed $\mu = 0.38$ D (Given)

$$\therefore \ \% \text{ ionic character} = \frac{0.38}{7.68} \times 100 = \mathbf{4.95\%}$$

Q. 20. The bond length of H–I bond is 1.60 Å. Which of the following is the correct value for its dipole moment (in Debye units), if it were present in the completely ionic form (charge on the electron = 4.8 × 10^{-10} esu)? 3.00, 3.20, 7.68, 0.33, 6.40, 2.77, 9.28.

Ans. For H—I, we are given that

$$d = 1.60 \text{ Å} = 1.60 \times 10^{-8} \text{ cm}$$

$$q = 4.8 \times 10^{-10} \text{ esu}$$

$$\therefore \quad \mu = q \times d = (4.8 \times 10^{-10}) \times (1.60 \times 10^{-8})$$

$$= 7.68 \times 13^{-18} \text{ esu cm} = \mathbf{7.68\ D.}$$

Q. 21. How will you arrive at the structure of (*a*) Carbon dioxide (*b*) Water from dipole moment studies ?

Ans. (*a*) Carbondioxide. The dipole moment of carbondioxide (μ_{CO_2}) is zero, although C = O bond moment ($\mu_{c=o}$) is 2.5 D. This implies that the two C = O bond moments nullify each other, acting in opposite directions.

$$:\overset{..}{O} \overset{\leftarrow +}{=} C \overset{+ \rightarrow}{=} \overset{..}{O}:$$

Using the parallelogram law of bond moments, the bond angle θ for a triatomic molecule is calculated as follows :

$$\mu^2 = \mu_1^2 + \mu_2^2 + 2\mu_1\mu_2 \cos\theta$$

or

$$\cos\theta = \frac{\mu^2 - (\mu_1^2 + \mu_2^2)}{2\mu_1\mu_2}$$

where, μ_1, μ_2 are the bond moments of bonds and μ is the dipole moment of the molecule. For arbon dioxide the above equation becomes

$$\cos\theta = \frac{\mu^2 co_2 - 2\mu^2 co}{2\mu_{co}^2} \qquad (\because \mu_1 = \mu_2)$$

$$\cos\theta = \frac{0 - 2\mu^2 co}{2\mu^2 co} = -1$$

$$\theta = \cos^{-1}(-1) = 180°$$

That is, carbon dioxide molecule is linear.

(*b*) Water. The bond moment, μ_{O-H} and the dipole moment μ_{H_2O} are found to be 1.505 D and 1.84 D respectively.

Again, using parallelogram law of bond moments, we can calculate the bond angle θ.

$$\cos\theta = \frac{\mu^2 H_2O}{2\mu^2 O\text{-}H} - 1$$

$$\cos\theta = \frac{(1.84)^2}{2(1.505)^2} - 1 = 0.2520$$

$$\theta = \cos^{-1}(-0.2520)$$

$$= 104°36'$$

Thus, water molecule has *bent* (*or V*) shape.

MAGNETIC SUSCEPTIBILITY

Q. 22. Explain the terms:

(*i*) Magnetic permeability. *(M. Dayanand 2000)*

(*ii*) Diamagnetic, paramagnetic and ferromagnetic substances

(*iii*) Magnetic susceptibility. *(M. Dayanand 2000)*

Ans. (*i*) Magnetic permeability: If two magnetic poles of strength m_1 and m_2 are separated by a distance *r*, then the force *F* acting between them is given by

$$F = \frac{1}{\mu}\frac{m_1 \times m_2}{r^2}$$

The quantity μ used in the above expression is called the magnetic permeability of the medium between the poles. *It is defined as the relative tendency of the magnetic lines of force to pass through the medium as compared to that in the vacuum.* For vacuum, μ is taken as equal to 1. Similarly, for air, μ is taken as approximately equal to 1.

(*ii*) **Diamagnetic, paramagnetic and ferromagnetic substances:** As mentioned above, for vacuum μ =1. For other media, can be μ less than or greater than 1. *If μ is less than 1, then the magnetic lines of force prefer to pass through the vacuum rather than through the substance comprising the medium.* In other words, *the magnetic lines of force are repelled away from the substance. Such substances are said to be* **diamagnetic**. On the other hand, *if μ is greater than 1, then the magnetic lines of force have greater tendency to pass through the substance rather than through the vacuum.* In other words, *the magnetic lines of force are attracted towards the substance. Such substances are called* **paramagnetic**. If μ *is very high* (*i.e.* of the order of 10^3 or more), *the tendency of the substance to attract the magnetic lines of force is very high and the substance is said to be* **ferromagnetic**. Ferromagnetism is exhibited mainly by iron, cobalt, nickel and their alloys. The behaviour of diamagnetic and paramagnetic substances in a magnetic field is shown diagrammatically in Fig. 18.8.

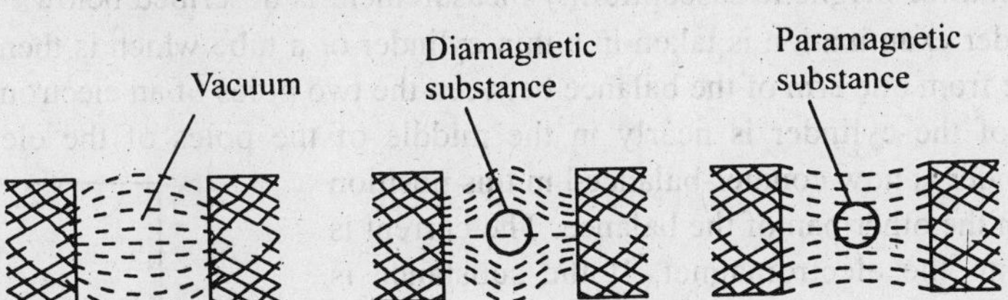

Fig. 18.8. (*a*) Magnetic lines of force passing through vacuum. (*b*) Magnetic lines of force pushed away by the diamagnetic substance. (*c*) Magnetic lines of force drawn in by the paramagnetic.

(*iii*) **Magnetic susceptibility :** If a material is placed in a magnetic field, the strength of the magnetic field in the material is different from that in the vacuum. The strength of the magnetic field thus present in the material is called **magnetic induction.**

If *H* is the strength of the applied magnetic field and *B is* the magnetic induction in the given material, the two are related according to the equation

$$B = H + 4\pi I \quad \text{... (i)}$$

where *I* represents the *magnetic moment per volume* in the material and is called **intensity of magnetisation.** It may be positive or negative depending upon whether the material is paramagnetic or diamagnetic.

The intensity of magnetisation produced per unit strength of the applied magnetic field is called **specific magnetic susceptibility**. It is usually represented by χ. Thus mathematically,

$$\chi = \frac{I}{H} \quad ...(ii)$$

It is this quantity which is usually measured experimentally in the study of magnetic properties of different materials.

Multiplying the specific magnetic susceptibility with the molar volume (M/ρ) of the material, we get what is called as *magnetic susceptibility per mole* or **molar magnetic susceptibility.** It is usually represented by χ_M. Mathematically,

$$\chi_M = \frac{M}{\rho}.\frac{1}{H} \quad ...(iii)$$

Thus **molar magnetic susceptibility** *may be defined as the intensity of magnetization induced per mole of the material by the unit field strength.*

Quite often it is convenient to deal with the quantity χ_M in place of χ.

Further, the magnetic induction B is related to the strength of the applied magnetic field H according to the equation

$$B = \mu H \quad ...(iv)$$

where μ is the magnetic permeability of the medium (*i.e.*, the material placed between the poles).

Also, from equation (*ii*)

$$I = \chi H \quad ...(v)$$

Substituting the values of B and I from equations (*iv*) and (*v*) in equation (*i*), we get

$$\mu H = H + 4\pi\, \chi H$$

or

$$\mu = 1 + 4\pi\chi \quad ...(vi)$$

Q. 23. Explain Gouy method for the determination of magnetic susceptibility.

(*Guwahati 2002*)

Ans. Gouy method for magnetic susceptibility measurement is described below:

The material under examination is taken in a thin cylinder or a tube which is then suspended vertically with a wire from one arm of the balance between the two poles of an electromagnet such that the lower end of the cylinder is nearly in the middle of the poles of the electromagnet (Fig. 18.9). The cylinder is now counter-balanced in this position by placing weights in the other pan of the balance. The current is now switched on into the electromagnet. If the substance is paramagnetic, the cylinder will be pulled down (so that more magnetic lines of force can pass through it). On the other hand if the substance is diamagnetic, the cylinder will be pushed upward (so that less magnetic lines of force may pass through it). In the first case, some extra weights will have to be added into the pan, whereas in the second case, some weights will have to be removed from the pan to restore the balance.

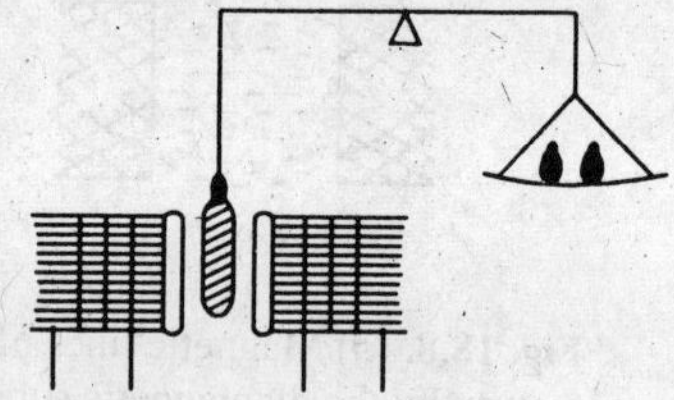

Fig. 18.9. Gouy's method for measurement of magnetic susceptibilities.

Since the change in weight is equal to the magnetic force acting on the material, hence if Δm is the change in mass, then we have

$$\Delta mg = \frac{1}{2}(\chi - \chi_a)AH^2 \quad ...(i)$$

where g is the acceleration due to gravity so that Δmg gives the change in weight. The right hand side is the magnetic force acting on the material of volume susceptibility χ when A is the area of cross-section of the specimen, H is the field strength and χ_a is the volume susceptibility of the air

Value of $\chi_a = 0.03 \times 10^{-6}$. Thus using equation (*i*), the value of χ can be calculated.

In the above calculations, the absolute value of the field strength H is also required. However, the necessity of measuring H and A can be eliminated if the experiment is first performed with substance of known susceptibility using the same magnetic field and the same cylinder for taking this substance. Usually water is taken as the standard substance for this purpose. Then evidently we will have

$$\frac{\Delta m_s}{\Delta m_w} = \frac{\chi_s - \chi_a}{\chi_w - \chi_a} \quad ...(ii)$$

where Δm_s is the change in mass when the sample under examination is placed in the magnetic field and Δm_w is the change in mass when water is taken in the same cylinder and placed in the same magnetic field. χ_s is the magnetic susceptibility (volume susceptibility) of the sample and χ_w and χ_a are those of the water and air respectively.

Q. 24. How does magnetic susceptibility help in elucidating molecular structure.

(Nagpur 2002)

Ans. It is found that molar magnetic susceptibility (χ_M) is an **additive** as well as a **constitutive** property. The values assigned to the various atoms and to the structural factors (*i.e.*, double bond, triple bond, rings etc.) can be obtained from the experimentally determined values for a large number of elements and compounds. The values of magnetic susceptibility for some of the atoms and structural groups are given in Table below.

Table. Atomic and structural susceptibilities

Atom	Magnetic *susceptibility* $\times 10^{-6}$	*Structural factor*	*Magnetic susceptibility* $\times 10^{-6}$
H	–2.93	C = C double bond	+ 5.4
C	–6.00	N= N double bond	+ 1.9
O (in alcohols and ethers)	–4.61	N= N double bond	+ 1.9
O (in ketones)	+1.73	C = N double bond	+ 8.2
O (in esters and acids)	–3.36	C ≡ C triple bond	+ 0.8
N (in open chain)	–5.57	C ≡ N triple bond	– 0.8
N (in ring)	–4.61	Benzene ring	– 1.5
N (in amines)	–1.54	**Note:** When for an aromatic compound, the value of benzene ring has been added, no value has been added for C = C double bond.	
F	–11.5		
Cl	–20.1		
Br	–30.6		

The calculation of molar magnetic susceptibility of a compound for the atomic and structural values help us to decide between the alternate structures for a compound or to confirm the structure of the compound by comparing the calculated value with the experimental value.

Q. 25. Establish that the structure of methyl acetate is

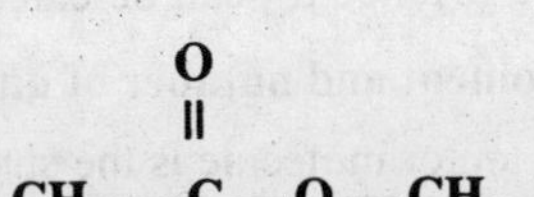

Given that the magnetic susceptibility of the compound is found to be -70.3×10^{-6}

Ans. The value for χ_M for this structure can be calculated as follows:

$$3\text{ C atom} = 3 \times (-6.00 - 10^{-6}) = -18.00 \times 10^{-6}$$
$$6\text{ H atom} = 6 \times (-2.93 \times 10^{6}) = -17.58 \times 10^{-6}$$
$$1\text{ O atom} = 1 \times (-4.61 \times 10^{-6}) = -4.61 \times 10^{-6}$$
$$1\text{O (ester)} = 1 \times (\times 3.36 \times 10^{-6}) = -3.36 \times 10^{-6}$$
$$\text{Total calculated value} = -43.55 \times 10^{-6}$$
$$\text{Observed value} = -43.6 \times 10^{-6}$$

Calculated value agrees well with the observed value. Hence the given structure is correct.

Q. 26. Confirm that the structure of benzoic acid is

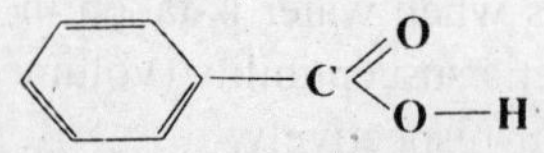

Given that the magnetic susceptibility of the compound is measured as –70.3 × 10⁻⁶.

Ans. The magnetic susceptibility contributions due to various atoms and structural factors and hence the total value for χ_M can be calculated as follows:

$$7\text{ C atom} = 7 \times (-6.00 \times 10^{-6}) = -42.00 \times 10^{-6}$$
$$6\text{ H atom} = 6 \times (-2.93 \times 10^{-6}) = -17.58 \times 10^{-6}$$
$$\mathbf{1\text{ O atom (alcohol)} = 1 \times (-4.61 \times 10^{-6}) = -4.61 \times 10^{-6}}$$
$$1\text{O (acid)} = 1 \times (-3.36 \times 10^{-6}) = -3.36 \times 10^{-6}$$
$$1\text{ benzene ring} = 1 \times (-1.5 \times 10^{-6}) = -1.50 \times 10^{-6}$$
$$\text{Total calculated value} = -69.05 \times 10^{-6}$$
$$\text{Observed value} = -70.3 \times 10^{-6}$$

As the calculated value is very close to the observed value, the given structure is correct.

Q. 27. How will you determine the magnetic moment of a paramagnetic substance?

Ans. Each atom, ion or molecule of a paramagnetic substance behaves like a micromagnet with a definite inherent magnetic moment of magnitude μ_m. When such a substance is placed in the external magnetic field, micromagnets tend to arrange themselves parallel to the applied field. However, this tendency is opposed by the thermal motion of the particles. Hence the actual arrangement taken up will depend upon the temperature. Langevin in 1905 put forward the following relationship between molar magnetic susceptibility χ_M, magnetic moment μ_m and temperature T:

$$\chi_M = \alpha_m \times \frac{\mu_m^2 N}{3kT} \qquad \text{... (i)}$$

where N is Avogadro's number, k is Boltzmann's constant and α_m is a constant for the given substance. It represents the diamagnetic molar susceptibility of the substance induced by the applied magnetic field.

In eqn. (*i*), all quantities are constant except χ_M and T. Thus determining χ_M for the given substance at different temperatures T, and then plotting χ_M *vs* $\frac{1}{T}$, a straight line graph will be obtained, the slope of which will be equal to $\frac{\mu_m^2 N}{3k}$. Hence μ_m can be calculated.

Q. 28. Derive a relation between magnetic moment and number of unpaired electrons.

Ans. The total magnetic moment of any atom, ion or molecule is the sum of the following two values:

(*i*) Magnetic moment due to the *orbital motion* of the electron. It is equal to $\sqrt{l(l+1)}$ B.M. where l is the azimuthal or orbital quantum number for that electron.

(*ii*) Magnetic moment due to the *spin motion* of the electron. It is equal to $2\sqrt{s(s+1)}$ B.M. where *s* is the spin quantum number having value equal to $\frac{1}{2}$.

If ions or molecules are considered instead of free atoms, it is expected that the orbital motions of the electrons are tied into the nuclear configuration of the molecule or the ion so tightly that they are unaffected by applied magnetic field. Hence the magnetic moment contribution due to the orbital motion is negligible. Thus the magnetic moment of the molecule or the ion is mainly due to the spin motion of the electrons. (Hence it is called spin magnetic moment). Thus we have

$$\mu_m = 2\sqrt{s(s+1)}\ B.M. \qquad ...(i)$$

Here *s* represented spin quantum number for one unpaired electron. For a molecule or ion containing a number of unpaired electrons, the expression is usually written as

$$\mu_m = 2\sqrt{S(S+1)}\ B.M. \qquad ...(ii)$$

where *S* representes the total spin quantum number

Thus for one unpaired electron, $S = \frac{1}{2}$

for two unpaired electrons, $S = \frac{2}{2}$

for three unpaired electron, $S = \frac{3}{2}$ and so on.

In general, for *n* unpaired electron, $S = \frac{n}{2}$ hence eqn. (*ii*) becomes

$$\mu_m = 2\sqrt{\frac{n}{2}\left(\frac{n}{2}+1\right)}$$

$$\boxed{\mu_m = \sqrt{n(n+2)}}\ \text{B.M.} \qquad ...(iii)$$

Thus knowing the number of unpaired electrons present, the magnetic moment (μ_m) can be calculated or *vice versa.*

Q. 29. What is the relation between molar susceptibility and the number of unpaired electrons?

Ans. Taking α_m as constant, eqn. (*i*) of *Q.* 27 can be written as:

$$\chi_M = \frac{\mu_m^2 N}{3kT} \qquad ...(i)$$

where χ_M and μ_m are magnetic susceptibility and magnetic moment respectively.

Substituting the value of μ_m in eqn. (*i*), we get

$$\boxed{\chi_M = \frac{N}{3kT}[n(n+2)]} \qquad ...(ii)$$

Q. 30. Discuss applications of magnetic susceptibility measurement.

Ans. Some important applications of magnetic susceptibility measurement are given below:

(1) **In confirming the structure of a given molecule:** The use of molar magnetic susceptibilities in confirming the structure of a given molecule is based upon the fact that magnetic susceptibility is additive as well constitutive property.

(2) **In calculation of the number of unpaired electrons in a molecule or ion:** Knowing the molar magnetic susceptibility χ_M which can be easily determined experimentally, the number of

unpaired electrons present in a molecule or an ion can be calculated. This has helped in understanding the electronic structures of a number of molecules. For example, the molar magnetic susceptibility of oxygen has been found to be 3400×10^{-6} which shows that it must contain two unpaired electrons. Thus the structure of oxygen molecule should be represented as

$$:\dot{\underset{\cdot\cdot}{O}}:\dot{\underset{\cdot\cdot}{O}}:$$

It is a paramagnetic molecule.

(3) **In the study of coordination compounds:** An important application of magnetic susceptibility in chemistry has been in the study of coordination compounds of transition elements. Measurement of magnetic susceptibility helps us to know the number of unpaired electrons present in the complex which in turn helps to assign the correct structure to the complex ion in some cases. For example, the magnetic susceptibility values of the complexes $[Fe(CN)_6]^{3-}$ and $[FeF_6]^{3-}$ (though each has the same coordination number *i.e.*, 6) suggest that there is one unpaired electron present in the former whereas there are five unpaired electrons present in the latter. These **are explained by suggesting the following electronic structures to these complexes: (Fig 18.10**

$$Fe = 26 = 1s^2\ 2s^2\ 2p^6\ 3s^2\ 3p^6\ 3d^6\ 4s^2$$

$$Fe^{3+} = 1s^2\ 2s^2\ 2p^6\ 3s^2\ 3p^6\ 3d^5$$

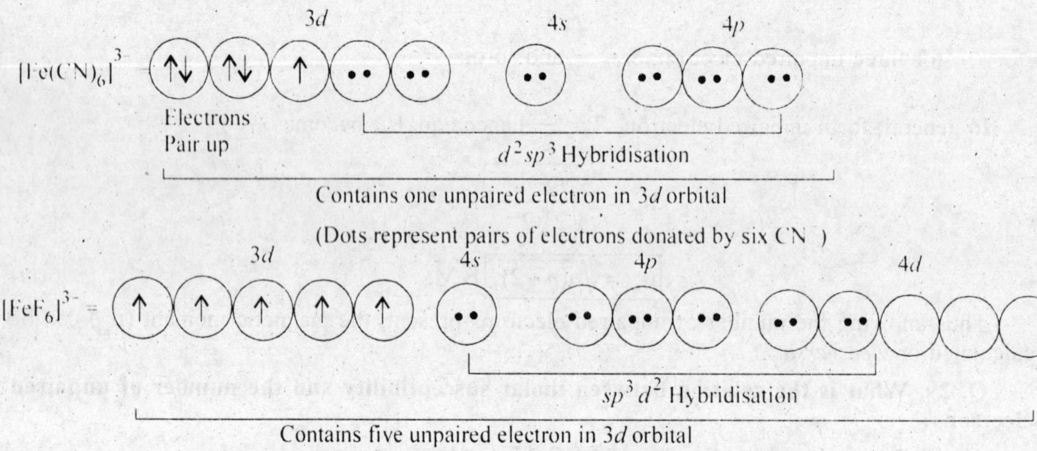

Fig. 18.10. Structure of (*i*) $[Fe(CN)_6]^{3-}$ and (*ii*) $[FeF_6]^{3-}$

(4) **In establishing the formulas of certain compounds:** For example, it has been shown that the formula of hypophosphoric acid is $H_4P_2O_6$ and not H_2PO_3. This is because if the latter formula were correct, sodium and silver salts, $NaHPO_3$ and Ag_2PO_3, would contain an odd number of electrons and should be paramagnetic while actually they are diamagnetic. Hence their salts must **have the formulas $Na_2H_2P_2O_6$ and $Ag_4P_2O_6$ respectively. Similarly the formula of mercurous chloride is Hg_2Cl_2 and not HgCl. Dimerisation leads to spin neutralization of electrons.**

(5) **In explaining the existence of odd electron molecules:** The compounds containing odd electrons are expected to be paramagnetic. This is actually found to be so for a few gaseous compounds, namely nitric oxide, nitrogen dioxide and chlorine dioxide. Moreover, their magnetic susceptibilities show that they contain one unpaired electron in each case.

Q. 31. Give molecular interpretation of (*a*) diamagnetism and (*b*) paramagnetism.

Ans. Molecular Interpretation of Diamagnetism.

The classical theory of diamagnetism and paramagnetism was put forward Langevin in 1905.

According to Langevin, *diamagnetism results from the orbital motion of the atoms and molecules.* According to Lenz's law which states that when a current flows through a wire in a magnetic field, the magnetic field induces current in the coil in such a manner that it opposes the applied field. Similarly, the applied magnetic field perturbs the orbital motion thereby inducing a magnetic field which opposes its own direction. In multi-electron atom, the orbits are spatially oriented at random relative to the field.

The molar diamagnetic susceptibility χ_M^d is given by

$$\chi_M^d = -\frac{Z\,Ne^2}{6\,mc^2}\Sigma r_i^2 = -2.832 \times 10^{10}\,\overline{r_i^2}$$

Where $\overline{r_i^2} \equiv$ the mean square distance of the *i*th orbit from the nucleus, Z = atomic number, e = charge on the electron, m = mass of electron and c = velocity of light.

Diamagnetism is a common property of all atoms of all matter whether it possesses a permanent dipole moment or not.

Molecular Interpretation of Paramagnetism

According to Langevins's theory of paramagnetism each atom or molecule behaves as a small bar magnet with magnetic moment μ. The magnetic moment may either be due to orbital motion of an electron or due to unpaired electron spin. The *molar paramagnetic susceptibility* due to permanent magnetic moment is given by

$$\chi_m^p = \frac{N\mu_M^2}{3kT} = \frac{\text{constant}}{T}$$

i.e. χ_m^p varies inversely with absolute temperature T. This is known as *Curie's law.* which holds over a limited range of temperature.

In case, the system is distorted by the external field, the molar paramagnetic susceptibility is more correctly given by

$$\chi_m^p = \frac{N\mu_M^2}{3kT} = \alpha_M$$

where α_M is magnetic polarizability of the system.

As all substances exhibit diamagnetic behaviour, the total molar magnetic susceptibility is the sum of χ_m^d and χ_m^p. However, the diamagnetic contribution is negligibly small as compared to paramagnetic contribution.

MOLECULAR SPECTROSCOPY

Q. 32. What is meant by electromagnetic spectrum ? Explain the terms : Frequency, wavelength, wave number. Derive relations between them.

Ans. Visible light, X-rays, microwaves, radio waves, etc., are all electromagnetic radiations. Collectively, they make up the **Electromagnetic spectrum** (Fig. 18.11).

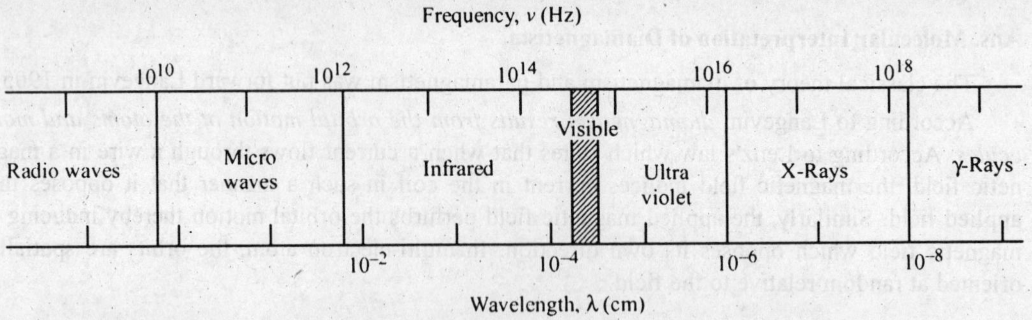

Fig. 18.10. The Electromagnetic Spectrum.

An electromagnetic wave may be illustrated by Fig. 18.11.

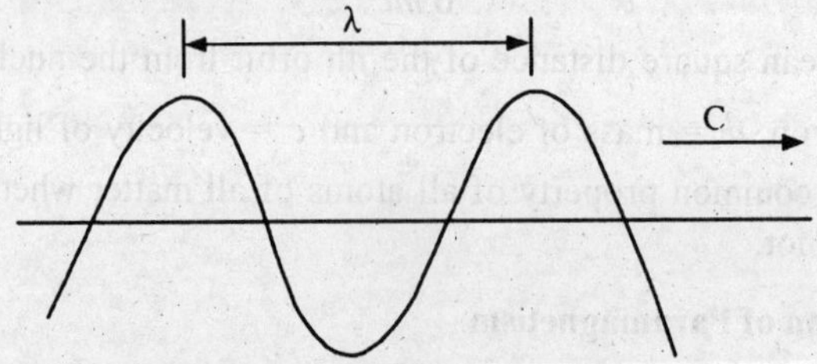

Fig. 18.11. A single electromagnetic wave.

Electromagnetic radiations consist of electrical and magnetic waves oscillating at right angles to each other. They travel away from the source with the velocity of light (c) in vacuum. Different terms used in the study of electromagnetic waves are discussed as under :

1. Frequency (ν) : It is the number of successive crests (or troughs) which pass a stationary point in one second. The unit is **hertz;** $1\text{Hz} = 1\text{s}^{-1}$.

2. Wavelength (λ) : It is the distance between successive crests (or troughs). λ is expressed in centimeters (cm), meters (m), or nanometers ($1 \text{ nm} = 10^{-9}\text{m}$).

3. Wave number ($\bar{\nu}$) : It is the reciprocal to wavelength. Its unit is cm^{-1}.

$$\bar{\nu} = \frac{1}{\lambda}$$

Relation between Frequency, Wavelength and Wave number

Frequency and wavelength of an electromagnetic radiation are related by the equation

$$\nu\lambda = c \quad \ldots(i)$$

Or

$$\nu = \frac{c}{\lambda} \quad \ldots(ii)$$

where c is the velocity of light. It may be noted that wavelength and frequency are inversely proportional. That is, **higher the wavelength lower is the frequency;** lower the wavelength higher is the frequency.

Further

$$\bar{\nu} = \frac{1}{\lambda}\ \text{cm}^{-1} \quad \ldots(iii)$$

From (*ii*) and (*iii*)

$$\nu = c\bar{\nu}$$

Energy of Electromagnetic Radiation

Electromagnetic radiation behaves as consisting of discrete wave-like particles called **quanta** or **photons.** Photons possess the characteristics of a wave and travel with the velocity of light in the direction of the beam. The amount of energy corresponding to 1 photon is expressed by **Planck's equation.**

$$E = h\nu = hc/\lambda$$

where E is the energy of 1 photon (or quantum), h is Planck's constant (6.62×10^{-27} ergs-sec.); ν is frequency in hertz; and λ is wavelength in centimeters.

If N is the Avogadro number, the energy of 1 mole photons can be expressed as

$$E = \frac{Nhc}{\lambda} = \frac{2.85 \times 10^{-3}}{\lambda} \text{ kcal/mol}$$

Q. 33. What is meant by : (*a*) Rotational energy; (*b*) Vibrational energy; and (*c*) Electronic energy of molecules.

Ans. Rotational Energy

It involves the rotation of molecules or of parts of molecules about the centre of gravity (Fig. 18.12).

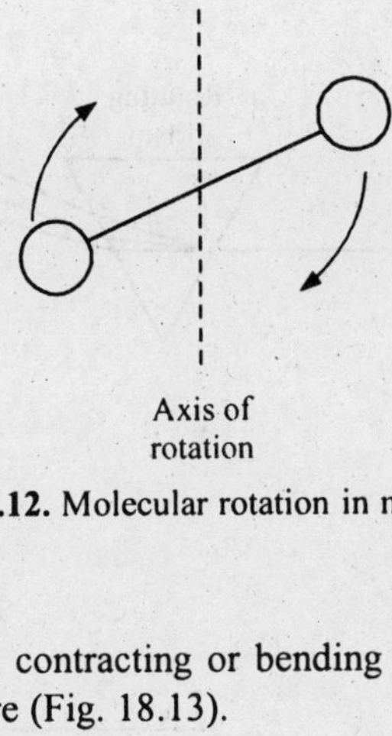

Fig. 18.12. Molecular rotation in molecules

Vibrational Energy

It is associated with stretching, contracting or bending of covalent bonds in molecules. The bonds behave as spirals made of wire (Fig. 18.13).

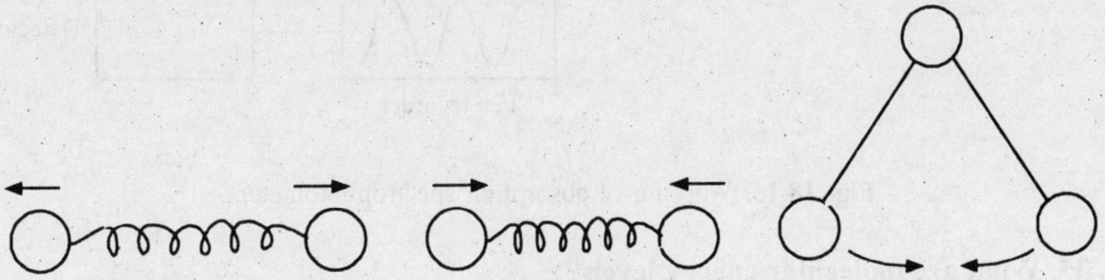

Fig. 18.13. Vibrations within molecules

Electronic Energy

It involves changes in the distribution of electrons by splitting of bonds or the promotion of electrons into higher energy levels (Fig. 18.14).

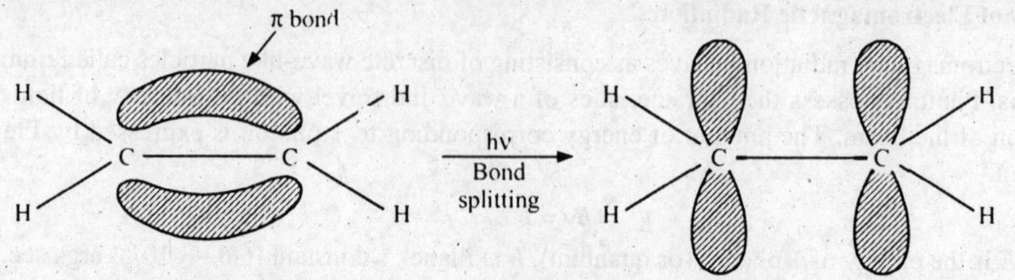

Fig. 18.14. Splitting of pi bond by absorption of energy.

Q. 34. Give the experimental set-up of absorption spectrophotometer.

Ans. Absorption spectrum of a given sample is obtained experimentally with the help of an apparatus called **absorption spectrophotometer** (Fig. 18.15). Light of a range of wavelengths from the source is passed through the sample. The wavelengths corresponding to allowed molecular transitions are absorbed. The transmitted light passes through a prism which resolves it into various wavelengths. It is then reflected from the mirror onto a detector. The prism is rotated so that light of each given wavelength is focussed on the detector. The pen recorder records the intensity of radiation as a function of frequency and gives the absorption spectrum of the sample.

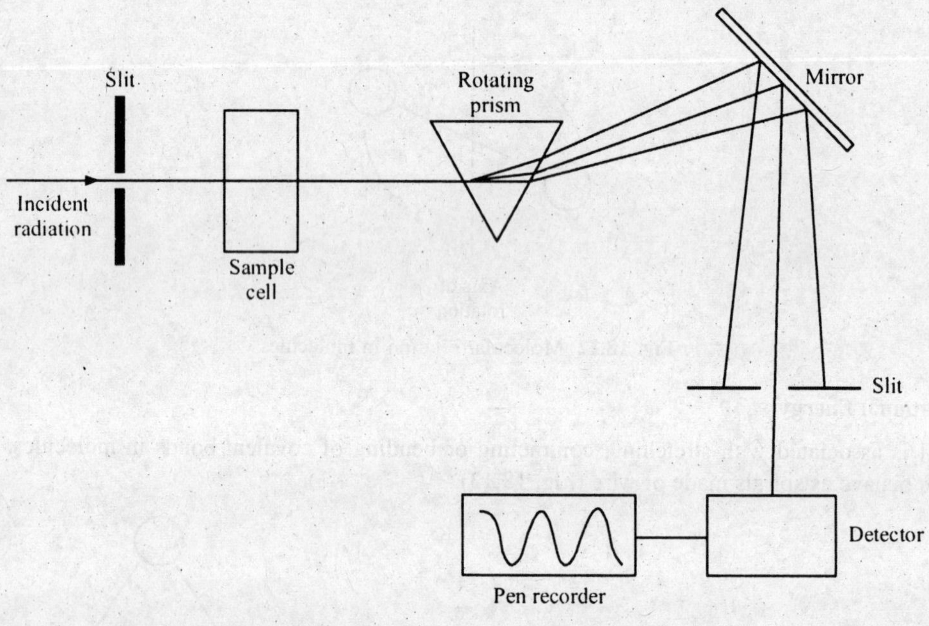

Fig. 18.15. Working of absorption spectrophotometer.

Q. 35. What are molecular energy levels?

Or

What is band spectrum ? Discuss electronic, vibrational and rotational transitions occurring in molecules with a diagram. (Nagpur 2003)

Ans. In case of molecules, when the energy is absorbed, it may result into rotation, vibration or electronic transition depending upon the amount of energy absorbed. Just as electronic energy is quantized *i.e.*, there are only discrete electronic energy levels in an atom or a molecule, the rotational and vibrational energies are also quantized *i.e.*, there are only discrete rotational and vibrational energy levels in a molecule. The rotational, vibrational and electronic energy levels of a molecule are collectively called **molecular energy levels.** The transitions of energies can take place only between these levels. The result is a **molecular spectra.** Just as the study of atomic spectra is helpful in the study of structure of atoms, the study of molecular spectra is one of the best experimental methods of studying the structure of molecules.

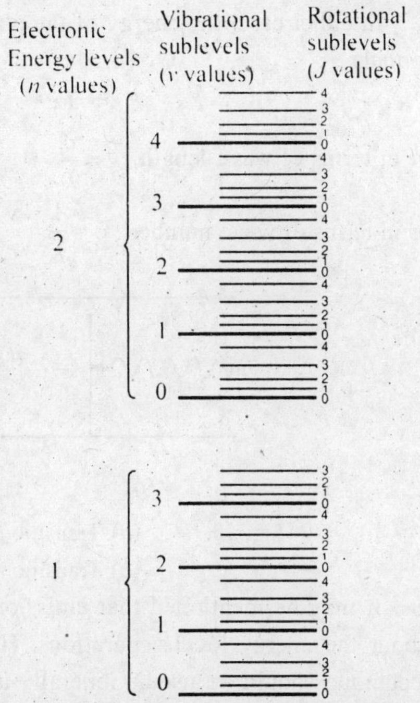

Fig. 18.16. Molecular energy levels.

When the energy absorbed by a molecule is so large that an electron can jump from one electronic energy level to another (say $n = 1$ to $n = 2$) then as the energy required for transition from one vibrational energy level to another is much less and for one rotational level to another is still less, hence the electronic transition is also accompanied by the transition between the vibrational levels as well as the rotational levels. The complete picture of the various molecular energy levels for the first two electronic levels is shown in Fig. 18.16.

It is evident that each electronic level consists of a number of vibrational sublevels represented by $v = 0, 1, 2, 3$ etc. and further each vibrational level consists of a number of rotational sublevels represented by $J= 0, 1, 2, 3$ etc. v and J are called *vibrational and rotational quantum numbers respectively.*

Because of a large number of energy level involved in transition, the molecular spectra are much more complex than atomic spectra. However, just as atomic spectra help in the study of the structure of the atoms, the study of molecular spectra gives even more information about the structure and properties of molecules such as bond lengths, bond angles, bond strength, shapes, dipole moments *etc.*

Q. 36. What are absorption spectrum and emission spectrum?

Ans. An atom or a molecule has a number of energy levels which are quantized. The transitions take place only between these energy levels according to certain rules, called *selection rules.* Transition between any two energy levels can take place in either of the following two ways:

(*i*) The transition may take place from lower energy level to higher energy level by absorbing energy Fig. 18.17(*a*). It is then called **absorption spectroscopy** and the result obtained as a result of number of such transitions is called "absorption spectrum". Absorption spectrum consists of **dark** lines.

(*ii*) The transition may take place from higher energy level to a lower energy level Fig.18.17(*b*) thereby emitting the excess energy as a photon. It is then called **emission spectroscopy** and the result obtained as a result of number of such transitions is called "emission spectrum". Emission spectrum consists of **bright** lines.

In fact, a number of groups of closely spaced lines are observed. Each such group of closely spaced lines is called a **band**. Thus a number of bands are observed. We can, therefore, conclude that whereas atoms give line spectra, molecules give band spectra.

In either case, the energy of the photon $h\nu$ emitted or absorbed is given by Bohr's frequency formula

$$E_2 - E_1 = h\nu$$

or in terms of wave length, $\nu = \frac{c}{\lambda}$

or in terms of wave number, $\bar{\nu} = \frac{1}{\lambda} = \frac{\nu}{c}$ (in cm^{-1} if λ is in cm or velocity c is in cm s^{-1})

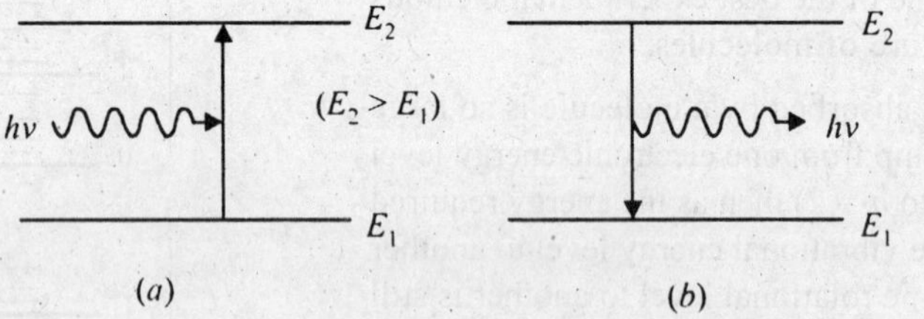

Fig. 18.7. (*a*) Transition involving absorption of photon.
(*b*) Transition involving emission of photon.

It may be mentioned that emission and absorption spectroscopy give the same information about the energy level separations. However, the practical considerations decide as to which technique should be used. Generally absorption spectroscopy is used as absorption spectra are easier to interpret than emission spectra.

Q. 37. What are different types of molecular spectra?. **(*Panjab 2000; Bhopal 2004*)**

Ans. Different types of molecular spectra are described as under:

(*i*) **Pure rotational (Microwave) spectra:** If the energy absorbed by the molecules is so low that it can cause transition only from one rotational level to another within the same vibrational level, the result obtained is called the rotational spectrum. These spectra are, therefore, observed in the **far infra-red region** or in **the microwave region** whose energies are extremely small ($\bar{\nu} = 1 - 100$ cm^{-1}). The spectra obtained is, therefore, also called **micro-wave spectra.**

(*ii*) **Vibrational rotational spectra:** If the exciting energy is sufficiently large so that it can cause a transition from one vibrational level to another within the same electronic level then as the energies required for the transitions between the rotational levels are still smaller, both types of transitions *viz.,* vibrational and rotational will take place. The result is, therefore, a vibration-rotation spectrum. Since such energies are available in the near infra-red region, these spectra are observed in this region ($\bar{\nu} = 500 - 4000$ cm^{-1}) and are called **infra-red spectra.**

(*iii*) **Electronic Band spectra:** If the exciting energy is higher such that it can result in a transition from one electronic level to another, then this will also be accompanied by vibrational level changes and each of these is further accompanied by rotational level changes. For each vibrational change, a set of closely spaced lines is observed due to rotational level changes. Such a group of closely spaced rotational lines is called a **band.** Thus for a given electronic transition, a set of bands is observed. This set of bands is called a **band group** or a **band system.** Each electronic transition gives a band system. The complete set of band systems obtained due to different electronic transitions gives the electronic band spectrum of the gaseous molecule. As such high excitation energies are obtained from the visible and ultraviolet regions, these spectra are observed in the **visible region** (12,500 – 25,000 cm^{-1}) and **ultraviolet region** (25,000 – 70,000 cm^{-1}).

(*iv*) **Raman spectra:** This is also a type of vibrational-rotational spectrum but is based on scattering of radiation and not on the absorption of radiation by the sample. It is based upon the principle that when a sample is hit by monochromatic radiation of the visible region and scattering is observed at *right angles to the direction of the incident beam,* the scattered radiation has frequency equal to that of the incident beam (called *Rayleigh scattering*) as well as frequencies different (higher as well as lower) from that of the incident beam (called *Raman scattering*). The difference in the frequencies of the incident beam and that of the scattered beam (called *Raman frequencies*) are similar to those observed for the vibrational and rotational transitions. However, by suitably adjusting the frequency of the incident radiation, Raman spectra are observed in the visible region (12,500 – 25,000 cm^{-1}).

(*v*) **Nuclear Magnetic Resonance (NMR) spectra:** This type of spectrum arises from the transitions between the *nuclear spin energy levels* of the molecule when an external magnetic field is applied on it. The energies involved in these transitions are very high which lie in the *radio frequency region* (5-100 MHz). The method is based upon applying such frequencies on the sample so that it resonates with the applied frequency.

(*vi*) **Electron Spin Resonance (ESR) spectra:** This type of spectrum arises from the transitions between the *electron spin energy levels* of the molecule when an external magnetic field is applied on it. These involve frequencies corresponding to microwave region (2000 – 9600 MHz). Frequencies of this range are applied on the sample to bring the sample in resonance condition.

ROTATIONAL SPECTRA

Q. 38. What are selection rules for rotational and vibrational spectra?

(Garhwal 2000; Lucknow 2002; Nagpur 2008)

Ans. Atomic spectrum as well as molecular spectrum are obtained due to transition taking place between energy levels. However, such transitions can take place only between definite energy levels not between just any two energy levels. *The restrictions thus applied on the transition are governed by certain rules which are called the selection rules.* If these rules are followed, the transition can take place and it is called an **allowed transition.** If these are not followed, the transition cannot take place and it is called a **forbidden transition.** The selection rules to be followed depend upon the type of transition.

The **selection rules** are generally expressed in terms of changes in quantum numbers for the allowed transitions. For example, for the pure rotational transition, the selection rule is $\Delta J = \pm 1$ where J represents rotational quantum number ($\Delta J = +1$ corresponds to **absorption** and $\Delta J = -1$ corresponds to **emission**). Similarly, for pure vibrational transition, the selection rule is $\Delta v = \pm 1$ where v represents the vibrational quantum number.

Q. 39. Derive an expression for the rotational energy of a diatomic molecule taking it as a rigid rotator. Draw the rotational energy level diagram for such a molecule.

(Bangalore 2004; Himachal 2010)

Ans. On the basis of wave mechanics, it can be seen that if *diatomic* molecule is considered to be a rigid rotator, *i.e.*, like a rigid dumbbell joined along its line of centres by a bond equal in length to the distance r_0 between the two nuclei (Fig. 18.18) then the allowed rotational energies of the molecule around the axis passing through the centre of gravity and perpendicular to the line joined the nuclei are given by

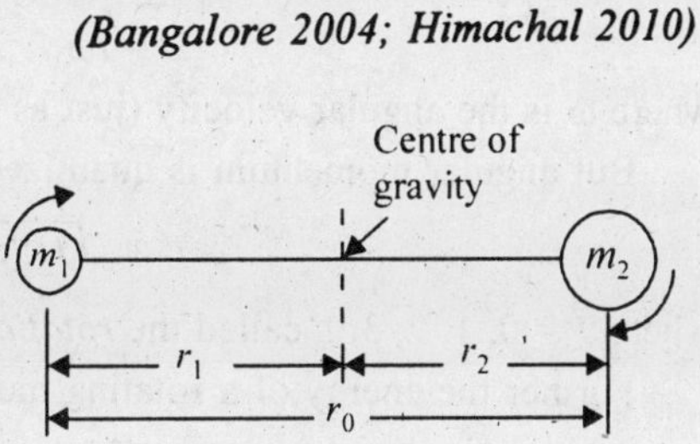

Fig. 18.18. Rigid rotator.

$$E_r = \frac{h^2}{8\pi^2 I} J(J+1)$$

where J is the rotational quantum number that can have values 0, 1, 2, 3 etc., and I is the moment of inertia of the molecule about the axis of rotation, *i.e.*,

$$I = \left(\frac{m_1 m_2}{m_1 + m_2}\right) r_0^2 = \mu r_0^2$$

where m_1 and m_2 are the atomic masses of the two atoms of the diatomic molecule and

$$\frac{m_1 m_2}{m_1 + m_2} = \mu$$

is called the *reduced mass.*

Derivation of the expression for rotational energy: Centre of gravity of a diatomic molecule is the point which satisfies the following conditions

$$m_1 r_1 = m_2 r_2 \qquad ...(i)$$

The moment of inertia of the diatomic molecule is given by

$$I = m_1 r_1^2 + m_2 r_2^2 \qquad ...(ii)$$

$$= m_2 r_2 r_1 + m_1 r_1 r_2$$

$$= r_1 r_2 (m_1 + m_2) \qquad ...(iii)$$

But $$r_1 + r_2 = r_0 \qquad ...(iv)$$

$\therefore$ From eqn. (i), $$m_1 r_1 = m_2 (r_0 - r_1)$$

or $$r_1 = \frac{m_2 r_0}{m_1 + m_2} \qquad ...(v)$$

Similarly, $$r_2 = \frac{m_1 r_0}{m_1 + m_2} \qquad ...(vi)$$

Substituting these values in eqn. (*ii*), we get

$$I = \frac{m_1 m_2^2}{(m_1 + m_2)^2} r_0^2 + \frac{m_1^2 m_2}{(m_1 + m_2)^2} r_0^2$$

$$= \frac{m_1 m_2 (m_1 + m_2)}{(m_1 + m_2)^2} r_0^2 = \frac{m_1 m_2}{m_1 + m_2} r_0^2$$

$$= \mu r_0^2 \left(\text{where } \mu = \frac{m_1 m_2}{m_1 + m_2} \text{ is reduced mass}\right) \qquad ...(vii)$$

Now, by definition, the angular momentum of a rotating molecule is given by

$$L = I\omega \qquad ...(viii)$$

where ω is the angular velocity (just as linear momentum is mass × velocity).

But angular momentum is quantized whose values are given by

$$L = \sqrt{J(J+1)} \frac{h}{2\pi} \qquad ...(ix)$$

where J = 0, 1, 2, 3..., called the *rotational quantum numbers.*

Further the energy of a rotating molecule is given by

$$E = \frac{1}{2} I\omega^2$$

$\therefore$ The quantized value of the rotational energy will be given by

$$E_r = \frac{1}{2} I\omega^2 = \frac{(I\omega)^2}{2I} = \frac{L^2}{2I} \qquad ...(x)$$

Substituting the value of L From eqn. (*ix*), we get

$$E_r = J(J+1)\frac{h^2}{4\pi^2} \times \frac{1}{2I} \quad \text{or} \quad E_r = \frac{h^2}{8\pi^2 I} J(J+1) \qquad ...(xi)$$

Putting $J = 0, 1, 2, 3$ etc. in equation (*xi*), pattern of the rotational energy levels obtained will be as shown in Fig. 18.19.

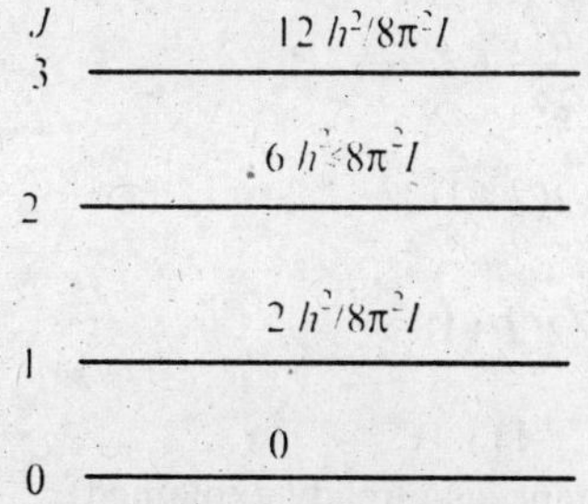

Fig. 18.19. Rotational energy levels of a diatomic molecule treating it as a rigid rotator.

Evidently, the spacing between the energy levels increases as J increases, because of the factor $J(J + 1)$ in eq. (*xi*).

Q. 40. Derive an expression for the frequency and wave number of lines in the rotational spectrum.

Ans. Allowed rotational energies are given by the expression

$$E_r = \frac{h^2}{8\pi^2 I} J(J+1) \qquad ...(i)$$

As $E = h\nu$, therefore, in terms of frequency, we can write

$$\nu = \frac{h}{8\pi^2 I} J(J+1) \qquad ...(ii)$$

Further as $c = \nu\lambda = \nu / \bar{\nu}$ (because wave number $\bar{\nu} = 1/\lambda$), therefore, in term of wave numbers, eqn. (*ii*) can be written as

$$\bar{\nu} = \frac{h}{8\pi^2 Ic} J(J+1) \qquad ...(iii)$$

$$= BJ(J+1) \qquad ...(iv)$$

where $B = h/8\pi^2 Ic$ is called *rotational constant.*

Putting $J = 0, 1, 2, 3$ etc in eqn. (*iv*), *the wave numbers of the different rotational levels* will be

$$0, 2B, 6B, 12B, 20B, 30B, \text{...... so on.}$$

When a transition takes place from a lower rotational level with rotational quantum number J to a higher rotational level with rotational quantum number J', the energy absorbed will be given by

$$\Delta E_r = E_r' - E_r \qquad (\text{or } E_{J'} - E_J)$$

$$= \frac{h^2}{8\pi^2 I} J'(J'+1) - \frac{h^2}{8\pi^2 I} J(J+1)$$

$$= \frac{h^2}{8\pi^2 I}[J'(J'+1)-J(J+1)] \qquad ...(v)$$

However, according to the rotational selection rules, only those rotational transitions are allowed for which $\Delta J = \pm 1$. Hence in the present case $J' = J + 1$.

Substituting this value in eqn. (v), we get

$$\Delta E_r = \frac{h^2}{8\pi^2 I}[(J+1)(J+2) - J(J+1)]$$

$$= \frac{h^2}{8\pi^2 I} 2(J+1) \qquad ...(vi)$$

To express in terms of wave numbers, we put

$$\Delta E_r = h\nu = h\frac{c}{\lambda} = hc\bar{\nu}$$

Hence $$hc\bar{\nu} = \frac{h^2}{8\pi^2 I} \times 2(J+1)$$

or $$\bar{\nu} = \frac{h}{8\pi^2 Ic} \times 2(J+1)$$

$$= 2B\,(J+1) \qquad ...(vii)$$

where $B = \dfrac{h}{8\pi^2 Ic}$ is rotational constant (as already explained).

$\bar{\nu}$ in eqn. (vii) represents the wave numbers of the spectral lines which will be obtained as a result of the transitions between the rotational levels. Putting $J = 0, 1, 2, 3$ etc. (*i.e.*, for the transitions $J = 0$ to $J' = 1$; $J = 1$ to $J' = 2$ etc.) or in general from J to $J + 1$ (involving absorption of energy), the wave numbers of the lines obtained will be

$$2B, 4B, 6B, 8B, ... \text{ and so on.}$$

We find that the most important feature of the pure rotational spectrum is that every two successive lines have a constant difference of wave number equal to $2B$. This is called **frequency separation** (or more strictly **wave number separation**). Thus the various lines in the rotational spectra will be equally spaced as shown in Fig. 18.20.

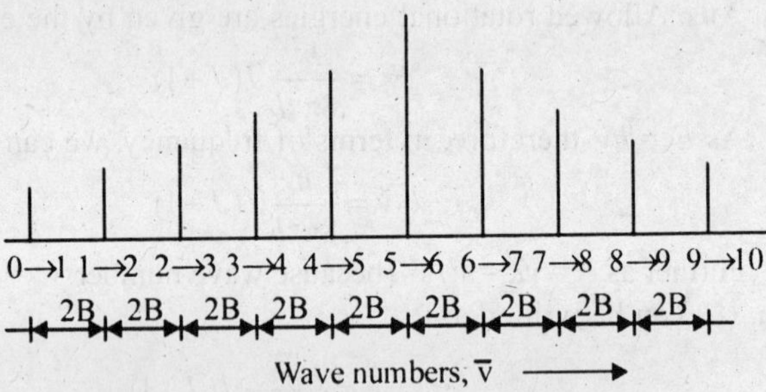

Fig. 18.20. Appearance of a rotational spectrum.

Any two successive lines will have the difference in wave number given by

$$\Delta\bar{\nu} = 2B = \frac{h}{4\pi^2 Ic}$$

Another important feature of spectra is that the intensities of different transitions are not equal. The intensities increase with increasing J, pass through a maximum and then decrease as J increases further.

Q. 41. Derive an expression for the intensities of rotational spectral lines.

Ans. It is expected that the relative intensities of the spectral lines will depend upon the relative populations of the energy levels. Greater the population of an energy level, greater is the number of molecules that can be promoted to the next higher level and hence greater is the intensity of absorption. But the population of a rotational energy level with quantum number J relative to that of the ground level for which $J = 0$ is given according to Boltzmann's distribution law by equation.

$$\frac{N_J}{N_0} = e^{-E_J/kT} \qquad ...(i)$$

However, rotational energy levels are degenerate and the degeneracy of any rotational level with quantum number J is given by

$$g_j = 2J + 1 \qquad ...(ii)$$

It we take into consideration degeneracy factor, equation (i) should be replaced by

$$\frac{N_J}{N_0} = g_j\, e^{-E_J/kT} \qquad ...(iii)$$

From (ii) and (iii)

$$\frac{N_J}{N_0} = (2J+1)e^{-E_J/kT} \qquad ...(iv)$$

$$\text{Intensity} \propto \frac{N_J}{N_0} = (2J+1)e^{-E_J/kT}$$

Putting $E_j = h\nu = h\frac{c}{\lambda} = hc\bar{\nu} = hcBJ(J+1)$ (refer to eqn. (iv) Q. 40).

We have $\dfrac{N_J}{N_0} = (2J+1)e^{-hcBJ(J+1)/kT}$

As J increases, the first factor *viz.* $(2J + 1)$ increases whereas the exponential factor decreases slowly in the beginning and then rapidly as the values of J become higher and higher. Hence plot of N_J/N_0 versus J for a rigid diatomic molecule is obtained as shown in Fig. 18.21.

The graph clearly shows that relative population and hence the intensity of transitions first increases, reaches a maximum value and then decreases.

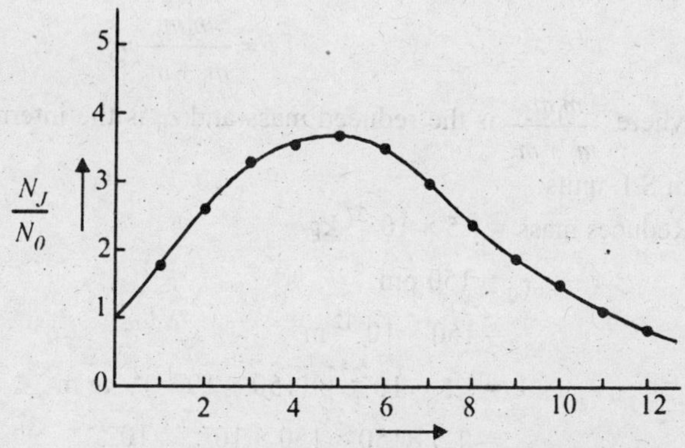

Fig. 18.21. Plot relative population of rotational energy levels versus rotational quantum number.

Q. 42. What type of molecules give rotational spectra? State which of the following molecules give rotational spectra: H_2, HCl, CH_4, CH_3Cl. ***(Nagpur 2008; Himachal 2010)***

Ans. The condition for obtaining a pure rotational spectrum is that the molecule must have a permanent dipole moment so that it can interact with the electromagnetic radiation and absorb or emit a photon. Thus *pure rotational spectrum is given only by polar molecules.* Consequently, homonuclear diatomic molecules (like H_2, N_2 etc.) and symmetrical linear molecules like CO_2 (or symmetrical molecules like C_6H_6) do not give rotational spectra. Thus polar molecules like H_2O, NO, N_2O *etc.*, give pure rotational spectra.

Out of H_2, HCl, CH_4 and CH_3Cl, HCl and CH_3Cl possess a permanent dipole moment and will give rotational spectra. H_2 and CH_4 being non-polar will not show rotational spectra.

Q. 43. Which of the following molecules show rotational spectra? HCl, CO, H_2 and O_2.

Ans. Out of HCl, CO, H_2 and O_2, the first two namely HCl and CO will give rotational spectra as they possess a permanent dipole moment. This is the necessary condition for a molecule to show rotational spectra.

Q. 44. How can you determine bond length of a molecule from its rotational spectra? How is the rotational spectrum of a substance obtained experimentally?

Ans. By determining the frequency separation, $\Delta\bar{\nu}$, it is possible to calculate the moment of inertia I of the molecule, using equation $B = h/8\pi^2 Ic$. Then knowing the masses of the atoms present in the diatomic molecule, it is possible to calculate the distance between the atoms, r_0 (*i.e.*, internuclear distance or bond length) using equation

$$I = \left(\frac{m_1 m_2}{m_1 + m_2}\right) r_0^2$$

The source of radiations is a special type of electronic oscillator (called klystron). It produces radiations of frequency corresponding to wavelength range of 0.1 to a few hundred centimetres. The energies corresponding to these wavelengths are same as required for rotational transitions. The procedure consists in passing the monochromatic energy of the microwaves through the gaseous sample of the substance being studied and measuring the intensity of the transmitted radiation. By varying the frequency of the oscillator and observing the intensity of the transmitted beams, the frequencies at which absorption takes place can be determined from the plot of transmitted intensity versus frequency.

Q. 45. What is the moment of inertia of a diatomic molecule whose internuclear distance is 150 pm and the reduced mass is 1.5 × 10⁻²⁷ kg.

Ans. Moment of inertia I of a diatomic molecule is given by the relation

$$I = \frac{m_1 m_2}{m_1 + m_2} \cdot r_0^2$$

where $\frac{m_1 m_2}{m_1 + m_2}$ is the reduced mass and r_0 is the internuclear distance.

In S.I. units

Reduces mass = 1.5×10^{-27} kg

and $r_0 = 150$ pm

$= 150 \times 10^{-12}$ m

$$I = 1.5 \times 10^{-27} \times (150 \times 10^{-12})^2 \text{ kg m}^2$$

$$= 1.5 \times 150 \times 150 \times 10^{-27} \times 10^{-24}$$

$$= 33750 \times 10^{-51}$$

$$= 3.375 \times 10^{-47} \text{ kg m}^2$$

Q. 46. The far infrared spectrum of HI consists of series of equally spaced lines with $\Delta\bar{\nu}$ = 12.8 cm⁻¹. What is (*a*) the moment of inertia and (*b*) the internuclear distance?

Ans. (*a*) Here we are given

$$\Delta\bar{\nu} = 12.8 \text{ cm}^{-1}$$

$$\therefore \quad I = \frac{h}{4\pi^2 c.\Delta\bar{\nu}} = \frac{6.62 \times 10^{-27}}{4 \times \left[\frac{22}{7}\right]^2 \times (3 \times 10^{10})(12.8)}$$

$$= \mathbf{4.36 \times 10^{-40} \text{ g cm}^2}$$

(*b*)

$$I = \left(\frac{m_1 m_2}{m_1 + m_2}\right) r_0^2$$

$$\text{Here } m_1 \text{ (for H-atom)} = \frac{1.008}{6.022 \times 10^{23}} g = 1.6739 \times 10^{-24} \text{ g}$$

$$m_2 \text{ (for I-atom)} = \frac{127}{6.022 \times 10^{23}} g = 2.1089 \times 10^{-22} \text{ g}$$

$$\therefore \quad r_0^2 = \left(\frac{m_1 + m_2}{m_1 m_2}\right) \times I$$

$$= \frac{1.6739 \times 10^{-24} + 2.1089 \times 10^{-22} g}{1.6739 \times 10^{-24} \times 2.1089 \times 10^{-22}} \times 4.36 \times 10^{-40}$$

$$= \frac{(1.6739 + 210.89) \times 10^{-24} \times 4.36 \times 10^{-40}}{1.6739 \times 2.1089 \times 10^{-46}}$$

$$= 2.6254 \times 10^{-16} \text{ cm}^2$$

$$\therefore \quad r_0 = 1.62 \times 10^{-8} \text{ cm}^2 = 1.62 \text{Å}$$

In terms of S.I.units

(*a*)

$$\Delta \bar{\nu} = 12.8 \text{ cm}^{-1} = 1280 \text{ m}^{-1}$$

$$\therefore \quad I = \frac{h}{4\pi^2 c \cdot \Delta \bar{\nu}}$$

$$= \frac{6.62 \times 10^{-34}}{4 \times \left(\frac{22}{7}\right)^2 \times (3 \times 10^8) \times 1280}$$

$$= \mathbf{4.36 \times 10^{-47} \text{ kg m}^2}$$

(*b*)

$$m_1 = 1.6739 \times 10^{-27} \text{ kg}$$

$$m_2 = 2.1089 \times 10^{-25} \text{ kg}$$

$$r_0^2 = \left(\frac{m_1 + m_2}{m_1 m_2}\right) \times I$$

$$= \frac{1.6739 \times 10^{-27} + 2.1089 \times 10^{-25}}{1.6739 \times 10^{-27} \times 2.1089 \times 10^{-25}} \times 4.36 \times 10^{-47} = 2.6254 \times 10^{-20}$$

$$\therefore \quad r = 1.62 \times 10^{-10} \text{ m} = \mathbf{0.162 \text{ nm.}}$$

Q. 47. The internuclear distance of carbonmonoxide molecule is 1.13 ×10^{-10} m. Calculate the energy (in joules and in eV) of this molecule in the first excited rotational level. The atomic masses of carbon and oxygen are 1.99 × 10^{-26} kg and 2.66 × 10^{-26} kg respectively.

Ans. We are given, $r = 1.13 \times 10^{-10}$ m

The reduced mass μ of CO is given by the equation

$$\mu = \frac{m_1 m_2}{m_1 + m_2}$$

$$= \frac{(1.99 \times 10^{-26} \text{ kg})\,(2.66 \times 10^{-26} \text{ kg})}{(1.99 + 2.66) \times 10^{-26} \text{ kg}} = 1.14 \times 10^{-26} \text{ kg}$$

Moment of inertia, $I = \mu r^2$

$$= (1.14 \times 10^{-26} \text{kg}) \times (1.13 \times 10^{-10} \text{m})^2$$

$$= 1.46 \times 10^{-46} \text{ kgm}^2$$

The rotational energy levels of a rigid diatomic molecule are given by

$$E_J = \frac{h^2}{8\pi^2 I} J(J+1) \text{ joule}$$

For the first excited rotational level, $J = 1$. Here

$$E_1 = \frac{h^2 \times 1(1+1)}{8\pi^2 I} = \frac{h^2}{8\pi^2 I} \text{ joule}$$

$$= \frac{(6.626 \times 10^{-34}\ \text{Js})^2}{4\pi(1.46 \times 10^{-46}\text{kg m}^2)} = 7.61 \times 10^{-23} \text{ joule}$$

Since, $1\text{eV} = 1.602 \times 10^{-19}$ J

$$E_1 = \frac{7.61 \times 10^{-23} J}{1.6023 \times 10^{-19} J(eV)^{-1}} = 4.76 \times 10^{-4} \text{eV}$$

Q. 48. The pure rotational spectrum of gaseous HCl contains a series of equally spaced lines separated by 20.80 cm^{-1}. Calculate the internuclear distance of the molecule. The atomic masses of H and Cl are 1.673 × 10^{-27} kg and 58.06 ×10^{-27} kg respectively.

Ans. The spacing between the lines $(2B) = 20.80\ \text{cm}^{-1}$

$\therefore$ $B = 10.40\text{cm}^{-1}$

$$B = \frac{h}{8\pi^2 IC}\text{cm}^{-1} \text{ or } I = \frac{h}{8\pi^2 BC}$$

substituting the values

$$I = \frac{6.626 \times 10^{-34}\text{ Js}}{(8\pi^2)(10.40\text{cm}^{-1})(3 \times 10^{10}\text{cms}^{-1})}$$

$$= 0.2689 \times 10^{-46}\ \text{kg}^2\text{m}^2 \qquad (\text{J} = \text{kgm}^2\ \text{s}^{-2})$$

The reduced mass is given by the equation

$$\mu = \frac{m_1 m_2}{m_1 + m_2}$$

where m_1 and m_2 are the atomic mass of H and Cl respectively.

Thus,
$$\mu = \frac{(1.673 \times 10^{27}\text{ kg})(58.06 \times 10^{27}\text{ kg})}{(1.673 + 58.06) \times 10^{27}\text{ kg}}$$

$$= \frac{(1.673 \times 58.06) \times 10^{-27}\text{ kg}}{59.74}$$

$$= 1.626 \times 10^{-27}\text{ kg}$$

Since, $I = \mu r^2$

$$r = \left(\frac{I}{\mu}\right)^{1/2} = \left(\frac{0.2689 \times 10^{-46}\text{kg m}^2}{1.626 \times 10^{-27}\text{ kg}}\right)^{1/2}$$

$$r = 1.29 \times 10^{-10}\text{ m} = 129 \text{ pm}$$

Q. 49. The pure rotational spectrum of the gaseous molecule CN has a series of equally spaced lines separated by 3.7978 cm^{-1}. Calculate the internuclear distance of the molecule. The molar masses of C and N are 12.011 and 14.007 g/mol respectively.

Ans. The spacing between the lines $(2B) = 3.7978\ cm^{-1}$

$$B = \frac{3.7978}{2} = 1.8989\ cm^{-1} = 189.89\ m^{-1}$$

$$B = \frac{h}{8\pi^2 Ic} cm^{-1}$$ [Refer to eq. (*iv*) Q. 40]

$$I = \frac{h}{8\pi^2 Bc} = \frac{6.626 \times 10^{-34}\ Js}{8\pi^2 (189.89\ m^{-1})(3 \times 10^8\ ms^{-1})}$$

$$= 1.4742 \times 10^{-46}\ kg\ m^2 \qquad (J = kg\ m^2\ s^{-2})$$

The reduced mass is given by the equation

$$\mu = \frac{m_1 m_2}{(m_1 + m_2)}$$

where N_A is the Avogadro's no. and m_1, m_2 are the atomic masses of C and N, respectively.

$$\mu = \frac{(12.011 g\ mol^{-1})(14.007 g\ mol^{-1})}{(26.018\ g\ mol^{-1})(6.022 \times 10^{23}\ mol^{-1})}$$

$$= 1.0737 \times 10^{-23}\ g$$

$$= 1.0737 \times 10^{-26}\ kg$$

Thus,

$$I = \mu r^2$$

$$r = \left(\frac{I}{\mu}\right)^{1/2} = \left(\frac{1.4742 \times 10^{-46}\ kgm^2}{1.0737 \times 10^{-26}\ kg}\right)^{1/2}$$

$$\therefore \quad r = 1.1717 \times 10^{-10}\ m = 117\ pm$$

VIBRATIONAL SPECTRA

Q. 50. What type of potential energy curve is obtained for a simple harmonic oscillator and why ?

Ans. A simple harmonic oscillator is the one in which the restoring force is proportional to the displacement in accordance with Hook's law *i.e.*,

$$F = -kx \qquad ...(i)$$

For a diatomic molecule, $x = R - R_e$, where R is distance to which the atoms have been stretched and R_e is the equilibrium distance between the two atoms.

k is called the force constant.

For such an oscillator, the potential energy is given by

$$V = \frac{1}{2}kx^2 \qquad ...(ii)$$

This is the equation for a parabola. Hence a *parabolic potential energy curve* is obtained as shown in Fig. 18.22.

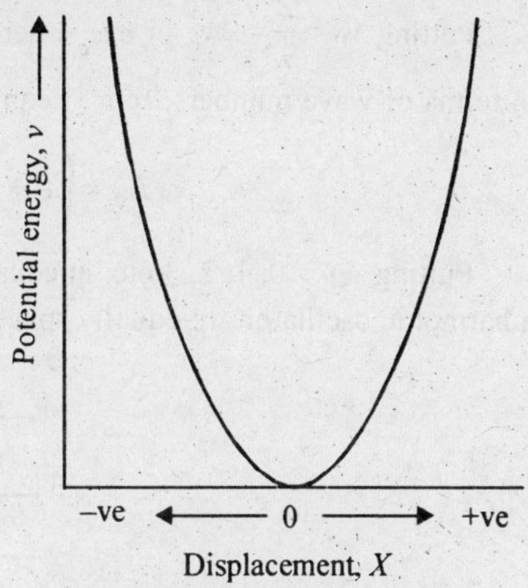

Fig. 18.22. Parabolic potential energy curve of a harmonic oscillator.

Q. 51. Write an expression for vibrational energy of a diatomic molecule taking it as a simple harmonic oscillator. Sketch the vibrational energy levels of such a molecule.

(Bangalore, 2004; Nagpur 2003)

Ans. Using the concepts of quantum mechanics, it can be seen that if the vibratory motion of the nuclei of diatomic molecule (Fig. 18.23) is taken as equivalent to that of **simple harmonic oscillator,** then the vibrational energy is obtained using quantum mechanics in the form of eqn. (*i*)

$$E_\vartheta = \left(\vartheta + \frac{1}{2}\right) h\nu_0 \qquad ...(i)$$

Fig. 18.23. Vibratory motion of a diatomic molecule.

where v_0 is the frequency of vibration and ϑ is the vibrational quantum number with allowed values of 0, 1, 2, 3, etc.

The energy absorbed in promoting a molecule from its lowest level to the next, that is, from v_0 to v_1 is given by the relation

$$\Delta E = \left(1+\frac{1}{2}\right) h\, v_0 - \frac{1}{2} h\, v_0$$

or $$\Delta E = h\, v_0 \qquad ...(ii)$$

Also for a diatomic system executing a harmonic motion, we have

$$v_0 = \frac{1}{2\pi}\sqrt{\frac{k}{\mu}} \qquad ...(iii)$$

where μ = Reduced mass

k = Force constant

Force constant is a measure of the strength of the bond between two atoms. Knowing the value of v_0 from absorption spectrum, the value of k can be obtained.

Combining (*ii*) and (*iii*), energy difference ΔE between two adjacent vibrational levels for a diatomic molecule is given by the equation

$$\Delta E = \frac{h}{2\pi}\sqrt{\frac{k}{\mu}}$$

Putting $v_0 = \frac{c}{\lambda} = c\,\overline{v}_e$ or $c\,\omega_e$ where $\overline{v}_e$ or ω_e, represents the equilibrium vibrational frequency in terms of wave numbers, (cm^{-1}) eqn. (*i*) can be written as

$$E_\vartheta = \left(\vartheta + \frac{1}{2}\right) hc\omega_e \qquad ...(iv)$$

Putting ϑ = 0, 1, 2, 3 etc. in equation (*ii*), it may be seen that the vibrational energy levels of a harmonic oscillator are equally spaced. These are shown diagrammatically in Fig. 18.24.

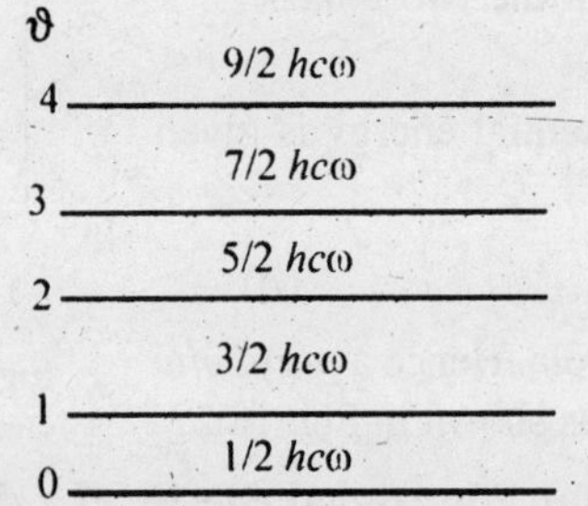

Fig. 18.24. Equally spread vibrational energy levels of a harmonic oscillator.

For the lowest vibrational level, ϑ = 0. The energy for the level will, therefore, be

$$E_0 = \frac{1}{2} hc\omega_e \qquad ...(v)$$

This is called **zero point energy.** It implies that even *at absolute zero when all translational and rotational motion ceases in a crystal, the residual energy of vibration* E_0 *still remains, i.e., the vibrational motion still exists.*

Q. 52. Write selection rules for vibrational transitions in a simple harmonic oscillator.
(Kolhapur 2000; Bangalore 2002)

Ans. The gross selection rule for a molecular vibration is that the *electric dipole moment must change when the atoms are displaced.* Here note that the molecule need not have a permanent dipole. Based on this, the specific selection rule for vibrational transitions is

$$\Delta\vartheta = \pm 1$$

i.e., change is vibrational quantum number should be unity. Transitions for which $\Delta\vartheta = +1$ correspond to absorption and those with $\Delta\vartheta = -1$ correspond to emission.

Q. 53. What type of a vibrational spectrum is expected for simple harmonic oscillator in the form of diatomic molecule?

Ans. Applying the selection rule, as the transitions can take place between vibrational levels whose vibrational quantum numbers differ by unity (*i.e.* $\Delta\vartheta = \pm 1$), the energy absorbed by a diatomic molecule, taking it as a simple harmonic oscillator, when the transition takes place from ϑ to $\vartheta + 1$ level will be given by

$$\Delta E_{\vartheta} = \left(\vartheta + 1 + \frac{1}{2}\right) hc\omega_e - \left(\vartheta + \frac{1}{2}\right) hc\omega_e = hc\omega_e$$

or $$\frac{\Delta E}{hc} = \omega_e$$

i. e., $$\Delta\overline{\nu}_e = \omega_e$$

Thus *only one absorption line* will be obtained in the vibrational spectrum whose wave number is equal to the equilibrium vibrational frequency of the diatomic molecule. Otherwise also, as the vibrational levels are equally spaced (Fig. 18.24), any transition from any ϑ to $\vartheta + 1$ will give rise to the same energy change and hence only one absorption line is expected. Further, at room temperature, as most of the molecules are in the lowest (ground state) vibrational level with v=0, therefore, the transitions take place from $\vartheta = 0$ to $\vartheta = 1$. The frequency thus observed is called the **fundamental frequency.** These spectra are observed in the infrared (IR) region.

Q. 54. What type of molecules give vibrational spectra?

Ans. The gross selection rule for a diatomic molecule to give a vibrational spectrum is that the *dipole moment of the molecule must change when the atoms are displaced due to vibration.* The condition required is only the change in dipole moment and the molecule need not have a permanent dipole moment. For example, in case of homonuclear diatomic molecules like H_2, O_2, N_2 etc., which have only stretching motion/vibrations and no bending motion/vibrations, the dipole moment does not change during vibration. Hence these molecules do not give vibrational spectra *i.e.*, they are said to be **infrared-inactive.** On the other hand, heteronuclear diatomic molecules like HCl, CO, NO etc., and polyatomic molecules like CO_2, H_2O, CH_4, C_2H_4 etc., which show change in dipole moment in some mode of vibration give vibrational spectra and are said to be **infrared-active.**

Q. 55. Why a diatomic molecule should be considered as an harmonic oscillator? Write Morse equation for the energy of anharmonic oscillator. ***(Nagpur 2008)***

Ans. When a molecule is undergoing simple harmonic oscillations, the restoring force of the harmonic vibrations is directly proportional to the displacement and the energy of the harmonic oscillator is given by

$$E_{\vartheta} = \left(\vartheta + \frac{1}{2}\right) hc\omega_e \qquad ...(i)$$

where ϑ is the vibrational quantum number, which may be zero or an integer and ω_e is the equilibrium frequency of vibration of the oscillator *i.e.*, for small displacements.

Strictly speaking the movement of a real oscillator is not perfectly harmonic. This is because as the displacement increases, the restoring force becomes weaker and for large amplitude of vibration, the atoms of the molecule must fall apart *i.e.*, the molecule must dissociate into atoms. Such a real oscillator is said to be an *anharmonic* oscillator. P.M. Morse in 1929 suggested that the equation (*i*) has to be modified for an anharmonic oscillator by including one additional term as follows:

$$E_{\vartheta} = \left(\vartheta + \frac{1}{2}\right) hc\omega_e - \left(\vartheta + \frac{1}{2}\right)^2 hcx_e\omega_e \quad ...(ii)$$

where x_e is called the *anharmonicity constant*.

Putting ϑ = 0, 1, 2.... in equations (*i*) and (*ii*), the energy levels of the harmonic oscillator and the anharmonic oscillator may be obtained. It may be seen that whereas the energy levels of a harmonic oscillator are equally spaced, those of the anharmonic oscillator are not equally spaced but fall more closely together as the quantum number increases. Further, whereas the potential energy curve for a harmonic oscillator is a parabola 18.25 (*a*), that of the anharmonic oscillator is as shown in Fig. 18.25 (*b*).

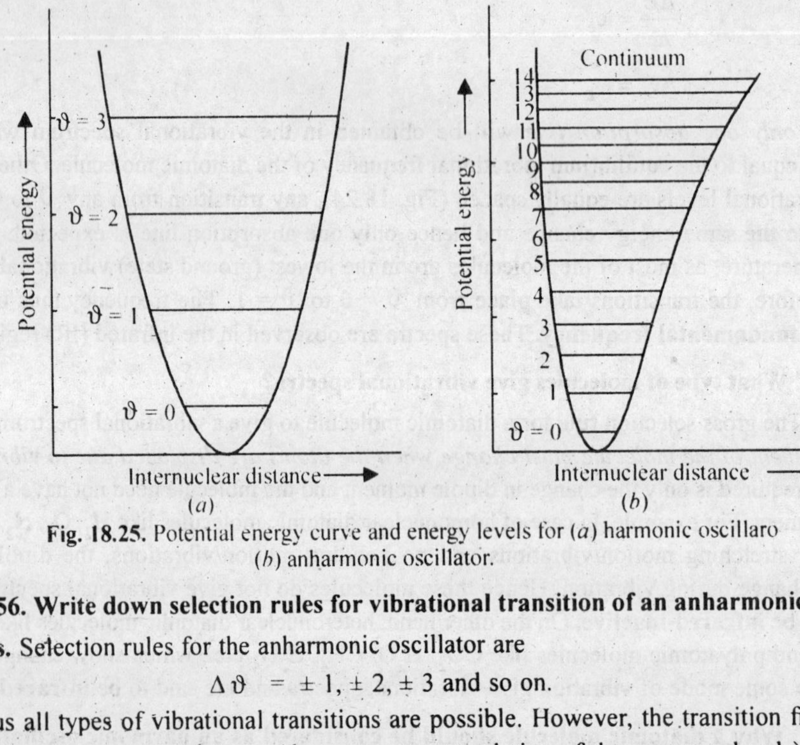

Fig. 18.25. Potential energy curve and energy levels for (*a*) harmonic oscillaro (*b*) anharmonic oscillator.

Q. 56. Write down selection rules for vibrational transition of an anharmonic oscillator.

Ans. Selection rules for the anharmonic oscillator are

$$\Delta\vartheta = \pm 1, \pm 2, \pm 3 \text{ and so on.}$$

Thus all types of vibrational transitions are possible. However, the transition from ϑ = 0 to ϑ =1 is found to be most intense (due to greater population of these energy levels) and is called the **fundamental absorption.** The transition from ϑ = 0 to ϑ = 2 has a very weak intensity and is called the **first overtone.** The transition from ϑ = 0 to ϑ = 3 has still weaker intensity and is called the second overtone and so on.

Q. 57. What type of vibrational-rotational spectra is obtained for a diatomic molecule taking it as anharmonic oscillator? *(Himachal 2000)*

Ans. As the energy required for rotation is much less than that required for vibration, the vibrational motion is always accompanied by rotational motion and hence we do not have a pure vibrational spectra; instead we have a vibrational-rotational spectra lying in the infrared region.

As rotational motion and vibrational motion take place independent of each other, the total energy of the molecule may be taken as the sum of the rotational energy and the vibrational energy. Thus

$$E_{\text{total}} = E_r + E_{\vartheta} \qquad ...(i)$$

or can be written as

$$E_{\vartheta,J} = E_J + E_{\vartheta} \qquad ...(ii)$$

Expressing in terms of wave numbers, it can be written as

$$\bar{v}_{\vartheta,J} = \bar{v}_J + \bar{v}_{\vartheta} \qquad ...(iii)$$

But the rotational energy in terms of wave number $\bar{v}_J$ and vibrational energy $\bar{v}_{\vartheta}$ are given by the following two equations

$$\bar{v}_J = BJ(J+1)$$

$$\bar{v}_{\vartheta} = \left(\vartheta + \frac{1}{2}\right)\omega_e - \left(\vartheta + \frac{1}{2}\right)^2 x_e\omega_e$$

Substituting these values in eqn. (*iii*), we get

$$\bar{v}_{\vartheta,J} = BJ(J+1) + \left(\vartheta + \frac{1}{2}\right)\omega_e - \left(\vartheta + \frac{1}{2}\right)^2 x_e\omega_e \qquad ...(iv)$$

Here $\bar{v}_{\vartheta,J}$ represents the wave number of a rotational level with quantum number J in the vibrational level with quantum number ϑ.

From eqn. (*iv*), it is clear that in the same vibrational level (so that ϑ = constant), there will be a number of rotational levels corresponding to $J = 0, 1, 2, 3,.....$

When a transition takes place from a level with quantum number ϑ and J to a level with quantum numbers ϑ' and J', the energy change expressed in terms of wave numbers will be given by

$$\Delta\bar{v} = \left[BJ'(J'+1) + \left(\vartheta' + \frac{1}{2}\right)\omega_e - \left(\vartheta' + \frac{1}{2}\right)^2 x_e\omega_e\right]$$

$$-\left[BJ(J+1)\left(\vartheta + \frac{1}{2}\right)\omega_e - \left(\vartheta + \frac{1}{2}\right)^2 x_e\omega_e\right] \qquad ...(v)$$

Restricting to the transition from $\vartheta = 0$ to $\vartheta' = 1$, we get

$$\Delta\bar{v} = \left[BJ'(J'+1) + \frac{3}{2}\omega_e - \frac{9}{4}x_e\omega_e\right]$$

$$-\left[BJ(J+1) + \frac{1}{2}\omega_e - \frac{1}{4}x_e\omega_e\right]$$

$$= B[J'(J'+1) - J(J+1)] + \omega_e - 2x_e\omega_e$$

$$= B[(J'-J)(J+J'+1)] + (1-2x_e)\omega_e \qquad ...(vi)$$

Putting $(1-2x_e)\omega_e = \omega_0$, we can write

$$\Delta \bar{v} = \omega_0 + B(J' - J)(J + J' + 1) \qquad ...(vii)$$

(*i*) For transition $\Delta J = +1$, *i.e.*,

$$J' - J = +1 \quad \text{or} \quad J' = J + 1$$

$$\Delta \bar{v} = \omega_0 + 2B(J+1) \quad \text{where } J = 0, 1, 2, 3.... \qquad ...(viii)$$

(*ii*) For transition, $\Delta J = -1$, *i.e.*,

$$J' - J = -1 \quad \text{or} \quad J = J' + 1$$

$$\Delta \bar{v} = \omega_0 - 2B(J' + 1) \quad \text{where } J' = 0, 1, 2, 3... \qquad ...(ix)$$

Combining equations (*viii*) and (*ix*), we get

$$\Delta \bar{v} = \omega_0 \pm 2Bm \qquad ...(x)$$

where $m = J + 1$ in eqn. (*viii*) and $m = -(J' + 1)$ in eqn. (*ix*). As $J = 0, 1, 2, 3,....$, this means that in eqn. (*x*), $m = \pm 1, \pm 2, \pm 3,...$ (+ sign for $\Delta J = +1$ and – sign for $\Delta J = -1$).

The nature of the vibration-rotation spectrum can be predicted from eqn. (*x*) as follows:

(*i*) As $m \neq 0$, therefore, line corresponding to the frequency ω_0 will not appear in the spectrum. This wave number is called the **band centre** or **band origin.**

(*ii*) As $m = \pm 1, \pm 2, \pm 3,....$, it follows that the spectrum will consist of equally spaced lines on each side of the band centre, with spacing equal to $2B$ between any two adjacent lines.

Keeping in view the selection rule, $\Delta J = \pm 1$, the absorption transitions that can take place **from level $\vartheta = 0$ to $\vartheta = 1$ and the corresponding lines obtained in the spectrum may be represented as shown in Fig. 18.26.**

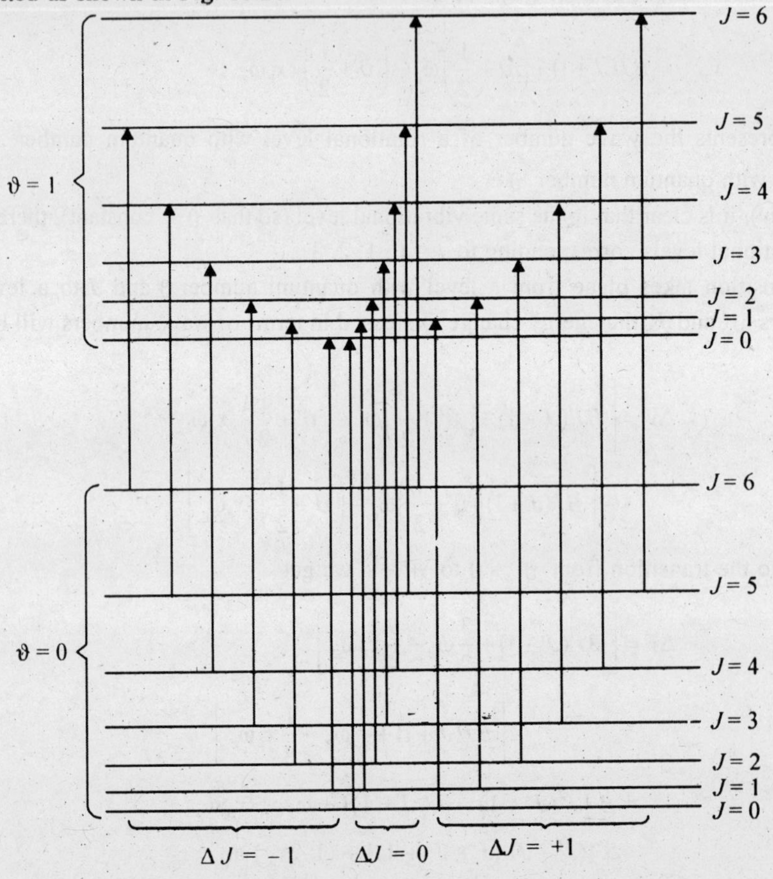

Fig. 18.26. Vibration-rotation spectrum of a diatomic molecule.

Q. 58. What are *P*, *Q* and *R* branches of rotation-vibration spectrum. (*Mumbai 2010*)

Ans. In the case of rotation-vibration spectrum:

(*i*) For $\Delta J = -1$, lines with frequency lower than the fundamental frequency are obtained. These lines are called ***P*-branch** of the spectrum (Fig. 18.27).

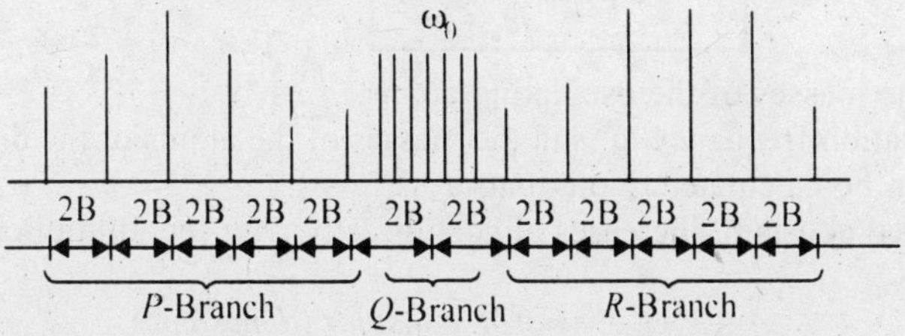

Fig. 18.27. *P*, *Q* and *R* branches of rotation-vibration spectrum.

(*ii*) For $\Delta J = +1$, lines with frequency greater than the fundamental frequency are obtained. These lines are called the ***R*-branch** of the spectrum.

(*iii*) For $\Delta J = 0$ *i.e.*, when a vibrational transition occurs without being accompanied by rotational transition (*e.g.*, in case of NO),

$$\Delta \bar{\nu} = \omega_0$$

i.e., a single line is expected at the centre. In actual practice, as the rotational constants of the two vibrational levels are slightly different, a cluster of closely spaced lines appears at the centre. This group of lines is called the ***Q*-branch** of the spectrum. When this type of transition is forbidden (*e.g.*, in case of HCl), a gap appears at the centre.

Q. 59. Give the applications of study of infrared spectrum.

(***Garhwal 2000; Nagpur 2002***)

Ans. Various applications of infrared spectrum studies are discussed as under:

(1) **Calculation of moment of inertia and bond length**

The value of B is given by

$$B = \frac{h}{8\pi^2 Ic}$$

$\therefore$ Frequency separation, $\Delta \bar{\nu} = 2B = \dfrac{h}{4\pi^2 Ic}$

\ **Thus knowing the frequency separation, the moment of inertia and hence the intemuclear distance *i.e.*, bond length of a diatomic molecule can be calculated.**

(2) **Calculation of force constant:** Taking a simple case where a diatomic molecule may be considered as a simple harmonic oscillator *i.e.*, an oscillator in which the restoring force (F) is directly proportional to the displacement, in accordance with Hook's law, we have

$$F = -kx \qquad ...(i)$$

where x is the displacement and is equal to the distance to which the atoms have been stretched (R) minus equilibrium distance between the atoms (R_e) *i.e.*, $x = R - R_e$. k in equation (*i*) is called force constant. Thus if $x = 1$ cm, $k = -F$. Hence **force constant** *may be defined as the restoring force per unit displacement* (*or per cm*) *of a harmonic oscillator.* It is found to be related to the equilibrium vibrational frequency ω_e (in s^{-1}) according to equation

$$\omega_e = \frac{1}{2\pi}\left(\frac{k}{\mu}\right)^{1/2} s^{-1} \qquad ...(ii)$$

where μ is the reduced mass of the system.

To convert the frequency ω_e from s^{-1} to cm^{-1}, we divide by velocity of light, c. Thus

$$\omega_e = \frac{1}{2\pi c}\left(\frac{k}{\mu}\right) cm^{-1} \qquad ...(iii)$$

To calculate *k*, equation (*iii*) can be rewritten as follows :

$$k = 4\pi^2 \omega_e^2 \mu$$

or

$$\boxed{k = 4\pi^2\omega_e^2 \frac{m_1 m_2}{m_1 + m_2}} \qquad ...(iv)$$

where m_1 and m_2 are the masses of the oscillating atoms.

Thus knowing vibration frequency ω_e and the masses of the atoms of the diatomic molecule, the force constant *k* can be calculated. Note that ω_e is in s^{-1}.

It is found that force constant, increases almost directly with the multiplicity of the bond is given below:

Bond	*Force constant (dynes/cm)*
C – C	4.6×10^5
C = C	9.5×10^5
C ≡ C	15.8×10^5

Thus multiplicity of the bond can be predicted from the value of the force constant. For example, the force constant for C ≡ O in carbon monoxide is 18.6×10^5 dynes/cm, whereas the force constant for C = O in carbon dioxide is 15.2×10^5 dynes/cm. This result confirms that CO_2 has resonating structures containing both carbon oxygen double and triple bonds *i.e.*, we write

$$O^+ — C \equiv O^- \leftrightarrow O = C = O \leftrightarrow O^- \equiv C — O^+$$

(3) **Calculation of equilibrium dissociation energy:** The equilibrium dissociation energy of a diatomic molecule is given by

$$D_e = \frac{\omega_e}{4x_e} \qquad ...(v)$$

This is the dissociation energy when it is measured from the bottom of the Morse potential energy curve. The true dissociation energy is, however, measured from the ground level ($v = 0$) and is given by

$$D_0 = \frac{\omega_e}{4x_e} - \frac{1}{2}\omega_e\left(1 - \frac{1}{2}x_e\right) \qquad ...(vi)$$

Q. 60. Calculate the force constant for the bond in HCl from the fact that the fundamental vibration frequency is $8.667 \times 10^{13}\ s^{-1}$. ***(Nagpur 2008)***

Ans. $m_1(\text{for H-atom}) = \frac{1.008}{6.022 \times 10^{23}} g = 1.6739 \times 10^{-27}$ kg

$m_1(\text{for Cl-atom}) = \frac{35.5}{6.022 \times 10^{23}} g = 5.8951 \times 10^{-26}$ kg

∴ Reduced mass $\mu = \frac{m_1 \times m_2}{m_1 + m_2} = 1.6277 \times 10^{-27}$ kg

Also it is given that

$$\omega = 8.667 \times 10^{13}\ s^{-1}$$

$$k = 4\pi^2\omega^2\mu$$

$$= 4 \times \left(\frac{22}{7}\right)^2 \times (8.667 \times 10^{13})^2 \times (1.6277 \times 10^{-22}\ kg)$$

$$= \mathbf{483.1\ Nm^{-1}}.$$

RAMAN SPECTRA

Q. 61. What is Raman spectra? ***(Gujarat 2003; Bangalore, 2004; Raipur 2010)***

Ans. Prof C.V. Raman of Calcutta University, observed in 1928 that when a substance (gaseous, liquid or solid) is irradiated with monochromatic light of a definite frequency *v*, the light scattered at right angles to the incident light contained lines not only of the incident frequency but also of lower frequency and sometimes of higher frequency as well. The lines with *lower frequency* are called **Stokes' lines** whereas lines with *higher frequency* are called **anti-Stokes' lines.** Line with the same frequency as the incident light is called **Rayleigh line.** Raman further observed that the difference between the frequency of the incident light and that of a particular scattered line was constant depending only **upon the nature** of the substance being irradiated and was completely independent of the frequency of the incident light. If v_i is the frequency of the incident light and v_s, that of a particular scattered line, the difference $\Delta v = v_i - v_s$ is called **Raman frequency** or **Raman shift.** Thus the Raman frequencies observed for a particular substance are characteristic of that substance. The various observations made by Raman are called **Raman effect** and the spectrum observed is called **Raman spectrum.** Raman spectrum is represented by Fig. 18.28.

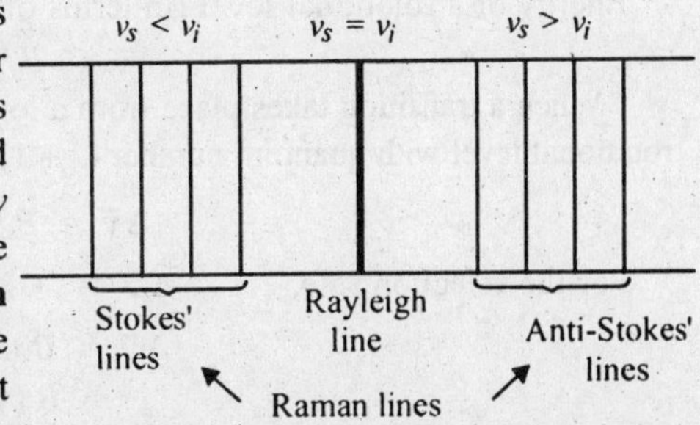

Fig. 18.28. A simplified reprensentation of Raman spectrum.

Q. 62. How do you explain Rayleigh's line, Stokes' lines and anti-Stokes' lines in Rama spectra? ***(Raipur 2010; Mumbai 2010)***

Ans. When the photon strikes the molecule, the energy is absorbed by the molecule and it gets excited to some higher energy level. Now if it returns to the original level, it will emit the same energy as absorbed and thus we have Rayleigh scattering. However, in most of the cases, the excited molecule does not return to the original level. It may return to a level higher than the original level, thereby emitting less energy than absorbed. This explains the occurrence of Stokes's lines. Thus a part of the energy of the incident photon remains absorbed by the molecule (so that molecule has higher energy than before). Alternatively, the excited molecule may return to a level lower than the original level. Thus more energy is emitted than absorbed. This explain the occurrence of anti-Stokes' lines. In this case, the molecule has less energy than before. The different cases may be represented diagrammatically by Fig. 18.29.

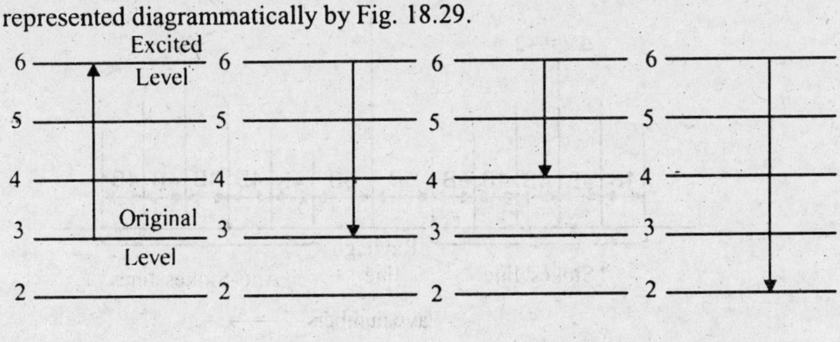

Fig. 18.29. (*a*) Energy absorbed by the molecule. (*b*) Rayleigh scattering. (*c*) Formation of Stokes's lines (*d*) Formation of anti-Stokes' line

Q. 63. Explain with mathematical equations the type of pure rotational Raman spectrum expected for diatomic molecule.

Ans. The selection rules for pure rotational Raman spectra of diatomic molecules are

$$\Delta J = 0, \pm 2$$

The selection rule $\Delta J = 0$ corresponds to Rayleigh scattering whereas selection rule $\Delta J = \pm 2$ gives rise to Raman lines as explained below :

Energy of a rotational level (in terms of wave number) with quantum number J is given by

$$\bar{\nu} = BJ(J+1) \quad ...(i)$$

When a transition takes place from a lower rotational level with quantum number J to a higher rotational level with quantum number $(J + 1)$ the energy absorbed in terms of wave numbers will be

$$\Delta\bar{\nu} = BJ'(J'+1) - BJ(J+1) \quad ...(ii)$$

For the selection rule, $\Delta J = +2$ *i.e.*, $J' - J = 2$, we get

$$\Delta\bar{\nu} = B(J+2)(J+3) - BJ(J+1)$$

$$= B(4J+6) \text{ where } J = 0, 1, 2, 3 \quad ...(iii)$$

Again for the selection rule, $\Delta J = -2$ *i.e.*, $J' - J = -2$, we get

$$\Delta\bar{\nu} = BJ'(J'+1) - B(J'+2)(J'+3)$$

$$= -B(4J'+6) \text{ where } J' = 0, 1, 2, 3... \quad ...(iv)$$

Combining the results of eqns. (*iii*) and (*iv*), the wave numbers of the lines obtained in the Raman will be given by

$$\bar{\nu} = \bar{\nu}_i \pm \Delta\bar{\nu}$$

or

$$\boxed{\bar{\nu} = \bar{\nu}_i \pm B(4J+6)} \quad ...(v)$$

where $J = 0, 1, 2, 3,....$

Here $\bar{\nu}_i$ represents the wave number of the Rayleigh line, plus sign gives lines with higher wave numbers called anti-Stokes' lines and minus sign gives lines with lower wave numbers, called Stokes' lines.

From eqn. (*v*), it may be seen that for $J = 0$, $\bar{\nu} = \bar{\nu}_i + 6B$ *i.e.*, the first Stokes' and anti-Stokes' line will be at a separation of $6B$ from the Rayleigh line. Further putting $J = 1, 2, 3, ...,$ the separation between any two adjacent Stokes' lines or anti-Stokes' lines will be $4B$. Thus, the pure rotational Raman spectrum expected for a diatomic molecule may be represented by Fig. 18.30.

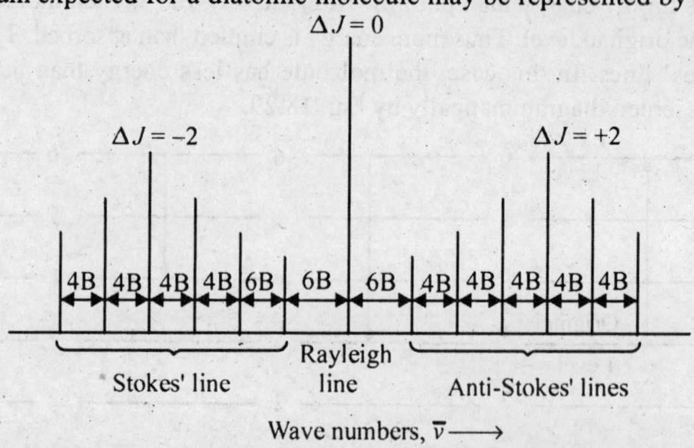

Fig. 18.30. Pure rotational Raman spectrum expected for a diatomic molecule.

Q. 64. What are the selection rules for rotation-vibration Raman spectra of diatomic molecules? Describe rotation-vibration Raman spectrum obtained for a diatomic molecule.

Ans. Diatomic gaseous molecules give rotational-vibrational Raman spectra which are governed by the following selection rules:

$$\Delta\vartheta = \pm 1$$

$$\Delta J = 0, \pm 2$$

However at room temperature, as most of the molecules are in the lowest vibrational level ($\vartheta = 0$), therefore, the significant vibrational transition is from $\vartheta = 0$ to $\vartheta = 1$ (called fundamental vibration transition). Restricting to this vibrational transition only, the results obtained are as follows:

For $\Delta J = 0$, $\quad \Delta \bar{v} = \omega_e (1 - 2x_e)$ (called ***Q*-branch**).

For $\Delta J = +2$, $\Delta \bar{v} = \omega_e (1 - 2x_e) + B(4J + 6)$ (called ***S*-branch**).

For $\Delta J = -2$, $\Delta \bar{v} = \omega_e (1 - 2x_e) - B(4J' + 6)$ (called ***O*-branch**).

Representing the value for *Q*, *S* and *O* branch by $\Delta\bar{v}_Q, \Delta\bar{v}_S$ and $\Delta\bar{v}_O$ respectively, and the wave number of the exciting radiation by $\bar{v}_i$ the wave numbers of the Stokes lines will be as follows:

$$\bar{v}_Q = \bar{v}_i - \Delta\bar{v}_Q$$

$$\bar{v}_S = \bar{v}_i - \Delta\bar{v}_S$$

$$\bar{v}_O = \bar{v}_i - \Delta\bar{v}_O$$

We have considered the case for $\Delta\vartheta = +1$ which represents Stokes lines. The anti-Stokes lines correspond to $\Delta\vartheta = -1$.

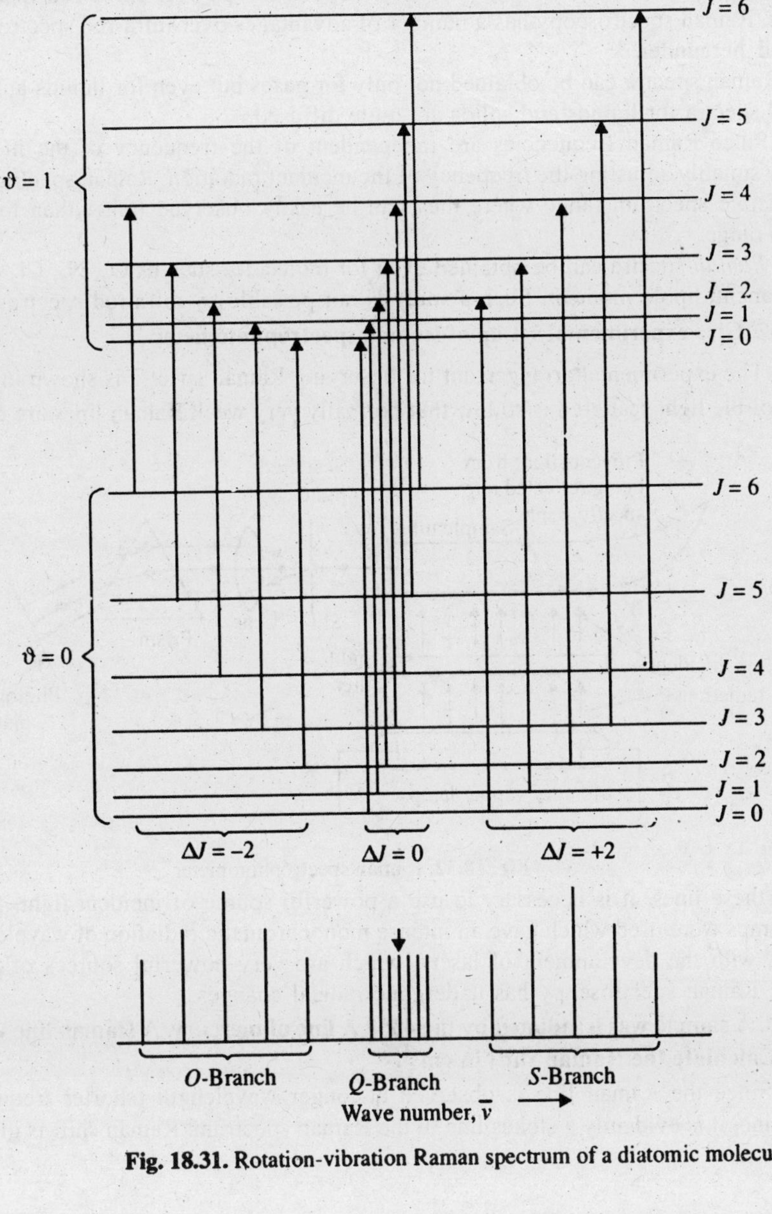

Fig. 18.31. Rotation-vibration Raman spectrum of a diatomic molecule.

The rotational transitions accompanying the vibrational transition from $\vartheta = 0$ to $\vartheta = 1$ and the rotation-vibration Raman spectrum obtained may be represented as shown in Fig. 18.31.

Q. 65. What type of molecules give rotation-vibration Raman spectra?

(*Nagpur 2008; Himachal 2010*)

Ans. The selection rule for a molecule to give rotation-vibration Raman spectrum is that the *polarization of the molecule must change as the molecule vibrates.* In case of both types of diatomic molecules *i.e.*, homonuclear as well as heteronuclear, as the molecule vibrates, the control of the nuclei over the electrons varies and hence there is a change in polarizability. That is why both types of diatomic molecules give rotation-vibration Raman spectra *i.e.*, they are *vibrational Raman-active.* This is one of the main advantages of Raman spectra over infrared spectra because homonuclear diatomic molecules do not give pure vibrational or rotational spectra because they do not possess a permanent dipole moment.

Q. 66. What are the advantages of Raman Spectroscopy over Infra-red spectroscopy ?

Ans. Raman spectroscopy has a number of advantages over infra-red spectroscopy which are discussed hereunder.

(*i*) Raman spectra can be obtained not only for gases but even for liquids and solids whereas infra-red spectra for liquids and solids are quite diffused.

(*ii*) Since Raman frequencies are independent of the frequency of the incident radiation, hence by suitably adjusting the frequency of the incident radiation, Raman spectra can be obtained in the visible spectrum range where they can be easily observed rather than the more difficult infra-red range.

(*iii*) *Raman* spectra can be obtained even for molecules such as O_2, N_2, Cl_2 etc., which have no permanent dipole moment. Such a study is not possible by infra-red spectroscopy.

Q. 67. Give experimental set-up of Raman spectrophotometer.

Ans. The experimental arrangement for observing Raman spectra is shown in Fig. 18.32. The fraction of the light scattered is so low that normally very weak Raman lines are observed. Hence

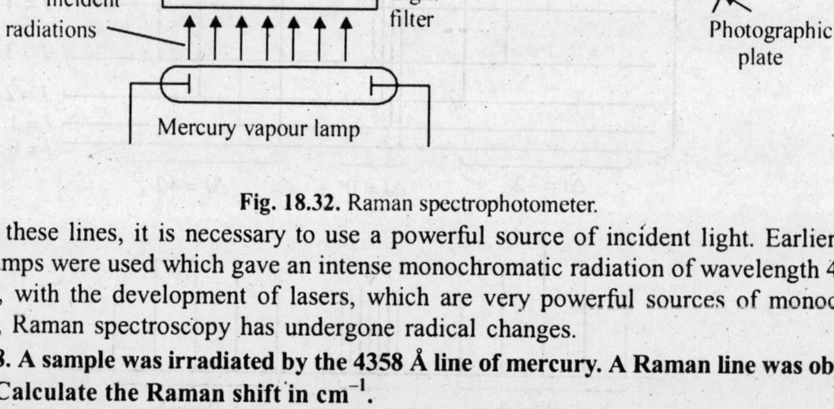

Fig. 18.32. Raman spectrophotometer.

to detect these lines, it is necessary to use a powerful source of incident light. Earlier mercury vapour lamps were used which gave an intense monochromatic radiation of wavelength 435.8 nm. However, with the development of lasers, which are very powerful sources of monochromatic radiation, Raman spectroscopy has undergone radical changes.

Q. 68. A sample was irradiated by the 4358 Å line of mercury. A Raman line was observed at 4447 Å. Calculate the Raman shift in cm^{-1}.

Ans. Since the Raman line is observed at longer wavelength (shorter frequency) than the exciting line, it is evidently a stokes line in the Raman spectrum. Raman shift is given by the eqn.

$$\Delta \bar{V}_{\text{Raman}}(\text{cm}^{-1}) = \frac{10^8}{\lambda_{\text{exc.}}(\text{Å})} - \frac{10^8}{\lambda_{\text{Raman.}}(\text{Å})}$$

$$= \frac{10^8}{4.358\times10^3} - \frac{10^8}{4.447\times10^3} = (2.295 - 2.249)\times10^4$$

$$= 0.046 \times 10^4$$

$$\Delta \bar{V}_{\text{Raman}}(\text{cm}^{-1}) = 460\,\text{cm}$$

ELECTRONIC SPECTRA

Q. 69. Write a note on electronic spectra

Ans. Spectra which arises due to electronic transition in a molecule by absorption of radiation falling in the visible and UV range is called electronic spectra. The range of visible region is 12500–25000 cm^{-1} while that of UV region is 25000–70000 cm^{-1}. It is observed that electronic transitions in a molecule are invariably accompanied by vibrational and rotational transitions, thus, making electronic spectra quite complex. Electronic spectra help in the determination of ionization energies of molecules.

It the light radiation incident on a molecule possesses very high energy, it causes removal of electron from the molecule *i.e.* it causes ionization. A beam of photons of known energy is allowed to fall on a sample and the kinetic energy of the ejected electron is measured. The difference between the photon energy and the excess kinetic energy gives the binding energy of the molecule.

Q. 70. What is natural linewidth (or life-time broadening)? What is meant by width and intensity of spectral lines? ***(Himachal 2010)***

Ans. Natural linewidth, $\Delta\nu$, is determined by Heisenberg uncertainty principle,

$$\Delta \text{E} \,.\, \Delta t = \frac{h}{4\pi} \qquad \ldots(i)$$

where ΔE = uncertainty in energy

and Δt = uncertainty in lifetime

For photon $\text{E} = h\nu$

or $\Delta \text{E} = h\,\Delta\nu$

From (*i*) $$h\,\Delta\nu \,.\, \Delta t = \frac{h}{4\pi}$$

Or Natural linewidth, $$\Delta\nu = \frac{1}{4\pi\Delta t}$$

Width and intensity (strength) are two characteristics of the spectral line obtained in the spectrum of a molecule. These two quantities are common to all branches of spectroscopy. Fig. 18.33(*a*) represents a sharp spectral line while Fig. 18.33(*b*) represents a line with a width ΔE at half height. It is much more convenient to interpret a sharp and intense line.

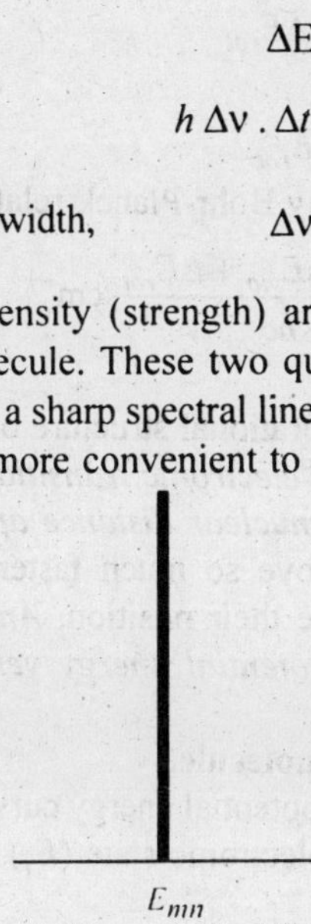

Fig. 18.33. (*a*) Sharp spectral line

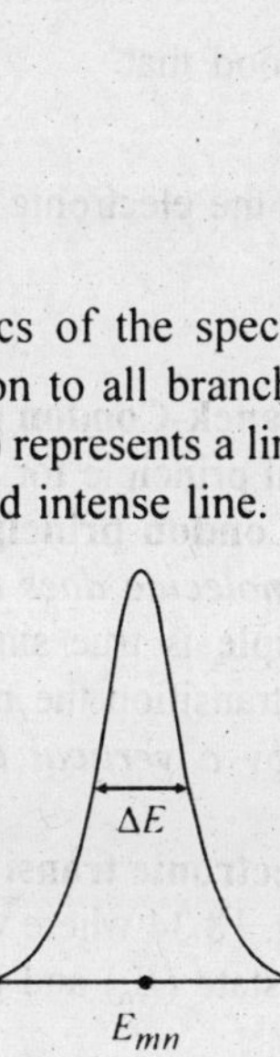

Fig. 18.33. (*b*) Spectral line having a width

The factors that contribute to the broadening of a spectral line are :

1. Collision broadening
2. Doppler broadening

Intensity of a spectral line is determined by :

1. Boltzmann population of the energy levels
2. Transition probability between the energy levels

According to Boltzmann, if N_0 is the number of molecules in ground state, then the number of molecules N in the excited state is given by

$$N = N_0 e^{-\Delta E / kT}$$

Where ΔE = Energy difference between the ground state and excited states. The relative population at equilibrium is given by

$$N / N_0 = e^{-\Delta E / kT}$$

Transition probability means the probability of transition between two given energy levels. A molecule upon absorption of light goes to an energy level allowed by selection rules. Selection rules decide the **allowed transitions** and **forbidden transitions.** Evidently, the allowed transitions yield spectral lines which are more intense than those involving forbidden transitions.

Q. 71. Explain Born-Oppenheimer approximation in relation to electronic spectra of molecules. ***(Raipur 2010)***

Ans. Electronic spectra of molecules are observed in ultraviolet and visible regions of the electromagnetic spectrum. A transition between two electronic states is almost invariably accompanied by simultaneous transitions between the vibrational and rotational energy levels as well. This is expressed by saying that electronic spectra have *vibrational fine structure* and *rotational fine structure*. According to the Born-Oppenheimer approximation, the total energy of a molecule in the ground state is given by

$$E'' = E''_{el} + E''_{vib} + E''_{rot}$$

Here, E''_{el}, E''_{vib} and E''_{rot} are the electronic, vibrational and rotational energies respectively. Assuming that the Born-Oppenheimer approximation is valid in the excited state as well, the excited state energy E' is given by

$$E' = E'_{el} + E'_{vib} + E'_{rot}$$

The energy change for an electronic transition is given by

$$\Delta E = E' - E'' = (E'_{el} - E''_{el}) + (E'_{vib} - E''_{vib}) + (E'_{rot} - E''_{rot})$$

$$= \Delta E_{el} + \Delta E_{vib} + \Delta E_{rot}$$

It may be understood that

$$\Delta E_{el} >> \Delta E_{vib} >> \Delta E_{rot}$$

The frequency for the electronic transition is given by Bohr-Planck relation.

$$\therefore \quad \bar{\nu} = \frac{\Delta E}{hc} = \frac{\Delta E_{el} + \Delta E_{vib} + \Delta E_{rot}}{hc} \text{ cm}^{-1}$$

Q. 72. What is Franck-Condon principle? ***(Raipur 2010)***

Ans. A very useful principle for investigating the vibrational structure of electronic spectra is provided by **Franck-Condon principle.** It states that *an electronic transition takes place so rapidly that a vibrating molecule does not change its internuclear distance appreciably during the transition.* This principle is true since the electrons move so much faster than the nuclei that during the electronic transition the nuclei do not change their position. *An electronic transition may be represented by a vertical line on a plot of potential energy versus the internuclear distance.*

Q. 73. Discuss electronic transition in a diatomic molecule.

Ans. Consider Fig. 18.34 where we have shown two potential energy curves for the molecule in the ground electronic state (E_0) and in the first excited electronic state (E_1). Since the bonding in the excited state is weaker than in the ground state, the minimum in the potential energy curve for the excited state occurs at a slightly greater internuclear distance than the corresponding minimum in the ground electronic state. Also quantum mechanically it is known that the molecule is in the *centre* of the ground vibrational level of the ground electronic state. When a photon falls on the molecule, the most probable electronic transition, according to the Franck-Condon principle, takes place from $\vartheta'' = 0$ to $\vartheta' = 2$ (written schematically as $0 \rightarrow 2$). Transitions to other vibrational levels

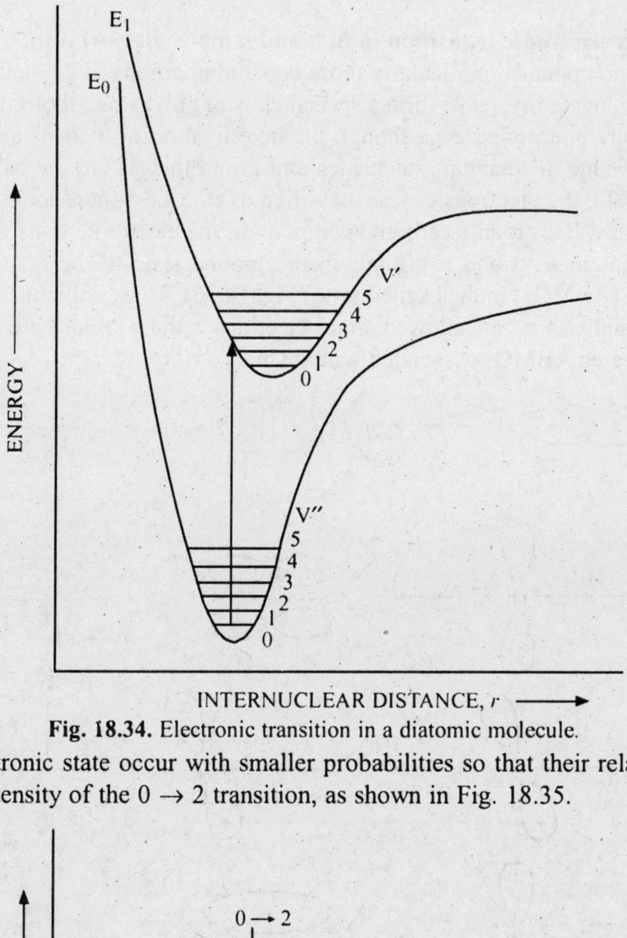

Fig. 18.34. Electronic transition in a diatomic molecule.

of the excited electronic state occur with smaller probabilities so that their relative intensities are smaller than the intensity of the $0 \rightarrow 2$ transition, as shown in Fig. 18.35.

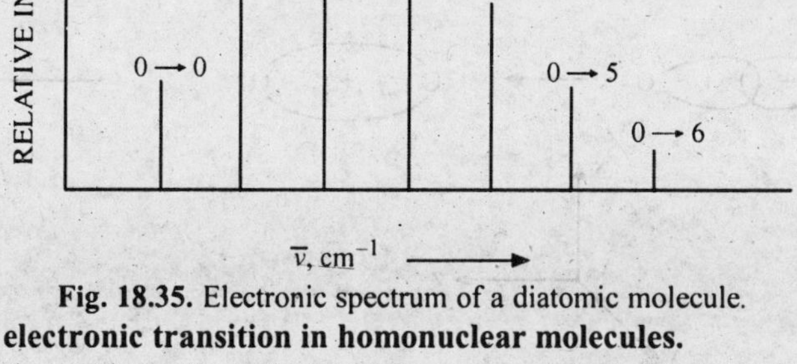

Fig. 18.35. Electronic spectrum of a diatomic molecule.

Q. 74. Discuss electronic transition in homonuclear molecules.

Ans. In molecules such as H_2 and N_2, highest occupied molecular orbital (HOMO) in the ground state is a bonding molecular orbital (BMO) whereas the lowest unoccupied molecular orbital (LUMO) is an antibonding molecular orbital (ABMO). The HOMO and LUMO orbitals are collectively referred to as *Frontier Molecular Orbitals* (FMOs). The electronic transition, HOMO $\rightarrow$ LUMO takes place, *when the electron absorbs a photon.*. This lies in the UV range. Oxygen molecule O_2 is an exception. It has two unpaired electrons in the ground state which is thus a *triplet state.* The electronic transition occurs from the triplet ground state to the triplet excited state (rather than to the singlet excited state). The electronic spectrum of O_2 molecule is, of course, complex.

Q. 75. Discuss electronic transition in σ, π and *n* molecular orbitals.

Ans. Organic compounds, particularly those containing groups like C = C, C = O, – N = N– and extensively conjugated systems, form a special class of polyatomic molecules whose electronic spectra can be easily interpreted even though the investigation of their detailed spectral features may require knowledge of quantum mechanics and group theory. On the basis of the molecular orbital theory (MOT), the electrons can be classified as σ, π or *n* (non-bonding) depending upon the MOs they occupy. For organic carbonyl compounds, the electronic transitions involve promotion of the electrons in *n*, σ and π orbitals in the ground state to the σ* and π* anti-bonding molecular orbitals. (ABMOs) in the excited state (Fig. 18.36). Thus, only the transitions of the type $\sigma \rightarrow \sigma^*$, $\pi \rightarrow \pi^*$, and $n \rightarrow \pi^*$ are allowed. Since electrons in the *n* orbitals are not involved in bond formation, there are no ABMOs associated with them.

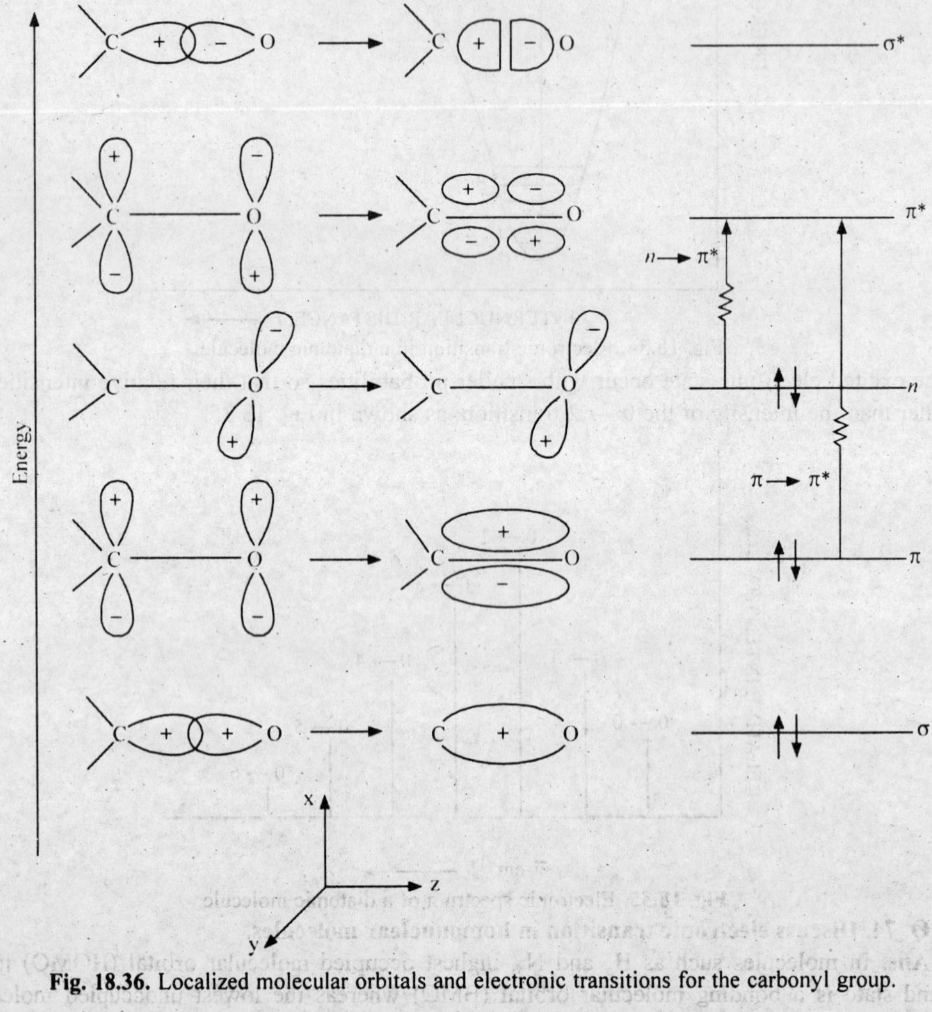

Fig. 18.36. Localized molecular orbitals and electronic transitions for the carbonyl group.

The $\sigma \rightarrow \sigma^*$ transitions occurring in saturated hydrocarbons and other types of compounds in which all valence shell electrons are involved in single bonds, are found in the vacuum ultraviolet region since they involve very high energy. The $\pi \rightarrow \pi^*$ and the $n \rightarrow \pi^*$ transitions, on the other hand, are found either in the UV or visible regions. The unsaturated molecules containing C = C and C = O groups (such as aldehydes and ketones) show $n \rightarrow \pi^*$ and $\pi \rightarrow \pi^*$ transitions. For aldehydes and ketones the more intense band near 180 nm is due to $\pi \rightarrow \pi^*$ transition and the weaker band around 285 nm is due to $n \rightarrow \pi^*$ transition. Olefinic hydrocarbons show $\pi \rightarrow \pi^*$ transition in the wave length 160 – 170 nm. Acetylene shows an absorption near 180 nm.

Molecules such as methylamine and methyl iodide which do not contain a π-orbital, show $n \rightarrow \sigma^*$ transitions.

In highly conjugated systems, the π-electrons are delocalized over the entire skeletal framework. It is found that *the absorption bands shift to longer wave lengths as the extent of conjugation increases.* Thus, in the compound $C_6H_5 - (CH = CH)_n - C_6H_5$, the $\pi \rightarrow \pi^*$ transition lies in the UV region when $n = 1$ or 2. As n increases, the electronic transition shifts to the visible region. This phenomenon is shown schematically in Fig. 18.37.

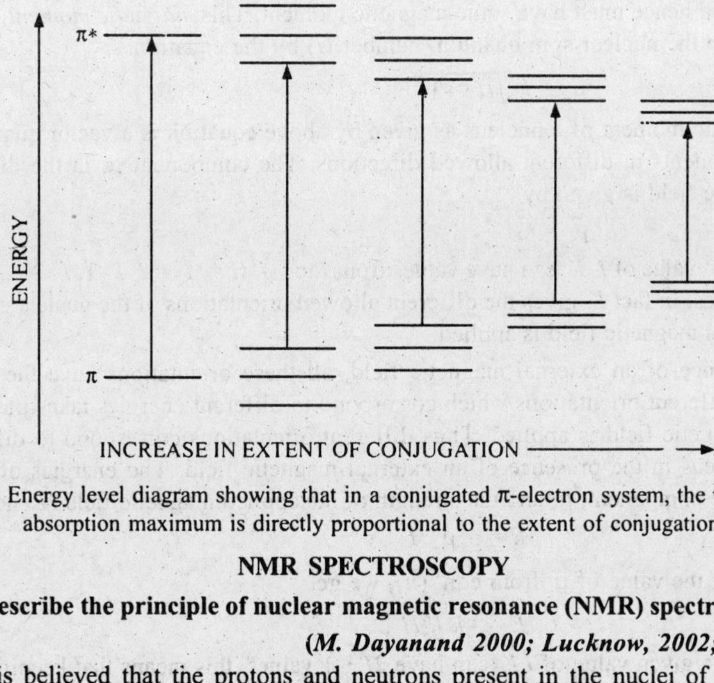

Fig. 18.37. Energy level diagram showing that in a conjugated π-electron system, the wave length of absorption maximum is directly proportional to the extent of conjugation.

NMR SPECTROSCOPY

Q. 76. Describe the principle of nuclear magnetic resonance (NMR) spectroscopy.

(M. Dayanand 2000; Lucknow, 2002; Bombay 2004)

Ans. It is believed that the protons and neutrons present in the nuclei of atoms also spin about their axes. Each proton or neutron is assigned a spin quantum number equal to 1/2. The *spin quantum number for the nucleus* can then have values 0, 1/2, 1, $1\frac{1}{2}$, 2 etc., depending upon the number of neutrons and protons having parallel and antiparallel spins. It is usually represented by I. For example, deuterium nucleus which contains one proton and one neutron is expected to have *nuclear spin quantum number* equal to 1 or 0 depending upon whether the neutron and proton have parallel spins (↑↑) or antiparallel spins (↑↓). Deuterium nucleus is actually found to have $I = 1$ in the ground state showing that the proton and neutron have parallel spins. It is observed that

(*i*) For nuclei in which the sum of protons and neutrons is even as well as the number of each of them is also even, $I = 0$ because there is pairing. For example, I is zero for ${}^{4}_{2}He$, ${}^{12}_{6}C$, ${}^{16}_{8}O$ and ${}^{32}_{16}S$.

(*ii*) For nuclei in which the sum of protons and neutrons is even but the number of protons is odd, I = 1, 2, 3 etc. For example, for $^{2}_{1}D$ and $^{14}_{7}N$, I = 1.

(*iii*) For nuclei in which the sum of protons and neutrons is odd, I is always odd integral multiple of $\frac{1}{2}$ *i.e.*, $I = \frac{1}{2}, \frac{3}{2}, \frac{5}{2}$.... For example, for $^{1}_{1}H$ and $^{19}_{9}F$, $I = \frac{1}{2}$; for $^{23}_{11}Na$, $I = \frac{3}{2}$; for $^{35}_{17}Cl$, $I = \frac{5}{2}$.

Nuclear spin angular momentum or simply called nuclear spin is given by $\sqrt{I(I+1)}\frac{h}{2\pi}$. where I is the nuclear spin quantum number, as explained above. Thus just as the electron spin angular momentum is quantized, so is the nuclear spin angular momentum *i. e.,* they can have only certain discrete values for the angular momentum.

The spinning of the various particles in the nucleus leads to net spinning of the nucleus about an axis. Now as the nucleus is a charged particle, when it spins about its axis, it acts as a small magnet and hence must have some magnetic moment. This *magnetic moment* of the nucleus (μ_n) is related to the nuclear spin quantum number (I) by the equation.

$$\mu_n = k\sqrt{I(I+1)} \qquad ...(i)$$

The magnetic moment of a nucleus as given by above equation is a vector quantity. Hence it can have components in different allowed directions. The component μ, in the direction of the applied magnetic field is given by

$$\mu_z = kI' \qquad ...(ii)$$

where for a given value of I, I' can have values from I to $-I'$ (*i.e.*, $I' = I, I-1, I-2, ...-I$) *i.e.*, I' has total $2I + 1$ values. In fact I' gives the different allowed orientations of the nuclear spin directions when an external magnetic field is applied.

In the absence of an external magnetic field, all these orientations have the same energy. Splitting into different orientations which correspond to different energies takes place only when an external magnetic field is applied. Thus different orientations correspond to different energy levels of a nucleus in the presence of an external magnetic field. The energies of these levels depend upon the component, μ_z and the strength of the applied magnetic field H and are given by

$$E = -\mu_z H \qquad ...(iii)$$

Substituting the value of μ_z from eqn. (*ii*), we get

$$E = -k\,I'\,H \qquad ...(iv)$$

Now as for a given value of I, I' can have $2I + 1$ values, this means that keeping H constant, a nucleus with a spin has $2I + 1$ energy levels in the presence of a magnetic field. This is further explained below with the help of a few examples:

(*i*) When $I = 0$, then $I' = 0$ and hence $E = 0$. This means that nuclei with zero nuclear spin show no splitting in a magnetic field.

(*ii*) When $I = 1/2$, then $I' = 1/2, -1/2$. The energies of the levels will be $-1/2\ kH$ and $+ 1/2\ kH$.

(*iii*) When $I' = \frac{3}{2}$, then $I = \frac{3}{2}, \frac{1}{2}, -\frac{1}{2}, -\frac{3}{2}$. The energies of the levels will be $-\frac{3}{2}kH, -\frac{1}{2}kH, \frac{1}{2}kH$ and $+\frac{3}{2}kH$.

Now in order that a transition may take place from one energy level (having $E = E_1$ and $I' = I'_1$) to another energy level (having $E_1 = E_2$ and $I' = I'_2$), the change in energy involved will be given by

$$\Delta E = E_2 = E_1 - kH(I'_2 - I'_1) \qquad ...(v)$$

However as the transitions can take place only between the adjacent levels, we must have $I'_2 - I'_1 = \pm 1$. Hence

$$\Delta E = \pm kH \qquad ...(vi)$$

where plus sign refers to absorption and minus sign to emission of energy.

Further, as $\Delta E = h\nu$, the frequency involved in the transition will be given by

$$\nu = \frac{\Delta E}{h} = \pm\frac{kH}{h} \qquad ...(vii)$$

Q. 77. Describe the set-up of NMR spectrometer. ***(Mumbai, 2005)***

Ans. Different ways to bring about the nuclear transition are:

(*i*) Keeping H constant, ν may be varied gradually. No energy is absorbed till ν reaches a value at which the energy supplied is equal to the nuclear energy level spacing. The sample is then said to be in resonance with the incident frequency.

(*ii*) Keeping ν constant, H may be varied.

The second procedure is found to be experimentally more satisfactory and is briefly explained under **procedure.**

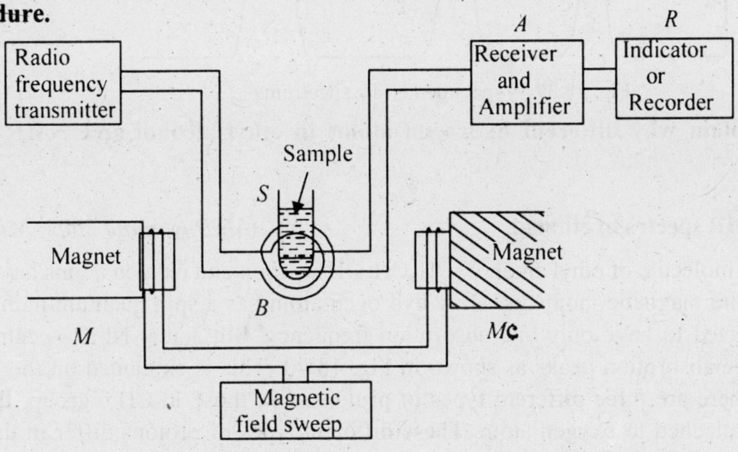

Fig. 18.38. A schematic arrangement of the NMR spectrometer.

The schematic arrangement of the apparatus used is shown in Fig. 18.38.

It consists of the following parts:

(*i*) *An electromagnet* with poles M and M' whose strength field can be varied by varying the feeding voltage.

(*ii*) *A sample tube S* placed between the poles of the electromagnet and containing the sample under study.

(*iii*) *A radio frequency transmitter T* which sends out constant frequency around the sample tube through the coil B.

(*iv*) *A receiver coil C* to pick up the frequency emitted by the sample.

(*v*) *A receiver and amplifier A* to receive and amplify the frequency picked up by the receiver coil C.

(*vi*) *A cathode ray oscilloscope or a recorder R* to record the amplified signals.

Procedure : The procedure consists in supplying some constant frequency (usually 60 megacycles/sec) to the sample from the radio frequency transmitter. A sweeping magnetic field is then applied on the sample. No absorption of energy by the sample takes place till the magnetic field strength is such that right hand side of equation

$$\nu = \pm \frac{kH}{k}$$

becomes equal to the frequency being supplied by the transmitter. When this energy is absorbed, the nuclei are said to be in the excited state. They then start re-emitting the energy. The energy thus emitted is picked up by the receiver coil *C*, amplified by the amplifier *A* and recorded on the recorded *R*. At this stage the sample is said to be in resonance with the applied frequency. When *H* is increased further, no absorption of energy takes place till it again reaches a value at which the sample is again in resonance with the applied frequency. In this way, a plot of the type shown in Fig. 18.39 is obtained. Such a plot is called an *NMR spectrum.* Each peak corresponds to a value of *H* at which the sample is in resonance with applied frequency.

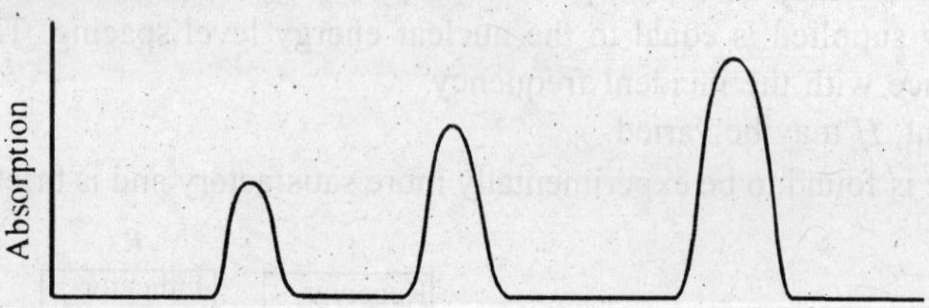

Fig. 18.39. A specimen NMR spectrum.

Q. 78. Explain why different hydrogen atoms in ethyl alcohol give NMR absorption frequencies.

Or

Explain NMR spectra of ethanol. **(*M. Dayanand 2002; Mumbai 2005*)**

Ans. In the molecule of ethyl alcohol (CH_3CH_2OH) carbon and oxygen atoms have no nuclear spin and hence no magnetic moments. Only hydrogen atom has a spin quantum number equal to 1/2 and is expected to have only one absorption frequency. But actual NMR spectrum of ethyl alcohol has three absorption peaks as shown in Fig. 18.40. This is explained on the basis that in ethyl alcohol, there are three different types of protons, *i.e.*, those in CH_3 group, those in CH_2 group and one attached to oxygen atom. These different types of protons differ in the sense that they are screened differently by the surrounding electrons *i.e.*, they have different electronic environment. As a result, different types of hydrogen atoms interact differently with the applied magnetic field and hence possess different absorption frequencies.

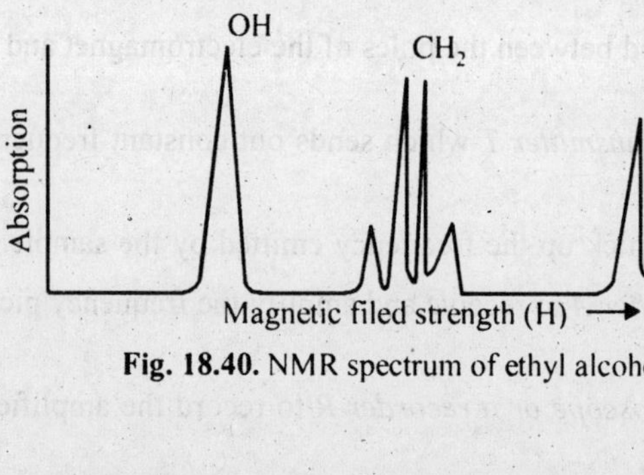

Fig. 18.40. NMR spectrum of ethyl alcohol.

Q. 79. What is meant by Chemical shift? *(M. Dayanand 2002)*

Ans. The environmental effect on any type of hydrogen atom is expressed quantitatively in terms of a quantity called *chemical shift.* It is generally represented by the letter δ and is defined by the equation,

$$\delta = \frac{H_z - H_r}{H_r} \times 10^6 \qquad ...(i)$$

where H_z is the magnetic field strength at which the maximum (peak) appears for a particular type of protons while H_r is the field strength at which absorption occurs for some reference substance such as water. For example, in the case of ethyl alcohol, for protons in CH_3 group, δ = 4.1, whereas for the proton in the OH group, δ = – 0.1.

As $H \propto \nu$, equation (*i*) can also be written as

$$\delta = \frac{\nu_s - \nu_r}{\nu_r} \times 10^6 \qquad ...(ii)$$

Q. 80. Why is tetramethylsilane used as a reference in NMR spectroscopy ?

Ans. Tetramethylislane is used as a reference substance because it has many advantages. **Firstly,** from its structure we see that all the 12 protons present in it are equivalent *i.e.* they have the same environment. Hence it gives a *single well defined absorption peak* in the NMR spectrum.

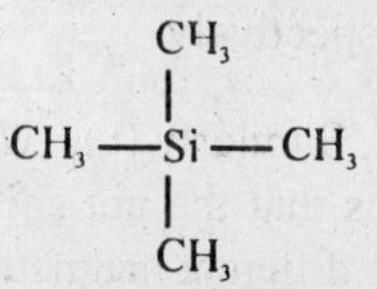

Tetramethylsilane

Secondly, the protons are highly shielded by their electrons from the external magnetic field so that it shows NMR at a very high external magnetic field strength as compared to other protons. **Thirdly,** it is chemically, inert. **Fourthly,** its boiling point is low (27°C) so that it evaporates off quickly and the sample compound can be easily recovered.

Q. 81. What is meant by shielding and deshielding of protons? *(Mumbai 2010*

Ans. Different protons have different electronic environment *i.e.*, the electrons present in the bond close to the proton and the electrons present in the neighbouring bonds show different effects. These electrons induce their own magnetic field which may reinforce the applied magnetic field or oppose it. Now for NMR to occur the proton has to flip its spin from lower energy level to higher energy level *i.e.*, it has to change its spin from ↑ to ↓. If the induced magnetic field reinforces the applied magnetic field, a smaller external magnetic field will be required to flip the spin of the proton. Such a proton is said to be **deshielded** and the absorption is said to be **downfield.** On the other hand, if the induced magnetic field opposes the external magnetic field, stronger magnetic field will be required for the flip. The proton is then said to be **shielded** and the absorption is said to be **upfield.** The situation may be represented as shown in Fig. 18.41.

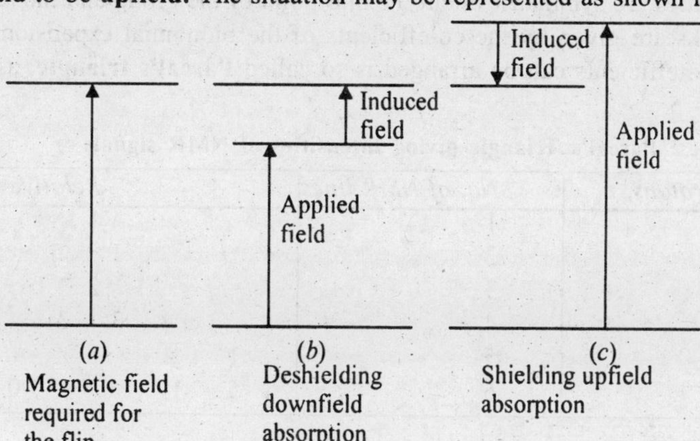

Fig. 18.41. Shielding and deshielding of protons.

Q. 82. Discuss nuclear spin-spin interactions taking the example of ethyl alcohol.

Ans. With respect to a proton of the CH_3 group the two protons of the CH_2 group can align themselves in the following four different ways:

↑↑	↑↓	↓↑	↓↓
(*i*)	(*ii*)	(*iii*)	(*iv*)

Thus their total spin values are 1, 0, 0 and – 1 respectively. As (*ii*) and (*iii*) have same total spin, they are energetically equivalent. Therefore, the net spin combinations of CH_2 group are three. These three different combination induce different magnetic field on the protons of the CH_3 group. As a result, CH_3 group gives rise to three different absorption peaks (triplet) having the relative intensities in the ratio 1 : 2 : 1.

Similarly with respect to a proton of the CH_2 group, the three protons of the CH_3 group can align themselves in 8 different possible ways as follows:

↑↑↑	↑↑↓	↑↓↑	↓↑↑	↓↓↑	↓↑↓	↑↓↓	↓↓↓
(*i*)	(*ii*)	(*iii*)	(*iv*)	(*v*)	(*vi*)	(*vii*)	(*viii*)

Their total spin values are $+\frac{3}{2}, +\frac{1}{2}, +\frac{1}{2}, +\frac{1}{2}, -\frac{1}{2}, -\frac{1}{2}, -\frac{1}{2}, -\frac{1}{2}$ and $-\frac{3}{2}$ respectively.

Thus (*ii*), (*iii*) and (*iv*) having same total spin, are energetically equivalent. Similarly (*v*), (*vi*) and (*vii*), having same total spin are also energetically equivalent. This means that the net spin combination of the CH_3 group are four. These four different combinations induce different magnetic field on the protons of the CH_2 group. As a result, the CH_2 group gives rise to four different absorption peaks (quartlet) having the relative intensities in the ratio 1 : 3 : 3 : 1.

Q. 83. Describe general rules about spin–spin interactions. What formation is provided by NMR spectrum ?

Ans. Some general rules about spin-spin interactions which are helpful in calculating the number of lines expected in the NMR spectra are as under :

1. Protons of the same group do not interact among themselves to cause observable splitting, (*e.g.*, two protons in CH_2 group do not interact with each other).

2. The multiplicity of the peak of a group of equivalent protons is determined by the neighbouring protons. In general, if *n* equivalent protons interact with the protons on an adjacent carbon atom, the peak is split into $(n + 1)$ peaks.

3. The intensities are symmetric about the mid-point of the group and the relative intensities of the $n + 1$ peaks are given by the co-efficients of the bionomial expansion of order *n*, *i.e.*, $(1 + x)^n$. These co-efficients can be arranged is so called **Pascal's triangle,** as shown in Table below for upto $n = 5$.

Table : Pascal's Triangle giving intensities of NMR signals

Number of protons, n	*No. of NMR lines*	*Relative intensities*
0	1	1
1	2	1 1
2	3	1 2 1
3	4	1 3 3 1
4	5	1 4 6 4 1
5	6	1 5 10 10 5 1

Spin-spin interaction is independent of the applied magnetic field strength whereas the chemical shift is a function of field strength. The NMR spectrum of a molecule gives the following information :

(*a*) The number of signals or peaks which let us know about the kinds of proton present in a molecule.

(*b*) The positions of the peaks which tells us about the electronic environment of each kind of proton.

(*c*) The intensities of the peaks which tell us about the no. of protons of each kind that are present and

(*d*) The splitting of a peak into several peaks which tells us about the environment of a proton relative to other nearby protons in the molecule.

Q. 84. Which of the following nuclei do not show nuclear magnetic resonance :

$$^{1}H,\ ^{2}H,\ ^{12}C,\ ^{13}C,\ ^{14}N,\ ^{15}N,\ ^{16}O,\ ^{19}F,\ ^{31}P\ ?$$

Ans. Nuclei that contain odd number of protons and/or neutrons show NMR spectra, since they have half-integral spin. Nuclei that have integral spin (in particular zero spin) are those which contains even number of protons or neutrons. These nuclei do not show NMR. Thus, the nuclei ^{12}C (6*p*, 6*n*), ^{12}O (8*p*, 8*n*) do not show NMR.

Q. 85. Predict the proton-NMR spectrum of ethylacetate using the three general rules.

Ans. The molecule has the structure

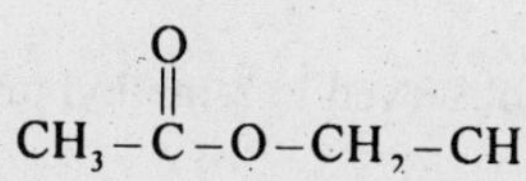

(*i*) There is no splitting due to spin-spin interaction of the protons in either of the two methyl groups or in the methylene group according to Rule 1.

(*ii*) CH_2 proton split the CH_3 peak into three lines (*i.e.*, a triplet) and the CH_3 protons split the CH_2 peak into four line. (*i.e.*, a quartet) according to Rule 2.

(*iii*) Intensities of CH_3 triplet are given by the co-efficients of the expansion $(1 + x^2) = 1 + 2x + x^2$, *i.e.*, by the ratio of 1 : 2 : 1 and the intensities of the CH_3 quartet are given by the co-efficients of the expansion $(1 + x)^3 = 1 + x^3 + 3x^2 + 3x$, *i.e.*,by the ratio 1 :3:3:1 according to Rule 3.

Q. 86. Determine which of the following molecules will show spin- spin coupling in their NMR spectra. If the coupling or splitting is observed, give the multiplicity of each kind of proton, *i.e.*, give the number of lines arising from each proton.

Ans. Spin-spin splitting is not observed in (*b*) and (*f*) which have only equivalent protons.

(*a*) $ClCH_2CH_2I$

(*b*) 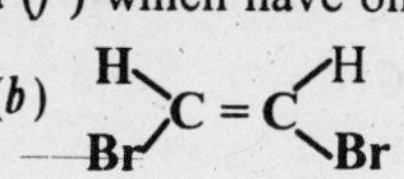

(*c*) 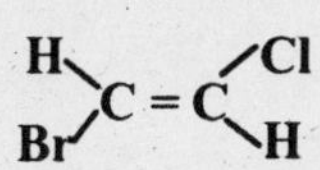

(*d*) 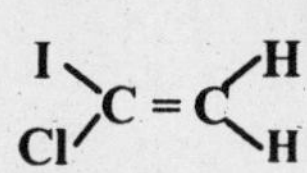

(*e*) $CH_3-CH=O$

(*f*) 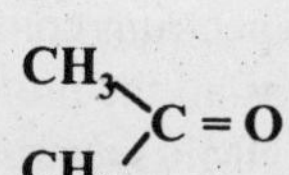

The protons of the CH_2 groups in (*a*) are non-equivalent each being split into triplet. In (*c*) the two protons are not equivalent and each is split into a doublet. The vinyl protons in (*d*) are not equivalent since one is *cis* to Cl and the other is *cis* to *I*, so that each is split into a doublet.

Notice that in this case the protons undergoing spin- spin interaction are on the same carbon. In (*e*) the protons of CH_3 and the proton of the – CH = O group are not equivalent; the CH_3 resonance is split into a doublet and the – CH = O resonance is split into a quartet.

Q. 87. How many kinds of protons are there in

(*i*) CH_3CH_3 (*ii*) $CH_3CH_2CH_3$ (*iii*) $(CH_3)_2CHCH_2CH_3$

(*iv*) $C_6H_5CH_3$ (*v*) $CH_2=CH_2$ (*vi*) $CH_3CH=CH_2$

Ans. (*i*) $CH_3{}^a\,CH_3{}^a$ has one kind of protons.

(*ii*) $CH_3{}^a\,CH_2{}^b\,CH_3{}^a$ has two kinds of protons.

(*iii*) $(CH_3{}^a)_2\,CH_2{}^b\,CH_2{}^c\,CH_3{}^d$ has 4 kinds of protons.

(*iv*) $C_6H_5{}^aCH_3{}^b$ has 2 kinds of protons.

Theoretically, there are three kinds of aromatic protons, *viz.*, ortho, meta and para. However, the ring protons are little affected by alkyl groups and are thus nearly equivalent.

(*v*) One kind. All the four ethylenic hydrogens are equivalent. $CH_2{}^a = CH_2{}^a$.

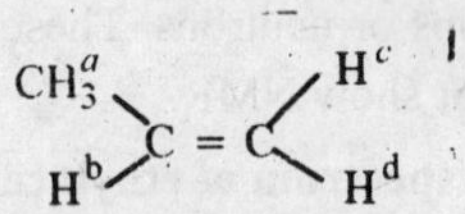

(*vi*) Four kinds.

Q. 88. Why is nuclear spin-spin splitting observed in 2-methyl propene but not in 1-chloro -2, 2-dimethyl propane ?

Ans. In 2-methyl propane

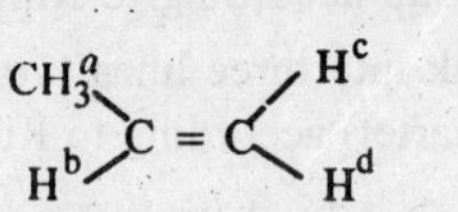

H^a and H^b although at a distance undergo spin-spin slitting because the coupling is transmitted through the μ-electrons. In $(CH_3{}^a)_2C - CH_2{}^b\,Cl$, H^a and H^b do not couple because they are intervened by atom; they are too far away from one another.

Q. 89. In organic compound C_3H_6O contains a carbonyl group. How can its NMR spectrum decide whether it is an aldehyde or a ketone ?

Ans. The two possible structures corresponding to aldehyde and ketone are

$CH_3{}^aCH_2{}^bCH^c = O$ and 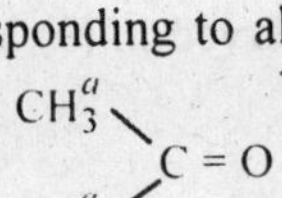

(A) (B)

For aldehyde, NMR spectrum consists of a triplet for H^a, a complex multiplet for H^b and a triplet for H^c. For ketone, as all the protons are equivalent, we get only a singlet, *i.e.*, no spin-spin splitting occurs. Thus, the NMR spectrum can readily decide whether the compound is an aldehyde or a ketone.

Q. 90. How can the NMR spectrum distinguish between the isomers p-xylene and ethyl benzene ?

Ans. NMR can help distinguish between the two as under :

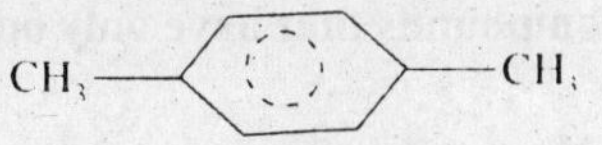 *i.e.*, $p\text{-}CH_3^a\,C_6H_4^b\,CH_3^a$

p-xylene

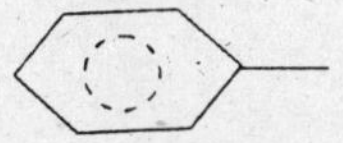 *i.e.*, $C_6H_5^a\,CH_2^b\,CH_3^c$

Ethyl benzene

$p\text{-}CH_3{}^a\,C_6H_4{}^bCH_3{}^a$ has two kinds of protons and its NMR spectrum consists of two singlets. $C_6H_5{}^aCH_2{}^bCH_3{}^c$ has three kinds of protons and its NMR spectrum consists of a singlet for H^a, a quartet for H^b and triplet for H^c. This is how distinction can be made.

Q. 91. Predict the NMR-spectrum of the compound $CH_3COCH_2C \equiv CCH_3$.

Ans. In the compound $CH_3^a - \overset{\overset{\displaystyle O}{\|}}{C} - CH_2^b - C \equiv C - CH_3^c$ the three groups of protons (*a, b* and *c*) are separated by carbon atoms. Hence, no spin- spin coupling will be observed. Only the effects of chemical shifts of the three kinds of protons will be observed. H^a, H^b and H^c each will give rise to a singlet with relative intensities of 3 : 2 : 3 respectively.

Q. 92. Below are given the molecular formulae of some organic compounds that give 2 singlets in their proton. NMR spectra are given below :

(*i*) C_2H_5OCl (*ii*) $C_3H_6O_2$ (*iii*) $C_3H_5Cl_3$

(*iv*) $C_5H_{10}Cl_2$ (*v*) $C_3H_8O_2$

Draw structural formula for each of them.

Ans. Since there is no spin-spin interaction, equivalent protons are situated on adjacent carbons.

(*i*) The compound has the structure

$$ClCH_2{}^a\,OCH_3{}^a$$

(*ii*) The compound has the structure

$$CH_3^a - \overset{\overset{\displaystyle O}{\|}}{C} - O - CH_3^b$$

(*iii*) The compound has the structure

$$CH_3{}^a\,CCl_2CH_2{}^bCl$$

(*iv*) The compound has the structure

$$ClCH_2{}^a\,C\,(CH_3{}^b)_2CH_2{}^aCl$$

(*v*) The compound has the structure

$$CH_3{}^aOCH_2{}^bOCH_3{}^a$$

Q. 93. Predict the proton-NMR spectra of CH_3CH_2Cl and CH_3CHDCl.

Ans. Deuterium (*D*) does not give a signal in the proton-NMR spectrum and does not split the nearby protons. Thus, we can ignore its presence in the molecule. The NMR spectrum of $CH_3{}^aCH_2{}^bCl$ consists of triplet for H^a and a quartet for H^b. The NMR spectrum of $CH_3{}^aCH_2{}^bCl$ has a doublet for H^a and a quartet for H^b.

Q. 94. Molecular formulae of some organic compounds that have only one NMR-signal are given below :

(*a*) C_5H_{12} (*b*) C_3H_6 (*c*) C_3H_4

(*d*) C_8H_{18} (*e*) C_2H_6O

Draw structural formula for each of them.

Ans. One NMR-signal indicates that all the protons must be equivalent.

(*a*) The compound is $(CH_3)_4C$ in which all 12 protons are present in 4 equivalent CH_3 groups.

(*b*) The compound is cyclopropane with structure

which has 3 equivalent CH_2 groups.

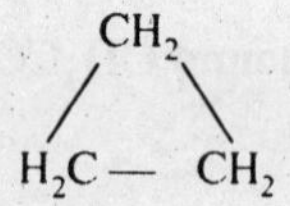

(*c*) The compound has the structure

$$H_2C = C = CH_2$$

which has 2 equivalent CH_2 groups and a C-atom with no protons,

(*d*) The compound has the structure

$$(CH_3)_3C - C\,(CH_3)_3 \text{ (allene)}$$

which has 6 equivalent CH_3 groups and 2 other carbons with no protons.

(*e*) The compound is dimethyl ether, $CH_3 - O - CH_3$ which has 2 equivalent CH_3 groups.

Q. 95. Predict the proton-NMR spectra of

H_2, CH_4, C_2H_6 and C_6H_6.

Ans. All the four molecules contain equivalent protons which do not show spin-spin coupling. Hence, their proton NMR-spectra consist of only a singlet.

19

CHEMICAL EQUILIBRIUM

Q. 1. What is meant by chemical equilibrium? Explain giving some examples.

Ans. Most of the chemical reactions, when carried out in closed vessels do not go to completion. A process starts of its own or on initiation, continues for sometime with falling rate and ultimately appears to have stopped. Although, reactants are still there, but they don't appear to react anymore. The process or the reaction is said to be in a state of *equilibrium*. The composition of the reaction mixture at equilibrium at a given temperature is the same irrespective of whether we start from the reactant side or product side. This may be illustrated with the help of the following examples.

1. Decomposition of calcium carbonate: Solid calcium carbonate on heating in a closed vessel starts decomposing into solid calcium oxide and gaseous carbon dioxide. Pressure built up due to carbon dioxide can be recorded on a manometer attached to the vessel. The pressure goes on increasing and then becomes constant and remains so after that. Calcium carbonate is still present in the reaction mixture but does not appear to decompose further. A state of equilibrium, as represented by the following equation, is attained.

$$CaCO_3\,(s) \rightleftharpoons CaO\,(s) + CO_2\,(g)$$

2. Evaporation of water in a closed vessel: If a small amount of water is placed in a closed vessel at room temperature, evaporation starts. A pressure due to water vapours is built up in the vessel, which can be recorded by the attached manometer. After some time, we find that the pressure of water vapours remains constant, if the temperature does not change. Sufficient water is still present, but it does not appear to evaporate further. We say that a state of equilibrium between evaporation of water and condensation of vapours has been established. This can be represented as

$$H_2O\,(l) \rightleftharpoons H_2O\,(g)$$

Q.2. Discuss some characteristics of chemical equilibrium.

Ans. 1. Chemical equilibrium at a given temperature, is characterised by constant values of certain observable properties such as pressure, concentration or colour. It is well known that hydrogen gas and iodine vapours react at higher temperature to give hydrogen iodide gas as per the following equation

$$H_2\,(g) + I_2\,(g) \rightleftharpoons 2HI\,(g)$$

It is also known that hydrogen iodide gas is not very stable and it dessociates into hydrogen and iodine as follows:

$$2HI\,(g) \rightleftharpoons H_2(g) + I_2(g)$$

The composition of the reaction mixture at equilibrium, is *1.56 moles of* HI, *0.22 mole of* H_2 *and 0.22 mole of* I_2 whether we carry out the reaction by taking one mole each of H_2 and Cl_2 as the reactants or two moles of HI as the reactant at 448°C.

This shows that *when equilibrium is attained at a given temperature, each reactant and each product has a fixed concentration and this is independent of the fact whether we start the reaction with the reactants or with the products.* Thus, *equilibrium can be attained from either side.*

2. *A catalyst can hasten the approach of equilibrium but does not alter the state of equilibrium.* In other words, *the relative concentrations of the products and the reactants remain the same irrespective of the presence or absence of a catalyst.*

3. *Chemical equilibrium is* **dynamic** *in nature.* Equilibrium involves two reactions proceeding in opposite directions. One of these reactions proceeds *from the reactants towards the products* and is known as the *forward reaction.* The other proceeds *from the products towards the reactants* and is known as the *reverse reaction.* When equilibrium is attained, there is no further change in the concentrations of the products or the reactants. This gives the impression that the reaction has come to a stand-still. But this is *not* the case. Actually, the two opposing reactions, the forward reaction and the reverse reaction, are proceeding simultaneously at *equal rates.*

Q.3. Given an experimental proof of dynamic nature of chemical equilibrium.

Ans. Consider the following reaction

$$H_2(g) + I_2(g) \rightleftharpoons 2HI(g)$$

When the equilibrium is attained at 448°C, a small amount of radioactive isotope of iodine is introduced into the reaction mixture. It is well known that radioactive isotopes have the same *chemical* properties as their non-radioactive counterparts. After sometime if the mixture is examined, it will be seen that HI *contains radioactive iodine.* There is, however, no change in the relative amounts of HI, H_2 and I_2. This shows that although there has been no change in the relative amounts of the reactants and the products, *chemical reaction has been taking place from left to right.* But since there has been no increase whatsoever in the amount of HI, it is evident that *the reaction has been taking place from the right to the left as well, at the same rate.* Thus, the two opposing reactions have been proceeding. This proves the dynamic nature of equilibrium.

Q.4. State and explain law of mass action.

(Lucknow 2000; Awadh 2000)

Ans. According to the law of mass action, *the rate at which a substance reacts is proportional to its 'active mass' and the rate of a chemical reaction is proportional to the product of the active masses of the reacting substances.*

The term *active mass* used in the above statement implies *activity* which, *for the sake of simplicity,* may be taken as equal to *molar concentration.*

Consider the following reversible reaction, taking place at a constant temperature:

$$A + B \rightleftharpoons C + D$$

According to the law of mass action, the rate (r_1) at which A and B react is given by the equation

$$r_1 = k_1 [A] [B] \qquad ...(i)$$

where k_1 is the proportionality constant, known as the **rate constant** of the reaction and the square brackets indicate molar concentrations. As the reaction proceeds further, the molar concentrations of A and B continue to decrease. Consequently, the velocity of the forward reaction goes on decreasing with time. The rate of the back reaction (r_2) is given by the equation

$$r_2 = k_2 [C][D] \qquad ...(ii)$$

where k_2 is the rate constant of the reverse reaction.

Thus, ultimately, a dynamic equilibrium is attained when the rate of the forward reaction becomes equal to that of the reverse reaction, *i.e.*, $r_1 = r_2$.

Q. 5. Deduce the law of chemical equilibrium. What is meant by equilibrium constant? ***(Lucknow 2000)***

Ans. Consider the reaction $A + B \rightleftharpoons C + D$

Rate of forward reaction, $r_1 = k_1 [A][B]$

Rate of backward reaction, $r_2 = k_2 [C][D]$

Since at equilibrium, $r_1 = r_2$, it follows that

$$k [A][B] = k_2 [C][D]$$

or

$$\frac{k_1}{k_2} = K_c = \frac{[C][D]}{[A][B]} \qquad ...(i)$$

K_c is known as the **equilibrium constant.** It is simply the *ratio of the rate constants of two opposing reactions.* It is *constant at a given temperature.* It has the same value at a given temperature *irrespective of the direction* from which the equilibrium is reached. It has also the same value *irrespective* of the *pressure* on the system or the *concentrations* of the reactants and products.

In a general reaction represented by the equation

$$aA + bB + cC + \cdots \rightleftharpoons lL + mM + nN + \cdots,$$

the equilibrium constant is given by

$$K_c = \frac{[L]^l [M]^m [N]^n \ldots}{[A]^a [B]^b [C]^c \ldots} \qquad ...(ii)$$

K_c, evidently, has the same units as concentration.

For gaseous reactions, it is more convenient to use *partial pressures* instead of concentrations. The equilibrium constant in that case is represented by K_p. Thus, for the general gaseous reaction

$$aA + bB + \cdots \rightleftharpoons lL + mM + \cdots,$$

$$K_P = \frac{[p_L]^l \times [p_M]^m \ldots}{[p_A]^a \times [p_B]^b \ldots} \qquad ...(iii)$$

K_p, evidently, has the same units as pressure.

Q.6. Deduce Vant Hoff reaction isotherm. *(Coimbatore 2000)*

Ans. Consider the following reaction

$$aA + bB + \cdots \rightleftharpoons lL + mM + \cdots$$

taking place *under any conditions* of temperature, pressure and composition. The free energy change of the reaction is given by the expression

$$\Delta G = (G)_{products} - (G)_{reactants} \qquad ...(i)$$

$$= (l\mu_L + m\mu_M + \cdots) - (a\mu_A + b\mu_B + \cdots)$$

where μ_A, μ_B, μ_L, μ_M, etc., represent chemical potentials of the species concerned.

Let us suppose that various reactants and products are in gaseous state. Chemical potential of a gaseous substance in *any state* is given by the equation

$$\mu_{i(p)} = \mu^\circ_{i(p)} + RT \ln p_i \qquad ...(ii)$$

where $\mu^\circ_{i(p)}$ is the chemical potential of a gaseous component i, in the *standard state* (partial pressure = 1 atmosphere) and $\mu_{i(p)}$ is the chemical potential of the gaseous component i in any state of partial pressure p_i.

Substituting the values of chemical potentials of various species at their partial pressures from eq. (*ii*) in (*i*)

$$\Delta G = \{\mu^\circ_{L(p)} + RT \ln p_L) + m\,(\mu^\circ_{M(p)} + RT \ln p_M) + \cdots\cdots \}$$

$$- \{a(\mu^\circ_{A(p)} + RT \ln p_A) + b\,(\mu^\circ_{B(p)} + RT \ln p_B) + \cdots\cdots\} \qquad ...(iii)$$

Re-arranging, we have

$$\Delta G = \{l\mu^\circ_{L(p)} + m\mu^\circ_{M(p)} + \cdots\cdots) - (a\mu^\circ_{A(p)} + b\mu^\circ_{B(p)} + \cdots\cdots)\}$$

$$+ RT \ln \frac{(p_L)^l\,(p_M)^m \ldots}{(p_A)^a\,(p_B)^b \ldots} \qquad ...(iv)$$

The first expression on the right hand side, evidently, is the tree energy change of the reaction when the products and the reactants are all in their respective standard states. This expression may be substituted by ΔG°. Hence,

$$\boldsymbol{\Delta G = \Delta G^\circ + RT \ln K} \qquad ...(v)$$

where K stands for the reaction quotient of partial pressures of the products and the reactants,

viz. $$\frac{(P_L)^l\,(p_M)^m \ldots}{(p_A)^a\,(p_B)^b \ldots}$$

Eq. (*v*) is known as **van't Hoff reaction isotherm.** This gives the *free energy change of a reaction* at any given temperature, pressure and composition of the reacting system.

Q.7. At 25°C, the ΔG° for the reaction

$$\mathbf{H_2(g) + I_2(g) \longrightarrow 2HI\ (g),}$$

is –3.1 kcal. Calculate ΔG when hydrogen gas at 0.1 atm and iodine vapour at 0.1 atm react to form hydrogen iodide at 10 atm, and 25°C.

Ans. For the given reaction,

$$K = \frac{p^2(\text{HI})}{p(\text{H}_2)\,p(\text{I}_2)}$$

where p terms represent partial pressures of the various species in the reaction mixture.

$$\Delta G = \Delta G^\circ + RT \ln K = \Delta G^\circ + 2.303\ RT \log \frac{p^2(\text{HI})}{p(\text{H}_2)\,p(\text{I}_2)}$$

$$= -3100 \text{ cal} + 2.303\ (1.987 \text{ cal K}^{-1}\text{ mol}^{-1})\,(298 \text{ K}) \log \frac{(10)^2}{(0.1)(0.1)}$$

$$= -3100 \text{ cal} + 1363.7 \log (10^4) \text{ cal} = -3100 \text{ cal} + 4\,(1363.7) \text{ cal}$$

$$= +2354.8 \text{ cal} = \mathbf{2.355\ kcal}$$

Q.8. Give the thermodynamic derivation of law of chemical equilibrium.

(G. Nanakdev 2000; Gujarat 2003; Kalyani 2004)

Ans. When a reaction is in a state of equilibrium, the driving force of the reaction is zero, *i.e.*, $\Delta G = 0$. Hence the van't Hoff reaction isotherm reduces to

$$\Delta G^\circ = -RT \ln K_{eq}$$

i.e., $$\Delta G^\circ = -\ RT \ln \left[\frac{(p_L)^l\,(p_M)^m \ldots}{(p_A)^a\,(p_B)^b \ldots}\right]_{eq} \qquad \ldots(i)$$

The subscript *eq* is put to indicate that the partial pressures of the products and reactants are those *existing at equilibrium.*

Since ΔG° represents *standard state free energy change* of a reaction, *i.e.*, free energy change when the reactants as well as products are in their standard states of unit partial pressure, *it must be constant at a given temperature.* Hence, the expression on the right hand side of Eq. (*i*) must also be *constant.* Further, since *R* is the gas constant. It follows that, if temperature *T* is constant, then

$$\left(\frac{(p_L)^l\,(p_M)^m \ldots}{(p_A)^a\,(p_B)^b \ldots}\right)_{eq} = \text{constant} = K_p \qquad \ldots(ii)$$

where K_p is called the **equilibrium constant** of the reaction.

Eq. (*ii*) is known as the equation of the **law of chemical equilibrium.**

It follows easily from Eqs. (*i*) and (*ii*) that

$$\boldsymbol{\Delta G^\circ = -\ RT \ln K_p} \qquad \ldots(iii)$$

The above relationship permits calculations of A G° of a reaction from the known value of its equilibrium constant and *vice versa.*

Q.9 At 2000 K, the standard state free energy change (ΔG°) for the reaction $N_2 + O_2 \rightleftharpoons 2NO$ is given by $\Delta G^\circ = 92048 - 10.46$ T J. Calculate K_p for the reaction at 2000 K ($R = 8.314$ J K^{-1} mol^{-1}).

Ans. $$\Delta G^\circ = -\ RT \ln K_p = -\ 2.303\ RT \log K_p$$

$$\log K_p = -\frac{\Delta G^\circ}{2.303\,RT}$$

$$= \frac{(-92048)\ +10.46\ \times\ 2000\ \text{Jmol}^{-1}}{(2.303)(8.314\,\text{J K}^{-1}\ \text{mol}^{-1})(2000\,\text{K})} = 1.857$$

$$\boldsymbol{Kp = 1.39 \times 10^{-2}\ \text{atm}}$$

Q.10. The value of K_p, for the water gas reaction

$$\mathbf{CO + H_2O \rightleftharpoons CO_2 + H_2}$$

is 1.06×10^5 at 25°C. Calculate the standard state free energy change (ΔG°) of the reaction at 25°C. ($R = 8.314$ J K^{-1} $mo1^{-1}$) *(Kerala 2000)*

Ans. $$\Delta G^\circ = -\ RT \ln K_p = -\ 2.303\ RT \log K_p$$

$$= -\ (2.303)\ (8.314\ \text{J K}^{-1})\ (298\ \text{K}) \log (1.06 \times 10^{-5})$$

$$= \mathbf{-28673.6\ J = -28.674\ kJ}$$

Q.11. Calculate the standard state free energy change (ΔG°) of the reaction

$$\mathbf{2H_2O(g) \rightleftharpoons 2H_2(g) + O_2(g)}$$

given that at 1000 K, water vapour at 1 atmosphere pressure has a degree of dissociation $= 3 \times 10^{-5}$ per cent ($R = 8.314$ J K^{-1} mol^{-1}).

Ans. The partial pressures of the various reactants in the equilibrium mixture will be evidently as below :

$$pH_2 = 3 \times 10^{-7} \text{ atm}, pO_2 = \frac{3}{2} \times 10^{-7} \text{ atm}$$

$$K_p = \frac{p^2_{H_2} \times pO_2}{p_{H_2O}} = \frac{9\times10^{-14} \times \frac{3}{2}\times10^{-7}}{1}$$

$$= 13.5 \times 10^{-21} \text{ atm}$$

$$\Delta G° = - RT \text{ In } K_p = - 2.303\ RT \log K_p$$

$$= - 2.303\ (8.314 \text{ J K}^{-1} \text{ mol}^{-1})\ (1000 \text{ K}) \log (13.5\times10^{-21})$$

$$\mathbf{= 380448 \text{ J} = 380.45 \text{ kJ}}$$

Q.12. Derive a relation between K_p and K_c.

(Nagpur, 2002; Delhi, 2002; Gulberga 2004)

Ans. The value of K_p for the reaction

$$aA + bB + cC + \cdots \rightleftharpoons lL + mM + \cdots$$

is given by the expression

$$K_p = \left[\frac{(p_L)^l\ (p_M)^m \cdots}{(p_A)^a\ (p_B)^b \cdots}\right]_{eq} \quad \ldots(i)$$

The value of K_c for the above reaction would be given by the expression

$$K_c = \left(\frac{(c_L)^l\ (c_M)^m \cdots}{(c_A)^a\ (c_B)^b \cdots}\right)_{eq} \quad \ldots(ii)$$

where p and c stand for partial pressure and concentration, respectively. Now, for an ideal gas,

$$pV = nRT$$

or $$p = (n/V)\ RT = cRT$$

or $$c = p/RT \quad \ldots(iii)$$

where c is the concentration. Thus, for any component i whose partial pressure is p_i,

$$c_i = p_i / RT \quad \ldots(iv)$$

Substituting the concentrations of various species in terms of their partial pressures in Eq. (*ii*) we have

$$K_c = \left\{\frac{\left(\frac{p_L}{RT}\right)^l \left(\frac{p_M}{RT}\right)^m \cdots}{\left(\frac{p_A}{RT}\right)^a \left(\frac{p_B}{RT}\right)^b \cdots}\right\}_{eq}$$

$$= \left(\frac{(p_L)^l\ (p_M)^m \cdots}{(p_A)^a\ (p_B)^b \cdots}\right)_{eq} \times \left(\frac{1}{RT}\right)^{\{(l+m+\cdots) - (a+b+\cdots)\}}$$

$$= K_p \left(\frac{1}{RT}\right)^{\Delta n} \quad \ldots(v)$$

where Δn = Number of moles of products — Number of moles of reactants

or $$\mathbf{K_p = K_c\ (RT)^{\Delta n}} \quad \ldots(vi)$$

Q.13. The value of *Kp* for the equilibrium

$$2H_2O\,(g) + 2Cl_2(g) \rightleftharpoons 4HCl\,(g) + O_2(g)$$

is 0.035 atm at 400°C when the partial pressures are expressed in atmospheres. Calculate the value of K_c for the same reaction.

Ans. $\Delta n = 4 + 1 - 2 - 2 = 1$

∴ We shall use the relation $K_p = K_c\,(RT)^{\Delta n}$

or $$K_c = \frac{K_p}{(RT)^{\Delta n}}$$

Substitute the values

$$K_c = \frac{0.035 \text{ atm}}{(0.0820\, l \text{ atm K}^{-1}\text{mol}^{-1} \times 673\text{ K})}$$

$$= \mathbf{6.342 \times 10^{-4}\ mol\ \mathit{l}^{-1}}$$

Q.14. Derive van't Hoff equation for the temperature dependence of equilibrium constant.
(Arunachal 2003; Delhi 2003; Punjab 2003; Himachal 2010)

Ans. The equation for reaction isotherm when the reactants as well as the products are *gaseous* and are also in their *standard states,* is represented as

$$\Delta G° = -RT \ln K_p$$

Differentiating with respect to temperature at constant pressure, we have

$$\left\{\frac{\partial(\Delta G°)}{\partial T}\right\}_p = -R \ln K_p - RT\frac{d\,(\ln K_p)}{dt}$$

Multiplying throughout by *T,* we get

$$T\left\{\frac{\partial(\Delta G°)}{\partial T}\right\}_p = -RT \ln K_p - RT^2\frac{d\,(\ln K_p)}{dT}$$

Substituting $\Delta G°$ for $-RT \ln K_p$, we have

$$T\left\{\frac{\partial(\Delta G°)}{\partial T}\right\}_p = \Delta G° - RT^2 \ln K_p - RT^2\frac{d\,(\ln K_p)}{dT} \quad ...(iii)$$

The well-known Gibbs-Helmholtz equation for substances in *standard states* may be written as

$$\Delta G° = \Delta H° + T\left\{\frac{\partial(\Delta G°)}{\partial T}\right\}_p \quad ...(iv)$$

Comparing Eqs. (*iii*) and (*iv*), we get

$$RT^2\frac{d\,(\ln K_p)}{dT} = \Delta H°$$

or $$\frac{d\,(\ln K_p)}{dT} = \frac{\Delta H°}{RT^2} \quad ...(vi)$$

Eq.(*vi*) is known as the **van't Hoff's equation.** $\Delta H°$ is the enthalpy change for the reaction at constant pressure when the reactants as well as the products are in their standard states.

It is known from experiments that the enthalpy change, ΔH, accompanying a chemical reaction does not vary appreciably with change in partial pressures of the reactants or products. Therefore, *we may take $\Delta H°$ as equal to ΔH, where ΔH* is the enthalpy change of the reaction *whatever may be the partial pressures of the reactants or products.* Hence, the van't Hoff equation may be written as

$$\frac{d(\ln K_p)}{dT} = \frac{\Delta H}{RT^2} \quad ...(vii)$$

Integrating Eq. (*vii*) between temperatures T_1 and T_2 at which the equilibrium constants are K_p' and K_p'', respectively and assuming that ΔH remains constant over this range of temperature, we get

$$\int_{K_p'}^{K_p''} d(\ln K_p) = \frac{\Delta H}{R}\int_{T_1}^{T_2} \frac{dT}{T^2}$$

or

$$\ln Kp'' - \ln Kp' = \frac{-\Delta H}{R}\left(\frac{1}{T_1} - \frac{1}{T_2}\right)$$

$$= \frac{\Delta H}{R}\left[\frac{T_2 - T_1}{T_1 T_2}\right] \quad ...(viii)$$

or

$$\log Kp'' - \log Kp' = \log \frac{Kp''}{Kp'} = \frac{\Delta H}{2.303}\left[\frac{T_2 - T_1}{T_1 T_2}\right] \quad ...(ix)$$

Knowing the equilibrium constant at a temperature it is possible to calculate the equilibrium constant at another temperature provided the heat of reaction (ΔH) is known.

Alternatively, knowing the equilibrium constants of a reaction at two temperatures, the heat of reaction (ΔH) can be calculated.

Q.15. The equilibrium constant K_p, for the reaction $H_2(g) + S(g) \rightleftharpoons H_2S(g)$ is 20.2 atm at 945°C and 9.21 atm at 1065°C. Calculate the heat of reaction.

Ans. Applying van't Hoff equation

$$\log Kp'' - \log Kp' = \frac{\Delta H}{2.303R}\left[\frac{T_2 - T_1}{T_1 T_2}\right]$$

$$T_1 = 945 + 273 = 1218 \text{ K}$$
$$T_2 = 1065 + 273 = 1338 \text{ K}$$
$$K_p' = 20.2 \text{ atm}$$
$$K_p'' = 9.21 \text{ atm}$$

$$\log 9.21 - \log 20.2 = \frac{\Delta H}{(2.303)(8.314\,\text{JK}^{-1}\text{mol}^{-1})}\;\frac{1338 - 1218}{(1218\text{K})(1338\text{K})}$$

$$\mathbf{\Delta H = -\ 88126.3\ J = -\ 88.826\ kJ}$$

Q.16. The equilibrium constant K_p for the reaction $N_2 + 3H_2 \rightleftharpoons 2NH_3$ is 1.64×10^{-4} atm at 400°C. What will be the equilibrium constant at 500°C if the heat of reaction in this temperature range is – 105185.8 J ? *(Delhi 2003)*

$$\log Kp'' - \log Kp' = \frac{\Delta H}{2.303R}\left(\frac{T_2 - T_1}{T_1 T_2}\right)$$

$$T_1 = 400 + 273 = 673 \text{ K}$$

$$T_2 = 500 + 273 = 773 \text{ K}$$

$$\Delta H = -105185.8 \text{ J}$$

$$\log K_p'' = \log (1.64 \times 10^{-4}) + \frac{-105185.8 \text{ J mol}^{-1}}{(2.303)(8.314 \text{J K}^{-1}\text{mol}^{-1})}\left(\frac{773\text{K} - 673\text{K}}{673\text{K} \times 773\text{K}}\right)$$

$$\therefore \qquad K_p' = 0.144 \times 10^{-4} \text{ atm}$$

Q.17. Deduce van't Hoff's equation for temperature dependence of equilibrium constant in terms of K_c. ***(Punjab 2003; Himachal 2010)***

Ans. K_p and K_c are related by the expression

$$K_p = K_c\, (RT)^{\Delta n} \qquad ...(i)$$

Taking logarithms of the above equation, we have

$$\ln K = \ln K_c + \Delta n \ln RT \qquad ...(ii)$$

Differentiating Eq. (*ii*) with respect to temperature, we get

$$\frac{d(\ln K_p)}{dT} = \frac{d(\ln K_c)}{dT} + \frac{\Delta n}{T} \qquad ...(iii)$$

or

$$\frac{d(\ln K_c)}{dT} = \frac{d(\ln K_p)}{dT} - \frac{\Delta n}{T}$$

$$= \frac{\Delta H^\circ}{RT^2} - \frac{\Delta n}{T} \qquad \text{Refer eq. } (vii), \text{ Q. 14.}$$

$$= \frac{\Delta H^\circ}{RT^2} - \frac{\Delta nRT}{TRT}$$

$$= \frac{\Delta H^\circ - \Delta nRT}{RT^2}$$

But, $\Delta H - \Delta nRT = \Delta E$ (from thermodynamics)

$$\therefore \qquad \left(\frac{d \ln K_c}{dT}\right) = \frac{\Delta E^\circ}{RT^2} \qquad ... (iv)$$

Q. 18. Explain the terms

(*i*) Homogeneous equilibrium

(*ii*) Heterogeneous equilibrium.

Ans. (*i*) An equilibrium which has all the reactants and products in the same phase is called homogeneous equilibrium. For example

$$H_2(g) + I_2(g) \rightleftharpoons 2HI\ (g)$$

Here H_2, I_2 and HI all exist in the same gaseous phase.

similarly $\quad CH_3COOH\ (l) + C_2H_5OH\ (l) \rightleftharpoons CH_3COOC_2H_5(l) + H_2O\ (l)$

is also an example of homogeneous equilibria as all the reactants and products are in the liquid phase.

(*ii*) An equilibrium which has the reactants and products in two or more phases is called heterogeneous equilibrium. For example

$$CaCO_3(s) \rightleftharpoons CaO\ (s) + CO_2(g)$$

Q. 19. Discuss the equilibrium of the reaction

$$\mathbf{H_2(g) + I_2(g) \rightleftharpoons 2HI\ (g)}$$

Ans. Suppose, *a* moles of hydrogen and *b* moles of iodine, occupying a volume *V* litres, react and that on the attainment of equilibrium, *x* moles of each have combined to form hydrogen iodide.

$$\begin{array}{lcccc} \text{Initial molar conc.} \rightarrow & a/V & b/V & & 0 \\ & H_2 & + \; I_2 & \rightleftharpoons & 2HI \\ \text{Eqbm. molar conc.} \rightarrow & \dfrac{a-x}{V} & \dfrac{b-x}{V} & & \dfrac{2x}{V} \end{array}$$

Applying the law of chemical equilibrium,

$$K_c = \frac{[HI]^2}{[H_2][I_2]} = \frac{(2x/V)^2}{\dfrac{a-x}{V} \times \dfrac{b-x}{V}}$$

$$= \frac{4x^2}{(a-x)\,(b-x)}$$

The equilibrium constant in this equation is written as K to indicate that active masses are expressed in terms of molar concentrations.

We see that Eq. (*i*) does not include the volume term, V. Thus, the equilibrium is independent of volume and, therefore, of pressure. In other words, *change of pressure or volume has no effect on the equilibrium state.*

The equilibrium constant may also be expressed in terms of partial pressures of the products and reactants when it is denoted by K_p, as already mentioned.

$$\begin{array}{lcccc} \text{Initial number of moles} \rightarrow & a & b & & 0 \\ & H_2 & + \; I_2 & \rightleftharpoons & 2HI \\ \text{No. of moles at eqbm.} \rightarrow & a-x & b-x & & 2x \end{array}$$

Total number of moles at equilibrium

$$= (a-x) + (b-x) + 2x = a + b$$

Suppose, the total pressure of the system is P atmospheres.

Then,

$$\text{Partial pressure of hydrogen, } (pH_2) = \frac{P(a-x)}{a+b}$$

$$\text{Partial pressure of iodine, } (pI_2) = \frac{P(b-x)}{a+b}$$

$$\text{Partial pressure of hydrogen iodide, } (pHI) = P\left(\frac{2x}{a+b}\right)$$

$$K_p = \frac{p^2HI}{p_{H_2} \times p_{I_2}}$$

$$= \frac{4P^2x^2}{(a+b)^2} \times \frac{a+b}{P(a-x)} \times \frac{a+b}{P(b-x)}$$

$$= \frac{4x^2}{(a-x)(b-x)}$$

If follows from Eqs. (*i*) and (*ii*) that $K_p = K_c$.

K_p and K_c are identical in the case of gaseous reactions where the number of molecules on the reactant and product sides is the same.

Q.20. A mixture containing 8.07 moles of hydrogen and 9.08 moles of iodine was heated at 448°C till equilibrium was attained when 13.38 moles of hydrogen iodide was obtained. How much hydrogen iodide will be obtained if we start with the same number

of moles of hydrogen mixed with 33.10 moles of iodine and the mixture is brought to equilibrium at the same temperature as before? Also calculate the degree of dissociation of hydrogen iodide at 448°C.

Ans. $H_2 + I_2 \rightleftharpoons 2HI$

If we sart with a moles of H_2 and b moles of I_2 and if $2y$ moles of HI are produced on the attainment of equilibrium, then the number of moles of various species at equilibrium will be as indicated below:

$$\underset{(a-y)}{H_2} + \underset{(b-y)}{I_2} \rightleftharpoons \underset{2y}{2HI}$$

$$a = 8.07,\ b = 9.08 \text{ and } 2y = 13.38$$

$$\therefore \quad K_c = \frac{(2y)^2}{(a-y)(b-y)} = \frac{(13.38)^2}{(1.38)\times(2.39)} = 54.29$$

(*ii*) Let $2z$ moles of HI be formed when we start with same moles of hydrogen but 33.10 moles of iodine. Then

$$54.29 = \frac{(2z)^2}{(8.07-z)(33.10-z)}$$

$$\therefore \quad z = 36.53 \text{ or } 7.90$$

Evidently, 36.53 cannot be a solution of the problem. Hence, $z = 7.9$

$\therefore$ Number of moles of HI = $\mathbf{2z = 15.8}$

(*iii*) The dissociation of hydrogen iodide is represented as

$$2HI \rightleftharpoons H_2 + I_2$$

If x is the degree of dissociation, the number of moles of various species at equilibrium will be as indicated below:

$$\underset{(1-x)}{2HI} \rightleftharpoons \underset{x/2}{H_2} + \underset{x/2}{I_2}$$

$$\therefore \quad \frac{x/2 \times x/2}{(1-x)^2} = \frac{1}{K_c} = \frac{1}{54.20}$$

Hence, $\mathbf{x = 0.213}$

Q. 21. Study the equilibrium of the reaction:

$$\mathbf{PCl_5 \rightleftharpoons PCl_3 + Cl_2}$$

Ans. If one mole of phosphorus pentachloride (vapour) is taken in a closed vessel of volume V and if x is its degree of dissociation, the concentrations of the various reactants and products will be as indicated in the equation given below:

$$\underset{(1-x)/V}{PCl_5} \rightleftharpoons \underset{x/V}{PCl_3} + \underset{x/V}{Cl_2}$$

The equilibrium constant, K_c, will be given by the expression

$$K_c = \frac{x/V \times x/V}{(1-x)/V} = \frac{x^2}{(1-x)V} \quad \text{...(}i\text{)}$$

We see that in this equation, V occurs in the denominator. Therefore, if the volume increases, the quantity $x^2/(1-x)$ increases correspondingly so as to maintain the constant value of K_c. This means that with increase in volume, the degree of dissociation x will increase. Hence, *decrease of pressure* (which will cause increase in volume) *in this system will lead to an increase in the degree of dissociation* and *vice versa.*

If the degree of dissociation x is small, Eq. (*i*) reduces to

$$K_c = x^2/V$$

$\therefore$ $$x^2 \propto V \qquad (\because K_c \text{ is constant})$$

or $$x \propto \sqrt{V} \propto \sqrt{1/P}$$

Thus, *the degree of dissociation* (when it is small as compared to unity) *varies inversely as the square root of the pressure of the system.*

The Value of K_p. Assuming that the system contains initially one mole of phosphorus pentachloride and that x is its degree of dissociation into phosphorus trichloride and chlorine when equilibrium is attained, the number of moles of the various substances will be as indicated below:

$$\begin{array}{ccccc} PCl_5 & \rightleftharpoons & PCl_3 & + & Cl_2 \\ 1-x & & x & & x \end{array}$$

$$K_p = \frac{p_{PCl_3} \times p_{Cl_2}}{p_{PCl_5}} \qquad ...(ii)$$

Since the total number of moles is $1 - x + 2x = 1 + x$, at equilibrium, the various partial pressures are

$$pPCl_3 = \frac{Px}{1+x}$$

$$pCl_2 = \frac{Px}{1+x}$$

$$pPCl_5 = \frac{P(1-x)}{1+x}$$

where P is the total pressure.

$$\therefore K_p = \frac{\left(P\frac{x}{1+x}\right)\left(P\frac{x}{1+x}\right)}{\left(P\frac{1-x}{1+x}\right)}$$

or $$K_p = \frac{x^2 P}{1-x^2} \qquad ... (iii)$$

Since K_p is constant at a given temperature, decrease of pressure of dissociation and *vice versa.*

If x is small in comparison to unity,

$$K_p = x^2 P$$

$\therefore$ $$x^2 \propto 1/P$$

or $$x \propto \sqrt{1/P}$$

It is evident from Eq. (*iii*) that if the value of K_p is known, the degree of dissociation of phosphorus pentachloride at any given pressure can easily be calculated.

Q. 22. Assuming that the degree of dissociation of PCl_5 at a certain temperature and under atmospheric pressure is 0.2, calculate the pressure at which this substance will be half dissociated at the same temperature.

Ans. The dissociation of PCl_5 is represented as

$$\begin{array}{ccccc} PCl_5 & \rightleftharpoons & PCl_3 & + & Cl_2 \\ 1-x & & x & & x \end{array}$$

where x is the degree of dissociation.

$$K_p = \frac{x^2 P}{1 - x^2}$$

In the present case, when $P = 1$, $x = 0.2$

$$\therefore \quad K_p = \frac{0.2 \times 0.2 \times 1}{(1 - 0.04)} = 0.041$$

Let P' be the pressure at which PCl_5 is half dissociated, *i.e.*, $x = 0.5$

$$K_p = \frac{x^2 P'}{1 - x^2}$$

$$0.041 = \frac{0.5 \times 0.5 \times P'}{(1 - 0.25)}$$

$$\therefore \quad \mathbf{P' = 0.123 \text{ atm}}$$

Q. 23. The equilibrium constant K_c for the reaction

$$2SO_2(g) + O_2(g) \rightleftharpoons 2SO_3(g)$$

is 800 at 527°C. Calculate the equilibrium constant for the reaction

$$SO_3(g) \rightleftharpoons SO_2(g) + \frac{1}{2}O_2(g)$$

at the same temperature.

Ans. For the first reaction,

$$K_c = \frac{[SO_3]^2}{[SO_2]^2[O_2]} = 800 \quad \text{... } (i)$$

For the second reaction,

$$K_c' = \frac{[SO_2][O_2]^{1/2}}{[SO_3]} \quad \text{...}(ii)$$

Taking the square root of both sides of Eq. (*i*), we have

$$\frac{[SO_3]}{[SO_2][O_2]^{1/2}} = \sqrt{800} = 28.2 \quad \text{...}(iii)$$

Taking the reciprocal of each side of Eq. (*iii*), we obtain the value of K_c' as given in Eq. (*ii*).

$$\therefore \quad \mathbf{K_c' = \frac{1}{28.2} = 0.0354}$$

Q. 24. Discuss the equilibrium of esterification of acetic acid.

Ans. The interaction of ethyl alcohol and acetic acid to form ethyl acetate and water is an example of a homogeneous equilibrium in the *liquid phase*.

$$CH_3COOH + C_2H_5OH \rightleftharpoons CH_3COOC_2H_5 + H_2O$$

If ethyl alcohol and acetic acid are mixed in equimolar proportions at 25°C, it is found that on the attainment of equilibrium, two-thirds of the acid has been converted into the ester. Therefore, two-thirds of the alcohol will also be consumed. If we start with one mole each of the acid and alcohol the number of gram moles of the various reactants at equilibrium will be as indicated below:

	CH_3COOH +	C_2H_5OH ⇌	$CH_3COOC_2H_5$ +	H_2O
Number of moles at eqbm.]→	$\frac{1}{3}$	$\frac{1}{3}$	$\frac{2}{3}$	$\frac{2}{3}$

$$K_C = \frac{\frac{2}{3} \times \frac{2}{3}}{\frac{1}{3} \times \frac{1}{3}} = 4$$

Knowing the value of K_c, the compositions of equilibrium mixtures when different amounts of acetic acid and ethyl alcohol are taken for the reaction at the same temperature, can be calculated.

Suppose, a moles of acetic acid and b moles of ethyl alcohol are mixed together at 25°C and that x moles of ethyl acetate are obtained at equilibrium. If V is the total volume of solution, the equilibrium concentrations will be as shown below:

$$CH_3COOH + C_2H_5OH \rightleftharpoons CH_3COOC_2H_5 + H_2O$$

Equilibrium concentrations]→ $\quad \frac{a-x}{V} \quad \frac{b-x}{V} \quad \frac{x}{V} \quad \frac{x}{V}$

$$\text{Then, } K_c = 4 = \frac{\frac{x}{V} \times \frac{x}{V}}{\frac{a-x}{V} \times \frac{b-x}{V}} = \frac{x^2}{(a-x)(b-x)}$$

The volume terms cancel out. The equilibrium state is, therefore, not influenced by the change of volume.

Q. 25. The standard free energies of formation for the esterification reaction

$$\underset{n\text{-Butyric acid}}{C_3H_7COOH\,(l)} + \underset{\text{Ethanol}}{C_2H_5OH(l)} \longrightarrow \underset{n\text{-Ethyl butyrate}}{C_3H_7COOC_2H_5\,(l)} + H_2O\,(l)\text{, at 25°C, are as}$$

follows:

$$C_3H_7COOH(l) : -91.5 \text{ kcal};\ C_3H_7COOC_2H_6 : -76.0 \text{ kcal}$$
$$H_2O(l) : -56.69 \text{ kcal};\ C_2H_5OH(l) : -41.63 \text{ kcal}$$

Calculate the amount of ester formed when 0.5 mole of *n*-butyric acid and 2 moles of ethanol are mixed and allowed to attain equilibrium at 25°C.

Ans. $\Delta G° = -76.0 - 56.69 - (-91.5 - 41.63)$

$= +0.44 \text{ kcal} = +440 \text{ cal mol}^{-1}$

Since $\Delta G° = -RT \ln K_p$

$$\therefore \quad \log K = -\frac{\Delta G°}{2.303\,RT} = -\frac{440 \text{ cal mol}^{-1}}{(2.303)(1.987 \text{ cal K}^{-1})(298 \text{ K})}$$

$$= -0.323 = \bar{1}.677$$

$\therefore \quad K = 0.48$

Let x be the number of moles of the ester formed at equilibrium. Then, evidently,

$$\underset{(0.5-x)}{C_3H_7COOH(l)} + \underset{(2-x)}{C_2H_5OH\,(l)} \longrightarrow \underset{x}{C_3H_7COOC_2H_5\,(l)} + \underset{x}{H_2O\,(l)}$$

$$K = \frac{x^2}{(0.5-x)(2-x)} = 0.48$$

or $\quad 0.52x^2 + 1.20x - 0.48 = 0$

On solving the equation, we get the positive value of $x = 0.35$. Thus, the amount of the ester formed = 0.35 mol.

Q. 26. The equilibrium between triiodide ions, iodine and iodide ions in aqueous solution is given by the following equation:

$$I_3^-\,(aq) \rightleftharpoons I_2(aq) + I^-\,(aq)$$

(*a*) In a particular equilibrium mixture the various concentrations are as follows:

$$[I_3^-] = 0.0067 \text{ M};\ [I_2] = 0.001 \text{ M};\ [I^-] = 0.01 \text{ M}$$

What is the equilibrium constant?

(*b*) In another equilibrium mixture at the same temperature,

$[I_2] = 0.0067$ **M and** $[I^-] = 0.20$ **M**

What is the value of $[I_3^-]$?

Ans. (*a*) $$K = \frac{[I_2][I^-]}{[I_3^-]} = \frac{(0.001 \text{ mol } l^{-1})(0.01 \text{ mol } l^{-1})}{0.0067 \text{ mol } l^{-1}}$$

$$= \mathbf{0.0015 \text{ mol } l^-}$$

(*b*) Since $$\frac{[I_2][I^-]}{[I_3^-]} = 0.0015 \text{ mol } l^{-1}$$

$$\therefore \quad [I_3^-] = \frac{[I_2][I^-]}{0.0015} = \frac{(0.001 \text{ mol} l^{-1})(0.20 \text{ mol} l^{-1})}{(0.0015 \text{ mol} l^{-1})}$$

$$= 0.13 \text{ mol } l^{-1} = \mathbf{0.13 \text{ M}}$$

Q. 27. Discuss the equilibrium of the following reaction:

$$\mathbf{CaCO_3(s) \rightleftharpoons CaO\ (s) + CO_2(g)}$$

Ans. In this reaction, there are three phases, two solid and one gaseous, in the equilibrium mixture. The equilibrium law equation may thus be represented as

$$K_p = \frac{p\text{CaO} \times p\text{CO}_2}{p\text{CaCO}_3} \quad \text{... } (i)$$

In applying the law of chemical equilibrium to heterogeneous system, the active mass of a solid substance is, ordinarily, taken as unity.

The equilibrium law equation, therefore, reduces to

$$K_p = p\text{CO}_2 \quad \text{...}(ii)$$

Thus, when calcium carbonate is heated in a closed vessel at a definite temperature, *the pressure of carbon dioxide produced becomes constant irrespective of the amount of calcium carbonate taken.* This is supported by experiment.

Q. 28. Deduce the value of K_p for the following reaction equilibriums:

(*a*) Reaction of steam on heated iron.

(*b*) Production of water gas.

Ans. (*a*) If the reaction takes place in a closed vessel, the following equilibrium is set up:

$$3\text{Fe } (s) + 4\text{H}_2\text{O } (g) \rightleftharpoons \text{Fe}_3\text{O}_4 (s) + 4\text{H}_2 (g)$$

Applying the law of chemical equilibrium,

$$K_p = \frac{p_{\text{H}_2}^4 \times p_{\text{Fe}_3\text{O}_4}}{p_{\text{Fe}}^3 \times p_{\text{H}_2\text{O}}^4}$$

Since the active masses of solid phases are taken as unity, $p_{\text{Fe}_3\text{O}_4} = p_{\text{Fe}} = 1$

$$\therefore \quad K_p = \frac{p_{\text{H}_2}^4}{p_{\text{H}_2\text{O}}^4}$$

or $$\sqrt[4]{K_p} = \frac{p_{\text{H}_2}}{p_{\text{H}_2\text{O}}}$$

Thus, according to the law of chemical equilibrium, the ratio of partial pressure of hydrogen to that of steam at equilibrium should be constant.

(*b*) The equilibrium involved in the water gas reaction is:

$$C\,(s) + H_2O\,(g) \rightleftharpoons CO\,(g) + H_2\,(g)$$

$$K_p = \frac{p_{CO} \times p_{H_2}}{p_C \times p_{H_2O}}$$

Since partial pressure of carbon is taken as unity, the equilibrium constant is given by the expression

$$K_p = p_{CO} \times \frac{p_{H_2}}{p_{H_2O}}$$

Q. 29. Discuss the equilibrium of the following reaction

$$\mathbf{NH_4HS\,(s) \rightleftharpoons NH_3\,(g) + H_2S\,(g)}$$

Ans. Applying the law of chemical equilibrium,

$$K_p = \frac{pNH_3 \times pH_2S}{pNH_4HS}$$

the active mass of ammonium bisulphide, being a solid, is taken as unity. Hence,

$$K = pNH_3 \times pH_2S$$

If the reaction is carried out in a *closed* vessel at 25°C, the total gas pressure at equilibrium is found to be 0·63 atmosphere. Since ammonia and hydrogen sulphide are present in equimolar amounts, as shown in the chemical equation given above, it follows that

$$pNH_3 = pH_2S = 0.315 \text{ atm}$$

Hence, $$K = 0.315 \times 0.315 = 9.92 \times 10^{-2} \text{ atm}$$

The equilibrium constant will have the same value even if some ammonia or hydrogen sulphide is already present in the reaction vessel.

Suppose ammonia is present initially in the vessel and its pressure is *p*. If *x* is the pressure of ammonia and hydrogen sulphide resulting from the decomposition of ammonium hydrogen sulphide at equilibrium, then,

$$pNH_3 = p + x\,;\; pH_2S = x$$

$$K_p = pNH_3 \times pH_2S = (p + x)x \qquad (i)$$

The total pressure (*P*) of the system at equilibrium, being equal to the sum of the pressures of ammonia and hydrogen sulphide, is given by

$$P = p + x + x = p + 2x$$

Since the total pressure *P* of the system can be measured by connecting the reaction vessel to a manometer and *p*, the initial pressure of ammonia is known, *x* can easily be calculated. Substituting the value of *x* in Eq. (*i*), the value of K_p can be obtained.

Q. 30. Discuss the equilibrium of dissociation of ammonium carbamate.

Ans. This reaction proceeds as under:

$$NH_4COONH_2(s) \rightleftharpoons 2NH_2\,(g) + CO_2(g)$$

Applying the law of chemical equilibrium,

$$K = p^2NH_3 \times pCO_2$$

If P is the total pressure of the system at equilibrium, then

$$pNH_3 = \frac{2}{3}P \text{ and } pCO_2 = \frac{1}{3}P$$

Hence, $$K_p = \frac{4}{9}P^2 \times \frac{1}{3}P = \frac{4}{27}P^2$$

Q.31. What is Le Chatelier's principle? Explain giving examples.

(*Purvanchal 2007; Himachal 2010*)

Ans. Le Chatelier, a noted French chemist, studied the effect of *concentration, temperature and pressure* on a large number of chemical equilibria. He summed up his observations in the form of a generalisation known as **Le Chatelier's principle,** which states as follows:

If an equilibrium is subjected to a stress, the equilibrium shifts in such a way as to reduce the stress.

According to this principle, *if a system at equilibrium is subjected to a change of concentration, pressure or temperature, the equilibrium shifts in the direction that tends to* **undo** *the effect of the change.*

Effect of change of concentration. Consider the following equilibrium:

$$\underset{\text{Pale yellow}}{Fe^{3+}(aq)} + \underset{\text{Colourless}}{(SCN)^-(aq)} \rightleftharpoons \underset{\text{Dark brown}}{[Fe(SCN)]^{2+}(aq)}$$

Suppose, some ferric salt is added to this equilibrium. The colour of the solution will darken immediately showing that there is increase in the concentration of the coloured complex ion $[Fe(SCN)]^{2+}$. This change is in accordance with Le Chatelier's principle. Addition of more of Fe^{3+} ions has resulted in increasing the concentration of the complex ferri-sulphocyanide.

Now, suppose a small amount of potassium ferri-sulphocyanide capable of giving the complex ion, $[Fe(SCN)]^{2+}$ is added to the equilibrium. The solution will be seen to become less dark showing that the dark coloured $[(Fe(SCN)]^{2+}$ ion has been changed into Fe^{3+} and $(SCN)^-$ ions. Thus, increasing the concentration of the *product* shifts the equilibrium in favour of the *reactants.*

Effect of change of temperature

Consider the equilibrium

$$N_2O_4(g) \rightleftharpoons 2NO_2(g); \; \Delta H = +59.0 \text{ kJ}$$

In this equilibrium, the reaction favouring the product (NO_2) is seen to be endothermic. Therefore, the opposing reaction favouring the reactant (N_2O_4) must be exothermic. Now suppose the system is *heated* and its temperature is allowed to rise. According to Le Chatelier's principle, the equilibrium will shift in the direction which tends to undo the effect of heat, *i.e.*, which tends to produce *cooling.* Therefore, the equilibrium will shift in favour of NO_2, *i.e.*, the dissociation of N_2O_4 into NO_2 will increase. If the system is cooled, on the other hand, the equilibrium will shift in the direction which tends to produce *heat.* Therefore, the equilibrium will shift in the reverse direction, *i.e.*, in favour of N_2O_4. The dissociation of N_2O_4 will decrease.

The combination of nitrogen and hydrogen to give ammonia is represented by the following thermochemical equation :

$$N_2(g) + 3H_2(g) \rightleftharpoons 2NH_3(g) \; ; \; \Delta H = -92.38 \text{ kJ}$$

The forward reaction is *exothermic.* Therefore, the back reaction must be endothermic. By Le Chatelier's principle, the increase of temperature will favour the back reaction, *i.e.*, the dissociation of ammonia. Therefore, in order to get a better yield of ammonia, the temperature at equilibrium should be low. But since the reaction rate varies with the temperature, the time taken to reach the equilibrium state becomes very long if the temperature is kept low. Therefore, a temperature close to 500°C, which is neither too low nor too high, is maintained.

Effect of change of pressure

Consider another gaseous reaction involving *combination of nitrogen and hydrogen to give ammonia:*

$$\underbrace{\underset{1\text{ mole}}{N_2(g)} + \underset{3\text{ moles}}{3H_2(g)}}_{4\text{ moles}} \rightleftharpoons \underset{2\text{ moles}}{2NH_3}$$

As is evident from the above equation, the forward reaction is accompanied by a decrease in the number of moles. If the pressure is *increased,* the volume will decrease and, therefore, *the number of moles per unit volume will increase.* According to Le Chatelier's principle, therefore, the equilibrium will shift in the direction in which there is *decrease* in the number of moles, *i.e.*, in favour of the formation of ammonia. Thus, *the higher the pressure, the greater would be the yield of ammonia.* It is on account of this consideration that a pressure of about 200 atmospheres is maintained in the Haber's process for the manufacture of ammonia.

The combination of sulphur dioxide and oxygen to give sulphur trioxide:

$$2SO_2(g) + O_2(g) \rightleftharpoons 2SO_3(g)$$

is also accompanied by a *decrease* in the number of moles. According to Le Chatelier's principle, therefore, a high pressure would give a better yield of sulphur trioxide.

The dissociation of calcium carbonate

$$CaCO_3(s) \rightleftharpoons CaO(s) + CO_2(g)$$

is accompanied by an *increase* in the number of gaseous moles. The dissociation of calcium carbonate would, therefore, be *suppressed* by an *increase* of pressure.

Q. 32. State Le Chatelier's principle and apply it for the following equilibrium :

$$\mathbf{PCl_5\ (g) \rightleftharpoons PCl_3\ (g) + Cl_2\ (g) - 17k\ cal}$$ (*Calcutta 2007*)

Ans. For the statement of Le Chatelier's principle, see Q 31

$$PCl_5(g) \rightleftharpoons PCl_3\ (g) + Cl_2\ (g) - 17\text{ k cal}$$

The above reaction is endothermic and takes place with increase of pressure (number of gaseous product molecules is greater than the number of gaseous reactant molecules) in the forward direction.

Effect of pressure. On increasing the pressure, the reaction will shift in a direction where the effect can be neutralized i.e. in a direction where the pressure is decreased. Thus, the reaction will move in the backward direction. On decreasing the pressure, the reverse effect will be observed.

Effect of temperature. On increasing the temperature, the reaction will shift in a direction where it is endothermic i.e. heat can be absorbed. Thus, the reaction will shift in the forward direction. On decreasing the temperature, however, the reverse action will take place.

Q. 33. State Le-Chatelier's principle. Explain the effect of heat on ice $\rightleftharpoons$ water equilibrium. (*Vidyasagar 2007*)

Ans. For statement of Le-chatelier's principle, See Q. 31

$$\text{Ice} \rightleftharpoons \text{Water} - \text{Heat}$$

Thus, it is an endothermic reaction,

If we heat the ice - water system, it will shift in a direction where the heat can be absorbed i.e. in the forward direction. Hence, more ice will melt if we raise the temperature.

Q. 34. Differentiate between K_p and K_c. How are they interrelated ?

(G.Nanakdev 2000; Awadh 2000)

Ans. See Q.5 and 12.

Q. 35. For a chemical reaction involving no change in volume the equilibrium constant K_p and K_c are equal. Explain. *(Lucknow 2000)*

Ans. $K_p = K_c\,(RT)^{\Delta n}$

For no change in the volume of the reactants and the products, $\Delta n = 0$

Therefore, $K_p = K_c\,(RT)^o$ or $K_p = K_C$

Q. 36. Write the expression for K_p

$$2SO_2\,(g) + O_2(g) \rightleftharpoons 2SO_3\,(g)$$ *(G. Nanakdev 2000)*

Ans. See Q.21

Q. 37. Write the relation between K_p and K_c for the reaction :

$$N_2\,(g) + 3\,H_2\,(g) \rightleftharpoons 2\,NH_3\,(g)$$ *(Vidyasagar 2007)*

Ans. Calculation of Kp

Initial	a		b		0
no. of moles	N_2	+	$3H_2$	$\rightleftharpoons$	$2\,NH_3$
Moles at Equil	$a - x$		$b - 3x$		$2\,x$

Total No. of moles at equilibrium $= (a - x) + (b - 3x) + 2x$

$= a + b - 2x$

Suppose, the total pressure of the system = P atmosphere

Partial pressure of N_2 $= \dfrac{P(a-x)}{a+b-2x}$

Partial pressure of H_2 $= \dfrac{P(b-3x)}{a+b-2x}$

Partial pressure of NH_3 $= \dfrac{P.2x}{a+b-2x}$

$$\text{Kp} = \frac{p_{NH_3}^2}{p_{N_2} \times p_{H_2}^3} = \frac{\dfrac{p^2.4x^2}{(a+b-2x)^2}}{\dfrac{p(a-x)}{a+b-2x} \times \dfrac{p^3(b-3x)^3}{(a+b-2x)^3}}$$

$$= \frac{4x^2(a+b-2x)^2}{(a-x)(b-3x)^3\,p^2}$$

Calculation of K_c

Let the total volume of the mixture at equilibrium = V

$$[N_2] = \frac{a-x}{V},\ [H_2] = \frac{a-3x}{V},\quad NH_3 = \frac{2x}{V}$$

$$K_c = \frac{[NH_3]^2}{[N_2]\,[H_2]^3} = \frac{\left(\dfrac{2x}{V}\right)^2}{\dfrac{(a-x)}{V}\left(\dfrac{b-3x}{V}\right)^3}$$

or $$K_c = \frac{4x^2V^2}{(a-x)\,(b-3x)^3}$$

20

COLLOIDAL STATE

Q.1. What is the difference between colloidal and true solutions?

(Garhwal 2000)

Ans. When water soluble substances like sodium chloride, copper sulphate, sugar etc., are put into water they dissolve and a homogeneous solution is obtained. The particles of the solute are not visible and their size is molecule size. Such mixtures are called *molecular solutions* or *true solutions.*

Now suppose we take muddy river water or an insoluble substance like lead sulphate or calcium sulphate is put into water. The particles of the solutes are visible even with the naked eye because their size is large. On keeping for some time, particles settle down. These mixtures are called *suspension.*

In between these two extremes, there are particles which are bigger than molecules but are too small to be seen even by a microscope. The colloidal state can thus be regarded as the intermediate stage between molecules and particles of a coarse suspension.

Q.2. Give in a tabular form the distinguishing characteristics of true solution, colloidal solution and suspension. *(Mizoram 2002)*

Ans. The distinguishing characters of the three types of solutions are given in table 20.1.

Table 201. Difference between a true Solution, colloidal Solution and suspension.

True Solution (1)	*Colloidal Solution* (2)	*Suspension* (3)
(*i*) The solute molecules in a true solution are not visible, even with a powerful microscope.	(*i*) The molecules of the dispersed phase cannot be seen with naked eye. However they can be made visible through a powerful microscope called Ultramicroscope.	(*i*) The particles are so big that they are visible even with naked eye. They sometimes settle at the bottom of the container, when the mixture is allowed to stand.
(*ii*) The particle size is below 1 nm in diameter.	(*ii*) The particle size is 1-100 nm in diameter.	(*ii*) The particle size is above 100 nm in diameter.
(*iii*) A true solution forms a homogeneous mixture.	(*iii*) A colloidal solution is heterogeneous.	(*iii*) Suspension is completely a heterogeneous mixture.

True Solution (1)	*Colloidal Solution* (2)	*Suspension* (3)
(*iv*) The particles of the solute in a true solution can pass through a parchment membrane.	(*iv*) The particles of a colloidal solution can not pass through a parchment membrane.	(*iv*) The suspension particle can not pass through the parchment membrane.
(*v*) The solute molecules in true solution can pass through even fine filter papers.	(*v*) The colloidal particles can also pass through ordinary filter papers, but can not pass through filter papers called Ultrafilters.	(*v*) The suspended partic les are retained even by ordinary filter papers.
(*vi*) The osmotic pressure exerted by the solution is relatively more.	(*vi*) Colloidal solutions exert a very low osmotic pressure.	(*vi*) No osmotic pressure is observed.
(*vii*) When a strong beam of light is thrown on a true solution, the path of light is not illuminated.	(*vii*) The path of light is visible due to scattering of light, when a strong beam of light is passed through a colloidal solution (Tyndall effect).	(*vii*) No Tyndall effect is observed.
(*viii*) The solutions do not possess the property of adsorption (a surface phenomenon).	(*viii*) Colloidal solutions possess remarkable property of adsorption.	(*viii*) The phenomenon of adsorption is much less marked.
(*ix*) No electrical properties such as electroosmosis, cataphoresis and coagulation are observed.	(*ix*) Colloidal solutions exhibit these electrical properties to a great extent.	(*ix*) No such electrical properties are exhibited.

Q.3. What are various phases in a colloidal solution? *(Kerala 2000)*

Ans. A colloidal solution is a heterogeneous system consisting of two phases viz., the dispersed phase and the dispersion medium.

(*a*) Dispersed phase. It is just like the solute in a solution. It can be said that the component which is present in smaller proportion is the dispersed phase. For example in a colloidal solution of starch in water, starch is the dispersed phase.

(*b*) Dispersion medium. It is just like the solvent in a true solution. It is the component which is present in greater proportion. In the example of colloidal solution of starch in water, water is the dispersion medium.

It should be noted that particles of dispersed phase (colloidal particles) are bigger than those of dispersion medium (molecules or ions).

Q.4. Give the classification of colloids on the basis of physical state of dispersed phase and dispersion medium.

Ans. Each of the dispersed phase and the dispersion medium can be a solid or liquid or gas. There are eight different combinations possible. Table 20.2 describes the various colloids in which the dispersed phase and dispersion medium may have the same or different physical states. A colloidal solution with both the components in the gaseous state is not possible as according to Graham's Law of diffusion, they form a homogeneous mixture.

Table 20.2. Classification of colloids based on physical state of the phases.

S. No.	*Dispersed phase*	*Dispersion medium*	*Name*	*Examples*
1.	Solid	Solid	Solid sol	Coloured glass, alloys, gems.
2.	Solid	Liquid	Sol	Paints, inks, white of egg,colloidal Au, colloidal Pt.
3.	Solid	Gas	Aerosol	Smoke, dust.
4.	Liquid	Solid	Gel	Curds, cheese, jelly, pudding.
5.	Liquid	Liquid	Emulsion	Milk, butter, oil in water.
6.	Liquid	Gas	Aerosol	Mist, fog, clouds.
7.	Gas	Solid	Solid foam	Cake, bread, pumice stone.
8.	Gas	Liquid	Foam	Soap lather, aerated water.

Q. 5. Give the classification of colloids on the basis of appearance.

Ans. On this basis, the colloids are divided into the following two main categories :

(*i*) **Sols.** When a colloidal solution appears a fluid, it is termed as sol. Sols are named after the dispersion medium. For example when dispersion medium is water, they are called **hydrosols**, if dispersion medium is alcohol they are called **alcosols** and if it is benzene, they are called **benzosols.**

(*ii*) **Gels.** When a colloid has a solid like rigid appearance it is called gel. The rigidity of a gel varies from substance to substance. Some substances may occur both as sols as well as **gels**. This depends upon the relative concentrations of the dispersed phase and dispersion medium.

The colloidal system in which the dispersed phase is a liquid and dispersion medium is solid is known as gel. A gel is formed by interlocking of the dispersed phase particles. Gelatin dissolves in warm water to form a colloidal solution which sets into a gel.

Q.6. Classify colloidal solutions on the basis of solvent affinity.

Or

What are lyophilic and lyophobic colloids? What are the points of difference between the two?

(Awadh 2000; Meerut 2000; Lucknow 2000; Kerala 2000; Nagpur; 2003; Arunachal 2003)

Ans. When the dispersion medium exerts an attraction on the dispersed phase then we get **lyophilic Sol** (solvent loving). When the attraction between the dispersion medium and the dispersed phase is very little then the sol is called **lyophobic sol (solvent hating).**

Typical examples of lyophobic sols (non-reversible): Metals, sulphur, sulphides, silver halides, egg albumen, silicic acid and ferric hydroxide give non-reversible colloidal systems with water.

Examples of lyophilic (reversible) sols: Protein, starch, glue, gelatin and agar-agar give reversible colloidal system with water. Rubber in benzene is also a reversible colloidal system. Main points of difference between the two are given in Table 20.3.

Table 20.3. Difference between lyophilic and lyophobic colloids.

Lyophilic (Solvent loving)or Hydrophilic Colloids (water as solvent)	*Lyophobic (Solvent hating) or Hydrophobic Colloids (water as solvent)*
(*i*) Lyophilic colloids pass into the colloidal form readily when brought in contact with solvent.	(*i*) Lyophobic colloids do not form the colloidal solutions easily when treated with solvent.

Lyophilic (Solvent loving) or Hydrophilic Colloids (water as solvent)	Lyophobic (Solvent hating) or Hydrophobic Colloids (water as solvent)
(*ii*) They are called the reversible sols because they can be recovered from the colloidal solutions and can be converted into the colloidal form when desired.	(*ii*) They are called irreversible sols, because they can not be recovered from their colloidal form.
(*iii*) Small quantities of electrolytes do not effect precipitation, but only large quantities of electrolytes cause precipitation.	(*iii*) Even small quantities of the electrolytes can cause precipitation.
(*iv*) The particles are not easily detected in the ultra-microscope.	(*iv*) The particles are easily detected by an ultramicroscope.
(*v*) The particles may or may not migrate in an electric field. The migration may be in any direction.	(*v*) Particles migrate only in one direction in the presence of electric field.
(*vi*) Substances like starch, proteins, gums, soaps, and metasilicic acid are common examples.	(*vi*) Colloids of metals (Au, Pt), sulphur, assenius sulphide and silver iodide are common examples of this class.

Q. 7. Write notes on the following

(*i*) Dispersion colloids

(*ii*) Multimolecular or association colloids

(*iii*) Macromolecular or molecular colloids.

Ans. (*i*) **Dispersion colloids.** In case of colloidal solutions in which the dispersed phase has little affinity for the dispersion medium (*i.e.* solid substance is insoluble in the dispersion medium), the size of the solid particles (dispersed phase) lies in the colloidal range. This system is called a dispersion colloid. For example, gold sol, hydrated ferric oxide sol, colloidal sulphur, a colloidal dispersion of As_2O_3 in water etc. Generally this type belongs to **lyophobic sols.**

(*ii*) **Multimolecular or association colloids.** Here an individual colloidal particle consists of an aggregate of a large number of small atoms or molecules. The size of such species lies in the colloidal range. The species thus formed are called associated colloids or multimolecular colloids, for example, soap solution, dyes, surface active agents etc. Generally in this type, there is occurrence of lyophobic and lyophilic portions in the same molecules. Thus the **aggregates of ions or molecules with lyophilic as well as lyophobic parts are called micelles.** Micelles are of colloidal dimension though the individual ions or molecules may not be of the colloidal dimensions.

(*iii*) **Macromolecular or molecular colloids.** In these systems each particle of the dispersed phase is of colloidal dimensions. Mostly this type consists of lyophilic sols. This resembles true solutions in some respects. Examples of this type include proteins, cellulose, the solutions of rubber and high polymers in organic solvents etc.

Q. 8. Where are micelles?

Ans. There are certain strong electrolytes which give a normal solution at low concentration but show colloidal nature at higher concentrations. These are called **micelles.** These substances may also be referred to as *Associated Colloids.* Soap is an example of this class. It forms a miscelle. These substance are commonly named as *surface active substances.* They consist of lyophilic as well as lyophobic parts in the same molecule. A long hydrocarbon chain serves as lyophobic and an ionic group like carboxy or sulphonic serves as lyophilic. A micelle may contain as many as 100 molecules.

Q. 9. Describe the cleansing action of soaps on the basis of micelle formation.

Ans. Water is not a good medium for cleaning our body or clothes. This is because of the high surface tension of water. It does not spread evenly over the dirt particularly grease in the clothes or on the body. The exceptionally high value of surface tension of water could be explained in terms of hydrogen bonding. Without complete wetting of grease by water, we can not expect efficient cleaning. The function of soap or other synthetic detergents is to decrease the surface tension of water. This is made possible by micelle formation as shown in Fig. 20.1.

Ordinary soap is sodium stearate $C_{17}H_{35}COONa$. It gives $C_{17}H_{35}COO^-$ in solution. The hydrocarbon chain consisting of 17 carbon atoms is hydrophobic in nature. A number of such hydrocarbon chains are directed towards grease as shown in Fig. 20.1. the negative poles of the stearate ions then attract the positive ends of water (water is a polar molecule) forming a miscelle.

The micelles are then washed away with more water. Thus soap helps in the cleaning of clothes or body. The function of soap is to reduce grease or oil to colloidal particle. therefore, soaps

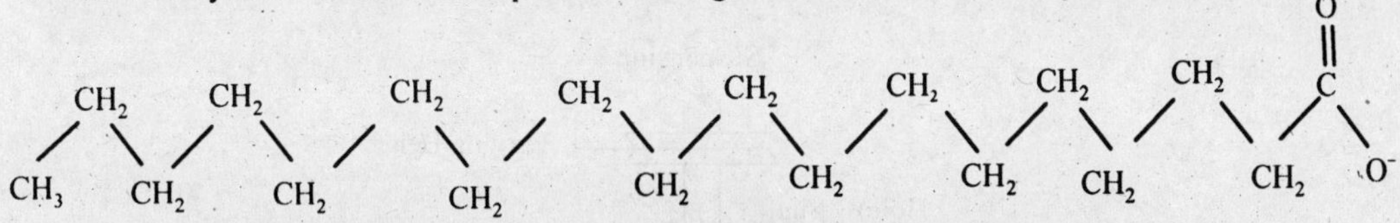

$(C_{17}H_{35}COO^-)$

THE ANION OF A SOAP IS REPRESENTED AS FOLLOWS

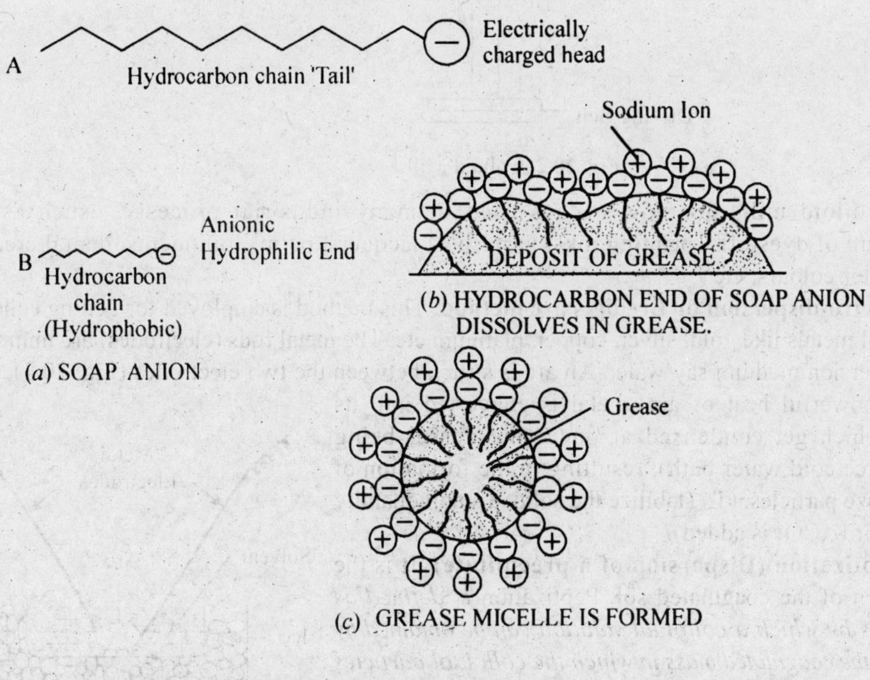

Fig. 20.1 A. Representation of a soap molecule,
B. Formation of a micelle.

are emulsifying agents. Grease particles are prevented from coming closer because of the covering by soap molecules. Any synthetic detergent like sodium lauryl sulphate $C_{17}H_{25}SO^-_4\ Na^+$ would form a miscelle by a similar mechanism.

Q. 10. Describe dispersion (size reduction) methods for the preparation of colloidal solution. *(Lucknow 2002; Venkateswara 2002)*

Ans. Dispersion methods for the preparation of colloidal solution are described as under:

1. Mechanical dispersion or disintegration. It involves the following steps:

(*a*) First of all, the substance is powdered throughly.

(*b*) The finely ground substance is brought in the form of coarse suspension, by shaking with the dispersion medium.

(*c*) The suspension is then made to pass through a colloid mill (Disc Mill), which consists of two metallic discs almost touching each other and rotating with a terrific speed in opposite directions (Fig. 20.2). The suspension is thus subjected to a powerful shearing force, resulting in the breaking of suspended particles to form particles of colloidal dimensions.

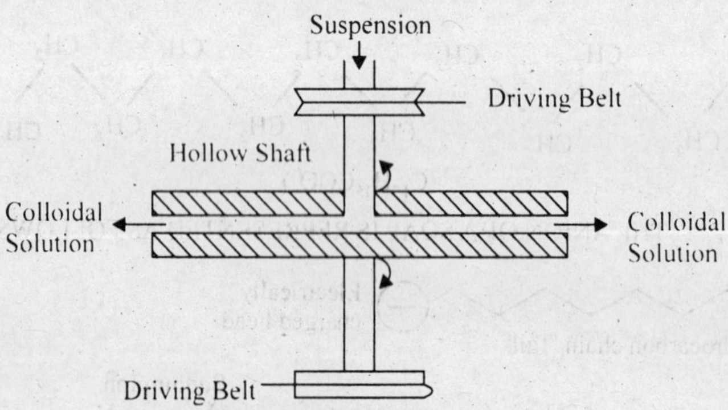

Fig. 20.2. Colloid Mill.

The colloid mills are now a days used in many industrial processes, such as the manufacture of dyestuffs, printing inks, varnishes, lacquers, paints, ointments, dental creams, oil and water colours, etc.

2. Electro-dispersion or Bredig's arc method. This method is employed for getting colloidal solutions of metals like gold, silver, copper, platinum etc. The metal rods (electrodes) are immersed in the dispersion medium say water. An arc is struck between the two electrodes (Fig. 20.3). As a result of powerful heat of arc, metal is converted into its vapours which get condensed at once (the vessel being placed in ice cold water bath), resulting in the formation of colloidal size particles. (To stabilize the solid, a small quantity of NaOH or K_2CO_3 is added).

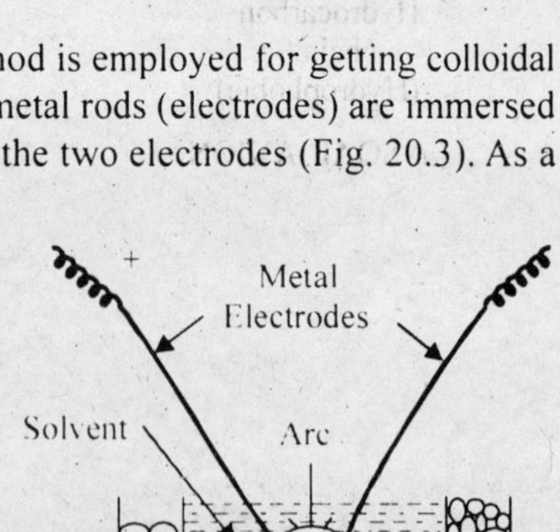

Fig. 20.3. Bredig's arc Method.

3. Peptization (Dispersion of a precipitate). It is the redispersion of the coagulated sol. Peptization *is defined as the process by which a colloidal solution can be obtained by dispersing the coagulated mass in which the colloidal particles pre-exist.* For example, ferric hydroxide is a coagulated sol and is a precipitate. On adding ferric chloride solution (Peptizing agent) to it, a reddish brown solution containing ferric hydroxide sol is obtained.

The excess of ferric chloride, if any, can be removed by dialysis. Similarly a precipitate of silver iodide can be peptised by shaking with a dilute solution of either KI or $AgNO_3$ when very dilute HCl acid is added to the freshly precipitated aluminium hydroxide, $Al(OH)_3$, a colloidal solution of aluminium hydroxide is obtained. The acid should be in small quantity.

Q. 11. Describe condensation or aggregation methods for the preparation of colloidal solutions.

Ans. Condensation methods for the preparation of colloidal solutions are described as under:

1. Excessive cooling. This method is employed to get colloidal solution of ice in organic solvent like ether and chloroform. This is done by freezing a solution of water in the given solvent. The molecules of water can no longer be held in solution. They collapse together to yield colloidal sized particles.

2. Exchange of solvent. There are certain substances which are more soluble in one solvent as compared to another. Colloidal solution of such substances can be obtained by pouring the solution of the substance in which it is more soluble to the other solvent, in which it is soluble to a smaller degree *e.g.*,

(*a*) Sulphur and phosphorus are more soluble in alcohol than in water. If their alcoholic solution is added to small quantity of water, hydrosol results.

(*b*) Phenolphthalein is soluble in alcohol but not in water. If its alcoholic solution is added to water, milky solution is obtained which is a sol of phenolphthalein.

Q.12. Describe chemical methods for the preparation of colloidal solutions.

Ans. Chemical methods for preparing colloidal solutions are given below.

(*i*) **Double decomposition.** A colloidal solution of arsenious sulphide is obtained by adding water saturated with H_2S gas, to a solution of arsenius oxide (one gm. in 500 ml of water).

$$As_2O_3 + 3H_2S \longrightarrow As_2S_3 + 3H_2O$$

(*ii*) **Hydrolysis.** A colloidal solution of **ferric hydroxide** is obtained by adding 2 or 3 ml of 30% solution of ferric chloride to about 500 ml of hot boiling water.

$$FeCl_3 + 3H_2O \rightleftharpoons \underset{\text{(deep red)}}{Fe(OH)_3} + 3HCl$$

(*iii*) **Oxidation.** A colloidal solution of sulphur is obtained by passing H_2S gas through a solution containing some oxidising agent, say dilute HNO_3 or by passing H_2S into a solution of sulphur dioxide in water.

$$H_2S + [O] \longrightarrow H_2O + S$$

$$SO_3 + 2H_2S \longrightarrow 3S + 2H_2O$$

(*iv*) **Reduction.** A colloidal solution of gold (violet coloured) may be obtained by adding stannous chloride solution (reducing agent) to a solution of gold chloride.

$$2AuCl_3 + 25nCl_2 \longrightarrow 3SnCl_4 + \underset{\text{(violet)}}{2Au}$$

Q.13. How will you purify colloidal solutions?

Or

What is dialysis? *(Garhwal 2000; Lucknow 2002; Calcutta 2007)*

Ans. A sol contains some electrolyte during its preparation. A small amount of electrolyte is essential for stability of the sol, but excess of it can cause coagulation. Therefore, it is essential to remove excess of electrolyte.

The purification of the sol is done by the following process.

Dialysis. The process of removing the dissolved electrolytes from the sol by means of a membrane is called dialysis.

A dialyser consists of a bag made of parchment or cellophane and suspended in running water in a trough. The mixture of the colloid and the crystalloid is taken in the bag and apparatus fitted as shown in Fig. 20.4 (a). After sometime, whole of the crystalloid in solution passes out while the colloid remains in the bag. It is further observed that the above process of separation can be accelerated if an electric field is applied around the parchment bag. The process is then called **electrodialysis.** This is shown in Fig. 20.4 (*b*).

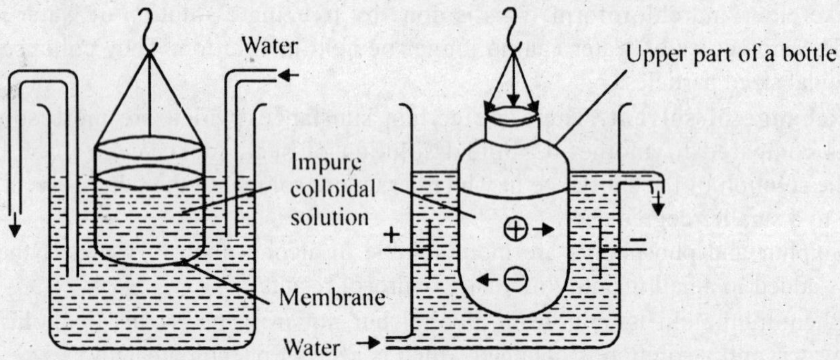

Fig. 20.4. (*a*) Dialysis (*b*) Electrodialysis

Explanation. Because of their bigger size, colloidal particles are unable to pass through the parchment bag while ions of the crystalloid are carried away by the running water in the dialyser. Further on applying the electric field, the movement of the ions is accelerated.

An important application of dialysis is in the purification of blood in the artificial kidney machine. The dialysis membrane permits small particles to pass through but colloidal haemoglobin does not pass through.

Q. 14. What is Tyndall effect ?

(Nagpur, 2003; Purvanchal 2007; Himachal 2010; Madras 2011)

Ans. It is an optical property of colloidal solutions. A beam of light entering a dark room lights up the dust particles floating in air. Similarly *when a strong beam of light is concentrated on a colloidal solution the path of the beam is illuminated by a bluish light* (and becomes visible when observed from the side). This phenomenon is known as *Tyndall effect* (Fig. 20.5).

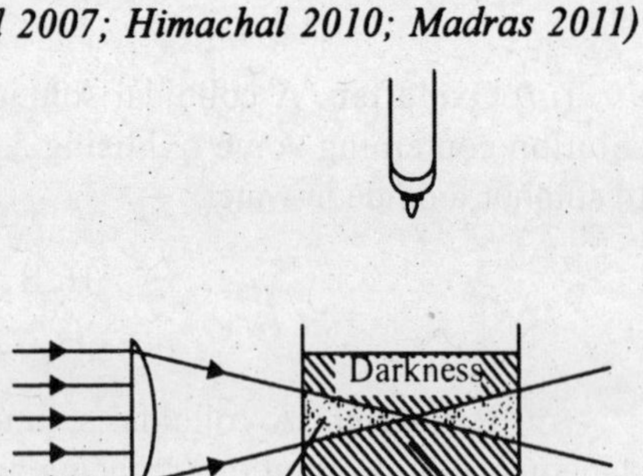

Fig. 20.5. Tyndall Effect.

The scattering of light by the colloidal particle is responsible for this phenomenon. The intensity of scattered light depends upon the difference between refractive indices of the dispersed phase and dispersion medium. Since this difference is quite appreciable in case of lyophobic colloids, they show well-defined Tyndall effect. This phenomenon has been employed as the basic principle for the construction of *ultra microscope.*

It is this effect which is made use of in the 'test box' in testing the purity of SO_2 for the manufacture of sulphuric acid by the Contact process.

Q.15. Discuss the following mechanical properties of colloidal solutions:
(*i*) Brownian movement;
(*ii*) Diffusion; (*Lucknow 2000; Nagpur 2008; Madras 2011*
(*iii*) Sedimentation.

Ans. (*i*) **Brownian movement.** If we examine the colloidal solutions with an ultramicroscope, we find that suspended particles are in a constant rapid *zig-zag motion* (Fig. 20.6). This zig-zag motion is called *Brownian movement.* Robert Brown first noticed this movement while examining pollen grains suspended in water under a microscope.

The kinetic theory gives a nice explanation of the phenomenon. The motion is due to unequal bombardment by the molecules of dispersion medium. With increase in size of particles, the

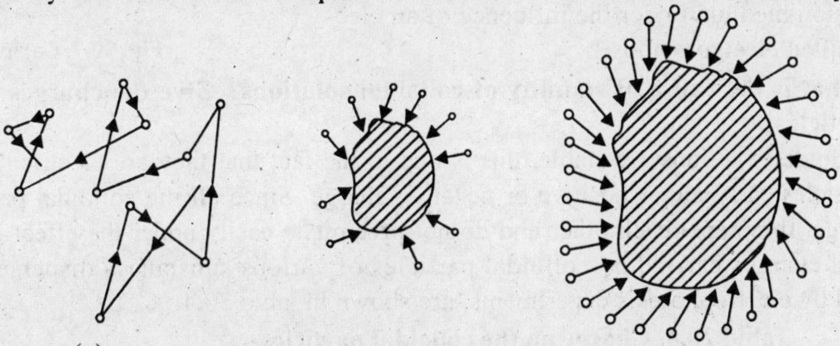

(*a*) (*b*) (*c*)

Fig. 20.6. The Brownian Movement.

probability of unequal bombardment diminishes and so Brownian movement decreases with increase of particle size.

This motion counteracts the force of gravity acting on the colloidal particles and is thus responsible to a certain extent for the stability of colloids.

(*ii*) **Diffusion.** Since the rates of diffusion decrease very rapidly with increase in the mass of the particles (Graham's law), colloidal solutions would diffuse much more slowly than ordinary molecular solutions. This property has been made use of in determining molecular weights of colloids. Since the equation to be used for the purpose also has a term for the *Avogadro's number* (the number of molecules in a gram molecule), the latter can also be determined from it, using solutes of known molecular masses.

(*iii*) **Sedimentation of suspensoids.** Although colloidal solutions are found to be fairly stable over considerable periods of time, they do very slowy settle down under the influence of gravity—even though some of them may take years to do so. The rates of settling [or sedimentation, as it is called] can be readily determined because the boundary line between the clear medium and the sol can be easily made out.

Q. 16. Write notes on
(*i*) Cataphoresis or electrophoresis (*Awadh 2000; Meerut 2000; Kerala 2000*)
(*ii*) Electro-osmosis. (*Lucknow 2002; Himachal 2010*

Ans. (*i*) Cataphoresis or Electrophoresis. Colloidal particles carry an electric charge and therefore, move towards the oppositely charged electrode when the solution is placed under the influence of an electric field (Fig. 20.7). This *migration of colloidal particles under the influence of electric field is called cataphoresis.*

The type of charge on sol particle can be determined by noting the direction of their motion. The important applications of cataphoresis are:

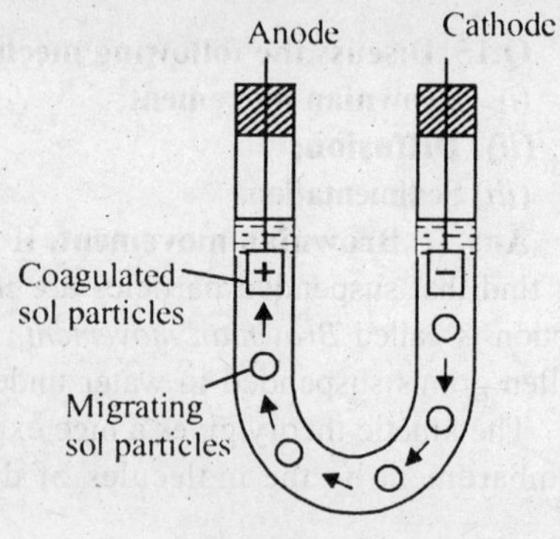

Fig. 20.7. Cataphoresis.

(a) Removal of colloidal gases.

(b) Removal of dirt particles.

(c) Electro-deposition of rubber.

(ii) **Electro-osmosis.** A sol is electrically neutral. This reveals that dispersion medium carries an equal but opposite charge to the dispersed particles. If the dispersed phase is kept stationary by some suitable means, it is observed that the dispersion medium itself begins to move under the influence of an electric field. This motion of the dispersion medium under the influence of an electric field is **called electro-osmosis.**

Q.17. What is the cause of stability of colloidal solutions? Give the charges on various colloidal particles.

Ans. Colloidal solutions are stable, this is due to the fact that they are electrically charged. Colloidal particles carry either positive or negative charge. Since all the colloidal particles bear identical charge, they repel each other and do not precipitate easily under the effect of gravity. The electrical charges carried by colloidal particles of various substances dispersed in water, as determined by electrophoresis experiments, are shown in table 20.4.

Tabld 20.4. Charge on the colloidal particles.

Positively Charged	*Negatively Charged*
1. Haemoglobin	1. Clay
2. Colloidal particles of Fe $(OH)_3$, Cr $(OH)_3$ and Al $(OH)_3$,	2. Metallic particles like those of gold, silver and platinum
3. Basic dye stuffs.	3. Sulphides of Cu, Pb, As, Sb, and Cd.
	4. Silicic acid, stannic acid and vanadium pentoxide.
	5. Gum-Arabic and soluble starch.
	6. Acid dye stuffs.

Q.18. Discuss origin of charge on colloidal particles.

(Arunachal 2002; Nagpur 2003; Garhwal 2010; Mumbai 2010)

Ans. The electric charge is acquired by sol particles by one or more of the following means:

(i) **Friction.** The charge on the sol particles appears as a result of rubbing of the particles of dispersed phase against those of dispersion medium.

(ii) **Selective adsorption.** The sol particles have a tendency to adsorb anions or cations from the solution. For example, the colloidal particles of As_2S_3 can adsorb either As^{3+} or S^{2-} ions. Since during the precipitation of this sol, only sulphide ions from H_2S are in excess, therefore on a colloidal particle of As_2S_3, the S^{-2} ions get adsorbed and the particles become negatively charged.

Similarly, the sol of Fe $(OH)_3$ obtained by the hydrolysis of $FeCl_3$ is positively charged. In the medium, are present $Fe(OH)_3$ particles, Fe^{3+} and Cl^- ions. Fe^{3+} ions get adsorbed on the surface of Fe $(OH)_3$ and particles become positively charged.

The AgCl particles can adsorb Cl^- ions from chloride solution and Ag^+ ions from solution containing Ag^+ ions. The colloidal particles becomes negatively charged in the first case and positively charged in the second case.

(*iii*) **Electron capture.** During electro-dispersion of the metals in Bredig's arc method.

(*iv*) **Dissociation of surface molecules.** Soap molecules dissociate to give ions. The cations pass into solution whereas the anions ($C_nH_{2n+1}COO^-$) aggregate together to form a negatively charged big particles called micelle.

(*v*) **Presence of acid or basic group.** Protein molecules yield colloidal solutions. Since a protein molecule contains an acidic group—COOH and also a basic group—NH_2, it will give a positive colloidal particle in acidic medium and a negative particle in alkaline medium.

In Acid :

$$H_2N-X-COOH+HCl \longrightarrow H^+{}_3N-X-COOH+Cl^-$$

In Alkali :

$$H_2N-X-COOH + NaOH \longrightarrow H_2N-XCOO^- + Na^+ + H_2O.$$

Q.19. Describe the theory of electrical double layer to account for the charge on the colloidal particles. *(Guwahati 2000; Meerut 2002)*

Or

What is zeta potential? How does it explain the stability of colloidal particles.

Ans. The electrical properties of colloids can be explained satisfactorily by postulating the existence of an electrical double layer of opposite charges at the surface of separation between a solid and a liquid, *i.e.*, at a *solid-liquid interface.*

According to the modern views, when a solid is in contact with a liquid, a double layer of ions appears at the surface of separation, as shown in Fig. 20.8. One part of the double layer is fixed on the surface of the solid. This is known as the *fixed part* of the double layer. It consists of either positive ions or negative ions. The second part consists of a *diffuse* or a *mobile layer* of ions which extends into the liquid phase. This layer consists of ions of both the signs but its net charge is equal and opposite to that on the fixed part of the double layer. Such an arrangement of charges is shown in Fig. 20.8. In one case (Fig. 20.8 *a*), the fixed part of the double layer carries positive charge while in the second case (Fig. 20.8 *b*), the fixed part carries negative charge. It will be seen that in each case the net charge on the diffuse layer is equal and opposite to that on the fixed layer.

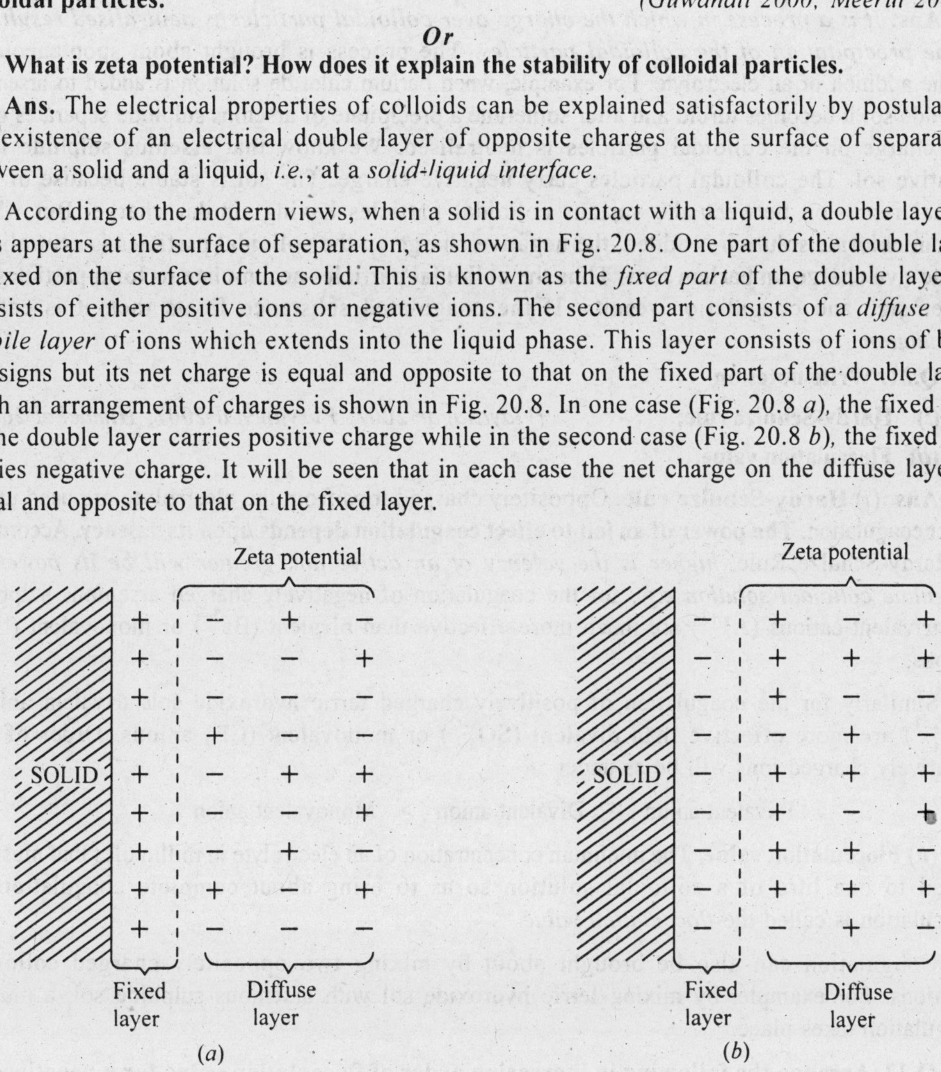

Fig. 20.8. The electrical double layer.

equal and opposite to that on the fixed layer.

This theory is applicable to colloidal systems also. The ions which are preferentially adsorbed by the colloidal particles, are held in the fixed part of the double layer. These ions give the characteristic charge to the colloidal particles. The ions carrying the opposite charge, *i.e.*, the counter ions, are present more in the diffuse portion of the double layer, giving a net opposite charge to this layer.

The existence of charges of opposite signs on the fixed and diffuse parts of the double layer leads to the appearance of a difference of potential between the two layers, as shown in Fig. 20.8. This difference of potential between the two layers is known as **zeta potential** or **electrokinetic potential.** Thus this theory explains the charge and hence the stability of the colloidal particles.

Q. 20. What is meant by coagulation? Which salt is the best for the coagulation of arsenius sulphide? *(Raipur 2010)*

Ans. *It is a process in which the charge over colloidal particles is neutralised resulting in the precipitation of the colloidal particles.* The process is brought about spontaneously by the addition of an electrolyte. For example, when barium chloride solution is added to arsenius sulphide sol it becomes turbid and after sometime a precipitate of arsenius sulphide separates out. The charge on the colloidal particles is neutralised. We know that arsenius sulphide is a negative sol. The colloidal particles carry negative charge. The sol is stable because of the mutual repulsion between the various colloidal particles having similar charge. But when barium chloride solution is added, the negative charge on the colloidal particles is neutralised by the +ve charge on barium ions. Thus the colloidal particles accumulate to form particles of larger size and coagulation results. If the coagulated substance floats then it is called *flocculation.*

Q.21. Write notes on

(*i*) Hardy-Schulze rule; *(Vidyasagar 2007; Purvanchal 2007; Himachal 2010)*

(*ii*) Flocculation value.

Ans. (*i*) **Hardy-Schulze rule.** Oppositely charged ions from the electrolyte are used up to effect coagulation. **The power of an ion to effect coagulation depends upon its valency.** According to Hardy-Schulze Rule, *higher is the valency of an active ion, greater will be its power to coagulate colloidal solution e.g.,* for the coagulation of negatively charged arsenious sulphide sol, trivalent cations (Al^{+3}) are much more effective than bivalent (Ba^{2+}) or monovalent (Na^{+}) cations.

Similarly for the coagulation of positively charged ferric hydroxide sols trivalent anions (PO_4^{-3}) are more effective than divalent (SO_4^{-2}) or monovalent (Cl^{-}) anions. Order of the negatively charged ions will be as under :

Trivalent anion > Divalent anion > Monovalent anion

(*ii*) **Flocculation value.** The minimum concentration of an electrolyte in millimoles that must be added to one litre of a colloidal solution so as to bring about complete coagulation or flocculation is called the *flocculation value.*

Coagulation can also be brought about by mixing two oppositely charged colloidal solutions. For example, by mixing ferric hydroxide sol with arsenious sulphide sol, a mutual coagulation takes place.

Q.22. Arrange the following in increasing order of flocculation value for a negative sol.

$MgCl_2, Al_2(SO_4)_3, Na_3PO_4$

Electrolyte	*Cation*	*Charge*
$MgCl_2$	Mg^{2+}	+2
$Al_2(SO_4)_3$	Al^{3+}	+3
Na_3PO_4	Na^+	+1

Hence the flocculation value of the electrolytes will be in the order

$$Al_2(SO_4)_3 > MgCl_2 > Na_3PO_4$$

Q. 23. Write notes on

(*i*) Protective colloids. *(Garhwa 2010)*

(*ii*) Gold number. *(Vidyasagar 2007; Calcutta 2007; Nagpur 2008)*

Ans. (*i*) **Protective Colloids.** Lyophobic colloids are sensitive and they are likely to coagulate in the presence of electrolytes. But they can be stabilized if a small amount of a lyophillic sol like soap, gum, gelatine etc. is added to them. This is called *protective action.* Such colloids which prevent the coagulation of other colloids are called protective colloids. It is believed that increase in the stability of the lyophobic colloids is due to the coating of lyophillic colloids on the lyophobic colloids.

(*ii*) **Gold number.** The protective action of the colloid is measured in terns of gold number. Gold number is defined *as the number of milligram of the protective colloid which just prevents the coagulation of 10 ml of gold solution when 1 ml of a 10 % solution of sodium chloride is added to it.* Gold number of a few protective colloids are given below :

Protective/colloid	*Gold number*
Gelatin	0.005
Albumin	0.08
Starch	25

Q. 24. The coagulation of 100 ml of a colloidal solution of gold is completely prevented by addition of 0.25 g of starch to it after adding 1ml of 10% NaCl solution. Final out the gold number of starch . *(Purvanchal 2007)*

Ans. As per the definition of gold number, it is the number of milligrams of protective colloid (starch in this case), which prevents the coagulation of 10 ml of gold solution when 1ml of 10% NaCl solution is added to it.

In the present case

100 ml of colloidal solution requires to prevent coagulation = 0.25 g starch

= 250 mg starch

10 ml of colloided solution requires to prevent coagulation = 25 mg starch

Hence. gold number of starch = 25

Q. 25. I efine gold number. The gold numbers of A, B, C and D are 0.005, 0.05, 0.5 and 5 respectively. Which of these has the greatest protective action? Explain your answer. *(Mysore 2004)*

Ans. For the definition of gold number, See Q. 23. *A* has the greatest protective action as we require the minimum quantity of *A* to prevent the coagulation of the gold solution when sodium chloride solution is added to it.

Q. 26. What is an emulsion? What are different types of emulsions? How do you differentiate between them? *(Purvanchal 2002; Garhwal 2010)*

Ans. Colloidal solutions in which the dispersed phase as well as dispersion medium are liquids, are called emulsions. Emulsions are of two types.

1. **Oil in water emulsions** are those in which oil is the dispersed phase and water is the dispersion medium *e.g.*, milk is an emulsion of liquid fat in water.

2. **Water in oil emulsions** are those in which water is the dispersed phase and oil is the dispersion medium *e.g.*, butter, cod liver oil.

Distinction between oil-in-water and water-in-oil emulsions.

The two types of emulsions can be distinguished from each other by the following tests:

1. **Dye-test.** If on heating the emulsion with an oil soluble dye, emulsion gains the colour of the dye, it is a water-in-oil emulsion. If the emulsion does not catch the colour of the dye, it must be oil-in-water emulsion.

2. **Conductance method.** A small amount of an electrolyte is added to the emulsion. If the conductance increase, the emulsion is oil-in-water type. If there is no appreciable change, it is water-in-oil type.

Q. 27. What are emulsifiers? Discuss their acton.

Ans. Normally emulsions are not stable colloidal systems. On standing, dispersed phase particles assemble together and form a separate layer. To stabilize emulsions, certain substances known as emulsifying agents or emulsifiers are added.

It is believed that the emulsifier surrounds the oil droplets, thus preventing them to come closer. It prevents them from coagulating. The function of the emulsifiers is to act as a protective layer around the oil droplets of the emulsion thus increasing its stability as shown in Figure 20.9.

Emulsions tend to coagulate because of the interfacial tension between the components of the emulsions. This effect is pronounced when there is a big difference between the surface tensions of the two components. In a bid to minimise the interfacial tension, they minimise the area of contact between the two components and thus the globules of the oil (*e.g.*, in oil-in-water emulsions) come together, thus destabilising the emulsion. An emulsifier prevents such a situation to occur.

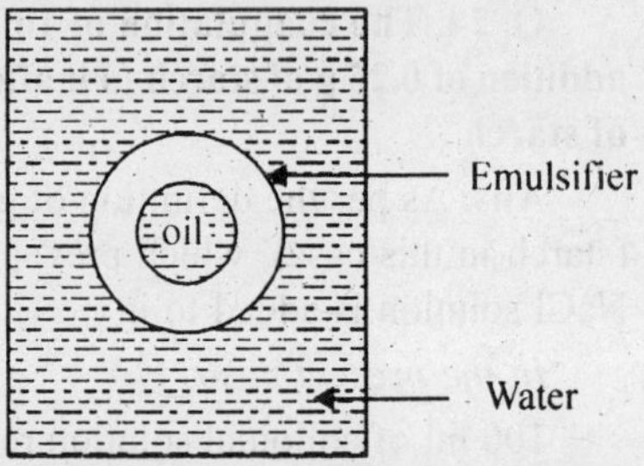

Fig. 20.9. Action of emulsifier.

Most commonly used emusifying agents are soaps and detergents. Other stabilizing agents are proteins and gums. Digestion of fats in the intestines is aided by emulsification. A little of the fat forms a sodium soap with the alkaline solution of the intestines and it emulsifies the rest of the fat thus making it easier for the digestive enzymes to do their work. Many of the lotions, creams and ointments are emulsions of oil-in-water or water-in-oil type. Emulsions find application in the concentration of ores also.

Q.28. What is demulsification?

Ans. Emulsion can be broken to recover the constituents by means of heating, centrifuging or by adding large amount of electrolytes to precipitate out the dispersed phase. This can also be achieved by chemical destruction of the emulsifying agent

The process of breaking emulsion to yield the constituents liquids is called demulsification. Separation of cream from milk is an example of demulsification by means of centrifugation.

Q. 29. What are gels? How are they formed? *(Madras 2011*

Ans. A gel is a colloidal system in which the liquid is the dispersed phase and solid is the dispersion medium. A gel is jelly-like. The process of formation of a gel is called *gelation.*

Gels are formed as a result of partial coagulation of a sol. The coagulating particles combine together to form thread like chains. Interlocking of these chains gives rise to a solid frame work. The liquid part of the sol gets trapped in the solid framework. We obtain a semi-solid porous mass called gel as shown in Fig. 20.10.

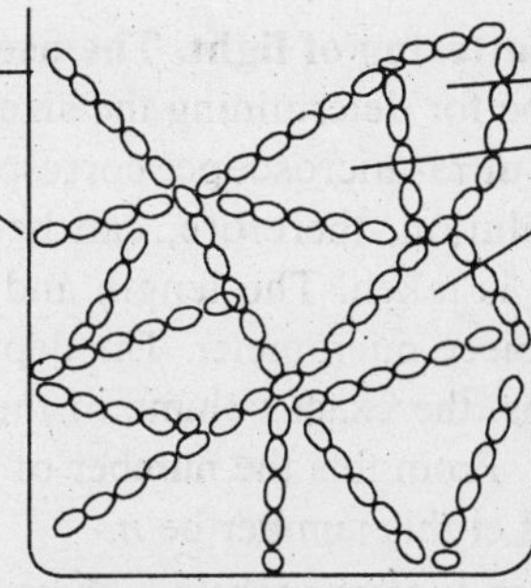

Fig. 20.10. Structure of a gel.

Q. 30. Give an account of classification of gels ? *(Purvanchal 2007)*

Ans. Gels may be classified into two types:

(*a*) Elastic gels: Gels which are elastic in nature are called elastic gels. Their shape changes when a force is applied on them. On releasing the force, the elastic gels regain their original shape. Gelatin, starch and soaps are examples of substances that can form elastic gels.

Such gels are formed by cooling concentrated solutions of lyophilic sols. There are no rigid forces of attraction between the particles of elastic gels.

(*b*) Non-elastic gels: Gel which are rigid and do not show elasticity are called non-elastic gels. Silica gel is an example of non-elastic gel. Such gels are obtained by a specific chemical reaction. Silica gel, for example, is obtained by the addition of conc. hydrochloric acid to sodium silicate solution of a particular concentration, silicic acid is formed which polymerises to form silica gel. Particles of such a gel are linked by covalent bonds which give the gel a rigid structure.

Q.31. Describe various methods to determine the size of colloidal particles.

Ans. Various methods are described as under :

1. By using ultrafilters. An approximate idea about the size of particles in a colloidal system can be obtained by the use of ultrafilters. These are prepared by impregnating filter papers with collodion or gelatin which are subsequently hardened by immersing in formaldehyde. The pores can be made small enough to retain particles of colloidal dimensions. The size of the pores depends on the particular filter paper employed and the concentration of the collodion or gelatin solution used for impregnating it. It is thus possible to obtain a series of graded ultrafilters by means of which a colloidal solution may be separated into fractions containing particles of different sizes. An approximate estimate of the size of particles can be obtained from a knowledge of the dimensions of the pores of the ultrafilters. The latter parameter is determined from the pressure required to force air or water through the pores.

2. From Brownian movement. Colloidal particles suspended in a liquid medium are subjected to Brownian movement. They also tend to settle down due to gravitation. Under the influence of both these effects, the particles distribute themselves in a vertical column in accordance with the equation

$$\frac{RT}{N_0}\ln\frac{n_1}{n_2} = \frac{4}{3}\pi r^3 (h_2 - h_1)(\rho - \rho')$$

Since N_0, the Avogadro's number, is known and since the number of particles n_1 and n_2 at two depths h_1 and h_2 of the vertical column can be counted with the help of an ultramicroscope and densities of the particles and of the liquid medium, ρ and ρ', can be determined by the usual methods, the radius of the particles r can be easily calculated.

3. **From scattering of light. The use of ultra-microscope.** Zsigmondy and others used the ultra-microscope for determining the size of the particles of colloidal dimensions. Each spot of light viewed in ultra-microscope, corresponds to a particle. The number of particles in a given volume of a solution, therefore, can be counted. The observation is repeated several times and an average is taken. The length and breadth of the field of vision are measured with the help of an eye-piece micrometer. The depth is determined by rotating the slit through 90°. From these dimensions, the exact volume of the solution containing the observed number of particles can be obtained. From this the number of particles contained per unit volume of the solution can be determined. Let this number be *n*.

The next step is to evaporate a known volume of the colloidal solution to dryness. From the weight of the residue, the mass of colloidal particles per unit volume can easily be obtained. Let this be *m* gram. The particles are supposed to be spherical and the density ρ of the colloidal particles is supposed to be the same as that in the bulk state. Thus the volume of colloidal phase is m/ρ.

$$\therefore \quad m/\rho = \frac{4}{3}\pi r^3 \times n$$

or

$$r = \left(\frac{3m}{4\pi\rho n}\right)^{\frac{1}{3}}$$

Q. 32. Describe various applications of colloids.

Ans. Colloidal solution and the charge on colloids find extensive applications in our daily life and various other processes. A few important applications of colloids are summarised below.

1. **Purification of drinking water.** Ordinary water contains suspended impurities of colloidal size which do not settle down due to their charged nature. On adding a little of potash alum, the positive charge on aluminium ions neutralises the charge on colloidal particles, they settle down and water is cleaned for drinking.

2. **Pollution control.** Chimneys of various industries bring out lot of smoke which acts as a health hazard. Smoke is a colloidal solution of carbon particles in air. This smoke is brought in contact with a conductor of opposite charge near the exit of the chimney. Smoke thus gets coagulated. Carbon particles settle down and carbonless fumes or air comes out of chimneys. The design is known as *Cottrell dust precipitator.*

3. **Medicines.** A large number of medicines are being produced in the form of sols these days. Such medicines are more effective as they have greater assimilation and adsorption qualities. It is found that ferric chloride solution checks excessive bleeding from a cut. It coagulates the blood due to it charged nature.

4. **Sewage disposal.** The sewage disposal carries dirt particles suspended in water. Separation of these particles from water is normally a difficult process. This difficulty is overcome by making the disposal water to come in contact with an electrode of opposite charge in a big tank. Dirt particles are precipitated and separated. They serve as a manure.

5. **Photography.** The photographic film contains a colloidal solution of silver bromide in gelatin. A photographic film can also be obtained on a glass plate by depositing a mixture of silver bromide and gelatin.

6. **Formation of deltas.** Delta is formed at a place where the river enters the sea. Sea water is rich in minerals whereas river water contains colloidal particles of clay. At the point of junction of the two waters, clay is deposited. This clay is fertile and is known as delta.

7. **Blue colour of the Sky.** Colour of a substance is the colour of the light transmitted by it. Colloidal particles of dust etc. scatter only the blue light and absorb the rest of the light. Hence blue light reaches us and the sky appears to be blue.

8. **Chemical warfare.** Smoke and mist screens are formed by the explosion of bombs. They are harmful to humans and gas masks which contain colloidal charcoal come to the rescue. They adsorb the toxic smoke.

Q. 33. Explain

(*i*) A colloidal solution is not precipitated on the addition of an electrolyte in the presence of gelatin;

(*ii*) A solution of ferric chloride is applied to a cut to stop bleeding;

(*iii*) Sky appears to be blue;

(*iv*) Alum is used in town water supply;

(*v*) $MgCl_2$ is a better coagulant than KCl for As_2S_3 sol.

Ans. For (*i*) See Q. 23.

For (*ii*) See Q. 32 under medicines.

For (*iii*) See Q. 32 under blue colour of sky.

For (*iv*) See Q. 32 under purification of drinking water.

For (*v*) See Q. 21.

Q. 34. Write short notes on 'Emulsion and Gels'.

(Kumaon 2000; Meerut 2000; Kerala 2000)

Ans. See Q.25, Q 29 and Q 30

Q.35. Explain why

(*a*) Alum is used in town water supply

(*b*) Alum is used is shaving

(*c*) Trudall cone is formed when a beam of light is concentrated on a colloidal solution

(Agra 2004)

Ans. For (*a*), See Q 32 under purification of water

For (*b*), See Q 32 under medicines. Feric chloride and alum have similar effect.

For (*c*), see Q. 14

Q.36. What is peptisation ? *(Calcutta 2007)*

(*a*) See Q 10 under peptisation

21

MOLECULAR SYMMETRY AND GROUP THEORY

Q. 1. What is meant by symmetry elements and symmetry operations ?

Ans. Symmetry Element. A symmetry element is an entity about which a symmetry operation is performed. A point, a line, or a plane are examples of symmetry elements.

Symmetry Operations. A symmetry operation is the movement of molecule about symmetry element in such a manner that the same configuration of molecule is obtained as the original. The molecule may assume an equivalent configuration or an identical configuration.

There are fine types of symmetry elements; E, C_n, σ, S_n and i.

1. E is called identity.
2. C_n is called n-fold proper axis of rotation.
3. σ_n is called plane of symmetry.
4. S_n is called n-fold improper axis of rotation.
5. i is called centre of symmetry. (or the inversion centre)

These symmetry elements and corresponding symmetry operations are listed in Table 21.1.

Table 21.1

Symmetry Elements and Symmetry Operations

Symmetry Elements	*Symmetry Operations*
1. Identity (E)	1. The operation leaves the molecule unchanged.
2. Proper axis of rotation (C_n)	2. Rotation by an angle $\theta = 2\pi/n$ about the axis.
3. Plane of symmetry (σ)	3. Reflection in the plane.
4. Improper axis of rotation or reflection axis, S_n,	4. Rotation about the axis followed by reflection in a plane perpendicular to the axis.
5. Inversion centre or centre of symmetry (i)	5. Inversion of all molecule through the inversion centre.

Q. 2. Explain the following with diagrams :

1. Proper Axis of Rotation (C_n)

2. Plane of Symmetry (σ_n) *(M. Dayanand 2000)*

3. Improper Axis of Rotation (S_n)

4. Centre of Symmetry (i)

5. Identity Element (E) *(Madras 2011*

Ans. 1. Proper Axis of Rotation C_n. A molecule is said to possess a proper axis of rotation of order n, if rotation about the axis by plane by angle $\theta = 2\pi/n$ leaves the molecule in a configuration which is appears same as original one. Consider, for instance, BF_3 molecule in which there is C_3 axis, *i.e.*, an axis of order 3, which is perpendicular to plane containing the three atoms and passing through the boron atom [Fig. 21.1].

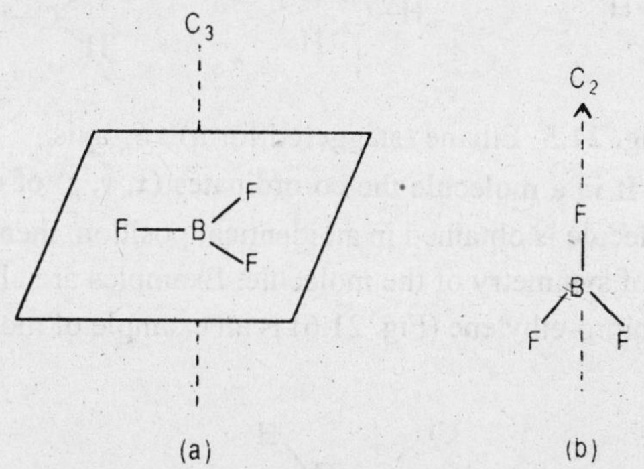

Fig. 21.1 (*a*) The C_3 axis of symmetry in BF_3 molecule.
(*b*) The C_2 axis of symmetry in BF_3 molecule.

The BF_3 molecule also possesses three C_2 axes of symmetry in addition to the C_3 axis. One such axis is shown in Fig. 21.1 (*b*). The C_3 axis is called the Principal Axis. In general, if molecule possesses C_n, axis of different orders, the axis of highest order is known as PRINCIPAL AXIS. The Principal Axis of benzene molecule is the six-fold axis, C_6 perpendicular to the regular hexagonal ring (Fig. 21.2).

Fig. 21.2. The 6-fold axis C_6 in benzene.

2. Plane of Symmetry (σ_n). A molecule possesses a plane of symmetry if reflection through the plane leaves the molecule unchanged. The plane of symmetry dividing the molecule into two equal halves, one part of which is the minor image of the others. For instance, BF_3 molecule possesses two plane of symmetry.

(*a*) The vertical plane (σ_v). Plane containing highest order rotation axis.

(*b*) The horizontal plane (σ_h). σ_h is perpendicular to C_3-axis (Fig. 21.3).

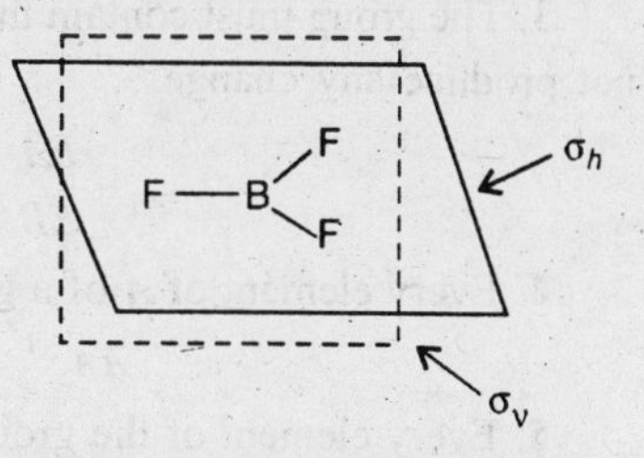

Fig. 21.3 The vertical and horizontal plane is FB_3 molecule.

3. Improper Axis of Rotation (S_n). A molecule possesses an improper axis of rotation of order n, if rotation about the axis by $2\pi/n$ followed by reflection in plane perpendicular to axis brings the molecule in an identical position. Examples are allene which possesses an S_2 and the staggered form of ethene which possesses an S_6 axis (Figs. 21.4 and 21.5).

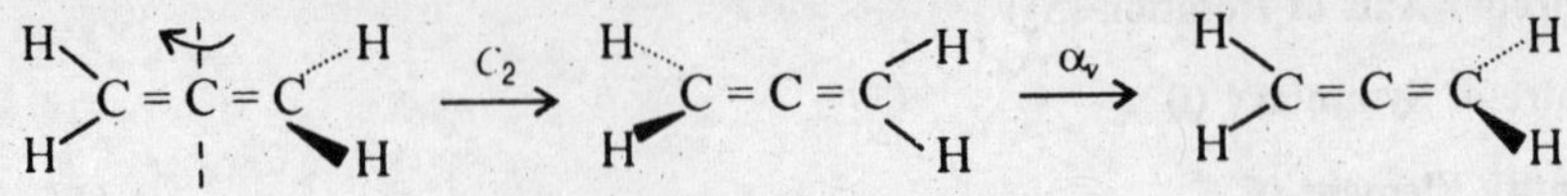

Fig. 21.4 Allene : S_2 axis.

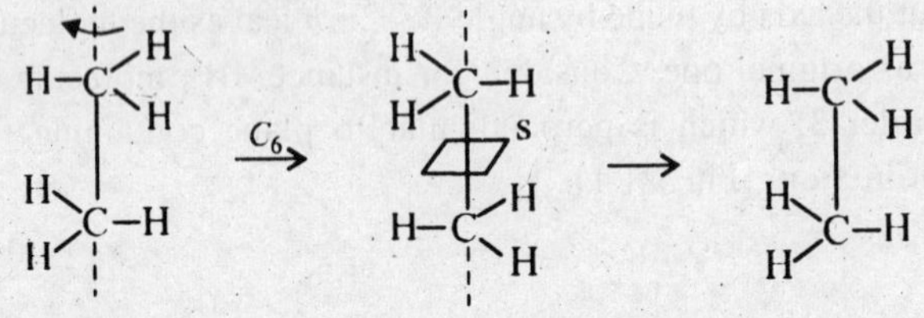

Fig. 21.5 Ethane (staggered form) : S_6 axis.

Centre of Symmetry (*i*). If in a molecule the co-ordinates (*x*, *y*, *z*) of every atom are changed into $(-x, -y, -z)$ and the molecule is obtained in an identical position, then the point of origin, *i.e.*, (0, 0, 0) is called the centre of symmetry of the molecule. Examples are all homonuclear diatomic molecules and *trans* 1, 2-dichloro-ethylene (Fig. 21.6) is an example of molecule possessing centre of symmetry.

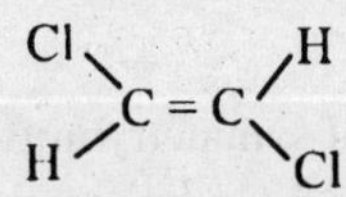

Fig. 21.6. Centre of symmetry in *trans* 1, 2-dichloro-ethylene.

5. Identity Element (E). All molecule possess an identity element which does not do anything to molecule.

Q. 3. Write the conditions (Group Postulates) for the symmetry elements to form a mathematical group G.

Ans. 1. Two elements *A* and *B* of a group combine to give the third element *C* which is also an element of the group :

$$AB = C$$

It means the applications of *B* followed by *A* is equivalent to the application of *C*. If $AB = BA$, the element of the groups A and B are said to commute.

2. An element combines with itself to form another element of a group.

3. The group must contain the identity element *E* which commutes with all elements and does not produce any change.

$$EA = AE = A$$
$$EB = BE = B \text{ etc.}$$

4. Every element of *A* of a group has an inverse A^{-1} which is also an element of the group:

$$AA^{-1} = A^{-1}A = E$$

5. Every element of the group obeys the associative law of combination :

$$A\,(BC) = (AB)\,C.$$

Q. 4. Define the following terms :

1. Abelian and Non-abelian Groups

2. Cyclic Groups

3. Order of the Group

4. Sub-group
5. Multiplication Table
6. Similarity Transformation
7. Class.

Ans. 1. Abelian and Non-abelian Groups. A group is said to be abelian if all elements commute and non-abelian if all the symmetry elements do not commute with one-another. H_2O molecule belongs to an abelian group and NH_3 to a non-abelian group.

2. Cyclic Groups. A group is said to be cyclic, if all its elements can be generated from the symmetry element. Thus, A, A^2, A^3, A^n form the elements of the cyclic group; here $A^n = E$, the identity element. In general, the roots of the equation $x^n - 1 = 0$ form a cyclic group.

3. Order of the Group. The total no. of elements of a group is called the order (h) of the group.

4. Sub-group. This is a smaller group within a group. If any selection or subset of the element of a group satisfies the definition of a group, then this subset of element is called a sub-group. Identity element is essentially a part of sub-group.

5. Multiplication Table. A water molecule has four symmetry elements, *viz.*, E, $C_2(z)$, $\sigma_v(xz)$, and $\sigma_v(yz)$ (Fig. 21.7).

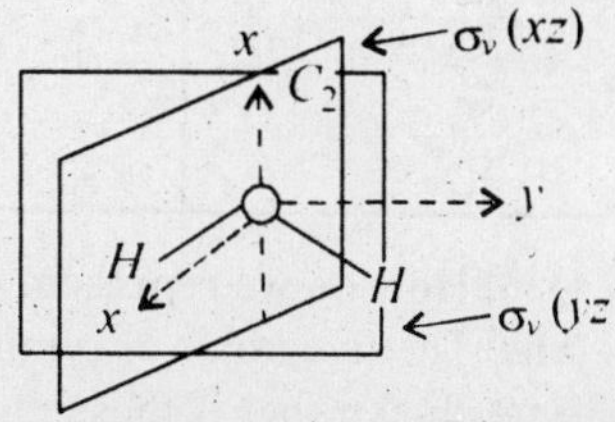

Fig. 21.7 The four-symmetry elements of H_2O molecule.

It can be seen that the product of any two symmetry elements is one of the four elements of a group. For instance, $C_2((z), \sigma_v(xz) = \sigma_v'(yz)$. Proceeding this way the symmetry operations of H_2O can be listed in a Group Multiplication Table (GMT) Table 21.2.

Table 21.2

Group Multiplication Table of the Symmetry Operations of H_2O Molecule

	E	$C_2(z)$	$\sigma_v(xz)$	$\sigma_v'(yz)$
E	E	$C_2(z)$	$\sigma_v(xz)$	$\sigma_v'(yz)$
$C_2(z)$	$C_2(z)$	E	$\sigma_v'(yz)$	$\sigma_v(xz)$
$\sigma_v(xz)$	$\sigma_v(xz)$	$\sigma_v'(yz)$	E	$C_2(z)$
$\sigma_v'(yz)$	$\sigma_v'(yz)$	$\sigma_v(xz)$	$C_2(z)$	E

6. Similarity Transformation. If A and X are two elements of group obeying the relation

$$X^{-1}AX = B$$

where B is also the same group, then B is called the similarity transformation of A by X and A and B are said to be conjugate to each other.

7.Class. A set of elements in a group which are conjugate to one another is said to form class.

Similarity transformation of the C_2 operation of H_2O molecule with all the other operations can be performed as under :

$$E^{-1}C_2E = C_2; \quad \sigma_v^{-1}C_2\sigma_v = C_2$$

$$(\sigma_v')^{-1}C_2\sigma_v' = C_2; (C_2)^{-1}C_2C_2 = C_2$$

We see that C_2 belongs to the a separate class since the similarity transformation of C_2 with all the operations generates C_2. A symmetry which commutes with all the symmetry operations is in a separate class. Thus, E, σ_h and i belong to the separate classes.

Q. 5. What is meant by Point Groups ? Illustrate with examples .

Ans. All the symmetry operations in a molecule can be combined to form a molecular group. This group is called *point group*, since all the symmetry elements of the molecule intersect at common point which remains fixed under all the symmetry operations. Table 21.3 gives molecular point groups of some molecules.

Table 21.3
Molecular Point Groups

Point Group	*Symmetry Elements*	*Molecules*
C_1	$E,\ C_1$	CHFClBr
C_2	$E,\ C_2$	H_2O_2
C_3	$E,\ C_3$	C_2H_6
C_s	$E,\ \sigma_v$	NOCl
C_{2v}	$E,\ C_2,\ 2\sigma_v$	H_2O
C_{3v}	$E,\ C_3,\ 3\sigma_v$	NH_3, PH_3
$C_{\infty v}$	$E,\ C_\infty,\ \infty\sigma_v$	HCl, NO, CO
C_{2h}	$E,\ C_2,\ \sigma_h,\ i$	*trans* CHCl = CHCl

Q. 6. How do we represent molecular point group ?

Ans. The five kinds of symmetry elements, *i.e.*, E, C_n, σ, S_n, i can be represented by matrices. It can be shown that if these elements operates on two dimensional position vector

$$\vec{r} = x\hat{i} + y\hat{j}$$

where $\hat{i}$ and $\hat{j}$ are unit vectors along the x and y axis respectively, they produce five 2×2 matrices

$$\begin{bmatrix} x' \\ y' \end{bmatrix} = \underset{2\times 2\,\text{matrix}}{\begin{bmatrix} \quad & \quad \\ \quad & \quad \end{bmatrix}} \begin{bmatrix} x \\ y \end{bmatrix}$$

Here, position vector is written as a column vector and the transformed vector as a primed column vector. The matrices are :

$$E = \begin{bmatrix} 1 & 0 \\ 1 & 0 \end{bmatrix};\ C_2 = \begin{bmatrix} \cos\theta & \sin\theta \\ -\sin\theta & \cos\theta \end{bmatrix};\ \sigma_{yz} = \begin{bmatrix} -1 & 0 \\ 0 & 1 \end{bmatrix};\quad \text{etc.}$$

On the other hand, if the same operation are carried on a 3-dimensional position vector

$$\vec{r} = x\hat{i} + y\hat{j} + z\hat{k}$$

they generate five 3×3 matrices.

$$E = \begin{bmatrix} 1 & 0 & 0 \\ 0 & 1 & 0 \\ 0 & 0 & 1 \end{bmatrix};\ C_n(z) = \begin{bmatrix} \cos\theta & \sin\theta & 0 \\ -\sin\theta & \cos\theta & 0 \\ 0 & 0 & 1 \end{bmatrix};\ \sigma_{yz} = \begin{bmatrix} -1 & 0 & 0 \\ 0 & 1 & 0 \\ 0 & 0 & 1 \end{bmatrix}$$

$$S_n = \sigma_{xy},\ C_n(z) = \begin{bmatrix} \cos\theta & \sin\theta & 0 \\ -\sin\theta & \cos\theta & 0 \\ 0 & 0 & 1 \end{bmatrix};\ i = \begin{bmatrix} -1 & 0 & 0 \\ 0 & -1 & 0 \\ 0 & 0 & -1 \end{bmatrix};\ \sigma_{xy} = \begin{bmatrix} -1 & 0 & 0 \\ 0 & 1 & 0 \\ 0 & 0 & -1 \end{bmatrix}$$

The sum of the diagonal element of the matrix is called the trace or character χ of the matrix. This characters of the various matrices corresponding to the symmetry operations given above are presented in Table 21.4.

Table 21.4
Characters of the Symmerty Operations

Symmetry operations (R)	*Character of R, χ (R)*
E	$\chi(E) = 3$
C_n	$\chi(C_n) = 3\cos\theta + 1$
σ	$\chi(\sigma) = 1$
S_n	$\chi(S_n) = 2\cos\theta - 1$
i	$\chi(i) = -3$

These matrices combine like the symmetry operations themselves as illustrated below:

$$\sigma_{xy}\sigma_{yz} = \begin{bmatrix} 1 & 0 & 0 \\ 0 & -1 & 0 \\ 0 & 0 & 1 \end{bmatrix}\begin{bmatrix} -1 & 0 & 0 \\ 0 & 1 & 0 \\ 0 & 0 & 1 \end{bmatrix} = \begin{bmatrix} -1 & 0 & 0 \\ 0 & -1 & 0 \\ 0 & 0 & 1 \end{bmatrix} = C_2(z)$$

The set of the matrices corresponding to the symmetry operations of a group is called *Representation.*

Q. 7. Explain (*a*) Reducible Representation (reps.)
and (*b*) Irreducible Representation (irreps.)

Ans. (*a*) **Reps.** Let A, B, C, ... be the matrices which form the representation of group and le X be similarity transformation matrix of this group.

$$X^{-1}AX = A'$$
$$X^{-1}BX = B'$$
$$X^{-1}CX = C'$$

$$X^{-1}AX = A' = \begin{bmatrix} \boxed{a_1'} & & & 0 \\ & \boxed{a_2'} & & \\ & & \boxed{a_3'} & \\ 0 & & & \boxed{a_4'} \end{bmatrix}$$

Then, if X is the proper transformation of matrix, we have

$$X^{-1}BX = B' = \begin{bmatrix} \boxed{b_1'} & & & 0 \\ & \boxed{b_2'} & & \\ & & \boxed{b_3'} & \\ 0 & & & \boxed{b_4'} \end{bmatrix}$$

The new matrix A' is now blocked out along the diagonal into smaller matrices a_1', a_2', a_3', a_4' etc., with the off-diagonal elements equal to zero. Similarly, we have

We express it by saying that the given sets of matrices form a reducible representation abbreviated as (*rep.*).

(*b*) **Irreps.** It is not possible to find a similarity transformation which will reduce the matrices *A*, *B*, *C*, ... to block-diagonalized form, the representation is called an irreducible representation abbreviated as *irrep*.

Q. 8. Explain Grand/Great Orthogonality Theorem (G.O.T.).

Ans. This is the most important theorem of group theory. It deals with matrix elements (elements of the matrices) which constitutes the irreps. of a point group. Mathematically it may be expressed as

$$\sum_R \Gamma(R)_{mn}\Gamma_j(R)^*_{m'n'} = \frac{h}{\sqrt{l_i l_j}}\delta_{ij}\delta_{mm'}\delta_{nn'}$$

Here

Γ_i = *i*th irreps. of point group of order *h* having dimension l_i,

Γ_j = *j*th irreps. of point group of order *h* having dimension l_j

$\Gamma_i(R)_{mn}$ is the *mn*th matrix element corresponding to symmetry element *R* belonging to *i*th irrep.

$\Gamma_j(R)^*_{m'n'}$ complex conjugate of the *m'n'* the matrix element corresponding to symmetry element *R* belonging to the *j*th irrep. The δ'*s* are the well-known *KRONECKER DELTAS* associated with the following properties.

$$\delta_{ij} = \begin{cases}1, & i=j \\ 0, & i \neq j\end{cases}; \quad \delta'_{mm} = \begin{cases}1, & m=m' \\ 0, & m \neq m'\end{cases}; \quad \delta'_{nn} = \begin{cases}1, & n=n' \\ 0, & n \neq n'\end{cases}$$

The summation is performed over all the symmetry operations *R* of the molecule. If the matrix element are real, then we have

$$\Gamma_j(R)^*_{m'n'} = \Gamma_j(R)_{m'n'}$$

The following three cases arises assuming that the matrix elements are real.

1. For two diff. irreps $i \neq j$, $m = m'$ and $n = n'$

$$\sum_R \Gamma_j(R)_{mn} = \Gamma_j(R)_{mn} = 0$$

2. For the same irrep. $i = j$, $m \neq m'$ and $n \neq n'$

$$\sum_R \Gamma_i(R)_{mn} = \Gamma_j(R)_{m'n'} = 0$$

3. For a irrep. *i* and for $m = m'$, $n = n'$

$$\sum_R [\Gamma_i(R)_{mn}]^2 = h / l_i$$

Q. 9. Describe important properties of irreducible representations ?

Ans. The properties of irreps. can be deduced from G.O.T. Consider a point group of order *h*. Recall that *h* is the number of symmetry operations of point group. Let the *h* symmetry operations be divided into *k* classes. Let the irreps. be denoted by $\Gamma_1, \Gamma_2, \Gamma_3, \Gamma_4, ..., \Gamma_k$ and their dimensions by $l_1, l_2, l_3, ..., l_k$. Using the G.O.T. of the following properties of the irreps. can be deduced.

1. The number of the irreps. in a point group is equal to the number of symmetry classes.
2. The sum of the squares of the dimensions of the irreps. of a group is equal to *h*, the order of the group. Thus,

$$\sum_{i=1}^{k} l_i^2 = h.$$

3. The sum of the squares of the characters of the identity operations in the irreps is equal to *h*, the order of the group,

$$\sum_{i=1}^{k} \left[\chi(E)^2\right] = h$$

4. The characters of the symmetry operations in two different irreps. obey the relation

$$\sum_{p=1}^{k} g_p \chi_i(R_p)\chi_j(R_p) = h\delta_{ij}$$

where δ_{ij} is the Kronecker delta with the following properties :

$$\delta_{ij} = \begin{cases} 1 & i=j \\ 0 & i \neq j \end{cases}$$

For $i \neq j$ this relation reduces to

$$\sum_{p=1}^{k} g_p \chi_i(R_p)\chi_j(R_p) = 0$$

and for $i = j$, it becomes

$$\sum_{p=1}^{k} g_p \left[\chi_i(R_p)\right]^2 = h$$

where g_p = Number of symmetry operations in pth class.

R_p = Symmetry operation of the pth class.

$\chi_i(R_p)$ = These are the characters of the matrix and $\chi_j(R_p)$ for the symmetry operation R_p in the i and j irreps. respectively.

5. The characters of all matrices belonging to the symmetry operations in the same class resemble one another.

Q. 10. How can we use the G.O.T. to construct the characters table for the C_{3v} point group?

Ans. This point group has three symmetry elements E, C_3 and $3\sigma_v$; the corresponding six symmetry operations can be divided into the following three classes :

$$E, (C_3^1, C_3^2), 3\,\sigma_v$$

We shall use rules mentioned in Q. 9 to arrive at the result.

According to the rule 1, the three symmetry classes give rise to three irreps.

From rule 2,

$$\sum_{i=1}^{3} l_i^2 = l_1^2 + l_2^2 + l_3^2 = 6$$

The only positive values of l_i that satisfy above equation are 1, 1 and 2. Now, every point group posseses a totally symmetric irrep. in which all the symmetry operations have character equal to 1.

From rule 3,

$$\sum_i \left[\chi_i(E)^2\right] = \left[\chi_1(E)^2\right] + \left[\chi_2(E)^2\right] + \left[\chi_3(E)^2\right] = 6,$$

which requires

$$\chi_1(E) = 1, \ \chi_2(E) = 1 \ \text{ and } \ \chi_3(E) = 2$$

The following table for the characters of different representations can be drawn.

Table 21.5
Characters of Different Representations

Irrep	$\chi(E)$	$\chi(E_3)$	$\chi(S_v)$
Γ_1	1	1	1
Γ_2	1	p	q
Γ_3	2	r	s

We have to determine the values of the unknown characters p, q, r, s in the irreps. Γ_2 and Γ_3. Applying rule 4, we have

$$\sum_{p=1}^{k} g_p\, \chi_i(R_p)\, \chi_j(R_p) = 0$$

$$\sum_{p=1}^{3} g_p \chi_1(R_p)\chi_2(R_p) = g_1\chi_1(E)\chi_2(E) + g_2\chi_1(C_3)\chi_2(C_3) + g_3\chi_1(\sigma_v)\chi_2(\sigma_v) = 0$$

From the table 21.5 and using $g_1 = 1$, $g_2 = 2$ and $g_3 = 3$ we obtain

$$1 + 2p + 3q = 0 \qquad ...(i)$$

Applying the following equation [Eq. (*i*), Q. 9]

$$\sum_{p=1}^{k} g_p\left[\chi_1(R_p)\right]^2 = h$$

to the irrep. Γ_2, we obtain

$$1 + 2p^2 + 3q^2 = 6 \qquad ...(ii)$$

Solving (*i*) and (*ii*), we obtain

$$p = 1,\ q = -1$$

Similarly, we obtain the following equations from the irreps. Γ_1 and Γ_3

$$2 + 2r + 3s = 0 \qquad ...(iii)$$

$$4 + 2r^2 + 3s^2 = 6 \qquad ...(iv)$$

Solving eq. (*iii*) and (*iv*), we get

$$r = -1,\ s = 0$$

From the values of p, q, r and s in Table 21.5 we obtain the desired character table for C_{3v} point group as given in Table 21.6.

Table 21.6
Character Table for C_{3v} Point Group

C_{3v}	E	$2C_3$	$3\sigma_v$
Γ_1 or A_1	1	1	1
Γ_2 or A_2	1	1	–1
Γ_3 or E	2	–1	0

The symbol Γ_1, Γ_2 and Γ_3 are the Bethe notations for the irreps. and A_1, A_2 and E are Mulliken notation for the irreps.

Q. 11. Verify that characters given in C_{3v} point group obey the properties of the irreps.?

Ans. Refers to the characters table for the C_{3v} point group.

1. The number of the irreps. is equal to the number of class (in this case 3) of the group.

2. Two irreps. A_1 and A_2 are one-dimensional, *i.e.*, $l_1 = l_2 = 1$ and the third is two-dimensional, *i.e.*, $l_3 = 2$. Thus,

$$\sum_{i=1}^{3} l_i^2 = l_1^2 + l_2^2 + l_3^2 = (1)^2 + (1)^2 + (2)^2 = 6$$

which is the order of the group.

3. $$\sum_{i=1}^{3} [\chi_i(E)]^2 = [\chi_{A1}(E)]^2 + [\chi_{A2}(E)]^2 + [\chi_E(E)]^2$$

$$= (1)^2 + (1)^2 + (2)^2 = 6$$

4. $$\sum_{p=1}^{3} g_p \chi(R_p)\chi_j(R_p) = h\delta_{ij}$$

For the irreps. A_2 and E (say), this relation gives (1) (1) (2) + (2) (1) (– 1) + 3 (– 1) (0) = 0.

For the irreps. A_1 and A_2 (say) it gives the (1) (1) (2) + (2) (1) (1) + 3 (1) (– 1) = 0 and for the irreps. E, (1) $(2)^2$ + 2 $(-1)^2$ + 3 $(0)^2$ = 6.

Q. 12. Explain diagrammatically that H_2O molecule is Abelian whereas NH_3 molecule is non-Abelian.

Ans. A molecule is said to be Abelian of all the symmetry elements commute with one another and non-Abelian if they do not commute with one-another.

We see that applications of C_2' followed by σ_v gives the same configuration as the

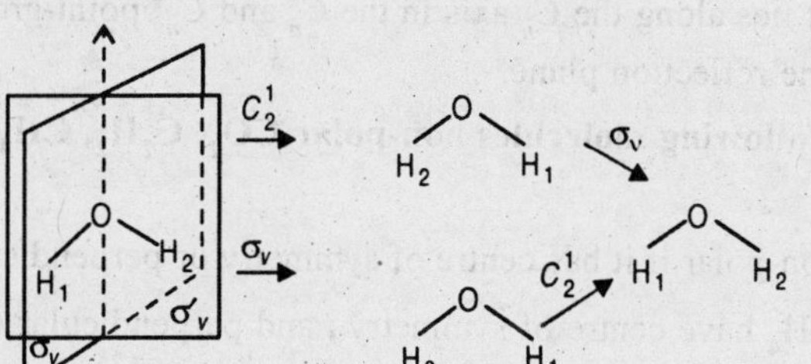

Fig. 21.8. Illustration of an Abelian group (H_2O molecule).

application of σ_v followed by C_2'.

Hence, C_2' and σ_v commute, *i.e.*, $\sigma_v C_2' = C_2'\sigma_v$. Thus, H_2O molecule belongs to an Abelian Point Group (C_{2v}) Fig. 21.8.

In the case of NH_3 molecule (Fig. 21.9) we find that applications of C_3' followed by σ_v' gives a different configuration of NH_3 molecule than the application of σ_v' followed by C_3', *i.e.*,

$$\sigma_v' C_3' \neq C_3'\sigma_v'$$

so that the two symmetry operations do not commute. Thus, NH_3 molecule belongs to non-Abelian Point Group (C_{3v}).

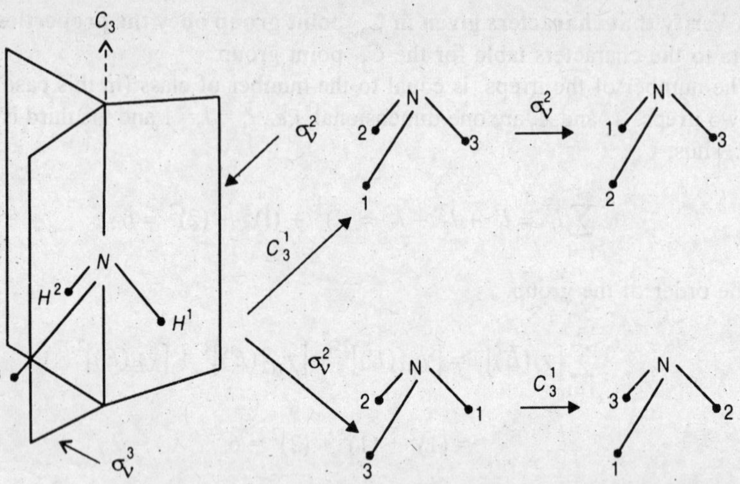

Fig. 21.9. Illustration of a non-Abelian group.

Q. 13. How can we study the following properties of a substance on the basis of group theory.

(a) Optical activity of molecule.

(b) Polar and non-polar nature of molecule ?

Ans. (a) A molecule will be optically active, if it does not possess the S_n-axis of any-order. Thus, it should not have the reflection plane (S_1-axis), the centre of inversions (S_2-axis) and any improper axis of order n. A molecule is not chiral if it belongs to the point group D_{nh} or D_{nd} or to cubic point groups T and O.

(b) A molecule cannot be polar if it has a centre of inversion and belongs to any of the D groups including D_n, D_{nh} and D_{nd}. It will be polar if it belongs to C_n, C_s, C_{nv} or the infinite $C_{\infty v}$ point group. The dipole moment lies along the C_n axis in the C_n and C_{nv} point-group. In the case of the C_s. point group it lies in the reflection plane.

Q. 14. Why are the following molecules non-polar CO_2, C_2H_2, CH_4 and staggered form C_2H_6?

Ans. A molecule is non-polar if it has centre of symmetry or perpendicular C_2 axis.

1. Both CO_2 and C_2H_4 have centre of symmetry i and perpendicular C_2 axis.
2. CH_4 has perpendicular C_2 axis.
3. The staggered form of C_2H_6 has the centre of symmetry i and perpendicular C_2 axis.

Thus, all the above molecules are non-polar.

Q. 15. Why are the following molecules not chiral (a) NH_3 (b) $CH_2 = C = CH_2$ (allene).

Ans. A molecule is not chiral if it has an S_n axis of any order or a centre of inversion or plane of symmetry.

(a) NH_3 has three vertical planes, hence, it is not chiral.

(b) Allene has an S_4 axis; hence, it is not chiral.

22

MATHEMATICAL CONCEPTS AND COMPUTERS

MATHEMATICAL CONCEPTS

Logarithmic Relations

Definition of Logarithm. Logarithm of a number with respect to a given base is the power to which base must be raised to represent the number.

For example $1000 = 10^3$

$\log_{10} 1000 = 3$

In general $N = a^x$

then $\log_a N = x$

As $a^0 = 1$ It follows that $\log_a 1 = 0$

$a^1 = a$ It follows that $\log_a a = 1$.

Four standard formulae of logarithm. These four formulae are commonly used. We are mentioning them without proof.

$\log_a mn = \log_a m + \log_a n$ (Product formula)

$\log_a \dfrac{m}{n} = \log_a m - \log_a n$ (Quotient formula)

$\log_a m^n = n \log_a m$ (Power formula)

$\log_a m = \log_b m \times \log_a b$ (Base changing formula)

Example 1. **Express $\log_a \dfrac{x^3 \sqrt[5]{y^2}}{\sqrt{z^3}}$ in terms of $\log_a x$, $\log_a y$ and $\log_a z$.**

Solution.

$$\log_a \frac{x^3 \sqrt[5]{y^2}}{\sqrt{z^3}} = \log_a \frac{x^3 y^{2/5}}{z^{3/2}}$$

$$= \log_a \left(x^3 y^{2/5}\right) - \log_a z^{3/2} \quad \text{(Quotient formula)}$$

$$= \log_a x^3 + \log_a y^{2/5} - \log_a z^{3/2} \quad \text{(Product formula)}$$

$$= 3\log_a x + \frac{2}{5}\log_a y - \frac{3}{2}\log_a z \quad \text{(Power formula)}$$

which is the result.

***Example 2.* If *a, b, c, d* are four positive numbers, prove that $\log_b a \times \log_c b \times \log_d c = \log_d a$**

Solution.

$$\begin{aligned} \text{L.H.S.} &= \log_b a \times \log_c b \times \log_d c \\ &= \frac{\log_m a}{\log_m b} \times \frac{\log_m b}{\log_m c} \times \frac{\log_m c}{\log_m d} \quad \text{(by Base changing formula)} \\ &= \frac{\log_m a}{\log_m d} \\ &= \log_d a \\ &= \text{R.H.S.} \end{aligned}$$

Hence the result.

EXERCISES

1. Express in terms of log *x*, log *y* and log *z*.

(*i*) $\log_a \frac{x^3}{y^2 z^4}$ (*ii*) $\log_a \frac{\sqrt[n]{x^m}\, y^r}{3\sqrt{z}}$

2. Prove that $\log \frac{a^2}{bc} + \log \frac{b^2}{ca} + \log \frac{c^2}{ab} = 0$.

3. Prove that $\log_a mnp = \log_a m + \log_a n + \log_a p$.

4. $\log \frac{42}{55} = \log 2 + \log 3 + \log 7 - \log 11 - \log 5$. Prove it.

ANSWERS

1. (*i*) $3 \log_a x - 2 \log_a y - 4 \log_a z$

(*ii*) $\frac{m}{n} \log_a x + r \log_a y - \frac{1}{3} \log_a z$

TWO SYSTEMS OF LOGARITHM:

The two systems of logarithm in common use are:

(i) Neperian log or Natural log. Here the base is *e*, where $e = 1 + \frac{1}{1!} + \frac{1}{2!} + \frac{1}{3!} + ...$ In all theoretical calculations, we use natural logarithm.

(ii) Common logarithm. Here, the base of the log is 10. In all numerical calculations, we use common log. We can convert natural log into common log as follows.

$$\begin{aligned} \log_e m &= \log_{10} m \times \log_e 10 \\ \text{and} \quad \log_e 10 &= 2.3026 \times \log_{10} 10 \\ &= 2.3026 \times 1 = 2.3026 \qquad | \because \log_{10} 10 = 1 \\ \therefore \quad \log_e m &= 2.3026 \log_{10} m \end{aligned}$$

Log tables. Log tables are the standard common log tables which are available to us for calculations. There are usually four digit tables and are given at the end of the book. The logarithm of any given number can be read from these tables. The logarithm of a number consists of two parts :

(*i*) Integral part called *Characteristic* (*ii*) Decimal part called *Mantissa*.

(i) Characteristic. It can be zero, positive or negative, we use the following two rules for finding the characteristic.

Rule I : If the given number > 1, its characteristic is +ve. Count the number of digits to the left of the decimal point. Subtract one from this number. This is the characteristic of the log of given number.

For example,

Characteristic of log of 1798.3 is $4 - 1 = 3$

Characteristic of log of 28.78 is $2 - 1 = 1$

Characteristic of log of 1.923 is $1 - 1 = 0$

Rule II : If the given number is less than one, its characteristic is negative. It is put under the bar. Count the number of zeros immediately after the decimal point. Add 1 to this number. It gives us the characteristic of the log of the given number.

For example,

Characteristic of log 0.5828 $= -(0 + 1) = \bar{1}$

Characteristic of log 0.0628 $= -(1 + 1) = \bar{2}$

Characteristic of log 0.0008 $= -(3 + 1) = \bar{4}$

(ii) Mantissa. As mentioned above, it is the decimal part of log of a given number. It is always positive, and can be read from the log tables as follows.

(*i*) In the given number, omit the decimal point and zeros in the beginning and at the end. For example in the log of 0.001175, we have to look up the mantissa of 1175.

(*ii*) Take first two digits *i.e.*, 11 and locate it in the first vertical column of the four figure log table.

(*iii*) Go through the horizontal row beginning with 11 and look up the value under the column headed by the third digit *i.e.* 7. From the log tables, we find this number as 0682.

(*iv*) Continue moving in the same horizontal line and note the number in the small differences column headed by the fourth digit, i.e., 5. This number is 18.

(*v*) Add 18 to 0682, we got 0700. ∴ The mantissa is 0.0700.

Note: If the number consists of a single digit, put zeros at the end of the number to complete four digits. Read mantissa as explained above.

For example log 7 = log 7.000 = 0.8451

Mantissa for 7, 70, 700, 0.7 etc. is the same.

Anti-logarithm: Antilogarithm is the reverse process of logarithm. Thus antilogarithm to logarithm is as division to multiplication.

For example, if log 7 = 0.8451 then Antilog 0.8451 = 7.

Standard Antilog tables are also given at the end of the book for calculations. We follow the rules given below for reading antilog tables.

(*i*) Read the antilog of mantissa only (i.e. of decimal part). Locate the four digits the way we did in reading mantissa from log tables.

(*ii*) If the characteristic is positive say n, the decimal is placed after $(n + 1)$ digits in the value read.

(*iii*) If the characteristic is –ve, say $\bar{n}$, then $(n - 1)$ zeros are placed before the left side of the number read; then decimal point is placed.

For example antilog of $\bar{2}.3478 = 0.02227$

antilog of 1.7192 = 52.38

antilog of 0 = 1.000

Use of logarithm. Logarithm is practically very useful in simplifying the complicated calculations. Logarithm means a rule to shorten arithmetic. It was invented by John Napier in 1614.

Following examples will enable the students to use the log tables for that purpose.

***Example 1.* Find the fifth root of 0.0076.**

Solution.

Let $x = (0.0076)^{1/5}$

Taking log on both sides, we get

$$\log x = \frac{1}{5} \log (0.0076)$$

$$= \frac{1}{5}(\bar{3}.8808) \qquad \text{(From log tables)}$$

$$= \frac{1}{5}(\bar{3}.8808 + 2 - 2)$$

$$= \frac{1}{5}(\bar{5}.8808 + 2)$$

$$= \frac{1}{5}(\bar{5} + 2.8808)$$

$$\log x = \bar{1}.5762$$

Taking antilog on both sides, we get

$$x = 0.3769$$

***Example 2.* Simplify $\dfrac{1.792 \times 77.4}{(129.7)^{2/3}}$.**

Solution.

$$\text{Let } x = \frac{1.792 \times 77.4}{(129.7)^{2/3}}$$

Taking log on both sides, we get

$$\log x = \log 1.792 + \log 77.4 - \frac{2}{3} \log (129.7)$$

$$= 0.2534 + 1.8887 - \frac{2}{3}(2.1130)$$

$$= 0.2534 + 1.887 - 1.4087$$

$$= 2.1421 - 1.4087$$

$$= 0.7334$$

$$x = \text{Antilog } 0.7334$$

$$= 1.5413 \qquad \text{which is the result.}$$

EXERCISES

1. Simplify the following
   (*i*) $(0.0036)^{1/5}$
   (*ii*) $(0.056)^{2/3}$
   (*iii*) $10^{-1/5}$
2. Simplify the following
   (*i*) $(0.05246)^{1/8} - 2.6055$
   (*ii*) $\dfrac{3.142 \times (80.2)^{1/2}}{(9.8)^{1/2}}$
3. Find the antilog of
   (*i*) –2.6674 [Hint. ($-2.6674 = \bar{3}.3326$)]
   (*ii*) –3.01

ANSWERS

1. (*i*) 0.3245
   (*ii*) 0.1464
   (*iii*) 0.6310
2. (*i*) –1.9137
   (*ii*) 8.989
3. (*i*) 0.002151
   (*ii*) 0.0009772

Curve Sketching

When a variable point $P(x, y)$ moves under a given set of conditions, then the path traced by P is called graph or locus of P. The equation which relates the variables x and y is called the equation of the graph of P.

Every point which is on the graph satisfies the given set of conditions and any point satisfying the given set of conditions lies on the graph.

To draw the graph or to sketch the curve we take y as the dependent variable and x is the independent variable. To draw the graph the following points are followed:

(i) Write the given equation equal to y.

(ii) Now give three, four values of x and find the corresponding values of y.

(iii) Plot these points in the graph by taking two perpendicular lines. Horizontal line as x-axis and vertical line as y-axis, we get the required curve.

***Example 1.* Sketch the curve $y = x^2 + 2x - 3$.**

Solution.

First of all, we get table of values for $y = x^2 + 2x - 3$ as under.

x	–2	–1	0	2	–4
y	–3	–4	–3	5	5

Plot these points in the graph and join it with free hand as illustrated below.

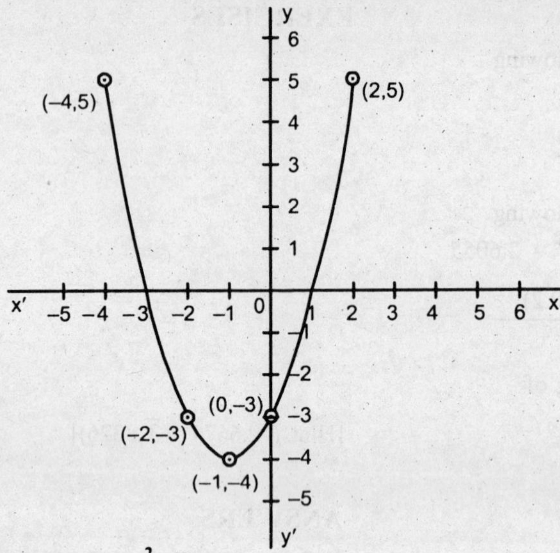

Example 2. **Sketch the curve $y = x^2 - 9$.**

Solution.

First of all we get table of values for $y = x^2 - 9$ by taking x as independent and y as dependent variable. We get the table of values as under.

x	–3	–2	–1	0	1	2	3
y	0	–5	–8	9	–8	–5	0

Plot these points in the graph and join it with free hand as shown below.

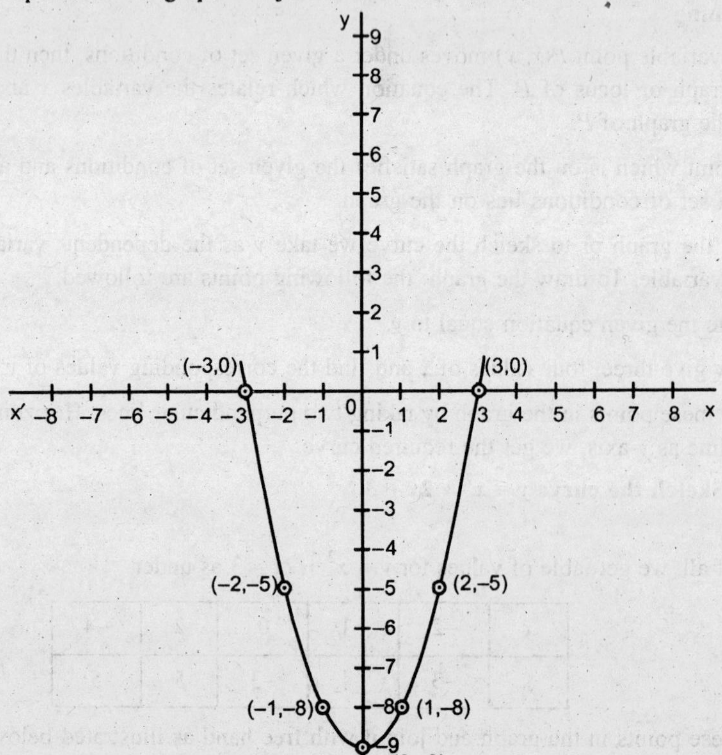

Example 3. **Sketch the curve $y = x^3 + 1$.**

Solution:

First of all we get table of values for $y = x^3 + 1$ by taking x as independent variable and y as dependent variable, we get the table of values as under.

x	–2	–1	0	1	2
y	–7	0	1	2	9

Plot these points in the graph and join it with the free hand as shown below.

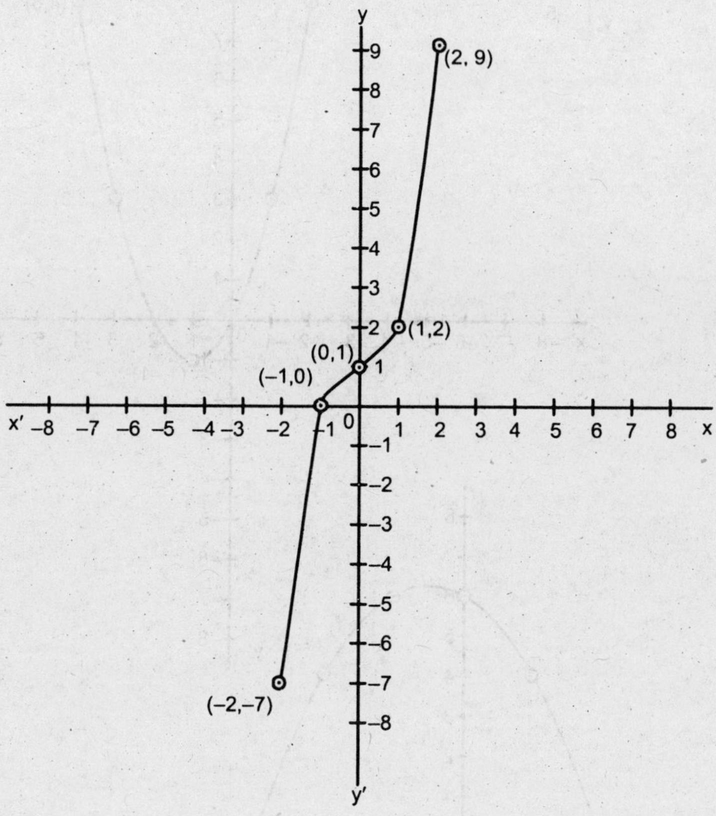

EXERCISES

1. Sketch the graph $y = x^4 - 1$.
2. Sketch the curve $y = x^2 - 2x$.
3. Sketch the curve $y = x^2 + x + 6$.

ANSWERS

1.

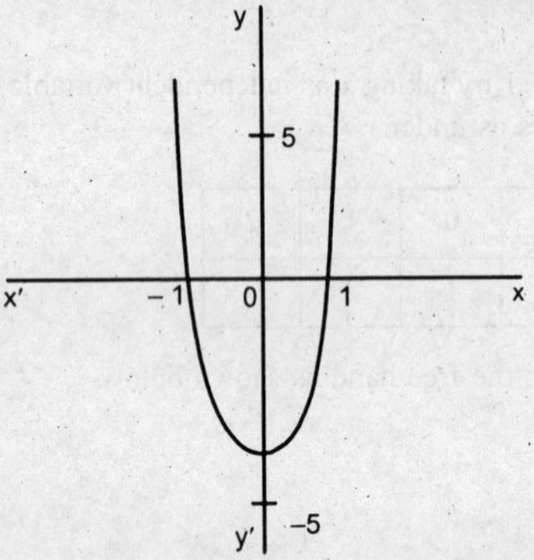
y
5
x'
– 1
0
1
x
y'
–5

2.

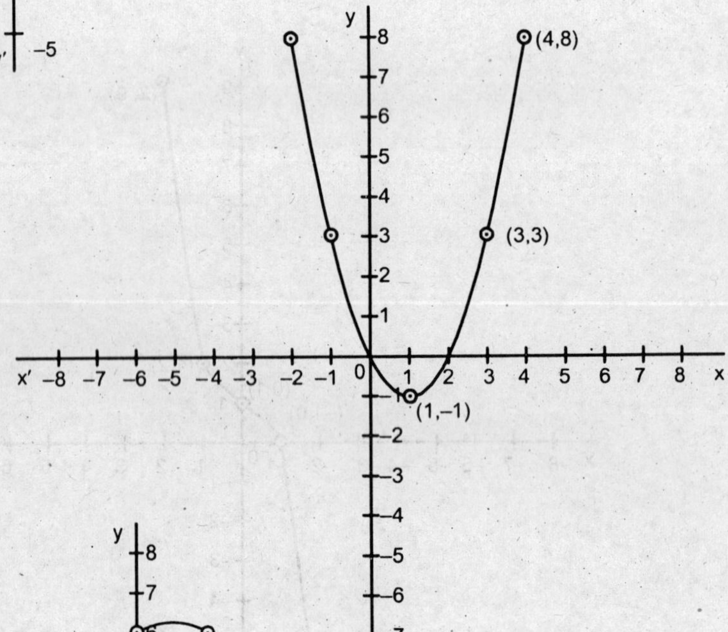
y
8
7
6
5
4
3
2
1
(4,8)
(3,3)
x'
–8
–7
–6
–5
–4
–3
–2
–1
0
1
2
3
4
5
6
7
8
x
(1,–1)
–1
–2
–3
–4
–5
–6
–7
–8
y'

3.

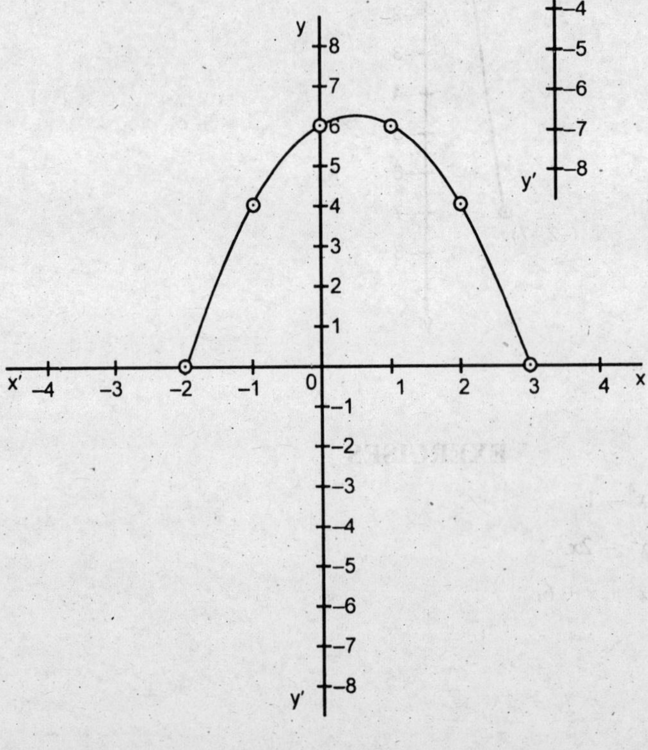
y
8
7
6
5
4
3
2
1
x'
–4
–3
–2
–1
0
1
2
3
4
x
–1
–2
–3
–4
–5
–6
–7
–8
y'

LINEAR GRAPHS AND CALCULATIONS OF SLOPES

Linear graph. The graph of a linear equation is called the linear graph.

Linear Equation. An equation of the form $ax + by + c = 0$, where a, b, c are constants and x and y are two variables, is called a linear equation of two variables. Every linear equation in x and y represents a straight line. The various cases are discussed below one by one.

(i) The equation $x = a$ represents a straight line parallel to y-axis and at a distance a from it.

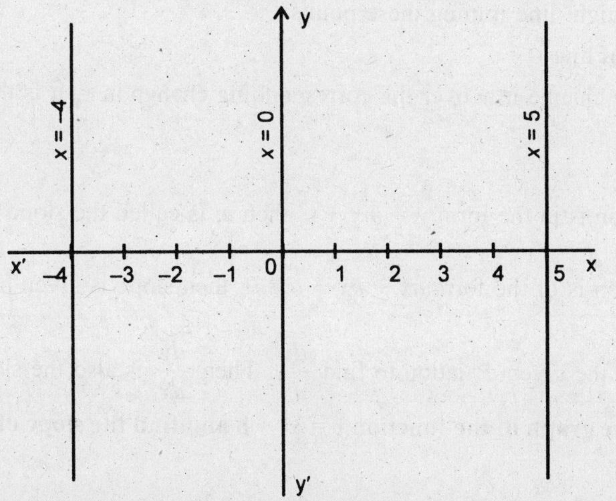

If the line is to the right of y-axis, a is +ve.

If the line is to the left of y-axis, a is –ve.

If the line coincides with y-axis, $a = 0$, the equation of the y-axis is $a = 0$.

The graph of $x = 5$, $x = 0$ and $x = -4$ is shown in the above figure.

(ii) The equation $y = b$ represents a straight line parallel to x-axis and at a distance b from it.

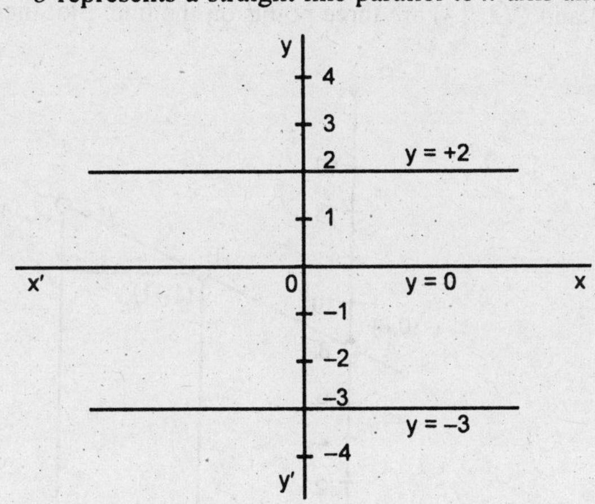

If the line is above x-axis, b is +ve.

If the line is below x-axis, b is –ve.

If the line coincides with x-axis, $b = 0$.

The equation of x-axis is $y = 0$.

The graphs of $y = 2$, $y = 0$ and $y = -3$ are shown in the figure above.

General method to draw the graph of linear equation $ax + bx + c = 0$ and to find the slope of the line.

1. Write the given equation in the form showing one variable in terms of the other variable (generally in the form $y = mx + c$)
2. Now give three, four values to x and find the corresponding values of y.
3. Plot these points named by ordered pairs.
4. Draw the straight line joining these points.

To find slope of line

1. Calculate the change in y over the corresponding change in x. It is the slope of the line *i.e.*, $\frac{\text{Change in } y}{\text{Change in } x}$.
2. If the equation is in the form $y = mx + c$, then m is called the slope of the line.
3. If the equation is of the form $ax + bx + c = 0$, then slope is given by $-\frac{\text{Coeff. of } x}{\text{Coeff. of } y} = -\frac{a}{b}$.
4. Differentiate the given equation to find $\frac{dy}{dx}$. Then $\frac{dy}{dx}$ is also the slope of the line.

***Example.* Draw the graph of the function $y = 3x + 8$ and find the slope of the line.**

Solution.

The equation of the line is

$$y = 3x + 8.$$

Putting $x = 0$ we get $y = 8$
Putting $x = 1$ we get $y = 11$
Putting $x = 2$ we get $y = 14$

$A(0, 8)$, $B(1, 11)$ and $C(2, 14)$ are three points on the line. Plot these points and join them by a foot ruler.

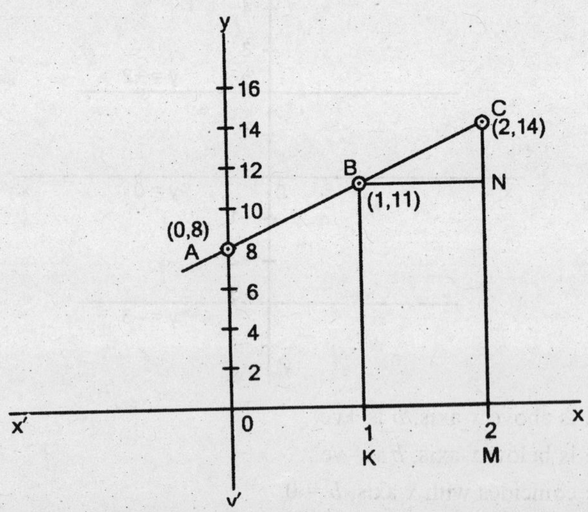

To find the slope draw BK, CM perpendicular to x-axis and BN perpendicular to CM.

The slope is given by $\frac{CN}{BN} = \frac{\text{Change in } y}{\text{Change in } x}$

Now

$$CN = CM - NM = 14 - 11 = 3$$
$$BN = KM = OM - OK = 2 - 1 = 1$$

$$\text{Slope} = \frac{3}{1} = 3. \text{ Hence the result.}$$

EXERCISE

Draw the graphs and find the slopes.

(i) $x + y = 0$
(ii) $2x - y = 3$
(iii) $x - y = 2$
(iv) $4x + y = 3$

ANSWERS

(i) 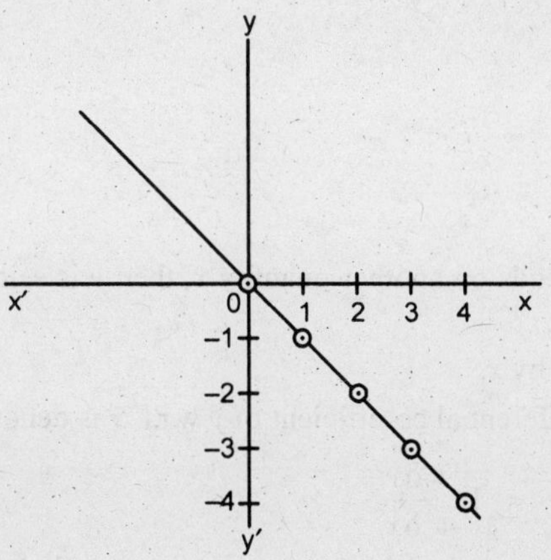

Slope = −1

(ii) 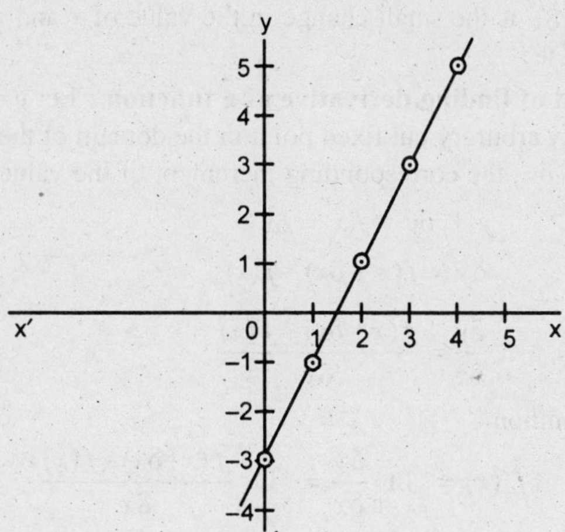

Slope = 2

(iii) 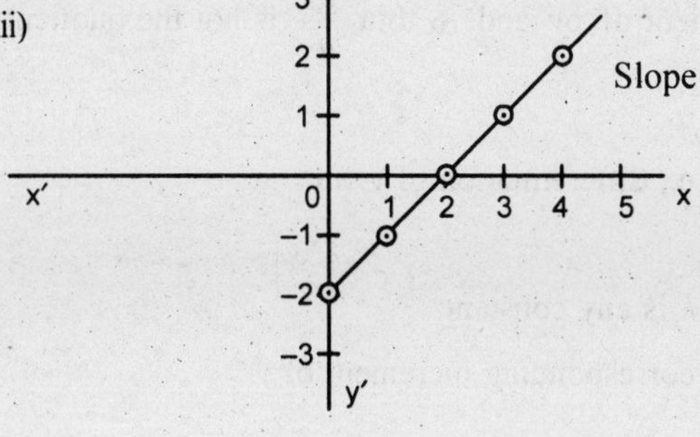

Slope = +1

(iv) 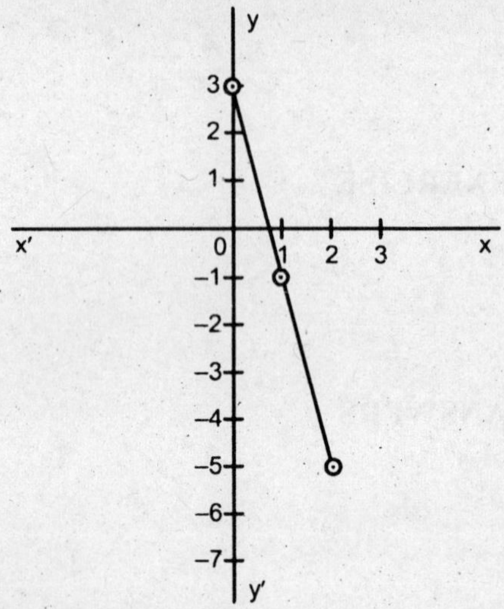 Slope = –4

DIFFERENTIATION :

Function. When the value of a quantity y depends on another quantity x, then y is said to be a function of x. We represent it as $y = f(x)$.

Note that $f(x)$ is a symbol. It is not f multiplied by x.

Differential co-efficient : The derivative or differential co-efficient of y w.r.t. x is denoted by

$$\frac{d}{dx}(y) = \frac{dy}{dx} = \underset{\delta x \to 0}{\mathrm{Lt}} \frac{\delta y}{\delta x}$$

Where δx is the small change in the value of x and δy is the corresponding small change in the value of y.

Method of finding derivative of a function : Let $y = f(x)$ be a derivable function of x. Let x represent any arbitrary but fixed point in the domain of the function. Let δx be the small increment of x and let δy, the corresponding increment of the value of y.

$\therefore \quad y + \delta y = f(x + \delta x)$

$\Rightarrow \quad \delta y = f(x + \delta x) - f(x)$

$\Rightarrow \quad \dfrac{\delta y}{\delta x} = \dfrac{f(x+\delta x) - f(x)}{\delta x}$

By definition

$$f'(x) = \underset{\delta x \to 0}{\mathrm{Lt}} \frac{\delta y}{\delta x} = \underset{\delta x \to 0}{\mathrm{Lt}} \frac{f(x+\delta x) - f(x)}{\delta x}$$

It is worthwhile to note that $\dfrac{\delta y}{\delta x}$ is the quotient of δy and δx but $\dfrac{dy}{dx}$ is not the quotient of dy and dx.

Instead, $\dfrac{dy}{dx}$ is to be interpreted as $\dfrac{d}{dx}(y)$ i.e., differentiation of y w.r.t. x.

1. Derivative of kx w.r.t. x

Let $\quad y = kx \quad$ .... (1) where k is any constant.

Let δx be an increment of x and δy, the corresponding increment of y.

$$\therefore \quad y + \delta y = k(x + \delta x) \quad \text{.... (2)}$$

$$(2) - (1) \Rightarrow \quad y + \delta y - y = k(x + \delta x) - kx$$

$$\delta y = k\,\delta x$$

$$\Rightarrow \quad \delta y = k\,\delta x$$

$$\Rightarrow \quad \frac{\delta y}{\delta x} = k$$

Taking limits $\delta x \to 0$

$$\underset{\delta x \to 0}{\text{Lt}} \frac{\delta y}{\delta x} = \underset{\delta x \to 0}{\text{Lt}}\ k$$

$$\frac{dy}{dx} = k$$

$$\frac{d}{dx}(kx) = k$$

2. Derivative of e^x w.r.t. x

Let $\quad y = e^x \quad$ .... (1)

Let δx be an increment of x and δy by the corresponding increment of y.

$$y + \delta y = e^{x+\delta x}$$

$$(2) - (1) \Rightarrow \delta y = e^{x+\delta x} - e^x$$

$$\Rightarrow \delta y = e^x . e^{\delta x} - e^x$$

$$\Rightarrow \frac{\delta y}{\delta x} = \frac{e^x \cdot e^{\delta x} - e^x}{\delta x}$$

$$= \frac{e^x (e^{\delta x} - 1)}{\delta x}$$

Taking limits $\delta x \to 0$

$$\underset{\delta x \to 0}{\text{Lt}} \frac{\delta y}{\delta x} = \underset{\delta x \to 0}{\text{Lt}} \frac{e^x (e^{\delta x} - 1)}{\delta x}$$

$$\frac{dy}{dx} = e^x \log e \qquad \left| \because \underset{\delta x \to 0}{\text{Lt}} \frac{e^{\delta x} - 1}{\delta x} = \log e \right.$$

$$\frac{dy}{dx} = e^x \qquad | \because \log e = 1$$

$$\frac{dy}{dx}(e^x) = e^x$$

3. Derivative of x^n w.r.t. x ***(Raipur 2010***

Let $\quad y = x^n \quad$ .... (1) where n is any natural number.

Let δx be an increment of x and δy, the corresponding increment of y.

$$y + \delta y = (x + \delta x)^n \, . \quad \text{... (2)}$$

$$(2) - (1) \Rightarrow \delta y = (x + \delta x)^n - x^n$$

$$\Rightarrow \quad \delta y = x^n + n_{C_1} x^{n-1}\delta x + n_{C_2} x^{n-2}(\delta x)^2 + ... + n_{C_n}(\delta x)^n - x^n$$

$$\Rightarrow \quad \delta y = n x^{n-1}\delta x + \frac{n(n-1)}{\lfloor 2} x^{n-2}(\delta x)^2 + ... + (\delta x)^n$$

$$\Rightarrow \quad \delta y = \delta x\left[nx^{n-1} + \frac{n(n-1)}{\lfloor 2} x^{n-2}\delta x + ... + (\delta x)^{n-1}\right]$$

Dividing both sides by δx

$$\Rightarrow \quad \frac{\delta y}{\delta x} = \frac{\delta x\left[nx^{n-1} + \frac{n(n-1)}{\lfloor 2} x^{n-2}\delta x + ... + (\delta x)^{n-1}\right]}{\delta x}$$

$$\Rightarrow \quad \frac{\delta y}{\delta x} = nx^{n-1} + \frac{n(n-1)}{\lfloor 2} x^{n-2}\delta x + ... + (\delta x)^{n-1}$$

$$\Rightarrow \quad \underset{\delta x \to 0}{\text{Lt}} \frac{\delta y}{\delta x} = \underset{\delta x \to 0}{\text{Lt}}\ nx^{n-1} + \frac{n(n-1)}{\lfloor 2} x^{n-2}\delta x + ... + (\delta x)^{n-1}$$

$$\frac{\delta y}{\delta x}(x^n) = nx^{n-1}$$

$$\therefore \quad \frac{d}{dx}(x^n) = nx^{n-1}$$

4. Derivative of sin *x* w.r.t. *x*

Let $\quad y = \sin x \quad .... (1)$

Let δx be an increment of x and δy, the corresponding increment of y.

$$\therefore \quad y + \delta y = \sin(x + \delta x) \quad .... (2)$$

$$(2) - (1) \quad \Rightarrow \delta y = \sin(x + \delta x) - \sin x$$

$$= 2\cos\left(x + \frac{1}{2}\delta x\right)\sin\frac{\delta x}{2}$$

$$\frac{\delta y}{\delta x} = \frac{2\cos\left(x + \frac{\delta x}{2}\right)\sin\frac{\delta x}{2}}{\delta x} \qquad \left|\ \because \sin C - \sin D = 2\cos\frac{C+D}{2}\sin\frac{C-D}{2}\right.$$

$$\Rightarrow \quad \frac{\delta y}{\delta x} = \frac{\cos\left(x + \frac{\delta x}{2}\right)\sin\frac{\delta x}{2}}{\frac{\delta x}{2}}$$

Taking limits

$$\underset{\delta x \to 0}{\text{Lt}} \frac{\delta y}{\delta x} = \underset{\delta x \to 0}{\text{Lt}} \cos\left(x + \frac{\delta x}{2}\right) \cdot \underset{\delta x \to 0}{\text{Lt}} \frac{\sin\frac{\delta x}{2}}{\frac{\delta x}{2}}$$

$$= \cos x \cdot 1 \qquad \left|\ \because \underset{x \to 0}{\text{Lt}} \frac{\sin x}{x} = 1\right.$$

$$\therefore \quad \frac{dy}{dx} = \cos x$$

i.e., $\frac{d}{dx}(\sin x) = \cos x$

On the lines of sin x we can find the following derivatives:

$$\frac{d}{dx}(\cos x) = -\sin x \quad \frac{d}{dx}(\tan x) = \sec^2 x$$

$$\frac{d}{dx}(\cot x) = -\operatorname{cosec}^2 x \quad \frac{d}{dx}(\sec x) = \sec x \tan x$$

$$\frac{d}{dx}(\text{cosec}) = -\operatorname{cosec} x \cot x \quad \frac{d}{dx}(a^x) = a^x \log a \quad (\text{on the lines of } e^x)$$

5. Derivative of log x w.r.t. x

Let $y = \log x$ .... (1)

Let δx be the small increment in x, then the corresponding increment in y be δy.

Then $y + \delta y = \log(x + \delta x)$ ....(2)

(2) – (1) $\Rightarrow \delta y = \log(x + \delta x) - \log x$

$$\Rightarrow \delta y = \log\frac{(x+\delta x)}{x}$$

$$\Rightarrow \frac{\delta y}{\delta x} = \frac{1}{\delta x}\log\left(1+\frac{\delta x}{x}\right)$$

$$\Rightarrow \frac{\delta y}{\delta x} = \frac{1}{x}\cdot\frac{x}{\delta x}\log\left(1+\frac{\delta x}{x}\right)$$

$$\Rightarrow \frac{\delta y}{\delta x} = \frac{1}{x}\cdot\log\left(1+\frac{\delta x}{x}\right)^{\frac{x}{\delta x}}$$

Taking limits

$$\underset{\delta x\to 0}{\text{Lt}}\frac{\delta y}{\delta x} = \underset{\delta x\to 0}{\text{Lt}}\frac{1}{x}\log\left(1+\frac{\delta x}{x}\right)^{\frac{x}{\delta x}}$$

| when $\delta x \to 0$ then $\frac{\delta x}{x} \to 0$

$$\frac{dy}{dx} = \frac{1}{x}\cdot\underset{\delta x\to 0}{\text{Lt}}\log\left(1+\frac{\delta x}{x}\right)^{\frac{x}{\delta x}}$$

| $\Rightarrow \underset{\frac{\delta x}{x}\to 0}{\text{Lt}}\left(1+\frac{\delta x}{x}\right)^{\frac{x}{\delta x}} = e$

$$\Rightarrow \frac{dy}{dx} = \frac{1}{x}\log e$$

$$= \frac{1}{x} \quad \left|\because \log e = 1\right.$$

$$\Rightarrow \frac{d}{dx}(\log x) = \frac{1}{x}\cdot$$

Example 1.

(i) $\frac{d}{dx}(x^7) = 7x^6$ $\quad \left|\because \frac{d}{x}(x^n) = nx^{n-1}\right.$

(ii) $\frac{d}{dx}\left(\sqrt{x}\right) = \frac{d}{dx}\left(x^{1/2}\right) = \frac{1}{2}x^{1/2-1} = \frac{1}{2}x^{-\frac{1}{2}} = \frac{1}{2}\sqrt{x}$

(iii) $\frac{d}{dx}(x^2+3) = \frac{d}{dx}(x^2) + \frac{d}{dx}(3) = 2x + 0 = 2x$ $\quad \left| \because \frac{d}{dx}(C) = 0 \right.$

(iv) $\frac{d}{dx}(C \cdot v) = C \cdot \frac{d}{dx}(v) = C \cdot \frac{dv}{dx} = C$ $\quad | \because v$ is the function of x

(v) $\frac{d}{dx}(5x^2) = 5 \cdot \frac{d}{dx}x^2 = 5 \cdot 2x = 10x$

(vi) $\frac{d}{dx}(\sin 7x) = \cos 7x \frac{d}{dx}(7x) = 7\cos 7x$

(vii) $\frac{d}{dx}\left[(2x+3)^5\right] = 5(2x+3)^4 \frac{d}{dx}(2x+3) = 5(2x+3)^4 \cdot 2 = 10(2x+3)^4$

(viii) $\frac{d}{dx}(\log x^2) = \frac{d}{dx}(2\log x) = 2\frac{d}{dx}\log x = \frac{2}{x}$

(ix) $\frac{d}{dx}(e^{5x}) = e^{5x}\frac{d}{dx}5x = 5e^{5x}$

EXERCISES

Find the derivative of following functions.

1. (*i*) $x^2 + 2$ (*ii*) $\frac{1}{x^2}$ (*iii*) $x + \frac{1}{x}$

2. (*i*) $x^{3/2}$ (*ii*) $x^{-2} - \frac{x^3}{x^4}$ (*iii*) $(ax + b)^n$

3. (*i*) $a^{\sin x}$ (*ii*) e^{x^2+5} (*iii*) $5x^{2/3}$

4. (*i*) $\sin 5x$ (*ii*) $\tan 5x$ (*iii*) $\cos 5x$

5. If $x = a(\theta + \sin\theta)$, $y = a(1 - \cos\theta)$

find $\frac{dy}{dx}$.

(Hint. $\frac{dx}{d\theta} = a(1 + \cos\theta)$

$\frac{dy}{d\theta} = a\sin\theta \quad \frac{dy}{dx} = \frac{\frac{dy}{d\theta}}{\frac{dx}{d\theta}}$)

ANSWERS

1. (i) $2x$ (ii) $-\frac{2}{x^3}$ (iii) $1 - \frac{1}{x^2}$

2. (*i*) $\frac{3}{2}x^{1/2}$ (*ii*) $-\frac{2}{x^3}+\frac{1}{x^2}$ (*iii*) $an(ax + b)^{n-1}$

3. (*i*) $a^{\sin x} \log a \cos x$ (*ii*) $2x\, e^{x^2+5}$ (*iii*) $5 \times \frac{2}{3}x^{-1/3}$

4. (*i*) $5 \cos 5x$ (*ii*) $5 \sec^2 5x$ (*iii*) $-5 \sin 5x$

5. $\tan\frac{\theta}{2}$

Some important results of Trigonometry which are used in differentiation and integration

1. $\sin^2\theta + \cos^2\theta = 1$
2. $\sec^2\theta = 1 + \tan^2\theta$
3. $\operatorname{cosec}^2\theta = 1 + \cot^2\theta$
4. $\sin(A \pm B) = \sin A \cos B \pm \cos A \sin B$ and $\sin 2A = 2 \sin A \cos A$
5. $\cos(A \pm B) = \cos A \cos B \mp \sin A \sin B$ and $\cos 2A = \cos^2 A - \sin^2 A$ or $1 - 2\sin^2 A$ or $2\cos^2 A - 1$
6. $\sin C + \sin D = 2\sin\frac{C+D}{2}\cos\frac{C-D}{2}$
7. $\sin C - \sin D = 2\cos\frac{C+D}{2}\sin\frac{C-D}{2}$
8. $\cos C + \cos D = 2\cos\frac{C+D}{2}\cos\frac{C-D}{2}$
9. $\cos C - \cos D = 2\sin\frac{C+D}{2}\sin\frac{D-C}{2}$
10. $1 + \cos A = 2\cos^2\frac{A}{2}$
11. $1 - \cos A = 2\sin^2\frac{A}{2}$
12. $\tan(A + B) = \frac{\tan A + \tan B}{1 - \tan A \tan B}$ and $\tan 2A = \frac{2\tan A}{1-\tan^2 A}$

θ	0	$\frac{\pi}{6}$	$\frac{\pi}{4}$	$\frac{\pi}{3}$	$\frac{\pi}{2}$
$\sin\theta$	$\sqrt{\frac{0}{4}}$	$\sqrt{\frac{1}{4}}=\frac{1}{2}$	$\sqrt{\frac{2}{4}}=\frac{1}{\sqrt{2}}$	$\sqrt{\frac{3}{4}}=\frac{\sqrt{3}}{2}$	$\sqrt{\frac{4}{4}}=1$
$\cos\theta$	$\sqrt{\frac{4}{4}}=1$	$\sqrt{\frac{3}{4}}=\frac{\sqrt{3}}{2}$	$\sqrt{\frac{2}{4}}=\frac{1}{\sqrt{2}}$	$\sqrt{\frac{1}{4}}=\frac{1}{2}$	$\sqrt{\frac{0}{4}}=0$

$\sin(-\theta) = -\sin\theta$	$\cos(-\theta) = \cos\theta$	$\tan(-\theta) = -\tan\theta$
$\sin(90° + \theta) = \cos\theta$	$\cos(90° + \theta) = -\sin\theta$	$\tan(90° + \theta) = -\cot\theta$
$\sin(90° - \theta) = \cos\theta$	$\cos(90° - \theta) = \sin\theta$	$\tan(90° - \theta) = \cot\theta$
$\sin(180° + \theta) = -\sin\theta$	$\cos(180° + \theta) = -\cos\theta$	$\tan(180° + \theta) = \tan\theta$
$\sin(180° - \theta) = \sin\theta$	$\cos(180° - \theta) = -\cos\theta$	$\tan(180° - \theta) = -\tan\theta$
$\sin(270° + \theta) = -\cos\theta$	$\cos(270° + \theta) = \sin\theta$	$\tan(270° + \theta) = -\cot\theta$
$\sin(270° - \theta) = -\cos\theta$	$\cos(270° - \theta) = -\sin\theta$	$\tan(270° - \theta) = \cot\theta$
$\sin(360° + \theta) = \sin\theta$	$\cos(360° + \theta) = \cos\theta$	$\tan(360° + \theta) = \tan\theta$
$\sin(360° - \theta) = -\sin\theta$	$\cos(360° - \theta) = \cos\theta$	$\tan(360° - \theta) = -\tan\theta$

MAXIMA AND MINIMA

Neighbourhood of a point : If a is any real number, then any open interval containing a is called a neighbourhood of a. For example (–2, 2), (–2, 1), (–0.6, 0.7) are all neighbourhoods of real number $\frac{1}{2}$.

Local maximum value. A function $f(x)$ is said to have a local maximum value at $x = a$ if $f(a)$ is greatest of all its values of x lying in same neighbourhood of a. A local maximum value is also called a relative maximum or simply a maximum value.

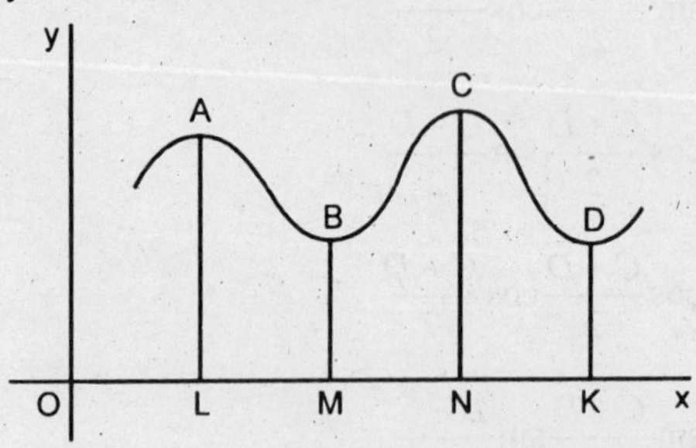

The above figure shows the graph of same function $f(x)$. It is seen that at A and C. The ordinate ceases to increase and if at A, $x = a$, we find that $AL = f(a)$ and it is greater than all other ordinates i.e., values of $f(x)$ in the neighbourhood of A i.e., $x = a$.

Thus from the above definition $f(x)$ is said to have a maximum value of A, C etc.

Local minimum value : A function $f(x)$ is said to have a local minimum value at $x = a$ if $f(a)$ is the smallest of all its values for values of x lying in same neighbourhood of a. A local minimum value is also called a relative minimum value or simply minimum value.

In the above figure B, D are points where the ordinates are less than those in the small neighbourhood of these points and as such we find that $f(x)$ has minimum values at B, D etc.

Local extremum value : A local maximum or a local minimum of a function is called a local extremum value.

Working rule for solving problems

Step I : Differentiate the given function and put $f'(x) = 0$ to get the critical values of $f(x)$.

Step II : Arrange the critical values in the ascending order of magnitude.

Step III : For a particular critical values at $x = a$

(i) If $f''(a) < 0 \Rightarrow f(a)$ is a local maximum value.

(ii) If $f''(a) > 0 \Rightarrow f(a)$ is a local minimum value.

(iii) If $f''(a) = 0 \Rightarrow f'''(a)$ must be studied.

If $f'''(a) \neq 0$ then $(a, f(a))$ is called a point of inflexion. If $f'''(a) = 0$ then $f^{iv}(a)$ is studied at $x = a$.

Example 1. **Show that the following functions do not have local maxima or minima.**

(i) $2x + 7$ **(ii) $\log x$**

Solution.

(i) Let $f(x) = 2x + 7$

$\therefore$ $f'(x) = 2$

$f''(x) = 0 \Rightarrow 2 = 0$ which is impossible.

$\therefore$ Given function does not have any critical values.

$\therefore$ Given function cannot have any local maximum or minimum value.

(ii) Let $f(x) = \log x$

$$\therefore \quad f'(x) = \frac{1}{x}$$

Put $f'(x) = 0 \Rightarrow \frac{1}{x} = 0$ which is impossible.

$\therefore$ Given function does not have any critical value.

$\therefore$ Given function cannot have any local maximum or local minimum value.

Example 2. **Find the local maximum and local minimum values of the following functions.**

(i) $2x^2 - 3x + 5$ **(ii) $-5x^2 + 20x + 9$**

Solution:

(i) Let $f(x) = 2x^2 - 3x + 5$

$\therefore$ $f'(x) = 4x - 3$ and $f''(x) = 4$

Put $f'(x) = 0 \Rightarrow 4x - 3 = 0 \Rightarrow x = \frac{3}{4}$

The critical value for $f(x)$ is $x = \frac{3}{4}$

$$f''\left(\frac{3}{4}\right) = 4 \text{ which is} > 0$$

$\therefore$ $x = \frac{3}{4}$ is a local minimum and the corresponding local minimum value is

$$f\left(\frac{3}{4}\right) = 2\left(\frac{3}{4}\right)^2 - 3\left(\frac{3}{4}\right) + 5$$

$$= 2\left(\frac{9}{16}\right) - \frac{9}{4} + 5$$

$$= \frac{9}{8} - \frac{9}{4} + 5$$

$$= \frac{31}{8}$$

Local minimum value at $x = \frac{3}{4}$ is $\frac{31}{8}$.

(ii) Let $f(x) = -5x^2 + 20x + 9$

$f'(x) = -10x + 20 \quad f''(x) = -10$

Put $f'(x) = 0 \Rightarrow -10x + 20 = 0 \Rightarrow x = 2$

The critical value for $f(x)$ is $x = 2$.

Now $f''(x) = -10$ which is < 0

$x = 2$ is a local maximum and the corresponding local maximum value is

$f(2) = -5(2)^2 + 20 \times 2 + 9 = 29$

$\therefore$ Local maximum value = 29 at $x = 2$.

Example 3. **Find the local maxima and local minima, if any, of the function $f(x) = x^3 - 6x^2 + 9x + 15$. Find also the local maximum and local minimum values.**

Solution.

We have $f(x) = x^3 - 6x^2 + 9x + 15$

$f'(x) = 3x^2 - 12x + 9 = 3(x^2 - 4x + 3)$

$= 3(x - 1)(x - 3)$

Put $f'(x) = 0 \Rightarrow 3(x - 1)(x - 3) = 0$

$\Rightarrow x = 1, 3$

The critical values for $f(x)$ in the ascending order are $x = 1, 3$.

Also $f''(x) = 6x - 12$

Critical value at $x = 1$

$f''(1) = 6(1) - 12 = -6$ which is < 0

$x = 1$ is a local maximum and the corresponding local maximum value is

$f(1) = (1)^3 - 6(1)^3 + 9(1) + 15 = 19$

Critical value at $x = 3$

$f''(3) = 6(3) - 12 = 6$ which is > 0

$x = 3$ is a local minimum and the corresponding local minimum value is

$f(3) = (3)^3 - 6(3)^2 + 9(3) + 15 = 15$

Hence local maximum value = 19 at $x = 1$

and local minimum value = 15 at $x = 3$.

EXERCISES

1. Find the critical values for the following functions.

(i) $x^2 - 2x + 4$ (ii) $x^3 - 1$

2. Show that the following functions do not have local maxima or minima.

(i) $x + 5$ (ii) $\log(x + 4)$

3. Find the points of local maxima and local minima and also find the local maximum and local minimum values of the following functions.

(i) x^2 (ii) $2x^3 - 21x^2 + 36x - 20$

ANSWERS

1. (i) $x = 1$ (ii) $x = 0$

3. (i) Local minimum value = 0 at $x = 0$

(ii) Local maximum value = -3 at $x = 1$

Local minimum value = -128 at $x = 6$

PARTIAL DIFFERENTIATION

Partial Derivatives of first order : Let $z = f(x, y)$ be a real valued function of two independent variables x and y, then $\underset{\delta x \to 0}{\text{Lt}} \frac{f(x+\delta x, y) - f(x, y)}{\delta x}$, if it exists, is called partial derivative of z (or f) w.r.t. x and it is generally denoted by $\frac{\partial f}{\partial x}$ or $\frac{\partial z}{\partial x}$ or $f\,x(x, y)$ or fx or zx. $\frac{\partial f}{\partial x}$ is read it as curly f by curly x.

Similarly $\underset{\delta x \to 0}{\text{Lt}} \frac{f(x, y+\delta y) - f(x, y)}{\delta y}$ if it exists is called partial derivative of z(or f) w.r.t. y and it is generally denoted by $\frac{\partial f}{\partial y}$ or $\frac{\partial z}{\partial y}$ or $f\,y(x, y)$ or $f\,y$ or zy.

From the above definition we conclude that if $\mu = f(x, y)$ is a function of two variables x and y, then the partial derivative of μ w.r.t. x is the differentiation of μ w.r.t. x taking y as constant.

***Example 1.* Find fx and fy for the function $f(x, y) = y^x$.**

Solution.

$$f(x, y) = y^x$$

Differentiating partially w.r.t. x, we get

$$fx = \frac{\partial f}{\partial x} = y^x \log y \qquad \Big| \because \frac{d}{dx}(a^x) = a^x \log a$$

Now differentiating partially w.r.t. y we get

$$fy = \frac{\partial f}{\partial y} = xy^{x-1} \qquad \Big| \because \frac{d}{dx}(x^n) = nx^{n-1}$$

***Example 2.* If $f(x, y) = \log(x^2 + y^2)$, find fx and fy.**

Solution.

$$f(x, y) = \log(x^2 + y^2)$$

Differentiating partially w.r.t. x we get

$$fx = \frac{\partial f}{\partial x} = \frac{1}{x^2 + y^2} 2x$$

$$= \frac{2x}{x^2 + y^2}$$

Differentiating partially w.r.t. y we get

$$\frac{\partial f}{\partial y} = \frac{1}{x^2 + y^2} \times 2y = \frac{2y}{x^2 + y^2}$$

EXERCISE

1. Find the first order derivatives of the following functions.

(i) $\frac{1}{\sqrt{x^2 + y^2}}$ (ii) $ax^2 + 2hxy + by^2$

2. Find $\frac{\partial \mu}{\partial x}, \frac{\partial \mu}{\partial y}, \frac{\partial \mu}{\partial z}$ if it is given that $\mu = x^2y + y^2z + z^2x$.

ANSWERS

1. (i) $\dfrac{-x}{(x^2+y^2)^{3/2}}$; $\dfrac{-y}{(x^2+y^2)^{3/2}}$ (ii) $2(ax + hy)$; $2(hx + by)$

2. $\dfrac{\partial \mu}{\partial x} = 2xy + z^2$ $\dfrac{\partial \mu}{\partial y} = x^2 + 2yz$ $\dfrac{\partial \mu}{\partial z} = y^2 + 2zx$

PARTIAL DERIVATIVES OF HIGHER ORDER

Since each of the partial derivatives $\dfrac{\partial f}{\partial x}, \dfrac{\partial f}{\partial y}$, in general, is a function of x and y, so each of them possesses partial derivatives w.r.t. the independent variables x and y.

(i) The partial derivative of $\dfrac{\partial f}{\partial x}$ w.r.t. x is given by $\dfrac{\partial}{\partial x}\left(\dfrac{\partial f}{\partial x}\right) = \dfrac{\partial^2 f}{\partial x^2}$ or fxx

(ii) The partial derivative of $\dfrac{\partial f}{\partial y}$ w.r.t. y is given by $\dfrac{\partial}{\partial y}\left(\dfrac{\partial f}{\partial y}\right) = \dfrac{\partial^2 f}{\partial y^2} = fyy$

(iii) The partial derivative of $\dfrac{\partial f}{\partial x}$ w.r.t. y is given by $\dfrac{\partial}{\partial y}\left(\dfrac{\partial f}{\partial x}\right) = \dfrac{\partial^2 f}{\partial y \partial x}$ or fyx

(iv) The partial derivative of $\dfrac{\partial f}{\partial y}$ w.r.t. x is given by $\dfrac{\partial}{\partial x}\left(\dfrac{\partial f}{\partial y}\right) = \dfrac{\partial^2 f}{\partial x \partial y}$ or fxy

$\therefore \dfrac{\partial^2 f}{\partial x^2}, \dfrac{\partial^2 f}{\partial y^2}, \dfrac{\partial^2 f}{\partial x \partial y}, \dfrac{\partial^2 f}{\partial y \partial x}$ are all called second order partial derivatives.

Example 1. **Verify that** $\dfrac{\partial^2 f}{\partial x \partial y} = \dfrac{\partial^2 f}{\partial y \partial x}$ **for the function** $\log \dfrac{x^2+y^2}{xy}$.

Solution.

$$\text{Let} \quad f(x, y) = \log \frac{x^2+y^2}{xy} = \log(x^2 + y^2) - \log xy$$

$$\frac{\partial f}{\partial x} = \frac{1}{x^2+y^2} \cdot 2x - \frac{1}{xy} \cdot y = \frac{2x}{x^2+y^2} - \frac{1}{x}$$

$$\frac{\partial^2 f}{\partial y \partial x} = 2x \cdot \frac{-2y}{(x^2+y^2)^2} - 0 = \frac{-4xy}{(x^2+y^2)^2} \quad \text{..... (i)}$$

$$\text{Now} \quad \frac{\partial f}{\partial y} = \frac{1}{x^2+y^2} \times 2y - \frac{1}{xy} \times x = \frac{2y}{x^2+y^2} - \frac{1}{y}$$

$$\frac{\partial^2 f}{\partial x \partial y} = 2y\left(\frac{-2x}{(x^2+y^2)^2}\right) - 0 = \frac{-4xy}{(x^2+y^2)^2} \quad \text{..... (ii)}$$

From (i) and (ii) we see that

$$\frac{\partial^2 f}{\partial x \partial y} = \frac{\partial^2 f}{\partial y \partial x}. \quad \text{Hence the result.}$$

Example 2. $f = \tan^{-1}\left(\dfrac{y}{x}\right)$ **verify that** $\dfrac{\partial^2 f}{\partial y\,\partial x} = \dfrac{\partial^2 f}{\partial x\,\partial y}$.

Solution.

Given $f = \tan^{-1}\left(\dfrac{y}{x}\right)$ ..... (i)

Differentiating (i) partially w.r.t. x, we get

$$\frac{\partial f}{\partial x} = \frac{1}{1+\left(\frac{y}{x}\right)^2}\cdot\left(\frac{-y}{x^2}\right) = \frac{-y}{x^2+y^2} \quad \text{..... (ii)}$$

Differentiating (i) partially w.r.t. y, we get

$$\frac{\partial f}{\partial y} = \frac{1}{1+\left(\frac{y}{x}\right)^2}\cdot\frac{1}{x} = \frac{x}{x^2+y^2} \quad \text{..... (iii)}$$

Differentiating (ii) partially w.r.t. y, we get

$$\frac{\partial^2 f}{\partial y\partial x} = \frac{\partial}{\partial y}\left(\frac{-y}{x^2+y^2}\right)$$

$$= \frac{(x^2+y^2)(-1)-(-y)(2y)}{(x^2+y^2)^2}$$

$$= \frac{y^2-x^2}{(x^2+y^2)^2} \quad \text{..... (iv)}$$

Differentiating (iii) partially w.r.t. x, we get

$$\frac{\partial^2 f}{\partial x\partial y} = \frac{\partial}{\partial x}\left(\frac{x}{x^2+y^2}\right) = \frac{(x^2+y^2)(1)-x(2x)}{(x^2+y^2)^2}$$

$$= \frac{y^2-x^2}{(x^2+y^2)^2} \quad \text{..... (v)}$$

From (iv) and (v), we get $\dfrac{\partial^2 f}{\partial y\partial x} = \dfrac{\partial^2 f}{\partial x\partial y}$. Hence the result.

EXERCISE – 2

1. Verify that $\dfrac{\partial^2 z}{\partial y\partial x} = \dfrac{\partial^2 z}{\partial x\partial y}$ when $z = \log\dfrac{x^2+y^2}{xy}$.

2. Verify that $\dfrac{\partial^2 z}{\partial y\partial x} = \dfrac{\partial^2 z}{\partial x\partial y}$ when $z = ax^2 + 2hxy + by^2$.

INTEGRATION

Integration is just the reverse process of differentiation.

The purpose of integration is to find out the function whose differential co-efficient is given.

The symbol $\int dx$ represents the operation of finding integral of the function w.r.t. x. Thus

$$\int \cos x \, dx = \sin x + c \quad \therefore \quad \frac{d}{dx}(\sin x) = \cos x$$

where c is called the constant of integration.

Important integration formulae

1. $\int x^n dx = \frac{x^{n+1}}{n+1} + C$ provided $n \neq -1$ as $\frac{d}{dx}\left(\frac{x^{n+1}}{n+1}\right) = x^n$

2. $\int 1 dx = x + C$ as $\frac{d}{dx}(x) = 1$

3. $\int (\mu + v) dx = \int \mu dx + \int v dx$

4. $\int \frac{1}{x} dx = \log x + C$ as $\frac{d}{dx}(\log x) = \frac{1}{x}$

5. $\int a^x dx = \frac{a^x}{\log_e a} + C$ as $\frac{d}{dx}(a^x) = a^x \log a$

6. $\int e^x dx = e^x + C$ as $\frac{d}{dx}(e^x) = e^x$

7. $\int \cos x \, dx = \sin x + C$ as $\frac{d}{dx}(\sin x) = \cos x$

8. $\int \sin x \, dx = -\cos x + C$ as $\frac{d}{dx}(\cos x) = -\sin x$

9. $\int \sec^2 x \, dx = \tan x + C$ as $\frac{d}{dx}(\tan x) = \sec^2 x$

10. $\int \text{cosec}^2 x dx = -\cot x + C$ as $\frac{d}{dx}(\cot x) = -\text{cosec}^2 x$

11. $\int \sec x \tan x \, dx = \sec x + C$ as $\frac{d}{dx}(\sec x) = \sec x \tan x$

12. $\int \text{cosec} x \cot x \, dx = -\text{cosec } x + C$ as $\frac{d}{dx}(\text{cosec } x) = -\text{cosec } x \cot x$

DEFINITE INTEGRAL

A definite integral is one whose lower and upper limits are given.

It is denoted by $\int_a^b f(x)dx$, and is read as integral of $f(x)$ w.r.t. x between $x = a$ and $x = b$.

If $\int f(x)dx = g(x)$

then, $\int_a^b f(x)dx = [g(x)]_a^b = g(b) - g(a)$.

Example 1. **Evaluate** $\int_1^2 x^2 dx$

Solution.

$$\int_1^2 x^2 dx = \left[\frac{x^3}{3}\right] = \frac{2^3}{3} - \frac{(1)^3}{3} = \frac{8}{3} - \frac{1}{3} = \frac{7}{3}$$

Example 2. $\int_0^{\pi/4} \sin x \cos x \, dx$

Solution. $\int_0^{\pi/4} \sin x \cos x dx = \int_0^{\frac{\pi}{4}} \frac{1}{2}(2 \sin x \cos x) dx = \frac{1}{2} \int_0^{\frac{\pi}{4}} \sin 2x \, dx = \frac{1}{2}\left[\frac{-\cos 2x}{2}\right]_0^{\frac{\pi}{4}}$

$$= -\frac{1}{4}[\cos 2x]_0^{\frac{\pi}{4}}$$

$$= -\frac{1}{4}\left[\cos 2\frac{\pi}{4} - \cos 0\right]$$

$$= -\frac{1}{4}\left[\cos\frac{\pi}{2} - \cos 0\right]$$

$$= -\frac{1}{4}[0 - 1]$$

$$= \frac{1}{4}.$$

EXERCISES

1. Integrate the following functions w.r.t. x and find within the limits $x = 2$ to $x = 4$.

(*i*) x^3 (*ii*) $\sqrt{x}$

2. Evaluate the following integrals

(*i*) $\int_0^{30°} \sin 7x \, dx$ (*ii*) $\int_{15°}^{30°} \cos(4x - 3) \, dx$

3. Evaluate the following integrals

(*i*) $\int_0^5 \frac{dx}{5^x}$ (*ii*) $\int \sqrt{1 + x} dx$

4. Evaluate $\int (1 + \cos x)^{1/2} dx$ Hint. $\int (1 + \cos x) = 2\cos^2 \frac{x}{2}$

ANSWERS

1. (*i*) 60 (*ii*) 3.448

2. (*i*) 0.2666 (*ii*) $\frac{1}{4}$ [sin 117° – sin 57°]

3. (*i*) $\frac{1}{5}\log\frac{5}{2}$ (*ii*) $\frac{2}{3}(1 + x)^{3/2} + C$

4. $2\sqrt{2}\cos\frac{x}{2}$

FACTORIALS

Factorial Notation : The continued product of first n natural numbers i.e., product of all natural numbers beginning with 1 and ending with n is written as $\underline{|n}$ or $n!$ and read as factorial n. Thus

$$n! = 1 \cdot 2 \cdot 3 \cdot .... n$$

or writing the factors in reverse order

$$n! = n(n-1)(n-2) .... 3 \cdot 2 \cdot 1$$

For example $4! = 4 \cdot 3 \cdot 2 \cdot 1 = 24$

$$3! = 3 \cdot 2 \cdot 1 = 6$$

and $0!$ is always 1 (i.e., $0! = 1$)

A property of factorials :

$$\begin{aligned} 6! &= 6 \cdot 5 \cdot 4 \cdot 3 \cdot 2 \cdot 1 = 6 \cdot 5! \\ &= 6 \cdot 5 \cdot 4! = 6 \cdot 5 \cdot 4 \cdot 3! \\ &= 6 \cdot 5 \cdot 4 \cdot 3 \cdot 2! \end{aligned}$$

Also we know that $n! = n(n-1)(n-2) .... 3 \cdot 2 \cdot 1$

which can be written as $n = n(n-1)!$

$$\text{or } n = n(n-1)(n-2)!$$

$$\text{or } n = n(n-1)(n-2)(n-3)! \text{ and so on.}$$

Example 1. **Find the value of $\frac{11!}{7!\,4!}$.**

Solution.

$$\frac{11!}{7!\,4!} = \frac{11 \cdot 10 \cdot 9 \cdot 8 \cdot 7!}{7!\,4!}$$

$$= \frac{11 \cdot 10 \cdot 9 \cdot 8}{4 \cdot 3 \cdot 2 \cdot 1} = 330$$

Example 2. **Prove that $(2n)! = 2^n n! \, [1 \cdot 3 \cdot 5 \cdot ...... (2n-1)]$.**

Solution.

$$\begin{aligned} (2n)! &= 1 \cdot 2 \cdot 3 \cdot 4 \cdot .... (2n-1)\, 2n \\ &= (2 \cdot 4 \cdot 6 \cdot .... 2n)\,[1 \cdot 3 \cdot 5 \cdot .... (2n-1)] \\ &= 2^n(1 \cdot 2 \cdot 3 \cdot .... n)\,[1 \cdot 3 \cdot 5 \cdot .... (2n-1)] \\ &= 2^n \; n! \; [1 \cdot 3 \cdot 5 \cdot .... (2n-1)] \end{aligned}$$

Hence the result.

EXERCISE

Find the value of

1. (a) $\frac{8!}{6!}$ (b) $\frac{10! - 9!}{9!}$
2. Prove that $12! = 2^6 \cdot 6!(1 \cdot 3 \cdot 5 \cdot 7 \cdot 9 \cdot 11)$
3. $\frac{(n+2)!}{n!} = n^2 + 3n + 2$. Prove it.
4. $(2n)!\,(n-1)! = 2(n!)\,(2n-1)!$. Prove it.

ANSWERS

1. (a) 56 (b) 9

PERMUTATIONS AND COMBINATIONS

Permutations. The different arrangements that can be made with a given number of things taking some or all of them at a time are called permutations.

The symbol ^nP_r is used to denote the number of permutations of n things taken r at a time.

To find the value of ^nP_r

The number of permutations of n things taken r at a time is the same as the number of ways in which r places in a row can be filled with n different things.

Suppose that we have n things. The first place can be filled up by any one of these n things. Thus there are n ways of filling up the first place.

When the first place has been filled up by any one of the n ways, we are left with $(n-1)$ things and any one of them can be put in the second place. Therefore the second place can be filled up in $(n-1)$ ways. So by the principle of Association, the number of ways of filling up the first two places is $n(n-1)$.

When the first two places have been filled up in any one of the $n(n-1)$ ways, then we are left with $(n-2)$ things. So the third place can be filled up in $(n-2)$ ways.

By the principle of Association, the number of ways of filling up the first three places is $n(n-1)(n-2)$.

Proceeding in this way, the number of ways in which r places can be filled up is

$$\begin{aligned} ^nP_r &= n(n-1)(n-2) \text{ .... upto } r \text{ factors} \\ &= n(n-1)(n-2) \text{ .... } (n-(r-1)) \\ &= n(n-1)(n-2) \text{ .... } (n-r+1) \end{aligned}$$

Therefore $^nP_r = n(n-1) \text{ .... } (n-r+1)$

For example $^7P_4 = 7 \times 6 \times 5 \times 4 = 840$

$^5P_4 = 5 \times 4 \times 3 \times 2 = 120$

When $r = n$,
$$\begin{aligned} ^nP_r &= {^nP_n} = n(n-1)(n-2) \text{ .... to } n \text{ factors} \\ &= n(n-1)(n-2) \text{ .... } (n-(n-1)) \\ &= n(n-1)(n-2) \text{ .... } 1 \\ &= n! \end{aligned}$$

So $^nP_n = n!$

Now to prove that $^nP_r = n \times {^{n-1}P_{r-1}} = \dfrac{n!}{(n-r)!}$

$$\begin{aligned} \text{L.H.S.} = {^nP_r} &= n(n-1)(n-2) \text{ ....} (n-r+1) \\ &= \frac{\{n(n-1)(n-2)\text{.......}(n-r+1)\}\{(n-r)(n-r-1)\text{.....}3\cdot 2\cdot 1\}}{(n-r)(n-r-1)\text{.....}3\cdot 2\cdot 1} \\ &= \frac{n!}{(n-r)!} \end{aligned}$$

$$\begin{aligned} \text{R.H.S.} &= n \times {^{n-1}P_{r-1}} \\ &= n \times \frac{(n-1)!}{((n-r)\cdot(r-1))!} \end{aligned}$$

$\because$ We have proved above $^nP_r = \dfrac{n!}{(n-r)!}$

$$= \frac{n.(n-1)!}{(n-r)!} = \frac{n!}{(n-r)!}$$

$$\therefore \quad {}^{n}P_r = n \times {}^{n-1}P_{r-1} = \frac{n!}{(n-r)!}$$

Example 1. **Find the value of $^{12}P_4$.**

Solution.

$$^{12}P_4 = 12 \times 11 \times 10 \times 9$$
$$= 11880$$

Example 2. If $^{n-1}P_3 : {}^{n+1}P_3 = 5 : 12$, **find the value of n.**

Solution.

$$\frac{^{n-1}P_3}{^{n+1}P_3} = \frac{5}{12}$$

or $$\frac{(n-1)(n-2)(n-3)}{(n+1)\,n(n-1)} = \frac{5}{12}$$

Simplifying $$\frac{(n-2)(n-3)}{(n+1)n} = \frac{5}{12}$$

Cross-multiplying , $12(n^2 - 5n + 6) = 5n\,(n+1)$

$$\Rightarrow \quad 12n^2 - 60n + 72 = 5n^2 + 5n$$

$$\Rightarrow \quad 7n^2 - 65n + 72 = 0$$

$$\Rightarrow \quad n = \frac{65 \pm \sqrt{(-65)^2 - 4(7)(72)}}{2 \times 7}$$

$$= \frac{65 \pm 47}{14} = 8 \quad \text{or} \quad \frac{9}{7}$$

Rejecting $\frac{9}{7}$ as n is a +ve integer.

Hence $n = 8$.

Example 3. **In how many ways can the letters of 'Lahore' be arranged ?**

Solution.

Total number of letters in the word Lahore = 6

These are all different letters.

$\therefore$ Number of arrangements of these letters taken all at a time

$$^{6}P_6 = 6! = 720.$$

Example 4. **How many words can be formed from the letters of the word DAUGHTER so that the vowels are never together ?**

Solution.

Total number of letters in the word DAUGHTER is 8. These 8 letters are all different.

First of all, let us find those arrangements in which vowels are always together.

When the vowels A, E, U are always together, they can be supposed to be put in a bracket and treated as one letter (A, E, U). So the number of letters becomes 6.

i.e., (8 – 3 + 1) (D, G, H, T, R (A, E, U))

These 6 letters can be arranged in $^6P_6$ ways = 6! ways

Now the three letters in the group (A, E, U) can be arranged among themselves in $^3P_3 = 3!$ ways.

∴ Required numbers of such words = 6! 3! = 720 × 6 = 4320.

Now total numbers of arrangements of letters of the word DAUGHTER is 8!

∴ Number of words in which vowels are never together

= Total number of words – Numbers of words in which vowels are always together

= 8! – 4320

= 40320 – 4320 = 36000.

EXERCISES

1. How many different words can be formed with the letters of the word 'LUCKNOW'?
2. Prove that $^{2n}P_n = 1 \cdot 3 \cdot 5 \cdot .... (2n - 1) \cdot 2^n$.
3. Find n if $^{n+3}P_6 : {}^{n+2}P_4 = 14 : 1$.
4. In how many ways 5 books on English and 4 books on Hindi can be placed on a shelf so that the books on the same subject always remain together?
5. How many words can be formed out of the letters of the word 'ORIENTAL' so that no two of the vowels are together?

ANSWERS

1. 7! = 5040
3. 4
4. 5! 4! 2!
5. $4!\ {}^5P_4 = 2880$.

Combinations : The different selections or groups that can be made out of a given number of things taking some or all of them at a time are called combinations.

The symbol ^nC_r is used to denote the number of combinations of n things taken r at a time.

To find the value of ^nC_r :

The number of combinations of n things taken r at a time is denoted by ^nC_r. Let these combinations be x. Because each combination contains r things and these r things in any one of the combinations can be arranged among themselves in $r!$ ways (as we know by permutations). Hence one combination will give rise to $r!$ permutations. So x combinations will give rise to $x \cdot r!$ permutations. But the number of permutation of n things taken r at a time is ^nP_r.

$$^nP_r = x \cdot r! = {}^nC_r \cdot r!$$

$$^nC_r = \frac{^nP_r}{r!} = \frac{n!}{(n-r)!\,r!}$$

$$\therefore \quad ^nC_r = \frac{n!}{r!(n-r)!}$$

To prove that $^nC_r = {}^nC_{n-r}$.

We know that $^nC_r = \dfrac{n!}{r!(n-r)!}$

Changing r to $n - r$ we get

$$^nC_{n-r} = \frac{n!}{(n-r)!(n-(n-r))}.$$

$$= \frac{n!}{(n-r)!\,r!}$$

$$^nC_r = {}^nC_{n-r}.\quad \text{Hence the result.}$$

Example 1. **Evaluate $^{50}C_{47}$.**

Solution.

$$^{50}C_{47} = {}^{50}C_{50-47} = {}^{50}C_3 = \frac{\lfloor 50}{\lfloor 3 \lfloor 47} = \frac{50{\cdot}49{\cdot}48{\cdot}\lfloor 47}{\lfloor 3{\cdot}\lfloor 47}$$

$$= \frac{50\cdot 49\cdot 48}{3\cdot 2\cdot 1} = 19600.$$

Example 2. **If $^nC_{10} = {}^nC_{15}$, find the value of $^{27}C_n$.**

Solution:

Given $\quad ^nC_{10} = {}^nC_{15}$

Either $10 = 15$ or $10 + 15 = n$ $\quad | \because {}^nC_r = {}^nC_{n-r}$

But $10 = 15$ is impossible

So $n = 10 + 15$

$\therefore \quad n = 25$

Now we have to calculate $^{27}C_n$ i.e., $^{27}C_{25}$ $\quad | \because n = 25$

$$^{27}C_{25} = \frac{27!}{25!\,2!} = \frac{27\cdot 26\cdot 25!}{25!\ 2!}$$

$$= \frac{27\cdot 26}{2\cdot 1} = 351$$

So $\quad ^{27}C_{25} = 351.$

Example 3. **If $^{n+1}C_{r+1} : {}^nC_r = 11 : 6$ and $^nC_r : {}^{n-1}C_{r-1} = 6 : 3$, find the values of n and r.**

Solution.

$$\frac{^{n+1}C_{r+1}}{^nC_r} = \frac{11}{6} \text{ and } \frac{^nC_r}{^{n-1}C_{r-1}} = \frac{6}{3}$$

$$\Rightarrow \frac{\dfrac{(n+1)!}{(r+1)!\{(n+1)-(r+1)\}!}}{\dfrac{n!}{r!(n-r)!}} = \frac{11}{6} \text{ and } \frac{\dfrac{\lfloor n}{\lfloor r \lfloor n-r}}{\dfrac{(n-1)!}{(r-1)!\{(n-1)-(r-1)\}!}} = \frac{6}{3} = 2$$

$$\Rightarrow \frac{(n+1)!}{(r+1)!(n-r)!} \times \frac{r!(n-r)!}{n!} = \frac{11}{6} \text{ and } \frac{\lfloor n}{\lfloor r \lfloor n-r} \times \frac{(r-1)!(n-r)!}{(n-1)!} = 2$$

$$\Rightarrow \frac{(n+1)!\ r!}{(r+1)!\,n!} = \frac{11}{6} \text{ and } \frac{n!(r-1)!}{r!(n+1)!} = 2$$

$$\Rightarrow \frac{(n+1)n!\ r!}{(r+1)\,r!\,n!} = \frac{11}{6} \text{ and } \frac{n(n-1)!(r-1)!}{r(r-1)!\ (n-1)!} = 2$$

$$\Rightarrow \frac{n+1}{r+1} = \frac{11}{6} \text{ and } \frac{n}{r} = 2$$

$\Rightarrow \quad 6n + 6 = 11r + 11 \quad$ and $\quad n = 2r$

$\Rightarrow \quad 6(2r) + 6 = 11r + 11 \quad\quad n = 2 \times 5$

$\Rightarrow \quad 12r + 6 = 11r + 11 \quad\quad = 10$

$r = 5.$

Hence $n = 10$, $\quad r = 5.$

EXERCISES

1. Evaluate (i) $^{51}C_{49}$ (ii) $^{100}C_{96}$
2. If $^nP_r = {}^nP_{r+1}$ and $^nC_r = {}^nC_{r-1}$, find n and r.
3. Prove that $^{2n}C_n = \dfrac{2^n \ 1\cdot3\cdot5\cdot......(2n-1)}{n!}$.
4. If $^{18}C_r = {}^{18}C_{r+2}$, evaluate $^rC_5$ and $^{11}C_r$.

ANSWERS

1. (i) 1275 (ii) 3921225
4. 56, 165

PROBABILITY

Probability means chance or possibility occurrence of some event.

Experiment. An experiment is defined as a process of well-defined outcomes.

Random experiment. A random experiment is defined as an experiment in which all possible outcomes are known in advance.

For example, If we toss a coin we will either get head or tail.

or, in throwing of a diçe there are 6 possibilities 1 or 2 or 3 or 4 or 5 or 6.

Sample space. The sample space of a random experiment is the set of all possible outcomes. For example,

	Random Experiment	*Sample Experiment*
1.	Throwing of a fair dice	$S = \{1, 2, 3, 4, 5, 6\}$
2.	Tossing of a coin	$S = \{H, T\}$
3.	Tossing of two coins	$S = \{HH, HT, TH, TT\}$
4.	A family of two children	$S = \{BB, BG, GB, GG]$

Event. An event is a subset of sample space. An event is called simple event, if it contains only one sample point. In the experiment of throwing a die, the event A of getting 3 is a simple event.

Equal likely outcomes. The outcomes cf a random experiment are called equally likely, if all of these have equal preferences. In the experiment of tossing of a coin, the outcomes head and tail are equally likely.

Exhaustive outcomes. The outcomes of random experiment are called exhaustive if they cover all the possible outcomes of the experiment. In throwing of a dice, the outcomes 1, 2, 3, 4, 5, 6 are exhaustive.

Probability. Suppose in a random experiment there are n equally likely, exhaustive outcomes. Let A be an event and there are m outcomes favourable to the happening of it.

Then the probability $P(A)$ of the happening of the event A is defined as

$$P(A) = \frac{m}{n}$$

$$\text{or } P(A) = \frac{\text{Total no. of favourable cases in the happening of } A}{\text{Total no. of equally likely exhaustive cases}}$$

It is clear from the definition that

$$0 \le m \le n \Rightarrow \quad 0 \le \frac{m}{n} \le 1$$

$$\Rightarrow 0 \le P(A) \le 1$$

The number of cases favourable to the non-happening of the event A is $n - m$.

$$\therefore \quad P(\text{not } A) = \frac{n-m}{n}$$

$$= \frac{n}{n} - \frac{m}{n} = 1 - \frac{m}{n}$$

$$= 1 - P(A)$$

$$P(\overline{A}) = 1 - P(A) \qquad (\overline{A} = \text{not } A)$$

$$\Rightarrow P(\overline{A}) + P(A) = 1$$

***Example 1.* What is the probability of getting an even number in the throw of an unbiased dice?**

Solution.

In this experiment, there are 6 equally likely possible outcomes, i.e., {1, 2, 3, 4, 5, 6} and let A be the event of getting even number.

A = {2, 4, 6}

Favourable cases = 3

Total no. of cases = 6

$$P(A) = \frac{3}{6} = \frac{1}{2}$$

***Example 2.* Find the probability of getting the sum 10 in a single throw of two dice.**

Solution.

Here S = {(1, 1), (1, 2), (1, 3) .... (6, 5), (6, 6)}

No. of possible outcomes are $6 \times 6 = 36$

Let A be the event of getting sum 10

A = {(4, 6), (5, 5), (6, 4)}

Favourable cases = 3

$$P(A) = \frac{3}{36} = \frac{1}{12}.$$

***Example 3.* Find the probability of getting a 'King' or 'Queen' in a single draw from a well shuffled pack of playing cards.**

Solution.

Let A be the event of getting a King or a Queen in the draw.

No. of favourable cases for happening of the event A is $4 + 4 = 8$

Total number of cases = 52

$$P(A) = \frac{8}{52} = \frac{2}{13}$$

EXERCISES

1. Find the probability of getting an odd number in a single throw of a fair dice.
2. Find the probability of getting a number less than 2 in a single throw of a fair dice.
3. In a simultaneous toss of two coins, find the probability of getting exactly two heads.
4. Find the probability of getting no head in a single toss of three coins.
5. A ball is drawn at random from a box containing 4 white, 7 red and 12 black balls. Determine the probability that the ball drawn is red.

ANSWERS

1. $\frac{1}{2}$ **2.** $\frac{1}{6}$ **3.** $\frac{1}{4}$ **4.** $\frac{1}{8}$ **5.** $\frac{7}{23}$

MUTUALLY EXCLUSIVE EVENTS

Two events are said to be mutually exclusive events if both cannot occur together with same trial. In experiment of rolling a die, the events $A = \{1, 3\}$ and $B = \{2, 4, 6\}$ are mutually exclusive events. In the same experiment, the events $A = \{1, 3\}$ and $C = \{1, 3, 5, 6\}$ are not mutually exclusive events because if 3 appears on the die it is favourable to both A and C.

Addition Theorem (For mutually exclusive events). If A and B are two mutually exclusive events associated with a random experiment, then

$$P(A \text{ or } B) = P(A) + P(B).$$

Proof: Let n be the total number of exhaustive, equally likely cases of the experiment.

Let m_1 and m_2 be the number of cases favourable to the happening of the events A and B respectively.

$$P(A) = \frac{m_1}{n}$$

and $$P(B) = \frac{m_2}{n}$$

As the events are given to be mutually exclusive, so there will be no sample point common to both events A and B.

$\therefore$ The event A or B can happen in exactly $m_1 + m_2$ ways

$$\therefore \quad P(A \text{ or } B) = \frac{m_1 + m_2}{n} = \frac{m_1}{n} + \frac{m_2}{n} = P(A) + P(B)$$

Hence $P(A \text{ or } B) = P(A) + P(B)$

Addition Theorem (For non-mutually events). If A and B are two non-mutually exclusive events associated with a random experiment, then

$$P(A \text{ or } B) = P(A) + P(B) - P(AB)$$

Proof. Let n be the total number of exhaustive and equally likely cases of the experiment.

Let m_1 and m_2 be the number of cases favourable to the happening of the events A and B respectively.

$$\therefore \quad P(A) = \frac{m_1}{n}$$

$$P(B) = \frac{m_2}{n}$$

As the events are given to be non-mutually exclusive, there will be some sample points common to both events A and B.

Let m_3 be the number of common sample points

$$P(AB) = \frac{m_3}{n}$$

These m_3 sample points are also included in the events A and B separately.

Total number of sample points in A or B

$$= m_1 + m_2 - m_3$$

$$P(A \text{ or } B) = \frac{m_1 + m_2 - m_3}{n} = \frac{m_1}{n} + \frac{m_2}{n} - \frac{m_3}{n} = P(A) + P(B) - P(AB)$$

***Example 1.* The probability that a company executive will travel by bus is $\frac{2}{3}$ and that he will travel by train is $\frac{1}{5}$. Find the probability of his travelling by bus or train.**

Solution.

Let A = the event that the company executive travels by bus

$\therefore \quad P(A) = \frac{2}{3}$

Let B = the event that the company executive travels by train

$P(B) = \frac{1}{5}$

Events in this question are mutually exclusive because either he can travel by bus or by train.

$\therefore$ The probability of his travelling by bus or by train

$$P(A \text{ or } B) = P(A) + P(B) = \frac{2}{3} + \frac{1}{5} = \frac{13}{15}.$$

Example 2. **One number is drawn from numbers 1 to 150. Find the probability that it is either divisible by 3 or 5.**

Solution.

Here $S = \{1, 2, 3, ..., 149, 150\}$

Let A = the event that number is divisible by 3

$A = \{3, 6, 9, ..., 147, 150\}$

$$P(A) = \frac{50}{150}$$

B = The event that the number is divisible by 5

$B = \{5, 10, 15, ..., 145, 150\}$

$$P(B) = \frac{30}{150}$$

The events are not mutually exclusive because some points are common to both A and B. The common points are 15, 30, 45, ..., 150.

So $AB = \{15, 30, 45, ..., 135, 150\}$

$$P(AB) = \frac{10}{150}$$

So by the Addition theorem

$$P(A \text{ or } B) = P(A) + P(B) - P(AB)$$

$$= \frac{50}{150} + \frac{30}{150} - \frac{10}{150} = \frac{70}{150} = \frac{7}{15}$$

Hence the result.

EXERCISES

1. A bag contains 30 balls numbered from 1 to 30. One ball is drawn at random. Find the probability that number of the ball is a multiple of 5 or 6.
2. Find the probability that a card drawn from a pack of playing cards is either King or Spade.
3. A box contains 4 red balls, 4 green balls and 7 white balls. What is the probability that a ball drawn is either red or white?

ANSWERS

1. $\frac{1}{3}$ **2.** $\frac{4}{52} + \frac{13}{52} - \frac{1}{52}$ **3.** $\frac{11}{15}$

INTRODUCTION TO COMPUTERS

Definition: Computer is an electronic device which is able to process unmeaningful raw data into meaningful information.

This raw data is called **Input** and the generated information is called **Output** in computer terminology.

Analog Vs Digital

An **Analog** device deals with the continuous measurement, recording or processing of information.

A speedometer is an analog device which measures the speed of the vehicle continuously. Similarly a slide-rule is an analog computing device.

A **Digital** device deals with discrete samples of information.

Both types of information can be converted to another with sufficient accuracy using either ADC (Analog to Digital converter) or DAC (Digital to Analog converter).

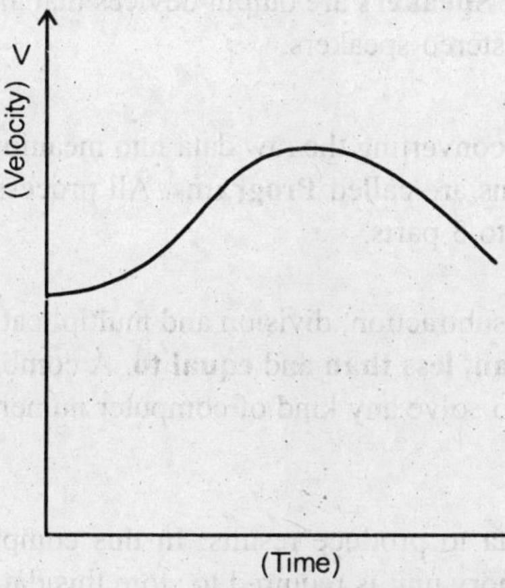

Analog Measurement

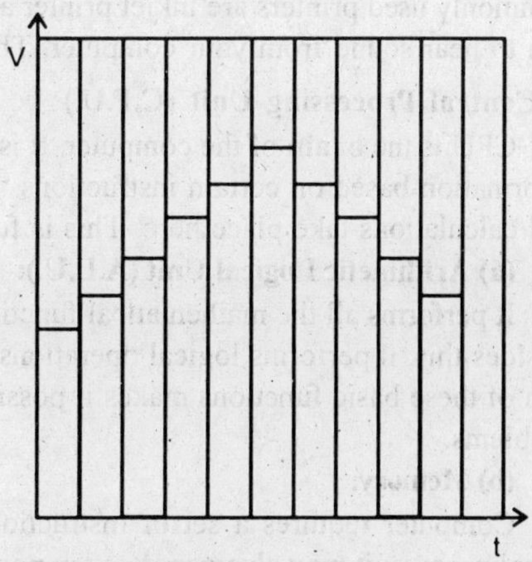

Digital Measurement

All modern computers are Digital in nature. They require Digital Output and can only generate Digital Output. Any interfacing with an Analog Input device requires use of ADC circuits. Similarly to produce an Analog output, DAC circuits are required.

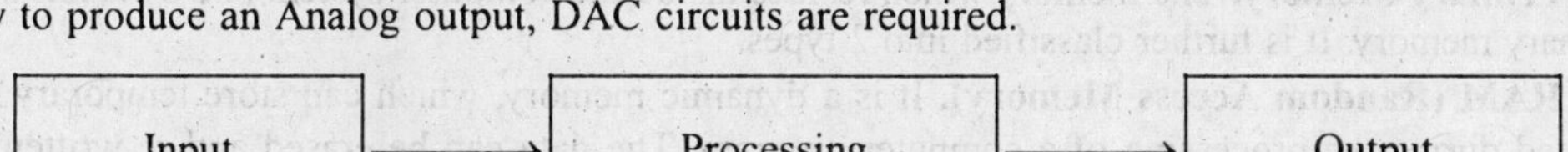

(The basic computational process)

COMPONENTS OF A COMPUTER

A computer system is made up of three basic components.

1. Input Devices
2. Output Devices
3. Central Processing Unit (CPU)

1. Input Devices provide data for the computer to operate upon. Some commonly used input devices are

(a) Keyboard (to enter data manually)

(b) Mouse (to re-position the cursor on the screen)

The **Keyboard** allows the entry of textual information. It has keys for characters (letters, numbers and punctuation marks) and special commands. Pressing the keys tells the computer what to do or what to write. The **Mouse** has a special ball that allows the selection of a point on the screen by moving a screen cursor to the point and pressing a mouse button. Alternative devices to the mouse are tracker balls, light pens and touch sensitive screens. Other types of input devices allow you to put images into the computer like **Scanner** and **Digital Camera**.

Besides these, there could be other specialised input devices, providing information about Industrial Automation System, Control Systems to the Computer, etc.

2. Output Devices are used to view or store the generated information. Some commonly used output devices are:

(a) Monitor (a T.V. screen to view results)

(b) Printer (to print the results)

(c) Speakers (outputs the sound)

The **Monitor** is a Cathode Ray Tube device like a T.V. which displays text and graphics. It allows you to 'see' what you and the computer are doing together. A **Printer** is another common part of a computer system. It takes computer data and generates a hard copy of that data. The most commonly used printers are inkjet printer and Laser printer. **Speakers** are output devices that allow you to hear sound from your computer. They are just like stereo speakers.

3. Central Processing Unit (C.P.U.)

CPU is the **brain** of the computer. It is responsible for converting the raw data into meaningful information based on certain instructions. These instructions are called **Programs**. All processing and calculations take place here. This is further divided into 3 parts.

(a) Arithmetic Logical Unit (A.L.U.):

It performs all the mathematical functions of addition, subtraction, division and multiplication. Besides this, it performs logical operations like **greater than, less than** and **equal to**. A combination of these basic functions makes it possible for A.L.U. to solve any kind of computer numerical problems.

(b) Memory:

Computer requires a set of instructions and input data to produce results. In this computational process, it may also generate temporary data. A memory unit is required to store this data as well as the generated result until it can be passed on to the output device.

Memory is further classified into 2 types.

Primary Memory. The memory which resides inside the computer system (CPU) is known as primary memory. It is further classified into 2 types.

RAM (Random Access Memory). It is a dynamic memory, which can store temporary data created during the processing of a computer program. The data can be erased and re-written any number of times. RAM is volatile in nature. As soon as the power supply of the computer is turned off, the content stored in this type of memory will be erased. Essentially, more RAM in a computer system means faster processing.

ROM (Read Only Memory) – This type of memory is used to store data, which is permanent in nature. This includes configuration data used to start a computer (BIOS—Basic Input Output System). Such information is related to hardware present in the system and does not change. Data stored in ROM cannot be erased or re-written.

Secondary Memory – Secondary memory resides outside the computer system (CPU). It is used to permanently store large amounts of data. Large amount of input data and the programs/ instruction required to process this data into information, both reside on the secondary memory and can be accessed any time later. Data on secondary memory has to be transferred to primary memory (RAM), for the computer to utilize it.

The various types of secondary memory are: — Floppy Disk, Hard Disk, Compact Disk (CD). Some of the devices which were earlier in use were — punched cards, magnetic tape and magnetic drums.

(c) Control Unit. It controls the functioning of the CPU. The flow of data between the ALU and Memory is controlled by the Control Unit.

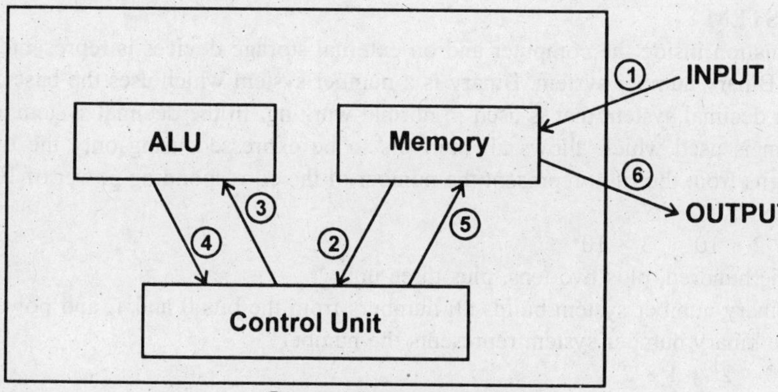

Central Processing Unit

It passes the input data present in the memory to ALU for processing. Once the data is processed, it sends the generated information to the memory so that it can be passed on to the output devices.

HARDWARE AND SOFTWARE

Hardware is a collective term used for all the physical parts of a computer which we can see and touch. Hardware consists of the monitor, keyboard, mouse, internal parts of the computer like Processor Chips, Memory Chips and other Integrated Circuits (I.Cs).

Software is a set of programs/instructions required by the hardware of the computer to perform the desired data-processing function. Software can be classified further into:

1. System Software
2. Application Software
3. Utility Programs

1. System Software consists of programs that are required for the functioning of the computer system. The following can be the examples of system software.

(a) Operating System. This software controls the loading of instructions stored in the Disk (Secondary memory) into the primary memory. The other components of O.S. help in controlling the overall flow of data to various parts of the computer. It also helps in the proper functioning of various peripheral devices like Printers, Modems, etc. Most popular operating systems are: Windows, Unix, Linux and MS-DOS.

(b) Instruction Interpreters (Languages): These sets of programs understand instructions written in English like language and convert them into Machine Language (Binary language) which the computer understands.

If they convert programs written in English like languages (called High Level Languages) into machine language, line by line, they are called **Interpreters**. If they convert the whole program together and then execute it, they are called **Compilers**. Some of the Interpreted Languages are Basic, Java. Some of the Compiled Languages are C, C^{++}, Pascal.

Internally, all the programs are converted to an intermediate Language called **Assembly Language** before getting converted to machine language. The program that converts Assembly Language programs to machine language are called **Assembler**.

2. Application Software. These are a set of programs written to solve a specific problem. These programs are specially written for such Applications like Railway Reservation, Library Management, Weather forecasting, etc.

3. Utility Programs are general purpose programs written to solve commonly faced problems/ functions. These could be word processing software (for creating a combination of text and graphics), photo-editing software, software for playing music on computer, anti-virus software, etc.

BINARY SYSTEM

All information inside the computer and on external storage devices is represented in a form related to the Binary number system. Binary is a number system which uses the base 2 instead of the base 10 in decimal system that is used in normal working. In the decimal system, a positional number system is used which allows all numbers to be expressed using only the digits 0 – 9. Successive digits from the right represent the number of the corresponding power of 10. Thus 123 in decimal is

$1 \times 10^2 + 2 \times 10^1 + 3 \times 10^0$

that is, one hundred, plus two tens, plus three units.

But the binary number system builds all numbers from the bits 0 and 1, and power of 2.

101 in the binary number system represents the number

$1 \times 2^2 + 0 \times 2^1 + 1 \times 2^0$

having value equal to 5 in the decimal system. Binary system allows all computation inside the computer to take place using cheap two-state electronic devices.

Modern computers organise information into small units called **bytes** which hold eight **bits,** each bit representing a 0 or 1. Each byte may hold a single character of text or a small integer. Larger numbers, computer instructions and character strings occupy several bytes.

Binary Arithmetic

All digital computers represent data as a collection of bits. A bit is the smallest possible unit of information. It can be in one of two states—**off** or **on,** 0 or **1.**

1 Bit [ ]

can be [0] or [1]

If one bit has two different states, how many states do two bits have? The answer is four. Likewise, three bits have eight states as represented below.

2 Bits = 4 States

[0 0] [0 1] [1 0] [1 1]

3 Bits = 8 States

[0 0 0] [0 0 1] [0 1 0] [0 1 1]

[1 0 0] [1 0 1] [1 1 0] [1 1 1]

A common and convenient grouping of bits is the **byte** or **Octet**, composed of eight bits. If two bits have four combinations, and three bits have eight combinations, how many combinations do eight bits have? If you don't want to write out all possible byte patterns, just multiply eight twos together – one two for each bit i.e.,

$$2^8 = 256$$

So, a byte has 256 possible states.

1 Byte = 8 Bits 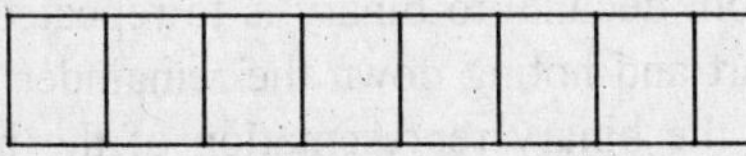

If a byte has 256 possible states, its exact state can be represented by a number from 1 to 256. However, since zero is a very important number, a byte is more typically represented by a number from 0 to 255. This is very common with bit pattern 00000000 representing zero, and bit pattern 11111111 representing 255. The numbers matching these two patterns, and everything in between, can be computed by assigning a weight to each bit, multiplying each bit's value (0 or 1) by its weight, and then adding the totals. For example, here's how 217 is represented as 11011001 in binary:

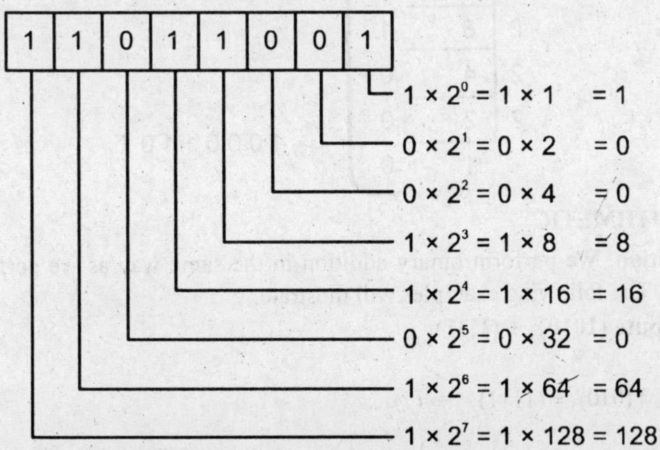

$$1 + 8 + 16 + 64 + 128 = 217$$

To convert a number from decimal to binary, begin at left most bit position (128). If the number is larger than or equal to the bit's weight, write a 1 in the bit position, subtract the bit's weight from the number, and continue with the difference. If the number is less than the bit's weight, write a 0 in the bit position and continue without any subtraction. Here's an illustration of converting 141 to binary:

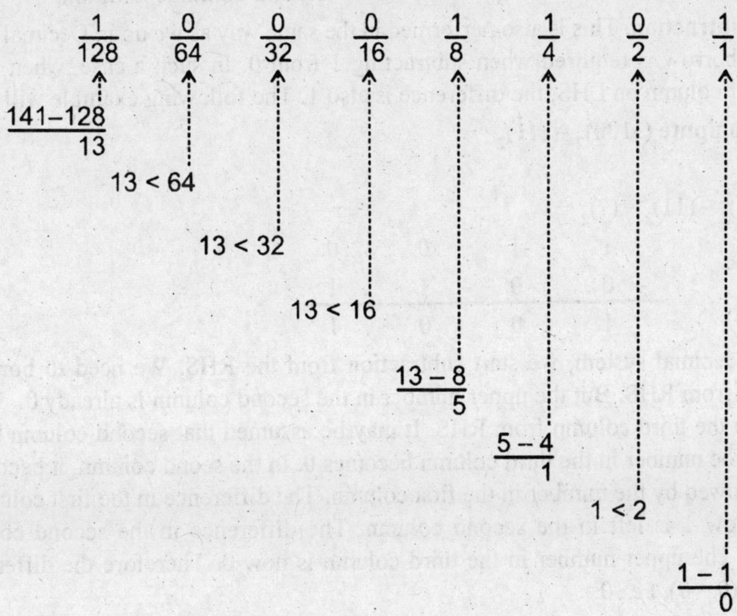

Another way of doing conversion from decimal to binary is to repeatedly divide the decimal number by 2, keeping only the whole part and noting down the remainder at each step, until we reach zero. At the end of this, we have the binary representation of the integer by reading the remainders in reverse order.

2	141	
2	70	1
2	35	0
2	17	1
2	8	1
2	4	0
2	2	0
	1	0

10001101

1.19 BINARY ARITHMETIC

Binary Addition. We perform binary addition in the same way as we perform addition with Decimal numbers. The following examples will illustrate.

Example 1. **Compute $(1010)_2 + (1111)_2$**

Solution.

$$(1010)_2 + (1111)_2 = (?)_2$$

	1	1	1	⟶ carry	
Augend		1	0	1	0
Addend	+	1	1	1	1
	1	1	0	0	1

We start the addition from the RHS as we do in Decimal system. As soon as the sum becomes two or more, we carry 1 to the next position on the left and continue addition.

Binary Subtraction. This is also performed in the same way as we do in Decimal system. While subtracting, a borrow is required when subtracting 1 from 0. In such a case, when 1 is borrowed from the higher column on LHS, the difference is also 1. The following example will illustrate.

Example 2. **Compute $(1100)_2 - (11)_2$**

Solution.

$$(1100)_2 - (11)_2 = (?)_2$$

1	1	0	0
− 0	0	1	1
1	0	0	1

As in the decimal system, we start subtraction from the RHS. We need to borrow from the second column from RHS. But the upper number in the second column is already 0. Thus, it needs to borrow from the third column from RHS. It may be assumed that second column borrows from third column. The number in the third column becomes 0. In the seond column, it becomes 2, out of which 1 is borrowed by the number in the first column. The difference in the first column will be (2 – 1), *i.e.*, 1. Now 1 is left in the second column. The difference in the second column will be (1 – 1), *i.e.*, 0. The upper number in the third column is now 0. Therefore the difference in third column will be (0 – 0), *i.e.*, 0.

Binary Multiplication

Binary multiplication is also similar to multiplication in decimal system. The following example will illustrate.

Example 3. **Compute $(1101)_2 \times (101)_2$**

Solution.

$$
\begin{array}{rrrrrrr}
 & & & 1 & 1 & 0 & 1 \\
 & & \times & & 1 & 0 & 1 \\
\hline
 & & & 1 & 1 & 0 & 1 \\
 & & 0 & 0 & 0 & 0 & \times \\
 & 1 & 1 & 0 & 1 & \times & \times \\
\hline
1 & 0 & 0 & 0 & 0 & 0 & 1 \\
\hline
\end{array}
$$

Binary Division. Binary division is performed by successive subtraction of the divisor from the dividend. The quotient bits are thus framed. The following rules are obeyed :

$$0 \div 1 = 0$$
$$1 \div 1 = 1$$

Division by 0 has no meaning.

The following example will illustrate.

Example 4. **Compute $(110)_2 \div (10)_2$**

Solution.

```
                 11     ←Quotient
Divisor ──→ 10 ) 1 1 0   ←Dividend
                 1 0
                 ───
                   1 0
                   1 0
                   ───
                   0 0  Remainder
```

Thus, $(110)_2 \div (10)_2 = (11)_2$

OPERATING SYSTEM

An operating system is a layer of software which takes care of technical aspects of a computer's operation. It shields the user of the machine from the low-level details of the machines operation and provides frequently needed facilities. There is no universal definition of what an operating system is. You can think of it as being the software which is already installed on a machine, before you add anything of your own. Normally the operating system has a number of key elements:-

(i) A technical layer of software for driving the hardware of the computer, like disk drives, the keyboard and the screen.

(ii) A file system which provides a way of organising files logically.

(iii) A simple command language which enables users to run their own programs and to manipulate their files in a simple way.

(iv) Some operating systems also provide text editors, compilers, debuggers and a variety of other tools. Since the Operating System (OS) is in charge of a computer, all requests to use its resources and devices need to go through the OS. An OS therefore provides legal entry points into its code for performing basic operations like writing to devices.

Operating systems may be classified by how many tasks they can perform 'simultaneously' and by how many users can be using the system 'simultaneously'. That is: Single-user or multi-user and single-tasking or multi-tasking.

Single-user operating system: We are all familiar with the concept of sitting down at a computer system and writing documents or performing some task. Operating systems such as Windows 95, Windows NT Workstation and Windows 2000 Professional are essentially single user operating systems. They provide you the capability to perform tasks on the computer system such as writing programs and documents, printing and accessing files. In essence, a single-user operating system provides access to the computer system by a single user at a time. If another user needs access to the computer system, he must wait till the current user finishes what he is doing and leaves.

Students in Computer Labs at Colleges or University often experience this. You might also have experienced this at home, where you want to use the computer but some one else is currently using it. You have to wait for them to finish before you can use the computer system.

Multi-user operating system: A multi-user operating system lets more than one user access the computer system at one time. Access to the computer system is normally provided *via* a network, so that users access the computer remotely using a terminal or other computer.

In the early days of large multi-user computers, multiple terminals (keyboards and associated monitors) were provided. These terminals send their commands to the main multi-user computer for processing, and the results were then displayed on the associated terminal monitor screen. Terminals were hand-wired directly to the multi-user computer system.

Today, these terminals are generally personal computers and use a network to send and receive information to the multi-user computer system. Examples of multi-user operating systems are Unix, Linux (A Unix clone) and main frames such as the IBM A5400.

The Operating System for a large multi-user computer system with many terminals is much more complex than a single-user operating system. It must manage and run all user requests, ensuring they do not interfere with each other.

Advantages and Disadvantages of Multi-user Operating Systems. The advantage of having a multi-user operating system is that it lets a number of users share the very expensive hardware resource. This means the cost is divided amongst the users. It also makes better use of the resources. Since the resources are shared, they are more likely to be in use than sitting idle and being unproductive.

One problem with multi-user computer systems is that as more people access it, the performance becomes slower and slower. Another disadvantage is the increased cost of memory devices, as a multi-user operating system requires a lot of disk space and memory. In addition, the actual software for multi-user operating systems tend to cost more than single-user operating systems.

Multi-tasking Operating System. A multi-tasking operating system provides the ability to run more than one Program at a time. For example, a user could be running a word processing package, printing a document, copying files to the floppy disk and backing up selected files to a tape unit. Each of these tasks the user is doing appears to be running at the same time.

A multi-tasking operating system has the advantage of letting the user do more than one task at a time, so this leads to increased productivity. The disadvantage is that more the programs that are run by the user, more the memory that is required.

COMPUTER PROGRAMMING

Writing software or computer programs, is a lot like writing down the steps it takes to do something. But, if what you need to do is not obvious, programming will challenge you like when you are solving puzzles.

A computer is very dumb. It does exactly what you tell it to do, which is not necessarily what you wanted. Programming will help you to learn the importance of clarity of expression.

These days, most people don't need to know how a computer works. Most people can simply turn on a computer and point the mouse at some little graphical image on the display, double-click the left mouse button, and a computer program starts up (like a Web browser); and, this is all they need to know about their computer.

To learn how to write computer programs, you need to know a little bit about how a computer works. Your job will be to instruct the computer to do things. These lists of instructions that you will write are computer programs, and the stuff that these instructions manipulate are different types of objects.

Basically, computer perform operations on objects. A microprocessor, which is the heart of a computer, is really primitive but very fast. It takes group of binary numbers representing parts of objects and moves them around, adds pairs together, subtracts one from another, compares a pair, etc.

Computers manipulate numbers, symbolic information (think characters), visual things (images), sound (MP3), and sets of instruction (the computer's native language). As an example, computer displays consist of a bunch of coloured points called pixels. A pixel is an object. It has a position which consists of the row and column it is in. Its colour is specified as three numbers—called RGB (Red, Green, Blue) values. So, like pixels are used to compose graphics and images, there are also standard representations for the other things: numbers, characters, sound, and the computer's instructions. At the lowest level, everything is a bunch of ones and zeros (binary number).

Programming Languages (The Microprocessor's Language)

Just like we have many different languages in use throughout the world for us humans to communicate with, there are different languages that we use to communicate with a computer. Computer languages are used to tell the computer what to do, you instruct it.

If you instructed a computer in its native language (machine language), you would have to write instructions in the form of binary numbers. This is very, very hard to do. Although the pioneers of computers did this, no one does this these days.

One step above machine language is assembler language. In assembler, the operations that the microprocessor knows how to do are given names. Address in memory can also be given meaningful names. This is a big step over binary, but still very tedious to do any large software program with. It still has its place for little pieces of software that need to interact directly with the microprocessor and/or those that are executed many times. Here is an example of DEC system–10 assembler language that:

1. adds two numbers together,
2. adds in a third number if it is greater than zero, and
3. stores the results in a third location.

Op. Code	Accumulator	Operand
Move	T_1	Number 1
Add	T_1	Number 2
Skiple		Number 3
Add	T_1	Number 3
Move	T_1	Result

Programming Languages (High-Level Languages)

Most software written today is in high-level languages. There are many and some are quite old. COBOL, FORTRAN, and Lisp were written in the 1950s. Higher-level languages make it easier to describe the pieces of the Program you are creating. They abstract away the specifics of the microprocessor in your computer. Most come with large sets of common stuff you need to do.

High Level Language Programming doesn't require an intimate knowledge of the internal structure of the computer and can be ported over different computer architectures. High level language programs are processed by a program, known as a **Compiler**, which translates the high level language program into an equivalent machine language program. Thus, if a compiler was available for each machine architecture, then a program could be run on any computer without change.

At the moment there are several internationally standardised languages as given in the table below which are popular.

Language	Application Area
COBOL	Business Applications
FORTRAN	Engineering & Scientific Calculations
C	General Purpose Language
Prolog	Artificial Intelligence
Lisp	Artificial Intelligence
C++	Object Oriented version of C
Java	Object Oriented Language widely used on Web

Most popular high-level languages have been standardised by one or the other international body. Thus if Programs are written to conform strictly to the standard then they should execute correctly on any architecture for which there is a standard conforming compiler.

APPENDIX – 1

LAWS OF LOGARITHMS

[I] First Law (Product law):

The logarithm of a product is equal to the sum of logarithms of its factors.

$$\therefore \quad \log_a (m \times n) = \log_a m + \log_a n$$

$$\log_a (m \times n \times p) = \log_a m + \log_a n + \log_a p$$

Remember:

$$\log_a (m + n) \neq \log_a m + \log_a n$$

[II] Second Law (Quotient law):

The logarithm of a fraction is equal to the difference between the logarithm of numerator and the logarithm of denominator.

$$\log_a \frac{m}{n} = \log_a m - \log_a n$$

Remember:

$$\frac{\log_a m}{\log_a n} \neq \log_a m - \log_a n$$

[III] Third Law (Power law):

The logarithm of a power of a number is equal to the logarithm of the number multiplied by the power.

$$\log_a (m)^n = n \log_a m$$

Corollary : Since $\sqrt[n]{m} = m^{\frac{1}{n}}$

$$\therefore \quad \log_a \sqrt[n]{m} = \log_a m^{\frac{1}{n}} = \frac{1}{n} \log_a m.$$

Note:

(*a*) Logarithms to the base 10 are known as common logarithms.

(*b*) If no base is given, the base is always taken as 10.

(*c*) Logarithm of a number to the same base is always one, *i.e.*,

$$\log_a a = 1; \log_{10} 10 = 1 \text{ and so on.}$$

(*d*) The logarithm of 1 to any base is zero.

i.e., $\log_a 1 = 0; \log_5 1 = 0; \log_{10} = 1 = 0$ and so on.

APPENDIX – 2

MATHEMATICAL RELATIONS
MATHEMATICAL FORMULAE

1. Algebra

(a) Exponents

$$a^x = a \times a \times a \times a \times \ldots \text{ to } x \text{ factor}$$
$$2^3 = 2 \times 2 \times 2 = 8$$
$$a^{-x} = \frac{1}{a^x} : 2^{-3} = \frac{1}{2^3} = \frac{1}{8}$$
$$x^{1/a} = \sqrt[a]{x};\ a^x \times a^y = a^{x+y}$$
$$2^3 \times 2^6 = 2^{3+6} = 2^9 = 512$$
$$\frac{a^x}{a^y} = a^{x-y};\ \frac{2^5}{2^2} = 2^{5-2} = 2^3 = 8$$
$$\frac{2^3}{2^5} = 2^{3-5} = 2^{-2} = \frac{1}{2^2} = \frac{1}{4}$$
$$(a^x)^y = a^{xy};\ (2^2)^3 = 2^6 = 64;\ a^0 = 1.$$

(b) Simple Quadratic Equation

If $ax^2 + bc + c = 0$, then

$$x = \frac{-b \pm \sqrt{(b^2 - 4ac)}}{2a}$$

When $b^2 - 4ac < 0$, the roots are imaginary.

(c) Approximations

When x and y are small quantities, then

$$(1 + x)^n = 1 + nx \text{ (approximately)}$$

If θ is very small angle expressed in radius, then $\sin \theta = \theta$, $\cos \theta = 1$, $\tan \theta = 0$ (approximately)

2. Trigonometry

(a) Notations

$$\text{sine (sin)} = \frac{\text{ordinate}}{\text{hypotenuse}},\ \text{cosecant (cosec)} = \frac{\text{hypotenuse}}{\text{ordinate}}$$
$$\text{cosine (cos)} = \frac{\text{abscissa}}{\text{hypotenuse}},\ \text{secant (sec)} = \frac{\text{hypotenuse}}{\text{abscissa}}$$
$$\text{tangent (tan)} = \frac{\text{orbinate}}{\text{abscissa}},\ \text{cotangent (cot)} = \frac{\text{abscissa}}{\text{ordinate}}.$$

(b) Numerical Values of Important Angles

	0°	30°	45°	60°	90°	180°
sin	0	$\frac{1}{2}$	$\frac{1}{\sqrt{2}}$	$\frac{\sqrt{3}}{2}$	1	0
cos	1	$\frac{\sqrt{3}}{2}$	$\frac{1}{\sqrt{2}}$	$\frac{1}{2}$	0	–1
tan	0	$\frac{1}{\sqrt{3}}$	1	$\sqrt{3}$	∞	0

(*c*) **Formulae**

$$\sin^2\theta + \cos^2\theta = 1;\ \sin 2\theta = 2\sin\theta\cos\theta$$

$$\cos^2\theta = \cos^2\theta - \sin^2\theta = 2\cos^2\theta - 1 = 1 - 2\sin^2\theta$$

$$\tan 2\theta = \frac{2\tan\theta}{1-\tan^2\theta} : \sin\theta = \frac{1}{\operatorname{cosec}\theta}$$

$$\cos\theta = \frac{1}{\sec\theta} : \tan\theta = \frac{1}{\cot\theta}$$

3. DIFFERENTIAL CALCULUS

If $y = x^n,\ \frac{dy}{dx} = nx^{n-1}$.

If $y = f_1(x) + f_2(x)$,

$$\frac{dy}{dx} = \frac{df_1(x)}{dx} + \frac{df_2(x)}{dx}$$

If $y = k$ (constant), $\frac{dy}{dx} = 0.$

If $y = kx^n \cdot \frac{dy}{dx} = kn.x^{n-1}$

If $y = uv,\ \frac{dy}{dx} = u\frac{dv}{dx} + v\frac{du}{dx}$

If $y = \frac{u}{v},\ \frac{dy}{dx} = -\frac{v\frac{du}{dx} - u\frac{dv}{dx}}{v^2}$

If $y = e^x,\ \frac{dy}{dx} = e^x.$

If $y = \sin x,\ \frac{dy}{dx} = \cos x.$

If $y = \cos x \cdot \frac{dy}{dx} = -\sin x.$

If $y = \tan x \cdot \frac{dy}{dx} = \sec^2 x.$

If $y = \log x,\ \frac{dy}{dx} = \frac{1}{x}.$

INTEGRAL CALCULUS

$$\int dx = x; \int a\,dx = a\int dx = ax$$

$$\int x^n dx = \frac{x^{n+1}}{n+1}; \int \frac{dx}{x^n} = \frac{x^{-n+1}}{(-n+1)}$$

$$\int \frac{dx}{x} = \log_e x; \int e^x dx = e^x; \int a^x dx = \frac{a^x}{\log_e a}$$

$$\int \sin x\,dx = -\cos x; \int \cos x\ dx = \sin x$$

$$\int \tan x\,dx = -\log\cos x; \int_a^b dx = [x]_a^b = (b-a)$$

$$\int_a^b x\,dx = \left[\frac{x^2}{2}\right]_a^b = \frac{b^2}{2} - \frac{a^2}{2}$$

$$\int_a^b \frac{dx}{x} = [\log x]_a^b = \log b - \log a = \log\frac{b}{2}$$

$$\int u\,dv = uv - \int v\,du$$ (Intergration by parts)

SOME USEFUL TABLES

Table 1. Seven base units

Physical Quantity	*Name of Unit*	*Official S.I. symbol*
Length	metre	m
Mass	kilogram	kg
Time	second	s
Electric current	ampere	A
Temperature	kelvin	K
Luminous intensity	candela	cd
Amount of substance	mole	mol

Table 2. Some commonly used derived units

Physical Quantity	*Name of Unit*	*Symbol*
Area	Square metre	m^2
Volume	Cubic metre	m^3
Density	Kg per cubic metre	$kg\ m^{-3}$
Velocity	Metre per second	ms^{-1}
Acceleration	Metre per sec per sec	ms^{-2}
Force	Newton	$N = kg\ ms^{-2}$*
Power	Watt	$W = Js^{-1}$
Pressure	Pascals (Newtons per sq. meter)	$Pa = Nm^2 = kg\ m^{-1}\ s^{-2}$
Work or Energy (Heat etc)	Joule	$J = Nm = kg\ m^2\ s^{-2}$**
Magnetic flux density	Tesla	$T = kg\ s^{-2}\ A^{-1}$
Frequency	Hertz	$Hz = s^{-1}$
Electric charge	Coulomb	$C = As$
Potential difference	Volt	V
Electric resistance	Ohms	$\Omega = VA^{-1}$
Electric conductance	$Ohms^{-1}$	$\Omega^{-1} = AV^{-1}$

Table 3. Some useful conversion factors

$1\ Å = 10^{-10}$ m	1 atm = 101325 pa or Nm^{-2} = 1.013×10^6 dynes/cm^2
1 a.m.u. = 1.66053×10^{-27} kg	1 mm or torr = 133.32 Pa or Nm^{-2}
$t° = t + 273.16\ K \simeq t + 273\ K$	1 calorie = 4.184 J
1 litre = $10^{-3}\ m^3 = 1\ dm^3$	1 erg = 10^{-7} J
1 dyne = 10^{-5} N	1 electron volt (eV) = 1.6022×10^{-19} J

* Newton is defined as the force that gives a mass of 1 kg an acceleration of 1 m sec^{-2} so that $f = ma$ = (1 kg) (1 m sec^{-2}) = 1 kg m sec^{-2} = 1 N.

** Joule is the work done when a displacement of 1 metre takes place by a force of 1 newton so that $w = fd$ = (1 N) (1 m) = Nm = (1 kg m sec^{-2}) (1 m) = 1 kg m^2 sec^{-2} 1 J

Table 4. Some common prefixes used for S.I. units

Prefix	*Symbol*	*Multiplying factor*	*Example*
deci-	d	10^{-1}	1 decimetre (1 dm) = 10^{-1} m = 0.1 m
centi-	c	10^{-2}	1 centimetre (1 cm) = 10^{-2} m = 0.01m
milli-	m	10^{-3}	1 millimetre (1 mm) = 10^{-3} m = 0.001 m
micro-	μ	10^{-6}	1 micrometre (μm) = 10^{-6} m
nano-	n	10^{-9}	1 nanometre (nm) = 10^{-9} m
pico-	p	10^{-12}	1 picometre (pm) = 10^{-12} m
femto-	f	10^{-15}	1 femtometre (fm) = 10^{-15} m
deka-	da	10^{1}	1 dekametre (dam) = 10^{1} m
hecto-	h	10^{2}	1 hectometre (hm) = 100 m
kilo-	k	10^{3}	1 kilometre (km) = 10^{3} m = 1000 m
mega-	M	10^{6}	1 megametre (Mm) = 10^{6} m
giga-	G	10^{9}	1 gigametre (Gm) = 10^{9} m
tera-	T	10^{12}	1 terametre = 10^{12} m
peta-	P	10^{15}	1 petametre = 10^{15} m

Table 5. Values of some common physical and chemical constants

Quantity	*Symbol*	*Values in CGS units*	*Values in SI units*
Velocity of light (in vacuum)	*c*	2.997925×10^{10} cm/sec $\simeq 3.0 \times 10^{10}$ cm/sec	2.997925×10^{8} m sec^{-1} $\simeq 3.0 \times 10^{8}$ m sec
Planck's constant	*h*	6.6262×10^{-27} erg sec $\simeq 6.62 \times 10^{-27}$ erg sec	6.6262×10^{-34} J sec $\simeq 6.62 \times 10^{-34}$ J sec
Avogadro number	N_0 or N	6.022169×10^{23} mole^{-1} $\simeq 6.022 \times 10^{23}$ mole^{-1}	6.022169×10^{23} mol^{-1} $\simeq 6.022 \times 10^{23}$ mol^{-1}
Gas constant	R	0.082053 lit atm K^{-1} mole^{-1} $\simeq$ 0.0821 lit atm K^{-1} mole^{-1} or 8.3144×10^{7} ergs K^{-1} mole^{-1} $\simeq$ 0.0821 lit atm K^{-} mole^{-1}	8.3144 JK^{-1} mol^{-1} $\simeq$ 8.314 JK^{-1} mol^{-1}
Molar volume at NTP	V	22.414 litres $\simeq$ 22.4 litres	0.022414 m^{3} $\simeq$ 0.0224 m^{3}
Faraday's constant	F	96487 coulombs/equivalent $\simeq$ 96500 coulombs/equivalent	96487 C mol^{-1} $\simeq$ 96500 C mol^{-1}
Electronic charge	*e*	1.6022×10^{-19} coulombs	1.6022×10^{-19} C
Electron rest mass	m_e	9.109558×10^{-28} g $\simeq 9.11 \times 10^{-28}$ g	9.109558×10^{-31} kg $\simeq 9.11 \times 10^{-31}$ kg
Proton rest mass	m_p	1.672614×10^{-24} g $\simeq 1.67 \times 10^{-24}$ g	1.672614×10^{-27} kg $\simeq 1.67 \times 10^{-27}$ kg
Neutron rest mass	m_n	1.67492×10^{-24} g $\simeq 1.67 \times 10^{-24}$ g	1.67492×10^{-27} kg $\simeq 1.67 \times 10^{-27}$ kg

GREEK ALPHABET TABLE

Capital	*Small*	***Pronunciation***	*English equivalent*
Α	α	alpha	a
Β	β	beta	b
G	γ	gamma	g, n
Δ	δ	delta	d
Ε	ε	epsilon	e
Ζ	ζ	zeta	z
Η	η	eta	$\bar{e}$
Θ	θ	theta	th
Ι	ι	iota	i
Κ	κ	kappa	k
Λ	λ	lambda	l
Μ	μ	mu	m
Ν	ν	nu	n
Ξ	ξ	xi	x
Ο	ο	omricon	o
Π	π	pi	p
Ρ	ρ	rho	r, rh
Σ	σ	sigma	s
Τ	τ	tau	t
Υ	υ	upsilon	y, u
Φ	ϕ	phi	ph
Χ	χ	chi	ch
Ψ	ψ	psi	ps
Ω	ω	omega	$\bar{o}$